CAMBRIDGE LIBRARY COLL

Books of enduring scholarly value

Botany and Horticulture

Until the nineteenth century, the investigation of natural phenomena, plants and animals was considered either the preserve of elite scholars or a pastime for the leisured upper classes. As increasing academic rigour and systematisation was brought to the study of 'natural history', its subdisciplines were adopted into university curricula, and learned societies (such as the Royal Horticultural Society, founded in 1804) were established to support research in these areas. A related development was strong enthusiasm for exotic garden plants, which resulted in plant collecting expeditions to every corner of the globe, sometimes with tragic consequences. This series includes accounts of some of those expeditions, detailed reference works on the flora of different regions, and practical advice for amateur and professional gardeners.

A Dictionary of the Economic Products of India

A Scottish doctor and botanist, George Watt (1851–1930) had studied the flora of India for more than a decade before he took on the task of compiling this monumental work. Assisted by numerous contributors, he set about organising vast amounts of information on India's commercial plants and produce, including scientific and vernacular names, properties, domestic and medical uses, trade statistics, and published sources. Watt hoped that the dictionary, 'though not a strictly scientific publication', would be found 'sufficiently accurate in its scientific details for all practical and commercial purposes'. First published in six volumes between 1889 and 1893, with an index volume completed in 1896, the whole work is now reissued in nine separate parts. Volume 6, Part 2 (1893) contains entries from *Sabadilla* (an imported plant, the seeds of which produce a neurotoxin) to *silica* (used in the production of glass).

A Dictionary of the Economic Products of India

Volume 6 – Part 2: Sabadilla to Silica

George Watt

CAMBRIDGE
UNIVERSITY PRESS

University Printing House, Cambridge, CB2 8BS, United Kingdom

Published in the United States of America by Cambridge University Press, New York

Cambridge University Press is part of the University of Cambridge.
It furthers the University's mission by disseminating knowledge in the pursuit of education, learning and research at the highest international levels of excellence.

www.cambridge.org
Information on this title: www.cambridge.org/9781108068796

This edition first published 1893
This digitally printed version 2014

ISBN 978-1-108-06879-6 Paperback

A

DICTIONARY

OF

THE ECONOMIC PRODUCTS OF INDIA.

BY

GEORGE WATT, M.B., C.M., C.I.E.,

REPORTER ON ECONOMIC PRODUCTS WITH THE GOVERNMENT OF INDIA.
OFFICIER D'ACADEMIE; FELLOW OF THE LINNEAN SOCIETY; CORRESPONDING MEMBER OF THE ROYAL HORTICULTURAL SOCIETY, &c., &c.

(ASSISTED BY NUMEROUS CONTRIBUTORS.)
IN SIX VOLUMES.

VOLUME VI, PART II.

[Sabadilla to Silica.]

Published under the Authority of the Government of India,
Department of Revenue and Agriculture.

LONDON:
W. H. ALLEN & Co., 13, WATERLOO PLACE, S.W., PUBLISHERS TO THE INDIA OFFICE.

CALCUTTA:
OFFICE OF THE SUPERINTENDENT OF GOVERNMENT PRINTING, INDIA,
8, HASTINGS STREET.

1893

CALCUTTA
GOVERNMENT OF INDIA CENTRAL PRINTING OFFICE,
8, HASTINGS STREET.

DICTIONARY

OF

THE ECONOMIC PRODUCTS OF INDIA.

(G. Watt.)

SABADILLA.

Sabadilla, see **Asagræa officinalis,** *Linn.;* LILIACEÆ; Vol. I., 336.

SABIA, *Colebr.; Gen. Pl., I., 414.* 1

A genus of scandent shrubs, which comprises about ten species, natives of tropical and temperate India. Of these the most noticeable are **Sabia campanulata,** *Wall.* (*Bakal pata,* KUMAON), **S. limonacea,** *Wall.,* **S. leptandra,** *Hook. f* (*Simali,* NEPAL; *Payengrik,* LEPCHA), **S. paniculata,** *Edgw.,* and **S. viridissima,** *Kurz.*

TIMBER. 2

With the exception of the last named, which is an inhabitant of the upper mixed forests of the Andaman Islands, the above species are found in the lower ranges of the Eastern Himálaya, the Khásia hills, and Assam. **S. campanulata** is the most westerly species, being diffused along the Himálaya to Simla. They have a soft wood with large pores and broad medullary rays (*Fl. Br. Ind., II., 1-3*).

SACCHARUM, *Linn.; Gen. Pl., III., 1125.* 3

A genus of grasses, which belongs to the tribe ANDROPOGONEÆ. They are tall plants with compound, often dense, panicles, covered with long silky hairs. The spikelets are very small, and there are no awns to the flowering glumes, as in the majority of the tribe. Twelve species are described, including sugar-cane (**S. officinarum**), *munj* grass (**S. ciliare**), and *káns* (**S. spontaneum**).

[*Northern India, 26;* GRAMINEÆ.

Saccharum arundinaceum, *Retz.; Duthie, Fodder Grasses of* 4

Syn.—S. BENGALENSE, *Retz.;* S. PROCERUM, *Roxb.;* S. EXALTATUM, *Roxb.*

Vern.—*Teng,* BENG.; *Sarkandá,* PB.; *Sarpat,* RAJ.; *Adavi cheruku, konda-kanamoo* (*Roxb.*), TEL.; *Phoung ga,* BURM.; *Rambuk,* SING.

References.—*Roxb., Fl. Ind., Ed. C.B.C., 81, 82; Voigt, Hort. Sub. Cal., 705; Elliot, Fl. Andhr., 10; Trimen, Sys. Cat. Cey. Pl., 106; Hackel in DC., Monogr. Phan., VI., 117; Atkinson, Him. Dist., 321; Drury, U. Pl., 371; Liotard, Paper-making Mat., 66, 68; Balfour, Cyclop., III., 467; Rep. Bot. Gar. Ganeshkhind, Poona, 1883-84, 12; Gaz., N.-W. P., IV., lxxx.; Gaz., Panjáb, (Delhi), 20; Journ. Agri.-Horti. Soc. Ind., X., 358.*

Habitat.—A handsome perennial species, with stems 10 to 20 feet high; found in Bengal, Sikkim, and Southern India. **Roxburgh** (under **S. procerum**) says: "By far the most beautiful of the genus I have met with. It comes nearest in appearance to **S. officinarum,** but is a taller and much more elegant plant."

DOMESTIC. Culms. 5

Domestic Uses.—The CULMS are strong and straight, and are employed by the Natives for screens and various other economical purposes (*Roxburgh l. c.*).

6 **Saccharum ciliare,** *Anders.; Duthie, Fodder Grasses, 23.*

Syn.—S. SARA, *Roxb.*; S. MUNJA, *Roxb.?*

Vern.—*Sara, sarkanda, sarkara, sarpat, sarpatta, rámsar, múnja,* HIND.; *Sar, sara, shar,* BENG.; *Sar,* SANTAL; *Sarkanda, sarhar, ikar* (W. Districts), *patáwar* (E. Districts), N.-W. P; *Palwa,* OUDH; *Sarkara, sarjbar, kharkána, kánda,* PB.; *Sara, sarpat,* AJMÍR; *Dargá, karre,* TRANS-INDUS; *Sar,* SIND; *Gundra, ponika,* TEL.; *Gúndra, tejanaka, shará,* SANS.

The following names are also given to certain portions of the plant in different localities:—*Munj* leaf-sheaths, *Sar* leaves (Panjáb); *Bind* or *vind,* culm or flowering stem (Doab); *Sararhi* (E. Districts of N.-W. Prov.); *Sentha, kána,* lower portion of flowering stem; *Sirki, til,* upper portion of flowering stem; *Thili,* upper portion of flowering stem (Lahore); *Majori,* the entire flowering stem; *Tilak, tilon,* the flowers (Panjáb); *Ghua,* the flowers (E. Districts, N.-W. Prov.).

References.—*Hackel, in DC., Monogr. Phan., VI., 118; Roxb., Fl. Ind., Ed. C.B.C., 82; Voigt, Hort. Sub. Cal., 705; Brandis, For. Fl., 548; Stewart, Pb. Pl., 261; Aitchison, Cat. Pb. and Sind Pl., 172; Sir W. Elliot, Fl. Andhr., 65, 119, 155; Sir W. Jones, Treat. Pl. Ind., V., 76; U. C. Dutt, Mat. Med. Hind., 293, 310, 316; Murray, Pl. & Drugs, Sind., 12; Baden Powell, Pb. Pr., 517, 520; Atkinson, Him. Dist. (X., N.-W. P. Gaz.), 321; Useful Pl. Bomb. (XXV., Bomb. Gaz.), 238; Econ. Prod. N.-W. Prov., Pt. V. (Vegetables, Spices, and Fruits), 91, 100-101; Royle, Ill. Him. Bot., 416; Liotard, Mem. Paper-making Mat., 24, 28, 66, 67, 68; Aín-i-Akbari (Blochmann's Trans.) I., 395; Settlement Reports:—Panjáb, Dera Ismail Khán, 345; Lahore, 13; Jhang, 23; Gazetteers:—Panjáb, Dera Ismail Khán, 11; Hoshiarpur, 14; Muzaffargarh, 26; Jhelam, 33; Montgomery, 18, 19; Karnal, 19; Ludhiana, 10; Jhang, 18; Jalandhar, 5; N.-W. P., I., 85; IV., lxxx.; Mysore and Coorg, I., 68; Agri.-Horti. Soc. Ind.:—XII., 331; XIII., 175, 315; XIV., 87; New Series, I., 108; VII., 6; Ind. Forester:—V., 31; VII., 179; VIII., 177; XII., 32; Append., 23; Balfour, Cyclop. Ind., III., 466, 467.*

Habitat.—A tall handsome grass, 8 to 12 feet high, abundant over the greater part of North-West India, where, especially in the Panjáb, it covers large tracts of country. It is sometimes also planted in lines as a boundary hedge, more particularly in low-lying localities subject to periodical inundation. It varies considerably in height, in the size and shape of the inflorescence, as well as in the quality of the fibre yielded by the leaf-sheaths. It flowers after the rains are over, and a little later than **Erianthus Ravennæ,** a tall grass of similar habit of growth, and with which it is often confounded.

MEDICINE. Root. 7

Medicine.—The ROOT is officinal in the Panjáb, under the name *garba ganda.* It is burned near women after delivery, and near burns and scalds, its smoke being considered beneficial (**Dr. Stewart**).

FIBRE. 8 String. 9 Paper. 10 Matting. 11 Ropes. 12 Thatch. 13 Flowering Stem. 14 Chairs. 15 Tables. 16

Fibre.—The *munj* or FIBRE is much valued on account of its strength, elasticity, and power of resisting moisture, and is extensively employed in the manufacture of rope, STRING, mats, baskets, and PAPER. *Munj* MATTING is said to be proof against the attacks of white ants. In some of the Panjáb Districts the *máls* or ROPES with which the earthen pots in wells are fastened are composed of *munj.* The *munj* is burned at one end, then beaten with a mallet, and, finally, wisted into a rope. *Munj* fibre, according to **Baden Powell,** sells at R2 or R3 a maund in October and November; *Sirki* is the light THATCH used for covering carts in wet weather, and is composed of the *til* or upper portion of the FLOWERING STEM, the lower and thicker parts called *kána* are used in the manufacture of CHAIRS, TABLES,

BASKETS, and SCREENS; also for roofing, for lining *kachha* wells, and for covering stores of grain. In the Jhelum District, when wood is scarce, *kána* is used for RAFTERS. FIBRE. Baskets. 17 Screens. 18 Rafters. 19

Fodder.—This GRASS is of too coarse a nature to be used for fodder except when quite young. In some of the Panjáb Districts, however, it is stated that during the cold weather the LEAVES of this grass are often the only pasturage for the cattle. They are also chopped up and mixed with *bhusa*, gram, oil-cake, or green stuff. In the early spring the grass is fired, and the young SHOOTS afford fine fodder for cows and buffaloes. In the Jhang District only the inferior patches are thus treated, as the plant seldom produces *munj-khána* after being burnt. According to **Coldstream** the young FLOWERING TOPS are regarded as good fodder for milch cows. FODDER. Grass. 20 Leaves. 21 Shoots. 22 Flowering Tops. 23

Domestic & Sacred.—Considerable confusion still prevails as to the particular species of **Saccharum** which should be regarded as having afforded the sacerdotal girdle. **Sir W. Jones** discusses the respective properties of the present species and of **S. spontaneum**, in a brief paragraph which will be found in the account below of that species. DOMESTIC. 24

Saccharum fuscum, *Roxb.; Duthie, Fodder Grasses of Northern India, 23.* 25

Syn.—ERIOCHRYSIS FUSCA, *Trin.*; MISCANTHUS FUSCUS, *Benth.*

Vern.—*Kilut, tilluk*, HIND.; *Khuri, pati-khori*, BENG.; *Kilik, tat, neja*, N.-W. P.; *Kandu-rellu gaddi*, TEL.; *Ikshwálika*, SANS. The Sanskrit name *ikshu* seems undoubtedly to denote the cultivated sugar-cane. It is somewhat curious therefore that this species should be called "the Sugar-cane—*wúlika* or *válika*," that is, thatch.

References.—*Roxb., Fl. Ind., Ed. C.B.C., 79; Voigt, Hort. Sub. Cal., 705; Elliot, Fl. Andhr., 81; Hackel in DC., Monogr. Phan., VI., 121; Drury, U. Pl., 371; Balfour, Cyclop., III., 466; Ind. Forester, VII., 179; Agri.-Horti. Soc. Jour. Ind., X., 358.*

Habitat.—Frequent on moist ground in Bengal and along the base of the Himálaya, as far as Kashmír. The flowering stems are 5 to 8 feet high.

Fibre.—The CULMS are used in the manufacture of pens, screens, and light fences; the LEAVES and REEDS, for thatch; and the LEAF-SHEATHS, like those of most wild species of this genus, may be used to supply the fibre from which the sacrificial thread is prepared. **Elliot**, in *Flora Andhrica, l. c.*, says:—"The best dark-coloured reeds with which the natives write are made from this species; *kandu* means black, scorched." FIBRE. Culms. 26 Leaves. 27 Reeds. 28 Leaf-Sheaths. 29

S. officinarum, *Linn.; Hackel, in DC. Monogr. Phanerog., VI., 112.* 30

THE SUGAR-CANE.

NOTE.—The reader may as well be warned that in the following attempt to give in this place the names that denote the plant as distinct from those for sugar and molasses, the author is conscious of the numerous mistakes that doubtless exist. Some of the names signify preparations of sugar, but they are often used by authors to denote the plant, and may, therefore, have both meanings.

Vern.—*Ukh, ganná, úk, ikh, nai shakar, rikhú, kumad*, HIND.; *Ik, ák, úk, kúshiar, púri, kullúa, kajúli, ganna*, BENG.; *Akh, ikshu*, SANTAL; *Tú*, NEWAR; *Ghenra*, PARBUTTIAH; *Uk, kakali chaku*, NEPAL; *Ukh, ukhi, ketári, khusiyár, katári*, BEHAR; *Aku*, URIYA; *Ikh, úkh, ikhari, úkhari, rikhu, ganna, puna-rikhu, kanthi-rikhu*, N.-W. P.; *Rikhú*, KUMÁON; *Shakar surkh, khand, ganna, kamánd, paunda, ikh*, PB.; *Kamand*, SIND; *Gándá, ús*, DECCAN; *Serdi, ús, gol*, BOMB.; *Usa, aos, ús, kabbu*, MAR.; *Sheradi, sherdi, sérdi, nai-sakar, úns*, GUZ.; *Karúmbú*, TAM.; *Cheruku, charki, kanupula-cheruku, lavu-cheruku* (a thick cane), *tillac-heruku, árukanupula-kránuga, cherukulo-bhedam, pottikanapu* (a short jointed cane), TEL.; *Khabbu, basari-mara*, KAN.; *Karinpa, tebu, karimba, tibu mirá*, MALAY.; *Keyán, kyán*, BURM.; *Uk*, SING.; *Ikshu, rusala, púndra* (a special variety of sugar-cane is denoted by the name *pundarika* which is the *cherukulo-bhedam* of the Telegus), *kanguruku*, SANS.; *Qasabus-sakar, kasib-shakar, kasabi-shakar*, ARAB.; *Nai-shakar*, PERS.; *Kánsia*, JAPANESE; *Tébu*, JAVA; *Fary*, MADAGASCAR; *Kan-ché* (S.-W. & Central), *chah-ché, tih-ché* (Canton), *Shih-mih, sha-t'ang* (sugar), CHINESE.

The reader will find a further enumeration of vernacular names under SUGAR below, and it need only be here repeated that, although most of the above denote the SUGAR-CANE, some of the names given mean simply SUGAR.

References.—*Roxb., Fl. Ind., Ed. C.B.C., 79; Voigt, Hort. Sub. Cal., 705; Kurz, For. Fl. Burm., II., 548; Dalz. & Gibs., Bomb. Fl., Supp., 99; Stewart, Pb. Pl., 260-262; Aitchison, Cat. Pb. and Sind Pl., 173; DC., Orig. Cult. Pl., 154-159; Graham, Cat. Bomb. Pl., 239; Mason, Burma and Its People, 505, 817; Sir W. Elliot, Fl. Andhr., 17, 37, 83, 107, 156, 175; Rumphius, Amb., 5, t., 74; Linn. Soc. Jour., XIX., 65; XXVIII., 197; Pharm. Ind., 252; Flück. & Hanb., Pharmacog., 649-655; U. S. Dispens., 15th Ed., 1254; Ainslie, Mat. Ind., I., 407; II., 460; O'Shaughnessy, Beng. Dispens., 638; Moodeen Sheriff, Supp. Pharm. Ind., 219; U. C. Dutt, Mat. Med. Hindus, 265, 300; Sakharam Arjun, Cat. Bomb. Drugs, 154; K. L. De, Indig. Drugs, Ind., 102, 103; Murray, Pl. & Drugs, Sind., 12; Bidie, Cat. Raw. Pr., Paris Exh., 93-94; Bent. & Trim. Med. Pl., 298; Smith, Contr., Mat. Med. and Natural Hist., China, 188, 207; Year-Book Pharm., 1871, 150, 169, 173, 200, 201, 204, 208, 212, 213, 214, 215, 261; 1872, 151, 152, 153, 156, 349; 1873, 465; 1874, 214, 252; 1875, 41, 49; 1876, 63; 1877, 139, 532; 1878, 42, 96, 97, 131, 158, 176; 1879, 77; 1880, 74; 1881, 116, 117, 118; 1882, 110, 111, 117; 1884, 107, 177, 208; 1885, 118; 1886, 42, 113, 114; 1887, 104, 107, 108, 141, 315; 1888, 33, 36, 106, 112; 1889, 31, 99; Watts, Dict. Chemistry, Vol. V., 464-474; VI., 1043-1046; VII., 1103-1110; VIII., Pt. II., 1833-1841; Bell, Chemistry of Foods, 97-114; Johnston (Church Ed., Chemistry of Common Life, 177-211; Agricultural Chemistry, 48, 388, 407; Johnson, How Crops Grow, 75, 76, 77, 78, 154, 156, 158, 338, 389; Birdwood, Bomb. Prod., 214, 250-253; Baden-Powell, Pb. Pr., 304-309, 383; Drury, U. Pl. Ind., 371-375; Atkinson, Him. Dist. (Vol. X., N.-W. P. Gaz.), 321, 692; Aitchison, Products of W. Afghanistan and N.-E. Persia, 199; Duthie & Fuller, Field and Garden Crops, 55-63; Fodder Grasses, N. Ind., 24; Useful Pl. Bomb. (Vol. XXV., Bomb. Gaz.), 185, 212, 277; Forbes Watson, Industrial Survey of India, 15, 86, 87; Royle, Prod. Res., 13, 67, 75, 85-94, 220, 231, 381, 393; Manual and Guide, Saidapet Farm, Madras, 36, 40; Liotard, Mem. Paper-making Mat., 14-15; Church, Food-Grains, Ind., 76; Kew Bulletin, 1888, 23, 294; 1891, 10, 35; Wallace, India in 1887, 223-230; Tropical Agriculture, 128-219; Symonds, Grasses, Ind. Pen., 50; Schrottky, Principles of Rational Agriculture applied to India, 230; Ordinances of Manu (Ed. Burnell & Hopkins), Lecture, VIII., 326, 341; IX., 39; X., 88; XI., 167; XII., 64; Ayeen-Akbary, Gladwin's Trans., I., 68-71, 79, 84, 89, 140, 341-342, 346-348, &c.; II., 27, 41, 44; Blochmann's, Trans., Vol. I., 63, 69-70; Linschoten, Voyage to East Indies (Transl. Burnell, Tiele & Yule), Vol. I., 95, 130; II., 56, 266; Milburn, Orient. Commerce (1813), Vol. II., 262-275; (1825), 305-307, 497; Buchanan-Hamilton, Jour. Mysore, Canara, etc.; also Statistics of Dinajpur, and Account of the Kingdom of Nepal; Hove's Tour in Bombay (1787), 17, 99, &c.; Colebrooke's Husbandry of Bengal, 126; Kirsop, some account of Cochin China, 1750; Observations on the Trade and Navigation of Great Britain, 1729; Stavorinus, Voyages to the East Indies, 1768; Staunton, Account of Lord Macartney's Embassay to China, 1793; M. de Guigne, Voyage to Pekin; Manilla and the Isle of France, 17[illegible]4; Henchman, Observations on the Report of the Directors of the East India Company respecting the Trade between India & Europe, 1801; Macpherson, Annuals of Commerce, 1805; Macpherson, History of the European Commerce with India, 1812; Edwards, History of the British West Indies, 1819; Young, West India Common-Place Book, 1807; Crawfurd, History of the Indian Archipelago, 1820; Botham, Observations on the mode of cultivating a Sugar Plantation in the East Indies; Marsden, History of Sumatra, 1811; Raffles, History of Java, 1817; Heyne, Historical and Statistical Tracts of India, 1814; Abel, Narrative of a Journey in the interior of China, 1816; Barrow, Travels in China; McCulloh, Sugar and Hydrometers; Porter, Tropical Agriculturist;*

Porter, Nature and Properties of the Sugar-cane; Pereira, Treatise on Food and Diet; Hassall, Food Adulterations; I. Bell, Culture of the Sugar-cane and Distillation of Rum (Calcutta, 1831); W. I. Evans, Sugar-planter's Manual; I. A. Lion, Manufacture and Refining of Sugar; T. Kerr, Cultivation of the Sugar-cane and Manufacture of Sugar; H. S. Olcott, Sorgho & Imphae, the Chinese and African Sugar-canes; D. M. Cook, Culture and Manufacture of Sugar from Sorghum; R. Niccol, Sugar and Sugar Refining; W. Crookes, Beetroot Sugar in England and Ireland; A. Voelcker, Cultivation and uses of Sugar-beet in England (Journ. Soc. Arts, XIX., 1871); F. Kohn, Methods of Extracting Sugar from Beet-root and Cane (Journ. Soc. Arts, XIX.; C. H. Gill, Manufacture and Refining of Sugar (Cantor Lecture Soc. Arts, 1872); Duncan and Newlands, the Alum process for purifying Sugar; J. Shier, Testing Cane-juice and the process of clarification; V. Drummond, Report on Production of Sugar from Sorghum; L. S. Ware, the Sugar-beet; W. G. Le Duc, Sorghum Sugar; I. H. Tucker, Manual of Sugar Analysis; R. H. Harland, Manufacture of Sugar from Sugar-cane; Grierson, Bihar Peasant Life, 232-237; Reports of the various Agricultural Departments, Experimental Farms, and Botanic Gardens; Indian Forester, 9th January 1886, 31st July 1886 9th October 1886; 1st October 1887; 9th June 1888; 5th January 1889; 16th February 1889, 24th March 1889, 22nd June 1889, 19th October 1889; Indian Agriculturist, numerous passages; Tropical Agriculturist, numerous passages; Produce Markets' Review; The Sugar-cane; Indian Agricultural Gazette (July 1885); South Indian Observer; Spons' Encyclopædia, II., 1830-1977; Encycl. Brit., XXII., 623; Balfour, Cyclop. Ind., III., 754-756; Morton, Cycl. Agri., II., 925-931; Ure, Dict. Indust. Arts and Manuf., III., 883; IV., 844; Smith's Dict. Econ. Pl., 396-397; Sugar Growing and Refining by G. Warford Lock and G. W. Wigner and R. H. Harland (1885); Sugar—A Hand-book for Planters and Refiners by G. Warford Lock, B. E. R. Newlands and I. A. R. Newlands (1888); A voluminous Official Correspondence from the Proceedings of the East India Company in the 18th Century down to 1891; Selection of Parliamentary Papers and Reports issued by the Board of Trade, etc., etc.

Many of the above works deal with sugar more than sugar-cane; but it has been thought desirable to give in this place the reference to all works of a general nature, and to reserve those of a more specific character for the various chapters of this article to which they more especially belong.

Habitat.—A strong cane-stemmed grass, from 8 to 12 feet high, which produces a large feathery plume of flowers; cultivated throughout tropical and sub-tropical Asia and the Islands of the Indian and Pacific Oceans. It is principally grown for its sugar; the expressed juice is boiled down, crystallised, and refined. The only mention of this plant having been found in a "wild" state in India is in the Transactions of the Agri.-Horticultural Society (*VI., Proc. 7*) where **Dr. H. H. Spry** is represented as having sent to **Dr. Wallich** "a small supply of sugar-cane procured from Car-Nicobar, where it grows in a wild state." This most interesting subject seems to have been overlooked. No modern botanist has recorded the occurrence of this plant in the Nicobars or anywhere else in India as an indigenous plant. (For an account of the cultivation see article **Sugar**, pp. 41-252.)

Sugar cane possibly wild in India. *Conf. with pp. 31, 32-33, 34, 49 57, 73-74, 76, 80, also regarding paunda pp. 7, 52, 64—66.*

Fibre.—The REFUSE of the sugar-cane mill has been recommended as a paper material (*Liotard*), and is said by **Stewart** to be sometimes made into well ropes and on the Chenab to be twisted into the rough CORDAGE used for tying the logs into rafts. The destruction of the fibre is one of the reasons why the Natives of many parts of India object to the improved iron rollers now very generally employed in the expression of the juice. It is somewhat surprising that the dried fibrous refuse is not universally employed as fuel in boiling the juice. In India this may be said to be only very occasionally utilised, the valuable fuel obtained from the sugar mills being thrown away as useless and what is even more surprising, it is in many cases not even used as manure.

FIBRE. Refuse. 31

Cordage. 32

Conf. with pp. 7, 8, 78, 110, 127.

SACCHARUM officinarum. The Sugar-cane.

MEDICINE. Sugar. **Medicine.**—In the *Materia Medica of the Hindus* compiled from Sanskrit authors, SUGAR and TREACLE are said to have been largely used from
33 **Treacle.** a very early age, principally for the purpose of disguising unpleasantly-
34 **Root.** tasted medicines. For medicinal purposes, old treacle is preferred to new. The ROOT also of the sugar-cane is said to have been employed in medi-
35 cine, and to have been considered demulcent and diuretic (*U. C. Dutt*). In Arabian works on Materia Medica, sugar is described as detergent and emollient, and is prescribed in doses of twenty direms. Many writers speak of it as attenuant and pectoral. It has also been supposed to have virtues in calculous complaints (*Ainslie*). In the Panjáb, **Baden Powell** says, sugar is considered by the Natives to be "heavy, tonic, and aperient, useful in heat delirium and disorders of the bile and wind." In another part of his work he remarks: "In cases of poisoning, by copper, arsenic or corrosive sublimate, sugar has been successfully employed as an antidote, and white sugar finely pulverised is occasionally sprinkled upon ulcers with unhealthy granulations. The Hindus set a great value upon sugar, and in medicine it is considered by them as nutritious, pectoral, and anthelmintic." The use of sugar as an antidote for arsenical poisoning is alluded to by many writers (*Chisholm, Voigt, etc.*).

Syrups. In European medicine sugar is employed for making SYRUPS, ELECTUA-
36 **Electuares.** RIES, and LOZENGES, and is regarded as useful not only for disguising the unpleasant taste of drugs, but also on account of the preserving influence it
37 **Lozonzes.** exerts over their active constituents. In India it is frequently employed in the preparation of PILLS The following statement of the European uses
38 **Pills.** of sugar in pharmacy may be reprinted here, since it summarises the facts generally given in works on Materia Medica:—

39 **Medical Uses.** MEDICAL AND PHARMACEUTICAL USES.—"The uses of sugar as an aliment and condiment are numerous. It is nutritious, but not capable of
40 supporting life when taken exclusively as aliment, on account of the absence of nitrogen in its composition. It is a powerful antiseptic, and is used for preserving meat and fish; for which purpose it possesses the advantage of acting in a much less quantity than is requisite of common salt, and of not altering the taste or impairing the nutritious qualities of the aliment. **Professor Marchand** has ascertained that a solution of sugar has no action on the teeth out of the body. It may hence be inferred that the popular notion that sugar is injurious to the teeth is founded solely upon the fact that the excessive use of sugar has a tendency to cause acid dyspepsia.

"The medical properties of sugar are those of a demulcent; and as such it is much used in catarrhal affections, in the form of candy, syrup, etc. According to **M. Provencal**, it acts as a powerful antaphrodisiac, when taken in the quantity of a pound or more daily, dissolved in a quart of cold water. For an account of the supposed therapeutic power of the vapour of boiling cane-juice, in bronchitis and incipient consumption, applied by living in a sugar-house, the reader is referred to the papers of **Dr. S. A. Cartwright**, of New Orleans, contained in the 47th and 51st volumes of the *Boston Med. and Surg. Journal*. In pharmacy, sugar is employed to render oils miscible with water, to cover the taste of medicines, to give them consistency, to preserve them from change, and to protect certain ferruginous preparations from oxidation. Accordingly, it enters into the composition of the compound infusion of roses, of several mixtures, pills, and powders, of many fluid extracts, syrups, confections, and of all the troches. Molasses is used for forming pills, for which it is well fitted, preserving them soft and free from mouldiness, on account of its retentiveness of moisture and antiseptic qualities.

"The influence of sugar in preventing changes in organic substances may be ascribed to an extraordinary osmotic power in its solutions, by

MEDICINE.

which infusoria and all other of the lower forms of life, to which fermentative processes are now generally ascribed, are almost instantly destroyed, the organism collapsing through the rapid exosmose of its fluids into the saccharine medium. All the different kinds of sugar susceptible of the alcoholic fermentation have this power" (*Dr. Louis, Mandl., Archives Gén. de Méd., 5e sér. XVI., 49, Juillet, 1860*). (*United States Dispensatory, p. 1261.*)

FOOD. Juicy variety eaten. 41 *Conf. with p. 52.*

Food & Fodder.—For an account of the extraction of sugar and of the by-products in the manufacture of that article, see **Sugar Manufacture** below. A thick JUICY VARIETY of sugar-cane is grown over almost the whole of India, which is largely used in the raw state as a sweetmeat. It is stript of its leaves, cut up into lengths of about 1 to 2 inches, and thus prepared may be seen exposed for sale in most of the bazárs throughout the country. The extent to which the cane is eaten does not appear to have been sufficiently taken into consideration in the estimates of yield of sugar from the acreage of cane. Indeed, in many parts of India, it may be almost said that cane is exclusively cultivated as a fruit. Thus of Montgomery it is stated that sugar-cane is very little cultivated for sugar-making, but is used simply as a pleasant article of food. Of Coimbatore and a few other districts the estimate has been made that the edible canes and seed-canes absorb about 10 to 15 per cent. of the total crop. It is probable that some such figure should be allowed for the whole of India; in other words, the area of sugar production should be accepted at 10 per cent. less than the actual area of the sugar-cane crop.

FODDER. Leaves. 42 *Conf. with pp. 5, 128.*

The LEAVES of the sugar-cane are employed as fodder. **Stewart** mentions that sugar-cane is occasionally grown without irrigation, the crop being used as *chari* for feeding elephants. The Financial Commissioner of the Panjáb (in a report dated 1883) says that in Sialkot the inferior crops are sometimes sold for fodder at R50 to R70 per acre, and in Multan at R100. **Mr. T. D. Macpherson**, writing of Bengal, says that the leaves, stripped from the canes, mixed with the crushed refuse obtained after the extraction of the juice, are given as fodder to cattle. A very similar statement is made of one district and another throughout India. Thus in the Karnal (Panjáb) we read that the "cane is cut down and dressed on the spot by stripping the leaves and cutting off the crown (*ganla*). These are given to the cattle to eat. In Ludhiana the flag which remains after cutting off the seed-joints is either given to the cattle to eat or is used as fuel for the boiling of the juice. But more direct references occur to the use of sugar-cane as a fodder. Thus, for example, of Gujranwala it is stated that a red coloured cane known as *chinkha* is "sometimes grown only as a fodder." The tops known as *bhadyas* are at Khandesh used to feed the cattle employed at the sugar-mill.

Mr. Benson (of the Saidapet Farm, Madras) furnishes the following instructive notice regarding the value of sugar-cane as a fodder :—

"In order to test the capabilities of the crop as a fodder-producer, an average row of canes was cut down in November; the canes weighed 1,162lb and the loppings 392lb; or, together, 1,554lb, equal to 131,624lb per acre, worth at least R290, so that it would have been far more profitable to have treated the crop as a fodder one, whilst, if the whole had been cut for fodder at the time the single row of canes was harvested, there would also have been a large saving in the cost of watering and weeding, and a large second crop would have been obtained. There can be no doubt but that sugar-cane as a fodder-producer is almost unequalled by any crop. Our municipalities, with their abundant supplies of manure, might find it worth while to grow sugar-cane as a fodder-crop; they might produce it in all favourable localities at R5 per ton, at which price it should meet

FODDER.

with a large demand for feeding milch cows and draught cattle" (*Saidapet Experimental Farm Manual and Guide*).

As having a bearing on the subject of the extended employment of sugar-cane as a fodder, it may be stated that many writers on the subject of the advantages of sugar or molasses as a fattening article of diet maintain that it has another property, and one highly injurious, *viz.*, it tends to render the breed sterile, both male and female. The reader will find interesting particulars on this subject in the Journals of the Royal Agricultural Society of England.

DOMESTIC. Torches. 43 Ropes. 44 Mats. 45 Chairs. 46

Domestic and Sacred.—The refuse cane (after expression of the juice) is sometimes dried and utilised as TORCHES by the Natives of the central parts of the Panjáb where the strips are called *pachchian.* At other times they are twisted and made into ropes, mats, or chairs. Owing to these uses of the refuse, objection is sometimes raised to the iron roller mills as breaking up the cane to such an extent that the fibre is valueless. The refuse or "megass" is very generally used as fuel to boil the juice, and all too rarely is it employed as a manure.

Votive offering. 47

In its unrefined state sugar is used as a VOTIVE OFFERING by the Hindus at the shrine of their gods. It is given by inferiors to superiors as a mark of respect. The cultivated plant cannot be said, however, to be held in the same veneration as the wild *Sara* or *Kasa.* While the plant is not worshipped as an emblem of the gods, every operation in cultivation and manufacture is governed by very pronounced religious observances, and the ultimate product holds a high place in the esteem of the Hindu. The bow of *Kámadeva* (the Indian Cupid) is sometimes represented as made of sugar-cane, at other times of sweet-smelling flowers. In either case the string is composed of bees. His five arrows are each tipped with the blossom presented to *Kámadeva* by *Vasanta* (Spring). **Sir W. Jones** translates a passage on this poetic conception as follows:—

"He bends the luscious cane, and twists the string
"With bees; how sweet! but ah! how keen their string,
"He, with five flow'rets tips the ruthless darts,
"Which through five senses pierce enraptured hearts."

The intimate association of sugar-cane and sugar with the Hindu religion has been urged (in the historic chapter below) as justifying the belief that the cane, if not a native of India, has at least been cultivated in this country for a longer period than can be shown in connection with any other part of the globe. The Institutes of **Manu** make undoubted allusion to sugar-cane as well as to palm sugar, honey, and other saccharine substances. There is, therefore, no room for the suggestion that sugar-cane has recently been substituted in the religious observances of Hinduism for manna or honey. Such substitution, if it took place, must have been some 2,000 or 3,000 years ago. It has to be admitted, however, that the earliest allusions in the classic literature of the Hindus to sweet substances are such that it is impossible to determine what is actually meant. An interesting feature of some of the religious practices have obviously been inculcated with the object of regulating and guiding the cultivator of cane. Thus, for example, the almost childish superstition against the flowering of the cane has doubtless its origin in the observation that when allowed to flower the cane loses its sweetness and degenerates until such stems would probably prove valueless for the purpose of propagation. This would lead to the supposition also, that it was early found that propagation by means of seed was of no value in preserving the saccharine property of the stems. The flowering of the cane was therefore pronounced a very ominous occurrence. It was a funereal flower, foreboding death to

Flowering & seeding. *Conf. with pp. 9, 11, 44, 47, 61, 83-88, 109.*

SACRED.

whomever might chance to look on it. It is impossible in this place to find space for the very extensive series of passages that might be here quoted regarding the religious observances connected with cane culture and sugar manufacture. The two which here follow may be accepted as representative. In 1792 the Political Agent at Banares furnished a long and most instructive report on sugar-cane, from which the following may be specially given as indicating the religious observances:—

"The attachment of the Natives to their established customs and usages is well known, and on the present occasion it may not be improper to state some of the superstitious notions of the *ryots* respecting the cane, as it will tend to show that any improvement which may be attempted in the culture thereof, can only be effected by gradual steps and the most encouraging and lenient measures."

"The *ryots* consider the sugar-cane (and also the betel plant) in a sacred and superior light; they even class it amongst the number of their *deutahs*. The first fifteen days of *Koar* (or September), termed *Pitereputch*, are devoted by the Hindus to religious ceremonies and offerings, on account of their deceased parents, relations, and friends; such of them as have been bereft of their parents refrain from every indulgence during the said period, as being the season of mourning and mortification; and as they deem the performance of the higher rites of their religion (such as making offerings of sweetmeats, cloths, jewels, etc., in the temples of their several deities, and also the sacrifices denominated *Howm-Jugg*, etc.) a pleasure and enjoyment, those are likewise carefully avoided."

"The sacred appellation of the cane amongst the *ryots* is *Nag'bele*, and hence, for the reasons above stated, the immediate owners of the cane plantations sedulously refrain from repairing to or even beholding them, during the continuance of the *Pitereputch*. On the 26th of *Cateeck* (or October), termed by the *ryots Deauthan*, they proceed to the fields, and having sacrificed to *Nag'bele*, a few canes are afterwards cut and distributed to the Bramins. Until these ceremonies are performed according to the rules of established usage and custom, no persuasion or inducement can prevail upon any of them to taste the cane or to make any use whatever of it."

"On the 25th of *Jeyte* (or May), termed the *Desharah*, another usage is strictly adhered to. As it is usual with the *ryots* to reserve a certain portion of the canes of the preceding year, to serve as plants for their new cultivation, it very frequently happens that inconsiderable portions of cane remain unexpended after the said cultivation has been brought to a conclusion. Wherever this happens to be the case, the proprietor repairs to the spot, and having sacrificed to *Nag'bele* (as before stated) he immediately sets fire to the whole, and is exceedingly careful to have this operation executed in as complete and efficacious a manner as possible."

Conf. with pp. 8, 11, 44, 47, 61, 83-88, 109.

"The cause of this extraordinary practice proceeds from a superstitious notion of a very singular kind. The act is committed from an apprehension that if the old canes were allowed to remain in the ground beyond the 25th of *Jeyte*, they would, in all probability, produce flowers and seed; for the appearance of these flowers they consider as one of the greatest misfortunes that can befal them."

"They unanimously assert that if the proprietor of a plantation happens to view even a single cane therein which is in flower, the greatest calamities will befal himself, his parents, his children, and his property; in short, that death will sweep away most of the members, or, indeed, the whole, of his family, within a short period of time after his having seen the cane thus in flower."

SACCHARUM officinarum.

Religious observances.

SACRED.

"If the proprietor's servant happens to see the flower, and immediately pulls it from the stalk, buries it in the earth, and never reveals the circumstance to his master, in this case they believe that it will not be productive of any evil consequences. But should the matter reach the proprietor's knowledge, the calamities before stated must, according to their ideas, infallibly happen."

"In support of this belief, many of the most aged *zemindars* and *ryats* in this province recited several instances of the above nature, which they affirmed to have actually happened during their own time, and, moreover, that they had been personal witnesses to the evils and misfortunes which befel the unhappy victims of the discription alluded to. These superstitious ideas must have originated at a very distant period since they are now so firmly rooted in the minds of the *ryots* in this part of the country."

"As the new cane is in the strength of vegetation during the rains, or in the months of *Saween* and *Bhadoon* (July and August), the proprietors in many parts of this province carefully avoid repairing to, or viewing, their plantations during those months, lest a cane flower should accidentally strike their sight, and thus entail upon them those miseries, which they are fully persuaded, must speedily follow such a circumstance."

"The *ryots* have several other singular notions in regard to the cane; but the particulars I have already taken the liberty to enumerate will sufficiently show, that any measures which may be adopted for future improvement, in respect to the cultivation, etc., must be introduced with circumspection and care, and must hold out a more than ordinary degree of encouragement, otherwise it will be extremely difficult to overcome those prejudices and opinions, which have acquired so absolute an ascendancy over their minds, and which appear to have been entertained in this part of the country for ages past."

Improvement difficult. 48

The following passage may be given as illustrative of the agricultural practices of the people of the present day:—

"Rites and sacrifices are performed on the germination of the cuttings, at the *Naudurga* festival in September-October, and in the following month, to avert a disease (*sundi*) which affects the crop. But the most important ceremonial connected with its growth is the *Deothán* in the end of October. This, which celebrates the awaking of Vishnu after his slumber in the infernal regions, is to sugar-cane what the *Arwan* is to other crops—a sort of harvest-home. Before this day no Hindu will eat the cane, and even jackals are said to avoid it. But on the *Deothan* several stalks are cut, five being reserved by the owner of the crop, and five each distributed to the village priests and craftsmen. On a board named the *sáligrám* are daubed, with cowdung and clarified butter, the figures of Vishnu and his consort. On the same receptacle are set *úrd*, cotton, and other vegetable offerings; while around it, tied together by their tops, the farmer places his five cane stalks. A burnt sacrifice and prayers are followed by the elevation of the *sáligrám*. During this last process the women of the household repeat five times the following incantation:—

Conf. with p. 171, 172.

"Arise, Oh God! Be seated, Oh Lord!
Spread thy carpets, God of Gaya Gajadhar:
Sit on them, highest Rama of Kampil.
Arise, God, a thousand times arise."

All present then move round the *sáligrám*. The tops (*juri*) of the five cane stalks around it are severed, hung up to the roof-tree, and burnt on the arrival of the *Holi* festival some months later. At the moment declared auspicious by the presiding Brahman the reaping of the crop begins. "The whole village is a scene of festivity, and dancing and

Conf. with p, 124.

singing go on frantically. Houses are set in order, and marriages which have been suspended during the rains recommence" (*Bareilly Gaz.*)

Saccharum spontaneum, *Linn.; Duthie, Fodder Grasses, 25.* 49

Syn.—IMPERATA SPONTANEA, *Beauv.;* S. SEMIDECUMBENS, *Roxb.;* S. CANALICULATUM, *Roxb.*

Vern.—*Káns, kagara, kosa, kus, kás,* HIND.; *Kash, kás, khágrá, kashiya,* BENG.; *Káns, kánsa, kánsi,* N.-W. P.; *Rara, khagar,* OUDH; *Kásh, jasha, jhánsh,* KUMAON; *Káhi, káns, sarkara, kánh,* PB.; *Kash, kashi, káns,* RAJ.; *Khán, káhu, kháu,* SIND; *Khán, káns, padar,* C. P.; *Kagara,* MAR.; *Rellu-gaddi, verri cheruku, kaki veduru, kore-gadi, billu gaddi,* TEL.; *Thetkia kyn, thek-kay-gyee,* BURM.; *Kasá, kásha, khaggara,* SANS. **Roxburgh** gives *khurí* as the Bengalí name for his **S. semidecumbens,** and *kagara* for **S. spontaneum.**

References.—*Roxb. Fl. Ind., Ed. C.B.C., 79; Voigt, Hort. Sub. Cal., 705; Trimen, Sys. Cat. Cey. Pl., 106; Dalz. & Gibs., Bomb. Fl., 304; Stewart, Pb. Pl., 261; Aitchison, Cat. Pb. & Sind Pl., 172; Mason, Burma and Its People, 524, 816; Sir W. Elliot, Fl. Andhr., 27, 77, 164, 191; Sir W. Jones, As. Res., IV., 248; U. C. Dutt, Mat. Med. Hindus, 266, 304, 305; Murray, Pl. & Drugs of Sind, 12; Birdwood, Bomb. Prod., 320; Baden Powell, Pb. Pr., 513; Drury, U. Pl. Ind., 376; Atkinson, Him. Dist. (Vol. X., N.-W. P. Gaz.), 321; Useful Pl. Bomb. (Vol. XXV., Bomb. Gaz.), 237; Econ. Prod., N.-W. Prov., Pt. V. (Vegetables, Spices, and Fruits), 101; Liotard, Mem. Paper-making Mat., 10, 19, 66; Settlement Reports:—Panjáb, Jhang, 24; Central Provinces, Upper Godavery, 40; Jubbulpore, 85; Nursingpore, 57; Gazetteers:—Panjáb, Karnal, 19; Hoshiarpur, 14; Muzaffargarh, 26; Dera Ismail Khan, 11; Jhang, 19; N.-W. P., I., 85, 153; IV., lxxx; Mysore & Coorg, I., 68; Agri.-Horti. Soc. Ind.:—Jour., X., 110, 358; XIII., 315; XIV., 87; New Series, V., Pro. (1875), 56; Ind. Forester, IV., 168; V., 31; VII., 179; IX., 245; XII., 565, App., 23; Balfour, Cyclop. Ind., III., 467.*

Habitat.—A coarse, perennial grass, with long creeping roots; abundant throughout India and up to about 6,000 feet on the Himálaya. It varies in height according to the nature of the soil, and appears to be most at home in damp low-lying ground, where it throws up flowering stems often 12 feet in height. Being a gregarious grass, the snowy-white pubescence, which surrounds the base of the spikelets, renders it a conspicuous feature when in flower; this usually takes place soon after the rainy season is over. Owing to its vigorous root-growth it is a most difficult plant to eradicate from cultivated land. In many districts of Northern India, and especially in Bundelkhand, it has given much trouble to the farmers by its encroachment on arable lands. The best known remedy is to plough up the land and smother the roots with a vigorous rainy season crop. On the other hand, it is said that *káns* grass, after a certain number of years, will wear itself out and disappear. It is somewhat curious that, as with **Saccharum officinarum,** **Roxburgh** had not seen the ripe seed of this species.

Flowering. *Conf. with pp. 8, 9, 44, 47, 61, 83, 88 109.*

FIBRE. **Thatching Material.** 50

Fibre.—This grass is largely used as a THATCHING MATERIAL, and the LEAVES are manufactured into ropes, mats, etc.

Leaves. 51

FODDER. 52

Fodder.—*Káns* is a favourite fodder of buffaloes, and is also, when young, given to elephants. In the Jhang District this grass is very plentiful in the moist land adjoining the rivers, when it affords valuable pasturage for buffaloes, so much so, that the zamindárs of those parts affirm that if there were no *káns* there would be no buffaloes, and they consider it to be too valuable to be used for thatching. In the Rohtak District it is said to be good for horses, hence the proverb:—"*Káns* grass for the horse, a staff for a man," and it is also said to be relished by camels and goats (*Gaz., p. 15*). **Roxburgh,** however, says that **S. spontaneum** (a very common Bengal grass) "is so very coarse that cattle do not eat it, except while very young."

SACRED. **Sacred & Domestic Uses.—Sir William Jones** (in *Asiat. Res. l.c.*) says:—"This beautiful and superb grass is highly celebrated in the *Puranas*, the Indian God of war having been born in a grove of it which burst into a flame. It is often described with praise by the Hindu poets for the whiteness of its blossoms, which gives a large plain at some distance the appearance of a broad river." **Atkinson** (*Him. Districts*) states that in Kumáon the long-rooting curculi are substituted for the *kusha* grass in religious ceremonies by the local Brahmans. Native PENS are made from
Pens. 53 the flowering stems. In the Dacca District, **Taylor** (*Topog.*, 59) says, it is
Fuel. 54 one of the earliest plants to appear on newly-formed *churs*, and is chiefly used for fuel.

55

SUGAR AND SUGAR-CANE.

SUGAR, *Eng.*; SUCRE, *Fr.*; LUCKER, *Germ.*; LUCCHERO, *It.*; AZUCAR, *Span.*; ASSUCAR, *Port.*; SAKHAR, *Rus.*; LUKIER, *Polish*; LUKUR, *Hung.*; Σάκχαρ or Σάκχαρον, *Greek*; SACCHARUM, *Latin*; SHAKKAR, *Pers.*; SUKKAR or AS-SUKKAR, *Arab.*; SAKKARA, *Sans.*

SOURCES of SUGAR. 56

SOURCES OF SUGAR.

Ainslie very justly remarks that "the Hindus value sugar very highly; in its unrefined state it is offered at the shrines of their gods; it is presented by inferiors to superiors as a mark of respect; and is considered by the *Vytians* as extremely nutritious, pectoral, and anthelmintic." It may, however, be remarked that the sugar of sugar-cane is not alone the material so used, nor is it the only saccharine substance known to the Sanskrit authors. The allusions that exist, in the classic literature of the Hindus, to sugar often clearly distinguish the special forms; but it may be said that far greater detail is given regarding sugar-cane than any of the other sugar-yielding plants. It may serve a useful purpose to give here a brief enumeration of the chief Indian sugar-yielding plants. Fuller details regarding these will be found in this work, however, under their respective
SUGAR-YIELDING PLANTS. 57 headings, so that the most superficial account is all that need be here given. The following enumeration will suffice to convey some idea of the relative value of each plant in the sugar supply of India:—

1. **Acer Negundo.**
   The Sugar-maple of Nebraska.

2. **A. rubrum.**
   The Swamp-maple of Pennsylvania.

3. **A. saccharinum.**
   The Sugar-maple of the Northern States and of Canada.

   India possesses some 13 species of maple, but up to the present date none of these have been found to afford sugar. In Vol I., 67, will be found a brief note on the subject of maple-sugar, in which the recommendation is made that it might be of value to the inhabitants of some of the Alpine tracts of India to ascertain whether some of the better known sugar-yielding species could be cultivated on the Himálaya. **Dr. Aitchison** urged this subject to the favourable consideration of Government many years ago, but apparently the experiment has never been tried in real earnest.

4. **Agave americana.**
   The American aloe may be said to have been completely acclimatised in India, but the juice obtainable from it in its native country, is either not yielded in India in sufficient abundance, or the Natives of this country, have not been made acquainted with the full properties of the plant. Neither liquor nor sugar are made from it in India. See Vol. I., 135-136.

SUGAR-YIELDING PLANTS.

5. **Arenga saccharifera.**

This is the Sago Palm of the Malaya, etc., but it is reported to be found in Burma and in Orissa. In Java it yields from its sap a large amount of cane-sugar. The process of preparation pursued in that country was described in detail by **Dr. J. E. deVry**, and his account of it will be found in Vol. I., 303-304, of this work Apparently the plant is too scarce in India to be regarded as a source of sugar.

6. **Beta vulgaris.**

The Beet, though largely grown as a vegetable, to meet the European demand, is not utilized as a source of sugar in India. A special form, recognised by **Roxburgh** as distinct from the European species, under the name **B. bengalensis**, must have been early introduced into this country. It is grown by the Natives fairly extensively as a garden vegetable, the leaves being eaten. See the article in Vol. I., 448-450.

7. **Borassus flabelliformis.**

The Palmyra Palm (the Barl tree or *tal*) is largely cultivated in India. **Rheede** alludes to its being tapped in his day on the Malabar coast. The juice, *ras*, obtained on tapping the flower-stalk, yields a large amount of sugar. This palm affords much of the jaggery sugar of Madras, particularly in Tinnevelly; but in Western India it is more extensively employed in the preparation of a fermented beverage. **Buchanan-Hamilton** wrote in 1807 that in Mysore the jaggery from this plam was more esteemed than that from the date. In Bengal it is rarely if ever tapped at all, or at least it is not utilized as a source of sugar. It is apparently largely used, however, as a source of sugar in Burma. The reader will find much useful information on Burmese palm sugar in the *Journal Agri.-Hort. Soc. India, X. (Old Series), 43-50.* (*Conf.* with the account given in Vol. I., 497-500, of this work.)

8. **Caryota urens.**

This is the Sago Palm of India. It is tapped for its juice very much after the same fashion as is pursued with the palmyra, and is the chief source of palm-sugar in Southern Ceylon. Though sugar can be, and doubtless is, made from the tree, wherever it occurs in India (such as in Orissa), the extent to which it is so utilised is relatively unimportant. In the Bombay Presidency, however, sugar is more extensively used from it than from the palmyra. (*Conf.* with the article in Vol. II., 208.)

9. **Cocos nucifera.**

The Cocoa-nut Palm is, perhaps, more extensively employed in Madras, as a source of sugar, than in other parts of India. In this respect it may be said (conjointly with **Borassus flabelliformis**) to take the place in South India (except Mysore) of the date palm in Bengal. (*Conf.* with the article in Vol. II., 452-454.)

10. **Manna.**

In Vol. V., 165-167, will be found a brief review of some 13 or 14 plants which in India are known to exude saccharine matters. These cannot well be viewed as sources of sugar, but they enter like honey perhaps more extensively than sugar itself into the pharmaceutic preparations of the Hindus, and from that point of view are important.

11. **Melia Azadirachta.**

The Neem Tree is known to afford a saccharine juice, from which sugar may be prepared; but it appears to be employed medicinally only, and cannot, therefore, be viewed as a source of sugar.

SUGAR-YIELDING PLANTS.

12. Phœnix sylvestris.

The Common Indian Date Palm is perhaps the most important source of Palm sugar in India. It is very extensively grown in Eastern Bengal for that purpose, and is also to be met with in Madras and Bombay. In Mysore it appears to be more important than either the palmyra or the cocoa-nut. **Robinson** wrote a prize essay on palm-sugar for the Agri.-Hort. Soc. Ind.—*see* their Journal, Vol. X., 243-274. For further information consult Vol. VI., p. 209.

13. Saccharum officinarum.

This is the Common Sugar-cane of which there are in India many very distinct varieties, each with well-recognised properties.

14. Sorghum saccharatum.

This form of sugar-cane (commonly called Sorgho or Chinese Sorghum) has been introduced into India, but does not appear to be very extensively grown. It is, in fact, perhaps more largely cultivated as a fodder than as a source of sugar. It seems probable, however, that it may have afforded the sweet canes which were eaten in China prior to the introduction of the manufacture of sugar; but, as opposed to that idea, it may be added that botanists seem to think that Sorgho was originally a native of Africa. The more distinctly African form of the plant (**S. kaffrarium**), the Imphee cane, **Hackel** has at all events reduced to **S. saccharatum**. By agriculturalists they are, however, regarded as different, and it may be reiterated that **S. saccharatum** is the Sorgho or Chinese Cane and **S. kaffrarium** the Imphee or African Sugar-cane. Both are extensively grown in America, and the latter was introduced into India at a much earlier date than the former. *Conf.* with the article **Sorghum saccharatum** in this volume.

15. Zea Mays.

Many writers affirm that it was from the stems of this plant that the ancient Mexicans made their crude sugar. As partly supporting that view it may be added that sugar has been made from the stems of the Indian-corn. Thus, for example, **Mr. C. B. Taylor** of Palamow describes the method he pursued in 1843 to prepare molasses from it. He remarks, however, that he failed to crystallize any of the juice, but distilled some of it into rum (*Jour. Agri.-Hort. Soc. Ind., Vol. II (Old Series), 541*). More recently the subject has been discussed in India; and in America, at the present time, it is attracting considerable attention, but the difficulty to crystallize the sugar seems to be insurmountable.

58

OTHER SACCHARINE-YIELDING PLANTS.

A much more extensive list of Indian plants, that afford saccharine substances, might be drawn up than that given above—the sources of the chemical substance, cane-sugar. Most of the articles which might, however, be here dealt with are of greater interest as materials from which alcohol is or might be prepared than as sources of sugar. Some of them are by-products from other industries, and if utilizable could be obtained cheap and in great abundance. Foremost among these should be mentioned indigo sugar. The reader might find it useful to peruse the remarks on that subject (article **Indigo**) in Vol. IV., pp 444-446. It will be seen that the method of manufacturing indigo, presently pursued, is to cause fermentation in the steeping vat, whereby the *Indican*, extracted from the plant, splits into indigo-blue and indigo-sugar, the former substance being, by the fermentation, reduced to indigo-white, only to have, by an expensive

OTHER SACHARINE PLANTS.

process of oxidation, to be reconverted into blue indigo. The indigo-sugar is rejected as useless. So, in a like manner, a large amount of sugar is annually thrown away by the coffee planters, who fail to utilize the pulp of the berry. Jute cuttings may be reduced to a form of sugar and fermented and distilled into a spirit or whisky. The flowers of the *mauha* (**Bassia latifolia**) contain a large quantity of saccharine matter which, in the region where the tree occurs plentifully, is taken advantage of in the preparation of an alcoholic beverage. A writer in the *Transaction of the Agri.-Horticultural Society of India* [Vol III., 173 (1836)], recorded his having failed to produce crystallizable sugar from the *mauha*; but he urged that since beet was being so used in Europe, the question of utilization of *mauha* flowers as a source of sugar should be investigated by the chemist. This subject is again dealt with in the Journal, Vol. VII. (New Series), Proc., 1884, p. lxxxv. Later on **Messrs. Turner, Morrison & Co.** furnished an analysis of *mauha* sugar. It contains 67·2 per cent., so they found, of glucose, but not even a trace of crystallizable sugar. In addition to the Manna-yielding plants (which need not be here dealt with) there are other sources of saccharine matter obtained through the instrumentality of insects. The Indian traffic in honey, for example, is very extensive. *Halwa*—a sweetmeat of camels' milk and honey—is largely imported and sold throughout the country. Many fruits also are well known to afford peculiar sweet beverages (*sherbets*), or to be capable of distillation into liquors. Among the latter may be mentioned the Pine-apple, the **Eugenia, Grewia, Opuntia,** and many others. The *Soma* or *Homa* of the ancient Sanskrit writers was very probably a flavouring ingredient, in its action similar to hops, which was used along with a maltose fermentation (see **Ephedra**), *Vol. III., 246-251.*)

THEORY of FORMATION. 59

THEORY OF THE FORMATION OF SUGAR.

The above enumeration is by no means an exhaustive one of all the plants which in India are known to afford sugar or saccharine matter. The formation of sugar-cane within the tissues of plants is one of the most obscure problems of vegetable economy. Sugar is a carbo-hydrate, that is to say, a compound formed of carbon, hydrogen, and oxygen. The last two elements are present in the proportions in which they exist in water, hence the term a carbon hydrate. Starch, from the botanical point of view, might be said to be the simplest or primitive carbo-hydrate, because all the known members of the series of such compounds can be expressed as derivatives from starch. The departures in their composition from that of normal starch are recognised as having been brought about by obscure functional changes which have hitherto escaped the chemist's methods of experimental determination They are vital modifications, the starch being reduced to one condition or raised to another, according to the varying requirements of life. It is, perhaps, scarcely necessary to explain that all the starches formed by plants, are elaborated in the leaves. The carbonic acid absorbed from the air may, roughly speaking, be said to enter into chemical union with the water brought up from the roots, oxygen being eliminated. The assimilation of carbonic acid and the formation of starch is thus so far effected independent of the food materials drawn from the soil by the roots. But of course the activity of the process will depend greatly upon the vigour of the plant, and hence a starch-yielding crop, such as the potato or the sugar-cane, may be a very exhausting one, although the product for which it is grown is primarily derived from the air. From starch all the solid materials of the plant are built up,—cellulose is simply a special modification of starch. But starch, as it is known to the chemist and as it is presumed to be formed through the assimilation that takes place in the leaves,

FORMATION of SUGAR.

is a substance of only partial solubility. To be carried from the leaves to the growing parts of the plant, it must, therefore, be rendered soluble. The exact method by which that change is effected has never been satisfactorily explained. A somewhat similar phenomenon occurs in germination. The starch stored up in the seed is rendered soluble through the action of the ferment termed *diastase*. The reader will find certain particulars of the changes that take place in germination under the article **Malt Liquors**, *Vol. V., 131-136*. The process of malting may be said to be an arrested germination. The action is allowed to proceed so far only as to effect the change of the starch into a form of sugar which is often designated *maltose*. By further fermentation and destillation alcohol is obtained from the *maltose*, an additional change taking place, which could not be accomplished on starch direct. It is believed that *diastase* is present in the sap of many plants, that is to say, that it is not confined to the germination of the seed. But *maltose* is by no means frequently found in plants. The soluble carbo-hydrate most abundant is *glucose*, or as it is often called *dextrose*, or grape-sugar. It thus seems necessary to presume the existence, in the sap of the plant, of some other ferment than *diastase*, or some reagent, so to speak, which would produce grape-sugar from the starch formed in the leaves. This is met by the common belief that the action of dilute acids, on maltose (starch-sugar), is to convert that substance into glucose Dilute acids abound in plants, and if this supposed action could be confirmed by direct chemical experiment, the presence of glucose would be readily accountable. In the *Kew Bulletin* (February 1891), the formation of sugar, in the sugar-cane, is discussed, the object being to draw attention to the subject of possible improvement of yield. In that article the following passages occur:—" Leaving glucose for a moment, we may turn our attention to cane-sugar. While the former is a migratory product, destined to afford material for the building up of tissues, the latter, as **Sachs** correctly points out, is a " reserve-material " stored up for some future effort of growth on a large scale, such as the process of flowering. Yet it is singular that it is twice as soluble as glucose. Nevertheless glucose seems to be what may be called the sugar 'currency' of the plant economy, and cane-sugar only the 'bullion' or banking reserve. The botanist is quite clear as to what happens in a cane-sugar plant. This is **Sachs'** account—" Starch is assimilated in the leaves of the Beet; in the petioles it is found again in the form of glucose. This glucose now enters the growing and swelling root, and is transformed into cane-sugar in its parenchymas."

In the Journal of the Agri.-Horticultural Society of India, July 1890, **Mr. Criper** published the results of his analyses of various parts of the sugar-cane stems. These exhibit, as **Mr. Criper** says, "the gradual formation of sugar in the sugar-cane, at different periods of its growth." Summarising his results **Mr. Criper** adds, " From the above analyses it will be noticed—

1st.—That the top joints contain no cane-sugar in November when nearly ripe.

2nd.—The glucose is invariably present, being highest in September in the top joints, and lowest in November in the bottom joints, *i.e.*, when the cane is about ripe.

3rd.—The top joints contain about 10 per cent. more water than the bottom ones, and this ratio does not appear to alter during ripening.

4th.—The amount of water present is from 8 to 11 per cent. more in July than in November.

FORMATION of SUGAR.

Analyses of cane-juice at different periods gave the following results:—

	1st Analysis, August 31st.	2nd Analysis, September 29th.	3rd Analysis, December 10th.
Height of canes to commencement of leaves . . .	4½ feet.	5½ feet.	5½ feet.
To end of leaves . . .	9 „	10½ „	10½ „
Specific gravity of juice . .	1·037	1·04	1·071
Cane-sugar	4·25	8·00	16·00
Glucose	1·27	2·00	·31
Ash	·73	·78	·73
Albuminous matter . .	1·51	·89	3·25
Acidity	·16	...	...
Water	92·08	88·33	79·71
	100	100	100

"The rise in the amount of albuminous matter, and decrease in the glucose, between September and December, is particularly noticed. It appears probable that the plant organism effects the conversion of the glucose into cane-sugar by combination with the elements of water."

This may be so; but, so far as the writer can discover, the formation of cane-sugar is even more obscure than the conversion of starch into glucose. Its formation has never been practically demonstrated; it would be necessary to find out in what part of the plant and by what agency starch reduced to glucose was made again to combine with water, in order to form cane-sugar. **Icery** was of opinion that at first uncrystallizable sugar is formed and that *its* subsequent transformation into cane-sugar is due to the force of vegetation, and especially to the influence of light. In the *Kew Bulletin*, from which some of the foregoing remarks have been taken, this difficult problem is briefly touched upon. The experiments performed by **Brown & Morris** lead to the supposition that maltose is directly converted into cane-sugar. "We cannot avoid the conclusion," these authors say, "that transformed starch is absorbed from the endosperm by the columnar epithelium of the embryo, in the form of maltose, and that this maltose, by the more or less complicated metabolic processes of the living cells of the embryo, is rapidly converted into cane-sugar. We have been able to demonstrate in a very striking manner the ability of the growing tissues of the embryo to convert maltose into cane-sugar. This was done by cultivating the excised embryos of barley upon a solution of maltose, and determining the cane-sugar in the plantlets after such cultivation. Although under these circumstances cane-sugar may be found within the embryo, not a trace can be discovered in the culture medium itself, which we should expect if the maltose were converted by the action of any secreted ferment. When, on the other hand, embryos are grown upon solutions of *dextrose* (glucose) instead of *maltose*, no cane-sugar is formed in their tissue." **Brown & Morris** continue their discussion of this subject in order to demonstrate the ultimate destination of cane-sugar:—"The intimate connection, they say, between cane-sugar and starch in plants has been clearly shown of late years by several chemists. In the case of the tuber of the potato, the dependence of its reserve starch upon the previous existence of cane-sugar

FORMATION of SUGAR.

in the juices of the plant has been very well shown by **Aime Girard** (*Compt. rend*, *108* (*1889*), *602*). The same has been done for maize by **H. Lepley** (*Compt rend.*, *94* (*1882*), *1033*), and for wheat by **Balland** (*Compt. rend.*, *106* (*1888*), *1610*). In a series of experiments which we conducted a few years ago upon the barley plant, taken from the fields at various stages of its growth, we were able to satisfy ourselves that cane-sugar forms a large proportion of the sugars existing in the sap of the plant, and that this cane sugar disappears *pari passu* with the formation and accumulation of starch in the seed. It is doubtless in the form of cane-sugar and its products of inversion that the transference of carbo-hydrates in the grasses mainly takes place."

Conf. with pp 69, 269.

The article in the *Kew Bulletin* concludes the discussion, of the formation of cane-sugar (briefly reviewed above), by giving a practical turn to the investigation. "Cane-sugar in the sugar-cane," we are told, "as in the beet, is, as will be seen, the derivative of starch. This substance is the result of the putting together under the constraining action of solar activity of the materials of carbonic acid and water. In the field of nature the process will be most effectively carried on, and the result for the same expenditure in cultivation must be largest where the supply of solar activity is most abundant. All things being equal, the formation of sugar as a product of solar activity ought, in the tropics, to be more easily and cheaply accomplished than in temperate countries." But it may be assumed that the beet sugar-producing sections of the Continent of Europe will not much longer persist in the effort to foster certain of their agricultural interests by taxing their home consumption of sugar, seeing that by so doing they make a gift of cheap sugar to England and other countries. The beet sugar-producing nations may now be regarded as having gained the chief object they aimed at, namely, local production of sugar and a new branch of agricultural enterprise. The extraneous aid of bounties has doubtless forced beet sugar into markets it could not otherwise have reached and thus precipitated the expansion of the trade. The removal of the bounties would doubtless in certain directions curtail the beet sugar demand and thus restore to the cane industry a portion of its lost ground. The question is not, however, so much as might at first sight be supposed, one of greater yield. Even were it the case that beet afforded considerably less sugar to the acre of land than cane, it will very probably, as matters now stand, always pay to produce sugar in Europe, rather than to import it from foreign and distant countries. An entire restoration of the monopoly of sugar supply to the cane planters would, indeed, be undesirable and need scarcely be looked for. Beet root, it may be said, has educated Europe to the use of sugar; it has, in fact, expanded the demand as it has increased the facilities of production. Any serious tendency to return to the former price of sugar would be in favour of beet production, and therefore, were there no other considerations than these, it may safely be predicted that beet will very likely preserve the position it has attained, even were the bounties entirely removed. Until some new source of supply, or fresh discoveries in chemical science, disturb once more the balance of the sugar market, there will continue to be a large demand for, and a remunerative trade in, beet sugar. The advantages of tropical environment might be admissible as holding good were the plants grown in the cold and warm countries identical in their chemical and physiological properties, but it breaks down when a comparison is made between a temperate-loving and a tropical plant, regardless of their individual characteristics. But it may be asked—Has it been demonstrated that the difference in yield is so very great as to lend direct support to the argument advanced? Is the average yield of the forms of

FORMATION of SUGAR.

cane in the tropics very considerably greater than the average of the various kinds of beet in temperate countries? Most writers affirm that cane gives from 16 to 20 per cent., beet from 10 to 20 per cent. The yield from cane in Barbados has been returned at 2½ tons (5,600lb) of sugar, but "grey-neck" beet is spoken of as having given 8,333lb. These figures, even if correct, are, however, only individual returns—results always open to the charge of being luxurious cultures or garden not field produce The yields in one country or with any one particular form of cane or of beet are, however, of less consequence than the average of all the sources from which the world draws its supply. For example, the average commonly quoted for cane would be considerably lowered were the yield obtained in India and China to be returned along with the Colonial figures.

The future of the sugar trade is, however, very much more obscure than the hackneyed controversy of sugar bounties, or even the advantages or disadvantages of cane and beet. The formation of sugar in the living tissue of the plant is a problem regarding which the chemist is not likely to much longer rest satisfied with the assurance that it is due to solar activity. Indeed, it may be affirmed that the conversion of non-crystallizable into crystallizable sugar (when accomplished) will exercise a far greater influence on the sugar trade than was produced by the abolition of slavery, or has been attained by the beet manufacturers. And there are features in the cultivation of both cane and beet that point to the possibility of advances in this direction being attainable even by the cultivator of the stock. The progress made with beet cultivation has, in fact, prepared the way for further advances. Even in cane planting there are certain well-ascertained facts, such as the observation that the different races of sugar-cane grown on the same field and therefore under the same degree of solar activity yield different amounts of sugar. The fact that a given variety of sugar-cane will not produce the same amount of sugar when grown on different soils or under different systems of cultivation (such as the degree and nature of manure or the abundance or scarcity of water) should also be borne in mind. Similar observations have been recorded in regard to beet cultivation. Thus, for example, it is very generally stated that "the nature of the season exercises much influence on the composition of sugar beet, especially on its richness in sugar which may range from 10 to 20 per cent. (*Encyclop. Brit.*). Then, again, "the formation of the sugar is favoured not so much by a hot summer as by dry weather and unclouded sky during autumn; hence the root succeeds better in North France and North Germany, than in Central France and South Germany: hence also the prospects of remunerative culture in Canada and New Zealand, and the failure in Australia. Nothing is so conducive to heavy crops as an abundance of rain during the first two months' growth of the plant" (*Spons' Encycl., p. 1832*). Many Indian writers affirm that the canes of the sub-temperate tracts of India are richer in crystallizable sugar than the canes of the tropical, a fact opposed apparently to the theory of greater yield under higher solar activity. In this connection the reader might consult the remarks regarding the canes of Nepal and of Kangra (*pp. 66, 185*). It may be said that the crop cannot endure severe droughts, but too much water makes the juice thin and deficient in sugar. A saline soil produces the same result. It is thus probable the lower yield of India, as compared with other sugar-cane-growing countries, is largely due to these causes, *viz.*, overmuch water and saline soils. The writer has, however, failed to discover any very definite statements regarding the behaviour of Indian sugar-cane, but it would seem safe to conclude these remarks by the affirmation that the future success of cane, as opposed to beet sugar, must be towards the lowering of the price at which the article can be placed in the market. That object

FORMATION of GAR.

will best be attained by two separate series of improvements—the one directed towards increasing the yield of crystallizable sugar in the cane, and the other towards chemical and mechanical improvements to facilitate and cheapen the production of sugar from the juice. It may be said that much chemical and engineering skill has, for many years past, been bestowed on the subject of beet sugar production. The new facilities brought to light from time to time have been tardily adopted by the sugar-cane planters; and in perfect fairness it may be said that the sugar-cane manufacturers of the world as a whole have, relatively to the beet root producers, done little or nothing to better themselves. The successes of their rivals have been solely attributed to "the pernicious system of bounties" granted to beet sugar production. Without desiring to add another view to the voluminous controversy that has been thrust on the public, it may safely be said of the actual growers of beet root and the manufacturers of beet sugar, that, in certain respects, they deserve their success, and have almost earned rather than received the bounties they now enjoy The political aspects of the question are, however, entirely different. It is for the countries that issue bounty-protected sugar to decide whether the gain of a new branch of agricultural enterprise more than compensates for the taxation imposed on their own consumption of sugar. The operation of the bounty system has been briefly and pointedly stated, by the author of the article "Sugar," in the *Encyclopædia Britanica*, thus:—

Beet Sugar Bounties.

Conf. with pp. 39-40, 316

"The efforts of growers have been largely directed to the development of roots yielding juice rich in sugar; and especially in Germany these efforts have been stimulated by the circumstance that excise duty on inland sugar is there calculated on the root. The duty is based on the assumption that from 12½ parts of beet 1 part of grain sugar is obtained; but in actual practice 1 part of raw sugar is now yielded by 9·27 parts of root. Moreover, when the sugar is exported a drawback is paid for that on which no duty was actually levied, and hence indirectly comes the so-called bounty on German sugar. In 1836 for 1 part of sugar 18 parts of beet were used, in 1850 13·8 parts, in 1860 12·7 parts, and now (1887) about 9·25 parts only are required. In France, until recently, the inland duty was calculated on the raw sugar; hence the French grower devoted himself to the production of roots of a large size yielding great weight per acre, and had no motive to aim at rich juice and economical production. Many processes, therefore, have come into use in German factories which are not admitable under the French methods of working. But since 1884 the French manufacturers have had the power to elect whether duty shall be levied on the roots they use or on the raw sugar they make, and a large proportion have already chosen the former." It will thus be seen that with beet sugar production prosperity meant essentially progression both chemically and agriculturally.

60

MICROSCOPICAL STRUCTURE AND RATIONALE OF EXTRACTION OF JUICE.

Fluckiger & Hanbury give the following practical observations on these subjects:—"No crystals are found in the parenchyma of the cane, the sugar existing as an aqueous solution, chiefly within the cells of the centre of the stem. The transverse section of the cane exhibits numerous fibro-vascular bundles, scattered through the tissue, as in other monocotyledonous stems; yet these bundles are most abundant towards the exterior, where they form a dense ring covered with a thin epidermis, which is very hard by reason of the silica which is deposited in it. In the centre of the stem the vascular bundles are few in number; the parenchyma is far more abundant, and contains in its thin-walled cells an almost clear solution of sugar, with a few small starch granules and a little soluble albuminous

MICROSCOPICAL STRUCTURE.

matter. This last is met with in larger quantity in the cambial portion of the vascular bundles. Pectic principles are combined with the walls of the medullary cells, which, however, do not swell much in water (*Wiesner*)."

"From these glances at the microscopical structure of the cane, the process to be followed for obtaining the largest possible quantity of sugar becomes evident. This would consist in simply macerating thin slices of the cane in water, which would at once penetrate the parenchyma loaded with sugar, without much attacking the fibro-vascular bundles containing more of albuminous than of saccharine matter. By this method, the epidermal layer of the cane would not become saturated with sugar, nor would it impede its extraction—results which necessarily follow when the cane is crushed and pressed.

"The process hitherto generally practised in the colonies—that of extracting the juice of the cane by crushing and pressing – has been elaborately described and criticized by **Dr. Icery** of Mauritius. In that island, the cane, six varieties of which are cultivated, is, when mature, composed of *Cellulose*, 8 to 12 per cent.; *Sugar*, 18 to 21; *Water*, including albuminous matter and salts, 67 to 73. Of the entire quantity of juice in the cane from 70 to 84 per cent. is extracted for evaporation, and yields in a crystalline state about three-fifths of the sugar which the cane originally contained. This juice, called in French ***veson***, has on an average the following composition:—

Albuminous matters	0·03
Granular matter (starch)?	0·10
Mucilage containing nitrogen	0·22
Salts	0·29
Sugar	18·36
Water	81·00
	100·00

"The first two classes of substances render the juice turbid, and greatly promote its fermentation, but they easily separate by boiling, and the juice may then be kept a short time without undergoing change. In many colonies the yield is said to be far inferior to what it should be; yet the juice is obtained in a state allowing of easier purification. when its extraction is not carried to the furthest limit.

"In beet root as well as in the sugar-cane, cane sugar was only said to be present; Icery, however, has proved that in the cane, some uncrystallizable (inverted) sugar is always present. Its quantity varies much according to the places where the cane grows, and its age. The tops of quick-growing young canes yield a ***veson*** containing 2·4 per cent. of uncrystallizable sugar; 3·6 of cane-sugar, and 9·4 of water. Moist and shady situations greatly promote the formation of the former kind of sugar which also prevails in the tops, chiefly when immature. Hence that observer concludes that at first the uncrystallizable variety of sugar is formed and subsequently transformed into cane-sugar by the force of vegetation, and especially by the influence of light. Perfectly ripened cane contains only $\frac{1}{75}$ to $\frac{1}{80}$ of all their sugar, in the uncrystallizable state" (*Pharmacographia by Flückiger & Hanbury*).

The writer has preferred to republish the above brief abstract, from the pen of one of the most eminent of authors, rather than to attempt a compilation of the extensive literature that exists on the subject. The technical reader who may desire more details would do well to consult the very able and elaborate article on sugar which will be found in ***Spons' Encyclopædia.*** Under the paragraph of references many works of special interest have also been mentioned. For more strictly scientific discussions on the microscopical structure of the cane, and the formation of sugar, a library of botanical and chemical works may be readily obtained.

CHEMISTRY. 61

CHEMICAL COMPOSITION OF CANE & CANE-SUGAR.

To give even the most elementary sketch of the chemistry of the saccharine substances, especially the practical bearings of the study on the sugar industry, would take many more pages than can be here devoted to the subject. In amplification, however, of the remarks which have already been made on the theory of the formation of sugar and the microscopical structure of the cane, the following brief passages may be republished:—"Formerly chemists called everything a 'sugar' which had a sweet taste, and acetate of lead to this day is known as 'sugar of lead' in commerce and familiar chemical parlance; but the term in its scientific sense soon came to be restricted to the sweet principles in vegetable and animal juices Only one of these—cane sugar—was known as a pure substance until 1619, when **Fabrizio Bartoletti** isolated the sugar of milk and proved its individuality. In regard to all other 'sugars,' besides these two, the knowledge of chemists was in the highest degree indefinite, and remained so until about the middle of the eighteenth century, when **Marggraf** made the important discovery that the sugars of the juices of beet, carrots, and certain other fleshy roots are identical with one another and with sugar of the cane. **Lowitz** subsequently showed that the granular part of honey is something different from sugar; this was confirmed by **Proust**, who found also that **Lowitz's** honey sugar is identical with a crystallizable sugar present largely in the juice of the grape. **Proust's** investigations extended to other sweet vegetable juices also. All those investigated by him owe their sweetness to one or more of only three species: (1) cane-sugar, (2) grape-sugar, (3) (amorphous) fruit-sugar. **Proust's** results obtain substantially to this day; a number of new sugars, strictly similar to these three, have been discovered since, but none are at all widely diffused throughout the organic kingdom.

"The quantitative elementary composition of cane-sugar was determined early in the nineteenth century by **Gay-Lussac & Thénard**, who may be said to have virtually established our present formula, $C_{12} H_{22} O_{11}$. Under FERMENTATION it has been explained how **Gay-Lussac** came to miscorrect his numbers so as to bring them into accordance with what we now express by $C_6 H_{12} O_6 = \frac{1}{2} C_{12} H_{24} O_{12}$. **Dumas & Boullay**, some years later, found that cane-sugar is what **Gay-Lussac & Thénard's** analysis make it out to be, while the 'corrected' numbers happen to be correct for grape-sugar. **Dumas & Boullay's** researches completed the foundations of our present science of the subject. 'Sugar' is now a collective term for two chemical genera named *saccharoses* (all $C_{12} H_{22} O_{11}$) and *glucoses* (all $C_6 H_{12} O_6$). Sugars are colourless, non-volatile solids, soluble in water and also (though less largely) in aqueous alcohol; from either solvent they can in general be obtained in the form of crystals. The aqueous solution exhibits a sweet taste, which, however, is only very feebly developed in certain species." "All sugars are liable to fermentative changes; a special character of the three principal vegetable sugars is that, when brought into contact as solutions with yeast (living cells of saccharomyces), under suitable conditions, they suffer vinous fermentation, *i.e.*, break up substantially into carbonic acid and alcohol. Dextrose and lævulose break up directly, thus $C_6 H_{12} O_6 = 2 C_2 H_6 O + 2 C O_2$. Cane-sugar first, under the influence of the soluble ferment in the yeast, gets inverted, and the invert sugar then ferments, the dextrose disappearing at a greater rate than the lævulose" (*Encycl. Brit.*).

CHEMISTRY.

The great property of the glucose carbo-hydrates as compared with the saccharose is their power of resisting the action of acids. If a saccharose such as cane-sugar (commonly called *sucrose*) be boiled with very dilute hydrochloric or sulphuric acids for example, it takes up water, and each molecule thereafter splits up into a molecule of *dextrose* and one of lævulose thus:—

$C_{12} H_{22} O_{11} + H_2 O = C_6 H_{12} O_6 + C_6 H_{12} O_6$ Cane-sugar + water = Dextrose + Lævulose.

For some years past the aid of the polariscope has been embraced in the analysis of sugar. Since the principle of that instrument and its application to the sugar industry exemplifies at once the difference between certain sugars, and manifests the properties of one of the most objectionable constituents of cane-juice, it may be here very briefly explained. If a polarised ray of light be made to pass through a medium, such as a solution of cane-sugar, the plane of incidence is seen to be different from that leaving the medium. A polarised ray may, in fact, be viewed as consisting of two circularly polarised rays, one of which becomes retarded in passing through a dense medium. That is to say, it is rotated to right or left. With the sugars the rotation varies both in regard to the angle and the direction. Cane-sugar, for example, turns the plane of the polarised light to the right, but if lævulose be employed, it is turned to the left—hence the name lævulose, or left hand rotatory glucose. The mixture of glucoses produced by the above decomposition of cane-sugar possesses the lævo-rotatory power or polarised light—hence the mixture has come to be called *invert sugar*.

The glucose group of sugars are not similarly acted on by acids; but, as has been already said, they break up directly into alcohol and carbonic acid with yeast. Before this result can be attained with cane-sugar it must first be inverted; but as acids are always present in the juice of the cane, the great danger in the sugar industry is the ease with which this inversion takes place. Invert sugar is the uncrystallizable portion of the saccharine juice. It is separated as molasses in the 'raw sugar' manufacture and as treacle in the hands of the refiner. So far as the production of crystalline sugar is concerned, it is a waste material which is most profitably disposed of by its conversion into rum. Chemistry has hitherto failed to effect the transformation of uncrystallizable to crystallizable sugar, though, as already briefly explained, this object is accomplished within the tissue of the plant. In the early state of the cane the percentage of glucose is very high; but as it matures, the lower portions of the stem get more and more sucrose and less and less glucose. But even within the living plant, the danger exists of invertion taking place. If the outer wall of the cane be injured, such as by being eaten by rats, ants, jackals, etc., and air be admitted, the heat of the atmosphere is sufficient to set up invertion, and fermentation rapidly follows. This danger is all the greater when the mature cane has been cut and is ready for the mill. Delay means imminent danger of the production of more invert sugar. The loss is not merely the loss of the quantity thereby reduced, but it has been ascertained that for every proportion of invert sugar present in a juice, a corresponding amount of cane-sugar may be said to be retained or cannot be entirely crystallized out of the mixture. All these dangers increases, therefore, tenfold when the expressed juice is retained for any time before removing from it the crystallizable sugar. From what has been said of the liability of cane-sugar to invert or become non-crystallizable through the instrumentality of dilute acids in the presence of heat, the advantage of processes of boiling at low temperatures will be readily appreciated. The most general method of preventing this evil is, however,

CHEMISTRY.

to add at once to the juice an amount of milk of lime sufficient to neutralize the acids it contains. In amplification of what has been said, the more direct chemical and physical characteristics of cane-sugar may be learned from the following brief statement :—

"Cane-sugar is the type of a numerous class of well-defined organic compounds, of frequent occurrence, throughout the vegetable and animal kingdoms, or artificially obtained by decomposing certain other substances; in the latter case, however, glucose or some other sugar than cane-sugar is obtained. Cane-sugar $C_{12} H_{22} O_{11}$ or $C_{12} H_{14} (O H)_8 O_3$ melts, without change of composition, at 160° C., several other kinds of sugar giving off water, with which they form crystallized compounds of the ordinary temperatures.

"Cane-sugar forms hard crystals of the oblique rhombic system, having a sp. gr. of 1·59. Two parts are dissolved by one part of cold water, and by much less at an elevated temperature; a slight depression of the thermometer is observable in the former case. One part of sugar dissolved in one of water forms a liquid of sp. gr. 1·33. Sugar requires 65 parts of spirit of wine (sp. gr., 0·84) or 80 parts of anhydrous alcohol for solution; ether does not act upon it"

"Cane-sugar is of a purer and sweeter taste than most other sugars. Though it does not alter litmus paper, yet with alkalis it forms compounds some of which are crystallizable. From an alkaline solution of tartrate of copper, cane-sugar throws down no protoxide. unless after boiling.

"If sugar is kept a short time in a state of fusion at 160° C., it is converted into one equivalent of *grape sugar* and one of *levulosan*; the former can be either isolated by crystallization or destroyed by fermentation, the latter being incapable of crystallizing or of undergoing fermentation. Cane-sugar, which has been melted at 160° C., is deliquescent and readily soluble in anhydrous alcohol, and its rotary power is diminished or entirely destroyed. It is no longer crystallizable. and its fusing point has become reduced to about 93° C., yet, before undergoing these evident alterations, it assumes an amorphous condition, if allowed to melt with a third of its weight of water, becoming always a little coloured by pyrogenous products. In the course of time, however, this amorphous sugar loses its transparency and reassumes the crystalline form. Like sulphur and arsenicous acid, it is capable of existing either in a crystallized or an amorphous state."

"If sugar is heated to about 190° C., water is evolved, and we obtain the dark-brown products commonly called *caramel* or *burnt sugar*. They are of a peculiar sharp flavour, of a bitter taste, incapable of fermenting and deliquescent. One of the constituents of caramel, *caramelane*, $C_{12} H_{18} O_9$, has been obtained by Gelis (1862) perfectly colourless. When the heat is augmented, the sugar at last suffers a decomposition, resembling, that which produces tar, its pyrogenous products being the same or very analogous to those of the dry distillation of wood" (*Flückiger & Hanbury*).

The briefest statement only of the chemical properties of sugar has been attempted above, the object having been kept in view to exhibit the features of the subject that have a direct bearing on the sugar industry. Of the substances that have been lightly touched upon, it may be said that *dextro-glucose* is manufactured on a large scale by the hydration of starch under the influence of diluted acids. It exists badly formed in cane-sugar and is found in fruits mixed with lævulose. Maltose is produced by the action of malt-extract or starch. Lævulose is formed from cane-sugar along with *dextro-glucose* by the action of diluted acids, the two sugars existing in equal proportions in what is known as invert-sugar.

Action of Acids. 62

ACTION OF ACIDS AND ALKALIES, ETC., ON SUGAR.

The main facts regarding the action of acids in decomposing cane-sugar and of the alkalies in forming definite compounds with it while they decompose grape-sugar have been incidentally alluded to above in more than one place. It seems, however, necessary to deal with this property of saccharoses and glucoses a little more fully. The mineral acids act differently on cane-sugar according as they are concentrated or dilute. Strong nitric acid, with the assistance of heat, converts it into oxalic acid. The same acid, when weak, converts it into *saccharic acid*, confounded by Scheele with malic acid. Concentrated sulphuric acid clears it. Diluted hydrochloric acid, when boiled with cane-sugar, converts it into a solid, brown gelatinous mass. Weak

CHEMISTRY. Action of Acids.

sulphuric acid, by a prolonged action at a high temperature, converts cane-sugar first into uncrystallizable sugar, afterwards into grape-sugar, and finally into *ulmin* and *ulmic acid*. Vegetable acids are supposed to act in a similar way. **Maumene** has found that cane-sugar undergoes the change into uncrystallizable sugar when kept for a long time in aqueous solution, as well as when heated with acids.

When the boiling with acids is prolonged for several days in open vessels, oxygen is absorbed, and, besides ulmin and ulmic acid, formic acid is generated. **Soubeiran** admits the change of the uncrystallizable into grape-sugar, but attributes it to a molecular transformation of the sugar, independently of the action of the acid, as, according to his observation, the conversion takes place only after rest. In confirmation of his views, this chemist states that he found the same changes to be produced by boiling sugar with water alone. Not only does cane-sugar change into the uncrystallizable when boiled with water, but, as clearly shown by an experiment of **M. E. M. Rault** in aqueous solution, under the influence of light, at ordinary temperatures it slowly changes into glucose, but this alteration does not take place in the dark (*P. J. Tr., Jan. 1872, p. 643*).

"Cane-sugar unites with the alkalies and some of the alkaline earths, forming definite combination which render the sugar less liable to change. It also unites with lead monoxide. Boiled for a long time with aqueous solutions of potassa, lime, or baryta, the liquid becomes brown, formic acid is produced, and two new acids are generated; one brown or black and insoluble in water, called *melassic acid*, the other colourless and very soluble, named *glucic acid*. Alkalies and alkaline earths are said to lessen the rotatory power of sugar in relation to polarised light; but the sugar recovers its power when the alkali is saturated (*Journ. de Pharm., 4 esér., IV., 314*).

"The account above given of the action of acids and alkalies on cane-sugar explains the way in which lime acts in the manufacture and refining of sugar. The acids naturally existing in the saccharine juice have the effect of converting the cane-sugar into uncrystallizable sugar, by which a loss of the former is sustained. The lime, by neutralizing these acids, prevents that result. An excess of lime, however, must be carefully avoided, as it injures the product of cane-sugar both in quantity and quality. The change in sugar which precedes fermentation, namely, the conversion of cane-sugar into the uncrystallizable kind, points to the necessity of operating on the juice before that process sets in; and hence the advantage of grinding canes immediately after they are cut, and boiling the juice with the least possible delay.

"Molasses is of two kinds, the West India and sugar-house. *West India molasses* is a black ropy liquid, of a peculiar odour, and sweet empyreumatic taste. When mixed with water and with the skimmings of the vessels used in the manufacture of sugar, it forms a liquor, which, when fermented and distilled, yields *rum*. *Sugar-house molasses* has the same general appearance as the West India, but is thicker and has a different flavour. Its sp. gr. is about 1·4, and it contains about 75 per cent. of solid matter. Both kinds of molasses consist of uncrystallizable sugar, and more or less cane-sugar which has escaped separation in the process of manufacture or refining, and gummy and colouring matter. When the molasses from cane-sugar is treated with a boiling, concentrated solution of bichromate of potassium and boiled, a violent re-action takes place, and the liquid becomes green; but if it be adulterated with only an eighth of starch-sugar molasses, the re-action is prevented, and the colour is not changed" (*Dr. G. Reich*) (*United States Dispensatory*).

Tests. 63

TESTS FOR THE PRESENCE OF SUGAR.—"Neither an aqueous nor an alcoholic solution of sugar, kept in large, well-closed and completely filled bottles, should deposit a sediment on prolonged standing (abs. of insoluble salts, foreign matters, ultramarine, Prussian blue, etc.). If a portion of about 1 gm. of sugar be dissolved in 10 c.c. of boiling water. then mixed with 4 or 5 drops of test-solution of nitrate of silver and about 2 c.c. of water of ammonia, and quickly heated until the liquid begins to boil, not more than a slight coloration, but no black precipitate, should appear in the liquid after standing at rest for five minutes (abs. of grape-sugar, and of more than a slight amount of inverted sugar)." "Cane-sugar may be distinguished from grape-sugar by **Trommer's** test, which consists in the use of sulphate of copper and caustic potassa. If a solution of cane-sugar be mixed with a solution of sulphate of copper, and potassa be added in excess, a deep blue liquid is obtained, which, on being heated,

CHEMISTRY.

Tests.

lets fall, after a time, a little red powder. A solution of grape-sugar, similarly treated, yields, by heat, a copious greenish precipitate, which rapidly changes to scarlet, and eventually to dark red. **Prof. Bottger** finds that, when a liquid containing grape-sugar is boiled with carbonate of sodium and basic nitrate of bismuth, a gray coloration or blackening of reduced bismuth is produced. Cane-sugar, similarly treated, has no effect on the test. **Dr. Donaldson's** test for sugar in the ***animal fluids*** is formed of 5 parts of carbonate of sodium, 5 of caustic potassa, 6 of bitartrate of potassium, 4 of sulphate of copper and 32 of distilled water. A few drops of this solution being added to an animal fluid, and the mixture heated over a spirit-lamp, a yellowish green colour is developed, if sugar be present. **J. Horsley's** test for sugar in diabetic urine is an alkaline solution of chromate of potassium, a few drops of which, boiled with the urine, will make it assume a deep sap-green colour. **Mr. J. Nickles** points out in the tetrachloride of carbon, obtained by decomposing carbon disulphide by chlorine and aqueous vapour, a new test for distinguishing glucose and cane-sugar. This test mixed with cane-sugar in a glass tube, kept for some time near 100°C. (212°F.), causes a darkening of the sugar, gradually increasing till it becomes black. Glucose undergoes no such change" (*Journ. de Pharm*, *4 sér.*, *III.*, *119*) (*United States Dispensatory*).

Estimation.
64

Estimation of Sugar.

"Cane-sugar does not precipitate the sub-oxide of copper from alkaline solutions of cupric tartrate, but it is very readily converted by boiling with dilute acid into invert-sugar, which does possess that property. Advantage is taken of this fact in what is generally called '**Fehling's** test.' A solution is made by dissolving 86 grammes of tartaric acid in crystals with 104 grams. of caustic soda. To this is added 29 grammes of sulphate of copper dissolved in water. The bulk is then made by additional water to 1 litre. This is **Fehling's** solution, and in its application for the estimation of sugar it may be used either 'volumetrically' or 'gravimetrically,' in either case, it is necessary, in the first place, to have a standard. In the volumetric process, which is the easier, ·625 gram. of pure cane-sugar is for this purpose boiled for ten minutes with about four ounces of water acidulated with 5 drops of concentrated sulphuric acid. The solution is then cooled, neutralised with solution of caustic soda, and made up to a bulk of 250 cubic centimetres. Twenty-five cubic centimetres of the copper solution are then heated in a white glass flask to the boiling point, and the sugar solution is run off into it from a burette, care being taken not to add more than will reduce the whole of the copper. It will generally be found that 40 c.c. of the sugar solution, which correspond to ·1 gram. cane-sugar or ·105 gram. glucose, will be required to reduce the copper or decolourise 25 c.c. of copper solution. If more or less than 40 c.c. are required, a corresponding difference will have to be made in the quantities of cane-sugar and glucose represented, respectively. This result is applied in the examination of saccharine substances or solutions in the following way: If a known weight—say 8 grams.—of a liquid which contains glucose and cane-sugar be taken and made up to 250 c.c., and if it be found that 45 c.c. of this diluted solution are required to reduce the copper in 25 c.c. of **Fehling's** solution, the per centage of glucose is thus found:—

$$\frac{250 \times 100 \times \cdot 1}{45 \times 8} = 6\cdot 94$$ **per cent., the cane-sugar equivalent, or 7·30 per cent. glucose.**

CHEMISTRY.

Estimation.

It is then necessary to make a second experiment to find the total amount of sugar present. A less weight than before—say 4 grams.—is taken and boiled for four minutes with about 4 ounces of water and 5 c.c. of normal sulphuric acid to invert the cane-sugar. It is then neutralised with soda and made up as before to 250 c.c. at 60° F. (15·5° C.), and if it be then found that 50 c.c. of this solution are necessary to reduce the copper in 25 c.c. of **Fehling's** solution, the total sugar in the liquid, calculated as cane-sugar, is as follows:—

$\frac{250 \times 100 \times \cdot 1}{50 \times 4} = 12\cdot5$; and $12\cdot5 - 6\cdot94 = 5\cdot56$, the percentage of cane-sugar present" (*Bell, Chemistry of Foods*).

It does no seem necessary to here detail the gravimetric method, but it may be said that the standard is the quantity of cuprous oxide precipitated by a given quantity of sugar-solution. A further chemical method of determining the quantity of sugar present in a solution is based on the production of alcohol from it as compared with the similar results with a standard. The quantity of alcohol formed, or the loss of carbonic acid being either or both of them resorted to for the determination of the sugar. The formula for the conversion of cane-sugar into invert-sugar as also that for the further reduction of the mixed glucose to alcohol and carbonic acid have been repeatedly given so that the reader should find no difficulty in applying this method. In working this system of analysis much labour is saved by using **Gilpin's** tables. The most ready method and the one now very largely employed, however, is that based upon the behaviour of a polarised ray of light on being made to pass through a solution of sugar. The principle of this analysis has already been briefly explained, but it may be further exemplified by giving an example. "If a tube 1 decimetre long be filled with a solution of pure cane-sugar, containing 1 gram in every c.c. of fluid, it will rotate the plane of polarisation of 73·8 degrees to the right, and this is called the specific rotatory power of pure cane-sugar. Rotation is in proportion to the length of the tube, and the mass of substance possessing the rotatory power, water being quite neutral.

It follows, therefore, that if we take a solution, containing a decigram of pure cane-sugar in every cubic centimetre of fluid, the tube being the same length as before, we obtain a rotation of 7·38°. If we then take an impure cane-sugar, and make a solution such that it shall contain 1 decigram in every cubic centimetre of liquid, fill a tube, 1 decimetre in length, with such solution, and find the rotation to be 6·3°. We should, supposing no invert-sugar to be present, find the percentage of sugar by the following proportion: as 7·38 : 6·3 :: 100 : *x*. The rule for finding the specific rotation from the observed rotation is: Divide the observed rotation by the length of the tube, multiplied by the weight of sugar in each c.c. of liquid, 1 gram. being the unit of weight, and 1 decimetre the unit of length. Thus, if a solution, containing 0·150 gram. of sugar in every c.c. of fluid, has an observed rotatory power of 16° in a tube 2 decimetres long, the specific rotatory power would be—

$$(1)\ \frac{16}{2 \times 0\cdot150} = 53\cdot33^\circ$$

and if this were a cane-syrup, the percentage of sugar would be 73·8 : 53·33 :: 100 : *x*. But raw sugars generally contain more or less invert-sugar; and as glucose has a specific rotatory power of 56° to the right, while lævulose, at a temperature of 57·2 F. (14° C.), rotates 106° to the left. the specific rotation of invert-sugar at 57·2° F. must consequently be—

$$(2)\ \frac{106 - 56}{2} = 25^\circ$$

CHEMISTRY.

to left. If therefore, at the temperature of 57 2' F. we obtain a solution of sugar which produces a specific rotatory power of 67°, and we find by **Fehling's** test that it contains 4 per cent. of invert-sugar, we have the data necessary for estimating the cane-sugar. Let *a*=the percentage of invert-sugar by **Fehling's** test, *b* the specific rotatory power of the sugar examined, and *x* the percentage of crystallized cane-sugar,

$$(3) \text{ Then } \frac{100\,b + 25\,a}{73{\cdot}8} = x.$$

In the trade the percentage of crystallizable sugar is not regarded as the sole criterion of value. The percentage corresponding to the angle, given by the mixed sugars, is, what is called by sugar merchants, the percentage of crystallized sugar; and the percentage of ash, as well as the appearance of the sugar, is taken into account along with this indication in fixing the price. It should also be remarked that beet sugar contains very little invert-sugar, so little indeed that it is disregarded on the continent" (*Bell, Chemistry of Food*).

It may be here added that a rough and ready mode of ascertaining the relative amounts of crystallizable sugar and molasses' present in *khar* or *rab* —the raw material sold in India to the sugar manufacturer—was proposed by **Mr. J W. Laidlay** in the *Journ. Agri.-Hort. Soc. of India*. IV., 147-151. This was based on the ascertained specific gravities of pure sugar and molasses. The former he accepted as having the sp gr. of 1·2299 at 84° F. the latter 1·37. Mixtures of these two, he thought, could be ascertained by a set of scales and a small bottle to hold say 500 or 1,000 grains of water. Having ascertained exactly the weight of the water contained in the bottle, he directed that the bottle should now be filled with *khar* and the contents thrown into the water. When dissolved the bottle should next be filled with the solution, "when the weight of the solution divided by the weight of pure water which the phial will contain will give the specific gr vity." The factor thus obtained was next to be used with the tables drawn up by him in the nearest figure to which would show the amount of crystallizable sugar and molasses present. This system, **Mr. Laidlay** affirmed, would be useful when more accurate chemical methods were not attainable.

HISTORY. 65

HISTORY OF SUGAR.

Vernacular Names for the Preparations of Sugar.—U. C. Dutt informs us that "twelve varieties of sugar-cane are mentioned by Sanskrit writers, but these cannot be all identified at present. The products or preparations of the sugar-cane, as described by Sanskrit writers, are as follows:—

1. *Ikshu rasa*, or sugar-cane juice.

2. *Phánita*, or sugar-cane juice boiled down to one-fourth. It can be drawn out in threads.

3. *Guda*, or sugar-cane juice boiled to a thick consistence, that is, treacle.

4. *Matsyandiká* is sugar-cane juice boiled down to a solid consistence, but which still exudes a little fluid on draining.

5. *Khanda* is treacle partially dried or candied into white sand-like grains.

6. *Sarkara*, or white sugar.

7. *Sitopalá*, or sugar-candy.

8. *Gaudi*, or fermented liquor obtained from treacle.

9. *Sidhu*, or fermented liquor obtained from sugar-cane juice.

The extent of the knowledge possessed by the early Sanskrit writers on the subject of sugar-cane is thus abundantly exemplified, and it will be seen that some of the above names have accompanied the diffusion of the

HISTORY.

knowledge in this most important article of food pretty well over the whole civilized world. On one point only would it have been desirable to have had more precise information, *viz.*, the separate recognition of molasses and treacle which are not the equivalents of *guda* or *gúr*, but the liquid which drains from *gúr* as it cools or is isolated from *ráb* by the refiner. In other words, these syrups are the uncrystallizable sugar removed from the crystallized substance at two stages of its manufacture. This distinction has not been made, so far as the writer can discover, by any of the classic Indian writers

For the vernacular and other names of the various forms of sugar and preparations therefrom, the reader might perhaps consult the chapter on MANUFACTURES, p.

Philological Evidence. 66

PHILOLOGICAL EVIDENCE.—The persistency with which certain Sanskrit names appear and reappear in the various languages of India, argues for the knowledge in sugar having proceeded from a common centre. The Arabic *kand* is apparently derived from the Sanskrit *khanda* (candied sugar). The Bengali *gura* comes from the Sanskrit *guda*,—a word which **Dutt** says is mentioned by **Charaka and Susruta.** The English candy is, in a like manner, derived from *kand* or *khanda*, and in both languages it means the same thing, *viz.*, sugar crystallized into large pieces. **Flückiger & Hanbury** very properly point out that *gura* is an old classic name for Central Bengal, and they add "whence is derived the word *gula*, meaning raw sugar, a term for sugar universally employed in the Malayan Archipelago, where, on the other hand, they have their own names for the sugar-cane, although not for sugar." It is significant that in the great Bengal sugar-producing district—Jessor—there are to this day towns with large sugar refineries that bear names highly suggestive of their chief industry, such as Khajura (which might be rendered the town of date-palms) and Mágurá (the town of *gúr*) (*Conf.* with the detailed article below on the sugar manufactures and trade of Jessor, *p. 270*). So, in a like manner, a town near Broach, in Bombay, came to be known as Sakarpur on account of the good quality of sugar made there. **Dutt** views, however, the fact of the early Sanskrit medical writers having described crystalline sugar, as proving that the manufactured article, *at least*, took its origin in Northern India and not in Gaura. Modern historic records would seem to confirm this view, since Bengal appears to have learned much from China, and pure white sugar is to this day known as *chíní*. But the most striking name in the enumeration given above is perhaps the Sanskrit *sarkara*, which originally signified "grit" or "gravel," hence crystallized sugar. The root of the name *khanda* means to crush, and it may therefore be accepted that when first used, it denoted the expression of the cane-stems and not, as by modern usage, candied sugar. So also the Sanskrit name *guda* indicated "a mass" or "ball," and was applied to the thickening of the juice by boiling; in its purest meaning, therefore, it conveyed the idea of a sweet syrup rather than sugar. In later Sanskrit *gula* was used for raw or unrefined sugar, and *guda* became associated with superior qualities or with sweetmeats, thus:—*Guda-trina*, sugarcane: *guda-pishta*, a sweet-meat; *guda-misra*, a sweetmeat-cake; *guda sarkara*, refined sugar; *ganda*, prepared from sugar or distilled, *e.g.*, rum *Sarkara* appears to have given origin not only to the Arabic, Persian, Greek, and Latin classic names, but to the extensive assortment of words in the modern languages of India and Europe which are very nearly the direct equivalents of the English word sugar. The qualifying additions to the Sanskrit root often met with, indicate for the most part the form or colour, and the country from which obtained or the plant which yielded the saccharine fluid. It is significant that this should

HISTORY.

so uniformly be the case, for, had sugar been manufactured prior to this discovery of sugar cane, it seems probable that a greater number of specific names would have even now survived, in which little or no trace could have been found to Sanskrit roots. But that the Sanskrit classic authors could have been acquainted with the sugar-cane plant, as indigenous to their ancestral country, is quite as improbable as that they should have possessed definite information regarding the sugar-yielding palms. The Sanskrit medical works were all written in India, and that, too, during the closing centuries of the classic epoch.

Charaka wrote the oldest treatise extant on Hindu medicine, apparently in the Panjáb; and at about the beginning of the Christian era. **Susruta** compiled in Benares the first great Hindu work on surgery, during the time of the early Muhammadan conquests. But it seems probable that both these writers derived their knowledge of medical science from the Greeks, and their value as original Indian observers is thereby greatly minimised. But even allowing for such imperfections it may safely be accepted that the glossary of vernacular names* in this work denotes an intimate knowledge with the various forms and preparations of sugar which is considerably more than 2,000 years old. In the references to sweet substances scattered throughout the older Vedas, it is possible the allusion is more to honey and manna than to sugar. And this idea seems to have descended to about the middle ages, sugar being confused with honey. Even the Chinese first spoke of it as *shi-mi* or stone-honey.

* *Conf. with pp. 3, 12, 28, 252—255.*

Between the very earliest Sanskrit works and the medical treatises, however, there are several sources of information that carry the Hindu knowledge in cane-sugar considerably further back than can perhaps be shown for any other country. To take but one example – the *Institutes of Manu*: There are numerous passages that clearly deal with sugar and indeed mostly cane-sugar. Thus in Lecture VIII., No. 341, we read: "If a twice-born man, being on a journey, finds his provisions are exhausted and takes two sugar-canes or two roots from the field of another man, he ought not to pay a fine." The punishment, however, for the theft of *gúr* is laid down, as also that for stealing "stuff to cause fermentation." The stealer of *guda* would, we are told, be punished hereafter by becoming a *vagguda* (a species of bat). So again in Lecture X. mention is made of the crystalline form of sugar and of sweetmeats. The *Institutes of Manu* are older than the Sanskrit medical works, but unfortunately a difference of opinion prevails as to when they were framed **Sir William Jones** assigned them to the period 1,250 to 500 B.C., but so fabulous an antiquity is now universally rejected. **Burnell** appears to have gone to the other extreme when he attributed them to the period 1 to 500 A.D. The more generally accepted view is that they were begun two or three conturies before the advent of Christ, but that the first and twelfth lectures were not added until considerably after that event. It may thus be safely accepted that the reference to a field of sugar-cane, in the passage quoted, is fully 2,000 years old. The picture of the weary traveller helping himself with impunity to sugar-cane from the wayside field has a reality and vividness about it that recalls the associations and scenes of modern India. The lesson of the degree of appropriation of a neighbour's goods that would amount to punishable theft might have been taught by other examples. But the illustration justifies the inference that sugar-cane was as well understood as any other crop, and, indeed, that it was perhaps the one from which such petty thefts were likely to be made. It may thus be accepted that sugar-cane was very generally cultivated in India, during the time the Institutes were written. The author of that great work is believed to have been a

Sugar-cane in India 2,000 years ago.
67

HISTORY.

Panjábí, who wrote in and for the Deccan. If, therefore, sugar-cane cultivation originated in Bengal it would be necessary to allow at least 1,000 years for its perfection and diffusion to Northern and Western India. It is thus probable that sugar-cane has been cultivated in India for something like 3,000 years.

But in this connection, it may be added, that the early European writers who speak of Indian sugar deal with it as a product of certain palms. For the most part they visited the Western and Southern coast, so that it is probably safe to assume sugar-cane cultivation did not take a very prominent place in the agriculture of India until almost modern times.

Yule & Burnell, perhaps, gave to palm-sugar an undue prominence. They wrote, "It is possible indeed, and not improbable, that palm-sugar is a much older product than that of the cane." The writer was for some time of that opinion, but unless *guda* meant originally date-palm jaggery there is no other Sanskrit nor any specific ancient vernacular name or names for palm-sugar. The evidence deducible from language will be seen to mostly favour an Indian origin for sugar-cane, if, indeed, its home might not be justifiably narrowed into the eastern division. One Indian author speaks of having found sugar-cane in a wild state. Plants were obtained by **Dr. H. H. Spry** in 1837 from Car-Nicobar and handed over to **Dr. Wallich**. The writer is not aware of any subsequent allusion to this interesting collection, so that it cannot be said **Dr. Wallich** gave any authority for the report that the samples were those of a truly wild plant. On the contrary, neither **Dr. Kurz** nor **Dr. Prain** (who have botanised in the Andaman and Nicobar Islands) make any mention of having seen sugar-cane in these islands except as a cultivated plant. **Kurz**, however, specially alludes to **Saccharum spontaneum** as covering large tracts of the northern side of Car-Nicobar. This may, therefore, have been the wild sugar-cane alluded to by **Dr. Spry**. (*Asiatic Soc. Beng. Jour. 1876, p. 162.*) **DeCandolle** seeks, however, to extend the area of sugar-cane eastward from India to Cochin-China. He observes, for example, that **Loureiro's** allusion to it would support the belief that sugar-cane was possibly indigenous to Cochin-China. This botanical opinion **DeCandolle** regards as obtaining support from the fact that a Chinese author of the fourth century speaks of it as a sweet bamboo which "grows" in Cochin-China. **DeCandolle**, on the authority of **Karl Ritter**, further tries to strengthen that opinion by demonstrating that the vernacular names for it become diversified east and south of Bengal. Having first pointed out that forms of the Sanskrit names exist in Bengal, he adds: "But in other languages, beyond the Indus, we find a singular variety of names, at least when they are not akin to that of the Aryans; for instance: *panchadara*, in Telinga; *kyam* in Burmese; *mia* in the dialect of Cochin-China; *kan* and *tsche* or *tche*, in Chinese; and further south, among the Malays, *tubu* or *tabu* for the plant and *gula* for the product." It will, however, be observed that very nearly as striking a diversity exists in the vernacular names currently used within India itself; besides which the name given by **DeCandolle** as Telegu for the plant is mentioned by **Ainslie** and **Roxburgh** as that of a peculiar form of sugar. The word is probably derived from the circumstance that it denotes sugar refined by means of the aquatic weed (**Hydrilla verticillata***)—the *panchadub*—of Ganjam. *See p. 311.*

Sugar-cane possibly wild in India. *Conf. with pp. 5, 32-34, 49, 57, 73, 76, 80.*

Sir W. Elliot, who devoted much patient study to the Telegu names of plants, makes no mention of *panchadara*, nor does that name appear in **Dr. Moodeen Sheriff's** enumeration of the synonymy of sugar and sugar-

* In the provinces of India the following plants are all used for this purpose: **Hydrilla verticillata, Lagarosiphon Roxburghii** and **Vallisneria spiralis.**

HISTORY.

cane in Southern India. It may, therefore, be rejected as a word of minor importance, since in every district of India almost, special names exist for the forms of sugar locally manufactured. Such names are of necessity of less importance to those that have a generic significance—the direct synonyms of the English sugar cane or sugar. It has already been shown that the Malay name *gula* is very probably taken directly from the Sanskrit *guda* and the Bengal *gura* or *gúr*. But may it not be the case also that the Burmese *kyan* (or *kyam*), the Chinese *kan che*, and the Japanese *kansia*, are but survivals of the Sanskrit root *khand*, to cut or crush, from which comes *khanda*, candied sugar. **Dr. Montgomerie** (*Trans. Agri.-Horti. Soc. Ind., X, 15*) tells us that the Malay name *tebú* or *leah* means flexible, and that *tebú* or rather *tebu kahor* denotes the white mealy incrustation upon the joints. They are names for cultivated forms, and being descriptive, are probably modern. He further tells us (*p. 71*) that the Natives recognise the purple or red cane as *tubú malacca*, thus pointing out Malacca as the place of origin, but that they think the three light-coloured varieties, *viz.*, *tubú leah*, *tubú tilor* (=egg-cane) and *tubú kapur* have been introduced by the Buggese traders from the eastern islands, in which case, he adds, they may be varieties of the Otaheite cane. An expert who reported on the samples furnished by **Dr. Montgomerie** to the Agri.-Horticultural Society had previously affirmed that they were forms of Otaheite cane. *See p. 55.*

Sugar-cane in the Malay Peninsula
68

Here and there throughout India there are often, however, remarkable similarities in the names given to sugar-yielding plants. These may be nothing more than coincidences, but they are at least quite as striking as the dissimilarities occasionally noticeable in the names given in the languages and dialects of India for sugar-cane itself. Thus, for example, while the Sanskrit *ikshú* has survived in Bengal as *ák*, and in Hindustan generally as *úk*, the date-palm is in Tamil known as *ichan*. or *ishan*. and the jaggery prepared from its juice *ich-cha-vellam*. In Telegu the date-palm is known as *itá* and in Malayal as *inite*. But a more curious case may be mentioned in the very general name, throughout India, for **Calotropis gigantea,** the *ák* or *ákanda*—a plant which yields manna. It may also be added that it is noteworthy that some of the Sanskrit names, now restricted in their signification to the cultivated forms of cane or to the preparation of sugar, should reappear as the vernacular names of some of the wild species of **Saccharum.** Thus, for example, *sar* or *sarkara* is given to almost any species, and that, too, even amongst the aboriginal tribes, such as the Santals. It is common also to find a combination of the two chief names such as *sarkanda*, and *kanda* itself is by no means infrequently given as the vernacular of a wild species. There is one singular point in connection with the Sanskrit names, *viz.*, that *guda* is never given to any form of **Saccharum,** wild or cultivated. It is restricted in its meaning to the inspissated juice, or to sweetmeats.

wild species eaten.
69

The culms of all the wild species, more specially **S. arundinaceum** (the *adave cheruku* of the Telegu people) and **S. spontaneum,** yield a certain amount of saccharine fluid, and, in consequence, boys in Bengal may be seen eating them. There are, indeed, many considerations that might be viewed as giving importance to the idea which would naturally be opposed by botanists that it may be possible that one of these wild forms might have given origin to the sugar-cane, the cultivation of 3,000 years having destroyed its original specific characteristics.

The writer is not, however, prepared to place much faith in the value of philological evidence by itself, but, in all fairness to the opinions that have been advanced, it must be admitted that with no other Indian product can there be shown a greater uniformity in the modern Indian names nor a more extensive distribution for the Sanskrit ones, than is the case with sugar. Taking into consideration at the same time the fact that the classic literature of India furnishes a more ancient historic record of sugar-cane cultivation than has as yet been shown for any other country, there is little room for doubt but that the world is indebted to India, or at most to Southern Asia, for sugar-cane. Still further, it might be contended that the Indian methods of refining sugar, though primitive as compared with the modern European improvements, are purely indigenous and are conducted on the same principles as the more skilled systems now followed by the great sugar manufacturers and refiners. What might be characterised as the superior indigenous methods may have come to Bengal at the time the name *chíní* was given to pure white sugar. But the Natives of India, even to the present day, prefer a dirty crudely refined sugar to the purer article. **HISTORY.**

Botanical Evidence is by no means backward in lending confirmation to the idea of India having been the original home of the sugar-cane. The genus embraces some 8 or 10 well recognisable species. These are all natives of the tropical and sub-tropical regions of the Old World. India possesses fully half the total number, and these too are, in many parts of the country, remarkably abundant. The suggestion, has already been made that, perhaps, one of these wild species may prove the ancestor of the cultivated stock. Be that as it may, confirmation of the opinion that, India is the home of the sugar-cane plant, may be seen in the abundance of wild species of **Saccharum** in the country, which philology and history agree in pointing out as the possible birth place of the sugar-cane industry. It has, moreover, been spoken of by certain writers as "wild" in Car-Nicobar and as "spontaneous" in Malabar. **Botanical Evidence. 70** **Sugar cane possibly wild in India.** *Conf. with pp. 5, 30, 32, 33, 34, 49, 57, 74, 80.*

Historic Records.—Incidentally it has been remarked that the majority of the early European writers who speak of Indian sugar seem to be alluding to the jaggery of palm-juice. The word jaggery comes, however, from the Sanskrit *sarkara*, and in the part of India with which the early European travellers were most familiar (South India) palm-sugar is relatively, even at the present day, more important than in the other sugar-producing provinces. But the term jaggery is equally applicable to cane-juice, and there are many undoubted references both to palm and cane jaggery that carry the history of that substance back to the classic periods of Greece and Rome. The conquests of **Alexander the Great** seem to have facilitated and extended the knowledge of sugar. **Eratosthenes** (223 B.C.) speaks of the roots of a cane that were sweet to the taste both when eaten raw and boiled. **Lucan** (A.D. 65) refers to the sweet juice expressed from reeds. The *Periplus of the Erythrean Sea* (A.D. 54-68) tells us that the "honey from canes," the *σακχαρι* of the early classic writers (**Herodotus, Theophrastus, Seneca, Strabo,** etc.) was exported from Barygaza, in the Gulf of Cambay, to ports of the Red Sea opposite Aden. Unmistakeable reference is made to sugar-cane as cultivated on the shores of the Persian Gulf in the ninth century. The crusaders found it in Syria. One of the historians of that remarkable period (1108) says: "The crusaders found sweet-honeyed reeds in great quantity, in the meadows about Tripoli, which reeds were called *sucra*." **Sanutus,** who wrote of 1306, says that in the countries subject to the Sultan, sugar-cane was produced in large quantities, and that it was likewise carried to Cyprus, Rhodes, Sicily, and other places belonging to the Christians. Europe was thus indebted to the Saracens for the introduction of sugar-cane cultivation. Refined sugar is recorded by the Chamberlain of Scotland "to have sold then (1329) at about one ounce of standard silver by the pound." **Marco Polo** gives us, in the thirteenth century, particulars regarding the sugar of Bengal, the art of **Historical Evidence. 71** **Classic Period. 72**

HISTORY.

manufacturing which had long before been the object of an emissary to Bengal from the Chinese Emperor **Taitsung** (627-650). **Dr. Smith** tells us that in consequence of the information obtained thereby, sugar was designated *sha-t'ang*, the name of the dynasty being combined with the radical for food. **Ramusio** gives a different account. Speaking of **Fu-chau**, he says, "The people knew not how to make fine sugar (*succhero*); they only used to boil and skim the juice, which, when cold, left a black paste. But after they came under the Great Can some men of Babylonia" (*i.e.*, of Cairo) "who happened to be at the court proceeded to this city and taught the people to refine sugar with the ashes of certain trees" (*Yule & Burnell*). In 1516 **Barbosa** wrote of "the sugar of palms, which they call Yagara." Of cane-sugar (speaking very probably of Chittagong) he says that the people make "much and good white cane-sugar, but they do not know how to consolidate it and make loaves of it, so they wrap up the powder in certain wrappers of raw hide." It would appear that the art of making loaf-sugar was discovered by a Venetian in the fifteenth century. *Jagara* is mentioned by **Barros** as an export in 1553 from the Maldives. **Garcia de Orta** described in 1563 the preparation of palm "*jagra*" in Goa. Speaking of Cochin, **Cæsar Frederike,** four years later, alluded to the *giagra* then prepared from the cocoa-nut. **Linschoten** (1598) says, "Of the aforesaid *sura* they likewise make sugar, which is called *jagra*; they seeth the water, and set it in the sun, whereof it becometh sugar, but it is little esteemed because it is of a browne colour" (*p. 102*). May not the *sura* alluded to be sugar-cane? **Yule & Burnell** quote the above passage regarding *sura* among an enumeration of references to jaggery—mostly palm-sugar. But in several other places **Linschoten** deals with the subject of sugar. Thus (*in Vol. I., 95*) he says of Bengal the "sugar and other wares were to be had in abundance." The much vexed question of the confusion between sugar and *tábáshir,* made by early writers, is exemplified by **Linschoten.** Thus he says, "All along the coast of Malabar* there are many thicke rheeds, especially on the coast of Choramandel, which rheeds by the Indians are called *mambu,* and by the Portingales *bambú.* These *mambus* have a certain matter within them, which is (as it were) the pith, such as quilles have within them, which men take out when they make them pens to write: the Indians call it *sacar-mambu,* which is as much to say as sugar of *bambu,* and is a very medicinable thing, much esteemed and much sought for by the Arabians, Persians, and Moores that call it *Tabaxur.*" There can be no doubt from the above passage that *Tabashír,* and not sugar, is indicated. **Linschoten** was, however, perfectly familiar with the sugar-cane proper. He gives, for example (*Vol. II., 266*), a detailed account of it in connection with the Canary Islands, and he says in another place: "There are also over all India many sugar-canes in all places and in great numbers, but not much esteemed." The sugar-cane of the Canary Islands was in his time regarded as the best.

Conf. with p. 65.

But that the Natives of India were thoroughly familiar with every feature of the cane-sugar industry, at the very time **Linschoten** and certain other European travellers were publishing their observations and dwelling more fully on the subject of palm than cane-sugar, there cannot be two opinions. **Barbosa's** account, abstracted briefly above, shows this to have been the case; but fortunately we possess in the *Ain-i-Akbari* so detailed a description of the methods of cultivation, manufacture of all forms of sugar,

* May the confusion of *tabashir* with sugar have given origin to the report that sugar-cane was "spontaneous" on the coast of Malabar? *Conf. with p. 5.*

HISTORY.

and the distillation of spirits from it, as to place the subject beyond doubt. **Abul Fazl**, the historian of the reign of the **Emperor Akbar** (A. D. 1590), says: "Sugar-cane, which the Persians call *Naishakar*, is of various kinds; one species is so tender and so full of juice, that a sparrow can make it flow by pecking it; and it would break to pieces, if let fall. Sugar-cane is either soft or hard. The latter is used for the preparation of brown sugarcandy, common sugar, white candy, and refined sugar, and thus becomes useful for all kinds of sweetmeats. It is cultivated as follows: They put some healthy sugar-cane in a cool place, and sprinkle it daily with water. When the sun enters the sign of Aquarius, they cut off pieces, a cubit and upwards in length, put them into soft ground, and cover them up with earth. The harder the sugar-cane is, the deeper they put it. Constant irrigation is required. After seven or eight months it will come up." The above extract will suffice to illustrate the fact that about the time **Linschoten** discussed the subject of sugar-cane in so obscure terms as to leave room for doubt as to whether he alluded to palm or sugar-cane jaggery, every detail of the cultivation and manufacture was fully known, at least in Upper India, and we have reason to believe that this was the case in Bengal also. **Early records of Cultivation in India. 73**

The Spaniards carried the cultivation and manufacture of sugar-cane to the Canary Islands in the fifteenth century, but prior to that (1420) the Portuguese had conveyed it from Sicily to Madeira and to St. Thomas' Island. In 1506 it was taken from the Canary Islands to San Domingo. The Dutch first established sugar works in Brazil in 1580, but on being expelled from that country by the Portuguese, they carried the art of sugar manufacture (1655) to the West Indies. Sugar was manufactured by the English in Barbadoes in 1643 and in Jamaica in 1664. A spirited competition soon took place between the British and the French and Portuguese manufacturers. The British, by greatly improving and cheapening the manufacture, were able to undersell the Portuguese in Brazil. The trade was at that time free, but on the restoration of Charles II., importation into Great Britain was by various Acts, restricted to British subjects. By 1726 the French had so vastly improved their manufacture in San Domingo that they began to compete with the British in the supply of Europe, and a serious decline in the British imports from the West Indies accordingly took place. **Conveyed to West Indies. 74**

It will thus be seen that sugar production had spread from India to Europe, but more especially to the West Indies. The record of subsequent events recalls the similar migration, but return again of indigo. Towards the close of the eighteenth century civil disturbances in San Domingo ruined the French planters. A greatly increased demand arose in British West Indian sugar with a corresponding rise in the price. Raw West Indian sugar, of the worst description, then sold in Britain at nine-pence per ℔, and a memorial was accordingly addressed by the public to the East India Company, to lower the price by bringing Indian sugar to Britain in competition with that of the West Indies. In Milburn's *Oriental Commerce* the following passage occurs regarding this critical stage of the British sugar trade, a stage which may be viewed as the starting point of India's foreign traffic in the article. "The East India Company, from these considerations, as well as from having been publicly called upon to lend their assistance towards effecting a reduction of the price of sugar, gave every encouragement to the importation of it from the East Indies; and the vigorous efforts they made to relieve the public necessity, increased the cultivation of the sugar-cane in India to an amazing degree, and secured to the Bengal Provinces a participation in this important article of trade." In 1792 the English Legislature, with the object of guarding **French plantations ruined. 75**

HISTORY.

against a further rise in price, imposed restrictions on exportation. But this state of affairs did not last long. Apparently the exports from India had some time previously begun to tell powerfully, and an increased production in the West Indies had also been brought about. Accordingly, in 1807, a Committee of the House of Commons had actually to be appointed to consider the depressed state of the West Indian trade. It was shown that an alarming fall in the price of sugar had taken place (since 1799), and it was anticipated that unless some "efficient remedy" was early thought of, ruin to the West Indian planters would rapidly supervene. Various measures were considered, but none apparently put into force. The sugar trade and the West Indian interest were left to shift for themselves and be adjusted by natural causes. Among the suggestions offered by the Committee it was proposed to increase the consumption of sugar by introducing its use into distilleries. The imposition of a heavy duty on the Indian sugar had not the desired effect. Indian sugar had to pay an import duty in 1792 of £37-16-3*d*. per cent., while the West Indian sugar paid only £0-15-5 per cwt. Far from contemplating the removal of the entire duty on West Indian sugar, however, the Committee deplored its threatened loss, though they heartily sympathised with the West Indian planters in the ruin which then seemed about to overtake them. The Committee thus recommended no practicable cure for the distressing problem they were convened to solve. The duty on West Indian sugar amounted in 1807 to £3,000,000 and the return export trade was valued at £6,000,000. The Committee could not, they seemed to think, recommend the sacrifice of so important an item of the English revenue. Popular feeling was strong in England against sugar manufactured by slaves. Preference was given to the inferior article from India, because it was made by freemen. The position was a critical one, but greater dangers were foreseen than those connected with sugar.

Prohibitive duty on Indian sugar. 76

History of Indian Sugar Trade. 77

Turning now to the more immediate history of the Indian sugar trade, it may be said that its progress, no more than that of the West Indies, was devoid of fluctuations. Indeed, in 1776 the merchants of Calcutta memorialized the Government on the decline of trade and the consequent losses that had been sustained. "Even so late, the memorial explained, as the period immediately preceding the capture of Calcutta, in 1756, the annual exportation was about 50,000 maunds, which yielded a profit of about 50 per cent., and the returns for which were generally in specie; so that in the 20 years immediately preceding the capture, it may be estimated that there flowed into Bengal for this article no less than R60,00,000, which was all clear gain to the country, and of the most eligible kind, the production of the ground, manufactured by the Natives. And this flow was regular, always feeding, but never overcharging the circulation. During the past 20 years the price of sugar has been gradually increasing, and the exportation and growth diminishing in the same proportion, so that the price is now 50 per cent. more than it was before that period. The charge of transportation is also greater; and the price at foreign markets not having risen in the same proportion, the export is so trifling and casual, that the sugar-trade of Bengal is, in fact, annihilated. It may be even doubted if Bengal produces enough for its own consumption, since there is annually an importation from Benares, and of candied sugar from China, the amount of which will be found equal to that of the trifling export which yet continues."

Establishment of European Plantations in India. 78

"Supposing the recovery of the trade to be an object deserving attention, we submit to your consideration whether it be attainable by any other means than by encouraging Europeans, distinguished by their property, situation, and credit from ordinary adventurers, to undertake the cultiva-

HISTORY.

tion and manufacture of sugar after the method practised in the West Indies, by grants of unoccupied lands and other reasonable privileges. We admit that much will depend on the conduct of the first undertakers; but with proper management on their part, and a reasonable support from Government, we think success would be infallible, and that in a few years the Natives would follow the new method, which would thence soon become general throughout the country, as the Italian mode of winding raw silk lately introduced, now is" (*East India Sugar, Papers respecting the Culture and Manufacture, etc., 1822, pp. 12-16*). It may be remarked that this same proposal has been made on more occasions than one. Indeed, it will be seen, from the pages below, that the self-same suggestion has been offered by a London mercantile firm, and that it is at the present moment before Government for consideration. What was the result of the Calcutta merchants' effort to establish European sugar cultivation and manufacture in 1776? The Governor-General, we are told, readily complied with the request preferred, and "a grant of land was accordingly allotted, in which a sugar plantation was afterwards set on foot, but after repeated experiments upon the soil, it was found so universally infested with white ants, that the Society were obliged to drop their scheme." Before abandoning the effort, however, they gave up cultivation and endeavoured to manufacture sugar from the cane they could purchase from the neighbouring Native growers, and "they produced both refined sugar and rum; thus evidencing the practicability of their plan, though that mode of producing sugar, with other circumstances, made it inconvenient for them to persevere in it." But the Society (or Company as we should now call it) that tried and failed to introduce into Bengal the sugar industry on the plan of the West Indian plantations made certain important discoveries. They found that sugar-cane was the worst of all crops to put on newly cultivated land, because of the fact that white ants were very much more severe a plague than on old cultivations. It was, at the period here alluded to, believed by the refiners of Europe that Indian sugar-cane contained too little crystallizable sugar to be of any value for the market by the refiner. The experiments performed by the first Indian Sugar Company demonstrated what has been confirmed over and over again since, that certain soils or certain cultivated races of the sugar-cane, grown in India, were quite as rich in crystallizable sugar as the West Indian forms.

European Plantations. 79

Conf. with pp. 39, 48, 62, 63, 91, 93-94, 103, 161, 162, 212, 306, 309.

Destruction by white-ants.

Conf. with pp. 101, 125, 161.

80

Little progress seems to have been made in the Bengal sugar trade down to 1790; but in that year the duty which had hitherto been levied on the coastwise exports of Bengal to other parts of India was cancelled. It was thought that by doing so Bengal might be able to compete with China, Manilla, Batavia, etc., in the Bombay market. The Mahrattas are spoken of, in the records of the trade, towards the close of the eighteenth century, as great consumers of sugar. Bombay was recognised as the province in which cotton should be encouraged, and Bengal that for sugar. The interchange of these commodities was accordingly viewed as the natural course of trade between these provinces (*Papers on Culture and Manufacture of Sugar in British India, 1822*). **Mr. Bebb's** very enlightened action, therefore, in removing the sugar duty had an immediate effect. Cultivation of sugar-cane was greatly extended, through the profitable export market thus opened up within India itself, and once started, the popular turn in favour of Bengal extended. Exports were made to Flanders, America, and some other countries. At this period, also, the East India Company had recieved the memorial from the British public (above alluded to), and the prosperous state of the Bengal and North-West cultivation enabled the Company to assure the English Government, that if the heavy import duty, which they held had, by accident more than intention, been

Removal of the Indian transit duty. 81

HISTORY.

imposed on Indian sugar, were withdrawn and India placed on the same footing with the West Indies, they could "permanently supply a considerable quantity of sugar for the relief of Great Britain." But their request was not granted, and the high duty continued to be charged on Indian sugar until 1836. The cheaper rate, however, at which sugar could be produced in India as compared with the West Indies, enabled the Company to compete in the European market in spite of the heavy import duty their sugar had to bear. But a more remarkable feature soon became manifest. It paid foreign traders to purchase the Bengal sugar and to ship it to the European market, in competition with the West Indian re-exports from Great Britain to the Continent. Sugar was also conveyed from Bengal to America, and even to the West Indies, to be consigned from these countries to Europe as colonial sugar, and was thus admitted into England itself on payment of the lower import duty. It will thus be seen that the West Indian sugar received for many years a distinct bounty at the expense of India. As might naturally have been anticipated, the action of the British Government very seriously retarded, indeed curtailed, the Indian trade in sugar-cane. It is not to be wondered at, therefore, that many writers spoke, at the beginning of the present century, of its being completely ruined. Thus, for example, **Captain Thomas Williamson** (in his *East India Vade Mecum,* 1810) says: "Although the sugar-cane is supposed by many to be indigenous in India, yet it has only been, within the last 50 years, that it has been cultivated to any great extent . . . Strange to say, the only sugar-candy used until that time" (20 years before the date of his book) "was received from China; latterly, however, many gentlemen have speculated deeply in the manufacture. We now see sugar-candy of the finest quality manufactured in various places of Bengal, and I believe it is at least admitted that the raw sugars from that quarter are eminently good."

Sugar Trade of East India Company.
82 But the East India Company could not have been expected to long remain silent. The Court of Directors accordingly published in March 1822 a voluminous and comprehensive report on the whole question. From that report the following tables may be here extracted:—

YEARS.	IMPORTED BY GREAT BRITAIN.			HOME CONSUMPTION.	
	From British Plantations.	From Foreign Plantations.	From East Indies.	From West Indies.	From East Indies.
	Cwt.	Cwt.	Cwt.	Cwt.	Cwt.
1817 . . .	3,440,565	192,780	127,203	3,220,595	33,131
1818 . . .	3,563,741	105,916	125,893	4,151,239	27,059
1819 . . .	3,665,520	138,032	162,395	2,672,226	24,775
1820 . . .	3,785,434	86,048	205,527	3,283,059	99,440
1821 . . .	3,623,319	162,994	277,228	3,661,731	83,232

The report alluded to, also gives the exports from Calcutta by private traders, during the above years as follows:—

YEARS.	To England.	To other countries.
	Cwt.	Cwt.
1817	129,858	199,288
1818	129,195	254,930
1819	157,957	258,746
1820	134,613	146,234
1821	112,830	132,137

HISTORY.

The Company's share in the export traffic in sugar from India must, therefore, have been very small; and, indeed, during the period covered by the above tables, they are further shown to have IMPORTED an average of 720,000 cwt., from which they sustained a loss of £12,107.

Effect of abolition of slavery. 83

Incidentally mention has been made of the opposition which arose in England to the consumption of slave labour sugar, as also to the heavy import duty levied by Britain on Indian and certain other sugars. The effects of the controversy which then raged, and of the strong prejudices that arose in connection with free and slave-made sugar may, in fact, be said to have governed to a large extent the ultimate development of the markets which have since become the world's chief sources of supply.

EUROPEAN PLANTATIONS.

Conf. with pp. 37, 48, 62, 63, 91, 93-94, 103, 111, 161-162, 212, 306, 309.

When slavery was in time abolished in British colonies, many West Indian planters sold their possessions and removed to India. They were not long, however, in discovering, to their utter ruin, that there existed in India other circumstances opposed to sugar planting on the European pattern, and circumstances, too, that were, if anything, more inimical to success than the prohibitive import duty which, prior to 1836, had been imposed by Great Britain. At one time that duty was alone regarded as the obstacle that existed in the path of a future great trade. When, therefore, it had been removed, and later on slavery prohibited, it was but natural that some West Indian planters should have turned their attention to India as a hopeful field of future enterprise. At the same time the duty on foreign sugars, not manufactured by slave labour, was also lowered, though it was preserved on the sugars of all slave countries with which Britain did not chance to have commercial treaties. There were many considerations doubtless which actuated the Government of the time to resist the equalization of duty, claimed by the East India Company, and to attempt a protective policy in favour of colonial non-slave made sugar. For a time the measures adopted were popular, but when the effects of the emancipation of the slaves reduced the average annnal sugar supply by one-half, and thus doubled the cost of the article, British philanthropy gave place to more rational considerations. Many writers of that period were then found willing to openly condemn the popular opinions, and they had little difficulty in showing the weakness of the national policy. They demonstrated conclusively that annually large quantities of British produce were exported to Cuba and Brazil, but that the sugar of these countries (the principal article they could return to England) being excluded, could not be brought to British ports. It was accordingly conveyed to Continental ports, sold, and the produce of slave labour converted into wool flax, silk, and other goods which could be freely admitted into England. The support thus given to slavery would have been the same had the sugar, purchased by English goods, been thrown into the sea.

Removal of Sugar Duty. 84

The scarcity of sugar that arose raised the price of that all-important article, until it could be clearly shown that Britain had lost in its sugar purchases £3,440,000 in one year, or £10,327,125 in three years. Accordingly, in 1846, the duty on all sugars, whether foreign or British, was equalized, and, as will be found in another chapter, it was in time entirely removed.

Discovery of Beet Sugar. 85

But a more serious difficulty was destined soon to oppose itself to the growth of an Indian foreign sugar trade—a difficulty that may be said to have paralized the sugar-cane trade of the world. In 1747 **Andreas Sigismund Marggraf**, Director of the Physical Classes in the Academy of Science in Berlin, discovered the existence of common sugar in beetroot, and in many other such fleshy roots. No use was, however, made of this discovery until **Marggraf's** pupil and successor, **Franz Carl Achard**, established a factory in Silesia in 1801. Through the policy of

SACCHARUM: Sugar.

Cultivation of the Sugar-cane.

HISTORY.

Napoleon I. this new industry was, however, for a time ruined in Germany, but was able to struggle along in France, and in 1830 it had become firmly established. By 1840 it had grown to a national enterprise, especially in Germany, and has since controlled the sugar market of the world. The influences of this new manufacture have been all-powerful and wide-spread, bringing ruin or expensive reforms into the utmost corners of the sugar-cane producing area. India has, perhaps, felt the effects of this revolution fully as much as any other sugar-cane-producing country. It would be impossible or nearly so to expect the time-honoured systems of production and manufacture of crude sugar (the article which in India or when exported, is refined into superior sugars) to change in obedience to foreign necessities. The apparatus necessary for direct manufacture is beyond the means of the ordinary Indian sugar-producer. It was, therefore, only what might have been anticipated that, instead of attempting to compete, the industry of refining or of preparing the article required by the foreign refiners, should have declined, and the demand for crystallized sugars been allowed to be more and more supplied by imported sugars.

Conf. with p. 81.

Many of the modern methods, discovered in connection with the development of beet-sugar trade, or which have been brought out in the keen competition which has arisen between cane and beet sugars, have been taken up by the wealthy sugar planters of the colonies, and hence, as remarked, these cane-sugar producers have, in some respects, felt the struggle that has recently taken place less severely than has been experienced in India. Still sugar

Conf. with pp. 81-82, 95, 104, 113, 325.

cultivation has by no means declined. The trade has been almost revolutionized, but the price and supply of the crude substance used by the people of this country is more satisfactory than ever it has been. The consumption of crude sugar has greatly increased, as the exports of Indian refined sugar have declined, just as the imports of superior sugars have increased with the decline of the local refiner's trade. The Indian people do now and always have preferred a crude raw sugar or even molasses

Decline of Indian Exports.

86 to a refined or crystallized article, and the sugar they thus use can be produced at a price which not even beet has as yet been able to approach. It will thus be seen that it by no means follows that, because the foreign

Conf. with pp. 19-20, 95, 316, 329, 341-344, 346.

exports of India have for some years past shown a serious decline, that sugar production is ruined, nor that the people are eating less sugar. On the contrary, it seems probable that the beet sugar trade and direct cane-sugar manufacture have lowered the value of the article formerly prepared for the refiner, and thus cheapened the crude sugar used by the people. The derangement that has taken place within the past few years may be demonstrated by the figures of the foreign trade since 1874:—

Balance Sheet of Indian Foreign Trade.

87

YEARS.	Imports of sugar.	Exports of sugar.
	Cwt.	Cwt.
Annual average from 1874-75 to 1879-80	550,284	576,817
" " 1880-81 to 1885-86	988,429	1,106,557
" " 1886-87 to 1890-91	1,842,217	1,058,311
Last year's actual Trade	2,743,491	824,741

Conf. with pp. 135, 136, 341, 346.

These figures exhibit the trade in refined or crystallized sugar only; unrefined sugar during the period dealt with was imported in such small quantities as to be unimportant. It will thus be seen that India imported last year three cwt. for every cwt. exported, whereas formerly it used to export more than it imported.

For further particulars carrying this brief historic sketch to more detailed modern commercial returns, the reader should consult the chapters below on "CULTIVATION OF SUGAR IN INDIA," "BOUNTIES PAID TO, AND DUTIES LEVIED ON, SUGAR," and "THE INDIAN TRADE IN SUGAR."

CULTIVATION OF THE SUGAR-CANE IN INDIA.

CULTIVATION: VARIETIES.

VARIETIES OR RACES OF SUGAR-CANE GROWN IN INDIA.

A. INTRODUCED CANES.

Introduced Canes. 88

The reader will find repeated reference to this subject in the somewhat voluminous series of passages quoted below, from the district gazetteers and manuals. But so very elaborate, and, in some cases, conflicting, are the statements which have been published on this subject, that the writer has found it the preferable course to allow the local authorities to very largely speak for themselves, instead of his attempting to summarise the opinions that have been advanced, the more so, since, for the present, he is debarred from a personal investigation of the forms of cane met with in India. It may be here also explained that the present chapter is not intended to be an essay on all the forms of cane recognised by planters in other parts of the world. It has been conceived by the author that one of the possible directions of improvement is the more careful cultivation of the better canes actually in India. To secure particulars of these canes has, therefore, been the object with which this chapter has been compiled. In his *Sugar Planter's Companion* **Mr. L. Wray** furnished a brief sketch of the better qualities of introduced canes, but seems, when he wrote that essay, to have been strangely prejudiced against the Indian local forms, as he scarcely makes mention of them. In his later publication—*The Practical Sugar Planter*—he, however, has something to say of the native canes, more especially of the large red Assam form. The earliest Indian systematic classification of the introduced canes is, perhaps, that given by **Mr. Joseph D'Cruz**, the Agri.-Horticultural Society's Head Gardener. His paper appeared in Vol. VI. (1848) of the Society's Journal; it gives a brief statement of his experience in cultivating the various introduced canes. Like **Mr. Wray**, however, he also practically ignores the indigenous forms; but several subsequent writers, while adding certain particulars on the subject of the introduced canes, occasionally allude to the native kinds. It may be said that the writer is unable to relegate to a definite standard of classification the particulars given by the early writers regarding Mauritius, Otaheite, Bourbon, and Java canes, and he has, accordingly, given the facts here collated under the names used by the authors consulted. The following brief review may be accepted as conveying the chief facts which have been published by the Agri.-Horticultural Society and by **Roxburgh, Buchanan-Hamilton, Wray**, and other Indian authors.

I.—MAURITIUS (OFTEN CALLED OTAHEITE).

Mauritius Canes. 89

References.—*Agri.-Hort. Soc. Ind.:—Trans., III., 42-43 (introduction into Bombay in 1837); 55-56 (grown in the Deccan and causing considerable interest); (introduced throughout Saugor and Nerbudda) (Proc.) 29, 66, & 72; IV., 187 (promises to confer a considerable benefit on the Agriculture and Commerce of the Bombay Presidency); V., 186 (Proc., 1837) 38, 56 (Gold medal of Society awarded to Major Sleeman "for zealous exertions in bringing the Mauritius sugar-cane to this country, and ultimately successfully establishing the permanent cultivation of that cane on the banks of the Nerbudda"); Proc., 70 (Formation of Sugar-cane Nursery); VI., Proc., 25, 34, 93; VII., 94; Journal (Old Series), II., Sel., 90, 289-291; IV., Sel., 289; (Troops at Poona provided with sugar),*

CULTIVATION: VARIETIES. Introduced Canes: Mauritius.

143; VI., 56 (cultivated in Col. Sleeman's plantation at Jabalpur, 1838) VIII. (Proc., 1853), 166 (Cultivation at Bogra by J. Payter); IX., (Proc., 1854), 61 (Cultivated in Society's Garden); X., (Proc., 1858), 87 (Cultivated in Society's Garden); New Series, Jour., IV., 48; V., 240; VI., 99.

Mr. D'Cruz remarks that this cane was introduced into the Society's garden from Col. Sleeman's plantation at Jubbulpur in February 1838. He remarks that it had previously been obtained from Bombay, but that the stock had disappeared. It excels the red Bombay cane, both in size and quality, yields one rattoon crop, and sometimes two, on rich soils; but high moulding is necessary as the roots get considerably above ground. A rattoon crop, Mr. D'Cruz, however, points out, should be little suited for the mill, seeing that it becomes hard, close-jointed, and full of lateral shoots. It also gives much less juice than the first year's plant As grown in the Society's garden, Mr. D'Cruz tells us, it averaged 9 to 10 feet in length and from six to seven inches in circumference. Of so-called Mauritius cane the early Indian records of the introduction of foreign canes point to Western India as the province where the greatest progress was made. The importance placed on the subject may be gathered from the following extract from an official letter addressed to the Secretary, Agri-Horticultural Society, India, by Mr. Thomas Williamson, Revenue Commissioner, dated the 1st April 1836:—

"In the Deccan it is now grown to a considerable extent; great attention has been bestowed on its culture by Mr. Sundt, at his estate near Poona. Government have made several extensive purchases of canes from him; they have been distributed for cuttings in the Ahmednuggur and Poona districts, in several parts of which the cane now flourishes. At a village near which I was encamped a few days ago seven bighas were cultivated by one individual, and the specimens he brought me of the produce looked very good. They were about three times the size of the common cane. Several respectable patels have, during my present tour, expressed a wish to have cuttings, and I have taken measures to supply them.

"The superiority of this cane may now be considered as permanently established. In the Surat districts, I understand, its cultivation has been extended a good deal during last year, and I expect it will be further extended in the present season. From the cultivators in two purghannahs alone Government lately purchased upwards of 5,000 canes, which have been distributed gratis among the people.

"In the Southern Concan the cane finds a congenial soil, and the Acting Collector gives a very gratifying account of its rising estimation among the ryots. He reports, 'I am happy to be enabled to state that there is every reasonable prospect of the extension of the Mauritius sugar-cane throughout the Concan.' He mentions one instance in which 10,000 canes had produced 2 khundies and 18 maunds of 'gur;' and he says the result of the experiment 'so satisfied the growers and their tenants, that the cane immediately rose in general estimation.' It seems, indeed, to have now excited interest in all parts of the Rutnagherry Collectorate. 'Already' remarks the Acting Collector 'seed cane has been bespoken from the stock now growing by the surrounding cultivators in the vicinity; and individuals at a distance have expressed their willingness to plant it. It will be satisfactory to the worthy proprietor of Powey* to know that extensive benefits this exotic promises to diffuse over the whole country are to be traced to some plants obtained from that estate.'

"Next to Framjee Cowasjee, Esq., the persons who deserve most credit for this improvement, are Mr. Sundt and Hurrybhai Omerashankur, Mamlatdar of the Chowrassee Pergunna in Guzerat, and I would venture to propose that the Society present them with a medal, or some other small token of its approbation" (*Transactions, Agri. Horti. Soc., Ind., Vol. III., 55*).

Captain (afterwards Colonel) Sleeman may be said to have been the most energetic experimentor in the field of the introduction of foreign canes into Western India. For his labours he obtained the Agri.-Horticultural Society's gold medal, and he subsequently seems to have established a

Conf. with p. 68.

* In another chapter of this article it will be learned that the Powey Estate, in 1792, belonged to Dr. H. Scott, who, along with two other enterprising European gentlemen, founded it in that year.

CULTIVATION: VARIETIES. Introduced Canes: Mauritius.

sugar-cane plantation at Jabalpur. Speaking of the introduction into the Saugor and Nerbudda Territory of what is probably the same plant as above designated Mauritius cane, he wrote in 1835:—

"The results have been the proof by successful experiments that sugar of excellent quality can be made in the Valley of the Nerbudda, a thing never believed by the people before this plantation was established. The sugar made by the aid of men from the sugar districts in Oudh bore the same price in the bazars as that brought from Mirzapore.*

"That the sugar made from the Otaheite cane is rather better in quality than that made from the small straw-coloured cane of the country; and very far superior to that made from the large purple cane.

"The cane, after eight years' planting, was, last season, as fine in its beautiful straw colour, in its size, the quality of its juice, as when gathered for me in the Mauritius by the present Secretary to the Government of that Colony, **Captain Dick,** in 1827. The plants I brought with me were deposited in the Botanical Gardens in Calcutta in March 1827, and in the following cold season I was supplied, at Jubbulpore, with cuttings *from these plants.* These canes, now sent into the bazar as they are cut and sold as a fruit, fetch about four times as much as the largest cane of the country, being much longer, and the juice much finer.

"In planting, I have adhered to the practice which prevailed in the Mauritius, and which will, I think, be everywhere found good. This practice was described by me in one of the Calcutta Magazines of 1827, and does not, I presume, differ much from that of the West Indies" (*Transactions of the Agri. and Horti. Soc. of Ind., Vol. III., p. 73*).

The reader will discover the present position of the so-called Mauritius and of the other foreign canes of Western India, by consulting the passages quoted below from the Bombay Gazetteers. It may also be remarked that the paragraphs which follow, on Otaheite cane, should be consulted. It will there be seen that **Dr. Thomson** maintained that the so-called Mauritius cane of Indian writers is in reality the yellow-violet of Java, which, having been grown in Mauritius, took the name of that island, with it to the countries to which it was subsequently conveyed.

II.—OTAHEITE—THE YELLOW AND STRAW-COLOURED.

Otaheite. 90

References.—*Agri.-Hort. Soc. Ind.:—Trans., II., App. 18-19; III., 57 (Introduction into Oudh); 72-74, 97 (supplied to Saharanpur), 172; IV., 184-190 (Bombay Government asks Bengal for a supply); (Sleeman sent it to Meerut, Bhopal, and Kosah); V., 18,36,66; (Introduction into Azimghur); (Proposal to send small vessel to the Island of Otaheite for a supply of canes), 204-210; (Proc.) 2-4, 12, 37-38, 46, 50-54; (Premium offered for its cultivation in Bengal); 59-61, 86-90, 98 (cane sold at Lucknow in 1837 at R10 per 100); 104, 121; VI., 56-58, 90-95, 137 (Introduction into Tenasserim from Botanic Gardens, Calcutta); 242, 249 (successfully introduced into Dhera Dhún, 1838); VII., 101-109, 130-134 (cultivated in Amherst in 1839); (Proc.) 22, 38-39 78 (grown in Asimghur); 116 (grown at Saugar); 127 (grown at Secundra near Agra); VIII. (culture of it at Tipperah), 89; 455 (grown in Dacca in 1840); Journal (Old Series), I., 257 (Dr. Thomson regards yellow Otaheite as wrongly named; it is, in his opinion, yellow Batavian cane); II. (Proc.) 260; III. 87, 229-230 (grown for past five or six years in Tavoy, 1844); (Proc.) 75, (grown at Cuttack); 179 (taken up by rayats of Dacca, in 1844); IV., (Sel.), 28; (Proc.), xcii.; IV. (Sel.), 32; (Proc.), xl.; VI. (Proc.), lxxxv., lxxxix.; VIII. (Proc., 1853), 185 (Flowering of cane at Gowhatti in Assam); IX. (Proc.), xiii., lxi.; X. (Proc., 1858), 87; etc., etc., etc.*

It is, perhaps, unnecessary to refer the reader to the long list of modern authors who deal with the subject of Otaheite cane. The above will serve the purpose of demonstrating how thoroughly the effort to acclimatize the

* Mirzapore for many years figured prominently in the efforts made to establish sugar-cane as a European industry. The Honourable the East India Company had a plantation and factory there as also a rum distillery. **Mr. R. Carden** was Superintendent. *Conf.* with p. 53.

CULTIVATION: VARIETIES. Introduced Canes: Otaheite.

cane, generally recognised by the name, was prosecuted in. India. That many recent, and, more especially, scientific writers should now question the possibility of any such cane being recognisable from the countless series of cultivated races which have been specialized, is a matter of less importance than that canes of the class to which the Otaheite of the early writers belongs have been fairly tried and proved valueless in India. The reader will find the subject spasmodically brought before the public in the reports of experimental farms and in the technical press that caters for the agricultural and planting interests. It may, however, serve a useful purpose to review very briefly the chief peculiarities of the cane as brought to light on its introduction into India, as also the fate that overtook the effort to acclimatize it in this country.

Mr. Wray speaks apparently of this cane taken conjointly with the Mauritius. He says: "This variety of Cane and the Yellow Otaheite are so much alike in all respects, and have become so intermixed on West Indian plantations, that it is a matter of some difficulty to distinguish between them: the Bourbon, however, greatly predominates. Of Otaheite Cane, there are two varieties which I am acquainted with; these comprise the yellow, or straw-coloured, and the striped. This latter, which has broad, purple stripes, is a little inferior in size to the former. In appearance it is very similar to the Ribbon Cane of Batavia, the difference being, in its greater size and the colour of its stripes, which in the Batavian are of a blood-red, on a transparent straw-coloured ground." "When **Captain Cook** first visited the Island of Otaheite, he found these canes growing in the greatest abundance and luxuriance; but, whether they really are indigenous or not I leave to be argued elsewhere. From Otaheite they were taken to the West India Islands." Discussing the produce of sugar from these canes, **Mr. Wray** says 2½ tons per acre or even 3 tons are commonly obtained in Jamaica; but, he adds, the general calculation is 2 tons of *plant canes,* that is, canes of first year's growth. He accordingly affirms that an estate of 200 acres in cane would very probably run thus:—

	Tons.
50 acres plant canes =	100
50 do. 1st Ratoons =	50
50 do. 2nd do. =	30
50 do. 3rd do. =	20
200 acres total crop =	200

Concluding his notice he says: "These canes require a generous soil, careful fencing, and attentive management. Many soils which agree with *other* varieties, are unfit for their proper development, whilst, it is generally remarked, that they are more sensible of the injuries committed by the trespassing of cattle, etc. during their early growth, than other descriptions. The foliage of the Bourbon and Yellow Otaheite is of a pale green, leaves broad and drooping much, and on arriving at maturity frequently "arrows" or flowers, especially on estates having a sea aspect. This renders it, when in extensive fields, exceedingly ornamental and graceful in appearance. The striped cane is darker in the colour of its leaves, and with less droop" "The Bourbon and Otaheite have been introduced many years into India, but, from some strange cause, they are held in great disrepute. Many persons I am acquainted with, after having for some time cultivated them largely, have reverted to the native canes in despair." "They did well for a time, but a dry season came and they were literally eaten out of the soil and destroyed by white ants." **Mr. Wray** says that though he found his Otaheite to suffer from that pest it was not worse than the

Flowering. 91

Conf. with pp. 8, 9, 11, 47, 61, 83-88, 109.

CULTIVATION: VARIETIES. Introduced Canes: Otaheite.

ordinary native forms grown side by side. It will be seen by a comparison of the above passages with the descriptions given by **Mr. Wray**, a few years later (in his *Practical Sugar Planter*) that he must have seen cause to materially alter his views. In that work he repudiates any distinction between the Bourbon and the Otaheite. He accordingly speaks of these canes collectively, under the latter name, and remarks that there are two forms "the yellow or straw-coloured and the purple-striped or ribbon cane." But, if this be so, then it may fairly well be asked—'In what respect does the so-called ribbon Otaheite differ from the ribbon Batavian.' That they are one and the same, from the botanist's stand-point, goes without saying, but that they differ in the hands of the planter or in the respective countries where they are grown is equally true. It has, therefore, seemed to the author desirable to bring together the various opinions that have been published about these canes on the responsibility of the original writers. **Mr. D'Cruz**, for example, says of Otaheite—"This variety, the genuine Otaheite, was received in November 1840 together with some canes of a purple variety." "An experience of several years leads me to the conclusion that this cane is superior to any other cultivated in the Society's garden, or, indeed, any other that has come within my observation. It is easy of culture, hardy, and exceedingly prolific; which, of course, adds to its value. It needs less labour for watering, replacing dead cuttings, and pulverizing holes than other foreign sorts." **Mr. D'Cruz** further remarked that when first planted it gave a magnificent crop and four rattoon crops. Fully 95 per cent. of the cuttings put down in October sprouted. This variety (the straw-coloured Otaheite) **Mr. D'Cruz** regarded as somewhat inferior in size to the Mauritius cane, "but it gives more juice, and is altogether richer in saccharine matter." Another writer, whose opinions are of equal weight with those of **Wray** and **D'Cruz**, *viz.*, **Robinson**, says in his *Bengal Sugar Planter* (*p. 113*)—a work published in 1849—that "The Otaheite cane was, at its first introduction, highly prized, and the produce it yielded per bigha so far surpassed that of any other variety, as to establish a pretty general opinion that any extra expense incurred in its cultivation was more than compensated for by the results of its yield. A year or two after its introduction, however, its virtues were found to have much degenerated, and its greater liability to the ravages of white ants, and the high cultivation it required, as compared with other varieties, now lost for it its character as the favourite, and the China and Native kinds came more into request as being hardier and involving less risk in their returns."

The chief Indian historic facts regarding the Otaheite cane (or perhaps in some cases confused with it of Mauritius cane also) will be found briefly indicated in the paragraph above under **References**. A few of these, however, may be here still further elaborated, in order to show how widely it had been distributed, and to manifest the view entertained fifty years ago regarding its value. The Civil Surgeon at Tipperah had, in 1840, distributed 25,000 plants in his district and expected that it would expel all other kinds in two or three years. In Dacca, about the same time, **Mr. Dearman** reported that the cane produced from rattoons improved in quality. Accordingly he was of opinion that the Otaheite cane, on the Dacca highlands, if properly attended to, would yield crops for several successive years. "The other kinds, such as Manipur, Singapore, Batavian, and two indigenous sorts, appear to be mere annuals. **Mr. Dearman** feels sure that the many thousand *bighás* of high waste land, lying near Dacca and beyond the reach of inundation offer a mine of wealth to any one having the means and disposition to engage in the cultivation of the Otaheite cane." In 1837 **Mr. J. W. Payter** recommended the Agri-

Proposal to charter a ship to bring Otahiete canes.

SACCHARUM: Sugar.

Cultivation of the Sugar-cane.

CULTIVATION: VARIETIES. Introduced Canes: Otaheite.

Horticultural Society to charter a small ship and to send it for a supply of canes to the Island of Otaheite. He offered to take R1,000 share in the expenditure involved and was of opinion many more planters would be equally willing to do so. The Committee of the Society did not, however, approve of the scheme, but in their report on the proposal made the remark—'The Secretary has ever been, and still is, a strong advocate for the speedy introduction of the Otaheite cane, and for the extermination of the indigenous cane, and hesitates not to hazard an opinion that in ten years from this date this result will have been attained.' The story of the failure of the Otaheite cane will be found, as told by **Mr. Payter's** successor, in the remarks below in connection with the Bogra district of Bengal.* **Dr. J. V. Thomson** wrote in 1842 a paper entitled 'Remarks on the variety of cane termed the Otaheite, but which is supposed to be identical with the yellow Batavia Cane.' He came to the conclusion that the so-called Otaheite cultivated in India was, as stated, in reality the Yellow Batavia, and that the true Otaheite had only recently been introduced by **Mr. Pritchard** *viâ* Sydney. This error has been indicated above by giving as a synonym for the Mauritius cane the fact that it is sometimes called Otaheite; but **Dr. Thomson** goes further and maintains that Mauritius got its cane from Batavia, so that, if that opinion be correct, the information given above under Mauritius cane should be transferred bodily to the paragraph below on the Yellow (Violet) Batavia cane (see *Journal Agri.-Horti. Soc. Ind., Vol. I., 257-262*). In a further volume of the Journal (*Old Series, IV., 143-147*) **Dr. Thomson** returns to this subject and his remarks may be here given in full:—

"In my former paper on this subject, I brought forward proofs from **Mons. Cossigny's** work *Amelioration des Colonies*, that the cane now principally cultivated in Mauritius is not the Otaheitean but the Batavian, *Cannes blanches* of that gentleman, which he introduced together with the other Batavian canes direct from Java in 1782, and not only distributed the *Cannes blanches* (rather *jaunes*) to Bourbon, but sent them to Cayenne, Martinique, and Saint Domingo in 1788 and 1789.

"Having since directed my attention more particularly to the subject of the Mauritius canes, I find that when the French were expelled from Madagascar by the natives in 1657, they are stated to have carried with them to Bourbon, where they first established themselves, the sugar-cane of Madagascar, which was probably one or other of the two yellow varieties which stand at the head of the appended list; from Bourbon the French subsequently removed to the Mauritius, of which they possessed themselves in 1715, so that the Madagascar canes became in all probability the general stock of the two islands. I feel quite satisfied that with such fine canes the French would give themselves no trouble to introduce others from so great a distance as Otaheite.

"The two kinds at present cultivated there, *viz.*, the Madagascar and the Batavian yellow cane, although probably so much alike as to lead to their being confounded together, have no doubt characters sufficient to distinguish them from each other, which intelligent members of the Society can now do, as they have been abundantly introduced from both the Mauritius and Bourbon, and cultivated to a considerable extent in the Society's nursery grounds for many years, under the appellation of Mauritius, Bourbon, and Otaheite cane. They may now be further compared with the genuine Otaheite cane, which was successfully introduced from that island several years ago. Subsequently, I also received a box of canes from Otaheite, which I am happy to say are doing well, and consist of four different varieties, *viz.*:—

1. A large pale yellow cane (*Cannes blanches?*).
2. A large purple cane.
3. A large reddish yellow cane.
4. A good sized striped cane.

"I have received canes of the same description from Batavia, but I entertain doubts of their identity with the Otaheitean canes, or with any of the Madagascar canes of the unmixed list.

"Independent of the canes introduced by the French into their islands originally, there exist in the great island of Madagascar a very considerable number of other fine varieties of the sugar-cane, many of them very remarkable for their size and beauty,

* See p. 48.

CULTIVATION: VARIETIES. Introduced Canes: Otaheite. Seeding of Cane.

Conf. with pp. 8, 9, 11, 44, 61, 83-88.

and of all which the natives appear to know the respective qualities irrespective of sugar making; as they appear to differ much in precosity, product, sweetness, hardness, etc., etc., some being best grown on the alluvial banks of the rivers; others on the drier slopes of the mountains; others, again, in the wet and swampy flats. As the cane is only grown by the natives of Madagascar for eating and for making intoxicating drinks by fermentation or distillation, and consequently not upon any great or extended scale, it is most probable that the various kinds originate in seminal varieties * naturally produced, from many of the plants being neglected and allowed to run to seed.

"While Government Agent at Madagascar from 1813 to 1816, I was instructed to collect and forward to the Botanic Garden at the Mauritius, all the varieties of sugar-cane I could obtain in duplicate, of which the appended is a detailed list of those so procured, which I succeeded in conveying safely, and delivering in a healthy and growing state in 1815, to the Superintendent of the Mauritius Garden at Pamplemouse. As I left the island in 1816, I am unable to state the fate of these kinds, but suspect their value not being appreciated, they attracted little attention, and have probably been dispersed or lost."

List of Sugar-cane introduced into Mauritius from Madagascar in 1816, alluded to above.

A.—Yellow Canes.

Yellow canes. 92

No. 1.† ***Fary-baymayvow*** (=large yellow), a large yellowish cane, probably identical with the original Mauritius and Bourbon cane.

No. 2. ***Fary-andrafow***, a moderate sized cane, of a pale yellowish colour.

No. 3. ***Fary-corowh***, a moderate sized cane, of a beautiful bright orange colour when ripe, so called from its colour resembling that of the beak of the little ground-parrot.

No. 4. ***Fary-boubaya***, of a moderate size and of a yellowish colour, slightly tinged with red. *N.B.*—Like one of my Otaheite canes.

No. 5. ***Fary-boubaya mayna***, a variety of the above, more deeply tinged and with a brighter red. (Mayna—red in Malg.)

No. 6. ***Fary-vonlon*** (Malgache name of a bamboo)!! extremely large with long joints, and of a greenish yellow colour, three inches in diameter!!

B.—Red or Purple Canes.

Red or Purple 93

No. 7. ***Fary-carang*** (Prawn-coloured), a large cane, of red colour above, and of a dark reddish purple towards the root, so called from its predominating colour being like that of the boiled shrimp.

No. 8. ***Fary-androwfow mayna*** (red *andrayfow*, Malg.), a red variety of No. 1 by its Malgache name which I doubt, and consider it a distinct variety, is only of a moderate size, with long joints of a purplish red colour, deeper towards the root.

No. 9. ***Fary-maentee*** (black-coloured, Malg.), a large cane, of a very deep reddish purple colour. *N.B.*—Resembles the purple Batavian and Otaheitean canes.

C.—Striped Canes.

Striped Canes. 94

No. 10. ***Fary***—(distinguishing name obliterated in my manuscripts), a large cane, with reddish purple stripes on a dark purple ground.

No. 11. ***Fary-ahombee*** (Bullock's-horn, Malg.), a very large cane, next in point of size to No. 6, Bamboo cane, both in size and marks, resembling a bullock's-horn, colour mixed stripes and shades of a yellowish and reddish purple.

No. 12. ***Fary-mang-indavalan*** or ***Fary-Ginghan*** (Ginghan cane), rather a large cane, of a dark reddish purple below and striped above, with a yellowish red on a bright reddish purple ground.

No. 13. ***Fray-Feesweet*** (Comb-striped, Malg.) of a moderate size, more closely and regularly striped, with a yellowish and a purplish red colour.

There are doubtless many more varieties of which I saw two or three, but did not procure sets, being *en route* at the time of their offering themselves to my notice. [*Agri.-Horti. Soc. Ind. Journal, Vol. IV., 146.*]

* This is a very significant remark both in the light of the controversy regarding the seeding of the cane and in the recently renewed interest in the possibility of improvement by seminal selection.

† ***Fary-baymayvow***: ***Fary*** is the Matgache generic name of the sugar-cane, to which they join a distinctive appellation descriptive of size, colour, etc.

Cultivation of the Sugar-cane.

CULTIVATION: VARIETIES. Introduced Canes: Otaheite.

It may be here incidentally added that modern botanists do not regard the sugar-cane as indigenous to Madagascar. Thus, for example, the Rev. **R. Baron** in a paper on the Flora of Madagascar (*Linn. Soc. Jour., xxv, 294*) includes sugar-cane among his introduced plants).

Mr. E. O'Riley, in a passage which will be found in connection with sugar-cane cultivation in Tenasserim wrote, in 1844, that the Otaheite cane, in fact all canes, seemed to there enjoy a complete immunity from the attacks of white ants. About the same time **Major Jenkins** made a similar observation regarding the Otaheite cane in Assam. Thinking that fact might be due to the nature of the soil, **Mr. O'Riley** forwarded to the Agri-Horticultural Society three samples of Tenasserim and Tavoy soils. These were analysed by **Mr. J. G. Scott** (*Journal, Vol. III., 233-236*), who said of one—the Tenasserim—that it contained so many metallic oxides as might make it offensive to insects. The other two samples—Tavoy soils—where the Otaheite cane was, by **Major Macfarquhar**, also said to be free from the danger of destruction by white ants, were examined by **Mr. Scott** and reported as simply good siliceous soils which contain nothing more than such soils do in general. **Mr. D'Cruz**, in the paper quoted above on sugar-canes, concludes by a statement of the number of canes distributed by the Society in Bengal and Behar. From 1839 to 1847 these amounted to 208,430, and we are told these were mostly Otaheite cane.

Subsequent reports speak of the greater favour that sprang up for Chinese and Singapore canes; but the facts given above will suffice to demonstrate the activity displayed in the effort to introduce the Otaheite cane. A large share of the above distribution was made to the branch Societies all over Bengal, and these propagated the supply, and each issued, in its neighbourhood, enormous quantities, so that the above by no means represents the total amount actually issued to the *rayats* of Bengal and Behar. Thus, for example, we read of Tipperah alone that the local Society distributed 25,000 plants. But from Volume VI. of the Society's Journal (published in 1848) there occurs a remarkable interruption in the interest in Otaheite cane. It is next mentioned in Volume IX. (published 1857) and gradually disappears from these journals, the only trustworthy records which exist of the remarkable period of India's interest in sugar, when it was thought all that was necessary to place India among the foremost producing countries, was the establishment of large plantations of the superior qualities with central factories. It is, perhaps, unnecessary to produce further evidence of the extent to which the Otaheite or Bourbon cane was diffused over India. The present chapter may, therefore, be fittingly concluded by furnishing **Mr. Payter's** account of its disappearance from Bogra, one of the chief sugar-producing districts of Bengal, and it need only be added that a similar calamity befel these introduced canes in every province of India:—

European Plantations in India. *Conf. with pp. 37, 62.*
95

"My uncle introduced the Otaheitean and Bourbon varieties of cane into the Ságuna estates about the year 1840. He obtained the greater part of the supply from the Agricultural Society's gardens in Calcutta; and after increasing the quantity by propagating in nurseries, he ultimately distributed it amongst the *rayats* of the *khás mahals*, whence it became disseminated all over the country. At first the people were unwilling to take it on account of its novelty, assigning various reasons for their refusal. Some of the wisest, however, accepted; and when its superiority in yield and quality became known it was eagerly sought for. The yield per *bighá* was fully double that of the indigenous plant, and the *gur* made from it, so much superior in quality as to command an enhanced price in the market. In short, those who cultivated it in any quantity became comparatively rich. The species introduced consisted of several varieties of the white and purple Bourbon cane, but in the course of a few years it all became of a uniform purple colour, caused, I suppose, by some peculiarity of soil. In the season 1857-58 the cane manifested symptoms of decline and ultimate-

CULTIVATION: VARIETIES. Introduced Canes: Otaheite.

ly rotted in the fields, emitting a most offensive smell. Since 1858 it has entirely disappeared, so that at the present time (1861) not a single cane is to be found, and the *rayats* have reverted to the cultivation of the native cane, which, though of a fair kind, is not to be compared to the Bourbon. I am unable to suggest any reason for the failure which, in this district and Rangpur, has become complete. In the latter district the Bourbon cane was also much grown. The disease first showed itself in Rangpur two or three years previous to its appearance in Dinajpur; in fact, the progress of the disease was from north to south, the cane in pargana Gilábári dying off the year previous to the disease manifesting itself in Ságuna, which is 15 or 20 miles further south. It may have been worn out by high cultivation, or the soil and climate combined may have caused it to deteriorate and decay."

Disease. 96 *Conf. with pp. 52, 76, 87, 102, 121–127, 161.*

The reader will find below that the so-called Red Bombay cane, which had been introduced into, Bengal, suddenly died in each district after it had been cultivated for a certain number of years. The canes became attacked by a worm, and when in that state, they emitted so offensive a smell that the fields could not be approached. A similar observation is, however, recorded with many of the indigenous canes when too constantly cultivated in the same district, a fact which the Natives very generally recognise, and every now and again obtain their seed-canes from a distance.

Bourbon. 97

III.—BOURBON CANE.

References.—*Agri.-Horti. Soc. Ind.:—Trans., VI., Proc. 16, 20, 128; Jour., IV., 144 (Introduced from Mauritius and Bourbon; IX. (Proc., 1854); lxi. (Grown in the Society's Garden); ccli. (Grown in Burma); X., Proc. lxxxvii.; Wray, Practical Sugar Planter. 3; Edwards' History of the British West-India; Statistical Account of Bengal, VIII., 215-219.*

Mr. Wray says this cane 'was introduced into the West-India Islands, from the Isle de Bourbon (Réunion), but came originally from the Coast of Malabar, where it was found growing spontaneously.* When first taken to the Isle de Bourbon, it is stated to have been a small sized, but soft and juicy, cane. By cultivation it, however, increased wonderfully in size and richness of juice, which speedily caused it to be generally cultivated in preference to the old species, until, at length, it entirely superseded them throughout the Island. This, in fact, has been the case, in a great measure, wherever it has been introduced.'

"Its good qualities do not consist merely in its rich juice and large size, but it has a degree of hardihood in its nature which renders it extremely valuable; for instance, during seasons of long continued drought, if the soil in which it is planted be congenial, no species of cane (save the Otaheite) can long withstand its destructive influence." The above passage appeared in **Mr. Wray's** first paper (*in the Agri.-Hort. Soc of India's Journal 1843*), but in his later publication (*the Practical Sugar Planter, 1848*) it is slightly modified. He there says of Bourbon cane "From my own experience in Jamaica, I can pronounce it a most valuable cane: but I entertain a strong suspicion that it is in reality no other than the *Tibbu Lint* of Singapore (sometimes called Otaheite cane), somewhat altered by change of soil and climate. It will thus be seen that **Mr Wray** came ultimately to hold very nearly the same opinion as had been advanced by **Dr. Thomson**, but curiously enough he did not see the necessity of his withdrawing from the theory that it came originally from Malabar. The Malabar origin of Bourbon cane is unhesitatingly affirmed in Edwards' *History of the British West-India*. It is there stated that in 1794 **Sir John Laforey, Bart.**, introduced the cane to the Island of Antigua from the French Charaibean Islands. It was reported by the French to be the growth of the coast of Malabar.' It was also viewed as much alike to the Otaheite cane. **Sir John** is reported to have said, 'In the spring of 1794, a trial was made of the Malabar canes on one of my plantations; 160 bunches from holes of five feet square were cut, they produced upwards of 350lbs of very good sugar.' 'The produce was in the proportion of 3,500lbs to an acre.'"

Mr. D'Cruz, who wrote (1847) on the canes grown in the Calcutta Society's garden, some years after the date of **Mr. Wray's** first paper, makes no mention of Bourbon canes, nor apparently has any other Indian writer.

* Could this be read to mean indigenous? *Conf.* with p. 5 regarding cane found in a "wild state" in Car-Nicobar; also with pp. 34 and 57.

CULTIVATION: VARIETIES. Introduced Canes: Bourbon.

It is thus difficult to ascertain how far these forms—Otaheite, Bourbon, Mauritius—of the early authors are distinct races or may be but other names for one and the same thing. One point may be regarded as specially interesting, namely, the controversy as to the origin of these forms. The opinion has been advanced that the straw-coloured Otaheite was in reality yellow Batavian, and that the true Otaheite is quite distinct from the cane so designated in the East and West Indies. It has been contended by another writer that the Mauritius, the Bourbon (as also the straw-coloured Otaheite) came originally from Madagascar. By still another that the Bourbon was a native of Malabar where it was "found spontaneously." While a fourth writer has added greatly to the confusion by the opinion that the Bourbon was "identical" with the Singapore. Out of all this confliction but one feature remains constant, namely, the association of these superior qualities of cane with islands or insular influences. This fact might be viewed as adding a certain amount of confirmation to the idea of the cane having been originally a native of Southern Asia and the Malay Archipelago.

From Madagascar in the extreme west to Java and the Philippine Islands in the east, there is in some respects a greater diversity in the names given to the plant than is the case in the more continental tracts of Asia, particularly India and China, where the knowledge in the manufacture of sugar seems to have undergone the greatest development. **Crawfurd** tells us, however, that "from Sumatra to New Guinea and the Philippines, it is known by one name, which, with very slight variations easily accounted for. is *tabu*. This is a native term, unknown, so far as our information extends, to any language ancient or modern, beyond the pale of the Archipelago, and we can, therefore, from analogical reasoning, entertain no doubt but the sugar-cane is an indigenous product of these countries." He then adds that the art of manufacturing sugar from it is certainly foreign. There is no name for sugar except *gula* which is of Sanskrit origin. It is significant that the word *Kan* (sweet) in Chinese should bear so close a resemblance to the Sanskrit *Khanda*, which has very nearly the same meaning and which even denotes the plant in some of the Aryan languages of India. *Kanche* in Chinese (the name for the cane) literally means sweet-bamboo.

Batavian.
98

IV.—BATAVIAN CANES (S. Violaceum, *Tussac.*).

References. – *Agri.-Hort. Soc. Ind:—Trans., V., Proc., 38 (a letter from a correspondent at Mauritius indicating that the Otaheite cane was nearly exhausted and hinting that Batavia might supply the want); VI., Proc., 6 "(a promise to send 25 cases Batavian sugar-cane from the Isle de Bourbon)," 15 (a letter from Bourbon advising despatch of 18 cases Batavian sugar-canes, for the Society's Nursery, 1839) Jour., I., 257: (also known as "Otaheite") II., Part I; 45—46 (classed along with the Bourbon and Otaheite, the three being held in the highest rank); 143 (varieties of sugar-cane by J. C. Thomson, M.D.); Wray, the Practical Sugar Planter; 5; Sir John Laporey in Edwards' History of the British West Indies; Voigt, Hort. Sub. Calc., 705.*

Sir John Laforey wrote in 1794 of these canes, "The Batavian canes are a deep purple on the outside; they grow short-jointed and small in circumference, but bunch exceedingly, and vegetate so quick that they spring up from the plant in one-third the time those of our islands;" the joints, soon after they form, all burst longitudinally. They have the appearance of being very hardy, and bear dry weather well. A few bunches were cut and made into sugar at the same time the experiment was made with white canes. The report made me of them was that "they yielded a great deal of juice which seemed richer than that of the others, but the sugar was strongly tinged with the colour of the kind; and it was

CULTIVATION: VARIETIES. Introduced Canes: Batavian.

observed that upon expression of them at the mill, the juice was of a bright purple, but by the time it had reached through the spout to the clarifier, a very short distance, it becomes of a dingy iron-colour." Crawford (*History of the Indian Arhipelago, 1820*) mentions three "indigenous canes" on the Indian islands, one a large cane often 2 inches in diameter which has so dark-coloured a rind that it is unsuited for sugar manufacture "because it tinges the sugar."

"Mr. Wray remarks that the Batavian with which he was familiar were of three* descriptions, *viz.*—

"The yellow violet, the purple violet or Java cane, and the transparent or ribbon cane. The *yellow violet*, so denominated in the West Indies, differs from the Bourbon and Otaheite, in being smaller, less juicy, considerably harder, of slower growth and with foliage much darker and more erect. When ripe, it is usually of a straw-colour, its skin or rind is thick, and the pith hard; but its juice is rich and tolerably abundant. It is seldom that this cane 'arrows,' but when it does so, it emits a faint, but agreeable fragrance, especially in the evening after a slight shower of rain, at which time it is particularly pleasing, and may readily be smelt even at a distance. Many persons have pronounced it extremely similar to the perfume of a violet bank, from which circumstance, probably, it has derived its name. The yellow violet does not require so rich a soil as those already treated of (Bourbon and Otaheite), but contents itself with that of an inferior description. This renders it of much importance in planting out large tracts of land, some portions of which may be too poor for its superiors. In Jamaica, it is usual, in such places, to plant the violet. Thus, we often see large patches of it flourishing in the midst of a field of Bourbon." "The sugar manufactured from this cane is of a very fine quality, but by Jamaica planters it is commonly mixed with Bourbon plants, according to proportion, for the purpose of rectifying the juice of the latter. This mixture gives excellent sugar." "The PURPLE VIOLET or LARGE JAVA CANE is fully as thick as the Bourbon, with joints from three to six inches long. In height it ranges from eight to ten feet, and the upper parts of the stalk often exhibit faint streaks, which are imperceptible in the lower joints, which are of a pure purple colour. The leaves are of a darker green than the yellow violet, when ripe and in perfection; it yields a juice generally esteemed more sweet and rich than that of any other description of cane; but being hard, and comparatively dry, it is more difficult to grind, and affords only a small quantity of juice. It is very hardy, and thrives well in poor, dry soils, whilst it is often planted in the outer rows of the cane-fields, as a protection against stray cattle, which browsing along the roads, and at intervals, break through fences, and tear and trample down the canes. These ravages would be very serious, were the plants *less* hardy, but fortunately this injury it quickly recovers from, and shoots up again with astonishing rapidity. This cane was introduced into the West India Islands much about the same time as the Bourbon, and is still much cultivated there. It is, like the yellow violet, generally mixed with Bourbon plants." In his *Practical sugar Planter* Mr. Wray dds that in the Straits the Malays term it *Tibbu Etem* (*Etam*=black) and cultivate it much around their houses, for eating.

"The TRANSPARENT, or RIBBON CANE, is of a transparent bright yellow, with a number of blood-red streaks, varying in breadth from a quarter to a full inch, and being very clear withal in its tints, it presents a very pretty appearance. Its leaves are of a green like that of the yellow violet, but far more erect. Its joints are from six to eight inches long, four in circumference, and six to seven in height. In Jamaica the transparent is generally planted in light sandy soils, where no other cane will thrive; sometimes it is planted promiscuously with the yellow violet. Though its rind is thick, and general texture hard, yet it yields a good quality of juice of excellent quality, which is easily converted into fine fair sugar. The transparent is also mixed with the Bourbon. These descriptions of cane I consider are admirably adapted to the East Indies, more especially the first and last varieties (the yellow violet and the transparent)." Mr. D'Cruz says that the transparent Batavian was introduced into India by M. Richard in February 1838 from Bourbon, from which circumstance it came to be known in India as *Striped Bourbon*. "There has been, a less demand for this cane than the other varieties, and consequently it has only been cultivated on a small scale."

* In his "*Practical Sugar Planter*" Mr. Wray adds a fourth which he calls the Tibbu Battavee or Batavian cane.

CULTIVATION: VARIETIES. Introduced Canes: Batavian.

The fourth kind which **Mr. Wray** describes in his later publication (*The Practical Sugar Planter*) is called the "Tibbu Batavee or Batavian Cane." It is, he says, common in the Straits of Malacca, where it is cultivated by the Malays. In appearance it is much like the yellow violet, except in the peculiarity of its colour, which is rather greenish, with a pink shade, in parts; in some of the lower joints, this pink colour is very bright and pretty, whilst in the *upper*, it is more faint and delicate: the joints are seldom more than from three to six inches apart. In height, size, and foliage, it closely resembles the yellow violet; although it differs from it in being much softer, more juicy, and less hardy in habit. In a rich soil it is prolific and rattoons well,—its juice is rich, clarifies easily, and gives a fine sugar: but on the whole, it is inferior to Otaheite variety, yet requires an equally rich soil."

The most recent point of interest in Java canes may perhaps be regarded as very remarkable. In the Report of the Botanic Gardens of Saharanpur for the year ending March 1891, mention is made of the visit of **Mr. R. D. Kobus**, a Dutch gentleman, who had been sent to India by the Government of Java in order to secure stock of Indian varieties of sugar-cane. **Mr. Kobus** explained that the sugar-cane had been attacked in Java by a disease supposed to be of fungoid nature, which threatened to extinguish the sugar industry there. **Mr. Kobus** recognised all the *paunda* forms (the class grown in India to be eaten) as the same as the cane grown by the Dutch in Java for sugar manufacture. The *ek* or *ganna* canes (the class specially grown in India for sugar-making) were, he said, entirely new to him. These are very much more hardy than the *paunda* canes and accordingly **Mr. Kobus** took back with him a large supply of these and was hopeful that he had thus secured a stock that might prove able to resist the disease.

Paunda. *Conf. with pp. 7, 64-6, 73, 74, 232.* **99**

It is at least curious that Java should now come to India for a fresh stock of cane. If it got its original supply also from this country, Java had so improved its quality that, as shown above, Indian planters, nearly half a century ago, were very anxious to bring back the Java improved canes to this country. It is just possible the *paunda*, as also many of the canes which, like the *paunda*, are eaten, came from foreign countries (in their present improved condition), but, if that be so, it is remarkable that they should not be valued as sources of sugar at the present day.

Diseases. *Conf. with pp. 48, 76.* **100**

The reader had, perhaps, better consult the special chapter on the DISEASES OF THE SUGAR-CANE as amplifying the brief reference above to the disease which is giving so much cause for anxiety to the Dutch planters.

China. 101

V.—CHINA CANE (S. sinensis, *Roxb., Fl. Ind., Ed. C.B.C., 80-81*).

References.—*Proceedings of the Hon'ble the East India Company—Official notice of successful introduction into Botanic Gardens, Calcutta, 30th Dec. 1797; Correspondence, 1799; Agri.-Horti. Soc. Ind.:—Trans., III., 62 (Culture in Canton); V., (Proc.), 90 & 104; VI., (Proc.), 30, 44; VII., (Proc.), 78 (Cultivation in Goomsur and Azimghur); Jour. (Old Series), IV., (Sel.), 131, 132 (Cultivated in Buxar & Dhoba in 1845); (Proc.), 92; V. (Proc.), 31 (Favourable progress of at the Society's Garden), 40; VI., 60; IX. (Proc., 1854), 61; Roxb., Fl. Ind., Ed. C.B.C., 80-81; Wray, The Practical Sugar Planter, 10-13; Voigt, Hort. Sub. Calc., 705; Royle, Productive Resources of India, 92.*

The earliest account of this cane in India is that given by **Roxburgh.** That distinguished botanist regarded it as a distinct species, from the native sugar-canes of India, which he referred to **S. officinarum.** The distinction which he tried to establish has not, however, been maintained by modern botanists, but as enabling those interested in sugar-canes to recognise this form, it may be as well to mention that **Roxburgh** distinguished the Chinese cane from the Indian by its having much flatter leaves, the margins of which were also hispid. The flowering panicles, he tells us, are ovate in general outline with simple or compound verticelled

CULTIVATION: VARIETIES. Introduced Canes: China.

branches. The corolla, he adds, is of two valves on the same side. The Indian canes he separates from the Chinese by the following characters:— panicles spreading, the branches alternate decompound and the corolla one-valved. The most ready character, therefore, to separate the Indian from the Chinese would seem to be the more compound inflorescences, the branches of which were scattered instead of given off in whorls.

The Chinese cane, **Roxburgh** says, was introduced into the Botanic Garden at the close of the year 1796, in the hope of finding it in some respects better than the common cane cultivated in India. "It promises," he continues, "considerable advantage; particularly from its being so solid, and hard, as to resist the forceps of the white ants, and the teeth of the jackals, two great enemies to our East Indian sugar plantations. At the same time it bears drought much better than the sorts in general cultivation. It produces a profitable crop even to the third year; while the common cane of India must be renewed every year. It is also said to yield juice of a richer quality." A correspondent of **Roxburgh's** (**Mr. Richard Oarden** of Mirzapore, Culna, in Bengal) furnished him in 1801 with particulars of his experiments with the then newly introduced China cane. "With respect to the produce of the common Bengal sugar-canes," writes **Mr. Oarden,** "I have never been able to collect an account that can be depended upon; the natives generally manufacture the juice into jaggery in my neighbourhood, which yields them nearly 14 *katcha* maunds per *bíghá* on an average; and a profit of about 11 or 12 rupees the *bíghá*. Neither the white ants nor jackals have committed any depredations on the China canes that I have planted, although the latter have often been seen among them, which certainly gives these canes a decided preference to the Bengal sugar-canes. I do not think the China ones degenerate in the least, nor do they improve; they appear to me to remain nearly in the same state If planted at the same time the Natives put their canes into the ground, they will not make such good returns as the Bengal sort, but planted in the West India mode, in the month of September or October, and suffered to remain on the ground till December or January twelve months, they will then yield double the returns of what the Bengal canes do, which is owing in part to the length of time they are in the ground, and principally to the ants and jackals not destroying them, whereas if the Bengal cane was to remain so long on the ground, the Natives would have great difficulty to prevent the greatest part of them from being destroyed, and the young shoots would suffer very much from the hot winds and ants, which I witnessed the second year I came to Mirzapore; but the shoots from the China canes I cut last January stood the last hot season uncommonly well, and will, next January, I have reason to believe, from their present appearance, make half, or nearly three-fourths, the quantity of sugar they did last January; and that with the trifling expense of clearing the ground twice, cutting, and manufacturing the juice."

Many other writers deal with the subject of Chinese cane and in a singular uniformity as to the terms of appreciation. Thus, **Mr. Wray** says of this Chinese cane that he obtained his supply from the Agri.-Horticultural Society at Calcutta. "In its nature, he remarks, it is extremely hardy, and very prolific. During the last hot season it remained uninjured in every respect, whilst the other canes were *all* either burnt up or eaten out of the ground by white ants. As the rains came on, the China canes sprang up wonderfully, many roots having no less than THIRTY shoots, which by September had become fine canes, about twelve feet in height, three inches in circumference, and with joints from six to eight inches apart. These were cut in October and planted out, yet, although, we have had a toler-

Cultivation of the Sugar-cane.

CULTIVATION: VARIETIES. Introduced Canes: China.

ably severe winter, the cold appeared to have little or no effect in checking their growth, but NATIVE CANES planted at the same time were ENTIRELY *kept back*." "For their extreme hardiness in withstanding heat or cold, white ants, jackals, etc., I can myself vouch, and consider it a variety of cane which deserves every attention." Mr. D'Cruz writes much in the same strain :—

"It is, he says, the hardiest of all varieties; the white ant seldom or never touches it; its ability to stand all changes of season is also a great argument in favour of its cultivation. It yields several rattoon crops, and requires less care and trouble than any other sort with which I am acquainted; though if a small degree of culture be bestowed, it repays the owner by an increased length and thickness. I may here mention that, in consequence of the demand during 1845 for this variety, being usually great, and much more than could be met, the Garden Committee increased the cultivation to meet a probable large demand during the following season."

"Several correspondents of the Society have, I observe, lately borne witness to the capability of this cane to stand drought and heat. As the number of the Journal in which their experience is recorded may not be available to every reader of this paper, it may not be amiss if I transcribe *verbatim* the remarks of these gentlemen. Mr. F. Nicol, writing from Chandpore factory, Jessore, states,—when comparing this variety to certain others, which he had also cultivated on a small scale in 1884,—'the China cane thrived the best, and grew to a great height, quite overtopping all the others. Several plants measured nine feet high; the white ants did not touch it, though they attacked the Otaheite, Singapore, Bourbon, and *Dholee*, and it is certainly the best description for land at all infested by these destructive insects.' Mr. P. P. Carter, of Bhojipore factory, near Buxar, in a communication dated June, 1845, observes—'Of the five descriptions of sugarcane I obtained from the Society in March last, the China has succeeded so wonderfully in spite of white ants, heat, and every evil, from which the others (and even the country cane of the district) have suffered so severely, that I am very anxious to have some correct information of its qualities. Should it prove to be a good yielding cane, producing sugar of good quality, I would cultivate it in preference to Otaheite and every other description I know of. I am rather curious to know what height and thickness it attains at full growth, as from its present vigorous appearance it promises even to surpass the Otaheite, while the seeds were scarcely thicker than my little finger.' And Mr. S. H. Robinson, writing shortly after from Dhoba, near Culna, remarks—'Of all the varieties of cane I have tried, the China has proved by far the hardiest, in surmounting the attacks of white ants, heat, and drought, and it has yielded me a good crop at the rate of 202 bazar maunds of cleaned cane per *beegah* from the same situation in which Otaheite, and two varieties of blue cane were all but destroyed by the white ants; and in which common country cane yielded its usual average of 150 maunds per *beegah*. At the mill the China yielded 55 per cent. by weight of juice to 45 per cent. trash, the juice being of the gravity indicated by 11 per cent. of Baumé's saccharometer, which is equal to the average of the best cane juice I have seen produced in these parts; so that there is nothing in these premises to discourage the hope of its produce in sugar, proving inferior to the results it gives in the field. Its aspect when growing also seems to confirm its adaptability to this climate, for though the canes are only from three quarters of an inch to an inch in diameter, it grows to the height of ten and twelve feet with very ordinary cultivation, and I have counted as many as 18 and 20 canes spring up from one stole. It has a bountiful supply of long slender leaves, which keep their fresh green appearance far better than the other varieties. I had an October crop last year half with China and half with country cane plants; the latter barely survived through January, and were all cut off by the hot winds before March was over; while the China are now (June) fine looking plants, and I hope will be ready to cut by November next' (*Journ., Vol. IV., pp. 27-131-32*). (*Journ. Agri. & Horti. Soc. of Ind., Vol. VI., pt. I., page 61.*)

"Mr. Wray in his more recent publication (*The Practical Sugar Planter*) adds certain particulars which are of interest to Indian observers. One of his correspondents (a practical planter in Bengal) wrote: 'As you advised, I wrote to the Society for five hundred canes, which arrived quite fresh, I then cut them up, allowing only one joint to each piece, and planting them in lines, four feet asunder each way, delivered them up to the same chances as my Otaheite and native canes were exposed to. The result has been beyond my utmost hopes; and this, too, after a season of unusual severity, which has grievously affected my native cane; and as to my Otaheite, what with hot winds, the white ants, the long continued wet weather, and the detestable jackals,

CULTIVATION: VARIETIES. Introduced Canes: China.

I have saved but a very few; whilst nothing seemed to injure or affect the *China Cane.* Did you ever know the Otaheite cane-sprouts to be devoured by caterpillars? I forgot to put those depredators in the list of enemies to the Otaheite plant; although they certainly are very formidable ones, as the partial destruction of many of my plants testified. They attacked the plant when only a few inches above the ground, from which many never recovered.

'I understand that Indigo plants are often quite destroyed in the same way, so perhaps it may not happen again for some years. This hope determines me in trying the Otaheite once more; when if it does not succeed, I shall keep entirely to the China cane; which, by the bye, I am now extending, as far as my plants will allow me. I am disgusted with the native cane, and shall soon put them aside altogether.' Mr. Wray then comments: 'I think these accounts of the China cane are sufficient to establish the fact of its being a variety well suited to India: although I need not say that it is immeasurably inferior to the Otaheite, wherever that cane can be cultivated successfully.' 'It is now very common throughout Bengal, although the natives think it a native cane, from its having been so long amongst them. They have given it a native name, which I quite forget. I met it in several places and recognized it at once; yet I never met but one native who knew it to be otherwise than a native cane. Certainly, its neglected cultivation during 50 odd years in India, has caused it to degenerate very much; hence my advice to all persons desirous of trying it 'send to the Society's Garden, in Calcutta for it.' It is a very small sized cane, being rarely more than one and a quarter inch in diameter; but it is better adapted for sugar candy making, than any other cane."

VI.—SINGAPORE CANES.

Singapore. 102

References.—*Agri.-Hort. Soc. Ind.:—Trans., IV., (Proc.), 58-52 (arrival in 1837 of canes of this nature); V., 15 (Dr. Montgomerie's letter reporting & Dr. Wallich's acknowledgment of receipt), 71; (Proc.), 30, 66 (red cane said to be native of Malacca), 90, 104; VI., 96, 103 (Balestier on Manufacture of Sugar at Singapore); (Proc.), 16, 20, 44, 56, 93, 108, 128; Journ. (Old Series) IV., Proc., 92; IX., Proc. (1854), 51; Wray, Practical Sugar Planter, 13.*

Singapore canes, says Mr. D'Oruz, were first received by the Society from Dr. W. Montgomerie.

"In June 1887 Dr. Montgomerie sent the first supply, the second in the October following. In his communication advising despatch of the latter, Dr. Montgomerie observes that he has not been able to obtain anything satisfactory relative to the origin of the canes grown at Singapore; they form part of the *sea-stock* of almost all native vessels, and as we have communication with all the East by such means, we may have got them either from Siam, Borneo, Celebes, Java, or any other neighbouring country. The natives recognize the red or purple cane, as the *Tubú* Malacca, pointing out Malacca as the place of origin, but they think the three light coloured varities, *viz., Tubú Leah*, or *Leah Tubú*, and *Kapiur* have been introduced by the Buggese traders from the eastern islands, and in such case they may most probably be a variety of the Otaheite cane, modified by the Malay, which may have been cultivated by the natives" (*Transactions, Agri.-Hort. Soc. of India, Vol. V., p. 66, Appendix*).

Conf. with p. 32.

"There is a material difference between this cane and other sorts. It is more transparent, and perhaps handsomer looking, than most other kinds in the Society's garden, and is, I am aware, held in much esteem by several planters. It is of a light-yellow colour, averaging in height from eight to ten feet, and from four and a half to five inches in circumference. It has a light-coloured, short, broad leaf, with a broad white stripe down the middle of the leaf, which serves to distinguish it from other sorts. Its cultivation is, however, precarious. It suffers more from the ravages of jackals, who are extremely partial to it, than other sorts. It is also easily blown down by high winds, and when once prostated it is difficult to raise it again, its natural brittleness causing it to break into pieces. Other canes are, of course, more or less subject to the same casualty, but I have seldom experienced any difficulty in raising them up again, and securing the greater portion. It might be worth the while of parties desirous of growing this particular variety, to make an experiment for counteracting the effect of the wind by planting China cane very thickly in borders, eight or ten feet wide, all round the plots in which the Singapore sort is to be cultivated. I have never tested the efficacy of this experiment, as there has been no occasion for it, the quantity cultivated in the Society's garden being on a comparatively small scale, and merely for the purpose of distribution; but I have been induced to offer the sugges-

Cultivation of the Sugar-cane.

CULTIVATION: VARIETIES. Introduced Canes: Singapore.

tion as possibly an easy and simple mode of securing a really fine variety of cane, during the period of its growth, against a stiff breeze or sudden blast of wind. I should mention that the red or purple sort alluded to in the previous page, was also increased, but being afterwards recognized as identical with the red Bombay variety its culture was discounted" (*Journ. Agri. & Horti. Soc. Vol. VI., pt. I, II., p. 58*).

Mr. Wray gives much greater detail regarding the canes of the Straits Settlements of Penang, Singapore, and Malacca. "The principal", he says, are eight in number. Of these, first and foremost is 'the Salangore cane,' by the Malays of Province Wellesley termed *tibbú bittong berabú* (the powdery bark cane), but by the Malays of Singapore and Malacca, it is named *tibbú cappor* (the chalk cane), from its having sometimes a considerable quantity of a white resinous substance on the stalk. This is the finest description of cane in the Straits Settlements, and perhaps in the whole world. In Province Wellesley, it is universally cultivated on all the estates, and is only known to those planters as the China cane; from the simple circumstance of the Chinese cultivators in the province having been in the habit of cultivating it for years, before any European embarked in such speculations in those parts. I have cut as many as five of the larger canes from one stool; each cane from ten to fifteen feet long, without leaves, and seven and a half inches in circumference (round the lower joints): each cane weighed from 17 to 25lb. That of 25lb weight, I kept some weeks in my house, and numbers of people saw it; it was thirteen and a half feet long, and two and a half inches in diameter; yet it was not by any means so large a cane as I have seen. The place where I found it growing was a newly cleared piece of jungle land, whereon a Malay had 'squatted,' built a house, planted rice and some three acres of sugar-cane around it." "The Salangore cane is remarkable for the quantity of '*cane-itch*' (so termed in the West Indies) which is found on that portion of the leaf attached to the cane-stalk. Sometimes touching a growing cane incautiously, I have had my hand covered with it, and thousands penetrating deep into the flesh, caused great irritation and inflammation. The leaves are very broad and deep serrated at the edges, with a considerable droop; they are some shades darker in colour than the Otaheite, and have so good a hold on the stalk as very seldom to fall off when dry, as some canes do, but require to be taken off by the hand. They rattoon better than any canes in the Straits; and I have known them to yield forty *piculs* (a picul is 133⅓lb of granulated but undrained sugar per *orlong* (an orlong is one acre and a third) in third rattoons. From what I have seen, I am disposed to think that in the West Indies, Mauritius, and India, they would be found to rattoon better than any other cane." Mr. Wray then proceeds to give some information as to the yield of sugar afforded by Salangore cane. He has known 6,500lb per acre and was informed by a French gentleman in the Province of Wellesley that he has in some cases obtained as much as 7,200lb of sugar (undrained) per acre, from which he has secured 5,800lb of shipping sugar well dried in the sun. Mr. Wray next describes the other forms of Straits canes and these need not be more than briefly indicated:—

Tibú líut (clay cane).—This is the form which Mr. Wray identifies as the Otaheite of most writers. It has already been sufficiently indicated.

Tibbú tılúr (or egg-cane).—This is viewed as the form peculiar to Tanne, one of the New Hebrides. This is a very clean and elegant cane and is remarkable for the almost entire absence of *cane-itch* The leaves are smooth and the stems bulge between the joints to such an extent as to have obtained for it the name of egg-cane. The leaves are shed as they ripen and the structure of the stem is so fine and brittle as to cause it to break readily—the chief danger in the cultivation of this cane. It is very

CULTIVATION: VARIETIES. Introduced Canes: Singapore.

prolific and is quickly and easily cultivated, as every one of the large eyes disposed along the stem shoots forth vigorously. The stools have generally from 5 to 15 canes. The juice is copious, of rich quality, and can be converted into fine sugar of a good, strong, sparkling grain. It cannot, however, be cultivated in situations exposed to wind, nor in damp rich soils, owing to its tendency to snap off.

Tibbú clám or *Obat* is the Black or Medicine cane of the Malays. This is a small but clean cane, of a rich purple colour, which colour it imparts to the hands of those who handle it or the lips of those who eat it. One remarkable character, Mr. Wray adds, is the rich delicate pink gradually darkening with age to a fine purple and dark green.

Bombay Cane. *Conf. with pp. 5, 31, 33, 34, 49, 75, 76.* 103

The recognition by **Mr. D'Cruz** of the red or purple sort of Singapore cane as identical with the red Bombay variety is very interesting in the light of the other suggestions and opinions (reviewed in the foregoing paragraphs) which point to the Bourbon and straw-coloured Otaheite canes being identical, and to their having originally been derived from Malabar. **Dr Gibson**, in an article on the Agriculture of the Deccan (1843), deals fully with cane cultivation and tells us that Mauritius cane had then shown distinct signs of retrogression But he makes no mention of "spontaneous canes" nor of any feature of special excellence in the local stock such as might have suggested its conveyance from Malabar to the West Indies. **Mr Wray** came to recognize Bourbon cane as the same thing as Singapore, so that the conclusion (and a not unnatural one) is unavoidable that there appears to have been three or four sources of the canes from which the early European cultivators obtained their stocks, namely, India (possibly Malabar), Batavia, Madagascar, and the Straits. By high class cultivation in new countries one or other of these stocks appear to have given origin to all the superior canes, of which so much has been written. As in the case with the attempt to bring back to India the greatly improved Carolina rice, so the triumphs of the West Indian sugar plantations have, by no means, proved suitable to India or the other countries from which, there is every reason to believe, the stock was originally obtained. Whether all the canes of Southern Asia and the Archipelago had been derived from India in the first intance (long anterior to the attempt on the part of Europeans to cultivate cane and manufacture sugar) is a point on which we shall very probably never be able to arrive at any very definite opinion. That India possesses a sufficiently comprehensive series of what are popularly designated indigenous forms to have given origin to the canes of the Straits, Batavia, China, Japan, Madagascar, etc, is a fact that scarcely requires to be stated. The diversity recognised by the pioneers in European sugar planting between Straits, Batavian, Madagascar, etc., canes was by no means so great as can be now shown to exist in the modern races of cultivated canes. The conclusion arrived at may, therefore, be briefly stated that nothing has been discovered in philology, botany, or history that seriously upsets the hypothesis, that all the forms of cane emanated form a common species which was very probably originally a native of India.

B.—The so-called Indigenous Canes.

Indian Indigenous Canes. 104

VII.—INDIAN CANES.

References.—*Trans. Agri.-Horti. Soc. Ind., I.* (*Ed. 1836, Communication dated 1824*); (*Black cane grown in Burdwan said to yield a stong grained sugar*), *102*; (*in Gasipúr District there are said to have been three kinds, the best Khura, the second Búrli and the third an inferior form not specialised; the 1st was used for making refined sugar and the 2nd and 3rd gúr*), *121*; *II.*, (*very superior qualities grown in Assam, but no sugar, or*

CULTIVATION: VARIETIES. Indian Indigenous Canes.

spirits manufactured there in 1835), 164-165; Report of experiments conducted by the Society on the cultivations of canes at Akra (p. 392 of Ed. 1838) shows red Bombay white Bombay, and red country kinds as experimented with; III., (in Bengal and Assam, three forms, Púri, Kajúli, Kullúa), 62; (Proceedings) Bombay cane said to be three times the size of Bengal indigenous, 34; IV., (three forms grow in Nepal), 104; VI., (varieties grown in Mirzapur, viz., *Mhungo and Huowoh, the latter yields dark coloured juice); 6-7; (Major Sleeman mentions the indigenous canes of the Central Provinces,* viz., *a purple cane—'uchrunga—and three good straw-coloured canes—Kúsiar), 91-92; (Proc.) 25 (a specimen of Púri cane 18 feet long); Hamilton (Buchanan) Statistical Survey of Dinajpur, Rungpur, Purniya, Bhagalpur, Behar, Patna, Shahabad, and Gorakpur; Buchanan Hamilton's Journey through Mysore, etc.; The Honb. the East India Company published in 1822 a series of papers and official correspondence entitled "East India Sugar;" the following references to that work may be given as denoting the sugar canes found in Bengal prior to the effort to introduce foreign canes:—The Resident of Benares reported in 1792 that there were five canes grown,* viz., *Punsari, Reonda, Mungú, Nivar, Kiwahi—the two first were in greatest repute, the Reonda being a purple cane; In Radnagore there were two forms—a white and a purple; In Rungpore the best cane was said to have been recognised as Kadgelí ukhor or black-cane; In Santipore there were two canes in 1792—the Púri and the Kadjuli.*

In the numerous district accounts quoted in another chapter of this article below, frequent mention will be found of the forms of cane grown at the present day. So diversified are these, however, that the writer has found it impossible to attempt a classification, or to give a review even of the statements made. The following passages, arranged under provincial sections, may, however, be accepted as conveying some of the chief ideas that prevail or have prevailed regarding the so-called indigenous canes; but it will be observed, in the remarks offered in this section, the effort has been made to mainly exhibit the canes which were cultivated at about the time of (if not prior to) the rage for imported qualities. It is believed that the effect of the importation of foreign canes has been largely to destroy all trace of the canes which had been adapted to the climates and soils, in which they had been evolved after countless ages of cultivation:—

BENGAL. 105

(a) **BENGAL.**—The Honourable the East India Company, seeing the necessity that existed to obtain definite information regarding the canes of India, and their suitability for the purposes for which they were grown, called for special investigations from its local agent. The various reports procured were reviewed by the Board of Directors in 1792, and the following *précies* issued, accompanied with coloured illustrations:—

Kadjúlí.—" A purple-coloured cane, yields a sweeter, richer juice than the yellow or light-coloured, but in less quantities, and is harder to press. It grows in dry lands; scarcely any other sort in Beerbhoom, much grown in Radnagore; some about Santipore, mixed with light-coloured cane. Grows also near Calcutta, in some fields separate, in others mixed with *púri*, or light-coloured cane. When eaten raw, is more dry and pithy in the mouth, but esteemed better for sugar than the *púri*, and appears to be the superior sort of cane. Persons who have been West Indian planters do not know it as a West Indian cane."

Púri.—" A light-coloured cane, yellow, inclining to white; deeper yellow when ripe and on rich ground. West India planters say the same sort as grows in the West India islands. Softer, more juicy than the *kadjúlí*, but juice less rich, and produces sugar less strong. Requires seven maunds of *púri* juice to make as much *gur* or inspissated juice as is produced from six of the *kadjúlí*. Much of this kind is brought to Calcutta market, and eaten raw."

CULTIVATION: VARIETIES. Indigenous Canes: Bengal.

Kullerah.—"This cane grows in swampy lands, is light-coloured, and grows to a great height. Its juice is more watery, and yields a weaker sugar than the *kadjúlí.* However, as much of Bengal consists of low ground, and as the upland canes are liable to suffer from drought, as was the case last year, and in some degree with the present crop in May last, it may be advisable to encourage the cultivation of it, should the sugar it produces be approved, though in a less degree than other sugars, in order to guard against the effects of dry seasons. Experiments alone can determine how far the idea of encouraging this sort may answer."

Conf. with p. 134.

"*Punsarí, reonda, mungo* (? *mango*), *newar, kiwahi.*—Different sorts produced in Benares district, but not known to the Board under these names." "*Punsarí* and *reonda* appear to be the most productive and the most esteemed."

Shortly after the appearance of the above, **Roxburgh** studied the canes of Lower Bengal, and in his *Flora Indica* he proposes a classification which may be said to be identical with the above, namely, into the common yellow cane *púrí; second,* the purple cane *kajúlí* (said to yield a juice ⅛th richer than that of the *púrí,* though the sugar made of it has an objectionable colour); and *third,* a very large light-coloured cane called *kallúa,* which, **Roxburgh** says, grows in low swampy lands.

In the Transactions and Journals of the Agri-Horticultural Society (especially the earlier volumes) frequent mention is made of the indigenous canes. These have, perhaps, been sufficiently indicated by the abstract given under the paragraph of references. One writer lays special stress on the fact that the ordinary cane of the vicinity of Calcutta, when planted on newly cultured lands in the Sunderbans had vastly improved until it resembled the much prized Otaheite. It was further noted that the rattoons yielded a second and very considerably superior crop of canes to the first. Many of the early writers urge the suitability of large tracts of the delta of the Ganges for sugar cultivation, especially from Dacca to Chittagong. The great sugar-producing area of Bengal (during the time of the East India Company) may be said to have extended from Calcutta through Eastern and Northern Bengal to Behar and Benares. **Dr. Buchanan-Hamilton** was directed to conduct a statistical survey of that region, and his report may be said to afford a mine of knowledge on all aspects of the then pressing sugar question.

Ratooning. *Conf. with pp. 76, 77, 78, 128, 195.* 106

Dinagepur. 107

He tells us that (from 1809 to 1814) there were two kinds of sugar-cane grown in DINAJPUR, *viz., khagra* and *kajolí,* or *kajalí.* The former was a yellow cane, hard, and not thicker than the finger, and was only grown by indolent farmers in the northern parts of the district. The latter had a tolerably thick stem, deeply stained with purple: it often grew 12 to 15 feet in height. In RUNGPUR both the canes already mentioned were also found, but the *khagra* was the most prevalent. This was probably due to its being too hard for both the white-ant and the jackal, and, to the fact of its being sufficiently stout to resist the winds without necessitating either protection, or of requiring its leaves to be tied up. In PURNIYA, he says, sugar-cane was most wretched and was confined to the banks of the Kankayi. The canes cultivated were, 1st, a most inferior kind known as *nargou* (from its resemblance to a common reed); this gave almost no juice; 2nd, the major portion, the *khagri* cane; 3rd, a large kind called *bangsa*; the comparison to the bamboo, **Hamilton** remarked, holds good only because of its being contrasted with the *nargou;* 4th, a little *kajali.* The *bangsa* differs from the *kajali* of Dinajpur in its stem being entirely yellow. In BHAGALPUR cane was mainly cultivated, **Buchanan-Hamilton** observed, near the hill streams, where it is copiously watered and in Rajmahal where it attains considerable perfection. In the interior of the

CULTIVATION : VARIETIES Indigenous Canes : Bengal.

district cane was seen to be fairly extensively grown but of inferior kinds. BHAGALPUR thus afforded a greater variety in the forms of cane met with than was the case with any of the other districts of the survey. The *kajlí*, Dr. Hamilton says, was by far the best and was confined to the Rajmahal division. The *mango* cane he regarded as of equally good quality if it was not the same thing. It was chiefly used for eating without preparation. The *paungdi* and *raungda* were, he further tells us, tolerably large yellow canes, and one of them, at least, appeared to be the same with the *bangsa* of Purniya. The *kirnya* was a small poor cane. The Collector of Bhagalpur wrote in 1792 that the chief discouragement against sugar culture in that district was "a salt of the nature of epsom salts, with which many of the wells are, and, of course, much of the land is, in the district south of the Ganges, impregnated; and to which with great probability, is imputed the bitter taste that renders Bhagalpur sugar so much inferior to that made in other districts. This bad quality is not found in the sugar made in Rajmahal, but there, as in Bhagalpur, the cane is of inferior size."

Salt in Soil Injurious. *Conf. with p. 130 134, 161.*

108

In BEHAR (PATNA) Dr. Buchanan-Hamilton observed six kinds of cane being cultivated. These were *ketar, baruka, mango, shakarchina, raungda,* and *paunda*. The three latter, he remarked, were thick and their juice very sweet, on which account they were eaten but not used for expression. The *ketar* was a form with stems not thicker than the finger, and was the most common of all, being, probably, the same as the *keruya* of BHAGALPUR. All the three kinds in common use for the mill, he adds had yellow stems. In Shahabad, Dr. Hamilton remarks, there were during the time of his survey, seven kinds of cane grown, *viz.*, *reongra*, or *rioda*, *saroti, mango, barukka, bhoronga, kaiva,* and *bhurli*. They were small canes not thicker than the finger and had all yellow stems. The *mango* was reckoned the best. These various kinds, though he could not make out their distinctive characteristics, were adapted for different soils. The high land near the villages was alleged to produce a large cane filled with juice that gave little saccharine matter, it was, therefore, used as an edible cane. In GORAKHPUR, Dr. Hamilton found four kinds cultivated, *viz.*, *reongra, mango, sarotiyo* and *baruka*, all these, he says, were suitable for yielding extract.

Introduced Canes *Conf. with Paunda, pp. 52, 64-66.*

109

The more recent publications, such as the district manuals, while largely compiling from the old authors (more particularly Dr. Buchanan-Hamilton) have occasionally furnished new and interesting facts. A selection of a few of the more instructive passages will be found below in the provincial chapter on cultivation, but the following may be here given, since they specially refer to the varieties grown in remote parts of the province:—

CHAMPARAN.—The plant seems to have been introduced about the year 1805 from Azimgarh and Gorakhpur. The kinds of cane which are most valued are *mango, pánsáhi, lálgainra, sarawati,* and *painwará*, of which the two first are principally grown in *Parganas* Manpur and Batsara. From *mango* is produced a very good *rab* for fine sugar, while *pánsáhi* is generally grown for *chaki* or *gur*.

In DACCA "there is a number of different varieties of canes grown here, some of which are very fine and well suited to the soil on which they are cultivated. *First*—the *khagri* is a thin, hard cane that will grow on any land. It is generally grown on land that will grow no other cane, namely, the low alluvial land on which stands four to six feet of water during the rains. The juice is rich in sugar and capable of producing first-class *gur*, but the outturn is very poor compared with other varieties. *Second*—the *dhalsundar* is a white cane with a reddish tinge, thicker than the *khagri*, but cannot stand stagnant water nearly so well as the latter. This is the variety that was formerly most extensively grown, but it has been of late largely superseded by better varieties. *Third*—the *merkuli* is generally grown on the banks of the Meghna and the Brahmaputra. It seems to be the same as *dhalsundar*, but altered a little

CULTIVATION: VARIETIES. Indigenous Canes: Bengal.

from having been grown long on much drier soil. *Fourth*—the *kali* or *kajli* is a hard red cane, having few qualities to recommend itself. *Fifth*—the red Bombay is the same as the Bombay of West Bengal and the red Paunda of Saharanpur. This was the variety of cane that Captain Sleeman brought from the Mauritius. It is a very superior cane, suited to high land, but extremely delicate. It was for a time largely grown near Dacca, but it does not seem to tiller well on the red clay, and has almost gone out of cultivation. *Sixth*—the *sharang*, known in the north-eastern part of the district as the *sharang*, of Dhalbazar, is a white cane somewhat inferior to the *shamshara* or the Otaheite cane of West Bengal. *Seventh*—The white Bombay is perhaps the same as the white Paunda of Saharanpur. It is a thick cane, juicy, and rich in sugar, somewhat harder than the *shamshara*, but grows to a much longer size than the latter. It is a hardy cane that will stand wet soil much better than the red Bombay. Taking everything into consideration, this is, I believe, the best cane for Lower Bengal."

In LOHARDAGA it is stated "There are three varieties in cultivation, *viz.*, (1) the *bánsá*, (2) the *punri*, and (3) the *kájali*. The *bánsá* is a thin white cane, less juicy than the *punri*, and hard to press. The *punri* is a little thicker, but shorter, containing a larger proportion of juice and softer in texture. The third variety is so called from the purple colour of the cane; it is never grown by itself, but crops up here and there among the first two varieties. The *bánsá* yields a *gur* of a nicer colour than the *punri*, but having a bad smell. In Támár, the *punri* is grown in preference to the *bánsá*, which is, however, more largely cultivated in other parts of the Five Parganás."

In PALÁMAU "Five distinct varieties of sugarcane are in cultivation. These are—

(1) *Ketár*, a thin but very tall variety grown in the central tuppehs of Palámau: it corresponds to the *bánsá* cane of the Five Parganás.

(2) *Manigo*, a thin and very short cane 4 or 5 feet high, commonly grown in Deogán in the north-east of the sub-division. This variety is extensively grown all over south Behar and is valued for the large proportion of sugar to juice which it yields.

(3) *Newár*, resembling the *manigo*, but a little thicker; *newár* cane yields a large quantity of sugar area for area than *manigo*, but the *gur* has a saltish taste and is less valued.

(4) *Básin*, yielding a very bright coloured *gur*, but an inferior yield.

(5) *Bhunli*, allied to the *newár*, but with shorter joints, yielding a bright coloured *gur*.

The first two varieties are commonly grown in Palámau; the other three are but little known. The *ketár* is an early variety, being reaped in Pous and early Mágh; the other varieties are late, and their harvest does not take place before Fálgun and is continued to the first fortnight of Cheyt."

ASSAM. 110

(b) ASSAM.—Numerous writers (half a century ago) allude incidentally to the superior quality of canes found in Assam. In fact the large red canes of Assam and Bombay and the large canes of the lower Eastern Himálaya, such as those of Nepal, were by many persons held to be identical with the superior canes of the Straits. Thus, Mr. Wray wrote, "I have seen only three varieties of large canes on the continent of India, which are supposed to be peculiar to the country; one is the large red cane of Assam, specimens of which were kindly sent me by Dr. Keith Scott, the Honourable East India Company's Civil Surgeon, at Gowhatty in Assam. This gentleman had established a sugar estate at Gowhatty and made sugar; so that he had experience of the right sort, which lends to his opinion considerable weight." Dr. Scott wrote of these canes that they "were juicy and sweet; the sugar produced from them is of an exceedingly fine grain and good colour, they are, moreover, strong in growth, and much less apt to fall over than the Otaheite to which they are fully equal in size, as well as in quantity and quality of juice. I a also preparing for you some '*flowers* of this cane, in different stages; which I will despatch when quite dried. I have now (January) canes in flower, which were planted last May!!" It will be seen from Major Jenkins' report on Dr. Scott's flowering canes (*Chapter on Improvement by Seminal Selection*) that a mistake was apparently made either by Jenkins or Scott. The former gentleman speaks of the Gowhatty canes which were seen flowering and seeding as having been the Otaheite form, the latter of their

Flowering. *Conf. with pp. 8, 9, 11, 44, 47, 83-88, 109.* III

CULTIVATION: VARIETIES. Indigenous Canes: Assam.

having been a large Assam cane which was quite as good as the Otaheite. It seems probable that **Dr. Scott's** version of the story is the correct one, the more so since he appears to have been in correspondence subsequently with **Mr. Wray** on the subject, and that author accepted the Gowhatty cane as a superior quality peculiar to Assam. There are two features of this cane which on passing, it may be desirable to note, 1st, the fact that there was in 1844 a sugar plantation and factory in Gowhatty under European supervision; and, 2ndly, that, as in all the other Indian allusions to the seeding of the cane, the subject appears to have been allowed to be forgotten.

Plantation under European Management. *Conf. with pp. 37, 48, 63, 91, 103, 111, 161-62, 212, 306, 309*

112

Mr. Wray concludes his notice of the Assam red cane by alluding to a similar kind found in Bengal. "In Lower Bengal (he says) near Calcutta and in the Straits of Malacca, a large red cane abounds, which bears so exact a resemblance to **Dr. Scott's** Assam cane, that I conceive it to be the same identical variety, somewhat improved in the rich and fertile soil of Assam." "The red cane of Bengal is a large and fine cane, much used about Calcutta for sugar manufacture; and I have had brought to me by Natives, sugar made from it by themselves (in their own rough and primitive way), which exhibits a grain of immense size, strength, and brilliancy. They, however, say that it is not a good cane for sugar making, as the juice is very dirty, and the sugar always dark coloured. These assertions, however, have no weight with me; for I can easily detect the cause, and know that it can be avoided."

To bring the history of the Assam canes down to more recent times, it is, perhaps, only necessary to quote a few passages from a very able note on the subject written in 1883 by the late **Mr. E. Stack,** then Director of Agriculture in Assam. It may be here pointed out, however, that **Mr. Stack** speaks of the superior qualities of Bombay or Bengal canes. This would seem to indicate that the system of exchanging stock seems to have extended even to Assam, and the mystery of the true nature of the so-called Bombay cane remains as great as ever. Many writers are suspicious that it may be the acclimatised stock of some of the foreign canes; but, as opposed to that view, **Scott** and **Wray** may be pointed to as having accepted these large canes as indigenous (or rather local) forms. There seems no doubt of the fact that India possesses so extensive a series of canes that a process of selection and distribution of the superior qualities would (as urged in another chapter) greatly improve the stock of the country as a whole.

Mr. Stack refers his remarks on the canes of Assam into those of the two great divisions of the province, *viz.*, the Brahmaputra and the Surma valleys:—

The Brahmaputra Valley (Assam Proper.)

"The varieties of sugarcane in the valley of the Brahmaputra are not numerous, and may be ranked as follows in the order of their importance:—

1. *Bagi* (white) or *mugi* and *mag* (amber-coloured).
2. *Rangi* (red), *kali* (black), or *teliya* (*i.e.*, the colour of newly-expressed mustard-oil).
3. *Bengal* or *Bombay* cane.
4. *Malahá* and *magará* or *megalá.*

"*Mugi* and *teliya.*—The two first kinds are regarded by the Natives as indigenous. They are commonly grown together, either intermixed at random, or with the red cane disposed round the edges of the field as a protection to the more valuable yellow cane against the depredations of men and animals. A well-cultivated field of *mugi* stands about seven feet in height, and the canes measure a little more than an inch in diameter at the thickest part; the colour is an amber yellow, and the texture soft and juicy. The *teliya*, on the other hand, is hard and thin, of a deep red colour, often passing into a dark shade of purple, whence its name *kali*, or black; and the

CULTIVATION: VARIETIES. Indigenous Canes: Assam.

average dimensions of the stalk do not exceed five or six feet in length by three-quarters of an inch in diameter. These two varieties of cane are more largely grown than any other, the *mag* being recommended by its superior qualities as a sugar-producer, while the hardiness and unattractiveness of the *teliya* render it well adapted for the careless style of cultivation which is affected by the Assamese ryot.

Plantations under European.

"*Bengal or Bombay cane.*—The cane called *Bangála* appears to have been brought from Bengal, either at the time of **Captain Welsh's** expedition in 1793-94 (as is the tradition in Nowgong), or by European sugar-growers in Kámrúp some thirty years ago. In the Mangaldai sub-division, where it is said to be of very recent introduction, and also in North Lakhimpur, this cane is called by the alternative name of *Bambo* or *Bam*, implying a doubtful connection with Bombay. This foreign variety of cane greatly excels the Assamese in size and juiciness; but as a sugar-producer it generally ranks below the indigenous *mugi*. Like the country cane, it is divided into yellow (*pura*) and red (*teli*); the former, which is much the commoner, is a large soft cane, with stalks averaging eight feet in height and an inch and a half in thickness, while the latter is said in the Mangaldai sub-division to be even larger and more juicy. The Bengal cane is grown chiefly in the southern part of Kámrúp, in the Mangaldai sub-division, in Sibságar, and, it is said, in North Lakhimpur. Elsewhere it is cultivated as a garden plant, in tiny patches of *basti* land close by the ryot's dwelling, and is eaten in the raw state after being slightly heated to increase its sweetness, or the juice is used as a syrup in compounding medicinal pills. Though yielding much more juice than the country cane, the Bengal kind is apt to break into small pieces in crushing, and thus gives more trash to the mill. A degenerate variety is known by the name of *Asamiya puri* in Kámrúp, and *Keteki puri* in Upper Assam.

"*Other varieties.*—The *mahlá*, or *malahá* cane of Kámrúp and Darrang, so called from its resemblance to a kind of reed of the same name, and the *magará* or *megalá* of Upper Assam, are a hard and thin variety of the country *mugi*, and, where grown at all, they are planted round the edges of the field, or intermixed with the *mugi* by chance. This cane, like the *telyai*, is so hard and dry that it may safely be left to protect itself against man and beast.

"Two local varieties of cane appear to be peculiar to the Mangaldai sub-division, namely, the *bhábeli*, resembling *mugi*, but with shorter intervals between the joints, and the *kâmrângi*, a cross between *mugi* and *teliya*. The former is used for medicinal purposes only (chiefly in disorders of the kidneys), and the latter is not intentionally grown at all.

THE SURMA VALLEY (SYLHET AND CACHAR).

"Sugar-cane is cultivated in the Surmá Valley in much the same fashion as in the valley of the Brahmaputra, but the local names of the common kinds of cane are different. In Sylhet, besides the so-called *Bombay* cane, we find the *dhálí*, or white, and the *surang*, or red, cane; in Cachar the *Bombay* cane is highly esteemed as the largest and juiciest and the best sugar-producer, and is sometimes grown as a garden product and eaten in the raw state, while the *lángli* cane appears to correspond to the *mugi*, and the *shamsheri* or *kámrángi* to the *teli* of the Assam Valley. Both districts possess also a small hard species, called *khágari*, or reed cane, which may be compared to the *malahá* of the Assam Valley, and in Cachar this and the other inferior sorts are said to be most in favour, as requiring less care, and being less liable to disease or the attacks of grubs and wild animals. The site chosen is high land, in the vicinity of a village if possible (*chára*), or, failing room there, on the bank of a river. Oil-cake (*parkai*) is used as manure a couple of months after planting, if the cultivator happens to have had a mustard crop of his own, and in Sylhet it is even bought for the purpose."

Mr. Stack makes the following remarks regarding the possibility of improving the Assam sugar-cane cultivation. His statements regarding former factories are of considerable interest in reference to the remarks which have been offered on the history of the effort to establish European sugar-cane plantations in India:—

Management. *Conf. with pp. 37, 48, 62, 91, 93-94, 103.*

"These enterprises are by no means the first of their kind. A similar experiment was tried in the preceding generation by a **Mr. Herriot** in Gauháti, and a **Mr. Wood**, at Dobapara, in the Goálpára district, and ruins of old rum or sugar factories existed or still exist, near Jorhát and at Numáligarh (now a tea-garden) in the district of Nowgong. The end of all these speculations, whether from the dearness of labour in the Assam valley, or, as has been alleged in Gauháti, from mismanagement of the concern, was failure so complete that no record of them can now be obtained,

CULTIVATION: VARIETIES. Indigenous Canes: Assam.

and we do not know how far they depended on the produce of the country, or sought to supplement it by importations of *gur* from Bengal. It is probable that they all started in the hope of finding a new field for sugar-cane in Assam, and the efforts made by Mr. Herriot to introduce a better kind of cane were so far successful that the best cane of the present day in Kámrúp and Darrang traces its origin to them. The Native cane being so small, hard, and dry, one obvious means of improvement is the introduction of a better stock from Bengal or elsewhere; and experience has proved that the soil and climate are sufficiently congenial; but the cultivator will take no steps in this direction by himself, and in the large areas of thinly-settled country the Native canes will always be preferred, as needing less protection from wild animals, and entailing a smaller loss in their destruction where protection proves insufficient. The ravages of wild beasts are no trifling obstacle to the developement of cane cultivation in this part of India."

N.-W. PROVINCES. 113

(c) NORTH-WEST PROVINCES AND OUDH.—Of Azamgarh district it is stated that "A number of varieties of sugar-cane are known. Those which are most in use are *sarautiá, raksida, reonra, mango,* and *phatnaiyá*. The last is grown chiefly in the *kachhar* country. The people are not more particular about their selection of sugar-cane seed than about the seed of other crops." In Bareilly thirteen forms are mentioned, *viz.*, (1) the white and (2) black forms of *paunda*, (3) *thun*, (4) *pandia*, (5) *dantur*, (6) *bakri*, (7) *chun*, (8) *dhaur*, (9) *agholi*, (10) *mittan*, (11) *kaghazi*, (12) *neula*, and (13) *kutara*. The *paunda* forms are grown as edible canes. In Gorakhpur four kinds are grown (1) *mahgujur*, a very tall cane, (2) *saroti*, (3) *bhaunwarwár*, and (4) *barokha* or *katarha*. In Jaunpur the canes grown are all small, the largest being known as *nasganda*, the next *paundia*; the *serotia* cane is the thinest and *kawai* the most inferior form—a cane grown mostly around the margin of the fields to deceive the pilfering wayfarer. One of the most instructive accounts of the canes of the North-West Provinces is that given by Mr. Butt, regarding the district of Shahjahanpur. As that district has always held a high reputation for its sugar, and it possesses, at the same time, a large sugar factory—the Rosa Works—Mr. Butt's remarks on the forms of cane grown may be here given in full:—

Varieties cultivated 114

"The food canes cultivated in Shahjahanpur are the *paonda, katahra, kálagana,* and *thún*; they are seldom or never pressed for manufacture, and are cultivated for direct consumption as food canes. These varieties are chiefly cultivated as garden crops near the city, and the cultivation is most remunerative; they have rich sweet juice and soft fibre, and these qualities fit them for use as food canes, but also render them liable to damage from thieves and jackals or other animals, and the crop must be carefully watched; the canes are also reputed to be delicate. Of the other varieties, *dikchan, dhaunr, matná,* and *dhúr* are perhaps the best known.

Dikchan 115

Dikchan is said to be a new variety introduced within the last forty or fifty years, but it is now the cane usually grown throughout the district; it grows freely in any fairly good soil, and gives a large and very quick yield of juice: a *másha* of juice being expressed from *dikchan* in three-fourths the time required for most other varieties. The juice, on the other hand, is rather thin and gives a smaller proportion of *ráb*, but in this respect the cane is said to have undergone a marked improvement. *Dikchan* is a tall cane, commonly ten feet high, and having a very bulky appearance as a growing crop. It is said that advances were most freely made on a field of *dikchan*. *Dikchan* is now out of favour, as in the last two or three years it has suffered more from unfavourable weather than any other cane, and many cultivators are again returning to *dhaunr* or *matná*.

Dhaunr. 116

Dhaunr is a variety somewhat similar, but on the whole inferior to *dikchan*. It is said to require less careful cultivation, and the fibre being harder it is less exposed to injury from cane-eaters, biped or quadruped; it also is commonly planted, and in Pawáyan especially *dhaunr* is commonly planted by Thákurs and Brahmans, *dikchan* by Kúrmís and other more careful cultivators.

Matna. 117

Matná is in almost every respect the opposite of *dikchan*. It is a small stunted cane, only about five feet high, with very hard fibre and a small yield of juice, but in quality the juice is the best of all, and gives the largest proportion of *ráb*. The preparation is more laborious, and it is never sown in low-lying land, as ordinary floods cover the head of the cane and destroy it, while taller canes high enough to

CULTIVATION : VARIETIES. Indigenous Canes : N.-W. P.

keep the head above water, are not materially injured by floods subsiding in a few days. Matná appears to be less cultivated than formerly, as sometimes there is a difficulty in obtaining cuttings for planting. In explanation of this it may be added that *matná* is usually planted in cuttings from the entire cane (*sabbáta*), while with other canes only the top piece next the arrow is planted (*aguwa* or *aghwái*). *Matná* is said to degenerate at once when planted as *aguwa*. Some admirers of *matná* go so far as to claim for it a yield of *ráb* per bigha double, and from the *ráb* a quantity of *khand* (dry sugar) some 15 per cent. in excess of that from any other variety.

Dhani or dhur. 118

Dhani is very commonly planted in lowlands subject to inundation. It is an extremely tall cane, very thin, and with very hard fibre : a small yield of juice, but of good quality. Some cultivators assert that they would as soon grow or press the stalks of *senthi* grass, but others praise *dhani* as one of the most paying canes, and its hard rind and fibre protect it from jackals. Almost every variety has its admirers, and some prefer a mixed growth, such as *dikchan*, *dhani*, and *dhanur*, holding that the denser juice of the *dhani* and *dhanur* assist the ultimate working up. Generally it would appear that the varieties, with hard fibre and knots close together, are best suited for lowlands, and can best withstand the floods, but that they must be sufficiently tall to keep the heads above water; and the canes with softer fibre are best suited for upland cultivation. Other varieties are the *agauli*, somewhat like *dikchan*; the *rilerta*, *bharauka*, *nyúra*, *mandgah*, and *rairi*, grown generally in khádir land."

Having exhibited the chief North-West forms mentioned in district gazetteers, it is, perhaps, unnecessary to do more in this place than to republish **Messrs. Duthie & Fuller's** account of the varieties grown in these provinces as a whole, since the passage that appeared in the *Field and Garden Crops* on that subject practically reviews all that had been previously written :—

"The varieties of sugar-cane (grown at the present time) are very numerous, and as their names vary greatly in different districts, it is a matter of some difficulty to identify them. A broad sub-division may be made into edible and non-edible cane, the former being grown for human food in the raw state and eaten as sweetm at, while the latter is intended for the production of sugar. Edible cane is, as a rule, much the thicker, softer, and juicier of the two, and is grown with very high cultivation. Its principal variety is the one known as *paunda*,* which is supposed to be a recent introduction from Mauritius. In the Dehra Dun district *paunda* is used for sugar making, but elsewhere it is grown merely as a sweetmeat. The most distinct varieties of non-edible cane are (1), a tall soft cane growing as high as 10 feet, requiring good cultivation and yielding a large proportion of juice (*dikchan* in Rohilkhand, *barokha* in Kawnpore); (2), a shorter and rather harder cane not often more than 5 or 6 feet high, yielding less but richer juice than the above (*agholi*, *matna*); (3), a hard, tall, reddish cane of poor quality much grown in damp localities without irrigation (*chin*); (4), a dwarf white hard cane yielding more juice than *chin*, but resembling it in being grown on second-rate land (*dhor*). The two first varieties are delicate and require a rich well manu red and well irrigated soil, the two latter yield a crop with much less care and expenditure and suffer much less from flooding in the rainy season."

As having a possible bearing on the canes of the North-West Provinces, it may be pointed out that **Munshi Sabhan Rai** of Patiala published in Persian his *Khulasatu-'t-Tawárikh* — a work which furnishes interesting particulars regarding sugar-cane. It appeared in A.D. 1695 **Sher Ali Afsos** issued in 1804 a Hindustani work —the *Araish-i-Mahfil*. This, while claimed as an original work, is literally a translation of the Persian history, amplified and moderanized in minor details. Through the kindness of **Lieut. Wolsey Haig** the author has been furnished with the following translation from the *Arai-h-i-Mahfil* and it may be allowed that the information it contains represents the currently accepted classification of the canes of Upper India during the seventeenth century.

* It will be seen that there was a cane in Behar and Bhagalpur that bore that name, at the beginning of the present century, or at least 30 years prior to the introduction of Mauritius cane. *Conf. with pp. 57, 60, 66.*

SACCHARUM: Sugar.

Cultivation of the Sugar-cane.

CULTIVATION: VARIETIES. Indigenous Canes: N. W. P.

"The villagers and zemindars of Oudh, Lucknow and that district call it *úkh*, but among those of Delhí and the surrounding country it is known as *ikh*. There are many species of sugar-cane, and each has a distinct name, but the only names commonly used in Urdu are *ganná, katárá,* and *paundá*.* The first of these is a generic name under which may be classed all species of sugar-cane, but the other two are names of distinct species. Thus the *katárá* is a hard and slender cane, equal in height to the *paundá*, or perhaps a little longer, but very hard and with little juice—*khánd, misri,* etc., are made from it. There are two sorts of *paundá, viz.*, the black and the white. Although the black is superior to most sorts of sugar-cane in some points, yet its sweetness is combined with a bitter, and sometimes with a saltish flavour. In spite of this it is far from deficient in sweetness. However, it is, from its hardness, apt to injure the teeth and tongue of any one eating it.

Edible Cane. Paunda. *Conf. with pp. 52, 64, 73, 232.* 119

"The white *paundá* is in every way superior to any other kind of sugar-cane. Its flavour is principally in the knots, but the parts in between the knots are most pleasant in taste, and each of the knots is full of juice, moreover, it is so tender that a toothless man, or even a child at the breast can chew it without trouble."

NEPAL. 120

(d) NEPAL.—**Dr. Buchanan-Hamilton** published in 1819 his "*Account of the Kingdom of Nepal.*" In that work he says remarkably little on sugar-cane, though he had only just completed his survey of the sugar-cane of Eastern Bengal. "The Newars," he remarks, "make a very little extract, soft sugar, and sugar-candy; but a large proportion of the cane is eaten without preparation." "The juice is generally expressed by a lever." In 1837, however, **Mr. A. Campbell** furnished the Agri.-Horticultural Society with a highly instructive and detailed paper on *The Agriculture and Rural Economy of the Valley of Nepal.* In that work much interesting particulars occur regarding sugar-cane. There were three forms grown, *viz.*, the *chi-tu*, the *kusha-tu*, and *ghewora-tu*. These are the Newari names and in Parbutiah they were known as *sano ghenra* (small reed-like cane), *kalo ghenra* (purple cane), and *sheto ghenra* (white cane). **Mr. Campbell** remarks "there are three kinds cultivated the large white one, the large purple cane, and a small white reed-like cane. The latter is most common in the valley; its produce is poor compared with that of the others; but it is suited for the only descriptions of soil allotted to its growth here, *viz.*, a hardish clay or light sand." In a foot-note, **Mr. Campbell** adds, "The purple and large white varieties ought not, perhaps, strictly speaking, to be enumerated as agricultural products of the valley, as they are only grown in very small quantity in the gardens of the wealthy. The small white kind is the one usually grown as a crop." The large purple cane is considered the most productive of sugar. "The fresh sugar-cane is a very favourite food of the people, hence more than half the crop annually raised is consumed in this way. Almost all the purple, and large white cane grown, is eaten fresh; the small reed-like cane alone being reserved for sugar making." The production of the valley, **Mr. Campbell** adds, was not, however, sufficient to meet the demand and large quantities were imported from "the neighbouring valley of Noakoti," 20 miles distant. The cane of Noakoti is principally of the larger kind, the purple predominating, while in the great valley, the small reed-like cane is the most abundant. It is seldom much thicker than the little finger or higher than 6 to 8 feet and is hard and juiceless. The large purple cane, especially in the warmer valley of Noakoti, attains a height of 10 feet and 6 inches in circumference. **Mr. Wray** (*The Practical Sugar Planter*) refers to a large black and yellow Nepal cane which were fully equal to the superior cane of Assam.

---

* *Paunda* must, therefore, have been a name given in India to a certain edible cane long anterior to the time when foreign canes were introduced.

CULTIVATION: VARIETIES. Indigenous Canes: PANJAB. 121

(e) PANJAB.—Mr Baden Powell, in his *Panjáb Products*, gives the following account of the canes of this province:—

"SUGAR-CANES, AND SUGAR therefrom. Synonyms—*Kumád; nai shakar* (Pers.); *Ganná; ûkh* (Hindustán).

"The first thing to be done is to describe the culture of the sugar-cane, and the way in which the juice is extracted and converted into sugar.

"In the Lahore district I obtained five kinds of sugar-cane; some of these were merely varieties. There is a purple cane, called '*kumád kálá*;' a hard thin cane, called *kumád lahorí*; another called '*kátá*;' and others, the plants of which were obtained from Jálandhar and Sahéránpúr. The principal difference observable is in the size of the canes: one sort is very thick and succulent, and is principally used for eating: it is cut up into pieces, peeled, and sold in the streets: contrary to what one would suppose, the thin hard canes yield the greatest quantity and the best syrup: the succulent ones are two watery.

"In Gujranwala, Major (now Major-General) Clarke mentions three kinds of cane: *Daulá, treda*, and *chinkha. Daulá*, or white, is the best; *treda* is yellowish; *chinka*, which is reddish and small, produces good *kand* and *chíní*, moist sugar."

The above passage gives a fairly comprehensive account of the canes grown in the Panjáb, but it may be amplified by a few passages from more recent publications. In the *Delhi Gazetteer* it is stated three kinds of cane are recognized, *viz., lálrí, mírati, soratha*, and *paunda* or *gunna* (edible). The *lalri* or *lalsi*, though not very sweet, is rich in saccharine matter. In—

"HOSHIARPUR—Five kinds of sugar-cane are grown in this district:—

(1) *Chan.*—A thin reddish juicy cane with a thin peel.

(2) *Dhaulu.*—Whiter, thicker, and rather more easily peeled.

(3) *Ekar.*—Resembles *dhanlu*, only with dark coloured lines; the peel is harder and there is less juice.

(4) *Kanara.*—White, very soft and juicy.

(5) *Pona.*—Thickest and the most juicy variety.

The kind almost always sown, except in the *kolha* or stream-irrigated lands in the hills, is *chan*. Its juice is considered superior to that of any other kind for making sugar; it is also less liable to injury from frost than *dhanlu*; but the latter is sometimes to be found mixed with *chan*. *Ekar* is not much thought of, being the hardest and least juicy variety, and some cultivators cut it down directly they recognise it in a field. *Kanara* is generally only cultivated in the hills; it is very soft and juicy, and the people have a saying that very little of it reaches the sugar press, most being chewed by the men working in the fields; the quality of its juice also is inferior to that of *chan*. *Pona* is never pressed, and is only grown near towns for chewing. A new kind of cane called *kahú* has lately been introduced for experiment from the Gurdáspur district: it is thick and juicy; but it has not been tried long enough for any definite opinion to be formed of it. The people seem to think the *chan* is the best kind for sugar.

Of Gujranwala, it is stated that "three varieties of cane are grown, *viz.*, the *treru* and *chinkha* are most generally cultivated; the *dowlo* (*dhaula*) or white, a delicate variety grown in the Charkhari, is esteemed the best; but the objection to its more extended culture is the extra labour and attention it demands, for which agriculturists consider the superior crop does not sufficiently compensate. The *treru* is a yellow sort, and the cane is not so strong or straight. The *chinkha* is an inferior kind, and of red colour; the cane is very sweet, but gives little juice; this sort is sometimes grown only for fodder." Two or three varieties of cane are said to be grown at KANGRA, called *chum, eikur, kundiari*, and a kind, containing a lot of juice which is raised only for eating, called *púna*. It may be here added that the sub-temperate canes of Kangra are sweeter and richer in saccharine matter than those of the plains. The following account is given of the varieties grown in Karnal district.

"The principal varieties sown are *surta* or *sotha*, with a long, soft, thick, white cane; the best of all, but somewhat delicate, and especially fancied by jackals.

Lálri with a hard, thin, red cane, very hardy, and will not spoil even if the cutting be long delayed; but not very productive of juice.

CULTIVATION: VARIETIES. Indigenous Canes: Panjab.

Merati or *Merthi* with a thick, short, soft cane, and broad leaves: it is very productive, but requires high cultivation, and suffers from excess of rain; it is not much grown.

Pondá, a thick sweet variety; grown near the cities for eating only, as its juice is inferior.

"Around the city of LAHORE, a good deal of the large thick cane called *pona* is raised, but *gúr* or sugar is never extracted from this species, and it is merely grown for sale in the bazar." The sugar-cane grown at LUDHIANA is almost entirely grown for the manufacture of some saccharine product (called *kátha* cane) but in a few villages the *ponda* or eating variety is raised. There are three varieties—*chan*, a soft juicy cane which grows to a considerable height, has a red colour and long joints (*pori*); *dhaulu* does not grow so high, has small joints, and is of a green colour and less juicy, while *ghorru* or *g·aru* is an inferior sort with many joints and a great deal of leaf at the top, very hard and yielding much less juice than the others. It is said to be suitable to sandy soils. The first of these (*chan*) is the real cane and the other two are mere degenerations; no one ever keeps a *ghorru* stalk for seed; and *dhaulu* is only planted if there is not enough of *chan*.

In the Montgomery district "sugar-cane is called *ponda* or *paonda*. There are two kinds, the *sahárni* or Sahâranpuri, and the *desi* or Jallandari. The former is the coarser and larger of the two. The *desi* is sweeter, softer and more juicy. The *treru* cane of Siálkot is distinguished by having dark lines from joint to joint. This pecularity is also said to exist in the *ekar* cane of Hoshiarpur. The *mendku* of this district is said to have come from America, but it does not meet with much favour, as its juice though large in quantity is poor in quality.

In a report on the sugar-cane of the Panjáb, written by the Junior Secretary to Financial Commissioner, information similar to that given above (which has been taken from the Settlement Reports and Gazetteers) was published in 1883. The paragraphs in that report on the cultivated canes conclude as follows:—

"Distinct from all these kinds is the large succulent cane called *paunda* or *pona*, which is sold in towns for chewing. It is not used for the manufacture of sugar in the Panjáb proper, but in Peshâwar it is preferred to all other kinds for crushing, as it is said to give little trouble, and its use is also extending in Bannu. There seems to be a tendency among the cultivators to distrust the larger varieties of cane, as they are not only more tender and subject to injury from frost, but also, as a rule, the juice which they yield seems to be more watery and less rich in sugar. The cultivators would undoubtedly adopt new varieties if they were established by practical experiment to be really better sugar-producers and at the same time not much more difficult to cultivate than those which they already possess."

CENTRAL PROVINCES. 122

(f) CENTRAL PROVINCES.—The following passage from an able report by Mr. J. B. Fuller regarding sugar-cane cultivation of the Central Provinces gives the chief peculiarities of the forms of the plant grown in these provinces:—

Conf. with p. 42.

"The varieties of cane grown in the Jubbulpore Division and in the Nerbudda valley are thrown into two classes known as *Gawna* and *Barahi*. The first class includes soft thick canes eaten to a great extent as sweetmeats; the chief of these are the white Otaheite cane, said to have been introduced by Colonel Sleeman, the common white edible *paunda*, and the red striped cane called *pachrangi*. A number of varieties are classed as *barahi* which bear different names in different districts. Amongst them may be mentioned the *kutiar*, *sararu*, and *kansia*, all of which are short, thin, and hard, but yield a juice which is in some respects better than that of *ganna* for sugar making. The varieties grown in the Nagpur and Chhattisgárh country fall into the same two classes, the first comprising the kinds known as *bangla*, *dhaori* (with yellowish stalks) *mailagir* and *pachrangi* (stalks striped with red and purple) and *kala* (or *karia*) with stalks of a deep purple colour. The second or hard stemmed class includes the hard white *kathia* cane, and the reddish coloured *ledi*,

CULTIVATION: VARIETIES. Indigenous Canes. Central Provinces.

both of which yield juice with a strongly saline taste. These two latter are the only kinds ever grown without irrigation. Speaking generally, the cultivation of the short hard varieties occasions far less trouble and expense than that of the softer kinds. They require less manure, less water, and less expenditure on fencing, since they are not so liable to be devoured by pigs and jackals. Indeed, it is not uncommon to see borders of the *ledi* variety grown as a fence round a field of a more valuable kind of cane. I may mention here that sugar-cane is commonly called *santa* in the Nagpur districts and *kusiar* in Chhattisgarh."

CENTRAL INDIA. 123

(g) CENTRAL INDIA & RAJPUTANA.—Very little of a definite nature can be furnished regarding the sugar-canes of the great central tracts of India. This is to be regretted since it seems probable the forms met with in the warmer and drier areas would be peculiarly interesting. The multiplicity of the forms mentioned by some writers demonstrates only the extent of the field of future research. It is, in fact, essential that some standard of comparison should be established not for the canes of Central India & Rajputana only, but for the whole of India. Of no part of India, in fact, has the canes been reduced to a scientific standard and practically nothing has been determined as to the relation of the various canes to the climates and soils on which they are found. But in general terms it may be said that it would appear heat, beyond a certain maximum, like a superabundance of water, operates adversely to the production of crystallizable sugar. A feature of such importance, it might have been expected, should have early attracted attention, but apparently it has been entirely neglected. Many planters, it may almost be said, preferred to waste their fortunes in trying to cultivate cane on sites selected arbitrarily, rather than to spend an initial sum in testing the suitability of the crop for the selected locality. To arrive at some ideas on this subject it would seem that the extremes of climate should be first investigated and hence the importance of definite experiments being performed in the dry hot tracts (such as much of Central India and Rajputana) and the humid regions and damp and flooded soils of Bengal.

Conf. with pp. 79-83.

Conf. with remarks regarding solar activity, pp. 18-20, 269, etc.

A good many reports have appeared on the subject of the canes of Central India and Rajputana, but these for the most part mention them by name only, and thus furnish very little by which they can be recognized and classified with the canes of other parts of India. Thus, for example, of CENTRAL INDIA the following selection of passages may be given:

Major General W. Kincaid, Political Agent, Bhopal, wrote in 1882: "The chief varieties of sugar-cane planted in this part of the country are—

1. *Ponda*—a white, thick cane, very tender, which yields juice of a superior quality, and in larger quantity than other canes.
2. *Kansia*—white, with faint rusty-coloured stripes.
3. *Khajla*—white and tender, but thinner than *Ponda*.
4. *Munggee*—white, *very* hard and tall.
5. *Mootora*—of a greenish-white colour and very hard.
6. *Kala*—black.
7. *Nuggurwar*—does not require irrigation.
8. *Bhurree* ditto ditto thinner than *Nuggurwar*.

"Sugar-canes of the above vaieties, except the last two, grow in the three kinds of land named below:—

1. *Morun*, also called *Kulmut* and *Mar*, which is thick, black loam, free from kunkur.
2. *Kabur*—brownish colour, not quite free from kunkur.
3. *Siyar* land, which has stone very near the surface.

Nuggurwar and *Bhurree* canes, which require no irrigation, will not grow except in *morun* soil, or in low, moist localities.

Captain C. B. Cooke, Political Agent, Bundelkhand, furnished in the same year a report in which he speaks of *dhur, bonsi, mutua, munga kansi, badouka, kachhra, rakhoti, tunia, safaid, siah*, and *raishmi* canes. **Captain D. W. K. Barr**, Political Agent of Baghelkhand and Superintend-

CULTIVATION: VARIETIES. Indigenous Canes: Central India.

ent of Rewah, describes the methods of cultivation, but makes no mention of the kinds of cane grown. **Major F. H. Maitland** gave in 1882 some useful particulars regarding the canes of the Charkhari State. The kinds grown, he says, are *bansi*, *kansi*, and *dhaur*. **Colonel C. Martin, C.B.,** Political Agent, West Malwa, in the same year wrote:—

"There are five varieties of sugar-cane, called—

(1) *Ponda*—white stalk, 2 inches in diameter, 10 feet high, superior quality, and used principally for eating.

(2) *Kala*—as tall as first variety, not so thick, black stem, much cultivated, principally used for eating, not so good in quality as the first.

(3) *Sufaid*—or *Dhola*, thin stem, 9 or 10 feet high, principally manufactured into *gúr*.

(4) *Mutaira*—7 feet high, as thick as the former, not eaten, but superior to No. (3) for the purpose of manufacture into *gúr*.

(5) *Surri* very thin, 5 feet high, white stem, used for manufacture into gur, and superior to Nos. (3) and (4).

"Nos. (1) *Ponda* and (5) *Surri* thrive in black soil; the other varieties thrive in all soils. All require much water and are therefore grown in situations suitable for rice, and are sown after the rice crop is gathered, very low ground, where water remains excessively stagnant, being avoided. Sugar-cane is not grown two years successively in the same soil; in the second and third years rice and peas are sown, followed again by sugar-cane."

In Goona it was reported:—"The cane in this district is of eight different sorts, *viz.*, *ponda*, white sugar-cane, black sugar-cane, (called *bhar* in Goona district), *kansia*, *ledoo*, *thirri*, *munh*, *tora*, and *chairi*."

RAJPUTANA. 124

RAJPUTANA.—**Colonel T. Dennehy** reported in 1883 that in Dholpur—

"Three kinds of sugar-cane are cultivated in the Dholpur State (1) *chain*, (2) *sarota*, both hard, thin canes, containing comparatively but very little juice, (3) *dhori*, a thicker cane more flexible and more juicy but containing less saccharine than either the two former kinds. The *chain* and *sarota* can be grown in any soil, everywhere in the State where facilities for irrigation exist, *dhori* is best cultivated in *mattiar*, a mixture of clay and sand principally found in the two parganas of "Kalari" and "Basreri." *Dhori* requires great care in cultivation and is considered the most valuable crop, as its juice, although inferior in quality to that of the other canes, is in quantity nearly double as much per acre as they give. None of the three kinds are so good as the best cane produced in the North-Western Provinces, and this Durbar has at present under consideration a proposal for obtaining from Moradabad some specimens of the best cane on trial for planting in various soils in the State. There appears to be a general feeling that unless fresh seed canes are procured from the outside every year or nearly every year, the quality of the crops will soon degenerate and recede to the present general standard of cane in Dholpur. It is, however, well worth a trial, and the attempt will be made."

Colonel W. F. Prideaux wrote that in Jeypore there were two kinds of cane grown, known locally as *khausila* and *dhol*. In Bhurtpore, on the other hand, he says only one kind is cultivated, *viz.*, *surait*, while in Kerowlie there are three kinds, *viz.*, *dhaur*, *sarauti*, and *katara*, the last mentioned being most preferred. **Colonel H. P. Peacock** alludes to two classes of cane grown in Ulwar as follows:—

"There are two kinds of cane cultivated, (1) *saroda* or *kansla* of a red colour; (2) *china*, *dhola*, *purhea* or *kotarea*, of a white colour, and the juice of which is found to have less saccharine matter than the former.

The Political Agent of Kotah wrote of the canes of that State that the undermentioned kinds are cultivated—

(1) *Ponda*, a very thick cane; it is only used for chewing, and *jagri* is not made from it.

(2) *Kali Gond Girri*, a thick dark cane: as in the case of *ponda*, *jagri* is not made from it and it is only used for chewing.

(3) *Dholi Gond Girri*, a light coloured cane used for chewing only.

(4) *Dhola*, light coloured; about 7 feet long, gives plentiful juice; resembles No. 5, but is lighter in colour, is commonly used for making *jagri*.

(5) *Bansbarra* grows to about 6 feet in length, produces much juice and is more used than any other cane for making *jagri*, the colour of the *jagri* made from it is reddish.

CULTIVATION: VARIETIES. Indigenous Canes. Rajputana.

(6) ***Mouhtora*** a shortish cane with joints far apart, light in colour; the *jagri* made from it is dark and granulated.

(7) ***Katarya,*** dark in colour: about 6 feet long, joints about 5 inches apart; juice of a brown colour; *jagri* dark and granulated.

(8) ***Chareri*** about 7 feet long, light coloured cane; joints about 8 inches apart; *jagri* is generally manufactured from its juice.

(9) ***Kansiya*** slightly red in colour and about 6 feet long; joints about 8 inches apart; *jagri* reddish and granulated.

(10) ***Barli,*** a light coloured cane about 6 feet long; joints about 5 inches apart, *jagri* reddish and only slightly sweet.

(11) ***Mungia*** light coloured with dark marks between the joints which are about 6 inches apart, grows to a height of about 6 feet; produces much juice.

(12) ***Machal,*** a very long cane, brown colour; produces much juice; *jagri* light coloured and granulated.

"The only two kinds of soil (both requiring irrigation) upon which the cane is grown are:—

(*i*) *Sri Mal* (black soil) of every description.

(*ii*) *Pilia Mal* (lightly brown soil).

"All the above kinds of cane are said to grow equally well on either of these soils; if either has the preference it should be given to *pilia mal* for each description. The inhabitants of towns and larger villages being comparatively the more prosperous, can afford to purchase cane freely for chewing, therefore those cultivators who possess cane-growing land in their vicinity turn their attention principally to those classes of cane, which, though not so valuable for their *jagri*-producing qualities, are the favourite sorts for chewing. In the wilder and poorer parts of the country, on the other hand, where people cannot afford to chew, but require *jagri*, the *dhola* and *bansbarra* kinds are the most frequently selected. The rural cultivators bring in *jagri* for sale in the towns and large villages. They do not find it remunerative to grow the canes used merely for chewing, as there is little or no demand among their poorer neighbours, and the cost of transport to the towns and large villages would be prohibitory. The kinds of cane grown in this State are believed to be similar to the best grown elsewhere, but owing to the inferiority of the soil for producing sugar-cane an inferior cane is the result. The principal defect here seems to be good manure. If this could be procured, the soil could be, it is said, improved to the level of the best cane-growing soils in other parts of India, and an equally good cane could be produced. In some cases irrigation is difficult, the kind of plough used is defective, nor are the cattle strong as a rule. Under existing circumstances it is believed that the ignorance of the cultivators is the only obstacle to the production of a better cane.

Mr. A. Wingate wrote in 1883 that "there are not many kinds of sugar-cane in common use for planting, the chief distinction being between the cane grown for selling, mostly found in the neighbourhood of towns, and that grown for crushing. Of the former kind that chiefly in favour is known as *gondgadi* sometimes called *paunda,* and of the latter kind, the commonest are *bharar* and *bhanisiawarchota.* In appearance the canes are light yellow or green in colour, and slender, there is also a dark variety locally called 'black,' but it is not common in Central Meywar."

Major W. J. W. Muir, Political Agent, Harowtee and Tonk, furnished in 1882 a detailed report on Sugar-cane cultivation from which the following passage may be here abstracted as giving an account of the cultivated forms:—

"In the districts of Keshorae Patan of Bundi and of Nimbahera, Chabra, Parawa and Sironj of Tonk, the *kali* and *dhamni* soils predominate, while the *bhuri* and *pili* are found only in parts. In the remaining portion of the Bundi State, in the Tonk and Aligarh districts of Tonk, in the Phoolya or British pergunnah of Shahpura and in the Kherar villages of Jeypore and Meywar in the neighbourhood of Deoli, the reverse is the case, and while the *kali* and *dhamni* are found in a lesser degree the *bhuri* and *pili* are the principal soils. Those of the best quality situated near wells with sweet water are selected for raising sugar-cane, which is a rabi or cold weather crop and is grown principally on lands irrigated by wells, though it is also raised on lands watered by tanks and rivers. It is, however, nowhere cultivated on unirrigated lands in the parts mentioned above.

SACCHARUM: Sugar.

Cultivation of the Sugar-cane.

CULTIVATION: VARIETIES. Indigenous Canes: Rajputana.

"The following table exhibits the different kinds of cane cultivated; the principal ones being shown separately:—

No.	Name of district.	VARIETIES.									
		Gondari black and white.	Mutora or dhari.	Ledu.	Bharal thick and thin.	Mungiya.	Machal.	Katariya.	Sarri.	Kansiya.	Principal ones.
1	The Bundi State	do	do	...	...	do	do	...	do	do	Mungiya and kansiya.
2	Tonk and Aligarh.	do	do	...	...	do	do	...	do	do	Do. do.
3	Nimbahera .	do	...	...	...	do	...	...	...	do	Do. do.
4	Pirwa . .	...	do	...	do	...	...	...	...	...	Bharal.
5	Chabra . .	do	do	...	...	do	...	do	do	do	Katariya and Kansiya.
6	Sironj . .	do	do	do	...	...	...	...	do	do	Ledu and Kansiya.
7	Phooliya . .	...	...	...	do	...	...	...	...	do	Kansiya.
8	Kherar, Jeypore and Meywar.	...	do	...	do	...	...	...	do	do	Kansiya.

"The two *gondgari* species are cultivated principally on *bhuri* soil which contains a certain portion of sand. The *ledu mungiya* and *katariya* kinds on *kali*, the *bharal* on both the *kali* and *bhuri*, while the remaining kinds on all four soils.

"The *gondgari* cane is thick and solid of thin watery juice, sweet taste, soft, and dear in price. It is considered of a superior quality, is used only for eating, and is sown in a limited quantity.

"The *mutora* grows to a maximum diameter of $\frac{3}{4}$ of an inch, is hard, and has a hollow space running through its centre containing two or three fibres which can be pulled out. Its colour is somewhat greenish, its juice is thin and slightly bitter. It is also largely eaten and the reason why it is not manufactured into *gur* in any quantity is that the *gur* produced is wanting (1st) in weight and substance, (2nd) in sweetness, and (3rd) in the *dana* or grains.

"The *ledu* resembles the *mutora* but is soft and sweet in taste.

"The *bharal* is of two kinds, one with a thick, the other with a thin cane. It has a watery juice which is not very sweet and yields but little *gur*, but is not wanting in grains.

"The *mungiya* is of a greyish colour, with a cane about half an inch in diameter and is solid and soft to the touch. It yields a thick juice, the *gur* from which is slightly greenish and not wanting in grains.

"The *machal* is like the *mungiya* in all respects except that the space between the knots is longer.

"The *katariya* is yellowish and is about half an inch in diameter. It is solid but soft to the touch, and its juice is thick and sweet. The *gur* is yellowish and of a good quality.

"The *sarri* is a thin greenish cane but solid. Its juice is thick and sweet but the knots are not very far apart. The *gur* is reddish, soft, and not wanting in grains and weighs heavy.

"The *kansiya* is like the *sarri* but is hard and has a longer space between the knots. It has also a hollow centre with fibres."

AJMIR. 125

In AJMIR-MERWARA it has been stated that three forms of cane are grown, *viz.*, *sagari*, *gundgiri*, and *kansea*. The last mentioned is said to be cultivated for the purpose of its juice which is sweeter, clearer and more palatable than that of any other cane. Sugar manufacture is, however, confined to the villages belonging to the Chokla of Pushkar The *sagrai* and *gundgiri* canes are grown near the homesteads and are eaten fresh.

CULTIVATION: VARIETIES. Indigenous Canes: BOMBAY.

(i) **BOMBAY.**—Most of the early writers speak of three forms of cane as met with in Bombay and that these are much larger than the canes of Bengal. The amount of crystallizable sugar which they contain appears, however, to be considerably less, hence apparently the reason of the greater importance of sugar cultivation in Bengal.

Ahmadnagar. 126

In AHMADNAGAR, according to the *Bombay Gazetteer*, four chief kinds of sugar-cane are grown, *kala*, or black, *pundyábás*, or pale yellow, *bahmani*, white and purple, and *kadi* or white. *Kala* or black also called *tamboa* or red, is of a dark mulberry colour and grows six to ten feet high and one and a half to two and a half inches thick. It is very juicy and yields dark, brown, raw sugar or *gúr*. *Pundyábás*, also called *pandhra* or white, is pale yellow in colour and is thicker, but shorter than the black and yields a lighter coloured and higher priced raw sugar. *Bahmani*, a variegated white and pale cane, is soft in the bark and is chiefly sold for eating raw. *Kadi* also called *balkya* or *bet* is white, and is slender, shorter, and less juicy than the others. It is sown along the edges of fields of the other varieties, as it requires little water, manure or care. In damp lands the *kadi* or *bet* yields a second growth (a ratoon crop) from the original stalk.

Baroda.

BARODA.—Only two varieties of sugar-cane are known in the district, the white and the purple coloured. There are two varieties of the white sort, *viz.*, *vasaigari* or *malbari* and *vansi*. The latter is thinner than the former. In DHARWAR the chief varieties used are *kabbu*, *ramrasdali*, *gabrasdali* and *kara kabbu*. The *halkabhu* or grass cane, though the smallest variety is considered the finest. It is white and thin, about the thickness of a good sized millet stalk. It is sown in rice-fields and is considered a hardy plant. It is very largely grown because it has several advantages over the other varieties. It wants less water than the large white and red kinds. After it has once fairly taken root little watering is required, the rain alone proving nearly enough. Though the larger cane gives much more juice it has much less saccharine matter in proportion than the small cane, and requires far more boiling to make *gur* or coarse sugar. The *gur* made from the small cane is also considered of superior quality. The *gur* of the small cane is light and granulated, while that of the large cane is heavy, wiry and of a somewhat darker colour. On account of its hard bamboo-like texture the small cane is much less subject to the attacks of jackals and wild cats than the large cane. To sow an acre of *halkabbu* requires 2,500 to 3,000 cuttings at three cuttings a cane. The *ramras dali* cane is streaked white and red, and is sown in rice-fields as well as in gardens. It grows to a fair height and thickness, and an acre yields about ten loads of inferior *jagri*, from which no sugar is made. The *gabrasdali* is grown in small quantities in garden lands for local use, and wants care and water once a fortnight. The skin of this cane is remarkably thin, the knots are far apart and it is very juicy. It is much like the Mauritius cane. For an acre of *gabrasdali* or *rámrasdali* 5,000 cuttings at five cuttings a cane are required. The *karra kabbu* is the common red cane. The other four minor varieties are the Mauritius or *morishyáda-kabbu*, *dodiya*, *byatalldoda*, and the *bile kabbu*. The Mauritius cane yields juice superior to that of the common cane, but, as it wants more water, and is more liable to be gnawed and eaten by jackals and porcupines its growth is limited.

Dharwar. 127

Paunda Cana.

Conf. with pp. 52, 64, 74.

128

Kanara. 129

Of KANARA it is said:—"Sugar-cane, *kabbu*, is largely grown both above and below the Sahyádris. It is of three kinds, *rasal* or spotted, *kare* or black, and *bile* or white. *Das kabbu* grows about two inches thick and six to seven feet long, and yields more juice than either of the other kinds. *Kare kabbu* grows about an inch thick and four to five feet long, and *bile kabbu* about half an inch thick and three and a half to four and a half feet long. The *kare kabbu*, whose molasses are reckoned the best, is mostly grown on the coast, on river and stream banks, near ponds, and in other places where water is available."

Khandesh. 130

In KHANDESH the five chief kinds of cane are:—"A small cane, *khadya*; a black cane, *kála*; a white cane, *bundya* or *pándhra*; a striped cane, *bángdya*; and Mauritius, a yellow cane. The small *khadya* cane is the most widely grown as, though it yields inferior molasses, its hardness makes it stand storing and carrying from one market to another. The black, *kála*, cane, the best for eating, is usually grown for that purpose only. The white *pundya* or *pándhra*, and striped, *bángdya*, canes are both good croppers, but require to be well watered and freely manured. They are usually cut for market, but also yield very fair molasses. One variety of the white cane, a little stouter than the finger, hard and woody, contains apparently but little juice. What there is must be very sweet as the yield of molasses is very great. The Mauritius cane, introduced on the Government farm at Bhadgaon, is now rather widely grown. As to bringing it to perfection it wants rich manuring and

CULTIVATION: VARIETIES. Indigenous Canes: Bombay.

watering, it is usually found only in the fields of well-to-do ryots. The molasses is sugary and fine. but, as it carries badly, its price rules little above the small *khadya* cane molasses."

Kolhapur. 131

KOLHAPUR.—The sugar-cane crop is one of the most important in the State. There are five kinds of sugar-cane grown in Kolhapur: *bhonga, chimnápunda, khadkya, rámrasál*, and *támbdi* or red. Of these five kinds *bhonga* is streaked white and red and is grown in garden lands to a less extent than *rámrasál*. *Chimnápunda* seems to be a species of *rámrasál*. Its skin is thin and its joints are close. As it is considered inferior to the other kinds very little sugar is made from it. *Khadkya* is white, grows about the thickness of a good Indian millet stalk, and has very little saccharine matter. It is grown in the Panhála, Karvir, and Bhudargad Sub-divisions. It is hard and requires to be watered thoroughly, only once during the dry months. *Rámrasál*, a white variety, about five to eight feet high and an inch thick, is largely grown in the garden lands of the Alta, Karvir, and Shirol Sub-divisions. Its joints are far apart, and it is the most juicy of all varieties. *Támbdi* or red was once very common, but it has now given way everywhere to *bhonga* and *rámrasál*. Though less juicy it is sweeter than *ramrasál* and is much eaten. Of these five kinds the white and the striped kinds seem to have been introduced about thirty-five years ago, and they, if not the acclimatised varieties of Mauritius, very much resemble it.

Nasik. 132

NASIK.—Of the varieties grown here, there are four kinds called—"White *khadya*, striped *bángdya*, black *kála* or *támbda*, and Mauritius called *baso*. The last is grown only to a very limited extent near *Násik* and Deoláli. The white cane, *khadya*, though very hard and coarse for eating, yields the best molasses, and the crop requires less labour and care. It is found over almost the whole district. In Málegaon and part of Yeola, the striped *bángdya* cane is chiefly grown, but it is seldom pressed. Mauritius cane requires the greatest care as regards water and manure, and the molasses are generally inferior" (*Bombay Gaz.*).

Mr. Ozanne, *Director of Land Records and Agriculture*, attempted a classification of the Bombay canes in a *Note on the Cultivation of the Sugar-cane*, which he published in 1887. He referred all the forms mentioned by local officers to four sections as follows:—

Large white Cane. 133

1. THE LARGE WHITE CANE.—"Soft, juicy, tall, and thick. The *gul* is soft and does not carry well. But with abundant irrigation the large outturn makes this variety very popular where water is plentiful. Its softness renders it excessively liable to damage by cattle and jackals. In some districts it is supposed that there are two varieties of the white cane, one indigenous and the other imported from Mauritius. The supposition may be correct, but it is more probable that one is merely an earlier importation than the other."

This includes the following forms:—

Maráthi.

Pundia. *Conf. with pp. 52, 64.* 134

1. Pundia or Phundia.—This is the commonest name. The word probably means white, though colloquially it is used of a fat dumpy child, and may be applied to denote the thick growth of this variety.
2. Morisas or Moris.—Corruptions of Mauritius, reported from Khándesh (Sávda) and Ratnagiri.
3. Pándhra.—White.
4. Viláyati.—Foreign.

Gujaráti.

5. Dholi.—White.
6. Bhári.—Brownish white.
7. Pundi.—Mar. Pundia. Occasionally used.
8. Malbári.—*i.e.* From the Malabar coast, where Mauritius cane was first introduced.

Conf. with pp. 5, 31, 32-33, 49, 57.

Kánarese.

9. Bile Kabbu.—White cane.
10. Dodd or Dás Kabbu.—Large cane.
11. Rasavali or Rasadáli.—Juicy cane.
12. Hotti Kabbu.—Bellied cane. So when wheat is puffed out and swollen it is called bellied wheat.
13. Gubbarasdáli.—Knotty cane. This variety is alleged to be distinct. Its joints are short.

CULTIVATION: VARIETIES. Indigenous Canes: Bombay.

Sindhi.

14. Acho.—White.
15. Viláyati.—Foreign—introduced from the Panjáb.

Red or Black Cane. 135

2. "THE LARGE RED OR BLACK CANE.—The colour varies from red to dark purple. This variety is gradually giving place to the white cane. It is, however, sweeter and more liked for raw eating. On the other hand, the colour of the cane is imparted to the *gul*, thus depreciating its commercial value. A yellow golden is the best colour for *gul*. The red cane is soft, juicy and sweet, but not so large a cropper as the white.

Maráthi.

1. Kála.—Black.
2. Támb or Támbda.—Red.
3. Nila.—Purple.
4. Jámbhla.—Purple.

Gujaráti.

5. Ráti.—Red. There are several local names descriptive of the source from whence the cane was introduced; thus the Balsári, the Vasai, and the Songhadi cane, introduced from Balsár, Bassein, and Káthiáwád.
6. Lál.—Red. There are several local names descriptive of the source from whence the cane was introduced; thus the Balsáti, the Vasai, and the Songhadi cane, introduced from Balsár, Bassein, and Káthiáwád.

Kánarese.

7. Kare Kabbu.—Black cane.

Sindhi.

8. Garho.—Red.
9. Vangrae.—Purple.

Striped Cane. 136

3. "STRIPED OR STREAKED CANE.—The stalk is variegated with streaks or lines of purple and white. It is the favourite for raw eating but makes good *gul*.

Maráthi.

1. Bángdia.—Bangle cane—derived from a coloured kind of bangle.
2. Kábara.—Variegated. A common name in Sátára, and occasional in Poona.
3. Bharal.—A trough or tube. Due probably to a larger central pith in this than in other canes.
4. Bhonga or Bhongála.—Literally hollow. This name is common in Kolhápur and conveys a meaning similar to that conveyed by Bharal and Dhamni.

Kánarese.

5. Rambáli.—Meaning doubtful.
6. Rámrasdáli.—Rám is the name of the god, meaning intensive, "The prince of juicy canes."
7. Rudragánthi.—Rudiagánth is a figure worked into cloth on the loom, "Figured cane."
8. Dhamni.—A tube, equivalent to Bharal in Maráthi.

N.B.—This variety is apparently unknown in Gujarát and Sind.

Straw Cane. 137

4. "THE STRAW CANE.—This is the hard, slender variety. Its hardness makes it popular where pig and jackals are harmful. It is much grown, for it will mature with a very scant supply of water. It is even grown in tracts of heavy rainfall without water at all or else with a preliminary flooding only at planting time. When thus grown it is called the waterless (*nipani*), cane. It is hardy and produces *gul* of excellent colour and keeping qualities."

Maráthi.

1. Vára.—Vára = air, cane not artificially watered.
2. Khadia.—From khadi a stone, a name due to the hardness of this variety.
3. Bharad.—Bharad land is *hard*, hilly land.
4. Dongaria.—Dongaria is hilly. These last three names convey the same idea.
5. Khajuria.—Date-palm cane, perhaps from the taste of the juice or the *gul*.
6. Bás.—North Deccan. } Both these words mean bamboo.
7. Kalakia.—South Decan. }

SACCHARUM: Sugar. **Cultivation of the Sugar-cane.**

CULTIVATION: VARIETIES. Indigenous Canes: Bombay.

8. Bhonsa.—A reed, in Ratnágiri.
9. Káthi.—In Bijápur, káthi = a stick—"Stick cane."

Kánarese.

10. Hul Kabbu.—Grass cane.
11. Básar Kabbu.—Meaning doubtful.
12. Betta Kabbu.—The "Reed cane." Betta = a reed or rattan, equivalent to Bhonsa (*see* above).

Gujaráti.

13. Vánsi.—In Surat and Káthiáwád, vans = a bamboo, "Bamboo cane."

Bombay Cane. *Conf. with pp. 5, 31, 33, 34, 49, 57, 73, 87, 102, 123, 144.* 138

It is somewhat remarkable that a superior quality of cane, met with almost throughout Bengal and Assam, should be designated a Bombay cane. There appears to be no distinct record of its introduction into Bengal, but its existence may be viewed as deriving interest from the fact that the early writers speak of the Mauritius cane having come originally from the Malabar coast of Bombay. The Bombay cane was first brought to prominent attention in 1857 through **Babu Joykissen Mukerji** having drawn attention to the fact, that in that year it was severely attacked by the Sugar-cane Borer **Babu Joykissen** wrote that "about 25 years ago **Mr. McDowell** of the Kissorigunge Indigo Concern introduced the red canes in the district of Rungpore, hence the Natives of that place call these canes *shahiban kusar*. "On comparison, he says, the experimental cultivators were convinced that the new canes had more saccharine matter in them than the country ones, and that they grew larger and yielded more juice than the latter, so much so, that the pecuniary gain to the ryot was more than twice the product of the other. Thus in a very few years the neighbouring fields of Kissorigunge were covered with these canes. In about 8 or 10 years these canes were introduced into most of the northern parts of the district and from thence gradually spread over throughout the Southern parts too." "The canes when ripe are reaped and carried to the mills, where they are cut in small pieces for being pressed, and the fields cleared of grass, etc. A few days after, new shoots begin to make their appearance out of the roots, and the fields are then taken care of, weeded, and the earth loosened and manured, and the heads tied together as in the first instance In a similar manner a third crop is reaped from the same field. In the first and third years the produce of the *shahiban* canes were moderate, but in the second year they yielded a far superior crop. In the fourth year some of these fields are ploughed and manured, and some other crops are cultivated, but in some instances the lands are left uncultivated for renewing the fertility of the soil. For some years the *shahiban* canes were very luxuriant in this district, and the cultivation of the country canes decreased in the proportion the other was introduced." **Babu Joykissen** then proceeds to describe the appearance and progression of the disease, which soon ruined completely "one of the much esteemed and principal harvests of the district." The facts regarding the disease willbe found alluded to in the chapter on DISEASES OF THE SUGAR-CANE, and it need only be here added that the wonder is, that some such disease did not appear earlier, for the value of the crop only served to work its own ruin through a process of over-cultivation. But having said so much, the subject of the Bombay red canes may be viewed as exhausted, except that writers on Bombay canes seem ignorant of any special and peculiar cane of Bombay that would answer to the cane so often alluded to by writers on the Bengal sugar-cane industry. The suspicion may, therefore, be offered that the so-called Bombay red canes may be the acclimatized form of Mauritius cane which on being translated to Bengal survived there though it has very nearly died out in Bombay.

Ratooning. *Conf. with pp. 59, 77, 78, 128, 143, 151, 195.* 139

Disease. *Conf. with pp. 48, 52, 87, 121-127, 151, 161.* 140

CULTIVATION: VARIETIES. Indigenous Canes: MADRAS. Coimbatore.

(d) **MADRAS.**—"In the Coimbatore district, the chief varieties are the white (*vellei* or *rastâlei*) striped *namam* and the red or purple cane. The first appears to be the Mauritius cane, introduced by Government some forty years ago; it has quite ousted the country cane, which was a very poor variety. The *namam* cane is chiefly used for eating, the *rastalei* for jaggery and sugar" (*Man. Coimbatore Dist*). In 1843 sugar-cane cultivation was introduced into the KURNOOL district when cuttings of Mauritius cane were sent from Madras, and for several years the Government encouraged the industry by remitting the land tax where it was grown.

Kurnool. 141

The Madras forms of sugar-cane were so fully investigated by **Dr. Buchanan-Hamilton** in his Journey from Madras through Mysore and Kanara, that it seems only necessary to supplement the remarks that will be found below regarding sugar-cane cultivation in this Presidency by furnishing a few of **Dr. Hamilton's** passages that more especially deal with the varieties.

MYSORE. 142

MYSORE & COORG.—**Dr. Buchanan-Hamilton** in the report of his Journey from Madras through Mysore, Kanara and Malabar gives many passages regarding the canes he saw. Of Mysore he says:—

"A considerable quantity of sugar-cane is cultivated by the farmers of the Ashtagram. It is of two kinds: *restali* and *puttaputti*. Both yield *bella*, or *jagory*; but the Natives can extract sugar from the *puttaputti* alone. The *jagory* of the latter is also reckoned the best. The *restali* can only be planted in *Chaitra*, the *puttaputti* may also be planted in *Sravana* or *Magha*. The crop of *restali* is over in a year, that of *puttaputti* requires fourteen months, but may be followed by a second crop or, as is said in the West Indies, by a crop of *rattoons*, which require twelve months only to ripen. The *restali* will not survive for a second crop. This is the original sugar-cane of the country: the *puttaputti* was introduced from Arcot by **Mustaph Aly Khan**, who in the reign of **Hyder** was *Tosha Khany* or paymaster general. The cultivation of *restali* has ever since been gradually declining." So again he remarks of Chinapatam, in Madura, that "both *puttaputti* and *restalu* canes are cultivated, and of both the white sugar can be made; but cane that is raised on a rich soil will not answer for this purpose, as its cane can never be made to granulate." "Near Bangalore," I observed, he continues, "the cultivation of a kind of sugar-cane called *moracabo*, or stick cane. This kind never grows thicker than the finger, and is very hard and unproductive of juice, but it requires less water than the *restali*. It seems to have been the original sugar-cane of the Kolar district, of which all the country on this side, of the central chain of hills form a part. The farmers have lately introduced the *puttaputti* from the lower Karnatic, and are extending its cultivation as fast as they can procure cuttings.

Ratooning.

"The kinds of sugar-cane cultivated in the country round Kolar are four, which are esteemed in the following order, 1st *restali*; 2nd *puttaputti*; 3rd *moracabo*; 4th, *cuttaycabo*. The two last are very small, seldom exceeding the thickness of the little finger; yet the *cuttaycabo* is the one most commonly cultivated. This is owing to its requiring little water, for by means of the machine called *yatam* it may have a supply sufficient to bring it to maturity.

South Kanara. 143

In SOUTH KANARA, **Dr. Hamilton** found two canes commonly cultivated. These were known as the *bily* and the *cari cabbu* or white and black canes. "The former, he continues, is the *restali*, and the latter the *puttaputti* of the country above the Ghats. The same ground will not produce sugar-cane every year; between every two crops of cane there must be two crops of rice. A piece of land that sows one *moray* of rice, will prodnce 4,000 canes, which are about six feet long, and sell to the *jagory* boilers at from half to one rupee a hundred. The *moray* sowing of *betta* land is here about 30,000 square feet, so that, according to the price of sugar-cane, the acre produces from R58 to R29, or from about £5-17*s*. to £2-18-6*d*. The land tax is the same as when the field is cultivated for rice. The want of firewood is the greatest obstacle to this cultivation; the trash, or expressed stems, is not sufficient to boil the juice into *jagory*, while that operation is performed in earthen pots placed over an open fire. If all the land in Codeal Taluc (district) that is fit for the purpose, were employed to raise sugar cane, it would yearly produce 1,000 *pagodas* worth of cane; that is to say, there are about 1125 maunds sowing of land, that once in three years might be cultivated. The quantity in the neighbouring district on the south side of the river is much greater. The *jagory* made here is

CULTIVATION: VARIETIES. Indigenous Canes: Madras.

hard. but black, and of a bad quality. It sells at 3 maunds for the *pagoda*, or at 12*s*. 3½*d*. a hundred weight. Between the rows of sugar-cane are raised some cucurbitaceous plants, and some kitchen stuffs, that soon come to maturity."

North Kanara. 144

In NORTH KANARA "sugar-cane," **Dr. Hamilton** tells us, was at the beginning of the present century, "raised on *mackey* land; but four years must intervene between every two crops; and for the first two years after cane, the rice does not thrive. The kind of cane used here, **Dr. Hamilton** adds, is called *billy-kabo* which above the Ghats is known as *mara kabo*. Inland they cultivate the *cari-kabo* which above the Ghats is called *puttaputty*."

In Kellamangalam (in the Karnata) **Dr. Hamilton** found four kinds of cane, *viz*., *restali*, *puttaputty*, *mara-kabo*, and *chittuwasun*. The jaggery of the *restali*, he tells us, sold higher than that of any of the other forms and that the *puttaputty*, was valued as an edible cane. The *cari-kabo*, a fifth form allied to the *puttaputty*, like it requires garden cultivation, but the *mara-kabo*, and *chittuwasun* may be grown anywhere.

BURMA. 145

(i). BURMA.—In a report on the sugar-cane of Burma **Mr. J. E. Bridges** furnishes the following particulars regarding the forms met with:—

"In Burmese times there where small patches of black cane grown round Beelin and sugar was manufactured in small quantities. Shortly after the English took the country, the 'Madras cane' was introduced from Moulmein, and it is now the cane almost exclusively grown in this tract. It is of a yellowish colour and so flexible that it does not require any supports. It grows to a height of 10 to 12 feet. Various testings of the juice of this cane were made with the saccharometer and the results are given below, together with corresponding percentage of coarse sugar obtained by actual experiment:—

Name of village.	Percentage of sugar in juice according to saccharometer.	Percentage of coarse sugar in juice according to experiment.
1. Kadipoo	23·57	14·76
2. Ngetchoon	25·71	15 27
3. Dawoon (Thatone sub-division) *	20·00	12·52
4. Ditto (ditto)	22·85	14·31
5. Nyounpalin	25·71	15·69
6. Shwaylay	22·85	14·31
7. Pokwon	22·85	14·31
8. Beelin	24·25	15·76
9. Pokwon	24·25	15·36
10. Payasaik	24·25	14·45
11. Thehbyoo river	25·71	15·69
12. Ditto	27·14	17·00
13. Ditto	27·14	17·00

"The percentage of jaggery to juice is said to increase as the dry season advances and the testings made would confirm this fact, as the testings on the Thehbyoo which were made at the end of November, give a much higher percentage than those made on the Beelin river about a month earlier. An iron boiling-pan (*kyaw*) containing 14·16 gallons, or 141·60lb. of juice, yields in November 22·75lb. of jaggery. The cultivators state that in January and February an iron pan, full of juice, yields 26 to 28lb of jaggery, or an increase of two to four per cent. The percentage of coarse brown sugar to juice may, therefore, be taken as varying from 18 to 20 per cent.

"Next in importance to the Madras cane comes the *kaingyan*, so called from its resemblance to the *kaing* or elephant grass. It is whitish in colour and grows to the same height as the Madras cane, but is much thinner. It is also flexible and does not require supports. It is said to yield 20 to 25 canes to each stool and to ratoon for five years. I found a few Burmese cultivators trying this kind of cane, as they think it will take less labour to cultivate than the Madras cane; they have, however, as yet only planted enough of it to obtain seed for next year. The Shans state that this kind of cane is almost exclusively grown in their country; that it is

Ratooning. 146

* This plantation was injured by floods and the cane was very poor.

CULTIVATION: VARIETIES. Indigenous Canes: Burma.

easier to cultivate, but yields much less juice than the Madras cane. Its juice was found to contain 21·43 per cent. of sugar according to the saccharometer and 13·42 per cent. of coarse brown sugar according to actual experiment. The macerated rind or cane trash of both the Madras cane and *kaingyan* are used as fuel in boiling the juice.

"The other kinds of cane are only found in small quantities here and there in the different plantations. They are—

(*a*) the *anyagyan;*
(*b*) the *kyoukgoungyan;*
(*c*) the *kyannet.*

"The *anyagyan* or Upper Burman cane is of a reddish colour and has short thick joints; its juice contains, according to the saccharometer, 20 per cent. of sugar and according to actual experiment 12·52 per cent. of coarse brown sugar.

Canes not suited for Sugar-making. 147

"The *kyoukgoungyan* is a large cane of greenish colour much resembling the Upper Burman cane; it has also short thick joints. According to the saccharometer its juice contains 14·28 per cent. of sugar, and according to actual experiment 8·94 per cent. of coarse brown sugar. Loamy soil does not appear to suit this cane, which grows best in the alluvial clay of the tidal creeks in the Bassein and Thonegwa districts. Its juice in Bassein contained, according to the saccharometer, 22·85 per cent. of sugar, or according to actual experiment 14·31 per cent. of coarse brown sugar. The Upper Burman cane and the *kyoukgoungyan* are exceedingly brittle and both require supports. They are eaten as a sweetmeat and cannot be utilized for manufacturing sugar, as they break at the joints whilst passing through the mill. The cane trash of these canes cannot be used as fuel for boiling the juice.

"The *kyannet* or black cane is a thin cane of a dark purple colour; it has green leaves and yields but little juice. According to the saccharometer the juice of the *kyannet* contains 24·25 per cent. of sugar and according to actual experiment 15·19 per cent. of coarse brown sugar. There is another variety of black cane with purple leaves which is used by the Burmans as a cure for insanity."

In conclusion it may be remarked that, from the above brief review of the canes of India, it will be seen that the distinction into edible canes, grown specially for the local markets (where they are sold as fruits), and into canes grown for the preparation of sugar, is urged by many writers. Indeed, it is often said that the edible cane crop is more profitable than the sugar-producing. The further distinction made by **Duthie & Fuller** into canes suitable for low damp soils and those for high rich lands, where much irrigation is necessary will be observed to have its exact parallel in **Roxburgh's** forms. It is, besides, the almost universal classification and need not therefore be further dwelt on. It is worthy of special consideration, however, that in their unconscious natural selection, the Natives of India are not now, and never have been atcuated by the singleness of purpose that characterizes the European planter's operations. They select not only canes suitable to particular climates and soils but canes which are good for sugar making, good for distillation, and good for eating. The two last mentioned would be highly unsuited for the sugar maker as their special merit may be said to be a copious and sweet juice with a low percentage of crystallizable saccharine matter. It is ignorance of the fact that such a cane is not only of great value to the people of India, but even more profitable to the cultivator that has caused so many writers to fail to appreciate the true character of the Indian sugar market and trade.

Conf. with p. 82.

THE IMPROVEMENT OF SUGAR-CANE.

IMPROVEMENT. 148

It may be said that, among others, there are four possible methods of accomplishing this result: (1) by experimenting with all available canes to test their adaptability to a new environment; (2) by ascertaining the effect of peculiar methods of cultivation; (3) by selection and propagation of sports or buds, found advantageous to the object aimed at; and (4) by a similar selection from seedlings.

Suitability. 149

SUITABILITY TO ENVIRONMENT.—It is, perhaps, scarcely necessary to deal in great detail with (1) and (2), nor to treat them separately. The

IMPROVEMENT.

Suitability.

Conf. with p. 69.

150

results that have been attained by the European planters in the great sugar-producing countries will at once be brought to mind. The canes of this country and that, have been carried here and there and subjected to all sorts of experiments with, in consequence, the production of an extensive series of now widely different cultivated races that can scarcely be said to preserve any of their original characteristics. But even without securing exotic forms on which to operate, the planters have effected vast improvements by ascertained definite systems of cultivation and treatment, and the resulting states of the cane have been found more or less permanent so long as the required treatment has been adhered to. Continued cultivation under certain conditions of climate and soil or under the influence of special manures, etc., may thus be said to tend to produce or preserve the peculiarities of many of the canes that have received distinctive names. These improvements are generally, however, rapidly destroyed, or at least altered, on the canes being carried to still further countries or even in the same country on their being subjected to diversified influences. Thus, for example, several writers affirm that Bourbon cane was originally obtained from the coast of Malabar. Assuming this to be correct (but it is of no moment should it not be so), the translation of the Malabar cane to the insular conditions of its new home and to the improved systems of cultivation it there received, resulted ultimately in the production of what is known as Bourbon cane. A few years' return of that improved state sufficed, however, in India to reduce it to what is perhaps a worse condition than its original. The effect of altered environment may be said to be often so immediate that a much less severe translation than the one indicated may suffice to produce startling results. Thus, for example, it was at one time thought that a great improvement might be brought about, in the Bengal sugar industry, by the cultivation of one of the superior Bombay qualities. High expectations were entertained, the plant was largely grown (as, for example, in Hughly, Rungpore, and Burdwan) but unfortunately disease appeared and proved quite as fatal as that which swept away the labour and outlay which had been spent for nearly half a century in acclimatizing the Otaheite cane. It may be inferred, however, from what has been said, that, just as the Malabar cane in its new environment improved into Bourbon, so the Bombay might in Bengal have become a superior stock. It follows accordingly that continued experimenting with the cultivation not only of foreign canes but of the canes of the provinces and even of the districts of India interchanged may result in the production of a condition of high merit. And this result may be obtained as much by the varied methods of cultivation to which the plant is subjected as to peculiarities of climate and soil. The student of Indian agricultural and economic questions cannot fail to bring to mind an extensive list of parallel examples of the behaviour, or rather what might be called the eccentric behaviour of plants under slightly altered conditions. The whole mystery of the multitude of forms of rice may be said to be a manifestation of this principle. It is, perhaps, needless to cite special examples in connection with rice, but the reader may consult the remarks (*Vol. V., 613*) regarding *bara* rice. That highly prized form is grown on one or two fields only in the Peshawar district, and when tried on other fields or in other parts of India has hitherto reverted to an immensely inferior condition. The Indigo planters of Bengal are well aware of the advantages of obtaining their seed from certain parts of the North-West Provinces. A very extensive list might be drawn up in order to demonstrate that in India, with most cultivated plants, there exist many peculiarly local manifestations. In the case of sugar-cane we are practically ignorant of the value of these. But this much may be said that infinitely greater and

Conf. with p. 5.

IMPROVEMENT.

Suitability.

more lasting improvements might be looked for from an extensive investigation of the merits and peculiarities of indigenous sugar-canes, than from all the efforts at acclimatising the racial peculiarities of exotic forms that have or are ever likely to be put forth. It seems to be the prevailing evil tendency of agricultural reformers to look to countries outside India for new economic products or superior races of existing crops. A state of indebtedness in these matters must necessarily mean the absence of the vitality essential to progression. Witness the load laid by its pioneers on the tea industry through the importation of the Chinese plant. It was not until the so-called indigenous tea was taken in hand and the Chinese stock largely exterminated, that tea-planting gave indications of success. Witness also the extravagant waste of money in the attempts to bring back to India the Carolina development of rice. Improvement by insidious adaptation of the indigenous stock may be less rapid (and hence by no means so attractive to the individual reformer) than the importation of a perfected race, but the result is more certain and the accomplishments, however slight, are permanent and direct gains. The failure of the past attempts at establishing in this country sugar-cane plantations, at a time when India might (at least along the more direct routes of export) have had reasonable expectations of success, may to a large extent be attributed to the chief effort having been directed towards the vain pursuit of methods by which to perpetuate, under the vastly different conditions of India, the special peculiarities of certain races of cane which had been brought to their perfection in the West India Islands. The idea of using the Indian forms of cane was only embraced when the industry was on the eve of expiration, or at all events when it had wasted fruitlessly its best opportunities.

The abolition of slavery was by many thought to be the death-blow to the West Indian sugar plantations. Experienced planters, accordingly, removed to India as a more hopeful field for future enterprise. The internal communications of this country were then, however, in a very backward condition. The sugar manufactured could not find its way to the coast, except by having to bear ruinously heavy transport charges. The selection of sites for sugar plantations was, in many cases, about as ill-advised as possible and the energies of the planters were, as already explained, far too much directed towards the futile endeavour to acclimatize West Indian canes. Their capital had been expended on the construction of unnecessarily large buildings or invested in unsuitable machinery. It was early seen that they could purchase cane cheaper than they could grow it, and that even a greater field was open in refining Native crude sugar than in extracting the juice and direct manufacture. Their refined sugar found little or no sale in India, and it failed to compete in quality and price with that which, despite the altered state of the West Indian labour market, continued to pour into Europe from the English, French, and other colonies. Emigration of coolies from India saved the sugar-planting colonies. But many conflicting influences came to bear on the young sugar-planting enterprise of India and in consequence it gradually died out.

While facilities of transport have now been greatly improved, and sugar might be conveyed to the port towns at a comparatively cheap rate, from the very localities chosen half a century ago for sugar plantations, beet-sugar has effected a complete revolution on the position and possibilities of the foreign export trade in Indian sugar. At no time has it been very important, but at the present day it is less so than it was a few years ago. Beet-sugar is not only coming to India in yearly increasing quantities, but having closed many of the European markets for West Indian sugar, large quantities of

Conf. with pp. 40, 95, 113.

IMPROVEMENT.

Suitability.

foreign refined cane-sugar are being thrust upon India at a very low price. At present, therefore, improvement of sugar-cane may be said to largely mean the improvement of the indigenous forms for the Native market. This is a very different problem from the improvement of cane for the production of superior qualities of crystallized sugar. It seems likely that the Native preference of, what is called by European writers, "impure sugar," may have more to recommend it than is generally supposed. The crude and inexpensive process by which it is prepared allows of competion even with the cheapest beet-sugar. The thickened mixture of molasses and of crystallized cane-sugar known as ***gúr*** is sweeter than the refined article. The so-called adulterations (when not direct admixtures) are generally wholesome enough principles, being derived from the cane and many of these are substances which contain nitrogen—an element largely deficient in a vegetarian diet. There has, however, always been a certain market (specially in Western India) for refined sugar, and, as already remarked, the imports of foreign refined sugars are telling heavily upon the Native and European refineries of this country, but it will take many years before the desire for pure crystallized sugar begins to affect the cultivator of Indian cane, and his manufacture of the substances which **Messrs. Travers & Sons** compare to "manure." It may, in fact, be safely said that, at present, improvement of cane in India means essentially improvement for the existing local market, and not for a prospective foreign trade which may not unjustly be characterised as a hypothetical market.

Price of Indian & Foreign Sugars. *Conf. with pp. 323-328.*

151

Selection.

152

Conf. with pp. 9, 10, 44-46, 61, 134-136.

SELECTION OF BETTER CANES.—But to return to (3) and (4)—the remaining methods by which improvement may be effected, ***viz.***, *selection of sports or buds* and *selection of seedlings*, which possess desirable properties, it may be remarked that these are the natural processes which would be expected to suggest themselves from the dictates of personal advantage, alike to the ignorant and the educated cultivator. In India the principle of selection has been in operation for countless ages of sugar-cane cultivation, and nearly every district possesses slightly different forms that are not to be met with anywhere else. Speaking roughly, the canes of India might be referred to three great classes:—

1st—Edible canes, that is to say, canes which are eaten in their raw state as fruits.

2nd—Canes that yield a large quantity of juice, used by the people of India as an inspissated syrup—***gúr***—in place of sugar.

Conf. with p. 79.

3rd—Canes that yield a large quantity of crystallizable sugar. The inspissated juice of this nature is boiled longer than is the case with ***gúr*** and it is then called ***ráb***

Spirit (or rum) is prepared from ***gúr*** (the entire juice of the cane) or from the molasses obtained on draining ***ráb***.

In bringing the various canes that represent either of these classes to their present perfection, it may safely be said that far greater attention has been paid to Nos. 1 and 2 than to 3. The edible canes (in the vicinities of cities) pay the cultivator better than any others. The principles that have guided his selection have, therefore, been a soft pulpy stem with a profusion of sweet juice, conditions by no means characteristic of a high percentage of crystallizable sugar. Such a cane is necessarily delicate, being liable to the attacks of white ants, easily injured, by the winds, and a prey to the pilfering proclivities of the people. It could therefore, be only grown on the lands near the homesteads where the soil is richly manured and the fields carefully tended. A cane of this nature would be unserviceable and wasteful if used in the manufacture of ***gúr***. But while having been thus actuated by what may be called selfish motives in his natural selection, the Native cultivator has kept another consideration

IMPROVEMENT. Selection.

in view, *viz.*, suitability to the conditions of his surroundings. Not the least important factor in this aspect has been a desire to possess the power to resist the natural enemies or disadvantages of his cultivation, such as white ants, jackals, winds, severe drought, inundation, swampy soils, and high temperatures. The evil consequences of high temperature on the expressed juice he combats, as will be seen below, by manufacturing the juice into *gúr* at once, and during the cooler hours of night. But a soft cane with a thin bark would be more liable to the injurious depredations of ants and jackals, and would also be more easily broken by the sudden gusts of wind and even heavy storms that sweep over the cane-fields as they are nearing maturity. A thin bark (or rind, as it is popularly called) would not only allow the cane to be more easily cut through by ants and jackals, but would expose the juice within the cane to destructive changes in its chemical nature. Drought and high temperature check growth and tend to dry up the juice within the cane before it can be reaped. From all these considerations it has come about that the Natives of India have selected small hard canes for their sugar-producing crops, the larger, softer and more juicy kinds being reserved for garden cultivation, where they can be more carefully looked after; the poduce affords the edible canes. But in this gradual evolution or progression into suitable forms the Native cultivator does not at present resort, and there is nothing to show that he ever did resort, to seminal selection.

Seeding. 153 *Conf. wtih pp. 8, 9, 11, 44, 47, 61, 109.*

FLOWERING AND SEEDING OF THE CANE.—The flowering of the cane is viewed as an evil omen. It is a token of death. The "arrow"-forming cane (as the West India planter calls the terminal panicle of flowers) is at once removed by the servants of the owner, who thus take on themselves and save their master from the evil consequences portended. It is not to be wondered at, therefore, that the notion is prevalent in this country that the cane never flowers or rather never seeds. The period at which the cane is cut very probably precludes the possibility of its flowering, and it is only with the stock, left sometimes in the field for next year's seed-canes, that the phenomenon of flowering is observed. But in India as in Java, the West Indies, and elsewhere the cane may often be seen flowering, and certain forms (as, for example the violet-scented canes of Java) obtain special names from the peculiarity of their flowering panicles. Some of the early European writers say that in the Gangetic basin the cane was supposed to be sometimes raised from seed. If ever this was so, modern writers do not appear to have recorded the existence of the practice at the present day. Indeed, it might be almost affirmed that in India the cane seeds but rarely, in fact, it is only very occasionally that it is allowed to flower. But the information that exists on this subject by no means justifies the assertion that sugar-cane has never been known to seed in India nor even that it has never been raised from seed. During two or three isolated periods the subject has been discussed in India and various reports published, but it appears that the rage for imported canes together with the discredit thrown by **Mr. Wray** and other practical planters on the possibility of improvement from seminal selection, have tended to consign the enquiry to the position of a curiosity in plant-life. It has been urged that since flowering deprives the cane of its saccharine fluid the production of seed should be discouraged. The only possible advantage, writers on this subject have contended, should be looked for in the production of a hybrid between cane and some hardier grass. Thus, for example, **Mr. Wray** wrote, "Experience and much consideration had quite convinced me that it was entirely useless to hope for any good results from cane flowers, of whatever variety they might be, being brought into contact with cane flowers. I, therefore, determined to try the GUINEA

IMPROVEMENT.

Seeding.

CORN or *bajra* and the INDIAN CORN or *maize* (*buta*) with the cane plant. Now, both of these plants perfect their seed; and I ventured to hope that by planting them together, I might get the flowers of the Guinea corn and Indian-corn to impregnate and fructify those of the cane." **Mr. Wray** failed in this expectation and naturally so. He succeeded, however, to make the cane flower, along with the other plants with which it was cultivated, and he adds "notwithstanding all my care and attention, I had not the gratification of seeing any seed appear on the cane plants so treated," and, therefore, "the failure of this, my last hope, set the question at rest in my mind." **Mr. Wray's** position, therefore, was that because he failed to cross two widely remote genera of grasses it was impossible to cross the various forms of sugar-cane, or to cause any particular form to produce seed. The simple fact that seed of the cane has been produced and germinated too, shows the absurdity of **Mr. Wray's** contention. Whether or not any great improvement is possible in this direction (or more readily accomplishable than by other means) is quite another matter. The subject of the possibility of improving the cane by seminal selection is being warmly investigated at the present time both in Java and the West Indies. The plant has been made to seed, and the seeds have been germinated. A controversy has, in consequence, ensued in the *Kew Bul letin*, part of which will be found in the Linnæan Society's Journal (*Vol. XXVII., 197-201, Pl. 33*) as to whom the honour should be ascribed of having first figured and described the seed. It may be of some interest, therefore, to trace out here the historic records of this subject which have a bearing on India. In 1792 **Mr. P. Treves** of Benares, in a long and able paper on the sugar and sugar-cane of that district, says: "I have never observed the cane in this country to flower. I therefore conclude it is cut too soon. The cane, like other productions of the vegetable world, produces seed, and analogy warrants the conclusion, that in that condition it is fittest for the hook." So again after dealing with the religious objections to the flowering of the cane he remarks: "I am informed that there is a species of cane called *Kuthari*, cultivated in or near the district of Champarun, and upon the banks of the Gogra, which is not cut down by the cultivators thereof until all the canes are in flower." Subsequent writers who speak, of the sugar-cane being raised from seed in some part of the Gangetic basin have very probably derived their information from the above passages, the facts being distorted. **Roxburgh**, also in 1792, published a report in the Proceedings of the East India Company (frequently quoted by the author of this article) in which the following remark occurs: "The flowering is the last accident they reckon upon, although it scarcely deserves the name; for it rarely happens, and never but to a very small proportion of some very few fields. These canes that flower have very little juice left, and it is by no means so sweet as that of the rest." In 1844 the **Rev. Doctor Stewart**, Honorary Secretary of the Royal Horticultural Society of Jamaica, addressed a letter to **Mr. Henry Pinkard** desiring him to procure, if possible, some information on 'the mode of propagating the sugar-cane in the East Indies." In that letter the Reverend gentleman says: "The sugar-cane in the West Indies is culitivated from cuttings, and although the plant flowers yet the seeds it produces are of no avail for planting; if they were, the produce would nc doubt be new varieties of the cane and the usual results would follow. It is stated that this is not the case in the East Indies, and I am anxious not only to ascertain the fact but to procure a quality of the seed, such a quantity as may be convenient, with information as to the mode of sowing and managing it." **Mr. Pinkard** forwarded this letter to the Secretary of the Agri.-Horticultural Society of India and asked to be informed "whether the sugar-cane is cultivated

IMPROVEMENT.

Seeding.

from seed, either in the East Indies or in China" (*Jour., Old Series, Vol. III., Selections, 84-87*). **Mr. L. Wray** was apparently asked by the Secretary of the Indian Society if he could afford any information on the subject, and his reply appeared in Vol. III. of the Journal. He there draws attention to the fact that "**Porter**, in stating that canes may be raised from the seed in the East Indies, evidently takes his cue from **Bryan Edwards**, who writes (*see Vol. 2nd, book 5th, p. 240*). 'In Abyssinia and other parts of the East, it (the sugar-cane) is easily raised from the seed (*vide Bruce's Travels*).' Now in **Bruce's** *Travels* I have not perceived that he fixes the fact in Abyssinia; but in Vol. 1st, Chapter 4th, page 81, he makes Egypt the scene of such reproduction. He says, 'About four miles from this, is the village of *Nizelet el Arab*, consisting of miserable huts. Here begin large plantations of sugar-cane, the first we have yet seen: they were then loading boats with these canes to carry them to Cairo.' 'I apprehend they (canes) were originally a plant of the old continent, and transported to the new upon its first discovery, because, here in Egypt, they grow from seed. I do not know if they do so in Brazil, but they have been in all times the produce of Egypt." **Mr. Wray** continues after the above quotation:—"About six years since, whilst I was in the West Indies, I fell in with a French work on sugar-cane (the title and author of which I forget) and I distinctly recollect it asserted, that the cane was raised from seed in Egypt, Arabia, and I think, Malabar. It particularly described the arrow of the cane and the singular fact of only one in every three plants producing perfect seed." **Mr. Wray** concludes, "For my own part, I h ve never seen any cane seed, nor do I believe that it is perfected in India." But on the other hand, in the same volume of the Agri.-Horticultural Society's Journal, **Major Jenkins** (a writer whose observations are entitled to the greatest respect) says that in Assam (Gowhatti) "some hundreds of canes (Otaheite variety* may be seen in flower at once in **Dr. Scott's** plantation, but I think only in plants 3 or 4 years old, *i.e.*, canes which have been planted 3 or 4 years, and allowed to remain undisturbed as far as regards the roots or shoots. I have sown some of the seeds but got no canes, perhaps from being lost among the other grasses. The flowering of the canes is not very uncommon anywhere, but the Natives consider the circumstances very unfortunate." "When the Agricultural Society were first importing canes from the South Sea Islands, I suggested whether seed might not be procured." It will be found that in 1845 **Dr. Thompson**, in connection with his remarks on Madagascar canes, thought of the idea of multiplying the forms of cane by seminal selection; see the paragraphs which deal with Otaheite Canes. (*See p. 47.*)

In 1853 the subject of the seeding of canes in India again attracted attention. **Mr. W. Haworth** procured seed in Ceylon (Kandy). These, he gave to the Secretary of the Agri.-Horticultural Society by whom they were given to the Head Gardener for cultivation. About the same time **Mr. J. Thomson** of Cossipore wrote that he had seen the flowering in Bengal, but not very often. He added, "I do not think, however, that what you believe to be the seed of the sugar-cane would germinate. At least I am not aware of sugar-cane ever having been produced from seed. In the West Indies, where they have not the same variation of season as in Bengal, the sugar-cane is allowed to stand much longer on the ground sometimes from 13 to 16 months. If I remember rightly the season of 'arrowing' is about November or December in the West Indies." "I believe that it is from the short time which the cane is allowed to stand

* See **Dr. Scott's** statement in chapter on Varieties of Assam canes, p. 61 above.

SACCHARUM: Sugar.

Cultivation in India.

IMPROVEMENT.

Seeding.

on the ground that we never see the arrow in Bengal; but I have very little doubt that if sugar-cane were allowed to stand through one cold season and on to the next, we should see plenty of cane arrow here."

In 1854 **Mr. J. W. Payter**, an experienced sugar-cane planter in Bogra, said that he was amused to read in the Society's Journal that doubt seemed to exist as to sugar-cane flowering in Bengal. "Twelve years ago, I had whole fields in blossom; this was from the cane I got from the Society; being unable to break through the prejudices of the ryots that year to cultivate it, the crop remained mostly uncut and *all* ran into tufts; but, I regret to say, I took no steps to ascertain whether it contained anything like seed or not. I have seen country cane in blossom but very rarely, and only one or two here and there. The ratoons produce tufts more generally than the first crop."

In 1881 the subject of cane seed was again taken up in India. The Secretary of the Agri.-Horticultural Society (*Vol. VI., 216-218*) drew attention to the allusion to the seeding of the cane in **Mr. Walter Hill's** article on *Beet-root* versus *Sugar-cane** and he took the opportunity to review the papers and correspondence that had appeared on that subject in the Society's Journal. The statement is there made that "The enquiry elicited replies from residents in certain parts of India, but no satisfactory affirmative information was obtained, though it was shewn that the cane when allowed to attain full maturity seeded freely." With reference to the Ceylon cane seed (furnished in 1853 by **Mr. Haworth**) the Secretary adds: "This was carefully tried in the Society's Garden, but entirely failed to germinate." This announcement is doubtless made on the authority of unpublished records, to which the Secretary has had access since, so far as the writer can discover, the head gardener nowhere alludes (in his monthly reports) to the failure or success of these seeds.

The subject of the cultivation of cane by seed does not appear to have been taken up in the District Manuals and Gazetteers. It would thus seem that writers who speak of such cultivation as taking place in India are in error. It is, however, very generally admitted that the cane flowers occasionally, and certain forms more frequently than others. If allowed to grow the required time, the same percentage of flowering spikes would be found to seed in India as has been recently observed in Java and the West Indies. It is significant that the so-called Otaheite cane is the introduced form which flowers most frequently in this country. **Mr. Wray** appears to hold that in his day this was the case also in the West Indies, especially on estates with a sea aspect. **Mr. D. Morris**, however (*Linnæan Society's Journal XXVIII., 199*), says: "The experiments at Barbados, confirmed by observations at Trinidad, Demerara, and latterly at Kew, have now very clearly proved that the varieties of sugar-cane known as 'Purple Transparent' and 'White Transparent' periodically produce seed at Barbados; and that the Bourbon cane, known also as the 'Otaheite Cane,' does so very sparingly. From the remarks given above under Otaheite and also Bourbon canes, it will be seen that the greatest confusion seems to prevail as to whether these forms should be regarded as distinct. The so-called "Otaheite," originally introduced into India, appears to have been the form now known as "Mauritius." Of what he regards as the true "Otaheite" **Mr. Wray** described two forms which might be the "Purple Transparent" and "White Transparent" mentioned by **Mr. Morris** or what is more likely these are two kinds of Batavian cane. If this latter conjecture be correct, it is significant that (nearly half a century ago) **Mr. Wray** (an experienced Jamaica planter) should have written that the

* *Conf. with Journ. Agri. Hort. Soc. Ind., VI, Sel. 99—104.*

IMPROVEMENT.

Seeding.

"*Yellow Violet*" (which is probably **Mr. Morris'** "*White Transparent*") differs from the "Bourbon" and "Otaheite" in certain particulars which he details, and that it is "seldom that this cane *arrows.*" It would thus seem that the planters' names for the cultivated canes have got hopelessly intermixed in recent times or, that in the liability to flower, the various races have materially changed.

The above review of the leading Indian published facts regarding the flowering of canes cannot, however, be concluded without mention being made of the fact that within the past few months a start has been made in growing cane in India from seed. Following up apparently the interest awakened by the Kew Bulletin in the subject of the improvement of cane by seminal selection, the Superintendent of the Saharanpur Botanic Gardens secured several sets of sugar-cane seeds. In the annual report just issued (1891) mention is made of the successful germination of some of these. One set (which failed to germinate) had been procured from **Mr. T. H. Storey**, Superintendent of the Sajjan Newas Gardens, Oodeypore. The second from **Mr. C. Maries**, Superintendent, the State Gardens, Gwalior. Of the grasses which sprang up from the Gwalior sowing many seedlings have been identified as that of cane. A third supply was got from Perak though this failed to germinate. It would thus appear that a start has been made, but the Superintendent (**Mr. Gollan**) remarks that the chief difficulty is to get a sufficient supply of seed. The plant is rarely allowed to flower owing to the strong prejudices of the people against this. It is reported, in fact, that in some part of the country if a single plant in a field flowers, the whole produce has either to be given to the Brahmans or burned. Though this belief prevails, **Mr. Gollan** adds that his informants had not heard of a field that had been actually so disposed of. Thus if any doubt ever existed, as to the cane seeding in India, **Mr. Gollan's** report must be accepted as setting that matter at rest. All that remains now is to extend the experimental cultivations until better forms are found among the seedlings than we presently possess.

Red Bombay Cane. 154 *Conf. with pp. 484, 76, 102, 123, 161.*

It need, therefore, be only repeated by way of conclusion that the practical interest in the subject of the seeding of the cane lies in the possibility of producing improved sugar-yielding forms. It is admitted by all sugar-cane planters that continued propagation from cuttings grown, year after year, on the same soil, results in a serious degeneration. On this account planters at a distance periodically exchange seed-canes or special nurseries are resorted to for the purpose of producing seed-canes. This same fact is fully appreciated by the Native cultivators of India and the dangers of too continuous a cultivation of any particular form are quite understood. Thus, for example, a Native cultivator wrote, in the Agri.-Horticultural Society's Journal, on the destruction of the Red Bombay canes of Bengal. This was due to the appearance of a worm in the cane after it had been grown in the same district without intermission for a certain number of years. Fresh stock, grown side by side, remained free from disease. It seems highly probable that the degeneration of the imported canes was largely due to the same cause, and that nurseries for interchanging stock from one province to another or from district to district would, therefore, effect greater improvements in the Indian sugar industry than anything else that could for some time to come be undertaken. In such interchanges and nursery treatment the stock might not only be kept up but improved, and should the idea of seminal selection be found beneficial, this could, by nurseries in every province, be carried out on a large scale. The chances of improvement by selection, whether seminal or otherwise, depend entirely on the extent to which the experimental cultivations are prosecuted. They are, therefore, quite beyond the means of the

Necessity for Nurseries. 155

IMPROVEMENT. Seeding. ordinary Native sugar-cane growers. Good forms, when discovered, could be perpetuated and distributed by cuttings. The possibility of improvement needs scarcely to be urged. The arguments in favour of this can be illustrated by almost any one of the numerous agricultural products of India, but unless continuously maintained in sugar-cane, the result would be the same as may be learned by the perusal of the review given in another volume of this work, of the late **Mr. Scott's** experiments with the opium plant. Certain forms, which that accurate observer fostered from seminal sports, were seen to possess well recognised properties both in the yield of the alkaloids and in the freedom from disease (*Vol. VI.*, 55). After a time the experiments conducted by **Mr. Scott** were abandoned, and it is highly probable the superior forms he tried to distribute over the opium districts have, by now, completely degenerated or have been entirely lost. The want of private enterprise in nursery produce and in the supply of, or demand for, superior seed is one of the greatest defects of India's agricultural interests.

SUGAR PLANTATIONS. 156

HISTORY OF THE EFFORT TO ESTABLISH SUGAR-PLANTATIONS & FACTORIES IN INDIA.

A perusal of the extensive literature preserved in the Proceedings of the Honourable the East India Company can scarce escape the conviction that, little more than a century ago, Bengal was regarded in England as the peculiar property of the great Company of merchants who laid the foundation of the present British Indian Empire. In its relation to Great Britain Bengal was practically classed as a foreign country. It was accordingly debarred from many of the special privileges and protections granted to the British colonies of America and the West India Islands. With no branch of Bengal trade is this fact more powerfully exemplified than in that of sugar. Prior to 1789 Great Britain had for some time received its sugar
1698 to 1729. exclusively from the West Indies. From the year 1698 to 1729 the supply came almost entirely from the British colonies. The imports progressively increased from a valuation of £629,533 to £1,515,421. About the close of the period specified, France, becoming jealous of the British success, made strenuous efforts, however, to organise sugar plantations in St. Domingo.
1726 to 742 Accordingly, from 1726 to 1742 the Sugar production of that island expanded from 400,000 cwt. to 848,000 cwt. In 1742 England was at war with Spain—a fact which may have favoured the French and other foreign sugar-planters. A more direct fostering influence doubtless existed, however, in the law passed by France which allowed her colonies to send sugar direct to foreign purchasers. The corresponding law did not come into force with the British colonies for some twelve years later, so that consignments for America and the Continent of Europe had to sustain the delay and bear the extra charge of being re-exported from England. The evident advantage thus enjoyed by French traders told much in their favour, while to evade the British law many reprehensible practices crept into existence. To India the restrictions imposed by Britain in this instance proved advantageous however, for American and other foreign ships gradually came to her ports and carried away sugar, indigo, and other goods. In an official report under date 1791, for example, we learn that "the export trade to America and Flanders is rising very fast in sugar." Thus Indian sugar had found its way to Europe and America before it was appreciably made available to the English people.

By 1742 the demand for sugar in Europe had, in fact, been firmly established. It had very nearly become a necessity of life and its production could not be repressed by fiscal prescriptions. The observation was accordingly made that, relatively to the French supply, the sugar obtained by Great Britain from her colonies had declined. England was, in fact,

SUGAR PLANTATIONS.

more dependent on the French than on the British colonies for her sugar. This state of affairs will be apparent by the following returns:—

Production of sugar in the French Colonies for 1742—	Cwt.
In Martinico, Guaduloupe, etc.	622,500
In Aispaniola (St. Domingo)	848,000
TOTAL .	1,470,500
Sugar produced in the British West Indian Colonies for 1742	791,400
European and American Sugar supply for the year 1742 .	2,261,900 cwt.

By a similar comparison in the trade for the period here dealt with, it has been shown that the French colonies had increased their production from 30,000 to 120,000 hogsheads, while the British colonies had been able to advance their outturn from 45 to only 75,000 hogsheads. It seems likely that this state of affairs would probably have continued but for the calamity which overtook the French colonists in the mutiny of St. Domingo. The French sugar planters were not only ruined but a sugar famine took place which very greatly raised the retail price of the commodity. An outcry arose not only against the protective measures that favoured the colonies and debarred India from participating in the British supply, but against the slave labour of the colonial plantations. While matters were in England thus maturing in a direction likely to prove favourable to India, a similar movement had for some years taken place in India itself. It was seen by the merchants in Bengal that the colonial prosperity had destroyed the export trade that formerly took place from India, as also the re-export Chinese transactions, and that the internal restrictions imposed by the Indian Administration were rapidly depriving Bengal of the market it had long enjoyed in the supply of sugar to the Malabar coast. It was pointed out that Bengal production had been so depressed that that province had actually begun to look to the North-West Provinces and even to China and Batavia for the sugar required by its own people. The memorial (which was submitted to the Government of Bengal in June 1776) urged that the Malabar trade was the more desirable, since it afforded an exchange between Bengal sugar and Bombay cotton. As matters were transpiring, the memorialists maintained the Dutch were drawing from India a large amount of specie, since they no more brought their sugar to the shores of India and sold it in exchange for Indian goods, but, trusting to India's necessities in the matter of sugar, were able to compel Indian ships to go to Batavia for the sugar. We accordingly read "that the vending or procuring a cargo of sugar was even considered as a sort of favour conferred by the officers of the Dutch Government on the Bombay merchant." But in these transactions the Dutch absolutely refused to take merchandise in return, and thus India was deprived of ten lakhs of rupees annually and had her own internal commerce disarranged. The memorial above alluded to, received the most careful and immediate consideration of the Honourable Company, and we accordingly learn that soon after the Indian restrictions complained of were greatly mitigated and in time entirely removed. Matters in Bengal were accordingly greatly improved the more so, since, through the changes that were taking place in Europe and America it became advantageous for foreign ships to come to the shores of India in search of sugar. The loss of this shipping traffic was apparently deplored, however, for in the Proceedings of the Honourable Company there occurs the regret that "much sugar was being carried to Europe and America on foreign ships". English merchants in India (private and Company's) were unable to

Sugar and Cotton. 157 *June 1776.*

SUGAR PLANTATIONS.

participate in this new foreign trade owing to the heavy duty charged on the sugar by England, and the fact that they were compelled, if they traded in it at all, to convey it to England, in the first instance, before they could consign it to countries where the import dues were more favourable. But this state of affairs had better be exhibited by the historic records of actual transactions. *April 1789.* In April 1789 the East India Company directed its agents in India to forward to England a consignment of Bengal sugar. This was complied with, and in May 1791 the first sample of the East India Company's investments in Indian sugar was submitted for report to Messrs. **Travers & Bracebridge.** It was found, that although it differed in some respects from West Indian sugar, it could be dealt with by the English refiners and that the quality was satisfactory. The next consignment arrived by the ship ***Haughton*** but was sold subject to the same duty and drawbacks as in the case of West Indian sugar. The English Custom authorities, however, declined to recognize these conditions, and on delivery being desired it was charged £37-16-3*d.* per cent. (or say 8*s.* a cwt. more than the duty on West Indian Sugar) on the gross sale price—the West Indian sugar being charged at the rate of only 15*s.* per cwt. This led to a protracted controversy and the exhibition of the strongest opinions both for and against the new Indian trade. *1792.* In 1792 the Court of Directors of the Honourable the East India Company placed before the Lord Commissioners of the Treasury a Resolution on the subject of the exorbitant duty claimed by the Custom authorities on Indian sugar. It was there set forth that the Company having been called upon by the British public to endeavour to lower the price of sugar by bringing the Indian article into the market, had done so and were prepared to guarantee to meet the entire requirements of Great Britain in sugar, provided India were placed on the same favourable terms as had been granted to the colonies. The Resolution, while disclaiming any intention of calling into question the desirability of the protective measures that had been enacted in favour of the colonies, pointed out that the greater distance of India and consequent heavier freight charges were considerations that would be seen to secure to the colonies a full participation in the trade. The application was not granted though frequently repeated, and the *1836.* heavy import duty continued to be charged on Indian sugar till 1836. The cargo of the *Haughton* appears to have been sold at a loss as will be seen from the following account:—

	£	s.	d.
Prime cost and invoice charges of 96 cwt. @ C. R1,268 .	126	16	0
Custom @ £37-16-3 per cent. on sale	165	5	0
Freight @ £26-10 per ton and 20 per cent. kintlege. .	152	12	0
Charges merchandise, @ 5 per cent.	21	17	0
	466	10	0
Sale @ £4-12 per cwt, ...	437	0	0

The next consignment per the ***Princess Amelia*** realized a profit of £286, but in the item of charges the British import duty on 1,746 cwt. amounted to £3,302 and the freight to £2,776.

In spite of many discouragements and losses the possibility of ultimate success in the sugar trade was kept vigorously in view. The East India Company called for detailed information from its officers in India. Every aspect of the trade was carefully enquired into and the reports, which continued to appear not only regarding Bengal sugar, but that of the North-West Provinces, of Madras, and of Bombay afford a very trustworthy source of information, being quite as complete a statement of the methods

SUGAR PLANTATIONS:

of cultivation and manufacture as any that we possess of a more recent date. It would endanger too great a repetition of the facts dealt with in other chapters of this article to follow up the numerous issues that arise in the present connection. Suffice it to say that the Company, in its own interests, were wise in their resolution to avoid, as far as possible, direct ownership of plantations or the investment of much money in the machinery necessary to test the practicability of sugar manufacturing in India, on the West Indian methods, until more favourable terms had been obtained for the admission of their sugar into England. The Company, accordingly, contented itself with purchasing sugar from the Natives on the most advantageous terms possible, and for a series of years, it continued to bring to England from 1,000 to 3,000 tons of the various kinds of Native made sugar. These purchases were recommended to their agents to be made mostly in the better qualities and to be used in place of saltpetre as balast on ships with light cargoes.

It may thus, in all fairness, be said that the policy which the Government of England pursued, during the first half century of its colonial sugar trade, withheld from India the possibility of its being to-day a great sugar-supplying country. The prohibitive British import duty was removed in 1836, and from that date the effort was once more put forth to establish plantations and factories and to create a large foreign demand for Indian sugar, but the opportunity had passed and will probably never again return to India. The failures of the nineteenth century were in fact, if anything, more complete than those of the eighteenth. But the discovery of beet-sugar gave an entirely new aspect to the trade and destroyed, in their turn, the refineries of India which had very considerably prospered, even although, European production and manufacture had failed.

1811.

In 1811 the East India Company gave up all further effort to contend against their losses in sugar. They accordingly, in that year, issued an order to their Indian representatives that, except very occasionally or when unavoidable, sugar should no longer be included among the Company's investments. There were, however, two great periods when the idea dominated that large plantations on the West Indian pattern were likely to succeed. These were from, say, 1790 to 1820 and again from 1830 to 1860. It will be seen in another section of this article that this very idea has, within the past few years, been once more urged as worthy of careful consideration. That subject need, therefore, be no further dealt with, but it seems desirable to review very briefly some of the early efforts which were put forth to test the practicability of large plantations and sugar factories. In the Proceedings of the Honourable the East India Company for 1791 there occurs, what appears to be, the first mention of a European desiring permission to go out to India as a sugar-planter. We are there told that **Lieutenant John Paterson** of the Bengal establishment had (in 1787) shown that sugar could be grown in India with many superior advantages and at a much less expense than in the West Indies. He was, accordingly, granted permission to return to India and to take up land in Behar which he could procure from the Natives, on the distinct understanding that he did so at his own risk. The Company simply agreed to purchase all the sugar he might make at a certain fixed rate. On arrival, we are led to believe that, he preferred Benares to Behar and had permission to alter his location. We next find him spoken of as having ultimately secured land in Beerbhoom. The fact that he had not, however, for some years commenced the manufacture of sugar is viewed by the Board of Directors with disfavour. A loan of R25,000 is then recorded as having been made to enable him to procure from England the machinery he required. An assistant (**Mr. W. Fitzmaurice**) whom he was permitted

European Plantations.
158
Conf. with pp. 37, 39, 48, 62, 63, 93-94, 103, 111, 161-162, 212, 306, 309.

SUGAR PLANTATIONS

to take to India left his service shortly after and applied for permission to take up land on the same terms as has been granted to **Paterson** We are then told that **Paterson** died on the 26th September 1794 and that directions had been issued to recover the loan from his estate.

February 1793. In February 1793 **Mr. William Fitzmaurice** (a Jamaica planter) submitted a long detailed memorandum on the advantages to India of the introduction of the West Indian methods of cultivating the cane and manufacturing the sugar. It may be inferred from various entries that he started a plantation, for he is subsequently shown as having sold sugar to the Company. The writer has failed to discover where **Fitzmaurice's** plantation was or what came of it. About the time of the submission of **Lieutenant Paterson's** application, **Mr. Robert Heaven** was granted authority to proceed to Bengal as a sugar planter. In 1794, we are told **Mr. James Hauson Keene**, an experienced sugar manufacturer, was allowed to proceed to Bengal for the purpose of starting a plantation and factory. *1795.* In 1795, **Lieutenant Charles Maddison** obtained the authority of the Board of Directors (on quitting "His Majesty's Service") to remain in India "and to engage in the culture and manufacture of sugar and indigo upon his entering into the covenants usually executed by persons of that description." So, again, **Mr. J. Walker** was allowed "to proceed to India under free merchants indentures, with a view to introducing improvements in the cultivation and manufacture of sugar."

It is thus, perhaps, needless to enumerate more examples to show that a large influx of Europeans (mostly West Indian planters) took place to India, and that there were rocognised covenants as *those usually granted to such persons.* From all parts of the country we get glimpses of the existence of sugar planters. Thus **Captain Andrew Pringle** wrote from Lucknow (a country beyond the territories of the Company) that he was prepared to supply 1,000 tons of sugar on certain apparently favourable terms. A **Mr. James Paull**, also of Lucknow, was ready to supply 60,000 maunds, but the Company did not think it politic to purchase sugar from "the Vizier's country." The sugars of various up-country districts are frequently alluded to. They were the regions apparently from which such private firms as **Messrs. Cockerell & Co.** drew their supplies. The sugars of "Benares, Mhow, and Azimghar" are very specially mentioned. That manufactured by **Mr. Carden** at Mirzapore was held in high estimation. Frequent reference is also made to the sugar of Burdwan, Calcutta, Nuddea, Jessor, Rungpore, etc., etc., and in such terms as to lead to the supposition that there were (about the close of the last century) factories and refineries owned and worked by Europeans, if not sugar-cane plantations also all over Bengal. Sufficient may perhaps be accepted as shown, by the above special cases, to justify the statement that little was thus wanting in skill, capital, and enterprise to have made sugar-cane cultivation and sugar manufacture a success, had sucess been possible. And not in Bengal only, but in Madras and Bombay similar strenuous efforts were made. Thus, *1799.* for example, we read of Madras by date 1799 that "being desirous of promoting the culture of sugar in such of our possessions as may be suited to the growth of the cane, with a view of affording to the European market a more ample supply of a commodity now become, in a manner, an important necessity of life, we have permitted **Mr. Edward Campbell** to proceed to India, for the purpoe of establishing sugar works in such of the districts under our Madras Government as he may conceive most favourable for such an undertaking." We learn subsequently that that gentleman settled in Trichinopoly and held a lease for ten years on a rental of R2,062 per annum. Shortly after, however, he converted his sugar plantation into an indigo factory. The Collector of Vizagapatam, by date 21st April

SUGAR PLANTATIONS.

1797, wrote that "although the trials which had been formerly made had failed to answer expectation if undertaken by Native management, the issue might be attended by very different success, if established under proper regulation." He was there dealing with the proposals for European planters. Shortly after this the Board recommended "that encouragement should be given to individuals who might wish to engage in the manufacture for themselves." **Mr. Robert Campbell** (of whom we have no previous history) seems to have anticipated this authority, for the Collector of the district proceeds to argue that he must have made great progress since, while by the Native methods 5 maunds of jaggery costing R5 were necessary to make 1 maund of sugar, **Mr. Campbell** was able to offer a superior sugar at R3-8 a maund. It is, however, explained that **Mr. Campbell** imposed certain conditions that rendered his proposals of supply to the Company inadvisable. One of these may be here mentioned. **Mr. Campbell** required "that the inhabitants, cultivators of the cane, should be compelled to sell to him, exclusively, the produce of their present gardens and all others they might in future cultivate for a period of six years." The want of a guarantee of continuous supply, it will be recollected, caused the ruin of the indigo industry of many parts of India, and it seems but natural that if the owner of a factory does not grow his own cane he requires some sort of security that the capital he expends on a factory may not be thrown away through the subsequently discovered inimical interests, or it may be the perverse inclinations of his neighbours in simultaneous ceasing to grow cane. The owner of a mill intended to directly manufacture sugar from the cane must either grow his own supply or possess some proprietary right that enables him to stipulate for a certain percentage of cane cultivation. This great difficulty was, therefore, quite as fully realized a century ago, as it is at the present day. Indeed, it seems, that this difficulty is entirely overlooked or ignored by those who advocate the construction in India of the most improved modern mills to manufacture sugar direct from the locally-produced cane. But **Mr. Campbell** (or rather the two **Campbells**) were not the only sugar planters who figure in the records of the experiments conducted in Madras about the close of the last century. **Mr. W. J. Colley** occupies a distinctly more prominent place. We learn, for example, by a letter of February 1800 that "**Mr. Colley**, who, in consequence of the encouragement he received from Government, has established expensive sugar-works at Mynsurkotah." In one of **Mr. Colley's** long explanatory letters he points out that he is prepared to sell the sugar to Government practically at what it cost him, provided he be allowed to take the molasses to manufacture the rum required by the Government. The profit of the sugar trade, he emphatically declared was in the rum. This was no doubt the case a century ago, as it is to a large extent at the present time. All **Mr. Colley** asked was "a preference for his rum, should its quality be approved." In 1815 it is stated that **Mr. Colley** manufactured 300 tons of sugar, so that his factory, which cost him R44,000, must have been fairly large. The Company soon withdrew its contract and little or nothing is further heard of **Mr. Colley** or his factory. Another of the Madras planters is specially mentioned, because of his superior knowledge in the distillation of rum, *viz.*, **Mr. Parkinson.** That gentleman was originally in the service of Government and in charge of one of their experimental sugar factories and distilleries. He left the service and became a sugar planter.

Conf. with p. 283.

February 1800.

Rum. 159

Conf. with pp. 95, 96, 158, 175, 313, 320, 321.

Regarding sugar plantations in Bombay (established during the closing decade of the last century) it, perhaps, is unnecessary to dwell on more than one or two. The object aimed at in the present sketch is to

SUGAR PLANTATIONS.

exemplify the widespread interest which was taken in the subject, and the amount of enterprise and capital which was expended on the futile effort to establish sugar-cane planting as an European industry. **Captain Robert Crauford** of His Majesty's 75th Regiment applied for, and obtained, a concession of land on the Malabar coast. It was granted on a long lease, the chief condition being the improvement of the cultivation of sugar. In a communication of 1792 there occurs the following remark addressed to the Board of Directors: "We have great satisfaction to acquaint you that **Messrs. Helenus Scott, Robert Stewart, & John Twiss** have undertaken to introduce the cultivation of sugar and indigo on the Island of Salsette, with very sanguine expectations of success; and we beg leave to assure you, that we shall give proper encouragement to a scheme which is so likely to prove beneficial to your interests." In a further communication (1795) we learn of permission being granted for a sugar-mill being sent out to **Dr. Scott**. Next, in 1801, of his having submitted four hogsheads of his Poway Distillery arrack, one ditto of Muscavado sugar from Poway, and a specimen of Bandap cotton raised from Guzerat seed. In a still later communication we find a discussion between **Dr. Scott** and the Bombay authorities on the admission of sugar (the produce of the Poway estate) as dead weight on the Honourable Company's ships. The reader will find in the chapter below on the sugar manufactures of Bombay (Thana District), p. 308-9, particulars regarding two subsequent efforts to establish sugar-planting in Bassein, in which Government not only gave lands on nominal rents, but made large advances of money to assist in the purchase of machinery. It will thus be observed that in Bombay, as in Bengal and Madras, there were, at the beginning of the present century large plantations and factories owned and worked by Europeans. In another part of this work the reader will find numerous incidental allusions to plantations and factories (dating back to the closing years of the last century) found in Assam, in Burma, in the North-West Provinces, and even in the Panjáb. But within a very few years after their establishment these were all either converted into Indigo factories, or they survived as sugar refineries for purifying native-made sugar, or they ceased to exist entirely, their places being indicated at the present day by unsightly ruins. The following brief allusion to this subject occurs in the Famine Commissioners Report: "Attempts were made in the early days of the East India Company to promote the industry of sugar manufacture in Upper India. Very large advances were given for sugar growing, and the factors in charge of districts in the neighbourhood of Benares introduced sugar mills which (as **Sir H. M. Elliot** mentions) were found to be much less effective than the crude sugar-mill of the country. As in the case of cotton, records of these early enterprises have almost disappeared, but the still existing ruins of sugar-mills testify to many a complete failure and many a broken fortune. The chief mistake appears to have been the concentration of operations in large central establishments which led to the deterioration and evaporation of cane juice during the carriage of canes to the factory, for, it is now well known that juice ought to be expressed and boiled as soon as possible after canes are cut."

Bombay Plantations. **160** *Conf. with pp. 212, 306-307, 309, 320.*

Madras Plantations. *Conf. with pp. 93, 232, 240, 309.* **161**

Conf. with p. 101.

It has, however, been stated above that a second period of renewed activity occurred, when the same high expectations were entertained and a corresponding waste of capital and enterprise took place. This may be said to have extended from 1830 to 1860. The failure of India to compete in the foreign markets has spasmodically been attributed mainly to the want of capital in planting enterprise. Unfortunately, however, all past experience disproves that theory. The reader, who may be sufficiently

SUGAR PLANTATIONS.

interested in the subject to peruse the numerous passages given below will have sufficient proof that the second great period in the history of the effort to establish European sugar-plantations and factories in India was neither less intelligently nor less earnestly prosecuted than the first. There is, in fact, in connection with the records of this period a greater uniformity in the verdict of failure to compete with the Native in the cultivation of the cane than occurs regarding the experiments of the eighteenth century. From one corner of India to the other the piteous story is recorded of planters (many of them with extensive West Indian experience) having spent all they possessed in the vain effort to establish sugar-cane plantations. Started on grants of land, in some cases given rent free for long periods, and the buildings necessary having been even sometimes constructed on loans of money made by Government, they have passed through a protracted struggle for existence. The end has almost uniformly been that the estates have become waste lands, or have been converted into indigo plantations, or become the cause of fruitless litigation. The ruined factories and refineries that can be pointed to in almost every district of India do not, however, prove that success is impossible. They should naturally enter, however, into the serious consideration of persons who may contemplate repeating experiments which have proved gigantic failures during at least a century of earnest endeavours. But while planting and direct manufacture have proved very nearly hopeless undertakings for Europeans in India, much greater success has attended the effort to refine the crude Native sugars. The Indian refineries (both Native and European), have been able to largely supply the Indian market with a superior sugar, and they even created and held for many years a modern export trade. But they were soon doomed to have to face a serious reversion. In beetroot sugar they were confronted with a far greater disturbing element than they had hitherto contended against in the numerous vicissitudes through which their industry had passed. The sugar manufactories and refineries that had not enjoyed a large monopoly in the production of rum were rapidly reduced to the verge of ruin, and there would seem to be reason in thinking that they have not even now reached their lowest level. Indeed, in perfect fairness, it may be said that the European cane-sugar industry of India, nay not of India only but of the world, at the present moment, receives its bounty against beet-sugar in the rum traffic.

Rum. 162 *Conf. with p. 40.* *Conf. with pp. 93, 96, 158, 175, 320-321.*

Mr. Westland's remark that paradoxical though it may seem, the Native collectors of palm juice and the manufacturers of *gúr* (= coarse sugar) have increased in wealth, while the Native and European refiners have been ruined, is true not of Jessor and palm-sugar only, but of all India and of cane-sugar as well. The consumption of sugar has undoubtedly greatly increased, but the enhancement in the imports of foreign sugar take at present but a small share in this modern feature of the trade. They are in no way lessening the consumption of the crude sugar by the people of India, but are rather taking the place of the refined sugar which was formerly manufactured in this country. But leaving for the present the subject of the decline of the Indian refiner's trade it may be as well to more fully exemplify the important steps which have been taken since 1830 to improve Native sugar-cane cultivation and to establish European plantations. For the first 20 or 30 years of the Agri.-Horticultural Society of India, its Transactions and Journals were not only the means of making public the then stirring events in the Indian agricultural and commercial world, but the Society itself took a leading part in the efforts that were being put forth to improve the productive resources of the country. That Society was founded in Calcutta in 1820, but local branches of it were rapidly formed in every important town. It gave

Crude Sugar eaten in India. *Conf. with pp. 40, 81-82, 104, 113, 137, 325.* 163

SACCHARUM: Sugar. History of Establishment of Sugar

SUGAR PLANTATIONS.

birth to a temporary activity, of a kind never since witnessed in India. Large sums were subscribed towards its funds both by Government and the public. Its first publications consisted of eight volumes of Transactions which appeared from 1836 to 1841. These literally teem with papers describing the sugar-cane industry of Bengal, and of the North-West Provinces, of Burma, etc. Articles appeared, for example, on the great extension of sugar-cane cultivation of Azimghur: sugar-cane in Benares: immense improvements on the ordinary sugar-cane of Bengal by cultivation in Chittagong: the value of Amherst as a sugar-producing country: the forms of cane grown in the Malay Peninsula: the formation of the Dhoba sugar-works and on the introduction of the superior Otaheite and Mauritius canes, etc., etc. In 1837 the gold medal of the Society was awarded for "Zealous exertions in bringing the Mauritius sugar-cane to this country, and ultimately successfully establishing the permanent cultivation of that cane on the banks of the Nerbudda." Premiums were awarded for the best sugar and for sugar made of the imported canes In Volume II., of the Transactions we read, for example, of the allotment of money awards amounting to R6,750 for the year 1836-37.

1836 to 1841.

From 1836 to 1841 vas a period of great energy with the Society, when sugar-cane may be said to have been the subject that occupied its attention before all others. The first volume of the Society's new publication—the Journal—which came out in monthly numbers, was completed in 1842. It opens with instructive original papers on the improvement of Indian cotton by the introduction of American seed; the utilization of hemp; the mode of improving East Indian sugar; and the method of reeling silk followed in Bengal. Of these subjects it may be said that sugar at first occupied a far larger share of the attention of the Society than any of the others. Interest in flax was first dropped, then in silk, next in sugar, and cotton survived until its place was assumed by indigo, then by tea, by jute and last of all by wheat. One of the most fruitful actions taken by the Society, contributed greatly towards bringing about the equalization of the import duty charged in England on Indian and West Indies sugar. Papers, of course, continued to appear on all the above-mentioned subjects and have done so down to the present day, but it may be said the greatest degree of interest was taken in sugar between the years 1830 and 1860. After that date sugar practically disappeared from the Society's consideration, and the efforts to improve the stock of Indian cane had, if anything, an even shorter portion of public favour. One of the earliest detailed papers on the subject of sugar and, perhaps, for the period of Indian interest, the best that appeared, was one by a "West Indian Planter" entitled "*The present imperfect mode of manufacturing East Indian Sugar and its attendant evils.*" The following passage from that article may be read with interest, since it marks the character of the then manufacture and the high expectations held out of the future of the industry:—

Rum. *Conf. with pp. 93, 158, 175, 277, 320-321.*

164

"I am perfectly well aware, that the great difficulty to be contended against is the total abolition of the present system of producing sugar from that acid, fermented stuff denominated 'gour', also the difficulty of getting the canes to the steam mill for the purpose of grinding in some districts. If this could be made practicable (and I have no doubt it could) the sugar cultivation here would eventually be the most lucrative in which moderate capital could be invested, and the example of the West Indies is staring us in the face as a convincing proof; there, on a well-managed estate, the rum pays all, or very nearly all, the expenses, the sugar going into the pocket of the proprietor as net profit !!! This is a fact, not come to my knowledge by hearsay, but from eleven years' experience and ocular demonstration. And could the services of the West India Negro be now depended on, the profits would be increased, as the planter only pays those who work for him; whereas, under the slave system, he was obliged to support, clothe, and hospital superannuated negroes and children, besides paying a heavy capitation tax: yet, notwithstanding this enormous expense, the rum made paid the outlay.

SUGAR PLANTATIONS.

"I flatter myself there is no need of further argument to convince us here that the present system is not an advantageous one, or one conducive to the improvement of the article in question, or to the permanent benefit of the proprietor. Why then do not the East India planters avail themselves of these facts? Surely, it cannot proceed from ignorance of the circumstance, as there must be many here conversant with West India agriculture and its profitable returns.

"With respect to the refining process practised here I do not think it will fully answer the purpose intended, as it is a well known fact that India never can compete with England in the art of refining; consequently, the article to be encouraged here, is the rich, clear, strong-grained, amber coloured Muscovado sugar, manufactured so as to be able to stand any sea voyage, and which, on reaching the market *weigh well.* Such being the West India sugar, the inferior qualities alone being reserved for the purpose of refining.

"I cannot resist making a few observations relative to the *useless and ruinous practice* observed here by sugar planters in involving both themselves and property by an enormous outlay in the purchase of expensive machinery, and that before one pound of sugar-cane be made as a return. A fifteen horse-power steam mill for cane-crushing I look upon as only necessary at the onset coupled with well hung and fast boiling coppers; with these, I say, a crop of 800 or 1,000 hhds. of first quality sugar could be manufactured with every facility in the regular boiling season. I even do not think the boiling in vacuo a judicious system as connected with ultimate profit, as the article produced therefrom is divested of most of its sweets and gravity—and the sugar owner should bear in mind that *weight* is the great thing to be looked to, this is retained by the good old system, and the splendid West India sugar which we all have both seen and used in England is the result. Such I wish the East India to become.

"The profession of a practical planter is one demanding much care, attention, and experience; inasmuch as it combines three distinct occupations, *viz.*, the agricultural part, the manufacturing part, and finally, the last, though not the least in importance, the distilling part; and as the superior or inferior quality of Muscovado sugar chiefly depends on the agricultural part of the profession, I am certain the great majority of the managers of sugar estates in this country would prefer having the cane matured under their own guidance" (*Trans. Vol. I.*, 50-52).

The essay from which the above passage has been extracted met with such favour that numerous correspondents called upon the anonymous writer to continue his valuable contribution. Accordingly there appeared "*Further suggestions, etc.*," from which the following may be here usefully extracted:—

"The more I see of India, the stronger my conviction is, that those engaged in cane cultivation, are not following the course necessary to make it a great sugar country, and perhaps, when too late, they will discover the mistake they have committed by adopting a plan of operations of which they are ignorant. I allude now to the present system observed towards cane agriculture. I most confidently assert, that a sugar estate never can succeed conducted on the ryot system, a system very applicable to indigo; but the one and the other being different and distinct plants, so must be their method of culture. As, during my short residence in India, I have frequently been asked by interested parties, 'could we not grow sugar this year and Indigo next?' My answer invariably was, 'Yes, you *could*, but as regards the former it would not be attended with success.' I shall prove this hereafter. Consequently we must have no ryots, such a cultivation being too extended, and affording too many ways and means for plunder. No! *Concentration of cane land is the great object*, the great desideratum, which, once obtained, a judiciously managed estate must flourish; but the idea of planting cane in the same manner as indigo, in small patches here and there, owned by numerous different people, scattered over a large extent of land (I know some places where cane was cut 14 to 20 miles) and each patch the produce of a different seed and soil, the irregularity and difficulty of carriage to the mill, the awful and wilful sacrifice of money that must necessarily and consequently be incurred in the employment of superfluous labourers and carts, the quarrels among the land-owners and ryots render the entire measure a decided loss, of both time and funds, and though last, not the least, valuable consideration to a sugar planter, and what is entirely lost sight of, is the *ratoon cane*, without which his produce cannot be first quality, as plant liquor boils both red and soft, owing to its richness and other causes, requiring deep mixture with that of first, second, third and fourth ratoons. Failures, and consequent embarrassment, will bring conviction that cane cannot be planted, or a superior sugar made from plant cane alone. A thorough knowledge of the cultivation and manufacture thereof can be gained only

SUGAR PLANTATIONS.

by three or four years' hard work as overseer, added to a subsequent long experience as manager: be it remembered, the duties of a planter in charge of a sugar estate are many and onerous, particularly in this country, where, in the onset, everything would be up-hill work, owing to the prejudices of those with whom he has to deal. Such a man is averse to the word *impossibility* or *such and such a thing cannot be accomplished.* He is reared in a school which teaches him that what is necessary must be done, consequently, his first business would be to entirely revolutionize the present system of cane agriculture, *viz.*, ploughing, hoeing, planting, weeding, moulding, drawing, trashing, and cutting. Then his attention is turned to his indoor work—reorganization must also take place here as to boiling, skimming, tempering, striking, and potting. In fact native systems and prejudices must be extirpated root and branch; that gained, everything goes on smoothly and profitably. In a country such as this, blessed as it is with every natural advantage, a proper application only of the means within our power is requisite to ensure a supply of sugar unequalled in quantity and quality.

"It is lamentable to see those gifts of nature lie dormant, and completely under the ignorant agricultural guidance of the natives who are content with small profit and little labour. I here must be permitted to express my regret, that those gentlemen who have endeavoured to establish and encourage cane cultivation, and to whom great prais ise due, should in the first instance have followed a mistaken course—a course singular in itself and adopted in no place out of India.

"As I before observed, a Sugar Estate, to turn out a profitable speculation, must *be concentrated*; the works placed exactly in the centre, so that the mill can command its food from all parts equally; the consequence is a great saving of labour, time, and money. Now we will suppose the first great difficulty, the attainment of the land, overcome, I shall at once proceed to open to view the gross revenue to be derived therefrom, and as nearly as I possibly can, conjecture the expenses indispensably necessary for the cultivation of the estate, and the manufacture of the produce on the West Indian system. I cannot err very much on this side of the question, being already thoroughly acquainted with my subject. I shall not attempt to speak so confidently as to the correctness of my calculation, relative to the estimate of the system, of manufacturing sugar from *goor*.

"I throw myself on the indulgence of my readers, and entreat them to make allowances for any discrepancies that may appear, although from enquiries made, I have been informed that I am not very far from the mark. Therefore, I commence with the West Indian plan and its results:—

Expenses of conducting a sugar estate in India on the West Indian system—Open pan.

	£	s.	d.
Rent of land on which cane sufficient to produce 750 tons of sugar might be cultivated	400	0	0
Salaries—			
1 Engineer, £150; Manager, £500; Overseer, £120 .	770	0	0
Coals, 500 tons, at £1-5 per ton	625	0	0
Labourers—			
150 for the field, R3 per month, R5,400; 30 for works, R6 for 6 months, R1,080; 10 Jobbers, R6 for 12 months, R720; total R7,200 at 2*s.* per rupee . .	720	0	0
1 European Cooper, £150; 4 Native Coopers, R5 per month, for 12 months R240=£24	174	0	0
Tools, hoes, cane bills, etc., £120; incidentals, £1,500	1,620	0	0
TOTAL .	4,309	0	0

Add to this interest on block—

Revenue from above Cultivation.

	£	s.	d.
750 tons sugar—20,454 maunds, selling to nett—R10 per maund, R2,04,540, at 2*s.* per rupee. . . .	20,454	0	0
70,000 gallons rum to nett in Calcutta, Re. 1 per gallon, R70,000 at 2*s.*	7,000	0	0
	27,454	0	0
Expenses as above .	4,309	0	0
Net revenue to pay interest on block .	23,145	0	0

SUGAR PLANTATIONS.

Expenses of conducting a sugar factory on the East Indian system—Vacuum pan.

Khaur will yield on an average 50 per cent. of boiled sugar; *Vacuum principle.*

	£	s.	d.
To produce 750 tons sugar, we should require 40,909 maunds Khaur, which at R4 per maund, R1,63,636 at 2s. per rupee	16,363	12	0
Boat hire to factory, say, at R3 per 100 maunds, R1,227 at 2s. per rupee	122	14	0
Salaries—			
Manager, £500; 2 Agents to buy *goor* (R150 per month each), £360	860	0	0
Boiler, £300, coals, 1,000 tons, say, at £1-5 per ton, £1,250	1,550	0	0
25 labourers for 6 months at R3 per month, R45,025, for 6 months at R6, R900; 10 Jobbers for 6 months, at R6 per month, R360; total R1,710 at 2s. the rupee	171	0	0
European Cooper, £150; 4 Native Coopers, 12 months R1 per month, R240 at 2s. £24	174	0	0
Tools, £20; incidental expenses, £1,000 . . .	1,020	0	0
TOTAL .	20,261	6	0

To this must be added interest on block.

Revenue from above Factory.

	£	s.	d.
750 tons, or 20,454 maunds of first quality vacuum sugar, would fetch in Calcutta (deducting river freight and Agent's Commission) R11 per maund or R2,24,994 at 2s. the rupee	22,499	8	0
The above would yield stuff to make 50,000 gallons rum, say, as above at 12 annas per gallon, R37,500 at 2s. the rupee	3,750	0	0
	26,249	8	0
Expenses as above .	20,261	6	0
Nett to pay interest on block, etc. .	5,988	2	0

"On West Indian system nett revenue £23,145, on East Indian £5,988-2-0; difference in favour of former, £17,156-18-0.

"The present dangerous state of the sugar market deserves attention. It is a matter that concerns us all more or less, should this country fail in its produce, and, as is generally feared, an alteration of the sugar duties by the admission of foreign sugar, the growth of Slave Colonies with the mother country, the consequences would be most disastrous to all those whose capital is at stake. It is now generally understood that *goor* and *khaur* will be very limited next year, owing to the almost total destruction of the cane in Bengal, caused by storms and inundations. The date trees are also seriously injured and in many places literally torn up by the roots. Now, as regards the cane, were the cultivation under the management of *properly experienced* planters, the injury from inundation would be materially lessened by draining, etc., etc., etc.

"Another point that requires serious consideration is the inferior quality of the East Indian rum, which, under the present system of manufacture, never can compete with that of the West Indies owing to the want of the cane skimmings, without which, no machinery, experience, or care in the process, can produce a spirit equal in quality to that of our Western Colonies."

It seems highly probable that the expectations entertained by the writer of the above forecast (as it may be called) of the future of the sugar industry of India, stimulated very greatly the effort that burst forth here and there all over the country, in the establishment of sugar plantations with their accompanying sugar works and refineries. That few, if any, of these survive at the present day the reader will, from the remarks which follow, have ample opportunity of seeing. In the same volume of the Agri.-Horticultural Society's publications numerous other papers appeared on sugar, but the second and third volumes of the Journal might be said

SUGAR PLANTATIONS.

to have been mainly devoted to the publication of "*The Sugar Planter's Companion*," a work by Mr. L. Wray of Goruckpore (*See Vol. II., 41-51; 64-83; 107-128; 149-160; 161-192; 209-218; 229-336; Vol. III., 30-60; 99-118; 153-186*).

Mr. Wray was himself a West Indian (Jamaica) planter of ten years' experience before coming to India, so that his opinions are entitled to the greatest respect. In the preface to his *Sugar Planters' Companion* he says: "I have hitherto found a very strong prejudice to exist amongst Europeans in this country, against the West Indian mode of cultivation and manufacture of the cane. They seem to prefer the Native system, with all its faults." "An entire change, however, of the East Indian method would be as unreasonable and difficult, as prejudicial and unsuitable." Mr. Wray, therefore, proposed in his admirable book certain adaptations or combinations of the two systems. His scheme of a plantation was very much similar to that detailed above from the writings of another West Indian planter settled in India. The estate, Mr. Wray says should be as concentrated as possible and preferably should not be much more than 500 *bíghás* in extent and have "the works" right in the centre. If larger estates be desired he suggested that each 500 *bíghás* should have its own works and be so far independent. While admitting that there seemed some truth in the greater liability of the Otaheite, Bourbon, and Batavian canes to the attacks of white-ants than was the case with the native kinds, still, he thought, this was due to the greater prevalence of that pest in newly-opened out lands than in old cultivations. "I must," he says, "have repeated any signal failures before I can be brought to turn my back on such valuable canes as the Bourbon and Otaheite." The writer has been unable to trace out fully the subsequent career of the energetic planter whose words have been briefly quoted. His name disappears from the list of active members of the Agri.-Horticultural Society of India about 1848. He then apparently went to Penang and, in 1852, he seems to have been in Port Natal. In 1848 he re-published his work as a separate book under the name of "*The Practical Sugar Planter*." He would appear, however, to have left India and abandoned sugar-planting in this country after an earnest effort of some five or six years.

But while Mr. Wray's experiences as an Indian sugar-cane planter were obtained in Behar, Mr. S. H. Robinson—a planter, who also wrote in 1849 a valuable manual on the subject—*The Bengal Sugar Planter*—, appears to have laboured in Lower Bengal. In his special chapter on the subject, Mr. Robinson reviews some of the efforts that, prior to his time, had been made in the hope of establishing sugar-planting as a European industry in India. While satisfied of ultimate success in his own peculiar system, he was confident of being able to ascribe the cause of failure in others. But scarcely had a decade passed ere the writers who followed enumerated Robinson also in the rapidly swelling list of energetic planters who had patiently spent their all in the vain endeavour to contend against circumstances essentially inimical to their interests and expectations. But it may be said that if any one could have succeeded Robinson would have, for many of his remarks manifest, a deeper insight into the problems that have since determined the failure of the undertaking than was given to the majority of his contemporaries. It may serve a useful purpose, therefore, to furnish a few passages from Robinson's useful little work, since it is now unfortu-
1829. nately not very accessible:—"In 1829, Mr. C. H. Blake proceeded to Bengal, with the sole purpose of developing this branch" (the cultivation and manufacture of sugar) of the productive resources of India. "This enterprizing gentleman established the Dhoba sugar works in the Burdwan district, the first in India, in which steam power was applied for the purpose

SUGAR PLANTATIONS.

of extracting the juice from the cane. The plan of operations adopted by him was to advance to the Native ryots or other land-holders in the vicinity of his factory, for the cultivation of sugar-cane, under a contract from them to deliver a certain return of cane in weight per *bighá* advanced for, it being in fact the same system, with slight modifications, adopted by the indigo planters for the cultivation of their plant by their ryots. But here, again, failure attended the experiment and after the first two years the cane cultivation was abandoned, the mills were closed, and the factory was converted into a refinery for bringing the Native sugars into the finest descriptions admissible in the home markets."

1830 to 1832.

"About this period (1830-32) another enterprizing gentleman, **Mr. T. F. Henley**, also embarked in a somewhat similar experiment, on a minor scale, at Barripore, a locality bordering on the Sunderbunds, in the southern part of the district of the 24-Pargannahs. He cultivated the Native canes and manufactured sugar from them on the West India principle. The result, as usual, was failure:—The sugar produced was sold in the Calcutta market at R3 to 4 per bazar maund, and must have entailed considerable loss on the proprietors. The soil was pronounced to be not adapted for the growth of sugar-cane, and the works were soon after abandoned."

Durwany Sugar-works, Rungpore. See Vol. VI., Pt. I., 466.

Attention appears to have been next directed to Eastern Bengal and Tirhut as the best localities to attempt the development of a sugar industry.* The indigo planters, in many cases, took the matter up. Better qualities of cane were by them carefully cultivated in their vegetable gardens. Highly manured and carefully watched, these experimental plots yielded, as might have been expected, highly encouraging results. It was ascertained, says **Robinson**, that a *bighá* of ground in Tirhut, equal to about three Bengal *bighás* or an English acre, produced, on a low average, 26 maunds of dry sugar; and under favourable circumstances, and from Otaheite cane, as much as 60 maunds, being upwards of 2 tons from the same surface; and, of this, the cost of cultivation was estimated at R15. Accordingly, the value of indigo factories rose, lands were eagerly secured in the supposed fruitful districts, and associations were formed for developing their resources in this new and inviting branch of cultivation: capital was largely embarked and mills and machinery imported for carrying it out with skill and spirit. But the evil genius which had presided over all former attempts of the same nature seemed still to prevail, and sadly disappointing, to all concerned, have been the results of these sanguine speculations. Four years had not elapsed since the promulgation of the flattering estimates of profit, when all engaged in carrying out the new enterprise confessed their disappointment, and failure. The history of the sugar works that were established in Tirhut comprises one uninterrupted catalogue of losses and disasters:—Mills broke down in the midst of the crop. The white-ants dealt destruction to whole tracts of cane, preferring always the finest varieties. In one year drought stunted or entirely destroyed the crop; in the next an inundation, such as had not been known before for many years, swept the lower parts of the districts and buried at once the canes and the hopes of the planter. But to crown all such Tirhut sugar as arrived in Calcutta for sale during these years was generally of an inferior quality. From the results of some shipments of it made to England it acquired the character of deteriorating greatly on the voyage home: other parcels were bought by the Calcutta refiners and were pronounced by them to be weaker than Native sugars of a corresponding degree of refinement. The opinions now began to

Conf. with p. 94.

Conf. with Mr. Clarke's remarks regarding white-ants, p. 125.

* Plantation of Date-palms under European management. See p. 275: see also ruined sugar factory, p. 277.

SUGAR PLANTATIONS.

circulate, that Tirhut especially was not adapted, from the nature of its soil, for sugar-cane cultivation, owing to the abundance of nitrogenous and other salts in its composition, causing a weakness in the crystal of the sugar and a great tendency in it to deliquiesce, to which causes its inferiority of quality was in great measure attributed. **Robinson**, after detailing many other particulars of the failure of the Tirhut effort to establish a European sugar-planting enterprise, arrived at the conclusion that it was due to *want of calculation.* It was thought, he remarks, that if one *bíghá* could be caused to yield so much sugar, 1,000 *bíghás* would yield 1,000 times that amount. Such miscalculations, or rather mistatements, of the prospects of the industry, doubtless, were made by adventurers; but it is scarcely possible that practical planters could have been so blind as to believe that garden results could ever be obtained over a wide area. The evils and dangers detailed so graphically by **Mr. Robinson** had, doubtless, far more to do with the failure than he was prepared to admit at the time he penned his *Bengal Sugar Planter;* subsequently, at all events, they have, from one end of India to the other, been accepted as the cause of the all but universal failure that has hitherto attended the efforts of European sugar planters in India.

The history of the Goruckpore sugar plantations does not differ materially from those of Lower Bengal and Tirhut—complete failure. It was soon found that although by cultivating the introduced and highly valuable canes on land as free as possible from white-ants, these canes were far more liable to diseases and to pests than were the forms that may, for the purpose of comparison with the foreign races, be called the indigenous kinds. But even after these evils had been guarded against and, as far as possible, mitigated, there came over the imported canes a still greater calamity for, without any very marked prognostications, the whole of the introduced canes of one district suddenly died from disease* or exhaustion, and this gradually spread over all India, the calamity moving often by almost arbitrary stages, and along definite routes until it enveloped the whole country. The reader would do well to peruse in this connection **Mr. Payter's** account of this calamity which will be found in connection with the Bogra district in the section above devoted to the forms of Bengal cane (*p. 48—49*). In most of the district reports of Bombay, it will also

1836 to 1846. be found that the canes introduced from 1836 to 1846 (or thereby) have either died out or got into such disfavour that they are now rarely, if ever, cultivated. A cane much valued in Bengal and known as the red Bombay, became diseased in a like manner in 1857 and ultimately ceased to be cultivated although, at the present time, it is occasionally met with as a curiosity or garden crop. (*Conf. with p. 76.*)

The publications of the Agri.-Horticultural Society (as has been remarked) continued till about 1860 to contain numerous papers all, more

1860. or less, in a strain of high expectation. Many of these deal with the success that had attended the introduction of the better class canes. But gradually all interest seems to have died out and it is significant that not a single paper should have appeared on the subject of this decline of interest, or on the failure that had overtaken the endeavour to establish European plantations. Public interest may be said to have been next directed to the subject, through the various provincial Agricultural Departments,—and this may be said to constitute the third or present awakening of interest in the sugar trade of India. The annual reports of some of the Agricultural Departments (especially of the Experimental Farms) deal with the effort made to introduce (or rather re-introduce) superior qualities

* *Conf.* with the brief chapter on the Diseases of the Sugar-cane, pp. 121-127.

SUGAR PLANTATIONS.

of cane to those in the country, more especially the **Sorghum** sugar-canes No one seems, however, to have re-undertaken to establish regular sugar-cane plantations, and it may, therefore, be fittingly said that the declining struggle of the earlier efforts survived, and even now in some localities survives in the form of manufactories or refineries to prepare sugar of various qualities from the *gúr* or rather *ráb* purchased from Native growers. It has thus been fully recognized as the practical lesson of very nearly a century's labours that in India sugar-growing is essentially a Native enterprise. But as such it is an enterprise of no mean importance, and, accordingly, much valuable information has been collected into the official publications which have within the past ten or fifteen years been issued by the Supreme and Provincial Governments. For the purpose of carrying the historic record of sugar-cane cultivation and the production of sugar down to the present date, it seems only necessary to start with the note on *Sugar-plants and Sugar* issued in 1887 by the Revenue and Agricultural Department of the Government of India. That report was drawn up by **Mr. F. M. W. Schofield** from the records (in many cases very incomplete) that then existed in the Revenue and Agricultural Department. It was issued by the Government of India to all Local Governments and Administrations with the invitation that it might be made the basis of a more thorough investigation than had previously taken place. The imperfections and defects of the report were corrected, and, at the same time, valuable new information was communicated. The correspondence that ensued has afforded the writer of the present article many of the recent statistical and other facts that will be found dispersed through the succeeding pages. One point only need be here specially dealt with. Copies of the above mentioned official correspondence found their way to Europe, and, apparently, as the outcome of the new interest in the sugar trade of India, **Messrs. J. Travers & Sons** addressed (May 8th, 1889) Her Majesty's Secretary of State for India on the subject of the backward state of the Indian sugar industry. The practical suggestion offered by **Messrs. J. Travers & Sons** may be said to amount to a proposal to re-endeavour to establish sugar-cane plantations and manufactures on the most approved modern methods. After the somewhat elaborate review given above, of past experiments in that direction, it is unnecessary for the writer to do more than give such passages from **Messrs. J. Travers & Sons'** communication as may suffice to convey their meaning. Having done so, it would seem sufficient to give an abstract of the opinions since obtained from the numerous private and official persons who have been consulted

European Plantations.
165
Conf. with pp. 37, 48, 62, 63, 91, 93-94, 111, 161-162, 212, 306, 309.

Messrs. J. Travers & Sons say :—

"The average production of India is given as a ton of sugar per acre, and the produce (with the exce ption of the three modern mills in Madras) is of the most wretched character.

"In the West Indies (which are also backward) sugar-growers obtain two tons of sugar per acre, or double the Indian average, and, with modern machinery, properly crystallized sugar can be made direct from the cane juice at a cost on the spot (that is, without carriage) of 8*s.* to 10*s.* per cwt.

"It is no doubt the competition of such direct cane-sugar from Mauritius which is leading to the closing of refineries in Bengal, if, as we imagine, those refineries work, not from the sugar-cane, but from coarse native sugar.

"In all the statistics sent us, Mauritius and similar sugars are described as *refined*, but this is altogether misleading. There are no *refineries* in Mauritius, where sugar is remelted, and the produce of the island is *simply raw sugar properly made by modern processes*.

"It is such sugar that India ought to make, and the Empire, with sufficiently improved cultivation and machinery, might readily supply the world with sugar. Refining is a secondary process, likely to altogether die out, by slow degrees, as cane and beet manufacture becomes more perfect. The disappearance of refining in

SUGAR PLANTATIONS.

Bengal, though hard upon individuals, is really a sign that there is progress elsewhere, and progress which no country is better adapted than Bengal to share in.

"That modern sugar can be well made in India is shown by **Messrs. Minchin** at Aska, Madras, and it is simply absurd that India should have first to export the labour to Mauritius, and then to re-import sugar from that distant island, which could be as well made, and certainly, more cheaply at home. India is generally regarded as the home of the sugar-cane, and with its teeming population, its climate, and (in some districts) its plentiful water and coal supply, it should be a large exporter of fine sugar instead of an importer.

Indian Sugar compared to Manure.

Conf. with pp. 40, 81-82, 95, 113, 137, 325.

"The manufacture of modern (or, as it is called, vacuum pan) sugar to be profitable must be on a large scale, because it involves costly machinery and chemical and mechanical supervision impossible for ryots, who, probably, do not extract one-third of the sugar that might be extracted from their crops, and make that third in a shape that looks more like manure than sugar, and which appears to fetch in many parts of India as little as 6*s.* per cwt. on the spot, whereas Mauritius sugar in India must net double that to pay the grower.

"Vacuum pan sugar-making is, probably, only possible on a large scale in India through the Central factory system, where the raw canes are brought by the mill from the growers. A system similar to this already prevails in indigo and silk mills in Bengal.

"We do not know whether the Government of India would be able to start a few model factories in suitable districts, or whether they must confine their attempts to develope sugar manufacture to the collection of information and figures like those in the returns forwarded to us. In any case, the efforts of the Government in this direction for some years past cannot fail to be of great value."

The numerous replies received by the Government of India, on circulating **Messrs. J. Travers & Sons'** letter, contain much of value that will be found incorporated in the provincial chapters below. It seems, therefore, only necessary to give in this place the substance of the Despatch sent to Her Majesty's Secretary of State by the Government of India (dated December 24th, 1889), together with one or two practical observations which have been furnished by **Messrs. Thomson & Mylne,** on the subject of **Messrs. Travers & Sons'** recommendation. The Government of India's Despatch summarises the local reports and gives in concise language practically all that can be said against the idea of model factories in India. The more important paragraphs of the Despatch were as follow:—

Rum.

166

Conf. with pp. 93, 95, 96, 158, 175, 313, 320, 321.

"The improvement of sugar production and manufacture in this country has been the subject of attention both of the authorities and of capitalists since the beginning of the century, and various attempts have been made to establish factories, none of which appear to have been attended with any permanent success, unless supplemented by the sale of rum and liquors. Sugar refining alone has not proved sufficiently profitable to maintain a factory. If this had been the case, there appears to be no reason why the industry should not have been largely taken up by private capitalists.

"Some of the main difficulties against which the industry has to contend are believed to be these:—

(*a*) The cultivation of sugar-cane is limited by the supply not only of water for irrigation but also of manure.

(*b*) As cultivation in India is confined to small farms or holdings, each cultivator, who is able to grow the crop at all, can only find manure enough for a small area, generally less than half an acre, of sugar-cane The plots of sugar-cane are therefore geatly scattered even in a canal-irrigated tract.

(*c*) A central factory has accordingly to bring in its supplies of cane in small quantities over varying distances, in many cases the distance being great.

(*d*) The carriage of canes over a long distance, even in a climate like that of the Mauritius, is detrimental to the juice for purposes of sugar making. It is much more so in India, where the canes ripen at the season when the atmosphere is driest and suffer, therefore, the maximum of injury.

(*e*) The Mauritius system of growing large canes at intervals is not adapted to the greater part of India where, in order to prevent the ingress of dry air into the fields, small canes have to be grown in close contact.

(*f*) The amount of cane which can be grown, limited as it is by the supply of water and manure, barely suffices for the wants of the Indian population. It seems to be at present as profitable to produce coarse sugar for their use, as highly refined

SUGAR PLANTATIONS.

sugar for export. There is, therefore, no sufficient inducement to capital to embark on the more difficult and expensive system.

"A further obstacle to sugar refining in India exists in the high differential rate which the conditions of our excise system require to be placed upon spirits made on the European method, as compared with that levied on spirits manufactured by the indigenous process. The sugar refiner in India is thus placed at a disadvantage in respect to the utilization of molasses in the form of spirits.

"In view of the circumstances above noted we are unable to advocate any attempt being made at the cost of the State to establish model factories. We are inclined to attach much confidence to the views and conclusions formed by **Messrs. Thomson & Mylne,** who have paid, for many years, practical attention to the subject of sugar cultivation and manufacture by *ryots,* and were the first to introduce the portable sugar mills which have now spread over India. They advocate the gradual improvement of the *ryots'* method of manufacture rather than the introduction of more expensive and centralising systems. The Provincial Departments of Agriculture have, of recent years, directed attention to this question and may usefully be desired to continue to do so.

"We are also willing to advocate the establishment of agricultural experiments in those comparatively limited tracts of the country (such as Eastern Bengal, where there is a moist climate and a more or less abundant supply of manure) in which the Mauritius methods of cultivation have *primâ facie* some prospects of success, and we are prepared to advise our Local Governments and Administrations to give every reasonable support to sugar factories and refineries which may be established by private enterprise."

In the communication (dated 23rd April 1891) alluded to above, **Messrs. Thomson & Mylne** give, amongst many other weighty reasons, the following objections to the establishment in India of Central Factories for the manufacture of cane grown on the surrounding lands:—

"**Messrs. Travers** wrote:—'The average production of India is given as a ton of sugar per acre and the produce (with exception of three modern mills in Madras) is of the most wretched character. In the West Indies, which are also backward, sugar growers obtain two tons per acre.'

"There is no doubt that the quality produced per acre in India is much below the average of most other cane-growing countries, and the quality also of the first products is very low, but in making any comparison and in considering what should be aimed at, in endeavours to secure a larger yield per acre as well as improvement in quality, there are several points of essential importance which need to be kept in view.

"The first is that the great bulk of the sugar-cane grown in India, is not, and cannot, be planted in large blocks or 'plantations' by either Native or European 'Planters' under conditions which would render it possible to deal with large quantities of cane or juice at Central Factories and profitable for capitalists to invest in the expensive scientific appliances requisite for the 'modern processes' which **Messrs. Travers** referred to. Nearly the whole of the two and half or three million acres of sugar-cane planted in India is grown in small plots by native farmers who put in a patch of cane in their holdings, of such size as suits them, in rotation with other crops. To ensure success plans for improving either cultivation or manufacture should be arranged with reference to this important factor.

"Another material point is, that, in most districts, each farmer crushes his own cane in the field or village, and converts the juice, on the spot, into *gúr* or *ráb* for which he finds a ready market in the local bazar. In some districts the custom is that several cultivators join in the purchase or hire of a mill, evaporating pan, etc., sharing these and other expenses of crushing and making *gúr* or *ráb*, but each man arranges independently for the cutting and carrying of his own cane, as also for disposing of it as he pleases, just as they do with their other crops.

"Another point of importance is that the bulk of the sugar-cane, now planted in India, is grown and manufactured for local consumption, not for export, and the form or character given to it is that which (unless and until the preferences and prejudices of the people can be altered) renders it most readily saleable in the local bazar. There *are* districts which produce a considerable quantity in excess of what is consumed locally, but the surplus is required for other districts which do not grow sugar at all, or produce less than they consume. Seeing that India now exports to Europe less sugar than was sent out twenty or thirty years ago, many merchants, refiners, and others, imagine that less is grown and made in India now than was formerly, but the truth of the matter seems to be that a much *larger* quantity is now produced than at any time previously, and that it can now be sold in the local bazars at such rates for

SUGAR PLANTATIONS,

consumption *in India*, that it would'nt pay merchants to buy for export.* The increase in consumption arises from the improved circumstances of the people, and, notwithstanding that much more is produced, a considerable quantity is now imported from Mauritius and other places. One explanation of the increased consumption is, that a great deal more is taken by the millions who grow the cotton, jute, wheat, oil seeds, and other products for which outlets have been created by railways and steamers, and for which such large sums have been received by the cultivators, also the improved means of the large numbers who, during the last thirty years, have found employment in the jute mills and presses, cotton mills and presses, tea gardens, iron works, collieries, railways (construction, maintenance, and working), and other industries which have been established. These have been making much higher wages than the same classes could do previously. In Britain, America, and some other countries, when work is abundant and wages good, the masses consume more largely beef, mutton, tea, and such articles, as well as sugar, but in India it is extra sugar in various forms both for daily use by the family and at marriages, festivals, etc., which is chiefly used.

"Another point of importance is that the fine white crystallized sugar, with large crystals, so much appreciated in Europe is not at all in favour with, and is, in fact, avoided by the masses of India if they see any reason to suspect that bone, charcoal, blood, or any such articles (impure to them) have been used in making it. So strong is this feeling and objection, that dealers frequently find it pays to smash up the large crystals to a fine powder which they then sell as native made 'Benares cheenee.'

"Another thing to be noted is that (apart from Cossipore, Rosa, Aska, and one or two other places, in all of which exceptional conditions have existed), the profitable carrying on of Central Factories by purchasing a sufficient quantity of cane at reasonable rates from those who grow it in their small plots is not practicable. There are several cogent reasons for this, one being that the rates such factories could pay for cane, which must be carted several miles, would be considerably less than the cultivators would realize by crushing it themselves and making *gúr* or *ráb* on the spot; another is that in most of the cane-growing districts there are arrangements, customs (established 'dustoors,') with regard to crushing the cane, and evaporation of the juice, which entitle the local carpenters, blacksmiths, kandoos, and other recognized institutions in each village to a share of the produce, rendering it difficult for the cane-grower to dispose of his crop in any other than the usual way.

"When the attention of the undersigned was first drawn to this subject, twenty-five years ago, through seeing how much their own tenants on the Jugdispore estate (a large portion being cane-growers) were losing of the good sugar the cane could yield, they thought, that the best way of securing substantial improvement would be to set up a central factory with machinery and apparatus, such as are found most efficient in Mauritius and other sugar-growing countries, and they began by getting large machinery. It soon became manifest, however, that arrangements which were suitable for countries in which cane can be grown in considerable blocks by, or for, a factory, the cane being brought in to be crushed, and the juice dealt with according to the latest improved processes under European management, would not suit India so long at least as the existing preferences and customs prevail. It was found that if such a factory were built and fitted with expensive machinery it would be impossible, in most districts, to get to the factory the needful quantity of either cane or juice, at the rates, and the condition, which would be necessary to avoid inevitable loss, that to work with large appliances, and make large quantities of sugar, it would be necessary to work with the *ráb* and *gúr* which is made in the scattered cane fields or villages by the cuitivators, whose methods of treating cane and juice would have already destroyed or lost a large proportion of the available sugar. It was also seen that the rates at which *gúr* and *ráb* could be brought delivered at the factory, would involve positive loss, and it was resolved, therefore, to try what could be effected by endeavours to improve the methods and appliances then in use. The only cane mills then used by (or within reach of, the cultivators of India were crude wasteful applicances made of wood or stone, wasteful of time, power, and a considerable percentage of the sugar in the cane, losing both quantity and quality."

In bringing to a conclusion this review of the facts which have been brought to light for, and against the idea of, establishing sugar-cane plantations in India and Central Factories to manufacture the locally-expressed

* The removal (in 1874), of the restriction formerly imposed on sugar, by the abolition of the iuternal registration and taxation greatly facilitated the consumption of sugar in India by allowing it to be carried from district to district and province to province without paying any transit dues.

SUGAR PLANTATIONS.

juice into sugar (according to the most modern methods), it may be said that all the officials consulted were opposed to the scheme except two, *viz.*, **Mr. Finucane**, Director of the Department of Land Records and Agriculture, Bengal, and **Mr. J. P. Goodridge**, officiating Director of the Department of Land Records and Agriculture, Central Provinces.

Mr. Finucane's reply (*letter No. 1913, 9th September 1889*) was as follows:—

"I am inclined to agree with the opinions expressed by **Messrs. Travers & Sons** that 'India, with sufficiently improved cultivation and machinery, ought to be able to supply the Empire with sugar.' It is, *a priori*, unreasonable that with a soil and climate admirably suited to the production of sugar-cane, with a superabundant population, and the cheapest labour market perhaps in the world, India should be found exporting labour from Behar to the Mauritius, and then importing sugar made by that labour, out of expensive raw material from the Mauritius back to Bengal.

"The importance of the subject did not escape the notice of **Sir Rivers Thompson**, who, in **Mr. Macaulay's** letter No. 1145, dated 3rd March 1884, remarked that the area under cultivation with sugar-cane in Bengal is very large, and pointed out 'that an old emigrant who had returned from the Mauritius, where he had learnt new methods of cultivation, had succeeded in imparting the knowledge to the villagers living in the neighbourhood to their great advantage.'

"Since that time **Mr. Sen**, one of the Assistants, lately employed in this Department, had, under my instructions, given special attention to the question of improvement in the methods of cultivation, and made special enquiries on the subject in the districts of Burdwan and Dacca. He reported that the Mauritius system of cultivation is known to, and practised by, the cultivators on the banks of the Damoodar, and by market gardeners in the neighbourhood of Dacca and Calcutta; while the system is unknown in the Bhagulpore and in the greater part of the Patna Division.

"After having, during the past three years, made various experiments as to the best methods of cultivation, the measures which might be applied with advantage, and the most suitable varieties of cane, and come to conclusions on these points, **Mr. Sen** proposed to take some ryots from the particular tracts where the best methods are practised to districts which are more backward in this respect, and through them to show cultivators in the latter districts better methods of cultivation, by the instrumentality of fellow cultivators working under his own supervision and control; but just as he had made his proposal, it was found necessary to transfer him to the general line of the public service, and the experiments in this direction will now, I fear, have to be abandoned or postponed for some time for want of an officer in this department possessed of such knowledge and experience of the actual details of cultivation, as would warrant reliance on the success of measures of this kind undertaken under his supervision and control.

"As regards the question of improvements in manufacture suggested by **Messrs. Travers & Sons**, I would remark that it seems not unreasonable to suppose that such improvement is possible, and it is not improbable that the establishment of model factories, in suitable districts, whether by Government or by private individuals, encouraged or subsidised by Government, would yield beneficial results. **Messrs. Thomson & Mylne**, in their letter, dated 28th February 1880, to the address of the Collector of Shahabad, reported that they had for years been trying whether cane could be profitably purchased and worked off at a Central Factory, and the conclusion to which they came was, that the price demanded for cane by the growers, which price the growers realized by making it into *gur*, was so high, that the experiment was not deemed to be profitable and was discontinued. **Messrs. Thomson & Mylne** added that the Rosa Sugar Works at Shahjehanpore had not found it advisable to make arrangements for crushing cane and making refined sugar from the juice direct, and the inference would seem to be that Central Factories such as are suggested by **Messrs. Travers & Sons** will not pay. The reason given for this is, that the factory could not work at a profit, if it paid as high prices for the cane as the cultivators realize by making it into *gur*. But this is only stating the fact in another shape, and is no explanation of the problem—why is it that with cheap labour, cheap raw material, refined sugar cannot be manufactured in India at a lower price than that for which it can be imported from the Mauritius or England? A similar question may be asked as regards other products, for example, iron—why is it that with cheap labour and cheap iron ore at Ranigunge, it is found profitable to import manufactured iron articles from England? I am not at present in a position to furnish an answer. The question as regards sugar is one of enormous importance to Bengal, and I would,

SUGAR PLANTATIONS.

therefore, suggest that Mr. Sen be placed on special duty during the cold weather to make a full and careful enquiry into it.

"The Cossipore Sugar Factory in Calcutta is a refinery. The Mauritius system of manufacture advocated by Messrs. Travers & Sons is one by which the suger is made direct from the cane by one process. The system adopted by Messrs. Minchin Brothers, in Ganjam, is a combination of both the Mauritius and Cossipore systems. Messrs. Minchin Brothers' system is not able to compete, it is said, with the sugar imported from the Mauritius in the northern districts of India, because the cost of transit is cheaper from the Mauritius than Ganjam. But if such a factory as Messrs. Minchin Brothers' were established in one of the northern districts of India themselves, or in Behar, it is by no means certain that the sugar produced would not drive the Mauritius sugar out of the market."

Mr. Goodridge's reply (*letter No. 1783-89A, dated 29th August 1889*), while fully supporting the recommendations contained in Messrs. Travers & Sons' communication, deals in an able manner with the whole subject of the differences between the West Indian and Indian systems of cultivation and manufacture. His reply may, therefore, be reproduced in this place, though it might perhaps find a more logical position in the section of this article which deals with the sugar-cane cultivation of the Central Provinces. Mr. Goodridge's final proposal amounts to this, that the experiments discussed in the early paragraphs of this section should be re-performed, the Government making grants of land, rent free or on easy terms, to practical planters, and even advancing the money necessary to enable them to procure the machinery for their factories. The remark may perhaps be pardoned, however, that the pioneers of none of the other important industries have obtained, nor indeed asked for so much direct assistance as has already been given to sugar, so that Mr. Goodridge's recommendation may from that point of view be regarded as a questionably sound one.

"I may mention that I am interested in sugar plantations in the Island of Barbados, and have frequently visited that Island, the last occasion being in 1879.

"The production of sugar in the Mauritius, with which Messrs. Travers compare India for the purpose of showing how backward the industry is in this country, is, as I gather, from the information that I have been able to obtain, carried on under circumstances very similar to those which exist in the West Indies. In both countries there are found—

(1) Sugar plantations of considerable size managed by Europeans and persons of European descent, and cultivated by paid labour by negroes in Barbados, negroes and coolies in Trinidad, and by Indian coolies in Mauritius.

(2) The employment of a considerable capital in this industry and the application of steam and mechanical and latterly of chemical science in the manufacture of sugar.

(3) An abundant rainfall of over 40 * inches per annum well spread throughout the year (though there is a well defined rainy season) affording adequate moisture during the months in which the cane crop is on the ground. The occurrence of frequent showers falling on a naturally well-drained soil which rests on porous coral or coralline rock that prevents stagnation and water logging.

* The average rainfall of Barbados for the 25 years from 1847–1871 was 57·74. In the Mauritius it is now about 40 inches per annum, though formerly, before the destruction of the forests, it was much more.

Size of Cane Fields, 167

Conf. with pp. 124, 143, 256.

"To those acquainted with the present condition of Indian agriculture it is only necessary to state the above circumstances to explain the great difference in the cultivation of the cane, and the manufacture of sugar in the West Indies and the Mauritius and in this country. Here the great bulk of the sugar production is by *la petite culture*. Instead of an energetic race, who have devoted themselves for generations to the sole object of producing sugar, we have an ignorant peasantry wedded to their own primitive methods of cultivation and cultivating, perhaps, a few acres of cane in addition to their wheat, rice, and cotton crops. I think it would be difficult in these Provinces to find many cultivators who have more than 5 acres of land under sugar-cane. In the Sambalpur district, where most of the sugar of these Provinces is grown, the whole body of cultivators, in a village, club together and sow about 8 or 10 acres, the area being divided among them into small strips. The Indian ryot has neither the inclination nor the means of improving his style of cultivation. Instead of a steady and well-distributed rainfall, we have nearly all of our rain during four months of the

SUGAR PLANTATIONS.

year with an occasional shower at Christmas and a dry season during the rest of the year. Hence at one season the cane is water-logged if not well drained, while at another it suffers from drought. To grow a crop of cane, irrigation from canals and rivers, or from tanks and wells is necessary, whereas, in Barbados and, I believe, in the Mauritius also, irrigation is not required and is never practised.

"It would be difficult to say whether the differences between the Indian and West Indian methods are most marked in the cultivation of the cane or in the manufacture of sugar. In the West Indies the ground is well prepared with the hoe and manured with farm-yard manure, which is placed at the bottom of the cane holes where it is wanted by the young cane. The whole field is afterwards carefully 'thrashed' by which means the ground is covered with a bed of cane straw, a foot thick, which retains the moisture round the roots of the young plants and prevents the surrounding grounds from being baked by the sun. In this country the manure applied, whether it be in the form of cow-dung, the droppings of sheep, or the alluvial deposits of tanks, is spread broadcast over the surface of the field and is exposed to the atmosphere. In the West Indies and in the Mauritius large quantities of guano, *nitrophosphates*, and other mineral and artificial manures are used. This is applied to the plants after they have made considerable progress in their growth. In this country the cane rarely gets a fresh supply of manure after it is planted. It is grown from the mature cane cut up into short pieces and laid horizontally on the ground. This is a most wasteful method and entails a large expenditure of cane for seed, perhaps as much as 10 per cent. of the whole produce.

"The young plant instead of firmly establishing itself by striking its roots downwards in search of food spreads them over the manured surface. It consequently becomes weak and straggling, and at a later period falls to the ground and has then to be propped up by interlacing one cane with another, or by means of small bamboos. From the moment it is put down till the young plant has provided itself with roots, it is exposed to the ravages of white-ants which find a convenient *nidus* in the manurial substances used, and attack the plants before they can establish themselves. In some villages in which these pests abound, it is found impracticable to attempt sugar-cane cultivation, and it is not uncommon to find considerable vacant patches in a cane field, the work of this destructive termite. Some years ago I introduced in the Sambalpur district the West Indian method of planting the cane tops vertically in hollows and between 3½ feet square banks, instead of sowing pieces of the mature cane horizontally on the level ground. This resulted in more vigorous canes and in large clusters, but the system had one drawback compared with the native method. If the white-ants destroyed the cane tops before they could be converted into healthy plants, there was nothing left but a bare field or one with numerous empty patches in it. On the other hand, even if two-thirds of the seed cane laid closely on the surface of the ground were destroyed, the other one-third was left for a crop. These destructive insects not only eat up the cane seed but consume a good deal of the manure. To check their ravages the Indian peasant finds it necessary that his manure should be placed where it is wanted and weathered during the rains before it is used. The insect does not then attack it with the same vigour as it does fresh manure. This exposure to the atmosphere, of course, deprives the manure of much of its fertilizing power, but it is better that the cane should be stunted or dry than that the ryot should have half of his field lying in empty spaces. It is well known that the amount of saccharine in the cane is dependent entirely on the stage of its growth. Hence the West Indian planter closely watches his cane-fields and cuts them at the right moment. The delay of a week would most seriously affect the outturn of sugar. The postponement of a month might be ruinous in these days of keen competition with bounty-fed beet sugar and when the margin of profit is so small. The Indian peasant, on the other hand, considers the time for reaping his canes an unimportant matter, and they are allowed to remain standing and to flower until he finds a convenient moment for reaping them.

White-ants. *Conf. with pp. 101, 125, 161.* 168

Flowering. *Conf. with pp. 8, 9, 11, 44, 47, 61, 83-88.* 169

"The Indian method of manufacture of sugar is as wasteful and primitive as the system of growing the cane. In the West Indies the cane is crushed in powerful mills with cylindrical rollers 4½ feet long by 3½ feet diameter, driven by steam or wind, and with every mechanical contrivance to extract a maximum amount of juice from the cane. Even the powerful crushing apparatus which has hitherto been used has, in the present struggle with beet, been superseded in some estates by chemical methods by which the whole of the saccharine substance is extracted from the cane. But I will compare the Indian method with what may be called the old West Indian system, not with the scientific process of later years. The cane juice or 'liquor,' as it is called, is subjected as soon as it is extracted to a process of defecation and clarifying in large vats, and is at once passed through several large 'tayches' till the liquor is reduced to the condition of a thick syrup. It is boiled at a low temperature in

SUGAR PLANTATIONS.

vacuum pans by which means a more highly crystalline mass is obtained. It is then placed in a centrifugal, a rapidly turning machine, which separates the crystal from its parent syrup. The whole is cooled in large shallow vats and afterwards put into hogsheads perforated, so as to permit the molasses to percolate through the sugar. When the molassees has been drained off in the stanchions, the sugar is said to be 'cured' and is in the form of the fine large grained crystalline, whitish brown sugar, or grocery sugar of commerce.

"This process is very different from that adopted in this country; instead of the large boiling-house with its long line of enormous copper 'tayches', its vac uum pans, and ingenious and economical heating apparatus by means of which the megass or woody fibre of the cane alone suffices to make the sugar, its centrifugals and its curing room, we have rough and improvised huts formed of branches and twigs placed at the corner of a cane-field. Here is put up a small crushing apparatus generally of wood, consisting of two or three rollers of about 1½ feet high and 10 inches in diameter and worked by a lever, moved by a bullock or a pair of bullocks. The cane is cut up into small strips by the owner, his family and friends, who consume a good deal of cane juice in the process of *gur* boiling. Only a small percentage of the juice is extracted from the cane by these small and inferior mills so deficient in crushing power. The pressed liquor is placed in large earthen vats and exposed to a quick fire. It is boiled just as it comes from the mill and no effort is made to cleanse or clarify it. The whole is then reduced by heat to the proper consistency and is thrown into a hole in the earth specially prepared for it and cooled long before the process of crystallization has set in. The finished article is more like a mixture of sand and dough sweetened with molasses than the sugar of commerce.

"In late years the wooden mill rollers have been succeeded in some places by iron ones, the best known being the Beheea mill of Messrs. Thomson & Mylne. This, as far as the rollers are concerned, is a miniature of the vertical West Indian sugar mill. It is of course only intended for sugar-making on a small scale. In some districts of these Provinces these mills are used, but in many others the people do not buy them and declare that, on the whole, the old wooden mills are better suited to their wants. The reason probably is that the village carpenter and blacksmith have to be supported in any case whether they make the old-fashioned wooden mill or not, and the ryot, who never has much spare cash for improvements of this kind, considers it cheaper to use this than to pay R150 for an article which he will need only for a few weeks in the year. I have never known an instance of a village community clubbing together to purchase one or more of such mills. Attempts have been made to introduce flat iron vats for sugar-boiling, but they are expensive and are not much appreciated, and most of the *gur* of these Provinces is made in large earthen pots. Iron rollers and iron vats will no doubt in time supersed wooden rollers and earthen pots, but in these Provinces the industry is still carried on by primitive methods which were perhaps in vogue 500 years ago. In most places the megass or woody fibre of the cane is thrown away* as useless. Efforts are now being made to show the value of this substance for boiling sugar, but it is only in those districts in which a difficulty is felt in obtaining fuel that the people show any inclination to utilise their megass.

Yield. *Conf. with pp. 134-136, 139, 211.* 170

"Such being the facts, it seems a matter for surprise that the outturn per acre of sugar-cane, cultivated by the Indian method should, as shown by the statistics, be less than in the Mauritius by one ton only. As a matter of fact, however, the produce per acre in Barbados is from 2½ to 3 tons, while in this country the produce of the same area, while nominally one ton, consists of such an inferior substance that the actual sugar yielded is considerably less than that quantity.†

"I now proceed to consider the question whether anything can be done to improve the method of production in this country. It is obvious that but little improvement can be effected under present conditions. The first thing necessary is that sugar should be grown on a larger scale and its manufacture supervised by properly-trained and experienced persons working with an adequate machinery. For making sugar Messrs. Travers suggest the introduction of the Central sugar factory where the canes of several cultivators could be converted into sugar. It is doubtful, however,

* It will be seen from the numerous passages quoted in this article that the megass is by no means universally thrown away. Indeed, one could wish that it were more frequently returned to the sugar-cane fields than used as fuelby the sugar-boilers. *Conf. with pp. 5, 7, 8, 79, 128, 196, etc.*

† Most writers speak of the yield of the West Indies as two tons of sugar and of India as one ton of *gur*. It seems highly desirable that this point should be more precisely dealt with by future investigators, since one ton of *gur* would in round figures yield but one-third of a ton of refined sugar.—*Ed., Dict. Econ. Prod.*

SUGAR PLANTATIONS.

whether a Central factory would answer in this country. Even if the Indian cultivator could be induced to bring his canes there to be made into sugar, which is not likely, there would be other insuperable difficulties. Here the sugar-cane fields are spread over a large area and are in patches instead of being concentrated as in the West Indies, were cane-field touches cane-field. In some of the West India Islands, and specially, I believe, in the French colonies, where labour is scarce and proper supervision costly, 'usines' or sugar factories have been established. Instead of each plantation having its own boiling-house one 'usine' serves for several. But even in the West Indies this system is worked with some difficulty and necessitates the construction of roads leading from the cane fields to the factory. In India their establishment would be quite impracticable considering the present scattered nature of the cultivation. I doubt whether there are many villages in these Provinces which contain as much as 50 acres of cane. To enable a Central factory to work successfully, an area of at least 500 acres of cane would be needed. Speaking from my recollection of Barbados, where there are many small estates, a boiling-house for an estate of less than 100 acres is exceedingly rare. Persons who grow cane in a smaller way use their neighbours' boiling-houses, giving them a share of the manufactured article.

European Plantations.

Conf. with pp. 37, 39, 48, 62, 63, 88-114, 161-62, 212, 306, 809, etc.

"There is much scope for the establishment of large sugar plantations in this country in places where the soil is good, labour cheap, and an ample and certain supply of water available. Land in Northern India in the vicinity of the canals would, I should say, be admirably adapted for this purpose. There the soil is good with a perennial supply of water for irrigation and a redundant population. The soil and climate of certain portions of the Central Provinces where there is, or could be, considerable irrigation from tanks as in the Sambalpur and Bhandara districts, and in some of the Feudatory States of Chhattisgarh, would also be suitable. The former would probably be more suitable than the latter, for while the canes might occasionally suffer from frost in Northern India, in the Central Provinces the supply of tank water might fail in years of insufficient rainfall.

"For the formation of a plantation after the model of those in the Mauritius and in the West Indies the action of Government will, at any rate, in the first instance, be necessary. The small cultivators of India have neither the means nor the inclination for undertaking such a task. It would never occur to a large landholder in this country to make money by growing sugar on a large scale by new and improved methods and by the expenditure of a considerable capital. By the trading classes the whole thing would be regarded as entirely beyond their sphere of action. The only persons who would perhaps have the requisite enterprise and means to undertake such an industry on a large scale are European planters who can command the necessary land and capital, but they have already profitable crops like indigo, which do not involve the same expenditure and which can be carried on without extensive irrigation. It would be impossible for a West Indian planter, supposing he could command the necessary capital and was prepared to make the venture, to provide himself with the requisite land. There are, it is true, extensive waste lands in this country, but they are quite unsuited for such an undertaking. They are either far removed from inhabited tracts and are situated in unhealthy countries where no European could live, and even if accessible, they are rocky and barren. The natives of this country are only too ready to appropriate all land which is at the same time fit for cultivation and fairly accessible, and they have already absorbed all such land as is available or worth appropriating. No native will willingly part with the land he cultivates, and if the whole area of an ordinary village could be purchased, a large portion of it would be in the cultivation of ryots with occupancy and other beneficial interests in their fields who could not, in the ordinary course of law, be ejected to make room for sugar cultivators. Under these circumstances, a sugar-planter who, whether he came from Barbados or the Mauritius, would be a stranger in this country, would find it difficult, if not impossible, to make satisfactory arrangements for the establishment of a plantation.

It will be necessary, therefore, for Government to take the initiative in this matter, and by means of the Land Acquisition Act or other appropriate procedure to acquire land sufficient for the establishment of a sugar plantation of 500 or 600 acres. This might be offered rent-free or on easy terms to a practical planter under certain conditions for a term of years, and he might also be given a subvention to aid him in providing the necessary machinery for the manufacture of sugar. There must be many enterprising planters in the Mauritius accustomed to Indian coolies who would be glad to accept an offer of this kind. By making success dependent on the efforts of the person chiefly interested in the project, there would be a guarantee that everything would be done to make the scheme a success. But in the event of no practical sugar-planter being willing to undertake the responsibility of a sugar plantation on the above terms, it would be well for Government to establish a few model plantations of

SUGAR PLANTATIONS.

its own in different parts of India. I understand that some years ago the services of a sugar-planter were obtained from the West Indies for the daira lands of the Khedive, and that a vast improvement followed the introduction in that country of the West Indian method of growing and manufacturing sugar. With a plentiful supply of water, such as would be afforded by our canals and large tanks, a good soil and cheap labour, no great difficulties would be encountered in the establishment of a sugar plantation. If the scheme were once shown to be successful, it is probable that many persons who can command large areas suitable for sugar-cane cultivation and the necessary capital would adopt it. The greatest difficulty to be encountered would be the securing of an adequate supply of manure. Much of the cow-dung of this country is used for fuel and, consequently, good farm-yard manure in large quantities is not readily obtainable. But if sugar cultivation by the West Indian method were shown to be profitable, mineral and artificial manures would be available in India as they are in Barbados and in the Mauritius. The value of such a plantation would not be confined to improving the production of sugar. It has often occurred to me that in establishing model farms and placing at their head men trained in England, and having a practical knowledge of the agricultural methods only of countries with a temperate climate, that we have somewhat overlooked the fact that the conditions of agriculture, in the greater portion of India, resemble those of the West Indies or the Southern States of America much more closely than they do those of Europe, and that it is in these former countries that those Indian crops which are most susceptible of improvement, such as rice, cotton, tobacco, Indian-corn, sugar, tropical roots, vegetables and fodder crops, are cultivated with the greatest success.

"The West Indies, like Mauritius, import the greater portion of their food, but a good deal of Indian-corn and vegetables are also grown in these Islands. A plantation is generally divided into two portions, one is under cane and the other is under preparation for cane, and is in the interval used for growing short crops, sweet potatoes, yams, Indian and guinea-corn (juari)—the two latter with guinea-grass supplying the necessary fodder for the farm cattle. All of the above crops are capable of great improvement and extension in India. While in this country a few yams * are to be found in pân baris the plant is reared in the West Indies in large open fields. The difference between the sweet potato † of India and that of the West Indies is striking. The former is generally an elongated tuber 5 inches long and 3 inches in diameter and is grown on a flat surface. In the West Indies it is ordinarily an ellipsoid with axes of 10 and 7 inches, and grown in rows on banks and not on level ground. There are other striking differences in the systems pursued in rearing other crops in the West and in the East Indies. The establishment of a plantation on the West Indian model in this country could not, I think, fail to improve the cultivation of all tropical products and to instruct the people in methods of which they have no idea at present. Some of the return coolies from the West Indies and the Mauritius might also be induced to take service in such plantations, and by instructing their countrymen would be of use to the manager in starting the work.

"I might usefully recapitulate the above remarks as follows:—The improvement of sugar production in India is not possible under existing conditions of scattered cultivation by numerous small cultivators, and in view of the fact that it is nowhere a staple but merely a subsidiary crop. I have further endeavoured to show—

(1) that cultivation on a large scale is essential if the requisite supervision in growing the cane and the necessary machinery for manufacturing sugar are to be provided;

(2) that such a change cannot be brought about unless an adequate area of irrigable land, in a healthy and well-populated country, with cheap labour, is first secured;

(3) that private effort and enterprise are probably unequal to the task of securing the conditions necessary for successfully starting the work;

(4) that it will therefore be expedient, in the first instance, at all events, for Government to take the initiative and to establish a model sugar plantation;

(5) that the best method of working such a plantation would be to interest the manager in the success of the scheme by leaving the profits to him, Government assisting by finding the land and giving it rent-free or at a low rent on certain conditions, and, if necessary, by a subvention to aid in

* The reader might perhaps consult the article **Dioscorea** (Yams) in this work, Vol. III., 115-136.

† See under **Ipomæa Batatas**, Vol. IV., 478-482.

SUGAR PLANTATIONS.

the constructing of the necessary buildings and in supplying the machinery needed;

(6) that in the event of no properly qualified person being willing to undertake the establishment of a sugar plantation on the above terms, Government should itself arrange for the working of the scheme by a paid agency;

(7) that it would be absolutely essential for the success of any scheme of this kind that the manager should be a successful and practical sugar-planter, preferably from the West Indies or the Mauritius, and accustomed to deal with the Indian cooly;

(8) that the establishment of such a model plantation would not only prove the superiority of the West Indian over the Indian system of sugar production, but would bring to the notice of Indian agriculturists the advantages of other modes of cultivating many tropical crops which, though of great value, have hitherto been much neglected in this country."

It has been stated that the other replies received by the Government to its letter, by which it forwarded **Messrs. J. Travers & Sons'** recommendations, were unfavourable, and that the Despatch issued by the Government conveys the facts brought out by the numerous contributors to the official enquiry. Some of these replies will be found placed under contribution below in the provincial sections of this article.

European Plantations. 171

The writer, in taking leave of the subject of the formation in India of sugar-cane plantations with Central Factories, desires it to be distinctly understood, however, that although disposed personally to join issue with those who regard the proposal as futile, he has endeavoured to review the past history of the enquiry impartially. The great success that has been secured in tea-planting was not attained at once, and past failures in sugar-planting can hardly be held as disproving absolutely the possibility of ultimate success. They, however, call for more careful consideration than would seem to be bestowed on them by modern writers, and it is likely that success will be obtained, if at all obtainable, by combating or escaping from the adverse circumstances that ruined the early experimenters. It seems probable that the greatest difficulty of all lies in the social habits of the people of India. They prefer the dirty sugar which **Messrs. J. Travers & Sons** compare to "manure." The preparation of that substance is more profitable to the cultivator than the disposal of his cane to a manufacturer would be. On this subject **Mr. T. W. Holderness** (the Director of Land Records and Agriculture in the North-West Provinces) very justly remarks: "The memorandum refers in contemptuous terms to the quality of the common sugars consumed by the Indian public. But they have an almost unlimited and active market, which is at present closed to machine-made sugar; and even if superstitious prejudices could be overcome, there would still remain the question of national taste. The compost known as *gur* has a peculiar flavour which is absent from machine-made sugars, and the tastes of a most conservative people will require to be changed before the local markets of India really open to the European sugar manufacturer." It thus seems likely that the whole question hinges on the rise or fall in the price of superior sugars and on the education of the people of India to the advantages of obtaining a better quality of sugar than they at present consume. But it may fairly well be here stated that chemistry, by no means, supports the opinion that beautiful crystalline sugar is more wholesome or rather more nutritious than brown sugar. Indeed, it may be said that one of the distinctive features between the saccharine juice of beet and cane is that in the latter substance the additional materials, over and above pure cane-sugar, are less objectionable than in the former. It is worthy of note also that the purity, or rather the whiteness of beet-sugar, is so far considered a disadvantage to it, that an industry has actually arisen in staining beet sugar so as to

Indian Crude Sugar. *Conf. with pp. 40, 81-82, 95, 104, 137, 325.* 172

SUGAR PLANTATIONS.

make it resemble certain qualities of cane-sugar. It is accordingly true that in the present state of the trade the Indian (or home market) is of far greater improtance to India than the foreign. It is, therefore, with the home market that the Indian cultivator and manufacturer alike are mainly concerned. But apart from purely financial considerations, important though these are, it seems probable that success is more likely to be attained by the cultivation and improvement of the indigenous canes than by the necessarily expensive systems of agriculture essential to the preservation of the good qualities of exotic forms.

AREA & OUTTURN.

173

AREA, OUTTURN, AND CONSUMPTION OF SUGAR-CANE AND SUGAR IN INDIA.

In dealing with these subjects, it may be remarked there are many fruitful sources of misconception and error. The area of sugar-cane cultivation (even were it possible to obtain for all India thoroughly trustworthy returns) is by no means the area that actually yields the sugar annually produced in the country. Large tracts are regularly cultivated with sugar-yielding palms, and these afford a by no means inconsiderable share of the supply. These palms are rarely cultivated in such a manner as to allow of estimates of acreage. They are for the most part grown in lines along the borders of fields by road sides, etc., and the yield has accordingly to be ascertained per 100 trees. But what is of perhaps equal importance, a very large amount of the sugar-cane grown in India is eaten as a fruit, is used by the distiller or is made into a thickened syrup (*gúr*), an article rarely if ever converted into sugar. An average yield of sugar, from the total area, would thus be quite misleading. Most writers have made a provision for palm sugar, but apparently the error due to neglecting to reduce the area of sugar production by the acreage devoted to the cultivation of edible canes or of those so deficient in crystallizable sugar that they might not inappropriately be spoken of as affording a superior quality of molasses, but no sugar, has not been guarded against. In some provinces or districts the edible canes are of greater importance than in others, and wherever a large market exists it is universally admitted that edible cane cultivation is more profitable than *gûr* and still more so than *ráb* manufacture. An ascertained abnormally lower consumption of sugar per head of population, than might be inferred from the area of sugar-cane cultivation, is in some cases at least largely explainable by the facts here indicated. At the same time it seems likely that many of the returns of consumption per head, have been falsified through the want of precision in the terms employed. A consumption of 12 seers of *gúr* per head of population, would represent a consumption of 4 seers of refined sugar, or of 4·8 seers of Native unrefined sugar (that is to say, of the inspissated syrup, drained and sun-dried but not refined). It would, in fact, appear to the writer that some of the estimates of consumption, that have been published, are open to the suspicion that there has been a want of uniformity in the use of the word "sugar" by the local authorities who have furnished the data on which certain calculations have been made. As indicated above a given area would produce three times the amount of *gúr* that it would of refined sugar. It seems, however, probable that many writers have not only used the word *gúr* (or jaggery) as synonymous with unrefined sugar, but even with molasses.* Remarkably few have thought of distinguishing between the various forms of sugar, though of refined and of unrefined sugars there are various qualities which differ not only in degree of purity and character of grain, but in the amount of

* See Mr. Butt's remarks below in connection with Shahjahanpur, p. 285.

AREA & OUTTURN.

molasses or treacle which they hold mechanically. Such an error is, however, relatively less important than the confusion of *gúr* with unrefined Native sugar (*viz.*, *shakar* or *bhúra*).

Bombay and the Panjáb are shown by the official returns to consume considerably more sugar than Bengal. This is doubtless largely due to the higher civilization and greater opulence of the mass of the community, an opinion borne out by the observation that the people of the Panjáb and Bombay use individually a far larger amount of silk than do the inhabitants of Bengal. While this is so, it seems likely that the immense population of Bengal with its large city and manufacturing communities may have had its consumption of sugar depreciated by the assumption that the article there eaten is admissible on the standard of *gúr*. Such a reduction would be fairly safe for the North-West Provinces, the Central Provinces, and the rural parts of Bengal, but it would be very misleading for Calcutta, Bombay, or the other centres of manufacturing enterprise. It is believed by the writer that the consumption of Native refined (or perhaps only drained) sugar is far greater than is generally supposed, especially in Bengal, and if this opinion be confirmed by future investigators, the area of sugar-cane cultivation in the Lower Provinces will have to be considerably increased and the yield per acre raised from the estimates currently quoted, or the value of the date-palm in the supply of Bengal sugar will have to be greatly enhanced. In this connection it may be added that it is somewhat significant that in 1847-48 it should have been found that there were in Bengal 6,390,590 date-palms which, yielding on the average for every 100 trees, 16 maunds 9 seers $5\frac{3}{4}$ chittacks, furnished 10,37,445 maunds of *gúr*, while at the present day the 30,000 acres estimated to be under date-palms afford only 7,43,000 maunds. From these figures it would appear that the date-sugar traffic has considerably contracted, or that the yield of date-sugar has been seriously under-estimated. This subject will be found to be returned to further on, in the remarks regarding Bengal and Madras, so that it need only be here added that hesitation to accept the accuracy of the palm-sugar returns would seem justified through the fact that, as presently estimated, the palms of Madras yield nearly twice as much per acre as do those of Bengal.

Palm Sugar Traffic, 174

The following statement of the averages for the five years previous to 1888 was published by the Government as an appendix to its Resolution of the 20th March 1889. It will be seen that in the columns of yield the product is spoken of as coarse sugar, but the average rate there shown is only some 27·9 maunds an acre (for the British Provinces), a rate which would by no means be a high one were it that of refined sugar. From the special chapter below, on the yield of sugar in Bengal, it will be seen that the writer suspects that "coarse sugar" is not entirely the equivalent of *gúr*, in the returns which have appeared on this subject. If "coarse sugar," however, means drained, unrefined Native sugar, the estimate of 1 ton would be more nearly what might be expected. It seems probable, however, that the chief error of all such estimates lies in the fact that large portions of the sugar-cane area are cultivated with canes that are not intended to be used in the manufacture of sugar and are, in fact, never so used. Accordingly the records of actual production of sugar, when expressed to the total ascertained acreage of cane, give a very considerably lower yield than would be the case had the entire area been devoted to the cultivation of cane suitable for the manufacture of sugar. This argument does not hold good with the edible canes only, but with a very important series of canes which, while meeting certain Native requirements, afford little or no crystallizable sugar.

Conf. with pp. 129, 131, 134, 137, 227, 251-254, 284, 297

SACCHARUM: Sugar.

Area, Outturn, and Consumption

AREA & OUTTURN.

Statement showing the supposed normal area under sugar plants, as well as the outturn of coarse sugar in India.

	AREA IN ACRES UNDER SUGAR-PLANTS.			OUTTURN IN LAKHS OF MAUNDS OF COARSE SUGAR PER ACRE.			Outturn in maunds of coarse sugar per acre.	Consumption in seers of coarse sugar per head of population.
	Sugar-cane.	Others.	Total.	Sugar-cane.	Others.	Total.		
British Provinces.								
Madras	62,000	29,800	91,800	27·42	22·22	49·64	54·1	4·30
Bombay (including Sind)	82,000	...	82,000	49·20	...	49·20	60·0	19·50
Bengal	282,000	30,000	312,000	82·32	7·43*	89·75	28·7	4·75
North-Western Provinces	788,000	...	788,000	180·50	...	180·50	22·9	17·00
Oudh	132,000	...	132,000	27·75	...	27·75	21·0	8·00
Panjáb	354,000	...	354,000	96·29	...	96·29	27·2	22·00
Central Provinces	54,000	...	54,000	15·40	...	15·40	28·5	9·00
Lower Burma	10,500	...	10,500	3·67	...	3·67	35·0	9·00
Assam	25,000	...	25,000	4·50	...	4·50	18·0	5·70
Berar	4,700	...	4,700	·54	...	·54	18·0	5·0
TOTAL	1,794,200	59,800	1,854,000	487·59	29·65	517·24	27·9	...
Native States.								
Hyderabad	57,000	...	57,000	12·30	...	12·30	18·0	5·0
Mysore	24,000	29,000	53,000	6·48	7·83	14·31	27·0	10·8
Central India	40,000	...	40,000	8·35	...	8·35	21·0	3·5
Rajputana	34,300	...	34,300	8·36	...	8·36	24·5	5·0
Bengal	5,000	...	5,000	·41	...	·41	8·2	Not stated.
North-Western Provinces (Rampur)	81,000	...	81,000	·50	...	·50	·6	
Panjáb	21,500	...	21,500	4·28	...	4·28	20·0	
Central Provinces	7,200	...	7,200	2·16	...	2·16	30·0	
Madras (Travancore)			Figures	not	furnished.			
Bombay and Sind	43,000	...	43,000	25·80	...	25·80	60·0	
TOTAL	313,000	29,000	342,000	68·64	7·83	76·47	22·4	...
GRAND TOTAL	2,107,200	88,800	2,196,000	556·23	37·48	593·71	27·0	...
			Reduced to Tons	1,986,535	133,857	2,120,392	Approximately 1	

* Conf. with p. 139.

AREA & OUTTURN.

By way of comparison with the figures shown in the foregoing table, it may be desirable to furnish here the statement of area and produce published in 1847-48 for three of the chief sugar-producing regions of India:—

Provinces for which statistics were collected in the year 1847-48.	Total area of land cultivated with cane in Bíghás: = 14,400 sq. ft.	Total produce of cane and date *gúr* in maunds of 80lb.	Total consumption of cane and date *gúr* in maunds of 80lb.	Remainder	Reduced to sugar at 13⅓ seers to one maund of *gúr*.
Bengal and the North-West Provinces . .	25,02,609	1,87,34,909	1,17,78 356	69,56,552	23,18,851
Madras	84,947	17,62,959	10,67,720	6,95,239	2,31,746
Bombay . . .	77,346	6,52,527	7,61,779*	*Nil.*	*Nil.*
	26,64,902	2,11,50,395	1,36,07,855	76,51,791	25,50,597
Reduced to	Cwt. .	15,107,425	9,719,896	5,465,565	1,821,855

* It will thus be seen that Bombay consumed in 1848, 1,09,252 maunds in excess of its local production. *Conf. with p. 210.*

The available surplus of production over consumption in 1847-48 thus amounted to, in round figures, 5½ million cwt., but Bombay must have drawn largely on the North-West Provinces and Bengal. The foreign exports amounted to 1,229,828 cwt. There was thus apparently sufficient sugar produced, from the great sugar-growing districts, to meet the home and foreign demands. The area of production in the above table, assigned to Bengal and the North-West Provinces is shown, by the more detailed tables, to have included 130 acres of Arracan, but otherwise it very closely corresponded to the country now embraced by these provinces. The tables furnished below, in connection with the provincial paragraphs, will be found to show the distribution of the sugar cultivation, and at the same time to exemplify the thorough manner in which the enquiry of 1848 was prosecuted. As justifying a degree of confidence in the accuracy of these returns, it may be pointed out that considerably greater interest was taken in the subject of sugar cultivation forty or fifty years ago than since. There were then a large number of European planters and manufacturers, who possessed an intimate knowledge of the sugar resources of their districts, and who could accordingly assist materially in the enquiry. It seems likely that had returns been preserved annually, a decline in the area and production would a few years subsequent to 1848 have been demonstrated until the industry recovered from the ruin of its European interests and settled down to the present form, as once more a purely Native branch of agricultural enterprise. Within more recent years, it is generally affirmed that it has greatly recovered and expanded; but it is significant that the relation of Bengal to the North-West Provinces has been preserved. Thus Bengal, in 1848, with its eight divisions (Jessor, Bhagalpur, Cuttack, Murshedabad, Dacca, Patna, Hazaribagh, and Chittagong) possessed 223,794 acres of sugar-cane (exclusive of date-palms) and the North-West Provinces, then referred to six divisions (Meerut, Kumaon, Rohilkhand, Agra, Allahabad, and Benares) had 595,441 acres. It will thus be seen that so far as Bengal and the North-West Provinces are concerned, the returns of 1848 when

AREA & OUTTURN.

contrasted with those of 1888 by no means manifest a very great expansion :—

	Area in acres of sugar-cane in 1847-48.	Area in acres of sugar-cane in 1887-88.	Yield of *gúr* from sugar-cane in 1847-48.	Yield of *gúr* from sugar-cane in 1887-88.
			Cwt.	Cwt.
Bengal	223,794	282,000	4,816,980	5,880,000
North-West Provinces . .	595,441	788,000	7,684,648	12,892,857

In the case of Bengal the sugar area may be said to have expanded 26 per cent. and that of the North-West Provinces 29 per cent. during the 40 years covered by the above returns. That is to say, sugar cultivation has increased in the former province by ·65 and in the latter by ·80 per cent. per annum, if it be admissible to assume a steady progression year by year instead of a fluctuation. With a record of an almost nominal expansion before us, it may be viewed as paradoxical to have to say that most writers are of opinion that there has, within recent years. been a great expansion of sugar cultivation in Bengal and the North-West Provinces. This anomaly seems to be attributable to either of two causes (*a*) there has been a greater production in India as a whole, or it may be that our statistics of large tracts, (formerly little known), have simply been perfected and our knowledge of the actual consumption thus made to approximate more nearly to the real state of the internal sugar trade of India ; or (*b*) subsequent to 1848 there must have been a serious decline of cultivation, so that the statement of recent expansion is contrasted with a very different state of affairs to what prevailed forty or fifty years ago. It is not, however, possible to verify this point, since, for large portions of India, only the merest approximations to a survey of the sugar-cane area have been made. The majority of writers, in fact, agree that, the expression of the ascertained area in Bengal by a uniform standard of production reduced to the head of population, gives a considerably lower figure than is believed to be the actual consumption. Thus, for example, the returns of Calcutta (which may be said to be accurately recorded) show an annual consumption by the 900,000 inhabitants of 2 seers and 4 chatáks of refined sugar (or the equivalent of 30 seers $9\frac{1}{2}$ chatáks of *gúr*), whereas the production of Bengal would represent but a consumption of $4\frac{3}{4}$ seers (of *gúr*) on the entire population of the province. The authorities best qualified to give an opinion, regard it as more likely that the actual consumption of Bengal is between 10 and 12 seers (of *gúr*) per head of population. But it may be pointed out that the figure shown in the table above (p. 116) against each province expresses in some cases apparently the consumption to the estimated local production. This, it will at once be seen, would not by any means be a fair standard, since many provinces grow largely for the purpose of exporting, while others draw supplies from foreign countries. Thus, for example, the average total exports from Bengal by all routes amounted (for the three years ending 31st March 1887) to 12,68,248 maunds (reduced to the standard of unrefined sugar), or say a little over one-seventh of the production. But during the same period the imports came to the average of 3,54,726 maunds, so that the net export was 9,13,522 maunds, or say one-tenth of the production. Then again these figures exclude from consideration the trade of Calcutta, and the population of that city should accordingly be deducted from that of the province in any calculation of the

Conf with p. 346.

AREA & OUTTURN.

consumption per head of the provincial population. The importance of this observation will be seen from the fact that the road traffic into Calcutta alone shows on the average a net import of over 3 lakhs of maunds of unrefined sugar for the three years ending March 1887. That amount must, therefore, have been drawn from the province, although it escaped registration until it reached the boundaries of the city. But these details, regarding Bengal and Calcutta, have been gone into in this place with the object of showing that in the case of Bengal at least there is abundant evidence in support of the belief that the area, production, and consumption, shown for that province, in the table at page 116, must be very seriously under-estimated. To local production must in every instance been added or substracted the net transactions. If the net import of foreign sugar to India be added to the estimated production and the total sugar supply be divided by the population (say, 210 millions), a figure is obtained which would represent the average consumption per head of population for the entire Empire. A calculation of this nature was framed by **Sir Charles Bernard** (and which appeared in the *Journal of the Society of Arts, May 1889*), based on the returns which had just then been issued by the Government of India. **Sir Charles'** estimates may be here corrected to the figures now available. The average shown in the table (page 116) of sugar-cane is 2,107,200 and with the average outturn of 1 ton of "coarse sugar" to the acre this should yield 2,107,200 tons, or by adding to that the sugar of palms, also the net import of foreign sugar, and allowing a margin for errors in the estimates, it might be put at 2,600,000 tons. Expressed to head of population that would be equal to about 14 seers (28℔), a figure which is very probably more nearly correct for all India than the consumption of 4·3 seers for Madras and 4·75 seers for Bengal, shown in the table. It may be accepted that the wholesale price of unrefined sugar in India is about R100 a ton,* so that the sugar consumed annually at the present time, costs 260 million rupees or between $1\frac{1}{4}$ and $1\frac{1}{8}$ of a rupee per head of population per annum. A slight error is involved in this estimate from the fact that a considerable amount of refined sugar is now used, the higher value of which should have to be provided for. But such an error affects the wealthy community only and may be therefore disregarded. Sugar (or rather *gúr*) is the great luxury of the poor in India, but it may not inaptly be here compared with salt—a necessity of life. The consumption of that article per head of population comes to about 12℔ and that amount costs the consumer eight annas or say nine pence From these figures it would, therefore, appear that the sugar and salt used by the people of India costs them less than three shillings per head a year. According to **Mr. A. E. Bateman** (of the Board of Trade) the yearly consumption of sugar in other countries comes to 70℔ per head in the United Kingdom; 60℔ in the United States; 27℔ in France; 19℔ in Germany; and 9℔ in Austria. By the estimate here given India consumes 28℔ of *gúr*; but reduced to the standard of the refined sugar used in European countries that would be equal to about 9℔. Tobacco, it might almost be said, is scarcely a luxury since it is very nearly universally used by men, women, and children. It is entirely free of duty and is sold at so low a price that any one who wishes it almost, can afford to procure a supply. Indeed, a large section of the Indian community (the cultivators) grow their own tobacco. The special preparations smoked by the well-to-do contain so much sugar (molasses) that the smoking of that article becomes a distinct item in the consumption that has to be provided for in all estimates of the sugar trade. But if doubt be admissible as to the

Conf. with pp. 40, 316, 331, 341—344, 346.

Consumption of Salt.

175

Conf. with p. 403, 428.

* *Conf. with prices shown on p. 323-328.*

SACCHARUM: Sugar. Area under and Consumption of Sugar.

AREA & OUTTURN.

accuracy of the yield per acre and consumption per head of population, a solution of at least some portion of the error may be looked for in the imperfection of the returns of area cultivated. The following passage appeared in Mr Schofield's *Note on Sugar* (issued by the Revenue and Agricultural Department of the Government of India, the publication of which led to further enquiries throughout the provinces of India.

"Summing up we find that the area under sugar-plants in India is returned at 2½ million acres, and the outturn of coarse sugar at 547·23 lakhs of maunds, or in round numbers 2 million tons. The greater portion of the sugar produced in India never goes beyond the unrefined stage, as the demand for sugar of this class is so large that there is little to spare for the refineries; and in estimating the average produce it is therefore usual to refer to coarse sugar. Messrs. Thomson & Mylne, in a note, dated 19th May 1883, stated that the total production of sugar in India might probably be estimated at 100 million cwts. (=5 million tons). This outturn would allow of 40℔ or 20 seers, as the consumption per head of population (250 millions) in India. But Messrs. Thomson & Mylne's estimate appears, from our statistics, to have been pitched too high. Allowing for the imperfection and incompleteness of the present statistics of sugar cultivation, I do not think we can give for India higher estimates than those subjoined:—

Area under sugar-plants.	Outturn of coarse sugar per acre.	Total outturn.
2½ million acres.	20 cwts. = 1 ton.	2½ million tons.

"To produce 100 million cwts. or 5 million tons, as estimated by Messrs. Thomson & Mylne, India would have to double either its present area under sugar-plants or its present outturn per acre. Doubling the area seems almost impracticable, for it must be remembered that the cultivation is limited by the supply of water and manure, also the unsuitableness of the soil in many parts of India, and that it could hardly be extended so largely without trenching seriously on the area under food-grains. As to doubling the present outturn, this seems possible in course of time by introducing improved methods of cultivation and of extracting the juice.

"We also find that, with the exception of the Panjáb, the Central Provinces, and Berar, the area under sugar-plants in the other British Provinces has increased during the last ten years: the area in the Panjáb and the Central Provinces appears to have contracted in consequence of the extension of railway communications: that in Berar has remained stationary.

"The following table, which has been compiled from the crop statements for 1885-86, professes to show the extent to which sugar-plant-growing prevails in each British Province and in Mysore:—

Province.	Area under cultivation in acres.	Area under sugar-cane Thousand acres.	Percentage of sugar-cane area, to total cultivated area.
North-Western Provinces	25,100	788	3·14
Oudh	8,800	142	1·61
Panjáb	20,500	331	1·61
Assam	1,600	20	1·25
Bengal	†54,500	312	·57
Mysore	4,700	24	·51
Central Provinces	13,500	48	·35
Bombay	25,400	70	·28
Lower Burma	4,200	10	·24
Madras	22,400	46	·20
Berar	6,500	5	·08

"These figures also indicate that the acreages given for Bengal, Madras, and Bombay have been under-estimated. The figures for Madras and Bombay are ad-

* Three o's omitted.

† Taken from Appendix I (page 39) to Famine Commissioner's Report.

AREA & OUTTURN.

mitted to be below the mark for the reasons already explained under the respective Provinces. As regards Bengal, it would probably be found (were actual measurement of the sugar-plant area undertaken) that 600,000 instead of 312,000 acres would be a nearer approach to the truth. In Appendix I (page 40) to the Famine Commissioner's Report, the area under sugar-cane is estimated at a million acres, but this again seems to be too high."

It will be seen from the above reference to **Messrs. Thomson & Mylne** that these gentlemen estimate the Indian production of sugar at exactly double what has been determined by the Government. This may be largely due to their providing more carefully for the mistake which some writers incur in regarding "*gúr*", ' unrefined sugar," and "coarse sugar" as synonymous terms, but it seems also probable that they allow for a larger area than has been ascertained by Government to be actually under the cane. **Mr. Schofield** suggests that the area in Bengal, when actually surveyed, would very likely be found to be more nearly 600,000 than 312,000 acres. The Bengal Government in its reply to **Mr. Schofield's** *Note*, however, reduced the original figure of the outturn in one or two districts, to that now shown in the table above (p. 116), but made no material alteration on **Mr. Schofield's** original calculations. The writer concurs in the opinion that Bengal at least of the provinces and Native States shown in the table above is likely to be considerably understated and from two reasons, *viz.*:—the yield per acre is abnormally low; the consumption per head of population is much less than all persons qualified to judge affirm it to be. This remark, however, applies to many other parts of India besides Bengal, so that in concluding the present chapter it may be added that very little of a trustworthy nature is known as to the production and consumption of sugar in India as a whole, though accurate returns exist of certain tracts or for certain features of the trade.

DISEASES. 176

DISEASES, PESTS, etc., TO WHICH THE SUGAR-CANE IS SUBJECT.

It is somewhat difficult to suggest a classification of this subject that would possess at once the advantages of scientific accurcay and brevity. It may, however, be admitted that there are three main groups: —Diseases proper, Pests, and Enemies. The first of these are manifested by fungoid growths which appear on the cane either as the cause or consequence of disease. The second and third may both be admitted as embracing insect pests, but a restriction to the section, which may be specially designated "pests." of all insects that live within or upon the texture of the cane and the relegation of ants to the same category with jackals, rats, cattle, etc., will, it is believed, be readily appreciated. Some of the disease, as also of the pests, may be said to be a consequence of weakness caused through defective cultivation or unsuitability to climate and soil. Indeed, the historic evidences of Indian sugar-cane planting favour the opinion that the most frequent cause of disease and pests is the over-profitableness (so to speak) of the undertaking, since it engenders a greed that neglects the most ordinary precautions. The continuous cultivation, on a certain tract of country, of a special race of cane ultimately results in weakness of the stock and its destruction either by fungoid disease or pests. *The most essential elements in sugar-cane cultivation are, therefore, rotation of crops and periodic renewal of stock* Where the former is disregarded, expensive manuring and high class cultivation have to be resorted to, but it is believed, renewal of stock cannot be neglected even by the most scientific planter. The calamity that has been repeatedly witnessed in India and which at the present moment is causing in Java and some of the British Colonies the gravest anxiety appears to universally supervene, *viz.*, a weakness from continuous cultivation which renders the crop an easy prey to

Conf. with pp. 48, 76, 126.

DISEASES.

disease and pests. Although several palms furnish a considerable amount of sugar, no writer appears to have made the diseases and pests of these palms the subject of any special inquiry. It is admitted, however, that if tapping of date-palms commences at too early an age, the yield of juice shows signs of decreasing vigour at a correspondingly early period. The tapping of palms is, however, so to speak, an injury to their growth, which must result in their destruction after a fixed number of years. The palm-sugar grower is, by his religion, required to set apart at least one tree in his plantation to the gods, and this, not being tapped, becomes much healthier and larger than the others, and produces annually a considerable amount of fruit, the seeds of which are used in rearing fresh plants to take the place of the exhausted juice-suppliers. Palm cultivation is, therefore, a very different undertaking from that of cane, and it does not appear that there are any special diseases attributed to the sugar-yielding trees as distinct from those grown for their fruits or fibre.

Diseases. 177 I. **Diseases of the Sugar-cane.**—Only one or two writers allude (and in the most general terms) to the existence of fungoid diseases on the Indian sugar-cane crop. Thus, for example, it will be seen below in the account of the Karnal district of the Panjáb (p. 187) that it is stated a

Smut. 178 *smut,* known as *al,* often makes its appearance when east winds prevail. The late **Dr. Barclay**, who devoted much patient study to the fungoid diseases of the crops of India, possibly never had the opportunity of examining sugar-cane. The subjects which he was able to accomplish were the diseases of the crops of the Himálaya, for, prior to his being located at Simla, he had not taken to that study. The reader will find under **Sorghum vulgare** a review of one of **Dr. Barclay's** latest papers, which will serve to indicate the immense importance of his researches, and the irreparable loss which the country at large sustained in his premature death. *Smut* is said to have been seen on the sugar-cane of Natal (and the fungus identified as **Ustilago sacchari**), but the opinion was formed by the investigators of the malady that it was caused through defective cultivation. The reader will find above (under the paragraph on Java Canes) allusion to the fact that a **Mr. R D Kobus** had recently visited India on deputation from the Java Government, to see whether he could secure fresh stock of cane, since that of Java had suffered so much from fungoid disease that it had been thought the preferable course to procure a fresh supply. Whether the disease alluded to is actually a fungoid malady or simply

Conf. with p. 52.

Rust. 179 "*rust*" or one of the numerous other insect pests that are known to overtake each country when cane has been cultivated for too long a period without renewal of stock, the writer is not in a position to say.

Pests. 180 II. **Pests.**—Various authors allude to insect pests as following mostly on the tract of defective cultivation and as often doing serious damage. It is difficult to discover how many pests there are in India of this nature. Some are spoken of as "caterpillars," others as "moths," etc. One only

Sugar-Cane Borer Moth. 181 has been hitherto made the subject of special inquiry—The SUGAR-CANE BORER MOTH (**dhosah*) The reader will find a brief abstract of the leading facts known regarding this pest in the article on **Pests**, Vol. VI., Pt. I, p. 152. **Mr. Cotes** (*Indian Museum Notes*) writes that for at least the last 100 years, the sugar-cane, in the different parts of the world, has been known to be subject to the attack, either of this pest or of others so closely allied, as to be scarcely distinguishable from it. The larva of the insect commits great depredations in sugar-cane fields by boring into the stalks, often thereby setting up putrefaction, so that the stalks become worthless. Incidental allusion to this pest will be found in other chapters of this article,

* **Dæatræa saccharalis,** *Fabr.*

DISEASES.

Sugar-Cane Borer Moth.

more especially that in connection with the destruction of the Red Bombay cane which was grown in Hughli, Rungpore, and Burdwan in 1857. Also the calamity that overtook the Otaheite cane in the Bogra district (about the same time), when that much-prized form was reported to have "rotted in the fields emitting a most offensive smell." Although the insect was not specially mentioned by the writers who described the destruction of the Otaheite cane, the symptoms of the disease that overtook the industry were precisely those given by subsequent observers to that of the "Sugar-cane Borer." **Mr. Cotes** remarks that this pest "is almost universally supposed to make its appearance only when moisture is deficient." **Babu Jaykissen Mukerji** (*Journal Agri.-Hort. Soc. India, IX., 355-358*) seems to have formed the opinion, however, that the worm appeared in the red Bombay cane (cultivated in Bengal) only after it had been continuously grown for a certain number of years. Many other writers allude to this same fact as observed with the ordinary canes. The continuous cultivation of a peculiar cane on a holding for more than a certain number of year's results, it is said, in its degeneration and very often complete extermination by the Borer Moth or some other equally potent disease. **Babu Jaykissen's** observations are pregnant with value, since doubtless a vast improvement in cane cultivation would be effected by encouraging an exchange of seed-stock from a distance. "In these places," **Babu Jaykissen** remarks, "where the red canes were planted earliest, *i.e.*, about twenty years, the disease appeared slightly about two years ago. Last year the decay increased, and this year total destruction has taken place. Where this cane has been introduced only lately or ten or fifteen years ago, there the crops, though they have somewhat suffered this year from excess of rains, yet they are free from the disease. In the lands of the Burdwan district, bordering on Hughli, a similar result has taken place." It would, therefore, seem highly probable that, although climatic conditions and peculiarities of soil may favour the growth of this peculiar disease, the most likely cause is the weakness engendered by a too continuous cultivation of a particular cane on the same fields. Although India very likely can never hope to grow the superior canes of the West Indies and other foreign countries, it seems probable that past failures were largely due to a too precipitate and greedy cultivation which neglected the most ordinary precautions against disease and exhaustion. Frequent exchange of stock from one province to another might have saved the superior qualities of cane, which half a century ago were highly appreciated and in great demand by the Native cultivators. It will be seen that a worm known as *kansua* is alluded to below (in the account of cane cultivation in the Karnal district of the Panjáb) as being common when the east winds prevail. It seems probable, as already suggested, that there are more insects than one that do damage to the Indian cane crops, but until these have been examined by an entomologist, it is impossible to form any definite opinions from the writing of unscientific observers. *The pon blanc* or "louse" (**Icerya sacchari & Pulvinaria gasteralpha**) which do so much damage in Mauritius and Bourbon do not appear to have been observed in India. In dry hot weather these insects frequent the roots of the cane and do much injury to the young rootlets. "The young run about on the green shoots and leaves, until they find a suitable spot where they may fix themselves for life. They are armed with a long sharp probe which they introduce into the new sap-wood, and suck away the juices of the plant, sometimes till they have quite destroyed it. They spread rapidly and are tenacious of life" (*Spons' Encyclop*).

Conf. with pp. 48, 52, 76. 87, 184, 220.

Kansua. 182

Pon blanc. 183

A disease or pest popularly called "RUST" has for some time now been determined to be due to minute mites which belong to the genus TARSONY-

Rust. 184

Codf. with 172.

DISEASES.

MUS. In the *Kew Bulletin* (*1890, 85-88*) the reader will find particulars of these creatures. They are extremely minute **Acari**, almost transparent, found chiefly in the axils of the leaves. Along with these several other species of **Acari** often occur, such as **Damœus** or **Notaspis** and several forms of GAMASIDÆ. The last mentioned are supposed to be predatory on the real pests—the TARSONYMUS. Rust has been noticed in Queensland, the Malay Archipelago, Mauritius, etc., but apparently no writer has discovered it in India.

ENEMISE.

The Money-lender.

185 **III. Enemies**—The MONEY-LENDER.—The poverty and social habits of the people of India should perhaps be ranked as the chief enemies of a greatly extended sugar-cane cultivation. The actual cultivator is everywhere the prey of a tyrant whose oppression is unequailed in the annals of the agriculture of any other part of the globe, namely, the money-lender. So very profitable is every stage, from the cultivation of cane to the refinement of sugar, that where loans of money are not required, through the indigence of the people, they are forced on them through ignorance of the dangers they are being lured into. Once in the hands of the money-lender, cultivator and manufacturer alike, become his slaves. No demands are made for a time, until the iniquitously high interest has raised the original loan to a sum which, by no chance, can ever be paid off. The various efforts made by Government to check this evil have, through the false reports, been distorted into arguments to strengthen the usurer's position. Thus the registration of loans, instead of giving the receiver the protection of law, on the legality of the giver's claims, has been represented as legalizing these claims. It would be beside the scope of the present article to review, however briefly, the various efforts that have been put forth to ascertain the extent of, and if possible to check, the indebtedness of the sugar-cane grower and manufacturer to the money-lender. Such questions fall more naturally into the field of the student of Political Economy, but it may fairly be said that no feature of the great problem of sugar cultivation in India calls more loudly for solution. Profitable though it be, the expense that it involves, deters the more thoughtful cultivator from attempting cane, since the experience of his neighbours warns him against the persecution of the money-lender. Without money he cannot cultivate cane.

Social Customs. *Conf. with pp. 8-11, 223, etc.*

186 SOCIAL CUSTOMS.—But the social habits of the people of India in other respects are opposed to extended sugar cultivation. The profits are inmensely minimised through the absured injunctions of religion and custom. The sugar-cane field becomes on the days of harvest the scene of universal jubilation. None who chose to demand a portion can be sent empty away, and every little service rendered by priest or artizan has to be paid for on a scale of remuneration quite disproportioned to the services rendered. The picture of the village wayfarer being, by the decrees of the Institute of Manu, permitted to take with impunity a certain number of canes, but exemplifies the antiquity of the social custom that has to admit petty theft as a necessary evil, best guarded against by surrounding the plot of cane by a hedge as it were of an inferior sort. In a paper which appeared in the *Kew Bulletin* (*1890, p. 72*) **Mr. C. B. Clarke** advances the very opposite opinion to that held by the writer, namely, that small holdings or plots of sugar-cane are not remunerative. **Mr. Clarke** says: "In Bengal, sugar-cane is often in half-acre plots; it does not pay the cultivator to watch so small a piece, therefore, every boy, every gharry-wallah* who passes, takes a few canes, and every elephant takes many. Gross robbery is also frequent. These small plots are very frequently thus

Size of Cane Fields.

187

Conf. with pp. 108, 143, 154, 157, 256.

* Cart-driver.

ENEMIES.

The Money-lender.

half destroyed before cut. I have seen them *wholly* destroyed. In plots of 100 acres the percentage of loss from this cause would be insignificant." Now Mr. Clarke's words might be true of the unfortunate cultivator whose half acre of cane chanced to be by the high trunk road. But let it be called to mind how few, how very few, roads there are in Bengal along which either children or carts, not the cultivator's own, ever pass, and the dangers thus exalted to be of primary importance disappear from consideration. The answer to Mr. Clarke's contention, however. is not far to find, and it is this that perhaps more than three-fourths of the cane grown in India is in plots that do not materially exceed the area which Mr. Clarke regards as unprofitable. A single cultivator, owning more than 2 or 3 acres of cane would, in Bengal, be a wealthy man *Conf. with p. 256*). The great bulk of the sugar-cane of India is grown near the homestead where it is not only well manured but easily tended. The depredations effected by man, in the way indicated by Mr. Clarke, are the least important of the losses sustained through countless centuries of social evolutions, but which still leave the half acre of cane the most profitable crop in the ordinary rayat's holding.

White-ants, Jackals, etc. 188

WHITE-ANTS, JACKALS, ETC.—These are, however, enemies against which the cultivator is ever on his guard, and against which at times he is helpless. Most writers, in fact, allude, as of serious moment, to the ravages of jackals, rats, and white-ants. These enemies of the cultivator have led to a long series of countless selections which have resulted in hard canes which "neither the teeth of the jackal nor the forceps of the white-ant" are able to break through. An inferior quality of cane, proof against these enemies, was found (and naturally so) to be more profitable in the end than a cane one half the produce of which would be removed by these depredators. Several of the reports here quoted show that in many parts of India, as, for example, in the great sugar cane area of Bengal, whole crops have been entirely destroyed by white-ants. This remark, it will be seen, has reference, however, more especially to the superior qualities and imported forms with thin barks. But Mr. C. B. Clarke in his paper which appeared in the *Kew Bulletin* (*1890, p. 72*) seems to greatly under-estimate the seriousness of white-ants as an enemy to sugar-cane planting. He says they are "fearful in Central India, troublesome in Chota Nagpore, and unimportant in Lengal." This opinion, the writer by no means finds borne out by the reports he has consulted White-ants would appear to be everywhere of equal moment, except with the canes grown in damp soils or submerged lands. Much has been written on the subject of the prevention of the pest of white-ants. The Natives in some parts of the country tie the leaves of some half a dozen canes into a tuft. This is said to give them strength against destructive wind storms and to allow of the admission of light which the white-ant and the rat both dislike. On the other hand, the practice is condemned as retarding the growth of the plant and the perfecting of crystallizable sugar. The individual canes thus deprived of their full share of light and air, by being tied together, become dirty on their stems and yield a foul juice—a disadvantage that more than counteracts any advantage that may be gained by letting the light reach the ground every here and there all over the field. The best protection against the jackal, and the one which saves the crop from the depredations of other animals as well, is careful fencing, but this is, as a rule, beyond the means of the ordinary cultivator who accordingly contends against the greater dangers of cane cultivation by contenting himself with the profit from a very small plot of cane grown near his homestead which thus gets the abundant manure due to human influence, and can be carefully tended and protected.

Conf. with pp. 37, 101, 161, 185.

ENEMIES. Cures. 189 *Conf. with p. 478.* White-ants, etc. 190

CURES FOR THE WHITE-ANT PEST.—To prevent the injury of white-ants the Natives often dip the ends of the seed canes in a fluid prepared with asafœtida, mustard-oil cake, and putrid fish, etc. Balls of flour or other grain, poisoned with arsenic, have also been recommended. These, placed in the field, are eaten by the first set of ants and the poison continued through the dead ants being eaten by their fellows. **Mr. Wray** recommends the use of petroleum as being more certain in its action. White-ants, he says, have a strong antipathy to the effluvia of petroleum, so much so, that if the ends of the seed canes be dipped in water impregnated with petroleum they will generally be found to be thereby protected from the attacks of these scourges. **Mr. Cotes** recommends as the best cure for the Borer worm that the diseased canes should be burned after being removed to some distant spot.

Parasitic. 191 *Conf. with p. 121.*

PARASITIC AND OTHER PLANTS INJURIOUS TO CANE.—**Mr. J. B. Fuller** in his report of sugar cultivation in the Central Provinces alludes to a pest, or rather enemy which often does much harm. It will be seen from **Mr. Fuller's** account below that the pest (a weed which belongs to the SCROPHULARINEÆ) only appears when land has been exhausted by over cultivation. This fact is of great significance since it lends support to the belief that to the same cause (as affirmed above) is due the plague of the sugar-cane borer. **Mr. Fuller** writes :—

Agia. 192

"White ants not uncommonly attack the cuttings. In the Betul district salt is reported to be used in this case, being tied up in a canvas bag and placed in the water channel. But the worst enemy which the cane has in these Provinces is a small parasitic plant called the "Agia" (**Striga euphrasioides,** *Benth.*) which grows on the roots of the cane and rapidly ruins it, producing an appearance in the crop as if it had been scorched. The character and effects of this weed have been carefully enquired into by the Deputy Commissioners of Narsinghpur and Chhindwara, and I make the following extract from the Narsinghpur report :—

"On my way from Birman to Sehora I could not help noticing the number of abandoned wells, even on the outlying fields near which there is now no sugar cultivation. On enquiring the cause I found nearly every one agreed to lay the chief blame on the *Agia,* a weed that appears in the month of 'Bhadon,' and lives till about 'Aghan' or 'Pus.' It appears not to injure rice or *jowari,* but to destory *kutki* and sugar-cane. The weed grows to a height of about 21 inches. When it touches a stalk of cane the latter seems to be blighted and scorched. In the cane-field it is said to appear very capriciously, so as almost to refute the generally accepted theory that, like *kans* grass, it appears seemingly spontaneously in exhausted soils. It cannot therefore be rooted up. It is now very common, and though known from of old is believed to have been formerly very scarce.

"'This points to the exhaustion theory. The men whom I consulted said that even when a field is untouched by *Agia,* or by two other maladies, *Kirohan* and *Durki,* of whose nature I am ignorant, the outturn is less than what it used to be. This may be true. The depth of water below the surface is what determines whether a field can or cannot be utilized for cane cultivation, and the quality of our land, which can be profitably used for this cultivation, is limited. I think that the main fact is that manure has not been sufficiently used to restore the original powers of their wonderful soil, a soil which had long rest previous to and for a good time after the beginning of the English rule in 1818, and but little rest for the last forty years beyond what is obtained by rotation. Manure is but little used even for sugar-cane.'

Conf. with p. 129.

"There can be no doubt that the *Agia* only attacks plants in poor soil, and it is for this reason that it has done so much harm in the Nerbudda valley, where manuring seems foreign to the habits of the people. Growing a crop of *san* hemp (**Crotalaria juncea**) and ploughing it in is occasionally used as a remedy. This is of course merely a form of manuring. The Deputy Commissioner of Chhindwara (**Mr. Tawney**) found that in a cane field manured with poudrette no *Agia* appeared save in a strip which had been used as a road by the sewage carts and had, therefore, escaped manuring. The *Agia* is therefore merely a concomitant of bad farming and is no cause of fear to a careful cultivator."

The determination given above of the *agia* of the Central Provinces may be quite correct, since that species might fairly well be found in Narsinghpur

ENEMIES.

and Chhindwara, but it may be added that the writer had the pleasure to receive from **Mr. R. D. Hare**, then Settlement Commissioner in Akola, Berar, samples of a weed which was found to effect a similar destruction of the *jauár* crop (**Sorghum vulgare**). These proved to be **Striga lutea**, *Lour.* **Mr. Hare's** account of this pest may be here quoted in support of **Mr. Fuller's** opinion that the *agia* denotes a soil impoverished to sugar-cane by too frequent cultivation of that crop. "The weed which chokes the *jowári* is called *taluk* by the Natives. It grows in the rains and commencement of the cold weather, and flowers in December. I do not think it is a root parasite, as it grows quite free from the *jowári* stems. I think it acts by taking all the nourishment or moisture out of the soil at the surface. *Jowári* and cotton are usually grown in rotation on the same lands. The *taluk* always makes its appearance among the *jowári* and practically never among the cotton plants. If *jowári* be grown two or three years running on the same field, the whole of it is overrun with *taluk;* but as soon as cotton is planted again it disappears entirely." It is somewhat significant that in the Central Provinces the *agia* should be regarded as not injurious to *juar*.

Taliuk. 193

194

Concluding Remarks on Diseases.

Many writers deal with the subject of the diseases of the sugar-cane plant. It is somewhat significant, however, that **Mr. Wray** should not have given a chapter on this subject, in his *Practical Sugar Planter*, and that the same oversight should have been made by **Mr. Robinson**, in his *Bengal Sugar Planter*. Both these expert writers, however, allude to the injury often done by ants, jackals, cattle, etc **Mr. Wray** referring to the effect of frost says that if planted in December, the cane will lie in the ground till February and March before it sprouts. Frost, as the canes are ripening, **Mr. Wray** adds, not only kills the plants but destroys the crystalline sugar present in the sap. But the writer must rest satisfied with what has been given above, together with the occasional allusions to the diseases that will be found in the quotations below regarding the chief sugar districts, since space cannot be afforded for further details. The general principles of sugar cultivation may be said to inculcate the theory that a wet season either during the early or late periods of the growth of the crop is very injurious. A cloudy closing season, causes the crop to be deficient in saccharine matter. A very dry season, immediately after planting, even if compensated for by artificial irrigation, results in a poor crop The young canes are more liable to be attacked by white-ants in a dry season than in a wet one, and if rain be deficient the canes have to be freely watered until they begin to sprout. New cultivations, deficient in vegetable mould, for the reason of their being exceptionally dry soils, are more liable to the destructive visitation of white-ants than old lands.

The reader should consult for further particulars on the subject of diseases, the district notices below as follows in Bengal, Bogra, p. 48; Lohardaga, p. 143; in Assam, p. 149; in the North-West Provinces (according to **Messrs. Duthie & Fuller**), p. 169; also Azamghur, p. 170; in the Panjáb, Hoshiarpur, p. 182; Jhang, p. 183; in Bombay, Khandesh, p. 218; etc., etc.

METHODS OF CULTIVATION, PLANTING, REAPING, AND EXPRESSION OF JUICE.

CULTIVATION. 195

So much has been said in the historic and other chapters of this article on the early records of the cultivation of sugar-cane in the provinces of India, more especially of Bengal, that it does not seem necessary to go over

CULTIVATION: Methods of.

these again, the more so since a fairly representative selection of passages from the gazetteers and district manuals will be found below. These convey the chief ideas of the various methods of cultivation and manufacture, and although the system pursued in any one district is very nearly the same in all, still the slight variations justify the publication of the selection given, in order that departures from the general practice may be clearly indicated. Besides which the constancy of the opinions and practices that prevail is better enforced by the publication of a selection of local reports than would be attained by a statement compiled by the author. For example, many peculiarities of sugar cultivation are, by writers on this subject said to be pursued by the West Indian planters, but not by the Indian.

Ratooning. 196 *Conf. with pp. 59, 76, 77, 78, 177, 184, 195, 215, 226, 247.*

RATOONING.—Thus it is frequently affirmed that ratooning the cane (that is, the production of a second or third crop off the same roots) is not understood in India. This is by no means correct, for the practice is alluded to more than a century ago, and is regularly followed at the present day in many parts of the country. A ratoon crop has even received distinctive names in the various provinces of India—names which can be shown to carry a knowledge of the subject considerably further back than the earliest records of European cane cultivation. Thus, for example, a ratoon cane crop in the Panjáb is known as *morda* or *mánda*, also as *muridaik* (in Delhi, etc.); in the North-West Provinces as *pairi* (*péri*) or *banjar;* in Bengal and the Central Provinces as *khunti* (*Conf. with Trans. Agri.-Hort. Soc. Ind., VI, 57; VII., 133*); in the Telegu country as *karsi;* and in Meywara (Rajputana) as *korbad*. In Bannu (see the passage quoted below, p. 180), the cane is ratooned for four or five years. The second crop is, by many writers, held to be richer in crystallized sugar. In Delhi the practice of ratooning, we are told, was formerly more extensively followed than at the present day. So again it is often said that the Indian cultivators throw the trash (or waste cane) away and neither use it for fuel nor manure. This statement is also scarcely correct, though neither of these practices are universally followed in India. One cultivator occasionally ratoons, but is ignorant of the great value of the trash as a manure for cane-fields; another ratoons and burns the trash as the fuel used in boiling the juice; whilst a third is ignorant of any advantage in ratooning, or even disputes that there is any advantage in that system, but manures his cane-fields with the trash. A fourth preserves the tops for next year's seed; while a fifth views these as useless for that purpose and accordingly gives them to his cattle.

Conf. with pp. 5, 7, 8, 79, 196, etc.

Conf. with pp. 140, 184, 186, 217, 240, 304.

Manures. 197 *Conf. with pp. 126, 140, 142, 144, 145, 149, 216, 225.*

MANURES.—It is therefore unnecessary to specialize any one feature of the system of sugar-cane cultivation pursued in India, excepting perhaps that of manuring. It is often said the Natives of India never manure their fields. While this may be true of certain crops or of certain tracts of country, it is certainly not true of sugar-cane. **Mr. Wray** enlarges on the fact that as in vine cultivation it is found one of the best of all manures is the prunings and decayed branches, etc., so with sugar-cane the most valuable manure is the cane itself. He, therefore, strongly condemns the practice of burning the trash obtained on the expression of the juice. It has been estimated, he says, by numerous planters and others, that not more than fifty per cent. of the weight of the cane is obtained from it as juice, by the ordinary mills used for crushing on estates in the West Indies; whereas it has been satisfactorily demonstrated that the plant consists of 90 parts of fluid and 10 parts of woody fibre.

"In the case of inefficient pressure, such as shows an amount of juice, not exceeding fifty per cent. of the whole weight of the cane, of course, the

CULTIVATION: Methods of. Manures.

remaining fifty per cent. is received again by the soil, under the system of manuring with cane-trash. Calculating, therefore, that the quantity of juice expressed amounts to 75 per cent. (with good and efficient mills), then the green cane-trash or *megass* available for manure will be 25 per cent. of the whole taken from the land. Now, plant-canes generally average from 30 to 35 tons an acre, which would give, if of the former weight, 7½ tons of green-trash; or if of the latter, then 8¾ tons as manure per acre, independent of the long tops and the dry leaves: the former being generally used as fodder for the cattle on the estate, and the latter not unfrequently burned either at the works or on the field." **Mr. Wray** estimates the value of the trash as fuel in comparison to coal, and arrives at the conclusion that it is more profitable to purchase coal or other fuel than to burn the trash. **Mr. Ozanne** furnished a very instructive though brief sketch of the manures generally used with cane in Bombay. The reader will find **Mr. Ozanne's** remarks in the special chapter below devoted to Bombay (p. 216).

Rotation. 198 *Conf. with pp. 145, 150, 170, 187, 215, 225.*

Many writers recommend green manuring with leguminous crops as highly beneficial to cane. For this purpose, beans, peas, lucerne, indigo, *san* hemp (*see p. 126*), may be grown between the rows of young canes and later on, if necessary, ploughed into the soil. The Natives so far recognise the value of leguminous crops in restoring the fertility of the soil that they very frequently follow cane in their simple rotation of crops with some of the plants mentioned. **Roxburgh**, in a paper published in 1792 (*see p. 225*), dwells on this subject, and **Buchanan-Hamilton** speaks of the usual rotation of cane lasting for four years, pulses or wheat being twice grown within that period. **Mr. Wray** extols the use of indigo as a manure for cane. He suggests that it should be sown in lines between the cane, and two cuttings obtained and ploughed in before it is rooted up. The indigo refuse, after removal of the dye, he says, is also highly valuable. The Chinese planters in the Straits, he adds, often obtain excellent crops of cane from a soil "so sandy and otherwise unfertile, that no European planter would for a moment think of planting canes in such lands." This result, **Mr. Wray** explains, is obtained by placing the stems and leaves fresh from the indigo vats over the roots of the cane and then moulding over them. The advantages of leguminous crop manuring, **Mr. Wray** urges, are entirely lost if the pulse be allowed to ripen its seed. The best time, he says, for ploughing in the manure is just before flowering when the green manure is quite green and succulent. **Mr. Wray** furnishes much useful information on the subject of manures, but it may fairly be said that his remarks are on general principles and have by no means any very special bearing on cane more than on any other crop. One point may be here alluded to, however, *viz.*, the utilization of the *dunder* or *redundar*, that is, the fermented wash after distillation. Of this substance he says that, as its name implies, it accumulates at the "works," and instead of there proving offensive and unwholesome, it should be carted off to the fields as a manure. An instructive paper on manures suitable for cane will be found in the *Journal of the Agri.-Horticultural Society of India, Vol. VI., pp. 61-91; also proceedings of that volume, p. 40.* A somewhat amusing statement occurs in the *Transactions of the Agri.-Horti.-Soc.* (*Vol. I., 116*) in which *gúr*, used as a manure, is said to hasten the fruiting of the mango.

Soils. 199

Soils.—But it may be said that **Mr. Wray's** opinion on the subject of manures for cane-fields amounts to this, that given a fairly suitable soil and a liberal supply of water, careful cultivation, repeated ploughing and manuring with the cane trash, is all that is needed. The defects of the soil may be combated by principles familiar to all cultivators up to a certain limit, but beyond that point sugar-cane cultivation must result in the production

CULTIVATION: Methods of. Soils.

Low Yield through Salt in Soil. 200

Conf. with p. 60, 161.

of a juice deficient in crystallizable sugar unless the favourable conditions of soil and climate be present. Many writers have pointed out that the presence of *reh* or other salts in the soil (beyond a certain proportion) invariably results in a watery juice deficient in crystallizable sugar. On this subject **Mr. Wray** wrote: "It often occurs in the Straits Settlements, Demerara, Louisiana, and other places, that lands are strongly impregnated with saline matter; which certainly causes the cane to grow most luxuriantly, but affects the juice (and, consequently, the sugar made from it) very prejudicially. In province Wellesley, I have known sugar that was quite salt produced the first year from such land; and in the Sunderbunds it was so very salt that the sugar estates had to be abandoned." It would appear that **Mr. Wray** regarded the most suitable soil for sugar-cane as one of granitic origin, but which possesses a fair amount of lime. The reader will find much useful information on the soils of India best suited for cane in the publications of the Agri-Horticultural Society of India, such as *Transactions, I., 121; III., 35; IV., 134; Journals, Vol. I., 126.*

It may, in concluding these introductory remarks, be said that the writer has thought it the preferable course not to attempt to give a review of the peculiar systems of cultivation pursued in India, but rather to furnish a fairly extensive series of passages from special and local publications. This, it is believed, may prove more useful than a compilation, since the works from which the writer has drawn, may fairly be said to constitute a library of books, many of which are not very generally accessible to persons not resident in India.

BENGAL. 201

I.—BENGAL.

References.—*Buchanan-Hamilton, Statistical Account, Dinajpur; Colebrooke, Husbandry of Bengal; Proceedings, Honourable East India Company, 1790 to 1822; also East India Produce, 1840; Sugar Statistics of 1848; Wray, Practical Sugar Planter; Robinson, Bengal Sugar Planter, also Agri.-Hort. Soc. Prize Essay on Date Palm; Agri.-Hort. Soc. Ind.:—Trans., I., 98-103; II., 188; III., 61-65; V., 184; VI., 46-47, 56-59, 239, Proc. 7, 41, 48, 94; VIII., Proc. 85, 128, 132, 200; VIII., 89, 157-860, Proc. 396, 410, 419, 426, 433, 455; Jour. I., 102, 147, 363-369; II., 345-348, Proc. 196, 260, 479, 541-544; III., 84, Proc. 179, 282, 293; IV., 61-91, 103, Sel. 32, 131-132, Proc. 55, 92; V. Sel, 33, 75-77, 105, Proc. 31, 40, 52; VI., 56-67, Proc. 26, 30, 85, 89; VIII., 1-12, 164-166, 181-82, Sel. 96; IX., 355-358 (Diseases), Proc. 271, Sel. 75; X., 243-274 (Prize Essay on Date), 358, Proc. 4, 87, XI., Proc., 42-44; XII., 109, 356-357, Proc. (1861) 45-46; New Series, VII., 162-176 (Ambu sorghum), 351, Proc. (1882) 27, 101-102, 141-143. 152, 161, 163; VIII., Proc. 16; An extensive official correspondence down to 1891; Hunter's Statistical Account of Bengal, numerous passages, etc., etc.*

Area and Outturn. 202

Area, Outturn, and Consumption in Bengal.—In 1846 the Chamber of Commerce of Bengal applied to the Government of India to procure for them a *Statistical Return of Lands*, cultivated in Bengal and the North-West Provinces, for the growth of Cane and Date *Gúr* and Sugar, and the probable consumption in each district. The reply, which was furnished by the Government in 1848, appears to deal with figures collected for the year previous. Whether the returns are actually those for 1846-47 or 1847-48 is, however, at this distance of time, of comparatively little importance. The small volume which was issued under the title of "*Statistics of British East India Sugar*," contains much of great value, and gives the data by which a comparison may be drawn between the sugar production and trade of India forty odd years ago and that of the present day. It is explained that the returns had all been reduced to one standard, namely, *bighás* of 14,400 square feet (or say, ⅓rd of an acre), and the

Error in Terms. 203

Conf. with pp. 114-115, 131, 252-255, 285, 298.

CULTIVATION in Bengal.

Area and Outturn.

produce expressed as *gúr* at the rate ⅓rd* of a maund of 82℔ as the equivalent in sugar. It may be pointed out that errors in the Agricultural returns of sugar are largely due to one set of figures being the freshly expressed juice, another the *gúr*, a third the raw or country sugar, and a fourth the refined sugar. A comparison of the yield, from such figures would obviously be misleading and fallacious. The following abstract statement of the area, produce, and consumption for the eight divisions of Bengal as recognised in the year 1847-48 may be here furnished from the *Statistics of British East India Sugar*:—

* Modern writers speak of the yield as 2½ maunds "of unrefined sugar to the maund of refined sugar." But what is meant by "unrefind sugar"? The refiner purchases *ráb*, *gúr*, *bura*, and other forms of unrefined sugar, and the yield from *gúr*, still more so from *bura* (country sugar) would be greater than from *ráb*. *Ráb* is the article chiefly used by the refiner. It has been drained of a certian amount of its molasses. *Bura*, as a rule, is too expensive for the refiner, but it is the unrefined sugar that may be said to be used by the middle classes.—*Ed., Dict. Econ. Prod.*

CULTIVATION in Bengal.

Area and Outturn.

Production and Consumption of Sugar in Bengal during 1847-48.

	Jessor Division	Bhagalpur Division.	Cuttack Division.	Murshedabad Division.	Dacca Division.	Patna Division.	Hazaribagh Division.	Chittagong Division.	Totals.
	Jessor, 24-Pergunnahs, Burdwan, Hughli, Nuddea, Bancura, and Barasat.	Bhagalpur, Dinajpur, Monghyr, Purneah, Tirhut, and Maldah.	Cuttack Balasore, Midnapur, and Kurdah.	Murshedabad, Bagurah, Rungpur, Rajshahye, Pubna and Beerbhum.	Dacca, Furridpur, Mymensing, Sylhet, Backergunge, and Cachar.	Shahabad, Patna, Behar and Sarun.	Hazaribagh, Lohardugga, and Manbhum.	Chittagong, Tipperah, and Bullúah.	
Cane Produce—									
Area in bighas	79,591	1,54,424	41,388	52,221	83,588	1,02,593	56,849	40,725	6,71,381 bighas=223,793 acres.
Gúr in maunds	7,56,118	14,21,073	4,20,412	5,28,661	5,65,698	26,29,268	2,09,929	2,06,438	67,73,600 mds.
Average per bigha	M. S. C. 9 20 0	M. S. C. 9 8 1	M. S. C. 10 6 5	M. S. C. 10 4 5½	M. S. 6 30	M. S. C. 16 6 13	M. S. C. 3 27 11	M. S. C. 5 2 12	8 mds. 38 s. (or say an average of 27 mds. an acre, or of 30 mds. an acre if total production be divided by the average).
Date Palm Produce—									
Number of Trees	37,83,000	8,03,200	44,663	44,414	15,56,010	Not returned	3,303	1,56,000	
Gúr in maunds*	6,25,367	not returned	25	10,903	3,74,954	ditto.	Not returned.	26,195	10,37,445 mds.
Average per 100 trees	M. S. C. 16 21 4	The returns are not sufficiently complete for the districts of the other divisions to allow of the average being worked out, but it has been estimated to have been in 1847-48 16 maunds 9 seers 5¾ chattaks.							
Total of Gúr obtained (both kinds) in maunds	13,81,485	14,21,073	4,20,438	5,39,564	9,40,652	26,29,268	2,09,929	2,32,633	77,75,045 mds.
Consumption of Gúr—									
Total in maunds	6,52,021	14,24,390	5,24,090	4,17,710	6,86,348	12,75,609	1,99,490	3,40,867	55,20,526 "
Average to head of population.	S. C. 4 14	S. 6	S. 7½	S. C. 2 8	S. 6¾	S. C. 7 5	S. C. 5 6	S. C. 5 10	S. C. 5 12⅞ (say 11·2lb).
Available for Export	7,29,464	None.	None.	1,21,854	2,54,304	13,53,658	10,439	None.	22,54,519 mds.

* The juice of the date, it is explained, was not always manufactured into *gúr*, but was, as at the present day, largely fermented and distilled.

CULTIVATION in Bengal.

Area and Outturn.

It will thus be seen that in 1847-48 there were 6,71,381 *bighás* (223,793 acres) under sugar-cane in Bengal, and that these were estimated to yield 67,37,600 maunds of *gúr* or, dividing the total production by the acreage 10 maunds 1 seer and 5¾ chatáks per *bíghá* (say 30 maunds an acre). This would represent the very low average outturn of about 8 cwt. of crystallizable sugar an acre. There were, however, found to be 6,390,590 date-palms which yielded *gúr*, as also 937,278 that afforded juice made into a beverage. These trees added to the Bengal supply of *gúr* 10,37,445 maunds and thus raised the total produce to 77,75,045 maunds. The consumption was said to have been 55,20,526 maunds of *gúr* or 5 seers 12½ chatáks on the population of 38,327,225. A balance was thus available for export which added to that shown in connection with the North-West Provinces, and that of Madras made a total of 1,821,855 cwt., expressed as refined sugar. The exports were for 1847-48, 1,229,828 cwt, so that there remained the ultimate balance of 592,027 cwt. of refined sugar (or its equivalent in *gúr*) as stock in hand to meet the details of internal trade. Thus, for example, Bhagalpur is shown to have consumed 3,317 maunds of *gúr*, Cuttack 1,03,651 maunds, and Chittagong, 1,08,233 maunds in excess of their production. A considerable trade took place from Calcutta to Bombay and Burma, so that there was *relatively* quite as extensive an interchange between district and district and between province and province forty or fifty years ago as at the present day.

It is to be regretted that detailed returns of the cultivation of sugar-cane cannot be obtained for each year for a number of years back. We are accordingly left to speculate on the fluctuations of production by the indications in the statistics of foreign exports. This remark is more peculiarly applicable to Bengal than to any of the other provinces of India. Indeed, as already remarked, fuller particulars, almost, are available regarding the sugar cultivation and trade of that province forty, fifty, or a hundred years ago than we possess at the present day. The *Statistics of Sugar*, which has been freely utilized above, is, for example, by no means the only publication that has attempted to set forth the area, outturn, and con-consumption of sugar in Bengal. The Honourable the East India Company published in 1792 a statement of the sugar cultivation of the Lower Provinces. There were (according to that statement) 1,59,732 *bighás* under cane in Sarun, Tirhut, Shahabad, Dacca, Jessore, Dinajpur, Ramghur. Murshidábád, Burdwan, Midnapur, Beerbhum, 24-Pergunnahs, Chittagong, Sylhet, Purnea, Calcutta, Rungpore, Tipperah, and Bhagalpur. The yield from these districts is said to have been 1,14,525 maunds of refined sugar and 4,75,824 maunds of *gúr*. But these figures, even if they could be believed to have any thing like expressed the real state of the Bengal sugar enterprise 100 years ago, are not in themselves of any very great interest. Abundant evidence exists in support of the opinion that the cultivation of cane greatly expanded in India, with the demand created in Europe through the action taken by the East India Company. There are, however, certain features of distinct interest, in the explanatory remarks made by the officers who furnished the returns for the year 1792. For example, of quite half the districts, remarks like the following occur:—no sugar is here manufactured; the juice afforded by the canes yields little or no crystallizable sugar; or again, the sugar-cane grown in this district is eaten fresh, the sugar used being imported. How far similar remarks are true of the sugar-cane grown in Bengal at the present day would seem a point that deserves careful consideration. Numerous writers dwell, for example, on the suitability of certain soils of Bengal for the production of cane, good for eating but bad for sugar-making. Others on the fact that owing to the unsuitability of

Yield. 204

Error in Terms. 205

Conf. with pp. 114-115, 134, 136, 183, 229, 252-255, 285, 298.

Methods of Cultivation

CULTIVATION in Bengal.

the soil the cane grown is of the most inferior kind. Upon the variability of yield, due to such canes may, therefore, be largely attributable the vast differences in the returns published and not of necessity to inaccuracies of compilation. Suitability of soil, superiority of cultivated race of cane, and greater facilities of expression of juice, and isolation of sugar may fairly well be admitted as doubling the outturn. The 223,793 acres of cane, recorded in 1848, yielded an average of 21·4 cwt. of *gúr* per acre. The figures of 1792 show a very different result. If these figures can be accepted as having fairly expressed the sugar-cane area of Bengal a century ago, as also the approximate outturn, it may be said that there were 53,244 acres under cane and that the yield amounted to 8,19,399 maunds of *gúr*, or an average of 11 cwt. per acre. That result would be very little more than half the acreage outturn of *gúr* shown for 1848, as also less than half the yield of *gúr* recorded at the present day. But this result shows conclusively the error of dividing the ascertained area by the total produce, in determining the yield of sugar per acre—a method, however, which has been pursued by all modern writers. The Honourable the East India Company realized a century ago that, to develope the sugar trade of India, there were certain subjects regarding which it was necessary that they should be possessed of detailed and accurate information. Amongst these may be mentioned the *area* devoted to the cane, the *yield of cane* per acre, the various *qualities of cane* grown, the *yield* of crystallizable *sugar* from the sugar-yielding varieties, the *consumption of cane* in its raw state; the *consumption of gúr*, deficient in crystallizable sugar; and the *consumption of sugar* itself. They fully appreciated the fact that the total sugar-cane in India can by no means be viewed as the raw material of the possible sugar supply. The Company were well aware that to the Natives of India a high percentage of sweetness (regardless of the yield of crystallizable sugar) was the chief criterion of cultivation. This made the area of sugar-cane cultivation by no means that on which estimates could be framed of the possible supply for the European market. They saw also that certain districts and certain cultivated races of cane were the chief sources of the sugar procurable in India, and accordingly their officers had instructions to draw up their forecasts with due regard to these considerations. Reviewing the numerous reports received the Board of Directors published in 1792 the following *précis*:—

Salt in Soil giving Low Yield.
206

Conf. with pp. 59-60, 130, 161.

"The Board observe that the highest produce of cane land in Benares is much less in quantity than the lowest above stated in Bengal. It is possible there is some mistake in the information received from the Resident; they will notice the circumstance to him. At the same time they remark that the natives reckon the Benares sugar to have less strength than the Bengal; and they understand that in experiments made in Calcutta on *gúr* from Benares and from different parts of Bengal, the former gave little or no sugar, the latter its due proportion. But these experiments cannot be considered as conclusive, without it were to be ascertained in what month the *gúr* was made, and in what repute the natives held it Were an experiment to be made on the *pateli gúr* from Rungpore, which is gathered in October, no sugar would be produced. The land about Santipore appears to yield less sugar than any other place mentioned. It is within the knowledge of the Board that the soil in that part of the country, and in general of the large *zemindarí* of Nuddea, is sandy, light, and poor. The Rungpore and Dinagepore lands stand at less produce than those of Burdwan and Beerbhum; but the computation is formed only upon the *awul khat*, or the sort of *chíni* which is more purified than the *ek bári* of Beerbhum and Burdwan, consequently, without the sugar is in the same state, an exact comparison cannot be formed. The same observation applies also to any comparison that may be endeavoured to be formed between the foregoing statements of the produce of an acre in these provinces and of an acre in the British West Indies, from whence, almost the whole of the sugars are exported in the Muscovado State. The Board understand that West Indian Muscovado sugar loses about ⅓rd of its weight by claying; and as the West

CULTIVATION in Bengal.

India clayed sugars are said the most to resemble the *chini* of this country some comparison may be formed between the produce of an acre in this country and in the West India Island. Many acres of plant-canes in the West India Islands are said to yield 5,000℔ of Muscovado sugar, deduct ⅓rd loss of weight by claying, *viz.*, 15 cwt. 44℔; and there will remain 29 cwt. 86℔. This, even allowing for the difference between clayed sugars and *chini*, is so much beyond the highest Bengal produce, that it marks something extremely favourable in the soil of these particular lands. But the general produce of the west Indian Islands is said to average not more than one hogshead of 15 cwt. to an acre: deduct ⅓rd loss of weight by claying, and the produce will be 10 cwt. per acre. This, allowing for the difference between the clayed sugar and *chini*, may be rated at about equal to the produce of the Rungpore and Dinagepur districts, and below those of Burdwan and Beerbhum, which are the chief districts in Bengal Proper that produce sugar for exportation."

Yield. 207 *Conf. with pp. 110, 134, 139, 211.*

It will thus be seen that the East India Company made its calculations on Native refined sugar (*chíní*) and purposely left out of calculation *gúr*—an article for which there was then little or no demand in Europe. The Board, accordingly, furnished the following estimates of the acreage yield of *chini*—

	HIGHEST PER ACRE IN *Chíní*			LOWEST PER ACRE IN *Chíní*.		
	Cwt.	qrs.	℔.	Cwt.	qrs.	℔.
Benares	4	1	3	2	1	22
Rungpore	9	1	11	6	1	5
Beerbum	(Average)			14	3	0
Radnagore	(Average)			12	2	6
Santipore	6	0	10	5	1	13
Burdwan	(Average)			14	0	0
Sulkí, Calcutta	(Average)			11	0	0

If it be of any value to state the average of all these records, we learn that in the districts above named the yield of refined sugar was 100 years ago, a little over 9 cwt. an acre, or, say, 27 cwt. of *gúr*. Thus, when in 1792 the total production in Bengal was divided by the area known to be under cane, the average yield was shown as 11 cwt., but, when the sugar producing acreage *alone* was taken into consideration, the yield was demonstrated to be 27 cwt. of *gúr*. That figure will be seen (by the table at page 116) to be the average yield of "coarse sugar" recorded by the Government of India in 1888-89 for all the provinces of British India, but there is strong presumptive evidence that from 1792 to 1848 a vast improvement took place in the yield of refined sugar. This opinion is arrived at by comparing the outturn recorded in 1848, in the chief sugar districts with that given for 1792. The fact that improvement can be thus shown to have taken place, during the first half of this century, led the writer to suspect that there may be some serious error in modern statistics, unless it be admitted that in this respect the sugar industry of India has retrogressed during the past fifty years. It is more than likely, however, that the low average in recent returns has been produced by including in the calculation large sugar-cane tracts that afford no sugar whatsoever. It seems, therefore, probable that a similar result would be obtained now, to that shown for 1792, were the area of Bengal classified according to importance in sugar supply. This would not, of course, affect the amount to be shown as the total production, but it would remove the false stigma from India that its cane affords less than half what is obtained in the other sugar-cane countries. Improvement and expansion are of necessity dependent upon a correct knowledge of actual sugar production. It has in more than one place been urged by the writer that considerable doubt exists as to the meaning that should be

Improvement 208

CULTIVATION in Bengal.

placed on the term, "coarse sugar," estimated for in the table at page 116. Comparisons with the West Indies and other advanced countries in sugar-cane production are, therefore, practically impossible. It is generally said, for example, that in the West Indies the yield of crystallizable sugar is 2 to 3 tons an acre. Certain writers compare India as affording 1 ton of crystallizable sugar; but the table alluded to gives the average for the British provinces as 27·9 cwt., or, say, 1 ton of "coarse sugar," and if this be *gúr* the actual yield of crystallizable sugar in India would be little more than one-third of a ton. The writer reiterates this feature of the Indian sugar trade because of the fact that, if the mistake of "coarse sugar" or *gúr* has been made with crystallizable sugar, it affects materially the estimates of consumption per head of population as well as the outturn per acre. One writer, however, whose communication has been submitted to Government under the seal of confidence, affirms that by the process adopted by him he regularly obtains from the Native cane a product equivalent to the Native *gúr*, which affords 14 per cent on the weight of cane with 16 per cent. of sweetness. This is equivalent to 2·44 tons of crystallized sugar an acre or to a total of 3·27 tons of sweetness. Such a result demonstrates not only the vast advantage of superior appliances and methods, but the possibility of the average yield in India being considerably nearer a ton of crystallized sugar, or 3 tons of *gúr* or "coarse sugar" *on that portion of the sugar-cane area which is specially grown for the production of sugar.* Were it the case that the entire area was grown for that purpose, the total production of India would, therefore, very probally be nearly three times as great as has been shown in the table at page 116. It will be seen below that **Mr. Basu** says that, while the cultivator in Palamau is happy to get 25 maunds of *gúr* per acre, the *rayats* of Hughlí and Burdwan would not consider the cultivation paying under 60 mauuds of *gúr* or, say, ¾ of a ton of crystallized sugar. An average of 28·7 maunds of *gúr* an acre for the entire province, thus very incorrectly represents its sugar-producing districts. But to revert to the returns of the Bengal sugar-trade of 1792, it may be said that a certain amount of confidence can be placed on the information procured by the East India Company, owing to the well known interest taken at that time in the subject of the development of the Indian supply of sugar. The Company enforced on its officers the greatest possible attention to the subject, and very elaborate and detailed reports were published which comprise several large volumes fully illustrated. If reliance may, therefore, be placed on the figures of area and outturn for 1792, we learn that from that date to 1848 the acreage of sugar-cane quadrupled itself, and the yield of *gúr* per acre on the total acreage was doubled. This latter fact seems deserving of special consideration, since it would justify the opinion that one result of the demand for crystallizable sugar from India seems to have been to improve the yield by selection, of superior qualities of cane and otherwise. This process of improvement has doubtless extended since 1848 to the present day, and if all be true that has been written of the inferior yield of India, there is ample room for still farther improvement. But this assumption involves the possible error that the people of Bengal, indeed of India, desire such improvement, just as the calculation of an average yield of crystallizable sugar to the returned acreage is erroneous without due deduction*

Error in Terms. **209** *Conf. with pp. 114-115.*

Yield. **210** *Conf. with pp. 110, 133, 211.*

Selection. **211** *Conf. with pp. 8, 9-10, 44-46, 58, 61-62, 82.*

* Madras, it will be seen from the remarks below, appears to have attempted to make the calculation correct, since, in the modern estimates of sugar-cane in that Presidency, the Government has referred the subject to three sections.—
(*a*) Edible canes. (*b*) *Gúr*-yielding canes.
(*c*) Sugar-yielding canes.

CULTIVATION in Bengal.

being made of the area devoted to edible and other canes which are never grown with any idea of being employed in the manufacture of sugar. It has already been remarked, but cannot be too prominently urged, that the Indian sugar-cane cultivation is by no means characterized by that singleness of purpose met with in the West Indies and other sugar-cane-planting countries, *viz.*, the yield of crystallizable sugar. The cane is grown mainly for local purposes and in small isolated patches, as an ordinary agricultural crop in rotation with many others. Central factories, such as have been proposed, would, in India, fail to draw their supplies of cane from the cultivators and accordingly would sink rapidly to the position all sugar factories in India occupy at the present day, namely, refineries, unless they grew their own cane. But even the refiners find it difficult to obtain their supplies at remunerative prices, since the local market for *gúr* and edible canes pays the cultivator better than the *ráb* required by the refiner. It is, therefore, just possible that the limits of improvement have very nearly been reached in relation to the existing nature of the Indian demand. To effect any very great further improvement it seems necessary, that the people should be educated to the advantages of using refined sugar in preference to the crude article they presently consume. But it may be contended that such education is rapidly taking place, and that one element in this tendency to change is the cheap rate at which foreign sugars can be landed in India. The yearly increasing imports have hitherto told heavily on the indigenous art of refining, but should any unforeseen accident disturb this state of affairs, such as the removal of the bounties or a favourable fluctuation in the rate of exchange, the imports would be checked and a greater demand arise for Indian refined sugar than ever existed before. It may fairly be said that certain communities of India are now using refined sugar in preference to *gúr* and crude sugar, and that as that demand expands, great improvements will be effected in the selection of cane, in the methods of cultivation, and in the appliances for the expression of juice and manufacture of sugar. It is in fact to some such reaction that the refiners of Bengal have to look for the restoration of their trade, for India itself must be their chief market in future, and the limit of expansion must accordingly be fixed by the rate of the social and material progression of the people.

Native Crude Sugar. 212 *Conf. with pp. 40, 81-82, 95, 104, 113, 325.*

The bulk of the evidence favours the assumption that the production and consumption of *gúr* is far greater in Bengal than has been shown by the estimated acreage devoted to the crop. It need scarcely be said that no actual survey of the sugar-cane area of Bengal has been made. The figures shown in the table at page 116, have been obtained as the result of the personal opinions of local officers. In most of the other provinces of India, where periodic settlements have to be accomplished, the area of sugar-cane, or of any other crop, is ascertained with very nearly as much accuracy as is the case in European countries of a like magnitude. The permanent settlement of Bengal has deprived the Government of that province of any trustworthy source of information, as to the present state of its agricultural prosperity. **Mr. Schofield**, in the *Note on Sugar*, which was issued by the Revenue and Agricultural Department in 1888, wrote that "it would probably be found (were actual measurement of the sugar-plant area undertaken) that 600,000 instead of 312,000 acres would be a nearer approach to the truth." In Appendix I. (p. 48) to the Famine Commissioner's Report, the area under sugar-cane is estimated at a million-acres, but this again seems to be too high. **Babu Addonath Banerjee** (of the *Statistical Department of the Government of Bengal*) in his review of **Mr. Schofield's** *Note on Sugar*, while lowering the acreage

CULTIVATION in Bengal.

for one or two districts of Bengal, accepts upon the whole **Mr. Schofield's** main conclusion, namely, that the area of sugar-cane in Bengal must either be understated or the outturn per acre incorrect, since "we must, perhaps, abide by the old estimates and accept the conclusions arrived at in the *Note* that the rate of consumption of *gúr* is 10 to 12 *seers* (20 or 24 lb) per head." **Babu Addonath**, however, offers no opinion as to the probable accuracy or otherwise of the suggestion that the area under sugar-cane and date-palms in Bengal is more nearly 600,000 than 312,000 acres. The writer of the present article, after the most careful perusal of all that has been written, is very much disposed to accept **Mr. Schofield's** conclusion, the more so, since the yield of date-palm sugar is by recent returns shown to be considerably less than it was recorded some forty-five years ago. Thus, in 1848, the 6,390,590 palms then registered were estimated to have afforded 10,37,445 maunds of *gúr*, whereas in 1888 only 7,43,000 maunds are credited to palms. And what is perhaps more significant still, if we accept as correct the acreage * in 1888, of the palms in Bengal and Madras, the former yielded only 24·7 maunds, while the latter gave 74·6 maunds an acre. If, however, we express the palms of 1848 to acreage (and accept 400 trees as the number to the acre) the yield in Bengal would in that year have been higher than in Madras, *viz.*, 64 8 maunds in the former and 40·8 maunds an acre in the latter province. It is practically impossible to believe that such changes in total outturn and acreage yield could be due to natural causes. We are practically driven to the other explanation, *viz.*, defects in one or other or in both sets of statistical returns. The writer in fact is strongly of opinion that any attempt to express acreage of palms is of necessity misleading, and that the East India Company's method of returning total production and outturn per 100 trees is preferable. It is obviously incorrect, at all events, to add together the acreage yield from palms and cane and to compare the figure thus derived with the ascertained acreage yield from canes alone. This is done whenever the returns of Madras or Bengal (see the table, p. 116) are, for example, compared with those of the North-West Provinces or of the Panjáb. If the palms be excluded from Madras and Bengal, the acreage yield of *gúr* becomes 44·2 maunds in the former and 29·2 maunds in the latter, and these figures may, if so desired, be contrasted with 22·9 maunds in the North-West Provinces and 27·9 maunds in the Panjáb.

Date-palm Sugar. *Conf. with pp 226-227, 231, 266, 270, 310. 352, 361, 370.* **213**

The reader will find much useful information regarding date-palm sugar below, in the abstract from **Mr. Westland's** Report on the Jessor District of Bengal (*Conf with p. 270; also with* **Phœnix**, *Vol. VI., Pt. I., 199-215*). In 1894, **Mr. S. H. Robinson** published his most suggestive work *The Bengal Sugar Planter*. Although that little book deals with the whole subject of the sugar of the Lower Provinces, **Mr. Robinson** has been able to devote two chapters to date-palm sugar. He defines the area of Bengal date-palm sugar production as extending due east and west from Kissengunge in Kishnagur to a little beyond Nollchiti in the Backergunge District; and north and south, from the vicinity of Comercolly in the Pabna District, to the borders of the Sunderbands. It thus covers a tract of country 130 miles long (east and west) by about 80 miles broad. Its principal districts are, therefore, Jessor, Furreedpore, and Backergunge with por-

* The writer has failed to discover the number of palms which has been allowed to the acre in the returns published by the Government of India. It very probably varys according to the nature of the palm grown and the province. Thus, **Robinson** allows 160 date palms to the Bengal *bighá* (480 trees to the acre), but, of the cocoanut, it is customary to estimate for only 100 trees to the acre. This matter is very important and it would be desirable if future reports furnished the rate adopted. *Conf. with* p. 144.

CULTIVATION in Bengal.

tions of Nuddea, Baraset, and Pabna. **Robinson** tells us that he had ascertained the annual produce of a full-grown date plantation to be equal to 78¾ maunds of *gúr* per Bengal *bíghá*, which, converted into *khaur*, might be taken as equivalent to a yield of about 5½ tons of Muscavado sugar per English acre. He then adds that "The calculation given in the subsequent chapter (on Native Sugar Manufacture) proved—*1st*, that date sugars could be produced at about two-thirds the cost of cane-sugar, of equal quality; *2nd*, that the date crop involved little or no risk, and a comparatively small outlay in the cultivation; and *lastly*, that good white sugar could be produced therefrom, by Native methods, at a cost of R4-10-7 per maund, and fine crystallized sugar at R6-13-9 per maund, delivered in Calcutta." **Robinson** assumed that the Bengal palm-sugar area might be accepted to embrace 10,400 square miles. "Let us suppose," he says, "only one-twentieth part of this surface to become, in the course of years, set apart for date-tree cultivation; and that the average produce be *one-half* of what has been calculated as the yield of trees in full bearing, which would allow 2¾ tons per acre per annum. The total annual production of such a tract of cultivation, we shall find, will amount to 915,200 tons of sugar, or more than sufficient for the wants of all Europe." This, however, is only an estimate of probable or rather possible production. **Robinson** gave the total outturn in Bengal of date-sugar, in 1848, as very considerably greater than what it is believed to be at the present day. If, however, we accept 2½ tons per acre as the yield of dry crystallized sugar, a figure below his lowest estimate, and apply that to the acreage returned for 1888, the outturn should have been 21,00,000 maunds of sugar (or 63,00,000 maunds *gúr*) instead of 7,43,000 maunds *gúr*. The possibility of such an error, existing in the modern statistics, renders it undesirable to accept the abnormally low consumption per head of population given for Bengal in the table at page 116. Indeed, it may be added that an increase of that magnitude would seem almost justified by the facts which will be found reviewed below, regarding the Madras palm-sugar production when viewed in the light of the estimates of production determined by **Mr. Westland.**

Errors in Terms.

Conf. with pp. 114-115, 130-131, 133, 136, 183, 229, 252-255, 285, 298.

The writer has repeatedly urged in this article that many of the misconceptions regarding the Indian sugar industry largely proceed from want of uniformity in the terms employed. There could be no more likely pitfall than the confliction of the statistics of yield of palm sugar from the reduction of numbers of trees to the acre being on different standards either due to necessity of the different kinds of palms or from local habit of cultivation. Thus, for example, if one hundred trees be taken as the number equivalent to an acre of land, the yield per acre would, of course, be just one-fourth what it would appear were the assumption made (as has been done above) that 400 trees are commonly grown on that space. So, in a like manner, much ambiguity has arisen from the use of the words *gúr*, coarse-sugar, sugar, and molasses as all synonymous. They each denote widely different products. Approximately it may be said that ⅓ the weight of *gúr* is the quantity of refined sugar that may be prepared therefrom. But most writers speak of the yield in India from cane as being nearly 1 ton of "*gúr* or coarse sugar," while others contrast India with the West Indies by affirming that its cane-fields afford only 1 ton of "crystallized sugar" as against 2 to 3 tons obtained in the colonies. Now it has been ascertained by actual experiment that 2½ tons of crystallizable sugar can, and often are obtained from the acre of Indian cane. Still the published returns for the country as a whole manifest a yield of less than a ton of coarse-sugar or, say, ⅓ of a ton of crystallized sugar. Such a state of affairs demonstrates powerfully the necessity of a throrough enquiry, as the first step towards improvement, for if it be the case that such diver-

CULTIVATION in Bengal.

sities in yield actually exist, a map of the country according to yield of sugar to the acre would prove of the greatest possible value. But it by no means follows that all that the cane possesses is removed by the Native manufacturer more than that his system of agriculture gives the maximum that the soil is capable of bearing. Defective appliances and methods of expression of juice and manufacture of sugar are largely accountable for the backwardness of India as a sugar-producing country. But it may safely be said, in conclusion, that the inferiority attributed to India is more apparent than real since it proceeds very largely from defective returns in (*a*) the acceptation of the total area of sugar-cane as the area of sugar production; (*b*) the confusion of returns of crystallizable sugar with those of *gúr*; (*c*) the probable underestimation of sugar derived from palms; and (*d*) the amalgamation of such returns as exists for palm-sugar with those for cane. The following passages may be now given (arranged alphabetically) in order to exhibit the methods of cultivation of cane and expression of juice as pursued in the chief districts of Bengal:—

Birbhum. 214

Conf. with pp. 128, 142, 144.

BIRBHUM.—"Sugar-cane is also grown on *do* land, but as a single crop. For this cultivation, which is a very exhausting one, a large quantity of manure is needed —150 maunds per *bíghá*, or about 60 tons an acre, being given if procurable. After manuring, the land is ploughed five times; meanwhile, a nursery is made on the muddy edge of a tank, which is kept well moistened, and planted with the top* shoots of the previous year's canes. When the cuttings begin to throw out shoots, they are taken up and put in another bed prepared with earth and rich manure, generally in cultivator's homestead land. Here they are carefully screened from the sun and watered morning and evening. In *Baisákh* or *Jaishtha* (April-June), the plants are put down in the field in furrows two feet apart, and at a distance of four or five feet in the furrows. They must be well watered and earthed up. Ten or twelve days afterwards, the earth between the rows should be dug up and heaped into ridges, channels for irrigation being cut across. This operation must be repeated twice, and the field hoed free from weeds. Towards the end of *Srában* (August), the trash is stripped off, and two or three plants tied together; the little bunch is called a *merá*. The stripping of trash must be repeated twice. By *Aswin* (September-October), the ridges should all be broken down and the soil levelled, trenches being cut three yards apart each way to allow the rain to run off. The beds thus formed, each three yards square, are called *gai*. In *Kártik* (October-November), the plant should be protected against storms by tying the heads of three or four *merás* or bunches together, and thus enabling them to resist the force of the wind. Monthly irrigation is now necessary. From *Phálgun* to *Chaitra* (February —April), the cane comes to maturity. The yield of one *bíghá* is about eight *palás*, a *pala* being the day's yield of one sugar mill or *sál*. The morning after the canes have been crushed, the juice is boiled into *gúr* or molasses. The refuse cane strips are used as fuel† to boil the *gúr*, and the ashes make a good manure. The use of the mill is charged at R2 or 4*s*. a day. The sugar-cane grown in Barwán *thánd* is of the variety known as *kájali*. It has a dark purple stem when stripped of trash, and grows about seven feet high, with a circumference of about three and a half inches. Sugar-cane is by comparison a capitalist's cultivation. The expense of tillage is retunred as follows:—Rent R4-8-0 a *bíghá*, or £1 7*s*. *od*. an acre; cost of cuttings, R5 a *bíghá*, or £1 10*s*. *od*. an acre; cultivation charges, such as labour, manure, irrigation, etc., R28-13 0 per *bíghá*, or £8 13*s*. *od*. an acre. A fair outturn is calculated to be 32 local maunds, equal to 23½ standard maunds, per *bíghá*, valued at R64, or about 55⅓ cwt. per acre, valued at £19 4*s*. *od*. The net profit, therefore, is about R25-11-0 per *bíghá* or £7 14*s*. *od*. an acre." (*Statistical Account of Bengal, Vol. IV., 353*).

Bogra. 215

BOGRA.—"The land selected for the cultivation of the sugar-cane is always raised above the level of inundation, either by nature or by excavating ditches all round it and using the excavated earth for the purpose. After lying fallow

* It is generally said that the Natives of India are wasteful and use the rich sugar-yielding stems, instead of the top shoots.

† It is often said the Natives of India throw this away, and do not use it as fuel. *Conf. with p. 217.*

CULTIVATION in Bengal.

Bogra.

for one or two years, the same sites are generally selected again. The ditches are re-dug, and the sediment taken from them used for manure. The cane is planted in straight furrows, having been cut into small pieces a few inches in length, which are placed obliquely in the ground five or six inches apart. It is planted in April and grows rapidly during the rains attaining the height of eight or ten feet by January, and is cut in February and March. The juice is extracted in a circular mill of tamarind wood, made by the village carpenter, which works on the principle of a pestle and mortar. One mill is often employed by several different parties, who may have cultivated the cane in the same or adjoining villages, and who share the expenses and assist with men and bullocks in the operation of pressing the cane and boiling the juice, in proportion to the quantity of cane grown by each party. The cost of a mill complete, including sheds for cutting up the cane and boiling the juice in earthen pots, the hire of an iron boiler and the rent paid for the jungle land that supplies the fuel required, called *jálkat*, amount to a total of from R25 to R30 or from £2-10*s.* to £3. The cultivation of the cane and the manufacture of the *gur* are regarded by the *rayats* as a profitable speculation.

"In 1846 Mr. Yule, the Collector, made the first attempt on record to estimate the extent of sugar-cane cultivation and the amount of the outturn. He adds, however, that his estimate is merely approximate and founded on 'data so vague that the statement cannot be considered by any means a guide to the capabilities of the district.' His return gives an estimated cultivation of 12,000 standard *bighás* of 120 feet square and an estimated produce in mans of *gúr*, of 80lb each, or 180,000 mans. He further calculated the entire yearly consumption of the district whether in sugar, *gúr*, or raw cane itself, the whole being reduced to their equivalents in *gúr* and 4lb being considered the average allowance for each individual of the population, at 45,000 mans." (*Statistical Account of Bengal, VIII., 215-219*).

Champarun. 216

CHAMPARUN.—"The cultivation of this plant is supposed to have been introduced into the district by immigrant rayats from Azimgarh and Gorakhpur about the year 1805. It is principally cultivated in the west and north-west of the Bettia sub-division, more especially in *parganas* Manpur, Batsara, and Patjarwa. The soil, though not liable to inundation, should be retentive of moisture. In order to obtain a good crop of sugar-cane very high cultivation is necessary. In the case of cultivated land, the field is ploughed altogether about sixteen times—four times in September, three times in October, twice in November, twice in December, and four or five times in January. Where fallow land is cultivated, the field must be ploughed five times a month during eight months from June to January. The soil is manured with cow-dung in November, just before the cuttings are planted. Sugar-cane is not irrigated, as the soil in which it is sown is generally moist; nor do the rayats tie the tops of the plants together to prevent them from being blown down, as in Bengal. From the time the cuttings strike till the *adha nischatra*, or June rainfall, five hoeings are necessary. The crop ripens from January to March, when it is cut with the *kodáli* or hoe. The roots are almost invariably dug. And a second crop or *khunti* (ratoon crop) is very rarely taken. The ground lies fallow till the next crop is planted.

"The following is an estimate of the cost of cultivating one local *bighá* or $1\frac{11}{16}$ acres:—Ploughing, R3 (6*s.*); manuring, R2 (4*s.*); planting, R2-8 (5*s.*); cane for cutting, R6 (12*s.*); hoeing, R7 (14*s.*); cutting, R3-8 (7*s.*); total, R24 or £2-8*s.*; to which must be added rent, at R4 or 8*s.* per *bighá*. The cost of conveying the cane to the mill has not been included, as the labourers obtain the green leaves of the plant instead of a money wage; but in the few cases, where this is not given, the carriage may be estimated to cost R1-8 (3*s.*) per *bighá*; so that the total expenses amount to R29-8 or £2-19*s.* per *bighá*. The rent is paid in four *kists* or instalments in the months of *Kartic* or October, *Magh* or January, *Phalgun* or February, and *Baisakh* or April. After the sugar-cane has been cut, it is taken to the pressing mill. This consists of (1) the *kolhu* or mortar of *kúsum* wood, in which the plant is pressed; (2) the *mohan* or pestle, which revolves inside the mill, (3) the *kathari* on which the driver sits; (4) the *parsá*, which joins the *mohan* and *kathari*. A bullock is yoked to the *kathari* and is driven round; the cane is crushed between the *mohan* and *kolhu*, and the juice escapes by a small hole at the bottom of the latter. The following men are employed during pressing:—two *náharwás* to clean the roots of the plant; one *gainri katwa* to cut the cane in pieces; one *murwá*, who feeds the mill; one *kathari hánkwá*, who sits on the *kathari* and drives the bullock; one carpenter for petty repairs, one *chulhá jhoankwá*, who prepares the *ráb* or *gúr*; two men to relieve the *murwa* and *kathari hankwa* and one for miscellaneous duties. Of these, the *chulha jhoankwa* is paid in *gur* according to the custom of the village, receiving in some places $\frac{1}{24}$th of the total outturn. The others are usually paid a

CULTIVATION in Bengal.

money wage, the total expenditure in crushing the produce of one *bighá* being not more than R25 or £2-10*s*. The proportion of juice to cane is as 1 to 3."

Dacca. 217 DACCA.—*Sugar-cane.*—" The cultivation of this important crop is of a very limited nature here, and is confined to a few localities. The cause of this is two-fold—*firstly*, there is very little land suitable for the cultivation of sugar-cane in the most populous part of the district; and, *secondly*, the part of the district which could be turned into the greatest sugar-cane growing tract in Bengal is a jungle and scarcely inhabited.

" Sugar-cane is grown here on the following classes of land and soil :—

1st—On the outskirts of the Madhupur jungle, namely, near the towns of Dacca and Mirpur and on the banks of the Lakhya. The soil here is red clay, mixed with more or less river silt and vegetable mould.

2nd—On the high banks of the Brahmaputra and the Meghna, containing soils more or less sandy.

3rd—In Rampal, where the soil is a fine loam.

4th—On the newly-formed alluvial land which is more or less flooded during the rains.

5th—On the artificially raised alluvial soil on the banks of the Dulai creek. There is not much sugar-cane land here, but what is grown is perhaps the best sugar-cane produced in any part of Bengal.

" *Tillage.*—The mode of preparing the land is different in different parts of the district, and the variation is due to the difference in the nature of the soil. *1st*—On the red clay the plough is very little used. The market gardeners near Dacca prepare the land with the sole aid of the hoe, and sometimes even the use of the hoe is as much as possible economised. The land here is covered with jungle, being in fact the southern extremity of the Madhupur jungle. When a plot of land is for the first time to be broken, the thorny plants and other bushes and long grasses are cut at the beginning of winter and allowed to dry for a month or so. They are then set fire to and burnt. The whole ground is then carefully hoed. The unburnt roots are gathered together and are either burnt again or used as fuel. A second hoeing is given about a month after, and a sort of rough tilth is obtained this time. The land is left in this condition till the beginning of the rains, when holes are dug all over the field exactly 36 inches apart each way. A few days after these holes are partially filled with well-rotten cow-dung bought of the *goalas* (milkmen) and *kolhus* (oilmen) of the town. The cow-dung is the only manure used here.

" While this preparation of the land is going on, the cuttings have to be obtained and seedlings raised from them. The market gardeners here obtain their cuttings from the entire plant, throwing away only about two feet or so from the lower end. The plant is cut into pieces about six or seven inches long, each containing two joints. These cuttings are then horizontally laid on a plot of ground well prepared for the purpose, and chosen near a tank or well. The cuttings are only half buried in the earth, and so arranged that the buds may be placed laterally. The ground is kept wet by wartering from time to time from the tank or well. The shoots soon come out and when they are about a foot high they are fit for transplantation. In places infested with white-ants the cuttings are laid on a bamboo platform covered with about two inches of earth. Near Dacca the plants that are unripe and thrown down by the storm and are rather thin-looking, are considered the best to obtain cuttings from. Advantage is now taken of a heavy shower of rain to transplant the seedlings. In each hole are placed two cuttings prepared as above, and the whole are then partially filled up by the earth formerly raised in making them. When sugar-cane is grown on fields already under cultivation, the time of transplanting the seedlings extend from Kartik to Jeith, but early planting necessitates the expenditure of much labour and money in watering the fields artificially.

"*After-treatment.*—All through the rains the land is kept clean by weeding it with the *pashuni*, and it is a noteworthy fact that the market gardeners here make very little use of the hoe after the seedlings have been transplanted. The old leaves are regularly stripped off, the cultivators here being under the impression that unless this is done the plants do not increase in length.

2nd—The land on both sides of the Dulai creek was artificially raised, while the creek was excavated. The soil is a fine sandy loam, rich in organic matter, and very retentive of moisture. The variety of cane grown here is the white Bombay, and it is sold entirely as sweetmeats in almost all the important hâts of the district from Teota to Narsingdee and from Kaoraid to Lahajang.

" *Tillage.*—Ploughing is commenced early in Aswin, each ploughing being followed by the rolling of the field twice with the ladder. Altogether about seven to eight ploughings and twice as many rollings are given. The ground is finely pulverized and stirred to a great depth and well cleaned. The land is this way got ready by

CULTIVATION in Bengal.

Dacca.

the middle of Kartik, when parallel lines are drawn all over the field, 36 inches apart, and along these lines holes are made at an interval of about 20 inches. The raising of the seedlings in the nursery and their transplantation are done exactly in the way described above.

"*After-treatment.*—When the seedlings have taken root in the soil, the field is weeded, and the soil near the plants loosened with the sickle. This loosening and ærating of the soil is absolutely needed, for otherwise the plants cease to grow. A few days after, the whole field is hoed, and the holes are again partially filled. About a fortnight after a second hoeing is given, and this time some well-rotten dung is placed around each clump and covered with earth, thus entirely filling up the holes. Another weeding and hoeing follow, and in the course of this latter operation the plants are earthed up. About the middle of Baisakh the canes begin to form *joints*, when the leaves are stripped off for the first time. Throughout the rains the field is kept clean and the soil loose by weeding and hoeing at intervals, and the canes are also stripped of the old leaves regularly. To prevent high winds and storms from throwing away the canes, all the plants in the same clump are tied together by means of old leaves.

"The plants become well ripe at the end of the rains, but the sale of the canes to the *beparis* begins early in Bhadra After the crop has been harvested, it is customary here to cover the fields with virgin earth taken from the bottom of the creek below and the shoots allowed to grow. This earth is the principal manure used here, and sometimes so much as R20 in this way spent per bíghá A ratooned crop thus obtained is generally heavier than the first crop. The operation is repeated at the end of the second year's harvest, and a second ratooned crop is taken, which is almost equal to the first crop. After this the field is ploughed and sown with *aus* paddy or *muskalai* The paddy or the *kalia* is followed by sugar-cane, which is again kept up for three years.

Ratooning. *Conf. with pp. 59, 76, 77, 78, 128, 177, 195, 215, 226, 247.*

218 *Conf. with Mr. Clarke's opinion on value of small holdings, pp. 124-25.*

"The profit derived from the cultivation of sugar-cane in this locality is generally very great, and sometimes simply enormous. In one instance a man having about 4 bíghás* of sugar-cane obtained the first year R350, the second year R400, and the third year R300, *i.e.*, R1,050 in all, while his expenditure for the three years did not amount to R500.

"Near Dacca cane seedlings raised in the way described above are sometimes sold by the cultivators, and bought by persons who are taking to sugar-cane cultivation for the first time. The usual price is about R5 per thousand.

"*3rd—Sugar-cane cultivation on the newly-formed alluvial soil of the district.*—The two varieties generally cultivated on such soils are *dhalsundar* and *khagri*, and those are partly sold as sweetmeats and partly made into *gur*.

"*Tillage., etc.*—Fallow land is generally selected for this crop. As soon as the rain-water has receded, the ground is deeply hoed by the *kodali*, and then ploughed and harrowed several times. In general, five to six ploughings and as many harrowings are given. Mustard seed is then sown at the end of Kartik at the rate of two seers per bíghá. As soon as the mustard is off the ground in Falgoon, the field is ploughed once or twice, and in Chait pond mud is spread over it as a manure. Not more than R4 to R5 is spent on manuring, lest if more mud be applied the cane may grow too luxuriantly and be blown away by storms. Three or four ploughings and harrowings more are then given, and the land is got ready early in Baisakh, and sometimes even at the end of Chait.

"*Seedlings*—Are raised in the way described above, but instead of using the whole canes, the tops only are used by the cultivators of these alluvial lands. The planting does not commence till the rains have set in earnest, which generally happens at the end of the first week of Jeith. Parallel furrows are now made all over the field, about 27 inches apart, the plough being drawn either by men or by a pair of steady bullocks. Cuttings, or rather the seedlings, are then planted in these furrows, about a foot apart. The after-operation consists in the hoeing and weeding the ground as often as the weeds appear, and the soil gets hardened by the rain. At the second hoeing the plants are slightly earthed up. The old leaves are neither stripped off nor wrapped round the plants.

"*4th—On the high banks of the Brahmaputra and Meghna.*—The varieties of cane under cultivation here are the *merkuli*, *kali*, and *sharang* of Dhalbazar.

"*Tillage.*—The tillage operation is the same as that described under *3rd*.

"*Cuttings.*—The tops only are used here, and in a year of sufficient rainfall the cuttings are planted without any previous treatment whatever, otherwise they are prepared as follows :— *a*) The cuttings of the *sharang* variety undergo the operation locally known as *baddi*. The *baddi* is the same as the putting in the *Hanpur* of West Bengal The tops are stripped off the leaves, cut into pieces, each containing a

* A bighá = ⅓rd acre.

CULTIVATION in Bengal.

couple of joints, and rubbed with some pond mud. They are then put in a hole and covered with straw or leaves, and water is daily applied to the cuttings to keep them slightly wet. For the tops of *merkuli* and *kali* a hole is dug close to a pond or creek, and the bottom of the hole is made into mud. The cane tops are driven into this mud in a compact mass for about 2 or 3 inches and kept in a standing position. Water is applied from time to time to keep the tops cool. Neither the cuttings nor the tops are planted till two or three heavy showers have fallen.

"*Planting.*—Paralled trenches are made by the *kodali*, 27 inches apart, and in these trenches cuttings are planted at intervals of about 18 inches.

Conf. with pp. 126, 128, 140, 142, 216, 225.

"*Manure.*—Cow-dung and mustard cake are generally used. The former is applied while the land is being prepared, and the latter before the plants are earthed up. These are always used in very small quantities. The after treatment is the same as that described under *3rd*.

"In some places on the banks of the Brahmaputra, when it is intended to grow sugar-cane on land that is too sandy for this crop, or that has been exhausted by repeated cropping, the fertility of the land is first restored by laying it down in a kind of long grass called *ulu*. How this is done will be described later on under the *ulu* crop. The *ulu* is harvested in Aghan, and immediately after the field is manured with dung and ploughed. The land is got ready by Falgoon, and suggar-cane cuttings are planted in Baisakh, and sometimes even in *Chait*" (*Sen, Rept. on Dacca District, 33-36*).

Faridpur. 219

FARIDPUR.—"Four kinds of sugar-cane are cultivated in Faridpur, *viz.*, *kájla*, *dhal sundar*, *khaila*, and *chuniá*. The two first varieties are sown on high, and the two last on low lying land; but all are planted and cut at the same season, being sown in January or February, and cut in February or March of the following year. Sugar-cane is largely cultivated in Faridpur, and its produce forms an important article of district trade. No attempt seems to have as yet been made to introduce any of the superior varieties of cane, and it is doubtful whether any innovation, either as regards the old plant, or the present modes of culture, would prove acceptable to the cultivators. The only manure used in the cultivation of sugar-cane is cow-dung. This is spread over the field during the rains preceding the cultivation, after which the land is allowed to remain fallow till sowing time. The crop is never grown on the same field for two successive years, and requires careful ploughing and pulverization of the soil before the young shoots are put into the ground. During the growth of the plant, the only care required is to tie or roll up, from time to time, the growing stalks with their own leaves, and sometimes to pick out a species of larvæ, which drills into the young cane, and does great damage if not timely guarded against."

Conf. with p. 138.

"The Date-Palm or *Khejur* tree (**Phœnix sylvestris**) is very largely cultivated in Faridpur, and the sugar produced from the juice of the tree forms the most important article of export from the district The trees are generally planted along the raised boundaries of fields, and throughout the village sites, about 8 or 9 feet apart and as a rule, are allowed to grow on the spot where they are first sown. But if the ground be low and subject to inundation for any length of time, the seedlings are first propagated in a nursery. They are transplanted from the nursery during the months of May and June, or soon after the commencement of the rains, a certain degree of moisture being absolutely necessary to ensure their flourishing in the new site chosen for them. The Natives generally prefer a deep, rich clay soil for date cultivation, if possible well above inundation limits" (*Statistical Account of Bengal, Vol. V., 308*).

An account of the Faridpur process of extracting the juice from the trees and its manufacture into sugar as also that of sugar-cane will be found below under the section of the article headed "manufactures," p. 267.

Hugli. 220

HUGLI.—"The cultivation of sugar-cane requires great care, and its production has been brought to a high state of perfection in Hugli district. The land at first receives several ploughings, and is afterwards plentifully manured with cow-dung and oil-cake. Cane cuttings are in the meanwhile nursed in a moist spot of ground near the homestead of the cultivator. After the cuttings have struck, they are transplanted in the months of April or May into the field specially prepared for their reception, which requires continual irrigation. As the plants grow, the leaves are folded round the cane, for the purpose of keeping off the attacks of insects The cane ripens and is cut in the months of January or February. It comprises three principal varieties,—1 Bombay,* 2 *sámsará*, and 3 *purá*. After the cane has been cut, the stumps left in the field throw out new shoots, and no new planting is required for two more years.

* *Conf.* with the remarks regarding the ruin that overtook the Otaheite (p. 48), also the Red Bombay (p. 75), and other canes in the chapter devoted to the varieties and races, as also that on the improvement of the Cultivated Canes and the Diseases of the cane.

In some fields, however, new cuttings are planted every year. Jackals and wild pigs occasionally do considerable damage to the sugar-cane crop" (*Statistical Account of Bengal III., 338*).

CULTIVATION in Bengal.

Lohardaga. 221

LOHARDAGA.—"*Extent of cultivation.*—The cultivation of sugar-cane is unknown on the central tabe-land of Chutiá Nágpur Proper, and is confined to the Five Parganás adjoining the district of Mánbhum. It is of recent growth, and said to be extending every year. The quantity of sugar-cane grown in the Five Parganás is as yet too small to make the crop sufficiently important. It is at present grown only in those villages in which exceptional facilities for irrigation exist."

Conf. with pp. 121-127.

"*Diseases.*—The only two pests to which sugar-cane is subject are—(1) *Nalipoká* (**Dæatræa saccharalis**), a species of caterpillar, eating into the soft growing part of the cane in early life. It is known as *dholsunrá* in Burdwan. (2) *Diyá* or white-ants; these cause considerable damage to the crop by eating away the cuttings and also infesting the young stems and eating into their pith. They are shaken off the canes when discovered.

Rotation. 222

Conf. with pp. 150, 170, 187, 215.

"*Rotation.*—In the Five Parganás, sugar-cane is usually grown for three successive seasons on the same land. After three years, some inferior crops like *gorá, sarguzá*, etc., are taken for a year with a view to enable the soil to recoup its lost fertility. Occasionally sugar-cane is grown every other year, alternating with wheat, tobacco, upland or lowland paddy, according to the position of the land, etc.

"*Soil.*—Sugar-cane is grown on any land which is within easy access of water for irrigation. It is usually grown on *bári* lands provided with wells, and on the banks of *bánds* and rivers. Alluvial soils on river banks are preferred to all the rest. Such soils are usually loamy in character and yield a heavy growth of canes. Besides the convenience of irrigation is a strong recommendation for choosing these soils for growing sugar-cane. *Nágrá* (*chite*) or heavy clay soils are also sought for sugar-cane; the *gur* made from canes grown on clay soils is said to be whiter and contains a larger proportion of crystallized sugar than that from any other cane. *Nágra* soils are, however, rarely found on uplands; the quantity of such soils available for sugar-cane is therefore very limited.

Conf. with pp. 126, 128, 140, 142, 144, 216, 225.

"*Manuring.*—Cow-dung, ashes, and mud from old tanks are the manures used for sugar-cane land. Raw cow-dung is avoided, as it encourages the attacks of white-ants, which occasionally do considerable damage to young shoots. Alluvial or *pánkuá* soil is seldom manured, being too rich to require artificial help.

"*Rent of sugar-cane land* —As a rule, sugar-cane land, although forming part of the ráyat's holding, is separately paid for at a much higher rate than is paid for upland. The rate of rent varies from 4 annas to R1 per *káhan* of cuttings planted. One *káhan* of cuttings is calculated to occupy about one-third of a *kát*, that is, about two-thirds of a bigha.

"*Cultivation.*—The land is first ploughed up in Mágh. If there has been no rain, it is irrigated before being broken up. Before the time of planting the cuttings, the land is ploughed five or six times in all, the larger clods broken by the *dhelphurá* and the soil harrowed and levelled by the *mher* or harrow. Before the last ploughing, cow-dung and ashes are applied to the field; these get mixed up with the soil by the ploughing which follows. Pond-mud, if used, is spread over the land before it is broken up in Mágh. When the ground is levelled and reduced to dust, it is dressed into ridges and furrows about 10 inches apart from one another. The cuttings are then laid down lengthwise along the hollows at intervals of about nine inches from end to end, and then lightly covered over with loose soil. A watering is given after the planting has been completed on the same day. The planting season extends all through the months of Fálgun and Cheyt; the earlier it is done, the better for the crop. Irrigation is repeated every three or four days in the beginning; the interval gradually increases up to seven to ten days according to the dryness of the weather.

"When the plants have become about a foot high, the ridges are hoed up with the *khrupi*, and the furrows slightly filled in. In Asár, after the rains have set in the land is levelled up with a view to facilitate the drainage. In this respect the practice is just the opposite of what is followed in the Bengal districts, where the land is drawn up into ridges and furrows at the approach of the rains, in order to prevent the stagnation of water in any part of the field. This difference of practice arises from the fact that Chutiá Nágpur Proper being an undulating country, the drainage is perfect at all times: in fact ridges, if allowed to remain, may stand in the way of free egress of water from the fields; while in the perfectly flat country of Bengal the only means of letting the surface water escape from a field lies in the running up of parallel water furrows across its face. During *Srávan* and *Bhádra*, one or two more hoeings are given to the field. About the end of *Bhádra*, when the plants have become six to eight feet high, short bamboo posts are stuck up at suitable intervals in the field. To these, four or more canes are tied up with the leaves with a view to prevent their

CULTIVATION in Bengal.

Lohardaga.

being blown down by high winds. Before so tying up the canes, they are wrapped up with their own leaves—a practice which is not so much insisted upon in the Five Parganás as is done in Burdwan, Hooghly, and many other districts of Bengal.

"*Harvesting.*—The sugar-cane harvest begins as early as the last half of *Pous*, and lasts up to the beginning of *Cheyt*. The canes, when ripe are simply cut at the base with the spade; the leaves stripped of and taken to the furnace for use as fuel. The tops of the canes are cut off and kept appart to supply the cuttings wanted for the next year's crop" (*Basu, Rept. on Lohardaga Dist.*, *79, 80*).

Palamau.

223 PALAMAU.—"*Extent of cultivation.*—Sugar-cane is cultivated to a small extent in the alluvial plains which intervene among the hills in the north and centre of the Palámau sub-division. The southern limit of its cultivation lies about 12 miles south of Daltonganj in tuppeh Bári. In the country about Hariharganj, in the extreme north-east corner of the sub-division, it is grown in most villages."

"*Rotation.*—The usual rotation of sugar-cane lands is the following:—

1st year . . .	Sugar-cane.
2nd " . . .	A *bhadoi* crop only and cold weather fallow.
3rd " . . .	Fallow.

"The rotation is one of three years, the third year's fallow being again followed by sugar-cane in the fourth year. The *bhadoi* crops taken in the second year may be *sáwán, máruá, kodo or gondli*; if the land is sufficiently low and moist, the *sáti* variety of early paddy may be taken in place of the millets. Commonly *sáwán* and *rahar* are made to follow sugar-cane, as the stubble of the former (*sáwán*) is believed to enrich the soil for the benefit of the succeeding crop.

"*Soil.*—In Palámau, sugar-cane is preferably grown on *pá-ur, i.e.*, light loamy soils. It will do just as well, and even better, on clay soils; but the produce is said to be of inferior quality. In the Five Parganás, on the other hand, clay soils are held best for sugar-cane, both in respect of the yield and colour of the *gur*. Among the five varieties of cane noticed above, the *newár* thrives best on clay soils.

Conf. with pp. 126, 128, 140, 142, 144, 145, 216, 225.

"*Manuring.*—Ashes produced by burning cakes of cow-dung, and every other description of wood ashes, are used for manuring sugar-cane. Sheep-dung obtained by folding a flock in the field for a night or two is also frequently used. Unburnt cow-dung is never used, as it is believed to stimulate the growth of the canes, which thus become liable to fall down; cow-dung also produces a larger quantity of cane-juice, which is, however, much less sweet in consequence. The objection to the use of raw cow-dung on the score of its encouraging the attacks of white-ants is not raised in Palámau. The manures are applied to the field some time before the planting takes place, and get mixed up with the soil by the ploughings which follow.

"*Cultivation.*—The ploughings commence in Asár, and are continued at intervals till Kártic. The frequency of ploughings during the rainy season depends a great deal on the leisure of the cultivator, whose hands remain more or less full of various other work during this time. During the three following months of *Aughrán, Pous,* and Mágh, the field is repeatedly ploughed and cross-ploughed, as many as 15 ploughings being often given during this period. Towards the end, frequent harrowings are also given in order to pulverize the soil and produce a fine tilth on the surface. If the weather before the planting takes place has been particularly dry, and in consequence large clods have been formed in the field, the latter is irrigated, harrowed, levelled, and pressed by the *hengá*.

"When the field has been thus prepared and levelled, the planting is done in the following manner. One plough goes in front of a second in the same furrow, which is thus made deeper. A man follows the ploughs with cane cuttings which he lays flat over the bed of the furrows, allowing a span breadth of ground between every two cuttings. He is immediately followed by a third plough, which makes a furrow a little on one side, and covers up the cuttings in the preceding furrow with a layer of soil about six inches deep. For the convenience of planting, the field is divided into several parcels, which are planted one after another. The three ploughs go round and round the parcel of land, the space allowed between two contiguous lines of cuttings being about 9 inches. After the entire field or a defined portion of it has been planted out it is smoothed and pressed by the *hengá* or *chowk* passing over it.

"Twelve thousand cuttings (*ponhrás*) are estimated to be required for planting one local bigha (roughly ⅝ths of an acre) of land. The tops of the canes of the preceding crop are cut into convenient lengths, each piece retaining 2 or 3, joints. For seed-cuttings of the shorter varieties of cane like the *manigo*, the entire canes are cut into pieces for the purpose, and the tops left off.

"On the fourth day after the cuttings have been planted, the upper two inches of soil are loosened by the *pháurá* or spade, and the loosened soil levelled and pressed by the *chowk*. Both these operations are repeated a second time on the 12th day.

CULTIVATION in Bengal.

Palamau.

"In three weeks time the shoots appear above the ground. The first watering is given about a week later on. Seven men working for 12 hours are required to irrigate 1 bigha of sugar-cane in a day. Four of these men work at 4 *láthás,* and are relieved by turns by two others; the seventh man guides the course of the water in the field. For the purpose of irrigation, the field is mapped out into small squares which are enclosed by low ridges; these squares are called *gánreris.* After one square has been flooded, the water is led into another, and so on, till the entire field is irrigated. Three or four days after the irrigation, when the surface soil has become sufficiently dry, it is loosened by a small spade or hoe called the *phâuri,* and then levelled by the feet. The watering followed duly by the hoeing and levelling is repeated once and sometimes twice at intervals of a month. Three waterings are usually required—the first in *Cheyt,* the second in *Bysák,* and the third in *Jeyt,*—but the number of waterings may be diminished by a seasonable fall of rain during these months. At the time of hoeing the crop in *Bysak* and *Jeyt,* the roots of the canes are earthed up and thus encouraged to tiller.

"During the rainy season, the intervals between the plants have to be dug over twice—once in *Asár* and a second time in *Srávan*—in order to loosen the soil and to remove all grasses and weeds that may have sprung up with the advent of the rains.

"Previous to planting, the sugarcane field is enclosed on all sides by trenches; and these latter are planted over with the thorny branches of *baer,* which serve as a rough sort of fence against pigs bears, and jackals.

"*Harvest.*—The harvest of sugar-cane commences in the last week of *Pous,* and is continued to the second week of *Cheyt.* The canes are cut down by the spade, and are then stripped of their leaves. The tops are also cut off to furnish cuttings for the next crop of suga--cane.

"The cultivation of sugar-cane by hired labour does not pay in Palámau, as indeed it would hardly do in other parts of Bengal. It involves a heavy strain upon the cultivator, and unless he has a sufficient number of hands in his own family, he never thinks of undertaking its cultivation. The usual plan is for several ráyats to combine and help each other by turns in cultivating parcels of land all lying close to each other. In this way hired labour can be mostly, if not entirely, dispensed with.

"The following is the cost of cultivating one local bigha of sugar-cane. The manures and the cane cuttings have not been charged for, as these are seldom bought. The wages of labour and hire of plough have been taken at higher figures than the average, first, because the rates of wages in the north of the sub-division where sugar-cane is grown are higher than the average for the sub-division, and secondly, because higher wages are always paid for all laborious work like irrigation.

Cost of cultivation.

	R	a.	p.
Twenty-five ploughings with harrowings at R0-5 per ploughing	7	13	0
Planting (three ploughs at R0-5 per diem and 6 men, *viz.*, 2 to lay cuttings, 4 to supply, at R0-1-9 each)	1	9	6
Four waterings (one watering takes seven men four days ∵ 7×4×4=112 men at R0-1-9)	12	4	0
Three hoeings following irrigation (one hoeing takes three men seven days ∴ 3×7×3=63 men at R0-1-9)	6	14	3
Two hoeings in the rainy season (three men for two days at R0-1-9) for each hoeing	1	5	0
Trenching (six men at R0-1-9)	0	10	6
Thorns (20 loads at 2 annas)	2	8	0
Rent of one local bigha	6	0	0
Total cost of cultivation	39	0	3

"*Remarks.*—The cultivation of sugar-cane, as it is now carried on in Palámau, is very negligent. The cuttings are planted, or rather sown at random and are covered with a very light covering of soil. During the rainy season it is found very difficult to hoe the field: the growth of leaves becomes so thick and close that the interior of the field may be said to become proof against air and light. Cane-fields are much better managed in the central districts of Bengal, where *samsera* and other larger kinds of cane are grown. In these districts the cane-fields are beautifully laid out in lines and furrows, and the canes are carefully wrapped up in leaves during the rainy season, thus admitting of plentiful air and light. The canes are, besides, plentifully manured with oil-cakes and hoed and earthed up at frequent intervals during the rainy

CULTIVATION in Bengal.

season. It is no wonder therefore that, while the Palámau ráyat considers himself happy to get 25 maunds of *gúr* per acre, the ráyats of Hooghly and Burdwan would not consider the cultivation paying under 60 maunds from the same area" (*Basu, Rept. on Lohardaga Dist., Palámau Sub-Division, 37-39*).

Rangpur. 224

RANGPUR.—"Sugar-cane requires a light dry soil. The crop is cultivated throughout Rangpur district, except in the eastern tracts. It is planted in February and March, and cut in the following January and February, being in the ground a period of about eleven months. The land requires eight or ten ploughings and as many harrowings and drillings. The seed-plants are sown on ridges or mounds of earth raised about a foot above the level of the field. Owing to the natural moisture, the crop does not require irrigation in Rangpur, as it does in other parts. When the young canes are three or four feet high, they are tied together in bunches of eight or ten to make them stand erect. The field requires careful weeding and manuring; and more care is taken of this crop than of any other. Four varieties of cane are grown, namely, *sarián, angá, handá, mukhí,* and *khári.* When the canes ripen they are cut into small pieces about six inches in length, and ground in a mill to express the juice. This is afterwards boiled into *gur* or molasses, which is sold and exported in its raw state without any attempt at refining. The outturn is estimated at from 9 to 10 maunds of *gur* per *bighá* or from 19$\frac{3}{4}$ to 22 cwt. per acre. The quantity of land under sugar-cane in Rangpur is estimated at 20,466 acres and the total net produce at 292,136 maunds, or 213,885 cwt. of *gur* (*Statistical Account of Bengal VII., 247*).

Santal Parganas. 225

SANTAL PARGANAS—"*Akh* or *ikshu* sugar-cane, is planted from cuttings in July and cut in February; three varieties are grown in the district, known as *bástá, kunri* and *kájali.* There is a fourth variety of sugar-cane, called *nárgari,* planted in September and cut in November and December of the following year. (*Statistical Account of Bengal, XIV., 337-338*).

Saran. 226

SARAN.—"This crop is grown on rich and high land, from cuttings which are planted about the month of March. The ground is thoroughly manured; and the cuttings are then inserted about eighteen inches apart. When they have struck, the field is irrigated about seven or eight times; the number of irrigations depending principally on the season, but also on the soil. In some places the cultivators tie the canes together at the top, to prevent them being blown down; but this is not generally the case. The plants are ready for cutting in the following February. Sugar-cane is considered to be a highly remunerative crop. The produce of a bighá of sugar-cane land is seldom worth less than R30 or more than R80" (*Statistical Account of Bengal, XI., 282*).

Tirhoot. 227

TIRHOOT.—"Sugar-cane (*akh*) is grown on first class high land. The soil is repeatedly ploughed and dug, until it is thoroughly pulverized. Cuttings are planted in the ground, eighteen inches apart, in the month of February. Irrigation is sometimes, but not often, adopted, and as the land is always of the best quality, no manure is required. The canes are cut in December or January; but sometimes the roots are not pulled up, when a crop is taken from them in the following year. The cultivators do not tie the cane into bundles as in Bengal and the North-Western Provinces; in fact, they take as little trouble as they can, and though the cane is of an inferior quality, the crop pays well in a good season. The juice is extracted by a mill," which is identical in its construction to that which will be found below in the account of the 24-Parganahs. "The juice is collected in earthenpots, and boiled down into *gúr;* for, without it, it would ferment and turn bad. The *gur* is largely used for sweetmeats and mixed with tobacco which is intended for smoking.

"Sometimes the *gur* is refined into sugar, but this is not often the case in Tirhoot where most of the local produce is exported to Barh, Patna, and other centres of trade, where there are sugar manufactories. According to the Collector's figures for 1871 there are nearly 20,000 acres under sugar-cane in Tirhoot district, the principal place of cultivation being the Darbangah sub-division" (*Statistical Account of Bengal, Vol. XIII., 86-87*).

24-Parganas. 228

24-PARGANAS.—The following account is quoted from Major Smyth's Revenue Survey Report:—

"A rich soil is selected high enough to be above the usual water-mark of the rainy season. The field is ploughed ten or twelve times, and manured. Cuttings of the cane are planted horizontally in the ground in March, about eighteen inches apart, which sprout up in about a month. In July or August, when the plants are about three feet high, they are tied up by three or four together with their own leaves, to prevent their being blown over. If there is no rain in September or October, it is necessary to water them. The canes are cut in January and February, and the juice is extracted by a mill, then boiled and made into *gur* or molasses. The mill acts on the principle of a pestle and mortar, the pestle rubbing the canes against the edge of the

CULTIVATION in Bengal.

24-Parganas.

mortar. To the end or the pestle is attached a beam from fifteen to eighteen feet long, which acts as a lever, and to this is attached another horizontal beam to which the bullocks are yoked. These walk round and so crush the cane between the pestle and the sides of the mortar. This last generally consists of the trunk of a tamarind tree, hollowed out, at the bottom of which is a small hole communicating with the outside, through which the juice escapes, and is received into an earthenware pot. The boiling is the next process; and this is done in a very similar method to that of the date juice before explained. The Revenue Surveyor estimates the yield of sugar-cane at £12 or upwards per acre; and the costs of cultivation £5 8s. od. **Mr. Westland** calculates the yield in the adjoining district of Jessore at £7 10s. od. an acre, which is stated to me to be too low. Another and more primitive method is to crush the canes between two revolving iron rollers, which are worked by hand" (*Statistical Account of Bengal, Vol. I., 145*).

II.—ASSAM AND CACHAR.

ASSAM.

References.—*Special Report by the late Dr. Stack, Director of Land Records and Agriculture: Agricultural Department Reports; Agri.-Hort. Soc. Ind.:—Trans., II., 164-167; III., 57, 61, 99; V., 22; VI., Proc. 60; VIII., 28; Journ. IX., 247-248.*

Area, Outturn, and Consumption.

Area & Outturn. 229 *Conf. with p. 211.*

The table given at page 116 for the normal area of Assam, devoted to sugar-cane cultivation, shows 25,000 acres, yielding 4,50,000 maunds of "coarse sugar." This is accepted as exhibiting an outturn of 18·0 maunds an acre, and the consumption was estimated to come to 5·70 seers (11·40℔) per head of population. It is perhaps unnecessary to go into these points very fully, since Assam (though it possesses some very good qualities of cane) cannot be ranked among the provinces of India largely interested in sugar production. Indeed, it imports very extensively from Bengal, so that its local production is by no means able to meet the consumption. It may, however, be said that during the past three years the area shown in the annual volumes of *Agricultural Statistics of British India* has been under the estimated normal area, thus:—Area in 1887-88, 17,756 acres; in 1888-89, 19,293 acres, and in 1889-90, 19,309 acres. The distribution of the acreage in the last-mentioned year was as follows:—Sibsagar 7,283 acres, Kamrup 5,109, Nowgong 2,406, Darrang 2,263, Lakhimpur 2,210, and Goalpara 38 acres. It will thus be observed that the areas in Cachar, Sylhet, the Naga Hills, Khasia, and Jaintia Hills, and in the Garo Hills have not been provided for, so that it is likely the normal area is not a high average.

The following account of the cultivation of cane and expression of the juice was written by the late **Dr. E. Stack**, while Director of Agriculture in Assam. It is, perhaps, as well to explain, however, that one or two paragraphs of **Dr. Stack's** article have been slightly altered to suit the arrangement followed in this paper. His remarks, for example, on the varieties of cane grown in Assam and Cachar have been carried to the chapter devoted to that purpose (*pp. 62—64*):—

A.—Brahmaputra Valley.

Brahmaputra Valley. Soil & Manuring. 230

"Soil and Manuring.—A light loamy soil, with a light admixture of sand, is the most suitable for sugar-cane. The Assamese name for this kind of soil, *mobulia*, denotes at once the waxy consistence of the loam (*mo* meaning wax) and the addition of sand (*báli*). The land must be high lying (*bám*) and beyond the reach of inundations. Favourite spots are the edges of a marsh, or the banks of rivers, which in an alluvial country tend to raise themselves above the level of the plain. In Nowgong and Kamrup the sloping plain at the foot of the southern hills furnishes good sites for cane, especially in the neighbourhood of streams, and it is in such places that the Bengal cane of Kamrup is chiefly grown. Gravelly or sandy soils will not produce sugar-cane, while rich alluvial land gives a luxuriant crop, but with watery juice. The degree of manuring depends entirely upon the ryot's means and inclination. Lands in the vicinity of stations are freely manured with cow-dung and crushed mustard seed,

Conf. with pp. 126, 128, 140, 142, 144, 145, 146, 216, 225.

Methods of Cultivation

CULTIVATION in Assam.

Brahmaputra Valley.

both before and after planting; on the other hand, a field in the jungle often receives no manure except the ashes of the grass and weeds raked out of the soil and burnt. On the whole, cane lands are not nearly so well manured in Assam as in Upper India. In Goalpara it is said that the spot usually selected is the site of an old cattle shed, but this can be true only of sugar-cane cultivation on *basti* or homestead land, which forms but a small portion of the whole.

Rotation. *Conf. with pp. 129, 145, 170, 187.* 231

"ROTATION.—The best cane is grown either on virgin soil or on old fallow; but land from which a crop of mustard, pulse, or summer rice (*áhu*) has been taken is often preferred as being of less laborious tillage. The exhausting nature of the crop is expressed by the proverbial saying that *áthia* (a kind of plantain), *kathia* (rice seedlings), and *gáthia* (the knotty crop, *i. e.*, cane) destroy the productive powers of the soil. A second crop of cane, unless ratooned, is never grown in the year next following the first, and though two or three years' fallowing is considered sufficient in the vicinity of stations or large villages, where manure is abundant and cultivation more than ordinarily careful, lands in outlying parts are not considered to have regained their vigour till they have lain six or seven years under a wild growth of grass. Hence such lands are not, as a rule, retained by the cultivator, but are relinquished and retaken at pleasure, whereas the patches near his homestead are usually kept in his own hands, to prevent their usurpation by others.

Tillage. 232

"TILLAGE.—Waste or fallow land is broken up in October. A good deep hoeing is the best treatment, and if the field be *káthani* (timber land) or *murháni* (stump land), that is to say, a forest clearing now for the first time brought under cultivation, this method is the only one possible; but it is not absolutely necessary in the case of a field reclaimed from reed jungle, while fallow land (*kiwári*) can usually be brought under the plough at once. Having thus been turned up, more or less thoroughly with the hoe or the plough, the land is then left till January or February, when the ryot, having gathered his crop of winter rice, is at liberty to recommence operations; at this time also previously cropped land (*jaháli*) is taken in hand, and ploughing, varied by harrowing, goes on with more or less diligence and frequency until the middle of April. The soil has now been thoroughly worked up, the weeds and grass raked out and burnt, and the clods which have escaped the harrow (*moyé*) are broken with the mallet (*dalimári*). The duration and number of these operations vary greatly according to the ryot's inducement or inclination towards careful tillage. The popular estimate of twenty ploughings at least is rather ideal than actual; but the ryot understands perfectly well that the value of his crop depends in a great measure upon the depth and thoroughness of tillage preparatory to planting. Then follows the partitioning of the field (*khandoá*) into strips of eight to twelve feet in width (*khand*), separated by drains communicating with the ditch (*kháwai*) which surrounds the field on the outside, and which is dug almost waist-deep. The field is now ready for planting as soon as the first showers fall.

Planting. 233

"PLANTING.—The layers from which sugar-cane is propagated in the Assam Valley consist invariably of the topmost joints, and are hence called *ág* (tops); they are sliced off pretty much at random, but are supposed to measure the length of the forearm with the fist closed, and usually comprise three or four joints. During the interval of two or three months, between cane-harvest and planting, the layers are kept in a cool and moist spot in the ryot's homestead, placed in a half-upright position in ground which has been turned up by the hoe, covered with rice-straw or plantain leaves to protect them from the sun, and watered occasionally if the weather be dry. When thus treated, they have already begun to throw out shoots (*gazáli*) before transplanting, but when cane-harvest has been prolonged till late in the year, the interval between the cutting and the planting of the layers is very much abridged, and a regular nursery is dispensed with, the bundles of layers being simply kept in a heap under damp straw until they are wanted; this is called *dhuliya* or 'dusty' planting. The day chosen for planting must be preceded by sufficient rain, and if drizzling rain lasts throughout the day, so much the better. It is seldom that the date fixed upon is later than the middle of May, though exceptional circumstances may cause it to be postponed till the end of that month, or even the beginning of June. The layers are placed at distances of about two feet from each other, in trenches three feet apart, which run at right angles to the drains (*khand*) dividing the field, and are thus cut up into lengths of eight to twelve feet. Thus calculated, the number of layers required to plant one bigha (120 feet × 120) would be 2,400. A carefully prepared estimate from the Nowgong district shows the number as 2,000 to 3,000. It is less in good land than in poor soil where losses from failure to germinate have to be made good. The rate at which layers are sold is liable, like everything else connected with the cultivation of sugar-cane in Assam, to great variations from year to year. The present selling price in the Darrang district is 400 to 500 the rupee, but it was

CULTIVATION in Assam.

Brahmaputra Valley.

600 the rupee a few years ago. After the layers have been planted, a little soil, often mixed with cow-dung, is lightly scattered over them, and they are left to themselves for ten days or a fortnight, until they have struck root.

Weeding. 234

"WEEDING, HOEING, AND EARTHING-UP.—The field is then weeded, and the soil around the young shoots lightly stirred with the spade (*khanti*) or hoe (*kuddli*). The latter process is one of great importance, and ought to be repeated at short intervals on sunny days throughout May and part of June, the earth being thoroughly stirred to the depth of six or eight inches, both around the shoots and also between the lines of canes. Manure also may be applied on these occasions, and one or more weedings are usually given. Later on, the earth from the ridges (*dilá*) between the trenches (*páti* or *khdli*) is heaped about the roots of the canes to strengthen their hold on the soil, and this process is repeated until the relative positions of trench and ridge are reversed, and the canes now stand on ridges with trenches between. This goes on till the middle of August, at intervals varying according to the leisure and industry of the cultivator; but the popular estimate is that the cane should be hoed once a fortnight until *Jeth* (ending on the 15th of June), and that the weeding and earthing-up should take place subsequently, at least once a month. Sunny days are always chosen, and in the earlier stages the prevalence of sunny weather is especially desired, as the earth about the young shoots cannot be stirred while it is wet without injurious effects.

Irrigation. 235

"IRRIGATION.—A prolonged break in the rains while the cane is yet young will occasionally compel the ryot to resort to irrigation to save his crop, but such cases are quite exceptional, and seem to be unknown in Upper Assam.

Tying-up. 236

"TYING-UP.—Working in the cane-field is usually at a standstill for about a month from the middle of August. During this time the juice of the cane is sweetening, and the ryot is said to feel a superstitious aversion from entering the field, lest the jackal should follow him. A final weeding and earthing-up are administered towards the end of September or the beginning of October, the canes being at the same time tied together in clusters as they grow by means of the leaves stripped off the lower part of the stalks, and bamboo props are sometimes added by way of support where the crop is exceptionally tall and valuable. The number of canes springing from a single layer may vary from three to ten, but is usually either four or five, and, where more numerous, the canes fall off in size. The person who undertakes the tying up must be completely clothed, with his hands protected by a covering of cloth, and his feet by sandals of leather or the bark of the betel-palm. It is a laborious process, and is often omitted. Indeed, the whole of the foregoing description must be taken as true only of the more careful style of cultivation practised in the immediate vicinity of villages, while in forest clearings or patches in the midst of reed jungle the cane is left pretty much to shift for itself.

Diseases. 237
Conf. with pp. 48, 76, 121-127, 161.

"DISEASES AND ENEMIES.—Nothing more has to be done now but to fence in the field securely with slips of bamboo intertwined, so as to form a continuous paling about three feet high, and strong enough to cost some trouble in pulling to pieces. Though of no avail against bears or wild elephants, this does save the crop to some extent from wild pigs, and from a still more mischievous enemy, the jackal, who, nevertheless, often contrives to find his way in and eat a large space clear in the centre of the field. The roots of the growing cane, especially if too freely manured are liable to be attacked by white-ants (*ui pok* , and in uncovering them, to rid them of this pest, injury is sometimes inflicted upon the plant. A rainy October, followed by a dry November and December, causes the top joints to wither and die. Apart from these calamities, however, sugar-cane in the Assam valley does not appear to be liable to any special disease. It does not suffer much from inundation, as the sites selected usually lie beyond the reach of any ordinary flood; while drought is a contingency that hardly comes into the cultivator's reckoning.

Ratooning. 238
Conf. with pp. 59, 76, 77 78, 128, 143, 177, 181, 195, 215, 226, 247.

"RATOONING.—A small proportion of the annual cane crop is ratooned, *i.e.*, grown from the roots of the last year's cane, instead of being propagated by layers. The stripped leaves of the previous crop are left lying on the field till April, when they are burnt, and a month later, when the young shoots begin to appear, a hoeing may be administered and some manure added. Such a crop is called *murha*, or stump cane; it receives little attention from the cultivator, ripens early, and yields only about half as much coarse sugar as an equal area of cane cultivated in the ordinary way. A peculiarity of *teliya* cane, and one of its recommendations to the Assamese ryot, is that it can be ratooned twice.

Harvesting. 239

"HARVESTING.—With the exception of *murha* cane, which is cut early in January, cane-harvest does not begin until the winter rice has been reaped and stored. The date is somewhat earlier in Goálpára and Kámrúp than in the upper districts, but generally it may be said that the festival of the *Mágh Bihú*, or harvest-home of the

CULTIVATION in Assam.

Brahmaputra Valley.

principal food-crop of the year corresponding in date more or less exactly to the 15th of January, is celebrated before the cultivator troubles himself with the labours of the cane-press. The operations of cutting, crushing, and boiling are carried on simultaneously from this date until the end of March, or even the first few days of April. The canes are cut off close to the root by a single stroke of the Assamese *dao* or bill-hook, the tops are lopped off for layers, and the stalks, stripped of their leaves, are bound in bundles (*palá*) weighing about half-a-maund, and carried to the mill. Where the crop is *pura* or good *mugi* cane, a small proportion is usually reserved for eating in the raw state, and is worth one or two pie per stalk in the village markets, while in the station bazars a single stalk is cut up into several pieces, each of which is worth a pice. Thus estimated, the value of a field of sugarcane depends on the number of stalks; and these vary greatly, according to the cultivator's ability to plant the field properly and to protect the growing crop. If we assume an average of one cane to every two square feet, the value of the canes on one rood of land, if sold for eating raw, would be about R50; but this is quite an imaginary case. The great bulk of the cane grown in the Assam Valley is destined for the mill. Unless of extraordinary length, the canes are not divided before crushing (in Goálpára, however, they are said to be cut up into lengths of 2½ feet), but are passed through entire; the average length of *teliya* cane, stripped and topped, is less than four feet, of *mugi* nearly five feet, and of Bengal *pura* about six feet. The cultivator is well aware of the importance of protecting the juice, while in the cane, from exposure to the air, and, therefore, he crushes his cane undivided, and only cuts it by parcels as he wants it for the mill, which is always set up in the immediate vicinity of the cane-field.

Sugar Mill. 240

"THE SUGAR-MILL.—This instrument (called *kherkha* in Goálpára, and *hál* in Upper Assam) is a rude but tolerably effective machine, and a quicker and less dangerous worker than the heavy beam-and-pestle arrangement of Upper India.* It consists of two vertical rollers (*bhim*) placed in juxta-position, with their lower ends resting in a flat trough (*bhorál*) scooped in solid and heavy block of wood (*toljoli*) resting on the ground, while their upper ends pass through a rectangular space cut in horizontal beam above (*borjoli*) supported by uprights (*hol khuta*) let through the lower block into the ground. The rollers are held in their places by vertical clamps (*ghará*), which grip them at the upper and lower ends, and are driven home by wedges (*khál*). The portions of the rollers which project above the upper beam (*borjoli*) are grooved so as to work into each other on the principle of an endless screw. The driving power is a horizontal beam (*kátari*), applied to the head of the taller or 'male' roller (*máta bhim*), upon which the shorter or 'female' roller (*maiki bhim*) revolves in the contrary direction. The 'male' roller is usually, if not invariably, that on the right hand as one faces the mill, and the direction of progress is from left to right, that is to say, the men at work walk round with the left shoulder inwards. Buffaloes are seldom yoked to an Assamese sugar-mill, and bullocks never. The whole machine is made entirely of wood, without a nail or a piece of iron in its composition, and its value varies according to the kind of wood used. A mill can be built of tamarind-wood for eight rupees but in *jam* wood (**Eugenia Jambolana**) it will cost twelve, and if *nahor* (**Mesua ferrea**) is used, as much as fifteen rupees.

Crushing. 241

"CRUSHING.—All being ready for crushing, the first thing the cultivator does is to bind two of the finest cane-stalks along the beam of the mill, as an offering to Viswakaráma, the god of artificers. The canes are then passed through the mill in batches (*aná* or *kaná*) of six or eight at a time, the juice falling into the trough, and thence through a hole on to a sloping wooden tray, which transmits it by a lip of plantain-leaf to the earthen vessel placed to receive it in a pit dug below. In some places the tray (*rasdhara*) is circular in shape, with a raised wooden edge and a funnel-shaped escapement for the juice, but usually a simple slab of wood, slightly concave, is considered sufficient. The working of the mill is accompanied by a loud and strident, noise, which is welcomed by the ryots as a sign that the rollers are biting well, and is, moreover, a cheerful and useful accompaniment while the work is carried on by night, as is the practice towards the end of the season, when the heat of the day would be injurious alike to the men and the cane-juice. Each handful of canes is passed through the mill three or four times, until they begin to yield mere foam, when they are thrown aside, and a fresh batch takes their place. *Mugi* and *pura* canes squirt

* **Messrs. Thomson & Mylne** claim for their Bihia mill the power of crushing thrice as much cane in a given time as can be done by the common *kolhu* of Behar or of the North-Western Provinces. Their calculations (which are supported by independent experimental evidence) make the average outturn of the *kolhu* about 100lb per hour. The Assamese mill works at least half as rapidly again.

CULTIVATION in Assam.

Brahmaputra Valley.

out their juice plentifully on the first compression, and give less afterwards, while the harder and tougher *teliya* passes through almost dry, and only begins to yield juice to the second squeeze. At the third and fourth crushings, the flattened canes are usually twisted into a rope, so as to present a bulkier body for compression. A boy sitting in front of the mill draws them out as they pass through the rollers, and hands them back to the man who sits behind and feeds the mill. Four or five men drive the machine, resting their hands on the beam, and pushing against it with the chest and shoulders. The force required to put the mill in motion was ascertained in one experiment made by Mr. R. T. Greer, sub-divisional officer of Golághát, to be 5 to 6lb without cane, and 40lb with *mugi* cane between the rollers, but 60lb with *teliya*. The rate of progress in crusing is about two maunds (165lb) per hour. A good deal of trash and impurity—earth from the imperfectly-cleaned canes, fragments of the stalk, dust carried by the wind, etc.,—enters the earthen pot along with the juice; in fact, after a couple of hours' work, mud can be plentifully scraped off the plantain-leaf lip of the tray, but the ryots seldom trouble themselves to clean it. When the pot is full, it is hanged for another. As the work proceeds, the wedges holding the clamps have usually to be driven home from time to time, to counteract the tendency of the rollers to work asunder.

Boiling. 242

"BOILING.—The juice is thrown into a boat (*náorá*) scooped out of a log. This stands at the edge of the boiling-house, a few yards removed from the mill, and sometimes contains leaves of the wild fig tree (*dimaru*), which are supposed to be useful in keeping the juice sweet. When some twelve or fifteen gallons have been collected, the boiling begins. The whole apparatus for this purpose is worth about two rupees, and consists of four earthen cauldrons (*tháli*), two ladles (*láokhola*) made of half-a-gourd attached to the end of a stick, one of which is usually perforated like a cullender, and a sieve or strainer (*jáki* or *chálani*) of plaited cane with a long handle. The furnace is excavated in the ground, and has four circular openings to receive the cauldrons; the first of these is set some three feet back from the furnace mouth, the second about as far behind the former, while the last two, which are placed side by side at much the same distance in the rear, lie almost beyond the reach of the fire, and are used merely as feeders in which the juice is heated before being transferred to the first or second cauldron for boiling. The cauldrons are invariably made of potter's clay, and in shape are almost exact hemispheres, with a diameter of eighteen to twenty-one inches; the first two, being somewhat larger and of superior quality, usually cost as much as seven or eight annas each, and must be procured from certain potteries where the clay is exceptionally good: Kokilamukh, for instance, enjoys this reputation in Upper Assam. The two feeders can be purchased for about four annas a piece. Before placing the cauldrons on the fire, their bottoms are smeared with clay tempered with cane juice, while a charm* is repeated to keep them sound and whole; in this way they can be made to last for one or two seasons, and though commonly cracked in all directions, the ryot continues to use them until the bottom falls out, when the fire is withdrawn and the spilt juice carefully scooped up from the floor of the furnace and strained through a cloth into the new cauldron, which is always held in reserve on such occasions, the whole apparatus being, at the same time, protected against the recurrence of such a malicious mischance by the sprinkling of water, over which charms have been muttered against the evil eye.

"The fuel consists of reeds (*khágari* or *ekra*), supplemented by the crushed cane-stalks (*jaban*) as the boiling proceeds. A man or boy feeds the fire, while two men mind the cauldrons, skimming the feeders with the sieve, and lifting the juice in the boilers with the ladle so as to prevent it from boiling over, while they replenish the second cauldron from the feeders, and thence transfer the heated juice to the first cauldron immediately over the fire. This latter operation is usually performed by the man who is entrusted with the duty of determining the exact point at which the juice has been boiled enough: he is always an experienced person, and must be fed well and treated with deference. Lime-water is said to be occasionally administered as the boiling goes on, but this is mentioned in the district of Nowgong alone, and is probably quite an exceptional precaution. In the latter stages of the boiling, care is

* A charm commonly used in the Nowgong district runs in this way:—"Sât patâlor máti, Anat kumâre ànile kâti; Khochi guli dilo châkat, Charu hol Brahmâr pâkot. Hari Har dâk. Phuta, phata, khola, khâpori, bâli jala, sahar kona, jora lági thâk."

That is to say, "Anat the potter cut and brought the earth of the seven worlds, kneaded and wet it, and put it on the wheel; it became a cauldron under Brahma's turning. Call on Hari (Vishnu) and Har (Siva). Breakage and cleavage, chip and potsherd, sand-leak and crevice, be joined and whole."

Methods of Cultivation

CULTIVATION in Assam.

Brahmaputra Valley.

taken by frequent interchanges of juice to keep the two boiling-cauldrons as nearly as possible at the same temperature. These stages are three in number, and are vulgarly known by the names of *o-phulia*, *babori-phulia*, and *temi-mulia* implying that the ebullient masses of liquor in the first stage are as large as the fruit of the O tree (**Dillenia indica**) that is, about three inches in diameter: in the second stage they are more frequent, and shrink to the size of the flower of the *babori* (an edible species of the *Compositæ*) in size about equal to the marigold; while in the final stage they present a hollow in the centre, and are thus compared to the little box (*temi*) in which the Assamese peasant carries his stock of lime for consumption with betelnut. On the appearance of this last sign, the boiling cauldrons are rapidly emptied with the ladle, and replenished again from the feeders without delay, while the fire, which had been slackened at the *o-phulia* stage, is again quickened by feeding it first with two reeds dipped in the fresh molasses, as an offering to the god Agni. The duration of operations depends of course upon the quantity of juice, but the ryot always reckons upon converting his *pál* of cane into sugar in a single day or night, that is, with eight to twelve hours' work. Reduced to an average rate, this means that about thirty gallons of juice can be boiled in five or six hours. When the last instalment of juice has been disposed of, the boiling cauldrons are rapidly rinsed with a little warm juice and lifted off the fire.

Beating. 243

"BEATING AND COOLING.—The liquid stuff ladled out of the cauldrons is received in a wooden vessel (*ghólani*) about six feet long, made and shaped in the same manner as the ordinary Assamese dug-out, but with one end cut square; where it is stirred with a Y-shaped instrument consisting of a triangle of bent bamboo fastened to the end of a stick (*hátbári* or *ghótanimári*). As the stirring continues, the liquid loses its dark brown colour, and assumes the hue and consistency of yellow mud. The process lasts half an hour. The *gur*, or compost, is then removed from the *ghólani* with the hand or a broad slip of bamboo, and put into earthen pots. This concludes the proceedings. The manufacture of refined sugar is an art which has yet to be introduced into the valley of the Brahmaputra.

Progress. 244

"RATE OF PROGRESS IN MANUFACTURE.—The word *pál* is used to denote the quantity of cane which is crushed and converted into sugar at a single spell of work, whether by day or by night. The quantity of cane in a *pál* depends a good deal upon whether the cultivator is or is not working against time. It usually consists of twenty bundles, which may be roughly assumed to weigh 10 maunds, or about 800℔, but twice as much can be disposed of towards the end of the season, when work begins after the evening meal (9 or 10 P.M.) and continues without intermission through the night and into the forenoon of the following day. The quantity of cane got through on such occasions is commonly reckoned as the produce of one *cottah* (one-fifth of a *bíghá*, or 320 square yards). When working by day, the cane is cut and brought to the mill as it is wanted, but for night work it must be cut and tacked before dark. Boiling begins when half the cane has heen crushed, and goes on for several hours after all other operations have been concluded. The usual custom is to boil the juice yielded by one *pál* of cane in two instalments, as nearly equal as can be guessed, neither of which, however, need fully test the capacity of the boiling apparatus which is capable of dealing with twenty gallons at once, if the ryot has so much to put into it. The relation between the weight and the volume of the juice has been determined by a series of experiments to be about 11℔* avoirdupois to a gallon: as compared with water, the weight, volume for volume, at a temperature of 75° F., was found in one experiment to be as 74 to 67.

Economical aspect. 245

"ECONOMICAL ASPECT OF THE INDUSTRY.—It will probably have been perceived from the foregoing description that the manufacture of sugar in the Assam Valley is a purely domestic industry. The ryot has no relations whatever with any manufacturer or money-lender. He grows his cane entirely on his own account, and converts it into sugar by the help of his neighbours, who work for him on the understanding that he will work for them when their turn comes. This system of mutual assistance relieves the ryot of a good deal of labour, and of almost all expense; nevertheless, the cultivation of sugar-cane is regarded as a most laborious undertaking, to be attempted only, as the proverb, by him who hath six sons and twelve grandsons.

Size of Fields. *Conf. with pp. 108, 124, 143, 157, 256.* 246

"The area planted by a single family rarely exceeds half a *bíghá* (800 square yards) and is often much less; and whenever a large field of cane is met with, it will be found to consist of several such plots belonging to different families, who have

* The Saidapet experiments give an average of 9¾℔ per gallon.

cultivated the whole, as they will crush and manufacture its produce, by their united labour applied fo each plot in turn.

CULTIVATION in Assam.

Brahmaputra Valley.

Estimate. 247

"ESTIMATE OF THE COST OF CULTIVATION.—In reference to an industry conducted on such conditions as these, the term 'cost of production' is apt to be misleading; and, in fact, on making the calculation at the ordinary rates of hired labour, the expenditure may easily prove to be greater than the value of the article obtained. It is difficult to form an estimate of the cost of cultivation and manufacture that can be relied on with any degree of confidence, but the following statement, compiled from returns furnished by district officers, may be regarded as not very far from the truth:—

Cost of cultivating, crushing, etc., the cane on half a bighá of land (800 square yards).

Cultivation.	R	a.
Hoeing in October	0	8
Ploughing and harrowing (eight times)	3	0
Draining and drilling	0	12
Price of 1,200 cane-tops	2	8
Planting	0	4
Weeding (twice)	0	8
Hoeing and earthing-up (four times)	1	0
Fencing	1	0
Watching	2	0
Revenue of land	0	4
TOTAL	11	12
Manufacture		
Cutting (wages of ten men)	2	8
Crushing (wages)	3	0
" (hire of mill)	0	8
Boiling (wages)	3	0
" (fuel)	2	4
" (one-half value of vessels)	0	12
TOTAL	12	0
GRAND TOTAL	23	12

Weight of Juice. 248

"WEIGHT OF JUICE AND COMPOST PER GIVEN WEIGHT OF CANE.—The quantity of sugar manufactured from a given weight of cane by the rude processes known to the Assamese ryot is considerably less than the cultivator obtains in other parts of India, and will not bear comparison at all with the produce of a West Indian factory. A large number of experiments have been made by various officers with a view to ascertain the actual proportions in weight, of the juice and the compost obtained from a given weight of cane. Where made by European or educated native officers, these experiments may generally be regarded as accurate, or as liable to error chiefly on account of the occasional reluctance of the ryots to assist heartily in operations, which they secretly regarded as the preliminaries of new taxation. A series of experiments, in a rougher fashion and on a larger scale, have been conducted by subordinate revenue officers (mauzadárs). Here the recorded weights represent the results arrived at by multiplying the average weight of a few bundles of cane, or vessels full of juice or compost, by the number of bundles crushed, and the number of vessels filled. Covering, however, comparatively so large an area, these experiments may be regarded as giving general results that are fairly trustworthy, especially when we consider their remarkable correspondence with the results obtained by superior officers. The two classes of experiments have been tabulated separately

SACCHARUM: Sugar.

Methods of Cultivation

CULTIVATION in Assam

Brahmaputra Valley.

Weight of Juice.

and in detail, at the end of this Note. Collating them by districts, we find results as follow:—

Experiments by District Officers.

District.	Number of experiments.	Weight of cane crushed. ℔	Weight of juice. ℔	Weight of *gúr.* ℔	Per 100 ℔ cane. ℔ juice.	℔ *gúr.*
Goálpára . .	5	10,613	4,959	707	46·7	6·6
Kámrúp . .	7	7,671	3,268	569	42·6	7·4
Darrang . .	6	13.567	5,722	1,115	42·2	8·2
Sibságar . .	11	7,218	2,781	367	38·7	5·1
Lakhimpur . .	3	2,837	1,205	223	42·5	7·8
General results .	32	41,906	17,935	2,981	42·8	7·1

Experiments by Mauzadárs.

District.	Number of experiments.	Weight of cane crushed. ℔	Weight of juice. ℔	Weight of *gúr.* ℔	Per 100 ℔ cane. ℔ juice.	℔ *gúr.*
Kámrúp . .	3	3,624	1,420	232	39·2	6·4
Darrang . .	9	18,201	8,137	1,442	44·7	7·9
Nowgong . .	2	2,592	1,326	234	·51	8·8
Sibságar . .	15	53,034	26,483	3,731	·50	·7
Lakhimpur . .	1	1,028	411	82	·40	·8
General . .	30	78,479	37,777	5,721	48·1	7·3

"We may reasonably conclude from these figures that the ordinary cane-crop of the Assam Valley cannot be counted on to yield more than 43 per cent. of its weight in juice, and 7 per cent. of its weight in coarse sugar. For an average struck upon all kinds of cane cultivated under all circumstances, even these figures are probably too high. Much better results may be obtained where special care has been bestowed upon the crop; the list of experiments by district officers shows that in several instances 50 and even 60 per cent. of juice, and 10 to 13 per cent. of *gur*, has been got from a given weight of cane; but these are exceptional cases, and do not represent the sugar-yielding capabilities of the common cane of the country.

Weight of Cane. 249

"WEIGHT OF CANE ON A GIVEN AREA OF LAND.—The weight of cane grown on a given area of land varies much more than the proportion between a given weight of cane and the weight of juice or *gur* obtainable from it. The species of the cane makes a considerable difference; *pura*, for instance, is a much heavier crop than *teli*. Speaking generally, a well-cultivated field will yield to the mill about one pound of cane to every square foot, while a field carelessly cultivated or insufficiently planted, or exposed to the depredations of animals, will hardly give one pound to every three square feet of its area: thus the limits vary from six to nineteen tons per acre; while an arithmetical mean, which is probably somewhat in excess of the actual average, may be deduced from the following statement, compiled out of the details given in the appendices:—

Experiments by District Officers (24 in number).

District.	Area cut (square feet).	Weight of cane ℔.	℔ per acre.
Goálpára	10,890	8,722	34,888
Kámrúp	14,537	6,067	18,180
Darrang	18,576	13,567	31,814
Sibságar	12,876	7,218	22,419
Lakhimpur	3,180	1,769	22,695
General result	60,059	37,343	27,083

Experiments by Mauzadárs (30 in number).

District.	Area cut (square feet).	Weight of cane ℔	℔ per area.
Kámrup	10,332	4,396	18,417
Darrang	29,916	18,201	26,500
Nowgong	4,284	2,592	26,356
Sibságar	79,920	53,703	29,270
Lakhimpur	1,440	1,028	31,097
General result	125,892	79,920	27,429

CULTIVATION in Assam.

Brahmaputra Valley.

Average Outturn.
250

"AVERAGE OUTTURN.—These figures, though without any pretension to absolute accuracy, may be accepted as representing the results of measurements and weighments made with as much care as would be taken in a wholesale commercial transaction. In using them for the purpose of educing general averages, it is necessary to remember that the most promising plots stand the best chance of being selected for experiment, that fields in the jungle must be rated far below those in the vicinity of villages, and that while the great majority of the experiments were made with *mugi* cane, it is the less productive *teliya* which the district reports would lead us to regard as the predominant species. Bearing these facts in mind, we may perhaps conclude that the average Assamese cane-field bears 10 to 11 tons per acre;* and such a weight of cane will yield about 1,400℔ of *gur*. Compared with other parts of India, these results are poor. In the North-Western Provinces the average yield per acre, irrigated and unirrigated, taking all the districts together, is estimated at 2,300℔ of *gur*,† and the *gur* of Upper India is better dried and more durable than that of Assam. Part of this superiority in yield of sugar is due to the greater quantity of juice expressed, for cane in the North-Western Provinces gives one-half of its weight in juice. If we look to Madras, it appears that the common country mill of the Bellary district, built on much the same principle as the Assam mill, but costing R72 for the rollers alone, can extract 66℔ of juice out of 100℔ of cane, and this will yield 12℔ of sugar,‡ or double as much as could be got from the same weight of cane in the Assam valley. The fault lies less in the Assamese mill than in the cane; for the Bihia mill extracts 67 per cent. of juice from Madras cane, while the best experiment with it in the Assam valley has not given more than 56 per cent. In Behar the average produce of *gur* per acre is estimated at the very high figure of 40 maunds, or 3,300℔; in Lower Bengal (the Rajshahye and Burdwan districts) at 2,500 to 1,800℔; § lastly in the Beelin cane tract in British Burma, the outturn of an acre well cultivated is estimated at 3,500℔ of *gur*.||

"VALUE OF OUTTURN.—We are now able to complete our calculation of the ryot's profits on sugar-cane. The cost of growing and converting into coarse sugar the cane on half a *bighá* of land (800 square yards) was estimated at R23-12. The produce will be some 4,000℔ of cane, which may be expected to yield about 240℔ of the compost called *gur*. The ryot will probably keep the greater part of this for domestic consumption; but on the supposition that he disposes of the whole of it by retail sale in the petty markets, it will fetch about 2½ annas per seer, or some R19

Size of Fields.
Conf. with p. 154.
251

* NOTE.—It is hardly necessary to repeat that a single field of one acre probably does not exist in the valley of the Brahmaputra.

Some additional statistics may here be quoted. Five experiments made last year in Sibságar, on an area of 1·13 acres altogether, gave an average outturn of 1,517℔ of *gur* per acre. The average assumed in the text is perhaps corroborated in some degree by the rough estimates of the ryots. In the southern part of the Kámrúp district 20 to 25 *kalsis* are estimated as a fair outturn for a *bighá* of land. The *kalsi* contains about 20℔ of *gur*, so that the outturn of *gur* per acre would be 1,200 to 1,500℔. Another estimate is 6 *kalsis* per *cottah*, or 1,800℔ per acre, as the produce of a good field. In some villages where cane-crushing was going on I measured up the area of cane cut for a single *pál*, and weighed the *gur* obtained, with results as follows:—

Square feet.	℔ *gúr.*	℔ per acre.
3,375	62	796
3,033	52	776
3,177	56	759
7,200	131	807
972	47	2,100
1,746	74	1,850

These very poor results obtained by ryots when working by themselves show that the estimate in the text is not too low.

† *Field and Garden Crops of the North-Western Provinces and Oudh*, Roorkee 1882.

‡ Saidapet Experimental Farm Report for 1881-82.

§ These figures are taken from papers published by Messrs. Thomson & Mylne.

|| Quoted from a Note by Mr. D. M. Smeaton, Director of Agriculture, dated the 9th October 1882.—These figures, however, seem small in comparison with some statistics of cane cultivation in Australia. I find it stated in the *Brisbane Courier* that the outturn per acre on one Queensland plantation is estimated at 37 to 40 tons of cane, and one ton of cane gives 150 gallons (about 1,500℔) of juice.

Methods of Cultivation

CULTIVATION in Assam. Brahmaputra Valley.

altogether, thus failing to cover the cost of cultivation and manufacture.* The mode of sale is in small earthen pots containing about 2lb each, and worth from two to three annas, or even as much as six annas in a dear year; or else in large earthen vessels (*kalsi* or *kalah*) holding some 20lb, and priced according to their weight; or he may sell by the maund, at the rate of R4 to R5. The conditions of production, however, are such that nothing like a fixed proportion exists between supply and demand. The market gets only the overplus from domestic needs, and the price rises and falls from year to year according as this happens to be little or much. In 1879-80, in the Nowgong district, a *tekli*, or small earthen jar, containing about two seers of *gur*, sold for 8 to 10 annas, or at the rate of about R10 per maund; the present year, on the other hand, is one of abundance, and *gur* was selling in April at R2½ per maund in Kámrúp and Darrang, while the price throughout last year in the vicinity of Dibrugarh ranged from R8 to R9 per maund. Assamese *gur* is never sold in the large balls or masses of hard compost which are so familiar in the bazars of Upper India. In the winter it barely attains a solid consistency, and shows a slight tendency to granular crystallisation, but as the weather grows hotter it liquifies, and if not speedily consumed often becomes sour and useless.

Area and Outturn. 252

"AREA AND OUTTURN.—Taking the total area under sugar-cane in the Assam valley (in 1882) to be 16,000 acres in round numbers, the average outturn as 1,400lb of *gur* per acre, and the price as R4-8 per maund of 82⅔lb, the whole weight of sugar produced would have been 10,000 tons, valued at R12,25,000.

Consumption. 253

"CONSUMPTION.—The whole of this is locally consumed, no portion being exported either to Bengal or to the fromtier tribes. It is not, in fact, sufficient by itself for the wants of the country. The import from Bengal during the last three years has been as follows:—

	1880-81. Maunds.	1881-82. Maunds.	1882-83. Maunds.
Refined	13,217	11,564	10,974
Unrefined	39,473	28,849	34,980
TOTAL	52,690	40,413	45,954

"Refined sugar is consumed almost exclusively by Europeans, well-to-do Bengalees, and Marwari traders, or is used at festivals in the great *Shattras*. If unrefined sugar alone be taken into consideration, we find that the average annual import during the last three years has been 2,833,426lb, and the local production of sugar being 10,000 tons, the sum of these two quantities, when divided by the population of the Assam valley (2,225,271), gives a yearly consumption of 11lb per head.† This calculation tends to show that the average outturn of sugar per acre has not been underestimated. In the Punjáb and the North-Western Provinces the estimate of the consumption of sugar made for the Famine Commission in 1879 was 30lb per head of the population, and when we remember how largely the Brahmaputra valley is peopled by races (Mech, Kachari, Mikir, Lalung, etc.) to whom the use of sugar is unfamiliar, besides the utter absence of large cities with their wealthy classes, it is difficult to believe that the average consumption in this part of Assam can exceed one-third of the figure estimated in Upper India.

Improvements. 254

"IMPROVEMENTS.—The first condition necessary to any improvement of the cultivation of sugar-cane in the Brahmaputra valley is a wider market. There is no present demand beyond domestic wants, if we except two small ventures in the Sibsagar and Lakhimpur Districts, which prove, in their limited way, that the producetion of sugar-cane can be stimulated without difficulty. There are two distilleries established by enterprising Europeans near Golaghát and Dibrugarh, where the *gur* of the country is converted into rum for consumption by tea-garden coolies. Situated in the centre of thickly-peopled tracts these factories have stimulated the production of sugar-cane considerably within the limited area on which they draw for their supplies. The Dibrugarh factory uses Bengal *gur* largely, while that of Golaghát depends entirely on local production. When the latter was first started in 1879, the

Rum. *Conf. with pp. 93, 96, 175, 320-321.* 255

* In confirmation of this estimate, which I believe rather underrates the loss which would follow cultivation *by hired labour*, I may mention that a European engaged in farming near Bishwanath in the Darrang district showed me a crop of *pura* cane which had already cost him so much that he doubted whether it would be worth his while to cut and crush it. It will be observed, moreover, that one of the ryots quoted in the note above got only 131lb of *gur* from his half *bighá* of cane.

† This will be seen to be the consumption according to more recent estimates. —*Ed., Dict. Econ. Prod.*

proprietor found some difficulty in procuring *gur* at all, but now he draws upon the cane crops within a radius of five miles, and cultivation in the neighbourhood has increased about 28 per cent. But neither the one concern nor the other is on a sufficiently large scale to affect seriously the general cultivation of cane, or to test the remunerativeness of such an enterprise if conducted with a larger capital. CULTIVATION in Assam. Brahmaputra Valley.

"It has already been stated that refined sugar is nowhere manufactured in the Assam valley. Even in the manufacture of *gur*, however, no one who has witnessed the rude processes employed by the ryot can doubt that a very great room remains for improvement. Reasons have already been given for believing that the country mill works more rapidly than the *kolhu* of Upper India, and perhaps it may, therefore, be somewhat less effective as a crusher; but, on the whole, it seems probable that the smaller proportion of juice obtained in these parts (42 per cent. against 50) must be attributed mainly to the inferiority of the cane. Recent experiments, however, with **Mr. Cantwell's** modified form of the Bihia mill show that Assamese cane can be made to yield as much as 56 per cent. of its weight in juice. It is in the boiling that the greatest loss occurs, 100lb of cane yielding only 5 or 6lb. in *gur*, against 15 to 18lb in the North-Western Provinces and 12lb in Madras. This difference, while, probably, arising in part from the poorer quality of the juice, is also due in great measure to carelessness in manufacture. In the vast majority of cases, no preventives of acidification are used in any stage of the process, and the boiling is often conducted by guess-work.

"The valley of the Brahmaputra is a country of peasant proprietors, in comfortable circumstances indeed, but without intelligence, enterprise, or capital, and any improvement, whether by the introduction of better kinds of cane or of a better mill, or by greater care in the manufacture of sugar, must be looked for from without.

"Land fitted for sugar-cane can be leased from Government at the yearly rent of 8 annas a *bigha*, or R1-11 per acre (including assessment to local rate), and there is the widest possible choice of sites. It is, however, more than doubtful whether cane-growing by hired labour could be made to pay: on the other hand, the central factory system, which has proved so successful in the West Indies and in Australia, can scarcely be introduced in the present defective state of communications and means of transport in the Assam valley. Where the commonest vehicle for loads is a bamboo carried on men's shoulders, there is obviously some difficulty in transporting the produce of an acre of sugar-cane to a mill situated at a greater distance than a few yards.

B.—The Surma Valley. SURMA VALLEY.

"Processes of manufacture.—The processes of manufacture are practically the same as those which have been described at such length as prevailing in the valley of the Brahmaputra. The mill, here called *kamrangi* or *ghani*, is sometimes driven by bullocks, and the Cachar ryot is said to cut his cane-stalks into pieces twenty inches long before crushing; in this district also iron cauldrons (*karhoi*) are occasionally employed. In some parts of Sylhet the cultivator boils the juice imperfectly, and sells the liquid product (or so much of it as he does not want) to men of the *Lowait* caste, who boil it down into solid compost (*bándha gur*). The liquid or *láli gur* is worth about R2, and the hardened compost some R4 to R5 per maund. Refined sugar is never made. 256 Process.

"Relations of cultivator with money-lender.—The cultivator of sugar-cane in the Surma valley is independent of the money-lender, unless he is beginning for the first time, and has not ready money to buy cane cuttings. In that case he takes an advance, repayable with interest when the crop is harvested. The rate at which cane-tops sell in Cachar is stated as 200 the rupee, but this seems exceptionally high. In Sylhet, again, the ryot is said to borrow money to buy oil-cake (*khoil*) for manure. The Money-lender. 257

"Area under sugar-cane.—There is no system of village records in the Surmá valley, and the estimates of the area under sugar-cane must, therefore, be regarded as conjectural. The method employed in Cachar was to require returns, through the officers in charge of police-stations, from the village policemen of their circles; these latter furnished lists of the sugar-cane fields within their beat, giving the length and breadth of each in "reeds" of 24 feet, and the station officers worked out the circle areas and sent them into head-quarters. This gave a total of 786 acres, but it was believed that their apprehensions of new taxation had induced the people to understate the facts, and on a comparison with the results obtained by actually measuring up the area under cane in three mauzas of each tahsil, the total extent of sugar-cane cultivation in the district has been estimated at 900 acres. For Sylhet no estimate that can be relied on with any degree of confidence is forthcoming, but the area under Area. 258

CULTIVATION in Assam. Surma Valley.

sugar-cane in this district has for some years past been shown in the annual administration reports as 8,000 acres.

"The average consumption of sugar per head of population is estimated at 4 chitaks a month in Cachar, *i.e.*, 6lb a year, but this seems low. Certainly Sylhet, with its large Muhammadan population, should not consume less *gur* per head than the Assam valley. Assuming, therefore, an average of 10lb, and dealing with coarse sugar only, we can make the following calculation:—

Population of Sylhet and Cachar	2,282,867
One year's consumption of sugar	22,828,670 lb
Deduct net imports of 1882-83	8,338,341 „
Remains to be provided by the produce of (say) 9,000 acres	14,490, 329 „
Thus the average produce per acre ought to be . .	1,600 „

"Whether sugar-cane is really more productive in the Surma valley than in Assam Proper, we have no means of judging with certainty. There is, however, nothing improbable in the supposition, considering the density of the population in parts of Sylhet, and the known fact that an acre of land yields more rice in Sylhet or Cachar than in the Assam valley districts. The custom of borrowing money to buy manure in Sylhet, if it prevails extensively, seems also to point to a more careful style of cultivation.

"The value of the sugar produced in 1882-83, as thus estimated in quantity, and taking the price at R4-8 the maund, would appear to be about eight lakhs of rupees; at the same rate, the sugar-cane crop is worth 80 rupees the acre; a sufficiently probable valuation, though evolved from data extensively coloured by conjecture.

N.-W. PROVINCES. 259

III.—NORTH-WEST PROVINCES AND OUDH.

References.—*Duthie & Fuller, Field and Garden Crops; Mr. G. Butt, Note on Sugar of Shahjahanpur; Note on Sugar Cultivation in N.-W. P. by Mr. Darrah; Report on Benares Sugar, in Proceedings, Hon'ble East India Company, 1792; Sugar Statistic in 1848; Agri.-Hort. Soc. Ind.:—Trans., Vol. I., 121-123; III., 57, 73, 124, 126, 211; IV., 104; 134-136, 144, 187, 203; V., 36, 66, 72, Proc., 51, 98; VI., 9, 95, 249; Proc., 17; VII., Proc., 38-39, 78, 117, 127; VIII., Proc., 408, 427; Journ., I., 130, 255, Sel., 402-411; II., Sel., 363-365; VII., 1-7, 230; XIII., Sel., 60; New Series, VI., 59-68; VII. (1883), 128, 140; Gazetteers:—I., 85, 115, 151, 153, 183, 252, 347, 433, 479; II., 160, 235, 266, 375, 479, 524, 555; III., 24, 162, 227-229, 305-307, 690; IV., 18, 19, 20, 28-29, 354, 525, 618; V., 28, 83-84, 267, 268, 334, 559-562; VI., 27, 30, 142, 146, 150, 151, 153, 238, 306, 325-326, 329-330, 647; VII., 39-40, app. 11, 12; VIII., Muttra Dist., 41, 182; Allahabad Dist., 116; Fatehpur Dist., 17; IX., Shahjahanpur, 46-49, 128-130; Moradabad Dist. 43, 45, 121, 127-128; Rampur Dist., 33.*

Area & Outturn. 260

Area, Outturn, and Consumption in the North-West Provinces and Oudh.

One of the most instructive and, at the same time, the earliest paper which treats of the sugar-cane of, at least, a very important part of these provinces (the Benares Division), is that to which frequent mention has already been made, *viz.*, a detailed article published by the Honourable the East India Company in 1793. The division then embraced the Sirkars of Benares, Chunar, Ghazípúr, Jaunpúr, and Terhar, as also the pergunnahs of Budhoí and Kera-Mungrowr. It is explained that, owing to the commission of sundry oppressions on the ryots, the Political Resident had, in the year 1788, fixed the *bíghá* to be used throughout the division at 3,136 square yards, or, say, ⅔rds of an acre. The report framed on that standard, it is further stated, was drawn up as the average of all the returns for the five years from 1787 to 1792. After detailing the proportions of cane-land held under each system of tenure, the writer adds: "Thus it appears that the total number of *bíghás* cultivated in five years amounted to 3,77,996, and the rental came to R17,32,310; so that the average number of *bíghás* for one year was 75,599." "It is further pointed out that this came to an average rate per *bíghá* (for the entire division of all classes of cane culti-

CULTIVATION in N.-W. Provinces.

Area & Outturn.

vation) of R4-9-6. The land, it is explained, was ploughed 16, 20, or 25 times from the month of June to September, and no grain was sown on it for the *kharif* crop. In the month of December the cultivator then placed a flock of sheep on the field and kept them there two, three, or four nights (according to his means, for they had to be hired from the shepherds), for the sake of their manure—such manure was deemed of singular value for sugar-cane. After this the land was watered from tanks or wells, and, when properly moistened, it was again ploughed two or three times. The cane was then planted in January, February, or March. Each *bíghá* generally required 8,000 canes, and these cost from R1-4 to R3 in proportion to the strength and maturity of the canes. The seed-canes were generally steeped in some tank of water for a whole night and were then cut into pieces 15 to 18 inches in length which contained from 4 to 5 eyes. The land was then furrowed at about 15 inches asunder and the seed-canes dropped into the furrows at a distance of 2 feet from each other. They were then covered with the earth and the field levelled. In two or three days the field was hoed to prevent the white-ants from lodging in the ground and again levelled. These operations were repeated two or three times till the canes appeared above ground. When the sprouts attained a height of 18 inches the field was manured, hoed, and then watered. This was repeated five or six times till the commencement of the rains. After this subsided in December, January, or February, the cane was cut down at leisure and when brought home was divided into small pieces, thrown into the mill and ground, and the juice expressed. The writer then explained the terms given to the various crops of cane, such as, 1st, *Chowmusskalis*, or lands cultivated in the first degree (*viz.*, that detailed above); 2nd, *Jerí*, or land which had previous to a cane-crop yielded a *kharíf* harvest, this was classed as cane-land second in value, since ploughing could only commence in October or November; 3rd, *Jownar*, land that had afforded a *rabí* crop and which therefore received still less preparation for cane, the seed-canes being planted in March; the fourth class was called *Muteri*, a term which had reference to the agreement. By this system the *rayat* secured so much land at a fixed *jumma* regardless of the crops he might choose to cultivate. The author of the report next discussed the character of the soils of the Benares lands as compared with the sugar-yielding tracts of Bengal, and added that "a striking difference between the soil of Bengal and that of the division of Benares, or, in general to the westward of the Gogra, as far as Delhi, is the immense quantity of calcareous matter it abounds with which, being precipitated by water between various strata of clay, forms an immense stratum of calcareous tufa, commonly called *kankar*. The depth of it varies much. Another remarkable difference is the numerous salt wells and grounds impregnated with salt. These are most common in the northern parts, and particularly in the north-west quarters. In some, the water is limpid, in others it is black and stinking like the ouze or mud of rivers near the sea. In many parts also, during the dry weather, an efflorescence is seen on the surface of the ground, this is termed flos-aseæ." The writer then proceeds to deal with the nature of the canes grown. Some, he says, were small but of considerable height, others an inch or one and a half inches in diameter and eight feet in height. The defective system of cultivation and the strong prejudices of the people he deplored, but deprecated at the same time the idea of improvement, especially in the direction of European ploughs and machinery; these were alike beyond the means of the people and their physical powers. An insect pest, he affirmed, often did great damage by lodging in the stems; it totally destroyed the plant, if not extracted. Speaking of the problem that was even then engaging

White-ants. *Conf. with pp. 37, 101, 125.* **261**

Soil. *Conf. with pp. 60, 130, 134.* **262**

Pests. *Conf. with pp. 48, 52, 76, 87, 102, 121-127, 151.* **263**

CULTIVATION in N W. Provinces. Area & Outturn. marked attention, namely, the possibility of European plantations, the Benares Political Agent, a century ago, arrived, on the whole, at an unfavourable conclusion. The European would have to give high daily wages to all employed, he would have to keep a large establishment of sirkars, peons, etc., besides being liable to numerous deceptions, so that it therefore became doubtful whether the higher yield obtained by a superior cultivation of an article that fetched so low a price as sugar would compensate for the greater expenditure. The prejudices of the cultivators, who would have to be employed, would entail a constant supervision, and in the Agent's opinions these prejudices were so strong that through "long custom and the practice of his forefathers" the cultivator "must literally be bribed to procure his own advantage." How far these conclusions have been justified the reader will be able to judge by the record given in this article of a century's endeavours toward the establishment of European plantations. The reader may, in fact, be disposed to comment that the cultivator's prejudices have not been materially changed, nor his systems of cultivation appreciably altered during the century that has lapsed. The author of the Benares report, here briefly summarised, dealt with the questions of the cost of production and outturn.

European Plantations. 264 *Conf. with pp. 37, 39, 48, 62, 63, 91, 93, 103, 111, 212, 306, 309.*

He gave the cost of cultivation of one *bighá* (approximately ⅝rds of an acre) of land with cane at R23 and the manufacture of the *gúr* therefrom at R9-15; the produce, 22 maunds of *gúr*, at R2 per maund (that is, R44), so that the profit came to R11-1. Other examples afforded very nearly the same result. Working out the total produce for the 3,77,996 *bighás* (251,997 acres) this was found to be (for the five years) 67,89,601 maunds (of 80℔), or an average of 17 maunds 38 seers 7 chattacks of *gúr* per *bighá*. The Agent then proceeded to estimate the amount of *packa chini* that quantity of *gúr* would have afforded assuming that it was all so manufactured. He allowed for a loss by evaporation in the first stage of boiling of 6 seers per maund, and further affirmed that as one maund of the syrup thus purified only afforded ¼ of its weight of *chini*, 14,42,790 maunds would have been obtained. The 17 maunds 38 seers and 7 chattacks of *gúr* produced per *bighá* was derived from maunds 98-31-10 of cane juice, or one maund of *gúr* from 5½ maunds of juice. The writer added, however, to this statement of the average yield that a superior quality of cane-juice would afford 1 maund of *gúr* from 3 maunds of the juice. It does not seem necessary to follow the Agent into every detail. Enough has perhaps been given to afford data upon which comparisons may be drawn with the modern systems and results. One further point may, however, be here mentioned. The Agent says that "in 1787-88 the exports of *chini* amounted to only maunds of sorts 1,70,352; hence the increase came to maunds 94,277." These exports went mainly to Calcutta, but a considerable amount went also to the Deccan. The consumption of sugar in Benares from 1787—92 was also dealt with by the writer whose report is here briefly reviewed. He arrived at the conclusion that the annual consumption of *chini* might be put at 1,20,000 maunds for the population of 2,911,556. This, it may be said, would come to 3·29℔ of *chini* or to three times that amount if expressed as *gúr*, or coarse sugar.

Yield. 265 *Conf. with pp. 110, 133-36, 178, 189, 198.*

In the *Statistics of Sugar* (published in 1848) it is explained that it was found desirable to refer these provinces to two great sections, *viz.*, the Benares Division (chiefly concerned in the Calcutta supply) and the upper districts that send little or no sugar to Calcutta, but which furnish the Panjáb, the Deccan, and the Central table-land with considerable amounts. The Benares Division, thus isolated, in some respects corresponds to the region which has already been dealt with, in connection with the returns of 1792. It may serve, therefore, a useful purpose to contrast the

CULTIVATION in N.-W. Provinces.

Area & Outturn.

figures which denote the trade, as it may be said, at the beginning and in the middle of the present century. The following table exhibits the position of the Benares Division in 1848:—

	Area cultivated with cane in *bighás* of 14,400 square feet (=⅓rd of an acre).	Produce of *gúr* in maunds of 80lb.	Consumption of *gúr* in maunds.	Surplus available for export in maunds.
Goruckpore	4,26,426	14,84,288	5,70,722	9,13,566
Azimghur	1,73,502	20,19,087	12,23,511	7,95,576
Jaunpur	70,053	3,57,764	85,747	2,72,017
Mirzapore	33,465	1,82,169	58,500	1,23,669
Benares	1,74,648	5,46,728	4,40,186	1,06,542
Gazípúr	1,11,293	9,71,605	4,10,088	5,61,517
TOTAL . { *Bighás* .	9,89,387	} 55,61,641	27,88,754	27,72,887
{ Acres .	329,795			

Field. 266 Conf. with *p. 211.*

The surplus thus shown when reduced to sugar would come to 9,24,296 maunds, "a quantity which approximates to that estimated to have been annually imported into Calcutta from Benares." It is explained that the produce of the soil varies from 3 maunds 5 seers and 3 chattacks to 11 maunds 25 seers and 7 chattacks per *bighá*. Taking the average of these two extremes and expressing the result in acres, the yield may be said to have been 22 maunds of *gúr* an acre. But there must have been more lands in 1848 that gave a lower return than the average of the two extremes than what exceeded that average, since the actual average yield of the division was found to be 5 maunds 24 seers 13¾ chattacks a *bighá*. The highest returns were in Azimghur 11 maunds 25 seers and 7 chattacks followed by Ghazípúr with 8 maunds 29 seers and 3 chattcaks. The lowest was in the Benares District itself, namely, 3 maunds 5 seers and 3 chattacks. We thus learn that the actual average outturn in the division was about 17 maunds of *gúr* an acre. It will accordingly be seen that the returns of 1792 manifest a higher yield than what was arrived at for 1848. This may have been due to the extension of sugar-cane cultivation (with the increasing demands) into portions of the Division which were, by no means, so well suited for the crop as were those to which its cultivation had been restricted at the time of the earlier estimates. The Division in 1792 did not embrace so large a country as that of 1848, but allowing for that error it may be said that there were under cane 251,997 acres as against 329,795 acres in 1844. These figures represent for the 56 years an average annual expansion of the sugar-cane area of the Division of ·55 per cent. Of one of the districts included in the returns of 1848, *viz.*, Azimghur, we have most instructive particulars for the year 1836. In that year **Mr. R. Montgomery** made a careful survey of the district, and his report on sugar-cane lends the strongest support to the possible accuracy of the returns of 1848. In 1836 there were 1,02,725 *bighás* under cane and these were estimated to have yielded 12,32,707 maunds *gúr*, or a little over 12 maunds a *bighá*. The exports to Calcutta came to 2,00,000 maunds of sugar, and a large trade also took place westward (see *Trans. Agri.-Hort. Soc., V.,* 72). From many other circumstances, besides results like the low expansion

CULTIVATION in N. W. Provinces.

Area & Outturn.

justified above by the statistical returns, the writer has had the conviction forced on him that the modern expansion spoken of at nearly every period of the Indian sugar trade has been the extension into tracts not formerly cultivated, until it may now be said sugar-cane is grown in every district throughout the length and breadth of the empire. The peace and prosperity which followed the advance of the British power in India may not only be viewed as having opened out a great foreign trade for the country, but as having vastly extended internal traffic. It became possible for the people to be scattered in small communities in the remotest corners of the empire and to there indulge in such a highly remunerative cultivation as cane, without fear of the loss of their labours from oppression and robbery. It is thus probable that did the data exist for such an enquiry it might be shown that, from year to year, the area of sugar-cane cultivation may have actually contracted in many of the once famous districts of production owing to their markets being cut away from them through new cultivations in tracts where cane was formerly unheard of. This idea is borne out by the traditions of cane cultivation, where it is believed that such and such a district obtained its stock from a neighbouring district and took to cane cultivation in a certain year, thereby being able not only to meet its own demands, but to actually enter in the lists of external supply. In this view it becomes therefore desirable to extend the enquiry of sugar-cane cultivation for the year 1848, so as to embrace the entire area of the North-West Provinces instead of confining observation to the Benares Division. The detailed table, furnished in 1848, included certain portions of what is now classed under the Panjáb, such as the Delhi Division*. This will, therefore, be excluded from present consideration. A column is also shown, in the original table (from which the form given below has been compiled), to exhibit the date-palms of these provinces. There were 258,071 palms in 1848, but as these do not appear to have afforded sugar, they may be omitted from consideration:—

* GHAZIPUR.—In the paging of this article the space below has been left unoccupied. It may, perhaps, be usefully utilized in drawing attention to a paper on the cane cultivation of the Ghazipur District of the Benares Division from the pen of P. Michea, Esq., which appears in the Journal, Agri-Horticultural Society, India, Vol. VI. (New Series) pp. 59-68. That article sets forth the chief defects of the system of sugar manufacture in the Benares Division and may be accepted as in some respects carrying the historic facts, here more especially dealt with, down to the year 1878. The opening sentence or two will be seen, however, to justify the inference that Mr. Michea was probably unaware of the European efforts that had, half a century and more prior to the date of his article, been undretaken to advance the Benares Industry to a position more nearly akin to that in the sugar producing colonies. Mr. Michea says:—"The production of sugar in the Benares Division, like all other industries essentially Native and on which no European improvement has been grafted, remains, to this day, governed by the same system if agriculture and manufacture which has existed for ages. The Natives have introduced no change whatever, and yet no other industry has been more thoroughly improved during the last 50 years than the surgar industry." While fully aware that no material improvement has been effected, the writer can hardly support the opinion that the present methods of cultivation and manufacture are essentially Native. They have been repeatedly modified and in the nature of the cultivated stock the modifications have resulted in a degeneration rather than an improvement.

CULTIVATION in N.-W. Provinces.

Area & Outturn.

Production and Consumption of Sugar in N.-W. Provinces during 1847-48.

	MEERUT DIVISION.	KUMAON DIVISION.	ROHILKHAND DIVISION.	AGRA DIVISION.	ALLAHABAD DIVISION.	BENARES DIVISION.	TOTAL.
	Dhera Dun, Saharanpur, Muzzaffernagar, Meerut, Bulandshahr, and Aligarh.	Kumaon and Garhwal.	Bignour, Moradabad, Budaon, Bareilly, and Shahjahanpur.	Muttra, Agra, Farruckabad, Mainpuri, and Etawah.	Cawnpore, Futtehpur, Hamirpore, Kalpí, Banda, and Allahabad.	Goruckpore, Azimghur, Jaunpur, Muzapore, Benares, and Ghazípur.	
(1) Area in *bighás* (=14,400 sq. ft.)	2,44,306	374	4,27,914	66,335	58,007	9,89,387	17,86,324
(1) Area in acres	81,435	124	142,638	22,112	19,335	329,795	595,441
(2) Production of *gúr* in maunds (=80lb)	15,03,167	167	28,85,178	3,02,389	2,71,326	55,61,641	1,05,23,868
	Mds. S. Ch.		Mds. S. Ch.	Mds. S. Ch.	Mds. S. Ch.	Mds. S. Ch.	Mds. S. Ch.
(3) Average yield of *gúr* per *bighá*	4 12 7	......	6 28 11	6 29 1	5 39 6	6 10 2	5 39 15 (Average, say, 18 maunds an acre) *Conf. with p. 209.*
(4) Consumption of raw canes, sugar, etc., expressed as *gúr* in maunds . .	11,83,012	14,000	10,26,016	3,98,051	4,09,285	27,88,754	58,19,118
(5) Average to head of population in lb	45·28	......	21·6	10·81	10·96	30·14	23·65 (Average)
(6) Available for export, maunds	7,16,525	*Nil.*	18,59,162	53,030	2,990	27,72,887	54,04,594
(7) Excess consumption over production, maunds.	Dhera Dun 59,125 Bulandshahr 2,21,559 Aligarh . 1,15,686	Kumaon 12,833 Garhwal 1,000	*Nil.*	Muttra . 93,687 Agra . 55,005	Cawnpore . 61,497 Futtehpur . 4,263 Kalpí . 5,224 Bánda . 27,000 Allahabad . 42,965	*Nil.*	6,99,844

NOTE.—The districts named (along the line No. 7) had no surplus, but imported the required balance to meet their consumption. The total of these imports must, therefore, be added to the production to show, by deducting the consumption, the net balance available for export from these Provinces.

CULTIVATION in N.-W. Provinces.

Area & Outturn.

It will thus be seen that there were in 1848, 17,86,324 *bighás* (or, say, 595,441 acres) under sugar-cane, and that these yielded of edible canes and canes used in the expression of their juice, a quantity which, when represented on the standard of *gúr*, came to 1,05,23,868 maunds. The local consumption in these provinces was, for the year under consideration, ascertained to have been equal to 58,19,118 maunds of *gúr*. Certain districts, however, consumed more than they produced and an interchange took place which resulted in an import of 6,99,844 maunds. That amount added, therefore, to the production, less the consumption, exhibits the net balance (47,04,750 maunds), which was available for export. The largest exporting division was Benares with an amount—27,72,887 maunds—which was almost entirely consigned to Calcutta. Commenting on this traffic (between Benares and Calcutta) the Government of India pointed out (in 1848) that that amount, *plus* the balance, shown by the province of Bengal as having been available for export (*viz.*, 22,99,523 maunds of 80℔) and expressed in sugar, would come to 16,90,803 maunds, a figure which very closely corresponded with the foreign exports from Calcutta for the year in question. It is, perhaps, scarcely necessary to urge that much faith should not be placed on the accuracy of the figures shown for consumption per head of population. The means of balancing imports and exports in production were not even so complete as they are at the present day, and it is doubtful if the census of the population for 1848 could be regarded as very trustworthy. It is probable, however, that any error due to imports would not have been so serious for the returns of that year as at the present time. The facilities of transport were nothing like so great as now, and the interchanges were, therefore, more from district to district than from province to province. Foreign sugar did not apparently penetrate to any very great extent from the coast. The determination of the centres of sugar production (intended for exportation to foreign countries) seems to have been regulated very largely by proximity to rivers and other routes of transmission towards the coast. Railways soon altered all this, however, just as canals afforded the water necessary for cultivation in fertile tracts which had hitherto not been brought under cane, if indeed under any crops. Expansion became equivalent more to diffusion than to increase of the original areas. Tracts of country not suited for cane culture were thereby apparently brought under the crop, with the result that the acreage yield was lowered. This peculiarity has already been indicated, but it seems necessary that the character of the expansion of sugar-cane cultivation should, as far as possible, be clearly appreciated. There is one feature of the figures here briefly reviewed that deserves special consideration. The officers who compiled the statistics of the North-West Provinces for 1848 were careful to avoid the error of regarding the area grown with cane as necessarily that which afforded sugar. They, however, did not carry this principle to its final issue by showing the amount of land devoted to edible canes, to canes used in the preparation of *gúr* and to canes grown exclusively for the sugar trade.

The consumption of the North-West Provinces will be seen to have amounted to only a little over half the production (table, p. 165), whereas in Bengal (table, p. 132) the consumption was quite two-thirds of the production. These provinces, therefore, grew far more largely for exportation than was the case in Bengal. But two other facts, of perhaps even greater significance, may be here alluded to as exemplified by the returns of 1848, *viz.*, while the consumption per head of population was double that for Bengal (*viz.*, 23·6℔ in the North-West Provinces and 11·2℔ in Bengal), the production of *gúr* per acre was considerably less, *viz.*, 18 maunds in these provinces and 27 maunds in Bengal. And these results will be found to

CULTIVATION in N.-W. Provinces.

Area & Outturn.

bear a distinct relation to the most recent corresponding returns—the averages for the five years ending 1888. Thus, for example (*p. 116*), consumption to head of population of coarse sugar, 34℔ in the North-West Provinces and 9·5℔ in Bengal; production 22·9 maunds an acre in the North-West Provinces and 28·7 maunds in Bengal.

Having thus briefly alluded to some of the more striking features of the statistical and other facts which have been published regarding the sugar trade of these provinces, one hundred years ago, and also during the middle of the present century, it becomes necessary to carry the inquiry down to the present date. Although in some respects superseded by more recent returns, the following table for the year 1881-82 may be here given. It deals with certain facts of the North-West Provinces sugar trade after the same plan as has been shown for the year 1848. A comparison of the results is thereby facilitated:—

Statement showing area under Sugar-cane and its outturn in the year 1881-82.

Division.	District.	Area under sugar-cane in acres.	Average outturn per acre of undrained sugar.		Full (or 16 annas) outturn of undrained sugar per acre in maunds.	Total outturn of undrained sugar in maunds.	Average price of undrained sugar per maund.	Estimated outturn of drained sugar.
			In annas.	In maunds.				
1	2	3	4	5	6	7	8	9
							R a. p	
Meerut.	Dehra Dun .	1,280	16	20·5	20·5	26,240	5 9 1	7,872
	Saharanpur .	36,258	11	14·1	20·5	511,237	5 1 3	153,371
	Muzaffarnagar	52,856	15	19·2	20·5	1,014,835	4 5 8	304,450
	Meerut . .	89,851	15	19·2	20·5	1,725,139	4 12 5	517,542
	Bulandshhr .	10,526	14	17·9	20·5	188,415	4 10 9	56,524
	Aligarh . .	1,735	15	19·2	20·5	33,312	4 13 0	9,994
	Total .	192,506	...	18·3	...	3,499,178	4 14 0	1,049,753
Rohilkhand.	Bijnor .	54,925	16	30·0	30·0	1,647,750	4 5 2	494,325
	Moradabad .	44,313	15	28·1	30·0	1,245,195	4 2 11	373,558
	Bareilly .	41,[illegible]40	12	22·5	30·0	930,150	5 7 6	279,045
	Badaun .	15,991	16	30·0	30·0	479,730	4 6 7	143,919
	Shajahanpur .	43,202	12	22·5	30·0	972,045	4 7 7	291,613
	Pilibhit .	29,961	12	22·5	30·0	674,122	4 8 1	202,237
	Total .	229,732	...	25·9	...	5,948,992	4 9 0	1,784,697
Agra.	Muttra .	804	12	22·5	30·0	18,090	4 11 10	5,427
	Agra .	2,523	8	15·0	30·0	37,845	5 1 3	11,353
	Mainpuri .	10,611	14	26·2	30·0	278,008	5 4 8	83,402
	Farukhabad .	16,231	16	30·0	30·0	486,930	4 14 2	146,079
	Etawah .	11,041	16	30·0	30·0	331,230	4 14 2	99,369
	Etah .	16,617	14	26·2	30·0	435,365	4 11 4	130,609
	Total .	57,827	...	25·0	...	1,587,468	4 14 11	476,239
Allahabad.	Cawnpore .	6,233	15	28·1	30·0	175,147	3 9 6	52,544
	Fatehpur .	3,021	13	24·4	30·0	73,712	5 0 8	22,114
	Banda .	4	...	...	20·5	...	4 7 7	...
	Allahabad .	10,253	14	26·2	30·0	268,628	4 13 0	80,588
	Hamirpur .	2,176	13	16·7	20·5	36,339	4 10 9	10,902
	Jaunpur .	52,340	12	18·4	24·5	963,056	3 7 8	288,917
	Total .	74,027	...	22·8	...	1,516,882	4 5 6	455,065

SACCHARUM: Sugar.

Methods of Cultivation

CULTIVATION in N.-W. Provinces.

Area & Outturn.

Divisions.	District.	Area under sugar-cane in acres.	Average outturn per acre of undrained sugar. In annas.	Average outturn per acre of undrained sugar. In maunds.	Full (or 16 annas) outturn of undrained sugar per acre in maunds	Total outturn of undrained sugar in maunds.	Acreage price of undrained sugar per maund.	Estimated outturn of drained sugar.
1	2	3	4	5	6	7	8	9
Benares.	Azamgarh .	75,310	13	19·9	24·5	1 498,669	2 11 2	449,601
	Mirzapur .	16,579	13	19·9	24·5	329,922	4 7 7	98,977
	Benares .	21,003	12	18·4	24·5	336,455	4 9 2	115,936
	Gorakhpur .	67,895	15	23·0	24·5	1,561,585	4 1 8	468,476
	Basti . .	29,448	14	21·4	24·5	630,187	3 5 7	189,055
	Ghazipur .	22,429	14	21·4	24·5	479,980	3 2 5	143,994
	Ballia . .	43,524	12	18·4	24·5	800,842	3 8 3	240,253
	Total .	276,188	...	20·3	...	5,687,640	3 11 1	1,706,292
Jhansi.	Jalaun . .	796	...	...	30·0	...	4 14 2	...
	Jhansi . .	282	13	24·4	30·0	6,880	5 6 1	2,064
	Lalitpur . .	516	14	26·2	30·0	13,519	6 5 5	4,056
	Total .	1,594	...	25·3	...	20,399	5 8 7	6,120
Kumaon.	Kumaon .	...	16	24·5	24·5	..	7 3 1	...
	Garhwal . .	...	...	...	24·5	...	7 14 5	...
	Tarai . .	4,150	14	21·4	24·5	88,810	4 4 9	26,643
	Total .	4,150	...	22·9	...	88,810	6 7 5	26,643
	GRAND TOTAL .	836,924	...	22·9	...	18,349,369	4 14 8	5,504,809

Column 6 shows the amount of maunds of *gúr* which has been ascertained to be the maximum outturn of a good crop. Column 7 shows the result in maunds of *gúr* of the crop of 1881-82 Column 8 shows the average price of *gúr* per maund during the year. Column 9 shows estimated outturn of refined sugar at the average rate of 30 per cent. of *gúr*.

It will thus be seen that in 1881-82 the area under sugar-cane had expanded very considerably in Meerut, Kumaon, Rohilkhand, Agra, and Allahabad, although it had contracted in Benares from that shown in the table for 1848. But, while the acreage in the last-mentioned division was less, the yield was considerably greater; the total outturn in Benares Division in 1881-82 was 56,87,640 maunds; in 1848, 55,61,641. The total returns of yield for the North-West Provinces in these years were in 1881-82, 1,83,49,3 69 maunds and in 1848, 1,05,23,868. Another striking feature may be noted. In the older returns the yield per acre came to 18 maunds, whereas in 1881-82 it was raised to 22·9 maunds. It is perhaps undesirable, however, to comment very specially on the figures for 1881-82, since more recent investigations have in some respects proved these to be defective. **Mr. Schofield's** *Note on Sugar* (to which frequent mention has already been made) gave a review of the information available up to 1887, and the report furnished by the Local Government on **Mr. Schofield's** *Note* added still more recent data. Thus, for example, on the subjects of yield per acre and consumption to head of population it was stated that—

"It is estimated that the net exports of unrefined sugar from the North-Western Provinces are on an average 39 lakhs maunds. This, when deducted from the total estimated produce (180½ lakhs), leaves 141½ lakhs maunds for consumption and an average rate of 17 seers per head of population in the North-Western Provinces. The total average produce in Oudh is now stated to be 27¾ lakhs of maunds (or 21 maunds per acre), which, after allowing for a net export of 5 lakhs maunds, leavees an amount sufficient to give a rate of 8 seers per head of population. As regards the

CULTIVATION in N.-W. Provinces.

Area & Outturn.

difference between the rate of consumption in the North-Western Provinces and Oudh, the Director writes as follows:—

"'Sugar in India is more an article of luxury than one of absolute necessity. Its consumption must, therefore, greatly depend on the extent of cultivation and on the means of the masses. It need not, therefore, be matter for surprise that the rate of consumption should be less in Oudh, where sugar-cane occupies 1·5 acres per 100 acres cultivated against 3 acres per 100 acres cultivated in the North-Western Provinces, and where the mass of the agricultural population are not as well-to-do as in the case of the North-Western Provinces. The agriculturists of Meerut, Rohilkhand, and Benares, the divisions most largely producing sugar, are believed to be on the whole better off than those of other divisions.'

"Large consignments of undrained sugar are said to be sent from Gorakhpore to refineries in Behar, whence it returns by rail in the form of refined sugar."

The table given at page 116 shows the area, outturn, and consumption in the provinces of India worked out to the average of five years. It may, however, by way of concluding this chapter be desirable to furnish the actual area returned for sugar-cane during the past three years:—

	1887-88. Acres.	1888-89. Acres.	1889-90. Acres.
North-West Provinces	960,693	990,219	871,008
Oudh	231,721	238,224	213,318
TOTAL	1,192,414	1,228,443	1,084,326

Oudh has not been very specially alluded to by the author, since to deal with every province of India upon a uniform plan and fairly completely would run to more than double the space that can be afforded in this publication. But it is believed that the reader will be able to learn all of very special interest regarding Oudh from the numerous quotations from district authors given below.

Perhaps the most instructive paper which has as yet appeared on the sugar-cane cultivation of these provinces is that by **Mr. H. Z. Darrah.** This was published by the Government of the North-Western Provinces in 1883, and the greater portion of it seems to have been directly utilized by **Messrs. Duthie & Fuller** in writing their *Field and Garden Crops.* The writer need offer no further apology for republishing the article referred to, than that he has taken the liberty to give one portion of it in this chapter and the remainder in that on sugar manufacture. To allow of this, it has been necessary in one or two instances to re-arrange slightly the paragraphs. It may also be added that **Mr. Moens'** description of the sugar-cane of Bareilly, which **Mr. Darrah** freely quotes, has been left in the form in which it appears in the Gazetteer and thus as a district account.

Distribution. 267

"DISTRIBUTION.—The total area under cane in the whole of the North-West Provinces and Oudh, may be assumed as 9½ lakhs of acres, amounting to 2·5 per cent. on the total cropped area, and 4·8 per cent. on the area under *kharif* crops. Its cultivation is greatly restricted to certain well-marked localities. The natural home, so to speak, of the cane, is the strip of damp country underlying the hills which comprises a large portion of Rohilkhand, Oudh, and the Benares Division. Here it is often grown without irrigation. But the increased facility for irrigation afforded by canals has led to a great extension of its cultivation in the drier districts of the Ganges-Jumna Doáb, notably in the upper portion of the Meerut Division, where it now forms one of the principal staples. It is also grown very largely in the districts of the Benares Division which lie between the Gogra and Ganges, where water is near the surface and irrigation from wells and tanks is much practised.

CULTIVATION in N.-W. Provinces.

Distribution.

South of the Jumna its cultivation is almost unknown, although the occurrence of numerous disused stone sugar-mills in the villages of this tract gives some ground for supposing that it was once one of the local crops. A striking fact in connection with the extension of sugar cultivation in the Meerut Division is its restriction to the three Northern districts of Saharanpur, Muzaffarnagar, and Meerut, although canal irrigation is equally abundant in the two southern districts of Bulandshahr and Aligarh. The explanation lies in the large extent of indigo cultivation in these two latter districts, which has as yet kept the sugar-cane completely in the back ground.

Seasons. 268

"SEASONS.—The sugar-cane season comprises, roughly speaking, a whole year. Sowing commences in February, and the harvesting of the previous year's cane is not concluded till very shortly before this. If, however, cane is to be classified with other crops, it must be ranked with those produced in the *kharif* season, since it is on the warmth of the summer months that its growth principally depends.

Rotation. 269 *Conf. with pp. 129, 145, 150, 187, 215.*

"ROTATION.—A cane crop is, as a rule, preceded by a whole year's fallow, the land not having been occupied in either *kharif* or *rabi* preceding. Occasionally, chiefly in the Sub-Himálayan tract, it follows a *kharif* crop of rice or pulse, when it is known as *kharik* as opposed to *pural*, or fallowed cane, and its produce is estimated to be decreased by ¼th to ⅓rd. Now and then it is even sown immediately after a crop of gram on land which has not been allowed even a half-year's fallow, but this is rare. The rents charged for cane in the Sitapur District are R10-12, R9-9, R8 and R6-12 per acre according as it is grown after a year's fallow (*purali*), after rice (*dhankeri*), after autumn pulse (*maseri*), or after gram (*charreri*). But these are exceptional cases, and the rule for the provinces is that cane requires a year's open fallow; land lying fallow for cane is known as *pándra*.

Mixtures. 270

"MIXTURES.—A crop of melons or onions is occasionally gathered off a cane field, being planted on the ridges of the irrigation beds, and being off the ground before the canes have made much progress. Hemp and castor are frequently grown as a border, but beyond this no subordinate crops are ever mixed with the cane.

Soils and Manuring. 271

"SOILS AND MANURING.—Sugar-cane land is usually good loam or light clay, and is invariably manured except in tracts such as the Himálayan Tarai and the old bed of the Ganges in the Etah District where the ground is saturated with moisture, which is made to supply the place of both manure and irrigation. The weight of manure applied per acre varies between 150 and 200 maunds. In the Shahjahanpur and Muzaffarnagar Districts it is the custom to apply the whole of the available manure to the cane-fields and the manured fields are therefore not collected in a belt round the village site, as is usually the case, but scattered at intervals over the village land. From Fatehpur the practice of herding cattle at night on cane-fields is reported. The manure is applied shortly before sowing and well intermingled with the soil by frequent ploughings.

Tillage. 272

"TILLAGE.—Ploughing commences with the rains and is continued in as opportunity offers till sowing time. During November the land is allowed a rest, it being considered unlucky to plough in that month (Bareilly), possibly because it may encourage the germination of weeds, many of which are seeding then. The number of times to which cane land is ploughed is occasionally as many as 25, and averages about 12 or 15.

Sowing. 273

"SOWING.—Cane is propagated by cuttings or layers and not from seed. The cuttings are made either from the upper portion of the cane, which is of but little use for sugar making, or from the whole cane, and must be always long enough to include two internodes, *i.e.*, three nodes or joints. The young canes are produced from buds which spring from the nodes under artificial stimulation, and with an eye to this, the seed canes are generally kept for some days buried in damp earth, and sometimes even soaked in water for 12 hours before sowing (Allahabad). The cuttings are covered with earth by a third plough following the sower, and, since the rows should be at least a foot apart, it is usual to strike two or three blank furrows between the one in which the seed has fallen and the one next sown. The amount of seed used per acre is about 20,000 cuttings, which represent some 3,000 to 5,000 canes. Cane is occasionally ratooned, *i.e.*, allowed to spring up from the roots of a previous crop, in which case the juice is said to be richer than in the first year, but only ½ to ⅔rds as much in quantity.

Irrigation. 274

"IRRIGATION.—On a comparatively small area cane can, as has already been noticed, be grown without irrigation at all, and over a great portion of Rohilkhand the ground often contains sufficient natural moisture in February to enable sowing to take place without a previous watering. But, as a general rule, this previous watering is required, and between sowing time and the commencement of the rains, waterings are necessary, which vary in number from three or four in the Meerut Division, to eight in the drier districts of the lower Doáb. Occasionally a watering is given in October or November if the rains have ceased early. It may be mentioned that *khari* water, *i.e.*, water impregnated with nitrates, is harmful to cane, seriously affecting the quality of the juice. In the few localities where cane is grown in Bundelkhand, a practice (called *palwár*) prevails of economizing water by covering the ground to a depth of 6 inches with grass and leaves so as to prevent the rapid evaporation of moisture.

CULTIVATION in N.-W. Provinces. Irrigation.

"WEEDING.—Two weedings are generally given, but they play an unimportant part compared with the frequent hoeings which are an essential feature in cane cultivation. The hoeing is performed with a small pickxae, the earth between the rows of canes being thoroughly stirred to the depth of 6 or 9 inches. The first hoeing should take place when the young shoots appear above ground, and from that time to the commencement of the rains, it should be hoed at least three times. When the rains have once set in the crop may be left to shift for itself, and will effectually stifle any weeds which may attempt to compete with it.

Weeding. 275

"HARVESTING.—Cane cutting nominally commences with the Deothan festival, which falls on a date varying in the solar calendar, but generally about the beginning of November. But practically it is generally delayed till a month later, and the cultivator has completely finished his *rabi* sowings. The delay is an advantage in one respect, since the juice of canes cut early in the season, though more abundant, is much less rich in crystallizable sugar than that of canes cut in January and February, and it is probable that it is due more to the slowness of the sugar-crushing process than to any other consideration that cane cutting commences so early as it does.

Harvesting. *Conf. with p. 10.* 276

"DISEASE AND INJURIES.—The most serious injury to cane grown on low lands results from being flooded in the rainy season, and large areas of cane may often be seen during the cold weather reduced to a mere snipe cover by the overflow of the tank or river on whose banks they are situated. Cane also suffers at times from the attacks of caterpillars, one kind called *kanswa* in the Meerut District, attacking the young shoots, and another known as *sildi*, the full grown plants. Jackals are also fond of sugar-cane, and do a great deal of injury, especially to the softer varieties unless the fields are watched at night.

Disease. 277 *Conf. with pp. 121-127.*

"COST OF CULTIVATION.—The average cost of growingl an acre of cane is shown below:—

Cost. 278

	R	*a.*	*p.*
Ploughing (twelve times)	9	0	0
Clod crushing (six times)	0	12	0
Seed (4,000 canes)	8	14	0
Sowing (three ploughings and three men) . . .	1	14	0
Weeding (twice)	4	0	0
Hoeing (three times)	5	8	0
Watching	2	0	0
Cutting	2	8	0
Total .	34	8	0
Manure (200 maunds)	6	0	0
Irrigation (seven times)	12	5	0
Rent	10	0	0
GRAND TOTAL .	62	13	0

The average cost of making a maund of *gúr* has been proved to be R1-6, so that, assuming an outturn of 30 maunds, the manufacturing expenses will amount to R41-4. Adding this to the cost of cultivation we obtain R104-1 as the cost of producing 30 maunds of *gúr*.

"OUTTURN.—The average outturn of irrigated cane calculated in semi-dried compost (or *gúr*) may be taken as 30 maunds per acre in the Meerut, Rohilkhand, Lucknow, Rai Bareli, and Benares Divisions; 24 maunds per acre in the Sitapur and Fyzabad Divisions, and 20 maunds per acre in the Agra and Allahabad Divisions. For the small amount of cane grown in Bundelkhand, an outturn of 18 maunds an acre would be a high average. If *ráb* is made instead of *gúr*, the outturn will be about 8 per cent. more than this, and if *shakar* be made about 3 per cent. less" (*Duthie and Fuller, Field and Garden Crops, N.-W. P.*).

Outturn. 279

A selection from the Gazetteers and other such works may now be given, as these in some respects amplify what has already been said.

CULTIVATION in N. W. Provinces.

Azamgarh. 280

Conf. with pp. 120-126.

AZAMGARH.—"Sugar-cane takes more of the time and labour of the Azamgarh agriculturist than any other crop." "The best soil for cane as a sugar-producer is a good clean clay, especially that known as ***karail***. The preparation of the land, the mode of sowing and the processes of hoeing, top-dressing and harrowing have been described more than once for other districts, and from the account given of them in the settlement report they seem to have no peculiarities in this district. Each root (***thán***) of strong plant should throw up from ten to twenty canes (***gohan***). An acre of fair crop should contain upwards of 90,000 canes. The crop suffers occasionally from blight (***kuswá*** or ***khairá***), which shows itself in the brown withered appearance of the leaves. But its chief enemy is a greenish caterpillar (***dhola***) which destroys the head of the young plant and prevents its growth. Canes attacked with ***dhola*** generally throw out side-shoots called ***pachkhís***, which grow from four to nine inches in length, but these never make up for the damage done to the head of the plant" (*Gaz. N.-W. P., XIII., 48-49*).

Bareilly. 281

Conf. with p. 10.

BAREILLY.—"Notwithstanding its large area, the rice crop yields in value and importance to that of sugar-cane. '***Ikh tak kheti, hathi tak banj***,' say the peasants—that is, sugar-cane is to tillage as the elephant to beasts.' There are thirteen recognised varieties, *viz.*, (1) white and (2) black ***paunda***, (3) ***thun***, (4) ***pándia***, (5) ***dantur***, (6) ***rakri***, (7) ***chun***, (8) ***dhaur***, (9) ***agholi***, (10) ***mittan***, (11) ***baghazi***, (12) ***neula***, and (13) ***katára***. The ***paunda*** varieties are grown only for chewing, others for both chewing and sugar, but most for sugar alone. The method of cultivation varies according to locality. In the uplands the field is prepared by a year's fallow, during which constant ploughings and manurings are administered. Sowings begin, as a rule, immediately after a watering in March-April. A consecrated plough, marked with a red stripe, is followed across the field by another of less hallowed character, bearing mould boards to widen the furrow. Immediately after the second plough walks the sower, or 'elephant' fresh from a feast of sweetmeats and clarified butter. He is adorned with a red frontal mark, with garlands, and silver. The bits of cane, which he throws crosswise (***tirchha***) into the furrow at every short pace, have been stored in a hole covered lightly with earth or moistened leaves. Behind the 'elephant' comes a man named 'the crow,' to adjust such cuttings as have not fallen right into place. The 'elephant' is sometimes accompanied by a third person, named 'the donkey,' who carries at his waist the basket containing the cuttings. The appearance of a horseman in the field, during the sowings, is hailed as a lucky omen. A feast of pulse-curry and other delicacies refreshes, on the completion of their labours, all those engaged in the process. Hemp and castor-oil plant (***andauwa***) are sometimes sown on the borders of the field, and ***urd*** and melons amongst the crop itself. The cost of cane-cuttings, when purchased, varies from R6 to R8 per acre.

"If rain falls in May-June the crop is watered once, and if not, twice; but in some moist tracts no irrigation is needed. From four to seven hoeings are administered in different months. That in June-July, known as the ***Asárhi-khod***, is considered the most important."

Conf. with pp. 10-11.

The sacred observances in connection with the reaping of the cane the reader will find detailed above in the special paragraph on that subject. The writer of the article in the Gazetteer, from which the above has been abstracted, continues:—

"In the Northern Parganahs the field destined for sugar-cane is not allowed a full year of preparatory fallows. The autumn harvest which precedes sowings finds it grown with rice millets (***kodon***, ***bájra***, etc.); but during the growth of the spring crops it at length enjoys a rest. Cane thus grown is named ***kharik***, and its outturn is rather less than that of ***purál***, or cane planted on lands fallowed for a whole year. Fields sown with a ***kharik*** crop after bearing autumn rice are sometimes called ***bartush***. In Aonla, Saneha, and parts of the Baheri ***tahsíls*** the crop is often suffered to sprout afresh after a first cutting as opposed to the ***naulaf***, or crop that is cut but once. Such cane is entitled ***pairi***. Its juice, though in quantity but a third or half that of ***purál*** and ***naulaf*** cane, is of better quality and better adapted for cicaring and concentration. The best sugar-cane is grown in Gurgaya of Richha, along the banks of the Deoha in Nawabganj and of the Katna in Bilaspur. Here the ***ráb*** syrup is finer and sells from ten to twelve per cent. higher than elsewhere. Local calculations show that the produce in juice of a ***purál*** crop is about 72, and of a ***kharik*** crop about 34 ***kacha maunds*** per ***kacha bighá***. The money value of good cane, such as grown in Nawabganj, is R13 per ***kacha bighá*** (R83 3 per acre); of medium cane R9 or 10 (R64 per acre); and of ***kharik***, ***baheri***, and ***khádir*** cane, R7 (R44 12 per acre).

The ***gur*** or ***rub*** prepared from the chopped cane is sold to the sugar-boiler (***khandsári***), who has in most cases advanced money on the crop. The increase during

CULTIVATION in N.-W. Provinces.

late years of sugar-boilers and agents points partly to an extension in this system of advance. In 1848 Bareilly proper possessed 174 *khandsáris* and 346 *árras;* in 1872 the numbers had risen to 561 and 948 respectively. Many landowners now engage in the business, which, owing to the ease of recovering at harvest the money advanced to their tenants, is to them peculiarly profitable. The amount but varies considerably, from R5 or 6 per *kacha bighá* in Baheri to R10 or even R18 in Bilaspur. A written engagement binds the borrower to sell the produce of the crop to the lender at a price fixed in the bond, and to pay on the advance a rate of interest, also specified therein. As the price is always fixed below market rates, and the interest ranges from 12 to 30 per cent. per annum, ruin is too often the result of taking such advances" (*N.-W. P. Gazetteer, Volume V., pp. 559, 560, 561, and 562*).

Benares. 282

BENARES.—"Sugar-cane is the principal agricultural product of the district. It is grown in every parganah, in every village, and by every class of cultivators. In Parganahs Pandrah and Kol Asla it is estimated that there is never less than one-fourth to one-third of the cultivated area taken up with it. In the *tari* lands along the banks of the rivers it is planted in February, and, although perfectly inundated, it does not suffer from this cause, so long as the tips of the leaves remain above the water. In these lands, although it is never irrigated, it grows with great vigour; but it does not yield *gur*, or unrefined sugar, to such an extent as the cane grown on the higher and artificially irrigated lands. In the latter description of lands it is sown between February and the middle of April and in the lighter soils is ready to cut in December, but in the better soils it is left in the ground till January or February. From the 15th of January to the 15th of March are reckoned by the natives the best months in which to manufacture *gur*. After February-March, although the produce is the same, the juice is thin, and the *gur* sticky and of an inferior quality. The lands to be sown with sugar-cane are either ploughed up and allowed to remain fallow from the commencement of the rains, or are sown with *san, urd,* or peas."

The above passage has been given in this place chiefly on account of the mention of a cane that is grown in land often quite inundated. The account of the Benares cane given a century ago in the proceedings of the Honourable the East India Company (and which has been freely drawn upon by the author of this article) is very much more complete than any thing which has since appeared on the subject. It is, in fact, an exhaustive and very able statement of the cultivation of the cane, the manufacture of sugar, and the trade in the product. The reader who may be interested in the Benares sugar production would do well to consult the original article which will be found in the volume of proceedings designated "East India Sugar" and which was published in 1822. The paper on Benares will be found from pp. 183 to 210.

Eta. 283

ETA.—"Ten species of sugar-cane are grown in the district: the *dhor, chin, barokha, paunda, manga, digilchin, gegla, agaul, rakhra,* and *kálaganna*. The cane for seed is cut into four or five pieces and stored until wanted in a place called bijhara. Mr. James writes: 'I saw in *parganah* Nidhpur a very curious arrangement for storing cane for seed. Just outside the village homestead was a square place, somewhat like a miniature cemetery divided off into twenty compartments or vaults. Each compartment has its respective owner, and here the cane is buried every year by the various sharers and taken up at seed-time. Each piece of cane so cut for seed is called a *painra*. It is sown in January, and is ready for cutting in November and December. When just sprouted sugar-cane is called *kulha;* when a little taller it is known as *ikh* or *ikhári*, and when the knots on the cane (*poi*) become distinct and developed the cane is termed *ganna*, and when ready for cutting *gánda*.' The cane is then cleaned (*chhol*) and gathered into bundles (*phándi*) of one hundred each. In this way they are carried to the *kolhú* (or press) where the cane is sliced into pieces (*gádíli*) about three inches long, and placed in the press, which is made of *shísham* or *babúl* wood, and rarely of stone. The refuse or pressed cane is here known as *páta* or *páti*. The juice pours out into an earthen vessel (*bogha*) below and is then taken off to the *karáhi* (or boiler), where it is made into *gur*, or undrained raw sugar. *Ráb* is made by putting the boiled juice into an earthen vessel called *karsi*, when after certain operations it becomes granulated (*rawa parjáta*). The *ráb* is then placed in a bag, and pressed and purified; the solid matter which remains in the bag after pressing is termed *choyanda*, and when dried is known as *khánd*, while the liquid which runs out of the bag is called *shíra*, and is used in making wine and in preparing tobacco for smoking. The scum which

Methods of Cultivation

CULTIVATION in N.-W. Provinces. Eta.

floats on the top during the process of boiling is called *laddoi,* and the whole juice when the boiling is just completed is known as *pág.* The first *bogha* of juice is usually distributed amongst the pressers, village carpenters, and blacksmiths during a ceremony termed *rasyáwal* or *raswái.* The next festival is the distribution of the first *gúr,* called *Jaláwan* by Hindus and *Sinni* by Musalmans when from two to five seers are given away. Sugar pressing work is known as *bhel,* and the large balls of *gur* are called *bhelis.* The large *bheli* weighing about seven seers and called *phúnka* is seldom made here" (*Gazetteer, Vol. IV.,* 28).

GHAZIPUR.—See foot-note page 164.

Gorakhpur. 284

GORAKHPUR.—This district, like that of Benares, figures prominently in the correspondence which took place towards the close of the last century. A little later on it came into even greater note, during the second effort which was put forth to establish European sugar plantations in India. This subject will be found fully dealt with in the special chapter above, which sets forth the historic facts connected with the effort to establish plantations. One point only need be here repeated, namely, that **Mr. Wray** (one of the most scientific planters of his day) owned or was in charge of a sugar factory and plantation in this district. The writer has failed to find definite records of the plantations and factories that were actually once upon a time worked, but **Mr. Wray** was not the only planter who spent many years in Gorakhpur in the anxious endeavour to make sugar manufacture a large and profitable undertaking.

"An extensive trade is carried on in coarse *chíní* (sugar), for whose preparation numerous factories have been built. The crop, which pays well, demands an immense amount of care and attention during the earlier stages of its cultivation. The process begins directly after the old crop is reaped. Cuttings of stalk, about 5 or 6 inches in length, are placed between layers of damp straw in a hole in the ground. This hole being closed up with a coating of earth forms a kind of hot-bed. The pieces of cane are called *porha,* and a bundle of one thousand an *anwala.* Some six of these bundles, costing from R1-8 to R3, are required for the *pakka b ghá.* After about eight days shoots sprout from the cuttings, which are dug up and planted in a field prepared with great care during the end of the rains and cold weather. It is necessary to plough the field some dozen times, besides taking a plank (*pahia*) over it to break up the clods. By March or April these preliminaries are complete, the shoots are planted lengthways in the furrow, about one inch apart and two inches below the surface; and the soil is smoothed down with an unweighted plank. Sometimes the cuttings are after three days extracted and replanted, the plank being passed over them; but this is not always done. Manure is spread over the surface, about 4 cart-loads to the *bighá* being sufficient. Partitions are then made in the field, which is carefully irrigated, the water being spread over the whole surface by means of a broad wooden shovel. From this time until the downfall of the rains the crop requires frequent watering, but it is of great importance that the soil should not be sodden by too much at a time. The labour required if the rains are late is extreme, as irrigation may be needed twenty times over; but when once the monsoon has broken, little remains to be done until the harvest in *Pús* or *Phálgun.* Fields in which rice or *kiráo* have been previously reaped are considered best for cane, unless land which has been a whole year fallow can be obtained. If rice has been cut, the field is ploughed up and the cane sowed at the end of *Phálgun;* if *kiráo,* at the middle or end of *Chait* (March-April). Two crops are often raised from the same plant, the stumps being left in the ground after harvest and frequently watered. New shoots sprout in May or June, and a fair crop is often secured. The more intelligent husband men assert, however, that this unrest is bad for the field. The name of the second crop is *peri* (or *banjar*). There are four kinds of sugar-cane:—

(1) *Mahgujur,* which grows to the greatest height.
(2) *Saroti* } both yielding *gur* syrup in abundance.
(3) *Bhaunwarwár* }
(4) *Barokha* or *katarha,* yielding little *gur* and used chiefly for eating."
(*Gaz. N.-W. P., VI.,* 325-326.)

Jaunpur. 285

JAUNPUR.—"One of the most important crops, to which the enterprising cultivator devotes his greatest time, labour, and capital, is sugar-cane. This is considered the most profitable of all agricultural products, but the extent cultivated is limited by the large outlay of money and labour which it requires. The kinds sown in this district are all small. The largest and best is called *nasganda;* the second *paunra.*

CULTIVATION in N.-W. Provinces

Jaunpur.

Serotia is the thinnest. *Kawai*, the worst kind, is sown along the edges of the field to disappoint and deceive the pilfering wayfarer.

"The cultivator who can afford it will leave fallow for six months or for an entire year the land in which he intends to sow sugar-cane. The land is previously prepared by three to five ploughings. Every kind of decayed vegetable and animal manure is applied. It is a favourite practice to fold sheep upon it, two rupees a hundred being paid to the sheep-owner. The season for sowing lasts from February to April. The lowest joint including the root is cut into pieces a foot in length; these are soaked in water and placed about a foot apart in furrows, also a foot distant from each other. After sowing the manuring is repeated, and the field is dug up by the hand with a hoe or pick, five or six times.

"The season for cutting lasts from November to January, varying with the time at which the cane was sown and the rain fall of the year. The juice of that first cut is whitest and clearest; of the last cut is reddish and contains most sugar. Men, women and children all turn out to cut the cane. It is then chopped into pieces three or four inches in length, called *gareri*, and is passed at once into the mill. It is a cylinder of stone fixed deep into the ground, the top of which is hollowed, to form a mortar, with a great pestle of wood turned in it by oxen, and weighted by the driver sitting on a board attached to it. The stone is often handsomely carved with figures of birds and elephants, and is worth from R40 to R100. As it is often owned in partnership by several cultivators, and also because the cane must be cut while fresh, the mill is kept working day and night. When nearly all the juice is expressed, water is added, and this last diluted juice, *pániwár*, is given to the labourers. The exhausted cane is used for boiling the sugar, and its ashes for manure" (*Gaz., XIV., 17-18*).

Mirzapur. 286

MIRZAPUR.—The early records of the endeavours to improve sugar cultivation in India are replete with passages bearing on this district. The Honourable the East India Company had an experimental plantation and factory at Mirzapur, the Superintendent of which, **Mr. Carden**, figures prominently in the efforts which were put forth, at the beginning of the century, to advance the cultivation of cane and the manufacture of sugar in India. **Mr. Carden** had charge also of the Government Rum distillery which was worked in conjunction with the sugar factory at Culna. In a letter dated 13th January 1804 we learn that "the Governor General in Council had authorised the manufacture in the present season of 100,000 gallons of rum, at the Honourable Company's distillery at Mirzapur; the estimated expense is stated by the Board of Trade to be R50,000, exclusive of charges of manufacture." But soon after the Company appears to have become dissatisfied with the working of their Mirzapur distillery, for in 1807 we read that the Honourable Company "were satisfied that rum could not be manufactured at so cheap a rate by the Company as it could be purchased from private European distilleries." Orders were accordingly issued for the disposal of the whole of the Company's property at Mirzapur. Whether this order embraced the disposal of their sugar factory and plantation as well as their distillery is not very clear, but although the Company early discontinued direct ownership of sugar factories, about the period mentioned above, large factories continued to be worked for some years later, and from the expression used above in connection with the purchase of rum it seems probable "the private European distilleries" were associated with the sugar factories. In this connection it may be added that one sugar-planter (**Mr. Colley** of Munsurcotah, Ganjam, *p. 93*) in a letter dated February 7th, 1800, admitted that the profit from sugar manufacture was derived from the rum prepared from the molasses.

Rum. 287

Conf. with pp. 93, 96, 158, 320-321, 306.

In the Gazetteer the following passage occurs regarding the sugar-cane cultivation of Mirzapur:—

"But of all sowings that of cane, the most prized and profitable of crops, is attended with the greatest ceremony. The day is kept as a sort of festival, and half a dozen canes and a day's wages are usually given to the labourers. After the cane slips have all been planted, an entire cane, called the '*raja*,' is buried in the centre of the

CULTIVATION in N. W. Provinces.

Mirzapur.

field. Then follows a scramble among the boys employed for the remaining cane slips, and a good deal of rough, good humoured horse-play. The same evening the women of the house, or hired labourers, if the farmer is of high caste, carry ash-manure to the fields, singing as they go, and on their return receive five pieces of sugar-cake each.

"The cutting of the cane is preceded by special ceremonies. The date chosen is always the *Deo-uthán ekádasi*, the 26th day of the month *Kartik* (October-November). The inevitable Brahman is called to the field, with rice—flour, turmeric, flowers—materials for a burnt offering (*hom*). After this the cane is adorned with the farmer's wife's silver collar (*hasuli*) and the burnt offering is made. A bundle is then cut, by way of first fruits, and carried home and eaten. The regular cutting then begins, and is carried on, at intervals, as the mill can work off the crop."

"Sugar is largely grown in the Gangetic valley, but there are no refineries worked according to European methods, and although the production of the various forms of country sugar is a flourishing industry at Náí Bázár near of Bhadohi the greater part of the produce of the cane is exported in the form of *gur*. Palm-sugar is made to limited extent from the *khajur* palm which is so abundant near Chunár. A good tree will produce a chhiták of *gur* every third or fourth day, and this *gur* fetches about three times the price of the corresponding produce of the cane.

Moradabad. 288

MORADABAD.—"Here, as in Sháhjahánpur, the manufacture of sugar in its various forms is a flourishing and highly profitable business. **Mr. Smeaton** writes: 'The demand for cane-juice has been all along on the increase. All who have a little capital embark it on sugar advances. Thrifty cultivators who have saved money—and these are numerous—are to be found in partnership with *banias* in the sugar business. *Zamindárs* themselves are finding how profitable it is, and many, among the wealthiest, have been lately taking to buying up the sugar of their villages. A regular competition has set in, and the tenantry have therefore found no difficulty in disposing of their juice to advantage. The influx of wealth formerly alluded to has of course greatly stimulated this competition. Many more persons now have capital than before: a great portion of these can afford to live more frugally, and therefore take a lower rate of profit than the old capitalists.'

"The measure by which the cane-juice (*ras*) is sold is almost always the *karda*, equal to a very little over 50 Government (or 100 *kachcha*) maunds. The system by which a sugar manufacturer obtains his supplies of juice, includes the giving of advances by him to the cultivator, and these are usually three in number. The price to be paid is fixed either on the first or second advance. The average produce of an acre may be put at 175 Government maunds, the value of which would be about R75 and the cost of cultivation and crushing R50, leaving the cultivator a profit of R25, though this varies enormously, according as the cultivator employs hired labour or not. The profits have increased since the Railway was opened by about R14 per acre. During the actual crushing operations, the hired labourer earns on an average R8 a month besides his food. He has to work hard, and runs some risk of having his hand crushed by the mill. The processes of manufacturing *gur*, *ráb*, and *khand* have been described in former notices.

"*Gúr* is made all over the district and is either made by *khandsális* (sugar manufacturers) or by the cultivators themselves. In the latter case it is usually sold to petty dealers at so many *bhelis* a rupee, a *bheli* being a ball of *gúr* weighing about 2½ Government, or two local seers. The purifying process by which *ráb* is turned into *khand* has been described elsewhere. The average percentage of *khand* to *ras* is about 7; **Mr. Butt** puts it at only 5·8, but *zamindárs* whom **Mr. Alexander** questioned on the subject put it as high as 8, and **Mr. Moens**, in his Barielly report, makes it seven. The manufacture is chiefly carried on at Sambhal, Belári, Kundarkhi, and Chandausi."

Shahjahanpur. 289

SHAHJAHANPUR.—It has already been explained that one of the ablest district reports on sugar that has hitherto appeared is that written by **Mr. Butt** regarding this district. If space could have been afforded it would have been useful to reprint **Mr. Butt's** report as it stands. The alternative course of giving the review of it that appeared in the Gazetteer was, however, thought preferable since the chief facts are there compressed into a third of the original space. The reader who may wish further details should, however, consult **Mr. Butt's** able report (*in the Revenue Reporter, North-West Provinces (1874), Vol. III., No. 1; see also below, pp. 282-292*), or the abstract of it as given by **Mr. Currie** in the Revenue Settlement Report of the Shahjahanpur District:—

CULTIVATION in N.-W. Provinces.

Shahjahanpur.

"Sugar-cane is cultivated all over the district, but chiefly within a radius of 15 to 20 miles round the city of Shahjahanpur, and least of all in the southernmost parganah, Jalálabad, for which, however, there is a special reason in the prejudice of Thákurs of that parganah against its cultivation. The percentage on the total cultivated of land under cane was found by Mr. Currie to be 5·6, and of land prepared area for the following year 3·9.

"The areas and percentages for each táhsil were in 1867-68 as follows:—

Tahsil.	Area in Acres.		Percentages.	
	Actual cane.	Prepared for next year.	Actual cane.	Prepared for next year.
Sháhjahánpur . . .	10,415	6,017	5·75	3·5
Jelálabad	984	*Nil.*	·75	*Nil.*
Tilhar	11,820	8,382	6·25	4·5
Pawáyan	18,245	15,006	7·5	6·
District Total .	41,464	29,405	5·6	3 9

"For the whole district the areas, in the three years for which crop areas have been furnished by Mr. Fuller, were in 1878-79, 63,680 acres; in 1879-80, 30,234; and in 1880-81, 35,266

"In river-valleys and low alluvial lands (*khádar*) the cultivation is much less careful than on uplands (*bángar*), the land is much less ploughed and worked and no irrigation is needed. The hardier and tougher kinds of sugar cane are grown, and the yield is comparatively less: and, besides this, the crop is liable to partial injury or total destruction by floods; so that the *khádar*-grown sugar cane bears about the same relation to *bángar*-grown, irrigated and manured sugar-cane that *bhúr*-grown barley does to irrigated wheat, as regards their culture and care respectively.

"So much has been written on the cultivation of sugar-cane that it seems unnecessary to detail the various processes which, except in a few minor points, are identical in this and the neighbouring districts of Bareilly and Farukhabad. The following account of the planting given by the late Mr. Currie may perhaps, however, be quoted without incurring much risk of repetition, as he alludes to differences observed in this district.

"The planting usually takes place in February and March, the time depending on the cultivators having leisure from the cutting, pressing, and boiling of the last crop.

The field is first ploughed, a man with a bundle of pieces of cane from eight to ten inches in length following the plough and dropping the pieces in lengthwise about a foot apart into the furrow; next the furrows are smoothed over and filled up with the clod-crusher (*patela*). Ordinarily the top part of the cane, from about a foot below the actual arrow or head, is used for seed, and only about 1½ to 2 feet of the cane.

"Some four or five of the immature stalks, which contain little or no expressible juice, are for this purpose cut from the full-grown canes. These cane-cuttings are tied up in bundles and earthed over to keep them from drying, till required for planting six weeks or two months later.

"The land lying fallow for cane is called *pandri*, and cane or any other crop sown after fallow is called *porach, polach* or *polcha*, in contradistinction to *khárag* or *khárik*. The reason why the *pandri* area is always less than the area actually under cane is because a large amount of cane is cultivated *khárag*, following rice, *bájra*, or *kodon* in the previous autumn; but even then the land is fallow for at least three months. It must not be supposed that rice and sugar alternate for several years in the same field, for of course this is never the case.

Ratooning.

290

Conf. with pp. 59, 76, 77, 78, 128, 143, 151, 177, 181, 195, 215, 226, 247.

"Ratooning (*peri rakhna*). *i.e.*, leaving the roots in the ground to sprout again and produce a second crop, is seldom resorted to except for food-canes and exceptionally even for them."

"The irrigating, hoeing, and cutting processes are the same here as elsewhere.

"The cultivator usually presses and boils his own canes, delivering the juice (*ráb*) to the manufacturer (*khandsáli*) who, as a rule, pays the cost of the removal. When the cultivator is in a position to work on his own capital and not on the advances made by the manufacturer, he frequently makes *gúr* (a coarse brown sugar), instead of *ráb*. The main difference between *gúr* and *ráb* is that the former is boiled rather longer over

CULTIVATION in N.-W. Provinces.

a hotter fire and is made up into moderately dry solid balls (*bheli*), whereas *ráb* is concentrated to only a little over crystallizing point, retains much more moisture than *gúr*, and is not intended for keeping, but for immediate conversion into manufactured sugar.

Shahjahanpur.

"Besides the system just described there is another called the *bel* system, prevailing chiefly along the western edge of the district adjoining Bareilly and Budaun, from one of which it seems to have been introduced. It consists in the manufacturer taking raw juice (*ras*) instead of concentrated (*ráb*) and boiling it himself. **Mr. Currie** writes:—

"The cultivator presses the juice all the same setting up his mill (*kolhu*) at the *bel*, which is merely a collection of mills and a boiling-house. There are usually from 12 to 20 mills at a *bel*, but sometimes as many as 30. Each jar (*matha*) of *ras*, as filled, is taken over at once by the manufacturer, who receives the refuse for fuel. The only expenses saved to the cultivator are the cost of one labourer (the boiler) and the hire of the boiling-pan. The real advantage to him is that the *ras* is taken over indiscriminately, without any tests as to whether it is good or bad, and he is relieved of the loss consequent on a small yield of *ráb* or of *ráb* of indifferent quality. The advantage to the *khandsári* is that *ráb* is prepared in larger quantities and on a more careful process, and as there remains no motive for fraud or deception as to the quality, it is, as the *ráb*, more uniform and superior to that purchased ready-made from the cultivators.

"The difference in the manufacture of *ráb* under the *bel* system consists in the boiling-pans being set up in sets of five over a furnace with a long flue, the largest pan into which the raw juice is first placed being furthest from the furnace over the far end of the flue, and the smallest, into which the heated juice is brought gradually, being immediately over the furnace. An experienced confectioner (*halwai*) is employed to conduct the boiling, and *sajji* (impure carbonate of soda) and other alkaline substances, with decoctions of bark and plants, are used to correct acidity and purify the syrup.'

"The *bel* system is said to have been extended rapidly since the Mutiny and to be likely to supplant the other method in which the cultivator himself manufactures the *ráb*.

"The manufacture of sugar-cane is, however, a subject which will be found treated of below in another chapter, and reverting to the cultivation of the plant, the following brief remarks on the cost of cultivation may be added to what has been stated already. Good sugar-cane lands have an average rental of about R15. There is little (if any) difference in the cost of cultivation of what turns out to be a good or an inferior crop. The net expenses of cultivation, omitting items which balance one another on the credit and debit side, *e.g.*, seed and cutting, amount to R43-7 per acre, made up as follows: rent R15, ploughing R8, carriage of manure R1-8, planting R1, irrigation R9-7, hoeing and tilling R6, carriage to the mill R2-8. The profits per acre vary from R36 to R115, the extremes being for the lightest and the best soils" (*Gaz., IX., 46-49*).

PANJAB.

IV.—PANJAB.

References.—*Baden Powell, Panjab Products, 304-308, 383; Sugar Statistics of 1848 (Delhi District); Agri.-Hort. Soc. Ind., Trans.:—V., Proc., 112; VIII., 157; Jour., VI., Proc., 116; VII., 231; VIII., Sel, 164; Gazetteers of each district too numerous to be separately quoted; A very extensive Official Correspondence and Reports down to 1891.*

Area & Outturn. 291

Area, Outturn, and Consumption.—It will be seen by the table given above (p. 116) that the Government of India views the normal area in this province (for the five years previous to 1888) to have been 354,000 acres. The yield came to 96,29,000 maunds of coarse sugar, or an outturn of 27·2 maunds an acre. The Panjáb imports, however, vary largely from the North-West Provinces and Karachi by rail and to a less extent from Sind by boat and by road from the North-West Provinces. The net imports during the past three years (by rail) were in 1887-88, 20,16,727 maunds, expressed as *gúr* or coarse sugar; in 1888-89, 15,76,311 maunds; and in 1889-90, 15,29,720 maunds. This may be taken as an average net import, during these years of 17,07,586 maunds. But no provision has been made for road and river traffic in that calculation. Allowing these sources of additional supply to cover errors, and net exports by transfrontier routes, we learn that the Panjáb had 1,13,36,586 maunds

CULTIVATION in Panjab.

Area & Outturn.

of coarse sugar in 1889-90. Reducing that amount to pounds, by allowing 80lb to the maund and accepting the population at 18¾ millions, it would appear that the consumption came to 48lb (or 24 seers) per head. This is a slightly higher figure than that given in the table at page 116, as that of the average of the five years preceding 1888, but **Mr. Schofield** in arriving at the consumption of 22 seers allowed for only a net import of 10 *lakhs* of maunds. It will be seen from the tables at pages 367-368 that the average net import of *gúr* was as stated above, during the three years from 1888 to 1890, 17,07,586 maunds on the rail traffic alone. The Native States within the Panjáb are said to have produced 4,28,000 maunds in the normal year, and by the census of 1881 they had a population of 3,861,683. It would be difficult to work out the proportion of the net imports that went to these Native States, but the allowance for the province must be reduced by that amount, so that a consumption of 22 seers per head is probably not far from correct. The area under sugar-cane has not materially increased during the past ten years:—

	In thousands of acres.
1880-81	386
1881-82	377
1882-83	401
1883-84	348
1884-85	335
1885-86	331
1886-87	354
1887-88	366
1888-89	391
1889-90	325

It may serve a useful purpose to exhibit in this place the distribution of the Panjáb sugar-cane cultivation by showing the amounts in all the districts that possessed during each of the past three years over 15,000 acres:—

	1887-88.	1888-89.	1889-90.
Delhi	26,702	29,403	17,387
Karnal	17,371	15,625	8,822
Amballa	23,592	27,601	23,736
Hoshiarpur	32,810	39,285	31,707
Jullandar	43,873	42,274	36,564
Lodhiana	15,327	14,905	11,311
Amritsar	29,559	26,521	21,153
Gurdaspur	48,861	57,035	54,565
Sialkot	39,644	44,805	41,981
Gujranwala	17,224	19,948	22,205
TOTAL ACREAGE IN THE PROVINCE	366,698	391,060	325,562

It will thus be seen that while the area has fluctuated to some extent, the decline in certain districts has been on the whole compensated for by the increase in others. The chief districts of sugar production in order of importance are Gurdaspur, Sialkot, Jallandar, and Hoshiarpur. The Financial Commissioner in a report on the sugar-cane of the Panjáb published in 1883 says that the sub-montane tracts from the Chenab to the Jumna constitute the chief area of the province. The reader will find so much of interest in the district accounts, which may now be given, that it does not

CULTIVATION in Panjab. Area & Outturn.

seem desirable that the author of this review should attempt a sketch of the sugar-cane cultivation of the province as a whole. The Financial Commissioner, reviewing the reports which had been obtained in 1883, gave a review, but had to admit that the diversity in the figures precluded the formation of averages that would have any value when applied to the province as a whole. Much useful information was then furnished, however, by the Financial Commissioner. The following passage may be given as an introduction to the series of district reports, as it furnishes a general sketch of the systems pursued in the province—:

"The mode of tillage and the times of the year in which the various processes are performed vary but little in the different districts. Sugar-cane is propagated from cuttings, each containing one or more of the joints of the cane, from which, when they are buried in the soil, several shoots are produced, and these grow into canes. It is absolutely necessary that the soil be very finely pulverised, or the shoots would not make their way to the surface. It is for this reason that the land is ploughed and re-ploughed for so long a time before the cane is planted. It is generally gone over not less than 10 or 12 times, the '*sohága*' also being used to break clods and reduce the earth to a fine and even condition.

"The Jat cultivators of the main sugar-producing districts repeat these processes an almost incredible number of times. In Hoshiarpur it is proverbial that sugar-cane requires 100 ploughings, and from 60 to 100 ploughings are stated to be the practice in Gurdaspur also. The amount of manure used in the Delhi district is estimated at about 11 tons an acre; but this quantity is exceeded in Gurdaspur and Amritsar, where 600 and 800 maunds, equal to 21½ and 28½ tons, respectively, are applied. The operation of planting takes place about the month of March. The cane cuttings are laid horizontally in a furrow only a few inches apart, the furrows themselves being also very close together; the quantity of cane planted is 20 or 25 maunds, equal to about three quarters of a ton. The ground is usually loosened with a hoe at the time the shoots should begin to appear. It is also constantly weeded while the crop is growing, and sometimes receives a further top-dressing of manure after it is above ground. Irrigation has to be almost incessantly continued during the heat of the summer until the commencement of the rains: about six waterings at intervals of from a week to a fortnight are the usual requirement, but if the rains are deficient, double that number are required. As a rule, the stiffer and closer soils require more frequent watering than those that are porous and absorbent: after the rains when the crop is ripening it is again watered to bring it to maturity. Cutting begins in October or November; but, as it can be done no faster than the operation of crushing proceeds, it often continues till the following February or March."

Bannu. 292

BANNU.—"Of highly remunerative crops two deserve special mention, sugar-cane and turmeric. Their cultivation is almost entirely confined to the richest parts of Bannu proper. The area under sugar-cane at settlement was a little over 4,000 acres. Both crops require large quantities of manure and repeated irrigation. The cane used in setting is cut into pieces about nine inches long so as to leave the knot or joint in the centre of each. It is then hand-planted, piece by piece, horizontally, in February or March, sometimes in prepared soil, but generally in the midst of a wheat or barley crop. About R12 worth of cane to the acre are so used. But fresh planting only occurs once every fourth and sometimes fifth year, as three or four crops are cut from the same root.* Those of the second and third years are the best. After the crop in which the cane has been set is removed, the soil is loosened and weeded, and if there were none such before, a low mud wall or hedge is run round each plot. The cane is of two sorts, red and white. The former is the superior, but it is also the more

* That the statement is incorrect that Ratooning is not practised in India the writer has repeatedly pointed out.—*Conf. with p. 128.*

delicate being very sensitive to frost. Both varieties are very thin in the stem, and grow in clumps very close together. The crop begins to ripen about the end of October and is cut by degrees between November and the end of March. The clumsy wasteful oil press of the Panjáb (*kohlú*, local *gauri* was till lately employed for extracting the juice; but within the last five * years fifty-six English presses have been imported, and are immensely popular, and the iron roller mills are now in almost universal use. The *gúr* produce is very inferior and dirty." **CULTIVATION in Panjab. Bannu.**

"From first to last the cultivation of the cane is careless. There is no division of labour. The juice is boiled down in iron pots to about one quarter its original bulk, by which time its consistency is that of treacle. It is then put to cool in wide-mouthed wooden or earthenware vessels, and when cooled, the stuff is made up into round balls of about 2¾ seers each. This is *gúr*. Little sugar is made, the people not having the skill to manufacture it, or perhaps the juice being in most *tappas* too poor to crystallize. The yield of *gúr* is very uncertain. Of the many causes which tend to diminish the supply of juice, frost in December and January is the most baneful, and most frequent. The average yield of *gúr* per acre is over 12 maunds, and the price current from ten to twelve seers the rupee, hence the average gross profit per acre may be set down at from R40 to R50. But the best lands in the best *tappas* (Suráni and Mítakhel) produce up to R32 maunds an acre, which would give a gross acreage profit of from R120 to R165. A little of the large thick cane known as *páunda*, and only used for chewing, is grown about Kálabágh, and yields enormous profits. It has lately been introduced in Bannu proper and in Miánwálí. Its cultivation is rapidly extending" (*Gazetteer, Bannu, p. 143*).

Delhi. 293

DELHI.—Sugar-cane is the most important and profitable crop of the *kharif* harvest in the Delhi and Sunipat Bangar tracts. The average acreage under cane in the district for the last ten † years is given as 4,347.‡ The land taken is the best in the village, that is to say, some of the best in the village is taken every year; it is a sign of weakness of resources when cane follows cane on the same ground. Nor without manuring is the cultivation profitable. It is not usual to try for a *rabi* crop when cane is to be planted in the spring; if this is done the latter will suffer by being planted late (*pachetr*). Ratooning (leaving the roots to produce a second crop in the succeeding year called *murıdaik*) is uncommon now, though in old times it was often practised. The change may be put down to the decreased fertility of the soil, or as the *zamíndárs* themselves say to the increase in the resources as shown in the greater power to buy seed, and the greater number of hands available for labour. There are three kinds of sugar-cane known in the district:— **Ratooning.** *Conf. with pp. 59, 76, 128, 195 215, 226, 247.*

"(1) *Lálri*, said to be the original kind and considered the best, as no insects attack it. This is the only kind actually used in the district.

"(2) *Mirati*, very productive and white, but if the *gúr* is kept long it gets worms and it is weak also in the rains and sometimes balls.

"(3) *Soratha*, white and productive. Good for sucking but sticky. Not so subject to worms as *mirati*.

Paunda or *ganna* is distinguished from the ordinary sugar-cane by its thickness. It requires more water for its cultivation and *gúr* is not made from it. Its only use in fact—often a very profitable one—for eating; it is sold in the bazar at prices varying from ½ to 1 or even 1½ anna the stick. The kind first sown is *mirati*, then *soratha*, and *lálri* last. *Mirati* is quickest in springing up. A speciality is said to exist in *lálri* that it can be reproduced from any knot of the stalk (*ganda*), whereas for *mirati* and *soratha* only the *top* knot of each stalk will do.

"Sugar-cane for seed is put in clump (*bijghara*) in *Phagan*, where the earth keeps it moist and fresh, a damp situation being considered good. What is kept in the house is for use; it does not keep long. The ploughing generally begins in June unless there is a crop tried for in the *kharif* preceding the cane crop. If a *zamindár* has enough ground he will avoid doing this. When the *kharif* crop is taken the ploughing for sugar-cane begins in December, and is continued at intervals according to

* The Settlement Report in which this statement was first made, appeared in 1878: the Gazetteer in which it is repeated was published in 1883-84.

† This statement was made by Mr. R. Maconachie in the Revised Report of Delhi Settlement originally conducted by Mr. O. Wood from 1872 to 1877 and completed by Mr. Maconachie from 1878-80.

‡ This is apparently a misprint for 30,447, but the tables for the year (or years?) of settlement give the total as 30,782 acres. The acreage returned in 1848 was 6,319, and the outturn 1,27,141 maunds of *gúr*, or 6 maunds 28 seers an acre.—*Ed., Dict. Econ. Prod.*

SACCHARUM: Sugar. Methods of Cultivation

CULTIVATION in Panjab.

Delhi.

Ratooning.

leisure and other circumstances, the number of times varying from 5 to 12. The first two ploughings may well be made one directly after the other, but the subsequent ploughings should come at intervals. For the first ploughing either rain or a watering (*palewá* or *paléó* is necessary. Sometimes the land is dug (with a *kasi* or *kahi*) for the first time and this is fully equal to two ploughings. No cash estimate of the cost of this can be usefully made as it is never done by hired labour."

"The quantity of manure used is very large; from 3 to 6 four-bullock waggon-loads go to a *kacha bighá*. This at the lowest estimate gives $3 \times 3 \times \frac{8}{5} \times 20$ maunds = 288 maunds = nearly 11 tons to an English acre. The *zamindárs* urge strongly that without such manuring the land will not be fairly productive. The time for putting in the manure begins in January-February and goes on to the end of February-March and sometimes even after planting. After manuring the land is ploughed unless of course it has been sown. Ploughing takes place in the end of February-March and may be continued through March-April, but the best time is the beginning of the latter month. Water is given before planting. Furrows are made regularly along the field, and a boy follows the plough putting in the seed pieces of cane (*gandiri*), which must have one or more joints in each piece, horizontally at regular distances, usually rather less than a foot along the furrow. The seed stalks are taken out of the clump; one or two men cut it up, as one cannot do it alone. Another man carries it to a place where it is put in, four or five are wanted to plant for one plough. There is, however, no lack of hands, as all the young boys of the family help in this, in order to get the holiday food which is given on planting day. The food consists of rice, sugar and *ghí* and mixtures of these, and such food giving is called *Mah Kálí* or *gúr bhata*; the work begins in the morning and goes on till it is done, three yoke of oxen can get through ten *kacha bighás* a day. One yoke ploughs and the other two follow with the *sohágá* (clod ch usher). Water is given a month after planting, and if the rains are good three subsequent waterings are enough, if they are not as many as five may be necessary at intervals of a month. Cultivation of cane by well-irrigation is not uncommon in the Khádar of Sunípat, but is not usually if ever met with in the Delhi *tahsíl*. In Ballabgarh there are three or four villages which have it. Delhi too has some in the Dáhar Circle from natural flooding. A fair well may water $\frac{3}{4}$ *bíghá* a day."

"Hoeing is carefully kept up; the number of times depends much on the character of the season, and varies from five to nine or ten. The first time comes a few days only after planting. A man's fair work per day at hoeing is put at three *biswas*. When the canes get high, they are generally tied together at the top. Cutting begins in October; it is a practice for Hindus not to begin till after the *Dashra* (September-October). Hired cutters get R3 a month and their food, the *zamindár*, unless lazy, does much himself in this. A two-ox waggon should cart one *bíghá's* cane in a month, but the animals do other work probably besides. Rent paid by *zabti* (special rental) is about R5 per *bíghá*, but in some villages it goes even up to R9. It is taken at the time the Government revenue falls due, and does not depend on the quality of the crop. No difference is made in the rent, whether in the previous *kharif* (season) another crop was taken, but when the land was left fallow it is called *tapar*.

The expenses of cultivation may be thus summed up:—

	R	a.	p.	Pakka bíghá. R	a.	p.
Planting (ten times)				10	0	0
Manure				5	0	0
Seed				5	0	0
Irrigation				4	2	0
Price of water	3	2	0			
Cleaning of water-course	1	0	0			
Hoeing				4	0	0
Tying of canes				2	0	0
Cutting and stripping				7	0	0
Rent				6	0	0
Carriage to the *kolhú*				3	0	0
Planting (estimated)				2	0	0

The *kolhú*, or sugar-mill, is made of four kinds of wood, first quality *sál* (**Shorea robusta**, second *kikar* (**Acacia arabica**), third *siris* (**Albizzia**), fourth *farásh* (**Tamarix**); *kikar* is the one most commonly used. The mechanism of the *kolhú* is the same as in Shahjehanpur."

S. 293

CULTIVATION in Panjab.

Delhi.

Ratooning,

"A *kolhú* complete costs R80 or R90, or even more, the work being made as durable and thorough in every respect as is possible to the not inconsiderable skill of the local carpenter. The *láth* (or pestle) often breaks, and must be replaced at the cost of a rupee. It is always made of *kikar*. The wages of the carpenter who looks after the *kolhú* are considerable. The produce of about 40 *bighás* of sugar-cane is pressed in one *kolhú;* a good many proprietors unite generally in working it. They bring their cane themselves from the field, and put it together, reckoning their shares by the number of oxen they each have. A *kolhú* lent on hire is said to cost R7 to the hirer, but it is often more than this. The men who own the cane, almost always own the oxen that work the *kolhú*. Four kinds of work are distinguished in the *kolhú*. Two *pindias* put the short cut pieces of cane (*gırariyán*) into the *kolhú*, and take out the cane straw *khói;* one man relieves the other at this arduous work, which is also rather dangerous for any but a left handed person. Wages R10 to R15 a month. Two *guriyas* who cook the *gúr*. Four *jhonknewalas* who keep up the fire, and dry the *khói*. Two *muthiyas*, who feed the *pindias* with cut up canes, put into a basket. The man who sits on the *páth*, driving the oxen, is not a hired labourer, but one of the proprietors. Two men are employed with each pair of oxen. The sugar-cane is generally cut by the proprietors, or by hired labourers at two annas a day each. The *kolhú* goes on day and night, but the workers are divided into day and night batches. A *matka* holding twenty *sers* is filled with the pressed juice in about an hour; and the oxen do this twice before they get taken off. The juice is thrown into the *kúnd*, a large earthen jar. From there it is put into the *karai*, or cooking cauldron, and is boiled slowly till it becomes pretty thick, and then it is conveyed into a second vessel smaller than the first, and the boiling process goes on till the *gúr* becomes thick and consistent enough to make the *bhélis* or *gúr*-balls. These are always four *sers* each. The place where the cooking goes on is called a *gurgói*. It is merely a thatched shed with a hollow floor to allow of the *kasais* being placed in it, and underneath them the cooking fires. Molasses (*ráb*),* and coarse sugar (*shakar*) are not made in this district, or if made, very rarely; and would of course be a more delicate process than the primitive one above described; yet this too requires care. If the boiling is too prolonged it spoils the *gúr* and diminishes its selling value. Delhi district *gúr* goes to Bághpat, Biwáni in Hisar, and Rewári and Fírozpur, Jhirka in Gúrgaon. The zamíndár generally manages his *gúr*-making himself, and there is no commonly received rate of sale, but Bághpat rates more or less influence the market. There is no custom of *katauti* † as in Shahjehanpur. The weight of juice turned out is commonly two-fifths of the sugar-cane. The straw is used for burning in the *gurgói;* it is good for nothing else, and from the juice one-fifth of its weight will turn out in *gúr*" (*Gazetteer, Delhi, 113-120*).

Confusion in names. *Conf. with pp. 114, 115, 131, 133, 136, 229, 252,-255, 285; 298.*

294

The above account is only slightly altered here and there from the original form, which appeared in the Settlement Report. The only serious departure is in acreage, and the writer has by the foot notes (*p. 179*) ventured to correct the figures as given in the Gazetteer to those of the Settlement Report. It will be observed that the above special report on the sugar interest of Delhi has been drawn up on the same plan as Mr. Butt's detailed report of the sugar culture and manufacture of Sháhjahánpur briefly reviewed in the remarks regarding that district under the section of the North-West Provinces.

Gujranwala.

295

GUJRANWALA.—"Sugar-cane is the most valuable crop of all for its acreage. It is grown chiefly on the river lands of Wázírábád and in the whole Charkhari mehal of parganahs Wázírábád and Gujránwála. Notwithstanding the manure, irrigation, and labour necessary to secure a good crop, it is the most remunerative of all produce. By the measurement papers, as compared with *patwáris'* yearly papers, it appears that the growth of sugar-cane has doubled within the last five years ‡ and the people are year by year more alive to the value of the crop. Sugar-

* Molasses is not synonymous with *ráb*, but is the crude treacle that drains from the raw sugar (*ráb*).

† Rents paid by contract rate for the whole cultivation.

‡ The account given in the Gujránwála Gazetteer (p. 52) is word for word the same as the above, which appeared nearly twenty years before the date of the Gazetteer. It is therefore not known whether the crop has continued to gain in popular favour to the same extent as that noted for the five years previous to 1866. The returns given in the annual publication of *Agricultural Statistics of British India* show the area occupied by sugar-cane to have been 19,782 acres on an average during the three years ending 1890.

CULTIVATION in Panjab.

cane is usually a *kharif* crop. After careful preparation of the land it is sown in February, and the crop ripens in November-December, in which months one or more sugar-mills will be found at work in nearly every village of parganahs Wázír-ábád and Gujránwála" (*Settlement Report for 1866-67, pp. 27-28*).

Hoshiarpur. 296

HOSHIARPUR.—"Sugar-cane requires a good soil but is seldom grown in the highest manured lands; the soils in which it is usually sown are *chhal rohi, jabar*, and *maira*. The greater part of the land under sugar-cane in this district is unirrigated; the rainfall is good, and the soil has an inherent moisture which precludes the necessity for irrigation; *chhal, jabar*, and *maira* will stand a little drought without much harm; *rohi* requires more rain, but with good rain or irrigation the outturn is splendid. The area recorded under sugar-cane is 29,117 acres,* of which only 3,553 were irrigated. There are two ways of preserving the seeds:—

(1) When the pressing begins the top joints of the canes are cut off to the length of four or five knots and tied up into bundles called *púla*, each sufficient for sowing one *marla* of land (about 23 square yards); these bundles are then buried upright in the ground till required. The top † joints are closer together, and the outturn in number of canes from such seed is probably greater than if the whole cane were cut up; but the size and the strength of the cane in the latter case are greater.

(2) The number of canes required for seed are left standing in the field till wanted, when the whole cane is cut up and sown.

Where sugar-cane is liable to injury from frost, the latter plan cannot be followed, and this appears to be the only reason in some parts of the district for the seed being cut early and buried in the ground. The *pona* cane seed is always buried, being most easily frost-bitten. The top shoots of the cane, called *ag*, form good fodder for cattle, and are considered the perquisite of those who cut and strip off the leaves from the canes. As a general rule, a cultivator keeps some of his best canes for seed. A Jat cultivator devotes a great deal of time and manual labour to the cultivation of this crop, and it is doubtful if his mode of tillage can be improved upon. Sugar-cane is generally sown upon land which has had wheat in it the previous year, so as to allow nine or ten months for preparation of the soil; but it sometimes follows an autumn crop of maize in dry lands, or of rice in marshy. In some special plots the old roots of the cane are taken up immediately after the crop is cut, and the same land immediately resown. When it follows wheat, ploughing is begun in May and continued at intervals, according to time and means available, through the rainy season, till the wheat sowings are commened in September and October. After an interval one or two more ploughings are given, and then all hands are required for working the sugar presses. Ploughing operations are begun again in January and February, and continued till the seed is sown in March. The *sohágá*, or clod-crusher, is used after every two or three ploughings. The people say land should be ploughed 100 times for sugar-cane, but it seldom gets more than 25 or 30 ploughings." There is a saying:—

"Seven ploughings for carrots,
A hundred ploughings for sugar-cane,
The more you plough for wheat,
The greater will be the gain."

"Great importance is attached to the pulverisation of the soil after the ploughings. The seed is sown in March in the following way: A furrow is made with a plough, and a man walking behind drops the seed in and presses it down with his foot at intervals of a foot between each seed. The furrows are made as close as possible to one another. Afterwards the *sohágá* is passed over the field to cover up the seed. The soil is then constantly loosened and weeded with a kind of trowel (*bagúri*), until the cane attains a height of two or three feet in the rains. This hoeing, called *godi*, is very important, and the more labour expended on it the better is the outturn of sugar-cane. After the canes are two or three feet high, nothing more is done until they ripen in November or December. Sugar-cane is always sown thick, and no attempt is made to strip off the lower leaves when it has grown up. The quantity of seed required is about two maunds per *kanál*, or 20 maunds an acre. The price of seed varies, but averages about R5 an acre. The cane is liable to various diseases and ravages of insects, the local account of which is as follows:—

Diseases. 297 *Conf. with pp. 121-127*

(1) White-ants attack the layers when first set, especially if the land is not well weeded at first. There are also destructive insects called *garuna* and *bhond*, the

* The average acreage for the past three years has been given as 34,601 acres.

† The editor has on more occasions than one drawn attention to the fact that it is incorrect to say the Natives of India do not use the tops as seed.

CULTIVATION in Panjab.

Hoshiarpur.

latter a kind of black beetle, which attacks the young shoots. The cane sown earliest is most liable to attacks of white-ants.

(2) *Tela*, a small insect, comes on the full grown canes in dry years.

(3) Frost is also destructive under the same conditions as *tela*. Sugar-cane is more liable to injury from frost in *chhal* land.

(4) Rats do much damage. For a remedy the tops of the full-grown canes are tied together in lots of 15 or so. This gives light below and checks the wandering instincts of the rats. The tying together of the canes is also a preventive against frost bite, and supports canes which have attained to any size. In good *chhal*, where fresh alluvial deposits can be depended on, the roots of the cane are sometimes left in the ground, and produce two or three and sometimes more years in succession.* This system is called *monda*. The outturn the second year is almost equal to that of the first; the third year a fourth less, and after that still less. *Monda* saves a great deal of trouble, but is only feasible in good alluvial lands. After the canes have been cut the land is ploughed a few times to loosen the earth round the roots, and the usual weeding and hoeing take place. As a rule, little or no fresh manure is applied. A not uncommon practice, when sugar-cane is quite young, is to cover the field with the leaves of *chhachra* (**Butea frondosa**) to keep the soil cool during the hot months of May and June. The leaves rot in the rains and add to the fertilization of the soil. Very little irrigation is required in this district. *Jabar* and *chhal* crops are not irrigated at all; in other soils, if available, water is applied first before sowing and afterwards three or four times until the rains set in. After that the land is only irrigated if the rains are deficient" (*Gazetteer, pp. 95-97.*)

Jhang. 298

JHANG.—*Sugar-cane* is grown for *gúr* in the Gilotar and adjoining villages of the Kálowál iláká in the Chiniot tahsil. In Chiniot itself and Maghiána it is grown to some extent and sold in the bazaars, but is not made into *gúr*. Sugar-cane grows best in a rich loam well manured, in or near the Hethár, where water is very near to the surface. If it is *once* flooded by river water so much the better, but floods are dangerous. Sugar-cane requires constant waterings, and if, as in Maghiána, the well is assisted by a *jhallár*, it is so much the better for this crop. Not only does a *jhallár* raise more water, but a change from well to river water seems to greatly benefit the cane. There is a good deal of uncertainty about this crop, and this, combined with the immense amount of labour needed and the long time that it occupies the ground, has brought it into some disrepute in Maghiána, where rice has, of late years, to a large extent taken its place. Sugar-cane is never grown near Maghiána as a sole crop. Vegetables and *chena*, one or other, sometimes both, always accompany it. Land cannot be ploughed too often for sugar-cane, and must be heavily manured. The cuttings are planted in trenches and lightly covered over with soil, and a watering is at once given when the cane plants are three months old, and about 2 or 2½ feet high, the trenches are filled up and manure put to their roots. At this time any other crop that may have been sown with the cane is pulled up. The cane is ready to cut about the middle of *Katik* (October-November), but it is often in the ground till *Phagan* (February). I have seen cane uncut in March. The crop is hoed four or five times. At first it is watered every fourth day up till the 1st *Jeth* (May-June) or later, and once a week from that time until it ripens. The worst enemy of sugar-cane is the white-ant, and constant waterings are needed to keep this pest away. Jackals are also extremely fond of cane. They 'chew,' but do not eat it. Frosts are injurious if they are early. A frost-bitten cane loses a large portion of its juice." (*Settlement Report of Jhang, 1874-80, p. 96*).

Conf. with pp. 121-127.

Kangra. 299

KANGRA.—"*Sugar-cane* is largely cultivated about Kangra, and the culture is gradually extending.† Some parts of the Palum valley, 3,200 feet above the sea, are famous for the cane they produce. In Noorpoor and Golier the plant is rarely met with. In *taluquas* Nadown and Rajgeeree, a portion of every holding will be devoted to sugar. There are two or three varieties, *chum, eikur, kundiari*, and a juicy kind called '*púna*,' raised only for eating. The quantity produced in different parts of the district is very unequal. Noorpoor and Hureepoor are dependent upon importations, while Palum and Nadown supply the neighbouring parts of the Mundee Principality.

Conf. with p. 19.

"*Peculiarities of Hill Canes.*—The cane, although not so thick and luxuriant in its growth as in the plains, contains a larger proportion of saccharine matter. The molasses of the hills is notoriously sweeter and more consistent than the produce below. The juice is expressed by means of cylindrical rollers revolving over each

* This is the West India practice known as ratooning.

† Relative to the area in other districts that of Kangra is, however, small. The average for the three years ending 1890 was only 4,594 acres.

CULTIVATION in Panjab.

Kangra.

other, and the motive power is usually a team of four bullocks. This process is universal over the Panjáb and is a great improvement on the mortar and pestle (*kolhu*) used in Hindustan. In the wilder hills, towards Dutwal and the Sutlej, a very rude and primitive method of extracting the juice is in force called '*Jhundur*.' I have not seen it, and scarcely understand the description; but the leading feature appears to be that no cattle are employed: strong active young men employ their force, and the cane is somehow compressed by the sudden closing of two frames of wood." (*Settlement Report of Kangra by G. Carnac Barnes* (*1850*), *p. 27*.)

Karnal. 300

KARNAL.—" Regarding the cultivation of cane in this district it is stated that it 'grows' best in fairly stiff loam, and worst in sandy soil. It likes abundant rain, and will stand a good deal of swamping, though too much makes the juice thin. It is occasionally grown in flooded land without irrigation; but the yield is precarious. Its cultivation is far more laborious than that of any other staple. The land must be ploughed at least ten times, and worked up to the finest possible condition. The more manure given the better the yield; and it is never sown without. If the soil is impregnated with *reh*, the juice becomes watery and yields but little sugar. The amount of seed is fixed in the following curious manner:—As many canes as will make up a total length of 21 hands is called a *panjá* or handful. Twenty-one *panjás* are a *púli* or bundle: and 30 bundles are sown in one acre. The word *panjá*, though common, generally, in the Panjab, is not used or known in the tract in any other connection than this. The seed cane will be worth R5 to R6 per acre. The seed cane is buried in the ground till wanted next year. Generally whole canes are buried; but a custom is growing in the Khádar of using only the top *18 inches or so of the cane for this purpose, as this is the piece which makes the best seed and gives the least juice. The seed cane is cut up into *pári* or slips with two knots in each and they are laid down a foot apart in the furrow by a man following the plough, who presses each in with his foot. The plough has a bundle of canes tied under the share to make a broad furrow. Nine men will sow an acre in a day. The *sohágga* is then passed over the field. On the first day of sowing sweetened rice is brought to the field, the women smear the outside of the vessel with it, and it is then distributed to the labourers. Next morning a woman puts on a necklace and walks round the field, winding thread on a spindle. This custom is now falling into disuse. Three days afterwards they hoe the field all over with *khodális* and follow with the *sohágga*. This operation is repeated four times at intervalls of 10 days. Ten men will work an acre in a day. The field is then watered.

"The *pachcha* is then given. They spread more manure, hoe it in, beat the ground to consolidate it, water, hoe, and beat again, and so on two or three times; it taking 20 men to do an acre once over in a day. A month after this they water again, and go on hoeing and watering till the rains set in. During the rains it must be weeded once *at least*; after the rains it is watered once or oftener according to the season, and if it shows any tendency to droop, tied up in bundles (jura) as it grows. As soon after *Diwálí* as the cane is ripe it is cut. If it is allowed to stand too long the flower (*nesari*) sometimes forms and it is then useless. Cane is occasionally grown a second year from the old roots, and is then called *mánda*. The cane is cut down and dressed (*chola*) on the spot, by stripping off the leaves and cutting off the crown (*gaula*). These are given to the cattle to eat This work and the crushing are done by the association of *lana*, there being one pair of bullocks for every acre of cane. When the cane is brought to the press, it is cut up into *ganderi* or pieces 6 to 8 inches long. The press is started on Sunday; and an altar called *makál* is built by it, where five *ganderis* and a little of the first juice (*ras*) expressed and 1¼ seers of the first *gúr* made are offered up, and then given to Brahmans on the spot. The press is tended by two *peria*, who feed the press with cane, opening out the canes in the press with an iron spike or kail, and driving new canes well in by beating them on the top with a leather glove faced with iron (*hatarki*); two *muthiás* who drive bullocks and hand the cane from a basket fastened on the beam to the *peria*; two *kárigars* who look after the boiling and make the *gúr*; and two *jhúkas* or firemen who feed the furnace. For each 24 hours the *perias* get 9 seers of *gúr*, and their food and tobacco; the *muthiás* get two seers and food, the *kárigars* 8 seers; and the firemen the same. The *kárigars* are generally *jhinwars*, and get 2½ seers on the first day in the name of Báwa Kálu, their *guru* or spiritual chief, a certain amount of juice and cane is also given to the workmen. The blacksmith gets ¾ of a seer, the carpenter 2 seers, and the potter ½ seer of *gúr* per diem. The hire of the iron pans is from R9 to R12 each per season.

Flowering. 301 *Conf. with pp. 8, 9, 11, 44, 47, 61, 83-88, 109.*

* European writers commonly speak of the Natives of India being ignorant of the possibility of using the tops for seed. *Conf. with pp. 125, 140, 184, 240.*

CULTIVATION in Panjab.

Karnal.

"As the juice runs out it is received in an earthen vessel (*báha kundi*) sunk in the ground and holding some 60 to 70 seers. A press will crush an acre of average cane in five days, working night and day. The juice is dipped out of a *kúndi* into a large pan called a *kúnd*. When the *kúnd* is full the juice is transferred to a *káraha, kárâhi* or *bel*, an iron evaporating pan let into the top of a furnace and is there boiled. After being similarly treated in a second evaporating pan, the inspissated juice is put to cool in a broad shallow earthen pan (*chák*) and worked about with a flat piece of wood (*háti hátwá*). When cool it is called *gúr*, and is ladled out with a wooden spoon (*dolera*) and scraper (*musad*), and made up into balls (*bheli*) weighing 4 seers each, of the shape of a cottage loaf. The first ball is given to the Bráhman at the *makál;* the others are taken to the *bania* and credited to the account. The crushed cane (*khói*) is used to feed the fire with. The cane saved for next year's feed is buried n a corner of the field. Young sugar-cane is attacked when about one foot high, by a worm called *kansua*, especially if the east wind blows. A smut called *ál* also attacks it under the same circumstances. Mice do much harm and also white-ants and frost." (*Gazetteer, pp. 173-175*).

Lahore. 302

LAHORE.—"Sugar-cane is but little grown at present in this district, and what is grown is generally sold in the larger cities or towns for eating purposes. It is the exception to see a *belan* or sugar-mill in any of the villages; the only parts of the district in which the cane is grown is to the north-east of the Sharakpúr parganah, or south of the Lahore tahsil. Around the city of Lahore, a good deal of the large thick cane called *pona* is raised, but *gúr* or sugar is never extracted from this species, and it is merely grown for sale in the bazár."

Ludhiana. 303

LUDHIANA.—"Sugar-cane is grown in an area of 13,213 acres, but its importance is much greater than is indicated by this, for the value of the yield is about 10 times that of an ordinary unirrigated crop, and the total annual value some R12,00,000. It is almost entirely grown for the manufacture of some saccharine product (called *kátha*, cane) but in a few villages the *ponda* or eating variety is raised. *Kátha* cane is grown in the unirrigated lands of the Samrála Bét (where it occupies 12 per cent. of the whole area), and of a few Ludhiana villages; and at the wells in the uplands of Samrála and the eastern portion of Ludhiana, the best crop being perhaps that raised about Malandh. The cultivation in the Dháia and Bét is much of the same description. Cane is sometimes the only crop in a field for two years, especially in outlying ones where the supply of manure is limited. It may also be grown with the aid of a great deal of manure on land just cleared of another crop of cane or of a *rabi* crop of wheat, but, as a rule, it occupies the land for three harvests following a *kharif* of cotton. Cane is not grown in the fields next to the site, but generally at a little distance. It is always planted, if possible, on land that has been cropped with cotton, and in the Upper Dháia Circle of Samrála, we find that the area under the two crops is nearly the same. The rotation is generally—

Rotation. *Conf. with p. 170, 215.* 304

YEAR.	RABI.	KHARIF.
First . . .	Ploughing.	Cotton.
Second . . .	Fodder, etc.	Ploughing.
Third . . .	Ploughing and cane sown.	Cane.

and back again to cotton, giving a cane, a cotton, and a fodder crop with, perhaps, a little grain in three years. The cane field is selected next to the well, as the crop has to be kept alive during the hottest months, and always gets more frequent waterings than any other. The land is ploughed not less than 7 or 8, and up to 20 times; the more ploughings the better. All the available manure has first been spread over the fields and is ploughed in. The planting is done from the middle of *Phagan* to the middle of *Chét* (March). The seed consists of joints (*pori*) cut from the last year's crop, which have been kept covered up in pits in the field. In planting them one man goes along with a plough, and another follows, laying down the joints at intervals of 6 or 8 inches in the furrow. The plough in making a new furrow covers up the former one; and the whole field is finally rolled. The canes spring from the eyes (*ankh*) of the joint. About 4 or 5 canes will come from one joint. Then follow waterings at intervals of 7 or 8 days in the uplands, and hoeings after each of the first fewer waterings. The fields are very carefully protected by stout hedges. In the Bét there are no waterings, and seldom any hoeings and the fields are quite open. The cane in the uplands grows to a height of 8 or 10 feet, and when it becomes heavy is protected by several stalks tied together. In the Bét the height is only 5 or 6 feet,

CULTIVATION in Panjab.

Ludhiana.

and this precaution is not necessary. There is altogether a great difference in the modes of cultivation in Dháia and Bét due chiefly to the difference of natural conditions, and partly to the different habits of the cultivators, those of the Dháia being industrious Játs, and those of the Bét apathetic Muhammadans, of the Rajpút and Gujar tribes principally.

"The method of extracting the juice is much the same both tracts. Cutting goes on all day in the field, each cane being stripped and the flag at the top, with the small joints immediately below it, being removed. In the evening the seed joints are separated from the flag (which is then used for fodder or for feeding the boiler furnace) and tied up in bundles for seed. The cane is carted to the *belna* or mill which stands trust outside the village site. The pressing is done in a *belna* or mill, the cane being passed between two horizontally wooden rollers, and the juice running into an earthenware jar set to catch it. In a corner of the enclosure of the mill stands the boiling shed, and the juice is taken into this and boiled in pans." "In the Dháia, the ját requires no assistance in the boiling and turns his juice into lumps (*bheli*) of *gúr* or into *shakar* which he may dispose of that very day. In the Bét the money.lender has invariably advanced money on the crop, and his man does the boiling. Here the produce when boiled assumes the semi-liquid form of *ráb* which is taken in part payment of the debt. Sugar cane is the crop invariably converted into cash may be said to be the revenue-paying one. It is very valuable, otherwise it could never have held its own so long, for it occupies the land the better part of two years; and in Dháia the labor of cultivation is incessant. Bullocks stand the work at the wells, and in the *belnas* for only a few years; and the cultivators are never tired of complaining of their hard life. These objections make it a dangerous crop to any but the most thrifty classes. The játs keep out of debt because it is in them to do so; but the Muhammadan of the Bét will tell one that he is a victim of the sugar-cane crop, and he is right to some-extent, for he has not the qualities which would enable him to subsist while his crop is growing" (*Gazetteer Ludhiana, 135-137*).

Muzaffargarh 305

MUZAFFARGARH.—"Sugar-cane is grown in every part of the district except the *thal* and the inundated tracts, but as it requires capital and abundant manure, it is mostly found in the neighbourhood of towns. The selection of land for the next year's sugar-cane is generally made in land which has just borne wheat. Beginning from May, the land is ploughed from four to five times during the summer. After each ploughing the land is rolled and levelled. It is then heavily manured. Between Septembe- and January a crop of turnips is taken off the land. The local theory is that turnips do not exhaust the land. The truth is that fresh unrotted manure is used, which requires the extra handling and watering caused by raising a crop of turnips to make it sufficiently decomposed to be beneficial for sugar cane. After the turnips have been removed the ground is ploughed eight times more and rolled. The sugar-cane is then sown in February and March. Canes for seed have been stored in mounds covered with earth called *tig*, since the last years harvest. These are now opened and the canes are cut into peices with one or two knots in each. A plough, which has a brick fastened about the sole, to make a wide furrow, is driven through the ground. A man follows, who places the pieces of sugar cane continuously in the furrow, presses them down with his feet, and covers them with earth. Then a log of wood called *ghíal* is dragged over the field. After planting the only care which sugar cane requires is constant watering and hoeing. Judging from the accounts of other countries hoeing is not done often enough.

"Two hoeings are considered sufficient. Sugar-cane is cut and crushed from the end of November to the end of January. The double-roller crusher is always used. In the mode of crushing and the management of the labour required, this district does not differ much from the rest of the Panjáb, but a few points may be noticed. There are ten attendants on the crusher and *gúr*-boiler. The crusher is worked from midnight to 10 A.M. This time is chosen as less severe on the animals than the day, and also because fewer visitors come at that time, it being *de-igneur* to give every caller as much juice as he can eat, drink, and carry away. It is very difficult to estimate the net profits of growing sugar-cane. Each owner extracts his own juice and makes his own *gúr*. The wages of the workmen are paid in every possible form. For instance, the *dhora*, or man who puts the cane into the crusher, gets one blanket and a pair of shoes; when crushing begins a quarter of a *seer* of *gúr* and a *chhiták* of tobacco every day, R4 and three seers of *gúr* per month; a present of R1 to R2 when the work is finished, and 15 seers of wheat under the name of *bijrái*. Then again some attendants are paid by the *kachcha* month, and some by the *pakka* month. A *kachcha* month is a calendar month. A *pakka* month is when a sugar-crusher has worked 30 times, and each time has extracted 10 maunds of *gúr*. A *pakka* month may occupy two calender months or more. We get into more certain ground when the owner of the cane has no sugar-crusher. He pays the owner of the crusher one-third of the outturn of *gúr*, the owner of the crusher supplying all attendants and animals required for working it. With the best knowledge that he had at his disposal Mr. O'Brien calculated for assessment purposes that the average net profits of sugar-cane per acre were R53. An intelligent Zaildár and sugar-grower of Jatoí, told him that the net profits of a successful crop were R200 per acre. The Extra Assistant Settlement Officer, who was a land owner and sugar grower estimated the outturn at 15 maunds of *gúr* per acre" (*Gazetteer, p. 92-93*). **CULTIVATION in Panjab.**

V.—CENTRAL PROVINCES.

CENTRAL PROVINCES 306

References.—*Special Report by Mr. J. B. Fuller; Agri.-Hort. Soc. Ind. Trans:—Vol. III., 72, 173; Proc., 91; IV., 190; Proc., 40; V., Proc., 51, 65; VI., 90, 95; Proc., 7, 28; VII., Proc., 116; ✓III., Proc.. 438; Journ., XIII., Proc., 9; New Series, Vol. VII., Proc., 179, 180 (Sorghum); Gazetteer, numerous pages; Official Correspondence, etc., etc.*

Area, Outturn, and Consumption. The table furnished at page 116 shows the average area devoted to sugar-cane, in these provinces, during the five years previous to 1888, as having been 54,000 acres. This yielded 15,40,000 maunds of coarse sugar, a quantity which gave an outturn of 28·5 maunds per acre and which, when corrected so far as possible by the trade returns, exhibited a consumption of 9·00 seers per head of population. Since 1888, the area has been returned as 48,524 acres in 1887-88; 49,650 acres in 1888-89; and 52,899 acres in 1889-90. It would thus appear that in the British portions of these provinces the sugar-cane area has slightly decreased. The Native States under these provinces manifested a normal area of 7,200 acres of cane with an outturn of 2,16,000 maunds, or 30 maunds an acre. **Area & Outturn. 307**

It is perhaps unnecessary to exhibit the distribution of sugar-cane cultivation in these provinces further than to indicate the shares taken by districts that have over 3,000 acres under the crop:—

	1887-88.	1888-89.	1889-90.
Saugor	3,498	3,421	5,500
Betul	8,069	7,836	7,987
Chhindwara	5,603	5,603	5,667
Chanda	3,232	3,333	3,141
Bhandara	4,819	5,434	5,640
Bilaspur	5,798	6,522	6,500
Sambalpur	5,282	4,353	4,353
Total acreage in these provinces	48,524	49,650	52,899

Having furnished the above brief abstract of the most recent figures that have appeared, it does not seem necessary to do more in this place than republish the main facts brought out by Mr. J. B. Fuller in a report on the sugar-cane cultivation of these provinces, published originally in 1883, since the methods of cultivation and manufacture have in no material respect changed:— *Conf. with pp. 108-113.*

"Excluding Feudatory States the total area under sugar-cane is returned as 53,937 acres. This is very greatly below the area which has been accepted in previous years. In a report submitted by this Administration to the Government of India in 1879, the total area under sugar-cane was given as 93,927 acres. The agricultural returns which were appended to the administration report for 1881-82 show it as

SACCHARUM: Sugar.

Methods of Cultivation

CULTIVATION in Central Provinces.

Area & Outturn.

87,084 acres. The area now returned is compared below division by division, with that returned in 1881-82:—

	1881-82.	1882-83.
	Acres.	Acres.
Jubbulpore Division	8,407	7,444
Nerbudda Division	18,969	13,921
Nagpur Division	17,798	15,027
Chhattisgarh Division	41,910	17,545
TOTAL .	87,084	53,937

"The decrease is large in every division, and in Chhattisgarh is enormous. It can only be explained by the assumption that the 1881-82 returns were absolutely incorrect. The present figures have been arrived at after special enquiry, and must be accepted as superior to the unchecked returns of *malguzars* and village accountants on which previous years' statistics have been founded. It is, however, at the same time, possible that the very fact of a special enquiry being held, led to a deliberate under-statement of area. The agricultural classes are notoriously suspicious of any attempts of Government to collect information on matters concerning them, and commonly believed that increased knowledge will certainly result in increased taxation. Sugar-cane cultivation has been undoubtedly very greatly falling off in the Jubbulpore, Nerbudda, and Nagpur divisions, since the import of sugar has been facilitated by railway communication, and I think that the decrease in the area now returned for these divisions is due to the returns for the last few years having been to a great extent each a mere copy of the one preceding it. The decrease now returned represents therefore the extent to which cultivation has fallen off in several years and not in a single season. The decrease in the districts of the Chhattisgarh division cannot, however, be thus accounted for. The area now returned as under sugar-cane in the Raipur and Bilaspur districts is compared below with that (1) recorded at Settlement, and (2) returned in 1881-82:—

Area under Sugar-cane

	At Settlement.	1881-82.	1882-83.
	Acres.	Acres.	Acres.
Raipur	3,390	18,618	2,349
Bilaspur	4,592	13,843	10,196

"Comparing the present returns with those collected at Settlement, it is seen that cane cultivation in Raipur has fallen off 37 per cent., but in Bilaspur has more than doubled. In the third Chhattisgarh district (Sambalpur) no land measurement has ever been made, and the area under cane has been up to the present year roughly estimated as being between 9,000 and 10,000 acres. The Deputy Commissioner (**Major Macdougall**) has now returned it as being only 1,558 acres, but I have not accepted this figure. Sugar-cane is more thickly grown in Sambalpur than in any other district, and there is hardly a village in the khalsa which does not contain some acres below the *bund* of the village tank. I have consequently increased the area to 5,000 acres, and I may add that in the opinion of **Major Bowie**, who is intimately acquainted with the district, this is considerably under the proper mark.

"Speaking roughly, therefore, the sugar-cane is cultivated most largely in the three following tracts: (1) the Satpura districts of Chhindwara and Betul, (2) the districts of Bhandara and Chanda, and (3) the Sambalpur and Bilaspur districts. This localization is the result of irrigation facilities which are afforded in the first tract by wells, and in the two latter tracts by artificial tanks, which are often of very large size. That in the village of Nawagaon in the Bhandara district is 24 miles in circumference.

"The great decline in cane cultivation since the opening of the Great Indian Peninsula Railway has already been noticed. I exemplify it by comparing the cane

CULTIVATION in Central Provinces.

Area & Outturn.

area now returned in certain typical districts with that ascertained at settlement (in 1866-68):—

	Area returned at Settlement. 1866-68.	Area now returned
	Acres.	Acres.
Saugor	5,106	3,217
Jubbulpore	4,056	1,980
Seoni	6,037	864
Hoshangabad	1,437	648
Nimar	420	199
Betul	7,000	6 412
Chhindwara	6,175	4,432
Bhandara	14,579	6,256

"The decline in cane cultivation has been made the subject of very careful enquiry by **Mr. Nicholls**, Deputy Commissioner of Narsinghpur, who has ascertained the area under cane in each of the last 12 years in a block of 31 villages in his district. His figures are epitomized as below:—

	Four years, 1871 to 1874.	Four years, 1875 to 1878.	Four years, 1879 to 1883.
	Acres.	Acres.	Acres.
Average area under cane	1,198	827	590

"The cost of raising sugar in these Provinces is considerably higher than that in the North-Western Provinces, and its cultivation is gradually receding before the large imports which the North-Western Provinces annually pour into these Provinces by rail. The most distinctive soil of the Central Provinces is that known as 'black cotton soil,' and there can be no doubt of the unfitness of this soil for sugar-cane. It is true that its great retentiveness of moisture enables it to produce a poor crop of cane without any irrigation whatever (as will be noticed further on), but its unsuitability for irrigation and manuring debars it from producing a good crop of cane without a disproportionate expenditure of time and trouble. The future of cane cultivation on black soil in these Provinces may be gathered from the Bundelkhand (black soil) districts of the North-Western Provinces where the only traces of sugar-cane which can now be discovered are the disused stone sugar-mills, which are still found lying here and there about the country.

"One of the features of the enquiry which forms the basis of this report, was to have been the ascertainment of the produce of certain selected fields in each district by actual experiment. I regret, however, to say that this appears to have been carefully effected in only four districts. The results are summarized below:—

District.	Plot experimentally cut.	Outturn of undrained sugar per acre.
	Acres.	℔
Damoh	·05	2,800
	·05	1,920
	·05	2,400
	·05	1,800
	·05	1,600
	·05	1,400
	·05	2,000
	Average of seven experiments	1,988
Mandla	·006	1,916
	·006	2,083
	Average of two experiments	1,999

SACCHARUM: Sugar.

Methods of Cultivation

CULTIVATION in Central Provinces.

Area & Outturn.

District.	Plot experimentally cut.	Outturn of undrained sugar per acre.
	Acres.	℔
Chhindwara . . .	3·0	2,733
	1·3	3,718
	2·5	2,952
	0·75	2,397
	0·8	3,659
	1·0	2,502
	0·9	2,802
	Average of seven experiments . .	2,966
Balaghat . . .	1·0 *	982
	0·6 *	1,400
	1·0 *	1,372
	1·0 *	1,194
	1·0 *	1,468
	0·75†	1,653
	0·5 ‡	720
	0·25§	296
	Average of eight experiments . .	1,165

"Experimental cuttings are also reported to have been made in the Jubbulpore and Betul districts, but the results obtained in the former district are discredited by the Deputy Commissioner as unreliable. No details are given of the experimental cuttings made in Betul, but the general average resulting from them is returned as 2,017℔ per acre. To these facts I may add (1) that the outturn of a good cane-field grown on black soil with manure and irrigation on the Nagpur Government Farm, was 4,720℔ per acre, and (2) that a plot of cane which was cut, pressed, and the juice boiled down into sugar in my presence in a village in the Sambalpur district yielded at the rate of 3,466℔ per acre. This plot had been selected with considerable care as being of average quality, and in the opinion of the village headman it was rather below than above the average.

"Judged by these facts the average outturns returned from some districts appear exceedingly ridiculous. As examples I give below the average outturn returned for the Jubbulpore, Hoshangabad, Bhandara, and Sambalpur districts:—

District.	Average outturn of undrained sugar per acre.	
	In tahsil in which outturn highest.	In tahsil in which outturn lowest.
	℔	℔
Jubbulpore	1,148	420
Hoshangabad	488	240
Bhandara	1,074	560
Sambalpur	698	101

"The Jubbulpore returns are based on the reports of Tahsildars, to which the Deputy Commissioner does not appear to attach credit. The Hoshangabad figures are not accepted by the Commissioner who considers them (as well as those given for Narsinghpur) much too low. The Bhandara figures are as reported by Tahsildars. The extraordinary discrepancy between the figures for the two tahsils of the Sambatpur district is ascribed by the Deputy Commissioner to one of the Tahsildars not having

* Kathia cane.
† Pachrangi „
‡ Kala cane.
§ Ledi „

CULTIVATION in Central Provinces.

Area & Outturn.

understood his instructions. Major Macdougall considers 720lb as the average outturn, and ascribes its lowness to the carelessness with which cane is cultivated in this district. On this point I quite disagree with him. What I saw of sugar-cane cultivation in Sambalpur gave me grounds for believing that the people make the best of the means at their disposal, and that the outturn is probably larger there than in any other district in the Provinces. I may add that the Deputy Commissioners of Narsinghpur and Bhandara admit that on their figures the sugar-cane involves a considerable annual loss to those who cultivate it.

"I have therefore not included the district estimates of outturn, but have preferred to frame estimates of my own on the facts which have been detailed above. There is nothing more certain that in respect to the outturn of his field a cultivator considers any deception legitimate when practised on a Government official, and any estimates which are merely founded on the statements of cultivators may be rejected as absolutely untrustworthy. For the purpose of my estimates I throw the 18 districts into three classes. The first includes those* in which the cane is sparsely cultivated, where black soil prevails and the average outturn is further lessened by a portion of the cane being grown without irrigation; the second includes the Betul and Chhindwara districts in which cane is grown on a reddish soil exceedingly well suited to it and watered from wells; and the third includes the districts of Chanda, Bhandara, Raipur, Bilaspur, and Sambalpur in which cane is, as a rule grown below artificial tanks and receives a plentiful supply of manure and irrigation. The average outturn in districts of the *first* class has been taken as 1,800lb per acre (=22½ mds. or 16 cwt.); in districts of the *second* class 2,600lb per acre (=32½ mds. or 23 cwt.); and in districts of the *third*, class 2,400lb per acre (=30 mds. or 21½ cwt.). The first estimate is based on the results of the Damoh and Mandla experiments, and the second on the experiments conducted by Mr. Tawney in Chhindwara, which are, I consider, entitled to much weight. Mr. Tawney considers that the fields experimented with were 'a little above the average, but not much,' and I have accordingly slightly reduced his estimate. For my third estimate I have no more solid foundation than the fact that the soil used for cane cultivation in the tank districts is, as a rule, not so well suited for it as that of Chhindwara and Betul, and that the outturn must be therefore rather lower. I should mention that in the two western tahsils of the Bilaspur district cane is grown on black soil occasionally without irrigation, and the average produce in these tahsils will be much lower than obtained in the third tahsil (Seerinarain), where cane is grown on the system followed in the adjoining district of Sambalpur.

* Saugor. Damoh. Jubbulpore. Mandla. Seoni. Narsinghpur. | Hoshangabad. Nimar. Wardha. Nagpur. Balaghat.

"The average outturn per acre assumed in the report submitted to the Government of India in 1879 was 4½ mds. or 360lb. The lowest of the estimates now adopted is five times this amount; but an additional justification for the present estimate can be obtained from the traffic returns showing the net imports of sugar by rail to Central Provinces railway stations (1) on the Bombay-Allahabad line, and (2) on the Nagpur branch line. It has been assumed that the former line serves the trade of all five distrcts of the Jubbulpore Division and of the Narsinghpur, Hoshangabad, and Nimar districts in the Nerbudda valley, whilst the latter line serves the trade of all five districts of the Nagpur Division and of the Betul and Chhindwara districts of the Nerbudda Division. The sugar imported has been all treated as *undrained*, and for this purpose the amount of *drained* sugar imported has been multiplied by 3, since 3lb of undrained sugar yield 1lb of drained sugar. The net imports during the year 1882-83 are shown below:—

	IMPORTS.			EXPORTS.	Net imports equivalent to undrained sugar.
	Drained sugar.	Undrained sugar.	Total equivalent to undrained sugar.	Undrained sugar.	
	lb	lb	lb	lb	lb
Stations Burhanpur to Jokhai on the Bombay-Allahabad line . .	9,309,000	22,634,000	50,560,000	3,060,000	47,500,000
Stations Pulgaon to Nagpur on the Nagpur branch line	662,000	136,000	2,123,000	11,000	2,112,000

SACCHARUM: Sugar.

Methods of Cultivation

CULTIVATION in Central Provinces.

Area & Outturn.

"The population of the eight districts served by the first strip of line is 3,280,508, and that of the seven districts served by the second strip of line 2,967,492. If the sugar imported by the latter districts stood in the same proportion to their population as that imported by the former districts, it would amount to 43,000,000 instead of 2,112,000lb. The difference (40,888,000lb) represents the amount by which the sugar produced in the latter districts exceeds that produced in the former.

"The area under cane in the Jubbulpore Division, and in the Narsinghpur, Hoshangabad, and Nimar districts is 10,519 acres, and that in the Nagpur Division and the Betul and Chhindwara districts 25,869 acres. There is therefore an excess of 15,350 acres in the latter districts to produce the 40,888,000lb sugar which they require. This allows for an average outturn of 2,658lb per acre.

"Calculating the gross produce at the assumed rate per acre, and adding it to the amount imported by rail the average annual consumption of undrained sugar per head of population comes to 20lb in the districts included in the first of the above classes, and to 21lb in the districts included in the second class.

'The price of sugar in the local market is given for each district in the sixth column of the appended statement, having been calculated for the cwt. as well as for the maund. Little or no refined sugar is produced in these Provinces, and it may be assumed that the whole of the produce is in the form of the mixture of molasses and sugar (containing roughly ⅔rds of the former and ⅓rd of the latter) known as *gurh.* The prices which are quoted refer, therefore, to *gurh* only.

"Sugar-cane is propagated by means of cuttings. Each cutting includes, as a rule, three joints or 'nodes,' but the Deputy Commissioner of Bilaspur reports that for an unirrigated crop the seed canes are planted whole. In making the cuttings it is usual to reject the upper 2 or 2½ feet of the cane. Selection of seed canes appears to be never practised, and, as a rule, a corner of a cane field is set aside for seed, and the canes growing in it are used as such, whether they are superior or inferior to those on the remainder of the field.

"Cane is grown on two entirely different systems, according as irrigation is or is not used. The only soil on which it is possible to grow cane without artificial watering, is that known as black cotton soil, and there is a certain amount of unirrigated cane in all districts in which this soil occurs. The Bilaspur district offers an illustration of the closeness with which methods of cultivation follow certain conditions. The Deputy Commissioner notices that in the west of the district (including the Mungeli tahsil and the Kawardha Feudatory State) almost the whole of the cane is unirrigated, whereas in Raipur and Seorinarain tahsils it is all artificially watered. He ascribes this to a difference in the varieties of cane grown, the hard, small kinds being grown in the west and the soft tall kinds in the east of the district. But the difference is entirely due to a difference in soil. In the Mungeli tahsil black soil forms 43 per cent. of the cultivated and culturable area, whereas in the Bilaspur tahsil it only forms 19, and in the Seorinarain tahsil only 14, per cent.

"Unirrigated cane is planted in November, December, and January on land which has, as a rule, enjoyed a year's fallow and has been ploughed again and again during the preceding 9 months. The field is manured with cowdung at the rate of from 50 to 200 maunds to the acre, it being often applied as a top dressing when the young shoots have appeared above ground. Pulverised oil-cake is also used in the Bilaspur district, being placed round the roots of the plants at the commencement of the rains. An important feature in this method of cultivating cane is the covering of the ground with leaves so soon as the young shoots have come up. This checks evaporation and renders the lack of irrigation less harmful than it would otherwise be. The field is hoed and weeded between the rows of cane three or four times during the rains and the crop is ready for cutting in November. Cane cultivation on this system is known as *Palwar* or *Nagarwa*.

"By far the largest and most productive portion of the cane area is however that to which irrigation is applied. With the aid of water and manure cane can be grown on almost any description of soil, but the kinds most preferred are clayey loams. The reddish loam of Chhindwara is perhaps the soil which best repays manure and irrigation, and is the one best suited to sugar-cane in the Provinces. The seed cuttings are planted from December to March in ground which has been moistened by irrigation. For this purpose shallow trenches are excavated throughout the field by a plough to which a broad wooden mould board has been attached in place of the ordinary pointed share. Water is allowed to run in these trenches, and they are planted with cane cuttings, laid horizontally, while the soil is still moist. The land has been, as a rule, well manured with cowdung before the cane is planted, but in some places it is customary to apply the manure in the form of a top dressing when the shoots have appeared above ground. The manure is laid on the surface of the soil round the plants, and water is then given so as to carry it down to their roots. A liquid manure is occasionally used consisting of cowdung mixed with water. From the sprouting of the plants up to the commencement of the rains constant irrigation, weeding and hoeing are required. The plants must be watered at least four times a month, and it is reported from the Bhandara district that cane planted in January will require in all 30 waterings, 24 of which are given before the rains commence, and 6 after their cessation. The field is, as a rule carefully fenced. In places where wood is easily obtainable scrubby bamboos are used, but if it is necessary to purchase them, fencing an acre is reported to cost as high as R7. In Chhattisgarh mud walls topped with thorns are commonly constructed round cane fields, and these entail a considerable amount of expense and trouble. During the rains it is customary to tie up together all the stems springing from a single cutting, using for this purpose the lower leaves of the plants themselves. In this way the plants are tied up in bundles of two or three, and the binding is continued with as the plants increase in height. This prevents the crop from being laid by high winds, and is further said to improve the quality of the juice by protecting the stems from the glare of the sun.

CULTIVATION in Central Provinces.

Area & Outturn.

Ratooning. 308

Conf. with pp. 59, 76, 77, 78, 128, 143, 143, 177, 181, 195, 215, 226, 247.

"The practice of 'ratooning' is reported from the Bhandara district, but is only adopted with the *kathia* variety of cane. After the first crop has been cut, the roots are manured with the ashes of the cane, leaves, and toppings, and copiously watered. A crop raised by ratooning is called *khunti*, and yields an outturn much inferior to that of a crop freshly raised from cuttings.

Irrigation. 309

"It has already been noted that both wells and tanks are used as sources of irrigation, the former in districts within or immediately below the Satpura range, and the latter in the south and east of the provinces. Wells are, as a rule, lined with stonework, but are sometimes mere temporary excavations made afresh each year. Water is most commonly raised from wells by means of the leather bucket drawn over a pully by bullocks. But the form of bucket which is used seems greatly superior to that used in Upper India. A leather tube is fitted to the bottom of the bucket, and a rope is attached to the mouth of the tube running parallel with that which carries the bucket. When descending or ascending this second rope keeps the tube doubled up with its mouth on a level with the mouth of the bucket, and no water consequently runs out. But when the bucket has been drawn up to the well's mouth, the rope attached to the tube (which runs over a pulley on a lower level to that which carries the main rope) draws it straight out and the contents of the bucket are at once discharged. The bucket is therefore self-emptying, and saves the labour of a man. In these Provinces a man is rarely, if ever, employed at the well's mouth to empty the bucket. In the Jubbulpore Division and the Nerbudda Valley the Persian wheel is not uncommonly used for raising water from wells. Water from tanks, as a rule, runs over the fields flush; when it is necessary to raise it the lever lift is commonly employed. No water rate is paid as the tank is, as a rule, considered the common property of the village, even when, as is often the case, it has been constructed by the malguzar.

"In both the Raipur and Bhandara estimates the only irrigation charge allowed for is the labour of coolies distributing water, and the cultivator is not debited with anything on account of interest on the capital expended on the construction of the irrigation tank. The Narsinghpur estimate is, I think, too low and omits several important items. To grow an acre of cane by means of well-irrigation cannot, I think, cost less than R60.

"Two distinct types of machines are used in these Provinces for the extraction of sugar-cane juice. The mortar and pestle mill or *kolhu*, and the vertical roller mill (*ghanra* or *charkhi*). The latter again has two different forms, consisting in one form of two and in another form of three rollers. The use of the *kolhu* appears restricted to the districts of the Jubbulpore Division and the Nerbudda Valley, down which it extends as far as Burhanpur. The two-roller sugar mill appears peculiar to the Satpura range, extending a short distance into the plains on either side of them. Thus it is reported to be used on the southern side of the Narsinghpur and Nimar districts, and on the northern side of the Bhandara district. The three-roller mill is used in the south of the Provinces and in the Chhattisgarh districts

"The *kolhu* consists of a stone mortar, round which a large wooden pestle revolves, drawn by a pair of bullocks. In order to press canes in it, it is necessary to cut them up into short pieces a few inches long, which are placed upright in the mortar and crushed by the pestle as it revolves over them and presses them against the side of the mortar. The two-roller sugar mill is, as a rule, fixed in a square pit excavated for the purpose, so that the tops of its rollers are just above ground level. One of the rollers is much bigger than the other, and is turned by a long cross beam fixed at its centre to the top of the spindle and drawn round by two pairs of bullocks. The smaller roller is grooved into the large roller by a screw at its upper end and is turned round by it in the reverse direction. The mill is fed by men sitting in the pit. There is not sufficient

CULTIVATION in Central Provinces.

Area & Outturn.

room in the pit to allow of whole canes being passed through and the canes are cut up into pieces of three or four feet. The three-roller mill is, as a rule, fixed above ground. The centre roller is the largest of the three, and turns the other two by means of a screw at its upper end, being itself turned by a long beam dragged round by two pairs of buffaloes. The size of the mill varies, but its rollers are often as long as four feet. There is a great deal of friction, and in spite of a liberal use of castor oil the mill cannot be worked without a most excruciating noise, too familiar to any one who has travelled through Chhattisgarh during the pressing season. The canes are passed through whole.

"No one of these machines extract all the juice in a single pressure. The *kolhu* pestle is allowed to revolve over the cane strips a large number of times, and the strips are wetted with water to soften them and make them yield their juice easier. In both the roller mills it is necessary to pass the cane through twice, and often three times, before the juice has been properly extracted.

"A great deal of trouble has been taken by some Deputy Commissioners (especially by Mr. Nicholls in Narsinghpur) to introduce the Bihia iron roller mill to the people. A considerable number have been purchased on account of Government, but few, if any, have as yet been sold to *bonâ fide* private purchasers. Cultivators readily admit its advantages, but will not invest their money in it. Its superiority to the *kolhu* has been demonstrated beyond doubt by its large sale in the Punjab, the North-Western Provinces, and Behar. It has been supposed that it was much less efficient than the large wooden roller mill, but I do not believe this. I found last camping season a Bihia 8" roller mill lying unused in the Sambalpur cutchery, having been condemned as much less efficient than the wooden implement. But I proved by actual experiment that there was very little difference in the rapidity with which it and the native mill respectively crushed cane. It was set up side by side with a three-roller wooden mill of large size, and in order to test the amount of labour expended in turning each I had them worked by coolies. The wooden mill required 9 men to turn it. The Bihia mill required only 3 men. The wooden mill required 3 men to feed it, since the cane as it emerged from between one pair of rollers had to be twisted by hand and pas ed back between the other pair. The Bihia mill was fed by one man. Three-and-a-half maunds of cane were given each mill, and were crushed by the wooden mill in 3 hours, and by the iron mill in 3¼ hours. The quantity of juice obtained was carefully measured, and found to be almost exactly the same in both cases. The Bihia mill therefore saved the labour of 8 men. The villagers admitted its superiority and asked to be allowed to keep it for a week or two and try it themselves. But as long as a malguzar can obtain labour practically free of cost, it will hardly pay him to purchase an iron mill. The Bihia mill cost R100, and the wooden mill had been made by the village carpenter for R2-12-0. Of course, it had in reality cost more than this since a large portion of the carpenter's annual village dues should be debited to it. But the village could not well dispense with its carpenter even though he were relieved of the labour of making a wooden mill by the purchase of an iron one.

Refuse rarely used as fuel. *Conf. with p. 110.*

310

"Over the greater part of the Provinces the juice is concentrated in a large iron vat fixed over a furnace sunk in the ground. The row of four boiling pans used in some parts of Upper India is unknown here. But in the districts of Chhattisgarh, where iron pans until lately could have been only acquired with great difficulty and expense, earthen pots are used for sugar boiling. Four pots are placed over a furnace, which may be circular in shape with a domed roof perforated with four holes for the pots to rest in, or may be a mere open trench in the ground. There is an immense consumption of fuel, and were not brush-wood easily obtainable sugar manufacture on this system would be a most expensive process. The crushed cane (megass) which furnishes in Upper India the entire amount of fuel used, would here go such a little way, that it is almost entirely disregarded and is very often not thrown into the furnace at all. Sugar boiling in these pots is a long process. In an experiment which I made it took 9 hours to concentrate 52lb of sugar, and the process very frequently ends in burning the sugar and turning it all into uncrystallisable molasses. The proportion of molasses in Sambalpur sugar is very large indeed, and represents an enormous annual waste of money, due to inperfections in the boiling process. There is probably no equal improvement which could be effected at so little cost as that which would result from the introduction of iron sugar boiling pans into the east of Chhattisgarh. They have already found their way into the western portions of the Raipur and Bilaspur districts.

"The cost of working a *kolhu* has been calculated by the Deputy Commissioner of Narsinghpur to be R1-10-0 per day as below:—

	R	a.	p.
Cutting cane in field	0	2	0
Stripping and cutting into pieces	0	4	0

CULTIVATION in Central Provinces.

Area & Outturn.

	R	a.	p.
Carting to mill with wages of driver	0	5	0
Feeding mill	0	2	0
Driving bullocks	0	2	0
Hire of two pairs of bullocks	0	6	0
Carpenter	0	2	0
Firing	0	2	0
Hire of vat	0	1	0
	1	10	0

But this should be increased by at least four annas on account of the wages of the sugar-boiler, making the total daily outlay R1-14-0. A *kolhu* will press some 14 maunds of cane in a day. To produce an outturn of 1,800℔ *gurh* per acre a weight of 250 maunds cane will be required. This will take about 18 days to press and will cost therefore in all (R1-14-0 × 18 =) R33-12-0. Assuming the produce to be 1,800℔ =(or 22½ maunds) sugar, the cost of manufacture comes to R1-8-0 a maund. It may be safely estimated as between R1-8-0 and R1-14-0.

"We have already seen that the cost of growing an acre of 250 maunds cane in Narsinghpur is about R50. Adding the cost of manufacture (R33-12-0) to this, the total expenditure per acre amounts to R83-12-0. Against this the cultivator obtains 22½ maunds of sugar, which at the rate of R5-5-0 per maund is worth R119-8-0. His profit per acre is therefore between R30 and R40.

"From what has been written above, it will have been seen that the production of sugar in these provinces is a domestic industry carried out by the cultivator himself without the intervention of a capitalist. The man who grows the cane himself as a rule makes the sugar, and although this system undoubtedly results in the manufacture of very bad sugar, it has its advantages in allowing the cultivator to remain his own master and not placing him at the mercy of a money-lender, which has had so disastrous an effect on his position in some parts of India. Speaking generally, the whole of the sugar which is produced is for local consumption, but it falls very far short of the wants of the population. Including the Feudatory States the amount of undrained sugar annually produced may be assumed to be 1,255,000 cwts. (= 17,57,000 maunds), and the amount annually imported is equivalent to some 458,000 cwts. (6,41,000 maunds) more. This allows 17℔ of undrained sugar as the annual consumption per head of a population (including hill tribes and feudatories) of 11½ millions.

"In conclusion, an interesting fact may be noticed concerning the cultivation of sugar-cane in Chhattisgarh. Its cultivation over the greater part of this tract should be impossible were it not for the supply of water afforded by the village tanks which were primarily constructed for the supply of drinking water and are held to be common property. A plot of ground immediately below the tank is reserved for cane cultivation, and is, as a rule, divided into two portions, one of which grows cane each year, while the other lies fallow or is put under a crop of pulse. Every cultivator in the village has a right to a small strip of land in this plot, which is divided off into a long series of small 'allotments.' In cultivating their allotments cultivators render mutual assistance to each other and in this way mutually reduce the cost of cultivation. A single sugar mill and boiling furnace suffice for the whole village and are used by the cultivators in turn. It is obvious that this system could not work well unless the cultivators agree to observe certain definite rules as to the apportionment of water and the use of the cane press. These rules are laid down by the village headman (or 'Gaontya') and seem to be generally carried out without difficulties arising. The idea of separate rights and interests, as opposed to possession in commonalty, is now, however, rapidly gaining ground in Chhattisgarh, and it will probably become each year more and more difficult to cultivate on terms which pre-suppose the existence of kindly feeling between the Gaontya and the cultivators and between one cultivator and another. In the Sambalpur district plots of cane cultivated by a ryot in his own field by means of a well sunk by himself are already of no uncommon occurrence."

VI.—CENTRAL INDIA AND RAJPUTANA.

CENTRAL INDIA & RAJPUTANA.

Reference—*Agri.-Hort. Soc. Ind. Trans:—VIII., Proc.,* 408.

Area & Outturn. 311

Area, Outturn, and Consumption.—It will be seen by the table given at page 116 that the Government of India has accepted the normal sugar-cane area for the five years previous to 1888 to have been in Central India 40,000

CULTIVATION in Central India & Rajputana.

Area & Outturn.

acres and in Rájputána 34,300 acres. The former is regarded as having yeild-ed 8,35,245 maunds of coarse sugar, say, 21 maunds an acre, and the latter 8,36,000 maunds, or 24·5 maunds an acre. The incidence of consumption on head of population was further found to have been 3·5 seers in Central India and 5·0 seers in Rájputána. But in the former calculation no provision has apparently been made for the imports, although 5,50,000 maunds were credited to the production of Rajputana. The rail-borne trade returns exhibit Central India and Rájputána conjointly, so that it is not possible to assign the exact shares taken respectively, but in 1888-89 Central India and Rájputána conjointly received 2,97,786 maunds of refined sugar (or, say, 8,93,358 maunds expressed as *gúr*) and 17,16,537 unrefined sugar. These two amounts may, therefore, be accepted as representing a grand total of 26,09,895 maunds; or deducting the exports a net import of 25,95,880 maunds of raw sugar. If, therefore, to cover all possible errors, we assume a net import of only 10 lakhs of maunds by each of the tracts of country here dealt with, it would appear that the average consumption per head of population (during the five years previous to 1888) was in Central India as near as possible 8 seers and in Rájputána 7 seers. And it is probable that these estimates very nearly represent the consumption. At all events, there is nothing to show that the people of Rájputána use more sugar than do those of the Central Indian Agency. It will be seen from the review of the internal trade of India (given in the chapter on that subject, *p. 366*) that these States draw very largely on the North-West Provinces, but that the demand for foreign refined sugar is rapidly increasing. The exports of refined sugar from Bombay port to Central India and Rájputána (*p. 364*), have been approximately doubled within the past four years. The supply from Bombay is, however, at present only about $\frac{1}{9}$th of that from the North-West Provinces. In the question of the sugar supply it may be said that so important are the Native States of Central India and Rájputána to the North-West Provinces that perhaps no better means exists of testing the progress that may in future be made by the imported, in competition with the locally produced sugars, than by watching the demands of these Native States.

CENTRAL INDIA. 312

A. Central India.—The methods of cultivation pursued in the various Native States of Central India and Rájputána are so similar that it is perhaps unnecessary to republish more than one or two of the reports which were furnished to Government. In the chapter on races of cane grown, a very complete selection of passages has already been given (*pp 59—72*) from these reports, owing to their being of more interest in that subject than the slight differences that exist in the observances and practices of the cultivators.

West Mlawa 313

Colonel C. Martin, C.B., Political Agent, West Malwa, wrote:—

"The land is well manured with cowdung, which should be two years old, and well rotted, or sometimes the cane suffers; about 20 cartloads of manure are required for one *bíghá*. The soil is dug $1\frac{1}{2}$ feet deep, well pulverized, manured and made up into beds suitable for opium, the cane is then layered 6 inches deep, being pressed in with the feet and opium seed sown over it at the same time. When the opium crop is gathered in March, the stalks are uprooted and the ground lightly ploughed in furrows, parallel to the layers of cane. The ground is well irrigated and weeded till the rainy season, after which, or in October, the crop is irrigated, the soil not being allowed to become dry, and continued at intervals until the cane is ready for cutting in December.

"Estimated cost of cultivation and manufacture so far as the latter is performed by the cultivators, and estimated net profit:—

"In one *bíghá* 3,000 stalks are required for layering: if *Ponda*, No. (1), is used the cost would be R100; (2) *Kalá*, the cost would be R60; (3) *Sufaid* and (4) *Mutaria*, the cost would be R20 each, and for (5) *Surri*, R25. The rent would be R25, weeding R12, labour of sowing R4, hedging R5, manufacturing the crop into *gúr* R40, total R106,

CULTIVATION in Central India.

West Malwa.

leaving a profit to the cultivator R14, R120 being the value of the average crop. In a very good season the profit may be R50. The above statistics regarding the expense of cultivation apply to all varieties, and the profit to the varieties (3) *Sufaid* or *Dhola*, (4) *Mutaria*, (5) *Surri*, (1) *Ponda*, and (2) *Kala* which are used for eating, are generally sold stand ing at a profit of R100 and R60 respectively.

"The juice is extracted in two ways in a *Kolhu*, and a Chirkhi *Kolhu* is of two kinds, both being blocks of hard stone, a cavity being excavated (1) sunk to the level of the ground, (2) 2 feet above ground, the stem of a babul or tamarind tree called *Lat* is introduced into the cavity of the stone, and the revolution given by a pair of bullocks moving round the mill expresses the juice. The Chirkhi consists of revolving rollers placed vertically, and the cane is pressed between them much as in machine of European manufacture.

"The juice is manufactured in a large iron vat (*kuras*), in which it is boiled to a thick consistency, cooled and tied in cloth, and thus becomes *gúr*. Before the juice is thick it is placed in earthen vessels and called *ráb*, and from this brown sugar (*shakar*) is made by placing it in bags of cloth and piled one on another. Small crystals crude, which are called *shira*, and are not used for eating, but for mixing with tobacco, or given to cattle; what remains in the bags is re-boiled and clarified with milk, and becomes *shakar*, or brown sugar.

"The people of this country mostly use *gúr*, and comparatively little sugar (*shakar*) is made, and it goes through no further refining process.

"*Gúr* and *ráb* are usually manufactured, the average market prices being for them from 8 to 16lb the rupee. The price of sugar (*shakar*) being from 6 to 8lb the rupee. I have applied to the States for statistics of area under sugar-cane, and will submit them when obtained; but as there will be delay in obtaining this information I give as much as I am able to obtain without greater delay."

Bhopal. 314

Major-Gen. W. Kincaid, Political Agent, Bhopal, furnished the following account of the cane cultivation of that State:—

"The mode of cultivation is as follows:—The fields are plouged seven or eight times, manure is then applied and square beds formed. Pieces of cane, and sometimes whole canes, are scattered in the beds, which are then flooded with water, and the canes are pressed in and buried in the mud with the feet. The process of planting the *Nuggurwar* and *Bhurree* canes is different. Furrows are made in the fields by means of a plough, and pieces of canes are put in them. They are then closed with earth and covered with the branches of *Khaukra* tree.

"The sugar-cane flourishes best in morun or the black soil before mentioned. The fields are generally watered twice a month in the cold weather, and once a week in the hot weather.

"The approximate cost of cultivation of cane, and manufacture of jaggery, is from R150 to R200 per acre, and the estimated net profit from R20 to R50.

"The estimated outturn of an acre is from 21 to 25 tons of sugar-cane. When fields of cane are sold (which is very rare), they fetch about R75 to R100 per acre.

"The juice is extracted by means of a stone-mill worked with two bullocks. The canes are cleared of leaves and cut in pieces about 1 foot each. A man puts these pieces into the mill, and takes them out when the juice has been expressed.

"The juice is usually converted into '*gúr*' or jaggery, but confectioners sometimes manufacture sugar from the juice in small quantities by the following process:—

"The juice being boiled down in a cauldron is poured into earthen vessels to cool and crystallize, after which it is filled in blanket-bags to allow the liquid part to drop through. After this draining process the raw sugar which remains in the bags is spread in a masonry cistern, in layers of the sugar and *choi* (an aquatic plant) alternately. By this method any moisture still remaining in the sugar oozes out, and the sugar becomes perfectly dry. The *choi* is then separated, and the sugar is trampled upon, by which means it gets refined.

"Sugar of local manufacture is not sold in the market, being manufactured only in small quantites by confectioners for making sweetmeat.

"The cultivator pays interest from 12 annas to R2 per cent. per mensem to his banker for money advanced, and sometimes repays him in jaggery at a cheaper rate than that current in the market, which constitutes the banker's profit.

"The demand for sugar is met by foreign importation and not by local manufacture."

Manpur. 315

Pundit Suroop Narain, C.I.E., Deputy Agent, Manpur (Malwa), supplied the Government of India with the following replies to the questions indicated by marginal notes:—

"Both the black and white kinds of cane are cultivated promiscuously in black,

Methods of Cultivation

CULTIVATION in Central India.

Manpur.

rich and Pandar soils. Depth of soils varies from 4 to 7 feet; patches of even ground capable of retaining moisture are preferred.

"The usual time for sowing is from November to January. Some begin sowing in February and continue it up to April, from the necessity, perhaps, of having to attend to opium fields in the earlier months, but late sowings of the cane give inferior results. The fields intended for cane sowing are ploughed up six or seven times, so that the soil gets thoroughly soft and even. Trenches of 1 to 1½ feet deep are then cut 1 foot apart in straight lines, across the fields. Village manure finely powdered to the amount of eight cart-loads per *bíghá* is then put over the surface. The field being watered, bits of cane, from 2 to 3 feet long, are then laid in the trenches horizontally, being so placed that the end of the last touches the head of the one coming next to it, as shown below.

"On the laying in of the bits being completed, the trench is watered, and as the watering goes on, a man stamps the bits down with his feet, causing them to sink deep into the soil. This operation is continued until the bits have got 12 to 15 inches deep under ground, and then they are left to germinate. The germination follows in from six to eight weeks in cold, and from 15 to 20 days in hot seasons.

"The black loam and Pandar soils are considered the best for sugar-cane. The answer as to the extent of irrigation is not uniform. One says that two waterings in cold and four in the hot seasons, with one in October, are sufficient. Another says four in cold and eight in hot season should be given. A third makes five in cold and six in hot season, with one in October; and the last has three in cold and six in the hot seasons.

"The first watering is done within 12 days after the sowing.

"There is again difference as to the cost of cultivation and manufacture.

"In Manpur the cultivation of one *bíghá* of cane in 1881-82 is returned as costing R82, and the manufacturing of cane-juice into *gúr* R62, or a total of R144. The yield is estimated at 30 maunds, which, sold at R180, leaves a profit of R36. The figures for Bagund and Jamnia are as follows:—

	Cultivation.	Manufacturing.	Total.	Yield.		Profit.
				Quantity of *gúr*.	Value.	
	R	R	R	Mds.	R	R
Bagund . .	109	50	159	21	172	13
Jamnia . .	120	48	168	30	190	22

"The cultivation of cane in Barwani has hitherto been rather unsuccessful, and loss to the cultivators has followed generally, such loss being estimated to have amounted to R42 per *bíghá* on the average in 1881-82.

"The juice is extracted either by means of stone-mills, in the manner in which oil is extracted from seeds—and in that case the cane is cut into small pieces, which are put in the mill like seeds—or by screw-mills—in which case entire canes are crushed between wooden rollers turned by screws. After the juice is extracted in either way, it is allowed to collect in large earthen vessels placed at the distance of 10 or 12 feet from the mills, and connected with them by wooden pipes. From these vessels the juice is transferred to large iron pans, which are placed over fire, and suffered to boil until it acquires consistency. In this state the juice is transferred to earthen pans to cool. When sufficiently congealed, the stuff is made into lumps varying from 2 to 3lb (in Malwa) to 60lb (in Nimar). The process of refining sugar is not followed anywhere in this part of the country.

"As stated above, the only process of manufacturing followed in this part of the country is that of making coarse sugar or *gúr*. The variety in the quality of the latter arises from the nature of the cane or the skill of those employed in manufacturing. The *gúr* made of the juice of the white cane is generally of light yellow colour, while that made from the black is brown or blackish. The first sells at the rate of from R6 to R7 per maund.

"There is not much cultivation of cane carried on in the territories under Manpur. The total quantity of land under this crop in 1881-82, as shown above, was only 107 *bighás* 12 *biswás*. This arises partly from the fact that Indore and Dhar districts coming under this Agency, are excluded from the return, and mostly from the cir-

CULTIVATION in Central India.

Manpur.

cumstance that, generally, cane cultivation in Malwa does not pay so well as that of poppy. It is only when the price of opium falls very much, reducing the profits arising from its cultivation below a certain limit, that cultivators are induced to turn to the cane cultivation. The best thing with poppy cultivation is that all labour and trouble connected with it is over in four months. Cane, on the other hand, takes a whole year before the labour and money spent in its cultivation is repaid to the cultivator. There is not much prospect, therefore, of the extension of cane cultivation in this part of the country. The largest extension, I believe, is going on in Holkar's territory: from the two-fold reason of His Highness extending irrigation works within his State, and the necessity of his subjects have to labour hard and utilise all means of raising crops to be able to pay the high rates of rent on land prevailing in this State:—

(1) In regard to the question of cultivator's obligations to the money-lender, it may be said that the normal condition of the class in Native States is indebtedness to the Sarcar; most of them have to draw upon him not only for the revenue they pay, but for money wherewith to pay for all their daily wants, excepting grain; and the consequence is that all the produce of their fields goes to the money-lender, who squares his accounts at stated periods, adding high rates of interest to the principal, and so the debt continues accumulating till it is wiped off by death, insolvency, emigration, or the like.

(2) The cultivation and manufacture is generally united. There is generally one mill, and the boiling establishment in a village, and each cultivator has to extract the juice and boil it into coarse sugar by turns. The latter process is said to be generally gone through with the assistance of Marwari Brahmins, who are called in, perhaps, both as particalarly skilled for the work, and as suited by their caste to make sugar which all people would use.

(3) As to comparative profits of the industry, the general impression is that cane answers better than grain, and worse than opium.

(4) There is not much capital engaged in cane cultivation, as will have appeared from the answers to the above queries.

(5) The production even of coarse sugar (*gúr*) does not equal the demand; much is imported from Khandesh and other parts of the country.

(6) Foreign competition does not seem to lead to the extension of cane cultivation, owing to the causes adverted to in the commencement of this query (9).

"So long as the cultivation of poppy is not diminished by natural causes and the necessity for the increase of that of cane is forced upon the people, no improvement in the modes of cultivation or manufacture is likely to take place."

RAJPUTANA. Dholpur. 316

A. Rajputana.—While discussing the sugar-cane of Dholpur and after having described the forms of the plant met with (see the chapter above on that subject), Colonel T. Dennehy wrote:—

"The area under cultivation of sugar-cane in Dholpur in 1882-83, is 2,443 acres, of which about 1,000 acres is *chain*, 800 acres *sarota*, and 600 acres *dhori*. This amount of cultivation is somewhat larger than it has been since the drought of 1876-77. For the last few years *zeera* (cumin seed) which is also a paying crop and requires less irrigation, has been largely grown. This year the price of *zeera* has fallen considerably and cane cultivation has been in many places resumed.

"The sugar-cane is generally sown in well irrigated fields which have grown cotton during the previous year.

"In January the land intended for cane is completely flooded from the adjacent tank or wells. A couple of days after this first irrigation it is thoroughly ploughed and then manured, and again ploughed five or six times so as to entirely break up the soil and disseminate the manure. In the meantime the cane intended for seed is buried under dried leaves which are kept moist for ten or twelve days. When the land is ready the cane is taken out, stripped of its leaves and cut into lengths of from 12 to 18 inches, each length containing one or more joints and eyes. Two ploughs are used for the sowing, the first makes a furrow 5 or 6 inches in depth in which the sower lays the seed, pieces of cane, horizontally, each at a distance of 8 or 9 inches from the other, a second plough taking the soil only 3 or 4 inches deep follows to cover in the earth in the furrow on the pieces just sown. After this last ploughing the land is harrowed with a harrow without teeth, a thick level plank, like a railway sleeper. The field is then surrounded by a mud wall or thorn fence as a protection against cattle and antelope. Sowings are generally made in January and February, all, even the late sowings, are completed by the end of March. The field is irrigated on the third or fourth day after the completion of the sowing, and from that time until the rains set in irrigation is almost unceasing; every portion of the field should be well moistened at least once in 8 days. The cutting and harvesting of the crop

Methods of Cultivation

CULTIVATION in Rajputana.

Dholpur.

takes place from the beginning of November to the end of December. It is a common practice in Dholpur to cut the sugar-cane close to the ground and leave the stumps and roots for another year. The leaves of the old canes are burnt and the ashes spread out over the roots. With frequent irrigation the roots are made to sprout again and a fresh crop is produced much inferior doubtless to the first, but which still repays the trouble and cost of irrigation. This second crop is harvested at the same period of the year and in the same manner as the first had been. A third inferior crop of *dhori* cane is sometimes thus obtained; *chain* and *sarota* can only produce at most two crops.

"Sugar-cane is cultivated in Dholpur in black soil, in *domat* (black soil and clay), in *pilia* (yellow clay), in *mattiar* (clay and sand), and in *bhoor* (sandy soil). The two first give the best results.

"The average cost of cultivation including hire for labour, ploughing, irrigation, etc., is about R85 per acre. The cost of manufacture of juice into *gúr* is about R20. The value of raw juice sold and of *gúr* manufactured is about R150. The average net profits would thus be R45 per acre. In reality the actual profits are much greater than this amount, as the hire of labour, although estimated for in this calculation, is scarcely ever actually disbursed by the cultivator, the work being in most instances done by him and by his family.

"The average outturn from an acre of sugar-cane in Dholpur, taking into consideration the proportion of the different kinds of cane cultivated, would be about 12 tons of raw juice. About one-fourth of this juice is either sold in its raw state or is consumed without any process of cooking or manufacture by the cultivator and his family. From the remaining 9 tons a little over 1 ton and 2 cwts. and a few pounds (or 30 maunds) of *gúr* is made.

"No saccharine product except *gúr* is manufactured in Dholpur. The process of manufacture offers no peculiar features. After expressing the juice in the *kolhú*, or stone cane-press, in the ordinary way, it is poured into large *karhaos*, or iron dots, in which it is boiled over temporary furnaces, *chulas* made in the immediate vicinity of the field. The refuse cane supplies the fuel. Several cultivators often combine to purchase one or more *karhoas* which they use in turn and which are let out for hire to others when not required by the owners. The juice of each field is boiled down separately; standing fields of cane are sometimes but rarely disposed of to speculators. Usually the produce is disposed of as manufactured *gúr*, and the selling price is usually on the average R4-8 per maund.

"The richer cultivators pay all the expenses of their sugar-cane cultivation from their own capital. Several others are able to obtain for the purpose loans from the State at the rate of 6 per cent. per annum interest; but many who neither have capital of their own nor can obtain a loan from the State, borrow from money-lenders about R40 per acre on their prospective crop, and for this accommodation they pay interest at the rate of 12 per cent. per annum. These loans are generally repaid in from 10 to 12 months.

"The profit realized by the cultivator on an acre of sugar-cane should be, allowing for interest paid to the money-lender, from R40 to R60. The total amount of capital employed in the growing of the sugar-cane and manufacture of *gúr* alone in the State would be about three lakhs.

"Putting the consumption of *gúr* at 12 seers per annum per head of population, the annual consumption in the Dholpur State would be 75,000 maunds or 2,754 tons, 9 cwts. and 32℔. The amount produced is 73,290 maunds or 2,691 tons, 13 cwts. and 28℔. As comparing production with consumption, therefore, we find that the State produces about 1,710 maunds or 62 tons, 16 cwts. and 4℔ less than its actual requirements. That is to say, that the quantity of *gúr* imported to Dholpur exceeds that exported from it by about 1,710 maunds or 62 tons, 16 cwts. and 4℔. If we consider the underpopulated condition of the State from which it is only beginning to emerge, and its inferiority of means of irrigation compared to districts in British territory, it is probable that for some time to come the cultivation of the sugar-cane (which necessitates considerable labour and exceptional irrigational facilities) will not increase to such an extent as to place Dholpur in the position of a considerable exportor of saccharine produce.

"A great deal has been done by the Durbar for the last ten years in the direction of increased means of irrigation by tanks and wells, a number of which are being made every year. A system of 'taccavi' advances established by the Durbar is also being largely taken advantage of by the people. Sugar-cane cultivation is undoubtedly increasing slowly and surely.

"The annual production of *gúr* averages about 73,290 maunds or 2,691 tons, 13 cwts. The imports exceed the exports by about 1,710 maunds or 62 tons 16 cwts. The consumption amounts to about 75,000 maunds or 2,754 tons, 9 cwts."

CULTIVATION in Rajputana.

Meywar.

317

Mr. A. Wingate furnished the following memorandum on the sugar-cane cultivation of Meywar State :—

"No special soil is devoted to sugar-cane, and it is found in clay, loam, or sandy soils wherever there are sufficient water and manure to permit its cultivation. The cane is finest in the rich black soils, but is most abundant in the sandy loams on the banks of rivers or in lands below tanks where the water-supply is constant and easily got. With a good water-supply and plenty of manure, the area under cane is sure to be large whatever the soil may be, but with the same advantages, a loamy would be preferred to a sandy soil. In black soil irrigation, cane will always be found in isolated plots of ½ of a *bighá* to a *bighá* or so, more if the water is near the surface, hardly any if the water is at a depth of 40 to 50 feet, each *asami* or cultivator growing his plot under his own well. Similarly under tanks with a full supply all the year round, cane will be found though the soil is very stiff; and in more compact areas, very refreshing to look upon in the hot weather. And along the banks of such a river as the Banas, where the soil varies from the richest loam, locally known as *dhamini* to the sandy or *retri* soils, and where water can be had by digging a few feet into the sub-soil, the planting of cane is only limited by the manure or by the price of its competitor, opium. The reply of the people to the question is that soil is by no means the important factor in successful sugar-cane cultivation, though in special cases it may hinder such cultivation as where the water is salt or the land *usar*.

"There is very little selection by the cultivators nor have they much opportunity of selecting. The bulk and weight of the cuttings required for planting prevent importation from a distance, and each village generally supplies itself. But were very good cane available, the villagers would readily buy it. The introduction of better varieties must originate with the Government. With his slender capital, a cultivator would never dream of making experiments; the experience of his forefathers indicates a certain, if a small profit, and any element of risk is avoided.

"I regret that under this head the information is very imperfect, first because there is none at all except for khalsa villages, and, secondly, because the papers of these are neither available in one place nor finally totalled. I have only been able to get figures for 251 villages and as for many of these the sugar-cane area has been hastily picked out for the purposes of this report, I am not inclined to place any reliance upon them. As far as the figures go, these 251 villages give an average of 11⅛th *bighás* of cane per village, so that the 850 villages measured ought to give 9,500 *bighás* under cane, and raising this (in the proportion of 3⅛ to 13½, *vide* **Mr. Smith's** report) so as to include jagri territory, there should be an area of nearly 37,000 *bighás* cultivated in Meywar. At 20 cwts. per acre, this area would produce 18,500 tons of *gúr* or 5,18,000 imperial maunds, which at 15 seers per head would supply a population of 1,381,000. The population is returned at 1,443,144, and doubtless this is below the mark. Taking the population at 1½ millions, 5,62,500 maunds of *gúr* at 15 seers per head per annum would be necessary, which at 20 cwts. per acre would require 40,000 *bighás* under cane; but there is some importation of *gúr* from the North-Western Provinces, and though there is some exports to Marwar, the imports on the whole exceed the exports. Still I think it extremely probable that an area of 18,000 to 20,000 acres may be under sugar-cane in Meywar, producing 18 to 20 thousand tons of *gúr*.

"Land intended for sugar-cane is very thoroughly ploughed, usually a crop of *makka* (Indian-corn) or *san* (flax) has been reaped in the rains, and during September and October the plot is ploughed up. At each ploughing the field is gone over twice, first one way and then across at right angles to the first furrows. After two or three such double ploughings, the clods are broken by a heavy log of wood usually a trunk of a tree roughly squared, or a felled palm, being dragged over by bullcoks, the driver standing on the log, another ploughing followed by the clod-crusher succeeds, and then the manure is carted out from the village and deposited in little heaps over the ground. Cane is an expensive crop to grow, for it occupies the ground for over a year and must therefore yield a return equivalent to two crops, *kharif* and *rabi*. To be successful it must be heavily manured, and few cultivators can afford manure for more than a plot of about ½ a *bighá*, or ¼ of an acre. The quantity of manure given varies with the soil, black soil requiring less and sandy soils more. For black soil about 250 to 300 imperial maunds per *bighá* (= say ½ acre) are necessary, and for river-side loams, which have a good deal of sand, about 350 to 400 maunds. The villagers do not reckon manure by weight but by cart-loads.

"The cart is drawn by 4 bullocks, and from a variety of replies appears to contain about 20 imperial maunds, and for sugar-cane usually from 15 to 20 cart-loads per *bighá* are allowed. From actual counting, about 10 to 12 heaps are turned out of a

CULTIVATION in Rajputana.

Meywar.

cart upon the field, and distributed at 10 to 12 feet apart. The manure is almost entirely collected where the cattle are penned during the night, often the court-yard of the dwelling-house, and is thus naturally stored at the village site. It consists of the cattle droppings and the refuse of the fodder they have been eating, bits of *makhi* and *jawar* stalks, etc., also of the ashes and sweepings about the house. By the time it is carted out, the lower part of the manure pit is very thoroughly decomposed to a rich dark-coloured mould. The droppings of the cattle while out grazing and of the bullocks at the wells are used for fuel, and so far as collected reach the manure pits in the form of ash, but as far as I have observed the women are not allowed to save themselves the trouble of collecting this fuel by taking from the supply accumulated during the night. As soon as the cattle have gone forth in the early morning, the manure and refuse are carefully scraped together and thrown on the manure heap, and there also go the droppings of any milch kine kept at home. The number and size of the manure pits about a Meywar village can hardly fail to be noticed, and the answer to every question about outturn is that it depends upon the manure.

"The manure is carted out during October and November, and it generally lies out in little heaps for some time before it is spread and ploughed and rolled in the clod-crusher. The field will then be left again to get another ploughing and clod-crushing just before planting operations begin. The lighter soils, though they take more manure, take much less preparation, and the bullocks can also plough more easily and quickly.

"Manure is seldom purchased except in the neighbourhood of towns. But a certain amount is always procurable in a village of any size from the cattle of mahajuns and others who do not possess land of their own. The price given varies from half an Oodeypuri rupee per cart to as much as two Oodeypuri rupees. But it may be taken that eight annas to R1 will generally procure a cart-load. The buyer has to cart it away himself. Carting manure is a tedious and laborious operation, the expense depending upon the distance of the field from the village site, and as many cultivators do not possess a cart they have to borrow one. Allowing five trips a day, it would take 3 to 4 days to manure a *bighá* for cane at a distance of about ¾ of a mile. From the time a cart entered a field with its load, to the time it left the field empty, was just twenty-five minutes, and they take about as long to load up. The ploughing and clod-crushing and manure carting are all done by the men, though when the cart returns a woman will often assist to load it. The current wages for this heavy labour are two Oodeypuri annas a day per man. Ploughing and clod-crushing require only one man, but for manure carting there are never less than two men employed and generally a woman or a boy besides to assist the loading. One man drives the bullocks and the other turns out the manure, while both help to load, one digging out the pit and the other throwing the manure with the aid of a basket into the cart.

"In a field thus carefully prepared, deep furrows are made about a foot apart, first by the plough and deepened and perfected by a man working with a wooden take with three or four broad wooden teeth. At each end of the perpendicular rows is a cross furrow running at right angles to the others. By this time it is nearly the middle of December from which period till the end of January planting goes on.

"In regard to procuring cuttings for planting, the practice varies, some wait till they are ready to cut the standing cane for crushing in the mill and then take from 1 foot to 18 inches from the top of each cane for seed, others select a sufficient number of whole canes and cut these up into pieces about 2 feet long. But, at all events, in Central Meywar, the most common plan is to get the planting as far as possible over before the crushing for juice begins, and also to plant as early as possible so as to be able to sow opium over the newly imbedded canes. As soon, therefore, as the season commences (in December), they open out a part of the protecting hedge or break down a corner of the mud wall and with the aid of a small hatchet a man chops and breaks off the canes close to the ground, leaving untouched any not fully grown. The canes are then carted or carried to the new plot where they are stacked under leaves and hay and occasionally watered for a few days. A woman then strips the leaves off and cuts off the tops. The next process is to measure the canes. If the canes are the cultivator's own property, the measurement may be omitted, but as they often prefer to purchase, and canes are sold, not by weight or number, but by length, the operation is carefully done. The buyer will cut down from the part of a field of sugar cane pointed out to him as many canes as he likes, and when the canes have been cleared of leaves and tops, he and the seller will sit down to watch the measuring. A man takes up a cane and applies to it his right elbow measuring as one *hath* from the elbow to the tip of the extended middle finger.*

"This point he marks by placing there the thumb of the left hand and measures

* Some add the breadth of the left hand.

CULTIVATION in Rajputana.

Meywar.

another *hath*. By the middle of the third *hath* the cane comes perhaps to an end, but another man supplies another cane and so a continuous measurement is kept up till 200 *hath* are reached. This constitutes a *bári* or bundle, and 20 such bundles are called 1,000 canes, really 4,000 *haths* of cane. I have measured several *haths* and find a *hath* averages 1 foot 10½ inches. So that they hold that a cane averages 7 to 7½ feet in length or 4 *haths*. The present price for seed canes is R8 to R10 Oodeypuri per 1,000 canes, *i.e.*, per 4,000 *haths*. They count 2,500 canes or 50 bundles to a *bighá*, which give 10,000 *haths* equal at my average for a *hath* to about 19,000 feet. They allow seven parallel rows to a *biswa*, which would give 140 rows per *bighá*, and as the canes are laid down in continuous lines in each row, and allowing 150 feet for the length of each row, they ought to use 21,300 feet of cane to a *bighá*, including the two cross-rows. But, as a matter of fact, the ground is not so accurately laid out, nor do the canes always touch, and I should say 20,000 feet of cane is about the quantity requisite for which they would pay R20 to R25 according to the market rate.

"After measurement, the bundles are conveniently distributed over the field, and a woman then lays the canes down in the dry furrows singly in one continuous line. Water is then let in and a man treads the cane with his feet into the soft mud. Another man re-dresses the ridges and arranges for the flow of water. The process is the same whether entire canes are planted or whether the canes have been cut into two or more lengths; but where only the top ends are planted, the planter is generally supplied by a woman and he puts a bit under each foot as he advances. In this case, the bits overlap and form two broken lines along the furrow. Over the great part of Meywar they plant with entire canes, sometimes cut into lengths and sometimes not.

"The newly planted cane is watered twice a month during the cold weather and four times a month during the hot weather. During the rains it gets no irrigation unless there is a long break. On the whole about 26 to 28 waterings will suffice. It is the constant watering during the hot season that makes cane cultivation such a labour. Where there are tanks or river-side wells with only a lift of a few feet, there is comparatively little trouble. But with ordinary wells of about 35 feet to 40 feet deep, the water gets very low and the bullocks have to drag to the full limit of their run, and where they could irrigate in the cold season 8 to 10 *biswas* a day, they in the hot months do not water half that area and are at work most of the week to keep the cane in vigorous growth. Not only is this a great strain on the cattle, robbing them and their drivers of the idlest part of the agricultural year, but as ploughing up the opium and barley fields begins in April and has to be frequently repeated, the bullocks employed on the sugar-cane are unable to assist in getting the farm ready for the *kharif* sowings. The people admit sugar is a paying crop, but say that no crop represents such wear and tear of well-gear and such a strain on the cattle. If asked how long a '*charas*' or water-bucket will last, they will say: 'perhaps 3 years, but if used for sugar-cane it will hardly get through half the second season' and so with every thing else. While to keep the bullocks working, green lucerne grass and plenty of fodder, with oil-cake are essential. Thus it happens that cane under wells which are not '*seja*' (that is river-side wells with short lifts and constant supply) is found only in very small patches.

"It is a very general practice to sow opium, broadcast, over newly planted cane, but this opium is always late sown and backward, and therefore the yield is poor, generally only 3 seers per *bighá* for which *half* the usual opium assessment is levied by the State. Some do not sow opium, but *methi* which they cut for cattle feeding and for which they have to pay nothing. The ground under cane must be broken up and weeded after each of the early waterings, to prevent caking, etc., and as opium requires this treatment it can be grown without any expense for watering, weeding or breaking up the ground. The only other expense is hedging or walling-in the sugar-cane and this has to be effectually done to keep out deer and pig. By the middle of December the cane is ready for cutting. Canes sold for eating are in the market in September and remain till January; but no crushing begins till quite the end of the year. With the 1st of January, mills are working all over the country, ceasing early in February. The busy season with the cane is thus December and January, as these two months see the new crop laid down and the old converted into sugar. The villagers are very fond of eating cane, and during the season, after the morning dose of opium, will always secure a cane or two to chew. They are very liberal with it among each other; and the consumption in this way must amount to several hundred canes per plot.

"As soon as the crop is off the ground, the stubble is generally burnt and then '*methi*' is sown for cattle-feeding without any ploughing. Or occasionally a crop of '*jaw*' (barley) or of '*rachka*' (lucerne grass) is taken, but always for green food

SACCHARUM : Sugar.

Methods of Cultivation

CULTIVATION in Rajputana.

Meywar.

for the bullocks, and as already said, these crops carry no assessment. Sometimes the plot is left till April, when it is ploughed up with the other wet lands in preparation for the *kharif*, and sometimes the cane is allowed to grow again. The local names for the first year's cane are '*sánta*' and '*bád*'; the second year's growth being called *korbád* pronounced *kolwad* generally. The cane is only allowed to shoot up again if the manure in the first instance has been very liberal and if the first crop has been decidedly good. A second crop can almost always be readily recognised by its stunted and rather straggling appearance. The outturn of a second year's crop is very inferior, but it seems to be in favour for furnishing canes for planting.

"When the water-supply is easily got as by flow or lift from a tank or from wells along river-sides with a never failing supply close to the surface, the assessment is about R10, Oodeypuri, per *bígáh*; otherwise R8. But there are all grades of privileged rates, and R4, R5, and R6 are not uncommon rates.

"As regards rotation, the only definite rule is that they never plant cane in a plot from which cane has just been cut. Rotations which have come under my own notice have been *makki* followed by opium, then san (flax) followed by sugar with opium sown over it, then *methi*. This is a period of three years.

"Another field has sugar, followed by *makki* and then *jaw* (barley); and in the rains following it was being prepared for cotton.

"One advantage of changing the plot yearly is that in the old plot, *makhi* followed by *jaw* can be grown without giving more manure; while they generally put the new canes down in an unexhausted field, as an opium field. But a plot of cane cut in January, might give a crop of *makki* in the rains and then be planted with sugar-cane again in the December following.

"It has been already said that either a deep black soil or a rich loam gives the largest gross outturn, but that lighter and sandy loams can often be more profitably worked if the water supply is abundant and easily available. Irrigation has been described under No. 3.

"I give below a calculation of the cost of cultivation and manufacture for one *bighá* of sugar-cane. If the average price of *gúr* is taken at R4, then the net profit per *bighá* comes to R25, and if the price is taken at R3½, then the profit is reduced to R11. But the 10 rupees set down for manure is really not expenditure. I do not think, therefore, that the net profit is less than R25 to R30 per *bighá*. As a matter of fact, it is very much more profitable, for very little of the labour is actually hired. The shares paid to village servants are not included as these are very trifling and very variable, and these as well as the amount eaten while crushing is going on, like the canes eaten before the crushing begins are deducted in estimating the gross outturn.

	EXPENSES.	R a. p.	R a. p.	R a. p.
PLOUGHING.	1st ploughing and clod breaking 3 times for 6 days, 1 man @ 2 annas		0 12 0	
	2nd ploughing and clod-breaking 2 times for 4 days		0 8 0	1 4
				1 4 0
MANURING.	20 carts manure @ 8 annas per cart . .		10 0 0	10 0 0
	3 days carting manure, 3 men for 3 days @ 2 annas		1 2 0	
	Cart-hire @ 12 annas per diem		2 4 0	3 10 0
	Spreading manure, etc., 2 men one day @ 2 annas		0 4 0	
				13 10 0
PLANTING.	Preparing field for planting, 2 men for 1 day @ 2 annas		0 4 0	
	Canes for planting, 2,500 canes @ R8 per thousand		20 0 0	
	Cutting, preparing and carting canes for planting:—			
	4 men @ 2 annas for one day .	0 8 0		
	1 cart for one day	0 12 0		
	Cutting of leaves @ R1 per thousand	2 8 0		
			3 12 0	
	Planting, 4 men @ 2 annas for 2 days . .		1 0 0	
				25 0 0

CULTIVATION in Rajputana.

Meywar.

		R a. p.	R a. p.
PICKING AND IRRIGATING.	Irrigation, 28 waterings @ R1¼ per watering	35 0 0	
			35 0 0
HEDGING.	Breaking up ground, etc., after watering 12 times with 6 men for 1 day @ 2 annas each	9 0 0	
	Hedging, 4 men @ 2 annas for 3 days . .	1 8 0	
			10 8 0
MANUFACTURING *Gúr*	Share of cost of mill, etc.	5 0 0	
	Digging holes, 2 men @ 2 annas for 2 days and setting up mill	0 8 0	
	Gathering fuel, 1 man for 16 days @ 2 annas .	2 0 0	
	Cutting and crushing canes, 5 men @ 4 annas for 8 days and 8 nights	10 0 0	
	3 women @ 1 anna for 8 days and 8 nights	3 0 0	
	2 pairs of bullocks for 8 days and 8 nights @ 12 annas	6 0 0	
			26 0 0
	GRAND TOTAL . .		86 14 0
ASSESSMENT.	Opium	3 0 0	
	Sugar-cane	9 0 0	
			12 0 0
			98 14 0

RECEIPTS.

	R a. p.	R a. p.
3 seers of opium @ R4 per seer . . .	12 0 0	
28 maunds of *gúr* at R4 per maund . . .	112 0 0	
	124 0 0	
Total receipts		124 0 0
Total expenditure		98 14 0
Net profit per *bighá* or ½ acre		25 2 0

"It is very unusual to dispose of canes in the manner indicated. About Oodeypur there is, I believe, a considerable sale in this way, but it is seldom met with in Meywar generally, and I have no information on the subject.

"When a field of cane is ready for cutting, a convenient spot close to it is chosen for setting up the mill. During November and December, the mill will be overhauled and put in thorough repair, or perhaps a large *babul* tree felled and a new mill made by the village carpenter. I append a rough sketch of the mill now in use, with the local name of each part of it. It is made entirely of wood and costs about 20 Oodeypuri rupees, the cultivator generally furnishing the wood; otherwise he will have to pay about R5 more for a suitable tree. They used to chop up the cane into little pieces and crush by turning an upright pole in a receptacle hollowed out in a large block of stone. The block was partially buried in the ground leaving the mouth exposed, and the crusher was turned by a bullock much as the oil mills are still worked. It was a tedious and wasteful process, while the stones cost over R100 to quarry and shape and convey by dragging or rolling, perhaps 10 or 20 miles to the village where required. Once arrived, they lasted for ever, and there are few villages where these stones may not be seen imbedded round the village site. In those days they used to bring the cane to the village site as they still do all other crops. But since the wooden roller has superseded the stone one they crush where the canes are cut. It is difficult to say when this new mill was introduced, but it was universal at least 40 years ago, and most say it came in with the British peace. Its advantages were its cheapness and simplicity; easily within the means of the cultivator and the capacity of the carpenter; while it turned out very much more juice.

CULTIVATION in Rajputana.

Meywar.

"The mill being ready to set up, an oblong hole is dug in the ground to receive the *chata,* or heavy solid plank on which the rollers rest and hold the whole mill firm. Another circular hole is dug in front of this to receive the *nánd,* or earthen large vessel into which the juice runs. A little distance off, a third pit is dug and divided into two compartments, one of which is the furnace over which the large iron pan *karáyi* rests, and the other is used for feeding the fire. Digging these holes is merely the work of a few hours. For some time previously they have been dragging large bundles of thorn to the spot accumulating *tilli* and cotton stalks, and anything else in the way of firewood as old hedging, etc., that is available.

"The mill is set up and a pair of bullocks yoked, and while a man cuts the canes with a hatchet close to the ground, a couple of women with sickles strip off the leaves, and another carries them to the mill and lays them down beside the mad feeding the mill. He passes two or three at a time between one pair of rollers, and a man on the opposite side catches the ends, twists them round and returns them through the other pair, so that the canes pass twice through the mill. The crushed canes are then useless except for platting into a thick mat which they lay by their wells to receive the rush of water as it issues from the leathern *charas* or bucket drawn up by the bullocks. The crushed canes are also spread in the sun and come in useful for fuel to the mill. The fault of the mill is that the crushing power is insufficient, and the straining appliances hardly worth the name. It will be noticed that both at the top and bottom where the rollers enter their sockets, wedges are drawn in to tighten the rollers. But the working is constantly loosening the mill and it would require an iron screw to secure and maintain the requisite pressure. For keeping the juice clean they put a few straws in the channel by which it flows into the earthen pot to catch the little bits of cane, etc., that fall, and they place a bit of cloth or thin matting over the mouth of the pot. But no juice that I have seen is perfectly clean or unmixed with particles of foreign matter, while the dark-dirty oil that assists the working of the mill finds to a greater or less extent its way to the juice.

"There are said to be 15 to 20 thousand canes to a *bíghá.* If the canes are juicy and of good length, then 500 canes would suffice to fill the pot or *nánd,* but as many as 1,500 will be used if the canes are stunted or the juice limited. One *nánd* full of juice suffices for the *karayi,* or iron pan. As a rule, the mill fills the *nánd* 4 times in the 48 hours, but if the juice is flowing slowly, then only 3 times. Two pairs of bullock suffice for this work, going one day and night and relieving each other every 6 hours, *i.e.,* each time the *nánd* is full. Meanwhile the juice is boiling in the open iron pan and occasionally skimmed with spoons called *chatılia* and stirred with long wooden scrapers. When boiled down to the proper consistency, the pan is taken off the fire and the *gúr* is pressed into holes made in the ground and lined with cloth, where it is cooled and moulded. Nothing remains but to weigh the outturn and examine the colour.

"The outturn of *gúr* is a fairly certain quantity. In a few favoured localities, as in some of the fertile little valleys with good black soil and abundance of water to the east of Chitorgarh, I have been told 50 local maunds per *bíghá,* equivalent to about 50 imperial maunds an acre or some 35 cwt. to 36 cwt., are produced. Fifty local maunds (the seer varies from 42 to 48 tolas) have been mentioned to me on several occasions at different times and places, by officials, patels, and cultivators. But few will admit 50 maunds in the case of their own village and all say it is a rare and extraordinary yield. Up to 40 maunds, however, the evidence is very distinct and abundant, and this would give about 28½ cwt. per acre. Forty maunds I consider a full outturn and know villages where probably the yield would average very little less. But for a general estimate that figure cannot be safely accepted. The cultivators talk very freely of 25 maunds local per *bíghá.* I have had replies as low as 10, 12, and 15 maunds, but the patels I have the most confidence in, have freely admitted 20 and 25 maunds, and for some time I have been of opinion that 25 maunds, nearly 18 cwt., is about the smallest outturn that an ordinary field should give. This is of *gúr* or the coarse brown unrefined sugar produced by merely boiling the juice. No further process takes place in Meywar, and all refined sugar is imported. I take it therefore that the produce ranges between 18 and 28 cwt. per acre. If manure has been spared or the hot weather waterings have been at all faulty, the outturn runs down at once. But, generally speaking, a sugar-cane plot is so carefully cultivated and looked after that the outturn may be regarded as fairly constant. One of the most intelligent and best informed Hakims of Central Meywar told me that 30 maunds per *bíghá,* equivalent to about 21½ cwt. per acre, was a fair average for his district. This is 3 or 4 cwt. more than the cultivators stated to me, and though I think him correct for the first class villages, when it comes to a general average for a large extent of country and taking one year with another, I do not consider more than 20 cwt. can be safely assumed. At the same time I am confident that 36 cwt. are

CULTIVATION in Rajputana.

Meywar.

got from the best manured fields in black soil. There is another way of looking at the outturn and that is in rupees, and I have never been given a reply of much less than R100 Oodeypuri per *bigha* as the value of the outturn of *gúr*, and replies have gone as high as R200 per *bighá*. At that date, the local price was R4 per local maund which would give from 25 maunds of *gúr* to 50 maunds as the outturn.

"As before remarked, cane is not grown solely for eating, but it is commonly stated that 15,000 to 20,000 canes can be cut per *bigha*. I feel pretty certain, however, that the speakers refer to 1,000 canes as meaning 4,000 *haths* of cane (as above described) though I was not then aware of this method of counting. Taking 20,000, however, and their estimate of the number of good canes required for a maund of *gúr* as 500 canes, then the outturn would still be 40 maunds local per *bighá* and 15,000 canes would reduce it to 30 maunds.

"Regarding the weight of canes my information is at present very little. Such as I have is to the effect that there are 10¾ tons of canes per acre, but I give this under reserve. The reason is that I was always told by the people that they never weighed canes and knew nothing of the weights, and since this report has been called for I have had no opportunity of weighing for myself.

"It is not enough, however, that the canes should yield a full weight of *gúr*, the colour of the *gúr* is quite as important, and it is surprising that the cultivators do not take more pains to keep the juice and boiling *gur* clean. A few puffs of wind will cover the boiling vessel with dust, and this going on all day must have a considerable effect. But they say that does not depreciate the *gúr* for local consumption. It is the colour of the *gúr* itself. This colour varies from a pale yellow to red and lastly to a blackish hue. The whitish yellow sells for about R5 Oodeypuri per local maund, and the darkest colours as low as R2. This condition of the *gúr* is stated to be quite independent of any efforts of the cultivator. The same field may give very fine *gúr* at one time and very black at another. Bad cultivation, they admit, will generally result in a dark-coloured *gúr* and good and careful tillage in a fine amber; but there is nothing certain in the matter and they attribute the colour to the season, *i.e.*, the character of the hot weather and rains. It is here there is the greatest and probably the likeliest room for improvement. Since I have been in Meywar the price of the best *gúr* has varied between Oodeypuri R3½ and R5 per local maund, the average appearing to be R4. But *gúr* is classed and sold under 3 heads, and in the following table are the quoted prices in Oodeypuri rupees (=12½ annas), per local maund (=½ an imperial maund) for yellow, red and black *gúr*, or 1st, 2nd, and 3rd quality:—

Sumwut	1st quality.	2nd quality.	3rd quality.
	R	R	R
1935	4	3	2
1936	4	3	2
1937	3¾	2¾	1¾
1938	3¾	2¾	1¾
1939	5¾	4¾	3¾

If the price of nearly all the *gúr* produced could be raised by more careful management by one or two rupees per maund so as to raise it all to first class *gúr*, the gain would be very great.

"Sugar-cane cultivation, though in favoured localities of considerable amount, is not extensively practised in Meywar and is generally confined to the well-to-do cultivators. There is no special dealing with the money-lender on account of the cultivation. But *gur* with wheat and opium, etc., goes to the *Asamis' Mahajuns* in liquidation of the current account between them. The relations between the *mahajan* and his clients in Meywar are very complex, but have nothing to do with the extent to which sugar-cane is cultivated and therefore need not be introduced here.

"The cultivator of the cane is invariably the manufacturer of the *gúr*, and refined sugar is nowhere made.

"Sugar-cane is rather more profitable than any other crop, but its cultivation is limited for reasons already detailed, and double cropping is preferred. To compete with opium, sugar ought to yield a profit of R40 to R50 per *bighă* for it represents land occupied for 1½ years. In the one case there would be *makki* and opium for 2 years running, in the other, *makki* and sugar with opium covering the same two years.

"There are no data for estimating the capital employed, seeing that the area under cane is very problematical.

CULTIVATION IN Rajputana. Meywar.

"*Gúr* is imported into Meywar as well as all the refined sugar used. The former comes from the North-Western Provinces and the latter both from there and Bombay.

"The effect of foreign competition is to keep up prices."

The official literature of Indian sugar-cane cultivation is so extensive that to reprint all the reports and correspondence, even regarding Central India and Rájputána would necessitate the allotment of a special volume of this work to sugar. The reader who may desire further information should consult **Major W. J. W. Muir's** report on the States of Harowtee and Tonk; the Commissioner of Ajmír-Merwara's communication; **Colonel W. F. Prideaux's** various memoranda on the Eastern States of Rájputána; **Mr. A. H. Martindale's** account of the sugar-cane of Kotah; **Major H. B. Abbott's** statement of the cane of Jhallawar, and many other similar replies which were received by the Government of India in 1882-83 in consequence of a circular letter which called for information regarding certain features of the cane industry.

BOMBAY & SIND

VII.—BOMBAY AND SIND.

References.—*Bombay Statistical Atlas; Various Special Reports by Mr. Ozanne, Director of Land Records and Agriculture; Reports Agricult. Dept. on Experimental Farms; Crop Experiments; Paper by Mr. Woodrow, in Indian Agricultural Gazette, June 1886; Agri.-Hort. Soc. Ind. Trans., II., App. 392-398, 417; III., 34, 42-43, 55-57; VI., Proc., 44; VII., 94, Proc., 151; Journ., I., Sel., 400; II., Sel., 87-89, 289-290; IX., 355-358 (Diseases), Proc., 271; New Series, VII., Proc.; 85; Gazetteers:—II., 41, 66, 408; III., 54, 233; VI., 39; VIII., 190; X., 146, 148; XI., 95, 425; XII., 167-169, 226-227, 352; XIII., pt. I., 290-291, 391-395; pt. II., 672, 675, 680, 682, 694, 697; XIV., 299-300; XV., pt. II., 19-20; XVI., 101-102; XVIII., pt. I., 516-517; pt. II., 51-55, 167, 168, 171; pt. III., 302, 303, 307; XIX., 166-168, 219; XXI., 251-252; XXII., 278-280; XXIV., 175-180; XXV., 185, 277.*

Area and Outturn. 318

Area, Outturn and Consumption.—In the small volume of *Statistics of Sugar* issued by the Government of Bengal in 1848 a brief notice occurs regarding sugar in Bombay. A table is furnished which shows the cultivation to have been 25,782 acres (= 77,346 *bíghás* of 14,400 square feet). The production was estimated at 6,31,192 maunds of *gúr*, the maund being 82℔ (or 6,52,311, the maunds being 80℔). This showed an average production 8 maunds 17 seers 5½ chataks (or 25 maunds 12 seers an acre). But there were found to be 250,063 date palms, and although these (as at the present day) appear to have been mostly used in the supply of a juice utilized in the preparation of certain beverages, a small amount of sugar seems, in 1848, to have been made from them and the province is accordingly credited with 216 maunds of palm *gúr* which thus raised the grand total of *gúr* (expressed in maunds of 80℔) to 6,52,527 maunds. The estimated consumption of "*sugar, khar, gúr* or raw canes, the whole reduced to their equivalent of *gúr* in maunds of 80℔," came to 7,61,779 maunds. The consumption to head of population was therefore 8℔, but to meet that consumption the Presidency imported (mostly by sea) 1,09,252 maunds. Owing to the production not being equal to the demand, the report explained that the law which then permitted Bengal to export sugar did not apply to Bombay.

Conf. with p. 117.

The small book from which the above information has been derived gives a table to show the comparative sugar productiveness of Bengal, Madras, and Bombay, but in republising that table the liberty may be taken to add that of the North-West Provinces (given in another part of the report) as also the corresponding figures for the year 1888:—

CULTIVATION in Bombay.

Yield of	Bengal per *bighá* of 14,400 sq. ft.	Madras per *bighá* of 14,400 sq. ft.	Bombay per *bighá* of 14,400 sq. ft.	North-West Provinces per *bighá* of 14,400 sq. ft.
	M. S. C.	M. S. C.	M. S. C.	M. S. C.
Gúr	10 1 5¾	12 37 2	8 17 5½	5 37 1¾*
Expressed per acre .	30 4 1¼	38 31 6	25 12 0½	16 32 1¼
Yield worked out for the average of 5 years ending 1888.	28·7	54·1	60·0	22·9

Yield. *Conf. with pp. 110, 133-134, 149, 162-63, 178, 189, 198, 226.*

319

It will thus be seen that in 1848 the yield in Bombay was viewed as very considerably less than in Bengal, whereas by the modern returns Bombay is shown to produce more than twice as much per acre as Bengal. In the reports which have been published within the past few years, one or two points seem anomalous and probably incorrect, upsetting as they do all previous notions, or they mark radical improvements in cultivation or what is more likely in tabulating results. The point already cited is specially significant. Sugar-cane cultivation in Western India must have either vastly improved or the careful crop experiments, conducted recently in that Presidency and with the view to obtain a definite knowledge of the yield from various crops, have demonstrated that all previously entertained notions of the outturn fell far short of the mark. So, again, the outturn of Madras (the other province which by modern investigations has had an exceptionally high yield assigned to its acreage) exhibits the equally anomalous feature of consuming, of all the British provinces of India, the least amount per head of population. It will also be seen from the statements quoted above that the consumption of sugar (*gúr*) in Bombay was in 1848 found to be 8℔ per head, whereas by the table given at page 116, it is now believed to be 19·5 seers (=39℔). We are not told the districts and the shares taken by each which in 1848 made up the cultivation of 25,782 acres, but the total area devoted to sugar in the British portions of Bombay and Sind, on the average of the five years previous to 1888, came to only 82,000 acres. The Native States of these provinces possessed 43,000 acres. Assuming, therefore, that the area treated of was the same in 1848 as in 1888, namely, the British territory, the above figures show an increase during the 40 years of 56,218 acres, and by a uniform expansion this would give an average annual increase of 1,405 acres. It may be contended, however, that, with the higher acreage yield now accepted and the greatly increased imports, an amount is estimated for sufficient to provide a consumption of 39℔ per head of population. But it seems probable that future investigations may slightly lower both the acreage outturn and the consumption per head in Bombay. If this be not found necessary, then it would appear highly probable the production and consumption in other provinces will have to be materially increased. It is hardly likely that so great a difference could actually exist between the productiveness of the North-West

*In the table given at page 165, the average has been corrected to 5 maunds 39 seers 15 chataks or, say, 18 maunds an acre. A slight mistake seems to have been made in the original calculation that gave the yield as 5 maunds 37 seers 1¾ chataks a *bighá*. The correction suggested would seem desirable though, to save the possibility of ambiguity, the original figures are left in the above table.

CULTIVATION in Bombay.

Provinces, for example, and Bombay as has been demonstrated (see p. 116) by a yield of 22·9 maunds in the one and 60·0 maunds in the other per acre. There must, it would appear to the writer, be some other explanation than superiority in systems of cultivation and better appliances for expressing the juice. So far as presently known the average Bombay cultivator is only slightly in advance of the *rayat* of the North-West Provinces. That the province which alone of all Indian provinces, during at least the past 200 years, has failed to meet its own wants and which at the present day should be consuming the greatest amount of foreign sugar, should nevertheless be the best sugar-producing province of India (viewed from the acreage yield), is a point which most persons are likely to believe requires to be supported by more than a tabular statement of yield worked out on admittedly imperfect data.

European Sugar-Plantations 320 *Conf. with pp. 37, 39, 48, 88-114, etc.*

The subject of the area under sugar-cane and the yield per acre in Bombay has recently (1889) been dealt with by the Officiating Director of Land Records and Agriculture in connection with a proposal to encourage European sugar plantations in India. A few passages from the Director's reply may be here given, as they confirm the exceptionally high production and consumption discussed above.

Area under cane. 321

"The subjoined table shows the area and estimated outturn of sugar-cane in the Bombay Presidency:—

PROVINCE.	Area.	Percentage on total cultivated area.	OUTTURN OF *Gul.*	
			Total.	Per head.
	Acres.		Cwt.	lb
Presidency Proper	80,000	0·32	3,440,000	27·4
Sind	2,000	0·12	86,000	4·0
TOTAL .	82,000	0·31	3,526,000	22·7
Native States	43,000	...	1,849,000	24·0
GRAND TOTAL .	125,000	...	5,375,000	23·5

"The area under sugar-cane is steadily increasing owing to the extension of canals, and it would increase at a more rapid rate but for the want of cheap manure.

"The chief sugar-cane growing districts in the Presidency Proper are shown below:—

DISTRICTS.	Area under sugar-cane, 1887-88.	Percentage on net cropped area.
	Acres.	
Satara	15,402	0·87
Ratnagiri	12,095	1·22
Nasik	10,541	0·54
Belgaum	9,940	0·57
Poona	8,860	0·43
Surat	7,089	1·54

"Compared with the other chief Provinces, sugar-cane is grown to a very small extent in the Bombay Presidency as will appear from the following table:—

CULTIVATION in Bombay.

PROVINCE.	Total cultivated area.	Area under sugar-cane.	Percentage of sugar-cane on total cultivated area.
	Acres in	thousands.	
North-Western Provinces	29,700	870	2·90
Oudh	11,200	190	1·70
Panjab	20,300	354	1·70
Bengal	54,500	287	0·53
Bombay	27,100	82	0·32
Madras	24,600	44	0·18

NOTE.—Madras has about 30,000 acres under cocoa and other palms from the juice of which much jaggery is made.

"The outturn of jaggery (*gul*) is estimated at an average of 60 maunds or 43 cwts. per acre. At this rate the total outturn in the Presidency (exclusive of Sind but inclusive of the Native States) amounts to 5,289,000 cwts., or 25·5lb per head. In Sind the outturn amounts to 86,000 cwts. **Outturn. 322**

"In 1887-88 the net imports, by all routes, of drained and undrained sugar, amounted to 1,122,158 cwts. according to the returns available for the Presidency Proper. The details are given below:— **Imports. 323**

Kind.	Imports.	Exports.	Net Imports.
	Cwts.	Cwts.	Cwts.
Drained Sugar	1,289,486	465,742	823,744
Undrained Sugar	359,624	61,210	298,414
TOTAL	1,649,110	526,952	*1,122,158

"The total consumption of drained and undrained sugar in the Presidency Proper (including the Native States) is estimated at 6,411,000 cwts.† (local outturn 5,289,000 cwts. net imports 1,122,000 cwts.) or 31 lb‡ per head of population. This estimate is liable to considerable reduction. Sugar-cane is largely used in its raw state. It is eaten as fruit and pressed for juice which is freely used as a cold drink, particularly by the Marathas. Sugar-cane is largely consumed in towns and cities. For instance, Bombay, in addition to its supplies of raw sugarcane received from Bassein, Ratnagiri and other places, takes almost the whole produce of the 400 acres under sugar-cane in the Mahin taluka, where hardly any *gul* is made. **Consumption. 324**

† 5,587,000 cwts. *gul.*
824,000 „ sugar.

‡ 27lb of *gul.*
4lb of sugar.

"The present area under sugar-cane is not large enough even to meet the local demand for *gul*. Besides as a substitute for sugar on account of its cheaper price, *gul* has a distinct demand, which is so great that, in addition to the local production, nearly 3 lakhs cwts. are imported chiefly from Northern India and Madras. *Gul* is considered to have a relish of its own and is on that account used in certain dishes. It is also believed to be sweeter than sugar and hence it is cheaper to use *gul* in sweetening a dish. The Gujaratis, whose partiality to sugar and ghee is proverbial, believed that a dish sweetened with *gul* requires less ghee than that required by a sugared one. Some poor and ignorant people have a prejudce against the use of sugar, because animal charcoal is used in the manufacture of crystallized sugar **Demand for Gul. 325**

"The import of cheap sugar, particularly from Mauritius, has dealt a death blow to the indigenous sugar-refining industry. Sugar of an inferior and unsavoury quality is made in Belgaum and Kolhapur. **Indigenous. Refineries. 326**

* More than three-fourths of the drained sugar was received from Mauritius and about one-fifth from Hong-Kong. Undrained sugar was mostly supplied by the North-Western Provinces and Oudh, Madras, and Bengal, the first-named Province having sent nearly half the quantity.

CULTIVATION in Bombay.

"Several attempts made between 1830 and 1861 at Bassein in the Thána district to manufacture sugar on modern and improved methods with pecuniary help from Government failed for some reason or other." The reader should consult the remarks at page 93 and again at page 308 for particulars of experiments at establishing sugar-cane plantations in Bombay.

Poona Factory. *Conf. with pp. 94, 306, 320.*
327 "The Poona factory was established in 1883. According to the information furnished by **Mr. Adarji**, the machinery was supplied by **Messrs. Manlone, Alliot, Fryer & Co.**, of Glasgow. The plant contains all the refining apparatus such as vacuum pan, centrifugal machines, crushing machines, with engines, bag filters, char filters, a reburning apparatus, coolers, steam-boilers, etc. The machinery including buildings cost nearly two lakhs of rupees. Sugar-cane is obtainable in the district for six months during the year. The factory is capable of daily manufacturing 2 tons of sugar, for which the produce of over an acre of sugar-cane is required. The Agricultural Returns show that in 1887-88 sugar-cane was cultivated to the extent of 8,860 acres in the Poona district, of which 4,400 acres were in the Haveli taluka, within a radius of 10 to 15 miles from the factory.

"The factory from its central situation can also be kept agoing during the other half of the year—by refining sugar from jaggery, which is obtainable in sufficient quantity both locally and by imports by rail from Satara and the Southern Maratha Country States, where jaggery is made in large quantities.

"Thus the Poona factory seems to be well situated so far as the supply of raw materials is concerned. Enquiry shows that the factory is not paying. **Mr. Adarji** mentions the following drawbacks:—

"'(1) That the crystallizable matter in the raw material in the Deccan is less than that found in the produce of Madras and Bengal. This is owing to the full use of poudrette as manure in growing sugar-cane.

"'(2) That the price of raw material is higher on this side than what it is in the other Presidencies.

"'(3) That at present the proprietor of the sugar factory could not dispose of the spirit that is made from the refuse or the non-crystallizable portion of the material to the best advantage.'

"It is a matter for enquiry and consideration whether the rum and sugar manufactured in **Mr. Adarji's** factory are of the standard quality required by the Commissariat Department, and whether assistance given to **Mr. Adarji** by the purchase of his rum and sugar for the Bombay army will help the factory out of its difficulties.

Baroda Factory.
328 "Under the patronage of His Highness the Gaekwar, a sugar factory was established in March 1887 at Gandevi in the Baroda State. It is situated in the heart of a sugar-cane-growing tract, the area under sugar-cane within a radius of 5 miles from the factory being estimated at about 1,200 acres. The raw sugar-cane is obtainable for three months, *i.e.*, from the middle of October to the middle of January. The appended copy of a report received from Baroda through the kindness of **Mr. Ozanne** shows that the machinery, including the buildings, were set up at a cost of about 2½ lakhs of rupees, that sugar is made direct from the juice, and that during the last two years of its working, the enterprise has worked at a considerable loss, a result attributed chiefly to a short working season and to the comparatively high price at which jaggery is locally sold. The average outturn of jaggery per acre in Gandevi is two tons per acre, while the yield of sugar, as judged by the trials, is from ⅜ to ⅛ ton.

"It would appear from the above that the factories already started at Poona and Baroda have many difficulties to contend with, and that the latter, in spite of all the support that it necessarily derives from the patronage of His Highness the Gaekwar, has worked during the last two years at considerable loss. Before any attempt is made to establish model factories as suggested by **Messrs. J. Travers & Sons**, it is advisable, with a view to guard against unnecessary expenditure of Government money, to encourage private enterprise by affording all possible facilities to secure the successful working of private factories."

Mr. E. O. Ozanne published in 1887 a *Note on the Cultivation of the Sugar-cane in Bombay*, which deals with most of the more interesting features of the enterprise. The paragraphs of that Note which discuss the forms of the plant met with will be found under the special chapter for that subject in the present article (pp. 73-76). Sugar-cane, **Mr. Ozanne** remarks, is grown in Bombay either for eating raw as a sweetmeat or for manufacture of *gul;* sugar is made in very few localities. Inferior sorts are made in Belgaum and Kolhápur. **Mr. Ozanne** published in connection with his report, figures which exhibited the averages for the ten years previous to 1885-86. Com-

CULTIVATION in Bombay.

menting on these he says: "The feature which is brought out by the figures is that cane is grown almost all over the Presidency, but no where is the crop very extensive. It is most important in Balsár, Jalálpor, and Chikhli tálukás of Surat; in Haveli and Junnar of Poona; in Devgad and Malvan of Ratnágiri; in Chikodi of Belgaum; in Hángul and Kod of Dhárwár; in Bassein of Thána; and, it may be added, in the Karvir and Alta tálukás of the Kolhapur State."

After furnishing brief notices regarding the sugar-cane cultivation in Poona, Khandesh, Konkan, Dhárwár and Gujarát, **Mr. Ozanne** deals with certain features of the Bombay sugar interests of a general nature. Such, for example, as—

Ratooning.

Conf. with pp. 59, 76, 77, 78, 128, 143, 151, 177, 181, 195, 226, 247, 278.

329

Rotation.—"Continuous cane is found only in the neighbourhood of Poona. It is always possible provided the ashes and skimmings are returned to the soil and provided manuring is liberal. Ratooning or growing a second crop from the stools of a former one is known everywhere, but is not largely practised. It does not pay. In Gujarát cane is grown at long intervals ranging from once in 4 to once in 20 years. So in Khándesh. The only regular rotation I am aware of is that noticed under the paragraph on irrigation in Násik and Kolhápur, in Máhím (North Thána) where cane is taken once in 7 years in a rotation in which the order of cropping is:—1, Betel vine; 2, Ginger; 3, Cane; 4, Plantain; or once in 5 years where betel vine is not grown; and in Bassein close by, where cane is followed by rice and a late crop of *udid* **Phaseolus radiatus** (for fodder), and in the third year by the same pulse as a substantive crop. In the Deccan and Karnátak except in rice land cane, if possible, follows chillies or tobacco or groundnut. But there is no very close adherence elsewhere to a particular crop to precede cane.

Sets for Planting.—"As a general rule the sets are pieces of cane from 9" to one foot long cut from the whole length. But frequently sets are taken only from the top of the cane, next the green unripe portion which is used as fodder. In places it is considered that such sets are superior. In others precisely the reverse opinion is maintained. Occasionally whole canes topped and stripped are planted. The plough is provided with a hole drilled through the body running backwards. Into this the cane is inserted. As the plough is driven the cane is deposited deep in the furrow. The depth at which it is deposited is the merit of this mode of planting. I have never seen it but write from descriptions given. It is not uncommon in the Karnatak to plant the sets in pairs at longer intervals than when planted singly. I shall notice further on the system of planting sprouting sets which is either resorted to when there is no irrigation available or when the farmer is not ready to plant when his sets are ready.

Irrigation.—"The extension of canals has naturally caused a considerable increase in the cultivation of cane. The most notable increase is that in the neighbourhood of Poona since the opening of the Khadakvásla Canal. The prejudices against the use of canal water have rapidly disappeared. The chief allegation was that it was too cold, which it no doubt was when used too freely.

"The principal sources of irrigation besides canals are wells, *dhekudis, páts* and tanks. Cane is grown largely under tanks in the Dhárwár and Belgaum districts of the Karnátak and in Ahmadabad of Gujarát. *Páts* or channels drawn from streams are most commonly used in the Deccan districts, and in them most largely in Khándesh, Násik, Sátára and Poona. An interesting practice is reported from North Násik. The land irrigable from a *pát* is divided into 3 or 4 blocks and cane is grown once in three or four years respectively in the whole block. The rotation is fairly constant. Rice precedes and wheat follows cane, or, where there are four blocks, peas or gram or sesamum intervene between the wheat and the rice. The block under cane may belong to many cultivators or to one alone. Even in the latter case there is a mutual agreement by which all join in the cultivation, arranging that those who cultivate another's land with cane should give that *rayat* a share of their land for other crops, so that each may apportion his cultivation to his needs. This plan effects much economy in irrigation and in the cost of watching the crop. A similar practice exists in the Kolhápur State. *Dhekudis* in Gujarát are water-lifts erected on the banks of rivers. They correspond to the *budkis* of the Deccan, but these latter are chiefly confined to the banks of small streams. Both are largely utilized for cane-growing. In Kolhápur the *rayats* club together to raise the water in stages from deep beds of streams or rivers till it can flow by gravitation. At each stage troughs are constructed and lifts placed on the troughs. On the whole, how-

CULTIVATION in Bombay.

ever, the reports received seem to indicate that the greater area of cane is watered from wells assisted in places by *páts*. By this means only can *páts* which do not afford a perennial supply be useful for cane, and economy in irrigation is effected by using the *pát* as long as its flow lasts.

"It may be noticed that the lifts used throughout the Presidency, except in parts of the Konkan and in Sind, are varieties of the leather bag. In the coast districts where the depth of water in wells is very small, and where the *mot* could not work, the Persian bucket wheel is used. Its use is general in Sind, even on river banks and on wells where the water is plentiful. In the Karnátak a hand lever and bucket-lift, consisting of an upright pole to the top of which is affixed a bamboo evenly balanced, is used in places where the water is close to the surface. To one end is attached a bucket, and to the other a stone or a lump of mud. There are no data for showing the area of cane under each variety of irrigation; but it may be stated generally that in the Deccan the largest area is under wells and *páts*, in the Gujarát districts under wells and *dhekudis*, and in the Karnátak under tanks and wells. In the Konkan, wells only are available when the cane is watered. Notice has already been made of the *nipáni* or unwatered cane. Cane is grown without water under different circumstances. In districts of very heavy rainfall, especially in bottom lands where the land retains its moisture after the cessation of rainy season, the *nipáni* cultivation is commonest. In this instance it is almost always taken in rice land after rice, but I have seen *nipáni* cane in land not low-lying and which would not grow rice. Where tanks exist the *nipáni* cane sometimes is planted after a single watering. Where no irrigation, even to moisten the soil into which the sets are deposited, is available, it is a common practice to place sets in layers separated by straw in a hole and to handwater till the buds sprout in about a week. The sets are then carefully planted. In parts of Gujarát cane is grown without irrigation in natural *bágáyat* (garden), *i.e.*, land generally alluvial, where the surface moisture is sufficient to bring the crop to maturity.

Mar:=*Mot*.
Guj:=*Kos*.

Mar:=*Rahát*.
Sindhi:=*Nár*.

Kan:=*Yala* or *Dotti*.

"It may be gathered from what has been stated above that the straw cane is the variety most generally chosen for *nipáni* cultivation, and, on the other hand, that the soft white cane demands the fullest supply of water throughout its period of growth.

Conf. with p. 129.

Manure.—"The dung of cattle mixed with house sweepings and all refuse by-products of crops is the chief manure for cane, as indeed for all other manured crops. There are great variations in the degree of care and success with which the muck heap is kept. The careful farmer of Gujarát pays the greatest attention to the proper preservation of the fertilizing ingredients *leaping* (*i.e.*, *daubing*) the surface of the pit with a thin layer of moist cowdung daily. In Kánara, too, the appreciation of the value of well-kept manure is conspicuous. It is a matter of regret that with these exceptions apathy or ignorance is everywhere to be observed.

"Poudrette or deodorized ordure is manufactured in the vicinity of some large cities, *e.g.*, Poona, Sholápur, and Ahmadnagar. The large extension of cane under the canals near Poona and Sholápur gave an incentive to the increase of the supply of manure. The manufacture of poudrette at Poona has been very successful. It met with much opposition, which was overcome by **Ráv Sáheb Narso R. Godbole**, the Secretary to the Municipality, whose foresight and energy met eventually with the fullest measure of success. Poudrette has, if anything, become too popular near Poona, and the cane cultivators would do well to use it rather as a supplementary fertilizer than as the sole manure. Where available, tank mud, especially in Gujarát, is used with profit.

"In the rich garden tracts near Bassein and Máhím in North Thána castor oilcake imported from Gujarát is the only manure. The climate is too moist to allow cattle dung to be well kept and it is all demanded for rice cultivation as *rab*, *i.e.* is burnt on the seed-bed for rice and *nágli* (a cereal) to kill weeds and give a readily assimilable food to the young seedling. Where sheep thrive they are folded on the land intended for cane, full use being thus made of the liquid as well as the solid excrement. Sheep do not thrive in districts of heavy rainfall.

"The ashes of the crushed cane and other material used as fuel to boil down the cane juice are returned to the soil with more or less care. But in this point, as well as in the matter of returning to the soil the skimmings of the juice, more attention would be very profitable.

"The use of green manures is well known and largely practised. The practice is increasing. The best crops for green manure are *tág* (Bombay hemp, **Crotalaria juncea**) and *guvár* (Field vetch, **Cyamopsis psoralioides**). These leguminous plants are rich in nitrogen and readily decay. They are sown in the early rains and

CULTIVATION in Bombay.

by September are well grown. As they are flowering they are cut down and after a short interval are ploughed into the soil. The decomposition is sufficiently rapid to make the manurial ingredients available for the cane crop planted in the hot weather following.

"Bones are not used, and this is one of the most melancholy features of Indian agriculture. It is especially to be regretted with regard to cane cultivation which more than any other crop requires a supply of phosphatic manure. The use of bones as manure will have to encounter the strongest opposition, but no opposition or prejudice could be greater than that manifested against the use of poudrette. Professor Cooke has patented a manure made chiefly of bones, but he has not yet followed up the experimental success which he secured. Crushed bones, *i.e.*, crushed by hand labour, would amply repay the cost. It is not necessary to dissolve bones with sulphuric acid or in other ways.

"Salt is by some supposed to be a manure for cane. It is rather used to prevent attacks of white ants or to drive them away. Salt and asafœtida (*hing*) are used throughout the Deccan and Karnátak, but, as far as I am aware, not in Gujarát. A small quantity is tied in a cloth and placed at the head of an irrigation channel and is gradually distributed by the running water.

Catch Coops.—"While the cane is young various catch crops are grown. They are not favourable to a maximum outturn of juice, but are principally the resort of poorer farmers, giving a quick return and thus easing the initial expenditure which cane cultivation demands. They are in cases of some value to the cane, providing shelter for the tender shoots till they have taken strong hold. In Gujárat pumpkins, cucumber, onions and other vegetables are thus grown. To these may be added ***guvár, vál,*** and ***bhendi.*** ***Guvár*** also serves as a sheltering crop, and for this purpose maize and bájri are grown between the rows. All are reaped early, the shelter crops providing an early and valuable supply of green fodder. The castor-oil bean and a shrub called ***shevri*** in Maráthi are also planted around the borders of cane to protect it from cattle. The leaves of the ***shevri*** are eaten readily by goats, and the stem is used as a rafter. Catch crops are less common in the Deccan and Southern Marátha Country. They are chiefly vegetables and pot herbs, but in Satára maize is grown as in Gujarát for early fodder. Castor-oil bean and ***shevri*** are also commonly grown. But it is more usual to plant around the borders of the valuable kinds of cane one or two rows of the straw-cane, which is hard and which serves a similar purpose.

Fuel for boiling. *Conf. with pp. 128, 140.* **330**

Fuel for Boiling the Juice.—"Except in the neighbourhood of Poona, in North Násik, and in parts of Belgaum it is reported that the crushed cane must be supplemented by other material as fuel. In Khándesh the crushed cane is not used. The potters who provide the earthenware pots claim the cane as their perquisite. They extract from it by lixiviation a small amount of juice from which they make inferior ***gul*** called 'Potters' gul,' and use the residue for their manufacture of pots and bricks. In Bassein also crushed cane, being fully utilized as a ***ráb*** material for rice nurseries, is not available for cane fuel. The quantity of fuel required depends principally on the shape of the boiling pan. The greatest economy is effected where it is shallow and wide as in Poona, and the greatest expenditure when it is deep and narrow as in Bassein. The quality of the fuel is governed by the facility of obtaining the supply free, at a rate below the market price, or at full market rates. In the Panch Mahals and in Bassein wood is used, because it is obtainable from the forests. The cost, till forest conservancy became an imperative necessity, was simply the cost of felling and carting. Where the Forest Department has become the guardian of the tree-growing areas a fee is charged. In places where free or cheap supplies of fuel are not available, the fuel used for boiling the juice consists of the stalks of cotton, ***túr***, safflower, etc., shrubs, thorny bushes and even branches of ***bábhul,*** tamarind, and mango grown on private lands or in waste places. In the localities where the crushed cane is the only fuel, a litttle extraneous material must nevertheless be provided to boil the first two or three panfuls till the crushed cane has dried sufficiently for use. It may be stated with confidence that where economy is forced by the dearness of fuel and where in consequence the most economical pattern of boiling pan is used, no other fuel than the crushed cane is absolutely necessary, except as just stated, for the first boilings, and it may be stated with equal confidence that the assertions that the crushed cane as fuel will not permit good lasting ***gul*** to be made are without foundation. When, however, it is used as a ***ráb*** material, or in other profitable ways, it would be bad policy to attempt to insist on the burning of the refuse cane."

The reader will be able to discover in the selection, which may now be given, of district accounts of sugar-cane cultivation, such additional information as he requires. For convenience these have been arranged

SACCHARUM: Sugar.

Methods of Cultivation

CULTIVATION in Bombay.

in alphabetical order. The selection has been made more with regard to diversities in the systems pursued than as denoting the chief sugar-producing districts of Bombay. The value of the districts in sugar production, it is believed, has abundantly been exemplified by what has already been said:—

Ahmadnagar. 331

AHMADNAGAR:—"Sugar-cane, *us*, which had in 1881-82 a tillage area of 2,801 acres, is one of the most important of watered crops. If the crop is good, in spite of the outlay on manure and water, the profit is very large. In growing sugar-cane the ground is several times ploughed in different directions and harrowed. Forty to seventy cart-loads of manure to the acre are spread over the field. The furrows are eighteen inches apart lengthways and four and a half to seven and a half feet apart crossways. The cane is propagated by means of layers which are cut in lengths of about a foot or a foot and a half. The planter takes a number of these pieces of cane in his hand, and, after a stream of water has been turned into the furrow, he walks along it dropping the pieces of cane one after the other lengthwise into the trench and treading them into the soft yielding earth. This cane requires watering every fourth, fifth or sixth day; shallow soils requiring water oftener than deep. During the hot season while the shoots are tender, to shade the young canes, in the spaces between the rows it is common to set some creeping plants, generally the *ghevdi*, which is cut as soon as the young canes have gained a certain height. As soon as the canes are planted the garden is surrounded with a thorn fence to keep out cattle. Growing sugar-cane wants constant watching, the jackal being its chief enemy from its fondness for biting the young stalks and sucking the juice. After about twelve months, the cane ripens and is cut down and carried in bundles to the sugar-mill. In the Akola *dángs* or hill lands a purple sugar-cane is grown without watering. As soon as the rice is off the ground in good level red soils in valley bottoms, the ground is ploughed and manured, and in January the cane joints are planted. They soon sprout, and next January the crop is fit for cutting without being watered in the hot season. This cane is said to take little out of the soil, and is followed by rice in the following rains" (*Bomb. Gaz., XVII., 273, 274*).

Baroda. 332

BARODA.—"Only two varieties of sugar-cane are known in the district, the white and the purple-coloured. The land requires to be repeatedly and deeply ploughed and manured before planting takes place. As the cultivation of the cane requires considerable moisture, it is not planted until after the latter part of October or the beginning of November, when the land is completely saturated with rain-water. It is planted either whole through the *nágar* or by the hand in pieces which are placed in a horizontal position and in rows at a distance from one another of from a half to three quarters of a foot. It takes full twelve months to grow. During this time it requires to be frequently and copiously watered. It is generally cut down after the rains, that is, in November or December. Each joint sends forth a full-grown cane. It grows to a height of from eight to ten feet. There are two varieties of the white sort of sugar-cane, *vasáigari* or *malbári*, and *vánsi*; the latter is thinner than the former. The Gandevi sub-division yields the largest crop of sugar-cane, an area of 846 *bighás* being covered by it. After the cane is harvested, the land is allowed to lie fallow for about six months, at the end of which period it is cultivated with *tuver* and *juvár*. These take six months before they are ready for the harvest. The land is then again allowed to lie fallow for a period of six months, when it is either planted with the same crop or with ginger. The ginger is dug out by October or November. The land is then again placed under sugar cane. It will thus be seen that the cane is planted every fourth year. As the cane ripens, it is dug out and removed to the *kolu* or crushing machine that the juice may be extracted for conversion into molasses." (*Bomb. Gaz., VII., 80*).

Dharwar. 333

DHARWAR.—"Sugar-cane *kabbu* (K.) or *us* (M.), which had 3,742 acres or 0·28 per cent. of the tillage area, is chiefly grown in the damp West or *malládu* and occasionally in gardens in the dry East. Except that when grown in fields, it is planted in a field from which rice has been reaped, the garden and field tillage of sugar-cane are much the same. The chief point is that the land must be damp enough. In December before the cane is planted the ground is prepared by breaking and levelling the rice field ridges. After a week the small plough or *ranti*, with two or more pairs of bullocks, is drawn three or four times across the ground. The clods are broken by the *korudu* or leveller, and in January the heavy hoe or *kunti* and the light hoe or *bal-lesal kunti* are used to powder and level the surface. Manure is laid in heaps, and towards the end of January, the large plough cuts the surface into furrows about eighteen inches apart. In February, and in some places in March, the cane cuttings are laid in the furrow and covered with manure. Sugar-cane wants more manure than

CULTIVATION in Bombay.

Dharwar.

any other crop; in fact, cane can hardly have too much manure. Six to nine cart-loads are generally given to the acre. After the cuttings are covered with manure the small plough or *ranti* is run along the side of the furrows and fills them with earth. The field is then once well watered and wants nothing more till the rains. Eight or ten days after the planting, when the surface is dry, the *korudu* is used to level it and break the clods. The small plough is again used to heap the earth on the cane and is again followed by the *korudu*. After a few days the surface is loosened by the smaller hoe or *ballesal kunti* to help out the young sprouts and destroy the weeds.

"Nothing further is done till the first showers fall, when the crop is a few inches above the surface and the field is weeded by the grubber or *yadi kunti*. Now, if not earlier, it is hedged, and weeded as often as wanted, at first with the *yadi kunti* and later with the *kurgi* or drill machine. The earth is heaped about the roots and the crop is ready for cutting in light porous soils in 11 months and in stiff soils in 13 or 14 months. Sugar-cane takes more out of the ground than any other crop. In fields sugar-cane is followed by rice and in gardens by pot-herbs. Unless the ground is richly manured, vegetables do not yield much during the first season after sugar-cane. It is not till the second or third year that sugar-cane can be again grown with advantage. In a fair season, on a rough estimate, an acre of sugar-cane will bring a net return of £1 12*s*. (R16)" (*Bomb. Gaz., XXII., 278-280*).

Kanara. 334

KANARA.—"In growing sugar-cane the ground is well dug, laid open to the sun for several days, and covered two or three feet deep with leaves and brushwood which, when dry, are set on fire. To the wood ashes old cowdung mixed with grass is added, and the ground is again turned and laid open to the sun for two or three days. Fresh cowdung ashes and leaves are again applied, and the ground is finally turned and divided lengthwise into beds two or three feet apart. Each bed has a trench a foot and a half wide and about half a foot deep for the water to run throughout the entire length. The trenches are joined at the ends, so that the water let into one of the trenches gradually finds its way into the rest and waters the whole garden. Except in some parts where it is as early as January or February, the season for planting sugar-cane is April or May. As soon as the beds are ready, the cuttings which for some days, or even for weeks, have been kept in a cool shady place dipped in cowdung water, are laid in the beds about five inches apart and watered. After it is planted the field is watered every morning by means of a palm-stem channel. In about fifteen days the cane begins to sprout and the watering is daily repeated. When the plants are about a foot high, cowdung manure is added and the ground is cleared of weeds and rank vegetation. This process is continued every month and the beds are raised as the plants grow. When the canes are three feet high each is tied up with its own leaves. This process which prevents the canes from breaking, is repeated till they reach their full height. Sugar-cane is ready for cutting eleven or twelve months after planting.

"Almost all husbandmen grow some little sugar-cane and make molasses. When the cane is cut, the roots, leaves, and dirt are carefully removed, and the juice is squeezed in a sugar-cane mill. The mill consists of three cylinders moved by a perpetual screw. The force is applied to the centre cylinder by two capstan bars which are worked by hand and require six to ten men at each end. The juice is boiled in iron, brass, copper, or earthen vessels. Lime is added during the process to harden and thicken the liquid. The thickened liquid is either stored in pots or cast into cubical masses by means of wooden moulds. The total cost of raising an acre of sugar-cane and of making the juice into molasses is estimated at about £22 (R220). The outturn of forty *mans* (24lb) of molasses is estimated to be worth about £20 (R200), and the value of eight thousand bundles of sugar-cane leaves about £3-4*s*. (R32) more, leaving a net profit of £1-4*s*. (R12) the acre. This cost of tillage is calculated on hired wages. If, as is generally the case, the land-owner himself works, he reaps a profit averaging £4 to £4-10*s*. (R40—R45) the acre. The details are £2 (R20) for seed canes; £3-10*s*. (R35) for preparing ground; 10*s*. (R5) for planting; £4-10*s*. (R45) for watering; 10*s*. (R5) for manure; 10*s*. (R5) for weeding; 16*s*. (R8) for fencing and hedging; £1 (R10) for cutting; £3-4*s*. (R32) for pressing; 10*s*. (R5) for boiling; £3 (R30) for fuel; and £2 (R20) for contingencies, giving a total of £22 (R220)" (*Bomb. Gaz., XV., ii., 19, 20*).

Kathiawar. 335

KÁTHIÁWÁR.—"Sugar-cane, *sherdi*, is an important crop all over Káthiáwár except in parts of Jhálávád. It grows in black soil and is planted in February and March and cut at the end of a year. The soil is ploughed ten times, broken up once, levelled twice, manured once at the rate of sixty cart-loads to the acre, weeded four times, and watered a hundred times. Two kinds of sugar-cane are grown, a reddish black and a white. The reddish black is the most generally cultivated, the white is found in Káthiáwár proper and in parts of Hálár and Porbandar. It is used locally

CULTIVATION in Bombay.

for making molasses and as fruit. The green tops are used as fodder. Sugar is not made in Káthiáwár" (*Bomb. Gaz., VIII., 190*).

Khandesh.

336 KHANDESH.—"Rich black loam is the best soil for sugar-cane, but highly manured light soils are also very productive. In growing sugar-cane, care is taken not to plant it on the same ground oftener than once in three years, and that the intervening sowings are ordinary dry crops, *jiráyat*. The ground is first ploughed crosswise and hoed to break the clods; manure, from 30 to 100 cart-loads the acre, is spread, and the field ploughed once or twice so as thoroughly to work in the manure. The surface is then smoothed, and any large clods are powdered with a wooden mallet. Then, after a final ploughing into parallel ridges one and a half feet apart, and letting water into channels between the ridges, the field is ready for planting. The seed canes are cut into short lengths, *kándis*, and the planter, filling a small basket and placing it under his left arm, drops, end to end and about six inches apart, the pieces of cane, along the channels, treading on each to settle it well into the mud. Every three or four planters have an attendant who keeps filling their baskets with cuttings. On the third day after planting, comes the first watering, *ambuni*, and on the seventh day the second *nimbuni*. After these follow regular eight-day waterings. A fortnight after planting, young shoots begin to sprout, and at the end of the first month, they are far enough on to allow the hoe, *kolpa*, to pass between the lines. This is done three times at interval of a month. After this it is weeded by hand.

"During the sixth month, or just before the *uttara nakshatra*, the latter half of September, the ground is, to help the after-growth, *háthbhar*, that comes thickly during the early rains, carefully loosened to a considerable depth by a small mattock, *kudal*. While rain is falling water is withheld. But as soon as rain ceases, a light watering, *veravni*, is given merely to wash in the rain water which is deemed cold and hurtful to surface roots.

Diseases & Pests. *Conf. with pp. 121-127.*

337 "The cane suffers from several enemies. The white-ant, *udhái*, may be kept in check by placing bags of pounded cowdung mixed with salt and blue vitriol, *morchut*, in the main water channels. Flowing over these bags, the water becomes salt enough to kill the ants without hurting the cane. *Alu*, a small grub which destroys the cane by boring numerous holes in it, is the larva of a large fly which lays its eggs in the axils of the leaves. No remedy for this pest is known. *Hamni*, a grub about four inches long, eats the young roots, and if not checked, works great havoc. It is got rid of by soaking dried *til* (**Sesamum indicum**) stems in the well until the water becomes light brown. Two or three doses of this water are usually enough. Nothing but fencing and watching can check the robberies of pigs and jackals. (*Conf. with p. 225.*)

"The cane is ready for cutting about the end of the eleventh month, if not, it is left until the thirteenth month, as the cultivators believe that, if cut in the twelfth month, the juice is much less sugary. When the canes begin to throw up flowering spikes, they are considered ready for crushing. As the root part is charged with particularly rich juice, the canes are cut over several inches below the ground. They are then stripped of all dry and loose leaves and carted to the mill. Here the tops, *bándyás*, are cut off, and used to feed the mill cattle. The crop is not at present so profitable as it might be made by improved machinery. A great deal of the sweet matter is wasted by the rude mode of extracting the juice. Besides, not acquainted with any method of refining sugar, the cultivator's only produce is raw molasses, *gul*. A large quantity of canes are also eaten by the people in their natural state.

"The crop is disposed of in three ways, by sale in the village markets to be eaten raw; by making cuttings, *bene*, for planting; and by crushing in mills for molasses. When sold to be eaten raw a good crop leaves a profit of from £10 to £12-10*s*. (R100—R125) an acre; when sold as cuttings for planting, it fetches from £20 to £30 (R200—R300) an acre; and when made into molasses, the acre yield is £5 (R50). Only the best and the largest canes are fit for cuttings. Smaller canes, if juicy and sweet, are set aside to be eaten raw, and those attacked by jackals, pigs, and white-ants are taken to the mill. The mill, *ghání*, made of *bábhul*, **Acacia arabica**, and kept under water in some well or reservoir, is generally the property of the cultivator. It costs about £2-10*s*. (R25) and lasts for two or three seasons. The boiling pan, *kadhai*, is hired from a Gujar or a Márvádi for 2*s*. to 4*s*. (R1 to R2) a day. The mill-workers are about twelve in number, seven of them *ghadles*, mostly of the Mhár caste, for removing the canes from the field and stripping them of their leaves; one *pertodya* to cut the canes into small two-feet pieces; two millers, *ghándárs*, one to feed the mill and one to take the canes from the other side; one fireman, *dastkuli*; and one boiler, *galva*. The boiler gets from 3*s*. to 4*s*. (R1½ to R2) a day besides an eighty-two pounds lump, *beli*, of molasses when the work is finished. The others get from 2½*d*. to 3*d*. (1½ to 2 annas) a day, and small quantities of molasses, cane, and juice.

CULTIVATION in Bombay.

Besides these, the village carpenter, potter, leather worker, washerman, and Mhár have their respective allowances. When cane is being crushed beggers infest the place night and day, and the *Kunbi* tries to please them expecting in this way to reap a good harvest. In the evening the mill is the resort of all the *pátils* and elders, and the owners distribute juice, cane, and bits of the new molasses, *gul*" (*Bomb. Gaz., XII., 167-169*).

Kolhapur. 338

KOLHAPUR.—"Sugar-cane, one of the most important crops in the State; it occupied in 1881-82, an unusually dry year, a tillage area of 9,900 acres. In ordinary years the tillage area under sugar-cane varies from 12,000 to 15,000 acres. As it requires a larger capital and a longer time to ripen than most other garden crops, it may be fairly presumed that the farmer who grows it is fairly prosperous. Sugar-cane is grown in three kinds of soil: black, red, and brown-red which is alluvial deposit on river-sides. The brown-red is considered the richest and best suited to sugar-cane. Sugar-cane requires much watering and heavy manuring. Sugar-cane takes much out of the soil. Unless he is satisfied with a poor return, the Kolhapur landholder does not grow sugar-cane oftener than once in three years. Still when the area of garden land is small, sugar-cane is grown alternately with either hemp, chillies, or spiked millet; but this soon impoverishes the soil and makes long rest necessary after a few years' cropping. In the plain country sugar-cane is followed in the second year after a heavy manuring by Indian millet, and in the third year either by hemp, chillies, ground-nuts, or spiked millet. In the western parts of Kolhapur sugar-cane alternates with rice or *náchni*. In garden lands and river-side lands which are flooded, as many as a thousand sheep are folded on one acre for five days and besides this about 35 to 50 cart-loads of ordinary manure are laid on the ground. Night-soil where procurable is preferred. It is considered superior and the quantity required is about half that of ordinary manure. In river-side alluvial deposits sheep urine and droppings are the only manure. When he cannot afford to manure the whole field a husbandman only covers the furrows in which the cuttings have been planted with ordinary manure.

"In parts near the Sahyádris sugar-cane cuttings are planted in December, and in the eastern subdivisions of Alta and Shirol between January and March. In the western parts the land is ploughed three to four times, the clods are broken down with the Kular, and furrows about eighteen inches apart are made by a heavy plough. The cuttings are then laid and are covered with manure. A small plough runs by the sides of the furrows and covers the cuttings. The field is then watered. After the cuttings have sprouted the field is weeded. Before the crop is ready the field is occasionally weeded and the plants are earthed up. The plantation is generally well hedged to protect it from jackals and wild pig. In garden lands and river-watered plots the field is ploughed crosswise in December and the clods are broken and the surface levelled with wooden mallets. Between January and February the field is manured with sheep urine and droppings and then with ordinary manure. The field is then thoroughly ploughed to work in the manure. Parallel ridges, or *sárs*, about eighteen inches apart, are made and water is let into the channels between the ridges. The field is ready for planting. Much care is taken in selecting cuttings. Cuttings are taken from the healthiest and biggest canes in the field. It is also seen that the canes have no *turás*, or flowery spikes at the top. Seed canes are cut into *kándis*, or pieces fifteen to eighteen inches long with three or four shoots. They are then dropped lengthwise into the furrows and pressed by the foot well into the ground. About ten thousand cuttings cover an acre. On the fourth day after planting comes the first watering or *ambavni*, and on the eight day the second watering, or *chimbavni*. After these waterings comes the regular irrigation after five to eight days according to the soil and sufficiency of water. A week after planting the cuttings begin to sprout; after three weeks when the plants have come a few inches above the ground the field is weeded by hand. During the first four months the field is weeded every month by hand. In four months the cane grows about four feet high and the *kulav* is run between the rows of plants to earth up their roots. In the fifth month the field is again weeded by hand. After this month till the crop is ripe no weeding takes place, but the field is watered at regular intervals. In the western parts where rainfall is heavier, sugar-cane does not want watering after the fifth of June; and in the eastern plains where rain is less heavy and falls at long intervals, it requires occasional watering even in the monsoon months. In the western parts sugar-cane is watered either by *páts* or by *budkis*, that is, wells built on the bank of a river or stream. In drawing water from *budkis* husbandmen club together. The water is raised from the *budki* to an intermediate receptacle and thence to another and so on to the level from which it can be distributed by gravitation. To draw up water from one place to another *mots*, or leather-bags, are used. There are generally three to four lifts, but sometimes as many as six. Considering the expense and labour thus required to raise the water, the land watered is taxed in proportion to the number of lifts. When more than four lifts are

CULTIVATION in Bombay.

Kolhapur.

used the land is assessed at the rate of full dry-crop assessment. In the eastern parts, like other garden crops, sugar-cane is watered by well-water raised by *mots* or leather-bags. Sometimes during the few months in the year, when the well-water supply is low, the field is watered by channels drawn from streams dammed at higher levels. While the crop is young pot-herbs are grown along the furrows. If the crop is stunted the ground is loosened with the hoe or *kudal*; and to give it a fresh start two to three inches of the roots of the plants are cut. Sugar-cane takes about eleven months to mature. When ripe it is heavy, its skin is smooth and brittle, and its juice sweet and sticky. If not cut in the eleventh month, it is kept till the thirteenth, as the husbandman believes that it yields much less juice when cut in the twelfth month. As it is believed that the root part contains particularly rich juice, sugar-cane is cut several inches below the ground. The dry and loose leaves are taken off and the canes are taken to the mills. Near large towns and market-places it often pays to take canes to markets to sell by retail for eating. But most of the cane goes to the mill" (*Bomb. Gaz., Vol. XXIV., 175-178*).

The continuation of the above passage which deals with the manufacture of sugar will be found in the chapter below on SUGAR MANUFACTURES.

Poona. 339

POONA.—"*Us*, sugar-cane, in 1881-82 covered 5,502 acres, 2,260 of which were in Haveli, 1,022 in Purandhar, 968 in Junnar, 428 in Khed, 378 in Sirur, 311 in Bhimthadi, 113 in Indápur, and 22 in Mával. Witn the help of water and manure sugar-cane is grown in deep black soils all over the district except in the extreme west; in the east it is one of the chief garden products. It is also much grown in Junnar, Khed, and Haveli, where, since the opening of the Mutha canals, the area under sugar-cane has considerably increased. In preparing land for sugar-cane the plough is driven across it seven or eight times; village manure is thrown on at the rate of about six tons (twenty large carts) to the acre; and the land is once more ploughed and flooded. When the surface is beginning to dry it is levelled with the beam-harrow and in December or March the sugar-cane is planted. The layers, which are species of matured cane about six inches long, are set in deep furrows drawn by the plough. Sugar-cane thus planted is called *nángria us*, or plough-cane, to distinguish it from *pávlya us*, or trodden cane, which is pressed on by the foot after the land has been ploughed, broken fine, and flooded. The treading system is usually followed with the poorer canes or in poor soil. Trodden cane, or *pávlya us*, is manured ten or twelve days after the layers are put down by folding sheep on the spot. Trodden cane sprouts a month after planting; plough-cane being deeper set takes a month and a half to show but suffers less from any chance stoppage of water and reaches greater perfection. Sugar-cane is either eaten raw or is made into raw sugar, or *gul*.

"The raw sugar, or *gul*, is extracted on the spot generally by the husbandmen themselves. A wooden press, or *gurhál*, worked by two or more pairs of bullocks is set up. The appliances used in making *gul* are: *chulvan*, a large fireplace; *pávde*, a wooden instrument like a hoe for skimming or for drawing the juice from the boiler into its receptacle; *shibi*, a stick with a bamboo bowl or basket for straining the liquid; *káhil* or *kadhai*, a boiling pan for thickening the juice; and *gurhál* or *charak*, the sugar-cane press. The press is made entirely of wood and is worked by two pairs of oxen. Two upright solid cylinders, eighteen or twenty inches across called *navrá-navri*, or husband and wife, whose upper parts work into each other with oblique cogs, are made to revolve by means of a horizontal beam fixed to the *navra* in the centre and yoked to the oxen at its ends. The cane, stripped of its leaves and cut into lengths of two or three feet, is thrice passed by the hand between the cylinders, and the juice is caught in a vessel below, which from time to time is emptied into the *káhil*, a shallow circular iron boiling pan.

"In 1881-82, in connection with sugar-cane experiments, **Mr. Woodrow**, the Superintendent of the Botanical Garden at Ganesh Khind, noticed that the soil of Poona had very little of the silica in combination with potash of soda and lime in the form known as soluble silicates. It was not difficult to reproduce these soluble silicates without which sugar-cane cannot grow; but it would be expensive in India and could not be done in a short time.

"To grow sugar-cane without wearing out the land it was necessary to manure with two tons an acre of quicklime and ten loads an acre of wood-ash, and to sow and plough in a green crop such as hemp or black mustard.

"After a crop of sugar-cane the land should be manured for four years as usual and such crops grown as the soil and the markets suit, preference as far as possible being given to pulses and cereals being avoided. In no case should more than one corn crop be grown. At the end of the four years if the ground is treated in the usual manner for sugar-cane, an average crop may be expected. Poona sugar-cane soil is usually rich in lime, in some cases lime is present in excess. It would often pay to

CULTIVATION in Bombay.

make a kiln and burn the calcareous earth on or near the field where lime was wanted" (*Bomb. Gaz., XVIII., ii., 51-55*).

Nasik. 340

NASIK.— Sugar-cane, *us*, which had in 1879-80 a tillage area of 7,449 acres, is one of the most paying of watered crops and very great care is taken in its growth. Four kinds of sugar-cane are grown,—white *khadya*, striped *bángdya*, black *kála* or *támbda*, and Mauritius called *baso*. The last is grown only to a very limited extent near Násik and Devláli. The ground is ploughed from corner to corner seven or eight times. Weeds, which are seldom found in watered lands, are carefully picked out as the ploughing goes on. The clods are broken and levelled, and a good deal of manure is spread over and mixed with the earth either by hand or by a light rake, *dáte*. Furrows, six inches deep and about 1½ feet apart, are cut by a deep plough, divided into small beds, and watered. Sugar-cane cuttings, about a foot long and three or four inches apart, are thrown into the furrows lengthwise, and pressed by the foot to drive them well into the ground. Planted in this way sugar-cane is called *pávlya us*. It is most suited to a shallowish soil. In the case of the white or *khadya* cane, the cuttings are thrown into the furrows without dividing the land into beds, and after levelling the furrows by a beam-harrow, the plantation is freely watered. Sugar-cane grown in this way is called *nángrya us*. The *nángrya us* being deeper set stands a scanty supply of water better than the *pávlya*, and if regularly watered comes to greater perfection than the other. The cuttings are planted in January or February, and more often in March, and begin to sprout after about fifteen or twenty days. Before it is five feet high the crop is twice or thrice carefully weeded. No further cleaning is wanted, as weeds do not thrive under the shade of grown canes. Before the rains set in, when the crop is not more than three feet high, except the white variety which wants only about half as much water, the cane requires a weekly watering, and after the rains, a watering every twelve or fifteen days. The crop takes full eleven months to ripen. The mill consists of two *bábhul* rollers called husband and wife, *navru, navri*, worked by two or four bullocks. A cane pipe joins the mill to the boiling pan which is under the charge of the owner of the cane or some other trustworthy person, as the work of choosing the proper time at which to take the pan off the fire requires much knowledge and care. As the fire must be kept burning fiercely, *bábhul* loppings are, as much as possible, used for fuel. Two men are required to feed the furnace, two to drive the bullocks and cut and supply the cane, one to feed the rollers, and one to see that the juice-pipe runs freely. The sugar mills are the resort of all the village when work time is over, and the smooth floor in which the moulds for the hot juice are built is pleasantly lit by the glow of the furnace. The white cane, *khadya*, though very hard and coarse for eating, yields the best molasses, and the crop requires less labour and care. It is found over almost the whole district. The Málegaon and part of Yeola, the striped, *bángdya*, cane is chiefly grown, but it is seldom pressed. Mauritius cane requires the greatest care as regards water and manure, and the molasses are generally inferior. Sugar-cane pressing usually goes on during the nights of the cold season, beginning with January. It employs a great number of hands. At the time of pressing, the owners never refuse cane or juice to any one, and crowds of beggars throng their fields. They even call passers by to take some of their sugar-cane and juice, believing that free-handed gifts are rewarded by a plentiful outturn" (*Bomb. Gaz., XVI., 101, 102*).

Satara. 341

SATARA.—"Sugar-cane, *us*, which had in 1881-82 a tillage area of 8,336 acres, is one of the most paying of watered crops. Very great care is taken in its growth, and it thrives best in shallowish soil. Three kinds of sugar-cane are grown,—white *khadya*, striped *bángdya*, and black *kála* or *támbda*. The ground is ploughed from corner to corner seven or eight times. Weeds, which are seldom found in watered land, are carefully picked out as the ploughing goes on. The clods are broken and levelled, and large quantities of manure are spread over and mixed with the earth either by hand or by a light rake called *dáta*. Furrows, six inches deep and about 1½ feet apart, are cut by a deep plough, divided into small beds, and watered. Sugar-cane cuttings, about a foot long and three or four inches apart, are dropped lengthwise into the furrows, and pressed by the foot well into the ground. When planted in this way sugar-cane is called *pálvya us*, or foot-pressed cane. In growing the white, or *khadya* cane, the cuttings are laid in the furrows without dividing the land into beds, and, after levelling the furrows by a beam-harrow, the plantation is freely watered. Sugar-cane grown in this way is called *nángrya us*, or ploughed cane. The *nángrya*, or ploughed cane, being deeper set, stands a scanty supply of water better than the *pávlya*, or foot-cane, and, if regularly watered, comes to greater perfection. The cuttings are planted sometimes in January and February, but more often in March, and begin to sprout after about fifteen or twenty days. Before it is five feet high the crop is twice or thrice weeded. No further cleaning is wanted, as weeds

CULTIVATION in Bombay.

do not thrive under the shade of grown canes. Before the rains set in, when the crop is not more than three feet high except the white variety which wants only about half as much water, the cane requires a weekly watering, and, after the rains, a watering once every twelve or fifteen days. The crop takes full eleven months to ripen. The sugar-cane mill consists of two *bábhul* rollers called husband and wife, or *navra, navri,* worked by two or four bullocks. A cane-pipe joins the mill to the boiling pan which is under the charge of the owner of the cane, or of some other trustworthy person, as to choose the proper time to take the pan off the fire requires much knowledge and care. As the fire must be kept burning fiercely, *bábhul* loppings are as much as possible used for fuel. Two men are required to feed the furnace, two to drive the bullocks and cut and supply the cane, one to feed the rollers, and one to see that the juice-pipe runs freely. The sugar-mills are the evening resort of all the village. The white cane, or *khadya,* is very hard and coarse for eating, but the crop requires less labour and care than the other kinds of cane. It is found over almost the whole district. The cane is usually pressed at night between January and March. It employs a great number of hands. At the time of pressing, the owners never refuse cane or juice to any one, and crowds of beggars throng the fields. They even call passers-by to take some of their sugar-cane and juice, believing free-handed gifts are rewarded by a plentiful outturn.

"In the year 1860 an experiment was made in the cultivation of *imphi,* **Sorghum saccharatum,** or Chinese sugar-cane. This plant which is grown in Europe as forage, has an advantage over the ordinary sugar-cane in the very short interval required between the sowing and ripening. In the case of *imphi* 100 days only are required. In Satara the result of the first experiment was so far satisfactory that the crop reached a height of eight feet and was much appreciated by cattle. Forty stalks made one pound of molasses. At present (1884) no Chinese sugar-cane is grown in the district" (*Bomb. Gaz., XIX., 167-168*).

Thana. 342

THANA.—"Sugar-cane, *us,* **Saccharum officinarum,** is, with the exception of Sháhápur, Kalyán, Bhiwndi, and Murbad, grown all over the district, especially in Bassein where sugar-cane and plantains are the chief watered crops. A loose, light, stoneless soil with at least one quarter of sand, is the best for sugar-cane. The ground should be slightly raised so that the water may readily drain off. A rice crop is first grown, and after the rains, when the rice has been cut (November), the land is thoroughly ploughed and cleaned and all the clods are broken. It is ploughed again twice every month for the next four months. In May, furrows are made six feet long, one and a half broad and one deep, with a space of about one foot between them. In these furrows, pieces of sugar-cane about $1\frac{1}{2}$ feet long are buried end to end, about two inches below the surface. If the land has been regularly ploughed since November, no manure is wanted. But if, as is sometimes the case, it has been ploughed only since March, oil-cake manure, *pend,* at the rate of fourteen pounds ($\frac{1}{2}$ *man*) to 100 furrows must be laid over the sugar-cane before it is covered with earth. On the day that cane is buried, the furrows should be filled with water, this soaking is repeated every third day for nine days, and afterwards every six days till the rains begin. From ten to fifteen days after the cane is buried, the young shoots begin to appear, and in about six weeks, when they have grown a foot or a foot and a half high, oil-cake manure (in Bassein called *dho* by the Christians and *khap* by others) is applied at the rate of about fifty-six pounds (2 *mans*) to every hundred furrows. In September, after this second dressing, a third supply of manure, *gádhni,* is given at the rate of eighty-four pounds (3 *mans*) for every hundred furrows. At the same time the earth between the furrows is gathered against the stems, its long leaves are wrapped round the cane, and water-courses are made ready. After another month (October) a fourth dressing, at the rate of twenty-eight pounds (1 *man*) for every hundred furrows, is given, and if the rains have ceased, the plants are watered every fourth or sixth day according to the moistness of the soil. In December, when the cane is about three feet high, the long leaves are again wrapped round the stems, and about the end of the month five or six plants are tied together. When the plants have grown five or six feet high, the long leaves must be again bound round the stems to preserve the flavour of the juice and prevent the plant being eaten. By May the cane is ready for cutting. The canes are bound in a bundle of six, and to the number of about 750,000 are yearly sent to Bombay, Surat, and Broach. The price is 2*s.* 6*d.* (R$1\frac{1}{4}$) the hundred" (*Bomb. Gaz., XIII., 290, 291*).

Sind. 343

SIND.—"For raising sugar-cane crops the land is richly manured, and ploughed over and over again, until the manure is well mixed with the soil. After the land has been carefully prepared and weeded the sowing commences in the month of March by small pieces of cane, each with an eye, being put into the ground at regular intervals. The field is then constantly irrigated, so as to be in a continual state of moisture. During the hot season it is perfectly saturated with water and kept free from weeds. In Upper Sind the sugar-cane is planted out in January or February and cut in November or December. The cane is usually sown standing, and is cut and manufactured by the purchaser. The expense of cultivating sugar-cane is heavy, owing to the long time the crop takes to mature, and the great quantity of water required for properly irrigating it. It is liable to injury at planting out from attacks of white-ants, and at different stages of its growth from jackals, rats, maggots, and frost" (*Gas., 11*). CULTIVATION in Sind.

VIII.—MADRAS. MADRAS. 344

References.—*Sugar Statistics in 1848; Numerous passages in Vol. of Proceedings of Hon'ble East India Company from 1790-1822; Madras Agri.-Horticultural Society; Agri.-Hort. Soc. Ind. Journ., II., Proc. 51, 52; Sugar-cane, Cultivation in Godavery District by R. E. Masters (Sel. Rec. Madras Gov. XXII., 1870); District Manuals:— Man. of Admn., Vol. I., 288, 363, II., 78; Man. Coimbatore Dist., 71, 95, 122, 123, 182, 183, 184, 189, 195, 196, 205, 235, 236, 237, 250, 251, 253, 288, 292, 394, 429, 449, 450, 451, 464, 476, 483, 489, 498, 518; Man. Kistna Dist., 365, 366; Man. Nellore Dist., 403, 624; Man. Salem Dist., Pt. I., 147, 149, 281, 354; Pt. II., 9, 61, 104, 105, 158, 159, 213, 214, 236, 237, 268, 299, 300; Man. Cuddapah Dist., 206, 207, 208, 209, 210-213, 251; Man. Kurnool Dist., 170, 179, 207; Man. Trichinopoly Dist., 4, 247; Man. Madura Dist., 100, 106; Man. North Arcot Dist., 156, 165, 167, 263, 323, 326, 327, 328.*

Area, Outturn, and Consumption.—The Proceedings of the Hon'ble the East India Company give many curious particulars of sugar cultivation in Madras from about the year 1792. One of the most useful papers that appeared was that by **Dr. Roxburgh** on "the Hindu Method of Cultivating the Sugar-cane and Manufacturing the Sugar Jaggery in the Rajahmundry Circar: also the Process observed by the Natives of the Ganjam District in making the Sugars of Berrampore." The most that can be done in this place, to convey an idea of the facts brought out by **Roxburgh**, regarding the sugar industry of the country indicated (100 years ago) is, to abstract a paragraph here and there from the leading sections of his paper Thus, for example, he says: "In the Northern Provinces or Circars, as well as in Bengal, Cadapah, etc., large quantities of sugar and jaggery are made. It is only in the Rajahmundry and Ganjam districts of these Northern Provinces where the cane is cultivated for making sugars." In the zemindaris of Peddapore and Pettapore (of the Northern Provinces), from 700 to 1,400 acres, **Roxburgh** tells us, were employed for rearing sugar-cane. "Besides these a third more should be added for the delta of the Godavery." "From the same spot they do not attempt to rear a second crop oftener than every third or fourth year. The cane impoverishes it so much that it must rest or be employed during the two or three intermediate years, for the growth of such plants as are found to improve the soil, of which the Indian farmer is a perfect judge. They find the leguminous tribe the best for that purpose." The juice may be boiled down to either of two forms of crude sugar. If when boiled to a certain extent the syrup is thrown on mats made of the leaves of the palm (**Borassus flabelliformis**) and stirred until cold, the sugar that forms is called *pansadarry*. But many persons prefer to make *bellum* or *jagary* because, although this sells for less, it keeps longer and may thus be retained till a favourable market is afforded. To make *jagary*, **Roxburgh** explains, a certain amount of quicklime is thrown into the boiler and the syrup is not, as in the preparation of *pansadarry*, scummed. When of a proper consistence some gingelly oil is added, and the syrup, when well mixed with the oil, is poured into shallow pits dug in the ground. The syrup as it cools solidifies and is then cut into cakes and these are wrapped up in dry leaves and put aside for sale. One acre, **Roxburgh**

Area & Outturn. 345

Rotation. *Conf. with p. 129.* 346

Use of Oil. 347 *Conf. with pp. 220, 234.*

CULTIVATION in Madras.

Area & Outturn.

adds, "in a tolerable season yields about ten candy of the above mentioned sugar or rather more if made into *jagary*." "Each candy weighs 500lb, and is worth on the spot from R16 to 24, according to demand. In the West Indies, the acre, so far as my information goes, (and it is chiefly from Mr. Beckfords's History of Jamaica) yields from 15 to 20 hundred-weight of their raw sugar, worth in the island from £15 to £20 currency. Here the produce is more than double: but on account of its inferior quality, and the low price it bears on the spot, the produce of the acre does not yield a great deal more money than in the West Indies." It would thus

Yield. *Conf. with p. 211.* 348

appear that in Roxburgh's time the accepted yield of unrefined sugar was about 44 cwt. per acre. Ratooning was sometimes practised, the second

Ratooning. 349 *Conf. with pp. 59, 76, 77, 78, 128, 143, 151, 177, 181, 195, 215, 247.*

crop being known as *karsni*, but it was so inferior that, Roxburgh says, when he asked the cultivators if they ever took a third, he got the reply that as the second crop was so inferior to the first there was no inducement for taking a third.

But having thus briefly reviewed some of the salient points of Roxburgh's paper (an observer whose statements carry such weight that many persons may be disposed to accept the yield of 44 cwt. an acre as likely to be correct, even at the present day), it does not seem necessary to deal, in this place, with any of the other authors who furnished the East India Company with reports on the Madras sugar industry, one hundred years ago. Passing therefore over a gap of some fifty years the information afforded in the *Statistics of Sugar* for 1848 may be next reviewed. The area shown to have that year been under cane was 84,947 *bighás* (or, say, 28,315 acres), and the yield 11,00,740 maunds (maunds of 80lb) of *jaggery*. The average yield was found to be 12 maunds 37 seers, 2 chataks a *bighá* (or, say, 39 maunds an acre). But it was found that there were 6,468,368

Date-palm Sugar. 350 *Conf. with pp. 138, 231, 266, 270, 310, 352, 361, 370.*

palms yielding sugar, and that these afforded 6,62,218 maunds of jaggery, so that the total amount of coarse sugar available in Madras, during 1848, came to 17,62,959 maunds; the consumption was estimated at 10,67,720 maunds, and the surplus available for export was therefore 6,95,239 maunds. It is explained that as the total population of the Presidency had not been determined, it was not possible to arrive at the consumption per head. The estimates of consumption for certain districts were, however, furnished, and it may be added that the highest of these quotations appears against Tinnevelly (18lb 12 oz.), next Madras (14lb), and the lowest Canara (1lb). Adding together these estimates and striking the average of all, the figure arrived at is 5lb 5 oz. No reliance can, however, be placed on that figure (as expressing the average of the Presidency, in 1848), except that it may be viewed as lending a certain amount of confirmation to the exceptionally low consumption shown for Madras in the table at page 116, namely, 4·3 seers (or, say, 8½lb).

The statistics of 1848 are, however, of more direct interest, in the view they afford of the to-day but imperfectly understood subject of the yield of sugar from palms. As already stated, it is recorded that there were in that year 6,468,368 palms in Madras and that these afforded 6,62,218 maunds of sugar. But no attempt was made to reduce the palm area to acres, and in the acreage of sugar production this source was accordingly kept quite distinct from that of cane. It is believed that the modern computation may be accepted as 400 trees to the acre. If, therefore, that standard be applied to the palms of Madras in 1848, the acreage yield would have been 40·8 maunds and similarly 64·8 maunds in Bengal. The average yield of the present day is said to be 24·7 maunds in Bengal and 74·49 maunds in Madras. The figures published for the palm sugar of 1888 thus reverse in every particular those of 1848. Not only would the total yield of palm sugar appear to be greater now in Madras, than in

CULTIVATION in Madras.

Area & Outturn.

Date-palm Sugar.

Bengal, but the yield per acre would seem to be also higher. It is practically impossible to believe that such radical changes could actually have taken place. The explanation must, as it seems to the writer, be sought in the defective nature of the returns. If the yield of palm sugar be lowered for Madras, the already abnormally low consumption to head of population in that Presidency would be rendered still lower. On the other hand, a material increase in the palm sugar, credited to Bengal, would not only seem to be justified by all the evidence the author has been able to bring to bear on the subject, but would raise the consumption per head of population much nearer to that which most writers think is actually used by the people of that province. But the statistics of the internal trade of Madras are admittedly imperfect (more so, in fact, than in Bengal), and it is, therefore, likely that even were the supply of palm sugar reduced by one-half, a more careful registration of trade would exhibit the province as obtaining from local production, and imports a quantity that would allow of a considerably higher consumption than is shown by the statistics hitherto published.

From what has been said it may have been inferred that the writer is strongly disposed to think that much of the ambiguity that exists, regarding the sugar trade of Madras and Bengal, is traceable to the fact of the palm supply being treated conjointly with that of cane. To exhibit this fact it is necessary to refer to the most recent official information. It has been explained (in other chapters of this article) that on the *Note on Sugar*, which was prepared by the Revenue and Agricultural Department in 1887, being issued, most of the Local Governments and Administrations furnished additional information, and, in some few instances, thus enabled the Government of India to modify the statistical returns that had appeared in the original *Note*. These corrections and amplifications were published in the form of a supplement to the *Note*, and from that supplement the following passage may be taken since it not only affords useful details regarding palm sugar (presently under special consideration) but exhibits the main facts of sugar production and consumption in the Presidency.

"Revised areas are given for the total cultivation of sugar-cane during the three years ending 1885-86, which are as follows:—

	1883-84.	1884-85.	1885-86.
	Acres.	Acres.	Acres.
Government	36,700	39,900	34,000
Inam	11,900	10,100	13,800
Zamindari	18,300	20,300	20,500
TOTAL	66,900	70,300	68,300

"These figures, which are believed to be approximately correct, are inclusive of areas the produce of which is eaten raw by the people instead of being manufactured into sugar or jaggery.

"The subjoined abstract details the area (from the produce of which sugar or jaggery was manufactured) and outturn for the same period:—

		Area.		Tons.	Maunds.	Maunds of coarse sugar.
		Acres.				
1883-84	Sugar-cane	6,500	Refined sugar	7,870	2,14,238	5,35,600
	Do.	53,500	Jaggery	100,100	...	27,25,000
	Cocoanuts, dates, etc.	30,000	Do.	87,200	...	23,74,000
	TOTAL	90,000		195,170	...	56,34,600

SACCHARUM: Sugar.

Methods of Cultivation

CULTIVATION in Madras.

Area & Outturn.

		Area.		Tons.	Maunds.	Maunds of coarse sugar.
		Acres.				
1884-85	Sugar-cane .	8,300	Refined sugar .	7,500	2,04,166	5,10,400
	Do. . .	55,000	Jaggery . .	73,600	...	20,03,600
	Cocoanuts, dates, and palmyra .	31,000	Do. . .	83,700	...	22,78,500
	TOTAL .	94.300		164,800	..	47,92,500
1885-86	Sugar-cane .	7,000	Refined sugar .	7,800	2,12,333	5,30,800
	Do. . .	55,800	Jaggery . .	70,500	...	19,19,200
	Cocoanuts, dates, etc. . .	28,500	Do. . .	74,000	...	20,14,500
	TOTAL .	91,300		152,300	...	44,64,500

"The average for these three years is as follows:—

	Area.	Outturn of coarse sugar.	Outturn per acre.
	Acres.	Mds.	Mds.
Sugar-cane	62,000	27,41,500	44·2
Cocoanuts, etc.	29,800	22,22,300	74·6
TOTAL .	91,800	49,63,800	54·1

"The average outturn of coarse sugar is 49⅝ lakhs of maunds and the net exports 17$\frac{1}{10}$ lakhs of maunds. This leaves 32½ lakhs of maunds for consumption in the Presidency, or about 4⅓ * seers per head of the population. As regards these figures of consumption, the Board of Revenue remarks as follows:—

* In Northern India the consumption per head is between 17 and 22 seers, and in Bombay 19½ seers: see p. 116.

"'The average seems certainly very low, but it must be remembered that a very large area under sugar-cane cultivation in zemindari and whole inam villages and under palm-trees is not brought to account, and the estimate of outturn given above is consequently much below the mark. It is further observed that the calculations do not take into account the traffic by road with the Native States and the adjacent Provinces. The land trade statistics compiled for some years show that the average imports of sugar from Hyderabad, Mysore, and the French Settlement amounted to 1,22,600 maunds or 22,400 maunds in excess of the exports to these countries. But the returns are obviously defective, being confined to a few stations on the frontier, and do not show the entire traffic. In determining the rate of consumption of sugar, it must also be borne in mind that in this Presidency a very large proportion of the rural population use sugar only on festive occasions, and not as a daily article of consumption.'

"The report sent up by the Government of Madras is silent as regards the statistics of Native States and the general trade of the Presidency in sugar."

There are several very instructive features in the returns thus furnished by the Madras Board of Revenue. The area that yielded edible canes, as also that which afforded refined sugar have been dealt with apart from the jaggery or *gúr* area. The average amount of land devoted to edible canes, during the three years, appears to have been 6,460 acres and that acreage has, therefore, been excluded from consideration. The Board appears to regard the yield of sugar as compared with jaggery at 2½ to 1: the writer in calculations of this nature has accepted 3 to 1 as more nearly correct for India as a whole. In Madras the term 'sugar' is commonly used for the unrefined but drained article which in Upper India is known as some of the forms of *khand* or *bura*. The provision made above shows, however, that the refined article is meant. But the most useful part of the figures given by the Board, in the above analysis of the

Madras sugar production, is the fact that the average area under sugar and jaggery, when reduced to the average production, shows a yield per acre of 44·2 maunds, whereas the yield of the fields employed specially for crystallized sugar (expressed as jaggery) came to 73·2 maunds and the jaggery area to only 40·5 maunds. These facts manifest the error of accepting an average production to total acreage, regardless of the relative shares of the land devoted to each purpose. It is this error, in the writer's opinion, that has caused much of the confusion that exists in the literature of Indian sugar—more especially when palm and cane sugars are discussed conjointly.

CULTIVATION in Madras.

Area & Outturn.

Conf. with pp. 114-115, 131, 133, 136, 183, 252-255, 285, 298.

Error in Terms.

It is, however, frequently stated that the area under sugar-cane in Madras has recently shown a tendency to expand. This may be so, but the writer, after perusing the fairly extensive series of publications available on the subject of Madras sugar, has been forced to the opinion that either serious mistakes were made in the returns that have appeared, within the past ten years, or the production of sugar has very probably contracted. Thus, for example, an official report on the sugar of Madras in 1881-82 contains the following tabular analysis:—

Statement showing the Area under Plantations and Trees used for the purpose of Sugar Manufacture in the Madras Presidency for 1881-82.

Whether Sugar-cane, Coconut, Palmyra, or Date, etc.	Total Area under				The Area from the Produce of which Sugar or Jaggery is manufactured.				Area from the Produce of which is manufactured		Average yield per Acre in Cwts.		Estimated total Produce in Tons.	
	Ryotwar.	Inam.	Zeminduri.	Total.	Ryotwar.	Inam.	Zemindari.	Total.	Sugar.	Jaggery.	Sugar in cwts.	Jaggery in cwts.	Sugar.	Jaggery.
1	2	3	4	5	6	7	8	9	10	11	12	13	14	15
	Acs.	Acs.	Acs.	Acs.	Acs.	Acs.	Acs.	Acs.	Acs.	Acs.	Cwts.	Cwts.	Tons.	Tons.
Sugar-cane	37,214	11,324	23,844	72,382	35,279	10,924	23,180	69,383	9,992	59,391	27	31	10,781	120,556
Cocoanut .	257,917	22,505	5,483	285,905	4,898	489	319	5,706	...	5,705	...	55	...	14,820
Palmyra .	97,670	24,959	33,737	156,366	17,715	2,258	4,911	24,884	...	24,884	...	108	...	75,279
Date .	22,789	7,465	7,824	38,078	925	477	192	1,594	...	1,594	...	61	...	3,204
Sago-palm	149	½	...	149	19	...	...	19	...	19	...	309	...	35

CULTIVATION in Madras.

Area & Outturn.

Commenting on the facts shown in that statement the report goes on to say:—

"If the average production be taken at 45 cwt. per acre, the total *jaggery* produced in this Presidency from cane would amount to about 150,000 tons. To this must be added the jaggery produced from about 25,000 acres of cocoanut trees, probably 12,500 tons; the jaggery produced from about 25,000 acres of palmyra trees, probably 125,000 tons; and also that produced from about 1,500 acres of date and sago-palms, probably 4,500 tons; giving a total estimate of 292,000 tons of saccharine matter for the whole Presidency. The imports from foreign countries are insignificant, seldom exceeding 2,000 cwt. per annum. The exports have increased rapidly since the famine, and in 1882-83 reached a total of 1,246,964 cwt., valued at R75,68,940. The details of this export trade will be found on pages 147 and 358 of the annual volume of trade of the Madras Presidency in 1882-83. If from the figures shown at page 358 is deducted the amount of sugar which was merely conveyed to some other port in this Presidency, the result is that 75,222 cwt. of refined sugar and 38,512 cwt. of unrefined sugar were exported to other ports in India, and the figures given on page 147 show that 13,219 cwt. of refined sugar and 1,119,930 cwt. of unrefined sugar were shipped to foreign countries, principally to the United Kingdom. This export of unrefined sugar includes palmyra jaggery, as no distinction is made between that and cane jaggery in the returns."

It will be observed that, according to recent returns, the average outturn of all kinds of sugar for the three years ending March 1886 came to only 170,756 tons, whereas it is apparently accepted that in round figures the production in 1881-82 came to 292,000 tons. **Mr. Schofield**, while alluding to the above report, points out that if the figures there given be correct, after making the deduction for net export, there would have remained in Madras an amount sufficient to have allowed the population a consumption 9½ seers per head. A consumption of 19lb would, in fact, be in keeping with the results worked out for the other provinces of India, but it is inadmissible until the area and production of sugar be regarded as something like that determined in 1881-82—now apparently viewed as an overstatement.

But to return to the subject of palm-sugar it will be discovered from the above passage that, according to the presently accepted view, the palm trees of Madras yield very nearly as much sugar to the Presidency as that obtained by cane cultivation. The averages for the past three years stood at 27,41,500 maunds sugar-cane and 22,22,300 maunds palm. What is still more remarkable the yield of jaggery from palms per acre is well on towards being double the average from cane, thus 74·6 maunds an acre from palm culture and 44·6 maunds an acre from cane. But if this be actually the case, the question naturally suggests itself is palm-sugar manufacture more profitable than cane? Surely the labour and expense of tapping the trees, for, say, four or five months a year, could never exceed that of the cultivation of cane. The area suitable for palm cultivation is, however, more limited than for cane, and the value of cane as an ordinary crop that may be grown at will in rotation with others must not be forgotten. The reader will find the subject of palm-sugar repeatedly dealt with in this work, as, for example, in the articles on **Borassus, Caryota, Cocos,** and **Phœnix**, and the general conclusion arrived at may be said to be that while very remunerative as a Native industry, for certain tracts of country, palm-sugar cultivation has not hitherto proved capable of serious expansion. Indeed, the most contradictory statements have appeared regarding the yield and profit of production, so that the subject seems to call for a thorough enquiry. It is probable, for example (see *Vol. II., 453*), that the fiscal restrictions imposed on the tapping of palms, owing to the very extensive employment of the juice in distillation, operate restrictively in the expansion of the trade in palm-sugar. It is equally probable, however, that even if the Madras yield of 74·6 maunds of *gúr* an acre can be confirmed by future investigations, there may be many objections

CULTIVATION in Madras.

Area & Outturn.

to palm-sugar manufacture which would render it undesirable that greater encouragement should in future be paid to this branch of the Indian sugar trade. The difficulty in forming a definite opinion regarding palm-sugar does not exist alone in the records regarding Madras. On the contrary, equally inexplicable statements regarding the Bengal sugar trade have been made by persons whose opinions are entitled to the greatest respect. Thus, or example, **Mr. Westland** (see *Vol. VI., 214*) speaks of the date-palm of Jessor as affording *nine tons of gúr an acre.* The rent of the land under date-palms, he tells us, is from R9 to R15, and the value of the produce R500 to R600. This would be a yield three times as great as that given for the Madras palms. So in a like manner **Robinson** (*Bengal Sugar Planter, p. 193*) says that "the annual produce of a full-grown date plantation was equal to 78¾ maunds of *gur* per Bengal *bíghá*, which converted into *khaur* may be taken as equivalent to a yield of about 5½ tons of Muscovado sugar per English acre." **Robinson** further estimates that the date-palm area of Bengal roughly measured is 130 miles long by 80 miles broad, or 10,400 square miles, and accordingly by accepting the produce at one-half the ascertained yield, say, at 2¾ tons an acre, the area in question might be estimated to be capable of yielding 915,200 tons a year. But the dilemma of palm-sugar is by no means solved by a verdict from these statements of its being a distinctly more productive cultivation (acre per acre) than cane. It has already been stated that, according to the most recent statistics of the date-palm industry of Bengal, there are at the present time 30,000 acres under that palm (devoted presumably to sugar production and distinct from the acreage assigned to the production of date-palm liquor) and, further, that, that area yielded on the average of the five years previous to 1888, 7,43,000 maunds of coarse sugar. This comes to 24·7 maunds (say, 17½ cwts.) of *gúr an acre,* and thus a little less than a third of the Madras yield. It surely cannot be the case that so great a difference exists between the palm-sugar yield in these two provinces. But we have still to deal with the fact that palm-sugar manufacture has hitherto proved a failure in Bombay. Some years ago the Government becoming alarmed at the increasing consumption of fermented palm juice, thought of diverting this by finding a better use for the produce of the palms. Jessor palm-sugar manufacturers were imported to teach the people, but, though everything was done that could be thought of, the industry failed to be established in that Presidency.* In order to combat the evil of intoxication, other alternatives had to be resorted to, namely, the destruction of large numbers of palms and the increase of the taxation on tapping. The reader will find information on this subject under the articles **Borassus** and **Cocos** (Vol. I., 499, and Vol. II., 452). In Bombay it has been estimated that there are 3,500,000 cocoanut palms, 47,810 palmyra palms, and 70,000 Caryota palms. Of these there are licensed to be tapped 50,000, 16,735, and 20,000, respectively, of the kinds named, but little or no palm-sugar is made in the Western Presidency from these trees. The owners of palm groves, we are told, would gladly hail a new utilization of their trees, since the fiscal restrictions have greatly lowered the value of their plantations.

Palm-sugar. 351 *Conf. with pp. 138, 226-227, 266, 270, 310, 352, 361, 370.*

Palm-sugar a failure in in Bombay. 352

The subject of the sugar supply of India derived from palms is one of so pressing importance that the writer cannot avoid recommending that it should receive the most careful consideration of all future investigators. The exact yield of each kind of palm should not only be thoroughly explored, but the effect of climate, soil, and systems of cultivation and tapping on the formation of crystallizable sugar should be looked to.

* See the passage quoted under Manufactures of Surat, p. 307.

CULTIVATION in Madras.

Area & Outturn.

The discrepancies briefly reviewed above may be found to exist only in the misleading nature of statistical returns unaccompanied by explanatory data. One feature of these remarks, it seems desirable to reiterate in conclusion, namely, that the comparison of acreage yield of palm with non-palm-sugar producing provinces is distorted through the returns of cane and palm-sugar being conjointly dealt with. Thus, for example, the apparent yield of cane sugar in Bengal is lowered, and that of Madras raised, by this process, in the table at page 116. Total production, were it to be worked out from total acreage by means of a previously determined yield per acre, would in such cases be more seriously wrong than is necessarily the case in agricultural calculations of this nature. Averages are, however, in most cases dangerous, especially when the relative values of the extremes have not been determined. Averages on distinct and conflicting data must of necessity be fallacious. The factor 54·1 may be correct as it stands (table, p. 116), but it can never be compared, for example, with 22·9, the ascertained production-rate in the North-West Provinces, since the one includes palm-sugar, the other does not. The Madras average yield of sugar-cane is 44·2 maunds an acre, a figure which might be viewed (assuming the returns of both provinces to be correct) as demonstrating that in Madras sugar-cane yields twice as much acre per acre as in the North-West Provinces.

Having thus briefly discussed the leading features of the available information regarding the area, outturn, and consumption of sugar in Madras, it remains only to give here a selection of passages illustrative of the methods of cultivation pursued in the Presidency. This purpose cannot be better served than by commencing with certain paragraphs from a review of information drawn up in 1881-82 by the Madras Board of Revenue:—

"*Particular kinds of Cane cultivated; their suitability to special soils; mode of selection by cultivators; possibility of introducing better kinds or better tillage.*—There are very numerous varieties of the cane quoted by the District Officers under local vernacular names, but **Mr. Robertson**, Agricultural Reporter to Government, states that many of these varieties cannot be distinguished, and that the distinguishing characteristics of other varieties arise from local conditions of soil and climate, and disappear when these conditions are absent. In popular parlance, the cane is divided into three varieties—the red cane, which grows on drier ground, the striped cane, which takes a richer soil, and the white cane, which succeeds in wet land unfavourable to the two other varieties. In the Madras Presidency, the cane is cultivated chiefly in the districts on the coast of the Bay of Bengal, and some inland districts which have a comparatively dry climate, while there is but little cane grown on the West Coast, where the climate is moist and resembles that of the Straits, the Mauritius, and the West Indies. The cane in those colonies attains to a luxuriant growth never equalled in this Presidency, and at the recent exhibition at Madras, a sample of cane from Penang was far superior to the sample of cane from Bellary district which gained the second prize. Many attempts have been made to introduce into this Presidency these larger varieties of cane. Otaheite and Bourbon canes along with the Minnesota Amber cane are now to be seen in the Godávari district, while at the Saidápet Experimental Farm successful trials have been made of the Chinese sugar-cane (**Sorghum saccharatum**) and other sugar-producing **Sorghums**. It has not yet been shown, however, that any of these foreign varieties will, in this climate, continue to produce more sugar than the country cane, and on this point the Board would quote the result of the experiment recorded in the Vizagapatam District Manual. It is there recorded that **Messrs. Arbuthnot & Co.**, the renters of the Pálkonda estate, brought a cane-planter from the West Indies to teach an improved method of cultivation, spent large sums in the introduction of the Mauritius cane and placed the experiment under the personal supervision of **Mr. John Young**, now Chairman of the Oriental Bank, but the result showed that the Native system of cultivation was more suited to the existing circumstances, and that the Mauritius cane was more precarious than the country varieties. It is not likely that any attempt to improve upon the tillage of the cane will ever be made more carefully or under conditions more favourable to success.

Sugar-cane Plantations. *Conf. with pp. 93-94.*

353

CULTIVATION in Madras.

Area & Cutturn.

"*Statistics of Area under Sugar-cane.*—The first year for which statistics are available is 1852-53, when the area under cane was acres 38,403. It remained almost stationary until 1869-70, when it was acres 37,805, and then increased steadily till 1875-76, when it was acres 52,094. The famine years show a great decrease, but in 1881-82, the area under cane in Government, Zemindari, and Inam lands in this Presidency was acres 72,382, the produce of acres 69,383 of which was manufactured into sugar and jaggery. The annexed statement gives the details of this area, and also of the area under cocoanut, palmyra, date and sago plams. These figures are derived from the special reports of Collectors in answer to this call, and, as far as Government land is concerned, may be accepted as tolerably accurate, but the statistics of Zemindari and Inam land must be regarded with less certainty. It is said that sugar or jaggery is manufactured from the produce of the following acreages:—

	Acres.
Under sugar-cane	69,383
Cocoanut palms	5,706
Palmyras	24,884
Date palms	1,575
Sago palms	19

so it is evident that the jaggery manufactured from the palmyra is the only considerable rival of the product of the cane.

"*Mode of Cultivation.*—It is not usual to cultivate the cane two years running upon the same land. In parts of Kurnool, Tinnevelly, and South Canara, however, the stumps of the cane are left in the ground to sprout and yield a crop the following year, and in the Nandyál *taluk* of the Kurnool district, the cane is left in the ground for three years, and in the Cumbum *taluk* for as long as ten years, the yield diminishing each year. These instances of slovenly agriculture are, however, exceptional. The cultivator usually permits land which has borne some other crop to lie fallow for a year, and then prepares it for the cane by several ploughings, or by breaking it up with crowbars which disturbs it to a depth of nearly a foot, and by heavily manuring the soil with whatever manure he can obtain, the most common manure being that obtained by picketing his herds or folding his flocks upon the land. The land having been manured, ploughed, and flooded, the cane is planted. The cane in India never bears seed, although it flowers. It is always propagated by cuttings. The top of the cane is commonly used, but some cultivators leave a few canes growing in the fields from the previous year and cut them up into lengths of one or two joints. These tops or cuttings are planted horizontally in the wet soil about eighteen inches from each other in rows about four feet apart. Six days afterwards the field is again watered, and about the twentieth day four or six shoots sprout from each cutting. In Ganjam and Vizagapatam, some ryots plants the cuttings in nurseries and afterwards plant out the shoots in the fields. After the shoots appear in the field, the ground is weeded and hoed, and when they are about a month old, chaff, weeds or some such manure is thrown around them. The soil is kept moist by occasional irrigation, and when about three months old the shoots ought to be a yard high. After this stage, it often becomes necessary to give the canes support, and this is done by bamboos or by a sapling stuck into the ground in the middle of each group of canes, the leaves being tied round so as to bind the canes together. This process requires constant care until the cane, at ten months from planting, is ready for cutting. It is then from four to six feet in length and about an inch-and-a-half in diameter. In the Vizagapatam district, it has attained a diameter of four inches.

"*The Classes of Soil best suited to Sugar-cane Cultivation, and the Extent to which Irrigation is required.*—The rich alluvial soils near the mouths of rivers are best adapted to the cane, but it is useless to attempt to grow cane upon land which cannot be irrigated during ten months of the year. The black soil (*regur*) which suits **Sorghum** does not suit sugar-cane, unless there is a considerable admixture of sand. It is remarkable that, although half the cane in the Presidency is grown in the districts of Ganjam, Vizagapatam, and Godávari, and although there is cane also grown in North Arcot, Nellore, and Kurnool districts, there is not a single acre under that crop in the Kistna. The black soil is not suitable, and the channels in the Kistna delta do not carry a sufficiently continuous supply of water. During the first month of cultivation, the field should be irrigated every week, and afterwards every fortnight but much depends upon the nature of the soil,—a garden rich in organic matter requiring water much less frequently than a sandy field. **Mr. Robertson**, at the Saidápet Farm, found it necessary to irrigate a crop of sugar-cane 114 times, but considers that usually forty or fifty times would suffice, giving the ground each time water equivalent to one inch of rainfall. Under a channel, the field of a ryot would

CULTIVATION in Madras.

Area & Outturn.

probably receive more than an inch at each irrigation, and probably twenty-five floodings would suffice. In this Presidency, fields on which sugar-cane is grown are charged with water-rate as if a double crop of rice had been produced on the land. It is true that sugar-cane is on the ground for ten months, and two crops of rice occupy only about seven months of the year, but it is not certain that the cane takes as much water as is taken for the two crops of rice. The Department of Public Works' estimate of the requirements of an acre of irrigated land is believed to be two cubic yards per acre per hour, which in seven months would amount to 75 inches of water. **Mr. Robertson**, Agricultural Reporter, considers that a crop of sugar-cane receives about forty-five floodings of one inch of water each.

"*Estimated Cost of Cultivation and of Manufacture and estimated Profit.*—It is difficult to ascertain the cost of the labour or of the manure necessary in this cultivation, for the cultivators do not keep accounts and are averse to give information. **Mr. R. E. Master**, Director of Revenue Settlement, estimated the cost of cultivation, at R145-8-0, and the cost of cultivation and manufacture at R182, per acre, while he estimated the outturn at 67 cwt. of jaggery. **Mr. Wilson**, the present Director, does not consider this outturn excessive for the Godávari district, and would lower the estimated cost of cultivation and manufacture to R150 or R125 per acre remarking that a profit of R200 per acre is the figure commonly quoted in the, Godávari. For the Presidency generally, **Mr. Wilson** would take the cost of production at R150 per acre, and estimates the outturn at 22½ tons of stripped cane, yielding 45 cwt. of jaggery, worth R250. This would give a profit of R100 an acre, but it must be remembered that the land lies fallow in the previous year, and this circumstance must be taken into account in any calculation of the profits.

"*Estimated Outturn of Cane and Value of Outturn if it is ever disposed of in this form.*—It is usual for the cultivator himself to manufacture jaggery, except in three localities where manufacturers purchase the cane, but it appears that there are 2,999 acres producing cane which is not crushed. The cane produced on these three thousand acres is sold retail for mastication at a price varying from one to six pies per cane. The average number of canes in an acre may be taken at 9,000 and their weight at 22½ tons. **Messrs. Parry & Co.** in South Arcot pay the cultivators R16 for each candy (℔ 500) of *jaggery* produced from their canes, or about R172 for the produce of an acre.

Conf. with p. 320.

"*Manufacturing Processes ordinarily employed.*—At Aska, in Ganjam, **Messrs. Minchin & Co.**, and in South Arcot **Messrs. Parry & Co.** and a Native capitalist, have European machinery. At Aska, the cane is sliced and the juice is extracted by the action of hot water, which is afterwards evaporated, the process requiring a large expenditure of fuel. In South Arcot, the process is that usual in the colonies, the cane being crushed in a three-roller mill, and the juice defecated with lime and passed through filters before being boiled *in vacuo*, the molasses being driven off by centrifugal action. The sugar prepared by either process is much the same in appearance, the grain is small and white. The ordinary process of manufacture of coarse *jaggery* does not differ from that in use in other parts of India. A wooden mill of two or three cylindrical upright rollers working into each other by endless screws at the top, the spirals being cut in opposite directions, is moved by a lever turned by bullocks. The canes, cut into pieces two or three feet long, after being soaked in water for a day, are passed between the rollers, and the juice flows down into a pit and thence by a channel into a tub or pot sunken in the earth. Near by is a boiler, and the crushed canes serve as fuel. The juice is poured into the boiler and a lump of lime is added. Sometimes gingelly-oil * (**Sesamum**) is also added. The juice is constantly stirred while boiling. To ascertain if it has arrived at the proper consistency, some is dropped into cold water, and if this solidifies, the boiling is poured into wooden vessels or bags and left to cool, when it becomes *jaggery*. In North Arcot and Cuddapah, there is a rude process of refining the jaggery. The boiling is stopped before the stage of crystallisation, and the juice is poured into pots with holes, through which the molasses drain for twenty days, leaving a crust of sugar, which is removed, boiled twice again and purified by means of milk and ghee. Sometimes when this crust of sugar is reboiled, thin slips of bamboo are left in the pot for forty days, and the syrup is allowed to drain off. The slips of bamboo are then found to be coated with sugarcandy.

* From the abstract given above of **Dr. Roxburgh's** description of sugar manufacture in 1792 it will be seen that gingelly oil was even then used. The action of the oil does not appear to have been investigated. It may have been to regulate ebullition. *Conf. with pp. 225, 254, 286.*

CULTIVATION in Madras.

Area & Outturn.

"*General Aspects of the Industry.*—The cultivation and manufacture of sugar are steadily increasing year by year in this Presidency. It is impossible to frame any estimate of the extent to which borrowed capital is used in this industry, but it is believed that the great majority of the cultivators of sugar are men of substance, who can afford to spend the requisite money and to wait the two years which must elapse before they can grow cane on the land where it was grown before. In the Godávari district especially, to embark in the cultivation of sugar is regarded as a certain sign of prosperous circumstances. As a rule, therefore, it is believed that the industry is not carried on by borrowed capital.

"The cultivation and manufacture are almost invariably united, except, as already mentioned, where the factories in Ganjam and South Arcot purchase the cane off the fields. The profits are doubtless much greater than that derived from any other cultivation. They amount to at least R70 per acre, while the profit from indigo does not ordinarily exceed R50 per acre. Only a rough estimate can be framed of the capital engaged in the industry. If the expenses amount to R150 per acre, the total expenditure must amount to more than a hundred lakhs, and this estimate of capital does not include the value of the land. The districts of Ganjam, Vizagapatam, Godávari, Cuddapah, and South Arcot export sugar or jaggery made from the cane. In North Arcot, Bellary, Salem, and Coimbatore, the supply appears to equal the demand. The other districts import it. The local consumption is not affected by foreign competition, as only refined sugar is imported, but **Messrs. Minchin & Co.** state that, since the import duty of 5 per cent. was removed, they have been unable to compete in the Bombay market with Mauritius sugar. If means could be taken to render the surf on the Ganjam coast passable, or if Ganjam were connected by canal with other communications, **Messrs. Minchin** could undersell the Mauritius sugar at Bombay.

"*Improvement.*—The improvements which may be effected are, no doubt, greater cleanliness in the mills and vessels used in the preparation of *jaggery;* some scientific method (such as the use of litmus paper) to ascertain the amount of lime required; closed boilers instead of open vessels; and the iron three-roller mill in place of the wooden-roller mills now used. **Messrs. Minchin & Co.** in Ganjam let out an iron mill, at R5 for each hiring, to the neighbouring cultivators, and a successful introduction of the Behea mill has been effected in the Bellary district.

Native States. 354

NATIVE STATES.—"The only Native States in this Presidency are Travancore, Cochin, Pudukota, Sandur, and Banganapalle. It appears that the area under sugar-cane in Travancore is comparatively limited; that jaggery and molasses are extensively manufactured from the juice of palmyra and cocoanut trees; Travancore is dependent on its imports for refined sugar. The improvement of the sugar industry is now engaging the attention of the Travancore Government, and experiments are being made with fair success with the amber sugar-cane obtained from the Saidapet Farm. Three sugar-cane crushing mills have been ordered out and sent to the sugar growing taluks for trial, and the services of an expert have also been engaged by the Government for the manufacture of sugar.

"In the Cochin territory, the cane is very sparsely cultivated, and what little is grown is sold for consumption as such, and not converted into sugar or jaggery. Some little jaggery is said to be manufactured from palm juice, but none is exported. During the past ten years only 2 cwts. were exported.

"Of the other three Native States, Banganapalle in the Kurnool district does not grow any sugar-cane. In Pudukota, attached to the Trichinopoly district, 22 acres are returned under sugar-cane, 161 acres under cocoanut, and 758 under palmyra. Only palmyra jaggery is manufactured, and that to a limited extent; less than 15 acres being utilised for the purpose, and the quantity manufactured averaging about 22 cwts. per acre. The sugar-cane is sold as such for consumption in Sandur in the Bellary district. About 65 acres were under sugar-cane cultivation in 1882-83. **Mr. Macartney**, the Agent to the Rajah, has furnished an interesting report regarding the cultivation of the cane in this small State. The outturn of jaggery per acre is given at over 53 cwts., and the net profit at about R160 an acre. **Mr. Macartney** appears, however, to have omitted to take into account the feeding charges of the bullocks, and the deduction required to be made for depreciation."

"Two kinds of sugar-cane are cultivated in Sandur State—the white and the dark coloured,—the former generally, the latter rarely, as, though it is said to produce a large quantity of juice, it is considered to be less rich in saccharine matter. The soil preferred for the cultivation of sugar-cane is a rich deep red loam. Some of the irrigated lands are well adapted for it; others are rather heavy and clayey. In their selection of the kinds of cane best suited for the soil, the ryots have been solely guided

CULTIVATION in Madras.

Area & Outturn.

Native States.

by experience. The plants or cuttings are often imported. The tillage is excellent, and leaves little to be desired, though probably some improvement might be effected by the introduction of ploughs and other implements of husbandry of a better description.

"In January the ground is well ploughed with four bullocks to each plough,—first in one direction, and then at right angles to the first ploughing. The process is done as effectively as possible, so as to expose the undersoil to the sun and air. The clods are then carefully broken up and cleared of roots and weeds. Manure is now applied to the extent of 30 cart-loads per acre. A well-to-do ryot will often expend as much as 50 cart-loads. It is also customary with the ryots to have large flocks of sheep and goats penned for several nights on their intended plantation, and for this they pay the shepherd at the rate of R2 and upwards per night, according to the size of the flock. This system is adopted in order to supplement any deficiency in the quantity of solid manure, or even as a substitute for it, when the ground is already in good condition. If tobacco cultivation has immediately preceded, the penning of a flock of 400 or 500 sheep for 3 or 4 nights per acre will often be thought sufficient. The field is again ploughed and reploughed in order to mix thoroughly the manure with the soil. Beds of about 3 yards square are made for convenience in watering, and the ground is well watered. The seeds or cuttings are then trodden in rows. This operation is usually carried out in April. The plantation should be watered twice a week in dry weather. During the south-west monsoon, however, it may sometimes be unnecessary to water it more than once or twice a month. The crop is 10 or 11 months in coming to maturity and during this period it will be necessary to weed it 4 or 5 times.

"In the first three months of the growth of a sugar-cane plantation, the ryots are accustomed to grow vegetables of various kinds among the young canes, and the proceeds of such crop assist in meeting the working expenses of the plantation.

"For a plantation of 3 acres in extent, the cost of cultivation and manufacture of jaggery comes to R570. The value of produce and assets amount to R1,072 so that a profit of R478 may be said to be obtained.

"It is unfortunately the exception, and not the rule, when a ryot is in the position to cultivate crops necessitating a considerable outlay, and it is to be feared that, what with high rates of interest and stipulations to dispose of the produce at a fixed rate to the money-lender or other conditions, the ryot enjoys but a moderate profit from his labour. The crop of sugar-cane is here never sold in bulk.

"The crushing mill is formed of two vertical cylindrical wooden rollers moved by an unending screw at the top. The canes are cut into two or three pieces for convenience in handling. Four bullocks are necessary for working each mill.

"It is usually necessary to pass the canes at least three times through the mill, as wooden rollers yield much more under high pressure than metal ones. Some years ago an iron crushing mill was purchased from the Collector of Bellary for experiment; but although it proved to the ryots that with two bullocks only it could extract more juice and do the work more efficiently than they with their mill, driven by four bullocks, could do, an offer for the mill could not be obtained and it was finally sold at a sacrifice. This mill was very portable, and could be taken down or set up in half an hour. It was a decided advantage over the native mill, as it could be carried easily to the crops, instead of having, in many instances, to carry them to it."

Coimbatore. 355

COIMBATORE.—"Sugar-cane (*Karumbu*) is cultivated chiefly in Coimbatore, Dhárápuram, and Udamalpet *taluks*, and but slightly in the others. It is usually grown on wet lands, which in Udamalpet and Dhárápuram require little or no aid from wells, as the channels run nearly the whole year; in Coimbatore, as the lands are under Noyil-fed tanks, wells are absolutely necessary. It is occasionally grown as a garden crop, and that under rain-fed tanks, such as *Puttúr pallapálaiyam* in Erode, is practically a garden crop.

"The chief varieties are the white (*vellei* or *rastálei*), striped (*náman*), and the red or purple cane; the first appears to be the Mauritius cane, introduced by Government some forty years ago; it has quite ousted the country cane, which was a very poor variety. The *námam* cane is chiefly used for eating, the *rastálei* for jaggery and sugar. It usually alternates with ragi and paddy, and in Coimbatore is said to follow betel well, probably because of the high manuring given to the latter. June is considered as the best season for planting, because of the abundant water for the next nine months. The land is heavily manured, usually by sheep folding at a rate equal to 6,000 to 8,000 sheep per acre for one night, at a cost of R15 to 20, ploughed six or eight times, ridged at about a cubit apart, and cuttings of three or four of the upper knots of the cane planted each about 1 or 1½ feet apart. From 15,000 to 20,000 cuttings, costing R23 to R30, are required per acre. It is watered twice in the first week and thereafter once a week for six months, and once a fortnight subsequently. Less water is used than for a five or six months' paddy crop, which requires a continuous

CULTIVATION in Madras.

Area & Outturn.

Coimbatore.

flow, but cane occupies the ground for a longer time. Five or six weedings are given each at the interval of a month; occasionally a compost of ashes, cow-dung, etc., is applied to the roots when the crop 3 or 4 feet high, and the earth ridged up over it. At eleven or twelve months old it is ready for market or for making *jaggery*; occasionally ratooning is practised, and this second crop is said to be nearly as good as the first, but this is doubtful, since, if so, it would be a general practice. The canes are all stout and strong, and being tied together when half grown, and surrounded by a strong hedge, they require no wooden props as in other districts. Garden cultivation is very similar.

"The total area in 1881-82 was 3,890 acres, of which 1,314 acres were in Coimbatore *taluk* within 5 or 6 miles of Coimbatore town. The yield of cane, numbering about 35,000, averages 25 to 30 tons; of juice 18¾ to 22½ tons, of jaggery 2 to 2½ tons, and of sugar * 2 tons per acre. The outturn of an acre will occupy a mill worked by two pairs of bullocks, one in the morning and one in the evening, for from 20 to 25 days. The value of the jaggery averages R210 to 240; the retail price is considerably higher than the price got by the ryot. When the canes are simply cut and sold for eating, R150 to R200 is about the price realised. The manufacturing process is as follows: the cane is cut into pieces about a cubit long, slightly beaten with a mallet, and then passed twice or thrice through the mill, which has two vertical wooden rollers of *karuvéla* (**Acacia arabica**) wood, about 5 feet high and 8 inches in diameter, geared at the top by a peculiar endless screw, and worked by a long lever attached to the head of one roller. The juice is received in pots and carried to the boiler, which is a simple large copper pan about 5 feet in diameter at the top, 1½ deep, and holding about 600lb of juice; it is placed over an open fire, fed with the cane trash. Two charges, each added in four successive instalments, are boiled each day, 18 *modas* of juice (=2,400lb at 1·1 specific gravity) being got through in that time. The yield of this is about 250lb of jaggery. A little lime temper is added to prevent acetous fermentation, and the scum is carefully cleared off. Owing to the rudeness of the process a great deal of the sugar becomes inverted, and is discoloured by partial burning, so that the sugary mass is a dark brown. When the juice has been inspissated to the consistency of thick treacle, the charge is struck by turning it out into moulds, which are small square holes cut in solid planks. Sugar is obtained exactly in the same way, except that it is more rapidly boiled until a minute sandy granulation appears, when it is turned into a shallow tub; it is then continually stirred with paddles, and repeatedly poured over the sloping sides of the tub until the minute crystals have somewhat developed by accretion. To secure a quick fire a little wood is used, this is the only occasion on which wood is required. The outturn of sugar is about 2 tons, value R230 to 250.

"The following table gives useful particulars:—

OUTTURN.						COST.				PROFITS ON		
Cane.		Jaggery.		Sugar.		Cultivation including assessment.	MANUFACTURE OF JAGGERY.			Cane.	Jaggery.	Sugar.
							Labour and wear and tear.	Bullocks.	Total.			
Tons.	R	Tons.	R	Tons.	R	R	R	R	R	R	R	R
25 to 30	150 to 200	2 to 2½	210 to 240	2	230 to 250	75 to 93	28	25	120 to 146	75 to 125	64 to 90	85 to 105

"The mill is clumsy, difficult to move about, requires powerful bullocks by reason of the friction of heavy, ill-cut screw gearing, and demands that the cane be twice squeezed; the services of a carpenter are frequently needed, as the rough threads of the gearing are apt to give way. The boiling is the process that most requires

* If the yield be 2 to 2½ tons of jaggery, by "sugar" must be meant *bura*; it certainly cannot be refined sugar, otherwise the jaggery would have to be more than double what is stated.

CULTIVATION in Madras.

Area & Outturn.

improvement; the ryots have recently adopted copper instead of iron for the pans, which is one step forward, but nearly everything is yet to be desired in the process.

"An immense area of the land is available for cane-growing; 11,000 acres of wet land (occupied), all of which would grow cane splendidly, are available within ten miles of Coimbatore town. The Erode wet lands are too wet for cane, which grows coarse, while the juice is very watery; a better irrigation system among the ryots would enable these to be utilised. Near Dhárápuram it is largely grown, and the area might be much extended. It is estimated that the cultivation of cane in Coimbatore, on 3,800 acres, employs a fixed capital of about 22 lakhs as the value of the land, together with a floating capital annually expended of 5 lakhs; the produce in jaggery and sugar alone is about 1,000 tons of sugar and 7,000 tons of jaggery, valued at 7½ lakhs. To this must be added the value of the cane used for eating and for supplying cuttings, which together absorb 10 to 15 per cent. of the gross outturn in cane. Actual profits, as in cotton and other crops, are somewhat greater than here shown, since much of the labour here charged in money is that of the owners and co-partners" (*Man. Coimbatore Dist.*, 235).

Cuddapah. 356

CUDDAPAH.—"Sugar-cane is principally produced in the sub-division, and is largely cultivated under the numerous small tanks. The cultivation of this crop lasts for the greater part of a year and a half, and it is in consequence always rated as a double crop. It is planted at two seasons, either just before the June rains, or else after the north-east monsoon, and pays accordingly, either two full assessments and one *fussaljasty* (charge for second crop on wet land), or else two *fussaljasty* rates and one full assessment. It requires a large and constant supply of water. The average rate per acre is R6-10, and the outturn is on an average 200 maunds per acre.

"Sugar is in considerable demand all over the south of India, and forms one of the principal articles of export from the sub-division. The cane is sweeter and more juicy than that raised in Tanjore or Trichinopoly, and a specimen of Madanapally sugar-cane gained a first prize in the Agricultural Exhibition at Madras in 1874. The common native mill used is made of *tumma* wood. It is formed of two screws fitting into each other, between which the sugar-cane is pushed with the fingers. The cane is crushed three times, and the juice, which has been caught in a pan below, is then boiled. There is a considerable amount of wastage in these mills, in labour for the three crushings, and because even then the whole of the juice is not extracted. I have endeavoured to induce the ryots to use the patent sugar mills with two iron cylinders, which press a greater quantity of juice with only two crushings, but have not as yet succeeded. The price of one of these machines (*viz.*, R225) is a fatal objection, and their weight prevents two or three ryots from clubbing together to purchase one, since the cane should be crushed as near the field and as soon after the crop is cut as possible, and there is considerable difficulty in the transport of so heavy a machine. A native mill is constructed for a few rupees and is easily carried. There is doubtless an opening for the employment of European capital in the manufacture of this article, and the favourable climate of the sub-division would render Madanapally a pleasant as well as a central station for a European agent. In 1872-73, the amount of acres under cultivation was estimated to be 2,384, which, calculating the average yield to be 150 maunds of jaggery per acre, would give an annual outturn of 3,57,600 maunds, which, at a valuation of R1½ per maund, would be worth R5,36,400. The cultivation of this product is on the increase, and the increase would be even more rapid if there were a more certain water-supply.

"Sugar-cane requires a constant supply of water for at least 18 months, and it is, therefore, seldom that it can be cultivated by means of tank water only. The numerous small tanks, however, though very ineffectual as a means of storing water, are of good in moistening the ground, and of thus supplying the wells with water. Of these there are a very large number indeed, and the majority are situated in the sub-division. Without these wells it would be almost impossible to carry on the wet cultivation, for by far the greater number of the tanks are dry before the second or Vaissaekam crop of paddy is cut (the first or Karticam crop is frequently never planted). Sugar-cane is therefore seldom planted, except where the water of a well is available, and as the competition for the ownership of the wells is very keen, and as the majority of the private wells have got into the possession of the wealthy ryots and village officials, it is they for the most part who grow it. Another reason which prevents the poorer ryots from cultivating this crop is the great expense attendant on the preparation of the ground. As this product is one peculiar to the district, and as the sugar produced ranks with the best in the Presidency (that of Astragram, which is raised under very similar circumstances of soil and climate not excepted), a description of its cultivation may not be out of place.

"Sugar-cane requires very deep ploughing, and the ground (say one acre) is gene-

CULTIVATION in Madras.

Area & Outturn.

Cuddapah.

rally ploughed over ten times with two or three pairs of bullocks, each separate ploughing lasting for one day. Between the ploughing the manure is placed on the ground, and subsequently ploughed in. For an acre of ground not less than 60 *bandies* of animal manure will be used and sometimes as many as 100. The refuse of the Kanooga nut, from which the oil has been pressed, is generally used in addition, when available, from 30 to 40 bandy loads per acre. Sheep are also penned on the ground for ten consecutive nights for the sake of the manure, and if the ryot does not own a flock of sheep, he has to pay a shepherd 8 annas per night for this privilege. Then comes the planting. This is done by cuttings, and the average number of cuttings per acre are as many as 8,000. The charge for such cuttings varies from rupees four to rupees four-eight per mille. The land is then prepared and portioned off into plots, intersected by deep trenches, fed by smaller ones, by means of which the water is carried round the whole plantation. The plants themselve s are banked up, so that the water acts upon their roots. The ground has to be continually kept moist, and water is generally flooded once in eight days, and allowed to stand in the trenches until it soaks in or evaporates. As the cuttings grow up, there succeeds the continual labour of tying them together. There are from five to six tyings before the crop matures. These tyings consist of binding together five or six plants covering about one square yard of ground. During the whole of this time the ground has to be continually kept free of weeds, which grow quickly in the moist earth sheltered by the growing canes. The field must also be carefully fenced in, as cattle are very fond of grazing on the sweet juicy stalks of the young plants. The fencing is generally done by thorn bushes, and is very effectual. After 18 months of this culture the crop is ready to be cut. Before doing so, a mill is generally set up as near as possible, and a hut with boiling pans. The canes are cut gradually, and in an acre of ground the cultivating will generally last for about 15 days. Each cane is passed three times through the mill, and the juice is at once taken to the boiling pans. There will be four such boilings in each day, and each boiling is calculated to produce 2½ maunds of jaggery. An acre of sugar-cane is, therefore, estimated to produce 60 boilings at 2½ maunds of each, or 150 maunds of jaggery. The average market price of jaggery is R1-4 to R1-12 per maund, so that the outturn is a valuable one. In order, however, to form an opinion of the net profit to the ryot, it will be as well to glance at the actual cost of production. For the details under this head, as well as for many of the particulars already given, I am indebted to the inquiries of the Sub-Division Serishtadar Teperumall Chetty. These inquiries I have also endeavoured to verify by personal investigation. The calculation is based upon the supposition that there are two or three members of the ryot's family, so that he has not to employ so much labour as a stranger would have to do.

Ryot debtor for one Acre of Sugar-cane.

	R	a.	p.
To 8,000 cuttings, at R4 per 1,000	32	0	0
To 60 *bandies* of manure, at 10 *bandies* per rupee .	6	0	0
To 30 *maunds* of oil-cakes, at 6 *maunds* per rupee .	5	0	0
To 10 nights of sheep penning, at 8 annas per night .	5	0	0
To rent for pan in which the juice is boiled, at ten annas per diem	9	6	0
To rent for mill, at five annas per diem . . .	4	11	0
To carpenter for regulating mill	3	0	0
TOTAL .	65	1	0

"The total expenses, irrespective of the value of the labour for weeding, tying, and ploughing, amount to R65-1. To this must be added the rent of the land, which, as the crop lasts for 18 months and receives a constant supply of water, will have to pay, say, R8 original assessment and two extra (*fassuljasty*) rates, or R8 in addition. The total cost of production is, therefore, R81-1, and the value of the product is R225, leaving a margin of profit of R143 for 18 months' labour to the ryot, or R95-15-4 per annum. This calculation, however, does not include the cost of agricultural stock and instruments. No ryot undertakes the cultivation of an acre of sugar-cane unless he has at least three pairs of bullocks, the original cost of which will not be less than R180. A bullock is supposed to last for ten years when used for wet ploughing, so that one-tenth of R180 should be added to the annual cost of production. In addition to this charge there is an annual expenditure on ploughs, ploughshares, ropes, buckets, etc., which I underrate when I give it at R20 per annum, so that these items will reduce the net profit from R95 to R58 per acre. Of course, the agricultural stock can also, during the time the cane is maturing, be employed in other labour;

CULTIVATION in Madras.

Area & Outturn.

Cuddapah.

but, as no ryot can undertake the cultivation of an acre of sugar-cane unless he possesses the stock already mentioned, the crop is only cultivated by wealthy ryots. If it happens that the ryot has to irrigate his cane from a well, his cost of production is materially increased, for additional expense is incurred in raising water, both in bullocks and servants. The result is that, though the outturn of sugar-cane is so good, it is generally only cultivated on inam or lowly assessed lands. In the complicated assessments which exist in this district much of the best land has been given away on rates lower than those usually charged. This was frequently done in former years out of consideration to the proprietor on account of services rendered. On lands such as these and on inam lands, therefore, ryots prefer to cultivate sugar-cane. On the fully assessed lands they are in the habit of cultivating paddy, cholum, or raggy, since far less expense is incurred in their agriculture, the crop is quickly matured, and in the event of a season failing they can count upon getting remission" (*Man. Cuddapah Dist.*, *206*, *210-213*).

Ganjam.
357

GANJAM.—"The sugar-cane grown in Ganjam is of excellent quality, and is said to be the best in India. It demands more care and attention, however, than any other crop, and is never grown for two years in succession on the same land. The ground requires to be well manured, with oil-cake or other suitable manure. Sugar-cane is estimated to require one-third more water than rice, and takes ten months before it reaches maturity. In spite of these drawbacks, however, the crop is one which is exceedingly profitable to the peasant who can afford to grow it. Sugar-cane is chiefly cultivated about Aska" (*Mad. Man. Admin.*, *II.*, *78*).

Conf. with pp. 93-94.

The account quoted above (*p. 225*) from **Dr. Roxburgh's** report now 100 years old may be accepted as giving the main facts regarding sugar-cane cultivation in this district. From the chapter on the History of the effort to establish Sugar Planting as a European Industry in India, it will be seen that attention was early directed to this district as one of the best in India. That interest may be said to have gradually matured into the Aska sugar mills—one of the largest and most successful works of this kind in India.

Godavery.
358

GODAVERY DISTRICT.—"The *masaka* or sandy *regada* soil is the best for this cultivation. Sandy soils also answer, if very well manured. The pure *regada* soil is the worst. Before planting a sugar-cane garden, the ground must be ploughed at intervals for a whole year, at least ten times in all. Twice or thrice it should be manured, and the sheep and cows may be picketed on it with advantage. During all this time the ground is left fallow. In the second year when the time of planting approaches, it is again ploughed and levelled, and small beds, each two yards square, prepared. Water is admitted, and the soil dug up with the *mamooty* to the depth of eight or ten inches, and kneaded till it acquires the consistency of mud. The heads* of the sugar-cane of the previous year, each about a foot long, and called *dawa*, are then planted or buried in the beds at the rate of ten in each — they are placed horizontally and well covered with the mud, which is then allowed to dry for ten or twelve days until it cracks, when sufficient water is again admitted to close up the cracks. Each cane head planted will have four or five joints, and from each joint a shoot springs, and makes its appearance from 20 to 30 days after planting. Other shoots spring from the sides of the first, some of which die. After the shoots appear, the beds should be weeded from this time till the cane is ready for cutting, fresh water must be admitted every four days. In three months from the time of planting, the shoots attain about the height of a yard, and at this stage, it is usual for the outside leaves, which decaying, have fallen down from the stalk, to be carefully wrapped and bound round it as a support and protection. This operation is called *topu suttu*. When the stalk gets to the height of four or five feet it requires still further support to prevent its being blown down. A bamboo, 15 feet, long, is placed between every two canes, which, together with their offshoots, are bound to it; eight or ten inches of the top of the cane are left loose, and any dead shoots are now removed. By the time the canes are six months old a stouter bamboo is inserted between every two of the groups of shoots tied to the smaller bamboos, that is, just half the number of large bamboos are stuck in, and the groups of canes tied to them. From this time the cane requires to be tied afresh to the bamboo every six weeks, as its height increases; altogether there are four separate tyings, besides the original *topu suttu*. The cane is ready for cutting in a year from the time of planting. In the Mogaltur and Nagaram

* See the frequent foot-notes on this subject, scattered throughout the series of quotations regarding district methods of cultivation as, for example, at *pp. 128*, *140*, *186*.

talúks, large expenditure is incurred in fencing the sugar-cane gardens to keep out jackals, which are very destructive; for this purpose, bamboos are placed close together and tied. Immense numbers of bamboos are required for these fences" (*Settlement Report, 1860, pp. 141-142*).

CULTIVATION in Madras.

Kurnool. 359

KURNOOL.—"Sugar-cane is cultivated in Nandyál, Kálva, Rámallakóta, Done, and Cambum. The method of cultivation is the same as everywhere else. Except in a few paces where sugar is not manufactured, the cane grown is the Mauritius sugar-cane introduced into this district in Fasli 1253 (1843) when cuttings were first sent from Madras. This cultivation was for several years encouraged by the remission of the Government tax on lands grown with it. The native cane is still grown in Cambum and a few other places, but sugar is not manufactured from it" (*Man. Kurnool Dist., 170*).

North Arcot. 360

NORTH ARCOT.—*Karumba*, Tamil; *Cheruku*, Telugu.—"This crop is always raised upon irrigated land, more often under a well than a tank, since the former affords a more certain supply, and the canes need constant watering for the ten or eleven months that they are growing. It also needs much manure, and is an expensive crop to raise, so that only the richer ryots attempt it; in many parts it is not the fashion, so that though there may be wealthy farmers, little or no sugar-cane is seen. It is chiefly raised above the Ghauts, where its cultivation is carried on more carefully and scientifically than elsewhere, but a good deal is also grown in Chittoor, Chandragiri, and the west of Kárvetnagar.

"There are several varieties of the cane, but the ordinary ones are callled *rasthâli, nâman, isar*, red and big, which only vary in the size or colour, and are cultivated in the same way and with much the same results. The crop is never sown on the same ground in consecutive years. A field which has carried paddy, rági, or other irrigated crop is therefore selected, and in December its preparation begins. Should it lie low a channel is dug all round to act as a drain, and the soil is daily ploughed across and across for several days until it has become thoroughly pulverised. To assist in producing this effect, men with heavy sticks beat the hard clods to pieces. When the tilth is fine enough, the surface is levelled and sheep are penned on it for several days. A great amount of farmyard manure, with faded *kánaga* flowers, is also brought (sometimes, it is said, as much as 75 tons to an acre) and worked in with ploughs. Then the field about April is divided into ridges separated by channels, and in the ridges are lightly placed cuttings of the previous season's cane, about a thousand to the acre, each set in a little powdered manure. For a fortnight the channels are flooded once in four or five days until the cuttings send out shoots; then the soil is loosened with a hoe or by a plough drawn by men, and no water is allowed for a week. After this the channels are cleared, the plants earthed up, and irrigation is carried on regularly twice a week (except in rainy weather) until the canes mature. When they are a foot high green leaves (above the Ghauts always of the *kánaga*) are buried in the trenches between each row. Above the Ghauts also when they have attained the height of a yard, ryots dig pits in the irrigating channels at the head of each trench, and in each pit place a maund or two of *kánaga* oil-cake mixed with fresh cowdung and water. This is allowed to ferment for four days, and at the end of that time, as the stream passes down each trench, a boy stirs up the mixture, a little of which mingle with the water and is carried along with it This mode of manuring has the very best effect, but is only repeated once more during the growth of the crop.

"When the canes are four feet high their sharp leaves begin to be troublesome and are therefore rolled round the canes and tied, thus protecting them from the sun and hot winds, preventing splitting and keeping them succulent. When 6 feet high upright posts are planted on the ground and bamboos tied to them horizontally, by which the canes are supported; a higher row is added as the crop increases in height. All this time weeds have to be carefully eradicated and the thorn hedge surrounding the field kept intact as cattle graze greedily upon the canes. In February or March the crop is cut close to the ground, except a portion left for cuttings, and a mill having been set up hard by, the juice is pressed out of each cane, after a foot or so of the top, which is sapless, has been cut off and thrown away. The mill is a rough-looking machine, made to take pieces, but in spite of its roughness, it performs its work fairly well. Two cylinders of Acacia wood are placed vertically side by side, having screws cut near their upper extremities which work into one another. One of the cylinders is slightly higher than its neighbour and has a horizontal arm at its summit which by means of ropes is dragged round and round by oxen. As the cylinders revolve, the canes are introduced between them and carried through, parting in the passage with their juice, which flows along a trough into a pot set on the ground. Each cane is thus pressed two or three times, and as soon as enough juice has been

CULTIVATION in Madras.

obtained to fill one of the broad shallow boilers the process is stopped for a time, and the liquid, before it has had time to ferment, is boiled, with some lime water, for about an hour, over a fire of wood and sugar-cane refuse, which burns with great heat. When sufficiently boiled it is poured into a tub or hollow in the ground faced with stone and is slowly worked about with a stick having a circular piece of wood at the end, until it stiffens and becomes jaggery. About 200 canes of the small, and 175 of the big, variety generally turn out a Madras maund worth of jaggery. The total yield of an acre is worth at the lowest R150, and often as much as R300 where it is carefully tended" (*Man. North Arcot Dist.*, *326*).

MYSORE. 361

MYSORE AND COORG.

References.—*Buchanan-Hamilton's Journey; Rice, Gazetteer, Mysore and Coorg; Agri.-Hort. Soc. Ind. Trans., VII., 94.*

The detailed report furnished by Dr. Buchanan-Hamilton on the sugar-cane of Mysore would, even if an abridgement of it were to be furnished, run to many pages. The writer is, therefore, compelled to allow the reader to consult the original essay, or to learn particulars from the Gazetteer—a work which should not be difficult to procure. The following brief statement of sugar-cane in Coorg may, however, be given, the more so since in many respects it is applicable to Mysore as well:—

COORG. 362

"The statistics of Coorg concerning the production of sugar are scarcely deserving of notice, as its cultivation is extremely limited, being confined to about 20 acres of land in the whole Province; and the plots seldom exceed one acre in size and are put down mostly at the head of the paddy flats, chiefly in the vicinity of Verajendrapett in the Yedenalknad taluk. No sugar whatever is extracted anywhere from the cane, but when ripe it is cut up into small pieces and offered for sale at the weekly markets, and none is cultivated in the taluks of Padinalknad and Kiggatnad. A few long stalks of cane are to be found growing in the back-yards of houses and in plantain gardens for home consumption.

"There are four varieties grown: (1) *Nili kabu*, or the bluish cane; (2) *Patta-patti*, the striped variety; (3) *Rosa dali*, a superior kind, and (4) *Cheni*, a small coarse variety. The first of these varieties, the *Nili kabu*, is raised mostly in black soil mixed with sand. *Patta-patti* thrives both in black soil and white clay, while *Rosa dali* and *Cheni* are found to be best suited for sandy soils and those mixed with pebbles. The Mauritius cane might be introduced with advantage, as it is of larger size and yields more saccharine matter than the common native varieties. It is, however, doubtful whether the natives themselves will take any great interest in the matter. For instance, some stalks were noticed by me growing well among a field of sugar-cane near Chickamaglur in the Mysore territory. The natives said it had been introduced many years previously by Mr. Stokes, and admitted that it was superior to the ordinary kind; notwithstanding this, they said that all that they ever did towards perpetuating the growth was to plant the top slips equally with those of the other kind just as they came to hand, and as an excuse for their indifference they complained of its requiring more pressure for the extraction of the juice. The present method of tillage seems best suited for the climate and requirements of the country.

"As no separate assessment is levied on lands cropped with sugar-cane, and the patches cultivated with it are so small, it is difficult to state with any degree of accuracy the exact area under cultivation. It is roughly estimated at within 20 acres.

"Trenches 1½ to 2 feet deep are dug 3 feet apart, over which weeds and other refuse are burnt; cuttings 1½ feet long, taken from the top of the cane, are planted 1½ feet apart, usually in the months of April and May. They require to be heavily manured first after being planted, a second time after four months, and sometimes a third and fourth time later on. The soil is dug up round the plants after they have formed one or two nodes. All dry leaves are removed from time to time. The crop requires to be irrigated once every 15 days during the hot season; in some places it has to be watered as often as once in two or three days. It generally takes 18 months to arrive at maturity, but the *Rosa dali* and *Patta-patti* variety is cut in 15 months when grown on good soil.

"The best soil for sugar-cane is black mixed with sand. It flourishes well also in a rich chocolate soil. The same extent of irrigation is not needed as on the plains, as Coorg possesses a damp climate.

"The cost of cultivation is roughly estimated at R300 the acre. As the cane is sold in the raw state, no charges are incurred in the manufacture of sugar. The proceeds are calculated to yield R400, or a net profit of R100 to the ryot. The profit, however, depends much on the fertility of the soil, the supply of irrigation and the quantity of manure used. It may in places be R100 more or R60 less than just given as being the average. It is a most exhaustive crop, and the soil requires to lie fallow for a year after to recover. **CULTIVATION in Mysore & Coorg.**

"It is difficult to estimate the production of sugar per acre, as the cane is not crushed in Coorg for that purpose. About 8,000 canes are grown on an acre plot, and if each be sold on an average at one anna in the market, they would yield R500, from which cost of carriage to the market and retail vendor's commission have to be deducted.

"Coorg draws its supply of sugar and jaggery from Mysore. The quantity of sugar said to be imported is 1,313⅛ maunds, valued at R26,264, or R20 the maund of 80lb. The quantity of jaggery imported is 1,050 maunds, of the value of R7,500, or on an average R7-2-3 the maund of 80lb.

"It is needless to enter into the further details called for in the remaining paragraphs of the memorandum, owing to the very limited extent of the enterprise in Coorg. The cultivation of the cane might be extended over a much larger area of land in Coorg which is suited for it, but the chief obstacles to such extension consist in the sparseness of the rural population, and the high wages paid to labourers, who have to be imported from Mysore and the Malabar Coast. Were large irrigation works constructed by throwing dams across the Cauvery and its tributaries, the Haringi and Latchmantirth, at points where the construction of such dams is said to be feasible, it might lead to this valuable product being cultivated on a larger and sufficiently remunerative scale in the low-lying eastern valleys bordering on Mysore, where the soil is rich and labour is cheaper and more abundant than in other parts of Coorg.

"At present the attention of capitalists in Coorg is directed almost entirely to the cultivation of coffee and cinchona" (*Col. Hill, Commissioner of Coorg, 1882*).

IX.—BURMA. **BURMA. 363**

References.—*Mason, Burma and Its People, 505; Gazetteer, Vol I., 423-427; Agri.-Hort. Soc. Ind., Trans.:—IV., 184; VI., 129-148, Proc., 6; VII., 129-134, Proc., 13, 142; VIII., 54-58, Proc., 443, 458; Journal, II., 252, Sel. 211-213, 271; III., 228-236, Proc., 162-163, 168, 282; IV., VI., Proc., 31; Proc., 25; X., 43-50 (palms); XII., Proc., (1861, Andaman) 3-4, (1862) 10, (1863) 45; An Extensive Official Correspondence down to 1891.*

Area, Outturn, and Consumption.—It does not appear that any additional information of much importance, applicable to the province as a whole, has been published since the date of **Mr. Schofield's** *Note on Sugar*. The following passage from that publication may therefore be here given:— **Area & Outturn. 364**

"The area under sugar-cane is small. It has, however, increased considerably, the figures for 1885-86 being 10,500 acres as compared with 6,500 acres in 1881-82. The outturn of coarse sugar per acre is reported to be 35 maunds. Accepting this rate, which seems rather high, the total outturn, calculated on the acreage of 1885-86, is 3·67 lakhs maunds. The net imports are five lakhs maunds, so that the total consumption is 8·67 lakhs maunds, or 9 seers per head of population (3¾ millions).

"Great efforts are being made by the Provincial Agricultural Department to extend the cultivation of sugar-cane in this Province, which at present imports largely from Upper Burma, Calcutta, Madras, and the Straits Settlements, though in some places its soil and climate are said to be well adapted to the cultivation. **Mr. D. M. Smeaton**, when Director of Agriculture in this Province in 1882, wrote as follows:—

"'There can be no doubt in my mind, after what I have seen, that the Bilin tract is in a remarkable degree suited to the cultivation of sugar-cane. The fact that the same land can go on from year to year producing cane at a constant and high rate is very striking, not to speak of the further fact that the land can bear, and bear well, other crops in the event of a rotation being expedient.

"The chief obstacle in the way of extension of cane cultivation in Lower Burma is the indebtedness of the cultivators. Advances were made to these cultivators commencing from 1883: this relief is said to have led to a slight reduction in the rates of interest and to an improvement in the condition of the cultivators. The following extract shows that this experiment was interrupted by the outbreak of rebellion among the Shans at the end of 1885:—

"'The efforts made in former years to encourage the cultivation of the sugar-cane and the manufacture of jaggery in the Shwegyin district were continued during the

CULTIVATION in Burma.

Area & Outturn.

year under report, and R17,820 were advanced to sugar-cane growers in the Bilin township of that district. The progress of cultivation was hindered and the recovery of advances was rendered difficult by the outbreak of rebellion among the Shans in December 1885. One of the most influential of the Shans in the Bilin valley joined the rebels and drew many after him. In consequence of these defections and of the prevailing disquiet, nearly R9,500 of the advances made during the past two seasons have yet to be recovered. In consequence of the unsettled state of the Shwegyin district no further advances have been made" (*Paragraph 70, General Administration Report, 1885-86*).

Mr. Edward O'Riley furnished many instructive papers to the Agri.-Horticultural Society of India on the subject of the cultivation of sugar-cane in Burma. Some of these deal with the efforts which had been made, about the year 1840, to introduce superior races of cane from the West Indies and other countries, but certain of his papers exhibit also the peculiarities of the indigenous Burmese cultivation. The following two passages may be given as an exemplification of these subjects. Speaking of one of the then most popular foreign canes—the Otaheite—Mr. O'Riley says:—

"Your remarks on the subject of the Otaheite cane in Tirhoot are very interesting: the same amount of produce, or even more on the average per acre of new soil, on this coast, may be obtained, with the very material point in its favour of *being entirely free from the attacks of white ants*, a subject which appears to oppose very serious obstacles to the extended cultivation of the Otaheite plant in the Upper Provinces. I have given this point a good deal of attention of late, and after a personal inspection of all the Otaheite cane grown in the vicinity, I have not found a single patch injured by the attacks of those insects, notwithstanding the fact of their abounding in the vicinity, and in many cases *literally in the cane-fields*, where they may be seen covering the stumps of the trees left after burning the new jungle. It is a singular fact, that I have never heard a complaint made by the native cultivators of the ravages of white ants, not only in regard to the Otaheite cane, but of all kinds in cultivation. To afford every information in regard to the nature and properties of the soil, the produce of which is exempt from this pest, I have the pleasure of forwarding in charge of Captain Russel of the *Ganges*, a box of the soil taken from one of the gardens, which is about the average of all the cultivation in this vicinity. In the same box are two paper parcels from Major Macfarquhar, at Tavoy, containing specimens of the soil of his garden, in which he states, 'that Otaheite cane has been growing for the last five or six years uninjured, although there are legions of white ants in every direction.' I trust that an examination of these soils may lead to some data, that may be useful in guiding future operations in this description of cultivation.

Diseases *Conf. with pp. 121-127.* 365

"I find that by placing a good layer of common charcoal from the furnaces, under all materials of wood, subjected to the ravages of white ants, they are well preserved and in no case have I discovered any damage when this precaution has been taken; perhaps the application of the refuse of the furnaces to the soil, when opening the furrows previous to planting, might be attended with success in this respect; at all events it is worth the trial" (*Journals, Agri.-Hort. Soc. Ind., Vol. III., 229.*)

Mr. Edward O'Riley, in one of his communications to the Agri.-Horticultural Society of India, furnishes the following particulars regarding sugar cane planting in Burma:—

"The descriptions cultivated in this province are the rattan and red canes, the former being in more general use on account of its extreme hardness; the latter, however, is superior in quality and generally attains the height of about 5 feet on an average with a diameter of 1½ inches.

"A site having been fixed upon for the purpose of planting the cane (always elevated above the level of the plains which are liable to inundation during the south-west monsoon) the cultivator commences cutting down the jungle, about two months previous to the rains, which is then burnt and allowed to remain until the first showers of the monsoon have penetrated the soil, rendering the previously hard surface soft and friable, and without any further process the stoles are planted perpendicularly in rows at the usual distance apart, covered over with the burnt soil and allowed to vegetate without any other attention being paid them. A plough is an article almost unknown to the cultivators, and in no instance ever applied to this species of cultivation; the only method they have of preventing noxious weeds from springing up is by planting cucumber, chillies, pumpkins, etc., between each row,

CULTIVATION in Burma.

Area & Outturn.

which coupled with the property the burnt soil possesses in this respect completely effects the object. Notwithstanding the heavy fall of rain during the season (about 200 inches) the cane with only this attention paid it, thrives progressively during the monsoon, and is at full maturity in nine months from the date of planting. The only labour attending a plantation of the above description besides that of clearing and planting is enclosing, which is done with the partially burnt stumps found on the ground.

"The cultivator having prepared a mill composed of two vertical wooden cylinders, supported by a frame and worked with a buffalo, the cane is manufactured on the spot. The juice undergoes no process of clarification, except that of removing the scum from the surface during the ebullition, and being sufficiently inspissated, it is thrown into a shallow frame where, from exposure to the air, it forms a heavy, hard mass; it is then cut into small squares and conveyed to the bazars for sale, where it is disposed of at about R20 per 100 viss of 365℔.

"The following statement is procured from a cultivator who, not taking into account the value of his labour, or that of his family, assured me that he had made R400 by his cultivation of one season; it will be seen, however, that his remuneration, taking every thing into account, and allowing 12 months for the completion of his undertaking, affords a much greater inducement to the natives to become cultivators than working as hired coolies, even at the high rates current here (½ a rupee per day), more especially as the Burmese cannot be convinced that they should take their own labour into account in their cultivation, and as the gross amount of returns appear to them the real profit. I am decidedly of opinion that a very large extent of this province would be soon brought into cultivation were the inhabitants more abundant; as it is, a number of persons who have become aware of the advantages derived from it have commenced clearing jungle to a considerable extent, and as they pay nothing for the ground, and the outlay required is only the trifling one for the plants, a yearly increase must necessarily ensue, an emulation will be created and improvements upon their present slovenly method of cultivating will obtain footing which, aided by the powerful stimulant of European enterprise and skill, will eventually give a name to these provinces hitherto not possessed.

"The following is the calculation above alluded to—

One man with a family of a wife and two children can clear ground and plant 10,000 plants which cost R14. If the man and his family were to employ their labour as hired coolies they could procure—

		R
The man R10 per month, 12 months . . .	=	120
The wife R5 per month, 12 months . . .	=	60
The children R4 per month, 12 months . . .	=	96
		276
The crop stands him at maturity		290
The average of each stole at the lowest is 4 canes, 10,000 stoles, 40,000 canes, which produce of coarse sugar 60 viss per maund, 2,400 viss sells in Moulmein, R20 per 100 viss		480
Leaving a profit to the cultivator more than he could procure as hired labourer for 12 months . .		170

"*N.B.*—It must be observed that very few Burmese will work even six months consecutively; hence the rate of R10 per month, which I have stated as the man's wages for 12 months.

"With the apathy peculiar to their character, the Burmese look no farther than present advantage dictates; the only labour required to produce a fair crop of ratoons would be nearly banking the plants during the rains and trashing them properly, neither of which is done; the consequence is that the old stems being left to themselves during the dry season are either killed by the heat or choaked by the parched surface through which the fresh roots must penetrate, and such straggling canes as make their appearance and come to maturity are little better than rattans in appearance" (*Trans. 1839, Vol. VII., 129-134*).

Mason says: "That Burma is well adapted for the cultivation of sugar-cane has been well tested by **Mr. O'Riley**, who made many tons of excellent sugar from cane that was raised at Amherst. In Hindustan the mucilage of the musk melon, **Abelmoschus (Hibiscus) moschatus**, is used to clarify sugar, and it is one of our most common indigenous plants. Both the Burmese and Karens grow sugar-cane,

CULTIVATION in Burma.

which they chew for its juice, and from which they make cake sugar. Considerable quantities are imported from the Shan States; and I have seen as fine looking sugar-cane on the mountains of Toungoo as I have ever beheld on the flats of New Orleans."

Area & Outturn.

Though dealing only with a section of Burma, the various papers which have appeared on the sugar-cane of the Beelin township may be mentioned as of special interest. This would appear to be the tract of Burma in which the greatest success has been attained and where expansion in the future is most likely to occur. **Mr. Bridges** defines this region as follows:—

"The cane-growing lands extend for about 33 miles on the banks of the Beelin river and for about 17 miles on those of the Thehbyoo. These two large rivers run nearly due south from the Yoma, and are separated by a tract of tree-jungle about 12 miles wide, which is watered by the Choungbya, Choungsouk, and Gonyinweh creeks.

"The available land is chiefly found on the banks of the Thehbyoo river, where cane-cultivation has only extended within the last four or five years.

"In addition to the land immediately available for cane-cultivation, there is a large amount of suitable land which remains uncultivated owing to various accidental causes. Thus the greater portion of the land now flooded could be reclaimed by connecting it by cuttings north and south with the Beelin river: these cuttings would not only drain the land, but also bring large quantities of alluvial deposit which would gradually raise the level of the land and rendered it suitable for cane-cultivation. In many instances the cultivators themselves are anxious to dig these cuttings: thus the villagers of Kadipoo wish to connect the Kanyinbin with the Beelin river; the cultivators of Nyoungpalin wish to make a cutting from the Belingyo to the Beelin: these two streams are, however, Government fisheries, and the lessees naturally object to any interference with their rights. There are other fisheries in the cane tract which are, I believe, of little value, and the abolition of which would improve the drainage of the country.

"On the Thatone side of the river, and below Ngetchoon in the Beelin township, there is still land available for cane-cultivation, but this land has either been taken up for paddy cultivation or is claimed by the paddy cultivators as grazing ground. Similar claims to grazing ground are made on the Beelingyo, near Beelin, and near Shwaynyo village on the Thehbyoo: these grazing grounds have not, I believe, been fixed by Government, and there would be no difficulty in fixing suitable grazing grounds in a tract where waste land is plentiful so as to allow of the more valuable land being taken up for cane-cultivation. At Ngetchoon the Beelin river is rapidly cutting through the paddy land in the bend, and this land will have to be thrown up by paddy cultivators and will be taken up for cane-cultivation.

"*Nature of Soil.*—The soil of the valleys of the Beelin and Thehbyoo is almost entirely deep grey loam, mixed here and there with light clay. The land is covered by the floods for a few days at intervals during the rains to a depth of two or three feet and a thick layer of alluvial soil deposited on it.

"*Modes of Cultivation.*—Where new land is cleared or land already cultivated has been left fallow, the cultivator turns up the soil with a hoe at the beginning of the rains (May or June): he then leaves the soil to rest until September, when he digs holes about 10 inches deep and one foot wide at intervals of one-and-a-half foot from each other. Three pieces of cane (agyoung) about five inches long are then placed in a slanting position in each of these holes, so that one end of each plant touches the ground and the other protrudes about an inch over the top of the hole. The cane-pieces are then partly covered up with loosened earth. There are generally three joints to each of the cane-pieces, and each joint has one eye. Many of the young shoots being however destroyed by the heat or other accidental causes, it is seldom that more than five or six canes are found to one stool.

"Some cultivators plough their land three times at the beginning of the rains instead of turning it up with the hoe; but the more general practice is simply to run deep furrows through the land in September and then place the cane-pieces longitudinally at the bottom of the breach, which is about 10 inches deep and one-and-a-half foot wide. The space left between the furrows varies from two to three feet according to the nature of the soil.

"Before the cane is planted, the land is cleared of grass and weeds. About ten days after the cane-pieces have been planted, the earth is loosened in the intervals between the holes and the cane-pieces further covered up with mould. In the month of Pyatho (beginning of January) the earth is again loosened and the plants further covered up. About the month of Kazon (May) the land is again cleared of weeds and grass and the plants are then left until the month of Wagoung or Tawthalin

CULTIVATION in Burma.

Area & Outturn.

(August and September), when they are stripped of the leaves that have become old and withered.

"There are generally three or four young shoots or ratoons which spring up from the old stool. Where these ratoons are sufficiently thick, no new plants are put down after the cane has been cut, but as a rule cane-tops (*kyanbya*) are planted (*phaseikthee*) in the intervals between the ratoons (*kyanngot*) after the land has been cleared of grass and weeds: these cane-tops are about five inches long and are planted from November to the beginning of January, when cane-pieces can no longer be planted. The cultivators state that one *ta* (0·28 acre) yields 3,500 cane-tops, whereas one *ta* of cane yields enough cane-pieces to plant out five *tas* (1·40 acres). Canes sell at R20 per 1,000, whereas cane tops sell at R2 per 1,000.

Ratooning. 366

Conf. with pp. 59, 76, 128, 143, 151, 177, 181, 195, 215, 226.

"After the second year's crop the land is either left fallow for a year or again replanted with cane-tops and then left fallow the following year. Some portion of each plantation, except in very small holdings, is left fallow every year, as only enough cane is cut daily to supply the amount of juice required for boiling, and part of the land is still uncropped in Pyatho (beginning of January), when the sun is too hot to allow of cane-tops being planted. In land thus left uncultivated, the cultivator often plants paddy for his food provision after leaving sufficient space between the rows of paddy to put down cane-pieces in September.

"*Cost of Cultivation.*—The cost of the different kinds of labour hired for cane-cultivation is as follows:—

	Per *ta* (0·28 acres). *R*	*a.*	*p.*
Ploughing	2	0	0
Clearing	1	12	0
Digging holes for plants	2	0	0
Planting	1	8	0
Loosening earth	2	0	0
Ditto (second time)	2	0	0
Clearing (second time)	2	0	0
Pruning	2	0	0
Cutting canes*	4	0	0
Seed	6	0	0
TOTAL	25	0	0

This hire includes the cost of keeping the labourers, which is by custom estimated at eight annas a man per *ta*.

"The cost of cultivation in a plantation worked entirely with hired labour would, therefore, amount to about R90 per acre. The greater number of cultivators, however, hire little or no labour and work the land with their families. Those who work very large holdings generally hire their labourers by the year and not by contract. I would estimate the average cost of cultivation per acre at R15 to R20.

"The instruments used in cultivation with their cost are as follows:—

	R	*a.*	*p.*
The plough or *teh*	3	0	0
The curved hoe (*dagouk*)	1	0	0
The straight hoe (*pauktoo*)	1	0	0
The da for cutting canes	0	8	0

"*Cost of starting cane-cultivation on new land.*—The only extra expenses incurred in cultivating new land are clearing the elephant grass and fencing in the land taken up. The cultivator, as a rule, cuts timber from the neighbouring forest to make a fence and clears the land with his family without any hired labour. This work is done in the dry weather, and at the beginning of the rains the cultivator generally plants paddy, leaving space enough between the rows of paddy to put down cane-pieces at the end of the rains. He then plants out one or two acres of cane if he has enough capital to buy seed and hire labour, but as a rule he only plants one-fourth or half an acre to obtain seed for the following year. In any case, he has to wait for about twenty months before he obtains any return from his plantation and to support himself and his family during this time.

* Two labourers are hired for cutting canes for each mill used; they are paid R15 each per mensem without food, and they cut about one acre of cane a month, one-half their hire being put down to cost of cultivation and one-half to cost of manufacture. This would give about R4 as the cost of cutting canes per *ta*.

SACCHARUM: Sugar.

Methods of Cultivation

CULTIVATION in Burma.

Area & Outturn.

"The thrifty Toungthoo manages to support himself and his family without incurring any heavy indebtedness by planting paddy and a little cane for two years, and working as a labourer in neighbouring plantations until he obtains a crop from his own land: he does not work more than one-fourth or half an acre, and gradually increases the area of his holding. Most of the Toungthoos, moreover, start with a small capital, as they have generally worked paddy land before taking to cane-cultivation, and their friends and relatives give them seed-stock or supply them with the small sums they require at a low interest (25 to 30 per cent.), so that altogether they work under exceptionally favourable circumstances.

"The Shans generally come straight from the Shan States to Beelin; they bring little or no money with them, and being an enterprising race they generally attempt to start relatively large plantations at once. They are not satisfied, like the Toungthoo, with cultivating one-fourth or half an acre, but plant out two to three acres. They require a considerable sum to buy seed, machinery, boiling-pans, and to keep themselves and their family, for with a plantation of two to three acres, they have to devote all their labour to working their own land and cannot, like the Toungthoos, eke out their small resources by working for others. They generally require about R300 before they obtain any return from their plantations, and as their friends are not generally rich enough to lend them such large sums, they are compelled to borrow from the Burmese and Chinese money-lenders at the rate of 48 to 60 per cent. The yield of the first year is seldom large enough to pay off the borrowed money with interest and leave them enough to keep their family until the following harvest. They consequently borrow again and gradually fall into a state of chronic indebtedness, unless they work very large holdings, the yield of which will leave them enough to keep their family after paying off their debt with interest.

"*Outturn per acre.*—The outturn per acre as obtained from crop statistics is as follows:—

Name of cultivator.	Canes.		Juice.		Jaggery.
	Number.	Weight.	Measure.	Weight.	Weight.
		lb	Gals.	lb	lb
Moung Pouk Gyain . .	15,185	38,750	2,734·78	25,316·85	3,991·25
Moung Sway	9,532	42,068·75	2,560·10	25,432·14	3,921·85
Moung Pan Noo . . .	10,646	42,762·50	2,580·35	25,818·75	3,966·25
Moung Lan	8,825	29,191·28	1,720·21	15,937·50	2,446·85
Moung San Dwa . . .	9,892	37,214·28	2,047·00	20,326·78	2,937·50
Average per acre .	10,809	37,377·67	2,328·43	22,566·48	3,452·74

"The average generally given by the cultivators is 875lb per *ta*, or 3,125lb of jaggery per acre. Other cultivators stated that the average was higher, and the experiments made would seem to confirm this view. I think a fair average would be 1,000 viss, or 3,500lb per acre.

"*Manufacture of jaggery.*—The canes bloom about the month of November; they are then severed with a *da* from the stool close to the ground. The branches at the top are given to the cattle for food, and the top, which is cut off where the hard cane ends, is preserved for planting. The canes are then divided with *da into* two pieces of about four feet each, tied up in bundles and carried by the cane-cutters to the mill, where they are bruised and the juice extracted.

"The mill consists of two heavy upright cylinders of about one foot in length and two feet in diameter. The cogs are circular and are cut in two rollers superposed to the crushing cylinders. A shaft, about 12 feet long, is fixed to one of the upper cylinders and is turned by a buffalo yoked to it.

"A few improved machines are now used in Beelin; they consist of three upright pyingado cylinders, and the middle cylinder, to which the shaft is attached, turns the the other two by means of short straight wooden cogs: these machines have been made by a Burman, Moung San Dwa, who imitated an English machine he had seen in Moulmein. The price of the machine with two cylinders is about R50 and of that with three cylinders about R70.

"Four of Messrs. Thomson & Mylne's mills were left by me in Beelin with the cultivators. These machines were considered a great improvement on the wooden

CULTIVATION in Burma.

Area & Outturn.

ones, but the cultivators objected to the small size of cylinders; they would have hollow cylinders with a diameter of 1½ foot and a length of one foot, thus increasing the speed of the mill without any additional strain on the cattle. They would also have the shaft ten instead of seven or eight feet long.

"Cane was crushed with a Burmese machine at the same time as with one of Messrs. Thomson & Mylne's mills, and the compared results are given in the following table:—

	Weight of cane crushed.	Time occupied.	Quantity of juice.	Quantity of jaggery.
	℔	Mins.	Gals.	℔
Messrs. Thomson & Mylne's mill (three converging cylinders)	845	125	59·16	80·49
Ordinary Burmese machine (two wooden cylinders)	845	110	47·16	64·18
Difference .	...	—15	+12	+16·31
Messrs. Thomson & Mylne's mill (three converging cylinders) .	700	100	42·75	61·42
New Burmese machine (three wooden cylinders)	700	75	39·58	56·87
Difference .	...	—25	+3·17	+4·55

As compared with the Burmese machine in general use, Messrs. Thomson & Mylne's mill was 12 per cent. slower, but yielded 20·28 per cent. more juice and jaggery; as compared with the improved Burmese it was 25 per cent. slower, but yielded 7·40 per cent. more juice and jaggery.

"The canes, which have been previously cut short, are passed three times between the wooden rollers, being handed back by a labourer who collects them as they come out of the machine. The cane juice is received in a large bamboo frame, and thence flows through a bamboo gutter into an iron pan in the boiling shed.

"The crushed cane (cane trash or megass) is dried in the sun for two days, and then used as fuel with branches of trees and dead wood brought from the neighbouring jungle.

"At the time I visited Beelin the cultivators were in a great state of anxiety about their supplies of fuel, as the Beelin forest goung had issued a general order forbidding the cutting of any kind of timber in the jungle. This order was not, I believe, issued from any dishonest motive, but from the mistaken idea that the old code and new code were to be taken together and that the Government intended to reserve every kind of timber. It is to be regretted, however, that very subordinate officials should thus be allowed to issue general orders and disturb the population of a whole township.

"The juice is carried in chatties from the receiving pan to the iron boiling pans, which are placed three in a row over a furnace dug in the ground. The fire is kindled at one end of the furnace and boils all the pans in succession on its way to the chimney. The liquor is at first placed in the pan furthest from the fire, and as it evaporates it is passed on by means of ladles to the next pan, and from this to the pan immediately over the fire. It gets thicker as it passes along, and the impurities are removed by means of a wooden skimmer. In each pan is placed a bamboo framework to prevent the boiling juice from escaping over the sides of the pan. After the juice in the third pan has become sufficiently thick, it is taken off the fire and poured into an iron pan, where it is allowed to cool for a few minutes. It is then poured on a bamboo mat, spread evenly with a piece of bone, and then divided with a piece of pointed bamboo into small squares. After it has become hard, it is broken into cakes and packed away in bamboo baskets covered with leaves. Each basket contains about 175℔ of jaggery."

CULTIVATION in Burma.

Area & Outturn.

"The cost per mensem of the labour employed in the manufacture of jaggery is as follows:—

	R
Two men cutting canes* at R10 each	10
Two buffaloes at R7 each	14
One herdsman	7
One labourer to boil the juice	20
Food of labourers	15
TOTAL .	66

"The cost of labour is somewhat lower on the Thehbyoo, where the hire of the labourer who boils the juice is R15 and the hire of a buffalo R4.

"The monthly outturn for one furnace is generally taken as 1,250 viss of jaggery, and if the outturn per acre is taken as 1,000 viss, this would give as the cost of manufacture per acre R53 (52·80). To this amount must be added the annual wear and tear of machinery, the cost of fuel, baskets, mats, etc., and the annual erection of crushing and boiling sheds.

"The crushing machine costs R50 and is stated to last about seven years; it works about three months annually, and takes one month to crush the cane of one acre. It moreover requires annual repairs which amount to R3 or R4. The annual wear and tear for the machine would therefore be about R3·50 per acre. The boiling pans cost about R21, and are said to last about three years; as one set is used for each mill, the annual wear and tear per acre for boiling pans would be R2·30; allowing R1·20 per acre for the share of other expenses, the cost of manufacture would amount to R60 per acre.

"The labourers who cut the canes also pass them through the mill. The herdsman not only grazes the buffaloes, but drives them whilst turning the mill and carries away the cane trash to dry. The labourer who boils the juice also breaks the jaggery into cakes and packs the cakes away in bamboo baskets.

"Although the hire of these labourers is always given by the month, they in reality work by contract. The headman has to boil daily seven *kyaws* (iron pan containing 99·12 gallons of juice, yielding from 150 to 196℔ of jaggery according to the season). The labourers cutting the canes have to supply and pass through the mill a sufficient number of canes to yield seven *kyaws* full of juice. If the full monthly outturn is not obtained, the pay of the labourers is reduced in proportion to the deficiency. Some headmen make 10 *kyaws* of jaggery a day; their pay then rises to R25 per mensem, and another cane-cutter and buffalo have to be hired.

"*Total outturn of the tract.*—We have found that the total area under cane, is 3,300 acres, of which one-third, or 1,100 acres, is uncultivated. The area yielding sugar in the present year would therefore be 2,200 acres, which, at the average found of 3,500℔ per acre, would yield 7,700,000℔ of sugar.

"*Selling price of jaggery.*—The selling price of jaggery at the beginning of the season varies from R33 to R30 per 100 viss; latter in the season the price falls to R25 and R20; and for two years it is said to have fallen as low as R9 per 100 viss. The cultivators generally agree in taking R25 per 100 viss as the average price, and this price may, I think, be taken as a fair average.

"*Average profits.*—In holdings worked entirely with hired labour, we have found that the cost of cultivation amounts to R90 and the cost of manufacture to R60 per acre. The average outturn per acre (3,500℔ at R25 per 878℔) is worth R250; the net profit per acre, not including cost of leaving, would therefore be R100. In small holdings the cost of cultivation has been estimated at R20 per acre, and the cost of manufacture being R60, the net profit would amount R170 per acre. In small holdings which do not measure more than one acre, the cultivator spends all the profits in supporting himself and his family, and in the larger holdings the profits of cultivation often go to pay the interest on money borrowed.

"*Jaggery how disposed of.*—The jaggery is carried from Beelin by four principal routes—

(*a*) by cart to Wimpadaw and thence by boat to Rangoon;
(*b*) by sea from the Beelin river to Rangoon;
(*c*) by cart from Dawoon to Thatone;
(*d*) by cart to Kyouksarit and thence by boat to Moulmein.

A small portion of the outturn is taken by boat to Pawata, and thence on elephants to Papoon and the Salween district, but the dangerous navigation of the Beelin river, owing to the numerous rapids below and above Wingalay and the difficulty of the

* One-half of this has been put down to cost of cultivation.

CULTIVATION in Burma.

Area & Outturn.

carriage across the hills, prevent large quantities of jaggery from being carried by this route.

"The bulk of the jaggery is carried by the Wimpadaw route, and the cost of carriage is as follows:—

by cart to Wimpadaw, R24 per 1,000 viss;
by boat from Wimpadaw to Rangoon, R12 to R15 per 1,000 viss.

"The cost of carriage from Beelin to Rangoon would therefore amount to R38 per 1,000 viss. At the beginning of the season, however, when the cultivators are most anxious, on account of the high price then prevailing, to send their sugar to Rangoon, the cost of carriage is much higher owing to the bad state of the road between Beelin and Kyiketo and to the absence of any road between Kyiketo and Wimpadaw; it then amounts to R24 per 1,000 viss between Beelin and Kyiketo and to R16 per 1,000 viss between Kyiketo and Wimpadaw. At the end of November this year the cart track between Kyiketo and Wimpadaw was not practicable, and the jaggery had to be dragged in small boats along a half dried up creek to Wimpadaw at a cost of R14 per 1,000 viss.

"A great deal was done for the Beelin road by Mr. Irwin, who had all the creeks and nullahs bridged, but nothing has yet been done to level this road. I would strongly recommend that this be done as early as possible; but the most urgent road is that from Kyiketo to Wimpadaw, as at the end and beginning of the rains there is no water in the cuttings for boats to pass and no road along which carts can travel, so that all communication is then stopped for some weeks. A branch railway connecting Beelin with Pegu would open out all the fertile country east of the Sittang, the produce of which can now only be brought to Rangoon during the rains, unless it is carted at great cost or brought round by sea at a great risk.

"The cane-cultivation, which is now rapidly extending along the banks of the Thehbyoo, will make it necessary to construct feeder roads connecting the different villages with the Beelin and Kyiketo road. These roads are absolutely necessary, as there is no sufficient water for boat traffic in the Thehbyoo at the end of November, and the sugar can only be brought to Kyiketo by cart.

"The boats that carry sugar by the sea route to Rangoon are from the villages of Zokekalee, Nimblay, Shwaylay, Kawkamay, and Zokethoke. It is stated that about five trips are made annually and that about 10 boats sail at each trip. These boats generally carry from 500 to 800 baskets, or 4,000 to 7,000 viss of sugar. Larger boats would be dangerous on account of the numerous sandbanks and of the large bore that sweeps up the river. The cost of carriage by this route only amounts to R25 per 1,000 viss, and the jaggery carried by sea sells in Rangoon at R25 to R30 more per 1,000 viss than sugar carried by other routes as the cakes are sent to Rangoon without getting broken.

"The cost of carriage from Dawoon to Thatone is said to amount to R2-8-0 per 1,000 viss.

"The cost of carriage from Beelin to Kyouksarit amounts to R3 per 1,000 viss, and by boat from Kyouksarit to Moulmein R8 per 1,000 viss. The want of a road between Kyouksarit and Beelin prevents much jaggery from being carried by this route.

"*Indebtedness of the cultivators.*—The indebtedness of the cultivators examined by me is shown in the following table:—

	Number examined.	Number indebted.	Amount of indebtedness.	Average amount of indebtedness.	Percentage of number indebted to number examined.
			R	R	
Burmans	23	10	1,750	175	43·47
Shans	75	56	17,020	303·92	74·66
Toungthoos	146	65	6,660	102·53	44·52
TOTAL	244	131	25,430	194·12	53·68

The money-lenders all state that the chief cause of indebtedness of the Shans is their propensity to gamble. The cultivators themselves attribute their indebtedness entirely to the high rates of interest charged, and to the extortions of the money-lenders. A certain number of Shans gamble as much if not more than Burmans, but,

CULTIVATION in Burma. Area & Outturn.

I believe, the greater number of them to be hard-working and enterprising cultivators. The system of cultivation adopted by them accounts, in my opinion, for their indebtedness. They attempt to cultivate large plantations as soon as they reach this country, and are not satisfied, like the Toungthoos, with plodding on slowly, clearing a small patch for themselves at the same time as they work for others. They consequently require large sums of money, which they can only obtain from the money-lender, and as the interest charged is never less than 48 per cent., the profits of the plantation are entirely swallowed up by the interest. Moreover, as the harvest approaches, the money-lender refuses to lend at money interest, and only advances money against payment in sugar at the rate of R12 to R16 per 100 viss, so that the interest before harvest-time rises to 100 and 150 per cent. per mensem.

"The documents which the cultivators sign are bonds mortgaging their land, cattle, plant, and crop to the money-lender for a sum to be repaid whenever called upon to do so. These bonds are registered, and from that moment the money-lender treats the cultivator as his serf: he does not allow him to sell his own sugar, but sells it in his name and allows him R1 or R2 less per 100 viss than he has himself received. The money-lender also pays all the labourers employed in the manufacture of sugar, and accounts are settled at the end of the dry season, when a new bond for the balance with interest is charged. If the cultivator refuses to sign the new bond, he is told that he will be sued in Court and all his property immediately sold up.

"This state of things is due, in my opinion, to the misconception of the people as to the effects of registration, which, in their opinion, renders enforceable any document however invalid and illegal its clauses may be, and to the fact that the civil courts, especially those of the Extra Assistant Commissioners, strictly enforce mortgage bonds without allowing any equity of redemption.

MANUFACTURE. 367 *Conf. with pp. 114-115, 285, 297.*

MANUFACTURE.

Vernacular Names of the Various Preparations of the Sugar-cane and Sugar.

Vernacular names, of Juice, 368

1.—JUICE OF THE CANE.

This is the liquid obtained from the sugar mill.

Vern.—*Ras*, HIND.; *Ras, kachras, kanchoras*, BEHAR.

of Refuse, 369

2.—REFUSE OR CANE AFTER EXPRESSION OF JUICE—THE MEGASS OF PLANTERS.

Vern.—*Pata, páti*, N.-W. P.

of Jaggery, 370

3.—JAGGERY OF SUGAR-CANE.

This is the sugar chiefly used by the poorer people of India; it is best known as *Gúr* or *Gul*. It might be defined as an impure muscovado sugar which contains, in addition to more molasses, a larger percentage of impurities and pulp of the cane. It is boiled to a greater extent than is the case in the preparation of the next article *Rab*, and in some parts of the country the boiling juice is clarified with lime. More frequently, however, *Gúr* is not clarified, the scum being boiled down with the sugar.

Vern.—*Gúr* or *gur, mithái*, HIND.; *Gúr*, BENG.; *Gurh* or *Mithái* (sold in cakes *bheli, chakki*), N.-W. P.; *Gúr* (the balls *bheli* or *rori*), PB.; *Gúr*, DEC.; *Vellam* (or *nulla vellam*), TAM.; *Bellamu, bellum*, TEL.; *Bella*, KAN.; *Vella, sharkkara*, MALAY.; *Akuru*, SING.; *Guda*, SANS.; *Qand*, ARAB.; *Kand*, PERS.

The jaggery or *gúr* of palms is generally distinguished by adding the name of the tree to the word for *gúr*. Thus *Tar-ka-gur* (HIND.) or *Panai-vellam* or *karaputi* (TAM.), Palmyra-palm-*gur*; *Nariyal-ka-gúr* (HIND.) and *Tenna-vellam* (TAM.), Cocoa-nut-*gúr*; *Sandólé-ka-gúr* (HIND.) and *Ich-cha-vellam* (TAM.), Date-palm-*gúr*; *Marí-ka-gúr* (HIND.) and *Kún-dar-panai vellam* (TAM.), Caryota-palm-*gúr*.

of Raw Sugar. 371

4.—SUGAR, RAW.

It is almost impossible to find an English word to express this substance. It is the first stage in the refinement of sugar. The *ras* is not boiled to the same extent as in the preparation of *gúr*, and the product is

MANUFACTURE of Sugar.

allowed to cool without being drained. The substance thus produced is never eaten, but is sold to the refiners.

Vern.—*Ráb*, HIND.; *Royadar, danadar, motki, hangra*, BENG.; *Ráb, ráwa*, BEHAR; *Rab*, N.-W. P.; *Rab*, PB.; *Ráb-shakkar, rab-ki-shakkar*, DEC.; *Ráp-sharukkarai*, TAM.; *Ráp-shakkara*, TEL.; *Gula, matsyandika*, SANS.

5.—MOLASSES AND TREACLE.

Vernacular Names of Molasses, 372

This is the uncrystallizable sugar removed or drained away from *gúr* (and to a less extent from *ráb*) in the preparation of various forms of common brown sugar. The finer qualities of this syrup (treacle) are isolated by the refiners. Both forms are largely used in the manufacture of sweetmeats, in the distillation of spirit (rum), and in the preparation of tobacco mixtures, etc., etc.

Vern.—*Shira, chhoa, lapta*, HIND.; *Máth, chitrah* (or *chitá*) *gúr, kotra* (treacle), BENG.; *Putri lát* (treacle), *chhoa, sira* (when used in the manufacture of tobacco it is called *sagár* in GAYA, *gariya* in PATNA, and *phánk* in SOUTH BHAGALPUR), BEHAR; *Shíra*, (ETA), *chota* (JAUNPUR), N.-W. P.; *Gul* (KHANDESH), BOMB.; *Svádu khanda*, SANS.

NOTE.—The boiled juice (*ras*) is sometimes designated *sira* (or *shira*) and may be used in the manufacture of rum of a superior quality to that made from molasses or *chhoa*.

6.—SUGAR, COUNTRY.

of Country Sugar, 373

This is the coarse brown sugar used by persons who can afford to pay more than is usually charged for *gúr*. In its preparation the cane juice is boiled a little longer than is the case with *gúr*, and on cooling it is stirred till it thickens. It is not, however, a refined sugar, though it is fairly well drained. It may, therefore, be regarded as the Indian equivalent of the muscovado sugars. There are many qualities of it, the inferior being scarcely different from *gúr* and the superior closely approaching to the next article *bura*, or once refined sugar. In the preparation of this article (*shakar*), however, clarification with lime and the scumming of the boiling *ras* seems to be very often followed, so that it is not only a drier substance than *gúr* but has been more carefully prepared. *Gur* and *shakar* are intended for human food without being purified any further. The object in their manufacture is, therefore, to obtain a certain colour rather than to crystallize any portion of the "compost" as it is often called.

Vern.—*Lal-shakar, shakar, khánr, bhúra*, HIND.; *Banglá-chini, sar* (BOGRA, etc etc.), BENG.; *Sakkar, sankar, khánr, bhúra* (when dry and of a brown colour), BEHAR; *Chóyanda, khand* (when dry), N.-W. P.; *Mulkácha-sakhar*, MAR.; *Gámni-skkar* (or *-chini,-búro* or *khánd*), GUZ.; *Makhtúmi-shakkar*, DEC.; *Náttu-sharukkarai*, TAM.; *Náttu-shakkara*, TEL.; *Nattu-panjasára* (or *-sharkkara*), MALAY.; *Kala-saghia, kalatigiya*, BURM.; *Nát-sakkare*, SING.; *Désha-sharkara, désha-panj asaram*, SANS.; *Sakkarul-hind*, ARAB.; *Shakare-hindí*, PERS.

7.—SUGAR PARTIALLY REFINED.

of Partially refined Sugar. 374

The sugar generally denoted by the names given below is *ráb*, once refined. That is to say, it has been washed and molasses (or it might now almost be called treacle) removed from it by pressing. The sugar is not, however, entirely crystallized and it has been bleached by exposure to the sun only. For this purpose the large lumps obtained from the pressing sacks are trodden out under feet and left on mats, or otherwise exposed to the sun. The same remark as already made applies to this and, indeed, to all the other forms of Indian sugar, namely, that, according to the variability of local practice, there are many grades or qualities of this article until the *bura* of one district may be inferior to the *shakar* of another or equal to the *chini* of a third. It will be seen in the account given below by Mr. Westland regarding Jessor sugar manufacture that

MANUFACTURE of Sugar.

an inferior form of *dhuluá* sugar is produced by pressing the *ráb* within sacks until the molasses is removed. The article thus produced would appear to answer to the *bura* sugar of other parts of India. The better class *dhuluá*, washed and partially crystallized with aquatic weeds would very probably be classed as an inferior quality of white sugar and differs from Mr. Westland's *paka* sugar in the fact that the *ráb* has not been re-boiled and skimmed before being crystallized. The Jessor *paka* sugar would, therefore, rank as a good quality of white sugar.

Vernacular Names.

Vern.—*Bura, bhura,* HIND.; *Dhuluá* (inferior qualities), BENG.; *Shakar* (*bura* in the Panjáb is used to designate a superior refined sugar), PB.; *Bura-shakkar,* DEC.; *Búrá-sharukkarai,* TAM.; *Búrá-shakkara,* TEL. *Gúla,* MALAY.; *Gula,* SANS.

8.—SUGAR (WHITE), OR ONCE REFINED SUGAR.

of White Sugar.
375

This is ordinary refined or crystallized sugar whether prepared by the Native or European methods. It does not seem necessary to enter in this place into a detailed description of the various methods of refinement. The reader will find much useful information on this subject in the passages which follow. Suffice it to say that the *ráb* purchased by the refiner in the preparation of this class of sugar is generally boiled, clarified, and the scum removed. The moist sugar first obtained on draining off the molasses is known as *putri* and this when dry is *khand.* The process is thus effected: the *pútri,* is placed in conically shaped vessels and washed and drained of its treacle by water passing through from a layer of aquatic weeds, placed on the top. The crystallized sugar, *khand,* formed below the weeds, is removed from the top and fresh weeds added till the whole has been crystallized. Various qualities of refined sugars are made by the Natives according to the extent of scumming, straining, and re-crystallizing. In the manufacture of the *dhuluá* sugar of Eastern Bengal, the *ráb* is apparently not dissolved in water and re-boiled. It is simply placed in the refining vessels and layer upon layer of aquatic weeds placed on the top until the molasses has been washed out and the mass crudedly crystallized. In the clarification of the *ráb* lime is, of course, very largely used, but various other salts sometimes take its place, such as the ashes of certain plants, impure carbonate of soda, etc. Milk is also often employed as also the mucilaginous substances obtained from certain plants (**Hibiscus, Kydia,** etc.), or more rarely oils are added to the boiling solution of *ráb* (*see p. 234*). The action of these mucilaginous or oily substances appears to be that on coagulating they mechanically remove impurities. Superiority depends on cleanliness, careful clarification, and extent of washing and crystallization.

In many parts of India different qualities of *khand* are recognised according as it is obtained from the layer immediately in contact with the aquatic leaves, lower down or at the bottom of the purified stratum. The top layer is in the Panjáb called *chitti khand* or white *khand,* below that *kkand* and the lowest of all *talauncha.* There is perhaps no problem that is more perplexing than that of the classification of the various names given to the forms of refined sugar. Indeed, it would seem that while in one province *bura* denotes a superior quality of unrefined, or at most partially refined sugar, in others, such as in the Panjáb, it denotes twice refined sugar, or *misri.* No classification can, therefore, be enforced for the whole of India, though possible when the lesser area of a single province is dealt with.

So much skill is required that the refinement of sugar has become a distinct branch of enterprise even in India. The cultivator never aspires to refine his own *ráb* and prefers accordingly to make *gúr.* Indeed, *ráb* is, as

MANUFACTURE of Sugar.

a rule, made from the cane juice by the money-lender, not by the cultivator. It will be seen, however, from the remarks below that gradually a superior system of manufacture is coming into use which more closely approximates to the European methods. Though to a very small extent as yet, still it may be said direct manufacture of sugar from the cane juice has been taken up by some of the better class cultivators, using for that purpose especially prepared apparatus, designed for the use of small manufacturers with hand labour.

Vernacular Names.

Vern.—*Sufėd-shakar,* (=white sugar), *chini* (=China sugar=*chini-shakkar*), HIND.; *Dhulua* (superior qualities), *paka* sugar, *dhóp-chini, badol,* BENG.; *Khand, chini* (or *kacha chini*), N.-W. P.; *Khand* when dry and *bed* when wet, PB.; *Pandhara-sakhar,* MAR.; *Saphéd-sakkar, ujlo-chini* (or *-búro,-khand*), GUZ.; *Vellai-sharukarai* [*páncha-dárá* in Ainslie or *pansadarry* in Roxburgh], TAM.; *Chiná-shakkara, tella-shakkara,* TEL.; *Bili-sakkare,* KAN.; *Ven-sharkkara* (or *-panja-sára*), MALAY. *Saghia-phiu, tagiyá-phiú,* BURM.; *Sarkara* or *shvéta-sharkará* (or *-panjasáram*), SANS.; *Sakkarul-abyaæ, sukkar* (*as-sukkar*), ARAB.; *Shakare-supéd, shakkar,* PERS.

9.—SUGAR-CANDY OR CRYSTALLIZED SUGAR, *e.g.*, TWICE REFINED SUGAR.

of Sugar-Candy, 376

The word *khand* in India does not correspond with the substance indicated by the English derivative from the same root, *viz.*, the Arabic *kand.* In Indian commerce, in fact, three widely different substances are generally returned as "sugar-candy," *viz., misri, kuza misri,* and *chini.* In the preparation of all three *khand* as defined above is taken and dissolved, boiled, and milk alone used in clarification. *Kuza misri* is the equivalent of sugur-candy. The specially prepared and clarified syrup is thrown into small vessels in which are suspended threads. The sugar crystallizes on these and on the sides of the vessel in large crystals.

The reader will find so much useful information regarding the preparation of the numerous forms of what may be called double refined Indian sugar (in the provincial chapters below), that it does not seem necessary to attempt a review of these in this place. Loaf sugar (*kand* or *qand*), as it is understood in Europe, is not, strictly speaking, made in India, though many of the qualities of both white sugar and sugar-candy are often formed in moulds and sold in blocks that in some respects resemble externally loaf sugar.

Vern.—*Misri* (or Egyptian sugar), *khand,* HIND.; *Misri,* BENG.; *Chini* (or *bura*), *misri* PB.; *Karkandu,* TAM.; *Mala-kanda,* TEL.; *Kalkanda,* KAN.; *Kulkantu,* MALAY.; *Sakari,* SING.; *Khanda, sitopalá,* SANS.; *Nabat, kand,* ARAB.; *Qande-suféd, kande-suped,* PERS.

10.—SCUM REMOVED BY THE REFINER FROM THE BOILING-PANS.

of Scum from Boiling-Pans. 377

(This is sometimes mixed with water and again boiled, it is then generally known as *pasawa.*) **Vern.**—*Mail, maila, mahiya,* BEHAR; *Laddoi* (Eta), N.-W. P.

Methods & Appliances of Manufacture.

METHODS & APPLIANCES. 378

So much has already been indicated (by the selection of provincial descriptions of cultivation) of the various methods of expression of juice and preparation therefrom of the coarse and refined sugars, made in India, that it is unnecessary to deal very fully with the MANUFACTURES. The reader may have discovered that the object has been kept in view (by what has already been written) to bring together in this article as much information as possible, regarding the peculiarities of the *Indian Sugar Industry* rather than to furnish an essay which, while bearing on India, would be a technical treatise on the modern advances and machin-

MANUFACTURE of Sugar.

Methods and Appliances.

ery used in the world's sugar-cane area. This remark must, therefore, be accepted as more especially applicable to the present chapter. To review, however briefly, the discoveries and inventions presently utilized, by the sugar-cane manufacturers alone, would necessitate the allotment to the present chapter of more space than can be afforded in this work for the entire article on sugar. The reader who may, therefore, desire information on these subjects should procure one or more of the special technical works which have appeared, and some of these will be found mentioned in the list of references to works consulted by the author. The article in **Spons'** *Encyclopœdia* may be accepted as a review of the chief inventions and chemical processes utilized by the sugar-makers and refiners, and the 90 odd pages, which have been there devoted to these subjects will be found highly instructive. The utmost that can be accomplished in this work will be to furnish a selection of descriptions of the methods and appliances employed by the Natives of India in the various provinces, accompanied with such information as can be procured regarding the better known European sugar factories and refineries that actually exist in the country. **Mr. S. H. Robinson**, who in 1849 published his little work, *The Bengal Sugar Planter*, enjoyed, as a Bengal planter of 16 years' standing, opportunities given to but few writers on this subject, to form definite opinions regarding the character of the Native industry and the possibility of its improvement. In the introduction to his work he wrote of the Native process that it "may not inaptly be characterised as a burlesque on the more scientific and comprehensive manufacture of the European planter and refiner. Yet rude and imperfect though it be when so contrasted, we should by no means despise it as unworthy of our notice, while seeking for the most beneficial modes of working in the same field with our larger capital and more scientific means and appliances. We cannot help admiring indeed how perfectly adapted the Native contrivances are, in every way, for the ends they are meant to compass; in giving the poor cultivator, as regards the first process of *gúr*-making, the most effectual, cheap and economical means of producing a saleable commodity from the small patch of cane his labour is limited to; and to the Native refiner similar advantages in cleaning and whitening for the market the limited quantity of sugar he is, with his small capital, restricted to working upon. This may be better understood perhaps by the consideration

Size of Cane Fields. *Conf. with pp. 108, 124-125, 143.*

379 that the quantity of sugar refined by one of the most substantial of these re-manufacturers for a whole year is about the same as an ordinary European refiner, with a single vacuum pan of medium size, can turn out in two days; and that the extent of cane cultivated, or owned, by any single *ryot*, seldom or ever reaches an acre in measurement, and more frequently occupies less than half of that space." As it seems to the writer a volume might be published on the subject of the present position and possible expansion of the sugar interests and capabilities of India and yet convey no more than has been thus pointedly indicated. The mistake of most would-be agricultural reformers may be said to be neglect to consider the conditions and requirements of India. Suggestions which in themselves are of the greatest value often lose entirely their merit when viewed from the stand-point of the Natives of India. With few subjects has this contention greater weight than with that of sugar. Improvements, to be of value to India, must be regulated by a due consideration of the necessities and capabilities of the people. An intimate acquaintance with the methods and appliances presently in use and with the dispositions and even prejudices of the cultivators and manufacturers are therefore essential to the invention of appliances that stand a chance of ready and extensive adoption. It is useless to tell the cultivator of less

MANUFACTURE of Sugar.

Methods and Appliances.

than an acre of cane that he does not extract more than a half the juice from his plant, if the apparatus which it is proposed to supply him with, costs so much (even were he capable of purchasing it) that the loss of interest on the capital invested would be greater than the value of the extra yield. And this principle may be exemplified with illustrations drawn from each subsequent stage up to the manufacture of the best qualities of refined sugar. Suffice it to say that the most superficial study of the Indian sugar question will reveal the fact that even in the present crude machinery and methods of manufacture vast improvements have been effected. Thus the clumsy presses by which a portion of the juice was removed through men standing on boards placed above and below the prepared cane, gave place to a pestle and mortar contrivance made of wood or stone and driven by men or cattle, This in time has in most districts been displaced by a primitive mill of two wooden rollers working either horizontally or vertically, and between which the canes are forced once or more times, until the juice is squeezed out. But within still more recent times an immense improvement has been effected by the substitution of iron for wooden rollers. An iron sugar-cane press is, however, too expensive for the ordinary cultivator, but his willingness to embrace inventions that at all come within his means could not be better exemplified than by the success that has attended the issue of the Bheea portable iron roller mills. The advantage of these having been recognised a few cultivators in a locality club together and purchase one. From the frequent allusion to these mills that will be found dispersed throughout this article, it will be seen that they are now being used in almost every district from one end of the country to the other. There could be no statement more unjust therefore than that the Indian cultivator or artizan is incapable of improvement. He has only to be shown that a departure from the time-honoured practices of his ancestors is in his own interests and can be accomplished by his limited resources, than he at once adopts the new method or appliance.

Bheea Iron Mill. **380** *Conf with p. 277.*

In a memorandum furnished by **Messrs. Thomson & Mylne** interesting particulars are given regarding the extent to which their portable iron roller sugar mill has been taken up. Having observed, these gentlemen say, the great losses sustained by the cultivators through the use of crude appliances—

"the efforts were made by the undersigned to contrive a crusher which, while suited as regards cost, weight, simplicity, etc., to the circumstances of small cultivators, would be a real improvement on the machines then within their reach and means. The aim was to produce a machine which would remedy the serious defects of the Native appliances; be suited to the means and wants of the cultivators; be so simple in construction, that the village carpenter or blacksmith might repair it; not liable to serious derangement by the blundering of people inexperienced in the management of exact machinery; and yet be of such size and form as to be easily portable. By novel contrivances and arrangements, a light portable mill was produced which proved to be so well adapted to the wants, means, and domestic arrangements of cultivators who grow cane in small plots, that it has in a few years been adopted in hundreds of districts, not less than 200,000 being now in the hands of the people

"So great an improvement did it prove to be, that in a village in which the greatest area of cane the cultivators could crush, previously, was about 30 acres, they planted a very few years after these mills were first placed within their reach, and with them worked off 250 acres, while last season they have grown and crushed about 600 acres. The completion of the Sone canals, and construction of village channels to convey the water to their fields for irrigation has been another main cause of this large increase, but without the improved mill for crushing, it would not have been possible for these cultivators to have worked off more than 50 acres at the outside.

"In the hope of finding a kind of cane which would yield more, or richer juice, and at the same time suit the soil and climate, seed cane was obtained from Lower Bengal, North-Western Provinces, Penang, Java, Mauritius, and other places. Portions of each were planted in the way usually adopted in Behar, and some according to the methods found most advantageous in Mauritius and other places, several kinds of

MANUFACTURE of Sugar.

Methods and Appliances.

Bheea Mill.

manure being also used. These trials appear to have started efforts to improve, but it has been found that none of the thick soft kinds would suit in Behar, and those which give the best results are what are locally known as *mongoo, pansaihi, barook,* and *Bhoorli*. The latter is being more largely planted lately, because it stands better than the others in wet soil, of which there has been a considerable increase in Shahabad since the opening of the Sone canals. Endeavours were next made to effect some improvement in the method of dealing with the juice. **Mr. Alfred Tryer,** an eminent authority on sugar, and sugar refining, has said that 'cane juice, from the moment it leaves the cells, should be treated with the same care and cleanliness as is new milk in a well ordered dairy,' but the practice of the cane-growers of India, and their helpers, was, and for the most part still is, the exact reverse of this.

"The tenants of the Judispore estate were urged to exercise more care, the reasons for and advantages of doing so being explained to them, and a few of the most enterprising were induced to go so far as to put a strainer over the mouth of the earthen vessel used to catch the juice as it comes from the mill, and so intercept trash, leaves, dust, etc., also to wash out (or rinse) the 'receiver' each time it was used, and to fumigate it by inverting it over a pinch of burning sulphur.

"The result was that *gur* and *rab* of a much higher quality were obtained, but here came a difficulty such as in India frequently occurs to hinder improvements. These men by taking trouble have obtained a superior article, but found they could get no more per maund or per cwt. for it than could have been obtained if all the dirt had been left in it, and if no care had been taken. It was not only an article unknown but was suspected, so that efforts had to be made to find purchasers who could appreciate it, and, amongst others, Marwaree dealers, from places 1,000 to 1,500 miles distant, were convinced that as it saved them paying rail freight for those long distances at the rate for sugar on so much trash and dirt, on this ground alone, apart from other advantages, it was well worth their while to give a higher rate for it. The Rosa refinery also, 500 miles away, found that even with rail freight for this distance to be added, they could pay more for this clarified *gur* than they could for the ordinary qualities in districts much nearer Shahjehanpur.

"A further step was taken to try and devise some simple, inexpensive apparatus and method by which cultivators of small plots who crush their own cane could produce sugar similar in character to that which **Messrs. Travers** describe as 'simply raw sugar properly made by modern processes.' In connection with these efforts, **Messrs. Manlove, Alliott, Tryer, & Co.** were consulted, and they, taking much trouble as well as interest in the matter, constructed for the experiment a novel form of Wetzel evaporator, with small steam boiler, fitted with special safety valve and other arrangements to admit of its being used by villagers having little experience in the management of such machinery or processes. They also made for the experiment a specially contrived portable centrifugal or spinner by which high speed could be obtained with hand power. Open evaporators, filters, and other appliances were made locally, and a second Wetzel was subsequently obtained from **Messrs. Manlove & Co.** These experiments (commenced in 1873) have been carried on from year to year since, and at an early stage proved quite successful as regards the quality of sugar produced, but as no good market could be found for the molasses, of which there was a considerable quantity, the indication for some time was, that the process must involve decided loss, unless a distillery were set up to work off and utilize the molasses. As making spirits was no part of the programme, and as there was no inclination to do this, the project of success seemed far from promising, when it was found that by carefully evaporating the molasses in the shallow pan used for the first evaporation of the juice in making the *rab*, a very saleable *gur* could be obtained which, being made from the molasses thrown off by the centrifugal, and strained through grain sugar (itself made from clarified *rab*), was speedily recognized as being specially clean and pure, and so is from year to year increasingly appreciated. Thus, a way was shown by which the millions of Indian cane-growers may secure greatly improved products and higher returns from their crop without any large, expensive, or complicated machinery, with only a small portable mill to crush the cane, an open shallow evaporating pan, a few *nánds* (cheap earthen vessels), in which the *rab* is placed for 8 or 10 days to let the crystals form or grow, and a portable centrifugal, any or all of which appliances the cultivators can hire or buy. The advantages of this process were found to be so real, the gain so substantial, that a demand arose for the small spinners which increased to a rush, and as with too rapid multiplication of cotton mills, jute mills, tea gardens, etc. so with these spinners, an excessive number were started within a few months, sufficient time not being given for the new and special products to become known over an area wide enough to admit of the whole quantity produced being sold at paying rates, and there was for a time the usual dis-

Wetzel Evaporator. 381

Utilization of Molasses. 382

MANUFACTURE of Sugar.

Methods and Appliances.

appointment, but the demand has overtaken production again, and there is every indication of another 'boom.'

"It will be seen that the chief aim of these arrangements and experiments has been (1) to secure a better cane crop, and (2) to put the cultivator in the way of getting more sugar from the cane.

"It was felt that, having regard to the conditions which prevail in India, it was best to begin with the cultivators, and the results obtained so far appear to justify this.

"**Sir Edward Buck** (now Secretary to the Government of India in the Department of Revenue and Agriculture), when Director of Agriculture in the North-West Provinces and Oudh, became aware of the extent to which the cane-growers of Behar were taking to these new machines, and in 1877 obtained Government sanction to the procuring of a number which were sent to different districts of his province. As in Behar and Lower Bengal so further west, it took time to convince the cultivators, but after several seasons of doubt, hesitation, and suspicion, numbers of cane-growers began to inquire for the new mills. and a rush for them followed. Depots for supplying them were opened in various North-Western Provinces districts, and in June 1888 **Mr. J. B. Fuller**, then Assistant Director of Agriculture, North-Western Provinces (now Commissioner of Agriculture and Settlements, Central Provinces), wrote regarding results obtained with the new machines as compared with the *kolnu*. 'If we may apply the result of this experiment to the total production of sugar in these provinces, it follows that by the substitution of the Beheea mills for the *kolhu* now used, the total annual produce would be increased by the value of nearly a crore and a quarter of rupees,' *i.e.*, a million and a quarter sterling. The benefit has been increasing year by year in the North-Western Provinces as in other parts of India. In the Panjáb also depots were opened, and a district committee of leading cultivators reported to the Director of Agriculture (**Colonel Wace**) after trials made in 1883, that the money gain per season, by using even the smallest size Beheea mill instead of the '*Kolhu*,' was about R360, and that it gave 'other important advantages.' Similar results were obtained in the Central Provinces, Madras, Lower Bengal, and other parts of India, and if only a third of R360 per season be taken as an average, to allow fully for mills which work only part of the season, as also for other deductions, the gain on the total number of mills in the hands of the cane-growers in each season from 1874 will be:—

			R
1874-75	from	800 mills, at R120 per mill	96,000
1875-76	,,	1,500 ,, ,, ,,	1,80,000
1876-77	,,	2,300 ,, ,, ,,	2,76,000
1877-78	,,	5,700 ,, ,, ,,	6,84,000
1878-79	,,	9,000 ,, ,, ,,	10,80,000
1879-80	,,	12,000 ,, ,, ,,	14,40,000
1880-81	,,	17,000 ,, ,, ,,	20,40,000
1881-82	,,	25,000 ,, ,, ,,	30,00,000
1882-83	,,	30,000 ,, ,, ,,	36,00,000
1883-84	,,	40,000 ,, ,, ,,	48,00,000
1884-85	,,	55,000 ,, ,, ,,	66,00,000
1885-86	,,	70,000 ,, ,, ,,	84,00,000
1886-87	,,	80,000 ,, ,, ,,	96,00,000
1887-88	,,	100,000 ,, ,, ,,	1,20,00,000
1888-89	,,	150,000 ,, ,, ,,	1,80,00,000
1889-90	,,	200,000 ,, ,, ,,	2,40,00,000
1890-91	,,	250,000 ,, ,, ,,	3,00,00,000
		TOTAL	12,57,96,000

"These figures are based on the experience of cane-growers in various parts of India, and on reports made by officers of Government Engineers, and others who have taken pains to make sure of the reckoning, and they indicate a small part only of the enormous loss sustained by cultivators in India through defective appliances and crude methods. It is to be noted that these figures show what was being lost by *some only* of those who grow sugar-cane in India, and that they represent only that part of the total which has been already recovered. What the total loss is in connection with this one crop can only be realized by those who have some knowledge of the careful treatment which is requisite to secure a full percentage of the sugar which cane or beet can yield, and who have also had opportunity of observing the crude wasteful appliances and methods which are used by Indian cane growers. It is also to be noted that

MANUFACTURE of Sugar.

similar losses are being sustained in connection with other crops such as oil-seeds, wheat, barley, maize, fibres, and dye-stuffs."

Methods and Appliances.

Having thus briefly indicated the advances that have already been accomplished, as also the fact that there exists in India a very extensive field for simple and cheap inventions, it remains only to sketch out the leading stages in the process of sugar isolation and refinement, as pursued by the European manufacturers, as also the more primitive systems of India. The following brief sketch of the European method of manufacture will be found to be compiled from *Spons' Encyclopædia*, the details of the machinery used to accomplish each stage or process having, from want of space, been omitted :—

Extraction of Juice.

383

I.—Extraction of Cane Juice.

"The juice in the cane exists in the plant enclosed in little cells, which are surrounded and protected by lignose (woody matter). the latter forming about $\frac{1}{10}$ of the total weight of the cane. The liberation of the juice may be effected.

(1) *by rupturing these cells, so that their contents flow out;*

(2) *by combining the crushing process with maceration in water;*

(3) *by utilising the membrane of the cells as a means of allowing the escape of the sugar and other 'salts' in solution by the process known as diffusion."*

(1) DISINTEGRATING.—"The imperfect liberation of the cane-juice by the crushing process of the ordinary mill has led to experiments in other directions. One result has been the invention of machines for effecting a more thorough mechanical disintegration of the cane-tissue. These may be conveniently considered under three sections:—(*a*) Defibrators, (*b*) Bessemer's press, and (*c*) Bonnefin's rasper." By the process (*a*) the cane is reduced to pulp, and by subsequent pressure 77 per cent. of juice is said to be separated. Some of the defibrators break the knots and joints and thus simply prepare the cane for the ordinary mill. By Bessemer's invention plungers were worked in cylinders, across whose path the canes were passed endwise and were thus crushed section by section. The result was unsatisfactory, and the invention never came into general use. By (*c*), Bonnefin's rasper, the cane is reduced to shreds by saws alternately moving through a cradle, then pulped by disintegrating apparatus, and the juice separated by pressure.

(2) MACERATION.—"It has been sought to facilitate the extraction of the juice by submitting the cane to the action of water or steam, either before the crushing operation in the roller mill, or at an intermediate stage between two such crushings. It seems to be undecided whether the saturation or the extra crushing should be credited with the increased yield of juice. Probably both assist, but it has been stated that the return of juice is raised from 60 per cent. to 75 per cent. by previously slicing the canes longitudinally, without any application of water or steam.'

(3) DIFFUSION.—"All the processes hitherto described for extracting the juice from the cane have depended for success upon the more or less complete *rupture* of the juice-containing cells. 'Diffusion' differs from them essentially, in dispensing with the breaking up of the cells and the machinery required therefor. The chief development of the diffusion process has been in the beet-sugar industry, but several methods of applying it to cane have been introduced. The cane is even said to possess an advantage over beet with regard to diffusion, in that the nitrogenous matters are so placed in the secondary cells that water at a high temperature can be used without injuring the membrane."

MANUFACTURE of Sugar.

The first operation is to slice the cane, and this is accomplished by various machines such as that patented by **A. Join & Co.**, by **Bouscaren**, or by **Fraz Rebicek**. There are various systems of diffusion and many patent machinery to accomplish the variations of the principle which they all manifest, namely, the removal of the saccharine substance by means of water at certain temperatures.

II.—Defecation and Clarification.

Clarification. 384

"Having, by any of the methods described, extracted as much as possible of the juice from the cane, the next operation is to eliminate from that juice all matters regarded as impurities from the sugar maker's point of view, *i.e.*, everything except the sugar and the water holding it in solution."

Straining. 385

(1) PRELIMINARY STRAINING.—"First of all, unless the juice has been extracted by diffusion, it is necessary to remove the gross impurities derived from the breaking-up of the canes. This may be done by a series of strainers, arranged so as to be easily removed, cleaned and replaced. One of the best contrivances is a modification of the endless wire-web strainer, not essentially different from that on which the rag-pulp of paper-works is agitated and filtered from a great part of its water. The wire gauze in common use has 40 to 60 threads per inch, but it can be obtained of 80 to 90: the finer the better, provided the web presents a clean surface as fast as necessary. The strained juice is received in a shallow tray placed immediately under the horizontal part of the straining web, and passes thence by a gutter to the clarifier. The chief means introduced for cleansing the juice are ***heat, chemicals,*** and ***filtration.***

Heat. 386

(2) HEAT.—"Heat alone will exercise beneficial effect both by checking acidity—scalding the juice prevents acetous fermentation setting in, probably by destroying the fungoid germs which are its necessary accompaniment (presumably its cause); and by evaporating a portion of the acids holding the alluminous matters in solution, whereby the albumen is coagulated and rendered insoluble. It is also a valuable aid to the action of chemicals upon the juice, increasing the energy of the reactions set up, and thus greatly reducing the duration of the operation. Hence heat is now universally availed of in recognized processes of defecation and clarification. But if the heat is applied injudiciously, much of the crystallizable sugar is inverted. As the degree of heat employed is a matter of vital importance, it is most conveniently applied in the form of steam, that being readily controlled.

"The use of the clarifier may be described in general terms as follows. The juice is raised to a temperature of 80° (176°F.), and sufficient milk of lime is added to neutralize the acid in the juice. The heat is then continued till a scum of impurities has risen to the surface, and commences to crack. The time occupied in this should be about 10 to 12 minutes from the commencement of the operation. The steam is then shut off and the liquor is allowed to subside for 15 to 20 minutes when the scum remains at the top; some heavy matter will have fallen to the bottom, and between them will be the clarified cane-juice, clear and of a pale straw-colour. The clarification being complete, the two-way cock is first turned on to the smaller aperture, until the top scum begins to appear; the cock is then turned to the large way, and the plug is taken out. The bottom sediment and top scum are conveyed to a cistern, whence they are placed in bags and any juice remaining in is squeezed out, leaving only a small portion of solid matter behind.

MANUFACTURE of Sugar.

Clarification.

Chemicals.
387

(3.) CHEMICALS.—"Of these the most important and most widely used is slaked lime; following it come bisulphide of lime, sulphurous acid, lead acetate, and sundry special compounds, as well as antiseptics.

Lime.—"The effects of heating are greatly augmented by the simultaneous application of a strong alkaline earth, such as lime, which combines with the liberated acids, and with any carbonates present, and thus forms an insoluble precipitate, which carries down much of the impurities. But any excess of lime beyond what is required to neutralize these acids will re-dissolve the coagulated albumen, and preserve it in a state of solution, until the excess of lime is again neutralized by addition of acid. The operation which is called 'tempering' is thus obviously one of extreme delicacy. The first point to ascertain is the exact amount of lime required by a given quantity of cane-juices." Various methods and apparatus exist for determining this, so that the matter is placed beyond the necessity of experience—often a very uncertain guide.

Bisulphide of lime has been used owing mainly to the bleaching and cleansing action of the sulphurous acid employed along with cream of lime. Other alkaline earths have also been proposed such as barium, strontium, etc. Their effect is more powerful than lime, but they cannot be said to have come generally into use. Sugar of lead (sub-acetate of lead) was also proposed as a defecating agent, and sulphur and chlorine compounds have similarly been recommended.

Filtration.
388

(4) FILTRATION.—"Filtration of the juice is a necessary adjunct to the defecation by heat and chemicals, its object being the removal of the matters rendered insoluble by these operations. The chief kinds used are bag, charcoal, and capillary filters." It is scarcely necessary to say more on this subject except that the use of bone and other animal charcoal filters, to which the Hindus object so strongly, are employed at this stage, but they are by no means indispensibly necessary. In fact many other processes exist by which the filtration is effected without the use of charcoal in any form. In the first process, mentioned above, cotton-twill filter bags are used, and in the third by capillary attraction along bundles of fibres, the saccharine juice is separated from the impurities. This is F. A. Bonnefin's method.

Granulation.
389

III.—Concentration and Granulation.

"The cane-juice, reduced to the condition of a clear solution of sugar (with some few salts as impurities) in water, has next to be deprived of so much of its water as will permit the sugar to assume a solid (usually crystalline) form. This operation, termed 'concentration' and 'granulation', has been described in principle. The inversion of sugar during concentration of cane-syrup is said to be prevented by the introduction of superphosphate of lime into the juice before boiling. There is no evidence as to the practical utility of this plan; but phosphoric acid appears rather to aid the crystallization of sugar, and the process would therefore seem to be based on good ground. Both *heat* and *cold* have been applied to the concentration of cane-syrup, but chiefly the former."

Heat.
390

(1) BY HEAT.—"The means by which heat is applied to the evaporation of cane-juice may be described under five separate heads according to their principles:– (*a*); *Pans heated by fire*; (*b*) *pans heated by steam;* (*c*) *film evaporators*; (*d*) *vacuum-pans*; (*e*) *bath evaporators*; (*f*) *Fryer's concretor.*

(*a*). *Pans heated by Fire.*—"The earliest and crudest system of evaporation was the 'copper-wall' or 'battery' of open pans called 'teaches' (taches, tayches, etc.). The first two pans of the series are the clarifiers; thence the juice flows into the teaches, sheet-copper pans set in masonry on a descending plane. As the juice concentrates, each lower

MANUFACTURE of Sugar.

Granulation by Heat.

pan fills up with liquor from the one immediately above it, until the density of the liquor in the 'striking-teach' permits granulation, when the mass is ladled into shallow wooden vessels, and conveyed away to be 'cured.' By the oldest method, the liquor was ladled throughout the series. More recently an improvement was introduced, consisting of a copper dipper, fitting inside the striking-teach, and having at the bottom a large valve opening upwards and worked by a lever. The dipper is attached to a crane, which commands the striking-teach and the gutter leading to the coolers. This greatly economises time. The furnace for heating the series is set under the striking-teach; the heat passes by flues to the chimney or to the boiler-flue. In working a battery, the difficulty is determining the exact moment when the boiling of the 'sling' in the stricking-teach must cease, *i.e.*, when to make a 'skip,' great skill and experience are required to suit each kind of juice. The main point is to bring about crystallization in the sling in as great mass as possible after it cools: if the sling be taken out too soon, there will be only a few large irregular crystals and a quantity of sugar will be left in the molasses; if the sling be boiled too long, a sticky mass of tiny crystals and syrup will result, from which the molasses can only be drained off with great difficulty, and from which it is impossible to obtain clean, dry, and hard crystals. An experienced 'wall-man' knows the approach of the striking-point, but a good test is the following. Pour a spoonful of the boiling sling into a glass of clear water; if, after a minute's cooling, the sling can be formed into a ball which does not stick to the fingers, and slightly flattens itself on the bottom of the glass on being dropped in, the correct period has arrived for striking. The continued use of the copper-wall is an illustration of the backwardness of the cane-sugar industry in many places. Its drawbacks are—(1) waste of fuel; (2) the amount of labour required and length of time occupied; (3) considerable waste of liquor in the sloppy manipulation; (4) the proportion of molasses produced is intensified by the churning-up of the liquor and consequent admixture of air, and by the irregular and uncontrollable action of the heat upon the surface of the metal with which the liquor is in contact."

(*b*) *Pans heated by Steam.*—"The simplest form of steam evaporating-pan consists of a rectangular wrought-iron tank, at the bottom of which is a series of copper steam-pipes, connected by gun-metal bands brazed to them, and carried on wrought-iron supports. The tank is fitted at the side with a steam-valve at one end of the steam-pipe range; at the other side, is a cast-iron box fitted with a wrought-iron pipe, for the escape of the condense-water to a condense-box. This form of evaporaton presents a large heating surface, with facility for cleaning. By passing the ends of the steam-pipe range through stuffing-boxes, the pipes can be turned up, and all parts of the interior of the tank be readily cleaned—a matter of great importance."

(*c*) *Film Evaporators.*—"Under this head are particularly included those evaporators which depend upon the principle of exposing thin films of liquid to the action of a heated surface in the open air. They are generally known as '**Wetzels**' among planters, and comprise the 'pans' bearing the names of **Gadsden, Wetzel, Schroeder**, and **Bour**, and many modifications, some of which, such as **Murdock's**, have steam-heated coils. The original form was **Aitchison's** simple cylinder, revolving with partial immersion in the liquid, and heated internally by steam. In its revolution, the cylinder carries on its surface a film of liquor, whose water is soon evaporated. In the **Gadsden** pan, the cylinder is replaced by a skeleton cylinder, consisting of two metallic discs, connected by a series of metallic rods fixed at short intervals around the periphery of each disc.

MANUFACTURE of Sugar.

Granulation. by Heat,

Here the drawbacks are the churning of the liquor (except at very low speeds), and the insufficiency of the heat derived from the steam-jacket of the pan. Wetzel's improvement upon this is the substitution of steam-pipes for the solid rods.

(*d*) *Vacuum-pans.*—"The principles which control the boiling of juices in *vacuo*, and the details of the construction of a vacuum-pan, need scarcely be gone into in this place. Briefly, it may be described as a closed iron vessel warmed by a worm or pipe passing through it, along which steam can be forced. The air-pump is started, and as soon as the vacuum reaches 26-27 inches, the feed-cock on the side of the pan is opened, and sufficient liquor is drawn in to completely cover the first coil; steam is next turned in, and the liquor rapidly concentrates; fresh supplies are admitted at short intervals, the feed-cock being opened, say, for 15 seconds at a time, until the mass commences to show 'grain.' The grain is fed carefully, the cock being opened frequently, and each time the quantity admitted is increased. As the amount of sugar in the pan continues to augment, steam is turned into the 2nd and 3rd coils, until, at the completion of the charge, the pan is nearly full, or just below the sight glass. In this way, the 'grain' 'grows' in size. On the conclusion of the boiling, the vacuum is destroyed, and the charge is run out into a tank, and allowed to stand for an hour or two, when a further crystallization takes place.

"The grain formed from syrups boiled in *vacuo* is larger and more solid than that from syrups simply concentrated to crystallizing-point in open batteries. A Cuban hogshead will contain only 1,600℔ of sugar made in a copper-wall, but 1,800℔ of vacuum-pan sugar. By the use of the vacuum-pan also, the planter is enabled to boil his molasses, and to extract from 1 gallon some 4-5℔ of sugar, still having a second molasses for the distillery.

(*e*) *Bath Evaporators.*—"The tempered juice, prior to evaporation, presses through a 'continuous preparator,' a metallic vessel 32 feet long and 18 feet broad, divided by partitions into four chambers of 2 feet in width; each chamber has a central partition not quite extending to one end, with holes for the inlet and outlet of a heating liquid, which therefore travels 36 feet in the chamber, on leaving which it is reheated. On the partitions, is a copper pan divided so as to form a continuous zig-zag channel about 1,100 to 1,700 feet long, the bottom being immersed in the heating liquid circulating in the chambers below The juice is admitted at one end, and issues at the other. Along one side of the pan, are hollows to collect the heavy bodies deposited during the flow of the liquid. The juice, introduced at 15° (59°F) being in contact during a travel of 1,100 feet or more with a liquid at about 99° (210°F.), leaves the further end of the pan at 80° to 90°, deprived of heavy organic and inorganic matters in suspension, and of light matters which become separated and rise to the surface. It successively fills capillary filters and is delivered in a pure state to be concentrated.

(*f*) *Fryer's Concretor.*—"In Fryer's concretor, no attempt is made to produce a crystalline article, but only to evaporate the liquor to such a point that, when cold it will assume a solid (concrete) state. The mass is removed as fast as formed, and being plastic while warm, it can be cast into blocks of any convenient shape and size, hardening as it cools. In this state, it can be shipped in bags or matting, suffering neither deliquescence nor drainage.

by Cold. 391

(2) BY COLD.—"More than 30 years ago, **Kneller** proposed to concentrate syrups by forcing cold air through them, and his plan was much improved by **Chevallier**. Sugar made in **Chevallier's** apparatus rivalled that of the vacuum-pan in every respect. A vessel holding 200 gallons of

MANUFACTURE of Sugar.

Granulation by Cold.

syrup (comprised of 3 parts of sugar to 1 of water) is estimated by **Wray** to turn out 12 tons of sugar daily. The cost of the apparatus is small; the power required is trifling; the ordinary air of the estate could be used at once in dry weather, and would entail an insignificant expense for drying in damp weather, and the quality of the sugar is unsurpassed. In 1865, **Alvaro Reynoso** proposed to rapidly cool the syrup in suitable machines, and thus form a confused mass of particles of frozen water (ice) and dense syrup. The mixture is afterwards separated in centrifugals, and the syrup deprived of ice is evaporated in vacuo ready for crystallization. It seems most singular that, in the face of the many drawbacks and great cost incurred by concentration by heat, and in presence of the many improvements introduced of late years into refrigerating and cold-producing apparatus, so little effort is made by sugar-growers to adapt the latter system to their needs. A similar crystalline product, namely, common salt, is obtained by hundreds of tons from sea-water by the effect of natural cold in favourable localities; and there would appear to be no valid reason why a modification of the plan should not succeed on an extensive scale with sugar solutions."

IV.—Curing.

Curing. 392

"'Curing' embraces the drying and whitening or bleaching of the sugar. The several plans will be discussed in succession.

(*a*) *Simple Drainage.*—"This is the oldest and crudest method. To remove a certain amount of the molasses and other impurities, the semi-liquid mass, dug out of the coolers as soon as sufficiently cold, is placed in casks with perforated bottoms; the holes in the casks are loosely filled with canes, twisted leaves, or rushes, (the latter long enough to reach above the contents of the casks), in such a manner as to form a rough strainer The casks stand meantime on rafters over an immense tank. Here the draining process slowly and imperfectly goes on, a portion of the molasses escaping into the tank below, but much still remaining in the mass of sugar, imprisoned between the minute crystals Even after months of standing, the separation of the molasses is so incomplete that very great leakage and waste continue while the sugar is on its way to European markets. Sugar cured in this way is termed 'muscovado,' and is the most impure form of 'raw' ('grocery,') 'moist.' or 'brown' sugar. It is nearly obsolete in the English and French colonies, and its manufacture is decreasing rapidly in Louisiana."

Muscovado Sugar. 393

(*b*) *Claying.*—"The first improvement introduced is based upon the fact that the impurities of muscovado sugar are much more soluble in water than the sugar itself: thus washing with water effects considerable purification. The earliest manner of carrying this out was by placing the sugar in inverted cones, with a minute aperture in the apex, stopped up during the filling and for about 12 hours afterwards; upon the mass of sugar in the cone, was placed a batter of clay and water (hence the term 'claying'), the object being to ensure a very gradual percolation of the water through the mass. This water carries with it the uncrystallizable sugar and colouring matters imbedded between the crystals. The resulting sugar is much lighter coloured than muscovado, but the grain is very soft, and the operation is most wasteful. In Bengal a wet rag* is sometimes substituted for the clay batter.

(*c*) *Spirit-washing.*—"The very slight solubility of sugar in alcohol, coupled with the ready solubility in that medium of many of its impurities, suggested the practice called 'spirit-washing.' This consists in substi-

* By the Natives aquatic weeds, are employed. *Conf. with p. 31.* A wet cloth is, however, referred to in the passage, pp. 311-312.

MANUFACTURE of Sugar.

Curing.

tuting cold alcohol or alcohol and water for simple water. The results are not perfect, however, and the costliness of the method soon caused its abandonment in this connection.

(*d*) *Vacuum-chest.*—"The vacuum-chest consists of an iron box with a tray of wire-gauze above, and connected with air-pump suction below. The sugar is spread on the tray, and the downward suction produced by working the air-pump creates a tendency in the fluid portion of the mass to separate itself. Effectual separation, however, can only be attained when the grain or crystal of the sugar dealt with is large, hard, and well-formed; with small or soft grain, the process is utterly inapplicable. This fault has restricted its use.

(*e*) *Centrifugals.*—"The preceding modes have been generally superseded by centrifugal machines or hydro-extractors. There are many varieties, but all consist essentially of a cylindrical basket revolving on a vertical shaft, its sides being of wire-gauze or perforated metal, for holding the sugar. The basket is surrounded by a casing at a distance of about 4 inches, the annular space thus left being for the reception of the molasses, which is expelled by centrifugal force through the sides of the basket, when the latter revolves at high speed. A spout conducts the molasses to a receiver."

From the above brief abstract of the various stages of sugar manufacture, as pursued at European Factories and Refineries, the reader may be able to follow the account of the crude methods practised in India. The possibilities of improvement will be indicated through the comparison thus rendered possible between the two systems

The following selection of passages, regarding the manufacture of sugar in the various provinces of India, may therefore be here given:—

MANUFACTURE in Bengal. 394

I. BENGAL.

The utmost that space can be afforded for in this place is to give two or three accounts illustrative of the manufacture of cane and palm sugar. It may, perhaps, be allowable to repeat that the modern reports consulted by the writer seem to greatly under-estimate the importance of the date-palm in the supply of Bengal sugar. It is customary, for example, to read of many advantages enjoyed by Madras owing to the very large amount of sugar which that presidency derives from palms. Such remarks imply that Bengal is placed at a disadvantage, because of its not having so much palm-sugar. Then, again, the palm-sugar of Madras is often spoken of as date-sugar. With the exception of Mysore, the major portion of the Madras palm-sugar is apparently, however, derived from the cocoanut and palmyra, not the date-palm. It seems worthy, therefore, of special consideration in future to ascerain whether Madras palm-sugar is in reality superior to that of Bengal, and whether that superiority is due to the particular palm used or to the system of manufacture. It is highly likely that Bengal has very nearly as much palm-sugar as Madras. The trade of Jessor and other districts of Eastern Bengal is mainly in palm-sugar. and the bulk of the manufactured article derived from these palms pours into Calcutta, so that it seems probable a much larger proportion of Calcutta sugar is derived from palms than is presently supposed.

Palm-sugar. *Conf. with pp. 138 226-227, 231, 270, 310, 352, 361, 370.* 395

Bogra. 396

BOGRA.—"The three police divisions of this district which formerly formed part of the district of Dinajpur were, during the greater part of the first half of this present century, the most important sugar-cane-producing tracts in this part of Bengal. In 1810, Dr. Buchanan-Hamilton, in his account of Dinajpur, speaking of Baddalgachi, says:—'The sugar made in this part of the country is called *bádal*, and is reckoned the best in the district. The observations of this accurate observer on the preparation of the inspissated juice or *gur*, and the subsequent process of refining are condensed below. These operations have since changed only in some minor particulars.

MANUFACTURE in Bengal.

Bogra.

"The boilers are of two sizes, one adapted for making, at each operation, about 540 Calcutta sers, or 1,105℔; the other boils 464 sers, or 950℔. The latter, which is most in use, weighs 490℔, and will contain about 2,672℔ of water, or about 42⅝ cubic feet, reckoning 1,000 ounces to the cubic foot. It is in shape a segment of a sphere 9 feet in diameter at the mouth. It is sunk into a cylindrical cavity in the ground, which serves as a fire-place, so that its edge is just above the floor of the boiling-house. Some manufacturers have only one boiler, others as many as four; but each boiler has a separate hut, in one end of which is some spare fuel, and in the other some bamboo stages, which support cloth strainers. This hut is about 36 feet long and 15 broad; has mud walls nine feet high and is raised about 18 inches above the ground. For each boiler are required two other houses. One, in which the extract of sugar-cane is separated from the molasses by being strained, is about 30 feet long by 15 wide. The other hut, which is about 45 feet long by 12 wide, is that in which, after the extract has been strained, boiled, and clarified. the treacle is separated from the sugar by an operation analogous to claying. Each sugar manufacturer has also a ware-house, the size of which is in proportion to the number of his boilers. The walls of these three last huts are of clay; and under the thatch, in order to diminish the risk from fire, they have a roof terraced with the same material. The floor of the ware-house is raised three feet above the soil and the whole premises is surrounded by a high wall of mud. The most simple process by which the sugar is procured from the pot extract, as performed at Badalgachhi, and by which the sugar called *bádal* in the neighbouring markets is produced as follows:—Take 960 maunds of pot extract, divide it into four parts, put each into a bag of coarse sack cloth (*chati*), hang these over an equal number of wide-mouthed earthen vessels and sprinkle a little water on them; there will drain from the bags 240℔ of substance called *máth* by the natives, and which is analogous to the molasses that flow from the hogsheads in a Jamaica curing-house. The remainder in the bags is called *sar*, and is a kind of coarse muscovado sugar, but it is far from being so well drained and freed from molasses as that which comes from the West Indies. Put the 720℔ of this substance into the boiler with 270℔ of water, and boil them briskly for 144 minutes. Then add 180℔ of water and boil 48 minutes more. In the meantime strain 90℔ of water through an earthen pot with some holes in its bottom lined with straw, and filled with ashes of the plantain tree (**Musa**). Four sers of this clear alkaline solution are added to the boiling sugar and occasion a thick scum which is removed. After twenty-four minutes 4½℔ of alkaline solution and three eighths of a pound of raw-milk are added, and the boiling and scumming are continued twenty-four minutes. This must be repeated from five to seven times until no more scum appears. Then add 240℔ of water, take out the liquor and put it into a number of strainers. These bags are of coarse cotton cloth, in the form of inverted quadrangular pyramids, each of which is suspended from a frame of wood about two feet square. The operation of straining occupies about ninety-six minutes. The strained liquor is divided into three parts One of these is put into the boiler with from three-eighths to one-and-a-half pounds of alkaline solution, $\frac{3}{8}$℔ of milk and 30½℔ of water. After having boiled for between forty-eight and seventy-two minutes, three-fourths of a pound of milk is added, and the liquor is poured in equal portions into four refining pots. These are wide at the mouth and pointed at the bottom, but are not conical, the sides being curved. The bottom is perforated, and the stem of the plantain leaf forms a plug for closing the aperture. When they have cooled a little the refining pots are removed to the curing-house and placed on the ground for twenty-four hours. Next day they are placed on a frame, which supports them at some distance from the ground. A wide-mouthed vessel is placed under each to receive the viscid liquor that drains off, which seems to be the same as the treacle of the European sugar-houses and by the natives is called *kotra*, *chitiyá*, and *rab*. In order to render the separation more complete, moist leaves of **Valisneria spiralis*** (*pata*) are placed over the mouth of the pot to the thickness of two inches. After remaining ten or twelve days these are removed, and a crust of sugar about half an inch in thickness is found on the surface of the boiled liquor. The crust is broken and removed and fresh leaves are repeatedly added until the whole sugar has formed which requires from seventy-five to ninety days. The sugar procured is usually 267℔; and the treacle 450℔, so that in scumming and straining the boiled liquor, very little is lost, or at least the loss is compensated by the water in the molasses and treacle; for the 240℔ of molasses, strained from the extract before it was boiled, must be also considered a part of the produce. When the cake extract is used it does not require to be strained before it is put into the boiler; but 720℔ of it are broken to pieces and put at once into the boiler with 120℔ of water, and are then treated exactly in the same manner as the *sar* or strained pot extract. The produce is reckoned to be usually 144℔ of sugar, 450℔ of treacle and nearly 91½℔ of scummings and strainings. It is not usual to carry the manufacture

* *Conf. with p. 31.*

MANUFACTURE in Bengal.

any further. The sugar and molasses are then exported by the *Jamúná* to different markets in Southern and Eastern Bengal.

"In 1863 Major Shirwell, the Revenue Surveyor, reported that the subsequent progress of this manufacture had been, from many causes, one of decline. It was supposed that the land had become less favourable for the growth of the sugar-cane, since the waters of the old Tísta river left this part of the country. However that may be, the deterioration of the cane was unquestionable. Mr. Payter, the farmer of the principal Government estates in Bogra," gave a detailed account of the introduction and decline of the Bourbon cane in that district. The reader will find Mr. Payter's remarks in the section above on the VARIETIES AND RACES OF SUGAR-CANE (*Otaheite*); *Conf. with pp. 45, 48, 140.*

Dacca. 397

DACCA.—"*Gur-making.*—The Behea sugar-cane mill, which has produced almost a revolution in the sugar-growing districts of Behar and West Bengal, is unknown in Dacca. The art of *gur*-making is also little known or practised, so that, though the extent of land capable of growing sugar-cane in this district is probably greater than in any part of Behar and Lower Bengal, with the exception of the neighbouring district of Mymensingh, yet the supply of sugar-cane for local consumption comes from such distant places as Ghazipur and Benares. The wooden mill, known as the *kerki*, is still in use here. It is worked by a pair of bullocks or six men, four working at a time. The cane is passed and repassed three times through the mill, and even then a large percentage of the juice is left in the *begass.*

"The juice is boiled in four large earthen pans arranged over a furnace in two rows, and as it gets thicker and thicker is gradually collected in one pan, fresh juice being put in the first pan from time to time.

"Two different preparations are made, corresponding to the *rab* and *gur* of Upper India. These are kept in earthen pots, each capable of containing from half to three-fourths maund. The yield per bigha varies between 7 and 20 maunds" (*A. C. Sen, Rept. on Dacca Dist.*, *36*).

Faridpur. 398

FARIDPUR.—The following account of the manufacture of crude sugar and of refining, as pursued in the district of Faridpur, gives the main particulars of the Bengal system. It will, however, be found to be greatly amplified by the more detailed description given below regarding the Jessor sugar manufactures. It should be recollected that in both Faridpur and Jessor date-palm sugar is more important than cane, and much of the information here furnished regarding these districts refer therefore mainly to palm-sugar:—

"The most important manufacture of Faridpur, and indeed the staple article of district trade, is sugar, prepared both from the juice of the date tree, and from the cane. The following description of the mode of extracting the date-juice, and the outturn of the produce, is taken from Colonel Gastrell's Revenue Survey Report, pp. 8, 9:—The trees should not be tapped to extract the sap until they are six or seven years old. But the Natives seldom permit them to attain that age, commencing the tapping ordinarily after the fourth, and sometimes as early as the third year. The evil consequence of this improvidence are small returns of sap, weak and sickly growth of trees, and finally their deterioration and destruction many years before they would otherwise have been exhausted. On the other hand, the advantages obtained by early tapping are quicker returns for the money laid out during the first years of the tree's growth; but these by no means compensate for the loss in after years. Tapping generally commences early in October, when the rainy season is passed, and continues until the middle of March following. Some persons continue to extract the juice still later, but the heat of the weather after that period generally causes it to ferment so rapidly that little or no *gur* (coarse crude sugar) can be obtained from it. The trees, moreover, require rest to recover themselves, after being deprived of so much sap for so long a period. Shortly before the regular process of tapping begins, the men employed in this work strip off the lower leaves of the tree, and make a horizontal incision close under the crown leaves, which are left untouched through the outer bark or skin, and well into the under-wood, about five or six inches in breadth by two or three inches in depth. Below this cut, the wood and bark is pared away to the length of ten or twelve inches, preserving a flat surface sloping outwards and down-wards from the inside of the top cut, and forming a deep notch in the tree, down the centre of which, and from both sides sloping downwards, small grooves are scooped out of about a quarter of an inch in depth, meeting at a point. These serve to conduct the sap to a small bamboo tube which the tapper inserts at

MANUFACTURE in Bengal.

Faridpur.

their point of junction, and below which an earthern pot is suspended to catch the juice. The sap runs all night, and is collected early in the morning in other pots by the same man who made the incision the previous night, aided by one or two boys. It is then carried away to the boiling-house, which is generally close at hand, and is at once boiled down. On the freshness of the juice, and its freedom from fermentation depends the return of *gur;* it is therefore essential to collect it early in the cool of the morning, and to convey it to the boiling-house as soon as possible. In the evening the tapper revisits the trees, scrapes the surface of the cut, cleans out the gr oves, and hangs up the pots that he left in the morning. He repeats this process for three days in succession, after which it is usual to give the trees a rest for three days before tapping again. In favourable weather this rule is followed throughout the season. But it is also usual to give the trees rest when fogs * are heavy, or rainy weather sets in; both states of the weather operating injuriously on the flow of sap, and rendering the tree liable to rot and die, if tapping be persisted in. As a rule, only one cut is annually made in the tree, but occasionally a second incision on the opposite side may be resorted to, although this is very rarely done. These cuts are made annually, and alternately on opposite sides of the tree, the age of which may be easily determined from the number of notches. One man, with the assistance of one or two boys or women, can efficiently look after and collect the sap of sixty trees. His wages would be, on an average, from 6*s.* to 7*s.* per month during the tapping season. He and his assistants receive their food daily; and at the close of his labours he is presented with a pair of waistcloths (*dhuti*) and one pair of shoes. The life is a hard one, and not free from danger. Serious accidents sometimes happen to these men from the breaking of the rope which they loop round their bodies and the tree to aid them, first in climbing the trees, and afterwards to support them. If the rope breaks or the knot slips, nothing can possibly save the man from falling head long backwards to the ground. Date-trees are usually rented by the score. Rates differ, but the general one appears to be three-half pence per tree, or 2*s.* 6*d.* the score. Young trees are said to yield about eight to ten pounds of juice per diem for the first few years; sixteen pounds when in full bearing, and again only eight to ten pounds when old or perhaps an average of about ten pounds throughout. The best and most productive, and at the same time the largest quantity of sap,† is collected during the cold season in the months of December, January, and February. The colder and drier the season is, the more favourable is it for the sap-grower. If the sap be of first rate quality and quite fresh, six pounds will boil down to about one pound of the coarsest kind of ungranulated brown sugar. But of sap of ordinary quality, from eight to ten pounds would probably be requisite to obtain that quantity of *gur;* seven pounds may therefore be taken as the average quantity of juice required to yield one pound of *gur.* The apparatus for boiling the juice into *gur* consists of a number of earthen pots, arranged in a circle over a fire in a cavity dug in the ground, and covered over with a clay roof or ceiling, having as many holes as there are pots to be inserted. The annual expenses for maintaining a hundred trees, such as rent of trees and land, wages, food and clothing of two men, pots, pans, and fuel, contingencies, etc., amount to about £8-16*s.* I have several rather inconsistent reports as to the profits and other details of the manufacture. The native sugar-boilers of the district informed Colonel Gastrell that a hundred trees would produce eighty-seven hundredweights of *gur,* with about £11-12*s.*, thus leaving a clear gain to the producer of £2-16*s.* per annum. Other data, given by Mr. S. H. Robinson in his prize essay on the cultivation of the date-trees, return the produce of a hundred trees at sixty-six hundredweights of *gur,* worth £81-8*s.*-9*d.* This calculation only leaves a profit of 2*s.* 9*d.* per annum on a hundred trees. I believe it to be below the truth especially at present prices."

"*Cane-sugar.*—The second kind of *gur,* or crude sugar, is called *kusuri* or *ákh gur,* and is obtained by boiling the juice of the sugar-cane. The process of extracting the juice is thus described in Colonel Gastrell's Report:—'The mill in common use ordinarily consists of two endless, coarse-threaded wooden screws, of about eight to ten inches in diameter, set vertically in two horizontal cross pieces, and firmly fixed to two uprights which are let well into the ground. These screws have their threads cut right and left, and play into each other. They are made of any hard, close-grained wood, tamarind being preferred. To the upper end of one of the screws,

* It will be seen that many writers say that the formation of sugar in cane is retarded by humidity still more so by flooding, beyond a certainextent.

† This remark is, it will be seen, opposed to the theory that greater solar activity favours the formation of sugar. See the remarks at pp. 18-20.

MANUFACTURE in Bengal.

Faridpur.

which projects above the horizontal bar, a long pole is attached, to which the bullocks that turn the mill are yoked. The cane is generally passed twice through the mill before being cast aside to dry for fuel. The expressed juice is received in a basin formed for the purpose below the screws. Women or boys are usually employed to feed the mill with canes and drive the bullocks.' The juice which collects in the basin is then boiled down into *gur*, the process of boiling being the same as for the sap of the date-tree.

"*Sugar-refining*.—The process of refining sugar is the same, whether it is obtained from the juice of the date or of the cane. The following description of the mode of manufacture is condensed from **Dr. Basu's** Report:—Two modes of manufacturing sugar from *gur* are reported on. By the first method, the boiled juice, in the form of *gur*, is placed in stout gunny or sackcloth bags. The molasses or refuse is squeezed out partly by twisting and tightening the mouths of the bags, and partly by laying weights upon them for additional pressure. The article thus produced is of a brownish colour. By far the largest quantity of sugar manufactured in the district, however, is prepared in a different way. The process, rather a cumbrous one, is as follows:—The *gúr* is at first boiled with a certain proportion of water in a large iron vessel, a quantity of diluted milk being added from time to time to separate the impurities, which are skimmed off as soon as they form on the surface. When no more skim appears, the thickened liquor is poured into a number of circular earthen pots or strainers, made wide at the top and pointed below, with a hole in the centre, called *bharnis*, and left for two or three days in the open air to cool. It is then removed to the refining house, where the final separation of the solid crystalline portion from the treacle is effected. The straining pots are generally arranged in rows on a bamboo frame at a certain height from the ground, and earthen pitchers are placed under each to receive the molasses as it slowly drains from the refining pot above. To complete the arrangement, as well as to quicken the operation, fresh moist leaves of a water weed called *pátá sáolá* are now placed on the top of the refining pot, and as soon as a layer of sugar from one to two inches thick is formed at the top, it is removed by scraping with the knife, fresh weed being laid on the remainder, and the same tedious process is repeated several times until the entire quantity of sugar is made. The native confectioner makes extensive use of this sugar for the purposes of his art; but before it is fit for use, it has to be clarified again by further boiling with the addition of a solution of milk as in the last process. When this is allowed to cool, it forms a hard crust, which requires to be broken and pounded before it can be employed. The molasses which drains off from the sugar in the process is employed for preparing *hookah* tobacco, inferior sorts of sweetmeats, etc., and the rest is sold for making country rum. **Dr. Basu** estimates the total quantity of *gur*, or raw sugar prepared in Farídpur District from the juice of the cane and date-tree, to be between two hundred thousand and three hundred thousand hundredweights, or from three to four *lákhs* of maunds per annum" (*Statistical Account of Bengal, V., 334*).

Jessor. 399

JESSOR.—Under the article **Phœnix sylvestris** (Vol. VI., pp. 209-215) will be found the first half of **Mr. Westland's** detailed account of the sugar manufactures and sugar trade of JESSOR. The writer feels that perhaps no better course could be found of conveying an idea of the sugar interests of a huge portion of Eastern Bengal than of completing **Mr. Westland's** account, even although to do so it has been found necessary to republish one or two paragraphs that have already been given under **Phœnix**. In passing the reader may be reminded that the chief facts dealt with in the quotation below refer more especially to the sugar of **Phœnix sylvestris**—the date-palm, but when once the saccharine fluid has been reduced by boiling to the crude syrup known in Bengal as *gúr*, the methods pursued are identical and little or no distinction is made whether the sugar bought and sold in the country has been derived from palm, or cane juice. The review of the sugar trade furnished by the concluding paragraphs may in fact be accepted, as conveying a vivid conception of the internal traffic of a large portion of Bengal, which centres in Calcutta, in this very important article of food.

Date-Palm Sugar. *Conf. with pp. 138, 226, 231, 266, 301, 352, 361, 370.* 400

"*Manufacture of Dhuluá sugar.*—"We have traced the *gur* into the hands of the refiners, and we shall now see what the process of manufacture is. But there are several methods of refining, and two or three sorts of sugar produced. We

MANUFACTURE in Bengal.

Jessor.

will take them in order, and describe first the method of manufacturing of *dhuluá* sugar—that soft, moist, non-granular, powdery-sugar used chiefly by natives, and especially in the manufacture of native sweetmeat.

"The pots of *gur* received by the refiner are broken up, and the *gur* tumbled out into baskets which hold about a maund each and are about fifteen inches deep. The surface is beaten down so as to be pretty level, and the baskets are placed over open pans. Left thus for eight days, the molasses passes through the basket, dropping into the open pan beneath, and leaving the more solid part of the *gur*—namely, the sugar in the basket. *Gur* is, in fact, a mixture of sugar and molasses, and the object of the refining is to drain off the molasses, which gives the dark colour to the *gur*.

"This eight days' standing allows a great deal of the molasses to drop out, but not all of it; and carry the process further, a certain river weed called *sáolá** which grows freely in the Kabadak, is placed in the baskets so as to rest on the top of the sugar. The effect of this weed is to keep up a continual moisture; and this moisture, descending through the sugar, carries the molasses with it, leaving the sugar comparatively white and free from molasses. After eight days' exposure with *sáolá* leaves, about four inches on the surface of the mass will be found purified; and these four inches are cut off, and *sáolá* applied on the newly exposed surface. This and one other application will be sufficient to purify the whole mass.

"The sugar thus collected is moist, and it is, therefore, put out to dry in the sun, being first chopped up so as to prevent its caking. When dry, it is a fair, lumpy raw sugar, and weighs about thirty per cent. of the original mass, the rest of the *gur* having passed off in molasses. Dishonest refiners can get more weight out of it by diminishing the exposure under *sáolá* weed, so as to leave it only five or six days instead of eight. The molasses is less perfectly driven out, and the sugar, therefore, weighs more. Of course, it has also a deeper colour, but that is in a measure remedied by pounding under a *dhenki*. There are also other dishonest means of increasing the weight; for example, the floors of the refineries are sometimes a foot or more beneath the level of the ground outside, the difference representing the amount of dust which has been carefully swept up with the sugar when it is collected after drying. It is also very easy so to break the pots that fragments of them remain among the sugar.

"*The Droppings.*—The 'first droppings' gathered in the open pans in the manner already described are rich in sugar, and are used, especially in the North-West, for mixing up with food. It entirely depends, therefore, upon the price, offered for them for this purpose, whether they are sold at once or reserved for a second process of sugar manufacture. In this second process the first droppings are first boiled, and then placed under ground in large earthenware pots to cool. Unless thus boiled they would ferment; but after being boiled in this fashion, they, on cooling, form into a mass somewhat like *gur*, but not so rich. After this the previous process is again gone through, and about ten per cent. more weight in sugar is obtained. The sugar is, however, coarser and darker in colour than the first.

"If the refiner is not very honest, and if he is sure of finding immediate sale, he will use a much more speedy process. Taking the cooled *gur*, he squeezes out the molasses by compressing the mass in a sack, and then, drying and breaking up the remainder, he it sells as sugar. It does not look very different from that prepared in the more elaborate way, but it will soon ferment, and hence the necessity of finding an immediate purchaser.

"The remainder, after all this sugar has been squeezed out, is molasses—*chitá gur*, as it is called. It forms a separate article of commerce, being exported to various places, as will be subsequently mentioned.

Manufacture of Paka Sugar.—"The sugar produced by the method above described is called *dhuluá*—a soft yellowish sugar. It can never be clean, because it is clear from the process used, that whatever impurity there may originally be in the *gur*, or whatever impurity may creep into the sugar during its somewhat rough process of manufacture, must always appear in the finished article. Another objection to it is, that it tends slightly to liquifaction, and cannot, therefore, be kept for any considerable time The *paká* sugar, whose manufacture I am now about to describe, is a much cleaner and more permanent article. It has also a granular structure, which the *dhuluá* has not. The manufacture of it is more expensive than the other, and the price of it, when finished, is about R10, whereas *dhuluá* costs only about R6 per maund.

"In this process, the *gur* is first cast upon flat platforms, and as much of the molasses as then flows off is collected as first droppings. The rest is collected, put into sacks and squeezed, and a great deal of the molasses is thus separated out. The

* This may be **Hydrilla verticillata,** the plant most extensively used in Bengal for this purpose. *Conf.* with footnote p. 31 also Vol. IV., p. 311 of this work.

MANUFACTURE in Bengal.

Jessor.

sugar, which remains behind, is then boiled with water in large open pans, and as it boils, all scum is taken off. It is then strained and boiled a second time, and left to cool in flat basins. When cooled, it is already sugar of a rough sort, and *sáolá* leaves are put over it, and it is left to drop. The result is good white sugar; and should any remain at the bottom of the vessel still unrefined, it is again treated with *sáolá*. The first droppings, and the droppings under the *sáolá* leaves, are collected, squeezed again in the sacks, and from the sugar left behind, a second small quantity of refined sugar is prepared in exactly the same way, by twice boiling. The droppings from the sacks are *chitá gur*, and are not used for further sugar manufacture. About thirty per cent. of the original weight of the *gur* is turned out in the form of *paká* sugar.

Kesabpur Method of Manufacture.—"There is another method of manufacture peculiar to Kesabpur, and slightly differing from that just described. The *gur* is first boiled in large open pots, and into each potful is put a handful of *bíchh*; it is then left to cool, and in doing so it coagulates, and is afterwards treated with *sáolá* leaf, and thus refined. The last droppings under the *sáolá* leaf are burnt; and this forms the *bíchh* used in the manufacture, the effect of which is apparently to make one boiling do instead of two. The droppings from this first process are collected, boiled with *bíchh* and cooled as before; then squeezed in sacks, mixed with water, boiled to drive off the water, and after cooling, purified with *sáolá* leaf. The droppings now are exhausted molasses, or *chitá gur*. The produce in sugar is twenty-five or thirty per cent. of the weight of the original *gur*.

English Process of Manufacture.—"There remains to be described the English process of refinement used in the factories at Kotchándpur and Chaugachha. In this the raw material is mixed with a certain amount of water and boiled in open cisterns, the boiling being accomplished, not by fire, but by the introduction of steam. The lighter filth now floats to the surface and is skimmed off, while the boiling solution is made to flow away through blanket-strainers into another cistern. After this, it is boiled to drive off the water. Now, if the mass were raised to boiling temperature, the result would be sugar, granular indeed in construction, but not differing in this respect from native *paká* sugar. But if the water be driven off without raising the mass to boiling point then we get the crisp and sparkling appearance which loaf-sugar always has. Whether there is any difference in the substances, I do not know; but so long as people prefer what *looks* pleasant and nice, sugar of this sparkling appearance will command a higher price in the market.

"The object is attained by boiling in a vacuum-pan, that is to say, a large closed cistern, from which a powerful pump exhausts the vapour as it rises. The lower the atmospheric pressure on the surface of the liquid, the lower the temperature at which the ebullition takes place. The pump is therefore regulated so as to diminish the pressure on the surface to such a point that the mass will boil at about 160° Fahrenheit; and the apparatus being kept regulated to the point, all the water is driven off by boiling by means of introduced steam, without the temperature becoming higher than 160°. It is out of place here to describe the mechanical devices for filling and keeping filled, and emptying, and watching and testing the liquid within the closed cistern, or for regulating the supply of heat and the action of the pump which is driven by steam It is sufficient to pass at once to the end of the vacuum-pan stage, which lasts eight hours, and to say that the mass in the pan is now run off into sugar-loaf moulds. It is already in a viscid state, and it is now left to cool in the moulds, which are placed upside down, having a hole in their vortex, placed above a pot. The molasses by its own weight drops out by this hole, and is caught in the earthenware pot beneath.

"The last of the molasses is wasted out in this way. The uppermost inch of the sugar in the mould is scraped off, moistened, and put back. The moisture sinks through the mass, and with it the molasses. This is done some three times, and then, the sugar having now been twelve days in the moulds, the purification is considered to be finished, and the loaves may be turned out of the moulds. If the raw material used was the *gur* as it comes from the cultivator, the result is a yellowish, sparkling, loaf-sugar; but if native-refined *dhulud* sugar is the raw material used, then the loaf is of brilliantly white sugar. The process used at Cossipur, near Calcutta, is similar to that last described. The principal difference consists in this, that the sugar is at one stage additionally purified by being passed through animal charcoal, and that the molasses, instead of being allowed to drop out by its own gravity from the moulds, is whirled out by the application of centrifugal force.

The Sugar Market.—"Although sugar is manufactured to some extent all over the district, the principal sugar country is the western part, which may be considered as included between these places—Kotchándpur, Chaugáchha, Jhingergachha,

MANUFACTURE in Bengal.

Jessor.

Trimohini, Kesabpur, Jessor and Khájura; and these are the principal marts for its production and export. There are two chief places to which export is made—Calcutta and Nalchití. Nalchiti is a place of great commercial importance in Bákarganj, a sort of central station for the commerce of the eastern districts. The demand there is for *dhuluá* sugar, as it is for local consumption; and except from Kotchándpur itself, almost all the *dhuluá* sugar produced in the district finds its way to Nalchití or Jhálkátí, which is near it. Kotchándpur also sends a great deal of *dhuluá* sugar there, but most of its produce goes to supply the local demand in Calcutta, as it is favourably situated for land carriage to Calcutta. Calcutta has, in fact, two demands, namely, a demand for *dhuluá* sugar for consumption in Calcutta and other places whither it sends the sugar, and a demand for *paká* sugar for export to Europe and other places. This last demand is met by Kesabpur, and by most of the other places in the southern half of the district. The former demand is, as stated, already met by Kotchándpur.

"The distribution of manufacture and export may therefore be shortly stated thus:—In the northern half of the sugar tract, *dhuluá* sugar is manufactured for native consumption, and sent either to Calcutta or to the eastern districts. In the southern half there are two manufactures: *dhuluá* is manufactured by the peasantry, and is brought up and exported to Nalchití and the eastern districts; and paká sugar is manufactured by professional refiners and exported to Calcutta.

State and Prospects of the Trade.—"The demand for *dhuluá* sugar increases every day, especially the demand from the eastern districts, while the paká sugar is decreasing. The increase of the former results from the increasing prosperity of the people, and the decrease of the latter is due to causes connected with the European market, for which most of the *paká* sugar sent down to Calcutta is intended. In the European trade there are, of course, several competitors with Calcutta. Mauritius especially is a close rival of Calcutta; and as the Mauritius cultivation is now extending and prospering, and as it has greater facilities for entering the European market than Calcutta, it necessarily results that exports from Calcutta are diminishing.

"The sugar trade is therefore less progressive in the southern half of the Jessor sugar tract, whence the export is chiefly to Calcutta, than in the northern half. Both at Trimohini and at Kesabpur there have been a large number of refineries closed. As for Kesabpur, the number of refineries has decreased in five years from about 120 to 40 or 50. Trimohini has for a long time been overshadowed by Kesabpur, being hardly more than an out-station of Kesabpur: it had some ten or twelve refineries about five years ago, and now it has not one. It must be remembered, however, that Kesabpur and Trimohini used to be not only refining, but also purchasing stations. I have stated that about these places a large number of husbandmen manufacture the sugar they produce; and as the sugar they make is all sold to merchants who have agencies at these places, it follows that a very large amount of sugar trade goes on apart from the refineries.

"While Kesabpur and the region near it have suffered especially from this cause, there is another cause for the decrease of the sugar trade, which has influenced equally every one of the sugar marts, the northern as well as the southern. A short time after European enterprise gave the first stimulus to the cultivation of the date, the native merchants began to step in and take away from the European manufacturers the fruits of their labour. The demand for native refined sugar was greater than for the first rate sugar manufactured by European means; and the consequence was, that the native merchants appropriated the trade to the exclusion of the English. But they came in too great a rush, and competed too keenly with each other for the produce. Since a date tree takes seven years to grow so as to produce *gur*, the demand cannot in this case produce supply till after the lapse of some time. The price of raw material rose, the merchants' profits became more limited, and the consequence was that a slight depression in the trade had the result of driving away many traders from it. The husbandmen, meantime, profited largely by these high prices, and there has been of recent years a great extension of cultivation. This will tend to reduce the price of *gur*, and to give the traders a large share of the profit; and if, as is most likely, the increase of demand from the eastern districts keeps pace with the increase of production, the sugar trade will soon recover from its present depression and extend even more widely than it did before.

The Cultivators.—"It should be noticed that the depression has been of such a nature that while it affects the merchants and refiners engaged in sugar traffic, it hardly, if at all, affects the cultivators. They have all along got high prices for their *gur*, and have prospered so much that, as already mentioned, new groves are starting up in all directions. Similarly, near Kesabpur and Trimohini, the many

MANUFACTURE in Bengal.

Jessor.

cultivators who manufacture their own *dhuluá* sugar have never felt the influence of the evil season that has caused so many merchants to withdraw from the trade. The demand from Nalchití for the *dhuluá* sugar has never fallen off, as has that for *paká* sugar from Calcutta; and thus the cultivators' manufacture has never diminished, as the merchants' has. It is thus that the apparent paradox is explained that while the sugar trade, so far as regards the cultivators, is in a most flourishing state, it is, as regards the merchants, in a somewhat depressed condition."

Description of a Sugar Mart.—"What I call depression is, of course, only comparatively so; for there can be few busier scenes than such places at Kotchándpur or Kesabpur display during the sugar season. For four or five months the produce is every day seen pouring in from every direction. At Kotchándpur alone two or three thousand maunds is the daily supply of *gur*, and at Kesabpur probably about one thousand. Carts laden with jars, cultivators bringing their own *gur*, fill the streets; the shops of the *bepáris* are crowded with sellers, and the business of weighing and receiving goes on without intermission. Larger transactions are going on at the doors of the refineries, where carts, full laden, stand to deliver their cargoes to the refiner. At Kotchándpur this occurs every day, more or less, though on the regular market days there is more business done than on others. At Kesabpur also there is a daily market, but at the other places the supplies are mostly timed so as to reach on the market day.

"Let us enter a refinery—a large open square, shut in with a fence, and having sheds on one or two sides of it, where part of the work, and specially the storing, is done. If it is a refinery of *paká* sugar, we find several furnaces within the yard and men busy at each, keeping up the fire, or skimming the pots, or preparing them. If it is *dhuluá* sugar, we see many rows of baskets with the sugar covered with *sáolá* leaf, standing to drop; rows of earthen pots, with *gur* or sugar, or molasses, according to the stage of manufacture, are seen on all sides; and in the same open yards all the different processes are at the same time going on.

"The manufacturing season extends from the middle of December to the middle of May. In December the merchants and the refiners all congregate at the sugar towns, and in May they finish their work and go home. Compared with their state during these five months, the appearance of such places as Kotchándpur and Kesabpur, during the rest of the year, is almost that of a deserted town. The refineries are shut up; no *gur* is coming in; nothing is going on. Many of the manufacturers belong to Santipur in Nadiya, and while they have their chief refineries in Kotchandpur or some other place, have also smaller ones in Santipur. Whether the Santipur factories derive any part of their raw produce from that part of the country, I do not know; but no inconsiderable quantity of *gur* is taken across from Kotchándpur, Jhingergáchhá, and Jádabpur to Santipur for manufacture there. The merchants of Kesabpur and Trimohini have their connection rather with Calcutta than with Santipur and places in Nadiya. Kotchándpur has, from its prominence, suffered more from the competition of the merchants than most other places, and it has got rather a bad name for the quality of its sugar. During that competition very many dishonest practices were introduced, some of which I have described before. The misfortune of such practices in this trade is, that as manufacturers have no distinguishing marks for their own sugar as indigo planters have for their indigo, a few dishonest men can cause a bad name to adhere to all the produce of the locality, and even honest men will find some difficulty in disposing of their wares. So much was this felt, that part of the *gur*, which otherwise would have been manufactured in Kotchandpur, was taken over to Santipur and manufactured there. Nay, in some cases, the same persons who manufactured dishonest sugar in Kotchándpur manufactured honest sugar in Santipur.

"It remains to give a view, in detail, of the chief sugar marts, so as to note matters which, in our general survey, have not found a place. I note first those places which are within, what I call, the chief sugar tract.

Kotchandpur. 401

KOTCHANDPUR.—"Is, by far the largest of the sugar marts, as both it and the adjacent village, Sulaimanpur, are covered with refineries. Of the sugar manufactured, most goes to Calcutta, but about a quarter or a third goes to Nalchiti and Jhálakáti in Bákarganj. The proportion of the latter is steadily increasing. From Kotchándpur to Calcutta there are two routes, by water and by land. The bulk appears to go by land to the Krishnaganj and Ramnugar stations of the Eastern Bengal Railway, going by it to Calcutta. The same carts that take away the sugar frequently collect *gúr* to bring back with them. The amount of sugar manufactured in and near Kotchándpur in each year must be near a hundred thousand maunds, worth about six lakhs of rupees. It is, perhaps, about a quarter of the whole sugar manufacture of the district. The principal merchants are Bangsi Badan, called

MANUFACTURE in Bengal.

Sadhu Khan by title, and **Guru Das Babu**, a great brass-ware manufacturer of Nadiyá. **Bangsi Badan**, now an old man, is, I believe, one of those men who, starting from a very small capital, become, by the application of extraordinary business qualifications, leading merchants in their country. He has several refineries all over the district, and an agency in Calcutta.

Jessor—Chaugachha. 402

CHAUGACHHA.—"Is, like Kotchándpur, on the bank of the Kabadah river. The *paká* sugar is manufactured here as well as the *dhuluá*. The refiners are chiefly residents of the place. Of the exports I have not obtained very much information, but apparently they are not very different from Kotchándpur. Part of the export goes by river, and part across country to Krishnaganj Railway Station. So far as sugar goes, the place has been made by the factory erected here by **Messrs. Gladstone, Wyllie & Co.**, a factory capable, I believe, of turning out a thousand maunds of sugar in one day, but which has not been worked for years. This factory cultivated the date very extensively, and Chaugáchhá is now surrounded by forests of date trees. *Gúr*, I am told, might have been bought at one anna a pot when the factory first came, a quarter of a century since, while now a pot is worth six or seven annas. The proprietor's revenue was then R118 from the whole bazár (probably about R5 per *bigha*), and it is now R4 per *bigha*.

Date-palm plantation. (Conf. with p. 101.) 403

JHINGERGACHHA.—"Still further south, is rather a place for the purchase of *gúr* than for the manufacture of sugar. There are three or four refineries in the place, but the greater part of the produce brought to market is bought up by *bepáris*, who take it across to Sántipur for manufacture there. This part of the district is, in fact, the part most accessible to Sántipur, being on the imperial road.

Jhingergachha. 404

JADABPUR.—"Is a little to the west of Jhingergáchhá, and, like it, supplies *gúr* to the Sántipur refiners rather than for local manufactures. It is simply a large *gúr* market, whither twice a week,—that is, on Mondays and Fridays—the sellers bring their *gúr* from all the places round about, and the *bepáris* come to meet them, purchase the produce, and carry it off to Sántipur.

Jadabpur. 405

KESABPUR.—"The business here consists in purchasing home-made *dhuluá*, and in refining *paká* sugar, most of the former going to the eastern districts, but partly also to Calcutta, and almost all the latter going to the Calcutta market. The purchasers are, for the most part, agents of Calcutta firms, and give their name to the chief street in Kesabpur, "CALCUTTA PATI." The export is either by the river from Kesabpur itself, or by carts to Trimohiní, and thence to Calcutta by river. There is a very large pottery manufacture at Kesabpur, the pottery being required for the sugar manufacture. Kesabpur has one advantage over the other places in the sugar tract, in its proximity to the Sundarbans. The river Bhadrá leads from it straight down towards the forest, and by this river large cargoes of firewood are brought up to be used in the manufacture of sugar. It is probably to this circumstance that it owes its prominence as a sugar manufacturing place, for it is the second largest in the district.

Kesabpur. 406

TRIMOHINI.—"Is now a sort of out-station of Kesabpur; for most of the merchants who have agencies here, have agencies also in Kesabpur. It is entirely a place for the purchase of sugar, and not for its manufacture; the *dhuluá* sugar manufactured by the husbandmen, and at the village factories round about, and also the sugar manufactured in and near Jhingergáchhá, are brought up here and exported to Calcutta and other places by river.

Trimohini. 407

TALA.—"Further south is another large sugar mart, also closely connected with Kesabpur.

Tala. 408

MANIRAMPUR.—"Has two or three factories, but which do little more than supply local consumption.

Manirampur. 409

KHAJURA.—"Is a place of a very large sugar trade, its name being derived from that of the date tree (*khajur*). I have not visited it, and cannot give details of its manufacture, but I believe I may say that its export trade goes to Nalchiti and Bákarganj.

Khajura. 410

KALIGANJ.—"Is farther up on the same river, and is only 8 miles from Kotchándpur. Most of the sugar which is exported from Kotchándpur to Nalchití is brought here to be shipped. Káliganj is not itself a large manufacturing place, but there are several refineries scattered in the villages round about it; for example, in Singhiá, Faráshpur, and others. The sugar manufactured is almost all exported to Nalchití and Jhálakáti.

Kaliganj. 411

"I have now enumerated all the marts which lie within the sugar tract proper, except one or two in the vicinity of Jessor itself, such as Rájáhát, Rupdiá, and Basantiá. These places and Nárikelbariá I have not had an opportunity to examine, but I believe I may state that their exports go to Nalchiti and Jhálakáti.

"A few of the manufacturing places on the outside of the sugar tract remain to be noticed. There is, first, the line of the road between Jhanidah and Mágurá, which

MANUFACTURE in Bengal. Jessor.

Kaliganj.

passes through a date-producing region. There are not any regular sugar-refining towns here, as the refineries are small ones, scattered and isolated. Ichákádá, a town upon the road, at a distance of 4 miles from Magura, is the principal place where the *gúr* is sold. The cultivators bring it there in considerable quantities upon the market days —Tuesdays and Fridays,—and sell it to refiners. Part of the *gúr* here produced is also carried farther east to Binodpur, 6 miles east of Magura, where there are one or two refineries established for the manufacture of the *gúr*, not very abundant, which grows about these parts. The export is almost entirely to Nalchití. Still farther east is Muhammadpur, where a little sugar is refined. The produce here is very scanty, but what is manufactured goes to Nalchiti.

"The Narál sub-division lies for the most part on a very low level, and is devoid of that high ground which is essential for the cultivation of the date. But at Lohágará there is some sugar manufacture, though of an abnornal sort. A few date trees grow near Lohágará, but on land so low that they produce no juice, and it is not from its vicinity that Lohágará derives its *gúr*. But the sugar tract proper is, as we shall afterwards see, deficient in rice cultivation; and as Lohágará, a low region, has some rice to spare, it sends a little, laden in ships, to Khajúrá and other places. The ships, which go laden with rice, bring back cargoes of *gúr*, and it is thus that the small amount of raw material required for the manufacture at Lohágará is supplied. The sugar manufactured in Lohágará is mostly *paká* sugar, and its export is principally to Calcutta; but some also goes to Bákarganj.

"We have another instance of this reciprocity between the sugar trade and the rice trade; for large quantities of rice pour up the Bhoirab river, conveying the rice from the great cultivating regions in the south to Náopárá, Basantiá, and Khajúrá, the inlets on eastern side into the sugar tract. From these places, but especially from Basantiá and Náopárá, the ships carry down *gúr* to be manufactured into sugar at Daus latpur, Senháti, Khulna, and Fakirhát. Near Fakirhát there is some high land, producing date trees, but for the most part it is dependent for its supply of raw material upon the cultivation further north. The places just mentioned, and also Phultalá (which is on the border land between the rice country and the sugar country, and can supply its down material for manufacture) produce for the most part, *paka* sugar. This is a natural consequence of their proximity to the Sundarban supply of firewood. Their export is chiefly to Calcutta.

Interchange of Sugar and Rice.—"I have already given instances of reciprocity of rice import and sugar export; but the principle extends further than I have stated. Throughout the delta there is a general westward movement of rice. Calcutta attracts most of the rice grown in the Jessor Sundarbans, and leaves the riceless districts in Jessor to be supplied from Bákarganj. All over the sugar tract the cultivation of rice is very deficient, and rice pours in from Nalchití all over Magura and the south of Jhanidah, and the head-quarters sub-division. The ships that come laden with rice, therefore, take back with them to Nalchití cargoes of sugar. So also, rice imported by the Kabadak from the south, and through Jhingergáchhá, Chaugáchhá, and Kotchándpur, is spread over the western part of the district, and the ships engaged in this import can carry away the sugar to the tracts whence they have come. From Calcutta itself the principal import is salt, and the salt-ships are employed in carrying back sugar to Calcutta.

"*Exporters.*—It remains to mention a few facts which should probably have found a place elsewhere. First as to the refiners. Professional refiners are for the most part themselves exporters; that is to say, those who buy sugar to refine it in large refineries, scarcely ever sell it to other merchants to export. In fact, they frequently combine, with their refining trade, the trade of purchasing from the smaller or village refiners for export. This latter, however, is also a separate trade; and, especially at Kesabpur and Trimohiní, there are merchants who, themselves doing nothing in the way of refining, purchase sugar locally refined, and export it to Calcutta or to Nalchití. Most of these are agents of Calcutta or Nalchití firms. In fact, according to the native system of trade, it will be found that the same firm, or firms, having in part at least the same partners, have establishments at many places, and carry on business at each place through different partners or agents. **Bangsi Badan Sadhu Khan**, for example, has refineries at all the large sugar marts, and has, besides that, a branch in Calcutta to receive and dispose of the sugar which he exports thither.

Chita or Refuse Gur.—"I have not yet said what becomes of the *chita gúr*, the refuse of the sugar-refining process. It is to a very small extent locally used for mixing up with tobacco to be smoked. By far the bulk of it is, however, exported to Calcutta, Nalchiti, and Sirájganj, but what ultimately becomes of it I do not know. An attempt has been made once or twice to utilize it by distilling it into rum at Tahir-

MANUFACTURE in Bengal. Jessor—Kallganj.

pur, where an old sugar * factory was converted into a rum distillery. The first attempts failed to produce any sufficient commercial return, and I do not know how the present attempt is prospering.

Sugar Trade a sign of Wealth.—"From what I have said it will be readily understood how great a source of wealth to the district lies in the sugar trade. The cultivation involves little labour, and it gives a productive return; and the manufacture also is such that many of the cultivating class can, and do, engage in it. I have roughly estimated the outturn of the district at about four lákhs of maunds, worth twenty-five or thirty lákhs of rupees; and I conclude from independent sources that this estimate is not far above the truth. In the Certificate Tax year, the sugar refiners were taxed upon an income of R3,24,000; and this excluded some of the largest firms (who were taxed in Calcutta), and all the small home refineries which fell under R500 profit. The whole trading profit distributed among the husbandman and professional trader amounts, I am pretty sure, to at least six or seven lakhs of rupees; and there is, throughout the sugar tract, an air of substantiality and comfort about the peasants and their homesteads, which testifies to the advantages they derive from engaging in sugar cultivation."

In **Sir W. W. Hunter's** Gazetteer of Jessor the above article is reprinted from the original, and the following brief paragraph added:—" Sugar is also manufactured by expressing the juice of the cane; but, as before stated, the manufacture is not carried on to a very large extent, in consequence of the greater expense. The process of manufacture is thus described in **Colonel J. E. Gastrell's** *Revenue Survey Report*:—The mill in common use ordinarily consists of two endless, coarse-threaded, wooden screws of about eight to ten inches diameter, set vertically in two horizontal cross pieces, and firmly fixed to two uprights, which are let well into the ground. These screws have their threads cut right and left, and play into each other. They are made of any hard, close-grained wood, tamarind being preferred. To the upper end of one of the screws, which projects above the horizontal bar, a long pole is attached, to which the bullocks that turn the mill are yoked. The cane is generally passed twice through the mill before being cast aside to dry for fuel. The expressed juice is received in a basin formed for the purpose below the screws. I was unable to procure any satisfactory returns of the expenses and profits of this cultivation" (*Statistical Account of Bengal, Vol. II, 285—98*).

Lohardaga. 412

LOHARDAGA.—*Pressing.*—"Sugar-cane is pressed in the Five Parganás by any of the following four kinds of machines:—(1) The *kalhu* or mortar and pestle—the same as is used by the Telís in pressing oil-seeds; (2) the *ráksi*, with two small horizontal rollers turning one over another; (3) the *choke ghání*, with two vertical wooden rollers turning one against another by means of a screw arrangement; this machine appears to me to be the prototype of the Beheea Mill; and (4) the Beheea mill. The use of the latter is as yet very limited, but is extending every year, and will no doubt in a few more years drive the old-fashioned native mills out of use. The *choke ghání* is in general use, and preferred to the first two native mills. The canes have to be passed twice through it to ensure thorough pressing.

Boiling of the juice to gúr.—"The furnace is made on some convenient piece of land near the ráyat's homestead. Its construction does not differ from that in Bengal. The top of the furnace contains four or more holes to accommodate the *báhánis* or boiling pans. These are oval-shaped earthen vessels of various sizes. A common-sized pan was found to measure 26 inches across and 15 inches deep. The pans hold from 30 to 50 seers of juice. During boiling, the scum that comes up to the surface is skimmed off at frequent intervals with a *jhánjri*, or perforated iron strainer. Beyond the removal of the scum, nothing else is done to purify the juice. A bamboo *birni* or brush, is placed in the liquid to prevent it overflowing.

" *Gur* is the only product of sugar-cane made in the Five Parganás. About five or six seers of cane-juice are calculated to yield one seer of *gur*. The actual proportion is variable. The juice of *bánsá* canes grown on *nágrá* soil is rich and yields as much as 25 per cent. of *gur*; while that of *punri* canes grown on alluvial soil is poor in quality and will not often yield more than half the amount the former gives.

" One *káhan* of sugar-cane land will yield at the best five maunds of *gur*. Taking the *káhan* as equivalent to two-thirds of a *bighá*, the produce will be at the outside about seven or eight maunds per *bighá*, valued at about R30.

Conf. with p. 257.

Suggestions for improvement.—" The use of the Beheea mill is on the increase; but the use of the large shallow evaporating basin is as yet unknown. The quality of *gur* turned out at present appears to me very inferior. There is much room for improvement in the direction of *gur* manufacture. As regards the cultivation of the canes, the manuring seems to me inadequate to a full yield of canes. Oil-cake, as a

* For other ruined sugar factories see remarks, p. 101.

MANUFACTURE in Bengal.

manure for sugar-cane and potatoes, is as yet unknown and may be usefully introduced. The *Bombay*, *Samsera* and other improved varieties may be tried, and unless they have been actually experimented with, it is difficult to pronounce upon their success or otherwise" (*Basu, Rept. on Lohardaga Dist., 80, 81*).

Palamau.

PALAMAU SUB-DIVISION.—"*Manufacture of concrete* (*gur*).—The cleaned canes are brought to the *kolsár* or gur-making yard. It is a small plot of ground close to the cane-field, and sheltered by trees, and has a small temporary thatch in one corner to serve as a store for the produce and shelter for the *gur*-makers.

"In one part of the *kolsár*, a temporary furnace or *chulhá* is made, and in another the mill is set up for grinding the canes. The furnace is of circular shape, 4 feet across, about 3 or 4 feet deep. It is enclosed by a mud wall, raised a cubit high above the ground. On one side of the *chulhá* is a large hole through which the fuel is fed; and on the opposite side is a longer gallery which leads away the smoke from the furnace, and thus serves as the flue. About half-way between the top and the bottom of the furnace is a shelf, made of twigs, plastered thickly over with mud; on this the fuel is placed while burning, and the ashes, as they accumulate, are pushed down by a long rake into the hollow of the *chulhá*.

"*Grinding.*—The only machine for pressing sugar-cane in Palámau is the Behea mill, which has completely driven the native *kalhu* out of use. The mill is driven either by a single bullock or by a pair. The manufacture of *gur* usually goes on by day and night. To keep one machine going for 24 hours, 8 to 12 country bullocks are required, each pair being relieved by another in regular rotation. One mill is estimated to press 32 *kundás* or earthen pots of juice in 24 hours, and the bullocks are changed at intervals of 2 *kundás* of juice.

"The shallow iron boiling pan is in use in Palámau Deep earthen pots for boiling the cane-juice, which are used in the Five Parganas, and in Bengal Proper, are quite unknown in Palamau, as they are in South Behar. The pan is 4 to 5 inches across and about 4 inches deep in the centre. It is capable of boiling 2 maunds of juice at one time. For a single furnace there are usually two boiling pans, particularly when *cháki gur* is to be made. For making this description of *gur*, it is necessary to stir it with a wooden rake for a length of time after the pan has been taken down from the *chulhá*: in the meantime the boiling of fresh juice may go on uninterruptedly in the second pan. When *ráb* is made, no such stirring is necessary, and the boiling may go on without any interruption, fresh juice being poured into the pan immediately after the preceding charge of *gur* has been ladled off.

"Cane-leaves and megass, that is the refuse of the canes after pressing, are the only fuel used in boiling the juice.

"The actual process of boiling the juice to *gur* is as follows:—The pan is set over the fire, and 4 *kundás* (roughly 2 maunds) of juice poured into the pan. The juice is seldom strained to remove chips of cane and other mechanical impurities. These are very probably retained on purpose in the juice in order to add to the weight of the outturn. After a short time the scum comes up and begins to accumulate on the sides of the pan. When sufficiently thick, it is skimmed off with an iron handle. The use of milk, lime-water, or any mucilaginous substance like castor seed emulsion, or the juice of *dheras* bark (**Hibiscus esculentus**) is either not understood or not known. The fact is that there is no demand for *gur* of superior quality and cleanliness, and the cultivator does not find it as yet worth his while to make first-class *gur* as he cannot get an adequate price for it: on the other hand, he is likely to suffer loss from the diminution of weight, which will no doubt result from the removal of the impurities. The boiling is continued for about 2 hours. Towards the end the syrup becomes thicker and apt to get burnt against the surface of the pan. A brisk stirring with a wooden rake is kept on all the time, until the *gur* has attained its proper consistency. The latter can be only guessed by the practised eye, and by feeling the thick syrup while yet hot between the fingers. When the *gur* is ready in the pan, the latter is taken down from the furnace and placed over a low earthen mound; the stirring with the rake is kept up until the *gur* gets thick enough on cooling. It is then ladled out into oblong forms of wood, about 16 inches long, 10 inches broad, and 4 inches thick, called *kátorás*, and allowed to gradually cool and solidify. In abount three hours' time the mass becomes quite hard, and is then taken out of the *kátorás* and becomes ready for the market. These square blocks of *gur* are known as *chákis*. These are very convenient for transport, as they can be readily packed in bags, and are not in the least subject to drainage of molasses like the *rab*.

"Sugar is not manufactured anywhere in or near Palámau, and consequently very little *ráb* is made. Its only use in Palámau is for making a cooling drink (*sharbet*) taken in hot weather. To make *ráb* the syrup is taken down from the pan in a

MANUFACTURE in Bengal.

slightly thinner condition than that of syrup intended to make *cháki gur*. The *ráb* syrup is poured into earthen vessels.

Palamau.

Outturn of gur.—"In Palámau 4 seers of juice are expected to yield 1 seer of *gur*. The juice must be very rich in saccharine matter to be able to yield this remarkably large proportion of sugar concrete. In Bengal the proportion of *gur* to juice seldom exceeds 1 to 5, and is very often as low as 1 to 7 or 8.

"One local bigha of cane is expected to yield from 25 to 30 local maunds of gur. This gives in standard measure 22½ to 27 maunds of *gur* to the acre. *Cháki gur* sells in Palámau on the average at 26 *kachchá* or 14½ *pákká seers* for the rupee. The outturn of one local bigha sold at this rate would be valued at from R48 to R57—

Cost of gur-making.

	R	a.	p.
Hire of machine and pan for 10 days at R0-8 *per diem*	5	0	0
One man to feed the machine for 10 days and 10 nights at R0-1-9	2	3	0
One man to drive bullocks for 10 days and 10 nights at R0-1-9	2	3	0
Five men to cut and clean canes for 10 days at R0-1-9 each *per diem*	5	7	6
One man to feed fuel for 10 days and 10 nights at R0-1-9	2	3	0
Total cost of gur-making .	17	0	6
Cost of cultivation	39	0	3
Cost of gur-making	17	0	6
TOTAL .	56	0	9
Average outturn of *gur* per local bigha = 28 local maunds, valued at 26 local seers per rupee	51	11	9

"It would thus appear that the cultivation of sugar-cane would not pay if every item of labour was charged for at the daily rates of wages. It is for this reason that no ráyat will attempt the cultivation of sugar-cance unless he has a sufficient number of hands in his own family to work in the field. To avoid the necessity of hiring labour, it is a common practice for several ráyats to combine and cultivate a field of sugar-cane either jointly or in separate parcels, all the men helping one another by turns" (*Basu, Rept. on Lohardaga Dist., Palamau Sub-Division, 39-41*).

Shahabad. 413

SHAHABAD.—In the brief notice above regarding SUGAR FACTORIES AND REFINERIES, it is mentioned that in this district there are 69 refineries. But throughout all the districts of the Patna division, sugar-cane cultivation is very important, and as Jessor and Faridpur may be spoken of as the great centres of date-palm sugar, Behar may be characterised as one of the most important tracts of Bengal from the sugar-cane stand-point. In the *Note on Sugar* (published by the Government of India) the outturn of Shahabad was estimated at 1,55,548 maunds; but Messrs. Thomson & Mylne give the area as 36,000 acres and the outturn 14,40,000 of maunds, representing a yield of 40 maunds of *gúr* an acre—

"The sugar industry is the most important of the district's manufactures, and the only prosperous one. It is carried on extensively in the Buxar and Sasseram sub-divisions, and elsewhere on a comparatively small scale. The outturn of the Buxar sub-division is estimated at 35,850 maunds against 23,880 maunds in the previous year, while in the Sasseram sub-division the quantity of sugar manufactured was 44,908 maunds. The principal feature of the year was the introduction of hand turbines at Nasirgunge, the seat of the industry in Sasseram, and elsewhere, which resulted in the increased manufacture of sugar. About seven-eighths of the total quantity of sugar manufactured in the district is exported to Cawnpore, Agra, and other places in the North-Western Provinces and to Bombay" (*Note on Sugar*).

Although Shahabad, Gya, Durbhunga, Champarun, and Sarun are most important districts in the production of cane and *gúr* and even Native refined sugars, the methods of manufacture do not differ materially from those detailed regarding Bogra, Jessor, etc. Babu Addonath Banerji,

MANUFACTURE in Bengal.

Shahabad.

while discussing the sugar trade of Behar, makes the following remarks regarding Shahabad:—

"In the year 1884-85 the sugar-cane crop was good in four divisions,* but not in the Patna and Dacca divisions. In the Durbhunga and Gya districts there was a partial failure of the crop, and in Furreedpore, another important sugar-growing district in Eastern Bengal, the crop was destroyed by wild pigs. After excluding these two years, it will be seen that, on the whole, the exports have not suffered, although the proportion of the *refined* to the *raw* material has leaned towards the latter. How far the sugar refineries in the interior are responsible for this state of things cannot be stated with any degree of confidence. Certain it is, however, that there has as yet been no deterioration in the prosperity of the refineries in Shahabad, where this industy is the most important of the district manufactures. The area of sugar-cane in Shahabad, irrigated in 1886-87, was 26, 16 acres. To what extent the facility of irrigation from canals has contributed to the extension of the cane cultivation is proved from the figures in respect of the export sugar trade of Shahabad. In a statement regarding the sugar trade of this district, recently compiled in the Statistical Department for the use of the late Sone Canal Commission, it was shown that 1,51,090 maunds of *refined* and 39,788 maunds of *unrefined* sugar were carried by rail during the calendar year 1876, while in the calendar year 1886, the exports of *refined* sugar amounted to 2,64,832 maunds, and of *unrefined* sugar to 3,52,062 maunds. The total trade in 1886 aggregated 6,16,894 maunds, which was more than three times the figures (1,90,878 maunds) of 1876. If expressed in *unrefined* sugar, the increase will be as follows:—

* Presidency. | Bhagulpore.
Rajshahye. | Chota Nagpore.

Year.	Exports from Shahabad by Rail (External and Internal).		
	Refined sugar.	Unrefined sugar.	Total in unrefined sugar.
	Mds.	Mds.	Mds.
1876	1,51,090	39,788	4,17,513
1886	2,64,832	3,52,062	10,14,142"

II.—ASSAM.

MANUFACTURE in Assam.
414

No sugar is manufactured in Assam. The province imports its supplies by river steamers from Bengal. The reader will find, under the chapter on cultivation, a detailed report of the crushing of the cane and the boiling of the juice into *gúr* as pursued in the province (pages 150-159). In the chapter on the History of the Effort to establish Sugar-planting as a European industry in India reference has been made to the experiments formerly undertaken to organize sugar plantations and factories, so that it does not seem desirable to say more in this place, except perhaps to add that, although land suitable for cane-culture doubtless exists in the province, there are many adverse influences, such as expensive labour, which preclude the possibility of Assam, for many years to come, at least from becoming a great sugar-producing country. It seems likely, however, that an inquiry (as already suggested) into the nature of the better qualities of cane found in the province might be productive of good results.

III.—NORTH-WEST PROVINCES AND OUDH.

MANUFACTURE in the N.-W. P. & Oudh.
415

So much has already been said regarding these provinces that it seems scarcely necessary to do more than to furnish three passages descriptive of the manufacture. The first passage, given below, completes **Messrs. Duthie & Fuller's** account of sugar in these provinces. It may be accepted as a review of the various methods which are pursued in the provinces as a whole. The second details the manufacture of sugar in Gorakhpur, a district which, in the early effort to extend the sugar trade of India, figured prominently. And the third supplies full information

MANUFACTURE in the N.-W. P. & Oudh.

regarding Shahjahanpur, the district which may be accepted as the centre of the present trade. That district not only possesses a large European refinery and distillery—the Rosa works—but is so important a locality for Native-made sugars that the prices determined annually at Baragaon, a village fourteen miles from Shahjahanpur, may be said to govern the sales of these provinces. The full extract from Mr. Butt's able report on the Shahjahanpur sugar will, it is hoped, meet the difficulties which most students of this subject experience. As his report is not very accessible, its reproduction seemed dsirable:—

I. "The boiling of the juice follows on the pressing with as little delay as possible, since fermentation rapidly sets in from exposure to the air. The process of boiling and concentration varies according as its result is to be *gúrh, shakar* or *ráb*. *Gûrh* is a compost of sugar crystals and uncrystallized syrup boiled till of a sufficient consistency to be made up into soft balls or cakes (*bheli* or *chakki*). *Shakar* is formed when the boiling is a little more prolonged and the mixture of crystals and syrup is violently stirred while cooling, when its colour becomes lighter and it crumbles into small pieces. In *ráb*-making the boiling is not so prolonged, and the result is syrup containing masses of crystallized sugar imbedded in it. *Gûrh* and *shakar* are for human consumption as they are, but *ráb* only represents the first stage in the manufacture of crystallized sugar. With *gûrh* and *shakar* the object is more to obtain a good colour than good crystallization, while the value of *ráb* entirely depends on the proportion of crystals which it contains. Hence the boiling process for *gurh* and *shakar* is, as a rule, much rougher than when *ráb* is manufactured. The boiling apparatus consists of a furnace excavated in the ground, over which one or more iron pans are set. If the boiler is supplied from only a single *kolhu*, as a rule one pan is used, while if two or more *kolhus* are used the number of pans is often increased to four or five, which are of different sizes and are placed in order, the largest one furthest from the feed end of the furnace, and the smallest one immediately over it. In this form the boiling apparatus is very similar to that formerly used in the West Indies. The use of a row of pans on this principle effects a great saving of time, and also perhaps enables the manufacture of better sugar, though this is by no means proved. The juice is collected in the large pan, where it is allowed to simmer slowly, scum rises to the surface, the formation of which is sometimes assisted by the addition of alkali (carbonate of soda) which promotes the coagulation of albuminous matter, or of milk, or the sticky juice of the edible Hibiscus, which, in becoming coagulated, collects and brings to the surface a good deal of impurity. From the large pan the juice is baled into the one next it, and so on from pan to pan down the series, becoming more concentrated in each transfer until it is finally worked up into sugar in the last and hottest pan. The preparation of sugar from *ráb* is not, properly speaking, an agricultural process, and needs therefore no notice in this account. It may be briefly mentioned that the process substantially consists in draining the uncrystallized molasses away from the sugar crystals. This is effected in the western districts by pouring the *ráb* into cloth bags and subjecting it to pressure, in which way about half of the molasses are strained off, and then placing the semi-pure result (called *putri* in the western, and *shakar* or *assára* in the eastern districts) in wicker crates, and allowing the molasses to filter slowly down, this filtration being assisted by a covering of the water weed known as *siwár* (**Hydrilla verticillata**), the moisture from which slowly filters downwards and washes the crystals clean. The European process of 'claying' was on exactly the same principle. The floury whitish sugar which results is known as *kacha chini* or *khand*, and is made over to the *halwais* for

Conf. with p. 295.

MANUFACTURE in the N.-W. P. & Oudh.

final refining. The following statement shows the average outturn per cent. of cane of each of the products mentioned above:—

"One hundred of cane yields 50 of juice, which latter may be divided into 18·0 of *gur, bheli* or compost; 17·5 of *shakar*; and 19·5 of *rab*. The *ráb* contains 13·0 *putri* or *assara* (semi-drained), and 6·5 *shira* or molasses. The *putri* may be also divided into two equal parts, *viz.* 6·5 *chini, hhand* or *shakar*; and 6·5 *shira* or molasses.

"Of the sugar exported from the Meerut division, 98 per cent. is in the form of *gurh* or *shakar*, but only 44 per cent. of that exported from Rohilkhand, the balance (56 per cent.), consisting in *chiní* or *khand*, the product of *ráb*. This difference illustrates something more important than a dissimilarity in local custom or even in equality of cane, for it represents a material difference in the distribution of the profits of sugar cultivation between cultivator, landlord, and capitalist. When a cultivator manufactures his own sugar he nearly always makes *gurh* or *shakar*, and *ráb* is, as a rule, only made by professional sugar-boilers or *khansaris*, with juice which they purchase from the cultivators. These purchases are all negotiated, like those of indigo factories and the Opium Department, by means of advances, and the system has an important bearing on the agricultural condition of a large portion of the provinces.

"In the sugar districts of the Meerut division, on the other hand, the rule is for the cultivator to boil his own cane juice, and add the profits of manufacture to those of cultivation. It is generally assumed that the cultivating classes of these districts are the most prosperous in the provinces, though their prosperity may be, perhaps, bought by a loss in the total value of the produce." (*Duthie and Fuller, Field and Garden Crops, North-West Provinces*)."

Gorakhpur. 416

II. GORAKHPUR.—" The manufactures of the district are few, and the only one of any great importance at present is that of sugar-boiling, extensively practised in the Hata, Padrauna, and neighbouring parts of the Deoria and Sadr tahsils. It is difficult to obtain any very accurate statistics of the number of sugar factories, but the following figures were furnished a few years ago by the tahsildars:—

PERGUNNAHS.	Number of factories.	Remarks.
Silhat . . .	28	Of which 5 are Melia village of tappa Indarpur.
Shahjahanpur . .	73	Of which 37 are said to be in tappa Patna, most of them being in Rampur Khanpur village, not far from Deoria.
Haveli . . .	37	Almost all in the tappas lying north-west and north of Silhat.
Salempur . .	65	Of which half are said to be in Barhaj.
Sidhua Jobna . .	...	The exact number is not stated, but is undoubtedly very large. Mr. Lumsden estimated that, in addition to the amount locally consumed, over 20,000 maunds of *chini* (sugar) were yearly exported from this pergunnah. Mr. Alexander thinks that the number cannot be far short of 100, as this is the pergunnah in which the cane seems to thrive best. Mr. Lumsden numbers 52 factories in his settlement report, but the number has since increased.

"The factory-owner does not, as a rule, cultivate his own sugar-cane. He makes money advances to a number of neighbouring villages, who grow the crop and usually also extract the juice (*ras*) in their own or hired mills. The *kolhu*, or sugar-mill, has already been described as 'a large drum-shaped mortar, in which an almost upright timber beam or pestle is made to turn by an arrangement attaching it to a pair

MANUFACTURE in the N.-W. P. & Oudh.

Gorakhpur.

of revolving bullocks.' The pestle is here called *játh*. The horizontal cross-beam which connects it with the bullocks is named *kátar*, and on the latter sits a man, partly to guide the bullocks, partly to give greater weight to the *jath*. Another man feeds the *kolhu* and pushes the cane against the *játh*. When seen for the first time this operation seems likely to end in crushing the hand of the operator, but accidents very rarely occur. The expressed juice trickles into a lower compartment of the mill, called *ghágu ;* and hence flows through a wooden spout or *parnáli* into the vessels set to catch it. In Gorakhpur, owing to the difficulty of obtaining stone, the *kolhus* are all of wood. When extracted, the juice is generally boiled at once in large iron vessels called *karáhi*, which are usually lent by the owner of the factory to which the boiled syrup (*gúr* or *ráb*) is to go, but are sometimes owned or hired by the cultivators. Occasionally, if the factory be very close, the juice is taken there at once. It makes of course a great difference to the cultivators whether he manufactures independently or on behalf of the factory-owner. The latter takes an ample return for the advances he makes and for the hire of the *karáhi*. But very few villagers grow cane altogether without advances, and one manufacturer informed Mr. Alexander that he did not care to deal with such persons. He had not, he explained, the same hold over them as over cultivators who had bound themselves, by taking his advances, to grow a certain amount of cane. In a year, however, when cane is at all scarce, an independent cultivator could command a very high price for his *gur* and obtain large profits. The clients of the factory who receive payment at a rate fixed behorehand, derive no additional profit from high prices. But where most of the cultivators must work on borrowed capital, this system of advances is perhaps the best way of supplying a useful want.

Conf. with p. 93.

"After its receipt at the factory the *ráb* syrup is again boiled twice and cleared of its scum. It is then allowed to harden and becomes *chini*, which finds a very large export towards the south. The sugar is sometimes refined by additional boiling and skimming, but is more often sent away in the rough state, packed in large earthen jars.

"No trustworthy statistics are available to show the average amount of *khand*, or dry sugar, produced yearly in a factory. But some establishments visited by Mr. Alexander at Pipraich confessedly turned out from 400 to 500 maunds of refined sugar (*chini*) each in a season. The average value was about R12 or R15 a maund, and as the cultivators get for their *ráb* about R3 to R4 only, the factories must make considerable profits. But they have usually, it must be remembered, to carry the *chini* some way before they can command a market.

"The principal places where the *khand* is collected for exportation are Captainganj, Pipraich, Gorakhpur, Sahibganj (in Sidhua Jobna), and Barhaj. From Captainganj a little is said to go up to Nepal, but by far the greater part of the trade finds its way by Gorakhpur, the Rapti or the little Gandak and Barhaj, to the Ghagra. A considerable amount also descends the Great Gandak to Calcutta. The Little Gandak is, as before mentioned, navigable only during the rainy season, but a large trade from along its banks travels by the Padruana and Barhaj road to the latter place" (*Gas., N.-W.P., VI, 411-14.*)

Shahjehanpur.
417

III. Shahjehanpur.—The following note on sugar-manufacture has been supplied by Mr. D. C. Baillie :—"The Native process was briefly described in the Budaun notice, but it may be interesting here to note the differences between the Native process and the European, as practised in Messrs. Carew & Co.'s works at Rosa. Messrs. Carew & Co., like the Native manufacturers of the district, work upon *ráb*, that is cane-juice boiled to such a viscidity that it crystallizes on being allowed to cool. The first operation in both the European and the Native process is the same : the *ráb* is tied up in the coarse cotton bags and subjected to pressure, in order to drain away the treacle from the pure sugar crystals. The treacle so drained away is in Rosa re-boiled, so as to make a lower quality of sugar ; by the Native sugar manufacturers it is made into an inferior quality of *gúr* and exported. The crystals left after the treacle has been drained away are termed *putri*. It is the raw sugar on which the English refiner works. It consists of grains of nearly pure sugar, coated on their surface with dark syrup, and generally contains some impurities, such as sand, vegetable fibre, and, in India, dried cow-dung. The last-named substance is usually employed as a cover for the vessel in which the *ráb* is kept.

MANUFACTURE in the N.-W. P. & Oudh.

Shahjehanpur

"In the English process the raw sugar is dissolved in hot water in certain proportions. The solution so formed is first filtered through cotton bags in order to remove the solid impurities above referred to, and then several times through a deep bed of charcoal, to remove colour and such impurities as escape the bag-filters. The decolourized liquid is concentrated by boiling off its water in a vacuum-pan till crystals have formed in proper quantity. Finally, in order to separate these crystals from the adhering 'mother liquor,' they are placed in the centrifugal machine. This consists essentially of a vertical metal drum, the curved walls of which are perforated by a great number of small holes, and which revolves with great speed round its axis. The centrifugal force produced by this revolution forces out the syrup through the pores of the drum, leaving the prepared sugar in the drum. The 'class' of the sugar depends on several matters:—(1) whether it is made entirely from *putri* or whether it contains a certain proportion of the crystals deposited after treatment (by the treacle being at first drained away); (2) on the number of times it has been passed through the charcoal beds; (3) on the amount of spinning it has undergone in the centrifugal machine.

"In the Native process the *putri* is not melted, and, consequently, impurities are not removed from it. The stages are two only. The treacle left adherent to the crystals in the *putri* is allowed to drain itself away under the force of gravitation. The *putri* is for this purpose placed in a large tank, the bottom of which is formed by a cloth placed over a bamboo frame and kept there for several weeks. The draining away of the treacle is aided by a partial fermentation which the sugar undergoes during this process In Shahjehanpur a layer of river weed (*siwár*) is laid over the top of the sugar, partly to aid fermentation (*sic*), partly because the moisture from the weed, slowly filtering through the sugar, aids the draining away of the treacle. The sugar, after having undergone this process. is technically termed *pachani*. This *pachani* is placed on a platform in the sun, and thoroughly trodden out by the feet. The product is *shakar*, or Native sugar ready for market. It is in colour rather whiter than the lowest quality of sugar turned out from the Rosa factory. Its crystals are much smaller: the great difference, however, is the presence in it of a large quantity of impurities, to which every stage of the process of manufacture from the expression of the juice to the final treading out has contributed its share, and towards the removal of which nothing has been done. The lower qualities of Rosa sugar, owing to the superior economy of the European process, and in spite of the expensive machinery and superintendence, can be sold cheaper than Native sugar is. It does not, however, in spite of its obvious advantages, make much progress amongst Native consumers. To Hindus the employment of animal charcoal during the process is a great stumbling-block, and has led to Rosa sugar being in the Panjab formerly cursed with bell and book" (*Gaz., N.-W. P., IX, 128-30*).

The account of the sugar manufactures of Shahjahanpur drawn up by Mr. Butt is so very instructive that it is difficult to abridge it and yet preserve its merit. Following after the passage just quoted, however, it is perhaps desirable to extract only such paragraphs as appear to amplify Mr. Baillie's statement. A *matha*, Mr. Butt explains, is an earthen vessel which holds on the average about 26 gallons of juice. To express a *matha* of juice may occupy from three to six hours—

"When the mill is worked by day only, seldom more than two *mathas* are pressed: one will be done before noon, when the men dine and rest, and press another in the afternoon. Working day and night the labourers are changed each *matha*, and

MANUFACTURE in the N.-W. P. & Oudh.

Shahjehanpur

generally four *mathas* are pressed in the day, but often five and occasionally even six *mathas* are pressed and boiled each day."

Preparation of Rab.—"Boiling takes about the same time as filling a *matha* of juice, and work keeps on evenly, one *matha* being boiled while the next is being filled with soft cane; the *matha* may be fully half an hour before the boiling pan is empty, and with very hard cane the boiling may be over some time before the *matha* is ready. At first starting the mill, some wood must be used as fuel, but as soon as the cane refuse (*khoe*) dries, the refuse from the mill and cane leaves (*patáhi*) from the field supply sufficient fuel. A fall of rain again necessitates the use of wood for a day or two; or if, as is often the case, no wood has been kept ready, causes a cessation of work, the fuel is supplied from outside through a hole in the end wall of the boiling shed. When the juice has been boiled nearly the usual time, the boiler now and then ladles up a little and judges the state of concentration by the appearance of the drippings as he pours the sample back; this eye test is probably universal, being that employed by the vacuum-pan boiler, as well as by the Indian cultivator.

"During boiling the scum, rising to the surface, is skimmed off by pushing a small board along the surface; the scum adheres to this and is scraped off with a potsherd. On being judged ready, the juice boiled for *ráb* is ladled from the boiling-pan into a vessel containing about 5 seers and used as a measure, and then poured into a large porous earthen jar (*kalsi*) containing about 3 *ráb* maunds, one boiling on another till the *kalsi* is full. It then cools, and is ready for removal to the sugar manufactory (*khandsár*). The cultivator almost always cultivates on advances from the manufacturer (*khandsári*), and the *ráb* is taken delivery of on the spot, and removed at the manufacturer's expense.

"*Ráb* is the product almost always made, as it is the product required by the manufacturer, and the cultivators are restricted to its manufacture.

Preparation of Gúr.—"When *gúr* (or *mitház*) is the product required, the juice is boiled somewhat longer and with greater heat, so that, on cooling, it can be made into hard balls; on removal from the boiling-pan the boiled syrup is poured into shallow pans, and there pounded and made into the *bhelis* (or round balls) of two to five sers each. The quality is injured by overboiling, but *gur* is a product at once saleable in the bazar, and can be stored or exported without injury. The manufacturers do not use *gur*, and it is not further manufactured. In order to make sugar from *gur*, it would first be necessary to boil it with water, and so bring it to a form resembling *ráb*; and elsewhere I believe sugar is made from *gur*, but never in Shahjahanpur. *Gúr* is exported and sold for direct consumption or for use by confectioners and tobacconists. Cultivators who cultivate without advances, and zemindars who are not manufacturers, commonly make *gur*, but generally *ráb* only is openly manufactured. As the cultivators are not under strict supervision, they, though under engagement to manufacture *rab* only, not uncommonly make one or two *mathas* into *gur* and sell it secretly. To prevent this, the money-lender often employs a servant to watch the mills in the village, but the supervision is difficult with scattered mills working day and night, and the watcher is often kept quiet by a small present. *Gur* is secretly removed to a friend's house and disposed of through him. Tilhar is the chief *gur* market in the district, and much *gur* is made by the Kurmis in the Tilhar Tahsil. The *gur* made in Tilhar is supposed to be the best, though Tilhar *ráb* is generally considered inferior to that made in either Pawáyan or Sháhjahanpur.

"It will be seen that the form of the product, whether *ráb* or *gur*, depends on the boiling. When *ráb* is the product required, the cane-juice is concentrated to a little over the crystallizing point, and consequently still retains much moisture. It is the product suited for the hotter fire, until, on cooling, it can be made into dry solid balls (*bhelís*). Here the excessive heat and burning destroys much more of the sugar present in the juice—*i.e.* renders it non-crystallizable.

Mistakes in Terms.—"In naming these products, concentrated cane-juice in the form of *ráb* or *gur* the most strange mistakes are usually made. 'Sugar,' 'molasses,' 'coarse molasses,' and 'treacle' are terms commonly used as equivalents for *ráb* or *gur*. Concentrated juice cannot correctly be called sugar, and denoting it as molasses or treacle is a gross mistake. Molasses (*shira*) is the syrup which drains from the *ráb* in the subsequent process at the manufacturer's, the remainder being raw sugar (*pútri*), from which again is obtained dry sugar (*khánd*). *Ráb* might perhaps be translated as undrained raw sugar. Treacle is the syrup that drains from refined sugar. 'Sugar' is a most indefinite term, as it may mean raw, dry or refined sugar. These errors in nomenclature lead to many mistakes. Thus, by the *Indian Economist*, in a comparison between East Indian and West Indian produce, a product presumably *gur* was taken to be the same form as West Indian raw sugar. The fact that from equal quantities of juice, produce, as raw sugar, will be less than half the weight of produce as

Conf. with pp. 114-15, 130-31, 252-55, 298.

MANUFACTURE in the N.-W. P. & Oudh.

Shahjehanpur

gur makes the comparison instituted of no practical value, and the difference in outturn was greatly under-rated.

"About the same time, a gentleman who had published in the *Economist* a series of estimates of agricultural profits in the North-West, wrote to the *Indian Statesman* in explanation of his tables:—' Up here the cultivator presses his own cane in rough wooden mills, boils down the juice and sells the coarse molasses to the dealer generally, as numerous enquiries have led me to believe, at R4 the maund. I particularly entered "sugar,"– that is, *gur*, not *ráb* or treacle, as I had not found it the custom to make the latter except from the refuse stalks, such treacle being obtained by pouring water over the already pressed cane, and the poor syrup being given to the children or the girl tenders at the boiler.' Here almost every term is a mistake, though the writer was a Settlement Officer who had paid some attention to the subject. In Madras and Bengal *jagari* is the term commonly employed for the first state of concentrated juice."

The bel system.—"In the greater part of the district the cultivators manufacture *ráb* and then deliver the product to the *khandsári*; but to the west, all along the Bareilly border, the custom is spreading of the manufacturers taking the fresh juice from the cultivators and themselves manufacturing the *ráb*. This custom has only been introduced from Bareilly since the Mutiny, but it is spreading fast, and will, in all probability, completely supplant the older system. In Bareilly, too, this custom is, I believe, of recent introduction. The change is not due to any change in the position of the cultivator, and under either system cultivation is carried on by advances, and the cultivator is bound to deliver his produce to the money-lending manufacturer. When the cultivator delivers *ráb*, he puts up his mill where he likes, usually close to his own house, and commonly the mill is worked day and night. When the manufacturer takes the juice, advances are made to a large number of tenants in the same or closely adjoining villages, and all these men put up their mills in some one place a little distance outside the village, where the manufacturer puts up his boiling-sheds. Commonly twelve or twenty mills may be seen working at one '*bel*,' and sometimes as many as thirty. Each *matha* of juice as filled is taken over by the manufacturer.

"The expenses of the cultivator are very little less than when he manufactures *ráb*. The manufacturer receives the cane refuse for fuel, and only the cost of one labourer (the boiler) and the hire of the boiling-pan is saved to the cultivator. The manufacturer has to build a boiling-house and to employ a writer and several servants to watch the mills, and also an establishment for boiling the juice. The mills are never worked at night, but even so, in the early morning, or when the overseer's attention is elsewhere attracted, the cultivator sometimes manages to pour some water into the juice. In the boiling-sheds the boiling-pans are put up in sets of five, the number of sets varying according to the number of mills, being generally one set per ten or twelve mills; the five pans are in a line, one directly over the fire, the others at intervals over apertures in the horizontal flue leading from the fire. The juice is first placed in the pan farthest from the fire, which is of a very large size, containing some 50 maunds of juice, and is called the *hauz* (or reservoir); the next pan, about half the size of the *hauz*, is called the *nikár*. In this some alkaline substance is added, generally *sajji* (impure carbonate of soda), but sometimes decoctions from the bark of various trees or plants are used. The next pan is the *phula*, and the juice is here heated nearly to the boiling-point. The fourth pan is the *phatka*. In this should the juice, now very much thickened, appear too viscid, some castor or mustard oil* is sprinkled as a corrective. The syrup is then moved to the fifth and last pan (the *cháshini*) directly over the fire, and a short boiling brings it to the proper *ráb* consistency. A *halwai* (confectioner) is always employed to conduct the boiling, and most of these *halwais* are men who come from the Mainpuri district each season. Whatever may be the scientific value of the substances added, there is here evidently a more careful process, and the *ráb* thus prepared on a larger scale must be of superior and more uniform quality, as the *ráb* being made by the sugar-manufacturer himself, there remains no motive for fraud or deception as to the quality.

Quality of the Juice.—"Shahjahanpur cane juice, when freshly extracted, seems equal in richness to that of most other cane-growing countries. Density, as shown by Baumé's saccharometer, is the test used; and our juice commonly shows a density above the average quotations of other countries. At the Rosa Distillery fresh juice, purchased from cultivators near and filtered, has often shown a density of 10°, and occasionally of 11°. Supposing a pure solution, the percentage of sugar present shown by each density is as follows:—

6° Baumé = 10·4 per cent. of sugar.
7° ditto = 12·4 ditto.

* Sesamum oil, it will be seen, is used in Madras, p. 234.—*Ed., Dict. Econ. Prod.*

MANUFACTURE in the N.-W. P. & Oudh.

Shahjehanpur.

8° Baumé	= 14·4 per cent. of sugar.		
9° ditto	= 16·3	ditto.	
10° ditto	= 18·2	ditto.	
11° ditto	= 20·1	ditto.	
12° ditto	= 22	ditto.	

Our fresh juice may be assumed to be of good quality, but destruction of the sugar commences at once. To prevent deterioration, cane-juice should be treated with the utmost rapidity and the most scrupulous cleanliness. Juice extracted at 9½ Baumé and allowed to stand an hour or two will only mark 8½ to 9°.

"The mill and receiving dish (*matha*) are seldom cleaned, and never thoroughly so; there is no appreciation of the loss and deterioration resulting from delay alone; the juice is insufficiently strained, and, as a rule, the juice is not treated with any alkali, to neutralize the natural or acquired acidity. The *sajji matti* (impure carbonate of soda) used in the *bel* system is but slightly alkaline: that used is always dirty, and in the after-processes there are no means of extracting the addition. The juice must then have undergone very great deterioration before reaching the boiling pan, and this acid and impure juice is then subjected for hours to the direct action of a strong fire.'

Mr. Butt estimates the total cost:—

	R. a. p.
Cultivation	43 7 0
Manufacture	31 0 0
TOTAL	74 7 0

Total cost to cultivator.—"This estimate has been drawn out on the principle usually followed if supposing all labour, even that of the bullocks, to be hired and good cultivation has been charged for. Of course, no such disbursement is ever made by any farmer. About R28 of the cost entered is payment for hired bullocks; a tenant, with the ordinary number of bullocks, need never hire cattle, as most of the labour is done by several tenants joining and working together till the work of all is done. A zamindár will have bullocks enough himself, and in either case the cost is the rateable share of the annual keep and original cost of the cattle—an amount, much under the cost of hired cattle. In the case of the cultivator very little labour is hired, and he commonly resorts to less careful cultivation in preference to paying for hired labour. A zamindár gets a good deal of labour at a very small cost, and his regular labourers receive a large part of their wages in kind. The cost of cultivation can be estimated without reference to the amount of produce, but that of manufacture depends on the produce, and may be reduced in proportion to reduction in the estimate of produce."

Produce in cane-juice.—"The produce is always estimated by the number of *mathas* of juice or *kalsis* of *ráb* produced per kucha bígha; generally by the amount of *ráb*; one, two, or three *kalsis*, as the case may be, estimating in juice the highest native estimate of produce, a rate often given as the maximum comes to exactly 100 *mathas* the acre. It is very generally considered that a good crop will give the series of thirteen *mathas* a bigha, equal to 81¼ *mathas*, or about 2,112 gallons, of juice per acre. Ten *mathas* is looked on as a fair average crop, and I think twelve *mathas* may be taken as a full outturn for good land in a cane tract. This gives 25 *mathas*, or 1,950 gallons of juice per acre, equal to about 21,200℔ of juice. The average West Indian produce per acre is generally estimated at from 30 to 35 tons of cane ready dressed for the mill. Taking 30 tons per acre, the produce in juice at 50 per cent. (Sháhjahánpur average) comes to 33,600℔; but at 70 per cent. (proportion extracted by good mills) amounts to 47,040℔. Taking what is there considered the low yield of 25 tons per acre, the juice at 50 per cent. should be 28,000℔, and at 70 per cent. 39,200℔. The actual results on a Queensland plantation in 1868, published in the first number of '*The Sugar-cane*,' give an average per acre of 4,170 gallons, or about 45,350℔ of juice, and one field gave an average of nearly 6,300 gallons per acre. The juice extracted appeared to be about 65 per cent.

MANUFACTURE in the N.-W P. & Oudh.

Shahjahanpur.

of the weight of cane and the density on this plantation as high as 10 Baumé, though on others ranging from 7½ to 9½ Baumé.

Experiments at the Rosa Factory.—"For some years the produce from several fields has been recorded at the Rosa Distillery. For the manufacture of cane-juice rum fresh juice is required, and the cane is pressed in the factory in Native mills. In 1870-71 the cane from three fields was pressed. One pressed before the cane was ripe gave an average of only 56¼ *mathas*, or 1,462 gallons per acre; the second gave an average of 75 *mathas*, or 1,950 gallons, and the third almost the same. In 1871-72 the average was 81¼ *mathas*, or 2,112 gallons. In 1872-73 some fields were cultivated by the factory, and three acres one rood were under plant canes, and five acres under ratoons. The ratoons gave an average of only 47½ *mathas*, or 1,225 gallons per acre, while the average for the plant canes was 94½ *mathas*, or 2,457 gallons per acre; this is equal to about 26,720lb of juice per acre. This actual result approaches that given commonly as the very maximum (100 *mathas*=2,600 gallons, or 28,275lb), and goes to show that with better cultivation and more manure our cane might equal that of other countries.

"Beside these fields cultivated by the factory in 1872-73 several standing fields close to the factory were purchased, but these fields were only sold by the owners on the cane commencing to dry and promising a very poor result; and **Mr. McAlister** considers that the produce should be viewed as a minimum. The aggregate area was nearly 18 acres, and the average outturn only 48½ *mathas*, or 1,261 gallons per acre. These crops were purchased at an average price of R44-8-0 per acre. The best of these fields (area little over one acre) gave an average of 83 *mathas*, or 2,158 gallons per acre. The canes were cut and carried by **Carew & Co.**, who also provided and kept in repair the mills (the ordinary native *kollu*); a native contractor provided cattle and labourers for pressing, and was paid in 1871-72 at the rate of 6½, and in 1872-73 at the rate of 7½ annas per *matha*. The cost of pressing here came to R30 for labour and cattle alone—a result somewhat above the cost assumed above. The contractor used eight bullocks with each mill, and five *mathas* were generally pressed in the twenty-four hours, the mills working day and night, and occassionally six. These Rosa experiments are the only practical experiments known, and they tend to show that the estimate of produce framed before these actual results had been received is not an extravagant one. In these experiments the juice extracted was always about 50 per cent. on the weight of dressed cane—an amount at least equal to that generally obtained by the native cultivator."

Produce in ráb.—"Estimating the produce in juice, the shares paid to the labourers, and customary dues to the revenue or village servants, are included, as well as any juce consumed by the cultivator or friends; but estimates of the produce in *ráb* include only the net amount handed over to the manufacturer, and hence estimates of produce appear lower when made in *ráb* than when made in juice. I have not any actual experiments as to the weight of *ráb* given by a *matha* of juice, but I have made constant enquiries from cultivators and manufacturers, and the received opinion is that, on the average, a *matha* should give rather over 20 sers *ráb*. Estimates range from 18 to 25 sers, and in assuming 20 as the average result, I am rather under than over the mark. Twenty *ráb* sers equal to 60·7lb, and the *matha* weighing about 283lb of juice, the *ráb* is about 21·5 per cent. of the weight of juice. I cannot compare this proportion with that obtained in other countries, as none of the returns I have seen give the weight of produce in any form corresponding to *ráb*. The produce in *ráb* then becomes 37½ *ráb* maunds per acre, equal to over 55 Government maunds, or to about 4,550lb per acre. This estimate is one of two *kalsís* of *ráb* per bígha, and so put would appear higher than an estimate of 12 *mathas*. Cultivators have spoken of an actual return of three *kalsís* of *ráb* per bigha, but two and a half is generally looked on as nearly the maximum; and many land-owners and manufacturers have wished me to believe in an average outturn of one *kalsí* per bígha, a rate of produce 20 per cent. below that looked on by **Mr. McAlister** as a minimum outturn."

"*Value of Produce.*—I assume an average price of R4 per *ráb* maund; the reasons for doing so will be explained when an account is given of the Baragaon *khataunti*. Taking the price, the value of the produce comes to R150 per acre, and deducting the expenses, amounting to R74-7, the profits remain R75-9 per acre. Remembering that the expense of manufacture varies directly with the amount of produce, the account for each rate of produce from one *kalsi* (three maunds) to three *kalsís* (nine maunds) per bígha becomes as follows; the expense of cultivation is

MANUFACTURE in the N.-W. P. & Oudh.

Shahjehanpur

supposed to remain constant, but, in fact, good cultivation is only practised where there is a hope of good produce:—

Rate of produce in *ráb* maunds per bigha.	Produce per Acre.			Value of produce in *ráb*.	Expense of cultivation and manufacture of *ráb* per acre.	Profits per acre.
	Juice.	In *ráb*. Government weight.				
	lb	Mds.	lb	R a. p.	R a. p.	R a. p.
3	10,600	27·7	2,277	75 0 0	58 15 0	16 1 0
4	14,134	36·9	3,036	100 0 0	64 1 8	35 14 4
5	17,668	46·1	3,795	125 0 0	69 4 4	55 11 8
6	21,202	55·3	4,554	150 0 0	74 7 0	75 9 0
7	24,736	64·5	5,313	175 0 0	79 9 8	95 6 4
8	28,270	73·8	6,072	200 0 0	84 12 4	115 3 8
9	31,804	83	6,831	225 0 0	98 15 0	135 1 0

418

"*Produce in gur.*—Supposing the cultivator to dispose of his produce in the form of *gur*, the expense of cultivation and manufacture remains practically unchanged. The produce in *gur* weighs less than in *ráb*, the extra boiling extracting more of the moisture; but the difference is generally estimated at only 10 per cent., produce in *gur* being to produce in *ráb* as 9 to 10, and the *gur*, consequently, 19·3 per cent. of the weight of juice: at this rate the produce in *gur* becomes nearly 50 imperial maunds per acre. *Gur* is sold by free competition in open market, and a fair valuation of the produce is less difficult than in the case of *ráb*. Using the price lists published in the *Government Gazette*, I take R3-8 per imperial maund as a fair average rate at the season when *gur* is cheapest; the price rises rapidly, but cultivators cannot afford to wait, and the rise is in great part due to the fact that considerable care is required for preservation of *gur* through the rainy season. The value of the produce and resulting profits, shown as for *ráb*, becomes as follows:—

Produce per acre.			Value of produce in *gur*.	Expenditure.	Profit.
Juice.	*Gur*.				
lb	Mds.	lb	R a. p.	R a. p.	R a. p.
10,600	24·9	2,050	87 0 0	58 15 0	28 1 0
14,134	33·2	2,733	116 0 0	64 1 8	51 14 4
17,668	41·5	3,416	145 0 0	69 4 4	75 11 8
21,202	49·8	4,099	174 0 0	74 7 0	99 9 0
24,736	58·1	4,782	203 0 0	79 9 8	123 6 4
28,270	66·4	5,465	232 0 0	84 12 4	147 3 8
31,804	74·7	6,148	261 0 0	89 15 0	171 1 0

"*Gur* then appears to pay much better than *ráb*, and this agrees with the general opinion that, even by free sale in both cases, a considerably greater profit is gained by sale of the produce in the form of *gur*. *Ráb* is sold exclusively to manufacturers, and competition has not free play. I had expected to find a greater difference in weight between the quantities of *ráb* and *gur* produced from equal weights of cane-juice, but the proportion given is that usually recognized.

"*Manufacture of raw and dry sugar.*—Before explaining the system on which advances and the price of *ráb* and cane-juice are settled between the cultivator and the manufacturer, it will be convenient to give a short account of the process followed in manufacturing dry sugar from the *ráb*. The *khandsári*, or manufacturer, takes delivery of the *ráb* at the sugar-mill and carts it home in the *kalsis*; on arriving at the *khandsár*, the *kalsis* are broken and the *ráb* is filled into woollen bags, each holding about half a maund; twelve of these bags are then placed, one on top of the other, in

MANUFACTURE in the N.-W. P. & Oudh.

Shahjehanpur

the *bojha* (also called *aráh* or *kuriá*), a high narrow chamber, some 2 feet square, or wide enough to take one very conveniently; a weight is placed on top, and a man occasionally lends his weight. The molasses (*shira*) drains out, and pressing through the *kundyer* (or drain at the bottom of the *bojha*) falls into the *mand* (or receiver).

"With good *ráb* the molasses should all pass off in one day, with inferior *ráb* in two days; what remains is *putrí* (or raw sugar), and 100 maunds of *ráb* (ser of R118) should give 50 maunds of molasses (ser of R107) and 50 maunds of *putrí* (ser also of R107). The *putrí* is then put into a closed room (*kanchi*). Large wooden pegs at close intervals are driven into the ground close to the wall round the room, the object being to keep the *putrí* from touching the wall.

"Then on the floor inside the pegs—thin bamboos raised slightly from the ground—are spread cotton twigs (*binaudhi, kapás*) and above this a white cotton cloth; on this is spread the *putrí* to the thickness of about three or four feet, and it is covered over with *siwár* grass. The *siwár* is a weed growing in deep slow streams, such as the Khanaut in Shahjahanpur. *Kahárs* collect and carry in the *siwár* The molasses that drain from the *putrí* in the *kanchi* at first drains into the same receiver as the molasses from the bags in the *aráh*, but later on the drainings become thin and clearer. It is then collected in a separate receiver for boiling, and this is considered to be the result of grains of sugar melting from the action of the *siwár* grass. This is called *ghaláwat shira*, and is priced and valued higher than the ordinary *shira*, but the quantity is small. On boiling it gives a small quantity of *ghaláwat putrí* and a large amount of black inferior molasses, also known as *ghaláwat shira*, there being no specific names to distinguish these two kinds of *ghaláwat shira*. They are very different in appearance at first. A thin covering is given, but the *siwár* is renewed several times, and at each renewal the fresh grass is placed next the sugar, and the old on the top; there thus becomes a thick covering, which must generate great heat. The converting *putrí* into *khand* takes some fifteen days. When the sugar on the top dries and hardens it is scraped off with an iron *kurpí*, and is spread out on sacking in the sun. The impurities (*banáwan*) are picked out, and the sugar is rubbed frequently and trampled on by men. It is then *khand* (dry sugar), and is stored in bags or store-rooms till sold to traders for export, most of the *khand* being purchased by traders for export westwards to Gwalior, Agra, Delhi, Biána, and intervening marts. Fifty maunds of *putri* (sers of 107 tolas) should give about 33 maunds of *khand* and 17 maunds of molasses. On the average of ten years the price of *khand* is R33-8 per *pulla* of three *khand* maunds. The price of molasses varies greatly, an average being perhaps about R1-8 per maund.

"The *khandsári* does not carry on the manufacture further; and on this account it is only necessary to mention the preparation of *louta*, as *khand*, *gur*, and *louta* are the exports. The *khandsári* disposes of the molasses; some is taken by confectioners, distillers, and others, but the greater part is sold to *gúráhas* who make *louta* from the molasses, and are generally, by caste, *Kulwárs*, *Halwaís*, or *Bharjis*,. They boil the molasses, adding some *ráb*, until it becomes of consistency like *gúr*; it is then put into vessels to cool, and as it cools is made into solid balls like *gúr bhelis*, but very much larger, being commonly over a maund each in weight.

"The balls are then exported under the name of *louta*. *Louta* is sold by Government weight, and now sells at about R6 or R7 per *gond* of three maunds. Molasses being sold by the *tikona* maund, the change in weight covers the loss in manufacture, and a maund of molasses gives, it is said, a full maund of *louta*. *Louta* is seldom consumed in Shahjahanpur, where it is disposed of; but I am told that in districts to which it is exported this *louta* is commonly sold under the name of *gur*, and used as *gur* is here. *Louta* is of dark colour, and being made from uncrystallizable refuse, contains no grains or crystals, and is in taste far inferior to proper *gur*. *Lapta* is *louta* less boiled, and too moist to be made into solid balls; *lapta* is exported in leather bags, and a large part of the *louta* and *lapta* is exported by water to Cawnpore. Manufactures from *khand* are only for local use, and export of refined sugar is confined to that made at the Rosa refinery; *gúr*, *khand*, *louta*, and *lapta* being the only forms in which native-made sugar is exported.

"*Sugar weights.*—The weights used in Shahjahanpur are—

1. The *ráb* maund of 40 sers, each ser R118; the maund equal to 1 maund 19 sers Government weight, or 121·4lb.
2. The *tikona* or *pukka* maund of 40 sers, each ser R107; the maund equal to 1 maund 13 sers 8 chataks Government weight, or 109lb.
3. The *khand* (dry sugar) maund of 40 sers, each ser R96; the maund equal to 1 maund 8 sers Government weight, or 98·8lb.
4. The *kucha* maund, which is half the *tikona* maund, and consequently equal to 26 sers 12 chataks Government weight, or 54·5lb.
5. The Government maund, equal to 82·3lb.

MANUFACTURE in the N.-W. P. & Oudh.

Shahjehanpur

"The *ráb* and *khand* maunds are special sugar weights, used only for these articles. *Ráb* sells by the maund, *khand* by the *pulla* of three maunds. The *tikona* maund is the ordinary weight of the district; and in sugar business is used for *putri* (raw sugar) and molasses. The *kucha* maund is subordinate to the *tikona*, and cane-juice is sold by the *kucha* maund—so much per 100 *kucha* maunds.

"The Government maund, as the rule, is little used except in markets in towns, but *gur* and *louta* are sold by Government weight.

"The sugar weights are supposed to have been framed with the view of covering the loss inherent to each process of manufacture, so that in writing out the account no entry need be made for loss.

"The received estimates of average outturn are: from each maund of *ráb*, half a maund of *putri* and half a maund of molasses, and from 50 maunds of *putri*, 33 maunds of *khand* and 17 maunds of molasses; the actual loss in each case is compensated for by the change in weights. To show the actual loss, I give the estimated results in local and English weights.

"Taking 100 maunds *ráb*, the figures are—

			lb
100 *ráb* maunds	. . . .	=	12,140
Should give	50 maunds *putri* . . .	=	5,450
	50 „ molasses . .	=	5,450
	Total 100 maunds	=	10,900

The actual loss here appears to be just over 10 per cent.

The return from the *putri* should then be—

			lb
50 maunds *putri*	. . . . . .	=	5,450
Should give	33 maunds *khand* . .	=	3,260
	17 „ molasses . .	=	1,853
	Total 50 maunds .	=	5,113

The loss here is over 6 per cent.

The final result from the *ráb* will be—

			lb
100 maunds *ráb*	. . . . . .	=	12,140
Give . .	33 maunds *hkand* . .	=	3,260
	67 „ molasses . .	=	7,303
	Total 100 maunds .	=	10,563

The total loss is 13 per cent. on the original weight.

"*The Baragaon khatauti.*—The price ot *ráb* is fixed in each year at a meeting held in Baragaon in the end of *Bhádon*. Baragaon is a large village in the parganah of the same name, distant fourteen miles from Sháhjahánpur and three from Pawáyan, and situated very fairly in the centre of the cane-producing country. It is a place of little trade, except in sugar; but the prices of all agricultural produce are commonly struck at Baragaon, and according to these prices all accounts between the cultivators and the Baniyas are settled in the greater part in the Sháhjahánpur district, and also in parts of the Bareilly, Kheri, and Hardui districts. A propitious day is settled by the pandits; notices are issued, and a *pancháyat* is held, composed of traders, zamindárs, and cultivators of the neighbourhood. Their duties are simple: the prices in the case of cereals, pulses, etc., being only the average of the Baragaon market prices during certain terms, and these market prices are invariably taken without question from the books of the leading firm in Baragaon. In the case of *ráb* there is no market price, and the *khatauti* price is derived from the average price of *khand* (dry sugar) for each of the three months of *Chait, Baisákh*, and *Jaith*. An

SACCHARUM: Sugar.

Manufacture of Sugar

MANUFACTURE in the N.-W. P. & Oudh.

Shahjehanpur

example will best illustrate the process. Suppose the average price of ***khand***, as ascertained from the entries of sales, to be as follows:—

		R	
Chait	. .	34	per *pulla*, *i.e.* 3 maunds of 96 weight.
Baisákh	. .	32	,, ,, ,,
Jaith	. .	30	,, ,, ,,
The total	.	96, divided by 3 gives an average of	R 32
		To this is added half	16
		Making a total of .	48

"This figure is now taken as annas, and 48 annas, or R3 per ***ráb*** maund is the ***khatauti*** price of ***ráb*** for the past season, the price according to which all transactions relating to the crops of the preceding year between the manufacturers and cultivators will be settled. The price of R3 will, however, only be allowed in the Pawáyan and Baragaon parganahs, and in Bísalpur (Bareilly district). The ***ráb*** here is supposed to be the best, and for the Sháhjahánpur and Tilhar tahsils the price is 2 annas a maund lower, and for parganah Khutár and the adjoining Oudh districts the price is 4 annas lower. Thus, with a price of R3 for Pawáyan, Baragaon, and Bísalpur, the price of Sháhjahánpur and Tilhar will be R2-14, and for Khutár and Oudh R2 12 per maund. The difference of 2 annas between Pawáyan and Sháhjahánpur is always admitted; but it is said that of late the Oudh cultivators have begun to object to the 4 annas reduction. Sugar sold at low rates in 1873, as in that year there were very few marriages among the Hindus; and a considerable part of the sugar being used in marriage entertainments, the demand was less than usual. The ***khatauti*** price of each for the past six years is shown, the average coming to just R3-4 per maund. The price of ***ráb***, though following the market price of sugar is not a market price, and is framed by an arbitrary process, so contrived as to give a price sufficiently below the value of the article to remunerate the manufacturer for his outlay and risk in advances—advances which do not bear interest.

Prices. *Conf. with p. 225.*

Year.	Price per *ráb* maund. R	a.	p.
1868	3	2	6
1869	3	4	0
1870	3	5	0
1871	3	5	9
1872	3	4	6
1873	3	2	0
6)	19	7	9
Average .	3	3	11½

Price of ráb.—"*Ráb* is always sold to manufacturers, and the few cultivators who have not received advances, and consequently are at liberty to dispose of their produce as they wish, generally make ***gúr***; but in parts of the district ***gúr*** is very seldom made and even tenants cultivating on their own capital prepare ***ráb***, and ***ráb*** is commonly sold by men who, for want of capital or other reasons, find it inconvenient to carry on the manufacture of sugar after having made advances to cultivators and arranged for a supply of ***ráb***. When thus sold by free sale, bargain is almost always made for a price so much above the ***khatauti*** price, and it is held that a cultivator can, at his own mill, readily obtain a price, 8 to 12 annas, or sometimes one rupee per maund above the ***hhatauti*** price.

"*Average prices of ráb and sugar.*—The price of ***ráb*** follows that of dry sugar, and chiefly depends on the demand for export. The rise in price is much less than in the case of the food grains, a result, probably, in part, due to extended cultivation of sugar in the Duáb. I give the average ***ráb khatautis*** for 30 years, with the price of dry sugar on which the ***khatauti*** is founded, and for comparison the average wheat ***khatauti***:—

TERMS.	*Ráb* price per *ráb* maund.			Dry sugar (*khand*) price per *pulla* (3 *khand* maunds).			Wheat *tikona* maund per rupee.		
	R	a.	p.	R	a.	p.	M.	S.	C.
1841-50	2	10	0	28	0	0	0	35	13
1851-60	2	8	0	26	10	8	1	2	14
1861-70	3	2	3	33	8	0	0	23	13

"*Sugar manufacturer's profits.*—The sugar-manufacturer's expenditure on his raw material is so complicated by the system of advances that it is almost impossible to frame any estimate of the profits of manufacture.

MANUFACTURE in the N.-W. P. & Oudh.

Shahjehanpur

"*Quality of sugar from Bel and Sárgar Ráb.*—It is always allowed that the *bel ráb* gives a much larger proportion of *khand* than ordinary or *sárgar ráb*, and that the *khand* is very good in appearance, and so sells well; but it is commonly stated that, in subsequent processes, this *bel khand* does not give good results, that it does not retain its whiteness, and that, confectionery made from it, very soon becomes bad, not keeping nearly so long as that made from other *khand*. The *sharbat* is also said to be very bad, and to burn the throat and stomach of any one taking it. These defects are very commonly imputed, and always charged to the *sajji*, etc., used in the boiling-house. These charges are probably well founded. Under the old system more of the sugar is destroyed, but what remains is of excellent quality. The carbonate of soda and other substances used under the *bel* system injures the quality of the sugar, as the native process has no means of removing them in any subsequent stage of the manufacture. In manufacture under European systems the lime mixed with the cane-juice is completely removed in a later stage; but the salts and alkalies added in the *bel* are never removed, and though much less of the sugar is destroyed, these additions injure the quality, and the effect is most apparent in refining processes after the sugar has left the *khandsári's* hands.

"Sháhjahánpur sugar (I speak of native-made sugar or *khand*) has at present a very high reputation, and, I believe, commands a higher price in the markets to which it is exported than sugar from any other district; but very probably the spread of the *bel* system may eventually lower the reputation and price of the sugar, and the increased quantity of sugar may, perhaps, hardly compensate for depreciation in value. At present but little of the sugar exported is made from *bel ráb*, and this *bel* sugar sells as well as sugar made from *sárgar ráb*. The *bel* system is not yet well established in the district, and my estimates are of outturn from *sárgar ráb*.

"*Comparison of final outturn with results in other cane countries.*—The following table shows the quantity per acre and relative proportions of the produce in each form according to the estimates of Sháhjahánpur produce given in the preceding paragraphs:—

PRODUCT.	lb	Percentage of juice.	Percentage of *ráb*.
Cane juice	21,202	...	...
Ráb	4,554	21·5	...
Raw sugar (*putri*)	2,040	9·6	44·8
Dry sugar (*khand*)	1,220	5·8	26·8
Molasses	2,739	12·9	60·1

"The final outturn in dry sugar is, thus, not very much over half a ton per acre. In other cane countries two to two and a half tons per acre is considered a fair average result."

IV.—PANJAB.

MANUFACTURE in the Panjab. 419

The reader, who may desire very special information regarding the sugar manufactures, has a very extensive series of publications to choose from. Mr. Baden Powell's *Panjáb Products* was one of the earliest, and, perhaps is still, one of the best. The more recently published Gazetteers also contain much of interest. There is only one sugar factory in the province, namely, that of Sujanpur, in the Gurdaspur district. (*See p. 319 below.*) The following two extracts may, however, be accepted as fairly representative of all —

Jullunder. 420

JULLUNDER.—Mr. W. E. Purser, in his interesting report on the sugar industry of this district, furnished, in 1884, the following particulars regarding the preparations of sugar:—

"*Method of crushing the cane.*—In working the *belna* two persons are required to drive the cattle. If there are three yokes, each will generally have one driver, but two in all are enough. Two yokes need two drivers, as, if there were only one, the second yoke would not work at all. For the actual work of the mill there

MANUFACTURE in the Panjab.

Jullunder.

are almost always three men; but in the Awán villages to the south of Jullundur city it is said that, as a rule, only two are employed. Two men sit in the pit at the side where the cane is put between the rollers; the third sits at the opposite side, where also is the pot into which the juice drains. The cane is brought from the field, trashed, and tied up in bundles about 9 inches thick. The drivers pass the bundles as needed to one of the two men sitting at the same side of the mill. He passes one bundle through the mill, and then another, and so on. If the cane is long, four bundles will be so passed; if short, six bundles. When the cane has gone through the mill, the man at the opposite side ties it up again and hands it to a driver who returns it to the feeders. When all the bundles have gone once through the mill, two bundles are tied together to form one crushing bundle, and when one of these double bundles has been pressed, it is not passed back by the drivers, but is shoved back along the boards under the juice tray. Of the two men who sit together, one is the captain of the mill. By means of certain wedges he can make the rollers work tight or loose, and it is his business to see to this. But he also helps in feeding the mill. The off-side man simply ties up the bundles and returns them. Where there are only two men at a mill, a feeder and a returner, the rollers are closer together, and the bundles crushed smaller in diameter than where there are two feeders. The cane is passed and repassed through the mill till it is considered fully crushed. The amount of juice extracted is about 54 per cent. of the weight of the uncrushed cane. Rather more than one-third of the same weight will be good cane fibre fit for rope-making, and the rest is broken fibre, good only for fuel.

"*Working mills in partnership.*—Generally several cultivators join together in working a *belna*. In Sikh times there was a tax, called *hunda*, of R5 on each *belna*. This tended to prevent the undue increase of mills; but now, if the men of a village do not pull together well, a needless number of *belnas* are found. When a partnership has been formed, it will consist of 3 or 5 *jogs*, each *jog* consisting of four bullocks. But the term is also applied to as much juice as four bullocks will press at one time, and this is 2 *matti*, that is, two of the pots into which the juice drains. So then 1 *jog* or *jori* = 4 bullocks = 2 *matti*. The partners then settle how many *jog* each man is to crush at a time, the usual number being 20 to 25, that is, 40 to 50 pots. Having done this, they cast lots who is to begin, and in what order the others are to follow. Then work begins, the men and cattle of the partnership working together. The man whose cane is being crushed supplies the stoker of the boiling furnace. He also arranges for the carriage of the cane from the field, and if *gur* is made, manages the manufacture himself. The partners merely help at the mill and supply one trasher per *jog*, who gets as his wages the arrow of the cane. In a year like this, when fodder is scarce, the whole country is only too glad to go out and trash for the sake of the arrows. It is a great advantage to begin; as the later ones turn, the more danger of frost, and in no case does the cane improve by delay. In some places it is the custom to have daily turns; but this is only possible when *gur* is made, or if *ráb* is made, when all the produce goes to one trader; as it would be impossible to arrange for daily despatches of *ráb* to different purchasers. It would take about eight days to press 20 to 25 *jogs* at 2 to 3 per diem. This would represent the produce of 13 to 14 poles per diem, or about 160℔ of *gur* (crude sugar).

"*The Boiling-house.*—When the juice has been extracted it may be made into *gur* or *ráb*. *Gur* is itself a completed product, and can be kept or sold at once in the market. But *ráb* is made only for the purpose of subsequent curing, and can be sold only to curers, directly or indirectly. It may be said that, as a rule, the cultivators themselves manufacture *gur*, while *ráb* is manufactured by their money-lenders. The methods employed are very simple. Near the village close to the mill is the boiling-house—one for each mill—made of thick mud walls and with a flat roof. It has a doorway, but no door. It is generally about 18 feet long and 8 broad. There is no aperture in the roof for the smoke to escape, and in consequence one has to sit as close to the floor as possible to see anything and escape being stifled. At one end of the room is the furnace. A hole about 5 feet deep and the same in diameter, but narrowing at the top, is dug inside the boiling-house. Over this the pan is placed and fixed in its place with mud plaster to prevent waste of heat. The pan may be on a platform of earth a foot high, or level with the ground. In the latter case, the stoke-hole, by which fuel is supplied to the furnace, is in a small excavation at one side of the pan, two or three feet square and a foot deep. Outside the house, at the opposite side of the wall to the furnace, a pit of the same depth and about six feet square is dug, and at the bottom a hole is made connecting this pit and the furnace. Through this hole the ashes are raked out of the furnace. On a level with the floor, on any side found convenient, another hole is made, and often a rough mud wall is built a couple of feet off facing the hole and connected by another wall with the side of the boiling-house. This forms a rough chimney with one side open. Through this

MANUFACTURE in the Panjab.

Jullunder.

hole most of the smoke escapes. About 3 feet above the ground over this orifice a small frame-work of branches of trees is fastened, and on it the *begass*, or crushed cane to be used for fuel, is smoke-dried. *Trash* and *begass* form the only fuel used. For boiling the juice a single pan of iron is used, about 4 feet in diameter and one foot deep. It is made now commonly of sheet iron, which is much cheaper than the hand-wrought iron that was formerly employed. The pan has two large iron handles.

"*Manufacture of ráb.*—The manufacture of *ráb* is carried on in this way:—When the earthen pot (*matti*) is full, it is brought into the boiling-house, and the broken pieces of cane, which are at the mouth, are picked out, and the pot filled full again with other juice, so that the trader may get his full measure. The *ráb*-maker then takes a large strainer of cotton cloth, and, placing it over the pan, has the contents of the pot emptied into the pan through the strainer. He then wrings out the strainer and ties it up to a looped rope made of cane fibre, suspended from the roof just over the pan. Any juice that remains in the strainer can drain out gradually into the pan. The furnace has been heated, and boiling goes on for an hour and a half to two hours. The *ráb*-maker regulates the firing, which is done by a boy who is supplied by the cultivator. In boiling three stages are recognized. The first stage is till the scum breaks. This takes place a few minutes after boiling begins, the time depending on the greater or less heat of the furnace. The scum is of a greenish grey colour, and when it begins to break, the fissures are white. At this point the boiler pours into the pan a couple of quarts of the extract of the bark of a tree, with the object of clearing the juice. The bark is called *sughlai* or *suklai*, and sells here at the rate of about 30 to 40℔ the rupee.* The extract is of a grey colour and viscous, and probably acts as the white of an egg does in clearing soup. The scum is next skimmed off with a round, almost flat, perforated ladle and thrown into a straining cloth, placed over a shallow rectangular basket made of cotton twigs. The baskets rest on the ground on one side, and on the other on a stick placed across a section of the pan. The skimmering is continued till the juice is clear, and during it *sughlai* extract is poured in twice more; and, finally, a quart or so of plain water is added. The second stage, during which the 'charge' is said to be rising, now begins, and continues almost to the end. During this stage the water of the juice is still in excess, and the boiling is less concentrated than at the final stage. While the *charge is rising* white scum forms in small quantities on the surface and adheres to the side of the pan; and, indeed, does so, till the boiling is completed. This scum is scraped off with the perforated ladle and put into a small earthen pot from which it is retransferred to the next charge. The scum in the basket strainer is subsequently tied up in the cloth (which is hung up to a peg in the wall), and any juice left drains into another earthen vessel whence it is poured into the next charge. The remaining scum, as well as that got from the straining before boiling, goes to Changars (a wandering tribe), if there are any about. If not, it is thrown away, or the very poorest of the poor may consume it.

Conf. with p. 281.

"At the third stage, the charge is said to be bubbling. Most of the water has evaporated, and the bubbles of the thick liquid are smaller than in the previous stage. At the proper moment, which seems to be determined by experience only, the *ráb*-maker pours a little oil rape-seed (*sarhon*) or sesamum (*til*), about half a wine glass-full, into the pan. This is said to check too rapid boiling. The effect is like that of oil on a troubled sea. The violence of the boiling at once decreases. The boiler now takes a large iron ladle containing, perhaps, a couple of quarts, and takes up some of the liquid and pours it out. He then rubs his finger on the ladle and feels the consistency of the syrup. Again he does so, till he is satisfied the boiling has proceeded far enough. He also judges of this by what is called the *efflorescence*. In the centre of the pan the juice gets whiter than on the sides, and the boiler professes to see something resembling flowers, principally there, but also on the sides. A third guide is the noise made by the boiling syrup when taken up in the ladle and poured against the side of the pan. When the proper consistency is attained, what remains in the pan is ladled into an earthen bowl buried in the earth close to the pan. A cloth is placed between the pan and the bowl to prevent the *ráb* which falls from the ladle being dirtied by contact with the earth. By this time another *matti* of juice is stand-

* *Suklai* is said to be got from the branches of various trees in the Hoshiárpur and Kángra lower hills. These trees are the *pula, bahal* or *dhâman, barna* and *tútri*; and in case of necessity the bark of the *fálsa* even may be used. But the *pula* and *bahal* are most prized. From **Stewart's** *Panjáb Plants* these trees would seem to be **Kydia calycina** (*pula*), **Grewia oppositifolia** (*bahal* or *dhâman*), **Cratæva religiosa** (*barna*), **Morus parvifolia** (*tútri*), and **Grewia asiatica** (*falsa*).

MANUFACTURE in the Panjab.

Jullunder.

ing ready, the pan is recharged, and the whole process goes on over again. The *ráb* in the meantime is taken out of the bowl into which it was ladled and is put into a large earthen jar, which rests on a pad made of cane-fibre, and which holds from 240 to 480℔. The country jars are small; but those got from Tanda, in the Hoshiárpur district, are large. A mark is made on the jar for every bowl of *ráb* put in. When the jar is full, it is closed with a cane-fibre plug which is tied to the mouth of the jar and plastered with mud. When the trader to whom the *ráb* belongs sees fit, he has it removed to the curing-house. It is considered an advantage to keep the jars when full some days in the boiling-house, as the heat is thought to help concentration.

"*Manufacture of gur.*—The manufacture of *gur* as ordinarily carried on is exceedingly simple. The juice is poured into the pan without any straining The broken pieces of cane and other impurities which float to the top are taken out by hand, and then the juice is boiled till the *efflorescence* appears. At this stage the *gur*-maker takes a short wooden crutch and stirs the boiling mass round and round and backwards and forwards with the arm of the crutch till the proper consistency is attained; that is till the *concrete* is a soft, very pasty mass, but not at all liquid. Next, the pan, which in the case of *gur* is not plastered on to the furnace, is lifted up by the two handles, and the charge poured and scraped out into a platter made of mud and chopped fibre of the false hemp plant (*sann*), sun-dried, about 3 feet in diameter and with a low raised rim. In this it is worked with a *ramba*, such as described above, till it becomes just about concentrated enough to retain the shape it is to have. A cloth is spread over one scale of a balance, and enough *gur* put in to equal the 2 seers (5 seers local weight, about 4℔) in the other scale. Then the cloth is closed so as to form a round ball of the *gur*, which is then taken out and kept till used or sold. During this keeping it gets gradually drier, till outwardly it appears quite dry. When the *ramba* gets clogged with the concrete, it is cleaned with a wooden wedge. In the Dhak and Sirwál, and rarely elsewhere, more attention is paid to the manufacture of *gur*, and impurities are strained out as in the case of *ráb*. The resulting *gur* is whiter than that ordinarily manufactured, and is made into small cakes about the size of a bun called *pesi*. The large round lumps are known as *bheli* or *rori*; these latter are mostly sold to men from the Malwa, who come across with their carts and are not particular about quality. *Pesi* are sold more by retail, where a finer kind is needed. They are of no fixed weight, but never exceed a pound-and-a-half.

"*Manufacture of shakar.*—From cane of superior quality, especially that grown on the sánwín system, *shakar* is made in the same way as the *pesi gur*. This product differs from *gur*, in that it cannot be formed into big lumps. It is consequently made up into small pieces, and these are rubbed by hand till they become a sort of powder.

"*Boiling of rab and gur.*—There appears to be no difference to speak of between the time *ráb* is boiled and that needed for *gur*; nor is there any difference in the way the furnace is heated.

"*The curing-house.*—A sugar curing-house consists of a room generally exceedingly dirty and with no apertures for ventilation. There may be one or more vats, and their size varies. The shape is usually rectangular. When possible, two walls of the room are utilized to form two sides of the vat, and the two other sides are formed of low walls made of brick and mud plaster. At the bottom of the vat are placed pieces of wood extending from one end to the other, or crossways, and spaces between them. On these a matting of *sarr* (**Saccharum Munja**) is placed, and this is covered with a cloth. The sides of the vat are faced with a matting made of river-flags. Under the floor of the room, but not under the vat, is a cellar, connected by a narrow well with the top of the floor. When the *ráb* comes from the mill, it is emptied out of the earthen jars into any convenient vessel by men, who get half an anna for each small jar, and from one anna to an anna and a quarter for each of the large jars. The *ráb* has to be extracted by means of a *ramba*, or trowel. It is carried into the curing-house in the vessel into which it has been emptied, and is there transferred to the vat. When the vat is full, the *ráb* is left in it for several days without further treatment, except that the room is kept full of smoke, especially at night, in order, it is said, to dry the *ráb*. After eight or ten days, the *ráb* is smoothed down and covered with a layer of *jálá*, a couple of inches thick. The *jálá* (**Hydrilla verticillata**) is a plant found in streams, and in this district is got mostly from the Bein river, but also from the Sutlej. It is collected by a caste called Jhiwars, and the price, which is fixed per 100 *mans* (of 33℔ each) of *ráb* to be cured, depends on the distance to which it has to be carried. Close to the Bein the price is about R1-8 per 100 *mans*. Two other aquatic plants which I cannot identify, *kareli* and *bhálú*, are also used, but not commonly, though *kareli* is said to be better than *jála*. Every third day the *jálá* is turned up, and a fresh supply put below it next to the sugar. The old *jálá*

Conf. with p. 31.

MANUFACTURE in the Panjab.

Jullunder.

is not removed, but is placed over the new layer. The effect of the *jálá* is to turn the upper crust of the *ráb* of a whitish colonr and to soften it, so that it becomes somewhat powdery. When the second supply f *jálá* is given, the soft white upper layer of the sugar is scraped off with a small curved iron scraper. This is done in the evening, and the next day the sugar, which has been thus got, is placed on coarse sacking made of false hemp, and trampled by men for several hours in the sun, after which it is of a very pale straw colour, and is known as *khand*, while before it was called *bed*. This process is repeated every time *jálá* is applied, till the curer thinks he has got as much out of his *ráb* as he can. In the meantime the *molasses*, or syrup, which was in the *ráb*, has beeen draining away through the straining cloth at the bottom of the vat into the spaces between the pieces of wood, and thence has flowed through an orifice in one corner of the vat into a bowl-shaped hole in the floor, and thence into the cellar. When needed it can be drawn up through the well. The time needed for curing a vat is from six weeks to two months, and the *jálá* will be renewed 10 to 15 times. Occasionally, when the vat is filled, a solution of about 1℔ of *sajji* (impure carbonate of soda) in 30 times its weight of water to every 100 *mans* (of 33℔ each) of *ráb* is poured over the *ráb*. But this system is going out of fashion. The *jálá* is not applied when wet, and so its effect seems hardly that obtained in the 'claying' process, the principle of which is the washing out the molasses from the crystals by means of the water draining away from the wet clay."

"*Grades of Cured Sugar.*—The cured sugar obtained by this process may be taken as three-tenths of the *ráb*. In the Dhak, one-third is said to be got, and further to the west only one-fourth. But three-tenths is a fair average, and these great differences are somewhat doubtful. A certain amount of wastage takes place, and there is also loss from evaporation. It is difficult to say what this amounts to, but probably it will not be far from one-tenth of the *ráb* used. The rest is *shira*, or molasses. The cured sugar is not of uniform quality. Some recognize more, some fewer classes, but there may be said to be three generally recognized grades. The top layers are called *chitti khand* or white *khand*; below them is second class *khand*; while the lowest portion is *talauncha*. *Chitti khand* is of a lighter colour than second class *khand*, while *talauncha* is of a dark-brown colour. The amount of each turned out depends entirely on the maker. If much *chitti khand* is made, there is said to be great loss of sugar which drains away with the molasses, owing to the repeated action of the *jálá*. If the manufacturer is making for the upper Panjáb market, he will produce much *chitti khand*, but if for the Malwa and Rájputána markets, he makes the lower qualities. In places where Europeans have penetrated, the first class *khand* is also called *chini*, but this name is applied by the people to a refined sugar, which will be noticed hereafter.

"*Molasses Sugar.*—Before going on to refined sugars, the manufacture of molasses sugar may be noticed. About 30 to 40℔ (15 to 20 seers) of molasses are boiled in an iron pan for about three-quarter of an hour, till the mass is reduced to three-fourths of its original weight. No straining or purifying takes place. The time of boiling may be less or more, depending on the extent to which the furnace is heated. The resulting *gúr* is made into large lumps, and shoved into bags, where it settles into a solid mass. Large but very variable quantities of molasses sugar are made. The amount depends entirely on the demand of the day.

"*Refined Sugar.*—The refined sugars usually made are *misri*, *kúzá misri* and *chini* or *búra*. *Misri* is made in this way. A certain amount of *khand* (say one *man* of 82℔) mixed with one-fourth its weight of water is boiled in a large pan. When the scum collects at the top as in the manufacture of *ráb*, about one pint of milk is mixed with a gallon of water and poured into the pan, afterwards a quart of water is added twice, and finally half a pint of milk in a quart of water is poured in. All this is to make the impurities in the *khand* rise, when they are repeatedly skimmed off. When the syrup has been thus cleared, it is removed from the fire, and as much as is needed is put into another pan, which is boiled again till the proper stage is arrived at, when the contents of the pan are poured into a flat iron tray about 2½ feet in diameter and with a raised edge of about an inch in height. Here it stands for a few hours. Occasionally a second boiling is poured over the first, in which case, when the *misri* is removed from the tray, it separates into two thin flakes. The tray is then placed in a slanting position, and the syrup, or treacle in the sugar, drains out of holes made in the lower edge of the mass.

"As this draining goes on only for one day, a good deal of syrup is left, and so *misri* is never really dry. It is of a damp yellowish white colour, and shows the crystals clearly. About three parts of *misri* are made out of four of *khand*, or a trifle less. In making *chini* or *búra*, the preliminary stage is the same as in the case of *misri*; the *khand* is mixed with water boiled, defecated with milk, skimmed and

 Manufacture of Sugar

MANUFACTURE in the Panjab.

Jullunder.

removed from the fire. From the syrup thus obtained half a dozen large ladle-fulls are boiled again till the proper consistency is reached, which is much the same as that of *ráb*. The contents of the pan are then poured into another pan, and rubbed against the sides with a small crutch till they set solid, which occurs in a few minutes. They are then scraped off with an iron instrument something like a chisel, and pounded with the crutch for about ten minutes. The result is a very soft, floury, almost white sugar, which is *chíní* or *búra.* There is no waste to speak of in making *chíní*—one in forty is what is usually reckoned. *Kúza misri* is simply sugar-candy. The syrup, after refining with milk and re-boiling as in the case of *misri*, is poured into small round earthen moulds, in which are suspended several threads. When it has set, the mould is turned upside down, and any superfluous molasses drains on along the threads. *Kúza misri* is of varying quality; some being quite white, and some yellow. In some the crystals are found only along the threads and in a thin layer on the sides of the mould, while in some the whole mould is filled up with crystals. The quality depends on the *khand* used, and also on the amount of clearing the syrup gets. In first-class candy, milk will be used half-a-dozen times, crystallization is completed in about four days, and the candy is ready in a week.

" *Quality of the sugars and average prices.—Relation of saccharine product to each other.*—The following table will show the chief saccharine products of the district and the relation they bear to each other. Vinegar is also made for home use; but further notice of it is not needed:—

Errors in Terms. *Conf. with pp. 114-115, 130-31, 131-134, 136, 183, 229, 252-255, 285.*

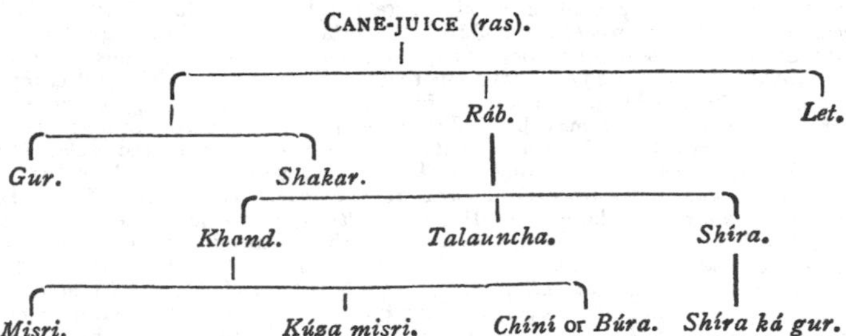

"*Gur* is also known as *kand siyáh*, and *shakar* is also called *shakar surkh*. But the second names are not used by ordinary country people, but by those who wish to show their learning. So, too, *khand* is also known as *shakar tari*. The word *kand* is Arabic, and the origin of the English word candy. *Khand* is the same as the Hindustani *khánd*, and the distinction between the two words *kand* and *khand* should be carefully kept in view. They may have the same root all the same. It is a question what English terms should be used to express these products: *gur* and *shakar* are apparently the same thing, the only difference being *shakar* is made from a superior quality of juice, and that its particles have not the same cohesiveness as those of *gur*. As far as I can see, *gur* and *shakar* are what is known as 'concrete sugar,' and in any case this is a very suitable name. For *ráb* nothing seems better than 'undrained raw sugar,' as suggested by the late **Mr. G. Butt, C.S.**, in his paper on 'Sugarcane cultivation, Shájahánpur District.' *Khand* is 'raw' [or 'moist' or 'brown' or 'grocery'] sugar, and *talauncha* is merely an inferior quality of *khand*. *Shira* is molasses, and the *gur* made from it may be called *molasses sugar*, a well-recognized term. *Misri* might be called 'crystallized sugar,' though the term lacks definiteness. *Kúza misri*, and similar products called simply *kand* and made in a mould, are 'sugar-candy.' *Chini* might be designated 'soft, refined sugar.' As to *let*, it is only 'boiled fermented cane-juice,' or 'uncrystallizable boiled cane-juice.' To the unlearned in sugar-making, the name 'loaf-sugar' or 'lump-sugar' would give the best idea of what *misri* is. But 'loaf-sugar' is already appropriated to the finest product of the most scientific refining, and to use the name might mislead those most interested in having accurate information. If a technical term is to be used, 'crystal sugar' would seem more appropriate; but for the same reason that leads to the rejection of 'loaf-sugar,' it is, perhaps, better to have a more indefinite name. When on the subject of names, the often-repeated derivation of *misri* from Misr (Egypt) and *chíní* from Chín (China) may be noticed."

MANUFACTURE in the Panjab.

Jullunder.

"*Prices.*—Sugar-candy is but little made, and not everywhere. *Misri* and *chini* are made by all confectioners and in large quantities, but only to supply local demands. *Shakar* also is consumed mostly locally.

"*Gur,* both ordinary and that made from molasses, *khand, talauncha* and molasses, are largely exported.* *Let,* as already said, is of no value. The average prices for a long series of years, from 1862-1881, may be taken as these:—

"Ráb 16; Gur 15; Shakar 12; and Khand 4 sers per rupee [the ser is 2lb nearly]. "Prices have been comparatively steady. The present prices are approximately—

Ráb 16; Gur (pesí) 14; Gur (bheli) 16; Gur (or molasses, 26; Khand (chitti) 4; Khand (2nd quality) 5; Khand (Talauncha) 8; Shíra 42; Misri 3; Chini 3½; and Kúza misri 2 sers the rupee. (This is the expected season's price, but it has not been fixed yet.)"

Hoshiarpur. 421

Hoshiarpur.—"Three pairs of bullocks are generally required to work it at one time, and, if worked night and day, nine pairs are necessary. There are, however, smaller *belnas* worked by only two pairs of bullocks. A *belna* costs R30, and lasts about seven years; but its rollers have to be constantly renewed. The village carpenter takes R2 for setting it up every year, as well as four canes a day while the pressing is going on, and a drink of the juice every third or fourth day. Another of his perquisites is half a ser (*kacha*) of *gúr* for every large vessel (*chati*) of juice expressed. The bullocks cost from R20 to R25 each and last five or six years. An iron boiling pan (*karah*) is also required, costing from R16 to R20; if hired it costs R4 a year. The number of hands required to work a sugar press are—(1) a man or boy to drive each pair of bullocks; (2) a man to put the bundles of canes between the rollers, called *dohra*; (3) another to pull out the canes on the other side and pass them back, called *mohra*.

"The canes are tied in bundles of 50 or 60, called *datha,* and are passed through the press 30 or 40 times until the juice is all extracted. The dry stalks or cane trash, called *pachhi*, are useful for making ropes and mats, and for tying sheaves of corn in the spring harvest. A *belna* is generally worked by partners, who help each other in stripping the leaves of the cut canes and preparing them for the press, and in providing bullocks to work it. The juice, as it exudes, flows into an earthen vessel called *kalari*, from which it is carried to the boiling pan.

"The next process is the boiling of the juice, and it differs according to the article required. The cultivator makes either—

"*Gúr*—Coarse undrained sugar, or compost.

"*Shakkar*—Coarse undrained sugar dried.

"*Mál ráb*—The sugar material from which drained sugar is made.

"For the first two the boiling process is the same. In making *gur* the boiled juice is emptied into a flat dish called *gand,* and allowed to cool, when it is worked up into round balls. For *shakkar* the cooled substance in the *gand* is well worked with the hands into a powder. *Gur* and *shakkar* will not generally keep good for more than a few months; they deteriorate in the damp weather of the rainy season and lose their colour, but are still saleable at a reduced price for a year or two. In making *mál ráb*, the cane-juice is not boiled so much as for *gúr* or *shakkar*, but during the process a material called (*suklái*), consisting of a gummy preparation of the bark of the *pola* (**Kydia calycina**) and sometimes of the *dhaman* (**Grewia oppositifolia**), is dropped into the boiling pan to clarify the juice. The scum is taken off as it rises, and when the juice has been boiled sufficiently it is emptied into open vessels, and when cool into large earthen jars called *mati*. The plan of using three or four separate boiling pans, as

Conf. with pp. 295, 305.

* Molasses are also extensively used locally in the preparation of tobacco for smoking purposes.

MANUFACTURE in the Panjab.

Hoshiarpur.

in the North-West Provinces, is not followed here, except in one village in Dasuah (Hardo Khandpur), where the method has been introduced by a man from the south. The leaves and refuse of the cane are used for feeding the fire, which is tended by a man called *jhoka*. The boiling and straining are superintended by one of the partners of the *belna;* if *gur* or *shakkar* are being manufactured, and in the case of *mál ráb*, by a servant of the trader who has agreed to purchase the *ráb*, called *rábia*.

"The making of drained raw sugar (*khand*) is generally carried out by a regular trader. The process requires a great deal of superintendence, and few cultivators proceed further than the making of the first crude substances above mentioned. In making *khand* the *mál ráb* is emptied into large vats (*khâchni*), lined with matting capable of holding from 80 to 400 maunds of *ráb*. At the bottom of the vat are a number of small channels leading to reservoirs outside, and on this flooring are placed pieces of wood on which is a reed mat, over that a piece of coarse cloth (*pal*), the sides of which are sewn to the side mats in the vat. After a time the molasses (*shira*) exudes through the cloth and matting at the bottom to the reservoirs outside, and is thence collected in earthen jars. After the *ráb* has been in the vat about 10 days, and the mass hardened sufficiently to bear a man's weight, it is worked up with an iron towel so as to break up all lumps, and smoothed with a flat dish previously rubbed with *ghi*. Then layers of *jála* (**Potamogeton**), a water-plant, are placed on the top, and after every few days the *jála* is rolled up, and the dry, white sugar at the top of the mass taken off, and fresh *jála* put next to the *ráb*, the old *jála* being placed over that, so that as the sugar is extracted the super-incumbent weight of *jála* increases. Towards the end, if it is found that the weight of *jála* is carrying sugar as well as molasses through the *pal*, some of the old *jála* is taken off. It takes three or four months to empty an ordinary vat by this process. If begun when the weather is cold, it is customary to light a fire in the room containing the vats before putting on the *jála*, in order to make the molasses drain off quicker. The sugar taken off is spread out on a piece of coarse canvas on a hard piece of ground in the sun, and well trodden with the feet until it has been reduced to a dry powder. This substance is called *khand*, and sometimes *chini*, and is the ordinary coarse drained sugar sold in the market. The other forms of sugar are:—*Búra*, made from *khand* boiled in water and clarified with milk. When the substance has become a sticky mass, it is taken off the fire, and well worked with a piece of wood until it becomes a dry powder. Another kind of inferior *búra* is made in the same way from the sugar which adheres to the *jála* in the vats. *Misri*, also made from *khand* mixed with water and boiled to evaporation. It is then put into a flat dish called *tawi*, and when set, placed in a slanting position for the moisture to drain off. *Kusa misri*, prepared as *misri* only with the best *khand*. After boiling, the preparation is poured into little round earthen vessels in which threads are placed, and when the sugar has set, the vessels are inverted. The crystals adhere to the sides of the vessels and the threads and the moisture drains off. The vessels are then broken, and the sugar taken out. This is the ordinary candied sugar. *Talaunchá*, coarse, moist, red sugar, being either that left at the end of the draining process in the vat, or molasses containing sugar and boiled and drained a second time, also called *dopak*. *Pepri*, the treacly sugar that adheres to the pieces of wood or the reed mat at the bottom of the vat.

Conf. with p. 31.

"It is difficult to put down the real cost of cultivation, as sugar-cane is only one of many crops grown by the cultivator, and nearly all the labour expended on it is that of his own hands and of his family and servants, but the marginal table is an average estimate for four acres of sugar-cane

MANUFACTURE in the Panjab.

Hoshiarpur.

which is about the amount that one *belna* can press. The results of experiments made as to the outturn of sugar-cane are given below:—

YEAR.	Detail.	Area under experiment.	Total outturn of *gur*.	Average outturn per acre.	Character of harvest.
		Acres.	Mds.	Mds.	
1879	Irrigated	6·	102·3	17·1	Average.
	Unirrigated	21·8	404·1	18·5	
1880	Irrigated	2·4	29.	12·1	Good.
	Unirrigated	33·4	643·5	19·3	
1881	Irrigated	6·2	159·5	25·7	Very good.
	Unirrigated	40·9	830·2	20·3	
1882	Irrigated	4·5	97·1	21·5	Average.
	Unirrigated	28·6	490 8	17·2	
TOTAL	Irrigated	19·1	387·7	20·3	
	Unirrigated	124·7	2,368·6	19·0	
	TOTAL	143·8	2,756·5	19·2	

	R
Seed	20
Manure	8
Field labour	30
Carpenter	2
Hire of boiling pan	4
Average annual cost of *belna*	4
Jhoka, or fireman	6
Other labour at the sugar-press	8
Government revenue	14
TOTAL	96

Or an average of R24 per acre.

"In every case the outturn of *gúr* has been taken, not boiled juice or *ráb*. It is curious that the average produce on unirrigated lands, on which the majority of experiments has been carried out, should be higher in two years than that on irrigated. The fact is, that scarcely any irrigation is required in this district; the great sugar-growing tracts have a naturally moist soil, and even where irrigation is available it is often not used. From the above statistics we are justified in taking 19 maunds of *gúr* as a good all-round average per acre. Assuming the price current to be 16 seers per rupee, the value of the outturn on four acres would be R190, or R47-8 per acre, and the net profit of the cultivator R23-8 per acre. The profit should be much the same if *mál ráb* is made, as the rather larger outturn of this commodity, as compared with *gúr* and the lower price, counterbalance each other. But, as a rule, *ráb* is more profitable, as the cultivator gets ready-money for it at once. In the case of *gúr* he has to consider the market in selling, and meanwhile some of it is eaten in the family, and some must generally be given to friends and relations. Captain Montgomery had an experiment carried out in order to show a statement as given in Appendix II, Government of India Resolution, No. 505 A., dated

SACCHARUM: Sugar. Manufacture of Sugar

MANUFACTURE in the Panjab.

Hoshiarpur

30th May 1882, Department of Revenue and Agriculture. The results are given in the margin. The outturn of *gur* here is much larger than the average given above for the whole district. Even so the outturn per acre is only about three-fifths of that given for the Sháhjehánpur district, though the relative percentages between the different manufactured commodities are much the same. Canes are never sold in the bulk, because the growing and pressing are done by the cultivator; only near towns are they sold separately for chewing. The estimated outturn per acre is about 300 maunds, equivalent to 10 tons and 14 cwt. The average market prices of the different kinds of sugar are as follows:—

	Outturn per acre.	Percentage on		
		Canes.	Juice.	*Ráb.*
	Maunds.			
Cane . .	296	...	...	...
Juice . .	149	50·3	...	...
Ráb . .	34	11·5	22·8	...
Gúr . .	25¾	8·7	17·2	...
Khand .	9½	3·2	6·4	27·9
Shira . .	20½	6·9	13·8	60·3

English Equivalent.	Native name.	Price per Rupee.	English Equivalent.	Native name.	Price per Rupee
Boiled cane-juice	*Mál ráb* .	19 seers.	Better sorts of drained sugar.	*Búra* . . *Misri* . .	3 seers. 2½ ,,
Undrained sugar	*Gúr* . *Shakkar* .	16 ,, 13 ,,	Candied sugar	*Kuza misri* .	1½ ,,
Common drained sugar .	*Khand* .	4 ,,	Very coarse red sugar.	*Pa;auncha* . *Pepri* . .	8 ,, Not ordinary.
			Molasses.	*Shira* . .	Sold 32 seers per rupee.

"As a rule, the cultivator is under no obligation to the money-lender during the period of cultivation; nor in the pressing, if only undrained sugars (*gúr* or *shakkar*) are made. If *mál ráb* is made, the trader often gives an advance when the pressing begins, calculated on the probable outturn, and accounts are settled after the whole has been delivered. Interest at 24 per cent. is usually charged only on the balance if the outturn has been over-estimated. The refinement of sugar is very seldom attempted by any but the most opulent cultivators. Probably not more than two or three per cent. of cultivators proceed further than the making of *gur, shakkar* or *mál ráb*" (*Gaz., Hoshiarpur*, 97-101).

MANUFACTURE in the CENTRAL PROVINCES. 422

V.—CENTRAL PROVINCES.

Nothing further need be said regarding the sugar manufactures of these provinces to what has already been given in Mr. J. B. Fuller's paper republished above (pp. 187-95) in connection with cultivation. The sugar manufactures of these provinces are relatively unimportant.

CENTRAL INDIA & RAJPUTANA. 423

VI.—CENTRAL INDIA AND RAJPUTANA.

So much has already been said regarding the sugar production of these states in the chapter on Cultivation (pp. 195-208) that it does not seem necessary to furnish a special chapter on the sugar manufactures. In some of the states sugar-candy is made, and obtains a high reputation, such as that of Bikanír.

BOMBAY & SIND. 424

VII.—BOMBAY AND SIND.

In 1887 the Director of Land Records and Agriculture, Bombay, furnished a note on sugar from which the following may be here given:—

"*Sugar-boiling Pans.*—These are either of copper or iron. The iron pan is

MANUFACTURE in Bombay & Sind.

far the most common. Pans again are of two descriptions. On the one hand is found a shallow wide pattern, and on the other a narrow deep pan. As far as I have been able to ascertain the Bassein pan is the only one of the latter description. It demands a far greater amount of fuel, but it is credited with making better and more lasting *gul*. There is no doubt that, as elsewhere, the deep narrow pan will in time disappear, for its consumption of fuel is greater than the scarcity and higher price of fuel will allow, without special concessions, such as have up to date been made from forests in Bassein. Even in Balsár close by the shallow pan is used. I know of no place, except Kánara, where the shallow pan is made of copper. The iron pans vary to some extent in size and also in depth. The shallowest and widest is that used in Poona, where fuel is dear. With it there is no difficulty in making *gul* with no other fuel than the crushed cane. The capacity of the Poona pan is ⅓rd greater than of the Bassein pan. The comparative requirement of fuel may thus be stated. The Bassein pan requires for an equal weight of juice as large a weight of *dry* wood as the Poona pan does of *freshly* squeezed cane. I have experimentally demonstrated this fact.

"The iron pan costs about R20 and lasts three or four years. In Bassein the copper pan costs much more (say R75), lasts much longer, and is sold for its full value as copper when no longer fit for use. But for its initial cost the shallow Kánara copper pan would probably be extensively used.

"*Cane Mills.*—Wooden mills are slowly, but gradually and surely giving place to the improved iron mills, of which several patterns are coming into use. These I shall describe separately. The objections to the iron mills are purely imaginary. The most common one is that the juice is discoloured. It is even asserted that the wooden mill extracts a larger percentage of juice, but this is so far from being the fact that the chief merit of the iron mill is its efficiency. On the Bhadgaon Farm iron mills have, after long and patient waiting, overcome all prejudices. The only complaint comes from the Kumbhárs (or potters), who by custom claim the crushed cane as their remuneration for supplying the pots and vessels required to receive the juice from the mill. The wooden mill was so imperfect that the Kumbhárs were able to extract a considerable amount of juice from the crushed cane and to make *Potters' gul*. They can get nothing from the cane which has passed through the iron mill. In Surat, Ahmadabad, Kaira, and the Pálanpur Agency a few iron mills are now in use, but the wooden mill still holds unimpaired sway in the rest of Gujarát.

"In the Deccan, iron mills are pushing their way. The inventor of a three-roller mill at Poona—whose mill will be described—has pushed his invention with great energy and is overcoming all objections. The ryots can seldom buy mills, but freely hire them.

"The cost of the wooden mill, made of babhul, is from R20 to R25, and that lasts ten years if well made and if kept under water in a well during the hot season.

"There are two patterns of indigenous wooden mills—one with two and the other with three upright rollers. In the former the upper portion of the rollers forms a male and female screw respectively, and is called the *navra-navri*—husband-and-wife—mill. The only difference in the latter is that there are two female screws, one on each side of the male, to the top of which is attached the lever, at each end of which the bullocks are yoked. The cane is passed and repassed as often as 6 times, but generally only 3 times till the juice has been extracted as far as possible. In the double squeeze pattern the cane is thrust through between the male and one female screw and back between the male and the other female screw. In the single squeeze it is thrust back as it came. I do not know the origin of the different patterns or their comparative merits.

"The following patterns of iron mills are in use:—

1. "Three-roller horizontal mills by McOnie of Glasgow or imitated from them of different sizes. The largest has rollers 16″ in diameter and 30″ long, and cost about R1,000. The smallest possesses rollers 8″ in diameter and 14″ long. Cost R500.

2. "Three-roller upright mill, patented by Mr. Subrav Chowhan of Poona, with iron frame. The rollers are smooth. This mill is made in sizes with rollers varying from 4″ to 13″, and costing from R25 to R300. The R200 size (rollers 11″) is the favourite. It is hired at R1 per diem.

3. "Single and double-squeeze Bihia mill. The former has rollers 8″ × 10″. The latter two rollers 7″ × 8″ and one break roll 4½″ × 8″. Cost R150 and R160 respectively. The rollers are slightly grooved.

"It is not necessary to discuss the respective merits of the iron mills. All are far superior to the wooden mills. They save in time and labour and extract a larger

SACCHARUM: Sugar.

Manufacture of Sugar

MANUFACTURE in Bombay & Sind.

percentage of juice The most efficient is the horizontal pattern by McOnie, but its price is very high. The Bihia mills are very portable. Wooden mills are superseded wherever any of the iron mills are available for hire. Few ryots can afford to purchase, and the village Bania is not educated to the point of encouraging agricultural improvement yet."

Khandesh. 425

KHANDESH.—"Sugar-making is carried on by all the better class of cultivators. Great stone sugar-mills, found in many of the Satpuda valleys, show that sugar-cane used to be more widely grown than it now is. The molasses is sold by the maker to the village shop-keeper at the rate of from 1⅛*d.* to 2¼*d.* (1—1½ annas) a pound. The dealer generally gathers a considerable quantity and forwards it to one of the district trade centres. Pimpalner and Ner in Dhulia are the chief producers of sugar, and the supply is gradually distributed among the district shop-keepers and travelling pedlars. The yearly outturn is estimated at about 1,100 tons. Almost all classes use it, and little leaves the district. Much is imported by rail. The ordinary retail price varies from 2¼*d.* to 3*d.* (1½—2 *annas*) a pound, with a slight rise during the marriage seasons. In preparing dainties the rich classes make use of refined sugar brought from Bombay and Benares.

"Sweetmeats are made in most large villages. The makers are chiefly Hindus of the *Pardeshi*, Gujarat Vani, and Bhatia castes. The industry supports about 100 families, the women helping the men. Their work is pretty constant, but they are specially busy in the marriage seasons and at fairs. They work from six to eight hours a day. They buy the sugar and spices, and offer the sweetmeats for sale in their shops or at fairs and markets. Sometimes materials are given them to be made up for a feast. The industry is fairly prosperous, the monthly earnings of a family varying from £1 to £3 (R10 to R30). The sweetmeats of Dhulia, Chopda, Jalgaon, and Bhusával, have a special local name. Very few leave the district"—(*Bomb. Gaz., XII, 226*).

Kolhapur. 426

KOLHAPUR.—The following account of sugar-manufacture completes the passage from the District Gazetteer, the first half of which has been given in the chapter of this article which deals with the cultivation of the cane:—

"The mill is set up in a corner of the field and employs about seventeen hands and sixteen bullocks. Five men called *phadkaris* are employed in cutting, topping, and stripping the cane. Fresh-cut canes give a larger percentage of juice, and so the cane is cut as required by the mill. One man called *molkya*, or the bundle-man, carries the cut canes to the mill. The *khándkya* chops the canes into pieces about a yard long. The tops* with one joint are kept for seed-cuttings, and the lower pieces are tied in bundles. Seven men work at the mill. The *bharkavlya* feeds the mill with the cut cane received from the *kándyághálnár*. The *lendkavlya* sits on the side of the mill opposite the feeder and thrusts back between the rollers the pieces of cane as they come through. Each piece passes three times between the rollers. The crushed cane or chipped is burnt with other fuel for boiling the juice. Two men called *pátkyás* drive the bullocks yoked to the mill. Two called *ádemodes* take the juice that falls into the *mándan*, an earthen pot large enough to hold about sixty gallons, to the boiling pan; and they also remove the boiled juice from the boiling pan or *káil*. The boiling pan, which is large enough to hold about 120 gallons, is placed on a stone and is heated by a long flue. When the scum rises in bubbles and breaks into white froth, the juice is sufficiently boiled. This takes about three to four hours. The impurities in the juice rise with the scum and are taken out with a bamboo sieve or *vávdi*. To cause impurities to rise, the juice is constantly stirred, and sometimes a handful of ashes of the myrobalam and milkbush† or *aghâda* (**Achyranthus aspera**)

* As pointed out in several other places the Natives of India, in many parts of the country, use the tops for their seed-cane. (*Conf. with pp. 128, 140, 184, 217, 240.*)

† If it is intended to give here "milk-bush" as the equivalent of *aghâda*, a mistake has been made, since **Achyranthus** has no milky sap: the *aghâda* yields an ash largely used, however, in dyeing, etc., and is likely enough to be used for the purpose indicated.

MANUFACTURE in Bombay & Sind.

Kolhapur.

are added to it. An expert styled the *gulrándhya*, from time to time takes a little juice, between his fore-finger and thumb to see whether the boiling has been carried on sufficiently. When he is satisfied, the juice is poured into a wooden trough to cool, and from the trough into regular holes made in the ground and lined with cloths to keep out dirt. At this stage the juice is called *kákvi* or molasses, which in the holes crystallizes into raw-sugar, are dark-brown in colour, and weigh thirty-six to forty pounds. The kindling of the fire and feeding it are entrusted to two men called *chuljálya* or hearth-burners. These are generally village Mhars. The burning cinders to light the fire must be brought from a Mhar's house. The labourers who work at the mill are paid in kind at the rate of three canes and 2¼ pounds of raw-sugar. The village servants or *balutedárs* are paid in proportion to the work they do. The carpenter or *sutár* has the largest share of work. He repairs the water-lifts and keeps the mill in good order. He receives six pounds of raw-sugar and eight canes a day while the pressing is going on. The leather-worker or *chámbhár* repairs the leather-bags and buckets and leather ropes and fastenings, and receives half as much as the carpenter. The blacksmith who mends the field tools, the Máng who supplies ropes and whips, the potter who supplies earthen pots, the barber who shaves the husbandman, and the washerman who washes his clothes, are entitled to three-fourths of a pound of raw-sugar and three canes a day so long as the mill is at work. The *tarál* sweeps the place where the mill works and gets three pounds of raw-sugar and five canes. The Brahman astrologer, the Jain Upádhya, and the *Lingáyat Jangam* fix the day for working the mill, and are granted two pounds of raw-sugar on the first day. The village *Gurar* prays to *Ganpati* to remove all difficulties that may come, and the *Mulláni* or Muhammadan priest extends the protection of his patron saint by distributing ashes of frank-incense burnt before the saint. These get one-fourth of a pound of raw-sugar, two canes, and a pot full of juice once only during the course of the pressing. When the pressing and boiling is over and the *gul* is being removed to the village, the village *balutedárs* receive half as much as they have already earned. Believing that retail sale of sugar-canes in the field will bring him ill-luck and free-handed gifts will be rewarded by a plentiful outturn, the husbandman freely gives cane, juice, and bits of new raw-sugar to any one who asks for them, and crowds of beggars throng the field. It is estimated that about twenty to twenty-five per cent. of the produce thus goes in wages and charity. As the juice easily ferments under the heat of the day, pressing and boiling take place at night. For home consumption the husbandman keeps a little molasses. The outturn of the molasses per acre is estimated at about 1,170 gallons, worth about £22-10s. (R225).

"Except in some of the villages of the Alta, Kagal, Karrir, and Shirol sub-divisions, no sugar is made in the State. The craft of sugar-making in Kolhapur is of late growth, and is wholly in the hands of Jains, Lingáyats, and Musalmans. Because it was first made at Yelgund in Alta by a Gujarat Musalman sugar-maker about thirty years ago, Kolhapur sugar is called *Yelgundi*. Of late it has improved in quality and quantity. Most of the sugar-cane juice in Yelgund and in the surrounding villages is made into sugar, and sugar of the present day is far superior in colour and taste to what it was about twenty years ago. The sugar refiner buys the juice off husbandman at 14*s*. (R7) a *can* of 120 gallons. Except that more care is taken to skim off the impurities, the juice is boiled in the same way as in raw sugar-making. To aid the rising of impurities to the surface, a handful of ashes of the *bhendi* (**Hibiscus esculentus**) is dropped into the boiling juice. The boiled juice is then poured into a wooden trough, and from it into earthen jars where it consolidates. After a week or ten days the lumps are put in a boiling pan, rubbed inside with salt water, and heated. The syrup is then poured into a bamboo basket six feet in circumference and two and a half to three feet in height, and placed on a stool nine inches high. Under the stool is dug a hole in which the treacle drains from the basket. For a week the basket is kept thus. Then the surface of the sugar in the basket is stirred to the depth of nine inches, two to three pounds of milk are poured into it, and the surface is smoothed with *pitáli* or platter rubbed with clarified butter. The surface is then covered with a thick layer of moss called *káju* in Hindustani, a piece of coarse cloth and a layer of sugar-cane leaves one over the other. The drainage in the hole below the stool goes on. Every third day the covering of the basket is taken off, the layer of refined sugar which has been formed is removed, and a fresh layer of the moss is laid. In this way all the refined sugar is gradually removed. The treacle which is collected in the hole is sold for making liquor. The average acre outturn of sugar-cane is 3,960 gallons of juice worth about £25 (R250). The same quantity of juice when made into sugar yields about 2,250 pounds of sugar worth £28 (R280), at the average rate of 6*s*. (R3) the *man* of twenty-four pounds. (*Bomb. Gaz., XXIV., 178-180*).

Conf. with pp. 295, 299.

Manufucture of Sugar

MANUFACTURE in Bombay & Sind.

Poona.

427

POONA.—The reader will find in the first half of the District Gazetteer an account of sugar-cane under the section of Cultivation. The following section has been separated, since it not only details the process of sugar manufacture pursued in Poona but in Ahmadnagar and other parts of the Bombay Presidency:—

"When the *káhil* or shallow circular iron boiling pan has been filled with juice, the fire beneath it is lighted and fed chiefly with the pressed canes. After eight to twelve hours' boiling and skimming, the juice is partially cooled in earthen pots and finally poured into round holes dug in the earth and lined with cloth, where, when it forms into lumps called *dhep* or *dhekul* it is fit for market. The pressing is done in the open air or in a light temporary shed, and goes on night and day till the whole crop is pressed. A sugar-cane press costs about £2-10*s*. (R25) and lasts three or four years. The boiling pan either belongs to the owner if he is well-to-do, or is hired at a daily or a monthly rate, according to the time for which it is wanted. The daily hire of a pan varies from 2*s*. to 4*s*. (R1 to R2) and the monthly hire from 10*s*. to £1 (R5 to R10). Each cane-mill employs about twelve workers. Seven remove the canes from the field and strip the leaves. One cuts the canes into pieces two feet long; two are at the mill, one feeding the mill, the other drawing out the pressed canes; one minds the fire, and another the boiling pan. The last is the *gulvia* or sugar-man. He is supposed to know exactly when the juice is sufficiently boiled and thickened to form lumps. As most sugar-cane-growers are without this knowledge, a sugar-man is hired at 6*d*. (4as.) a day or £1 (R10) a month. The two-feet long pieces of cane are passed between the upright cylinders two or three at a time. To stop any leaks the pan is smeared with *lodan*, a glazed preparation of *udid* or *náchni* flour. It is then put on the fire-place, and the hollow between the pan and the fire-place is closed with mud. About 600 pints (300 *shers*) of juice are poured into the pan and the fire is lighted. The boiling lasts six or seven hours during which the juice is constantly skimmed, and lime-water and *náchni* flour are thrown into the juice to keep it from being too much boiled. When the sugar-man thinks the proper time has come, the pan is taken off tne fire and the juice, with constant stirring, is allowed to cool for about an hour. When cool it is poured into cloth-lined holes in the ground two feet deep and a foot-and-a-half across. It is left in the holes for a couple of days until it has hardened into lumps or nodules weighing fifty to sixty pounds (29—30 *shers*). When the lumps are formed they are taken away. If the sugar-cane is of eighteen months' growth, it yields *gul* equal to one-fourth of the juice boiled, in other cases it yields about a sixth. If the juice is allowed to overboil, it cannot make the *gul*; it remains the boiled juice of sugar-cane which is called *kákavi*. The people believe that sugar-cane fed with well water yields one-fifth more *gul* than the same cane fed by channel water. The correctness of this belief is doubtful.

Sugar-cane Plantation.

Poona Factory.

428

Conf. with pp. 214, 320.

"As far back as 1839-40, the growth of Mauritius cane spread greatly in Junnar. The land was well suited to this cane, the supply of water was abundant, and the people were anxious to grow it. Mr. Dickinson, a planter of considerable experience in the West Indies, was employed in making sugar. But the produce did not find a ready market. He turned his refuse sugar and treacle to account by manufacturing rum. In 1841, besides fifty-seven acres planted by the people on their own account, about 100 acres were planted in Junnar under contract with Mr. Dickinson, the manager of the sugar factory at Hivra. The sugar was used only by the European inhabitants of Poona and Ahmadnagar. In 1842-43, the area under Mauritius sugar rose from 157 to 388 acres. The cultivation spread from Junnar to Khed and Pabáe. Sugar-works were started at Hivra by a Joint Stock Company, and were afterwards bought by Mr. Dickinson. In Bhimthadi a Musalman planted some cane in Chakar Bág with the view of making sugar, and some husbandmen turned out sugar equal in grain to Mr. Dickinson's, but not free from feculence. They also made *gul* which was sold at a higher price than that produced from the local cane. At first Mr. Dickinson was in the habit of contracting with the husbandmen to plant cane for him. He was afterwards able to obtain a sufficient supply at all times, chiefly from the gardens of Bráhmans, headmen, and well-to-do husbandmen In 1842, Mr. Dickinson made 87,000 pounds of sugar worth £1,500 (R15,000) more than the outturn of the previous year. Messrs. Sundt & Webbe also planted about three acres of land with Mauritius cane in their garden at Mundhue, about five miles north-east of Poona, and made about 2½ tons (2,826 *shers*) of *gul*, which was sold at 16*s*. (R8) the palla of 120 *shers*. In 1844, the area under Mauritius cane rose from 388 to 547 acres. Mr. Dickinson's farming continued successful, partly because he was able to dispose of his rum and sugar by Government contracts. Many

Rum.

Conf. with p. 175.

MANUFACTURE in Bombay & Sind.

husbandmen were willing to make sugar, but from want of capital and of local demand were obliged to content themselves by producing *gul.*

Poona.

"In 1849 **Mr. Dickinson's** sugar had a good year at Hivra. He made five tons (330 *mans*) of muscavado sugar and sold it to the families of the soldiers and other Europeans at Poona and Ahmadnagar. Among the Natives the demand was trifling, and this discouraged its more extended manufacture. The Natives, even in the immediate neighbourhood, preferred the soft blanched sugars sold by the shopkeepers; their objection to **Mr. Dickinson's** sugar was its colour, but to refine it would have caused a serious loss in quantity. In 1847 a committee which met in Poona to distribute prizes for the best specimens of superior field products, awarded a prize of £30 (R300) to two persons. One of the prize specimens was some grained muscavado sugar, the other was sugar made by evaporation. Before crystallization had set in, this sugar had been poured into pots with holes in the bottoms through which the treacle was allowed to pass. A prize of £20 (R200) was awarded to two other Natives for the best brown sugar; and a third prize of £10 (R100) to two others for the best specimens of *rási* or inferior sugar. All the prize specimens came from near Junnar, and were due to the exertions and influence of **Dr. Gibson.**"

Many particulars of the Poona sugar factory that exist at the present day will be found in other chapters of this article, but the above information regarding Mauritius cane and the early efforts to establish sugar factories may be viewed as of special interest. *(Conf. with pp. 94, 214.)*

Surat. 429

Conf. with p. 231.

Surat.—"With the double object of introducing a new industry and of checking the manufacture of liquor, the Government of Bombay, in 1874 (November 25), authorized the Collector of Surat to spend a sum of £150 (R1,500) in an attempt to introduce the Bengal system of manufacturing sugar from the juice of the wild date tree. Skilled workmen brought from Jessor, in Bengal, succeeded in making sugar of a marketable value. But the returns of the first set of experiments show that the juice of a date tree, which, sold as toddy, brings in a yearly profit of 3*s.* (R1-8), would, if manufactured into sugar, yield only 1*s.* 3*d.* (annas 10). The experiments have been repeated, and the results may be more satisfactory. But so far (1876) there would seem to be little reason to expect that the manufacture of sugar will take the place of the manufacture of toddy" (*Bomb. Gaz., II., 41*).

Thana. 430

"Thana.—Raw sugar is chiefly made in the Bassein sub-division by Pachkálshis, Malis, Native Christian, and Samvedi Brahmans. The sugar making season lasts from February to June. Women and children help by carrying the sugar-cane from the gardens to the sugar-mill or *ghâni.* Eight tools and appliances are used in making sugar. These are the *vila* or sickle for chopping the roots of the cane, worth from 1*s.* to 2*s.* (8 annas R1); the mill or *ghâni,* consisting of two or three rollers, each about a foot in diameter, plain and smooth in body, with the upper one-third cut into spiral ridges or screws into which the screws of the adjoining roller fit and move freely while the machine is working. The rollers fit into circular grooves on a thick horizontal plank supported by two strong uprights. These grooves communicate with each other, and while the cane is being crushed between the rollers, they carry the juice to an earthen pot which is buried below. On the top of the rollers there is another thick horizontal board with circular holes to allow the rollers to move freely round their axes. One of the rollers is longer than the other, and has a square top fitting into a corresponding groove in the yoke-beam. At the slightly tapering end of the yoke-beam, which is about eight feet long and six inches square, is the yoke. Including the uprights the cost of the mill ranges from £7 to £8 (R70-80). Besides the mill, there are required three or four boiling pans, *kadhais,* of copper, hemispherical in shape with two handles, worth from £3 to £4 (R30 to R40) each; five scumming sieves, *manichádivás,* copper saucer-like pans about a foot in diameter, with the bottom full of small holes except a belt near the sides. Over the sieve is a bamboo about three feet long whose lower end is split into three parts, which, by the elasticity of the cane, press tightly against the edge of the sieve and makes the upper part of the bamboo into a handle; five stirring ladles, saucer-shaped bamboo baskets, a foot and a half in diameter and provided with a long bamboo handle, worth 3*d* (2 annas) each; two broad-mouthed cylindrical earthen pots or *koṅdyás* brought from Virár at 1*s.* (8 annas) each; two to four dozen earthen pots, also called *kondyás*

MANUFACTURE in Bombay & Sind.

Thana.

but sloping at the lower end and not cylindrical, worth 3*d*. (2 annas) each; and half a dozen rods for stirring the juice after it is poured out of the boiling pan.

"Besides these appliances one cart, worth from £5 to £6 (R50 to R60), and four pairs of bullocks are required. But the carts and bullocks belong to the sugar-maker's garden rather than to his sugar-making establishment. The earthen pots with narrow mouths at 3*d*. (2 annas) each, which, as is described below, are required for storing such of the boiled juice as is intended to make crystallized sugar are generally supplied by the *váni* customers. Of late, instead of the hemispherical copper boiling-pan, some sugar-makers have introduced the Poona flat-bottomed iron boiling-pan. This is an improvement, as the large iron pan requires less fuel and is not so likely to overflow.

"When the cane is ripe it is pulled out, the tops and roots are cut off, and the canes are taken to the mill. The mill is worked by bullocks, and, as the rollers revolve, a man sits by and keeps feeding them with fresh cane. On the other side of the rollers a second man receives the squeezed canes and heaps them on plantain leaves ready to be again squeezed; for, to bring out the whole juice the cane has to be squeezed half dozen times. As the juice gathers in the earthen pot which is buried below the mill, it is removed to the boiling pan or *kadhai* in a small egg-shaped jar. As soon as enough juice is collected, the pan is moved to the fire-place and the juice is boiled, after mixing with it about a pound of shell-lime brought from Rangoon and Kalamb in Bassein. When the juice begins to boil, the scum is removed by the *manichádivi*, the saucer-like copper sieve which has already been described. If the juice begins to overflow, it is sharply stirred with the long-handled saucer-shaped ladles. The boiling goes on till the juice, if thrown into cold water, becomes as hard as stone. Then the juice is poured into a set of earthen pots or into a bamboo basket lined with a thick layer of dried plantain-leaves, stirred with a wooden rod, and left to cool. If the raw sugar or *gul* is to be made into crystallized sugar or *sákhar*, the juice is heated on a less violent fire and poured into earthen pots with narrow mouths..

"All the raw sugar or *gul* made in the district is sold to local and Márwár Vánis, to whom, in many cases, the sugar-makers are indebted. The price varies from £3 to £4 (R30 to R40) the *khandi* of 25 *mans* (700℔). Raw sugar is divided into three classes, yellow or *pivla*, red or *lál*, and black or *kála*. When the boiled juice fails to become hard enough to make sugar and remains a thick molasses-like fluid, it is known as *kákvi* and is sold for £1-5*s*. to £2 (R12½ to R20) a *khandi* of 25 *mans* (700℔). As is noticed later on, in crystallizing the raw sugar, the part that oozes through the bottom of the jar is also used as molasses. Labourers are seldom employed. When they are, they are paid 6*d*. (4 annas) a day in cash. If they work at night, they get about 6*d*. (4 annas) worth of raw sugar. Each sugar mill requires eight men, four for gathering and bringing the cane, two to watch the mill, and two to boil the sugar. The sugar-pan holds 168 pounds (6 *mans*) of juice and in the 24 hours, if worked night and day, six panfuls can be boiled.

"The owners of sugar-cane gardens, whether they are Mális or Brahmans, prefer to dispose of the sugar in its raw or uncrystallized state. The whole supply of raw sugar comes to be crystallized into the hands of Maráthas and Gujrat traders and Márwár Vánis. The crystallizing of sugar requires four appliances, a number of earthen pots to hold the raw sugar, worth 14*s*. to £1 (R7 to R10) a hundred; a few iron scrapers with wooden handles, worth 1*s*. (8 annas) each; some coarse cloth, worth about 6*s*. (R3); a stone mortar, worth from 6*s*. to 8*s*. (R3 to R4); wooden pestles with iron tips, worth from 1*s*. 4*d*. (8 to 12 annas); and sieves, worth from 1*s*. 6*d*. to 2*s*. (12 *annas to R1*). The work is done by Native Christian or Musalman labourers, who are employed by the Vánis at from 6*d*. to 7½*d*. (4—5 annas) a day. The Vánis buy the raw sugar in large earthen pots holding about 56℔ (2 *mans*). To crystallize the sugar, the first step is to bore a hole about the size of the little finger in the bottom of each of the earthen pots which contain the raw sugar. The sugar pot is then set on a broad-mouthed earthen jar called *hánd*. The cover on the mouth of the raw sugar is taken away and a layer of water plant, **Hydrilla verticillata**, locally called *sákhari shevál* or sugar-moss is laid on the top of the sugar. On the third or fourth day the plant is taken off and the surface of the sugar, which, by this time, has become crystallized, scraped with a curved notch-edged knife and put on one side. The top layer is called the flower or *phúl* and weighs about a pound. The second layer, which is a little duller in colour, is named *dána* or grain, and weighs about a couple of pounds. The sugar of both sorts is then laid in the sun on a coarse cloth 16 yards long and one yard broad. After lying in the sun for one or two days, it is pounded in a stone mortar or *ukhali* by iron-tipped wooden pestles. It is then passed through a sieve and is ready for sale. Within the last 30 years, competition from Mauritius is said to have reduced the production of crystallized sugar from 600 to 60 *khandis*.

Conf. with p. 31.

MANUFACTURE in Bombay & Sind. Thana. European Plantations. 431

Conf. with pp. 37, 48, 62, 63, 91, 93, 103 161-162, 212, 306.

"The great growth of sugar-cane in the neighbourhood of Bassein has on two occasions, about 1830 and in 1852* led to the opening of a sugar factory in Bassein. In 1829 a **Mr. Lingard** applied for land at Bassein to grow Mauritius sugar-cane and other superior produce, and to start a sugar factory. Government, anxious to encourage private enterprise, gave him a 40-years' rent-free lease of about 83 acres (100 *bighás*) of land on the esplanade of Bassein fort. They also advanced him £2,300 (R23,000). **Lingard's** mill was soon built and some sugar-cane was planted, but his death in 1832 checked the scheme. At his death he owed Government £2,300 (R23,000) the security being a mortgage on the building, worth £220 (R2,200), the land, and its crops. Government took temporary possession of the estate. When the Revenue Commissioner visited the place in 1833 he found the mill greatly out of repair. He suggested that it should be made over to some enterprising man, and a Hindu named **Narayan Krishna** was given a two-years' rent-free lease of the estate. In 1836 **Narayan's** tenancy expired. He had failed, as he could neither bring his sugar to perfection nor persuade other planters to press at his mill. Government, who were exceedingly anxious to extend the growth of Mauritius cane, engaged to remit the rent of all land under that crop and resolved to let the Bassein estate on favourable terms. In 1837 **Messrs. McGregor Brownrigg & Co.** were allowed a trial of the estate for three months, and, being satisfied with the result, they asked for a long lease. In 1841 they were granted in perpetual lease some 115 acres (136 *bighás*) near the travellers' bungalow on the esplanade. The lease began to run from 1839. For forty years they were to hold the land rent-free and were then to pay a yearly rent of £2-4s. the acre (R22 the *bighá*). They agreed to grow sugar-cane, but the promise was made binding for only seven years, as Government hoped that, by that time, the manufacture of sugar would be firmly established. This hope was disappointed. **Messrs. McGregor Brownrigg & Co.** continued to grow sugar-cane only so long as they were obliged to grow it. In 1843 they reported that from the poorness of the soil and the want of shelter, sugar-cane did not thrive and did not pay. They levelled the ground, dug wells, and grew other kinds of superior produce. In 1848 they sold the estate to a **Mr. Joseph**, who, in 1859, sold it to one **Dosabhai Jahangir**, and he, in the same year, sold it to a **Mr J. H. Littlewood.**

"In 1829 the land inside Bassein fort was leased to a **Mr. Cardoza** for thirty years at a yearly rent of £40 (R400). He died soon after, and in 1836, to help his widow, the rent was lowered by £10 (R100) with a further reduction of £2-18s. (R29) on account of excise payments. In 1852 **Mrs. Xavier**, a daughter of **Mr. Cardoza,** was allowed to repair the ruined church of St. de Vider and turn it into a sugar factory. **Mrs. Xavier** seems to have sublet the land to **Mr. Littlewood,** who, with a **Mr. Durand,** fitted up a building for making and refining sugar. The scheme proved a failure, and was for a time abandoned. Afterwards, with the help of fresh capital, a new start was made under the name of Bassein Sugar Company. New machinery was bought and an experienced manager and assistants were engaged. In 1857 **Mr. Macfarlane,** a Bombay solicitor, and **Mr. J. H. Littlewood** (that is, the Bassein Sugar Company) applied for a new lease on easy terms, as **Mrs. Xavier** was willing to forego the unexpired portion of her lease. On March 21st, 1860, **Messrs. Macfarlane and Littlewood** were granted a thirty-years' lease of certain lands in the fort of Bassein on a yearly rent of £27-2s. (R271). The lease was to be renewable at the end of the thirty years. **Messrs. Macfarlane and Littlewood** carried on business under the name of the Bassein Sugar Company until 1861, when the concern was sold to **Messrs. Lawrence & Co.** In 1868 **Messrs. J. H. Littlewood, H. Worthing,** and **Navroji Manekji** bought the estate. **Mr Littlewood** had the management, and, though the Sugar Company has long ceased to exist, he still (1881) lives in a small house in the fort" (*Bomb. Gaz., XIII., 391-395*).

VIII.—MADRAS.

MANUFACTURE in MADRAS. 432

In the volume of the Proceedings of the Honourable the East India Company (to which the writer has repeatedly referred), there occurs a long and detailed account of sugar manufacture in Ganjam as practised in 1792.

* Consult the chapter above on the History of the effort to start Sugar-planting as a European Industry in India. Particulars will there be found of a still earlier effort, *viz.*, in 1792. *Conf. with pp. 93, 94.*

MANUFACTURE in Madras.

This was furnished by **Mr. Alexander Anderson**, and appeared as an appendix to **Dr. Roxburgh's** very admirable paper on sugar-cane cultivation. That volume, at pages 245-275; again, appendix second, pages 22-26; pages 45-50; appendix third, pages 1-22, and at several other places, furnishes particulars of the Madras sugar manufactures and trade prior to 1822. The drawings of sugar mills (given in the volume) are peculiarly instructive, as they admit of comparison with those now in use. Although two or three large sugar factories or refineries have now for some years been in existence, the fittest perhaps surviving out of the many started towards the close of the last century, still the Native industry cannot be said to have materially changed. One point only seems worthy of very special remark. If the statistics of past and present trade can be depended upon, there has been recently a greater expansion of palm than of cane-sugar. By modern returns nearly half the sugar of Madras is derived from palms. No writer that the author has been able to discover deals with this fact anything like to a satisfactory extent. There are several palms that yield sugar, and while the area occupied by these has been determined, few authors have apparently considered it necessary to investigate the yield of these comparatively, nor to detail the methods of manufacture of sugar from them as followed by the people. It is often affirmed that one reason of the greater success recently of Madras as compared with Bengal, is due to the large amount of palm-sugar which it obtains annually. But the area of date-palms is quite as great in Bengal as the total area of the sugar-yielding palms of Madras (Mysore being excluded from consideration). If there be any such superiority it would seem to be due either to the fact that the palmyra (the chief sugar-palm of Madras) yields better sugar than the Bengal date-palm, or to the existence of a superior system of palm-sugar manufacture. In Mysore the chief sugar-yielding palm appears to be the date. **Dr. Buchanan-Hamilton** furnished, nearly a century ago, the only detailed account of the date-palm industry of that province, but **Roxburgh** seems to have regarded the palm-sugar of the portion of Madras of which he wrote, as scarcely deserving of special consideration. The reader will find many of the obscure questions of the Madras and Bengal palm-sugar industry, discussed in the chapters devoted above to Cultivation. It is only necessary, therefore, to add that the imperfect nature of the information available and the limited space that can be devoted to it, both combine to preclude the manufacture of sugar in Madras being here fully dealt with. The province of Mysore, has in fact, to be left out of consideration. But it may be repeated that we shall never obtain a definite knowledge of the Indian sugar question, until palm-sugar has not only received more careful consideration but been made the subject of independent investigation. While of Madras and Bengal, it may be said, palm-sugar is almost of equal importance with cane; the most that authors have considered necessary to say of palm-sugar has been that once the juice is obtained its subsequent treatment in the preparation of sugar is identical with that of cane. Most writers have thought necessary to go into every detail of cane culture, such as the yield, cost, profit, etc., but of palm-sugar they have deemed it sufficient to give the bald statement of area and production. Much of the apparent repetition in the present article is due to the fact that few writers deal with the same features of either cane or palm-sugar; a compilation of their various opinions into one article became therefore impossible. This remark is more peculiarly applicable to the Madras Presidency.

Palm-sugar. *Conf. with pp. 138, 226-227, 231, 266, 270, 352, 361, 370.* **433**

But to follow the usual course, the passages which may now be given will be found to amplify the information already furnished, in the special chapter on Sugar-cane Cultivation, more specially, with reference to manufacture:—

Presidency. **434**

"MADRAS PRESIDENCY AS A WHOLE.—"Sugar or *jaggery* is manufactured

MANUFACTURE in Madras.

from the produce of the following acreages; under cane, 69,383, cocoa palms, 5,706; palmyra, 24,884; date palms, 1,575; sago palms, 19. In 1882-83 the exports to foreign countries included 13,219 cwt. of refined sugar and 1,119,930 cwt. of unrefined sugar. The ordinary jaggery is made by a rough process, the canes being crushed in a wooden roller mill and the juice boiled in dirty utensils; but refined sugar of a superior quality is manufactured by **Messrs. Parry & Co.** and by a native manufacturer in South Arcot who follows the usual method of boiling in *Vacuo*, and also by **Messrs. Minchin Brothers & Co.** of Aska in Ganjam who have adopted the method of diffusion" (*Mad. Man. Admn., I., 363*).

Ganjam. 435

GANJAM.—**Dr. Roxburgh** in 1792 published the following special report on sugar manufacture :—

"After the cane is ready, it is cut in pieces of a foot or eighteen inches long, and on the same day it is cut, these pieces are put into a wooden mill* which is turned round by bullocks. On one side of the mill is a small hole, sufficient to let the juice pass through, which is received in an earthen pot placed for the purpose. The juice is then strained into other pots containing about twenty-four *puckar seers*, and to each pot of juice is added about three ounces of quick-lime. It is then boiled for a considerable time, till on taking out a little and rubbing it between the fingers it has a waxy feel, when it is taken off the fire and put into smaller pots with mouths six inches in diameter. The mass may now be kept in this state for six or eight months or more, and it is necessary, at any rate, to do so for a month or six weeks. When the process is intended to be continued, a small hole is made in the bottom through which the syrup drains off. It is then taken out of these pots and put into shallow bamboo baskets, that any remaining syrup may exude; after which it is put in a cloth, and the syrup is squeezed through the cloth, adding a little water to it occasionally, that it may be more perfectly removed. The sugar is then dissolved in water, and boiled a second time in wide-mouth pots containing only three seers with not too fierce a fire, adding from time to time a little milk and water, and stirring it frequently; which is used by these people to clarify it, instead of eggs, which their religion forbids them to touch. The scum is removed as it is thrown up, and when it resumes the waxy feel on rubbing a little of it between the fingers, the process is finished, and the sugar is put into small wide-mouth pots to cool and crystallize; after which a small hole is bored for the purpose of draining off any little quantity of syrup that may still exude. The outside of the pots are now covered with cow-dung, and for the purpose of making the sugar white, or removing any syrupy or blackish appearance, the creeping vine, called in Moors *pancha-dub*, and in Tellinga's *nectynas* growing in tanks and marshy places† it is put on the top of the sugar in the pots, and renewed every day for five or six days. Should the sugar on taking it out of the pots be blackish or less pure towards the bottom, the bottom of the loaf being set up on this plant and renewed daily will effectually remove that appearance. If it is wrapt in a wet cloth and renewed twice a day, the sugar will also become white; it must be then thoroughly dried and kept for use.

"To make sugar-candy, the sugar must be again dissolved in water, and boiled in the same manner as before, adding milk to it in small quantities : the proportion of three seers of sugar and half of milk, with water to dissolve the sugar. It is then put into other wide-mouth pots, with but three seers in each pot, putting thin slices of bamboo, or some dried date leaves, which prevents the sugar as it candies from running into large lumps."

Roxburgh commenting on the above adds :—

"Here we see a very superior sugar, and sugar-candy of the first quality, manufactured in a simple but tedious manner, and at a most trifling expense. A few earthen pots are the only vessels or boilers they require. But it is not to be imagined that such would succeed, if the work were carried on to any great extent. The iron

* A very large wooden mortar, the pestle of which rests obliquely towards the side, and is so moved round in a circular manner by means of a lever fixed at its top projecting eight or ten feet over the side, to end of which lever two bullocks are yoked. It is the common oil-press of the Hindus, but is exceedingly inconvenient for extracting the juice of the sugar-cane, and shews how far behindhand the natives of that district are in this part of the process, when compared with the small convenient mill employed hereabout. Nothing can shew more clearly how exceedingly adverse these people are to any change in their old customs.

† **Hydrilla verticillata** : it grows in great abundance, particularly in clear standing water near the sea. *Conf. with p. 31.*

MANUFACTURE in Madras.

Ganjam.

boilers employed hereabout might be laid aside for those of copper or of cast-iron from Europe or not, as they like themselves, for it seems of no great consequence; but by having a greater number of them for the liquor to pass through and be well clarified in, would render unnecessary the second process mentioned by Mr. Anderson, which, on account of its tediousness must become very inconvenient; consequently, all that seems to be wanted to render the sugars made hereabout, fit for any market, is a boiler, or two or three more in each set, with wooden coolers, instead of losing time to let it cool in the boiler, as is the practice here at preesnt, the addition of some quick lime, and probably alum, to the cane-juice, and the subsequent claying of it in conical pots, as is done in the West Indies, for which process the Natives of the Ganjam district substitute moist Hydrilla (or Vallisneria) for covering the sugar in the pots with, wrapping the loaves, when not sufficiently white, in wet cloth, to extract the molasses" (*Papers on Cult. & Manuf. of Sugar in British Iudia, published 1822, Appendix 3rd, pp. 1-8*).

Wet cloth. *Conf. with p. 265.*

North Arcot. 436

NORTH ARCOT.—"To produce coarse sugar the boiled juice is rapidly stirred about with a rolling pin until it has set. For fine sugar and sugar-candy the process is slightly different, the boiling being stopped earlier than for jaggery. When sufficiently boiled the juice is put into pots, which are covered, and also to stand for a fortnight, by which time their contents have become solid. A few holes are then made in each of the pots, which are placed upon empty ones, and in the course of three weeks, most of the molasses drips through, leaving behind a crust, some two or three inches deep of fine sugar, which is at once removed, the rest being allowed to drain for about a month or six weeks longer. The sugar thus produced is further purified by boiling. It is then strained and boiled again for another hour, towards the close of which a little milk and ghee are added. Finally, the syrup is moved from the fire and well stirred for a quarter of an hour. When dry the finest native sugar, called *búrá*, is produced.

"To make sugar-candy, the second of the above two boilings is slightly curtailed, and the syrup is poured into pots in which are placed thin spits of bamboo. Cloth is then tied over the mouths of the pots, and they are stood for forty days upon paddy husk. After that the fluid portion which remains, called '*kalkanda panakam*' is poured off and considered a very good and wholesome beverage. The bamboo spits coated with sugar crystals are separately secured. This manufacture is almost confined to the town of Baireddipalle in the Palmanir taluk.

"Sugar and jaggery are largely used by the natives, mixed with their food or spread upon cakes with ghee. Pieces of the cane are often bought by the poor, stripped of their bark and masticated" (*Man. North Arcot Dist., 327*).

Godavery. 437

GODAVERY DISTRICT.—"The mode in which the sugar or jaggery is made is as follows:—A large shed is prepared and arranged so as to admit the south wind, and a fire-place some eight feet in diameter is constructed, to hold a round iron boiler. The canes are brought direct from the field and at once passed through a press composed of two circular pieces of hard wood, which are made to revolve by rude machinery worked by bullocks. Under the press is a pit in which is placed a chatty, and into the chatty the juice falls as it is expressed from the canes. The canes are generally cut in two before being placed in the press, and the head of each (to be used as seed for the next year) is at the same time removed. The canes will not keep and must be passed through the press the day that they are cut. When about 20 *chatties* of juice have been expressed, their contents are poured into the boiler, and boiled for nearly an hour. To each boiler full of juice a viss of *chunam** is added. The juice as soon as boiled, is poured into an iron pot, and after being stirred for a while, is poured out again on a mat, where the sugar dries and becomes hard. It is then broken up and packed up in baskets, containing five maunds each:—

Expenses for Eight Acres.

	R
Khist of 1st year, during which the ground lies fallow . .	60
Do. 2nd year	60
Cost of manuring	60
Preparing beds and planting	50
Weeding	50
Tying up the canes five times at R40	200
Half value of 60,000 bamboos at R8 per 1,000, R480 . .	240
Making jaggery	132
One-sixth value of 24 bullocks at R15, R360	60
Cost of feeding do.	89

* Lime.

MANUFACTURE in Madras. Godavery.

	R
One-fifth value of boiler, at R100	20
Half value of press, at R32	16
Half value of sheds, at R16	8
Making fence	50
Watchers to keep off jackals and cooly for watering	270
Firewood	100
TOTAL	1,456
Returns.	
120 *pooties* jaggery at R20	2,400
Deduct expenses	1,456
Profit on eight acres in two years	944
Or per acre per annum	59

"This is an estimate of the expenses and returns for a *pooty* of eight acres. It applies to the Mogaltur and other southern *talúks* only—the system in force in the sugar-growing *talúk* of Peddapur, being quite different. The bamboos are calculated to last two years, and therefore, half their value is entered. The press and the sheds are supposed to last each two years, the boiler five, and the bullocks six years, and the proportions of their value have been entered accordingly" (*Settlement Rep. 1860, 142, 143*).

BOUNTIES PAID TO, AND DUTIES LEVIED ON, SUGAR.

BOUNTIES & DUTIES. 438

It will be seen from the historic chapter that, for many years, what amounted practically to a sugar bounty, paid by England to her West Indian Colonies and against Indian sugar, existed in the higher import duty levied by Great Britain on Indian sugar. This, as Robinson (*Bengal Sugar Planter*) remarks, amounted to an additional burden of 8*s*. per cwt. on Indian sugar. On rum "the difference," he adds, "was even more oppressive and acted as an effectual check to the application of British enterprise to the growth of these articles." The duty was ultimately, however, equalized (in 1836), but it is perhaps doubtful if India has to-day attained the position it might have enjoyed had so great opposition not existed against it in the earlier years of the creation of her present foreign trade. Then, again, a bounty, it may be said, was paid to Bengal as against Madras and Bombay. These two Presidencies were at first regarded by the Honourable the East India Company as undesirable regions from which sugar should be exported. This restriction was, however, early removed in the case of Madras, but survived for some years later with Bombay. It was thought by the Company that as the Western Presidency did not produce enough for its own consumption, it should be debarred from participating in the export traffic. So, in a like manner, a distinct advantage was gainsaid to foreign sugars, in competing in the Indian market, by the fact that heavy restrictions were imposed on the internal transit of sugar. This arose largely from what was known as the SALT LINE. It became necessary to protect the Company's salt interests. and a measure was gradually matured and which continued till 1874 whereby a large tract of country was regularly patrolled. This not only regulated the salt traffic, but to do so necessitated a complete registration of all goods that passed either way across the line. On sugar this was peculiarly injurious, as the line may be said to have crossed the numerous routes by which Bengal and North-West sugars could reach Bombay. It became actually cheaper to convey sugar from the Straits, China, Mauritius, and other foreign countries, than from Bengal. To contend against this state of affairs the exports and imports sent coastwise were

Rum. 439 *Conf. with pp. 93, 95, 96, 104, 158, 175, 320, 321.*

Salt Line. *Conf. with p. 420.* 440

BOUNTIES and Duties.

Salt Line.

allowed to pass free of duty. The salt difficulty was got over, however, by the Government of India purchasing up all the salt mines, lakes, etc. The duty on salt could thus once for all be levied before the article left the sources of supply, so that it became no more necessary to establish any internal registrations. As stated, therefore, the salt line was entirely removed, and from that date a considerable improvement in the Indian internal transactions took place. But an even greater impetus to the home production and consumption of sugar was given by the energetic efforts put forth to extend canal irrigation and to open up the country by the formation of great arteries of communication in main lines of railway. It has already been remarked that what has taken place with sugar, since India became a British Empire, has been the extension of cane cultivation into suitable regions where it was formerly little, if at all, grown owing to want either of water (irrigation) or means of export. The total production and consumption in India has, therefore, greatly extended, though it seems likely that some of the tracts formerly regarded as the sources of Indian supply have relatively become of less importance.

It does not fall within the natural scope of this work to review the measures which have been adopted by England and other Continental countries, from time to time, to obtain a revenue from sugar or to protect their sugar interests. The much talked of sugar bounties of the present day, it may be said, are not by any means, however, the first occasion when certain sugar interests have been fostered or protected by receiving national aid. Towards the close of the last century the English Government adopted a system of direct bounties and drawbacks of sugar duties. Thus, for example, we read that in 1766 "the merchants and traders of the city of Dublin represented to the Lord Lieutenant of Ireland, that four thousand families were supported by the trade of refining sugar, in which a capital of £340,000 was engaged; and they complained that the bounty given on the exportation of English refined sugar to Ireland was a hardship upon their trade, which it could not possibly bear. They, therefore, begged of him either to endeavour to get the bounty taken off from refined sugar shipped for Ireland or to promote a bill in the Irish Parliament, for laying a duty on the sugar when landed in Ireland which should be equivalent to it." It is probable that with no other article of trade could the ultimately injurious effects of protection and bounties be shown than with sugar. Reform after reform, when first contemplated, was opposed by the most powerful interests. Calamities were foretold that in their ultimate effects would bring ruin to every household, but scarcely had the measures thus opposed become law than they were admitted to have proved of the utmost value. The onward progress of the sugar trade has, it might almost be said, marked events of national importance in the history of modern times. In India sugar and saccharine substances, as also tea, bore an import duty of 7½ per cent. (Act XVII.) from March 1867 till 1875 In that year (Act XVI. of 5th August), the import duty was reduced to 5 per cent. On the 10th March 1882 (Act XI.) both sugar and tea became free of any import duty. The loss of this source of revenue was not, however, serious. The following exhibits the amounts collected in India :—

	1875-76.	1876-77.	1877-78.	1878-79.	1879-80.	1880-81.	1881-82.
	£	£	£	£	£	£	£
Sugar, all kinds .	44,79[illegible]	18,506	33,190	62,803	49,099	81,290	60,616
Tea . . .	14,813	7,753	10,603	7,906	11,238	14,320	10,969

S. 440

BOUNTIES and Duties.

An export duty of 2 per cent. was also imposed by Act XVII. of 1865, but this was repealed by Act XVIII. of 1866, since which date the export of all classes of sugar from India has been free.

Remission of Duty.

When the remission of the import duty was contemplated, the measure was opposed on the ground that it would be "disastrous to the trade in Indian grown sugar by enabling foreign sugar to undersell it." **Mr. J. E. O'Conor**, in his *Review of the Trade of British India for 1882-83*, alludes to this subject in the following passage:—"The dismal anticipations of the opponents of the measure have certainly not been realised, for in the first year following the remission the quantity of sugar imported has diminished by more than 13 per cent. as the figures in the margin will show. These figures refer to refined sugar, which, however, constitutes the bulk of the imports into India, the imports of unrefined sugar being a very small business."

	Cwt.	R
1878-79 .	918,202	1,47,75,653
1879-80 .	647,630	1,06,59,414
1880-81 .	982,262	1,60,96,243
1881-82 .	772,519	1,24,21,892
1882-83 .	669,348	1,08,56,003

In the *Review* for the following year **Mr. O'Conor** wrote: "The quantity of refined sugar exported last year was more than three times as great as in 1881-82, that of unrefined sugar was nearly 37 per cent. larger. At the same time, as pointed out in another part of this *Review*, the import of foreign sugar into India materially decreased. Bombay, in fact, was supplied with sugar from Calcutta to the extent of 67,440 cwt; in 1881-82 the quantity was only 29,464 cwt.; in 1880-81, 23,590 cwt. The cane crop was a very good one, and supplies for manufacture abundant and moderate in price." "Most of our unrefined sugar is exported from Madras, chiefly to England, for refining and for brewers' purposes. Refined sugar is still a comparatively small trade for export, though it has expanded very greatly during the year." In the *Review* for 1883-84 **Mr. O'Conor** again returned to this subject: "The quantity of sugar exported last year exceeded that of 1881-82 by 77 per cent., the value of the trade being 58 per cent. larger. This year (1881-82) is taken for comparison with last year because the duty of 5 per cent. on imported sugar was taken off, with other import duties, at the end of that year (in March 1882), and the remission was vehemently opposed by the representative in the Legislative Council of the mercantile community of Calcutta, on the ground that it would assuredly bring about the extinction of the sugar industry in Bengal. The prediction, so far, has been singularly falsified, and if the trade should collapse now, after having had two full years since the abolition of the duty, a far more flourishing existence than it had previously known, its decay must be attributed to other and wholly different causes than the removal of a protective duty. At present, however, there is no sign of decay, and this is all the more remarkable considering the condition of the sugar markets in England to which most of the Indian sugar is sent." In his very next *Review*, however, **Mr. O'Conor** had to comment on a radical change in the sugar trade of India. The imports were very nearly double those of the previous year, while the exports had seriously fallen off. Whether or not the removal of the Indian import duty favoured the admission of sugar may be accepted as an open question. **Messrs. Turner, Morrison & Co.**, of Calcutta, hold that the abolition of the import duty "has entirely killed the trade between Calcutta and Bombay." Beet-root production had by 1884-85 not only begun to flood the markets of Europe with a cheap sugar, thus depriving India very largely of her foreign outlets, but it had liberated large quantities of colonial sugar which thus sought among others an Indian market. To the removal of the import duty might, as it seems, be attributed some share in the facility with which this new import trade has developed. But if it be admitted that the

Conf. with p. 344.

BOUNTIES and Duties.

Remission of Duty.

existence of an import duty would have operated against the growth of this new Indian supply, it would only have thereby liberated a certain amount of sugar to compete with the Indian exports in the foreign markets. The imports might have been less, but the exports would very probably have fallen to an even lower ebb than they have as yet attained. The radical change of the Indian trade which sprang suddenly into existence in 1884-85 has since continued. In the historic chapter this has been already alluded to, but it may be demonstrated by the figures of the imports and exports for last year. These were, imports 2,743,491 cwt. and exports 824,741 cwt. Thus, far from India now exporting more than she imports, last year she received 3 cwt. of sugar for every cwt. of *gúr* she exported. And this revolution is admittedly the result of the Continental system of granting bounties on beet-sugar. The subject of the beet-sugar bounties has already been briefly indicated in the concluding paragraph of the chapter on the History of Sugar,* but it seems probable that a more detailed statement might prove useful. That object could not be better attained than by furnishing some of the leading facts so ably dealt with by Dr. Giffen, in his report on *The Progress of the Sugar Trade*, which appeared as a Parliamentary paper in May 27th 1889:—

Conf. with pp. 20, 40, 329, 341-342, 344, 346.

Amount exported from bounty-giving Countries.

441

"BOUNTIES, AND THE AMOUNT EXPORTED FROM COUNTRIES GIVING BOUNTIES.

"It is foreign to the purpose of this Report to deal with any controversial matter, but it is proposed to show what the facts are as to the amount of exports from the leading continental countries which have a system of duties and drawbacks in which bounties arise, and to indicate in what way calculations as to the amounts of the bounties may be connected with the facts as to the exports.

"It is agreed that the bounties here in question, for the most part, were not in their origin formally given as bounties, though in recent years the fact of bounties being given has been fully acknowledged, and in most cases laws have been passed with the full knowledge that bounties would arise, and with the intention that they should be given. They arise in the administration of duties upon sugar. Import and excise duties being levied, drawbacks are necessary in order to permit the countries levying such duties to send their produce into the general market of the world. Bounties arise because the drawbacks are in excess of the duties which have been levied previously on the sugar when produced or imported. The 'roots' or 'juice' or the like raw material from which sugar is made, when the duty is levied, are calculated to yield so much sugar, and the duty is assessed accordingly. The raw sugar, when the duty is levied, is calculated to yield so much refined sugar, and the duty is assessed accordingly. The raw and refined sugar actually yielded, respectively, prove to be in excess of the calculations, and the surplus either passes into consumption duty free, or, if exported, receives a drawback as if it had paid duty, The amount of revenue thus lost to the exchequer is spoken of as a bounty.

"It appears important to distinguish, however, between the amount of revenue thus lost on surplus sugar which is not exported, and surplus sugar which is exported. In either case the exchequer loses something it would have had if the duty had been strictly levied, but in the former case there may be no bounty, properly speaking, as the duty thus lost may sometimes be a mere reduction of tax to the consumer within the country; and in any case the matter may be considered as a purely internal affair of the countries concerned, and not affecting the trade of other countries in the same way as an actual bounty on export. At any rate, whatever may be the precise effect of a surplus of sugar which is not exported, the bonus on it cannot be spoken of as a bounty on export, whereas, if the drawback is given on surplus sugar actually exported, there is clearly an export bounty. In dealing with the statistics it is proposed to keep this distinction in mind.

"There are also several methods of calculating the amount of the bounty arising in this way. Naturally, foreign Finance Ministers, and financial authorities reckon the whole surplus of sugar escaping duty, whether an export drawback is paid or not, as receiving a bounty; and, dividing the aggregate production by the aggregate amount lost to the exchequer, they say the amount is so much per ton produced. It would be a different calculation, however, to reckon only the surplus sugar exported as receiving an export bounty equal to the amount of duty that sugar had paid; and the amount so calculated again may be divided either by the whole

* *Conf. with pp.* 19-20, 39-40.

BOUNTIES and Duties.

quantity of sugar exported, so as to shew the average bonus on the export, or it may be divided by the whole production obtained by means of export bounties. One of the two latter methods appears to be the more precise, but it seems necessary in stating the facts that the different methods of calculation should be stated and explained, and their exact bearing made clear.

Exports from Bounty-giving Countries.

"In the Appendix [1] a brief account is given of the present law as to duties and drawbacks on sugar in the principal countries of the continent, with references to the immediately previous legislation. Practically, in all the countries named, the fact of a bounty at the present time is officially acknowledged, and, whatever questions may arise as to the exact amount of surplus sugar in each case, these official statements, which are summarised in the Appendix, appear to be explicit enough for the practical consideration of the average effect of the bounties.

"In the following table, accordingly, an attempt is made to show in a condensed form on the authority of these official data the practical effect of the system of duties and drawbacks, on the average, in several of these principal countries. The principal point in each case is the surplus sugar which escapes duty, the other particulars being a deduction from that figure and from the rates of duty in connection with the figures of production and export.

Calculation as to Amount of Bounty on Sugar given by the undermentioned Countries.

	France (1887).	Germany (1888 Law).	Belgium (1887).	Holland.
1. Total production of sugar.	555,000 tons*	990,000 tons*	150,000 tons*	36,000 tons †
2. Estimated proportion of surplus sugar manufactured to total production as in 1.	36½ per cent.	25 per cent. .	20 per cent. .	16 per cent.
3. Amount of surplus sugar.	200,000 tons .	250,000 tons.	30,000 tons .	6,000 tons.
4. Rate of duty	50 fr. per 100 kilos., plus 10 fr. on all sugar. ‡	10*s.* 3*d.* per cwt., *viz.*, 6*s.* on all finished sugar and 4*s.* 3*d.* on the roots. §	45 fr. per 100 kilos.	27 fl. per 100 kilos.
5. Estimated total loss of revenue from surplus sugar escaping duty.	4,000,000*l.* ‖ .	1,000,000*l.* .	550,000*l.* .	162,000*l.* ¶
6. Estimated bonus on production, dividing total loss of revenue by quantity produced.	7*l.* 4*s.* per ton	1*l.* per ton .	3*l.* 13*s.* per ton.	4*l.* 10*s.* per ton.

[1] Dr. Giffen alludes to an appendix to his Report in which the information is given in Table XXVI.—*Ed., Dict. Econ. Prod.*

* Including imports of raw sugar.

† Not including imports of raw sugar as in this case bounty does not arise to the same extent, it is believed, as in connection with the home production.

‡ The 50 fr. is the ordinary sugar duty levied on the roots, or raw sugar, according to the legal yield, which is known to be less than the real yield; the 10 fr. is levied on the finished sugar and is paid on the surplus as well as the non-surplus sugar. The surplus sugar only escapes the 50 fr.

§ The 4*s.* 3*d.* on the roots is the part of the duty which the surplus sugar escapes. *See* above note as to France.

‖ The surplus sugar escapes the ordinary duty, but pays 10 fr. per 100 kilos. *See* note above.

¶ Dr. Giffen deals with this in the Appendix, page 66, of the original report.

SACCHARUM :
Sugar.

Sugar Bounties and Duties.

BOUNTIES and Duties.

Exports from Bounty giving Countries.

	France (1887).	Germany (1888 Law.)	Belgium (1887).	Holland.
7. Total sugar exported.	159,000 tons.	619,000 tons.	111,000 tons.	96,000 tons.
8. Apparent bonus on export, dividing total drawback on surplus sugar exported by total quantity exported.	3,180,000*l.*, or 20*l.* per ton.	1,000,000*l.*, or 1*l.* 12*s.* per ton.	550,000*l.*, or 5*l.* per ton.	162,000*l.*, or 1*l.* 14*s.* per ton.
9. Estimated bonus on production, dividing total drawback on surplus exported by quantity produced.	5*l.* 14*s.* per ton.	1*l.* per ton .	3*l.* 13*s.* per ton.	4*l.* 10*s.* per ton.

"Thus, in whatever way the bounty is calculated in the case of the countries named a large bonus is given to the producers and manufacturers in the sugar trade by the Governments concerned. The effective bounty on export may not always be as great as the loss of revenue through surplus sugar escaping duty; but in the case of Germany the whole surplus and more appears to be exported; in the case of France very nearly the whole surplus is exported; and in the case of the Netherlands and Belgium much more than the whole surplus is exported. This surplus then receives a bounty per cwt. or per ton equal to the duty in those countries, and even when the amount of the bounty thus received is divided up among the whole quantity exported and still more the whole quantity produced, it is still, in some cases at least, very large. The figures as to Russia and Austria cannot be treated in exactly the same way, the bounty arising in a somewhat different manner in those cases. In any case, however, it will be understood the table is inserted here for the sake of reference to show the different ways in which calculations may be made as to the amount of bounties.

"The farther observation may also be made that the effectiveness of the bounty will depend on the market for the surplus sugar obtainable. There must be an effective home market for the non-surplus sugar to begin with, and there must be an effective foreign market where the surplus sugar can be sold. Otherwise the drawback on export will not be available as a bounty, and in proportion as the price falls in the foreign market its effectiveness will diminish. But the precise consequences and effects of bounties given on sugar in connection with the system of duties and drawbacks are involved at this point, and farther explanations would bring in matters of controversy and argument.

"Putting all the figures which have been stated together, the amount of the total bounties on export, with the amount per ton of surplus sugar exported, and per ton produced, as regards the countries named may be stated as follows :—

Calculation of Bounties from data stated above and in Appendix.

	Total Bounty on Export.	Rate per Ton on Surplus Exported.	Rate per Ton on Production.*
	£	£ s.	£ s
France	3,180,000	20 0	5 14
Germany	1,000,000	1 12	1 0
Belgium	550,000	5 0	3 13
Holland	162,000	1 14	4 10
Austria-Hungary (maximum) . . .	500,000	†	†

* The rate per ton on production is on the production plus the imports in the case of all the countries except Holland, where it is on the home production only. *See* note to previous table.

† Bounty does not arise in connection with an export surplus.

BOUNTIES and Duties.

Exports from Bounty-giving Countries.

"It may be added without going into details, that bounties do not appear to have increased generally since the adoption of the law of 1884 by France, which was, however, a very great step in the direction of an increase. As to the laws now being passed or contemplated, no information can properly be given, as this question belongs to the Commission appointed under the Convention of last year.

"As to whether the increase in the production of sugar in the last few years is more in the kinds of sugar which receive bounties generally, than in sugar which does not receive bounties, reference may be made to the statements as to production in the early part of this Report. It is beyond question that there is an increase in all descriptions on sugar, even in the most recent years. It may also be pointed out, as was done in the Report of 1884, that the increase of the production of beet-sugar is enormously greater than the increase of the amount exported. It does not appear, in fact, that the surplus sugar escaping duty available for export can have increased greatly, though it has increased, whatever effect the existence of bounties may have had in stimulating production generally, a question outside the province of this Report. The facts as to the export from beet countries are given in the table on page 9 (*supra*), which shows that the overflow from beet countries in all, which was *nil* in 1868, there being in fact, in that year an import of 48,000 tons on balance, amounted to 353,000 tons in 1878, 520,000 tons in 1882, and 950,000 tons in 1886-87, but a large part of this increase was not of surplus sugar. It is to be noticed that the greatest increase in the exports of sugar in recent years, that is, since 1884, the date of my former Report, has not been from France, which increased her bounties so much in 1884, but from Germany, which exports a great deal more than the surplus sugar, and which gives a much smaller bounty, while there is also a large increase in the exports from Russia, which, it is claimed, does not now give bounties on any export by the European frontier. This large export from Germany and other countries, as well as the rapid growth of the exports in years immediately before 1884, may, of course, be traced back to causes operating before that year, among which temporarily higher bounties than those lately existing, as well as improvements in production might be included. The principal effect of the French bounties in stimulating production may also be felt more in future years that has yet been the case. It is not proposed, however, to make any comments, as the exports from bounty-giving countries are fully shown above, and the figures can easily be compared by those interested with the above information as to the bounties themselves.

"It may also be pointed out that the growth of some kinds of sugar may have been stimulated by protectionist measures other than bounties, such as import duties without corresponding excise duties, which most European countries and the United States appear to levy; and it is not suggested, therefore, that the whole increase of production above stated, which is not due to bounties is an increase under natural conditions. On the contrary, the existence of other causes of disturbance of the natural course of the sugar trade must be recognised, though it would be foreign to the purpose of this Report to discuss them.

SUGAR MILLS AND REFINERIES.

SUGAR MILLS AND REFINERIES. 442

According to the *Statistical Tables of British India* there are in India, at the present day, 12 large and 81 small sugar factories with, so far as is known, a capital of R28,26,000 and an outturn valued at R54,60,677. But since most of the factories and refineries are private concerns whose capital is not subject to registration, the above return by no means expresses the actual capital employed in the industry. Moreover, by far the major portion of the sugar transactions is in the hands of the cultivators or village artizans, so that the produce of their labours never reaches any person who could be called a manufacturer or refiner. This state of affairs is likely to prevail in India for many years to come, so that it may safely be said that neither the present nor any future quotation of the factories and refineries of this country can be accepted as representing, even approximately, the capital and outturn of the sugar production of India. Such as they are, however, the returns of registered sugar factories and refineries are instructive. The following may be specially mentioned: In Madras (1) the Aska Factory located in the Ganjam district. This is said to employ 496 hands permanently and 746 temporarily. (2) Two factories owned by Parry & Co., one at Nellikuppam and the other at Tiruvennanallur, both in the South Arcot district; the former of these gives employment to 560

Sugar Mills and Refineries.

SUGAR MILLS and Refineries.

and the latter to 169 persons. In Bengal there are, especially in Eastern Bengal, very many refineries largely concerned in palm-sugar. The reader will find much interesting particulars on this subject in the passages quoted above regarding the districts of Jessor and Furreedpore (*pp. 267-276*). The estimated outturn of the Jessor refineries is said to be 282,405 cwt. and of Furreedpore 8,852 cwt. The Jessor refineries employ 1,485 permanent and 4,033 temporary hands. So, again, in Khulna there are many Native factories and refineries which are said to turn out 57,976 cwt. Behar may be said to be exclusively concerned in cane-sugar. In Shahabad there are no less than 69 refineries turning out 37,537 cwt., and in Champarun there is a sugar factory shown to produce 2,204 cwt. The Shahabad refineries give employment to 115 permanent and 492 temporary hands. In the suburbs of Calcutta there is a large sugar factory and refinery with a capital of R10,00,000. In Bombay there is but one refinery—the works at Poona. This is said to have a capital of R2,26,000. In the North-West Provinces and Oudh the industry is almost entirely in the hands of the cultivators. There are no great centres of refining like those of Jessor and Shahabad in Bengal. There is, however, a very large and long established sugar factory, the Rosa Mills, in Shajahánpur. This is said to have a capital of R16,00,000 and to give employment to 1,015 persons. The outturn has been valued at R19,06,557. In the Panjáb the sugar industry is, as in the North-West, in the hands of small cultivators and village manufacturers. There is, however, a sugar factory and distillery—the Sujanpur works in the Gurdaspur district. This is said to give employment to 214 persons and to produce 3,456 maunds of sugar and 3,140 gallons of rum. In Mysore there is also a sugar factory which is said to produce 6,000 cwt. of sugar valued at R84,000. In Baroda a sugar factory has recently been started, but apparently it has met with but little success.

Conf. with pp. 175, 214, 234, 283, 288, 293-294, 306.

So far as the published particulars of these factories and refineries are concerned, then they may be said to give employment to 4,500 persons permanently and about 6,000 temporarily. It does not seem desirable to specialize in this place any one factory, and since full particulars of all are not available, it is therefore, perhaps, as well to say nothing regarding their present position and possible future prospects. It has been contended by many writers that the larger works are all pure and simple refineries and that they exist solely through having large contracts for rum. There would seem to be no manner of doubt that the refining trade of India has felt very keenly the effect of beet-sugar competition. Some of the writers quoted above will be found, in fact, to affirm that the small refiners who do not also own rum distilleries have been ruined. On the point urged by **Messrs. Travers & Sons**, *viz.*, the desirability of encouraging the extension of the system of central factories it may be said that, on the whole, an adverse opinion has been recorded. Some of the local Government reports have already been quoted, but as having a direct bearing on the subject of the present chapter, the following passages may be here given from the discussion raised by **Messrs. Travers & Sons**:—

Rum.

443

Conf. with pp. 93, 96, 158, 175, 321, 322.

The Director of Land Records and Agriculture, in the North-West Provinces, furnished many strong arguments opposed to the suggestion of State aid in the effort to extend sugar planting and manufacturing on the European pattern. The following passage may be here given from his reply:—

"The suggestions made by **Messrs. Travers & Sons** is that the Government of India might start a few model factories for the preparation of sugar by modern processes in suitable districts. This appears to be the only point of practical importance in the memorandum. In my opinion the Government would be ill-advised were it to act on the suggestion. I base my opinion on the general ground that

SUGAR MILLS and REFINERIES.

private enterprise in India is now sufficiently alert and well organized to undertake the business of sugar-refining on a large scale and with ample capital if there were a reasonable prospect of success. That sugar-refining companies, working on scientific principles, such as the Rosa Company and the Aska Factory, show no signs of multiplying in India is to my mind a clear proof that, under existing commercial conditions, the prospects of successful trade are small. Nor is the explanation why prospects are not encouraging far to seek. European sugar-refineries in India have two markets, and two only, open to them. They can manufactnre for export to Europe, in which case they have to contend with the bounty-aided sugars of the Continent, and are no more able than the Mauritius factors to make a reasonable profit on their capital in such a market. Or they can manufacture for local consumption in India, endeavouring to supplant sugars refined by Native or crude European processes, and sugars imported from the Mauritius. Here they are met with the great difficulty that the mass of the Native population regards with dogged suspicion all machine-made sugar, holding it to be impure aud contaminated with bones and blood. The market is thus a very small one, and the prices ruling in it are by no means improved by the quantities of similar sugar thrown in despair upon it by Mauritius planters. Assuming that the cost of producing a given amount of crystallized sugar by modern processes is about the same in India and in the Mauritius (and from such information as I have at hand, I do not think a sugar refinery in India could manufacture cheaper than the Mauritius planter), what are the probabilities of commercial success? They are bounded, it seems to me, by the actual success attained by the Mauritius planters, and as we are constantly told that sugar in Mauritius does pot pay, scientific sugar-refining in India is not a hopeful industry. The Rosa Factory in these Provinces depends more on its rum than on its sugar, and I believe this is the case with the few other similar concerns existing in other Provinces." Rum. 444

The Director of Land Records and Agriculture in the Panjab wrote (*August 20th, 1889*) :—

"It is pointed out that the manufacture of modern or vacuum pan sugar to be profitable must be on a large scale, because it involves costly machinery and chemical and mechanical supervision impossible for ryots, who probably do not extract more than one-third of their sugar, etc. This fact of itself renders it useless to discuss the subject further from a Panjab point of view.

"But, I may ask, if it be true that so much profit may be made out of the sugar of India, why is it that more English capital has not been invested in the undertaking? If the case be as stated by Travers & Sons, surely it would, with money almost a drug in the market as it is in London at present, be a very easy thing to get up a company to start sugar works in a sugar-producing district in this country to put down the necessary plant and to buy the canes as they stand from the growers. If the venture were so certain to succeed, private enterprise would soon provide the capital and would not wait for the establishment of model factories by Government. Such a suggestion amounts to a proposal that Government should first run all the risk of the experiment, the success of which might very much benefit capitalists without affording a corresponding advantage to the zamindars who would, however, probably derive benefit from a rise in price of produce.

"In regard to sugar, as to every other agricultural product in this country, established custom stands in the way of all improvement and a strong dislike to sacrifice any present advantage to a prospective future gain. For instance, we often find sugar-growers preferring their clumsy old '*belna*' which takes three pairs of oxen to work it, to the Beheea mill which takes one pair. This is partly due, no doubt, to obtuseness, but it is also due to the fact that the old mill does not break up the cane fibre which is much employed in making ropes, etc., whereas the new mill, which gives much more juice, destroys the fibre and renders it useless. They thus prefer to sacrifice some of the juice in order to save the fibre. It is the neglect of little facts of this kind, I think, more than anything else which has hindered introduction of improved methods in agricultural pursuits. Instances of this kind might be multiplied, but they are not necessary to show that only by the introduction of capital can the system proposed by Messrs. Travers be brought into use in this country."

The Board of Revenue in Madras issued the following resolution on the subject :—

"Messrs. Travers & Sons' views and suggestions are practically that—

(1) the ryots do not extract one-third of the available sugar from their cane;
(2) the product is more like manure than sugar and is worth only about

SUGAR MILLS and REFINERIES.

half what Mauritius 'modern sugar' must fetch to permit of its import into India at a profit;

(3) 'modern sugar' can be made well in India;

(4) to make 'modern sugar' in India, a system of central factories and manufacture on a large scale is necessary to profit;

(5) if such a system be adopted, India might readily supply the world with sugar;

(6) the Government might start model factories in suitable districts.

"There can be no doubt that most of the sugar (jaggery) at present manufactured by the ryots is coarse and dirty; but the Board believe that, though the ryots are quite aware that by taking a little extra trouble in its manufacture a much cleaner sample can be obtained, they do not in practice find that this is profitable. The jaggery is produced to meet the local demand, and that in the crude form at present turned out. Until quite recently, prices have been so low that the growth of cane-sugar has in most places left but little margin for profit. There is in most places practically no demand for export and, except under the influence of such demand, the ryot is not likely to change his customs. The introduction of iron-sugar-mills in most parts of the Presidency where sugar-cane is grown may lead to a superior outturn being obtained; but so long as the defective systems of evaporating at present in vogue are adhered to, there is not likely to be a marked improvement in the quality of the produce.

"Whether modern sugar can be made profitably in India, the Board are unable to say; but it is believed that **Messrs. Parry & Co.**'s works in South Arcot and **Mr. Minchin's** at Aska are worked as commercial undertakings, and therefore for profit, on the Central factory system recommended by **Messrs. Travers & Sons.** There are, however, but very few localities in the Presidency where cane is at present grown on a sufficiently wide, yet concentrated area, to be able to supply the requirements of a large factory, and it is believed that the success of both the factories named is due to their being able to find work for their expensive establishments and large plant during those parts of the year when no cane is obtainable for crushing on other * work than the manufacture of 'modern sugar.' Besides this, there does not appear to the Board to be any reasonable ground for anticipating that the area under sugar-cane can be largely increased on any such concentrated area as the working of a factory would demand. The two requisites—a good soil and a perennial supply of water—in a tract where the drainage is good, and manure is abundant, are not forthcoming in many places. Better prices will probably lead to slight extensions in many places, and this will be pushed on by the economy of production obtained by the use of superior mills, but it is neither possible nor probable that cane will be grown on any much larger area than it now is.

Rum or Arrack. 445

Conf. with pp. 93, 320.

"The Board are therefore of opinion that **Messrs. Travers & Sons'** suggestions as to large Central factories for the manufacture of 'modern sugar' are impracticable as far as this Presidency is concerned, and would deprecate most strongly the idea of model factories being established by the State, for if it were possible for the enterprise to succeed, there is no doubt that the commercial community would embark in it at once."

Messrs. Thomson & Mylne were also invited to favour the Government of India with their opinion on the subject of **Messrs. Travers & Sons'** suggestion. The following is their reply which was dated August 27th, 1889:—

"If the ryots planted and cultivated cane as the terms are understood in the West Indies and Mauritius, the average produce per acre would be very much increased, probably doubled in Behar, North-West Provinces, Oudh, and the Panjab, where small cane is grown for its extract.

"We believe the ryots get as large a percentage of juice from their cane even with their old wooden mills as is obtained by large steam-driven cane mills. The 'manure' referred to in paragraph 8 of **Messrs. Travers & Sons'** letter is the result of crude methods and appliances with a fatal ignorance of the sensitive nature of cane-juice, *viz.*, the small quantity of juice extracted by one mill in a given time, and its consequent long exposure, in contact with foul vessels and surfaces permeated by the germs of fermentation, and in a state of active fermentation transferred to evapora-

* At both these factories there are arrack distilleries and the molasses obtained in the manufacture of sugar are largely used in the manufacture of arrack. The latter is believed to be the more important and paying industry at these factories, but the Board have no statistics.

SUGAR MILLS AND REFINERIES.

ting pans so made, as if the intention is to reduce the contents to a sweet charcoal or to a soapy low class glucose. We know by experience that much of this might be remedied gradually. Though the ryot will not be able to produce a vacuum pan sugar of such quality direct from the cane as would increase his own returns considerably, and help to meet, if not to stop, the importation of foreign vacuum pan sugars.

"The conditions of the sugar industry of India being essentially domestic are not likely to be replaced by Central factories with vacuum pans evaporating a free flow of defecated juice at the lowest possible temperature, and producing a granulated sugar which needs no refining. We do not think that the sugar industry of India, as a whole, would benefit by any effort Government or private individuals might make by establishing model factories. By a judicious modification of the patent law rendering it suitable to the special condition and circumstances of the ryots, much might be done by private enterprise to prevent the odious "manure" being applied to the product of their cane fields, the result of twelve months' labour and payment of land rent. Government in effect leaves the ryot severely alone, ignorant and isolated, in his village to meet foreign competition, backed by the skill and science of Europe and America, which aims at supplanting him in the production of a valuable article of food for his own people. The growing demand for a better quality favours the importer even against caste prejudices; alone he is unable to maintain his ground, and must be assisted either directly by Government or by private enterprise. We have already represented that there is no encouragement given to the latter, to make a sustained effort or protection for money invested in doing so. The patent law is more suited to the self-reliant, progressive, manufacturing peoples of Europe and America, than to the extremely conservative agricultural population of India, whose circumstances and needs in the direction of any improvement are altogether different."

It will thus be seen that, did we not possess a record of past failures, which extend over fully a century, of patient endeavour to accomplish the very object aimed at by **Messrs. Travers & Sons**, the arguments adduced by the above passages would be quite sufficient to decide the question. Were there room in India for more sugar mills and refineries, private enterprise would not be long in meeting the necessity, the more so, since the Government of India in all its dealings with commercial undertakings has shown itself ever ready to afford whatever assistance may be required and which can consistently be given.

PRICES OF SUGAR.

PRICES. 446

With perhaps no other section of this article is it more difficult to furnish trustworthy information than that of the price of the so-called "coarse sugar"—the chief form used in India. Not only are there so many widely different classes of *gúr, ráb, bura, gurpatha, dulloah, chiní*, candy, etc., etc., all recognised as of different merit and classed at different rates, but the so-called sugar (under any one of these names) produced in the differand provinces and even districts have properties peculiar to themselves which occasion the greatest possible range in prices. The classification of trade returns into "drained" and "undrained" or "refined" and "unrefined" necessarily throws together under any one of these sections a series of forms of widely different values. Thus, for example, **Babu Addonath Banerjee** has very rightly pointed out that, if the fact be disregarded that the major portion of the exports of refined sugar from Madras go to Ceylon, whereas those from Bengal go mainly to the United Kingdom, the inference would be incorrectly drawn that the refined sugar of Bengal is R2-11-7 per cent. cheaper than that of Madras. The explanation lies in the fact that the Ceylon people desire a sugar of a different kind and of a more expensive quality than that which is exported to England, The so-called refined sugar exported from Bengal was valued at R6-12-8 a maund (R9-3-10 a cwt.) an inferior article as may be seen from the fact that the *gúr* or unrefined sugar shown in the returns of Internal Trade of Bengal was valued at R4-8 a maund. The refined sugar exported from Madras (as already stated) goes mainly to Ceylon, and that article is valued on the average at R10-3-11 a maund (R13-15-2 a cwt.), but Bengal also exports to Ceylon

PRICES.

a small quantity of sugar annually and the article recorded in the returns of this trade must be very similar to that sent from Madras, as it is valued on the average at R10-2-5 a maund (R13-13-2 a cwt.). It will thus be seen that unless the utmost care be taken in analysing the relative nature of the returns of trade, an average price for a province, still more so for all India, would be most misleading. The bulk of the Bengal exports to foreign countries in refined sugars) are very inferior qualities, not much higher classed than good unrefined sugars; this, therefore, lowers the average value, worked out for the total trade. Bearing this in mind it may be said, that the average value of the refined sugar exported from Bengal in the year 1885-86 was R8-3-9 a maund or R11-3-6 a cwt. The corresponding figure for Madras was R10-3-6 a maund or R13-15-1 a cwt. From the reason given above, the one figure would give a low average value of the refined sugars of Bengal, and the other a high average for Madras. It is probable that refined sugar of fairly good quality would fetch in both provinces about R9 to 10 a maund (say R13 to 14 a cwt.).

Turning now to the subject of unrefined sugars; the average value of the exports from Bengal in this article is shown as high and that of Madras as low. The Bengal average came to R7-5-2 a cwt. or R5-6-1 a maund against R4-4-3 a cwt. and R3-2 a maund Madras. Here, again, the difference of price is due to the quality of the article required by the chief markets to which the exports are usually consigned. In 1885-86 Madras, out of its total exports of *gúr* (1,126,794 cwt.), sent 791,217 cwt. to the United Kingdom, valued at R4-4-2 per cwt. or R3-2-1 a maund. Bengal, on the other hand, shipped out of its total exports (2,313 cwt.), the major quantity, *viz.*, 2,306 cwt. of a superior quality to the United Kingdom, valued at R7-5-3 per cwt. or R5-6-1 a maund. Far from these two valuations therefore representing the average prices of *gúr* in the provinces named, they would more correctly be the average prices of good and of indifferent qualities of *gúr* in India as a whole. It will thus be seen that in India refined sugar may be said to range in value (wholesale) from R6-8 a maund to R16 a maund, and unrefined from R3 to R6-8 or R8-8 a maund.*

How far the cheap imported sugars are now competing with the Native refined sugars and with the Indian European sugars is a point which admits of some difference of opinion. The imported sugars are doubtless directly competing with the sugars refined in India at factories worked on the European methods. And there is a large population, and a yearly increasing one, that has no religious scruples against refined sugar, however produced. The Hindus, of course, have the very strongest objection to European refined sugars owing to the apprehension of animal

Beet.-Sugar.
447
Conf. with pp. 19-20, 39-40, 81. 313, 326, 341. 361, 365.

charcoal having been used in its preparation. "Beet sugar has," says **Babu Addonath Banerjee**, "reached the shores of India, but its operation here cannot be expected to be sufficiently wide to compete successfully with Indian sugar, so far as its consumption among the Hindu population is concerned. It is only the ignorant among the Hindus who may be deceived into eating sugar manufactured in the European refineries." It is useless to assure persons with a strong religious prejudice like that of the average Hindu, that it is by no means necessary that European crystallized sugar should have been refined with the blood and bones of the most sacred of all animals—the cow, or with animal charcoal of any kind. The answer,

* The following may be given as the average prices of unrefined sugar that prevailed in 1890-91 per maund of 82lb:—Madras R5; Bombay R6-13-5; Bengal R4-14; N.-W. Provinces R4-11-4; Oudh R4-9-6; Panjáb R5-6-11; Central Provinces R6-14-10; Upper Burma R5-6-10; Lower Burma R5-11; Assam R8-6-2; and Berar R8-10-3. *Conf. with p. 353.*

PRICES.

Crude Native Sugar.

448

Conf. with pp. 40, 81-82, 95, 104, 113, 137.

and a powerful one, naturally arises, when not in its package, how are we to recognize sugar free from the polution of such refinement? There is also a flavour about Native refined sugar which to many makes it much preferable to the best European sugars. It is needless, therefore, to endeavour to further exemplify the reason why, for many years to come, European refined sugars stand a poor chance of finding in India a market of very great proportions. Refined sugar is, moreover, beyond the means of the vast majority of the people of India. The cheapest beet-root sugar appears, at the present rates, to be about 2½ times as expensive as the coarse sugar commonly eaten in India. This fact may be shown by the following table:—

	Average price a cwt.		
	1884-85.	1885-86.	1886-87.
	R a. p.	*R a. p*	*R a. p.*
Price of beet-sugar imported into Calcutta from the United Kingdom . . .	11 2 7	11 6 0	10 2 5
Price of unrefined sugar exported to the United Kingdom from Madras . .	4 8 1	4 4 2	4 1 5

Conf. with p. 292.

But when the cultivator manufactures his own *gúr*, it certainly never cost him anything like R4 a maund.

But an effort may here be made to exhibit the average prices of refined (Indian and Foreign) sugars in order to admit of a comparison with the average prices of *gúr* shown on page 324. Having found it difficult to obtain this information, from published reports, the writer recently suggested to the Government of India the desirability of inviting opinions from the provincial authorities. He has accordingly been favoured with returns for the year 1891 (from most of the provinces), in which the prices of Indian and Foreign sugars are given to maunds of 82℔. It may be added, however, that in few cases has the distinction been made into Native and European Indian refined sugars, with the not improbable result of raising slightly the average value of the Indian refined sugar which has been contrasted in these returns with the imported article. That is to say, the European refinery sugar of India is a much superior article to that eaten by the people of this country who use *a refined sugar*. One point of considerable importance has been made out by these reports, however, namely, that foreign sugar is procurable in few districts only. Thus, for example, of the 21 districts of the NORTH-WEST PROVINCES AND OUDH that have furnished prices of sugar, three mention foreign sugar; the others expressly say that it is not procurable. In these 21 districts the average price of Indian refined sugars appears to have been R10 a maund and foreign (three districts) R12. The cheapest refined Indian sugar is given for Allahabad, *viz.*, R7-3-8 and the most expensive for Aligarh, *viz.*, R13-8. The Director of Agriculture, commenting on the trade in foreign refined sugar, makes the significant remark that the imports from Bombay during 1890-91 came to 26,080 maunds of refined sugar (presumably foreign). The imports in the previous year from that port stood at 6,031 maunds (*see p. 366*), so that this fact may be regarded as denoting the growth of a demand for foreign refined sugar. In only one district of the CENTRAL PROVINCES has any mention been made of foreign sugar. The "drained" sugar used in these provinces showed an average value of R11. Of the PANJAB, returns have been furnished for seven of the chief districts, but of three of these it is expressly stated foreign sugar is not sold (*viz.*, Mooltan, Jallandar, and Delhi). The averages of all the returns show foreign refined

PRICES.

sugar to have sold at R11-10 a maund and Indian at R10-7. The most expensive Indian (Sujanpurí) refined sugar appears against Rawalpindi, *viz.*, R12 to 14 a maund and the cheapest in Delhi, *viz.*, Sujanpurí R8 and Shajihanpurí R10. Of the imported sugars, Amritsar is shown to have consumed Australian sugar valued at R12 to 14, German R11-12, and Mauritius R11-4 to R11-6 a maund. Statistics of 13 districts of BOMBAY and SIND have been supplied, but of six of these no information regarding foreign sugar has been furnished, or it has been stated that no foreign sugar is procurable. One feature of the Bombay sugar trade may be here specially alluded to, namely, the abnormally high rate of unrefined sugar and the correspondingly low valuation of foreign refined sugars. Unless some mistake has been made in framing the returns, it would appear that in many of the districts of Western India unrefined sugar fetches nearly as high a price as the refined article. During 1891-92 the average price of unrefined sugar (estimated from the figures furnished to the author) would appear to have been R7-8 a maund—a price considerably above that shown in the "Returns of Agricultural Statistics for 1890-91" quoted above on page 324 Refined sugars sold as follows:—Average of imported sugar R10-11 a maund; of Indian R10-15. It would thus appear that in Bombay foreign sugar is directly competing with local produce, but it should not be forgotten that this has been the case during at least the past century. The returns with which the writer has been favoured regarding MADRAS, show no foreign sugar, and it is further even affirmed that prices of imported sugars could not be procured. Aska refined sugar sold at R9 a maund and Berhampur at R9 to 10. It would thus seem that both the European and Native refined sugars of South India are considerably lower priced than the cheapest imported article. In ASSAM foreign refined sugar is mentioned in connection with four out of the seven districts reported on. The average of the prices of foreign sugar may be given at R14-10 a maund and of Indian at R12-8. With regard to BURMA, foreign sugar is quoted in connection with two out of the five districts of which prices have been furnished. These are Moulmein (where foreign sugar sold at R18 and Indian R9-12 a maund) and Rangoon (foreign R9-14 and Indian R9-10 a maund).

It may thus be safely stated that before cheap foreign sugar can be regarded as seriously affecting indigenous production, it becomes necessary to possess stronger evidence than has as yet been made out, of the growth of a demand for refined in preference to unrefined sugar. It would, in fact, seem that as long as *gúr* can be procured for less than half the price of refined sugar, and as long as poverty is characteristic of the vast majority of the hundreds of millions of the people of India, so long will local production and consumption remain unaffected by the fluctuations of foreign markets in refined sugar. But beet-root sugar has undoubtedly got a footing in Western India, a province in which foreign sugar of some sort is a necessity to meet local demands. The fall in the price of refined sugar and the high rate of unrefined sugars must have greatly contributed to expand the traffic in refined foreign sugar in Bombay. Indeed, it may fittingly be concluded that the only indication of India being abnormally affected by the depreciation in the value of sugar in the world generally, is in the depressed state of the refiner's trade. And, although this cannot be treated with indifference, it is of less moment to the national prosperity of India than a disturbance, if such existed, that threatened the value of the agricultural interests in sugar-cane.

The writer has failed to obtain, however, a sufficiently exhaustive series of prices for a sufficient number of years back to allow of a satisfactory review being offered. This difficulty exists more in judging of the internal

PRICES.

than the external trade. If the valuation given for the exports can be accepted as trustworthy, we possess in the returns of foreign trade full particulars. Thus, **Mr. O'Conor**, in his publication *Prices and Wages*, furnishes the rates at which three classes of Native refined sugars have been exported from Calcutta since 1843. But as Calcutta did not apparently export each of these classes every year, there are so many gaps in the returns that it is difficult to learn all that is required. **Mr. O'Conor** accepts the value in 1873 as a standard (which he expresses as 100) and thus exhibits the fluctuations in price. He also supplies the London prices of Manilla sugar during each year since 1873. But even had these returns been complete, more would have been required for the purpose here aimed at. It would be necessary, for example, to know the selling prices of Indian and imported sugars, in a selection of centres, such as Calcutta, Cawnpore, Lahore, Karachi, Bombay, Nagpur, Madras, Rangoon, etc. We learn, however, sufficient from the figures furnished by **Mr. O'Conor** to be able to affirm that nothing like the decline, shown in the London prices, has taken place in the Calcutta. In 1873, Manilla sugar sold in London at 23*s*. 3*d*. per cwt., and the price declined steadily till every fifth year it stood as follows 15*s*. 3*d*., 12*s*, and 10*s*. 3*d* Since 1860, **Mr. O'Conor** records the prices of 23 years' shipments of *dulloah* sugar. The average of these comes to R7-13 per mannd (82lb) and the shipments for the past five years were valued at R8-8 in 1886; R8-10 in 1887; R7-8 in 1888; R7-4, in 1889; and R10, in 1890, and the average of these would be R8-6 a maund. The standard of 100 in 1873 was equal to a valuation of R7-8 a maund, so that in only one of the past five years has the declared valuation fallen below the standard, and according to the evidence here adduced, therefore, *dulloah* sugar has not only preserved its price, but has even slightly risen, in the face of the decline of the price of sugar throughout the rest of the world. The inference might perhaps be admissible from these figures alone, that foreign sugars had not as yet materially lowered the value of *dulloah*, one of the most appreciated forms of Native refined sugars. But unfortunately such an opinion is opposed by the almost universal belief that the European and Native refiners alike have, within the past few years, felt to an alarming extent the effect of a keen competition, in the foreign markets, and in the Indian markets being flooded with foreign sugar. This position might be understood, in the case of the European refiners of India, since to a very large extent their Indian consumers could have little or no objection to using foreign sugars. The consumer of Native refined sugar, on the other hand, has an equally strong prejudice against the European sugars refined in India, and those imported from foreign countries. It is, therefore, as it would seem, the loss of the foreign markets which alone can have hitherto seriously injured the Native refiners. The value of these markets may be judged of from the statistical information furnished in the chapters on the Indian trade in sugar. Even could it be clearly established, therefore, that the imported sugars are actually displacing the Native refined sugars from the Indian makets, such a contention would be but on the threshold of the main problem, *viz.*, the effect of foreign sugars on the total production of sugar (*gûr*) in India. The first effect of increased importation would naturally be to lower prices. It may be admitted that *gúr* has become slightly cheaper in India, as a whole, but this has resulted in an increased consumption, and to such an extent as to encourage an expansion of the area of cultivation. Until a preferential demand arises for refined, instead of unrefined, sugars, the Indian cultivator, it would appear, is thus benefited by cheap prices. The produce of his labours is more generally used. So very profitable in fact is the cultivation of cane and the manufacture of *gúr*, that both branches of the industry are benefited by increased demand, even although a slightly lower price be offered. The increase of the acreage of production and greater consumption would in other words appear to have balanced the disturbance created by beet-sugar, so that it may be said India, as a whole, has been benefited by the low prices that have recently prevailed.

INDIAN TRADE IN SUGAR.

INDIAN TRADE.

449

It would be beyond the scope of the present article to attempt, however briefly, to review the complex problem of the *World's Supply and Consumption of Sugar*. While that is so, there are aspects of the Indian trade in sugar, the full force of which cannot be realized until some idea has been conveyed of the nature and value of the Indian markets of outlet and of supply. In the historic chapter above, it has been pointed out that Europe, during the classic period of Rome, received from India its knowledge of sugar as also for many ages its supplies of that commodity. Centuries intervened between the discovery of sugar and the time when it began to be a necessity of European life. Sugar, in fact, first assumed importance in Europe on cane cultivation being established as a European industry in the Colonies. The birth of the new Colonial, was the death of the old Indian, trade. But the Honourable the East India Company, becoming aware of the loss India had sustained, in its failure to create or even to participate in the greatly increased traffic, made (towards the close of the last century) strenuous efforts to awaken interest in the subject Although many obstacles were soon found to have been thrown across the path of progression, the Company succeeded to revive and greatly enlarge India's foreign interests in sugar. Heavy losses were for years patiently borne in the hope of ultimate success. East Indian sugar in time became regularly quoted, and the amounts that poured into England and other European countries improved in quality very greatly. The internal sugar trade of India also manifested distinct indications of expansion. The demands of India for superior qualities of sugar had grown so strong that the imported, refined, article gradually came to bear, in the various languages of the peninsula, names to denote apparently the two countries of chief external supply, *viz.*, *chiní* (China) and *misrí* (Egypt). There is abundant evidence in support of the belief that for many centuries the art of refining sugar was known in India to comparatively few, and was for the most part practised by the special artisans of the nobles. To a greater extent, relatively, therefore, than at the present day, the bulk of the people consumed (a century ago) a crude unrefined sugar—*gúr*. Those who desired a better article looked mainly to a foreign supply. The inhabitants of the coast were probably always, as they are now, a wealthier community than the people of the more interior agricultural tracts. Within a belt of country skirting the coast, the demand for sugar has always been greater than further inland. But that demand it was felt, could be more easily met by the merchants who traded with India, bringing, as a return cargo, sugar from China, the Straits, Batavia, and perhaps also from Egypt than by the Indian refiners. An import traffic had accordingly by the seventeenth and eighteenth centuries assumed considerable proportions. It was only natural that the East India Company should have recognised, therefore, that the subject of internal trade deserved quite as much attention as the encouragement of India's foreign export in sugar. The restrictions imposed by coastwise dues and internal transit charges were seen to operate prohibitively on Indian sugar conveyed from the chief sugar-producing districts to the great consuming centres. The supply of Bombay, for example, was regarded as a natural outlet for the

INDIAN TRADE.

surplus Bengal stock. The success of the Dutch traders, in meeting the demands of Western India, was viewed, and not unnaturally, with the strongest disfavour. To facilitate Bengal competition, the dues were removed from Indian sugar, but the tax on foreign imports retained. In spite of this *chini* and *misri* continued to be largely imported and to even find a market in Bengal itself. The unsatisfactory character of the Indian refiners' art and the insufficient supply was at last recognised as the chief reasons for the success of the Chinese and Batavian sugars. Effort was accordingly directed towards improving and extending the refining interests of India. The result for a time was distinctly satisfactory. The manufacturies and refineries started in Bassein, for example, at once told on the imports into Western India. The demand for Chinese, Straits and Dutch colonial sugars declined, and a large export from India in refined sugar was gradually developed. But in course of time, and coming down to comparative recent events, a more formidable rival appeared in the Bombay market in the supplies of sugar that yearly began to pour into India from Mauritius, the West Indies, and ultimately from Europe. The success of European planting in India had been but temporary. Most of the refineries fell into Native hands. The quality of the article degenerated, and thus, though the Chinese and Java trade had largely been ousted, the advantage was reaped by new-comers, and a greater import than ever was created. The exports also began to change from refined to unrefined sugar. But startling though many of the ups and downs of the Indian sugar trade have been during the century 1784 to 1884, the revolution that has since taken place surpasses in magnitude all previous experiences. The exports of refined sugar have ceased, the traffic is now practically in the unrefined article, and the trade has migrated from Bengal to Madras. And this is not all. India now imports 3 cwt. for every cwt. exported.* These imports are largely drawn from the very countries which were formerly the chief consumers of our exports. So very significant is this modern phase, that India must be regarded as taking no longer any part in the world's supply of this article, but rather as affording a market for the produce of other countries. One feature of importance still, however, remains to India. The advance of civilization has created new necessities and afforded the means of realizing greater luxuries. The people of India are able to afford a larger consumption of sugar than they ever did before. There is no evidence to support the inference that might be deduced from the expansion of the import traffic that foreign sugars are driving the Indian out of the home markets On the contrary, everything points to a greatly increased cultivation and a corresponding immensely enhanced consumption. The imports are undoubtedly, however, competing with the Indian refiners' trade, in the supply of certain sugars. A larger amount of crude sugar is thereby released and rendered available for the consumption of the vast majority of the people who never have and for many years to come are not likely to care for refined sugars, still less to eat such refined sugars as bear the stigma of religious prejudice against the process of their manufacture. So far it may safely be affirmed the loss of the markets to which India formerly exported sugar and the creation of a foreign supply have not told injuriously upon the production of sugar in India. It would be to trespass beyond the field of legitimate criticism and review to venture further by foretelling the probable future. Radical changes have already taken place in India. Caste, where opposed to the spirit of the age, has given place on many points, but it would seem highly improbable that for years to come foreign sugars should succeed

Conf. with p. 324.

* *Conf. with pp. 19-20, 39-40, 316, 341-344, 346.*

INDIAN TRADE.

to invade the domestic life of by far the most important section of the Indian community—the Hindus. This fact, therefore, greatly restricts the possibilities that are open to the importers of sugar. But there exists an even more potent consideration than the religious views of a section of the community, namely, the poverty of perhaps four-fifths of the entire population. The question of the effect of imports on production in this country has to be regarded not in the light of the influence on the refiners' and exporters' sections of the trade, but on that of the crude sugar manufacturer—the cultivator of the cane. Little has as yet transpired in support of the opinion that the value of sugar, in Europe and the colonies, can be lowered much below what it has already touched. On the contrary, it stands a greater chance to be augmented. If that be so, the cheapest foreign sugars hitherto landed in India cost at least twice as much as the article eaten by the vast majority of the people of this country. Were the colonial sugar manufacturers to attempt to produce an article that would directly compete in price, even with the more expensive qualities of *gúr*, the import trade would at once become of graver moment than it has as yet assumed. These remarks may therefore be said to tend towards the conclusion—and a not unnatural one—that the Indian internal trade in sugar—(the home consumption of that commodity)—is the feature of greatest importance to this country. In perhaps no other article of Indian commerce, therefore, is the want of definite particulars of production and consumption more keenly felt than with sugar. Where sufficient importance is denoted by a large foreign supply or demand, the prosperity or otherwise of the country as indicated by certain articles of trade may be learned from the returns of foreign transactions. The utmost reliance can be put on the accuracy of these returns, for the articles that leave or are brought to India are not only carefully registered, but from the records of other countries, in their dealings with India, it becomes possible to confirm every transaction. But in a vast empire like that of India where certain provinces have not as yet been even surveyed, agricultural resources become very largely a matter of speculation. The construction of railways has afforded the Government, however, one direct mode of gauging the extent of internal traffic, since the movement of articles are not only registered from province to province, but from district to district. The railways are the great arteries of certain tracts. Rivers and canals serve the same purpose, and a registration of the trade on these is also preserved so far as possible. But when the returns of rail, river and canal have been all tested, there has been left out of consideration the road traffic of which no registration whatever is kept, except in the imports and exports with certain large towns. It may safely be assumed that in a commodity which like sugar is grown mainly for local consumption, the interchanges from village to village along the roads throughout the entire country are very much greater than that shown in the returns of rail, river, and canal which tap but limited tracts of the total area of India. This statement will at once be borne out, for example, by an examination of the imports and exports by road into Calcutta. Along the coast of India and across its land frontier a record of the interchanges is also preserved, so that the returns of the shipments coastwise and of the goods carried across the land frontier come to bear on the problem of the annual production of sugar in India. But of many of the provinces the area under cane has been annually determined, and periodically surveyed. Experiments to ascertain the acreage yield have also been performed; so that allowing a margin for error not greater than would be necessary for countries of like magnitude, it may be accepted that the returns of the surveyed provinces can be

INDIAN TRADE.

accepted as fairly accurate. Thus the tendency of the defective returns of India would be to under rather than over estimate production.

By bearing every possible error in view and by bringing every available aid on the enquiry, it has been determined that within the past few years the normal area devoted to sugar-cane and sugar-yielding palms has not fallen far short of 2,500,000 acres, and that the yield has been about 2,500,000 tons of coarse sugar. It is probable that to compare that amount with the consumption of sugar in Europe it should be reduced to a little more than one-third as the "coarse sugar" of India, if refined into an article similar in quality and equal in value to the sugar used in Europe and America, would be in the ratio of 2⅛ of *gúr* or coarse sugar to 1 of refined sugar. But although a consideration of the nature indicated is perfectly right, still it must be added that as an article of food with the people of India, *gúr* serves the purpose, and it might almost be said is to them of corresponding value with the sugar of more civilized countries.

Conf. with p. 119.

In Dr. Giffen's paper on "*The Progress of The Sugar Trade*" published on the 24th of May 1889 a useful table is given of the consumption of sugar in the chief countries of the world. After showing the amounts used in the United States, in the United Kingdom, and in all the other European countries, Dr. Giffen allows a quantity for Australasia and then for "other countries." He does not show India, China, Japan etc., so that it is not permissible to affirm that the "other countries" include or exclude India. The total consumption of the world, according to Dr. Giffen, comes to 5,200,000 tons of sugar, of which the United States takes 29 per cent., the United Kingdom 21 per cent., all European countries (excluding the United Kingdom) 36 per cent., Australasia 2 per cent., and "other countries" 12 per cent. The amount shown against other countries is only 580,000 tons. India at the present day imports 3 cwt. of sugar for every cwt. exported. If, therefore, the net import be added to the estimated production and the figure just given for the production be reduced to the value of the figures dealt with by Dr. Giffen, it may safely be stated the consumption shown against "other countries" is less than half the actual amount consumed in India alone. In other words, it seems correct to say that expressed as sugar equal in quality with that consumed in Europe, the people of India use up annually fully one million tons or about the same amount as in Great Britain. Or, leaving Great Britain out of consideration, the people of India use a little more than half the amount consumed in the whole of the rest of Europe. Taking this view of the comparative value of the sugar used in the United Kingdom and in India, the subject assumes a greater importance than is given to it by an inspection of the dirty-looking masses exposed for sale on the village traders' stall. Dr. Giffen remarks that the 1,100,000 tons of sugar, as imported into the United Kingdom, is valued at £16,500,000. In 1884 sugar was estimated in the United Kingdom to have been about half the value of the wheat. The consumption of wheat came to 26,000,000 quarters, the value of which at 32*s.* per quarter would have been about £42,000,000. Sugar, for many years, kept pace with other articles of food in growing cheaper year by year, but recently it has begun to lead in that respect. Thus, the 8,000,000 cwt. consumed 30 to 35 years ago cost the British as much as the 22,000,000 cwt. taken annually within the past two or three years. But there are certain features of the British sugar trade that must be specially noticed. The value of that article is largely increased beyond its declared import value through the very great amount of it that is usually refined. About 739,000 tons are refined or prepared for special industries. This trade gives employment to 4,260 men at the 26 refineries that exist in the

Conf. with p. 346.

INDIAN TRADE.

country. This shows about 30 men for every 100 tons refined per week, or 5,000 tons per annum. The imports of refined sugar into England (mainly from Germany) have recently, however, been greatly increased and the business of refining for export has decreased. It will thus be observed that the refiners of Great Britain have felt the modern tendencies of the sugar trade quite as much, if not more, seriously than the Indian. But there are industries in England that use very large amounts of sugar which scarcely, if at all, exist in India. Thus, for example, Dr. Giffen tells us that certain leading confectionery and jam-making firms in London alone use annually 34,000 tons of sugar: similarly that the turn-over for Scotland of this nature comes to 40,000 tons. An actual estimate of the consumption of sugar for jams, confectionery, biscuits, brewing, mineral waters, etc., cannot be obtained, but the point of interest which it is desired to urge here is that a very large amount of the articles so prepared are exported, so that the actual cousumption of sugar in the United Kingdom has to be reduced by that amount. In India no such export trade exists. The sweetmeats made are articles of actual food, not sweetmeats in the sense of such luxuries in Europe. The extent to which the so-called sweetmeats of India are eaten both daily and at festivals, marriage ceremonies, etc., must impress, therefore, the argument here advanced, namely, that if the sugar used in the United Kingdom was valued at £16,500,000, a considerably larger sum may be accepted as representing the annual consumption in India. It is little to be wondered at, therefore, that in the present keen competition, the exporters of sugar should have earnestly turned their attention to India as a market of great importance. At the lowest possible estimate the industries of cultivation of cane and palms and manufacture and refinement of sugar in India must be admitted as equal to at least £20,000,000 annually.

The reader should consult the introductory remarks offered above, in each of the provincial chapters, on the subject of the area, outturn and consumption of sugar. The defective nature of the internal returns of trade, in some provinces, precludes, as explained. a detailed statement being prepared of the total trade for all India. The series of tables that may now be here given, although in some instances defective, are believed to be accurate so far as they go. Their value is only lessened when it is desired to obtain for each and every province returns of the exact same nature. The absence of such uniformity debars a total statement being prepared, but does not render inaccurate the statistical information furnished. The Indian sugar trade may be viewed under three great sections: I. Foreign, II. Internal. and III. Coastwise. The trans-frontier transactions (that is, the sugar carried by land routes to or from India) are unimportant and may, therefore, be placed under the section on Internal Trade.

FOREIGN TRADE. Exports. 450

I.—FOREIGN TRADE IN REFINED AND UNREFINED SUGAR.

A. **Exports from India.**—In various passages of this article the writer has endeavoured to exhibit the manner in which the possible growth of a great trade in Indian sugar was precluded through the prohibitively heavy import duty charged on Indian sugar by Great Britain. That duty amounted to about 8*s*. a hundredweight more than was charged on colonial sugar. It existed until 1836. It will, therefore, be instructive, as manifesting the early records of Indian sugar, to give tables of the trade, for some years prior to and after the removal of the duty. Column III of the table below shows the imports into Great Britain of Indian sugar from

FOREIGN TRADE. Exports.

1800 to 1821. The contrast with column I shows the proportion of the Indian to the total imports. Columns IV and V classify the actual sales made in London for each of the years, into the two sections, sales from Company's imports and from private imports. The decline of the Company's transactions and the growth of the private, is a feature of some importance as it shows the extent to which this branch of trade found it necessary to escape from the restrictions enjoined by the Company's regulations in the character and conditions of their investments.

Statement of the Sugar Trade of Great Britain for the first twenty-two years of the present century designed to specially exhibit the share taken by India.

YEARS.	Total of all sorts imported into Great Britain.	Total of all sorts exported; reduced to raw sugar.	Total of all sorts imported from India.	ANALYSIS OF THE ACTUAL SALES AT LONDON OF THE IMPORTS FROM INDIA.			
				Company's Imports.	Private Imports.	Total.	
	Cwt.	Cwt.	Cwt.	Cwt.	Cwt.	Cwt.	£
	I	II	III	IV	V	VI	VII
1800 . .	3,390,974	618,537	120,471	111,070	109,766	220,836	545,937
1801 . .	3,164,474	1,657,551	226,538	55,797	19,111	74,908	197,134
1802 . .	3,976,564	1,202,769	61,213	55,786	27,704	83,490	158,317
1803 . .	4,297,097	2,046,767	57,381	27,510	21,769	49,279	102,473
1804 . .	3,185,849	1,693,285	97,928	78,620	25,477	104,097	273,514
1805 . .	3,248,306	1,103,936	125,155	102,735	29	102,764	294,757
1806 . .	3,178,788	1,102,685	124,360	65,806	156	65,962	144,797
1807 . .	3,815,183	1,013,435	37,227	105,503	7,980	113,483	211,658
1808 . .	3,641,310	1,363,642	118,586	48,447	5,936	54,383	96,728
1809 . .	3,753,485	910,67?	72,587	31,618	119	31,737	68,990
1810 . .	4,100,198	1,496,691	26,200	40,534	8,088	48,622	113,410
1811 . .	4,808,663	1,319,349	49,240	1,824	12,059	13,883	28,550
1812 . .	3,917,543	690 869	20,322	67,610	9,646	77,256	177,433
1813 . .	3,763,423	1,158,162	72,886	47,559	22,632	70,191	216,600
1814 . .	4,000,000	1,615,500	50,000	42,548	13,394	55,942	216,608
1815 . .	4,035,323	2,002,109	49,849	3.312	121,032	124,344	383,610
1816 . .	3,984,782	1,906,711	125,639	18,951	90,770	109,721	247,108
1817 . .	3,760,548	1,663,617	127,203	2,774	73,050	75,824	180,757
1818 . .	3,795.550	1,671,740	125,893	19,086	98,249	17,335	263,071
1819 . .	3,965,947	1,695,627	162,395	20,754	114,649	135,403	237,356
1820 . .	4,077,009	1,302,179	205,527	18,318	154,553	172,871	263,530
1821 . .	4,063,541	1,659,556	277,228	39,731	141,653	181,384	243,726

Although the East India Company gave orders, about 1820, that sugar should no longer form a part of their commercial investments, it continued to be exported. Factories had, however, been formed by private persons which soon drifted into refineries and agencies to purchase Native sugar, and the result was that, by the time the heavy British import duty was equalized with that charged on other sugars, the exports from India to England stood at a little over 250,000 cwt. In 1840, or six years later, they had, however, increased by 1,000,000 cwt.

The following table compiled from the Proceedings of the Honourable

SACCHARUM:
Sugar.

Foreign Trade in Refined

FOREIGN TRADE.
Exports.

the East India Company (*Statistics of Sugar*) shows the trade from India from the years 1836 to 1848:—

The Exports of Sugar from India during the twelve years immediately following the equalization of the duty charged on the imports into Great Britain with that levied on Colonial Sugar.

YEARS.	Exported to Great Britain.	To other Ports	Total exports of sugar and *gúr*.	Value in rupees.	Average value per maund.
	I.	II.	III.	IV.	V.
	Cwt.	Cwt.	Cwt.		R a. p.
1836-37 . .	260,617 1/7	180,354 3/7	440,971 2/7	51,38,460	8 5 2
1837-38 . .	425,611 3/7	156,367 6/7	581,979 3/7	67,18,911	8 3 11 1/8
1838-39 . .	522,741 3/7	98,049 2/7	620,790 5/7	74,63,088	8 9 4 7/8
1839-40 . .	523,322 1/7	79,455 5/7	602,777 6/7	73,60,036	8 11 6 1/2
1840-41 . .	1,226,635 5/7	48,215	1,274,850 5/7	1,64,68,898	9 3 7 2/3
1841-42 . .	1,037,501 3/7	49,707 1/7	1,087,208 4/7	1,39,16,426	9 2 3 2/3
1842-43 . .	1,123,675	23,132 1/7	1,146,807 1/7	1,48,35,773	9 3 2 3/4
1843-44 . .	1,097,482 6/7	4,360	1,101,843 6/7	1,46,04,641	9 7 5 3/4
1844-45 . .	1,084,292 1/7	15,077 1/7	1,099,369 2/7	1,46,91,956	9 8 8 3/4
1845-46 . .	1,308,045	5,793 4/7	1,313,838 4/7	1,78,93,188	9 11 7
1846-47 . .	1,203,811 3/7	21,343 4/7	1,225,155	1,67,98,655	9 12 0 5/8
1847-48 . .	1,168,944 3/7	60,883 4/7	1,229,828	1,66,28,524	9 10 6 1/3

The very sudden rise in the exports, which is shown to have taken place in 1840-41, is to be accounted for by the material reduction in the production of sugar in the West Indies in consequence of the Emancipation Law which came into effect in 1838. A decline is also perceptible in the above table from 1841 to 1845, in which year a reduction took place in the duty charged by England on Bengal and all muscavado sugars, from 24*s*. to 14*s*.

But there is still another peculiarity in the figures shown in the above table. The grand totals in column III. (at least for the years 1843-1848) will be found by comparison with the totals in the table below to have represented the exports from Calcutta alone. Madras, which at the present day has become the chief exporting province, took no share (or practically no share) in the trade during very nearly the first half of this century. It enjoyed a small trade during the closing decades of the eighteenth century, but was not placed on the same favourable terms as Bengal until the end of 1839. In Wilkinson's *Commercial Annual* we are furnished with particulars of the External Trade of Bengal compiled from the Customs returns of that province, and the information there given should naturally be regarded as not embracing any portion of the Madras trade.

FOREIGN TRADE.

Exports.

The following analysis, prepared from **Wilkinson's** *Commercial Annual* —a tabular statement of the *External Commerce of Bengal*, during the years 1843 to 1850, may, therefore, be usefully given in this place:—

Exports of Sugar and Khaur from Calcutta.

Countries to which Exported.	1843-44.	1844-45.	1845-46.	1846-47.	1847-48.	1848-49.	1849-50.
	Mds.	Mds.	Mds	Mds.	Mds.	Mds.	Mds.
Great Britain	15,36,476½	15,18,009	18,31,263¼	16,85,336	16,^6,522¾	16,34,569	17,94,797⅔
Bombay	1,150½	9,403	567½	15,786	70,373¾	92,356¼	14,210
Gulfs (Arabian and Persian)	3,054½	1,449½	2,957¾	5,932½	6,101	5,954½	6,914⅔
Madras	70½	157¾	52¾	2,081	856	101	266
Ceylon	590½	5,859¾	768½	4,753½	3,776¾	2,827¼	4,535¼
Pegu	28	174¾	90½	136½	384¾	340¼	241¾
New Holland	390	2,888¾	2,017½	70½	993½	5,637½	49½
Cape and St. Helena	670½	730	1,019¾	456½	1,385½	701½	552¼
North America	...	...	...	...	130¼	385¼	535
Maldives	150½	446	487	664¾	1,038¾	462¼	703½
Mauritius	...	...	150	...	146	...	...
Balasore	...	...	...	...	50¾	...	...
Antwerp	...	...	...	...	...	115¾	...
TOTAL IN INDIAN MDS.	15,42,581	15,39,117	18,39,374	17,15,217	17,21,759	17,43,450	18,22,805
TOTAL IN CWT. (The maund being accepted as ⅘th cwt.)	1,101,843	1,099,369	1,313,838	1,225,155	1,229,828	1,245,321	1,302,003

It may be here pointed out that it is significant **Wilkinson** should not have shown Calcutta as importing sugar from Madras or from foreign countries. Bengal, during the years dealt with, may therefore be regarded as having held its own. The imports of foreign sugar forty years ago went almost entirely to Bombay, and Bengal exported largely to countries from which it now draws supplies. Another feature of some importance may also be alluded to, *viz.*, that the exports seem to have been almost entirely in refined sugar. The amounts of *ráb* or of *gúr* were so unimportant that they were viewed as involving no serious error by being treated along with sugar generally.

As bearing on the question of Madras exports, the following analysis of the Indian Export Trade may be furnished for the years 1851 to 1862. It will then be seen that Madras is exhibited as having had in 1851 a very considerable foreign export so that the table from **Wilkinson's** *Commercial Annual* should, as it is stated to be, be accepted as indicating the Bengal section only and the earlier tables (furnished above) which have been compiled from the Honourable the East India Company's Proceedings may be, therefore, regarded as dealing also with the Bengal Trade although stated to be "East India sugar," a definition which should, of course, have included Madras as well as Bengal. This is the only explanation by which the totals in certain years, which appear in both sets of returns, could be identical. The Act XXXII. of 1836 which equalized the duty in Bengal sugar imported into England was only extended to include Madras (Act XV.) in 1839. There is, however, abundant evidence in support of the opinion already advanced that for well on to the middle of this century the export trade of Bengal in sugar might almost be viewed as that of all India. There is no very serious error, therefore, involved in the acceptation of the early transactions shown in the above tables as expressing the total sugar export trade from India. From about 1850, however, we possess precise information as to the total trade and the shares taken by each province.

SACCHARUM: Sugar.

FOREIGN TRADE.

Exports.

Foreign Trade in Refined

Analytical Statement showing the quantities of Sugar exported annually of Europe, America, and other Ports Foreign to India, distin-

From Ports in the Presidency of	Places whither Exported.	1851-52.	1852-53.	1853-54.
		Cwt.	Cwt.	Cwt.
BENGAL	Great Britain	1,106,298	1,048,236	458,429
	Continent of Europe . . .	3,473	153	1,667
	America	1,735	2,876	1,004
	Other Ports Foreign to India .	16,578	31,999	39,218
	TOTAL .	1,128,084	1,083,264	500,318
	To Ports in India, but not in Bengal Presidency † . . .	59,291	116,376	9,175
	TOTAL .	1,187,375	1,199,640	509,493
MADRAS .	Great Britain	399,753	307,624	493,712
	Continent of Europe . . .	2,775	..	...
	America	...	461	...
	Other Ports Foreign to India * .	9,609	23,857	10,052
	TOTAL .	412,137	331,942	503,764
	To Ports in India, but not in Madras Presidency † . . .	11,988	8,113	30,709
	TOTAL .	424,125	340,055	534,473
BOMBAY .	Great Britain	...	770	22
	Continent of Europe . . .	...	...	...
	America	...	...	...
	Other Ports Foreign to India * .	67,287	61,671	83,950
	TOTAL .	67,287	62,441	83,972
	To Ports in India, but not in Bombay Presidency † . . .	4,653	4,290	4,261
	TOTAL .	71,940	66,731	88,233
BRITISH INDIA	Great Britain	1,506,051	1,356,630	952,163
	Continent of Europe . . .	6 248	153	1,667
	America	1,735	3,337	1,004
	Other Ports Foreign to India * .	93,474	117,527	133,220
	TOTAL .	1,607,508	1,477,647	1,088,054
	To Ports in India from one Presidency to another, but exclusive of the Port to Port Trade within each Presidency . . .	75,932	128,779	44,145
	TOTAL .	1,683,440	1,606,426	1,132,199

* Consisting of Aden, Africa, Arabian and Persian Gulfs, Australia, Cape of Good Saint Helena, Straits Settlements, Sonmíani and Meckran, Suez, Turkey, West

† Bengal to Ports on the Coromandel and Malabar and Canara Coasts, Madras Madras to Indian French Ports, Bombay, Cutch, Sind, Calcutta, Arracan, Chitta Bombay to Calcutta, and Ports on Malabar and Canara Coasts.

N.B.—The exports from Bombay Presidency are entirely Imports Re-exported.

FOREIGN TRADE.

Exports.

from each Presidency of British India, to Great Britain, the Continent guishing also the Port to Port Trade, from one Presidency to another.

1854-55.	1855-56.	1856-57.	1857-58.	1858-59.	1859-60.	1860-61.	1861-62.
Cwt.	Cwt.	Cwt.	Cwt.	Cwt.	Cwt.	Cwt.	Cwt.
520,431	607,117	755,982	419,808	662,528	381,611	380,904	288,595
14,027	19,799	104,683	23,043	4,484	16,508	16,328	11,862
14,903	21,340	50,265	33,405	15,519	11,756	19,711	39
79,876	81,802	77,294	41,687	87,609	76,114	63,228	78,147
629,237	730,058	988,224	517,943	770,140	485,989	480,171	378,643
274,753	160,708	not stated.	226,086	256,654	179,152	232,187	338,694
903,990	890,766	...	744,029	1,026,794	665,141	712,358	717,337
187,954	430,280	502,281	329,654	233,175	304,022	313,902	245,360
...	...	717	22,449	1,572	...	223	240
...	...	...	5,035	...	...	...	...
18,887	2,864	2,220	8,026	3,124	4,980	3,307	1,999
206,841	433,144	505,218	366,164	237,871	309,002	317,432	247,599
24,422	30,400	26,148	11,868	18,303	25,552	22,150	10,467
231,263	463,544	531,366	378,032	256,174	334,554	339,582	258,066
362	28,728	8,929	2,283	445	1,857	1,206	225
...	...	...	...	...	...	...	...
...	...	...	...	...	...	...	...
106,500	85,130	66,300	75,738	105,441	63,153	47,152	82,220
106,862	113,858	75,229	78,021	105,886	65,010	48,358	82,445
2,684	3,727	3,677	4,125	4,898	4,406	5,277	7,475
109,546	117,585	78,906	82,146	110,784	69,416	53,635	89,920
708,747	1,066,125	1,267,192	751,745	896,148	687,490	696,012	534,180
14,027	19,799	105,400	46,492	6,056	16,508	16,551	12,102
14,903	21,340	50,265	38,440	15,519	11,756	19,711	39
205,263	169,796	145,814	125,451	196,174	144,247	113,687	162,366
942,940	1,277,060	1,568,671	962,128	1,113,897	860,001	845,961	708,687
301,859	194,835	...	242,079	279,855	209,110	259,614	356,636
1,244,799	1,471,895	...	1,204,207	1,393,752	1,069,111	1,105,575	1,065,323

Hope, Ceylon, China, Java, Maldive Islands, Mauritius and Bourbon, New South Wales Indies
Pegu, Bombay, Karachí, and Indian French Ports.
gong, Goa, Moulmein, and Pegu.

SACCHARUM: Sugar.

Foreign Trade in Refined

FOREIGN TRADE.

Exports.

It may be again observed that in the returns furnished by the above tables no distinction was made into refined and unrefined sugar. In the early years of this trade such a distinction scarcely existed. The East India Company found that it only paid to export the purer article, and indeed sugar was treated as a ballast cargo to be used in place of saltpetre. It would not pay to export it as an ordinary cargo, the freight from India being too high. Even as a ballast cargo it was found that the better qualities were more profitable than the inferior. With the growth of refineries in England and through the vast improvements in shipping and consequent cheapening of freights, it not only became possible however, but was more profitable to export the crudely refined or unrefined sugars Accordingly, the distinction in the trade returns had to be made into these two great sections. In the tables which may now be given, the exports from India are shown from 1871 to 1891. The figures in column I, it will be observed, are the totals of refined and unrefined sugar from 1871 to 1875, and after that year columns I and III separate the returns into the classes named. Each year, therefore, as the relative proportions of the two classes changed, it would have become more and more incorrect to add together the quantities of refined and unrefined sugars. Before a figure to express the total can now be arrived at, the former has to be reduced to the value of the latter, in the ratio of 2⅔ to 1, and in some provinces 3 to 1 would be even safer. The table above, which gives the trade from India to England from 1851 to 1862, a decline is shown of the quantity from 1,506,051 cwt. to 534,180 cwt. In 1871-72 the exports had still further declined to 373,897 cwt. From that year to the present there has been an almost uniform falling off in the quantity of refined sugar, but a steady improvement in the unrefined article. The highest record of Indian exports appears to have been 1883-84 (if the year 1876-77 be excluded, an abnormal year), since which year it may be said a decline in the total of the two classes has been manifested:—

Exports by Sea of Indian Sugar to Foreign Countries

YEARS.	Refined or Crystallized, and Sugar-candy.		Unrefined Sugar, Molasses, Gúr, etc.		To allow of comparison with the early returns, the grand total may be shown by reducing refined to unrefined sugar.
	I. Cwt.	II. R	III. Cwt.	IV. R	V. Cwt.
1871-72 . .	372,897	28,80,482	...	These figures should not be accepted as correct, since the amount of refined sugar is not known.	372,897
1872-73 . .	630,938	49,27,432			630,938
1873-74 . .	294,818	22,78,227			294,818
1874-75 . .	498,054	31,92,383			498,054
1875-76 . .	107,208	11,04,274	313,554	14,35,100	581,574
1876-77 . .	674,627	72,57,281	418,998	19,94,680	2,105,565
1877-78 . .	477,128	49,74,679	366,997	24,83,834	1,559,817
1878-79 . .	51,043	6,96,792	228,713	13,46,808	356,320
1879-80 . .	44,963	5,91,652	279,616	14,67,061	392,023
1880-81 . .	18,915	3,24,562	515,259	27,92,946	562,546
1881-82 . .	34,010	5,05,854	883,483	54,76,463	968,508

FOREIGN TRADE.

Exports.

YEARS.	Refined or Crystallized, and Sugar-candy.		Unrefined Sugar, Molasses, Gur, etc.		To allow of comparison with the early returns, the grand total may be shown by reducing refined to unrefined sugar.
	I. Cwt.	II. R	III. Cwt.	IV. R	V. Cwt.
1882-83 . .	111,274	13,01,331	1,207,424	67,86,428	1,485,639
1883-84 . .	203,693	22,86,004	1,426,827	71,46,181	1,936,059
1884-85 . .	55,323	7,14,940	1,015,596	47,45,755	1,153,903
1885-86 . .	24,942	3,29,787	1,142,598	49,24,337	1,204,953
1886-87 . .	33,340	4,41,435	953,066	46,06,597	1,036,416
1887-88 . .	37,723	4,62,388	1,008,565	41,95,899	1,102,872
1888-89 . .	34,523	4,33,621	978,955	50,69,771	1,065,262
1889-90 . .	111,323	14,91,320	1,309,321	76,80,470	1,587,628
1890-91 . .	28,768	3,83,754	795,973	37,91,871	867,893

In order to demonstrate more fully the present state of the Indian sugar trade, the following analysis of the returns of each fifth year since 1875-76 may now be shown:—

Analysis of the Indian Foreign Exports for each fifth year since 1875-76.

REFINED OR CRYSTALLIZED SUGAR AND SUGAR-CANDY.

Years.	Provinces from whence exported.	Cwt.	R	Countries to which exported.	Cwt.	R
1875-76.	Bengal .	83,995	8,09,768	United Kingdom .	65,592	5,51,119
	Bombay .	13,161	1,99,031	Ceylon . .	17,633	2,30,094
	Sind . .	224	3,534	Arabia . .	8,794	1,28,348
	Madras .	9,826	99,917	Aden . . .	3,939	58,700
	Burma .	2	24	United States .	3,751	28,133
				Turkey in Asia .	2,816	44,282
	TOTAL .	107,208	11,04,274	Other countries the balance.		
1880-81.	Bengal .	9,460	1,63,200	Ceylon . .	11,656	2,09,169
	Bombay .	3,618	64,145	Arabia . .	3,299	51,024
	Sind . .	27	481	Aden . . .	1,131	18,369
	Madras .	5,810	96,736	Other countries the balance.		
	Burma .	*Nil.*	...			
	TOTAL .	18,915	3,24,562			
1885-86.	Bengal .	4,238	47,551	Ceylon . .	18,644	2,60,547
	Bombay .	2,913	34,089	United Kingdom .	2,400	22,181
	Sind . .	9	161	Arabia . .	1,480	14,198
	Madras .	17,782	2,47,986	Other countries the balance.		
	Burma .	*Nil.*	*Nil.*			
	TOTAL .	24,942	3,29,787			
1890-91.	Bengal .	1,803	24,206	Ceylon . .	16,216	1,95,571
	Bombay .	11,616	1,76,874	Aden . . .	3,137	40,752
	Sind . .	95	1,419	Persia . .	2,573	42,013
	Madras .	15,202	1,80,850	Turkey in Asia .	2,320	37,313
	Burma .	52	405	Eastern coast of Africa.	1,875	28,952
	TOTAL .	28,768	3,83,754	Other countries the balance.		

SACCHARUM:
Sugar.

Foreign Trade in Refined

FOREIGN TRADE.

Exports.

UNREFINED SUGAR, MOLASSES, GÚR, ETC.

Years.	Provinces from whence exported.	Cwt.	R	Countries to which exported.	Cwt.	R
1875-76.	Bengal .	121	1,560	United Kingdom .	273,128	12,20,290
	Bombay .	12,390	1,04,767	United States .	15,110	50,640
	Sind .	97	751	Arabia . .	13,238	1,06,480
	Madras .	299,807	13,24,175	Ceylon . .	8,140	29,417
	Burma .	1,139	3,849	Aden . . .	2,133	19,520
				Other countries the balance.		
	TOTAL .	313,554	14,35,100			
1880-81.	Bengal .	13,983	1,08,127	United Kingdom .	498,074	26,74,717
	Bombay .	6,140	58,821	Ceylon . .	8,602	41,654
	Sind . .	167	1,604	Arabia . .	6,051	57,890
	Madras .	494,375	26,22,939	Aden . . .	833	7,920
	Burma .	594	1,455	Other countries the balance.		
	TOTAL .	515,259	27,92,946			
1885-86.	Bengal .	2,313	16,937	United Kingdom .	793,525	33,88,536
	Bombay .	12,342	95,782	Egypt . .	145,931	6,39,103
	Sind . .	27	218	Spain . . .	91,172	3,90,651
	Madras .	1,126,794	48,08,833	St. Helena . .	84,277	3,24,436
	Burma .	1,122	2,567	Arabia . .	11,720	82,828
				Other countries the balance.		
	TOTAL .	1,142,598	49,24,337			
1890-91.	Bengal .	422	3,100	United Kingdom .	756,438	34,95,695
	Bombay .	13,767	1,23,268	Aden . . .	13,611	1,15,950
	Sind . .	212	1,870	Arabia . .	13,024	1,09,031
	Madras .	780,283	36,58,658	Ceylon . .	10,659	57,864
	Burma .	1,289	4,975	Straits Settlements	772	4,010
				Other countries the balance.		
	TOTAL .	795,973	37,91,871			

The reader, in contrasting the figures exhibited in the above table, with those for the years 1851 to 1862, will be able to discover the radical changes that have taken place. The Indian export traffic may be said to have changed from Bengal to Madras, and from refined to unrefined sugar, during the past twenty or thirty years. Indeed, it may safely be said that there has been a steady decline in the export of refined sugar from India since the year 1845.

Imports. 451

B. Imports into India.—If the change in the character of the Indian Foreign Export Trade in sugar be regarded as significant, many persons may be disposed to view the revolution of the import traffic as fraught with positive danger to the Indian cultivator. The writer has already tried to combat that position by showing that the immense and yearly increasing imports do not, so far, appear to have caused a decline of production. He does not wish it to be thought, however, that he regards it as impossible that foreign sugars may in the future effect that result but rather that there are no indications of immediate danger.

FOREIGN TRADE.

Imports.

Imports by sea of Sugar from Foreign Countries.

Years.	REFINED OR CYSTALLIZED SUGAR. I.	II.	UNREFINED SUGAR, MOLASSES, GÚR, ETC. III.	IV.
	Cwt.	R	Cwt.	R
1871-72	562,559	70,63,545	No information as to quantity available for these years.	34,240
1872-73	342,450	43,61,124		40,337
1873-74	435,570	55,55,169		34,612
1874-75	395,715	51,58,647		6,989
1875-76	610,524	89,39,283	2,627	19,989
1876-77	256,304	40,22,105	1,801	13,458
1877-78	473,332	79,67,329	1,773	13,029
1878-79	918,202	1,47,75,653	5,179	33,152
1879-80	647,630	1,06,59,414	4,379	28,467
1880-81	982,262	1,60,96,243	4,059	15,329
1881-82	772,519	1,24,21,892	3,463	15,688
1882-83	669,348	1,08,56,003	3,324	13,607
1883-84	729,321	1,14,61,689	7,588	22,012
1884-85	1,613,067	2,13,89,937	3,807	18,440
1885-86	1,164,056	1,45,58,063	7,130	22,909
1886-87	1,678,490	2,05,46,411	71,065	2,58,985
1887-88	1,715,002	2,08,03,360	93,477	3,32,810
1888-89	1,450,481	1,74,12,643	167,229	4,96,747
1889-90	1,623,621	2,16,91,047	99,492	3.09,441
1890-91	2,734,491	3.32,68,496	197,410	7,30 365

In the remarks offered above it has been the custom in discussing published returns both of production and export, to reduce the Indian sugars to the standard of *gúr*. This may be done by accepting the ratio at $2\frac{1}{2}$ *gúr* to 1 sugar or 3 *gúr* to 1 sugar. Some writers consider the former as sufficient; others the latter. In order to keep up this standard it becomes necessary to express the figures in column I as *gúr* and to add the result to column III, in order to compare the total imports with the exports (column V of table, page 338). It will be seen that the Foreign Imports are almost entirely in refined sugar, while the Foreign Exports are almost exclusively in unrefined sugar.

Conf. with pp. 19 20, 39-40, 316, 329, 343-44, 346.

The first direct effects of the beet-sugar production of Europe on India were (*a*) the closing of the markets to which India exported refined sugar, (*b*) the throwing on the market large quantities of colonial sugar which sought an outlet in India. From both of these influences, it will be observed, a larger amount of crude sugar must have become available in India. Consumption increased through the slight fall in price thereby occasioned, and, finding a demand, production extended. But in time beet-root sugar had to seek foreign markets. Within the past few years, therefore, large quantities have begun to pour into India, so that we have now not only cheap Colonial cane refined sugar but still cheaper beet-sugar being pressed on the Indian consumer. This fact explains the immense and sudden expansion of the imports shown above for 1890-91.

The following table may be now given in order to furnish an *Analysis of the Indian Imports of sugar during each fifth year since 1875-76*:—

SACCHARUM: Sugar.

Foreign Trade in Refined

FOREIGN TRADE.

Imports.

Refined Sugar and Sugar-candy

Years.	Provinces into which imported.	Cwt.	R	Countries from whence imported.	Cwt.	
	Bengal	150	2,715	Mauritius	518,202	74,45,312
	Bombay	585,844	86,10,151	China (Hong-Kong)	63,005	11,00,727
1875-76	Sind	162	3,226	Straits Settlements	28,009	3,71,367
	Madras	1,036	22,633	Madagascar	520	7,540
	Burma	23,332	3,00,558	United Kingdom	308	5,916
	Total	610,524	89,39,283	Other countries the balance.		
	Bengal	43,173	7,58,134	Mauritius	746,209	1,19,90,799
	Bombay	890,521	1,45,67,164	China (Hong-Kong)	140,956	25,29,771
1880-81	Sind	8	236	Straits Settlements	81,992	13,50,741
	Madras	1,038	25,369	Java	12,004	1,98,062
	Burma	47,522	7,45,340	Other countries the balance.		
	Total	982,262	1,60,96,243			
	Bengal	83,347	11,53,897	Mauritius	890,545	1,09,69,899
	Bombay	1,037,718	1,27,64,644	China (Hong-Kong)	190,556	24,18,976
1885-86	Sind	12,377	1,67,790	Straits Settlements	38,399	5,65,434
	Madras	2,251	40,079	United Kingdom	19,279	2,28,533
	Burma	28,363	4,31,653	Java	8,684	1,89,942
	Total	1,164,056	1,45,58,063	Other countries the balance.		
	Bengal	511,796	61,90,135	Mauritius	1,345,383	1,63,03,189
	Bombay	1,724,991	2,06,14,116	Germany	709,195	84,02,707
1890-91	Sind	360,964	46,54,293	United Kingdom	281,196	34,32,357
	Madras	7,561	1,06,097	China (Hong-Kong)	195,912	26,13,508
	Burma	129,179	17,03,855	Straits Settlements	114,467	15,00,124
	Total	2,734,491	3,32,68,496	Other countries the remainder.		

Unrefined Sugar, Molasses, Gúr, etc.

Years.	Provinces into which imported.	Cwt.	R	Countries from whence imported.	Cwt.	R
	Bengal	2	114			
	Bombay	958	8,340	Eastern Coast of Africa	975	7,007
1875-76	Sind	178	1,313	Straits Settlements	810	6,603
	Madras	680	3,649	Ceylon	573	2,594
	Burma	809	6,573	Arabia	126	1,377
				United Kingdom	116	1,276
	Total	2,627	19,989	Other countries the balance.		
	Bengal	*Nil.*	12			
	Bombay	2,388	7,897	Mauritius	2,229	6,921
1880-81	Sind	569	2,888	Ceylon	896	2,670
	Madras	988	3,806	Arabia	629	3,163
	Burma	114	726	Other countries the balance.		
	Total	4,059	15,329			
	Bengal	5,603	15,351	Mauritius	6,247	16,967
	Bombay	1,080	4,649	Straits Settlements	355	2,564
1885-86	Sind	289	1,929	Arabia	264	1,500
	Madras	74	302	Other countries the balance.		
	Burma	84	678			
	Total	7,130	22,909			
	Bengal	192,840	7,10,747	Mauritius	149,562	3,60,828
	Bombay	3,120	13,095	Java	40,053	3,12,916
1890-91	Sind	15	96	Straits Settlements	4,477	44,940
	Madras	1,370	5,583	Arabia	1,908	5,724
	Burma	65	844	Ceylon	1,364	5,546
	Total	197,410	7,30,365	Other countries the balance		

FOREIGN TRADE.

Imports.

But, to exemplify more fully the leading features of this modern import trade, the following analysis of the past five years may be furnished. The growth of the imports from the continent of Europe will perhaps be viewed with greater concern than the older traffic with Mauritius, since it represents the amounts of beet-sugar being used in India:—

Analysis of some of the chief items of the Imports of Foreign Sugar into India since 1885-86.

Countries from whence obtained.		1885-86.	1886-87.	1887-88.	1888-89.	1889-90.	1890-91.
		Cwt.	Cwt.	Cwt.	Cwt.	Cwt.	Cwt.
Mauritius .	Refined .	890,545	1,310,250	1,205,465	1,243,224	1,277,119	1,345,383
	Unrefined	6,247	66,791	91,183	162,991	89,240	149,562
Germany . .	Refined .	0	59	6,881	0	40,734	709,195
France . .	Refined .	272	812	150	757	3,979	9,356
Austria . .	Refined .	0	9,899	4,380	196	3,087	31,374
Belgium . .	Refined .	0	579	1,195	1,402	3,497	25,044
Italy . .	Refined .	0	347	234	6	0	51
United Kingdom	Refined .	19,279	37,340	47,297	617	25,102	281,196
Straits Settlements	Refined .	38,399	51,117	91,853	88,367	97,341	114,467
	Unrefined .	355	,725	773	1,560	8,763	4,477
China (Hong-Kong)	Refined .	190,556	244,859	316,035	115,814	152,252	195,912
Ceylon . .	Refined .	620	7	5,276	55	20,193	571
	Unrefined .	74	288	1,237	679	1,097	1,364
GRAND TOTAL OF ALL FOREIGN IMPORTS	Refined, cwt	1,164,05-	1,678,490	1,715,002	1,450,481	1,623,621	2,734,491
	Unrefined, ,,	7,130	71,065	93,477	167,229	99,492	197,410
	Refined, R	1,45,58,063	2,05,46,411	2,08,03,360	1,74,12,643	2,16,91,047	3,32,68,496
	Unrefined, ,,	22,909	2,58,985	3,32,810	4,96,747	3,09,441	7,30,365

Mr. J. E. O'Conor (*Review of the Trade of India for 1890-91*) says of the imports of sugar by India (from Foreign Countries) that the increase is a noticeable feature:—"Of refined sugar, which is mainly what India imports, the quantity imported was about 68 per cent. more than in 1889-90, the excess being chiefly beet-sugar imported from Germany. This is an immediate and direct result of the system of sugar bounties, aided by the development of direct steam communication between India and Germany, and by the course of exchange. It is worth while to draw attention here to the fact that, whereas in former years India exported more sugar than she imported, that feature has in the last few years been rapidly reversed" Last year India imported 2,734,491 cwt., while the exports came to 824,741 cwt. only, so that fully three cwt. were received by India for every cwt. that she furnished to the outer world.* **Mr. O'Conor**, commenting still further on this state of affairs, puts certain salient questions, to which, however, he hazards no answer:—"The question suggests itself: Is this feature of the trade the result of artificial encouragement of production in Europe? or is it the result of natural causes, Indian sugar being really dearer and, therefore, unable to compete, or has the limit of our production been reached? If it is the result of State encouragement in Europe, then after a time the imports will diminish, if they will not cease entirely, for the bounty system will probably terminate in a few years; but if it arises out of natural causes, we must expect imports to increase progressively with increase of population, while the exports diminish." A couple of years before **Mr. O'Conor** offered these suggestive

* *Conf.* with the chapter on History, p. 40; also the remarks (*pp. 341, 346*) on the Foreign, as also the Internal Trade, where these figures will be shown to be in one respect misleading. They do not express in its full bearing the altered nature of the Indian trade.

FOREIGN TRADE.

Imports.

questions, the Government of India, in a despatch which reviewed the information that had been then brought to light, arrived at the conclusion that the bounty system was not affecting India to any appreciable extent, but that, on the contrary, the cultivation of sugar-cane had recently been greatly extended, and that the consumption of sugar was greater than ever it had been, while the industry of growing the cane was highly remunerative. The despatch may be here quoted :—

"With the information now before the Government of India, it may be said that the consumption of sugar in India has increased to a great extent during the last thirty or thirty-five years. Not only is more sugar produced now, but the imports are larger, in fact almost as large as the exports, which have now considerably diminished, used to be thirty-five years ago. The imports, which comprise chiefly Mauritius sugar (refined). are mostly taken by the Bombay Presidency, where it appears to supply a distinct demand for crystallized sugar, and whence a portion is despatched inland. The present exports by sea consist chiefly of unrefined sugar, which is supplied almost exclusively by Madras. The trade of this Presidency is flourishing, and the same seems to be the case with regard to the industry of Bengal. The sugar, which in former years was sent away from Bengal in large quantities, is now said to be consumed in India.

"The Government of India considers that it may be said in general terms that the sugar industry of India is at the present day in a thriving condition, and that it has not been affected to any appreciable extent, like other sugar-producing countries, by the system of sugar-bounties prevailing in continental Europe. In the despatch of May, 1882, the Government of India made the following remarks, which may be taken to apply equally to the present condition of the sugar industry in India :—

"'The increasing import, and the decreasing export do not, we consider, indicate the decadence of the industry in India. The area under sugar-cane has largely increased in Upper India through the development of canal irrigation, and is reported to be extending from the same causes in Bombay. More sugar is produced in India than formerly; but the demand is much greater. Not only has the increasing prosperity of the people increased the average consumption, but sugar is now borne by rail into tracts where the cane is cultivated to a limited extent, and which were formerly very scantily supplied. The profitableness of the industry is seen in the high price which the Indian cultivator can obtain for his produce, and all the evidence before us leads to the belief that the capital invested in sugar cultivation in India is steadily increasing.'"

The imports of FOREIGN SUGAR into India first exceeded one million cwt. in the year 1884-85, since which date they have fluctuated, but, on the whole, shown an upward tendency until, as stated, they last year assumed the very considerable proportions of 2,734,491 cwt. In 1884-85 the Collector of Customs, Calcutta, reported that the low price of refined sugar in Europe caused large imports of beet-sugar, chiefly from the United Kingdom and Austria. At first, however, the effect of the beet-root trade may be said to have been the supply of markets formerly met by the West Indies, thus releasing a large quantity of cane-sugar which was poured into India from Mauritius. Gradually, however, the beet-sugar began to tell directly on India until the hitherto unprecedented state of affairs came to pass that India became no more a country to which Europe looked for sugar, but an outlet for its surplus production. Some writers hold that the remission of the import duty greatly favoured the importation of foreign sugar. (*Conf with. p. 315.*) Even were this admitted, it cannot be said India as a whole has thereby been injured, but it would be hard to prove that the disturbance of the sugar markets caused by beet would or would not have produced the same result whether India possessed or did not possess an import duty.

Re-exports. 452

C. Re-exports of Foreign Sugar from India.—The trade under this section is by no means a very important one. During the past ten years it has averaged about 150,000 cwt. of refined sugar, and practically no unrefined sugar. The following analysis of each fifth year, since 1875-76, may be accepted as fully representing the trade :—

FOREIGN TRADE.

Re-exports.

Refined or Crystallized Sugar and Sugar-candy Re-exported

YEAR.	Provinces from which exported.	Cwt.	R	Countries to which exported.	Cwt.	R
1875-76	Bengal .	1	8	Persia . . .	38,616	5,17,872
	Bombay .	86,247	12,27,160	Arabia . . .	27,629	4,00,299
	Sind .	152	2,484	Turkey in Asia .	11,067	1,72,504
	Madras .	241	4,844	Aden . . .	4,068	57,618
	Burma .	*Nil*	*Nil*	Ceylon . . .	2,812	43,315
				Other Countries the	balance.	
	TOTAL .	86,641	12,34,496			
1880-81	Bengal .	*Nil*	*Nil*	Persia . . .	56,314	9,93,211
	Bombay .	109,443	19,37,846	Turkey in Asia .	24,072	4,28,127
	Sind .	88	1,505	Arabia . . .	19,090	3,31,376
	Madras .	822	13,661	Ceylon . . .	3,640	63,869
	Burma .	*Nil*	*Nil*	Eastern Coast of Africa.	3,477	70,053
				Other Countries the	balance.	
	TOTAL .	110,353	19,53,012			
1885-86	Bengal .	*Nil*	*Nil*	Persia . . .	1 09,494	13,80,224
	Bombay .	161,815	20,31,188	Arabia . . .	20,571	2,46,567
	Sind .	1,066	14,674	Turkey in Asia .	19,553	2,54,935
	Madras .	252	3,804	Aden . . .	5,178	57,522
	Burma .	210	2,940	Other Countries the	balance.	
	TOTAL .	163,343	20,52,606			
1890-91	Bengal .	*Nil*	*Nil*	Persia . . .	78,156	9,66,470
	Bombay .	155,754	19,13,785	Arabia . . .	28,574	3,48,841
	Sind .	4,618	59,858	Aden . . .	19,876	2,44,493
	Madras .	195	2,932	Turkey in Asia .	17,582	2,14,024
	Burma .	1	12	Eastern Coast of Africa . .	7,860	97,134
				Other Countries the	balance.	
	TOTAL .	160,568	19,76,587			

It will thus be seen that the bulk of these re-exports are sent from Bombay and go mainly to Persia.

INTERNAL TRADE. 453

II.—INTERNAL TRADE OF INDIA IN REFINED AND UNREFINED SUGAR.

It has already been stated that the area under sugar-cane and sugar-yielding palms in India may be accepted as 2,500,000 acres. This has been estimated to produce 2,500,000 tons of coarse sugar. Last year, however, India imported (when expressed as coarse sugar or *gúr*) 7,033,637 cwt. and exported 867,893 cwt. A net import was, therefore, obtained by the country of 6,165,744 cwt. of *gúr*. It should be observed that the exports of India are almost exclusively in *gúr* or coarse sugar, while the imports are entirely, or very nearly so in refined sugar. **Mr. J. E. O'Conor** in the *Review of the Trade of India for 1890-91* (in the passage quoted at page 343), has not apparently thought it necessary to make this distinction. He has added together the exports of refined and unrefined sugar and compared the total thus obtained with the similar total of the imports. The result came to this, that last year India exported 824,741 cwt. and imported 2,734,491 cwt, or fully three cwt. received by India for every cwt. furnished to the world. This is, perhaps, sufficiently startling by itself without the further argument that the money spent in purchasing the imports would have procured very nearly three times as much *gúr*, so that, as stated above, it is quite fair to say that India imported approximately 7 cwt. for each cwt. exported. If this view be not accepted, the exports of *gúr* might be expressed in the quantity of refined sugar that they would have yielded in Europe, and that figure compared with the Indian imports. Some such consideration would seem necessary, since in stating the comparative consumption of sugar in India with that in European countries,* the reduction has been made of the 2,500,000 tons of *gúr* produced as equivalent to 1,000,000 tons of sugar of like value with that used in European countries. But as the imports were in refined sugar, and had to be consumed as such, it is probably the more correct consideration to credit India with a net import of only 1,909,750 cwt. (or 95,487 tons) instead of 6,165.944 cwt. (or 308,287 tons). It will thus be seen that a consumption for all India (including the net imports) of 1,000,000 tons of sugar, or 3,000,000 tons of *gúr*, is very considerably under than over the mark. In discussing the internal transactions of India in sugar that quantity had better, therefore, be accepted as the amount which estimates of local consumption and records of internal trade have to confirm.

Conf. with pp. 40, 118, 120, 316, 329-30, 340-41, 343.

The subject of consumption of sugar per head of population has been so fully discussed already that it seems sufficient to refer the reader to the paragraphs above (*pp. 117-18*), that deal with that subject, and to rest satisfied with furnishing in this place such particulars as are available regarding the movement of sugar on its railways, or otherwise.

BENGAL. 454

1.—Bengal.

It is extremely difficult to convey a clear conception of the internal sugar trade of this province. No statement of the road traffic, nor indeed of the river-borne trade, can be furnished, except for that comparatively small section of the transactions that passes by these routes to and from Calcutta. The registration of traffic on the railways (although even these tap but limited tracts) affords the only tangible conception of the provincial trade. In dealing with this subject it is essential that the trade of Calcutta, so far as possible, should be treated as distinct from that of the province. It is only by so doing that the chief modern aspects of trade can be understood, namely, (*a*) the loss of the foreign exports, (*b*) the existence of a large import from Madras and foreign countries, and (*c*) the admitted

* *See p. 119.*

INTERNAL TRADE OF BENGAL.

expansion of sugar production, the outlets for which are (*d*) increased consumption and (*e*) increased exports to the upper provinces of India.

Rail-borne. 455

Rail-borne Trade of Bengal.

The following table may be given of the rail-borne sugar traffic to and from Bengal:—

Intei-Provincial Trade of Bengal by Rail.

Years.		Imports into Bengal.		Exports from Bengal.	
		Quantity.	Total in Unrefined Sugar.	Quantity.	Total in Unrefined Sugar.
		Mds.	Mds.	Mds.	Mds.
1881-82	Refined	8,576	25,818	2,21,873	9,49,458
	Unrefined	4,378		3,94,774	
1882-83	Refined	10,227	33,144	1,92,526	8,60,552
	Unrefined	7,577		3,79,237	
1883-84	Refined	8,325	31,396	1,54,530	6,95,317
	Unrefined	10,584		3,08,992	
1884-85	Refined	9,453	29,841	1,00,735	5,58,274
	Unrefined	6,209		3,06,437	
1885-86	Refined	12,201	35,084	88,525	9,20,179
	Unrefined	4,582		6,98,867	
1886-87	Refined	10,657	32,895	1,33,808	9,78,538
	Unrefined	6,253		6,44,018	
1887-88	Refined	10,361	30,097	78,104	11,88,641
	Unrefined	4,195		4,44,215	
1888-89	Refined	10,501	45,443	58,216	6,95,927
	Unrefined	19,191		2,78,371	
1889-90	Refined	10,113	1,01,664	58,696	10,44,538
	Unrefined	76,382		3,94,337	

Babu Addonath Banerjee, commenting on the figures shown above for the years 1881-87, points out that, while the imports had remained stationary, the exports had fluctuated in a marked degree both in refined and unrefined sugar. Since then, however, it will be observed, the imports have vastly increased, while the exports have continued to fluctuate to exactly the same extent as formerly. It will further be noted that the increase in the imports is mainly in unrefined sugar. The chief item of this increase has been the very much larger supplies drawn from the North-West Provinces by Calcutta and Behar, two very important centres of sugar-refining. It is noteworthy in passing therefore that the increase in imports by rail is not in refined sugar, though the foreign imports shown in the table (page 342) manifest a considerable increase in that item. In 1875-76 Bengal received 150 cwt. and in 1890-91 511, 796 cwt. of refined sugar from foreign countries. It will however be seen (page 364) that Bombay is increasing its supply of foreign refined sugar to the North-West Provinces. (*Conf. with p 361.*)

The following tables analyse the returns of the Inter-Provincial (external) rail-borne trade of Bengal for the years 1887-90:—

SACCHARUM: Sugar.

Internal Trade in Refined

INTERNAL TRADE of Bengal.

Rail-borne.

Analysis of the Rail-borne (External) Bengal Trade in Refined Sugar.

Whence Imported or whither Exported.	Imports into Bengal.								Exports from Bengal.							
	To Behar.	To Western Bengal.	To Eastern Bengal.	To Northern Bengal.	To Dacca.	To Calcutta.	To Chota-Nagpur.	Total.	From Behar.	From Western Bengal.	From Eastern Bengal.	From Northern Bengal.	From Dacca.	From Calcutta.	From Chota-Nagpur.	Total.
1887-88.	Mds.	Mds.	Mds.	Mds.	Mds.	Mds.	Mds.	Mds.	Mds.	Mds.	Mds.	Mds.	Mds.	Mds.	Mds.	Mds.
Bombay	...	...	...	...	...	...	...	...	185	...	...	...	...	...	...	185
North-West Provinces and Oudh	7,511	137	...	199	...	2,454	..	10,301	20,198	22	107	...	..	25,885	...	46,212
Panjáb	12	2	...	1	...	34	..	49	1,536	...	287	...	..	4,749	..	6,572
Central Provinces	...	...	...	...	...	...	..	...	405	...	...	...	..	...	..	405
Berar	...	...	...	...	...	...	...	...	...	...	...	...	...	...	...	...
Rajputana and Central India	...	...	...	...	...	9	...	9	15,679	...	...	...	...	604	...	16,283
Bombay Port	...	...	...	...	...	2	...	2	71	...	...	...	...	...	...	71
Total	7,523	139	...	200	...	2,499	...	10,361	46,064	22	394	...	...	31,624	...	78,104
1888-89.																
Bombay	..	...	...	..	...	...	...	...	417	...	...	...	...	...	...	417
North-West Provinces and Oudh	6,903	44	16	905	...	2,603	...	10,471	6,366	758	3	...	...	27,223	...	34,350
Panjáb	...	...	...	1	...	24	...	25	2,546	161	...	...	...	5,533	...	8,240
Central Provinces	..	...	...	...	...	...	...	...	2,635	343	...	...	...	495	...	3,475
Berar	...	...	...	...	...	...	...	...	...	.	...	...	...	2	...	2
Rajputana and Central India	2	...	...	...	...	...	...	2	10,708	129	...	...	...	894	...	11,731
Bombay Port	3	..	...	...	...	...	...	3	3	...	...	...	...	...	...	3
Total	6,908	44	16	906	...	2,627	...	10,501	22,675	1,391	3	...	...	34,147	...	58,216
1889-90.																
Bombay	2	...	...	..	...	...	...	2	...	...	...	...	...	...	...	...
North-West Provinces and Oudh	7,223	143	99	139	...	2,461	39	10,104	16,999	7	...	...	...	23,262	...	40,178
Panjáb	1	1	...	1	...	...	...	3	2,030	...	...	...	...	1,906	...	3,936
Central Provinces	1	...	...	...	...	...	...	1	5,037	...	...	...	...	197	...	5,234
Nizam's Territory	...	...	...	...	...	...	...	...	...	...	...	...	...	12	...	12
Rajputana and Central India	2	...	...	...	...	...	...	2	9,162	...	...	...	...	174	...	9,336
Madras Ports	..	...	...	...	...	1	.	1	...	...	...	..	...	...	...	..
Total	7,229	144	99	140	...	2,462	39	10,113	33,138	7	...	...	...	25,551	...	58,696

INTERNAL TRADE of Bengal.

Re-exports.

Analysis of the Rail-borne (External) Bengal Trade in Unrefined Sugar.

Whence Imported or whither Exported.	Imports into Bengal.								Exports from Bengal.							
	To Behar.	To Western Bengal.	To Eastern Bengal.	To Northern Bengal.	To Dacca.	To Calcutta.	To Chota-Nagpur.	Total.	From Behar.	From Western Bengal.	From Eastern Bengal.	From Northern Bengal.	From Dacca.	From Calcutta.	From Chota-Nagpur.	Total.
1887-88.	Mds.	Mds.	Mds.	Mds.	Mds.	Mds.	Mds.	Mds.	Mds.	Mds.	Mds.	Mds.	Mds.	Mds.	Mds.	Mds.
Bombay	...	...	...	...	...	...	...	...	69,940	...	...	...	...	...	...	69,940
North-West Provinces and Oudh	1,975	189	...	75	...	1,934	...	4,173	1,38,136	22	34	...	...	16,789	...	1,54,981
Panjáb	22	...	...	...	...	...	...	22	3,120	34	2	...	...	7,149	...	10,305
Central Provinces	...	...	...	...	...	...	...	...	1,01,762	2	...	...	...	2,684	...	1,04,448
Berar	...	...	...	...	...	...	...	...	29,425	...	...	...	...	...	...	29,425
Rajputana and Central India	...	...	...	...	...	...	...	...	70,700	1	1	...	...	4,065	...	74,767
Karachi	...	...	...	...	...	...	...	...	76	...	...	...	...	...	...	76
Bombay Port	...	...	...	...	...	...	...	...	270	...	...	...	...	3	...	273
Total	1,997	189	...	75	...	1,934	...	4,195	4,13,429	59	37	...	...	30,690	...	4,44,215
1888-89.																
Bombay and Sind	...	...	...	...	...	...	...	...	17,319	...	...	...	...	...	...	17,319
North-West Provinces and Oudh	10,861	892	...	125	...	7,294	...	19,172	1,24 832	267	315	2	...	10,172	...	1,35,588
Panjáb	...	...	...	...	...	...	...	...	4,183	6	111	...	...	3,827	...	8,127
Central Provinces	...	...	...	...	...	...	...	...	78,882	389	...	...	...	943	...	80,214
Berar	...	...	...	...	...	...	...	...	3,865	...	...	...	...	40	...	3,905
Rajputana and Central India	14	5	...	...	...	...	...	19	29,558	48	...	...	...	3,443	...	33,049
Nizam's Territory	...	...	...	...	...	...	...	...	...	...	...	...	...	2	...	2
Bombay Port	...	...	...	...	...	...	...	...	167	...	...	...	...	...	...	167
Total	10,875	897	...	125	...	7,294	...	19,191	2,58,806	710	426	2	...	18,427	...	2,78,371
1889-90.																
Bombay and Sind	...	...	...	...	...	...	...	...	37,246	...	...	...	...	...	...	37,246
North-West Provinces and Oudh	25,065	2,094	554	425	...	47,783	...	75,921	1,73,429	42	335	...	..	2,595	...	1,76,401
Panjáb	5	17	...	...	...	305	62	389	7,920	7	817	880	...	677	...	10,301
Central Provinces	...	...	...	...	...	...	...	...	1,10,073	3	1	290	...	142	...	1,10,509
Berar	...	...	...	...	...	...	...	...	21,824	...	...	..	...	...	...	21,824
Rajputana and Central India	71	...	...	...	...	1	...	72	37,333	...	...	...	...	346	...	37,679
Bombay Port	...	...	...	...	...	...	...	...	370	...	...	...	...	7	...	377
Total	25,141	2,111	554	425	...	48,089	62	76,382	3,88,195	52	1,153	1,170	...	3,767	...	3,94,337

INTERNAL TRADE of Bengal.

The reader may have observed how very important Shahabad and Gya districts are in the supply of sugar. Behar may not incorrectly be described as the chief area of sugar-cane cultivation (and Shahabad its principal district) just as Eastern Bengal is the great region of date-palm sugar (and Jessor the chief district). In the above analytical tables of the Bengal external land traffic in sugar the importance of Behar will be fully realized. Thus, of the exports of unrefined sugar that left Bengal in the years 1887-88, 1888-89, and 1889-90, Behar furnished the entire amount, except about 20,000 to 30,000 maunds. What is, perhaps, of almost equal significance (if Calcutta be left out of consideration) the major portion of the imports of unrefined sugar are taken by Behar, a fact to be accounted for by the very extensive trade that exists in Shahabad in refining sugar. This idea is borne out by an inspection of the table for refined sugar where it is shown that fully half the total exports in that class go from Behar, the other half being from Calcutta.

The chief external provinces that draw on Bengal for sugar are the North-West Provinces and Oudh, the Central Provinces, Rajputana, and Central India. The trade with these provinces fluctuates often within wide limits, but the analyses of the three last years given above are in these respects quite normal, and manifest, if anything, a tendency (particularly in unrefined sugars) to improvement.

By Rail.
456 *The Intra-Provincial Trade of Bengal by Rail* may be now discussed. It may be observed that the movement of sugar from one part of Bengal to another is that alone referred to in this place, and which it is desired to recognize as distinct from the conveyance of sugar to and from Bengal and other provinces. On the subject of this trade Babu Addonath Banerji wrote:—

"I may mention that the trade in *refined* sugar is not chiefly between Behar and Calcutta. It has two distinct currents—one flowing downwards from Behar, and the other going upwards from Calcutta. The former loses volume in Calcutta, a small supply only going to Western Bengal, while the upward trade, which shows a steady development since 1883-84, has a wider distribution, the chief importers being Behar, Western Bengal, and Northern Bengal. The following statement shows the statistics of this trade for four years, commencing from 1883-84, since which year the block system of registration was extended to the Eastern Bengal State Railway and connected lines:—

Traffic in Refined Sugar in the Internal Blocks of Bengal Railways.

Year, and whence exported.	To Behar.	To Western Bengal.	To Eastern Bengal.	To Northern Bengal.	To Dacca.	To Calcutta.	To Chota-Nagpur.	Total Trade.
1883-84.	Mds.	Mds.	Mds.	Mds.	Mds.	Mds.	Mds.	Mds.
From Behar . .	...	5,923	...	16	...	69,885	...	75,824
,, Calcutta . .	5,801	5,551	1,475	6,202	...	...	...	19,029
,, Other places .	93	...	...	1,680	...	8,202	...	9,975
TOTAL .	5,894	11,474	1,475	7,898	...	78,087	...	1,04,828
1884-85.								
From Behar . .	...	2,879	29	178	...	22,720	...	25,806
,, Calcutta . .	10,899	5,096	2,800	4,317	...	...	...	23,112
,, Other places .	173	...	511	786	...	3,607	...	5,077
TOTAL .	11,072	7,975	3,340	5,281	...	26,327	...	53,995
1885-86.								
From Behar . . .	...	2,847	...	...	...	18,676	...	21,523
,, Calcutta . .	11,396	4,358	3,429	3,416	...	...	...	22,609
,, Other places .	62	6	6	231	...	2,579	...	2,884
TOTAL .	11,458	7,221	3,435	3,647	...	21,255	...	47,016

and Unrefined Sugar in Bengal. (*G. Watt.*)

INTERNAL TRADE of Bengal.

By rail.

Year, and whence exported.	To Behar.	To Western Bengal.	To Eastern Bengal.	To Northern Bengal.	To Dacca.	To Calcutta.	To Chota-Nagpur.	Total Trade.
1886-87.								
From Behar	...	7,234	14	4	...	11,346	...	18,598
,, Calcutta	11,403	12,537	2,366	5,646	346	...	...	32,298
,, Other places	704	53	...	286	6	1,169	...	1,218
TOTAL	12,107	19,824	2,380	5,936	352	12,515	...	53,114
1887-88.								
From Behar	...	2,881	...	456	...	9,523	...	12,860
,, Calcutta	10,502	13,891	6,051	11,301	485	...	...	42,230
,, Other places	17	134	16	625	59	2,018	...	1,869
TOTAL	10,519	16,906	6, 67	12,382	544	11,541	...	57,959
1888-89.								
From Behar	...	1,736	...	1,299	...	9,482	...	12,517
,, Calcutta	7,885	12,636	1,787	8,794	178	...	...	31,280
,, Other places	262	101	39	6,409	...	754	...	7,565
TOTAL	8,147	14,473	1,826	16,502	178	10,236	...	51,362
1889-90.								
From Behar	...	6,533	137	8,195	...	20,694	108	35,657
,, Calcutta	10,358	24,725	1,505	14,983	559	...	224	52,354
,, Other places	208	54	37	175	25	925	45	...
TOTAL	10,566	31,312	1,679	23,353	584	21,619	377	89,490

"The increase in the exports from Calcutta occurred simultaneously with the increase in its imports by sea and coast, and the figures given above show how the imported article is distributed in the interior. Western Bengal, which, previous to 1883-84, drew largely upon Behar, now gets the largest supply from Calcutta.

"As regards *unrefined* sugar, the exports to Calcutta from Eastern Bengal form the largest item, but the enormous supply of *gúr* which annually comes to Calcutta and Western Bengal from Behar should not, I think, be ignored in making a generalization of this trade." The figures of traffic for the four years 1883—87 are given below:—

Year, and whence exported.	To Behar.	To Western Bengal.	To Eastern Bengal.	To Northern Bengal.	To Dacca.	To Calcutta.	To Chota-Nagpur.	Total Trade.
	Mds.	Mds.	Mds.	Mds.	Mds.	Mds.	Mds.	Mds.
1883-84.								
From Behar	...	51,350	...	62	...	63,919	...	1,15,331
,, Eastern Bengal	...	...	...	5,635	...	3,23,233	...	3,28,868
,, Calcutta	1,627	2,004	5,807	3,628	...	...	...	13,066
,, Other places	371	...	248	...	...	480	...	1,099
TOTAL	1,998	53,354	6,055	9,325	...	3,87,632	...	4,58,364
1884-85.								
From Behar	...	52,787	...	252	...	33,365	...	86,404
,, Eastern Bengal	2	...	...	10,914	...	2,64,090	...	2,75,006
,, Calcutta	5,238	8,336	3,665	8,221	...	...	...	25,460
,, Other places	935	...	89	...	...	435	...	1,459
TOTAL	6,175	61,123	3,754	19,387	...	2,97,890	...	3,88,329
1885-86.								
From Behar	...	51,654	...	23	...	20,282	...	71,959
,, Eastern Bergal	71	...	...	15,844	...	2,75,469	...	2,91,384
,, Calcutta	4,907	16,555	4,815	10,392	...	...	...	36,669
,, Other places	1,656	...	16	...	...	2,070	...	3,742
TOTAL	6,634	68,209	4,831	26,259	...	2,97,821	...	4,03,754

SACCHARUM: Sugar.

Internal Trade in Refined

INTERNAL TRADE of Bengal.

By rail.

Year, and whence exported.	To Behar.	To Western Bengal.	To Eastern Bengal.	To Northern Bengal	To Dacca.	To Calcutta	To Chota-Nagpur.	Total Trade.
1886-87.								
From Behar . .	...	76,575	...	83	...	29,925	...	1,06,583
,, Eastern Bengal .	15	4	...	15,613	18,596	2,24,535	...	2,58,763
,, Calcutta . .	2,893	59,865	4,339	16,260	1,803	...	...	85,160
,, Other places .	849	25	2	2	...	315	...	1,193
TOTAL .	3,757	1,36,469	4,341	31,958	20,399	2,54,775	...	4,51,695
1887-88.								
From Behar . .	...	78,014	69	3,411	15	56,062	...	1,37,571
,, Eastern Bengal .	11	3,571	...	23,143	13,864	2,43,895	...	2,84,484
,, Calcutta . .	5,257	65,543	11,532	26,807	5,081	...	...	1,14,220
,, Other places .	1,314	26	154	20	...	697	...	2,211
TOTAL .	6,582	1,47,154	11,755	53,381	18,960	3,00,654	...	5,38,486
1888-89.								
From Behar . . .	..	51,392	119	23,640	...	46,550	...	1,21,701
,, Eastern Bengal .	145	6,954	...	22,106	20,667	2,95,159	...	3,45,031
,, Calcutta . .	4,583	62,566	7,220	29,668	2,249	...	...	1,06,286
,, Other places .	1,149	202	198	2,429	...	1,433	...	5,411
TOTAL .	5,877	1,21,114	7,537	77,843	22,916	3,43,142	...	5,78,429
1889-90.								
From Behar . . .	...	1,02,395	13,830	23,918	9,713	2,28,803	2,212	3,80,875
,, Eastern Bengal .	328	9,921	...	15,105	23,624	2,65,317	...	3,14,295
,, Calcutta . .	2,808	89,517	26,131	16,852	10,842	...	115	1,46,268
,, Other places .	1,328	264	685	2	79	772	1,054	4,141
TOTAL .	4,464	2,02,097	40,645	55,877	44,258	4,94,892	3,381	8,45,619

Conf. with remarks regarding Palm-sugar, pp. 138, 226-27, 231, 266, 270, 310, 361, 370.

The reader should very particularly observe the importance of Eastern Bengal in the supply of Calcutta in *gúr*. In connection with the external rail-borne trade, the value of Behar has been specially noted. That province cannot, however, be viewed as unimportant in the traffic here specially dealt with. It furnishes Calcutta with a very considerable quantity of refined sugar, and a fluctuating trade also exists in *gúr* from Behar. Last year the exports of *gúr* from Behar to Calcutta attained their highest recorded amount, *viz.* 2,28,803 maunds. It is, however, to Eastern Bengal that Calcutta looks for the bulk of its unrefined rail-borne sugar, as also a large portion of its refined sugar. In the passage below which deals with the Calcutta trade by itself, this subject will be found returned to, more especially in connection with the road and river traffic.

Summarizing the main facts learned regarding the external and internal rail-borne sugar trade of Bengal, Babu Addonath Banerji remarks:—

"Since the year 1883-84 the internal trade of the Lower Provinces carried by the East Indian Railway and the Eastern Bengal State Railway and connected lines has been registered under the 'block' system of registration, which, however, has no pretensions to register the entire trade of the Province, but only such portions of it as move from one trade block to another. The trade since that period has been well sustained, as the following statement shows:—

INTERNAL TRADE of Bengal.

By Rail.

Internal Sugar Trade of Bengal.

YEARS.	Refined sugar.	Unrefined sugar.	Total in unrefined sugar.
	Mds.	Mds.	Mds.
1883-84	1,04,828	4,58,364	7,20,434
1884-85	53,995	3,88,329	5,23,316
1885-86	47,016	4,03,754	5,21,294
1886-87	53,114	4,51,699	5,84,484
1887-88	57,959	5,38,486	6,83,383
1888-89	51,362	5,78,429	7,06,834
1889-90	89,490	8,45,615	10,69,340

"The year 1883-84 was, as already stated, one of deficient harvests, and consequently also of high prices of food-grains. The quantity of sugar exported during that year from Bengal to other provinces in India amounted to 6,95,317 maunds against 8,60,046 maunds in the previous year, and the large internal trade of 1883-84, as shown above, was carried on at the cost of the external trade, which could not compete with local demands. The year 1884-85 was also a bad one, and there were decreases both in the external and internal trade. In comparison with the previous year, the decline under the former head aggregated 1,37,043 maunds, and under the latter head 1,97,118 maunds.. In 1885-86 the external trade rose from 5,58,274 maunds to 9,20,179 maunds, but the internal trade remained stationary, while in the following year there was a satisfactory increase under both heads. The combined total of the external and internal export trade during 1883-84, 1886-87, and 1889-90 was as follows :—

Exports. 457

Exports from Bengal

YEARS.	To external blocks in India.	To internal blocks in Bengal.	TOTAL.
	Mds.	Mds.	Mds.
1883-84	6,95,317	7,20,434	14,15,751
1886-87	9,78,538	5,84,484	15,63,022
1889-90	10,44,538	10,69,340	21,13,878

"The figures given above do not, therefore, show that there has been any falling off in this section of the trade."

It may be observed that the author has deemed it desirable to preserve in many places above Babu Addonath Banerji's original criticisms of the figures of trade prior to 1886-87. In republishing some of his tables the modern figures, down to 1890, have, however, been added. The chief inference, Babu Addonath desired should be drawn, was, that, viewed from every aspect almost, the Bengal trade in sugar had manifested an expansion. The more recent returns will be seen to fully substantiate that opinion, for in many directions the traffic has within the past three years increased by 50 per cent. So far as the rail returns are concerned, also, the increase can in no way be attributed to a fall in price, for the wholesale recorded value of refined sugar (carried by rail) manifested an increase in 1889-90 on that of the previous year of 8·54 per cent., and unrefined sugar an increase of 24·14 per cent. The former stood at R10-4 a maund in 1888-89 and R11·2 in 1889-90, while the latter was R3-10 in 1888-89 and R4-8 in 1889-90.

SACCHARUM: Sugar.

Internal Trade in Refined

INTERNAL TRADE of Bengal.

By River & Canal.

458

River and Canal-borne Trade.—Turning now to the subject of the river-borne trade of Bengal, it may at once be explained that this practically consists of the transactions between Calcutta and Bengal with Assam. It is, therefore, an inter-provincial record. Of the intra-provincial transactions from district to district, along the rivers of the province, very little can be shown in a tabular form. Of the Assam trade **Babu Addonath Banerjee** wrote :—

"Under this head is shown the registered trade between Bengal and Assam carried along the Brahmaputra and Megna rivers by country boats and inland steamers. The total quantity of such traffic since the year 1881-82 is shown below :—

Inter-Provincial trade between Bengal and Assam by river.

Years.		Imports into Bengal from Assam.			Exports from Bengal to Assam.		
		Refined sugar.	Unrefined sugar.	Total in unrefined sugar.	Refined sugar.	Unrefined sugar.	Total in unrefined sugar.
		Mds.	Mds.	Mds.	Mds.	Mds.	Mds.
1881-82	By boat	2	1,526	1,531	13,165	1,05,537	1,38,449
	,, steamer	...	...	...	11,355	2,666	31,053
	Total	2	1,526	1,531	24,520	1,08,203	1,69,502
1882-83	By boat	...	1,375	1,375	22,783	1,33,334	1,90,291
	,, steamer	...	...	...	9,497	5,554	29,296
	Total	...	1,375	1,375	32,280	1,38,888	2,19,587
1883-84	By boat	...	150	150	30,279	1,15,911	1,91,608
	,, steamer	6	...	15	11,355	5,555	33,942
	Total	6	150	165	41,634	1,21,466	2,25,550
1884-85	By boat	...	51	51	37,202	1,60,515	2,53,520
	,, steamer	...	...	...	17,480	5,793	49,493
	Total	...	51	51	54,682	1,66,308	3,03,013
1885-86	By boat	...	64	64	44,185	1,91,803	3,02,265
	,, steamer	...	...	...	15,694	8,362	47,597
	Total	...	64	64	59,879	2,00,165	3,49,862
1886-87	By boat	2	14	19	44,592	1,71,987	2,83,467
	,, steamer	...	7	7	11,336	4,216	32,556
	Total	2	21	26	55,928	1,76,203	3,16,023

"The import trade is unimportant, but the exports show great development. Compared with 1881-82, the increase under *refined* sugar amounted during 1886-87 to 128·92 per cent., and under *unrefined* sugar to 64·84 per cent. The advance has been considerable since the year 1884-85, and is attributable to the steady growth of the trade under favourable conditions."

The traffic indicated by **Babu Addonath Banerjee** in the above table may now be brought down to the returns of last year.

INTERNAL TRADE of Bengal.

By River & Canal.

Traffic on the Brahmaputra and Megna rivers between Bengal and Assam, as carried by Inland Steamers aud Country Boats.

Years and Routes.		Imports into Bengal from Assam.			Exports from Bengal to Assam.		
		Refined sugar.	Unrefined sugar.	Total in unrefined sugar.	Refined sugar.	Unrefined sugar.	Total in unrefined sugar.
		Mds.	Mds.	Mds.	Mds.	Mds.	Mds.
1887-88	By boat	...	11	11	32,265	1,63,099	2,43,761
	,, steamer	...	4	4	10,760	9,503	36,403
	Total	...	15	15	43,025	1,72,602	2,80,164
1888-89	By boat	...	11	11	26,491	1,68,835	2,35,063
	,, steamer	...	...	...	17,291	13,101	56,328
	Total	...	11	11	43,782	1,81,936	2,91,391
1889-90	By boat	...	72	72	24,518	1,72,817	2,34,112
	,, steamer	2,353	402	6,284	22,053	15,573	70,705
	Total	2,353	474	6,356	46,571	1,88,390	3,04,817

It will thus be seen that the export-traffic to Assam has fluctuated slightly, but has not seriously increased since 1884-85, although it is double what it was in 1881-82.

But the traffic with Assam, though the chief item, is by no means the only river-borne sugar trade of Bengal. Sugar appears in the returns, for example, of the steamer traffic on the Nuddea rivers; on the Midnapore Canal; on the Hidgellee Canal; on the Orissa Canals; the Calcutta Canals; and on the Ganges and Hooghly rivers. It is somewhat difficult, however, to prepare a statement of the river traffic, as supplies of a commodity like sugar are often conveyed to certain marts, landed, sold, reshipped or sent by train, so that the same amount may appear more than once. This error is overcome by selecting important sections, such as the trade with Calcutta or with Assam. As records of actual transactions by water carriage the following may be cited:—

Years and Routes.		Down-Stream.		Up-Stream.	
		Refined sugar.	Unrefined sugar.	Refined sugar.	Unrefined sugar.
Nuddea Rivers.		Mds.	Mds.	Mds.	Mds.
More than half being to and from Calcutta.	1887-88	10,348	41,291	5,086	23,526
	1888-89	17,649	36,422	3,721	25,316
	1889-90	28,608	36,930	1,586	12,496
Midnapore Canal.					
Almost entirely to and from Calcutta.	1887-88	...	72,995	...	22,177
	1888-89	...	98,960	...	17,497
	1889-90	...	43,442	...	12,817

SACCHARUM: Sugar.

Internal Trade in Refined

INTERNAL TRADE of Bengal.

By River & Canal.

Years and Routes.		Down-Stream.		Up-Stream.	
		Refined sugar.	Unrefined sugar.	Refined sugar.	Unrefined sugar.
Hidgellee Canal.					
The down-stream in this case are exports mostly from Calcutta and the up-stream imports.	1887-88	...	10,426	...	855
	1888-89	...	7,015	350	310
	1889-90	...	14,991	...	40
Orissa Canals.					
This indicates the traffic of Cuttack, Balasor, and Puri. The imports are chiefly into Cuttack.	1887-88	...	37,057	...	715
	1888-89	...	41,014	...	520
	1889-90	...	22,515	...	1,510
Orissa Coast Canals.					
The down-stream traffic is mainly exports from Calcutta.	1887-88	713	12,135	20	80
	1888-89	1,720	11,798	...	21
	1889-90	2,270	26,054	...	...
Calcutta Canals.					
The figures shown as down-stream are imports into Calcutta and up-stream exports.	1887-88	26,999	2,77,121	4,339	34,134
	1888-89	33,465	2,73,009	3,290	47,633
	1889-90	18,050	2,84,119	2,979	34,404
The imports are mainly from Panspotta and Dhappa. The exports go to Kowrapookur and Dhappa.					
Brahmaputra and Megna.					
This trade has already been sufficiently indicated in the table above of Assam trade.					
Ganges, Bhagiruthee, Jellinghee, and Hooghly Rivers.					
Down-stream here means imports into Calcutta and up-stream exports.	1887-88	15,757	1,314	4,98	3,692
	1888-89	7,940	559	7,323	5,923
	1889-90	6,414	929	8,232	17,236

Having now discussed the Bengal sugar trade under the various headings of rail, river, and canal, as carried to and from both the internal and the external blocks of the province, it may serve a useful purpose to give here a review of all the figures that have been obtained, and to furnish a corresponding statement for Calcutta. Calcutta may be accepted as the chief, if not the only, seaport town to which foreign and coasting supplies are brought to the province, or from which exports are made by sea. The net balance of the marine transactions has, therefore, to be added to the supplies brought to Calcutta by land routes before either the consumption of the capital can be dealt with or the exports (in most cases re-exports) from Calcutta by land routes can be rightly understood. Owing to the complete overlapment of each and every item of the trade with all others, it becomes difficult to trace out transactions, and the totals in one table may at first sight seem to conflict with those shown in another, until the particulars of

INTERNAL TRADE of Bengal.

each table are critically examined. In reviewing the Bengal and Calcutta transactions by land routes, it need only be necessary to take the returns of one year, *viz.* 1889-90, but the subject requires to be broken into the two sections—Refined and the Unrefined.

Bengal land-route transactions. 459

Bengal Province Land-route Transactions.

A.—Refined Sugar.

The imports and exports *by rail* resulted in a net import by the province of 5,241 maunds and a net export *by rivee* of 8,066 maunds. The ultimate balance by the two routes was a net export of 825 maunds on a trade of 1,28,082 maunds imported and 1,30,907 maunds exported. The Imports *by rail* were 7,645 maunds from the North-West Provinces; and 52,354 maunds from Calcutta. *By river* 68,077 maunds from Calcutta The Exports *by rail* were 16,916 maunds to the North-West Provinces; 2,030 maunds to the Panjáb; 5,037 maunds to the Central Provinces; 9,162 maunds to Rajputana and Central India; and 21,619 maunds to Calcutta. *By river* 23,583 maunds to Assam, and 52,560 maunds to Calcutta.

B.—Unrefined Sugar.

The imports and exports of this class left a net export *by rail* of 7,10,904, and *by river* 3,57,164 maunds on a total trade of 3,75,191 maunds imported, and 14,43,259 maunds exported. Of the Imports *by rail* 28,138 maunds came from the North-West Provinces; 84 maunds from the Panjáb; 71 maunds from Rajputana and Central India; and 1,46,265 maunds from Calcutta. *By river*, 72 maunds came from Assam; and 2,00,561 maunds from Calcutta. Of the Exports *by rail*, 37,083 maunds were consigned to Bombay Presidency; 163 maunds to Sind; 1,73,806 maunds to the North-West Provinces; 624 maunds to the Panjáb; 1,10,367 maunds to the Central Provinces; 21,824 maunds to Berar; 37,333 maunds to Rajputana and Central India; 370 maunds to Bombay (port town); and 4,94,392 to Calcutta. *By river*, 1,64,526 maunds to Assam, and 3,93,271 maunds to Calcutta.

Calcutta land-route transactions. 460

Calcutta Land-route Transactions.

A.—Refined Sugar.

The transactions *by rail* resulted in a net export of 53,824 maunds and *by river* in a net export of 14,059 on a trade of 90,562 imported, and 1,58,445 exported. The Imports were *by rail*, 21,619 maunds from the province of Bengal; 2,461 maunds from the North-West Provinces; and 1 maund from Madras. *By river*, 52,560 from Bengal; and 13,921 maunds from the North-West Provinces. While the Exports were *by rail*, 52,354 maunds to Bengal; 23,262 maunds to the North-West Provinces; 1,906 maunds to the Panjáb; 197 maunds to the Central Provinces; 174 maunds to Rajputana and Central India; and 12 maunds to the Nizam's territory. *By river*, 68,077 maunds to Bengal; 112 maunds to the North-West Provinces; and 12,351 maunds sent to Assam.

B.—Unrefined Sugar.

The imports and exports left a net import *by rail* of 3,92,949 maunds and *by river* of 83,291 maunds in the city, out of a total 9,36,252 maunds imports and 3,60,012 maunds exports. The Imports *by rail* came from the Bengal Province 4,94,892 maunds; from the North-West Provinces 47,783 maunds; from the Panjáb 305 maunds; and from Rajputana and Central India 1 maund. *By river*, from the Bengal Province, 3,93,271 maunds. The Exports *by rail* went to Bengal 1,46,265 maunds; to the

INTERNAL TRADE of Bengal.

North-West Provinces 2,596 maunds; to the Panjáb 677 maunds; to the Central Provinces 142 maunds; to Rajputana and Central India 346 maunds; and to Bombay (port town) 7 maunds.

Calcutta trade.
461

CALCUTTA SUGAR TRADE.

Before concluding this notice of the Bengal Sugar Trade, it may be useful to bring together some of the main facts regarding the Calcutta sections. Some of these have already been exemplified, but there are others that seem to call for special consideration as, for example, the indication that is afforded of the Bengal road-traffic by the registration of transactions carried by carts into or out of Calcutta. **Babu Addonath** says that:—

"the bulk of the imported sugar is consumed by the me tropolitan population. In Calcutta there is a congregation of all nationalities, and the prejudice against this sugar is, therefore, not so general here as it is in the interior; hence almost all the foreign sugar that comes in goes to add to the luxury of the towns-people without, in any way interfering either with the condition of the sugar-cane cultivation, or the sugar trade in the mofussil. How the large imports of foreign sugar into Calcutta have passed into consumption may be seen from the statement below, which gives the grand total of traffic imported and exported by all routes, *i.e.* by rail, river, road, sea, and coast, and the quantity not exported before the close of the year."

Sugar Trade of Calcutta by all Routes.

		Imports.	Exports.	Surplus of imports over exports expressed in unrefined sugar.
		Mds.	Mds.	Mds.
1878-79	Refined	2,37,534	4,04,435	1,33,925
	Unrefined	7,55,092	2,03,915	
1879-80	Refined	3,10,560	4,07,047	3,98,724
	Unrefined	9,19,632	2,79,691	
1880-81	Refined	3,69,133	2,46,845	7,76,000
	Unrefined	7,89,603	3,19,323	
1881-82	Refined	5,10,591	2,55,350	10,58,135
	Unrefined	8,42,475	4,22,442	
1882-83	Refined	5,77,578	3,71,519	9,52,819
	Unrefined	8,17,931	3,80,260	
1883-84	Refined	5,09,044	4,23,187	9,16,281
	Unrefined	10,53,516	3,51,873	
1884-85	Refined	3,83,927	1,82,100	12,03,266
	Unrefined	9,92,036	2,93,337	
1885-86	Refined	3,79,529	1,63,415	11,75,846
	Unrefined	9,31,331	2,95,770	
1886-87	Refined	4,23,587	2,00,739	11,54,301
	Unrefined	8,90,459	2,93,278	
1887-88	Refined	5,46,053	2,29,311	14,90,629
	Unrefined	10,72,876	3,74,102	
1888-89	Refined	4,13,947	2,01,926	12,75,549
	Unrefined	11,83,888	4,38,391	
1889-90	Refined	4,87,837	2,85,312	11,44,780
	Unrefined	12,13,193	5,74,725	

NOTE.—To reduce the above figures of maunds to cwt. multiply by $\frac{4}{5}$.

In the Annual Report of the Inland Trade of Calcutta for 1889-90, it is pointed out that there was manifested in that year a recovery of the Calcutta trade both in refined and unrefined sugar. The advance under the former was 17·85 per cent. compared with the previous year, but in comparison with 1887-88 the present figures manifested a decrease of 10·66 per cent. In the case of unrefined sugar, the traffic was in excess of these

INTERNAL TRADE of Bengal.

years by 2·47 per cent. and 13·08 per cent., respectively. The boat traffic, however, showed a large falling off, which amounted to 27·76 per cent.

Calcutta trade imports 462

The following analysis of the Calcutta import trade may be here furnished for the past three years :—

Analysis of the Total Imports of both classes of Sugar into Calcutta.

Provinces and countries.	Refined sugar.			Unrefined sugar.		
	1887-88.	1888-89.	1889-90.	1887-88.	1888-89.	1889-90.
	Mds.	Mds.	Mds.	Mds.	Mds.	Mds.
Bengal	1,68,810	1,57,272	1,06,788	8,40,657	8,05,230	8,06,163
Behar	19,061	12,323	23,739	56,107	46,808	2,30,558
N.-W. Provinces and Oudh .	5,820	9,261	16,382	1,967	7,446	47,783
Madras	51,190	43,105	36,852	5	98	...
Bombay	1,049	30,999	25,363	...	...	...
Other countries . . .	3,00,123	1,60,987	2,78,713	1,24,140	2,24,306	1,28,689
Total in mds. .	5,46,053	4,13,947	4,87,837	10,72,876	11,83,888	12,13,193
Total in cwt. .	3,90,038	2,95,676	3,48,455	7,66,340	8,45,634	8,66,566

The quantities imported by sea showed a rise of 45·11 per cent. in refined sugar, but a decline of 42·79 per cent. in unrefined sugar. The table above exhibits the countries or provinces from which the supplies of Calcutta were drawn during each of the past three years.

The magnitude of the road traffic may be judged of by the transactions shown in the table below of the Calcutta trade by boat, steamer, and road :—

Years.		Imports into Calcutta.			Exports from Calcutta.		
		Sugar refined.	Sugar unrefined.	Total in sugar unrefined.	Sugar refined.	Sugar unrefined.	Total in sugar unrefined.
		Mds.	Mds.	Mds.	Mds.	Mds.	Mds.
1878-79	By boat .	1,11,611	4,75,296	7,54,324	67,095	96,303	2,64,040
	,, steamer .	...	...	...	7,703	266	19,523
	,, road .	67,217	87,253	2,55,295	68,641	55,373	2,26,976
	Total .	1,78,828	5,62,549	10,09,619	1,43,439	1,51,942	5,10,539
1879-80	By boat .	1,60,196	5,41,623	9,42,113	99,088	1,66,798	4,14,518
	,, steamer .	...	...	...	10,536	2,116	28,456
	,, road .	96,891	71,822	3,14,050	37,632	52,903	1,46,983
	Total .	2,57,087	6,13,445	12,56,163	1,47,256	2,21,817	5,89,957
1880-81	By boat .	1,12,354	4,37,405	7,18,290	70,597	1,94,353	3,70,846
	,, steamer .	...	66	66	14,947	3,934	41,301
	,, road .	1,46,262	1,66,083	5,31,738	39,813	61,974	1,61,506
	Total .	2,58,616	6,03,554	12,50,094	1,25,357	2,60,261	5,73,653
1881-82	By boat .	1,77,932	3,99,685	8,44,515	63,499	2,03,823	3,62,570
	,, steamer .	...	...	...	11,945	1,877	31,740
	,, road .	2,03,145	1,94,023	7,01,885	50,625	79,307	2,05,869
	Total .	3,81,077	15,93,708	15,46,400	1,26,069	2,85,007	6,00,179

SACCHARUM: Sugar.

INTERNAL TRADE of Bengal.

Calcutta trade imports.

Internal Trade in Refined

Years.		Imports into Calcutta.			Exports from Calcutta.		
		Sugar refined.	Sugar unrefined.	Total of sugar unrefined.	Sugar refined.	Sugar unrefined.	Total of sugar unrefined.
		Mds.*	Mds.	Mds.	Mds.	Mds.	Mds.
1882-83	By boat	2,18,069	4,17,821	9,62,994	56,152	1,54,250	2,94,630
	,, steamer	...	...	...	12,776	3,295	35,235
	,, road	1,93,923	1,07,224	5,92,031	39,401	96,261	1,94,763
	Total	4,11,992	5,25,045	15,55,025	1,08,329	2,53,806	5,24,628
1883-84	By boat	1,45,804	4,64,595	8,29,105	50,136	128,783	2,54,123
	,, steamer	2	...	5	12,306	3,327	34,092
	,, road	2,50,353	1,94,693	8,20,575	29,332	94,027	1,67,357
	Total	3,96,159	6,59,288	16,49,685	91,774	2,26,137	4,55,572
1884-85	By boat	71,867	4,94,302	6,73,970	54,684	1,16,427	2,53,137
	,, steamer	...	...	...	20,096	3,791	54,031
	,, road	1,44,213	1,96,724	5,57,256	28,383	61,925	1,32,882
	Total	2,16,080	6,91,026	12,31,226	1,03,163	1,82,143	4,40,050
1885-86	By boat	83,933	4,42,205	6,52,038	51,140	98,846	2,26,696
	,, steamer	...	...	..	15,355	5,460	43,847
	,, road	1,11,333	1,82,596	4,60,928	23,999	47,319	1,05,817
	Total	1,95,266	6,24,801	11,12,956	89,894	1,51,625	3,76,360
1886-87	By boat	58,464	3,54,219	5,00,379	78,287	1,00,629	2,96,347
	,, steamer	15,476	27	38,717	14,909	4,355	41,627
	,, road	96,059	1,88,493	4,28,641	24,456	49,860	1,11,000
	Total	1,69,999	5,42,739	9,67,737	1,17,652	1,54,844	4,48,974

It may now be useful to bring together into one table the Calcutta trade in both classes of sugar, picking out for that purpose the figures which, in the above tables, have been exhibited in various ways with the object of demonstrating the sources of total Bengal and Calcutta traffic:—

Sugar Trade of Calcutta as manifested by the returns of marine, rail, river and road traffic.

Routes.		Imports.			Exports.		
		1887-88.	1888-89.	1889-90.	1887-88.	1888-89.	1889-90.
		Mds.*	Mds.	Mds.	Mds.	Mds.	Mds.
By Boat	Refined	69,376	79,892	57,714	60,458	62,610	55,666
	Unrefined	3,95,697	4,06,517	3,91,940	1,69,479	1,65,929	1,81,892
,, Inland Steamer	Refined	15,757	7,940	8,767	8,483	14,178	28,374
	Unrefined	1,314	559	1,331	6,960	11,722	28,388
,, East Indian Ry.	Refined	12,044	12,109	23,162	57,093	55,530	61,499
	Unrefined	58,678	55,033	2,77,487	1,01,538	85,603	1,09,799
,, Eastern Bengal Ry.	Refined	1,996	754	919	16,761	9,897	16,406
	Unrefined	2,43,910	2,95,403	2,65,494	47,372	39,110	40,233
,, Road	Refined	94,563	78,304	56,348	24,544	22,821	21,583
	Unrefined	2,49,162	2,01,972	2,48,558	47,535	39,142	40,795
,, Sea	Refined	3,52,317	2 34,948	3,49,927	61,972	36,890	1,05,284
	Unrefined	1,24,115	2,24,401	1,28,383	5,218	96,885	1,73,918
Total	Refined	5,46,053	4,13,947	4,87,837	2,29,311	2,01,926	2,85,312
	Unrefined	10,72,876	11,83,888	12,13,193	3,74,102	4,38,391	5,74,725

* To express the figures of maunds to cwt. multiply by $\frac{5}{7}$.

INTERNAL TRADE of Bengal.

Calcutta trade.

There is, perhaps, only one feature of the above table that calls for special consideration. The importance of Behar in the internal and external supply of Bengal sugar has been fully exhibited in what has already been said. The East Indian Railway is that by which Calcutta draws its supplies of Behar sugar. The amounts, however, brought by that railway to Calcutta are very much less important than the traffic by the Eastern Bengal State Railway or by road.

Date-sugar. 463

Conf. with pp. 138, 226-27, 231, 266, 270, 310, 352, 370.

It may be said that the Calcutta supplies brought from the Eastern Districts are drawn mainly from the region of date-palm cultivation. It would not be safe to infer, however, that 2,43,910 maunds brought by the Eastern Bengal State Railway and the 2,49,162 maunds carried by road were entirely date-sugar. The districts tapped by these routes of transit have a considerable acreage devoted to cane, but, while that is so, it is admissible to assume that an important share of the amounts shown consists of date-sugar. It will be found that the spirit of the remarks below, regarding Madras, would lead to the inference that the chief reason of the Madras success in the recent sugar trade, is the large amount of palm-sugar which it turns out, and it has even been assumed that Bengal, being deficient in that respect, has failed in the competition with Madras. There is little that directly supports an opinion for or against such a conclusion, but, trusting to personal observation alone, the writer would be more disposed to arrive at the very opposite opinion, if indeed, it be necessary to seek for any external explanation of the modern phase of the sugar trade, by which Bengal has lost or found unprofitable the foreign market which it formerly held, while Madras has taken its place. It would seem far more likely that the admixture of date-palm sugar has lowered the reputation of Bengal sugar than that the absence of that form of the article has depreciated the value of the Bengal commodity. The subject seems deserving of careful investigation.

Conf. with the remarks in chapter on Prices, p. 323 et seq.; see also pp. 347, 365.

Beet-sugar began to be first imported into Calcutta in the year 1884-85, since which date, it will be observed, there has been a serious decline of the imports of Indian refined sugar by land routes, particularly by boat and road. This may be accepted as demonstrating largely the decline of the refining industry of Eastern Bengal—a direct result, therefore, of the traffic in cheap imported sugar. The exports by sea in refined sugar appear to be very largely re-exports of foreign sugars. This trade would seem to have developed in the same ratio with the imports of foreign sugar, so that an additional evidence is thereby obtained of the declining importance of the refining art of Lower Bengal. In Behar, however, that industry has not been in any way injured, but, on the contrary, has greatly improved, for the supplies of both classes of sugar drawn last year from Behar were higher than in any former year. It thus seems probable that the traffic in palm-sugar may have been more seriously affected than in cane by the modern tendencies of the trade.

Babu Addonath Banerjee (of the Statistical Department of the Bengal Government) in his very able statement of the sugar trade of Bengal, discusses the question of the Calcutta surplus of total imports over exports, in relation to the population, and arrives at the conclusion that the Calcutta consumption for the seven years previous to 1886-87 amounted to 12 seers 4 chattacks of refined sugar, or (if expressed in unrefined sugar) of 30 seers $9\frac{1}{2}$ chattacks per head. The figures by which this result was obtained may be here exhibited, since they are instructive in themselves apart from the subject of consumption per head of population. Babu Addonath Banerjee wrote as follows:—

"The surplus of imports over exports has considerably increased since 1878-79. After allowing a third of the annual surplus for stock in trade, there remained the following quantities for consumption during each year since 1880-81:—

SACCHARUM: Sugar.

Internal Trade in Refined

INTERNAL TRADE of Bengal.

Calcutta trade.

464

Consumption per head of population, consisting of 900,000 souls, expressed in refined and also in unrefined sugar.

Year.	Refined Sugar.			Unrefined Sugar.		
	Quantity.	Rate per head.		Quantity.	Rate per head.	
	Mds.	Sr.	ch.	Mds.	Sr.	ch.
1880-81	2,06,933	9	3	5,17,334	23	0
1881-82	2,82,169	12	8	7,05,342	31	5
1882-83	2,54,085	11	4	6,35,213	28	3
Average .	2,47,729	10	15⅜	6,19,323	27	8
1883-84	2,44,341	10	13	6,10,854	27	2
1884-85	3,20,871	14	4	8,02,177	35	10
1885-86	3,13,559	13	15	7,83,898	34	13
Average .	2,92,924	13	0	7,32,310	32	8
1886-87	3,07,814	13	10	7,69,534	34	2
Average for Seven Years	2,75,682	12	4	6,89,205	30	9½

"The high rate of consumption shown in the foregoing statement is not to be wondered at, considering the higher standard of life adopted by the great majority of the towns-people. A large quantity of sugar passes daily into consumption in Calcutta in the shape of sweetmeats, of which there are a hundred sorts, and confectionery shops have sprung up in considerable numbers, and there is scarcely any locality where there are not two or three such shops. Then again, a great number of people in Calcutta have taken to drinking tea, and this also necessitates the use of sugar.

"S milar information for the remaining forty-four districts in the Lower Provinces is not available, and we have, therefore, to depend for a general knowledge of the subject on the annual administration reports of Commissioners of Divisions. These reports do not show that any diminution in the cultivation of sugar-cane has taken place in producing districts. The great bulk of the sugar produced in Bengal is, it is well known, consumed in the shape of *gur*. The use of *refined* sugar or *chini*, on the other hand is confined by its dearness to the wealthy classes, and it is therefore chiefly in demand in head-quarters towns of district and sub-divisions for consumption by the well-to-do people. The sugar refineries in Bengal turn out *chini* according to the local demand in India, as well as in foreign countries. Whenever, therefore, there is a slackness in demand for this class of sugar, these refineries suffer, but, on the other hand, the rural and urban population enjoy the luxury of getting more *gur* to eat at a low price in the same way as the metropolitan population gets more *chini* to eat at a low price when it is imported in large quantities from foreign countries. The internal commerce in sugar is prodigious, and, with facilities afforded by roads, railways, and canals, is rapidly increasing, but its consumption is very unequally distributed. Sugar is still regarded as luxury among the poorer classes, who can barely afford to have two meals of rice a day."

In concluding this brief notice of the Bengal Sugar Trade, it may be remarked that, although we do not possess a definite statement of the area under cane and sugar-yielding palms, the registration of rail, river, and canal traffic is more exhaustively recorded in the Lower Provinces than, perhaps, in any other part of India. A more detailed review of the information thus available may help, therefore, to convey an idea of the very great importance of the internal traffic in sugar, and at the same time may throw light on the trade of other provinces regarding which the statistical information is less perfect. To complete this review of the Bengal traffic in sugar, the reader requires to consult the section below, which furnishes tabular statements of the coastwise trade.

INTERNAL TRADE of Bombay. 465

2.—BOMBAY.

In order to convey an idea of the chief items of the *Bombay Internal Trade in Sugar,* it is, perhaps, only necessary to furnish the tables which will be found below. The river traffic of the presidency is believed to be less important than in Bengal, and as to the road transactions it may at once be said nothing whatsoever is known. Should the reader wish to construct a statement of the total sugar trade of Bombay, it would be necessary to bring into one place the imports and exports shown in the section of FOREIGN TRADE, with those here given for the RAIL-BORNE and those which will be found below as COASTWISE TRAFFIC. The particulars of the trade with provinces for which separate chapters have not been furnished in this review of the sugar trade, such, for example, as the Central Provinces, can be worked out by the reader picking out the items of exports from Bombay, from the Panjáb, North-West Provinces, and Bengal, etc., into the Central Provinces. These items would represent the chief imports, and similarly the exports from these provinces appear as imports into Bombay, the Panjáb, the North-West Provinces, Bengal, etc.—

Analysis of the Bombay Presidency Sugar Trade for the years 1888-91.

Presidency. 466

FROM OR TO	REFINED SUGAR.					
	Imports into			Exports from		
	1888-89.	1889-90.	1890-91.	1888-89.	1889-90.	1890-91.
British Provinces.	Mds.	Mds.	Mds.	Mds.	Mds.	Mds.
Madras	23,934	19,681	12,429	249	1,675	170
Bombay	...	...	...	...	...	...
Sind	...	...	1	...	...	...
Bengal	417	...	...	...	2	17
North-West Provinces and Oudh .	3,787	6,099	10,373	2	...	10
Panjáb . . .	39	12	4	49	7	63
Central Provinces .	95	121	51	85	148	254
Berar . . .	39	90	41	153	1,975	250
Native States.						
Rajputana and Central India . . .	265	109	434	3,466	3,481	8,890
Nizam's Territory .	1,029	224	263	56	110	73
Mysore . . .	15,197	29,987	40,405	...	23	107
Seaports.						
Madras	113	18	276	...	...	...
Bombay	4,77,501	4,59,610	6,34,740	284	1,974	57
Calcutta	...	...	...	...	...	...
TOTAL .	5,22,416	5,15,951	6,99,013	4,344	9,399	9,891

SACCHARUM: Sugar.

Internal Trade in Refined

INTERNAL TRADE of Bombay. Presidency.

FROM OR TO	UNREFINED SUGAR.					
	Imports into			Exports from		
	1888-89.	1889-90.	1890-91.	1888-89.	1889-90.	1890-91.
British Provinces.	Mds.	Mds.	Mds.	Mds.	Mds.	Mds.
Madras . . .	59,145	44,164	2,44,820	24	2,434	15
Bombay . . .	...	...	...	...	...	...
Sind . . .	...	...	...	...	21	...
Bengal . . .	17,176	37,083	73,315	...	...	...
North-West Provinces and Oudh . .	2,15,170	2,83,278	4,86,279	7	7	...
Panjáb . . .	9,948	4,453	31,605	13	76	...
Central Provinces .	744	1,086	3,634	23,345	28,046	22,157
Berar . . .	189	77	478	61,789	91,885	98,904
Native States.						
Rajputana and Central India . . .	3,322	4,864	9,567	10,024	8,165	7,218
Nizam's Territory .	443	347	2,333	410	387	241
Mysore . . .	1,052	4,214	71,011	...	114	151
Seaports						
Madras . . .	4	1	1	...	...	...
Bombay . . .	2,49,314	2,47,157	1,71,242	65,780	60,469	34,837
Calcutta . . .	...	...	...	...	...	...
TOTAL .	5,56,507	6,26,724	10,94,285	1,61,392	1,91,604	1,63,523

Port Town.

467

Analysis of the Bombay Port Town Trade in Sugar during the years 1888-91.

FROM OR TO	REFINED SUGAR.					
	Imports into			Exports from		
	1888-89.	1889-90.	1890-91.	1888-89.	1889-90.	1890-91.
British Provinces.	Mds.	Mds.	Mds.	Mds.	Mds.	Mds.
Madras . . .	76	...	...	1,973	1,837	2,620
Bombay . . .	284	1,974	57	4,77,501	4,59,610	6,34,740
Sind	...	...	...	...	...	...
Bengal . . .	3	...	...	3	...	...
North-West Provinces and Oudh . .	50	238	285	5,538	6,031	26,080
Panjáb . . .	20	4	33	3,926	6,449	35,850
Central Provinces .	12	49	7	82,053	57,770	1,16,901
Berar	...	...	...	40,357	38,428	65,909
Native States.						
Rajputana and Central India . . .	8	26	8	93,818	70,375	1,66,432
Nizam's Territory .	...	...	...	23,010	22,421	25,414
Mysore . . .	...	892	197	...	68	91
Seaports.						
Madras . . .	...	...	...	...	...	...
Bombay . . .	...	...	...	...	...	...
Calcutta . . .	...	...	...	...	...	5
TOTAL .	453	3,183	587	7,28,179	6,62,989	10,74,072

INTERNAL TRADE of Bombay.

Port Town.

From or to	Unrefined Sugar.					
	Imports into			Exports from		
	1888-89.	1889-90.	1890-91.	1888-89.	1889-90.	1890-91.
British Provinces.	Mds.	Mds.	Mds.	Mds.	Mds.	Mds.
Madras	8	51	5,294	21	52	2
Bombay	65,780	60,469	34,837	2,49,314	2,47,157	1,71,242
Sind	...	...	...	...	...	...
Bengal						
North-West Provinces	167	370	174	...	...	...
and Oudh	2,468	3,399	1,027	183	111	33
Panjáb	568	352	526	99	141	106
Central Provinces	...	29	8	608	53	599
Berar	53	30	...	18	21	117
Native States.						
Rajputana and Central India	...	310	16	8,525	2,864	2,835
Nizam's Territory	...	...	25	9	52	1
Mysore	...	901	16	...	...	...
Seaports.						
Madras	...	...	...	...	...	...
Bombay	...	...	...	...	...	...
Calcutta	...	7	...	...	...	...
Total	69,044	65,918	41,923	2,58,777	2,50,451	1,74,935

It may be said that the most striking feature of these tables is the very large amounts of refined sugar consigned by rail from the Bombay Port Town to the Presidency (6,34,740 maunds), to Rajputana and Central India (1,66,432 maunds), and to the Central Provinces (1,16,901 maunds). These may be said to be the amounts of foreign sugars which were drained by the interior tracts in 1890-91 from that chief port of Western India. The manner in which these imports have increased, year by year, gives a better indication of the effect of the imported sugars than can be obtained from any other source of inquiry. The remarkable increase in the consumption in Rajputana and Central India is peculiarly instructive, since foreign sugar is there reaching a country where Hinduism is certainly all-powerful. While there is thus demonstrated to be a largely augmented consumption of foreign sugar in Western India and the provinces that draw supplies from Bombay, it cannot, however, be held that these imports have checked the trade in the Native article. On the contrary, it will be seen from the table of the Bombay Presidency imports of *gúr* from the North-West Provinces, that these have increased from 2,15,170 maunds in 1888-89 to 4,86,279 maunds in 1890-91, and an even more notable example may be cited in the supplies drawn by Bombay Presidency from Madras: the imports in 1888-89 stood at 59,145 maunds, whereas last year they had increased to 2,44,820 maunds.

Conf. with pp. 361, 369, 370.

Consumption by Hindus. *Conf. with pp. 324, 377.*

Increased consumption of sugar.

It will thus be seen that in Bombay, as in Bengal and all the other provinces of India, there has recently not only been a greatly increased supply of sugar from foreign countries, but the comsumption of Indian-grown sugar has also expanded to an even greater extent.

N.-W. P. & OUDH.

3.—NORTH-WEST PROVINCES AND OUDH.

It does not seem necessary to explain the table which follows on the *Rail-borne Sugar Trade* of these provinces. The chief item of in-

INTERNAL TRADE of N.-W. P. & Oudh.

terest is the very large quantity of *gúr* which these provinces export to the Panjáb. This trade will, however, be seen to indicate a very considerable contraction during the past three years from 13,32,749 maunds in 1887-88 to 9,35,620 maunds in 1889-90. The trade in that article with Bombay has, however, been greatly increased, and that to the Central Provinces greater than in 1887-88, though less than in 1888-89. The exports of refined sugar from these provinces to the Panjáb have also been greatly augmented during the three years under notice, so that, it will be added, the grand totals of the exports of both classes taken together have been fully maintained. The imports are, comparatively speaking, unimportant, and are drawn mainly from Bengal.

FROM AND TO	Refined Sugar, including Sugar-candy.					
	Imports.			Exports.		
	1887-88.	1888-89.	1889-90.	1887-88.	1888-89.	1889-90.
British Provinces.	Mds.	Mds.	Mds.	Mds.	Mds.	Mds.
Madras	...	...	7	8	...	...
Bombay	9,007*	2	...	7,627	3,787	6,099
Sind	43	...	...	1,185	603	633
Bengal	46,365	7,127	16,916	10,373	7,868	7,643
Panjáb	305	349	985	1,29,976	1,37,121	2,17,143
Central Provinces	...	49	31	37,138	8,577	19,543
Berar	...	...	...	13,749	1,420	6,879
Native States.						
Central India and Rajputana.	24	112	336	98,449	1,86,742	1,73,240
Nizam's Territory	...	...	...	42	21	78
Mysore	...	...	...	9	2	7
Chief Seaport Towns.						
Madras	...	...	3	...	...	2
Bombay	...	5,538	6,031†	...	50	238
Karachi	...	2	2	...	8	3
Calcutta	...	27,223	23,262	...	2,603	2,461
Total	55,744	40,402	47,573	2,98,556	3,48,803	4,33,969

FROM AND TO	Unrefined Sugar, Molasses, Gur.					
	Imports.			Exports.		
	1887-88.	1888-89.	1889-90.	1887-88.	1888-89.	1889-90.
British Provinces.	Mds.	Mds.	Mds.	Mds.	Mds.	Mds.
Madras	...	...	...	952	...	...
Bombay	601*	7	7	1,84,772	2,15,170	2,83,278
Sind	3	...	...	2,01,039	2,01,725	1,38,122
Bengal	1,58,313	1,25,416	1,73,806	4,174	11,878	28,138
Panjáb	7,915	26,628	39,364	13,32,749	13,15,115	9,35,620
Central Provinces	1,225	22	189	1,75,966	3,13,128	2,74,099
Berar	...	...	...	49,522	70,261	95,657

* In 1887-88 the distinction was not made into Presidencies and Port Towns.

† In a communication received since the above was sent to Press, the Director of Land Records and Agriculture draws the author's notice to the fact that in 1890-91 these provinces imported 26,080 maunds of refined sugar from Bombay. These appears to have been foreign sugar. It is given in the table of exports from Bombay, page 364. *Conf. with p. 378.*

INTERNAL TRADE of N.-W. P. & Oudh.

FROM AND TO	UNREFINED SUGAR, MOLASSES, GUR.					
	Imports into			Exports from		
	1887-88.	1888-89.	1889-90.	1887-88.	1888-89.	1889-90.
Native States.	Mds.	Mds.	Mds.	Mds.	Mds.	Mds.
Central India and Rajputana.	689	1,521	3,022	14,90,061	13,09,181	14,31,088
Nizam's Territory .	...	...	...	68	43	45
Mysore . . .	...	...	...	...	...	...
Chief Seaport Towns.						
Madras . . .	...	...	...	...	...	...
Bombay . . .	...	183	111	...	2,468	3,399
Karachi . . .	...	...	...	...	19	267
Calcutta . . .	...	10,172	2,595	...	7,294	47,783
TOTAL .	1,68,746	1,63,949	2,19,094	34,39,303	33,46,282	32,37,496

4.—PANJAB.

PANJAB. 469 *Conf. with p. 179.*

The table given below fully demonstrates the *Rail-borne Sugar Trade* of the Panjáb. It is unnecessary to repeat what has already been shown that this province depends mainly for its external supplies on the North-West Provinces and Karachi. The Panjáb, like Bombay and Sind, consumes a large amount of foreign sugar. These three provinces, in fact, may not inaptly be characterised as India's present market for that commodity—

FROM OR TO	REFINED SUGAR, INCLUDING SUGAR-CANDY.					
	Imports into			Exports from		
	1887-88.	1888-89.	1889-90.	1887-88.	1888-89.	1889-90.
British Provinces.	Mds.	Mds.	Mds.	Mds.	Mds.	Mds.
Madras . . .	...	..	...	...	...	26
Bombay . . .	5	49	7	5	39	12
Sind	2,347	1,598	1,263	29	1,211	260
Bengal . . .	1,823	2,707	2,030	15	1	3
North-West Provinces and Oudh.	1,23,764	1,37,121	2,17,143	290	349	985
Central Provinces	...	...	...	...	20	...
Native States.						
Rajputana and Central India.	1	171	103	310	980	1,002
Seaports.						
Bombay . . .	3,946	3,929	6,449	6	20	4
Karachi . . .	1,20,517	1,05,096	92,565	21	52	17
Calcutta . . .	4,709	5,533	1,906	34	24	...
TOTAL	2,57,112	2,56,204	3,21,466	710	2,696	2,309
Reduced to Gúr	7,71,336	7,68,612	9,64,398	2,130	8,088	6,927

SACCHARUM: Sugar.

Internal Trade in Refined

INTERNAL TRADE of Panjab.

From or to	Unrefined Sugar, Molasses, Gur, etc.					
	Imports into			Exports from		
	1887-88.	1888-89.	1889-90.	1887-88.	1888-89.	1889-90.
	Mds.	Mds.	Mds.	Mds.	Mds.	Mds.
British Provinces.						
Bombay . . .	26	13	76	793	9,948	4,453
Sind	605	135	103	58,496	52,367	1,12,053
Bengal . . .	3,156	4,300	9,624	22	...	84
North-West Provinces and Oudh.	14,37,555	12,15,115	9,35,620	7,849	26,628	39,364
Central Provinces .	...	...	2	31	208	441
Native States.						
Rajputana and Central India.	277	195	686	1,49,105	3,49,115	2,20,141
Nizam's Territory .	...	...	...	...	...	2
Seaports.						
Bombay . . .	71	99	141	65	568	352
Karachi . . .	15,389	31,037	2,919	346	100	404
Calcutta . . .	7,149	3,827	677	...	...	305
Total .	14,64,228	12,54,721	9,49,848	2,16,707	4,38,934	3,77,599

MADRAS. 470

5.—MADRAS.

The want of details of the rail-and canal-borne trade in sugar precludes the preparation of a statement of the Madras trade similar to what has been furnished for the other provinces. It does not, however, appear that any additional information of much moment has been brought to light since the appearance of Babu Addonath Banerjee's note on sugar, so that the following passage may be accepted as manifesting the chief points of interest in the Madras trade. The reader should, however, compare the statements made below with the returns of foreign trade above, and those of coastwise transactions on a further page—

Comparison of exports. 471

Export Trade of Madras compared with that of Bengal.—In connection with this subject it will be necessary, in the first instance, to examine in detail the condition of the sugar market in the Madras Presidency. The total export from that Presidency of *refined* or *unrefined* sugar by sea and coast since the years 1876-77 were as follows:—

INTERNAL TRADE of Madras.

YEARS.	REFINED SUGAR.			UNREFINED SUGAR.		
	To foreign ports.	To coast ports.	Total.	To foreign ports.	To coast ports.	Total.
	Cwts.	Cwts.	Cwts.	Cwts.	Cwts.	Cwts.
1876-77 . . .	20,997	15,003	36,000	410,184	892,52	499,436
1877-78 . . .	16,774	8,369	25,143	358,774	76,751	435,525
1878-79 . . .	802	9,707	10,509	216,604	84,610	301,214
1879-80 . . .	661	26,065	26,726	258,559	193,967	452,526
1880-81 . . .	5,810	28,178	33,988	494,375	194,002	688,377
1881-82 . . .	12,535	65,342	77,877	789,684	199,003	988,687
1882-83 . . .	13,290	79,528	92,818	1,119,930	238,836	1,358,766
1883-84 . . .	31,579	72,802	104,381	1,347,278	206,406	1,553,684
1884-85 . . .	34,780	81,248	116,028	980,884	194,966	1,175,850
1885-86 . . .	17,782	83,114	100,896	1,126,794	188,128	1,314,922
1886-87 . . .	23,834	113,559	137,393	938,966	224,819	1,163,785

"The above table shows that, in the case of *refined* sugar, by far the largest exports went to the coast ports in India, while, as regards shipments to foreign ports, *unrefined* sugar stands first. Both these sections of the trade will be separately considered below.

In the case of *refined* sugar, a marked increase in exports commenced in the year 1881-82. The bulk of the supply is consumed within British India, the chief consumers being Bengal, Burma, and Bombay. The development of the trade in *refined* sugar is of recent date, and is entirely due to the establishment of sugar refineries on an extended scale.

"Unlike Bengal, almost the whole of the exports of *unrefined* sugar from the Madras Presidency go to foreign countries. Its transactions with these ports have risen enormously since 1880-81, but its despatches to coast ports have been pretty steady since that year. The customs returns of that Presidency for the year 1880-81 contain the following remarks on that subject:—'The trade in sugar has shown a considerable expansion, the quantity exported during the year being the highest yet recorded. The increase is due to a brisk demand in the London markets, and to the good crops of sugar-cane in 1879-80, which led to large shipments at the beginning of the official year. Almost the whole sugar exported is unrefined.'

"The largest increase occurred during 1883-84, in which year the following full account of the progress of this trade is given in the Annual Volume of the Sea-borne Trade and Navigation of that Presidency:—"Nine years ago, the exports barely amounted to 400,000 cwts., valued at less than 19½ lakhs. In the year under report they rose to 1,478,600 cwts., valued at more than 80½ lakhs, showing an increase of 271·7 per cent. in quantity and of 315·8 per cent. in value. During the same period the cultivation of the cane has increased by about 19·5 per cent, and the exports of Indian products generally by 33·8 per cent. A large quantity of unrefined sugar (jaggery) exported from this Presidency is made from the juice of the palmyra and date trees; but there are no accounts to show the area covered by these trees growing on waste lands, much less the increase since 1875-76. The development of the sugar industry has been very marked since 1880-81, but the value has not increased in the same proportion as the quantity exported. In the year under report there was a fall in prices, owing, it is said, to good harvests of beet-root in Europe, and large supplies sent thither from America. The bulk of the exports (93·1 per cent.) consisted of unrefined sugar.

"Considerably more than two-thirds of the refined article was consumed within British India,* and the remainder by foreign countries. Almost the whole of the refined sugar is manufactured at the Aska Factory in Ganjam, owned by Messrs. Minchin & Co. and at two other factories established by Mssrs. Parry & Co. in the South Arcot district. Manufacture is also carried out to a small extent at a factory established by a native gentleman in the latter district at Iruvelipet.

	Cwts.	R
* British Burma	26,551	3,66,944
Bombay	22,051	3,31,714
Bengal	21,103	3,23,854

"In 1884-85 the exports fell off largely, owing to the 'increased production of beet-sugar in Europe, the sugar bounties, and the fall in prices the London in market.' There was some improvement in the following year, but the figures were still below those of 1883-84 and 1882-83, and the trade did no altogether escape the evil effects consequent upon the enormous supply of beet-sugar in England. The following

INTERNAL TRADE of Madras.

extract from the Trade and Navigation returns for 1885-86 gives the result of the transactions during the year:—"The condition of the sugar trade was somewhat better than in the previous year. 'The increase in quantity was 13·5 per cent, but value by only 1·4 per cent., owing to lower average prices than in 1884-85 consequent on the production of beet-root sugar in Europe. The exports of refined sugar amounted to 80,700 cwts., and were confined to Ceylon, Bengal, Bombay, and Burma. The bulk of the unrefined sugar, 791,217 cwts., out of 1,171,937 cwts., was shipped to England.'

"By far the largest supply of *unrefined* sugar goes to the London market for the purpose of being refined and for brewers' purposes; also for feeding and fattening cattle. In 1883-84 it was ascertained that the cultivation of the cane in Madras had increased by about 19·5 per cent., and it was further stated that a large quantity of *unrefined* sugar (jaggery) exported from that Presidency was made from the juice of the palmyra and date trees, but there are no statistics to show the area covered by these trees growing on waste lands. The juice is obtained from these plants by cutting off the male spadix when young, and from the cut portion there is for four of five months a continual flow. The liquid is at first clear, and is immediately boiled down to a thick syrup, which granulates on cooling, and constitutes, if not otherwise purified, the coarse, brown sugar called jaggery. If the juice is not immediately boiled it becomes turbid, and passing into the vinous fermentation, forms the intoxicating drink called toddy. This kind of sugar sells very cheap, and so answers the purposes of the London market, where 'coarse sugar was sold in 1884-85 at 1*d.* per pound to be used for fattening cattle in the place of linseed cake.' Now thes conditions do not exist in connection with the sugar exports from Bengal* which supplies refined cane-sugar, which is much dearer than the coarse date-sugar mentioned above. The following remarks were made on this point by the Officiating Director:—'With regard to the increased export of unrefined sugar from Madras, it appears that Madras obtains a very large proportion of its sugar supply from date and other palm-trees. The area under such trees is 31,000 acres, producing 22,00,000 maunds of coarse sugar, against 57,000 acres under sugar-cane, producing 26,00, 000 maunds of sugar. The superiority of Madras appears, therefore, to arise from the possession of this large area under sugar-producing trees, and the question of the relative condition of the sugar-cane industries in Madras and Bengal cannot be solved merely by a comparison of exports of sugar from those Provinces. Date-sugar is not distinguished in the Madras returns from that produced from cane.'

Palm Sugar. 472

"The sugar-cane crop is a troublesome one to grow, as it necessitates a considerable amount of care and expenditure, but the profits realized from it are large. Hence it is apparent that the decline in the exports of Bengal sugar, and the improvement in those of Madras sugar, are due to causes unconnected with one another. For instance, the Bengal trade has collapsed owing to the enormous production of beet-sugar in Europe, while the Madras trade has been influenced by the development of the natural resources of the country, coupled with the existence of a brisk demand for cheap coarse sugar in England."

The success of Madras sugar in Burma and in Bombay has been quite as marked as in Bengal. The explanation that it is due to palm-sugar will hardly, as the writer thinks, hold good. The fact that Madras sugar is cheaper, or rather that Madras has succeeded to prepare a cheap sugar that proves sufficient for the wants of certain markets is more likely to be the reason of the remarkable increase recently of the sugar trade of South India.

TRANS-FRONTIER. 473

6.—TRANS-FRONTIER (LAND) TRADE IN SUGAR TO AND FROM INDIA.

It will be seen from the tables below that this consists mainly in exports. Nepal obtains the largest amount of unrefined sugar, being followed by Kashmír and by the traffic along the Sind-Pishin Railway. In refined sugar the largest quantities are shown as conveyed by the railway just mentioned, and the next most important transactions are with Kashmír.

* Bengal possesses very nearly as great an area under sugar-yielding palms as Madras. The imports into Calcutta from Eastern Bengal (Jessor particularly) must be very largely in date-sugar. It is difficult to see how the argument advanced above gives Madras any very special advantage which Bengal does not, or could not, participate in.—*Conf. with pp. 138, 226–227, 231, 266, 270, 310, 352, 361.—Ed., Dict., Econ. Prod.*

TRANS-FRONTIER.

Refined Sugar including Sugar-candy.

	IMPORTS.			EXPORTS.		
	1888-89.	1889-90.	1890-91.	1888-89.	1889-90.	1890-91.
	Cwt.*	Cwt.	Cwt.	Cwt.	Cwt.	Cwt.
Kashmír	...	43	...	18,999	15,956	26,439
Nepal	...	...	80	11,061	9,083	10,219
Hill Tipperah	11	...	...	9	21	10
Sind-Pishin Railway	22	7	68	26,194	25,942	23,037
Lus Bela	...	...	...	946	563	1,080
Khelat	...	...	...	1,1 5	881	305
Kandahar	...	...	...	264	419	223
Sewestan	...	...	...	451	430	585
Tirah	...	...	...	18	124	21
Kabul	...	...	...	6,860	6,057	7,813
Bajaur	...	...	...	570	505	373
Ladakh	...	...	...	335	277	270
Thibet	...	...	...	51	72	31
Sikkim	...	...	...	108	48	...
Bhutan	...	...	...	...	53	...
Manipur	...	...	...	17	4	8
Lushai Hills	...	...	...	...	...	4
Siam	...	...	...	8	...	..
N. Shan States	..	...	...	90	105	42
S. Shān States	...	...	...	...	25	45
Karenne	...	...	...	40	16	55
TOTAL IN CWT.	33	50	148	67,136	60,581	70,560

Unrefined Sugar, Molasses, Gúr, etc.

	IMPORTS.			EXPORTS.		
	1888-89.	1889-90.	1890-91.	1888-89.	1889-90.	1890-91.
	Cwt.	Cwt.	Cwt.	Cwt.	Cwt.	Cwt.
Kabul	...	15	...	1,829	2,447	2,255
Kashmír	703	1,328	4,5*o	20,292	21,342	19.036
Nepal	400	443	260	52,862	48,103	50,505
Hill Tipperah	18	3	10	70	101	89
Western China	...	...	12	...	...	...
N. Shan States	...	59	16	103	80	125
S. Shan States	...	497	914	...	68	167
Karenne	...	3	...	...	48	19
Sind Pishin Railway	59	196	17	18,249	17,117	19,441
Lus Bela	...	...	...	356	140	290
Khelat	...	...	...	1,154	736	896
Kandahar	...	...	...	152	74	27
Sewestan	...	...	...	3,142	2,446	2,770
Tirah	...	...	...	769	757	628
Bajaur	...	...	...	2,524	1,309	802
Ladakh	...	...	...	27	27	15
Thibet	...	...	...	4,462	3,901	1,426
Sikkim	...	...	...	50	256	330
Bhutan	...	...	...	825	625	460
Naga and Mishmi Hills	...	...	...	2	6	3
Lushai	...	...	...	...	...	1
Siam	...	...	...	...	22	...
TOTAL IN CWT.	1,180	2,544	5,779	106,868	99,604	99,285

* To reduce cwt. to maunds so as to allow of comparison with the rail-borne traffic multiply by $\frac{7}{5}$.

COASTWISE TRADE.

COASTWISE TRADE.

It does not seem necessary to do more in this place than to furnish a series of tables in illustration of coastwise trade. The following four tables show the totals imported and exported coastwise by each of the provinces of India under the two sections, INDIA and FOREIGN sugars. The transactions in the latter class might almost be designated re-exports, that is to say, they are amounts of sugar imported in the first instance from foreign countries and exported again by coastwise steamers. The term "re-exports" has, however, been restricted to foreign imports subsequently exported to foreign countries, and which thus leave India entirely. The three tables first given below involve an error which it is necessary the reader should be guarded against. The figures shown are the totals of refined and unrefined added together, without the former having been first reduced to the standard of the latter. In the subsequent tables, given under the heading of the provinces concerned, these two classes have not only been separately dealt with, but inter-provincial transactions have been excluded from consideration. Thus, for example, exports from the port town of Bombay, say, to Surat, have not been regarded as exports, since the amounts thus conveyed from port to port are still within the presidency. The imports and exports shown are, therefore, the amounts actually brought into a province or which leave it.

Imports, India Grown Sugar. 474

Imports Coastwise of Indian-grown Sugar of all kinds.

YEARS.	Into Bengal.	Into Bombay	Into Sind.	Into Madras.	Into Burma.	TOTAL.	
	Cwt.	Cwt.	Cwt.	Cwt.	Cwt.	Cwt.	R
1870-71	...	...	...	...	...	...	...
1871-72	Quantities not returned for these years except in value which includes of foreign sugar.						39,29,895
1872-73							29,65,240
1873-74							44,55,271
1874-75							42,76,202
1875-76	12,796	170,186	23,858	63,763	43,323	313,926	33,56,386
1876-77	5,432	139,661	31,834	65,398	43,859	286,184	34,01,152
1877-78	7,398	283,251	32,104	88,720	65,049	476,522	55,72,703
1878-79	16,134	324,400	14,096	104,324	47,538	506,492	54,89,828
1879-80	13,999	388,209	27,010	173,354	40,520	643,092	62,21,478
1880-81	30,250	297,487	18,264	203,722	40,703	590,426	52,50,567
1881-82	52,685	308,237	18,436	173,082	58,484	610,924	57,08,896
1882-83	42,198	304,058	13,199	213,406	61,550	634,411	58,29,011
1883-84	28,872	373,977	8,847	211,719	66,332	689,747	58,43,054
1884-85	44,346	462,172	9,822	183,282	76,443	776,065	61,40,866
1885-86	45,580	484,437	12,602	231,186	65,865	839,670	64,59,396
1886-87	52,565	376,177	11,160	247,967	142,827	830,696	67,04,861
1887-88	48,058	329,400	15,153	206,538	107,425	706,574	55,87,367
1888-89	54,461	404,610	14,589	208,769	122,010	804,439	63,79,425
1889-90	48,995	462,256	12,495	159,316	92,943	776,015	65,36,625
1890-91	63,269	388,392	8,722	185,264	78,625	724,272	61,99,359

Coastwise Trade of Bengal. (*G. Watt.*)

SACCHARUM: Sugar.

COASTWISE TRADE.

Exports, Indian-grown Sugar. 475

Exports Coastwise of Indian-grown Sugar of all kinds.

YEARS.	From Bengal.	From Bombay.	From Sind.	From Madras.	From Burma.	TOTAL.	
	Cwt.	Cwt.	Cwt.	Cwt.	Cwt.	Cwt.	R
1870-71 . .	...	...	...	...	...	...	...
1871-72 . .	Quantities not returned for these years except in value which includes foreign sugar.						57,70,601
1872-73 . .							63,68,663
1873-74 . .							69,62,920
1874-75 . .							73,14,409
1875-76 . .	193,035	118,430	1,974	140,115	24,372	477,927	47,42,837
1876-77 . .	264,121	121,989	1,285	104,255	30,813	522,463	55,61,546
1877-78 . .	381,408	318,734	3,367	85,120	20,966	809,595	90,67,898
1878-79 . .	117,792	430,581	2,528	94,317	22,449	667,667	69,20,653
1879-80 . .	132,944	574,796	2,674	220,032	29,345	959,791	90,19,069
1880-81 . .	52,910	466,171	811	222,180	23,865	765,937	66,67,583
1881-82 . .	70,347	479,348	1,327	264,345	16,873	832,240	74,22,888
1882-83 . .	104,834	313,284	1,266	318,366	23,308	761,058	64,76,435
1883-84 . .	71,763	438,531	1,068	279,208	19,851	810,421	69,58,007
1884-85 . .	24,121	627,629	705	276,214	28,774	957,443	67,55,237
1885-86 . .	33,964	659,248	839	271,242	28,564	993,857	76,56,441
1886-87 . .	33,398	455,559	263	338,378	32,614	860,212	69,16,818
1887-88 . .	35,869	463,598	311	251,032	35,464	786,274	64,02,501
1888-89 . .	30,226	663,753	854	259,639	38,831	993,303	80,40,281
1889-90 . .	55,235	716,886	641	216,426	33,532	1,022,720	84,35,745
1890-91 . .	42,639	505,997	2,181	397,475	29,849	978,141	83,54,533

Imports, Foreign Sugar. 476

Imports Coastwise of Foreign Sugar of all kinds.

YEARS.	Into Bengal.	Into Bombay.	Into Sind.	Into Madras.	Into Burma.	TOTAL.	
	Cwt.	Cwt.	Cwt.	Cwt.	Cwt.	Cwt.	R
1870-71 . .	...	...	...	...	...	...	...
1871-72 . .	See remark for these years in the Imports Coastwise of Indian grown sugar.						
1872-73 . .							
1873-74 . .							
1874-75 . .							
1875-76 . .	3	33	75,699	8,709	642	85,086	11,75,589
1876-77 . .	14	1,345	41,165	4,767	3,251	50,542	7,44,184
1877-78 . .	6	19,550	52,465	7,973	375	80,369	13,41,512
1878-79 . .	22	33,137	104,079	9,786	384	147,408	24,31,893
1879-80 . .	1	23,844	99,550	6,944	40	130,379	22,79,960
1880-81 . .	9	30,178	170,415	10,716	34	211,352	37,51,217
1881-82 . .	7	27,214	111,590	6,872	422	146,105	25,56,655
1882-83 . .	6	30,061	87,207	4,380	46	121,700	20,59,476
1883-84 . .	13	34,812	83,380	5,077	27	123,309	20,66,893
1884-85 . .	22	45,021	221,877	8,949	112	275,981	40,33,524
1885-86 . .	148	40,812	208,023	10,312	38	259,333	34,38,519
1886-87 . .	20	53,500	150,259	15,912	899	220,590	28,72,926
1887-88 . .	1	53,090	79,733	19,396	1,087	153,307	19,20,247
1888-89 . .	8,976	50,393	91,268	23,824	624	175,080	21,75,202
1889-90 . .	16,535	42,122	52,981	15,991	1,218	128,847	17,66,237
1890-91 . .	5,721	57,811	88,446	17,453	538	169,969	21,19,134

SACCHARUM:
Sugar.

Coastwise Trade of Bengal.

COASTWISE TRADE.
Exports.
477

Exports Coastwise of Foreign Sugar of all kinds.

YEARS.	From Bengal.	From Bombay	From Sind.	From Madras.	From Burma.	TOTAL.	
	Cwt.	Cwt.	Cwt.	Cwt.	Cwt.	Cwt.	R
1870-71 . .	...	...	...	...	...	...	...
1871-72 . .	See the remark against these years in the table of Exports Coastwise of Indian grown sugar.						
1872-73 . .							
1873-74 . .							
1874-75 . .							
1875-76 . .	5	131,473	184	131	342	132,135	19,19,019
1876-77 . .	5	79,184	215	99	346	79,829	12,20,802
1877-78 . .	...	125,044	528	445	214	126,231	20,07,053
1878-79 . .	...	233,357	836	219	98	234,510	39,11,476
1879-80 . .	...	209,171	914	76	90	210,251	34,84,866
1880-81 . .	...	294,943	1,187	224	82	296,436	50,53,916
1881-82 . .	65	235,004	927	322	304	236,622	39,99,419
1882-83 . .	...	206,355	766	255	152	207,528	34,58,859
1883-85 . .	...	229,907	921	87	147	231,062	37,00,687
1884-84 . .	...	463,387	1,475	160	57	465,079	63,75,136
1885-86 . .	*Nil*	411,834	1,129	92	66	413,121	50,77,472
1886-87 . .	55	417,808	6,715	185	630	425,393	51,49,666
1887-88 . .	55	336,117	5,593	247	3,970	345,982	41,31,186
1888-89 . .	*Nil*	365,357	4,607	437	244	370,64	43,69,084
1889-90 . .	1	286,531	4,593	566	407	292,099	37,80,477
1890-91 . .	73	399,396	5,003	546	279	405,297	48,94,134

The main ideas to be learned regarding the coastwise trade in sugar will, perhaps, be best exemplified by analysing the trade in the three chief sea-board provinces, Bengal, Bombay and Madras.

Bengal.
478

I.—Bengal.

The following table exhibits the Imports and Exports of this province Coastwise during each fifth year since 1875-76:—

Imports.
479

Imports.

REFINED SUGAR INCLUDING SUGAR-CANDY.

FROM	1875-76.		1880-81.		1885-86.		1890-91.	
	Indian.	Foreign.	Indian.	Foreign.	Indian.	Foreign.	Indian.	Foreign.
	Cwt.	Cwt.	Cwt.	Cwt.	Cwt.	Cwt.	Cwt.	Cwt.
Bombay . . .	41	...	907	...	750	145	...	...
Sind . . .	...	...	...	...	...	...	...	...
Madras . . .	7,734	...	19,181	...	32,754	...	39,449	...
Burma . . .	3	2	514	2	37	...	99	5,631
TOTAL .	7,778	2	20,602	2	33,541	145	39,548	5,631

COASTWISE TRADE.

Bengal.

Imports.

Unrefined Sugar, Molasses, Gúr, etc.

From	1875-76.		1880-81.		1885-86.		1890-91.	
	Indian.	Foreign.	Indian.	Foreign.	Indian.	Foreign.	Indian.	Foreign.
	Cwt.	Cwt.	Cwt.	Cwt.	Cwt	Cwt.	Cwt.	Cwt.
Bombay . . .	...	...	...	...	...	...	...	...
Sind . . .	...	...	...	...	...	...	...	...
Madras . . .	...	...	122	...	4	...	4,177	...
Burma . . .	15	...	49	...	4	...	273	...
Total .	15	...	171	...	8	...	4,450	...

Exports. 480

Exports.

Refined Sugar including Sugar-candy.

To	1875-76.		1880-81.		1885-86.		1890-91.	
	Indian.	Foreign.	Indian.	Foreign.	Indian.	Foreign.	Indian.	Foreign.
	Cwt.	Cwt.	Cwt.	Cwt.	Cwt.	Cwt.	Cwt.	Cwt.
Bombay . . .	157,518	...	23,591	...	6,562	...	7,231	...
Sind . . .	...	...	...	...	...	...	312	...
Madras . . .	5,652	5	5,170	...	967	...	552	...
Burma . . .	23,033	...	10,268	...	11,498	...	6,117	...
Kattíwar . . .	...	...	...	...	...	...	27	...
Travancore . . .	...	...	122	...	25	...	16	...
Cochin . . .	...	...	60	...	...	...	...	...
Total .	186,203	5	39,211	...	19,052	...	14,255	...

Unrefined Sugar, Molasses, Gúr, etc.

To	1875-76.		1880-81.		1885-86.		1890-91.	
	Indian.	Foreign.	Indian.	Foreign.	Indian.	Foreign.	Indian.	Foreign.
	Cwt.	Cwt.	Cwt.	Cwt.	Cwt.	Cwt.	Cwt.	Cwt.
Bombay . . .	18	...	2	...	...	...	4	...
Sind . . .	...	...	...	...	...	...	...	...
Madras . . .	1	...	998	...	7	...	160	...
Burma . . .	2,049	...	3,048	...	1,997	...	3,764	...
Kattíwar . . .	...	...	...	...	...	...	12	...
Travancore . . .	...	...	...	...	...	...	...	...
Cochin . . .	...	...	...	...	...	...	...	...
Total .	2,068	...	4,048	...	2,004	...	3,940	...

It will be seen from the tables of the coastwise trade during the past twenty years, that the records of the Bengal transactions manifest a very remarkable change. The Imports of Indian sugar have increased from, say, 10,000 cwt. to over 60,000 cwt., while the Exports have contracted

COASTWISE TRADE.

Bengal.

from about 200,000 cwt. to 40,000 cwt. The trade in Foreign sugars brought to Bengal or exported therefrom by coasting steamers is much less important, but a somewhat remarkable feature in this section may be said to be the fact that these imports have greatly increased. The analysis of each fifth year, since 1870, shows that increase in the imports of Indian sugar is mainly due to the improved traffic with Madras in refined sugar. During the years named, the supply derived from Madras has increased from 7,734 cwt. in 1875 to 39,449 cwt. in 1890-91. The decline in the exports would appear to be due to the same cause, *viz.*, the success of Madras. The Bengal exports to Bombay have decreased from 157,518 cwt. of refined sugar to 7,231 cwt., so also to Burma they have contracted from 23,033 cwt. in 1875-76 to 6,117 cwt. in 1890-91, and it will be found that the exports from Madras to Bombay and Burma, during these years, have almost correspondingly improved. Madras is thus not only contesting in Bengal itself the trade in refined sugar, but it has practically ousted Bengal sugar from Bombay and Burma Formerly the Bengal exports of refined sugar to Madras were very considerable : last year they had almost ceased. It was these facts that mainly actuated the Government of India in the recent enquiry into the sugar trade, which resulted in the discovery, which the writer has, he ventures to think, abundantly confirmed, namely, that if Bengal has lost its foreign markets, the home consumption has greatly increased. The home market have, in fact, proved more profitable than the foreign, for some years back,—a state of affairs that can hardly be regarded as unsatisfactory.

BOMBAY. 481

2.—Bombay.

As already remarked, one of the chief features of the Bombay coastwise trade in sugar is the loss of the Bengal supply and the birth of a demand for Madras refined sugar. Unlike the tables given above for Bengal, those which follow for Bombay derive their chief interest, however, in the fact that the transactions in foreign sugar are very much more important than in Indian :—

Imports. 482

Imports.

From	Refined Sugar including Sugar-candy.							
	1875-76.		1880-81.		1885-86.		1890-91.	
	Indian.	Foreign.	Indian.	Foreign.	Indian.	Foreign.	Indian.	Foreign.
	Cwt.	Cwt.	Cwt.	Cwt.	Cwt.	Cwt.	Cwt.	Cwt.
Bengal	131,130	...	27,387	...	6,438	...	6,533	...
Sind	1	...	...	...	136	1	...	18
Madras	435	12	3,492	17	9,709	...	26,396	119
Burma	...	...	...	...	...	...	...	...
Goa	...	21	...	...	3	30	186	...
Pondicherry	...	...	...	...	...	...	...	...
Cambay	...	...	...	...	3	...	22	5
Cutch	...	...	...	...	...	3	...	3
Kattíwar	...	...	152	...	1,930	115	1,119	17
Koncan	...	...	...	...	...	2	8	3
Travancore	...	...	...	...	21	...	...	...
Gaekwar's Territory	...	...	...	...	...	...	45	...
Total	131,566	33	31,031	17	18,240	151	34,309	165

COASTWISE TRADE.

Bombay.

Imports.

Imports

FROM.	UNREFINED SUGAR, MOLASSES, GÚR, ETC.							
	1875-76.		1880-81.		1885-86.		1890-91.	
	Indian.	Foreign.	Indian.	Foreign.	Indian.	Foreign.	Indian.	Foreign.
	Cwt.	Cwt	Cwt.	Cwt.	Cwt.	Cwt.	Cwt	Cwt.
Bengal	1,445	...	2	...	398	...	4	...
Sind	...	...	...	...	...	...	166	112
Madras	37,168	...	2,456	...	32,036	...	84,053	...
Burma	...	...	...	...	...	...	44	...
Goa	5	...	...	...	19	...	310	...
Pondicherry	...	...	...	...	...	...	666	...
Cambay	...	...	...	...	28	...	51	...
Cutch	...	...	...	...	...	...	...	...
Kattíwar	1	...	...	...	115	...	125	...
Koncan	...	...	...	...	76	...	185	...
Travancore	1	...	...	...	...	...	123	...
Gaekwar's Territory	...	...	...	...	10,770	...	12,774	...
TOTAL	38,620	...	2,458	...	43,442	...	98,501	112

From the table below it will be seen that the Bombay coastwise exports in Foreign sugar have increased from 131,473 cwt. in 1875-76 to 340,649 cwt. in 1890-91. A somewhat striking feature may be here alluded to (though in point of value less important than many others manifested by the table), namely, the growth of an export of refined foreign sugar from Bombay to Madras.* Sind is, however, the country that drains coastwise the largest amount of foreign sugar from Bombay. The demand in Kattíwar has been steadily increasing

Conf. with pp. 365, 378.

Exports.

483

Exports.

To	REFINED SUGAR INCLUDING SUGAR-CANDY.							
	1875-76.		1880-81.		1885-86.		1890-91.	
	Indian.	Foreign.	Indian.	Foreign.	Indian.	Foreign.	Indian.	Foreign.
	Cwt.	Cwt.	Cwt.	Cwt.	Cwt.	Cwt.	Cwt.	Cwt.
Bengal	...	...	4	934	...	474	...	150
Sind	17,916	75,457	5,369	166,531	715	209,521	2,100	70,509
Madras	5,177	6,097	4,707	10,144	5,142	9,757	3,461	16,273
Burma	...	75	...	265	8	3,014	...	756
Damaun	...	...	...	...	2	142	118	647
Diu	46	146	38	407	12	395	77	521
Goa	207	3,124	186	3,895	414	7,858	1,363	16,660
Pondicherry	...	...	...	...	3	...	1	...
Cambay	...	...	46	191	36	3,391	641	3,377
Cutch	7,316	22,791	4,934	19,411	1,764	21,343	5,858	30,265
Kattíwar	4,990	23,281	4,781	57,877	3,690	114,000	8,358	191,212
Koncan	337	502	184	1,136	7	500	27	951
Travancore	...	...	61	136	59	65	34	148
Gaekwar's Territory	...	...	99	504	119	605	856	9,180
TOTAL	35,989	131,473	20,409	261,431	11,971	371,065	22,894	340,649

* In the chapter on Prices it will be seen that a quotation of the price of foreign sugar in Madras could not be procured.

SACCHARUM: Sugar.

Coastwise Trade of Madras.

COASTWISE TRADE. Bombay. Exports.

Exports.

UNREFINED SUGAR, MOLASSES, GÚR, ETC.

To	1875-76.		1880-81.		1885-86.		1890-91.	
	Indian.	Foreign.	Indian.	Foreign.	Indian.	Foreign.	Indian.	Foreign.
	Cwt.	Cwt.	Cwt.	Cwt.	Cwt.	Cwt.	Cwt.	Cwt.
Bengal . . .	...	...	...	...	...	...	...	...
Sind . . .	5,494	...	6,112	...	11,333	...	6,070	...
Madras . .	2,102	...	1,623	...	574	...	273	...
Burma . . .	...	...	...	...	...	...	...	...
Damaun . .	...	...	...	...	10	...	100	...
Diu . . .	144	...	620	...	420	...	603	...
Goa . . .	236	...	550	...	4,662	...	2,600	...
Pondicherry . .	...	...	...	...	...	...	...	...
Cambay . .	...	...	1,279	...	6,816	...	3,161	242
Cutch . . .	53,182	...	48,743	...	96,184	70	85,074	12
Kattíwar . .	19,544	...	50,631	45	100,211	65	120,403	968
Concan . .	1,739	...	1,191	8	968	...	1,977	...
Travancore . .	...	...	...	...	1	...	5	...
Gaekwar's Territory	...	...	23	...	167	...	891	...
TOTAL .	82,441	...	110,772	53	221,346	135	221,157	1,222

It has been remarked in connection with the rail-borne traffic, that the foreign sugars of Bombay are yearly succeeding to penetrate further and further into the country, and seem to find favour in such countries as Rajputana, where they would (viewed from the religious standpoint) have been least expected to have succeeded. The exports coastwise to Kattíwar may be cited as another example of this fact. The exports of foreign refined sugars from Bombay to that State were 23,281 cwt. in 1875-76; last year they had attained the very considerable proportions of 191,212 cwt. The traffic with Cutch, though less remarkable, shows a similar progression. Cutch, however, draws a very much larger quantity of Indian unrefined sugar than of refined sugar. It will be observed by the foot-note to page 366 that the exports of refined sugar from Bombay to the North-West Provinces and Oudh last year manifested a sudden and very considerable expansion. This is the more remarkable, since for many years the North-West Provinces have largely contributed to the Bombay supply. The demand for refined sugar from Bombay may be accepted as another indication of the successful competition of foreign against Indian refined sugar.

MADRAS. 484

3 — Madras.

After what has been said regarding Bengal and Bombay, it seems unnecessary to specialize any of the features of the Madras coastwise trade in sugar. The imports of Bengal refined sugar have greatly contracted, and a new trade has come into existence in the demand for foreign refined sugars from Bombay. The exports of unrefined sugar from Madras to Bombay are, however, very considerable, and the trade has been more than doubled within the past 20 years. It has also been explained that the exports of refined sugar from Madras to Bengal and Burma have been very considerably increased within the past few years.

COASTWISE TRADE.

Madras.

Imports.

485

Imports.

REFINED SUGAR INCLUDING SUGAR-CANDY.

FROM.	1875-76.		1880-81.		1885-86.		1890-91.	
	Indian.	Foreign.	Indian.	Foreign.	Indian.	Foreign.	Indian.	Foreign.
	Cwt.	Cwt.	Cwt.	Cwt.	Cwt.	Cwt.	Cwt.	Cwt.
Bengal	5,015	172	4,096	6	544	10	534	1
Bombay	3,635	5,771	3,682	7,568	3,283	9.467	3,398	16,654
Sind	...	...	...	...	...	...	...	...
Burma	1	...	...	1	...	...	25	...
Goa	...	...	...	...	3	...	27	...
Pondicherry	...	...	8	6	...	49	...	...
Kattíwar	...	...	...	...	...	...	28	...
Travancore	36	...	...	...	...	...	...	...
Cochin	...	...	...	...	...	...	1	...
TOTAL	8,687	5,943	7,786	7,581	3,830	9,526	4,013	16,655

Imports.

UNREFINED SUGAR, MOLASSES, GÚR, TEC.

FROM	1875-76.		1880-81.		1885-86.		1890-91.	
	Indian.	Foreign.	Indian.	Foreign.	Indian.	Foreign.	Indian.	Foreign.
	Cwt.	Cwt.	Cwt.	Cwt.	Cwt.	Cwt.	Cwt.	Cwt.
Bengal	12	...	2,476	...	37	...	54	...
Bombay	1,140	...	1,748	2,978	832	676	260	...
Sind	...	...	...	...	...	...	...	...
Burma	...	...	4	...	4	...	3	...
Goa	...	...	1	...	...	...	...	...
Pondicherry	...	...	...	...	...	...	...	...
Kattíwar	...	...	...	...	...	...	...	...
Travancore	75	...	...	...	56,125	...	180	...
Cochin	...	...	...	...	...	...	...	...
TOTAL	1,227	...	4,229	2,978	56,998	676	497	...

SACCHARUM: Sugar.

COASTWISE TRADE.

Madras.

Exports.

486

Coastwise Trade of Madras.

Exports.

REFINED SUGAR INCLUDING SUGAR-CANDY.

To	1875-76.		1880-81.		1885-86.		1890-91.	
	Indian.	Foreign.	Indian.	Foreign.	Indian.	Foreign.	Indian.	Foreign.
	Cwt.	Cwt.	Cwt.	Cwt.	Cwt.	Cwt.	Cwt.	Cwt.
Bengal . . .	7,061	...	19,582	...	33,649	...	37,488	...
Bombay . .	380	...	4,226	36	11,342	...	25,650	...
Sind . . .	...	...	1,251	...	...	...	...	...
Burma . . .	292	1	867	...	17,927	...	29,166	3
Goa . . .	...	...	...	...	...	...	...	...
Karical . . .	...	...	...	...	...	...	...	...
Mahe . . .	...	...	...	...	...	...	...	...
Pondicherry . .	8,250	...	7	...	...	...	...	...
Cutch . . .	...	...	...	...	...	...	...	...
Kattíwar . .	...	...	...	...	...	...	...	...
Koncan . .	...	...	...	...	...	...	...	...
Travancore . .	...	...	...	1	...	...	...	...
Cochin . . .	...	...	...	...	...	...	6,104	...
TOTAL .	15,983	1	25,933	37	62,918	...	98,408	3

Exports.

UNREFINED SUGAR, MOLASSES, GÚR, ETC.

To	1875-76.		1880-81.		1885-86.		1890-91.	
	Indian.	Foreign.	Indian.	Foreign.	Indian.	Foreign.	Indian.	Foreign.
	Cwt.	Cwt.	Cwt.	Cwt.	Cwt.	Cwt.	Cwt.	Cwt.
Bengal . . .	...	...	244	...	6	...	2,229	...
Bombay . .	40,084	...	2,834	...	37,196	...	92,446	...
Sind . . .	1,678	...	975	...	215	...	9,294	...
Burma . . .	24	...	58	...	992	...	2,442	...
Goa . . .	2	...	4	...	102	...	871	...
Karical . . .	...	...	801	...	...	...	...	...
Mahe . . .	...	...	2	2	2	...	..	...
Pondicherry . .	14,190	...	7,963	...	...	...	...	...
Cutch . . .	16,188	...	2,509	...	6,041	...	13,563	...
Kattíwar . .	...	...	...	...	...	...	9,000	...
Koncan . .	...	...	...	...	...	...	85	...
Travancore . .	...	...	...	...	589	...	...	...
Cochin . . .	...	...	...	...	...	...	...	...
TOTAL .	72,166	...	15,390	2	45,143	...	129,930	...

(*W. R. Clark*)

SACCOPETALUM, *Benn.; Gen. Pl., I., 28.*

[ANONACEÆ.

Saccopetalum tomentosum, *H. f. & T.; Fl. Br. Ind., I., 88;* 487

Syn.—UVARIA TOMENTOSA, *Roxb.*

Vern.—*Kirna, karri,* HIND; *Patmossu,* URIYA; *Omé, hake húmú,* KOL.; *Thoska,* GOND; *Humba,* KURKU; *Karri,* N.-W. P; *Umbi, umbia,* RAJ.; *Karí,* C. P.; *Kirna, karri, húm,* BOMB.; *Húmb, húm,* MAR.; *Chilkadúdú, pedda chil ka-dadúga,* TEL.; *Hessare,* KAN.

References.—*Roxb., Fl. Ind., Ed. C.B.C., 456; Brandis, For. Fl., 7; Beddome, Fl. Sylv., t. 39; Gamble, Man. Timb., i., 10; Dalz. & Gibs., Bomb. Fl., 4; Elliot, Fl. Andhr., 147; Lisboa, U. Pl. Bomb., 4, 277; Gazetteers:—Bombay, XIII., 24; XX., 427; N.-W. P., IV., lxvii.; Land Rev. Settlement, Seonee Dist., Cen. Prov., 10; Ind. Forester:—III., 200; X., 325; XII., App., 5; XIII., 119; For. Admin. Rep., Chutia Nagpur, 28.*

Habitat.—A large tree, with straight stem, found in Oudh, Nepál Tarai, Gorakhpur, Behar, Central India, and on the Western Gháts.

Gum.—It yields a gum which belongs to the false tragacanth or hog-gum series. GUM. 488

Food & Fodder.—The oval BERRIES are said to be eaten in some parts of Bombay. The LEAVES are used as fodder. FOOD. Berries. 489 FODDER. Leaves. 490

Structure of the Wood.—Olive-brown, moderately hard, smooth, close-grained; no heartwood. It is not apt to warp, but often cracks in seasoning. Weight 45lb per cubic foot. TIMBER. 491

Domestic Uses.—The TIMBER is used in Oudh for building huts and cattle sheds; in the Western Gháts it is reckoned a good timber, and is much used in house-building. DOMESTIC. Timber. 492

Sacred, see **Domestic and Sacred,** Vol. III., 191.

Saffron, see **Crocus sativus,** *Linn.;* IRIDEÆ, Vol. II., 592.

SAGAPENUM.

Sagapenum, *Cooke, Report on Gum Resins, 63.* 493

Vern.—*Kundel,* or *kundal* (?), HIND.; *Isus,* MAR.; *Kundel* (?), SANS.; *Sugbinuj, sakbinaj,* ARAB.; *Sagafiún, iskabínah,* PERS.

References.—*Flück. & Hanb., Pharm., 291; Ainslie, Mat. Ind., I., 357; O'Shaughnessy, Beng. Dispens., 363; Dymock, Mat. Med. W. Ind., 396; Baden Powell, Pb. Pr., 403.*

Habitat.—A gum-resin imported into India from the Persian Gulf and coasts of Arabia, and said to be produced from a species of **Ferula** which grows in Arabia and Persia.

SAGAPENUM consists of masses made up principally of brownish-yellow semi-transparent tears, resembling Galbanum, but having a darker colour and a more alliaceous odour. These tears are agglutinated together by a proportion of soft gum-resin, which varies considerably in amount; indeed, some specimens appear to be made up entirely of that substance and show no distinct tears. The analysis of Sagapenum shows that it contains—resin 50-54 per cent., gum 31-32, volatile oil 3-11, bassorin 1-4, malate and phosphate of lime, 0·40-1·12, and small amounts of sulphur water and impurities. It is distinguished from Galbanum by the presence of sulphur, and by the comparatively large amount of resinous residue it yields to petroleum spirit. 494

Medicine.—According to **Ainslie,** SAGAPENUM was known to the Greeks, and is described by **Dioscorides** as the produce of a **Ferula** growing in Media. **Dymock** remarks, "I see no reason to suppose that the ancient Hindus knew the drug, although *kun el* is in some books given as the Sanskrit and Hindí name for it." MEDICINE. 495

MEDICINE. Arabic writers seem to have been acquainted with the substance, probably from their intercourse with the Greeks. The *Makhzan-el-Adwiya* says the substance is found near Ispahan. Muhammadan writers describe it as a powerful attenuant and resolvent, and say that when combined with purgatives it exerts its resolvent power on every part of the body, removing noxious humours; they also value it as an anthelmintic and emmenagogue. A Sagapenum pill is often prescribed by them in flatulent dyspepsia; it contains equal quantities of Aloes, Sagapenum, Bdellium, and Agaric. The dose of the gum-resin is two or three dirhems taken with warm water (*Dymock*).

TRADE. 496 **Trade.**—Goden Powell mentions the difficulty that exists in the Panjáb to get *sakbínaj*. *Jaushír* or *ammoniacum* are usually sold in place of it. "The quantity annually imported into Bombay varies greatly, most of it going to London. It is seldom to be obtained in the retail shops. Value R½-R¾ per ℔" (*Dymock*).

SAGERETIA, *Brong.; Gen. Pl., I., 379.*

497 **Sageretia Brandrethiana,** *Aitch.; Fl. Br. Ind., I., 642;* RHAMNEÆ.

Vern.—*Ganger, goher, kunjar, kohér, kanger, bhándí, bajan,* PB.; *Mángri,* PUSHTU; *Maimúna, momanna, númáni,* AFG.; *Ganger,* SIND.

References.—*Brandis, For. Fl., 95; Gamble, Man. Timb., 92; Stewart, Pb. Pl., 42; Aitchison, Cat. Pb. and Sind Pl., 32; Boiss, Fl. Orient., II., 22; Murray, Pl. and Drugs, Sind, 147; Baden Powell, Pb. Pr., 596; Settle. Rep. Kohat Dist., Panjáb, 29; Agri.-Horti. Soc. Ind., XIV., 4.*

Habitat.—A deciduous, distorted shrub, met with in the Sulaiman and Salt Ranges and North-West Himálaya, between the Indus and the Jhelum. Distributed westwards to Persia and Arabia.

FOOD. Fruit. 498 **Food & Fodder.**—The FRUIT is small and black and has a sweet flavour not unlike that of the bilberry. It is a great favourite among the frontier tribes and Afgháns, and is regularly collected and exposed for sale in the bazárs of Peshawar. In the Salt Range, a *chatni* is made of it.

FODDER. Leaves. 499 Young Twigs. 500 The LEAVES and YOUNG TWIGS are much blowsed by sheep and goats (*Lace, Quetta Fl. in MSS.*).

TIMBER. 501 **Structure of the Wood.**—Yellow, very hard, and close-grained.

502 **S. oppositifolia,** *Brong.; Fl. Br. Ind., I., 641.*

Syn.—S. FILIFORMIS, *Don;* RHAMNUS FILIFORMIS, *Roth.;* R. RIGYNUS, *Don;* ZIZYPHUS OPPOSITIFOLIA, *Wall*

Vern.—*Aglaia,* KUMAON; *Kanak, gidardák,* KASHMIR; *Drange, girthan,* PB.; *Mmánrai,* PUSHTU.

References.—*Brandis, For. Fl., 95; Gamble, Man. Timb., 92; Stewart, Pb. Pl., 42; Aitchison, Cat. Pb. & Sind Pl., 32; Prod., II., 28; Baden Powell, Pb. Pr., 596; Atkinson, Him. Dist., 307; Gaz., N.-W. Provinces, IV., lxx.*

Habitat.—A large shrub, found in the North-West Himálaya (sub-tropical), from Peshawar to Nepál, also in Southern India from Konkan southwards. It is distributed to Java.

FOOD. Fruit. 503 **Food.**—The FRUIT of this species, like that of the preceding, is eaten.

DOMESTIC. 504 **Domestic Uses.**—The wood is used as fuel.

505 **S. theezans,** *Brong.; Fl. Br. Ind., I., 641.*

Syn.—RHAMNUS THEEZANS, *Linn.*

Vern.—*Dargola* (Simla), *drangu, ankol, kauli, karúr, phomphli, kánda, brinkol, chaunsh, katráin, thúm, kúm,* PB.

References.—*Brandis, For. Fl., 95; Kurz, For. Fl. Burm., I., 267; Gamble, Man. Timb., 92; Atkinson, Him. Dist., 307.*

Habitat.—A large, spinescent shrub, found in the Salt and Sulaiman Ranges, and in the Western Himálaya from Kashmír to Simla, from 3,000 to 8,000 feet. It is also met with in the forests of Ava (*Kurz*), and is distributed westwards to Balúchistán and eastwards to China.

Food.—The FRUIT is eaten in parts of the Himálaya and in China. In China the LEAVES are used by the poorer classes, as a substitute for tea. FOOD. Fruit. 506 Leaves. 507

Structure of the Wood.—Very hard, white, with irregular dark-coloured heartwood. Weight 56℔ per cubic foot. TIMBER. 508

SAGITTARIA, *Linn.; Gen. Pl., III., 1006.* 509

A genus of aquatic herbs, belonging to the water plantain family (ALISMACEÆ) three species of which are described by **Roxburgh** as common in the Indian Peninsula. Two of these, however, have been reduced in **DeCandolle's** Monograph on Phanerogams to kindred genera, and the remaining one—**Sagittaria sagittæfolia**—alone need be here separately noticed.

Sagittaria sagittæfolia, *Linn.; DC., Monog. Phan., III., 66,* ALISMACEÆ. 510

ARROWHEAD.

Vern.—*Muyá muyá, choto.-kut,* BENG.

References.—*Roxb., Fl. Ind., 675; Voigt, Hort. Sub. Cal., 680; Smith, Econ. Dict., 24; Hunter, Orissa, II., 187 (app. VI.); Gazetteers:—N.-W. P., I., 85; X, 319.*

Habitat.—A common aquatic herb, found on the borders of fresh-water lakes, tanks, and ditches throughout India, and distributed to Europe and Asia generally, and throughout North America.

Food.—In North America, the fleshy CORMS of this plant are used as an article of food by the Native population, and in China it is even cultivated as a food-plant. It contains a bitter milky JUICE which is expelled by boiling. Apparently the Natives of India are ignorant of its value as a food-plant, since no information on the subject is given in any of the books that deal with the known properties and uses of Indian plants. FOOD. Corms. 511 Juice. 512

SAGO. 513

This form of starch is obtained from several palms and a few other plants.

Sago. 514

SAGON, *Fr.;* SAGO, *Germ.*

Vern.—*Sagu-dana, sagu-chawul,* HIND.; *Sagu-dana,* BENG.; *Shanárisi,* TAM.; *Sa-uké-chawal,* DEC.; *Sagu,* MALAY.; *Sikumi,* CHIN.

References.—*Roxb., Fl. Ind., Ed. C.B.C., 299, 668, 723; Kurz, For. Fl. Br. Burm., II., 530, 533; Brandis, For. Fl., 550, 556, 560; Gamble, Man. Timb, 415. 419; 420, 421; Mason, Burma and Its People, 426, 506, 811, 812; Pharm. Ind., I., 248; U. S. Dispens., 15th Ed., 1743; Fleming, Med. Pl. & Drugs. (Asiatic Reser., XI.), 189; Ainslie, Mat. Ind., I., 361; O'Shaughnessy, Beng. Dispens., 622, 640; K. L. De. Indig. Drugs Ind., 103; Bidie, Cat. Raw. Pr., Paris Exh., 92; Birdwood, Bomb. Prod., 236, 238, 239; Drury, W. Pl. Ind., 118; Useful Pl. Bomb. (XXV., Bomb. Gaz.), 135, 178; Forbes Watson, Indust. Survey Ind., 45; Royle, Prod. Res., 230; Smith, Econ. Dict., 362; Balfour, Cyclop., III., 484; Encyclop. Brit., XXI., 148; Madras Mail, June 14th, 1889; Indian Agri., July 6th, 1889; July 13th, 1889.*

SOURCES OF SAGO - The chief source of the Sago of Commerce is a palm indigenous to the East Indian Archipelago, known as **Metroxylon Sagu.** It flourishes in low marshy situations, and seldom attains a height of 30 feet, but is low and thick-set in character. At the age of fifteen years it becomes mature as a starch-yielding plant, and then the whole interior of the stem is gorged with spongy medullary matter, around which is a rind of hard wood. If the plant be allowed to flower, and the fruit to ripen, all this medulla becomes absorbed, the stem is left a mere hollow shell, and the tree dies. Before this occurs, however, the trees SOURCES. 515

SOURCES.

are cut down, the stem is cut into lengths, split up, and the pith extracted and grated to a powder. This powder is then kneaded up with water and strained, the starch passes through the strainer, and the woody fibre remains behind. The starchy fluid is then floated into troughs, the starch settles to the bottom, and after one or two washings it is considered by the Natives fit for their domestic purposes. What is intended for exportation is made into a paste with water and rubbed into grains which are known according to their size as *Pearl Sago, Bullet Sago, Sago Meal,* etc. The great proportion of the Sago of Commerce comes from Borneo, where there are large forests of sago palms in the low-lying marshy lands along some parts of the coast.

The proper Sago palm (**Metroxylon Sagu**) is not indigenous to the Indian Peninsula, but large quantities of Sago, some of which is said to be quite as good as the Sago of Commerce, are obtained in India from other sources Such as –

1 **Arenga saccharifera,** I., 302.
2. **Borassus flabelliformis,** I., 502.
3 **Caryota urens,** II , 208.
4. **Corypha umbraculifera,** II., 575.
5. **Cycas circinalis,** II., 675.
6 **C. pectinata,** II., 675.
7 **Cycas Rumphii,** II., 675.
8. **Metroxylon** (several species), V., 239.
9. **Phœnix acaulis,** VI., 199.
10 **P. rupicola,** VI., 207.
11. **Tacca pinnatifida,** VI.

It will be observed that, in the enumeration given, Nos. 5, 6, 7 and 11 are not palms. The principal sago palm of India may be said to be **Caryota urens.**

MEDICINE. 516 **Medicine** —Sago is used exclusively as an article of diet. It is nutritive, easily digestible, wholly destitute of irritating properties, and is therefore often employed as a bland, innocent article of diet in febrile disorders, bowel complaints, and during the convalescence from acute disease.

Chemistry. 517 CHEMICAL COMPOSITION.—Chemically considered Sago has the characters of starch. Under the microscope, the granules appear oval or ovate, and often truncated. Many of them are broken, and in most the surface is irregular or tuberculated (*U. S. Dispensatory*).

DOMESTIC. 518 **Domestic Uses.**—In the East Indian Archipelago the Natives chiefly make the sago starch, which is intended for domestic use, into biscuits which, if kept dry, may be preserved for a very long time. It is also dried and made into sago meal, from which they prepare a variety of dishes.

In ndia, the Sago from **Caryota urens**, and from other indigenous trees, is used as an article of food, principally in the form of a gruel or thick soup, but the use of Sago by the Natives of India is not nearly so extensive as in the East Indian Archipelago, where it forms almost the staple diet.

Sajji, see **Barilla,** Vol. I., 394; see also **Carbonates of Potash** and **Soda,** Vol. II, 152, 154; and **Reh,** Vol. VI., Part I., 400 to 427.

519 **Salagrama.**—Fossil ammonites reverenced and worshipped by the Hindus, and supposed by some to be the ætiles or eagle stones of the ancients. Those used in the worship of the Hindus are black, mostly rounded, and commonly perforated in one or more places by worms, or, as the Hindus believe, by Vishnu in the form of a reptile. Some are supposed to represent gracious incarnations of Vishnu and are highly prized by their owners; others, which border a little on violet, denote a vindictive *avatara*, and are shunned. The possessor of a SALAGRAMA keeps it wrapped up in clean cloth, from which it is frequently taken out and bathed. The water used for the purpose is thought to have acquired a sin-expelling virtue.

Sal Ammoniac, see **Ammonium**, Vol. I., 219.

SALEP.

520

The name given to the dried tubers of numerous species of the genus Orchis, and in India of the genus Eulophia.

Salep.

SALEP. 521

Vern.—*Salab-misri*,(=*Egyptian Salep*), Hind.; *Chálé-michhri*, Beng.; *Sálib misri*, Pb.; *Salap, salab*, Afg. *Sálam-misri*, Bomb.; *Sálama-misri*, Mar.; *Sálammisri*, Guz.; *Shálá mishiri*, Tam.; *Sálá-misiri*, Tel.; *Sálá-mishri*, Malay.; *Sala-misri*, Burm.; *Saalab-misri, khusyus saalab, khusyatus-saalab*, Arab.; *Saalabmisri*, Pers.

References.—*Stewart, Pb. Pl., 236; Afgh. Del. Com., 113; Flück. & Hanb., Pharmacog., 654; U. S. Dispens., 15th Ed., 1744; Ainslie, Mat. Ind., I., 368; O'Shaughnessy, Beng. Dispens., 653; Moodeen Sheriff, Supp. Pharm. Ind., 221; S. Arjun, Cat. Bomb. Drugs, 137; K. L. De, Indig. Drugs, Ind., 81; Murray, Pl. & Drugs, Sind., 22; Dymock; Mat. Med. W. Ind., 2nd Ed., 789; Baden Powell, Pb. Pr., 261; Atkinson, Him. Dist. (X., N.-W. P. Gaz), 722; Prod. Res., 12, 226, 231; Davies, Trade and Resources, N.-W. Boundary, India, VI., cxxviii; ccclxxv; Ind. Forester, XIII., 91, 95; Smith, Dict. Econ. Pl., 363; Balfour, Cyclop. Ind., III., 500.*

SOURCES. 522

Sources of Salep.—Although most, if not all, the species of **Orchis**, found in Europe and Northern Asia, are furnished with tubers capable, if properly prepared, of yielding salep, only a few of them are actually in use as sources of the substance. The following are the most important of these:—**Orchis coriophora, O. latifolia, O. laxiflora, O. maculata, O. mascula, O. militaris, O. Morio, O. pyramidalis,** and **O. saccifera.**

The Salep of the Indian bazárs, which is highly esteemed by the Natives as a remedy for various diseases and for which high prices are often paid by them, is principally derived from the tubers of **Eulophia campestris** and **E. herbacea,** and probably also from the species of a few other genera. The variety of Salep known as Royal Salep or *badjah*, resembles a bulb more than a tuber, and was identified by **Mr. J. G. Baker** of Kew (in the discussion which followed the reading of **Dr. Aitchison's** paper before the British Pharmaceutical Society in December 1886) as being derived from **Ungernia trisphæra.** a plant belonging to the Natural Order of Amaryllideæ. **Dymock,** in his *Materia Medica of Western India*, had previously, however, on the authority of **Mr. N. M. Khansahib,** describes the Royal Salep as the *pseudo* bulb of an Orchidaceous plant **(Pholidota imbricata). Aitchison** (*Prod. of W. Afgh. & N. E., Persia, 215*) speaks of the bulbs of **Ungernia trisphæra** as collected and given to camels.

The Salep of European commerce is prepared chiefly in the Levant, and to some extent in Germany and other parts of Europe. German Salep is said to be more translucent and more carefully dried than the Levant variety. That of the Indian bazárs is produced on the hills of Afghánistán, Balúchistán, Persia, and Bokhara, but the Nilghiri Hills and even Ceylon are said to furnish a part of it. Besides these, imitation Salep is largely prepared for the Indian trade. This is said to be made up of pounded potatoes and gum.

PREPARATION. 523

Collection and Preparation.—The tubers are dug up after the plant has flowered, the plump firm ones are washed and set aside, those that are shrivelled and soft are thrown away. The selected tubers are then strung on threads, scalded to destroy their vitality, and dried in the sun, or by gentle artificial. heat. By drying, they become hard and horny, and lose their bitter taste and peculiar odour.

CHEMISTRY. 524

Characters and Chemical Composition.—The Salep of the Indian bazárs is met with is three varieties:—palmate, large ovoid, and

The Salep of European Commerce.

CHEMISTRY.

small ovoid masses threaded together into long strings, all of them more or less translucent and gum-like. They have very little odour and a slight, not unpleasant taste. After maceration in water for several hours, they regain their original shape and size. Salep contains a substance known to the chemist as Bassorine, which is reported to be more nutritive than any other vegetable produce. One ounce, it is reported, will suffice per day to support a man (*Smith*). The following account of the microscopic structure and chemical composition of the Salep tubers is taken almost *verbatim* from **Flückiger & Hanbury's** valuable work:—"On microscopic section the tuber is found to consist almost entirely of parenchymatous cells, containing starch and some acicular crystals of oxalate of lime. In the midst of these parenchymatous cells are numerous larger ones filled with homogeneous mucilage. Small fibro-vascular bundles are irregularly scattered through the tuber. The most important constituent of Salep is a sort of mucilage, which is constantly present in a varying large amount. Salep yields this mucilage to cold water, forming a solution which is turned blue by iodine, and which mixes without precipitation with a solution of neutral acetate of lead. Mucilage of Salep, if precipitated with alcohol and then dried, becomes violet, or blue in colour if moistened with a solution of iodine in iodide of potassium. The dry mucilage is readily soluble in ammoniacal solution of oxide of copper; when boiled with nitric acid, oxalic but not mucic acid is produced. In these two respects the mucilage of Salep agrees with cellulose rather than with gum arabic. In the large cells in which it is contained, it does not exhibit any stratification, so that its formation does not appear due to a metamorphosis of the cell wall. Mucilage of Salep contains some nitrogen and inorganic matter, of which it is with difficulty deprived by repeated precipitation with alcohol.

It is to the mucilage just described that Salep chiefly owes its power of forming, with even 40 parts of water, a thick jelly, which becomes still thicker on addition of magnesia or borax. The starch, however, assists in the formation of this jelly, yet its amount is small or even *nil* in the tuber which bears the flowering stem, whereas the young lateral tuber abounds in it. Salep also contains sugar and albumen, and when fresh, a trace of volatile oil. Dried at 110°C., it yields 2 per cent. of ash, consisting chiefly of phosphates and calcium" (*Dragendorff, Pharmacographia, 654*).

MEDICINE.

Medicine.—From ancient times, Salep has been considered to possess great invigorating virtues and has hence been extensively prescribed both in Europe and the East for diseases characterised by weakness or loss of the sexual powers. Under the superstitious influence of the *doctrine of signatures* this idea was no doubt strengthened by the supposed appear-

Tuber. 525

ance of the TUBER, which was thought to resemble the form of the testicles (hence the French name "testicle di chien"). In the East, the odour of the fresh root is supposed to resemble that of the seminal fluid, and is thought to have a powerful aphrodisiac effect when clasped in the hand. The dry tuber has an immense reputation as a nervine restorative and fattener. It is much prescribed in paralytic affections. The palmate tubers are most sought after. It is much used by Native practitioners in conjunction with other nervine tonics. It is also considered a very nourishing article of diet, and is given mixed with milk and flavoured with spices and sugar. All scientific opinion, however, concurs in the belief that not only is it devoid of medicinal virtue, but it is highly doubtful if its nourishing properties are so great as they are supposed to be.

TRADE. 526

Commerce.—A considerable trans-frontier trade in Salep from Afghánistán, Persia, and Balúchistán, and also from Bokhara *viá* Kashmír exists, and a little Salep is also prepared in India from indigenous species of orchids, but the great proportion of the ordinary article met with in Indian bazárs is imported by sea into Bombay from Persia and the Levant. **Dymock**, in his *Materia Medica of Western India*, gives the prices of Salep as follows:—"*Abushaheri* or *Lasaniya* R15 to R35 per maund of 41℔; *Panjábi* R2½ per ℔; *Panjáb-i-salab* (Palmate Salep, Persian), R5-10 to 10 per ℔." **TRADE.**

For further information on the sources of Salep in India, see **Curculigo**, II., 650; **Eulophia**, III., 290; and **Orchis**, V., 492.

SALICORNIA, *Linn.; Gen. Pl., III., 66.*

Salicornia brachiata, *Roxb.; Fl. Br. Ind., V., 12; Wight, Icon., t.* [*738;* CHENOPODIACEÆ. 527

Syn.—ARTHROCNEMUM INDICUM, *Thwaites.*

Vern.—*Oomarie keeray* ,TAM.; *Quoilú, koyalú*, TEL.

References.—*Roxb., Fl. Ind., Ed. C.B.C., 28; Voigt, Hort. Sub. Cal., 320; Thwaites, En. Ceylon Pl., 246; Elliot, Fl. Andhr., 100; Drury, U. Pl., 377; Royle, Prod. Res. Ind., 75; Gaz., Mysore and Coorg, I., 65; Ind. Forester, III., 238.*

Habitat.—A gregarious, herbaceous shrub, growing abundantly on the coasts of India and on the margins of salt lakes. It is found abundantly on the northern shores of the island of Ceylon. **Moquin** (in *DeCandolle's Prod., Vol. XIII., Pt. II., 145*) quotes **Wallich** as giving a Nepál habitat; but this, **Sir J. D. Hooker**, in the *Flora of British India*, says, is incorrect.

Medicine.—This is one of the numerous sources of the alkaline earth, *sajji*, used in medicine and also in the arts. (Compare list of plants given under **Barilla**, *I.*, 394.) **MEDICINE.** 528

Food & Fodder.—The Natives pickle the LEAVES and young SHOOTS of this plant (*Drury*), and in times of scarcity utilise them as greens (*Shortt, in Ind. For*). **FOOD.** **Leaves.** 529 **Shoots.**

S. indica, *Willd.*; see **Arthrocnemum indicum**, *Moq.*; Vol. I., 328. 530

SALIX, *Linn.; Gen. Pl., III., 411.*

A genus of deciduous, diœcious trees or shrubs, containing about 160 species, 26 of which are found in the Indian Peninsula. They are very rare in the tropics and southern hemisphere and absent in Australia and the Pacific Islands.

Salix acmophylla, *Boiss.; Fl. Br. Ind., V., 628;* SALICINEÆ. 531

Syn.—S. GLAUCA, *Anderss.*; S. OCTANDRA, *Del.*

Vern.—*Bedh*, AFG.; *Budha*, SIND; *Bísu, bada*, PB.

References.—*DC., Prodr., XVI., ii., 195; Brandis, For. Fl., 463; Aitchison, Cat. Pb. Pl., 140; Murray, Pl. & Drugs of Sind, 29; Aitchison, Botany of Afgh. Del. Com., 111.*

Habitat.—A moderate-sized, quite glabrous tree, found in North-West India, and distributed to Afghánistán, Balúchistán, and westward to Syria. In India, it occurs usually in a cultivated or semi-cultivated state.

Medicine.—A decoction of the BARK is used in Balúchistán as a febrifuge. **MEDICINE.** **Bark.** 532

Fodder.—The LEAVES are largely utilized as cattle fodder, for which purpose, in some localities, the tree is severely lopped. **FODDER.** **Leaves.** 533

Structure of the Wood.—Tough and elastic. Weight 37℔ per cubic foot. **TIMBER.** 534

Domestic Uses.—The WOOD is employed for small carpentry. The TWIGS afford good *buttalis* for binding purposes. **DOMESTIC.** **Wood.** 535 **Twigs.** 536

537 **Salix alba**, *Linn.; Fl. Br. Ind., V., 629.*

THE WHITE OR HUNTINGDON WILLOW; SAULE BLANC, *Fr.*; WEISSE WEIDE, *Germ.*

Vern.—*Vwír*, KASHMÍR; *Bís, yür, changma, málchang, chámmá, kalchan, chung, búshan, madánú*, PB.; *Bed-i-siah*, AFG.; *Kharwala*, TRANS-INDUS.

References.—*Brandis, For. Fl., 466; Gamble, Man. Timb., 375; Stewart, Pb. Pl., 206; Aitchison, Afgh. Del. Com., 111; Year Book Pharm., 1874, 629; Baden Powell, Pb. Pr., 596; Christy, Com. Pl. & Drugs, V., 43; Smith, Dict., 440; Ind. Forester:—I., 119; V., 478; IX., 170; XIII., 505.*

Habitat.—A large tree, cultivated in the North-West Himálaya and Western Thibet; distributed to Europe and Northern Asia. [**Stewart** and also **Baden Powell** give the above vernacular names for this species, but as these are not repeated by **Brandis** nor by **Gamble**, it may be inferred that they do not denote this species. Indeed, **Stewart** calls the plant he is dealing with **S. alba**?—*Ed*]

MEDICINE. Bark. 538 **Medicine**—[The BARK of the white willow yields the crystalline glucoside—*salicin*—a drug, by modern practice, largely used in the treatment of acute rheumatism. It is recognised as antiseptic, antipyretic, and anti-periodic. It will be observed that these properties have for many centuries been recognised as possessed by the barks of certain Indian willows, so that, although new (perhaps) to Europe, the drug can hardly be regarded as new to India. **Sir William O'Shaughnessy**, fifty years ago, followed up the Native reputation of willow bark by endeavouring to isolate from each species its salicin.—*Ed., Dict. Econ. Prod.*]

FODDER. Branches. 539 Shoots. 540 Bark. 541 **Fodder.**—The BRANCHES are severely lopped, and used as fodder. The young SHOOTS and BARK of the larger trees are removed by hand and used as fodder.

TIMBER. 542 **Structure of the Wood.**—Soft white near the circumference, yellow or brown towards the centre. Weight 26 to 33lb per cubic foot. "The tree reaches 8 and 9 feet in girth when well protected **Moorcroft** mentions one of 16 feet, but the largest trees are very often hollow" (*Stewart*).

DOMESTIC. Wood. 543 **Domestic Uses.**—In Kashmír the TWIGS of this and other species of willow are much used in basket-making. The WOOD is valued in Tibet, Spiti, and Afghánistán, where it is employed for boarding. In Afghánistán, willow-wood is the one most used for building, as insects do not attack it much. The wood of willow is also used in Tibet for making ploughs and other agricultural implements. On the Chenab, pails are, according to **Stewart**, cut from single blocks of the wood and fine combs are made of it.

544 **S. babylonica**, *Linn.; Fl. Br. Ind., V., 629.*

THE WEEPING WILLOW.

Vern.—*Tissi, bhosi*, NEPAL; *Giúr, bisa*, KASHMIR; *Bísa, bada, bed, katira, majnúm, bidái, bitsú bes, besu, wala, majnún, laila, bed majú, majnun*, PB.; *Mo-ma-kha*, BURM.

References.—*Roxb., Fl. Ind., Ed. C.B.C., 712; Brandis, For. Fl., 465; Gamble, Man. Timb., 376; Stewart, Pb. Pl., 207; DC., Prod., XVI., ii., 507; Mason, Burma and Its People, 778; Boiss., Fl. Orient., IV., 1185; Aitchison, Afgh. Del. Com., 111; Baden Powell, Pb. Pr., 385, 596; Atkinson, Him. Dist, 317; Smith, Dic., 439; Gazetteers:—Panjáb, Hoshiarpur Dist., 12; Sialkote, 11; N.-W. P, IV., lxxvii.; Mysore & Coorg, I., 65; Settle. Rep., Simla Dist., xliii., App. II.; Ind. Forester, V., 181, 186; X., 126; Agri.-Horti. Soc. Panjáb, Select Papers up to 1862, 155.*

Habitat.—A tree with pendant branches, commonly grown for ornament in North India, both in the plains and in the Himálaya up to an alti-

The Sallow—an Essential Oil. (*W. R. Clark.*) **SALIX Caprea.**

tude of 9,000 feet. Said by **Stewart** to be indigenous in the Sulaiman Range. It is distributed to Europe, North and West Asia.

Medicine.—The LEAVES and BARK contain a neutral principle, salicin **MEDICINE.**
and tannic acid. They were formerly officinal in India, and are still much **Leaves.**
used by Native practitioners as astringents and tonics, chiefly in the treat- **545**
ment of intermittent and remittent fevers (*Stewart*; *Baden Powell*). The **Bark.**
bark is also said to be anthelmintic. **546**

Structure of the Wood.—The tree sometimes attains a height of 50 **TIMBER.**
feet, with a girth at the base of 6-7 feet. The wood is white in colour, **547**
smooth, and even-grained and takes on a good polish.

Domestic Uses.—The tree grows rapidly and is easily raised in moist **DOMESTIC.**
places by cuttings of considerable size, which are often planted to consoli- **Twigs.**
date the banks of canals and watercuts (*Agri.-Horti., Soc. Panjáb*). The **548**
TWIGS and BRANCHES are much used for making baskets, wattles, weirs, **Branches.**
dams, fences, fire-wood, etc. For these purposes they are invaluable, **549**
Good cricket bats have been made also from the wood.

Salix Caprea, *Linn.*; *Fl. Br. Ind., V., 629.* 550

THE SALLOW, *Eng.*; SAULE MARCEAU, *Fr.*; SAHLWEIDE, *Germ.*

Vern.—*Bed-mushk*, HIND., & PB.; *Khwagawala*, PUSHTU; *Khiláf*, *Má-el-khilaf* (the distilled water), ARAB.; *Bede-mushk*, PERS.

References.—*DC., Prod., XVI., ii., 222; Brandis, For. Fl., 467; Stewart, Pb. Pl., 207; Boiss., Fl. Orient., IV., 1188; Pharm. Ind., 213; Dymock, Mat. Med. W. Ind., 731; Year-Book, Pharm., 1880, 251; Baden Powell, Pb. Pr., 596; Smith, Dic., 363; Kew Off. Guide to the Mus. of Ec. Bot., 129; Ind. Forester, X., 126; Agri.-Horti. Soc., Panjáb, Select Papers up to 1862, 271-275.*

Habitat.—Cultivated in North-West India and Rohilkhand, usually from cuttings. Distributed to Europe and Northern Asia. It is abundant at Peshawar and at Lahore, where it is said to have been introduced from Kashmír after the conquest of that country by Ranjít Singh.

Oil—In Kashmír, an essential oil or *attur* is obtained by distillation **OIL.**
from the FLOWERS of this tree, It is used in Native perfumery. **Flowers.**
551

Medicine.—On distillation with twice their weight of water, the FLOWERS **MEDICINE.**
yield a scented water, which is highly valued in native medicine, being con- **Flowers.**
sidered cordial, stimulant, and slightly aphrodisiac. It is used as an external **552**
application in headache and ophthalmia. The ASHES of the wood are taken **Ashes.**
in hæmoptysis, and, mixed with vinegar, are applied to hœmorrhoids. The **553**
STEM and LEAVES are considered astringent and resolvent, and the JUICE **Stem.**
and GUM are said to increase the visual powers (*Agri.-Horti. Soc, Panjáb*). **554**
In Europe, the BARK of this species of willow was at one time used as a **Leaves.**
substitute for cinchona. "This species of willow is frequently mentioned in **555**
Persian books as a popular and well-known remedy. The Persian settlers **Juice.**
in India have introduced the flowers and distilled water, but they are only **556**
used by the better classes of Muhammadans and Parsis" (*Dymock*). **Gum.**
557

SPECIAL OPINIONS.—§ "Leaves have been found useful in fevers in **Bark.**
the form of a decoction" (*Assistant Surgeon Bhagwan Das, (2nd), Civil* **558**
Hospital, Rawal Pindi, Panjab). "The distilled water from the flowers is usetul in palpitation of the heart, acting as a stimulant" (*Civil Surgeon F. F. Perry, Jullunder City, Panjáb*).

Domestic Uses.—The scented water, distilled from the flowers, is an in- **DOMESTIC.**
gredient of one of the *sherbets* in common use among the wealthier Muham- **559**
madans of India. The wood is, in Europe, much sought after for the manufacture of charcoal for powder factories.

560 **Salix daphnoides,** *Vill.; Fl. Br. Ind., V., 631.*

Syn.—S. ACUTIFOLIA, *Willd.*

Vern.—*Richang, roangching, chankar,* LAHOUL.; *Yúr,* KASHMIR; *Bedi, bidái, betsa, beli, bushan, bashal, bhail, bhéul, mudanu, shún, tháil,* PB.; *Changma, chámma, malchang, kalchang,* WEST TIBET.

References.—*DC., Prod., XVI., ii., 261; Brandis, For. Fl., 469; Gamble, Man. Timb., 377; Boiss., Fl. Orient., IV., 1191; Atkinson, Him. Dist., 317; Ind. Forester, IV., 198; XIII., 505.*

Habitat.—A shrub or tree of the temperate North-West Himálaya, both on the outer ranges and in the inner arid tract, at altitudes between 2,500 and 1,500 feet. It is distributed to North and West Asia and to Europe.

FODDER. Branches. 561 Leaves. 562 Twigs. 563 Bark. 564 **Fodder.**—The BRANCHES and LEAVES are used for cattle fodder. In Lahoul this willow is particularly abundant, and is much valued for that purpose. The trees are pollarded every third or fourth year, at high elevations every fifth year. This is done in spring, before the new leaves appear. The smaller TWIGS are given unstripped as fodder, together with the BARK of the larger branches (*Brandis*).

TIMBER. 565 **Structure of the Wood.**—Heartwood red, shining. Weight 33·5lb per cubic foot.

DOMESTIC. Twigs. 566 Wood. 567 Branches. 568 Leaves. 569 **Domestic Uses.**—The TWIGS are used for baskets, wattles, and twig bridges. The WOOD is employed for building and for making agricultural implements. The larger BRANCHES, which are cut off in pollarding the tree, when stripped of their bark, are employed as firewood. In Lahoul the LEAVES of this and other species are used as cattle litter.

570 **S. elegans,** *Wall.; Fl. Br. Ind., V., 630.*

THE INDIAN WEEPING WILLOW.

Syn.—S. KUMAONENSIS, *Lindl.;* S. DENTICULATA, *Anders.*

Vern.—*Bail, blail, bhains* (Simla), *bitsu, bed, bida, beli, yir* (Chenab), *bada* (Ravi), PB.

References.—*DC., Prod., XVI., II., 356; Brandis, For. Fl., 466; Gamble, Man. Timb., 377; Atkinson, Him. Dist., 317; Settle. Rep., Kohat Dist., 29; Ind. Forester, IV., 90.*

Habitat.—A shrub or small tree of the Himálaya, from Lahoul to Nepal, at altitudes ranging from 6,000 to 11,000 feet.

FODDER. Leaves. 571 Twigs. 572 **Fodder.**—The LEAVES and TWIGS are used as fodder for cattle and goats.

TIMBER. 573 **Structure of the Wood.**—Pinkish coloured; weight 33lb per cubic foot.

574 **S. fragilis,** *Linn.; Fl. Br. Ind., V., 630.*

THE CRACK OR RED WOOD WILLOW.

Vern.—*Tilchang,* LAHOUL.

References.—*DC., Prod., XVI., ii., 209; Brandis, For. Fl., 466; Gamble, Man. Timb., 376; Boiss., Fl. Orient., IV., 1184; Kew Off. Guide to the Mus. of Ec. Bot., 129.*

Habitat.—A fast-growing, moderate-sized bushy tree, cultivated in Ladak, Lahoul, and Western Tibet; distributed to North and West Asia and Europe.

TIMBER. 575 **Structure of the Wood.**—Heartwood yellowish red, supposed in Europe to be more durable than that of other willows. Weight 28lb per cubic foot.

DOMESTIC. Timber. 576 **Domestic Uses.**—The TIMBER is used in Lahoul and Ladak for much the same purposes as mentioned above under the other species of willow.

577 **Salix sp.** Vern.—*The manna bed khist,* PB.

MEDICINE. Manna. 578 **Medicine.**—From a species of dark-barked cultivated willow, met with in Turkistan, much of the MANNA found in the Indian bazárs is said to be produced (see **Manna**, V., 165).

Salix tetrasperma, *Roxb.; Fl. Br. Ind., V., 626; Wight, Ic., t. 1954.* 579

Syn.—S. DISPERMA, *Don.*; S. HORSFIELDIANA, *Miq.*; S. NILAGIRICA, *Miq.*

Vern.—*Bed, bent, baishi, bet,* HIND.; *Páni jamá,* BENG.; *Nachal,* KOL.; *Gada sigrik,'* SANTAL.; *Bhesh,* GARO; *Bhi,* ASSAM; *Laila, bains,* N.-W. P.; *Bilsa,* OUDH; *Yir,* KASHMIR; *Bís, beis, bitsa, bín, bidu, bakshel, magsher, safedar, badha, bidá, bed leila, bed,* PB.; *Bitsa, badha, sufaida,* SIND; *Dhanie,* C. P.; *Wallunj, bacha, bed, baishi,* BOMB.; *Boch, bach,* MAR.; *Atrupálai,* TAM.; *Eti pála,* TEL.; *Niranji,* KAN.; *Atrapála,* MALAY.; *Mo-ma-kha,* BURM.; *Burum,* SANS.

References.—*DC., Prod., XVI., ii., 192; Roxb., Fl. Ind., Ed. C.B.C., 712; Brandis, For. Fl., 462; Kurz, For. Fl. Burm., II., 493; Beddome, Fl. Sylv., t. 302; Gamble, Man. Timb., 375; Dalz. & Gibs., Bomb., Fl., 220; Stewart, Pb. Pl., 208; Elliot, Fl. Andh., 54; O'Shaughnessy, Beng. Dispens., 606; Pharm. Ind., 213; Dymock, Mat. Med. W. Ind., 732; S. Arjun, Bomb. Drugs, 130; Baden Powell, Pb. Pr., 596; Atkinson, Him. Dist., 317; 749; Drury, U. Pl., 377; Lisboa, U. Pl. Bomb., 133, 278; Gazetteers: N.-W. Prov., I., 84; IV., lxxvii.; Bomb., XV., 75; Panjáb, Gurdaspur, 55; Mysore & Coorg, I., 65; Burma, I., 138; Forest Admin. Rep., Chota Nagpur, 1885, 34; Ind., Forester, III., 204; X., 325.*

Habitat.—A moderate-sized, deciduous tree, found throughout India, on river banks and in moist places; in the Himálayan valleys it ascends to 6,000 feet. It occurs also in Burma and as far south as Singapore. It is distributed to Sumatra and Java.

Tan.—The BARK is used for tanning (*Kurz*). TAN. Bark. 580

Medicine.—The BARK is said to be febrifuge. **Sir W. O'Shaughnessy,** however, carefully examined it and failed to detect any trace of salicine in it (*Pharm. Ind.*). MEDICINE. Bark. 581

Fodder.—The LEAVES are lopped and given to cattle. FODDER. Leaves. 582

Structure of the Wood.—Sapwood large, whitish; heartwood distinct, of a dark brown colour (*Brandis*). Weight about 37lb per cubic foot. TIMBER. 583

Domestic Uses.—The TWIGS are made into baskets. The WOOD is rarely used, as it is soft and porous. It has, however, been employed for gunpowder charcoal. **Mann** says that in Assam it is used for posts and planks. DOMESTIC. Twigs. 584 Wood. 585

S viminalis, *Linn.; Fl. Br. Ind., V., 631.* 586

THE OSIER; OSIER BLANE, *Fr.*; KORBWEIDE, *Germ.*

Vern.—*Bitsu,* PB.; *Kumanta,* LAHOUL.

References.—*DC., Prod., XVI., ii., 264; Brandis, For. Fl., 470; Gamble, Man. Timb., 377; Boiss., Fl. Orient., IV., 1191; Smith, Dic., 304; Kew Off. Guide to the Mus. of Ec. Bot., 130; Ind. Forester, X., 126; XVIII., 505.*

Habitat.—A shrub or small tree of the Temperate Himálaya, from the Jhelam to Sikkim, at altitudes between 5,000 and 9,000 feet; distributed to North and West Asia and to Europe. Common throughout Europe in osier beds.

Structure of the Wood.—White and soft. TIMBER. 587

Domestic Uses—This plant forms the principal material used in Europe for basket-making. DOMESTIC. Plant. 588

S. Wallichiana, *And.; Fl. Br. Ind., V., 628.* 589

Vern.—*Bhains, bhangli, katgúli,* N.-W. P.; *Bwir,* PB.

References.—*Brandis, For. Fl., 468; Gamble, Man. Timb., 376.*

Habitat.—A large shrub, met with in Afghánistán, Kashmír, and the Temperate Himálaya, eastwards to Bhután, ascending to 9,000 feet. It occurs also in the plains of the Panjáb.

Structure of the Wood.—The white or pinkish white in colour. Weight 32lb per cubic foot. TIMBER. 590

Domestic Uses.—The BRANCHES are made into baskets. The TWIGS are used as tooth-brushes. DOMESTIC. Branches. 591 Twigs. 592

Salmalia malabarica, *Schott.;* see **Bombax malabaricum,** *DC.;* [MALVACEÆ; I., 487.

SALSOLA, *Linn.; Gen. Pl., III., 71.* [CHENOPODIACEÆ.

593 **Salsola arbuscula,** *Pall.; Boiss., Fl. Orient., IV., 960;*

Vern.—*Narruk, randuk, randu,* BALUCH.

References.—*Aitchison, Bot. Afgh. Del. Com., 103; also Prod. W. Afgh. and N. E. Persia, 181.*

Habitat.—A very characteristic shrub of the deserts of Northern Balúchistán and Persia, and distributed to similar tracts of country in Russian Turkistan.

TAN. 594 **Tan.**—The Natives employ it in preparing the skins for their water-bottles (*Aitchison*).

FODDER. 595 **Fodder.**—Camels are very fond of it (*Aitchison*).

596 **S. fœtida,** *Del.; Fl. Br. Ind., V., 18; Wight, Ic., t. 1795.*

Syn.—S. MOORCROFTIANA, *Wall.;* S. INDICA, *Herb. Royle;* S. SPINESCENS, *Wight;* CAROXYLON FŒTIDUM, *Moq.*

Vern.—*Motí láne, gorá láne, láná góra,* PB.; *Shora, shorag,* PUSHTU; *Lánan,* SIND; *Ella-kura,* TEL.

References.—*Stewart, Pb. Pl., 177; Aitchison, Cat. Pb. & Sind Pl., 127; Aitchison, Hand-book Trade Prod. Leh, 201, 234; also Afgh. Del. Com., 103; also Prod. W. Afgh. and N. E. Persia, 181; Ind. Forester, XIII., 93; XIV., 370.*

Habitat.—A large shrub, not uncommon in the Central and South Panjáb, the Trans-Indus region, and Sind. It is distributed to Balúchistán, Persia, Arabia, and North Africa. Its name is derived from the fact that in Egypt it has been observed to exhale an odour like rotten fish.

MEDICINE. Plant. 597 **Medicine.**—[Aitchison states that in the desert country from Quetta to the Hari-rud, this PLANT is burnt to obtain Barilla, *khár, ishkhár.* In the *Trade Products of Leh, p. 234,* the same writer states that with the Bhotes the process of manufacture is quite unknown, they employing in its place *Phúli*—a soda salt.

FOOD. Manna. 598 **Food.**—"At Sha-ishmail," writes **Aitchison,** "I obtained from the surface of its leaves a quantity of MANNA, which presented the appearance of drops of milk that had hardened on its foliage; this seemed to be well known to the Baluchi camel-drivers, who collected and ate it. The only name they had for the substance was *Shakar* (sugar)" (*W. Afgh. and N. E. Persia, 181*).—*Ed., Dict. Econ. Prod.*]

S. Griffithii, *Stewart,* see **Caroxylon Griffithii,** *Moq.,* Vol. II., 176; also [**Haloxylon,** Vol. IV., 199.

S. indica, *Willd.;* see **Suæda indica,** *Moq.;* Vol. VI.

599 **S. Kali,** *Linn.; Fl. Br. Ind., V., 17.*

References.—*DC., Prod., XIII., ii., 187; Stewart, Pb. Pl., 179; Aitchison, Bot. Afgh. Del. Com., 103; Boiss., Fl. Orient., IV., 954.*

Habitat.—A spreading bush, found in the North West Panjáb, common in Balúchistán. It also occurs in Western Tibet at altitudes ranging from 12,000 to 14,000 feet, and is distributed westward to the Atlantic and throughout North Asia, North and South Africa, Australia, and North America.

MEDICINE. Plant. 600 **Medicine.**—Stewart remarks that this may be the PLANT which, **Bellew** states, is used in the manufacture of *sajji* in the Peshawar Valley. **Stewart** is, however, disposed to think that *sajji* is imported into Peshawar.

(*G. Watt.*)

SALT, *Man. Geol. Ind., III. (Ball), 475-492; IV. (Mallet), 33-34.* 601

According to Indian writers there are two forms of Salt—the Common or White Salt, and the Medicinal or Black Salt. The last-mentioned is also known as *Bit-loban*, and is held in high esteem by the Natives on account of its reputed medicinal virtue. It is, however, only an impure preparation of sodium chloride made with that salt and certain other ingredients, its composition varying with the locality where it is made. It generally contains sulphuret of iron; but inferior forms are devoid of the odour of that salt and are prepared by boiling chloride and carbonate of sodium with **Phyllanthus Emblica** and **Terminalia Chebula**, etc. The present article is intended to deal chiefly with Common Salt—Sodium Chloride.

SALT—CHLORIDE of SODIUM. 602

COMMON SALT, SEA SALT, TABLE SALT, *Eng.*; SEL COMMUN, SEL DE CUISINE, SEL MARIN, *Fr.*; CHLORANTRIUM, KOCHSALZ, *Germ.*; SALT, *Dan. & Sw.*; CHLORURO DI SODIO, SAL COMMUNE, *It.*; SÁL, *Sp.*

Vern.—*Namak, lón,* HIND.; *Nimok, nun,* BENG.; *Namak, nimak,* BOMB.; *Mitha,* MAR.; *Mithú,* GUZ.; *Uppu,* TAM.; *Lavanam, uppu,* TEL.; *Uppu,* KAN.; *Uppa, lavanam,* MALAY.; *Sa,* BURM.; *Lunu,* SING.; *Lávana,* SANS.; *Milh, milhul-aajín,* ARAB.; *Namak, namake, khurdani, numake-taám,* PERS.; *Uyah,* JAVA; *Yen,* CHINA.

References.—*Mem. Geol. Surv. Ind., IV. (H. F. Blanford, 1865), 215; (King & Foote, 1865), 374; IX. (Wynne & Warth, 1875), 89, 299; X (Theabald, 1870), 351; XI. (Mallet, 1875), 91; XIV. (Wynne, 1878), 284; and XVII. (1880), 92; Rec. Geol. Surv. Ind., IV. (Dr. Oldham, 1871), 80; V. (Blanford, 1872), 42; VI. (Theobald, 1873), 67; X. (W. T. Blanford, 1877), 10; XIII. (Hacket, 1880), 19; Plowden's Rept. (1856), Pt. III. Finance & Revenue Accounts; Sel. from Rec. Govt. Beng., No. III. (A. I. M. Mills, 1851), 39; No. XXX. (H. Rickets), 60; Sel. from Rec. Govt. Bomb., XVII. (1855), 705; No. XLI. (1857), 178; Sel. Rec. Govt. of Madras No. 16 (1855), Memorandum on Salt; Mason, Burma & Its People, 577, 731; Pharm. Ind., 327; U. S. Dispens., 15th Ed., 1326; Ainslie, Mat. Ind., I., 370-71; O'Shaughnessy, Beng. Dispens., 60; Moodeen Sheriff, Supp. Pharm. Ind., 71, 231; U. C. Dutt, Mat. Med. Hind., 23, 84-87; Baden-Powell, Pb. Pr., 69-76, 81-82, 98, 371; Royle, Prod. Res., 18, 382; Crookes, Handbook, Dyeing, etc., 119; Simmonds, Science and Commerce, their Infl. on Manuf., 96; Commercial Products of the Sea (339-348); Hamilton, Account of the Kingdom of Nepal, 93, 214, 286, 301, 316; Crawfurd, History Indian Archipelago, I, 199; Colebrooke, Remarks on the Husbandry of Bengal, 181; Kirkpatrick, Account of Nepal, 207; Man. Madras Adm., I., 436-445; II., 40, 291; Man. of Kurnool, 179; Boswell, Man., Nellore, 67; Moore, Man., Trichinopoly, 68, 248; Nelson, Man., Madura, 25, 40; Gribble, Man., Cuddapah, 228; Mackenzie, Man., Kristna, 369; Assam, by W. Robinson (1841), 33; Gazetteers:—Bengal, I. (Midnapur, Hijili), 389; II. (Jessore), 300-301; III. (Midnapur), 150-152; XVIII. (Cuttack), 175; (Balasor), 249, 336; XIX. (Puri), 151, Bombay, V. (Cutch), 21; VIII. (Kathiawár), 92-93; X. (Ratnagiri), 190; XIII., Pt. I. (Thana), 363-378; XV. (Kánara), 72; Panjáb, Gurgaon Dist., 11; Shahpur Dist., 4; Rohtak Dist., 10; Dera Ghazi Khan Dist., 9; Kohat Dist., 18; Bunnu Dist., 18; N.-W. P., I., 294; II., 207, 416; III., 34-37, 249; VI., 20, 142, 146, 148, 151-152, 345, 363, 411, 418-420, 428, 466, 646, 647, 666, 696, 698, 702-704, 706, 755, 784, 793; IX. (Moradabad), 123-124; Burma, I., 40, 415, 418; Orissa, I., 41-44; II., 152-163; Mysore & Coorg, I., 451, 452, 454-456, III., 308; Journ. Asiatic Soc., Beng., I. (Sir A. Burnes, 1832), 145; II. (1833), 365; (Dr. J. Malcolmson), 77; III. (J. Stephenson, 1834), 36, 188; V. (J. Prinsep, 1836), 798; VII. (C. Gubbins, 1838), 363; XII. (Dr. Jameson, 1843), 183; XXII. (Dalton & Hannay, 1853), 518; Proceedings, 1880; (H. B. Medlicott), 123; Bomb. Branch Royal As. Soc., I. (Dr. Giraud, 1843), 303; XVII. (Pt. II.) (Dr. Fleming), 500; XXII., 229, 333, 444; Bomb. Geol. Soc., Trans., X. (Dr. Buist, 1852), 219; Madras, Monthly*

Jour. of Med. Sci. (*J. Nicholson, 1872*), *July, 1; Cal. Jour. Nat. Hist., II.* (*McClelland, 1842*), *251; VI.* (*Capt. Hutton, 1846*), *601; Aylwin, Pamphlet on Salt Trade, 1846; Burke, India Salt, Scinde versus Cheshire* (*1847*); *Wilbraham, The Salt Monopoly* (*1847*); *Hamilton, Notes on Manuf. Salt in the Tumlook Agency, 1852; DeLisle, on Bombay Salt Dept., 1851; Parliamentary Paper, Rep. Commissioner on Salt, 1856; Annals of Indian Adm., pt. I., 1856; Spons' Encycl., II., 1710-1740; Balfour, Cyclop. Ind., III., 504-507; Morton, Cycl. Agri., II., 791-796; Ure, Dict. Ind. Arts & Man., III., 602.*

HISTORY. 603

HISTORY OF SALT.

It is perhaps scarcely necessary to deal with the ancient history of Salt, since, in one word, it may be said to have been known to the Hindus from time immemorial. *Lávana*, its best Sanskrit name, has but few other meanings than salt or saltness, but such as it does possess show the high esteem in which salt was held. Various forms of the word were employed to denote loveliness, beauty, grace, or the private property of a married woman. Most writers regard the acquirement of the habit of using salt with human food, as marking the advance from a nomadic to an agricultural life. To a people who live on milk or on raw or at most only roasted meat, salt is not necessary, but the process of boiling removes the natural salts of the animal diet. On the other hand, to a people who live on cereals, salt is an absolute necessity, and hence it is not a matter of surprise to find the most ancient works of India treating of this mineral substance with as much detail as can be found in modern publications. **Dutt** tells us that **Susruta**, the great father of Indian medicine, describes eight different kinds of salt. Of these four were forms of salt identical with the corresponding kinds met with in present Indian commerce. The other forms were impure salts or special preparations which correspond with the various kinds of black salt already alluded to, or to the efflorescences that occur on the surface of the soil in many parts of India. **Susruta's** four pure forms of sodium chloride were as follows:—

FORMS OF SALT. Saindhava. 604

1st, Saindhava.—This was produced, we are told, in Sind or the country bordering on the Indus. "The term is applied to rock salt, which is regarded as the best of all salts. Three varieties of rock salt are recognised, *viz.*, white, red, and crystalline. The pure white crystalline salt is preferred for medicinal use. For alimentary purposes also rock-salt is considered superior to the other varieties. It is regarded as digestive, appetising, sweet, and agreeable, and is much used in dyspepsia and other abdominal diseases." The word 'Sind may have embraced a larger area in ancient times than at the present day. No rock salt occurs in Sind or further south than Kohat.

Samudra. 605

2nd, Sámudra.—This literally means produced from the sea. "The term is applied to sun-dried sea-salt, manufactured in the Madras Presidency. It is called *karkach* in the vernacular. Orthodox Natives, who consider common salt as impure from the circumstance of its having undergone the process of boiling, and who take only rock-salt, substitute *karkach* for rock-salt, if the latter is not available. Sun-dried sea-salt is described as somewhat bitter and laxative. In other respects its properties are said to resemble those of rock-salt."

Romaka. 606

3rd, Romaka, also called *Sákambari*, is the salt produced from the Sambar Lake near Ajmír. "The name *romaka* is said to be derived from a river called *Rumá*. It is obtained by the evaporation of salt water in the shape of clear rhomboidal crystals. It has a pungent taste, and is considered laxative and diuretic, in addition to possessing the other properties of salts. It is said to be the best and purest of evaporated salts."

Pansuja. 607

4th, Pánsuja or Ushasuta.—This literally means salt manufactured from

HISTORY.

saline earth, "*Panga* or common salt, manufactured from earth impregnated with salt water, would come under this head." It is prepared by boiling.

Forms of Salt. Vit lavana. 608

Sauvarchala. 609

The other four salts described by **Susruta** are (1) *Vit lávana*—a substance which, **Dutt** says, occurs in dark red shining granules, somewhat resembling coarsely powdered lac. This, **Dr. Fleming** said, was prepared from Sambar salt and **Phyllanths Emblica** fruits. (2) *Sauvarchala,* called also *sonchal,* or *kálánimak.* The description of this substance **Dutt** takes from **Baden Powell's** *Panjáb Products,* but it seems probable that **Moodeen Sheriff** is more nearly correct when he restricts the name *kálá nimak* to a preparation of sodium chloride and sulphuret of iron, which when fresh always smells strongly of sulphuretted hydrogen. (3) *Audbhid.*—This is the impure sodium chloride which forms the white efflorescence on *reh* lands. This form of common salt is often called *reha* or *kalar* in the vernacular. It consists chiefly of sulphate of sodium with a little chloride of sodium, and is described as alkaline, bitter, pungent, and nauseating. And (4) *Gutika.*—This form of salt, mentioned by **Susruta** and some of the later Sanskrit writers, **Dutt** says, cannot be identified. According to **Siva Dass** it would appear to be a form of salt produced by boiling. It is said to be stomachic, digestive, and laxative.

Audbhid. 610

Gutika. 611

In the pages which follow, the present system of administering the salt interests of India will be found dealt with in considerable detail, while at the same time the facts of historic interest will be incidentally mentioned. It would, therefore, only expose the present brief review of Salt and the Salt Question to the charge of needless repetition, were this chapter made to bring together all the historic features of the subject. In concluding these remarks, therefore, it need only be said, that ever since the time of **Alexander the Great** (*see* **Strabo, V.,** 2, 6, **XV.,** 1, 30) the salt mines of Northern India have been worked and the traffic in salt, by all the various rulers of this vast empire, has been recognised as one of the most effectual means of causing the teeming millions to contribute towards the expense of administration.

OCCURRENCE, AGE GEOLOGICALLY, AND DISTRIBUTION OF SALT.

OCCURRENCE. 612

Salt, or, as it is expressed chemically, Sodium chloride (NaCl), is perhaps to man one of the most valuable products of the mineral kingdom. By European writers it is customary to find it stated that there are two forms, *viz.,* Sea Salt and Rock Salt, but in India two other kinds have to be added—Marsh Salt and Earth Salt. In chemical nature, when obtained pure, these forms are identical, and it seems accepted by geologists that they have also been all derived from the same source, namely, the sea, though from seas of widely different geological epochs. They may, in fact, be viewed as differing chiefly in the relative age of their isolation from marine water. The majority of the beds of rock-salt bear abundant evidence of being only more ancient marsh deposits, that is to say, of being arms of the sea that in course of time became isolated, first as inland seas and then drying up became salt marshes or lakes, until ultimately they formed thick deposits of salt in the superficial structure of the dry earth's surface. "This is seen from their stratified nature, with their interposed beds of clay, which could only have been deposited from solution. The crystals of selenite (hydrated calcium sulphate), moreover, which they contain, can only have been formed in water and can never since have been subjected to any considerable amount of heat, otherwise their water of crystallisation would have been driven off. The beds also

OCCURRENCE.

of potassium and magnesium salts found at Starsfurt and other places, interposed between or overlying the rock deposits, are in just the position in which one would naturally expect to find them if deposited from salt water. Finally, the marine shells often occurring abundantly in the surrounding rocks of contemporary periods also testify to the former existence of large neighbouring masses of salt water" (*Encycl. Brit.*).

The above passage has been given here in order to exhibit the very generally accepted view regarding the formation of salt, but which may be said to be the outcome of European geological experience. In India difficulties arise on every hand in connection with the study of the rocks and soils of this vast continent. The unfortunate absence of indications of life from many of the geologic systems renders it difficult, if not impossible, to fix their corresponding ages with the rocks of Europe. The isolated salt wells that occur here and there all over India, in many cases tap strata in which there is little or no evidence of marine influence. Thus, for example, **Medlicott & Blanford** (*Man. Geol. Ind.*) say of the Púrna alluvial deposits of Berar: "Throughout an area more than 30 miles in length extending from the neighbourhood of Dhyanda, north of Akola, to within a few miles of Amraoti, wells for the purpose of obtaining brine are sunk in several places on both sides of the Púrna river. The deepest wells are about 120 feet deep; they traverse clay, sand, and gravel, and, finally, it is said, a band of gravelly clay, from which brine is obtained. No fossils are found in the clay and sand dug from the wells. The occurrence of salt in the alluvial deposits of India is not uncommon, and it is impossible to say, without further evidence, whether it indicates the presence of marine beds. The absence of marine fossils in all known cases is opposed to any such conclusion, but still it is not impossible that the land may have been 1,000 feet lower than it now is in late tertiary, or early post-tertiary, times, and this difference in elevation would depress the Púrna alluvial area beneath the sea-level." That large tracts of the salt-yielding area of India owe their salinity directly to the sea influence, however, these distinguished geologists freely admit. Thus, for example, they say of the Indus valley marshes and wells: "In the Indus valley some proof has lately been obtained, shewing that the sea may have occupied part of the area in post-tertiary times. East of the alluvial plain of the Indus near Umarkot (Omerkote) is a tract of blown sand, the depressions in which are filled by salt lakes. The lakes are supplied by water trickling through the soil from large marshes and pools supplied by the flood-waters of the rivers, and it is evident that the depressions amongst the sand-hills are at a lower level than the alluvial plain, and that the salt is derived from the soil beneath the sand. To the southward is a great flat salt tract known as the Ran of Cutch, marshy in parts, dry in others, throughout the greater part of the year, but covered by water when the level of the sea is raised by the south-west monsoon blowing into the Gulf of Cutch, and the old mouth of the Indus, and all water which runs off the land, is thus ponded back. The Lúni river, which flows into the Ran, is, except after rain, extremely salt, and salt is largely manufactured from the salt earth at Panchbhadra, close to the Lúni, more than 100 miles from the edge of the Ran, and nearly 300 from the sea. Both the present condition of the Ran and tradition point to the area having been covered by the sea in recent times, and having been filled up by deposits from the streams running into it; and the occurrence of salt lakes near Umarkot, 150 miles from the sea, of an estuarine mollusk **Potamides (Pirenella) Layardi,** common in the salt lagoons and back-waters of the Indian coast, seems to indicate that these lakes were formerly in communication with the sea. The enormous quantity of blown sand, also, which covers the Indian

OCCURRENCE.

desert, can only be satisfactorily explained by supposing that it was derived from a former coast line of the Ran and east of the Indus valley.

"It appears probable that in post-tertiary times an arm of the sea extended up the Indus valley at least as far as the salt lakes now exist, or to the neighbourhood of Rohri, and probably farther, and also up the Lúni valley to the neighbourbood of Jodhpur; the Ran of Cutch being of course an inland sea. The country to the westward has been raised by the deposits of the Indus, and the salt lakes have been isolated by ridges of blown sand.

"It is true that along the western margin of the Indus alluvium later tertiary rocks (Manchhar) are found containing remains of mammalia, and precisely resembling the Siwalik formation, and as there is nevertheless a probability that the lower Indus valley was an arm of the sea in post-tertiary times, it may fairly be argued that the existence of the sub-Himálayan Siwaliks is no proof that the Ganges valley was not an inland sea at the same epoch. But in the Indus region the representatives of the Siwaliks pass downwards into miocene marine beds; in lower Sind the Manchhar formation itself becomes interstratified with bands containing marine shells, and not very far to the westward, on the Baluchistán coast, there is a very thick marine pliocene formation, so that there is evidence in abundance of the sea having occupied portions of the area in later tertiary times, whilst there is no proof of any such marine conditions in the Ganges plain."

Having thus exhibited by these quotations the somewhat obscure nature of the indications regarding the formation and age of some of the Salt Waters and Saline Soils of India, we may turn now to the Rock Salts. Geologists appear to accept these as belonging to two widely distinct epochs, *viz.*, the eocene or nummulitic beds of Kohat and of Mandí, and the salt marl of the Salt Range which seems of Silurian age. Thus, while some of the brine wells and salt-impregnated soils are of post-tertiary, others are of later tertiary; the most recent rock-salts are early tertiary and the oldest of palæozoic age. Speaking of the Salt Range deposits **Medlicott & Blanford** say: "At the base of all the Salt Range sections throughout the range from East to West, there is found a great thickness of red marl, varying in colour from bright scarlet to dull purple, and containing thick bands of rock-salt and gypsum and a few layers of dolomite. The base of this group is nowhere seen, so that the thickness is unknown; all that can be ascertained is that it is not less than 1,500 feet.

"The beds of rock-salt to which the group owes its name are very rich, some separate bands being as much as 100 feet in thickness and there being frequently several thick beds at one locality. Thus, at the Mayo Mines of Khewra, there are altogether no less than 550 feet of pure and impure salt in the upper 1,000 feet of the salt marl; of this thickness, 275 feet, or one-half, consists of nearly pure salt; the other half, known as *kalar*, being too earthy and impure to be of marketable value without refining. The salt of the Panjáb, it should be noted, is transported and sold in the market as it is dug from the mine, without being refined. The beds of salt, so far as they are known, are most abundant in the upper portion of the group, and the principal bands of gypsum overlie the salt beds. The salt bands do not appear to be continuous over a large area; but owing to the manner in which the outcrops are usually dissolved by rain, and then covered up by the marl, it is impossible to trace the beds. The salt itself is white, grey or reddish, and is frequently composed of alternating white and reddish layers, differing in translucency as well as in colour. Some bands are almost pure, others contain small quantities of sulphate of lime and chlorides of calcium and magnesium." . . . .

SALT. **Occurrence, Age Geologically,**

OCCURRENCE

"There can be no reasonable doubt that the salt marl is a sedimentary rock, although its very peculiar appearance has induced some observers to suspect an igneous origin. The red colour, however, due to the occurrence of iron sesquioxide, is a normal character of beds containing salt. The absence of organic remains is also a common peculiarity of saliferous rocks. Whether such formations with their beds of rock-salt and gypsum have been deposited in salt lakes under process of desiccation, does not appear to be equally generally admitted.* The amount of salt in the beds of the Salt Range is so great, that successive supplies of salt water and repeated evaporation alone could produce the thickness of the mineral found in places.

"The geological age of the salt marl and of the next formation in ascending order, owing to the absence of fossils, is somewhat doubtful, but the presence of a bed, probably of Silurian age, at a higher horizon, shows that both must be of very ancient palæozoic date."

Of the Kohát rock-salt **Medlicott** & **Blanford** write that "The salt consists of a more or less crystalline mass, usually grey in colour with transparent patches, and never reddish, like the salt of the Salt Range." . . . "The quantity of salt is sometimes marvellous; in the anticlinal near Bahádur Khel alone rock-salt is seen for a distance of about eight miles, and the thickness exposed exceeds 1,000 feet, the width of the outcrop being sometimes more than a quarter of a mile. Hills 200 feet high are sometimes formed of pure rock-salt. As a rule, the salt contains sulphate of lime (gypsum), but none of the potassium and magnesium salts of the Salt Range beds." . . . . "It is by no means certain that the Kohát salt and gypsum are eocene, but, in the absence of any evidence to the contrary, it appears best to class them with the nummulitic beds immediately overlying them." For the purpose of the present article it does not seem necessary to deal further with the geological aspects of the Indian salt supply. Suffice it to add, in conclusion, therefore, that **Mr. Theobald**, after a very careful study of the geology of Kangra, arrived at the conclusion that it was safest to regard the Mandi rock-salt as, like the Kohát, referrable to the nummulitic age.

Having thus briefly conveyed some idea of the ages of the various kinds of salt met with in India, it may be as well to indicate the distribution of local supplies over the country. For this purpose an abstract of the detailed account given by **Mr. Ball** will perhaps suffice, since to a certain extent the same facts are dealt with again in the two concluding chapters of this article.

Madras. 613

Madras.—In this Presidency salt was formerly manufactured by two processes, *viz.*, by evaporation of sea-water and by lixiviation of saline earths. The latter process has, for some years, been abandoned, indeed prohibited, but the former is still carried out, and the locally-made salt may be said to entirely supply the wants of the Presidency; only small quantities are imported from Bombay and none from foreign countries. A most instructive and detailed statement of the salt industry of South India will be found in the *Madras Manual of Administration* published in 1885 (Vol. I., 436-445).

Bengal. 614

Bengal.—Formerly salt was manufactured in Midnapur and Jessor, but at the present day only a small quantity is made at Behar, Bhagalpur, Monghyr, and the neighbourhood of Calcutta as a bye-product in the saltpetre manufacture. In Orissa, however, salt is still manufactured from sea-water by solar evaporation. Formerly it was also made by artificial

* For a discussion of this question and references, see **Wynne**, *Mem. Geol. Surv. Ind.*, XI., 141, and XIV., 82.

OCCURRENCE.

heat (*panga* salt) but this has recently been prohibited. For particulars regarding this industry the reader should consult Sir W. W. Hunter's *Statistical Account of Bengal* (Vols. XVIII. and XIX.), and also his *Account of Orissa*.

Berar. 615

Berar.—Salt was formerly prepared in Berar to a considerable extent, the sources of supply having been drawn from the Lonar Lake and the numerous brine wells of Púrna (in Akola) which have already been fully discussed. Salt is not now made in Berar.

Rajputana. 616

Rájputana.—In the pages below full particulars will be found regarding salt manufacture from the numerous marshes and salt-impregnated soils of this portion of India. Suffice it to say that the chief sources of supply are the Sambar Lake, the Didwana Lake, and the Kachor-Rewasa Lake.

Bombay. 617

Bombay—Salt has been manufactured by solar evaporation of sea-water along the coast of Bombay for a great many years. So important is this industry that of Thána it is said to be second in importance only to agriculture, and gives employment to 20,000 persons. A full and detailed account of the Thána salt-works will be found in the *Gazetteer* (Vol. XIII., 363-378). Formerly similar works existed in Cambay, but these have been closed, the Nawab receiving a compensation of R40,000 a year. Salt was, some few years ago, also manufactured from the saline earths of the Deccan, but the industry, like all similar manufactures of salt, has declined in consequence of the better quality and lower price of the Government salt. Indeed, it was found that salt prepared from earth contained so many impurities as to give rise to disease among the people who ate it and, as an abundant supply of pure salt was available, the manufacture of earth salt was, in Bombay, as elsewhere, prohibited. There are numerous and large accumulations of salt in Sind, but these are not now utilised as sources of supply. The salt wells of the Indus basin and of the Ran of Cutch have already been fully discussed, and it need, therefore, be only added that, except perhaps as local supplies, little or no salt is now made from these sources.

Panjab. 618

Panjáb.—This Province may be said to differ from all the other provinces of India in its possession of rich and inexhaustible supplies of rock-salt. These have already been discussed from a geological point of view and need scarcely be further dealt with, since the methods of working the mines, the annual production, and revenue from these, constitutes the chief feature of the note which will be found below from the pen of Mr. G. F. Buckley—a gentleman who, from his extensive personal acquaintance with the subject, is highly qualified to deal with it. It may, however, be here remarked that the rock-salt of India is capable of a geographical as well as a geological classification. The Trans-Indus region embraces the deposits of Kohát and Kalabagh (in Bannu District), while the Cis-Indus corresponds to the Salt Range. The Kalabagh like the Salt Range deposits are supposed to belong to the Silurian age, while the Kohát, like the distant patches of salt in Kangra (the Mandi deposits), are of a much more recent formation. In the Gurgaon District of the Panjáb there are also extensive brine wells which were formerly of far greater value than at the present day. The salt prepared from them is of inferior quality to that of the Sambar Lake, and it is reported that the workers of these wells are generally so indebted that they are unable to produce an article at all capable of competing with the superior salt of the Sambar Lake.

N. W. Provinces. 619

North-West Provinces.—Salt was formerly manufactured to a considerable extent from the brine wells of these provinces. But the salt so prepared was inferior and expensive and always contained considerable impurities of sodium and magnesium sulphates, sodium carbonate, and nitre. The salt industry may be said to have been most active in the vicinity of the Jamna,

SALT. **Occurrence, Age Geologically,**

OCCURRENCE. especially in the districts of Bulandshahr and Muzaffarnagar. Full information will be found regarding the former salt manufacture of these provinces in the District Gazetteers, but it may be said that the prohibition against earth-salts, mentioned above (regarding the similar manufactures of Bombay and Sind), applies to the North-West Provinces and Oudh as well as to the rest of India. Earth-salt is not now permitted to be made anywhere in British India, except as a by-product in the saltpetre industry. A small amount of salt is annually brought into the Himálayan districts of Kumáon and Garhwál across the frontier from Tibet, but the trade is unimportant, rarely having exceeded 10,000 maunds.

Assam. 620 *Assam.*—Formerly salt was (and among the hill tribes is to a small extent still) manufactured from brine wells. The springs, for example, of Borhat and Sadiya were in 1809 said to have yielded 1,00,000 maunds. In Cachar and Manipur and in the Chittagong hill tracts, isolated localities are even to the present day famous because of their salt wells. The hill tribes in these localities used formerly, to boil down the brine in joints of the bamboo, and to some extent do so still.

Burma. 621 *Burma.*—In the tertiary rocks of Pegu numerous salt springs occur, but none of these are located in the western side of the Arakan range nor on the eastern side of the Pegu. Salt pans may be seen along the coast from Akyab to Mergui, in which sea-water is evaporated in earthen or iron vessels, but the trade in sea-salt, as in the Pegu brine salt, is not, comparatively speaking, very important. When once brought under proper supervision it will, as in Bengal and Assam, very probably give place to the superior article imported from foreign countries.

The above brief provincial notices may be accepted as manifesting the extensive sources of supply which exist in India, but the perusal of the succeeding pages on the trade of this commodity (which have been obligingly furnished by Mr. J. E. O'Conor) will reveal the fact that, within a certain radius around the eastern coast of India (more especially Bengal and Burma), English and other imported salts may be said to have driven, or to be driving, the local article out of the market. As already remarked, it was found necessary, in the interests of the public, to prohibit the manufacture of Earth and Well Salts within the limits of British India. This may be said to have been due to two reasons chiefly: (1) the stuff produced was most impure and unfit for human food, while the profits from its sale to the local traders, were such as to induce them not to import the superior article which could be furnished at the same or very nearly the same price; (2) it was impossible to control such manufactures owing to the wide area over which saline earths occur. But in considering the question of India's salt supply, the effect of heavy land transport, as compared with marine, has to be borne in mind. While the imported salt can never compete with Sambar salt in the North-West Provinces, Oudh and Rajputana, or with Panjáb rock salt west of Delhi, so neither of these can contest the markets near the coast of Bengal (that is to say within a zone from that coast) with imported salt. In the future, therefore, it may be said imported salt will hold the country from say Patna to the sea. When the locally-manufactured salt in Burma is brought under excise control, the expensive and wasteful process pursued must cause it to disappear before the competition of the imported salt, at least in every place to which roads, railways or navigable rivers reach. Elsewhere than in Bengal, Assam and Burma locally-made salt will always, however, hold the field. While the Panjáb alone might be said to possess a supply so extensive that were the world's requirements concentrated on it no anxiety need be occasioned, still, the rock-salt of Upper India in all probability never can meet the demands of a larger area in the future

OCCURRENCE.

than it is doing at the present. The control and equalisation of the supply of so important an article of human food, of necessity becomes the duty of the State, and the most marked result of the supervision of the present system, of the cheapening of the cost of production, and lowering of the duty, is the uniform and low rate at which the article can be had throughout the empire. Inferior or impure salt finds no market, while manufacture on a large scale has lowered the cost of production. The import duty may be regarded as the balancing power against cost of transit which prevents foreign salt from disarranging the local production, and together with the tax on Indian salt it affords what is perhaps the only means by which the working classes of India pay their quota towards the administration that has secured to them peaceful and profitable lives.

The Editor has to here acknowledge the great services rendered him in the preparation of this article. The major portion of the information from this place to page 415, has been derived from a Note on Salt prepared (under the orders of the Commissioner) from the records of the Commissioner's Office by G. F. BUCKLEY, Esq., *Superintendent, Northern India Salt Revenue.*

DISTRIBUTION.

In India salt has been lavishly provided by nature; it is dissolved in a wide expanse of sea which lashes the shores of the Peninsula; is stored up in mines; is spread out in salt-impregnated lakes and marshes; and is found to effloresce at many localities in the interior and on the sea-board.

Upper India. 622

In Upper India, with a population of over 100 millions (including the Panjáb, North-West Provinces, Oudh, Rájputana, and Central India), only local salt is consumed, of which there are practically inexhaustible sources in mines at the Panjáb and in salt lakes and marshes at Rájputána. Some salt from Thibet is imported into the Himálayan Districts of Kumáon and Garhwál and into Oudh. Earth-salt is made under treaty with the British Government in the feudatory States of Gwalior, Dattia, and Bikanir, and a little salt is also made in the Patiala State in the Panjáb. A certain amount of salt educed in the process of refining saltpetre is also used in Upper India.

SOURCES.

INDIA'S SOURCES OF SALT.

Imported Salts. Kinds of Salt Consumed. 623

The local and imported salts consumed in India may be here briefly enumerated :—

1. Panjáb rock-salt, of which there are three kinds—
   (*a*) From the cis-Indus salt range called *Lahori* and *Sendha*.
   (*b*) From the trans-Indus salt region called *Kohati* and *Nimak Sabaz*.
   (*c*) Rock-salt from the Mandi State in the Himálayas.
2. Pit brine salt from the Delhi salt sources called *Sultanpuri*.
3. Salt from the Sambhar Lake in Rájputána called *Sambhar*.
4. Pit brine salt from Didwana in Rájputána called *Dindu*.
5. Pit brine salt from Pachbadra in Rájputána called *Kausia* in Central India.
6. Pit brine salt from Falodi in Rájputána called *Falodi*.
7. Salt from Gujrat in the Bombay Presidency called *Baragara*.
8. Bombay coast salt called *Kokan*.
9. Madras coast salt called *Kirkatch* and *Banwar*.
10. Bengal coast salt called *Panga*.
11. Earth-salts called generically *Khari Nimak*.
12. Saltpetre salt called *Packwa* and *Nimak Shor*.

SALT.

Duty on Salt.

DISTRIBUTION.

Imported Salts.

Kinds of Salt Consumed.

13. European salts from—
   (*a*) England, (*b*) Germany, (*c*) France; these are called *Nefurfuli*, a name which is mostly applied to English or Liverpool salt.
14. Ceylon salt called *Suffri*.
15. Red Sea salt called *Ajudhiapuri*.
16. Aden salt. This is very important—about 33,000 tons imported last year and the same amount this year.
17. Persian Gulf salt called *Muscat* and *Muscat Sendha*.
18. Thibet salt called *Lencha*.
19. Minor local supplies, such as those of Manipur.

Cheshire salt shipped from Liverpool to Calcutta, Chittagong, Rangoon, and other rice ports of Burma forms the bulk of the imported salt.

DUTY.
624

Duty on Salt.—[Since 1882 it may be said that the duty levied on salt has been equalized throughout India (except Burma) to R2-8 a maund. The amounts which went into consumption or paid duty and the amount of revenue so collected, during the past ten years, were as follows:—

YEARS.	Quantities in maunds.	Duty in tens of rupees (=£).
1880-81	26,621,089	7,115,988
1881-82	29,620,715	7,375,620
1882-83	31,060,651	6,177,781
1883-84	31,574,426	6,145,413
1884-85	32,531,020	6,507,236
1885-86	32,064,822	6,345,128
1886-87	34,074,088	6,657,644
1887-88	33,216,615	6,670,728
1888-89	33,485,353	7,678,634
1889-90	33,480,141	8,187,739

The revenue, as indicated above in nominal pounds sterling, should be understood, however, to have been the gross receipts. The total charges in administering the Salt Department and collecting the revenue in 1889-90 came to R44,53,054, so that the net revenue in that year was (in nominal pounds sterling) £7,737,262, realized on a consumption of 23,914,386 cwt.

As appendices to this article the reader will find two elaborate tables, one showing an analysis of the consumption of salt in the various provinces, the sources from which derived, and the revenue thereform (*p. 429*); the other, the quantities of salt sold per rupee since 1861 in the provinces of India (*p 430*). It is only necessary to explain that one item of receipts has been omitted throughout, namely, miscellaneous. This usually amounts to a little more than half the expenditure of administering the department, so that, if about R25,00,000 (or Rx. 250,000) be deducted from the above gross receipts, the approximate net revenue would be indicated.—*Ed., Dict. Econ. Prod.*]

Prices of Salt.—The growth of the railway system and other improved facilities in communication generally have cheapened salt to the mass of the people:— PRICES. 625

Rates at which salt sold per British maund of 82⅞lb.

Province and Mart.		Rate in Indian Currency. R a. p.	Province and Mart.		Rate in Indian Currency. R a. p.
ASSAM	Sylhet	4 3 4	PANJAB	Lahore	3 5 4
	Kamrup	4 0 0		Multan	3 5 4
BENGAL	Calcutta	3 14 0	SIND	Karachi	3 1 6
	Cuttack	3 6 6		Sukkar	3 5 4
	Patna	3 8 0	BOMBAY	Bombay	3 8 9
NORTH-WEST PROVINCES.	Cawnpore	3 4 9		Surat	3 1 0
	Meerut	3 5 6	CENTRAL PROVINCES.	Hoshangabad.	4 7 0
RAJPUTANA	Jeypur	3 5 4		Jubbulpur	4 5 6
	Abu	3 8 0	BERARS	Akola	4 0 0
OUDH	Lucknow	3 5 0	NIZAM'S TERRITORIES.	Sekunderabad.	4 7 0
	Sitapur	3 8 0	MYSORE	Mysore	4 7 0
CENTRAL INDIA.	Indore	3 12 0		Shimoga	4 0 0
	Gwalior	3 14 0	MADRAS	Madras	2 12 6
				Bellary	3 5 4

[The reader will find by the table at the end of this article (*p. 430*) that, taking India as a whole, salt has materially cheapened since 1861.—*Ed., Dict. Econ. Prod.*]

Rate of Consumption.—The rate of consumption of salt varies greatly in different provinces, but on a general average is estimated at 5 seers or 10lb per head of population. In India, and especially in Upper India where patches of saline soil and saline herbs abound and brackish water is not uncommon, it has never been the general practice to give salt habitually to cattle, especially grazing cattle (which constitute the majority), except as a religious observance or as medicine when sick. (CONSUMPTION.—*Conf. with para. p 428.*) CONSUMPTION. *Conf. with p. 119.* Cattle. 626

Salt Tax.—Ever since the conquest of the country a tax on salt has been one of the chief sources of revenue to the Indian Government. During the Muhammadan rule a considerable revenue was derived from salt by farming the sources of production or imposing a duty on it, in towns and cities. Act 38 of 1803 was the earliest regulation under which salt was taxed by the British Government, according to quality at from 4 annas to one rupee a maund (82⅞lb). The highest rate to which the tax has ever risen was 3 rupees and 4 annas a maund. Until 1882 the duty varied in different provinces, being higher in Bengal than elsewhere; in that year it was equalised throughout India, and is now fixed at 2 rupees 8 annas per maund, except at the Kohat and Mandi Salt Mines. For fear of causing frontier difficulties were a higher rate of duty imposed, salt from the Kohat Mines on the Afghánistán border is taxed at the rate of 8 annas per Sikh maund (102lb). Himálayan salt from the Mandi Mines is taxed more heavily, but less than British Indian salt. The Raja gets a share—and the reason is not merely the inferior quality, but also the impossibility that it can compete in the plains with our salt. The total salt tax collected in India during the 12 months ending 31st March 1890 amounted to R7,93,06,523. (*Conf. with p. 429.*)

SALT TAX, etc. Historical. 627

First rate of tax. 628

Highest rate. 629

Present rate. 630

Total tax collections. 631

SALT TAX. Mode of collecting tax. 632

For the purpose of realising the duty on salt* produced in Native States and in British districts subject to a lower rate of duty, when imported into Upper India, the Customs line which was commenced in 1843 and which by 1870 stretched across the whole of India from a point north of Attock on the Indus river to the Mahanadi on the border of Madras, a length of 2,500 miles of an impenetrable hedge of thorny bushes and trees supplemented in places by a stone wall or a ditch and earth mound and which was guarded and patrolled by day and night by a force of 14,000 officers and men, was maintained. In 1869 the policy of collecting the tax at the sources of production was initiated, and in 1879 the old system and with it the Customs line disappeared. This was rendered possible (1) by agreement with the Native States under which the British Government obtained leases and control of all the important sources of salt in the Native States, and (2) by the equalisation of salt duties throughout India.

Conf. with pp. 314, 410, 420.

Liberal compensation is given to the Native Chiefs, to the extent of R27,85,000, exclusive of royalties amounting to about 2 lakhs of rupees (paid to the States of Jodhpore and Jeypur) on all salt sold over a fixed limit at the Sambhar Lake.

CHIEF CLASSES OF SALT.

CLASSES. Modes of mining, quarrying, and manufacturing salt. 633

Rock salt; 634

I. Rock Salt.—For methods of mining and quarrying see the account below of the Mayo mines, Kohat quarries, and Mandi quarries.

Lake, marsh, and pit salt; 635

II. Lake and Pit Salt.—See below Sambhar Lake, Didwana, Pachbadra, and Delhi salt sources.

Sea water salt. 636

III. Sea Salt.—Alluvial muddy flats on the coast liable to submersion at high tides are selected for the sites of the salt works. On the flat a reservoir to contain sea water is first made by means of a mud embankment, near it another rectangular embankment is constructed and carefully divided off into shallow rectangular crystallisation pans separated by ridges wide enough to work on. Between the enclosure of crystallising pans and the sea water reservoir, another reservoir, sometimes two, are made for concentrating the sea water before its admission into the crystallising pans. Levels are so adjusted that sea water may at high tides run into the main reservoir, and, as required, into the secondary reservoirs or condensers and crystallising pans, by gravitation. The processes are (1) 3 to 9 inches of brine are admitted into the pans and allowed to evaporate, and (2) before total evaporation takes place a fresh supply of brine is let into the pans, and so on. The result of (1) is a thin crust of salt, of (2) a heavier crop; in both the salt is scraped up, drained on the dividing ridges and then stacked for sale on suitable spots—8 to 30 days are required for a single crop.† The yield averages during the season 10 to 15 seers (20 to 30lb) per square foot of crystallising surface, which, on the Bombay coast, ranges from 50,000 square feet upwards. Twenty-thousand maunds (735 tons) may be considered a fair annual outturn of a salt-work of 200 crystallising pans, each of 270 square feet.

The mean rate of evaporation of sea water in salt-works is one vertical inch in three or four days or 8 inches per month An acre of sea water which contains 2·3 per cent. salt should, on evaporation, yield about 19 tons of salt; the ordinary produce of Indian sea coast salt-works (owing to defective working) seldom exceeds one-fifth of the estimate Indian sea salt

* Another purpose of the line was to tax the export of sugar from the North-West-Provinces, southwards.—*Conf. with p. 313.*—*Ed., Dict. Econ. Prod.*

† In Burma and Orissa artificial heat is or was used for the manufacture of sea salt. This is known as *Panga* salt.—*Ed., Dict. Econ. Prod.*

CLASSES OF SALT.

contains from 80 to 95 per cent. chloride of sodium or average 86·53 per cent.

Swamp Salt. 637

IV. SWAMP SALT.—This salt is due to the sea breaking in upon the low-lying lands upon the Indian coast in the shallow basins in which it is caught and evaporated naturally. Swamp salt is thus formed in extensive cakes about an inch thick upon the surface of littoral wastes, and needs only to be carefully taken up. Crystals of swamp salt are remarkable for their solidity and purity, equivalent to 97 per cent. chloride of sodium.

Saline Efflorescence. 638

V. SALINE EFFLORESCENCE.—In numerous places in India, after the rainy season, owing to the action of rain and capillary attraction, chloride, sulphate, and carbonate of sodium and potassium nitrate, effloresce upon the surface of the earth. Stretches of such efflorescence may be seen and are remarkable for, in many places, their total absence of vegetation. Though tracts and patches of the efflorescence are common, in which one of the salts named may largely preponderate, the other salts, possibly in minute quantity, will almost invariably be also found present. The preponderating salt gives the efflorescence a distinctive appearance and in the terminology of the salt industries a separate name, *e.g.*, sodic chloride efflorescence is usually of the colour of dirty chamois-leather and is known as *Lonha* in Upper India, and sodic sulphate and sodic carbonate efflorescence is more or less white and is called *Khariar* and *Reh* and *Kallar Shor* (see the article **Reh**, VI., Pt. I., pp 400-427).

Earth Salt. 639

VI. EARTH SALT (*khari nimak*).—In Upper India this is made in the Gwalior State, Central India, and in the Narnoul district of the Patiala State, the Panjáb, by a solar evaporation process known as *Abi*. The plant of a work consists of a rude filter, hollowed out of a mound, or built up on the ground and connected by a channel with a reservoir for brine and a few pans about 20′ × 20′ each. The pans and reservoirs are plastered with chunam or lime cement to render them watertight. A factory is usually tended by a single family with a few bullocks or donkeys to carry in salt soil from the neighbourhood. Brine, which is produced by lixiviating salt earth in the filter, is collected in the reservoir, whence it is passed into the pans to evaporate naturally. This it does in from a few days to three weeks, according to the state of the weather, and leaves a residue of salt which is scraped up and pitted to dry. This salt which rarely contains over 80 per cent. sodic chloride, sells in Gwalior and Patiala at R1 to R2-8 annas a maund. The average yield of a single work during the season is about 100 maunds (3·7 tons), and the total outturn of such salt in Upper India does not exceed 1,500 tons; none of it, since it is not taxed by the British Government, is allowed to enter British territory.

Oudh earth salt-works 640

Prior to the annexation of Oudh when, owing to the state of the country, trade in other salts was difficult, earth-salt was largely made and consumed in that province. In 1869-70 the British Government tried to revive the industry at Mallowna in the district of Unao and at Karor in the district of Jounpore. After a patient trial in which the Government spared no pains to make the experiment a success, the attempt had to be abandoned with a loss to the State of R50,000 in advances made to the salt manufacturers who were unable to manufacture salt which could compete with the superior salts imported into Oudh from Rájputana and elsewhere.

Saltpetre Salt. 641

Conf. with pp. 431-447.

VII. SALTPETRE SALT (*Puckwa* and *Nimak Shor*).—Nitrous efflorescences, from which crude or rough saltpetre is made, usually contain from one-fifth to one-third as much sodic chloride as nitrate. A little common salt is thus occasionally produced illicitly in making crude nitre which is manufactured by the process known as *Abi*, already described, under Earth salt, or by what is known in Upper India the *Jaria* process, a process in which the brine is concentrated by boiling and is then set out to cool and crystallise

CLASSES OF SALT.

Saltpetre.

in dishes. In the refinement of crude saltpetre, which contains from about 15 to 30 per cent. of sodic chloride, common salt is frequently educed. This salt, however, owing to the presence of nitre and other salts, is unfit for human use, though some samples may be procured which contain a high percentage of sodic chloride. The process of refining crude saltpetre may be briefly described. Crude saltpetre is dissolved in about twice its weight of boiling water or nitrous brine (obtained by lixiviating nitrous earth in a filter—see Earth salt), heat is applied to the boiler and the boiling continued until the solution is sufficiently concentrated when it is run off into a pan to clear it of suspended impurities, after which the clear liquor is set out to cool and crystallise in pans, dishes, or vats. In from three to ten days the crop of saltpetre crystals is extracted and the residual liquor is utilised instead of water and brine as the refining medium. Common salt is produced by boiling the residual liquor and by continuing the boiling of the solution of crude saltpetre in the residual liquor, when some of its contained sodic chloride will deposit in the boiler, as common salt, more or less pure, and this is extracted, washed, tied up in cloth, and placed on wood ash to dry.

If the residual liquor is sufficiently saturated, it will dissolve little or none of the chloride of sodium, contained in the crude nitre; in this case the chloride will remain as a mass at the bottom of the boiler from which it may be extracted and subsequently purified by being dissolved in water or weak brine and then boiled in solution until common salt deposits. Even this purified salt, however, is hardly fit for human use. The solubility of sodic chloride is practically unaffected by temperature, between the freezing and boiling points of water, but the solubility of nitre is enormously increased by heat; on these properties of common salt and nitre depends the eduction of common salt in saltpetre refineries. In Northern India (the Panjáb excepted) a saltpetre refiner who works under a license may pay duty on the salt he produces in his refinery and sell it or destroy it if unfit for sale. During the twelve months ending 31st March 1800 in the saltpetre refineries of Behar, Oudh, and the North-West Provinces 65,684 maunds of salt were prepared, of which 27,814 maunds were destroyed by the producers as unfit for sale and the balance sold at from R2-10 to R3 per maund (82$\frac{2}{7}$ lb). Saltpetre salt is consumed in the area east of Allahabad, and in Behar and parts of Oudh it is employed to adulterate Liverpool and Sultanpur salt. It is also used for preserving hides and skins.

SALT-MINING.

Sources of Supply of Salt in Northern India.

642

SALT-MINING.

THE SALT RANGE.

643

I.—The Salt Range.

The Salt Range extends from near 71° 30′ to beyond 70° 30′ E. long. and lies wholly between the parallels of 32° 23′ and 30° N. lat. forming part of the Kohistan or upland of the Sind Sagar Doab. One extremity touches the Jhelam river, the other rests upon the Indus. The Salt Range proper lies entirely cis-Indus and stretches away for about 152 miles. The enormous deposits of rock-salt make it one of the most important regions in India. As regards its geology, the following extract from Dr. Warth's report on the stratification of the Kheora hill which contains the most important mine must for want of space suffice:—

	Average thickness in feet.
Recent formation—	
Debris of gypsum	150
Limestone formation—	
Nummulitic limestone	200

SALT-MINING.
Salt Range.

	Average thickness in feet.
Coal formation—	
Coal alumshab marl	20
Sandstone formation—	
Green sandstone	600
Blue marl	125
Red sandstone	600
Salt formation—	
Upper layer of white gypsum	5
Brick red marl	130
Brown gypsum	140
Lower layer of white gypsum	200
Salt marl and salt	600

The salt formations on the right bank of the Indus river at Kalabagh in the Bannu District, owing to lithological resemblance and geographical position, are believed to be continuations of the Salt Range proper. The regularity with which the red marl, gypsum, and salt are overlaid by aqueous deposits, together with their interna stratification, point to evaporation of salt water as the origin of the salt. Although the thickness of the salt deposits, best known, is enormous, it is believed to have been accumulated in detached basins and not to extend in one vast sheet everywhere beneath the range. The exposed deposits at numerous places along the whole southern face of the range, however, show that the supply is practically inexhaustible. It has been estimated that there is probably not less than 10 cubic miles or 70 milliards of tons of rock-salt in the range. It is the oldest known salt deposit in the world and belongs to an epoch not later than the Silurian. The salt occurs in broad bands separated from each other by layers of red marl and gypsum which, especially the first, are characteristic of the occurrence of the salt which varies in colour, from pure white through all intermediate tints to brick red. The salt is of a purity such as few known mines can yield. Analyses of samples reveal the presence of 98 per cent. of sodic chloride. Beautiful crystals of salt several inches in diameter are often found.

Rock-salt is excavated in the Salt Range in—

(*a*) The Mayo mines.
(*b*) The Warcha mines.
(*c*) The Kalabagh quarries.
(*d*) The Nurpur mine.

The Mayo Mines.
644

(1) The Mayo Mines, so named to commemorate Lord Mayo's visit to them in April 1870, are the largest in the Salt Range. In these mines vast caverns have been left by the old Sikh workmen. Dr. Warth introduced great improvements and a scientific system of mining, in consequence of which the mines are now perfectly safe and thoroughly ventilated. Stalactitic masses are found in the abandoned workings which when lighted up have a most picturesque effect. Formerly entrance to the mines was gained down a slippery incline or through an adit; now the mines may be entered by a wide drift running at a low level with a tramway laid down for a distance of 1,700 feet which is prolonged upwards by a steep gradient of 1 in 8½ to a further length of 344 feet. From the head of the gradient is another spacious passage in which a tramway is laid down in connection with the lower tram line, and by these two tram lines the whole of the salt intended for sale is removed to the sale depôt outside at the mouth of the gorge. The mines are now worked in a regular series of galleries or chambers. The chambers are constructed across the strata from marl seam to marl seam enclosing the whole of the intermediate salt. The

SALT-MINING
Salt Range.
Mayo Mines.

chambers, which are each 45 feet wide, are separated by pillars or walls 25 feet thick, left in the salt seam to support the roof. The chambers and pillars run at a magnetic bearing of 330° at right angles to the strike of the seam. The dividing walls are never pierced except when a narrow passage is absolutely necessary to connect galleries on the same plane. Notwithstanding the magnitude of operations in the mines serious accidents are almost unknown. The method of excavating the salt is to carry on work in the chambers, from the roof downwards, for which purpose a forward working, called by the miners *Kutti*, as high as possible is commenced. After the *Kutti* is completed the roof of the chamber is blasted down until the crown of the roof is as high as is desired; this work is called by the miners *Chath*. The greatest care is taken to make the roof a parabolic arch. The floor is then worked down, called by the miners *Pur*, by blasting until the level of the gorge outside the mines is reached when the inflow of brine prevents further downward excavation, as pumping out the brine would have to be resorted to; this is quite unnecessary, as there are such enormous stores of salt at a higher level: some of the existing chambers are 250 feet long, 45 feet wide, and 200 feet high.

There are about 400 miners employed in these mines. They are a healthy and contented class and are paid at the rate of 8½ pies per cubic foot including excavation of the salt, separation of marl from the mass, removal of rejections, called *Kallar*, to appointed places in the mines, and carriage of the pure salt to the tramway loading stations in the mines, whence it is removed on trucks to the sale depôt, stacked there and as required filled into gunny bags, weighed and loaded into the Railway wagons for despatch to purchasers at a cost to the State of R1-15 per hundred maunds.

Output.
645

OUTPUT AND TRADE FROM MAYO MINES.—These mines with the rest of the Panjáb passed into the possession of the British after the overthrow of the Sikhs in 1849, since when over *a million and a half of tons* of rock-salt have been sold. The available supply of salt embedded in the Mayo mines alone is estimated at 8 millions of tons of salt. Owing to improvements at the mines and sale depôt, the growth of the Railway system, the cheapening of Railway rates, and to the Mayo mines being in direct railway communication with all the chief marts in Upper India, to which the salt can be sent without breaking bulk *en route*, the trade in rock-salt has greatly developed. The trade in rock-salt has grown from less than 5 lakhs of maunds to over 12 lakhs of maunds in 1870, and now amounts to over 20 lakhs of maunds (72,000 tons), bringing into the Treasury over 50 lakhs of rupees. About nine-tenths of the whole trade in rock-salt from the Salt Range is contributed by the Mayo mines, and over 96 per cent. of the Mayo mines trade is carried by the Railway. This salt is consumed all over the Panjáb east of the Indus river, in Upper Sind, in Kashmír, and in the upper districts of the North-West Provinces; it is used by pilgrims at Benares and Gya, and by ascetic Hindus all over Upper India as a pure or *pák* salt.

Warcha Mine
646

(*b*) The WARCHA or RUKLA mine is about 70 miles west-south-west of the Mayo mines and is accessible by the Sind-Sagar Railway. All the neighbouring heights about Warcha are composed of limestone, below which the salt formation crops out. The mine is large and at a considerable elevation; there are large remains of old Sikh workings and great natural vertical water-courses. The new workings are in a seam of salt 20 feet thick; the total seam is thicker but is not pure enough for commercial purposes. The miners make a cutting along the bottom of the seam and blast away the superincumbent salt until the roof is reached. They are

SALT-MINING.

Salt Range.

paid at the rate of R3-12 for every 100 maunds delivered outside the mine at the sale depôt.

Kalabagh Quarries. 647

(c) The SALT QUARRIES OF KALABAGH, 100 miles west-north-west of the Mayo mines, were famous long before the advent of British rule. In no other place in the Salt Range could salt be quarried in such quantities. The quarries are on the right bank of the Indus river above the town of Kalabagh. The hills and ranges about Kalabagh are tertiary ossiferous sandstone and conglomerates. The quarries are in workable seams from 4 to 20 feet in thickness in the midst of marl and small unworkable salt seams. The quarries extend for two miles. The salt is very pure, nearly all of it is red and homogeneous, some of it being as finely grained as alabaster and capable of being turned on the lathe.

Nurpur Mine. 648

(d) The NURPUR or NILAWAN salt mine is the smallest in the Salt Range and is kept open only to meet the requirements of local consumption. The mine is in a seam of salt about 30 feet thick and about 70 feet above the level of the gorge.

Output—all Mines. 649

OUTPUT FROM ALL MINES.

PERIOD.	QUANTITY IN MAUNDS (82⅞lb) OF SALT SOLD FROM EACH MINE.				Total sales.
	Mayo Mines.	Warcha Mines.	Kalabagh Quarries.	Nurpur Mines.	
From 1st April 1889 to 31st March 1890 . . .	1,968,466	139,391	28,682	4,442	2,140,981

THE KHOAT OF TRANS-INDUS SALT REGION. 650

II.—Kohat Salt Region.

The KOHAT SALT REGION lies to the west of the Indus river between 32°47′ and 33° 52′ N. lat. and 70° 35′ and 72° 18′ E. long. on the Afghánistán border between the Peshawar and Bannu valleys.

The area occupied by the salt region is about 1,000 square miles and contains the largest known exposures of salt upon the globe. The salt where covered is overlaid by white and grey gypsum and grey gypserous clay much confused as to stratification. The salt deposits are believed to belong to the early Eocene Age. The salt has in some places a visible thickness of over 1,000 feet. It is very prominent in localities, notably at Bahadur Kheyl, where the salt forms high detached hills and cliffs and where for a distance of 4 miles with a width of a quarter of a mile or more the salt is quite exposed. Throughout this region salt is seen in numerous places forming precipitous outcrops within the elliptical boundaries of nummulitic limestone. The exposures of salt vary in size from the enormous one at Bahadur Kheyl to others of a few feet. The area of exposed salt has been estimated at over four millions of square feet and the available salt supply at 40 milliards of maunds (over 1⅓ milliards of tons) sufficient at the present rate of consumption to last practically for ever.

In colour Kohat salt is of varying tints of grey with transparent blotches; in some places it is dark, smelling strongly of petroleum. Its texture varies from a crystalline mass to a somewhat earthy salt, intermingled with finely divided grey clay; the latter character, however, seldom prevails to the extent of interfering with the commercial value of salt, which is remarkably pure, containing often as much as 99 per cent. sodic chloride and no trace of associated salts of other kinds.

SALT.

Salt-mining.

SALT-MINING

Kohat Salt Region.

Though at one time or another salt has been excavated at fourteen places in the Kohat district the quarries now worked, enumerating them according to geographical position from the east, are :—

1. Jutta, 2. Malgin, 3. Narri, 4. Kharak, and 5. Bahadur Kheyl.

Jutta was first worked in 1850. Malgin and Bahadur Kheyl are older quarries. Narri and Kharak were first resorted to about the beginning of the present century.

Two methods of quarrying the salt are followed. At Jutta, Narri, and Malgin gunpowder is used and the salt is blasted out in irregular pieces, the salt being quarried in the shape of a vault sloping downwards at an angle of 60°. No artificial light is ever used. At Kharak and Bahadur Kheyl the salt is cut out in slabs called *tubbis* of uniform size and weighing 51℔, or half a Sikh maund, by means of pickaxe and wedge. The *tubbis* which are ingeniously cut from the sloping face of the salt are the most convenient shape in which salt can be carried on camels through the difficult passes and defiles which lead into and through Afghánistán.

The quarrymen, though under the supervision of the salt officer at the quarries, carry on work entirely on their own account; they make their own bargains with purchasers for excavation of the salt, which ranges from about 16 to 30 Sikh maunds per rupee. The only revenue derived by the State is the tax of 8 annas per Sikh maund (102℔) levied at the quarries. In consequence of the light duty imposed on this salt an establishment is maintained along nearly 500 miles* of the left bank of the Indus river to shut out Kohat salt from the cis-Indus districts, where it would, if allowed to cross the river, displace the fully-taxed salt of the Salt Range.

Kohat salt is consumed over an area of 60,000 square miles; it goes as far as Kandahar, Balkh, and Ghuzni in Afghánistán; it is used in Swat, Boneyr, Bajor, and the Afridi country, and is the only kind consumed in the British districts west of the Indus.

Sales. 651

SALES OF KOHAT SALT.

PERIOD.	QUANTITY OF SALT SOLD IN BRITISH MAUNDS (82$\frac{2}{7}$℔).					
	Jutta.	Narri.	Kharak.	Malgin.	Bahadur Kheyl.	Total.
12 months ending 31st March 1890 . .	298,526	48,976	85,259	161,599	103,059	697,419

III.—Mandi Salt Quarries.

MANDI SALT QUARRIES. 652

The MANDI SALT QUARRIES, 77° E. long. 32° N. lat., are situated at Guma and Drang in the Himálayan feudatory State of Mandi.

The existence of a considerable quantity of this salt is geologically indicated in the neighbourhood of the quarries; the extent of the salt deposit is, however, unknown. The salt is of a dark red colour mixed with quartzite, sandstone and limestone pebbles; it contains about one-fourth of insoluble impurities. Salt of excellent quality but in small quantity is known to occur in these quarries. Mandi salt is roughly refined for domestic uses by purchasers who dissolve it in sufficient water to make strong brine which they use to season food. At Guma, which is 5,000 feet above sea level, the salt is dug out of the side of the gorge in which it is found; at Drang, which is 2,000 feet lower than Guma, the salt is quarried in the open air; at both places blasting is resorted to. Sometimes fresh

* More correctly 325 miles. See pp. 404, 420.—*Editor.*

SALT-MINING.

Mandi Quarries.

water is led from the neighbouring stream to the salt over which it is made to trickle in thin streams a few feet apart, the water cuts the salt into bonds which are blasted and broken up for sale. Frequent interruptions occur owing to slips and falls of earth due to the haphazard way in which the salt is quarried. The quarries are the property of the Raja of Mandi, who charges to purchasers at the quarries, where they are bound to give a day's free labour, 10 annas 6 pies a maund, as price of the salt and 7½ annas as duty on it. Two-thirds of the duty, by virtue of a treaty between the Raja and the British Government, is credited to the latter on account of the Mandi salt which is consumed in British territory. Actual record of sales at the quarries shows that two-thirds of the salt sold at these quarries is consumed in British districts.

Mandi salt is used in the adjoining Himálayan Native States and in the British districts of Kangra, Kulu, Simla, and Hoshiarpur. In the twelve months ending 31st March 1890 sales at these quarries amounted to 1,30,716 maunds (4,700 tons).

SALT EVAPORATION.

SALT EVAPORATION.

IV.—Delhi Salt Works.

DELHI SALT WORKS. 653

The Delhi Salt Works, E. long. 77° 35′ and N. lat. 28° 39′, are situated in the Gurgaon and Rohtak districts of the Panjáb, about 30 miles south-west of Delhi. This salt tract called *Surr* occupies an area of about 1,500 square miles. The salt made on it is called *Sultanpuri* and is the product of natural sub-soil brine derived from wells sunk to depths of from 7 to 20 feet. The specific gravity of the brine in the wells ranges from 2° to 4° Beaumé.

Salt is obtained by evaporating the brine by solar heat in shallow lime-plastered pans which average about 200′ × 60′ and 10 to 12 inches deep; a set of 10 pans is attached to each well and so arranged that there is a slight fall from each pan into the one next beyond it. The highest pan is first filled with brine which is gradually passed from pan to pan, until on reaching the last pan it is so concentrated that salt is deposited. As the brine contains foreign salts, crystallisation of common salt has to be carefully watched so that it may be harvested before the other salts begin to fall. The average period occupied in harvesting a single crop is a fortnight. An average season's crop amounts per factory to 3,000 maunds (108 tons). The mass of manufacturers are agriculturists during the rains and part of the winter, and only turn to their salt-pans when their fields no longer need their labour.

Sultanpuri, which is a small-grained and not very pure salt, is in great favour with the people in Oudh and North-West Provinces, to which its use is wholly confined. Though the works are in railway communication with the railway system of Upper India, and though *Sultanpuri* is an old favourite, it is gradually, but steadily, being ousted by the superior Sambhar salt, sales having fallen from seven lakhs of maunds (25,000 tons) in 1870 to 3⅓ lakhs of maunds (11,800 tons) in the twelve months ending 31st March 1890, in which period it sold at 3⅓ annas per maund at the works.

V.—Sambhar Salt Lake.

SAMBHAR LAKE. 654

The Sambhar Salt Lake lies in lat. 26° 58′ and long. 75° 5′ on the east of the Aravali range of hills which runs through Rájputána in a north-westerly direction; in the height of the rains it covers an area of nearly 200 square miles, the greatest length being then 23 miles, its average breadth 4 miles, its circumference about 60 miles, and its average depth 2 feet. In the dry or hot weather its bed is much less in area, and in

SALT. **Salt Evaporation.**

EVAPORATION. Sambhar Lake.

seasons of exceptional drought the entire area dries up. The bed of the lake shelves very gradually from 9 inches at 100 yards from the edge to 2½ feet at 5,750 yards.

The lake bed is composed of 1½ feet of black fetid mud saturated with sulphuretted hydrogen; below the mud is a layer of quicksand overlying a stratum of micaceous schist decomposed on the top, but harder below. No rocks are found to a depth of 20 feet. The lake is believed to derive its salinity from the denudation of the rocks of the surrounding country which is supposed to belong to the Permian system, a system which abounds in limestone on salt. After the rainy season the specific gravity of the lake water is about 1·03 (3° Beaumé), about equal to sea water; it rises as the dry hot weather advances to 1·24 (29° Beaumé), a supersaturated solution, the specific gravity of a saturated salt solution being 1·204 (25° Beaumé). Salt forms in large crystals in the shape of truncated pyramids, and in May and June a layer of crystals 2 inches and more in thickness overlies the bed of the lake. In colour the crystals are white, grey (owing to the presence of finely divided clay in the fissures of the crystals), and shades of pink (due, it is believed, to **infusoria**). After the first fall of rain the lake teems with animulculæ upon which thousands of water-birds of kinds feed

Tradition ascribes the formation of the lake to the gift of Sakumbri Devi, a goddess, who in return for milk supplied her in A.D. 551, converted a forest into a vast plain of the precious metals which she subsequently transformed into salt. The lake is owned conjointly by the Maharajas of Jodhpur and Jeypur from whom the British Government lease it at 5½ lakhs of rupees (£55,000) per annum. The lake was worked by the Emperor Akbar and his successors up to Ahmed Shah, when it reverted to the Rajput Chiefs of Jodhpur and Jeypur from whom the British Government under treaty took it over on 1st February 1870, since when it has yielded over 62 millions of maunds (2¼ millions of tons) of salt. In 1870 the price of salt at the lake averaged 10 annas per maund; it is now 4½ annas. The cost of extraction and storage of salt is about 1 anna per maund. Extraction and storage of salt at the lake gives employment to a large number of men, carts and cattle.

The salt in the lake is believed to be practically inexhaustible. A recent assay of the water gave 8·81 per cent. of dry residue composed of—

Sodium chloride	87·6
,, sulphate	7·5
,, carbonate	4·6
Balance	0·3

Salt is not only held in solution in the lake but pervades in minute crystals the whole substance of the black mud which forms the bed. One-fifth of this mud is salt. Enormous quantities of sodic sulphate and sodic carbonate lie in the lake and on its shores from 1 to 2 inches in depth.

The salt is obtained from the lake in three ways:—

(1) As evaporation of the water of the lake proceeds, crystals of salt deposit in immense quantities all over the bed and are picked up and stored in large oblong pyramids sloped to an angle of 36° of one to two lakhs of maunds (3,600 to 7,300 tons). These heaps are beaten and smoothed as a protection from rain but are not covered; wastage of salt due to rain amounts to about two inches of surface salt per annum.

(2) Early in the season, and before the water in the lake is sufficiently concentrated to deposit, salt brine is run into shallow artificial pans on the lake edge; thus a large quantity of good salt is produced by solar evaporation before it could be obtained from the lake bed.

EVAPORATION.

(3) Twelve large deep permanent pans are maintained alongside of the railway which crosses the lake in which, by irrigating the pans with brine from the lake and allowing it to evaporate, naturally heavy crops of large crystalled white salt are produced for customers who desire a good-looking salt.

VI.—Didwana Salt Marsh.

DIDWANA SALT MARSH. 565

The DIDWANA SALT MARSH, which the British Government leases from the Maharajah of Jodhpur at an annual rental of R2,00,000, (£20,000) lies 40 miles from the Sambhar Lake (the nearest railway station) in an oval-shaped depression about 3 miles long by 1 mile wide which is covered in the rainy season with water 6 to 8 inches deep which, however, soon dries up, after which manufacture of salt commences.

The soil of this marsh is not unlike the bed of the Sambhar Lake. The mode of manufacture is as follows:—In the bed of the marsh, wells 6 feet wide and about 14 feet deep are sunk, the sides of the wells being supported by a wood lining. Brine from the wells is filled into solar evaporation pans, having a superficial area of about 2,000 square yards and about a foot in depth. In from 10 to 20 days the brine naturally evaporates and leaves a deposit of small grained pure salt which is scraped up, removed to the edge of the marsh and stored for sale. This salt is much esteemed in Shekawati and Jodhpur and the British districts of Hissar, Saharanpur, and Mozuffernagar, the area in which it is consumed. The trade in *dindu*, as this salt is called, has been nearly stationary, averaging about 3½ lakhs of maunds (12,500 tons) a year.

Actual cost of manufacture and storage amounts to 4½ pies per maund. It is issued to purchasers at 3 annas per maund *plus* duty.

VII.—Pachbadra Salt Works.

PACHBADRA SALT WORKS. 656

The PACHBADRA SALT WORKS are in a valley about 8 miles long and 2½ miles wide—evidently at some remote period the bed of a river. The salt-works are about 50 miles south-west of the city of Jodhpur with which and the general railway system of the country they are connected by railway. This salt tract is leased from the Jodhpur State by the British Government at an annual rental of R1,70,000 (£17,000).

Kausia, as the produce of these works is called in Central India, is a pure salt; samples have yielded 99·87 per cent. of sodic chloride, which forms in opaline cubes of from ¾ to 1½ inches; it bears carriage well, and of all Indian salts suffers least from exposure to damp. The method of manufacture is simple. Oblong pits 100 to 400 feet long, 60 to 100 feet wide, and 10 to 12 feet deep are dug in the valley and soon fill by percolation of subsoil brine (2° to 3° Beaume), which in from two to three years dries naturally when the crop is harvested. When the brine in the pits is sufficiently concentrated (20° to 25° Beaumé), the branches of a thorny shrub are sunk in it to help the growth of the crystals. To extract the salt men enter the pits, cut through the thorny branches with a crow, and draw the salt (the masses of which are broken up in the pit) to the sides, with a hoe and remove it to the top in baskets: a single crop may amount to from a few hundred maunds to 8,000 maunds (288 tons).

Kausia is sold to purchasers at 1½ annas per maund *plus* duty. It is consumed over an area of 100,000 square miles in Jodhpur and Central India. Of the total trade about one-third is carried by the railway, the balance being conveyed by camels and horned cattle owned by the Banjarahs, the carriers of trade in Rájputana and Central India, in areas not opened up by the Railway.

SALT.

Salt Evaporation.

EVAPORATION.

VIII.—Luni and Falodi Salt.

LUNI SALT TRACT. 657

In the neighbourhood of Pachbadra is a considerable saliferous tract known as the Lúni salt tract on which very good salt forms spontaneously. No sales of it are permitted, but about 5,000 maunds (180 tons) are issued free to the people of the villages near the salt deposit.

Falodi Salt-Works. 658

The Falodi salt source is in a depression about 5 miles long by 3 miles wide and 60 miles north of Jodhpur, from which State the British rent it at R4,500 per annum. Salt is made here in much the same way as at Didwana. This source is worked at a loss by the British Government, as it is the only source of salt supply to the population of the sandy deserts of north Jodhpur and Bikanir. The salt is sold to purchasers at 3 annas per maund.

Trade. 659

Trade in Evaporated Salt.

Period.	Quantity of salt sold in British maunds (82⅚lb).					
	Sambhar.	Didwana.	Pachbadra.	Falodi.	Delhi salt sources.	Total.
12 months ending 31st March 1890 . .	3,834,805	377,068	614,901	44,955	328,851	5,200,580

IX.—Tibet or Lencha Salt.

TIBET SALT. 660

Tibet Salt, called *lencha* by the Tibetans, is imported into the British Himálayan districts of Garhwal and Kumáon and also into the northern section of Oudh. During the twelve months ending 31st March 1890 33,000 maunds were imported. Tibet salt is said to be the produce of salt lakes and swamps in the region traversed by the Yaru river.

SODA SALTS.

OTHER SODIUM SALTS.

In India in very many districts sulphate and carbonate of soda are manufactured from the earth, and in the case of the carbonate by incineration also of the Salsola plants.

Sodic Sulphate. 661

I.—Sulphate of Soda (*khari, khari-nun,* and *chamra-khari*) is made by both the solar evaporation (*ábi*) process and by the use of artificial heat (known as the *jaria* process).

The solar evaporation work consists of a rude filter about 10 yards long by 1 yard wide and a lime-plastered evaporating and crystallising bed divided into two sections. Brine, which is produced by lixiviating in the filter the sulphate of soda efflorescence collected from the neighbourhood of the works, is admitted into the pan and allowed to evaporate, which it does in from 10 to 15 days according to state of the weather, and leaves a residue of brown crystalline soda which would give on refinement 40 to 50 per cent. of sodic sulphate. The average outturn of a season (*i.e.*, April to June) is about 500 maunds (18 tons), worth on the factory 8 annas to R1 per maund (82⅚lb). The cost of manufacture amounts to from 6 to 12 annas per maund

In the *jaria* process the brine is boiled until the soda deposits in the boiler. In the districts of Behar the sulphate is not manufactured by this process, but from brine produced by first burning the soda efflorescence built up in alternate layers with paddy straw in conical mounds, and then lixiviating the calcined mass. This process produces relatively pure sulphate of soda known as *patna-khari*. This soda is made in Upper

SODA SALTS.

India in Behar, in Fattehpur, Etah, and Bulandshahr, in the North-West Provinces, in Hurdui, in Oudh, and in the Patiala State near Umballa. It is used chiefly for hide-curing, as a cathartic for cattle, and by the people of the Himálayan as a prophylactic in goitre affections.

Carbonate of Soda. 662

II.—Mineral CARBONATE OF SODA (*sajji, papri-sajji*, and *sajji-nun*) is made as follows: 10 or 12 *kyaris*, or beds, each about 10 yards square, are constructed in an alkali tract by raising enclosure walls of the surrounding efflorescent soil about 6 inches high. Into these beds, water from a well at hand is admitted to a depth of 2 or 3 inches. The soda efflorescence of the neighbourhood is thrown in by hand until the water is absorbed and a pasty mass formed. This is smoothed over and within a week a crust of concentrated efflorescence forms on the surface of the beds, which crust, as moisture is dissipated by solar evaporation, breaks up into flakes. These are gathered by hand and constitute the *papri-sajji* of the bazárs used by tobacconists and dyers, etc. To make *sajji*, the *papri* is dissolved in water and the solution boiled in an earthen vessel until the impure carbonate forms into a hard mass.

Barilla. 663

III.—BARILLA (*sajji*) is made in Upper India in the districts of Ghazipur, Azamgarh, Benares, &c., in the North-West Provinces, and is worth about R1-1 per maund.

Vegetable *sajji*, the barilla of commerce, is made from various plants called *lani* in the Panjáb in this way. In December and January when the *lani*, a small green bush with tiny succulent leaves and branches, ripens, it is cut. Circular pits of from 2 feet and upwards in diameter are then dug at convenient distances, according to the requirements of the crop, which grows spontaneously. Into the pits sheaves of the half-dried plant are thrown and set on fire, fresh sheaves being added until the pit is full of ashes in a state of semi-fusion, when the contents of the pit is well stirred and allowed to cool; this occupies about 24 hours from the burning of the first sheaf. When sufficiently cool the pit is covered with a little dry earth, to prevent evaporation. Within a week the covering is removed when the contents of the pit are found as a hard cellular mass of *sajji*. This is broken up into pieces and sells at from R1-8 to R4 a maund. It is used extensively for the manufacture of glass, paper, soap, for bleaching purposes, and by the poorer classes as a substitute for soap. This *sajji* will give on refinement 25 to 40 per cent. sodic carbonate. (For further information the reader is referred to the article **Barilla**, Vol. I., 394-399, and to **Haloxylon**, Vol. IV., 199.)

Black salt. 664

IV.—BLACK SALT is prepared in Upper India chiefly at Bhewani in the Hissar district by heating together in a large earthen pot 82lb of common salt, one pound of the fruit of **Terminalia Chebula**, one pound of **Phyllanthus Emblica**, and one pound of *sajji*, impure-carbonate of soda, until by fusion of the salt the ingredients are well mixed, when the pot is removed from the fire and its contents allowed to cool and form a hard cellular mass. This preparation is used medicinally principally as a digestive.

Having thus indicated, as Mr. Buckley has done by the above *Note*, some of the leading features of the salt interests of India, more especially as affecting the working of the Northern India Salt Department, it becomes necessary to extend the enquiry all over India. The extensive papers placed by the Department of Finance and Commerce at the disposal of the Editor fortunately constitute an invaluable source of information. The chapters which here follow, under the headings of provinces, may be said to be directly drawn from these official papers. Although, n some instances, part of the ground has already been covered by Mr. Buckley's remarks, it is believed the *precis* below will be found instructive as tracing out not only the historic facts but the main features of the trade in salt.

HISTORY of BENGAL SALT. 665

I.—BENGAL (LOWER PROVINCES).

Historic Sketch and Regulations.

Under the Muhammadan rule a tax on the salt consumed by the people of Bengal was levied, by means of imposts, on the privilege of manufacture and duties on the transportation of salt from the places of manufacture to the interior of the country.

A monopoly for the manufacture and sale of salt was first established in Bengal by **Lord Clive** in 1765, the chief object being to provide fitting emoluments from the profits for the principal persons concerned in the Government, and thus to prevent their mixing in the intrigues and questionable transactions by which Civil Servants and others in those days often amassed enormous fortunes. Half the monopoly profits were to be distributed among the officers of Government, and the other half it was proposed to credit to the Company. In his Minute of the 3rd September 1766, **Lord Clive** assumed that this share would yield, "according to the present state of the salt trade, from 12 to 13 lakhs of rupees annually." The rate fixed for deliveries was R2 per maund. The existence of this monopoly was of but short duration, as the Court of Directors wholly disapproved the arrangements. At the same time, however, the Court stated that they did not object to the levy of the ancient duties on salt, which had always constituted part of the revenues of Bengal.

In the year 1772 the manufacture and wholesale trade were farmed out by Government, to private individuals; but this complicated farming system was never very productive and soon failed. In 1780 **Warren Hastings** introduced a system for manufacture and sale under the agency of the Company's civil servants. In accordance with this system, the *molunghis* (salt-makers) received advances from the agents at the beginning of the season, on the stipulation that they delivered their salt when made to the Government at a certain price, and the agents afterwards stored the salt and sold it to wholesale dealers at a price fixed from year to year by the Government. The difference between the price agreed upon with the *molunghis* and the price at which it was delivered from store to the merchants was thus in effect the duty levied upon the salt. In 1788 sales of salt by public auction, instead of at fixed rates to the dealers, was introduced by **Lord Cornwallis**. The revenue immediately rose, but the system was eventually abolished by the Court of Directors in 1837, as it was found to lead to the establishment of sub-monopolies, injurious to the interests of both the people and the Government. In their despatch of the 4th January 1837, the Court of Directors ordered that the price to be thenceforward paid by the purchasers of salt should be determined by the cost price of manufacture, added to a fixed rate of duty.

The rates of duty since fixed from time to time have varied from a maximum of R3-4 to a minimnm of R2 per maund; but the system for manufacture and supply, as introduced by **Mr. Hastings** in 1780, continued in force, with but few modifications, until the year 1862, when the several salt agencies were gradually abolished, leaving the supply, either by importation or excise manufacture, to private enterprise. The several agencies were situated in the province of Orissa and in the districts of Chittagong, 24-Pergunnahs, Jessor, and Midnapore. The full rate of duty was not, however, levied uniformly throughout Bengal until 1862. From the year 1810 a system of retail sales at reduced prices from shops, established on the part of Government, was introduced in districts and localities where salt was manufactured, or was capable of being easily produced; the object being, as stated at the time, to leave the people

HISTORY of Bengal salt.

residing in such tracts without excuse for violating the law under the temptation of a high rate of duty, and to obtain some revenue in a part of the country where, from the great facilities for smuggling, it had been found impracticable to realise full prices. From inquiries held during the years 1860 and 1861, the Government concluded that the loss of revenue entailed by the remission of a large proportion of the duty on the salt consumed within the saliferous tracts was, under existing circumstances, far larger than would arise were the full duty levied. The system of retail sales at reduced prices was therefore abolished from the year 1862.

From the commencement of the salt monopoly, a preventive establishment was employed for the protection of the revenue. This establishment was for the most part separate from, and independent of, the agency constituted for the manufacture and supply of salt, and was employed in Northern Behar to prevent the influx of lighter-taxed salt from the westward, and also within certain defined limits which included the saliferous tracts on the sea-board of Lower Bengal. By Regulation X of 1819 the general control of this preventive department was vested in the Board of Customs, Salt and Opium, established in that year. On the passing of Act XIV of 1843, imposing an additional duty of R1 per maund on salt passing from the North-Western Provinces to the eastward of Allahabad, the establishment in Behar was withdrawn. The one rupee was, in June 1847, reduced to 12 annas, and in April 1849 to 8 annas. In March 1861 the additional duty was abolished. In Lower Bengal the limits within which the preventive force were entertained have been narrowed from time to time so as to concentrate their operations on the salt-producing tracts only. As a further check against illicit manufacture within these limits, all salt under transport was required to be conveyed by certain specified routes and pass stations, and to be covered by protective documents, under penalty of confiscation. Merchants and dealers were also required to record all sales and losses from their stocks on the reverse of their protective documents. The law and rules on this subject have been modified from time to time. Those now in force are contained in Act VII of 1864 and the Government Notification issued under that Act. In 1863 the special preventive establishment was abolished, and all duties previously discharged by them were delegated to the regular Police Force. In addition to the sea-board and salt-producing tracts, the Police have also to guard the frontiers of Arrakan to prevent the ingress of the lighter-taxed salt from those districts.

In 1835-36 the excise manufacture of salt was first commenced by private individuals, but the continuance of the system was subsequently negatived by the Court of Directors in 1840. In 1847 the manufacture of salt under certain excise rules was again permitted, but the quantity produced is now very small and is limited to Orissa, the total quantities produced in the three years ending 1889 having been—

	Maunds.
1887	103,795
1888	244,507
1889	70,293

The reasons for this decline in local manufacture are twofold. In Northern Orissa the salt locally produced can no longer compete with Liverpool salt, which is cheaply brought to the province by steamers and sailing vessels to Balasore. Liverpool salt is vastly superior in quality to the locally-made article, and the conditions of manufacture in Orissa are so inefficient and costly that even if the salt were of much better quality it could not compete with imported salt. The manufacture was gradually declining when it became apparent to Government that its continued existence

SALT.

History of, and Trade in,

HISTORY of Bengal salt.

could be due only to evasion of the revenue which was easy in the circumstances. Proper control implied enormous expenditure, and it was therefore determined to suppress the manufacture of *panga* salt (salt made by artificial heat). This having been done, the only manufacture left is that of *kurkutch* salt (salt made by solar heat) in Southern Orissa, the conditions of such manufacture rendering control comparatively easy.

Foreign salt was first imported into Bengal in the year 1818-19. No large importations, however, occurred until the year 1835-36. At first, and until the fixed duty system was adopted, Customs dues were levied at such rates as were considered necessary to maintain the average prices of the Government sales. As the old stocks of salt manufactured at the Government agencies were exhausted in 1873-74, the consumption in the whole of Bengal, with the exception of Orissa, may be said to be now supplied by imported salt.

Production, Trade, and Duty in Lower Bengal.

Trade. 666

TRADE.—The following are the descriptions of salt commonly imported, the bulk of the importations being from Liverpool :—

Manufactured by solar evaporation.	Manufactured by boiling.
French. Red Sea. Aden. Muscat. Bombay.	Liverpool. Hamburg.

The importations of salt into Bengal were as follows, in the year 1889-90, all the salt being brought to either Calcutta or Chittagong. Calcutta is the centre of distribution for the province, except for that small tract which is more easily supplied from Chittagong :—

	Tons.
From United Kingdom	264,234
,, Germany	20,317
,, Egypt	2,746
,, Aden	33,782
,, Arabia	35,705
,, Persia	7,207
,, Madagascar	23

Liverpool salt does not penetrate further west than the western frontier of Behar, Sambhar salt meeting it somewhere in the vicinity of Zamania.

Duty

DUTY.—The duty levied on salt is now R2-8. Since 1837 the rate of duty has been frequently changed. The duties have been as follow: From 1837 to 1844 at the rate of R3-4 per maund. In October 1844 this rate was reduced to R3, in April 1847 to R2-12, and in April 1849 to R2-8. In December 1859 the duty was again raised to R3, and in March 1861 to R3-4. In January 1878 the rate was again reduced to R3-2, in August 1878 to R2-14-0, and in March 1882 to R2. On 19th January 1888 the rate was raised to R2-8, the rate now levied.

In 1815 a convention was made with the French Government, under the terms of which the East India Company agreed to supply sufficient salt for the consumption of the French Settlement of Chandernagore at prime cost from the Orissa and Midnapore agencies. The quantities of salt thus supplied free of duty varied from 4,000 to 12,000 maunds per annum. This arrangement held good until the year 1839, when the

Company entered into an engagement to pay annually a sum of R20,000 to the French Government on their agreeing to buy their salt in the open market at the same price paid for it by other inhabitants of Bengal: this payment is still continued.

HISTORY of salt in

II.—NORTHERN INDIA.

NORTHERN INDIA. 667

Historic Sketch and Regulations.

Under the Sikh Government, salt was one among forty-eight articles liable to customs, excise, town, or transit duties. The Sikh Government did not establish any systematic management for their salt revenue; no scale of duties was fixed. The cis-Indus mines were farmed out to individuals of rank and eminence. The farmer, as long as he paid in the amount of this contract, enjoyed a monopoly of the sale. He was under no restrictions as regards time, place, or price. He might sell, wholesale or retail, at the time, or at distant markets. He might regulate his proceedings by the state of prices and markets, by the briskness or sluggishness of the demand, or if he preferred he might hoard up the salt in depôts or entrepôts.

The trans-Indus mines were managed differently or rather were not managed at all. They were held by the fierce mountaineers of Kohat; no speculator would be rash enough to get up a concern there, and even the Government would have to collect its revenue with the sword; so the matter was compromised by surrendering the mines to some local chieftain on the payment of a small annual tribute, but the salt, when in transit, was liable to town duties at Peshawar and other cities.

When the Sikh Government passed under British control after the Sutlej campaign, the Lahore Council of Regency, acquiescing in arrangements proposed by the British Resident, abolished the duties on twenty-seven articles, chiefly the products of domestic industry, indigenous agriculture, or internal commerce, and also reduced the duties on nine articles. All the interior lines were swept away, and the town and transit duties were abolished. The three grand frontier lines were kept up—one along the Indus, to intercept goods coming from the west; one along the western bank of the Beas and the Sutlej for goods, chiefly British, coming from the east; and the third, running along the base of the Himalaya range, to meet the imports from Kashmír and Jummu. The province of Multan was excluded from these arrangements, which took effect during the year 1847.

To compensate for the deficiency in the revenue occasioned by these remissions and reductions of duties, amounting together to upwards of six lakhs of rupees, a moderate toll on ferries was introduced, the excise on drugs and spirituous liquors was improved by a system of licenses, and the salt revenue was reformed. A fixed duty of two rupees on the Panjábí maund was demanded on this article from the merchants at the cis-Indus mines. But these duties were levied by a new contractor, who bore the cost of management and collection and paid to the State an annual revenue of six lakhs of rupees, being an increase of two lakhs on the previous outturn. No alteration was made in the management of the trans-Indus mines.

After the annexation of the country to the British Empire in India, in March 1849, the Customs and excise duties levied in it under the reformed arrangements introduced by the Council of Regency in 1847 were taken into consideration. By those arrangements duties were still levied under twenty heads. They comprised duties of customs both of import and export; excise duties on spirituous liquors and drugs; fines, seignorage on mints, tolls on ferries; contract of the salt mines and other things.

SALT. | History of, and Trade in,

HISTORY of Northern India salt.

From the year 1850 the whole of these duties were abolished, excepting three, namely the ferry tolls, the spirit excise, and the salt excise; and one new tax was added, the stamp duty. An entire change was at the same time made in the system under which the revenue from salt was derived during the regency. Instead of letting the salt mines by contract, the Government took the management of the cis-Indus mines into its own hands, levying an excise duty at the mouth of the mine, of two rupees per Company's maund on the salt delivered, to cover all charges, and allowing the salt after this payment to pass free throughout the British dominions, subject, however, to the additional duty of 8 annas per maund levied on all salts passing the special line at Allahabad, for the protection of the Bengal duty, which was 2 rupees 8 annas per maund. The manufacture of alimentary salt in the Panjáb was at the same time prohibited.

With respect to the trans-Indus mines, it was resolved, on political and social considerations, to impose a light duty of two annas per maund at the Bahadur Kheyl mine, and four annas at the other mines, and to allow certain perquisites to the local Khuttuk chieftain with a view to reconcile the hill chiefs to the new system.

In 1851, in order to guard against the influx of this lightly-taxed salt across the Indus to the detriment of the revenue derived from the produce of the cis-Indus mines, a system of prevention, resembling that which obtained under the Sikh Government of watching the ferries of the Indus, was introduced. Under this system, parties were stationed at each ferry, on the eastern bank of the Indus, from Attok on the north to Leia on the south, controlled and watched by a roving party constantly moving up and down the line, the establishments on the upper portion of the line, between Kalabagh and Attock being superintended by one European officer, those on the lower portion, between Kalabagh and Leia, by another.

Production, Trade, and Duty in Northern India salt.

Production. 668

The revenue was formerly collected through the agency of the Inland Customs Line, which was formed in 1843-44 and which was extended at various times, as briefly indicated above, until it reached from Torbeila near Attock, on the Indus in the Panjáb, to the Mahanuddy in the Sumbulpur District of the Central Provinces. It was 2,472 miles in length, and was manned by 10,496 officers and men. In consequence of the development of railway communication 764 miles of this line were abolished in the Central Provinces in 1874-75, and eventually in 1878-79 the whole line from Leia in the Multan Division in the Panjáb to the Central Provinces was removed; there remain only 325* miles from Leia to Torbeila on the Indus maintained for the purpose of preventing the low-taxed Kohat salt from crossing into the Panjáb.

Conf. with p. 404.

EFFECT ON SUGAR. *Conf. with p. 313.*

Irrespective of this line, in February 1870 the Inland Customs Department assumed charge of the Sambhar Lake belonging conjointly to the States of Jaipur and Jodhpur, and in October 1878 they received charge of the salt sources at Pachbadra, Didwana, Phalodi, and Lúni from the Jodhpur State. In addition to the above, the salt-works at Sultanpur and Nuh in the Delhi Division are worked under the supervision of the Northern India Salt Revenue Department. There is also the charge of the salt mines in the Salt Range in the Panjáb, and a force, designated the Internal Branch, existing in Oudh, the North-West Provinces, and the Province of Behar in Bengal, for supervising the saltpetre trade, levying the duty on salt educed therefrom, and for the control of the manufacture of other saline substances, such as sulphate of soda, carbonate of soda, etc, administered by the above-mentioned Department.

* *Conf. with p. 410.*

HISTORY of Northern India salt.

The whole Inland Customs Department (now become the Department of Salt Revenue in Northern India) is administered by the Commissioner of Northern India Salt Revenue, formerly responsible to the respective Local Governments and Administrations within whose jurisdictions the Customs organization was established, but now immediately under the Government of India. The duties levied on the Customs line were on salt imported to the North of the line and sugar exported to the South of the line.

Conf. with pp. 313 et seq.

Duty. 669

DUTY.—The general rates of duty on salt were as follow :—

	Per maund on *Rajputána salt.*	
	R	a.
From 1843-44 to 1845-46	1	8 and R2,
From 1846-47 to end of 1859	2	0
From beginning of 1860 to March 1861	2	8
From March 1861 to end of December 1877 . . .	3	0
From January 1878 to July 1878	2	12
From July 1878 to 9th March 1882	2	8
From 10th March 1882 to 18th January 1888 . .	2	0
From 19th January 1888 to date	2	8

with the following exceptions, *viz.* in the Sirsa Division of the Panjáb Section, the duty was 8 annas per maund for a part of 1843-44, when it was raised to R1 per maund, at which figure it remained until 1846-47, when the duty was equalised with that generally levied. In the Saugor Division of the Central Provinces Section a rate of R1-8 per maund prevailed from the formation of the division in 1855-56, till the introduction of the general rate in 1859-60. In the Hoshangabad Division, also formed in 1855-56, R1 per maund was levied until a date in 1859-60, and thereafter R1-8 until the general rate was introduced in 1860-61.

The duty on Madras salts entering the Central Provinces was levied from the outset, until the abolition of the portion of the line across which it passed at R1-8 per maund. Bombay salt entering the Central Provinces paid from the outset at the rate of R1-8 per maund until 1st May 1874, when a mileage rate was introduced on all salt travelling by rail from Bombay, the object of the differential rate being to level up the duty to a uniform rate by the time it reached Jubbulpur, where it came in contact with Northern India salt paying R3 duty. This mileage rate was abolished in 1878, when the salt duty was made uniform throughout India (except in Bengal and Burma).

In addition to the above, all salt excavated in the Sind-Saugor Doab and the Kalabagh mines was subject to duty as follows :—

	Per maund.	
	R	a.
From 1849-50 to April 1860	2	0
From April 1860 to September 1861	2	2
From September 1861 to December 1877	3	0
From January 1878 to July 1878	2	12
From July 1878 to March 1882	2	8
From 10th March 1882 to 18th January 1888 . . .	2	0
From 19th January 1888 to date	2	8

Salt excavated at the Kohat mines on the frontier of Afghánistán was subject to a duty of 4 annas, 3 annas, and 2 annas per maund. A uniform rate of 8 annas per Sikh or Lahori maund (102lb) at all the mines was, however, ultimately introduced with effect from the 7th July 1883. Salt educed in the process of manufacturing saltpetre has always been subject to the duty prevailing in Northern India.

SALT.

History of, and Trade in,

HISTORY of Northern India salt.

Trade.
670

TRADE.—The production of salt in Northern India was as follows in the last three years (in Indian maunds of 82℔):—

	1887.	1888.	1889.
Panjáb Salt Mines	1,458,451	2,256,796	2,559,045
Kohat „ „	560,933	615,679	634,300
Mandi „ „	125,276	126,848	129,343
Sambhar Salt Source	4,512,661	5,118,542	2,916,916
Didwana „ „	458,878	584,304	477,433
Pachbhadra „	777,856	803,304	547,776
Falodi „ „	15,114	45,431	5,595
Sultanpur Salt Works	455,409	397,278	256,219
Saltpetre Refineries	51,608	51,826	50,079
TOTAL	8,416,186	10,000,008	7,576,706

There are not infrequently violent fluctuations in the quantity produced, due to climatic causes at Sambhar. The Government endeavours, as far as possible, to guard against the vicissitudes of the seasons by keeping large stocks in reserve ready for the market when the outturn falls to a low point. This practice tends to keep prices steady. Occasionally, when stocks have run abnormally low at Sambhar, and it has been found necessary to raise the price to prevent absolute exhaustion, the place of Sambhar salt is taken in the Central Provinces and Rájputána by salt from the Government works at Kharagora on the edge of the Ran of Cutch and in the western districts of the North-Western Provinces by rock-salt from the Panjáb mines.

BOMBAY.
671

III.—BOMBAY PRESIDENCY.

Historic Sketch and Regulations.

The salt produced in the Bombay Presidency is partially exported beyond the Presidency by land to Central India, including Rájputána and Malwa, the Central Provinces, the North-Western Provinces, and the Nizam's Territories, and by sea to British and Foreign Malabar in the Madras Presidency, and to Calcutta and the Straits Settlements. Under existing arrangements exports, both by land and sea, pay full excise duty in Bombay, with the exception of those to Foreign Malabar, on which a nominal charge of 3 pies per maund only, to cover cost of establishment at the salt pans, is levied, in pursuance of a trade convention concluded with Travancore and Cochin in 1865. Credit for periods varying from thirty to ninety days, according to destination, is allowed, however, in the case of full duty exports to the value of R10,000 and upwards, both by land and sea, on the exporter entering into an agreement and depositing Government securities blank endorsed of sufficient value to cover the duty. Down to 1874 exports to the Malabar Coast were free, subject to payment of customs duty on arrival, and the same system was followed temporarily in the case of exports to Calcutta from 1860 (up to which year excise duty had been levied subject to a month's credit under bonds) until 1874. In the latter year the system of pre-payment of excise duty was made general. The plan of giving credit on security of Government paper was introduced almost simultaneously, and these arrangements have since been maintained with a short interruption during 1876-77, when exports to British Malabar were again allowed for a few months free of excise duty. Exports to Calcutta and to certain fixed ports on the Malabar Coast are allowed a drawback of excise duty on actual wastage not exceeding 5 per cent.

HISTORY of Bombay salt.

Mr. Plowden remarks in his report on salt in British India, dated 24th May 1856, that "it was not until the 15th December 1837 that salt was erected, by Act No. XXVII. of that year, into a source of considerable revenue in the Presidency of Bombay; prior to that date it was one of many miscellaneous items, as under the Native Government." Under Act XXVII of 1837 the manufacture of salt in the Bombay Presidency was placed under restriction, and the produce, in common with the importations of salt by sea and land, was subjected to a duty of eight annas a maund. The object of the duty was to compensate partially the loss to the general revenue from the abolition of inland transit duties. In 1844, to set off a further loss of revenue from the abolition of the *moturpha*, or tax on trades and professions, the duty was raised from 1st September to one rupee a maund, but immediately after it was reduced in the same month, with effect from the same date, to 12 annas a maund, at which rate it continued until August 1859, when it was raised to one rupee a maund, and was again raised on 13th April 1861 to one rupee four annas, and on 20th January 1865 to one rupee and eight annas a maund By Act XXIV of 1869 the duty was raised to one rupee thirteen annas a maund, and by Act XVIII of 1877 to two rupees and eight annas a maund. By the notification of Government, dated 10th March 1882, the duty was reduced to two rupees a maund. By the notification of Government, dated 19th January 1888, the duty was raised to two rupees and eight annas a maund.

Except in Gujrat and at certain works in Goa Territory the excise system is followed, under which licenses are issued for private manufacture at places approved of by the officers of Salt Revenue, which are guarded, and from which no removals are allowed, except upon payment of the prescribed duty, or under the credit rules already referred to. In Gujrat, where all the works in British Territory are the property of Government, the monopoly system was introduced in 1873-74, and manufacture was concentrated at two places only—Kharaghora on the borders of the Ran and Balsar on the sea coast of the Surat Collectorate. At Kharaghora large crystal salt is made from brine wells under departmental supervision, bought from the manufacturers as it is ready and stored, and sold at the cost and risk of Government. At Balsar the salt is ordinary sea-salt, and Government merely fixes the price at which it is to be sold without taking it over or interfering directly with manufacturers. A small quantity of salt is issued annually free of duty from the works at Kharaghora to certain Native States and Chiefs, in pursuance of the arrangements for the establishment of the monopoly above referred to.

By a treaty concluded in 1880, the manufacture of salt in Portuguese Territory was placed under the control of the British Government for a term of twelve years. On the expiry of certain tentative arrangements made for the first three years, those in Diu have been handed back to the Portuguese authorities, and private manufacture has been stopped at all works in Damaun and Goa, except such of those in Goa as the owners were willing to work under the British excise system. Besides these, manufacture is carried on at a certain number of selected works, either by the Portuguese Government on their own account, or by contractors for the British Government, to supply the quantity (about 1,40,000 maunds) of duty-free or nominal duty salt required annually, under the terms of the treaty, for local consumption, fish-curing, and manure.

There were formerly certain works in Cambay Territory, at which the duty was shared between the Nawab, a territorial owner, and the British

SALT. **History of, and Trade in,**

HISTORY of Bombay salt.

Government as successor to the *Chouth* formerly levied by the Peshwa. This arrangement, however, ceased in 1878, when the works were finally closed by agreement with the Nawab. Besides the duty-free salt issued to certain Chiefs in Gujrat and to the Portuguese Government above noticed, the Nawab of Janjira is allowed, under a salt and customs convention concluded in 1884, to purchase and remove duty-free from the neighbouring salt-works in British Territory the quantity required for the use of himself and his subjects.

Production, Trade, and Duty in Bombay.

Production. 672

PRODUCTION.—The production has been as follows :—

	Maunds.
1887	220,501
1888	212,902
1889	270,864

Trade. 673

TRADE.—The production of salt in Bombay was as follows in the last three years (Indian maunds) :—

	1887.	1888.	1889.
Made by Government . .	2,436,413	2,681,926	1,774,668
Made on private account .	7,201 813	7,076,962	8,466,049
TOTAL .	9,638,226	9,758,888	10,240,717

In Sind salt is made exclusively by Government, all private manufacture, in a country where saline soil abounds in every district and control is practically impossible, being absolutely prohibited. The Government works are located on the Moach plain on the sea face a few miles from Karachi.

MADRAS. 674

IV.—MADRAS PRESIDENCY.

Historic Sketch and Regulations.

The question of introducing a salt monopoly in the Madras Presidency, similar to that which prevailed in Bengal, was first mooted in paragraphs 453 to 467 of a letter to the Government from the Board of Revenue, dated the 2nd of September 1799, the Board advocating its introduction. At that period the only salt-producing territories belonging to the Company were the Northern Circars, comprising the Districts of Ganjam, Vizagapatam, Rajahmundry, Masulipatam, and Guntoor, and the Jaghír comprising the district of Chingleput. Accordingly, in permanently assessing the land tax in 1802, the Company having, in the meantime, acquired the Carnatic and the Territory of the Nawab of Arcot by treaties, the exclusive right of manufacturing salt was reserved to the Government, but it was not until 1805 that the salt monopoly was established on its present footing to meet the expenses of the new judicial establishment. Previous to that year the manufacture was either farmed out, or managed by the officers of Government, upon what system the records do not clearly show. The gross revenue, according to the most authentic estimate extant, amounted, on an average of five years previous to the monopoly, to 80,000 star pagodas, or R2,80,000 exclusive of all charges, and the average sale price at the pans was 6 pagodas 6 fanams, or a little more than R21 per garce of 120 maunds. In the year preceding the monopoly the gross receipts amounted to R2,21,607 and the charges of establishment to R11,467.

The Board of Revenue, to whom the question was referred, declared the introduction of a monopoly impracticable, and advocated the imposition of a high duty on all salt, manufactured or imported, the Home manufacturers being required to take out permits and register their pans. The

HISTORY of Madras salt.

Trade.

collectors in general also preferred a fixed duty to a monopoly. All salt exported by sea, the Board thought, should be exempt from duty; and they forwarded the draft of a law framed in accordance with their views. One of the members (Mr. Falconer) dissented from these views of his colleagues; he considered the establishment of a close monopoly, as in Bengal, practicable and necessary, and recorded a minute on the policy of adopting that measure in preference to leaving the manufacture free. The Madras Government, without entering in the least into the merits of the question, upon the simple ground that the introduction of a monopoly of the salt on the part of the Company on the principles of that established in Bengal had been prescribed by the orders of the Governor General in Council, rejected the excise proposition of the Board of Revenue and directed such modifications to be introduced in the draft law as might be necessary to adapt it to the plan of a monopoly. The draft of a law for regulating the revenue derivable from salt on the plan of a monopoly was accordingly prepared and submitted to the Government of India, by whom its general principles were approved, and, ultimately, Regulation I of 1805 was passed on the 13th of September of that year, establishing the monopoly in all the provinces of the Presidency, except Malabar and Canara, to which it was afterwards extended by Regulation II of 1807. Under the monopoly system the private manufacturers, who occasionally received advances from Government, and who were paid at different rates varying with the locality, were prevented from selling the salt to any but Government. The salt was resold by Government at a price calculated so as to include the purchase money paid to the manufacturers and the expense of storage, transport, etc., reduced to an average for the whole Presidency. This price was independent of duty, and at first was two annas, but by Act XVIII of 1877 it was fixed at three annas per maund, and under the latest enactment, Act XII of 1882, three annas per maund is now the *minimum* rate at which monopoly salt can be sold *ex* duty. In the Eastern maritime districts these arrangements prevailed without modification up to 1881-82. In 1882-83 the Excise system was extended to the group of factories near Tuticorin. Under this system manufacture, storage, and sales are carried on under Government supervision on private account, subject only to the payment of the duty on removal from store *plus* a cess to cover interest on the capital cost of the works executed by Government under the old system, and made over to the licensees on the introduction of excise. The system has gradually been extended, so that out of the forty-six factories on the East Coast only six are worked under the Monopoly, and the remaining forty under the Excise system.

In Malabar no salt is manufactured. The district is supplied by imports from Bombay, Goa, etc. These imports were formerly made by Government, but Government has withdrawn from the trade since 1877, and now only charges import duty. In South Canara the excise system was substituted for the monopoly system in 1877, but the local manufacture of salt was abolished in 1883-84, as the salt produced was of inferior quality, and the pans were difficult and expensive to guard. Even when salt was manufactured locally, the supply fell short of the demand. The deficiency was made good by importations from the Bombay Presidency, as in Malabar. When these importations were made by the Government, or down to 1877-78 inclusive, the imported salt was sold at a price equal to the Madras duty, *i.e.* the general gross selling price (less three annas a maund) *plus* the supposed cost of the salt. Private imports on payment of duty were also permitted, but the trade was almost *nil* until Government withdrew from importation, as the cost of salt sold by Government had been improperly fixed so low as to forbid competition. Since

HISTORY of Madras salt.

Government ceased to import the trade has greatly increased. Act XVI of 1879, for regulating the transport of salt, renders its conveyance in any vessel, other than a vessel of not less than 300 tons burden, illegal on the West Coast within certain limits, unless under cover of passes which practically confine the privilege to duty-paid salt. Certain quantities of salt are annually supplied to the French Settlements in the Madras Presidency, under an old convention with the French Government, at cost price. A further quantity, generally less than one thousand maunds annually, is also supplied at prime cost from the Canara District for the Amindivi Islands. The greater part of Mysore and the Nizam's territory and of the southern and eastern parts of the Central Provinces are supplied with Madras salt.

The *general* selling price of salt in the Madras Presidency has been as follows, *viz.*—

9⅓ annas a maund of 82$\frac{2}{7}$℔ from 1805 to November 1809.
14 „ „ „ from November 1809 to a date in 1820.
9⅓ „ „ „ from 1820 to June 1828.
14 „ „ „ from June 1828 to 31st March 1844.
1 rupee 8 annas a maund, reduced in the same year to 1 rupee a maund, from April 1844 to July 1859.
1 rupee 2 annas a maund, from August 1859 to 2nd April 1861.
1 „ 6 „ „ from 3rd April to 3rd June 1861.
1 „ 8 „ „ from 24th June 1861 to a date in 1865-66.
1 rupee 11 annas a maund, from a date in 1865-66 to October 1869.
2 rupees a maund, from October 1869 to 27th December 1877.
2 „ 11 annas a maund, from 28th December 1877 to 9th March 1882.
2 rupees 3 annas a maund, from 10th March 1882 to 18th January 1888.
2 rupees 11 annas a maund, from 19th January 1888 to date.

These rates are inclusive of a cost price of two annas, or latterly of three annas per maund, and were formerly abated by 5 per cent. for purchase, without measurement or weighment, of a heap of 1,200 maunds.

Trade. 675 TRADE.—The production of salt in the Madras Presidency has been (in Indian maunds) in the last three years:

	1887.	1888.	1889.
Made by Government . .	868,447	927,312	1,220,969
Made on private account . .	7,990,341	7,976,312	8,196,729

V.—BURMA.

BURMA. 676 In Burma until January 1888 the duty levied on salt was only three annas a maund: it was in that year increased to one rupee. There is a considerable local manufacture by artificial heat, in most of the littoral districts of Arakan, Pegu, and Tenasserim, and salt is also obtained from brine wells in Upper Burma; but in this portion of the province such production is of small importance. What the exact quantity produced may be it is impossible to say, for the collection of the revenue is based upon an estimated production per pan or pot, and the estimates, as may easily be understood, are probably far from the truth. The local authorities give the following figures:—

	Maunds.
1887	337,646
1888	429,116
1889	414,119

No salt is made by Government in Burma.

HISTORY of Burma salt. Trade.

The local production was greatly stimulated by the increase of the duty in 1888, that increase falling upon imported salt. But the tax on locally-made salt having been raised subsequently to a level supposed to approximate to the rate of duty on imported salt, it may be imagined that the stimulus has ceased to operate. The quantity manufactured in the province is, however, entirely insufficient to supply the needs of the people, three-fourths or four-fifths of the consumption is met by imported salt from Liverpool and Germany.

TRADE IN, AND PRODUCTION OF, SALT IN INDIA.

TRADE. 677

The total quantity of salt produced in India in 1889 was just over 28 millions of maunds, including a small quantity produced in Gwalior. This quantity was supplemented by about 11,219,000 maunds of imported salt, the total quantity produced and imported amounting therefore to 39,225,276 maunds, about two-thirds of the whole being salt produced in India. Importation is practically limited to Bengal and Burma, which are the only two provinces where consumption is in the main met by imported supplies. The very small quantities imported into Bombay, Sind, and Madras are table salt for European consumption and rock salt from Muscat, supposed by orthodox Hindus to be specially pure and used by them in religious ceremonials. A small quantity is imported from Tibet into the Panjáb, the North-Western Provinces, and Assam, but though this salt is free of duty it is unable to compete with Indian-taxed salt except in the inner ranges of the Himálayas. It is mainly brought into Kumáon and Garhwál by Bhutia traders who import salt and borax and take away grain and other commodities. The total imports in the last three official years were as follow :—

Imports. 678

Imports of Salt by Sea and Land into British India.

Provinces.	1887-88.		1888-89.		1889-90.	
	Mds.	R	Mds.	R	Mds.	R
Bengal	9,563,082	63,32,741	10,159,805	81,01,819	9,909,474	76,71,316
Bombay	3,974	26,051	29,563	23,183	6,043	25,170
Sind	697	2,403	599	2,415	599	2,149
Madras	1,388	6,479	1,062	4,943	1,470	7,481
Burma	1,971,025	15,89,951	672,927	690,483	1,265,997	12,40,777
Northern India	33,330	1,34,467	52,280	2,14,267	34,623	1,39,817
Assam	785	3,885	331	1,711	513	2,517
TOTAL	11,574,281	80,95,977	10,916,567	90,38,821	11,218,719	90,89,227

In the last of these years the imports came from the undermentioned countries :—

	Tons.
United Kingdom	285,767
Germany	45,439
Egypt	2,746
Aden	33,782
Arabia	35,736
Persia	7,312
Other Countries	26

SALT.

Trade in, and Production of,

TRADE. in salt.

The following statement may be offered of the Imports of Foreign Salt placed alongside of Indian production, the figures in both cases being in maunds of 82lb :—

	1887-88.	1888-89.	1889-90.
Imports	11,574,281	10,916,567	11,218,719
Production	27,586,696	29,560,806	28,006,557
TOTAL .	39,160,977	40,477,373	39,225,276

These imports, it will be seen by the table already given, are distributed in the provinces, but they go practically to Bengal and Burma. In the table given at the end of this article a classification will be found of the sources of local salt, grouped under the various provinces in which that salt is ultimately consumed. It need scarcely be added that the difference that may appear in the amounts shown in the tables given in this review is due to the fact of the one set of figures being importation and production, and the other the actual consumption, or rather the amounts that paid duty on going into consumption.

CONSUMPTION. 679

The subjoined table shows the progress in consumption in recent years. In the five years ending with 1877-78, the increase, compared with the consumption in the preceding period, was very small, hardly exceeding an annual average of one per cent. In the next five years (ending 1882-83), the rate of increase was nearly 3 per cent. yearly. In the last five years (ending 1887-88), consumption again increased, the average annual rate being over 3 per cent.—a rate much in excess of the rate of increase in the population. The great advance in consumption in this last period must be attributed mainly to (1) the reduction of duty in March 1882 to the moderate rate of R2; (2) the reduction in cost of transit effected by the substitution of railways for pack-bullocks and carts; (3) the energetic reforms in the last five years in the administration of the salt revenue in the Madras Presidency, which practically stopped the illicit manufacture of salt there :—

YEARS.	Average annual quantity.	Average annual duty.
	Maunds.	R
1868-69 to 1872-73	22,973,432	5,75,19,725
1873-74 to 1877-78	24,183,707	5,97,97,641
1878-79 to 1882-83	27,790,576	6,59,39,847
1883-84 to 1887-88	32,475,600	6,15,68,427
1888-89	31,394,857	7,58,82,438
1889-90	33,086,400	7,93,06,523

Salt in India. (*G. Watt.*)

SALT.

CONSUMPTION of salt in India.

Statement of Consumption of Salt of all kinds in the various Provinces of India, and the Duty paid thereon.

Provinces and Sources of their Supply.	1887-88.		1888-89.		1889-90.	
	Quantity.	Duty.	Quantity.	Duty.	Quantity.	Duty.
	Mds.	R	Mds.	R	Mds.	R
Bengal.						
Foreign Imports . . .	9,068,649	1,83,28,906	9,181,775	2,25,02,373	8,940,088	2,23,50,336
Imports from Bombay and Madras	554,965	11,10,330	499,679	12,18,978	530,104	13,35,260
Excised Salt	182,860	3,82,854	128,835	3,22,087	79,082	1,97,705
Total .	9,806,474	1,98,22,090	9,810,289	2,40,43,438	9,549,274	2,38,73,301
Madras.						
Imports from other Provinces; paying duty at Madras .	1,881	3,882	1,136	2,859	1,475	3,671
" " Bombay .	1,039,305	21,60,096	1,079,953	26,99,882	1,184,738	29,61,845
" " Goa .	51,309	1,06,027	40,154	1,00,385	34,248	85,620
Local Production . . .	905,396	18,90,719	747,472	18,69,666	868,696	21,73,020
Excised Salt	5,706,634	1,19,06,247	5,883,068	1,47,04,191	5,933,367	1,48,33,413
Total ..	7,704,528	1,60,66,971	7,751,783	1,93,76,983	8,022,524	2,00,57,569
Bombay.						
Foreign and Indian Imports and removals from Stock less exports to Calcutta and Madras	5,848,728	1,23,16,282	5,529,986	1,38,27,150	5,846,108	1,46,18,081
Imports from Goa, Damaun, and Diu	211,185	4,54,325	187,063	4,68,637	249,563	6,23,913
Total .	6,059,913	1,27,70,609	5,717,049	1,42,95,787	6,095,671	1,52,41,994
Sind.						
Imports by Sea . . .	569	1,161	576	1,439	588	1,471
Excised Salt	233,934	4,89,033	227,946	5,65,553	243,404	6,05,607
Total .	234,503	4,90,194	228,522	5,66,992	243,992	6,07,078
Northern India.						
(*N.-W. Provinces, Oudh, Panjab, Rajputana, Central India) etc.*						
All classes consumed except the next section . . .	6,683,456	1,38,99,537	6,799,302	1,60,96,098	7,157,033	1,78,92,584
Mundi and Kohat Salt . .	697,731	2,56,675	773,759	2,93,530	828,137	3,14,347
Total .	7,381,187	1,41,56,212	7,573,061	1,72,89,628	7,985,170	1,82,06,931
Burma.						
Imports by Sea . . .	1,031,222	3,83,564	314,153	2,64,201	1,189,769	11,89,769
Excised Salt	Not returned.	26,563	Not returned.	45,409	Not returned.	1,29,881
Total .	1,931,222	4,10,127	314,153	3,09,610	1,189,769	13,19,650
Grand Total of Salt consumed in India . . .	33,117,827	6,37,16,203	31,394,857	7,58,82,438	33,086,400	7,93,06,523
Net Revenue from Salt .	—	6,18,81,597	—	7,39,62,027	—	7,73,72,623

Note.—This table omits from consideration the miscellaneous receipts which on the average come to about R25,00,000 per annum.

SALT.

Quantities of Salt sold.

PRICES of salt.
680

Statement showing the quantities of Salt sold per rupee in seers and decimals of a seer of 80 tolas.

Provinces and Divisions.	Average of the five years from 1861 to 1865.	1866 to 1870.	1871 to 1875.	1876 to 1880.	1881 to 1885.	1886.	1887.	1888.	1889.
Burma—									
Tenasserim	42·6	35·64	31·42	28·22	31·66	20·37	19·94	18·28	18·43
Pegu (deltaic)	24·21	21·43	24·82	23·46	23·38	28·27	24·21	17·47	14·56
Pegu (inland)	30 91	31·77	24·66	22·13	21·62	21 04	19·63	15·09	13·83
Upper Burma	...	...	...	...	...	...	21·34	21·39	16·73
Arakan	52·29	44·36	35·40	34·90	32·79	37·60	34·55	30·90	23·39
Assam—									
Surma	8·19	7·61	8·31	8·51	10·99	11·42	11·55	9·84	9·55
Brahmaputra	7·33	6·82	7·10	7·38	9·82	10·45	10·21	8·91	8·72
Bengal—									
Eastern	8·26	8·10	8·05	8 36	11·02	11·28	11·29	9·16	9·51
Deltaic	9·01	8·52	8·57	9·11	11·78	12 30	12·29	9·55	9 64
Central	8·55	8·24	8·21	8·88	11·50	11·97	12 14	10·02	9·74
Northern	7·80	7·63	7·38	8·01	11·08	11·93	11·06	9·87	9·48
Orissa	10·07	8·83	9·13	10·49	12·27	12·85	12·79	9·79	10·23
Chutia-Nagpur	5·98	6·35	6·64	7·27	9·26	9·79	9·88	8 07	8·01
Behar, South	8·31	8·31	7·75	8·46	11·16	11·84	11·90	10·01	9·63
Behar, North	7 83	7·68	7·35	8·03	10·69	11·49	11·74	9·73	9·88
N.-W. Provinces—									
Eastern	6·8	6·9	7·27	8·2	9·7	10·88	11·48	9·93	9·76
Central	6·76	7·01	9·06	9·51	11·31	12·1	12·68	11·08	10·91
Western	7 19	7·54	9·34	10·00	11·69	12·54	12·98	11·45	11·36
Sub-montane	7·22	6·95	8 14	8·85	10·92	11·78	12·14	10·69	10·62
Oudh—									
Southern	6·76	6·25	7·41	8·39	9·84	11·04	11·77	10·5	10·26
Northern	6 75	6·81	7·52	8·26	10·37	11·39	12·00	10·45	10·41
Rajputana—									
Eastern	...	...	27·52	19·45	12·76	12·94	13·26	11·68	11·41
Western	...	...	58·95	38·00	14·93	18 53	18·87	16 37	15·43
Central India	...	...	17·58	12·18	11·23	12·15	12·27	10·85	10·82
Panjab—									
Central	8·49	8·01	8·34	9·51	12·01	12·59	12·84	11·31	10·93
Southern	8·00	7·99	9·98	8·70	11·54	12·58	13·00	11·36	10·54
Sub-montane	9·94	9·06	9·44	10·33	11·67	14·27	14 78	12·92	12·67
North-Western	27·90	23·82	24·68	21·68	22·63	25·59	23·81	21·88	23·74
Western	10 79	10·56	10·66	19·42	23·00	23·30	20·16	19·22	20·17
Sind and Baluchistan	35·58	37·32	40·57	22·09	13·67	13·89	12·99	11·36	11·23
Bombay—									
Konkan	17· 1	15·80	15·88	12·52	12·44	12·40	12·81	10·68	11·12.
Deccan	17·38	14·52	13·76	11·67	11·73	12·40	12·49	10·80	11·04
Khandesh	14·71	13 30	13·49	12·31	13·62	13·89	13·92	11·64	11·78
Guzerát	19 85	16·04	17·44	15·59	13·99	15·12	15·05	12·34	12·30
Kattywar	...	...	41·96	59·42	40·63	37·50	40 00	40·00	40·00
Central Provinces—									
Western	7·41	6·77	8·09	9·55	10·06	10 84	10·96	9·38	9·30
Central	5·53	5·59	7·19	8·35	9·75	10·29	10·50	9·04	9·07
Eastern	7·72	3 71	6·85	8·32	8·98	10·34	9·55	7·07	8·75
Berar	10·45	5·03	8 28	9·43	10·56	11·03	10·96	9·62	9·55
Nizam's Territory	14·02	11·01	10·07	9·69	10·21	10·37	10·09	8·87	8·99
Madras—									
Malabar Coast	19·66	17·09	15·95	12·52	13·58	15·68	14·97	12·65	12·43
South, Central	17·22	14·51	15·81	12·72	14·66	15·00	14·35	12·65	12·70
Central	13·56	12·76	16·81	14·07	14·42	13·73	13·36	11·94	11·68
East Coast, North	20·33	17·31	16·23	12·99	13·45	13·37	13·00	10·70	12·01
East Coast, Central	20·87	17·27	17·27	14·60	14·38	14·47	14·50	12·50	12·76
East Coast, South	20·72	17·33	17·07	14·58	15·43	14·72	14 10	12·16	12·06
Southern	19·78	16·23	16·45	14·93	16·20	16·06	15·87	13·84	14·07
Mysore	14·15	12·65	11·92	9·73	10·27	10·80	10·36	8·84	8·98
Coorg	15·50	11·65	10 54	8·40	9·62	11·94	11·28	9·33	9·31

NOTE.—It will be observed that when the number of seers procurable for a rupee is shown to be greater, salt has grown cheaper, and the reverse when a fewer number is obtainable.

(*John Watt.*) [*499-501.*

SALTPETRE, *Man. Geol. Ind. Vol. III. (Econ. Geology by Ball),* 681

This term is used to designate various salts found in a natural state in many parts of the world, chiefly South America, Spain, Persia, Hungary, but more especially India. Chili saltpetre has sodium as the base; lime-saltpetre is often found on the walls of stables, while in various districts of India, potassium saltpetre, or nitre, is met with, either as an efflorescence upon the soil, or disseminated through the superficial stratum itself. It is also found in plants, such as tobacco, sunflower, borage, etc., in certain porous rocks, in spring and rain water, and is produced artificially as the result of the process called "nitrification." In addition, caves in Ceylon, Teneriffe, Kentucky, etc., which are resorted to by birds and other animals, are found to contain saltpetre, the birds, etc., providing the necessary organic matter by which the substance is formed when the other essential conditions are present.

Saltpetre occurs in thin, white, brittle, sub-transparent crusts or silky tufts, composed of delicate hexagonal or rhomboidal crystals, which have a saline taste. The salt is anhydrous, non-deliquescent, and sparingly soluble in cold, but readily so in hot water.

Wagner's "*Chemical Technology*" (edition by **Crookes**) expresses very briefly and pointedly the theory of the formation of saltpetre, and its distribution over the globe. "Although native saltpetre is met with under a variety of conditions, they all agree in this particular, that the salt is formed under the influence of organic matter. As already stated, the salt covers the soil, forming an efflorescence, which increases in abundance, and which, if removed, has its place supplied in a short time. In this manner saltpetre, or nitre as it is sometimes called, is obtained from the slimy mud deposited by the inundations of the Ganges, and in Spain from the lixiviation of the soil, which can be afterwards devoted to the raising of corn, or arranged in saltpetre beds for the regular production of the salt. The chief and main condition of the formation of saltpetre, which succeeds equally in open fields exposed to the strong sunlight, under the shade of trees in forests, or in caverns, is the presence of organic matter, *viz.* humus, inducing the nitre formation by its slow combustion; the collateral conditions are dry air, little or no rain, and the presence in the soil of a weathered crystalline rock containing felspar, the potassa of which favours the formation of the nitrate of that base. All the known localities where the formation of nitre takes place naturally, including the soil of Tacunga, formed by the weathering of trachyte and tufstone, are provided with felspar. The nitric acid is due to the slow combustion of nitrogenous organic matter present in the humus, it having been proved that the nitric acid constantly formed in the air in enormously large quantities by the action of electricity and ozone, as evidenced by the investigations of **MM. Boussingault, Millon, Zabelin, Schoubein, Frochde, Bottger,** and **Meissner,** has nothing whatever to do with the formation of nitre in the soil, a fact also supported by **Dr. Goppelsroder's** discovery of the presence of a small quantity of nitrous acid in native saltpetre."

Saltpetre, Nitre, OR Potassium nitrate. 682

NITRE, *Eng. & Fr.*; SALPETER, *Germ & Dut.*; NITRO, *It & Sp.* [& *Port*; SENITRA, *Russ.*

Vern.—*Suriakhar, shórá,* or *shorah, shora kalmi* (refined). HIND. GUZ.; *Sórá,* BENG.; *Shóra-mitha,* MAR.; *Sóro-khar,* GUZ.; *Potti-luppu,* or *potluppu,* TAM.; *Petluppu, shúrá-káram,* TEL.; *Petluppu,* KAN.; *Veti-uppa, sandawa,* MALAY.; *Yán-zin,* BURM.; *Vedi lúnú, pot-lunu,* SING.; *Yavakshra,* (?) SANS.; *Ubkir, abqar, malh-i-barut,* ARAB.; *Shora, shórah,* PERS.

References.—*Pharm. Ind., 315; Ainslie, Mat Ind., I, 375; U. C. Dutt, Mat. Med. Hind., 89; K. L. De, Ind. Drugs, 160; Baden-Powell, Pb. Pr., I, 79; Buck, Dyes, and Tans, N.-W. P., 42, 78; Liotard, Dyes, 76; Dana, System Mineralogy, 592; Spons', Encyclop, I., 273; Agri.-Hort. Soc. Trans., VII., 186-187; Journ., XII., 107 (old series); Stevenson, Jour. Asiatic Soc. Bengal, II, (1833), 23; Wagner, Chem. Tech. (Ed. Crookes), 134; O'Conor Review of Trade of India; Milburn's Oriental Commerce; Proc. Hon'ble East India Company, 1790 to 1820; Report of the Committee of Warehouses (on a memo. from the manufacturers of Gunpowder), date 1793.*

Occurrence.—This subject has already been briefly indicated in the introductory note above, and will be found to be still further elaborated OCCURRENCE. 627

below in the chapters devoted to Manufacture and Trade. Briefly stated, it may be said the saltpetre of Indian commerce is derived from Behar and, to a very much less extent, from several districts of the North-West Provinces, the Panjáb, Bombay, Madras, and Burma.

HISTORY. 684

History.—Previous to the invention of gunpowder and the resulting demand for nitre—one of the most important constituents of that article—little attention seems to have been given to this salt by the Natives of India. So much was this the case that in Sanskrit literature, it may be said, there is no specific name for it. **U. C Dutt** (*Materia Medica of the Hindus, 89*) writing on this subject, says, "Nitre was unknown to the ancient Hindus. There is no recognised name for it in Sanskrit. The **Bhávaprakása** mentions *Suvarchiká* as a variety of *Sarjika* or barilla, and gives *Sorá* as its vernacular equivalent. But *Suvarchiká*, according to the standard lexicons, is a synonym of *Sarjika* and not a separate article. Some recent Sanskrit formulæ for the preparation of mineral acids containing nitre mention this salt under the name of *Soraka*. This word, however, is not met with in any Sanskrit dictionary, and is evidently Sanskritized from the vernacular *Sorá*, a term of foreign origin. The manufacture of nitre was, therefore, most probably introduced into India after the adoption of gunpowder as a material of warfare. It is necessary to observe here that many writers have erroneously translated nitre into the Sanskrit term *Yavak-shára*. This last, however, is not a nitrate but an impure carbonate of potash, obtained by reducing to ashes the spikes of the barley." The lack of interest shown by the Native population in saltpetre was, to a certain extent, exemplified by the European traders, for we read in **Milburn's** "*Oriental Commerce*" (published in 1813) that "we have had no account of the manner in which it is prepared in the East Indies, no person on the spot having taken particular notice of the manufacture." For long the trade was carried on as a monopoly by the East India Company, who were bound under special restrictions and regulations which prevented an extended trade being carried on in this article. For over a century the Company were under obligation to supply to the British Government saltpetre to the amount of 500 tons (or 8,000 bags) annually, before being allowed to offer any for public sale. The trade in saltpetre has always been subject to extreme fluctuations, the course of political events, if likely to terminate in war, having a direct effect on it. In 1755 the quantity offered for sale was 14,747 bags, the whole of which, under the prospect of war with France (which took place early in 1756), was disposed of. From 1775, when disturbances with America commenced (which eventually produced war with France, Spain, and Holland), to the year 1783, there were sold 120,154 bags. During the greater part of that time "the putting up price" was £4 per cwt. In 1791 on account of the disturbed state of the continent, and especially the unsettled condition of the Provinces of Holland, the manufacturers of England were possessed of orders from abroad for considerable supplies of powder, and they wanted a larger quantity of nitre than the East India Company, consistently with their duty to the British Government, could furnish.

The manufacturers thereupon applied to, and obtained from, the Lords of the Privy Council, a license to import nitre from the continent. At the same time the manufacturers complained of the monopoly in nitre (held by the East India Company), and represented to their Lordships that prices were exacted for this raw material that were highly injurious to the gunpowder industry. In consequence, an Act was passed "by which, and from September 1791, the Company are required to put up at each of their half-yearly sales, such a quantity of saltpetre as shall be equal to 5,000 bags more than what shall appear to have been disposed of, upon

HISTORY.

an average of the four preceding sales at a price of £1-11 per cwt. or £31 a ton in peace, or £2 per cwt., or £40 a ton, in war. The Company are also required by the same Act to supply 500 tons annually for His Majesty's service, at the like specified rates." If the Company failed to carry out these conditions, they were to forfeit their monopoly (*Milburn*). Under the above Act the East India Company continued to export nitre to the extent of 60,000 bags (about 4,000 tons) annually; 500 tons of which were reserved for the British Government. Although allowed in the year 1808 an advance in price to £50 a ton by the Government, it appears that from the risky nature of the trade, the East India Company did not consider it advisable to export more than about 4,000 tons annually. This quantity, too, was directed by the Board to be exported as 'dead weight' for vessels returning light. Thus, in a letter, written at the beginning of this century, to their Representatives in India, instructions were given to "form proper assorted cargoes for eight or nine regular ships, a very considerable proportion of dead weight will be required. This may be effected by a provision of saltpetre, to the extent of 60,000 or 70,000 bags, and the deficiency may be made up with sugar." With the exception of occasional extra demands, this annual 'regular investment of saltpetre' continued for a number of years, the Company limiting their venture to little more than their contract supply to the English Government and standing orders. Sugar and saltpetre were considered so bad investments that "they must necessarily constitute a drawback upon the profit on the raw silk and on other fine goods, which cannot be conveyed to England without the aid of some kind of dead weight." In the Report of the Committee of Warehouses (published in 1793) it is shown that while the East India Company were required by the British Government to sell saltpetre at a dead loss—the price paid for the article having fallen from £4-0-6 (in 1783) to £1-18-6 (in 1789)—the manufacturers of gunpowder were making large fortunes since the price of that article had (during the period named) declined only from £4-7-6 (per barrel of 100℔) to £3-12-6. This represented a difference in value of the raw materials of 65 per cent. and of only 18 per cent. in the price of powder.

This state of affairs continued until discoveries in Europe and America established new, cheaper, and more regular sources of supply. Although the artificial manufacture of saltpetre may be said to have been the chief element that upset the old Indian trade, the discussion of the processes pursued would be beyond the scope of this article.

PURIFICATION AND MANUFACTURE OF SALTPETRE.

MANUFACTURE. 685 *Conf. with pp. 405-406.*

Ball (*Econ. Geology*) states that more than two-thirds of the saltpetre which is exported from Calcutta is derived from Tirhut, Saran, and Champaran in Behar. The districts of Cawnpore, Ghazipur, Allahabad, and Benares, however, also contribute, and so does the Pánjab to a limited extent. About the year 1868 the manufacture of saltpetre in the Madura district, Madras, was a monopoly in the hands of a European firm who were under contract to supply the Government with a fixed amount annually. Latterly this trade, **Ball** adds, was found not remunerative, and accordingly was discontinued.

Bengal. 686

Bengal (Behar).—As this province furnishes by far the largest supply of saltpetre, and is accordingly the locality where the industry may be best seen, it may be selected as a type of what is carried on in other parts of India, although to a much less extent than in Behar. The climate best suited for the production of nitre, is where dry weather follows the rains, and thus, by evaporation, allows the salt to effloresce on the surface. So

MANUFACTURE in Bengal.

long ago as 1833, the manufacture of this salt at Tirhut was fully explained (in a very able paper, by **Mr. J. Stevenson**, Superintendent of the Honourable Company's Saltpetre Factories, Behar, which appeared in the *Journal of the Asiatic Society*) and as a more recent author does not deal with the subject in such detail, no further apology need be offered for reproducing in this place the main facts brought out by **Stevenson**: "The soil of Tirhut almost everywhere contains a large proportion of saline matter, such as nitrate of potass (saltpetre), nitrate of lime, sulphate and muriate of soda, etc.;* but in general (the sulphate of soda is most abundant. The saltpetre (as well as the other salts) lies in patches as it were, some parts being more productive than others, according as carbonate of lime and sand alternately predominate. By analyzing the different soils I have found those places most productive of nitre to contain a redundancy of the former; and, on the contrary, where the soil was unproductive, I found a redundancy of the latter substance. I am, therefore, naturally led to the conclusion, that carbonate of lime is one of the principal agents in the formation of this article. This will also account for the district of Tirhut being more productive of nitre than any other place in India, for almost half of its soil is calcareous; an average sample of it, collected from various places where saltpetre abounds, and carefully analyzed, gave me the composition as follows, 100 parts being operated upon:—

Matter insoluble in the three mineral acids, Silex		50·0
Matter soluble in ditto	Carbonate of Lime	44·3
Matter soluble in water	Sulphate of Soda	2·7
	Muriate of do.	1·4
	Nitrate of Lime	0·9
	Nitrate of Potass...	0·7
		100·0

This analysis does not agree with **Dr. John Davy's**, but be it remembered that that scientific gentleman operated upon saltpetre earth from the factories, which, of course, contains more saline matter than the general soil. In the month of November the *loneahs*, or native manufacturers of saltpetre, commence their operations, by scraping the surface off from old mud heaps, mud buildings, waste grounds, *etc.*, where the saltpetre has developed itself in a thin white efflorescence, resembling frost rind. This saline earth being collected at the factories, the operator first subjects it to the processes of solution and filtration. This is effected by a large mud filter, lined on the inside with stiff clay. It is a round hollow basin, in shape resembling the top of a well, from six to eight inches in diameter. A false bottom is formed of pieces of bamboo laid close, and resting upon pieces of brick. This leaves an empty space of a few inches above the solid bottom for an outlet to the filtered liquor. Over these bamboos a covering of strong close-wrought grass-mats is laid, which completes this simple form of filter. The operation then proceeds with the process by spreading over the mats a thin layer of vegetable ashes, generally from the indigo plant, upon which the earth to be subjected to the filtering process is laid, and trodden down level, and to the desired solidity, by the operator's feet. This requires great attention on the part of the man who performs it: for if too solid, the water will pass through too slow; on the contrary, if too soft, the water will pass through too quick; for the solution of the saline matter to take place, and the full products would not be obtained. After

* I have not been able to ascertain whether the *sajji mati* (native carbonate of soda) is found in this district; as far as my own observations have extended, it does not form a part of the composition of the soil. I also could not detect any alumina, though it is likely some parts may contain it.

MANUFACTURE in Bengal.

this point has been adjusted, water is poured gently upon the earth to the depth of four or fives inches, according to the size of the filter and quantity of earth used (one of six feet diameter will filter 20 maunds of earth). The whole is then suffered to remain tranquil for several hours, during which time the water gradually passes through the earth, dissolving the saline matter in its passage, and filtering through the mats, drops into the empty space between the solid and false bottoms, and is conveyed by means of a spout of bamboo or a hollow tile, into an earthen receiver, made large enough to hold the full quantity of filtered liquor, and half sunk in the ground for the purpose. The saltpetre liquor thus obtained is more or less coloured with oxide of iron and decomposed vegetable matter. Its specific gravity also varies with the quality of the earth operated upon. An average from a great number of filters gave me 1·120. The second process is to evaporate the saltpetre liquor to a crystallizing state, which is effected in earthen pots fixed in two rows, over an oblong cavity dug in the ground, the interstices between the pots being filled up with clay. An aperture at one end of the cavity serves for an egress to the smoke; another, at the opposite end, is used for the introduction of fuel, which is generally dry fallen leaves, gathered from the *âm topes* (mango groves): such are the simple materials used in this part of the manufacture. The boiling is continued till the liquor is evaporated to the crystallizing point, which is ascertained by the operator taking, from time to time, a small portion of the liquor from the pots, and setting it aside to cool in small earthen dishes, like a common saucer. After the liquor has cooled, and the crystals formed agreeable to the practice of the operator, the fire is stayed, and the liquor removed to large shallow earthen dishes (which are used instead of crystallizing coolers), placed in rows, and sunk to the brim in soft earth. At the end of about 30 hours the process of crystallization is finished. The crystals of saltpetre are taken out of the coolers, and put into baskets to drain, after which they are removed to the store-house, ready for sale. During the operation of boiling, it occasionally happens that too much heat has been used, and the pots are in danger of boiling over. To prevent this the operator has a very simple remedy, which our more scientific operators might not be ashamed to take a lesson from. A bunch of dry jungle grass is fixed at a right angle to the end of a stick; this is dipped into the liquor, and held up over the pot, and the liquor, which it had absorbed, falls down in a shower (cooled by the air) into the vessel it had been taken from. The temperature being thus reduced, the evaporation proceeds more steadily and the accidental boiling over is prevented. The mother liquor remaining after the crystals of saltpetre have been removed, is returned to the evaporating pots, and mixed with a fresh portion of the liquor from the filters, for a second boiling, and crystallization. The extraneous salts, such as sulphate and muriate of soda, which the filtered liquor from the earth always contains, are partly found at the bottom of the pots (the muriate of soda in particular), and partly in the mother liquor, remaining after the process of crystallization. But to separate them more effectually, the manufacturer passes the liquor from the boilers through a piece of coarse cloth, placed in a basket; and when the liquor has drained through, the greater part of the extraneous salts are found on the cloth. To do this effectually, it is necessary that the liquor should be at the boiling-point, otherwise the saltpetre liquor would not leave the sulphates and muriates, but would form an anhydrous mass. The muriate of soda, or common salt, is rendered more pure by subsequent boiling. It is then called by the natives *pakwa nimak*, and is sold in the bazars as an article for culinary purposes. The remaining extraneous salts—sulphate of soda, nitrate of lime, etc.—are returned to the earth,

MANUFACTURE in Bengal.

to undergo a change by decomposition against another season. The *nitrate of lime* is decomposed by the *carbonate* of potass, which the vegetable ashes, used in the process, contain. When solutions of these salts come in contact with each other, a mutual decomposition takes place. The *nitric acid* of the lime combines with the potass, and the *carbonic acid* of the *potass* combines with the lime. Thus two new salts are formed, *viz. nitrate of potass* (saltpetre), and *carbonate of lime*. In this manner the old earth, which has already produced saltpetre is, regenerated, and rendered productive against other seasons. The native manufacturers are aware of this fact, but, not being able to account for it on scientific principles, they say that *saltpetre generates* or *developes saltpetre*; but I dare say that most scientific men will concur with me that the above idea of the natives is next to a physical impossibility. Owing to the porous nature of the earthen crystallizing vessels, a part of the saltpetre liquor oozes through the bottom, and is absorbed by the earth on which the utensils are placed; occasionally they are broken, and the contents, of course, falls into the earth below. This earth is again subjected to the process of filtration, and the practice of the manufacturer, in order to obtain what had been wasted in the above manner. Thus the *loneahs* proceed from season to season, without the least deviation or alteration in their manufacture. No persuasion, however reasonable, by way of improvement, will cause them to alter the plans which their forefathers had in practice; and it is probable that the methods used at present were the same three thousand years ago. The saltpetre obtained in the above manner, which I have attempted to describe, is a very impure article, termed by the natives *dhoah*, and is sold at the rate of from two to three rupees a maund. It generally contains from 45 to 70 per cent. of pure nitre. The following analysis was tried from an average of several hundred maunds of what was stated to be of good quality, and brought three rupees eight annas per maund: 100 grains operated upon:—

Insoluble matter, sand, and mud	5·0
Sulphate of soda	9·1
Muriate of do.	8·0
Total impurity	22·1
Nitre	77·9
	100·0

This may be taken as a fair sample of the quality that the *loneahs* produce in general, but when it passes from their hands to the saltpetre merchants it is frequently adulterated with sand, mud, and dirty salts of various kinds, to such a degree that it scarcely contains 50 per cent. of pure nitre. A sample of this adulterated article from 15,000 maunds gave me the following result:—

Insoluble matter, sand, and mud	22·7
Sulphate of soda	23·8
Muriate of do.	4·2
Total impurity	50·7
Nitre	49·3
	100·0

To produce the article called by the natives *kalmee* (crystallized in long prisms, meaning the best kind of saltpetre) the *dhoah* is re-dissolved and crystallized; the percentage of nitre will then amount from 85 to 95 pure,

MANUFACTURE in Bengal.

but this is only done by the opulent native merchants who supply the Calcutta bazar.

In conclusion, I have only to observe that the above methods of manufacturing saltpetre used by the Natives of this country, although rude, yet are very simple, and more effective than most of our scientific chemists, at first sight, would suppose. No manufacture in Europe can equal it in point of cheapness and simplicity; and when it is considered that these simple people have no knowledge whatever of chemistry as a science, it is surprising how well they manage to make the rough article. At least such were the ideas that struck me during the many hours (and I may add pleasant ones) that I have spent in observing the simple, but not altogether ineffective, plans and operations of this industrious manufacturing people.

The above notices claim no merit, except that of truth. They are the result of observations and notes taken on the spot, during a residence of two years in the district of Tirhut, province of Behar."

In a more recent article which appeared in the Journal of the Agricultural and Horticultural Society (*XII, p. 107, Old Series*) Mr. R. W. Bingham, Honorary Assistant Magistrate of Chynepore, describes the manufacture of saltpetre in Behar as follows: "Saltpetre is made extensively all over the district, particularly upon the sites of old towns and villages. It is all made by a peculiar caste called *núníah*, and, so far as my experience shows, is principally in the hands of Ghazipore and Patna *muhajuns*, who make yearly advances, charging at 12 per cent. for the same. The *núniahs* are a tolerably safe class, compared with the ordinary *ryot*, to deal with, and pay the *zemindars* a comparatively large price (if measured by the *bígah*) for the old walls and old sites in which they revel. The supply of saltpetre from these old sites appears to be practically inexhaustible; for we find the *núníah* very busy making up his piles of loose earth just after the setting in of the rains. This earth he exposes to the sun and the rain, and takes care, by erecting walls, etc., that the precious stuff is not wasted away. A casual visitor would not be able to understand what he is after, but when the hot suns of April, May, and June come on, then himself and his family boil away merrily, and eliminate saltpetre and salt from this apparently useless soil. Then the *mahajan* is on the look-out and secures the saltpetre as it is made, and carries it to his own refinery for final manipulation; while the salt which is always bitter, and I should say unwholesome, under the name of *kharí nimuck* is sold to the lowest classes of the community at a cheap rate. The business must be a profitable one, as the large bankers of Ghazipore, Patna and Benares are always ready to go into the trade, and to advance money to responsible middlemen of Sasseram, Bubbooah, and other local marts for the purpose of its extension. Sometimes these men experience considerable trouble in recovering their advances, but in that case they quietly walk off with the bullocks of the *núníah*, who, considering himself as the *assammí* of the man from whom he receives advances, never dreams of making any complaint, but begs or borrows from his comrades or friends till he has got money enough to release them by paying back principal and interest; well knowing he will get no more advances, and will, besides, be put out of caste by his castemates if he does not, at all events, pay the original advance. If, on the contrary, he makes more saltpetre than will cover his advance, and he has no particular ceremony going on, he will clandestinely sell his partially refined saltpetre to other petty purchasers, and get drunk while the money lasts, and ask contemptuously 'What, am I a poor man that I should work'? The trade is too hazardous a one, and the petty advances spread

MANUFACTURE in the over too wide an extent of country to make it worth the while of Europeans with capital to attend to; in consequence it is almost wholly in the hands of the large houses abovenamed (who are connected with Calcutta Native firms) and who in turn have their small branches in every petty town in the district."

Panjab. 687 **Panjáb.**—The exports of saltpetre from this province during the year 1889-90 came to 67,771 maunds valued at R3,38,855. Of that amount 46,552 maunds is in the rail-borne trade shown as having been drawn from Delhi City, 45,180 maunds from the Cis-Sutlej territory, and 17,147 maunds from the tract of country between the Jhelum and the Sutlej. These facts may, therefore, be accepted as denoting the region from which saltpetre is derivable in the Panjáb.

The following series of passages may be given as manifesting the methods pursued in the separation of the salt from the soil, as also its refinement.

Under the name of *Shora*, nitre is found as a natural efflorescence over the ground in many parts of the Panjáb, especially near old buildings. In **H. Baden-Powell's** (*Panjáb Prod.*, 79) will be found a full account of the manufacture, as carried on in that province, from which the following extracts may be sufficient: "Saltpetre is manufactured in two methods, first, by boiling; the other, by evaporating in shallow basins, termed '*ágar*.' The boiling pans pay R2 a year as their tax; the *ágars*, R8. The whole number of pans in the Panjáb appears to be 4,200, and 20 *ágars*. The annual yield of the pans is variously estimated at from 100 maunds in Hissar to 35 maunds in Múltán, but this latter is much too low. The chief expense of preparation is in fuel and wages of work-people. In Sialkót, it is calculated, the profit to the maker is about 65 per cent. on his outlay. **Mr. Roberts** thinks this about the average of the whole province."

"In the *Lahore Chronicle*, May 1855, will be found a good description of the process of manufacture as carried on in this Province, written by **Mr. Gardener**, and from which the following particulars may be taken: 'The saltpetre trade is still in its infancy, and should it receive the impulse of European capital and energy, the Panjáb is capable of producing from 4,000 to 5,000 tons yearly of this useful and necessary article, which would realize somewhat about £70,000 to £110,000 annually in the London market. The surface of the soil is scraped off with a small spade, (called *kai* or *kúdar*) to the depth of an inch or two, and collected into conical piles or heaps from two to four feet high, which are afterwards removed sometimes four to six miles, to attain a locality where fuel and water may be convenient; there the process of accumulation proceeds, affording employment to both the male and female members of numerous families, until a sufficient quantity of the earth is procured to insure to them the manufacture of saltpetre, at least for a season, or, say, five or six months. These people are generally solely dependent on this article for their subsistence. The accumulated earth is their whole stock-in-trade; it is usually left exposed to the full action of the weather, without, however, any perceptible detriment or change. The process of extracting the saltpetre next ensues. Large-mouthed earthen vessels, of the form of those used on the Persian wheel, are placed on an earthen tripod, each vessel having a small aperture at its bottom; first, a layer of straw and then of wood-ash is introduced, and on this the saltpetre earth is loosely placed to within a few inches of the mouth of the vessel. The straw acts as a filter, and, no doubt, experience has taught them the chemical and neutralizing property of the potass contained in the wood-ash; a line of such vessels is erected, with earthen empty cups beneath the orifice of each vessel, to receive the dripping liquid, the earth in the pot being kept well saturated with water until the whole of the saline matter contained in it is pretty well carried off. This simple, though wearisome, operation is continued day and night, to ensure a sufficient supply or quantity of the mother liquid for daily operations. This liquid usually contains but from two to five per cent. by weight of the required article. Oval iron pans, from one to two feet diameter, and six to nine inches in depth, are next filled with the liquid; and heat being applied evaporation commences; the diminishing quantity of liquid being, from time to time, replenished by additional supplies. This part of the process

MANUFACTURE in the Panjab.

requires care and experience, and occupies fromtwelve to eighteen hours of continued labour. The impurities, as they rise, are carefully skimmed off the surface of the boiling liquid, from which, on attaining a certain degree of concentration, the impure salt and other foreign matters are copiously precipitated.

"This results from the muriate of soda, or, properly speaking, the chloride of sodium, being equally soluble in cold as in hot water. The filthy sediment is scooped out of the bottom of the pan at intervals and heaped by the side of the boiler. The small pan, in the Upper Panjáb, after thirty to thirty-six hours' continued labour, usually yields 8 to 16℔ saltpetre, while the larger pans, of the lower country, in the same time, will yield from 15 to 30℔, the average yield being the medium figures of each. The quantity and even quality depend on the nature and richness of the earth used.

"The soil of the lower part of the Panjáb contains a much stronger impregnation of common salt than the upper. A line drawn from Pind Dádan Khán or from Kálábágh eastwards to Pakpatan on the Sutlej, would pretty well define the line of difference."

N.-W. Provinces and Oudh. 688

North-West Provinces and Oudh—Next to Bengal these provinces are the most important sources of saltpetre in India. According to the Annual Report of the Rail-borne Traffic, the exports in 1889-90 were 226,302 maunds, valued at R14,70,961. These were sent mostly to Calcutta (208,650 maunds), Bombay port (10,438 maunds), Rajputana and Central India (2,989 maunds), and the Central Provinces (2,383 maunds). That a very considerable quantity appears in the rail returns as having been derived from the Agra Division (117,299 maunds), Oudh (45,369 maunds), Allahabad (38,237 maunds), Meerut (17,879 maunds), and Benares (7,264 maunds). Although these are merely the blocks from which the supplies were drawn and not necessarily the districts in which the article was actually prepared; still, the figures given denote approximately the importance of the districts of, supply in these provinces.

Very little has been published on the subject of the saltpetre industry of these provinces, but the following extract may be accepted as exemplifying the method of preparation or isolation from the soil and refinement as practised:—

"Saltpetre, both crude and refined, is exported in considerable quantities. In the manufacture of crude saltpetre there are two processes, the *jariya*, or artificial-heat process, and the *aliya*, or solar-heat process. In both of these processes it is necessary to make the brine first, and, for this purpose, a shallow trough (*kariya*) is excavated in some mound or artificial eminence raised a few feet above the level of the surrounding country. The trough is usually from 18 inches to 2 feet broad, 7 to 10 feet long, and from 18 inches to 2 feet in depth. The bottom is lined with several rows of bricks, on which are laid twigs of cotton or stalks of *arhar*, and over these a layer of grass, so as to form a rough kind of filter, while the interstices left between the bricks allow the brine to flow. About twelve maunds of earth impregnated with saltpetre are then thrown in loosely and covered by about fifty to sixty *gharas* of water for eight to twelve hours, when it is allowed to run off into a reservoir (*kanda*), and yields about 25 to 30 *kharas* of brine. In the *jaria* process the brine is then boiled for about six or seven hours in a bowl-shaped boiler of iron (*karahi*) to crystallizing point. As soon as a drop of the liquid will solidify on a leaf, the fire is damped and the solution is removed to earthen vessels known as *nánds*, where, on cooling, the crystals form and yield about one-half the weight of saltpetre, the liquor left in the *nands*, on the removal of the crude saltpetre, is known as *tor*, and may be used for extracting alimentary salt, or be sprinkled again over the *kariya*. About four maunds of wood or five maunds of leaves are required as fuel for one operation, and the permission to gather and use the wood or leaves is usually included in each case. The leases vary from R10 to R100, but the average for the circle, including the Etáwa and Mainpuri districts, is R16 to R18 per factory maund. The *aliya* process allows the brine to flow into a large, shallow vat of masonry known as a *kari*. The vat is about six inches deep, and there are usually two or three in each factory, placed on different levels to allow of the brine flowing from one to the other until the saltpetre forms. This process occupies about two days in favourable weather, and can only be followed in the driest and hottest weather. The same quantity of saltpetre and twice the amount of alimentary salt is produced. To refine the saltpetre some thirty *gharas* of water are boiled in a large iron vessel, and to this are added some five maunds of crude

MANUFACTURE in the N.-W. Provinces and Oudh.

saltpetre, and the solution is allowed to boil for two hours. It is then drawn off into a large wooden trough or succession of troughs, and when the sediment falls to the bottom, the clear solution is drained off and allowed to cool. The operation takes from three to five days and gives about two maunds of refined saltpetre. The boiling is generally continued until five vats are filled, when the crystals are removed from the first vat, and the remaining *tor* or mother liquor is again worked up with four maunds of crude saltpetre and some water, so as to prevent the liquor from becoming too thick for crystallization. The alimentary salt produced pays duty at R3 per maund and sells for R4 per maund. The cost of manufacture varies in every parganah, as well as the cost of the case of the right to remove nitrous earth. As a rule, a crude factory can yield fifteen sers of crude saltpetre a day, or about a maund every three days during the working season, worth R3 8. Against this must be charged wages of licensee and labourer for three days, at two annas a day, 12 annas; fuel 7½ annas; lease, 3 annas; hire of boiler, 3 annas, or R1-9-6 per maund. For seven maunds of refined saltpetre valued at R47-4, or R6-12 per maund, the charges are for three labourers, 6 annas; fuel, R1-4; hire of boiler, 1¼ annas; lease, 4 annas; and value of 9½ maunds of crude saltpetre, R32-2 or a total of R35-1-3, leaving R12-2-9, profit on the operations out of which the license, amounting to R50 per annum, must be paid. Another refining process known as *ras-galai* is thus conducted. A certain quantity of crude brine is boiled to crystallizing point, and when incalescent, two or three maunds of raw material are added, and the boiling is continued for an hour longer. The solution is then removed to the vats as before, and when the sediment has fallen to the bottom, the supernatant liquor is retransferred to the boiler, and mixed with a portion of the *tor* or mother-liquor, is again boiled for two hours. The salt then precipitates and after its removal the solution is deposited in vats, and crystallization takes place. This process gives the superior saltpetre known as *ekbara*, and also a considerable quantity of good alimentary salt. There are several hundreds of crude factories, but only a few refineries, in the district. The *rasi* factories number about a hundred; each refinery employs six to ten workmen; *bhari* works, three labourers, and the license and crude saltpetre factories, two men. A crude factory with one boiler can turn out eleven maunds per mensem, or in the seven working months about 80 maunds of crude material, worth on an average about three rupees per maund" (*Mainpuri Gaz.*) 531-533.

Bombay. 689

Bombay.—Little or no mention of saltpetre manufacture occurs regarding the districts of this province, except that of Ahmedabad. The following passage, however, regarding that district may be accepted as conveying the chief ideas that prevail The method of isolation and purification pursued resembles more that of the Panjáb than Bengal:—

"At the beginning of the century, near certain villages in the Limbdi district, such as Thalavad and Patan, etc., saltpetre was made in large quantities. But on the introduction of British rule the widespread peace and the cheaper supply from Bengal put a stop to the Gujarat manufacture. The Vanias declared that, because of the murderous uses to which gunpowder was put, it was a sin to make saltpetre, and in 1825, except a little by Musalmáns of the Bohora class, none was produced. In 1830 the Revenue Commissioner, Mr. Dunlop, made an inquiry into the cause of the failure of the manufacture, and by the help of Mr. Vaupell, a gentleman of much intelligence and knowledge, supplied Government with a full account of the processes employed and of the state and prospects of the industry. The manufacture was then on a very small scale. The Vanias opposed the Bohras in their attempts to increase the production, and a Parsi, who had come to Dhollera with pots and other tools, failed from ignorance of the proper kind of earth. Still the Natives were willing to make saltpetre, and if a demand arose, at a shilling for five pounds (R4 the *man*), an unlimited quantity might be supplied. Bengal saltpetre, though a little dearer and inferior to the best local variety, was even in Limbdi able to compete with it. Unless Government came forward as a buyer, there seemed little hope of reviving the industry. The manufacture is carried on only during the cold season. The earth used, of a dark red, mixed with white, becomes whiter the deeper it is dug. The richest patches are near villages in places frequented by cattle. When one plot is exhausted the workers change to another and keep moving so long as the cold season lasts. Except the alkaline earth and pure water, nothing is used in the manufacture. The first process is to scrape off and gather the surface soil. When enough is gathered the earth is placed in large pierced earthen vessels called *gola*. Fresh water is then poured on the earth till the vessel fills. And as it strains through, the salt water is collected into smaller pots, *moria*, placed below. These are again emptied into deep iron pans, *karav*, holding from ten to twenty-five *morias* of the strained liquor.

These iron pans are set over a cow-dung fire and as the contents boil and evaporate, common sea-salt forms, and as it forms is taken away in pierced iron ladles. The boiling goes on till, as the water begins to crystallize, it thickens into a jelly. It is then, in the evening, poured into shallow earthen vessels, *kunda*, and allowed to stand all night. In the morning the crystallized nitre is taken away and put into bags. In this state, called *ekvara* single, or once washed from its large proportion of common salt and other impurities, it is of no use. To refine it, the saltpetre is again washed and purified in clear fresh water. It is then termed *bevda* or twice washed, and, though somewhat inferior to the Bengal variety is used for making gunpowder. After a third purifying, it is called *tevda*, or thrice washed. Thus the best saltpetre made in Gujarat is principally used in the manufacture of the finest gunpowder called *ranjki*, or priming powder. Besides these varieties at Sami and Munjpur a kind called *kalmi*, charged with alum is much used for fireworks "(*Bombay Gaz., Vol. IV, 125*). MANUFACTURE in Bombay.

Madras.—It has already been stated that formerly saltpetre was manufactured in the Madura district, but that the European firm who engaged in the trade had found it unremunerative, and had, accordingly, discontinued the manufacture (*Madura Dist. Man., 25*). Mention is also made of the industry in Nellore, where, in 1873, the estimated possible outturn was put at 556 maunds, valued at R3 a maund for single refined, and R4 a maund for double. Several other districts are also known to afford the salt, but the industry can scarcely be said to be one of any importance in South India. Madras. 690

Burma.—"Saltpetre is manufactured in several places in Upper Burma to about 50 tons annually. It is found in some of the caves of Tenasserim and is imported into Rangoon" (*Balfour, Cyclopædia of India*). Ball, however, says that "in Upper Burma there appears to be a large manufacture and the price realised is high" (*Indian Economist, Vol. V, 14*). Burma. 691

EUROPEAN MANUFACTURES AND USES. USES. 692

"The chief use of saltpetre is in the manufacture of explosives, fully five-sixths of the total consumption being applied to this purpose" (*Spons' Encycl.*). The readiness with which this nitrogen compound parts with its oxygen, as also the large percentage of that element which it contains, combined with its non-deliquescence, renders it of the highest value in the manufacture of explosives. In the manufacture of gunpowder, potassium saltpetre is exclusively used, the analogous sodium salt being unsuitable for that purpose on account of its slow combustion and tendency to absorb moisture from the air. On the other hand, in the manufacture of nitric acid and other uses, Chili, or sodium saltpetre, is preferred to nitrate of potash on the score of cheapness, and because it produces about 7 per cent. more nitric acid, weight for weight—60, as against 53.

To render the crude saltpetre fit for the manufactures of gunpowder, the impurities, chiefly the chlorides of potassium and sodium, have to be separated. The following extract of the process of refining, taken from Spons' *Encyclopædia*, is given *in extenso*, as it would appear that Indian saltpetre has been objected to in certain of the colonies on account of its colour, the colonial buyers preferring to secure the refined article from England. To refine saltpetre advantage is taken of the "different rates of solubility of the various salts at different temperatures, partly upon the mechanical action of animal gelatin upon the extractive matters contained, and partly upon the fact that crystals of saltpetre, being homogenous (that is, consisting of one salt alone), separate out without contamination from the solution containing the chlorides of potassium and sodium. The crude article is dissolved in boiling water, the salt being added to saturation and the heat gradually increased. A density of 1·5 or 1·6 should be attained. Small quantities of dissolved glue are introduced into the boiling solution,

USES. which separate out the various extractive matters. These partly rise to the surface, and form a scum which is removed from time to time, and possibly sink to the bottom of the pan. Sometimes the hot solution is further diluted with water to prevent the depositing of crystals of saltpetre and allow time for the insoluble matters to separate out. The liquors are then run off into flat copper crystallizing pans, and, while cooling, are kept thoroughly stirred up with wooden rakes to prevent the formation of large crystals, which are apt to contain appreciable quantities of the mother liquors in their interstices, and yield, when pulverized, a damp powder. The fine needles which are obtained, having the appearance of a white powder, are termed "saltpetre flour." This is fished out and thrown upon a wire-gauze strainer placed across the crystallizing pan, to drain, the mother liquor falling back into the pan. The saltpetre flour is almost pure, the mother liquors containing the chlorides and returning them into the pan. The flour is then removed to the wash-pans and treated with cold water, or a saturated solution of pure saltpetre. The wash-pans are usually about 10 feet long, 4 feet wide, and 3 feet deep, fitted with a false bottom, upon which the flour is placed. When thoroughly washed, and freed from all adhering mother liquor, it is dried at a gentle heat and sifted to separate out the lumps. The mother liquors are evaporated, a sufficient quantity of potash salt is added to decompose the nitrates of the earths contained, and worked over again as crude lye from the saltpetre earths" (*Spons' Encycl., I, 275-276*). It would seem that the methods followed by the Natives of India might be improved on the pattern of the European refinements without entailing any great extra expenditure. The use of glue at all events might be recommended. The complaints made against the Indian article by the Australian and other consumers might, as it seems, be thus easily removed by a very slight improvement on the Indian methods of manufacture.

DYE. 637 **Dye.**—Saltpetre is used as a mordant, especially in the case of wool-dyeing with animal colours, such as lac or cochineal. **Sir E. C. Buck** says that the peculiarity of the process adopted in India consists in the use of a preparation known as *tezab.* This "is a distillation from saltpetre, sal-ammoniac, sulphate of iron, and alum in the following proportions:—Saltpetre 8 parts, sulphate of iron 8 parts, sal-ammoniac 2 parts, and alum 2 parts. These are mixed in an earthen jar, used as a retort and placed over a fire, the distillation being conducted much in the same way as that of country spirits The vapour is conducted down a tube into a second jar, the condensation being effected in the tube, which is cooled by being wrapped in wet rags. The *tezab* is mixed with some lemon-juice, and the dyed cloth is boiled in it for about half an hour" (*Dyes of North-West Provinces, 76*). **Liotard** (*Dyes of India, 76*) alludes to the practice followed in Lahore with the lac-dyeing of wool. "The mordant used is an acid prepared from *kahi* (salts of iron), saltpetre, and sulphur. The *modus operandi* is thus explained; the woollen fabric is steeped in the dye equal to it in weight, and hot water is then poured on it The vessel is then closed up for three or four days, by which time the whole of the dye is absorbed in the fabric, which is then washed. Subsequently, the mordant is mixed with the dye and the fabric is then placed in the mixture and boiled over a fire. The result is a crimson colour which is fast." The writer is not aware of any very special mention of the use of saltpetre in dyeing in Europe, so that the few brief remarks offered above regarding its use by the Indian dyers are likely to be read with some degree of interest. Similar passages might be cited from other works, but it is, perhaps, enough to repeat that saltpetre as a mordant is used chiefly with wool or with animal tinctorial re-agents.

Medicine.—This salt is eliminated chiefly by the kidneys, and in its passage through them it acts as a stimulating diuretic, the urine containing it as a nitrate, thus unchanged. It has distinct diaphoretic powers and hence, by getting the skin to secrete freely, there is, under its use, a fall of the temperature in most febrile states. **Ainslie** (*Mat. Ind., Vol. I, 375*) says: "The Native doctors prescribe saltpetre for nearly the same purposes that we do; to cool the body when preternaturally heated, and in cases of *Neercuttoo* and *Kull-addypoo* (ischuria and gravel). They are also in the habit of cooling water with it (which it does by generating cold while dissolving) for the purpose of throwing over the head in cases of phrenitis. Given in repeated small doses, not exceeding ten or twelve grains, it abates heat and thirst, and lowers arterial action." MEDICINE. 693

"Under the Hindustani name *shora*, nitrate of potash may be obtained in most of the bazars of India, but often in a very impure state. To fit it for internal use it should be purified by dissolving it in boiling water, removing the scum, and, after the liquid has been allowed to settle, straining it through a hempen cloth, and setting it aside to crystallize" (*Pharm. Ind., 315*). **Sakharam Arjun** says that saltpetre is prescribed in Bombay as a diuretic in combination with milk.

Food.—From its antiseptic power it is used to preserve fish and meat, to the latter of which it gives a red colour (*Milburn*). For the purpose of being used as an antiseptic, nitre is said to be largely imported by the United States of America. The Natives of India cannot be said, however, to use saltpetre as an article of food except in the fact that the poorer classes eat the common salt (sodium chloride) isolated by the saltpetre manufacturers which always contains saltpetre and other impurities. The reader should consult the remarks that occur on this subject under the article **Salt**, p. 405. FOOD. 694

Domestic.—In certain parts of the North-West Provinces and Oudh saltpetre in the crude state is used as MANURE. **Mr. F. Ashton** (Assistant Commissioner, Northern India Salt Department) states that "the salt which appears in the streets and environs of villages is carefully scraped up and used as manure." The two crops generally treated with this manure are wheat and tobacco. In the case of wheat the saltpetre is taken out to the fields in baskets, and is then carefully scattered over the young plants when they are about 6-8 inches in height. In the case of tobacco the efflorescence is applied not only to the root of a plant, but is also carefully sprinkled over each leaf. This use of nitre as a manure occurs mainly in the Upper Doab of the North-West Provinces where the people are well-to-do; the price precludes, however, the general use of nitre for the growing of crops. DOMESTIC. Manure. 695

Nitre is also said to be used in India as a FLUX in glass-making. The most extensive use of saltpetre is, perhaps, the preparation of the fire-works and crude gunpowder used throughout India. The article employed in the ordinary village fire-works can hardly be called gunpowder, but if it be accepted as a crude form of that substance, it may be contended that the Natives of India knew of gunpowder long before it was discovered in Europe, although they never thought of using it in fire-arms until they saw these weapons in the hands of their European conquerors. Flux. 696 Fire-works. 642

TRADE IN SALTPETRE.

From the brief allusions offered above to the early records of this trade it may have been inferred that saltpetre engaged the attention of the mercantile governors of India for fully a hundred years. In the Proceedings of the Honourable the East India Company from 1784 to 1820, frequent mention is made of it, the exports being then viewed as the most profit- TRADE. 697

TRADE.

able form of ballast cargoes to be sent in ships that carried such light substances as silk and piece-goods. In 1846 the discovery was made that Chilian saltpetre (sodium nitrate) might be reduced to potassium nitrate (ordinary saltpetre) by the natural decomposition of the Chilian saltpetre. This new method rendered a practically limitless source available for the demands of commerce. It was, therefore, only what might have been expected that interest in the natural salt in India obtained by a crude process and which was constantly liable to adulteration through the impecuniosity of the manufacturers, or the criminality of the traders should have greatly decreased. It is commonly urged that the restrictions imposed through the salt monopoly have also raised the price at which Indian saltpetre can be produced, and have accordingly lessened the chances of competition. There may be a certain amount of truth in such a contention, but the historic records of the trade manifest a shrinking in the exports from India (or rather a check given to the development of the trade) prior to the existence of the more stringent regulations of the salt monopoly.

The following table exhibits the exports from Bengal for the five years ending with 1850 :—

Exports. 693

Exports of Saltpetre from Calcutta.

Years.	Cwt.	Rx.
1845-46	441,829	350,649
1846-47	404,740	325,615
1847-48	446,052	345,280
1848-49	464,293	362,769
1849-50	508,316	394,596

To allow of comparison with English valuations, the money value of the Indian saltpetre has been shown in tens of rupees—the nominal pound sterling.

It will thus be seen by comparison with the figures below that the Indian trade has practically fluctuated, and, if anything, manifested a tendency to shrink rather than expand during the past half century :—

The total exports from all India during the past ten years.

Years.	Cwt.	Rx. = tens of rupees.
1880-81	352,995	351,728
1881-82	354,860	359,437
1882-83	399,565	388,766
1883-84	491,668	464,410
1884-85	451,917	425,000
1885-86	402,174	370,200
1886-87	397,572	376,091
1887-88	386,396	364,016
1888-89	420,503	401,801
1889-90	422,229	411,276

It may be noted that the exports of Bengal alone were in 1850 in excess of the total from all India during 1890, but that the article had increased in value in almost an inverse ratio to the decrease in quantity. The following table analyses the total Indian exports during the past five

TRADE.

Exports.

years, so as to exhibit the more important countries to which the Indian saltpetre is usually exported :—

Countries to which exported.	1885-86.	1886-87.	1887-88.	1888-89.	1889-90.
	Cwt.	Cwt.	Cwt.	Cwt.	Cwt.
United Kingdom . .	195,206	158,503	176,470	192,059	167,052
United States . . .	90,882	86,045	61,111	61,382	90,981
Hong-Kong . . .	60,431	79,074	90,137	104,437	76,872
France	33,766	50,900	36,516	32,319	34,949
Straits Settlements . .	7,539	8,202	7,697	12,462	17,929
Australia	4,242	985	2,285	1,285	895
Belgium	3,443	4,333	1,002	1,143	...
All other countries . .	...	...	...	5	11,851
TOTAL .	402,174	397,572	386,396	420,503	422,229

The United Kingdom thus takes by far the largest amount and is followed by the United States, then by China, France, the Straits, Australia, etc. **Mr. J. E. O'Conor**, commenting on the fluctuations of the exports, remarks, "The trade is subject to extreme oscillations, the result partly of political influences. It reached its highest point during the Confederate struggle in the United States, the value of the exports being at that time: 1860-61, £661,614; 1861-62, £828,378; 1862-63, £896,808; 1863-64, £722,165; 1864-65, £542,461; and in 1865-66 (the impetus given to the trade continuing after the termination of the war) £605,376. It should be remarked, however, that the trade in this article is extremely sensitive to the fiscal action of Government. Thus, there is no doubt that the reduction of the export duty by one-half in 1865 tended to maintain the trade at a high level in that year, though the extraordinary demand of the previous years had ceased; and, on the other hand, when in the following year a heavy duty was imposed, the value of the trade fell at once to £297,713 (in 1866-67), and again to £256,301 (in 1867-68). The trade was freed in 1867, but the blow had been heavy and recovery was slow, other sources of supply having been sought by consuming countries. The French, for instance, commenced a local manufacture, and a chemical substitute also entered into competition with native saltpetre." **Mr. O'Conor** then gives the figures of the exports from 1868-69 to 1874-75, and shows a similar result to that demonstrated above, namely an export trade fluctuating from 390,000 cwt. to 550,000 cwt. **Mr. O'Conor** then concludes his review of the figures up to 1874-75 as follows :—"There would not seem to be much apprehension that, in these days of international conflicts, followed by internecine struggles,—days of military supremacy and bloated armaments,—the demand for 'Villanous saltpetre' will fall off materially, and as the article is increasingly required, moreover, in some of the useful arts, it is probable that the trade will remain tolerably steady, unless it be interfered with from time to time by political exigencies of the administration of the day."

The following classification of the imports into Calcutta exhibits the routes by which the saltpetre reaches the sea-board. The balance of imports

TRADE.

over exports would be the amount available for local consumption or retained in stock :—

Total quantity of Saltpetre Imported to, and Exported from, Calcutta by all routes:—

Specification of Routes.	Imports.			Exports.		
	1887-88.	1888-89.	1889-90.	1887-88.	1888-89.	1889-90.
	Mds	Mds.	Mds.	Mds.	Mds.	Mds.
By East Indian Railway.	631,422	679,680	712,658	913	832	990
By Eastern Bengal State Railway.	8	1	...	1,258	1,311	1,405
Boat . .	5,689	5,429	3,884	1,038	227	92
Inland steamer .	2,481	664	3,933	867	583	727
Road . .	...	12	...	5	...	...
Sea . .	...	34	...	520,230	566,393	573,432
Total .	639,600	685,820	720,475	524,304	569,346	576,646
Total expressed in cwt.	456,857	489,871	514,482	374,503	406,676	411,890

The difference between the totals shown as exported by sea from Calcutta and the totals given in the previous tables of the Indian foreign transactions is the amount exported by the other provinces of India. But the Calcutta transactions may be subjected to a further criticism in order to demonstrate the sources of supply from which the exports and local consumption are drawn :—

Sources of the Calcutta Supply.

	1887-88.	1888-89.	1889-90.
British Provinces.	Mds.	Mds.	Mds.
Behar	408,417	468,059	456,508
North-West Provinces and Oudh . .	170,462	165,940	208,650
Panjáb	59,368	51,001	53,110
All other Provinces	829	506	2
Native States.			
Rajputana and Central India . . .	524	314	2,205
Total .	639,600	685,820	720 475

N.-W. Provinces. 700

North-Western Provinces and Oudh.—It will thus be seen that after Bengal (Behar) the North-West Provinces and Oudh are the next most important. It is perhaps unnecessary to deal with these provinces in the same detail as has been done with Calcutta—the great emporium of the Indian saltpetre trade—but it may be remarked that the total exports from these provinces during the year 1889-90 came to 226,302 maunds. Of that amount 208,650 maunds went to Calcutta, mostly by rail, and of the remainder Bombay port town took 10,438 maunds; Bombay Presidency

TRADE.

644 maunds; the Central Provinces 2,383 maunds; Rajputana and Central India 2,989 maunds; and the balance went to Bengal province, the Panjáb, Berar and the Nizam's Dominions. The total value of these exports was R14,70,961. The imports into these provinces were very small, *viz.*, 1,944 maunds.

Panjab. 701

Panjáb.—The chief item of the Panjáb transactions was that shown as delivered at Calcutta. But the province also furnished Bombay port with 8,554 maunds in 1889-90, and during that year 864 maunds to the North-West Provinces.

Bombay Port Town. 702

Bombay Port Town.—The total imports by rail and road were in 1889-90 21,380 maunds, the net import being 13,438 maunds, of which 10,438 maunds (as already stated) were derived from the North-West Provinces; 8,554 maunds from the Panjáb; 371 maunds from Bombay Presidency; and 2,017 maunds from Central India. Of the exports, 463 maunds were consigned to Berar; 428 maunds to the Nizam's Dominions; 7,703 maunds to Bombay Presidency; 277 maunds to Central India; 58 maunds to Mysore; 11 maunds to Madras, and 2 maunds to the North-West Provinces. The total net imports into the Bombay Port Town and Presidency came to 23,876 maunds, and that amount, it will be noted, was drawn almost exclusively from the North-West Provinces and the Panjáb, little or none being derived from Behar, the chief seat of Indian production.

Coastwise. 703

An analysis of the coastwise transactions and trans-frontier trade would not materially disturb the impression conveyed by the facts already exhibited. A large local demand exists all over India which is met by the Indian article. Little or no foreign saltpetre comes into the country. It is chiefly used up in the preparation of the fire-works employed at festivals and ceremonials.

Gunpowder. 704

Practically no gunpowder, except of the crudest kind, is manufactured by the Natives. There is one factory in the country, however, in which the Government prepares largely its own gunpowder, namely, that at Dum-Dum near Calcutta. The gunpowder used in sport may be said to be entirely imported. There are, however, many refineries for saltpetre in Behar and the North-West Provinces, and one near Calcutta is owned and worked by a European. The value of the outturn of the better known Indian refineries has been given at R21,87,126; they give employment permanently to 1,942 persons, and temporarily to 3,248 additional hands. But it will be observed that these figures convey no more conception of the actual number of persons employed in the entire saltpetre trade than that the returns of foreign trade express the total transactions. The persons employed in the preparation of the crude article would in the one case be overlooked and, in the other, the local consumption would be ignored.

(*W. R. Clark.*)

SALVADORA, *Linn.; Gen. Pl., II, 681.*

[*t. 1621*; SALVADORACEÆ.

705

Salvadora oleoides, *Dcne.; Fl. Br. Ind., III, 620; Wight, Ic.,*

Syn.—S. STOCKSII, *Wight*; S. INDICA, *Royle*; S. PERSICA, *T. And.*

Vern.—*Jhal, jál* (the fruit=) *pílú* (or *bara-pílú*), HIND.; *Kubbur, diár jhal*, N.-W. P.; *Jál, ván, váni, kubbur, diar, jhal, pil, ták, wan, sál* (the fruit=) *mithi ván, pílú* (the dried fruit=) *khokar, pinju*, PB.; *Plewane*, TRANS-INDUS; *Miswak, plewan*, PUSHTU; *Jhal, jhár, diár, mithi diár, kabbar*, SIND.; *Kankhina, kakhan*, BOMB.; *Pílu, khakhan* (the oil=) *kinkanela*, MAR.; (The oil=) *Khakananutela*, GUZ.; *Ughai, koku*, TAM.; *Arák*, ARAB.; *Darákh i-misvák*, PERS.

References.—*DC., Prod., XVII, 28; Brandis, For. Fl., 316; Gamble, Man. Timb., 260; Stewart, Pb. Pl., 175; Pharm. Ind., 170; Moodeen Sheriff, Supp. Pharm. Ind., 223; O'Shaughnessy, Beng. Dispens., 527; Dymock, Mat. Med. W. Ind., 626; S. Arjun, Bomb. Drugs, 113; Murray, Pl. & Drugs, Sind, 170; Baden-Powell, Pb. Pr., 273,*

597 ; Drury, U. Pl., 378 ; Royle, Ill. Him. Bot., 319 ; Balfour, Cyclop., III, 509 ; Gazetteers :—Panjáb, Jhang Dist., 16 ; Muzaffargarh, 22 ; Mooltan, 102 ; Dera Ghazi Khan, 10 ; Delhi, 18 ; Rohtak, 14 ; Karnal, 16 ; N.-W. P., IV, lxxiv. ; Sind, 746 ; Settle. Rep. :—Lahore, 14 ; Montgomery, 17 ; Dera Ghazi Khan, 4; Delhi, cclxii., App. xxv ; Jhang, 21 ; Agri.-Horti. Soc. Panjáb, Select papers up to 1862, 50.

Habitat.—A large evergreen shrub or tree, of the arid tracts of Sind, the Panjáb, and Rájputana, often forming the greater part of the vegetation of the desert. It ascends to 3,000 feet in the Trans-Indus hills and to 2,400 feet in the Salt Range. It is distributed to Aden.

DYE. Galls. 706

Dye.—The GALLS found upon this plant are used in dyeing (*Stewart*).

OIL. Seeds. 707

Oil.—On expression, the SEEDS yield an oil of a bright green colour and with the consistence of butter. The yellowish substance sold in the bazárs under the name of *khárkhanela* is much adulterated (*Dymock*).

MEDICINE. Fruit. 708 Oil. 709 Root-bark 710 Leaves. 711

Medicine.—The FRUIT is sweet in taste, and is supposed by the Natives of the Panjáb to have aphrodisiac properties, but "this," Stewart says, "is to be attributed to the fact of crowds of both sexes wandering in the wilds at the ripening time." The fruits eaten singly are said to cause tingling and small ulcers of the mouth, hence people prefer to eat them by handfuls, seeds and all, and the latter are apt to accumulate in masses in the sigmoid flexure of the intestines and lead to disagreeable results (*Stewart*). The OIL expressed from the seed is used as a stimulating application in painful rheumatic affections and after child birth. In Sirsa and other parts of the Panjáb the ROOT-BARK is ground up and used as a vesicant. The LEAVES are made into a decoction and given as a purgative to horses.

FOOD. Fruit. 712 FODDER. Berries. 713 Leaves. 714

Food and Fodder.—The tree flowers in April and its fruit ripens at the beginning of the hot weather. The FRUIT is sweetish and is largely eaten by the Natives, large numbers of whom go out to collect it in the season; and so much do they depend on it that Coldstream states that a bad crop is considered a calamity. In Muzaffargarh the fruit is often dried for future use, and has then much the appearance and flavour of currants. The fruit of the *jal* is in fact supposed to be a very cooling diet. Cattle are fond of the BERRIES, and it is thought to increase both the quantity and the sweetness of their milk. The LEAVES are the favourite diet of camels during the first quarter of the hot weather, but other animals will not eat them.

TIMBER. 715

Structure of the Wood.—Light red, moderately hard, with small, irregular, purple heartwood. Weight about 54℔ per cubic foot.

DOMESTIC. Wood. 716

Domestic Uses.—The WOOD is sometimes used for building; also for agricultural implements, Persian-wheels and the knee timbers of boats (*Stewart*). It furnishes a bad fuel, as it smoulders, emits a disagreeable smoke, and leaves a very large quantity of ash (*Coldstream*). Being, however, almost the only wood available, it is much used for burning purposes in the Multan, Montgomery, and Jhang districts. Mixed with deodar and pine scrapwood, it has been found to answer well for burning bricks (*Stewart*).

The thick groves of these trees are much used by the cattle thieves of the Panjáb as places of concealment for stolen animals. The shade of the *jal* is esteemed by the agriculturists as being particularly cool and a good protection for cattle against the sun.

717 **Salvadora persica,** *Linn.; Fl. Br Ind., III, 619; Wight, Ill., t. 181.*

THE TOOTH-BRUSH TREE. Supposed by Royle to be the Mustard Tree of the Bible.

Syn.—S. WIGHTIANA, *Planch.;* S. INDICA, *Wight.;* CISSUS ARBOREA, *Forsk.*

Vern.—*Pílú* (or *chhota-pílú*), *jál*, HIND. & BENG.; *Jál*, N.-W. P.; *Jit, kauri ván, kaurijál, chhota ván, jhár, jhit, jhal, arak, pílú*, PB.; *Ple-wan*, PUSHTU; *Jhal*, RAJ.; *Kabbar, kharídjar, pilu* (the fruit=), *khari pírú, kússeer*, SIND; *Pilvu, kákhan*, BOMB.; *Pilu, rhakhan*, MAR.; *Piludi*, GUZ.; *Opa, ughai, kalarva, kárkol, ugá*, TAM.; *Waragu-wenki, ghunia, pinna-vara-gógu, pedda vara góki*, TEL.; *Pilu*, SANS.; *Arák, irak, pilu, kharjal, kabbar*, ARAB.; *Darakht-i-misvák*, PERS.

References.—*DC., Prod., XVII., 28; Roxb., Fl. Ind., Ed. C.B.C., 130; Brandis, For. Fl., 315; Beddome, Fl. Sylv., t. 247; Gamble, Man. Timb., 259; Thwaites, En. Ceylon Pl., 190; Dalz. & Gibs., Bomb. Fl., 312; Stewart, Pb. Pl., 174; Elliot, Fl. Andhr., 150, 153; Boiss., Fl. Orient., IV., 43; Pharm. Ind., 170; Moodeen Sheriff, Sup. Phar. Ind., 222; Ainslie, Mat. Ind., II., 266; O'Shaughnessy, Beng. Dispens., 526; U. C. Dutt, Mat. Med. Hind., 313; Dymock, Mat. Med. W. Ind., 624; S. Arjun, Bomb. Drugs, 113; Murray, Pl & Drugs, Sind, 170; Irvine, Med. Top. Ajmir, 178; Baden Powell, Pb. Pr., 597; Drury, U. Pl., 378; Lisboa, U. Pl. Bomb., 98, 401; Birdwood, Bomb. Pr., 167; Balfour, Cyclop., III., 509; Kew Off. Guide to the Mus. of Ec. Bot., 95; Gazetteers:—Panjáb, Bannu, 23; Peshawar, 27; Dera Ismail Khan, 19; Muzaffargarh, 22; Dera Ghazi Khan, 10; Bombay, V., 26, 285; VI., 14; N.-W. Prov., IV., lxxiv.; Sind, 603, 746; Manual of the Trichinopoly Dist., 79; Settle. Rep., Dera Ismail Khan, 7; Ind. Forester, IX., 174; XII., App. 1, 16.*

Habitat.—A small, evergreen tree, found in the drier parts of India, from the Panjáb and Sind to Patna, and extending southward to the Konkan, the Circars, and North Ceylon. It is distributed to Persia, Syria, Arabia, and East Africa.

Oil.—The oil appears to be similar in character to that of the preceding. The LEAVES and PEDUNCLES as well as the SEEDS contain a large amount of essential oil (*Trans., Med. Phys. Soc. of Bombay*).

OIL. Leaves. 718 Peduncles. 719 Seeds. 720

Medicine.—In Persian works on medicine, the FRUIT is described as deobstruent, carminative, and diuretic (*Dymock*). It is said to be administered in Sind with good effect in cases of snake-bite, and to be used both in the fresh and in the dried state, although in the latter it loses much of its efficacy, and has to be administered in considerably larger doses and combined with borax.(*Dr. Milach*). The fruit is also held to be purgative. Ainslie states that the BARK of the stem is a little warm and somewhat acrid, and is recommended by Native physicians to be used as a decoction in low fever, and as a stimulant and tonic in amenorrhœa. The dose of the decoction is half a teacupful twice daily (*Materia Medica*). The SHOOTS and LEAVES are pungent, and are considered by the Natives of the Panjáb as an antidote to poisons of all sorts (*Murray*). The JUICE of the leaves is given in scurvy. The leaves are used by the country-people in the south of Bombay as an external application in rheumatism; they are heated and tied up in thin cotton cloth (*Dymock*). The bruised bark of the ROOTS is acrid, and acts as a vesicant (*Ainslie*). It is "remarkably acrid; bruised and applied to the skin, soon raises blisters, for which purpose the Natives often use it. As a stimulant it promises to be a medicine possessed of very considerable powers" (*Roxburgh*).

MEDICINE. Fruit. 721 Bark. 722 Shoots. 723 Leaves. 724 Juice. 725 Roots. 726

The tree derives its Persian name (*darakht-i-miswák*, or tooth-brush tree) from the fact that the wood is much employed for the manufacture of tooth-brushes, and it is supposed by the Natives that tooth-brushes made of it strengthen the gums, keep them from becoming spongy, and improve digestion (*Stewart; Murray*).

Food and Fodder.—The FRUITS (or small red berries) are eaten by the Natives of India. They have a strong aromatic smell and a pungent taste like mustard or garden cresses, and are not very much appreciated as articles of diet. The pungent SHOOTS and LEAVES are eaten as a salad. They are also used as camel fodder.

FOOD & FODDER. Fruits. 727 Shoots. 728 Leaves. 729

TIMBER. 730 **Structure of the Wood.**—White and soft. It is easy to work and takes a good polish, but is little used on account of its small size. White-ants are not liable to attack it. Weight about 46℔ per cubic foot.

DOMESTIC. 731 **Domestic Uses.**—It is not a good fuel, since it possesses properties similar to those of **S. oleoides.**

SALVIA, *Linn.; Gen. Pl., II., 1194.*

732 **Salvia ægyptiaca,** *Linn.; Fl. Br. Ind., IV., 656;* LABIATÆ.

References.—*Benth. in DC., Prod., XII., 355; Boiss., Fl. Ort., IV., 631.*

Habitat.—A dwarf, scaberulous undershrub of the Panjáb plains, from Delhi westward; found also at altitudes up to 2,000 feet, and distributed to Afghánistán, Western Asia, and North Africa.

This species does not appear to be of any economic importance, or at least it is not distinguished economically from the following variety:—

733 **Var. pumila,** *Benth.*

Vern.—*Tukhm malanga,* PB.

References.—*Stewart, Pb. Pl., 172; Aitchison, Kuram Valley Rept., Pt. I., 183; Gazetteer, N.-W. P. (Bandelkhand), I., 83; Agri. Horti. Soc. Ind. Journal (Old Series), XIV., 6.*

Habitat.—A small undershrub, more scabrid and hispid than the preceding, found in the Panjáb plains and hills from Delhi westward, and distributed to Afghánistán and Balúchistán.

MEDICINE. Seeds. 734 **Medicine.**—The SEEDS are used in diarrhœa, gonorrhœa, and hæmorrhoids. Stewart thinks that from their vernacular name these seeds seem to be confounded in Native medicine with those of **Lallemantia Royleana.**

FODDER. Plant. 735 **Fodder.**—The PLANT is greedily browsed by sheep and goats in many parts of the Panjáb.

736 **S. lanata,** *Roxb.; Fl. Br. Ind., IV., 654.*

Habitat.—A herbaceous plant, found in the Western Himálaya at altitudes from 5,000 to 8,000 feet.

MEDICINE. 737 *Conf. with p. 482.* **Medicine.**—According to **Stewart** this species is often confused with **S. Moorcroftiana.** It may be used separately or as an adulterant, but the majority of writers who deal with this subject very probably refer to the more Alpine species.

738 **S. Moorcroftiana,** *Wall.; Fl. Br. Ind., IV., 654.*

Vern.—*Kalli jarri, shobri, thut, halu, gurgumna, laphra, papra,* PB.

References.—*DC., Prod., XII., 286; Stewart, Pb. Pl., 172; Gazetteer, N.W. P., X., 315; Agri.-Horti. Soc. Ind. Journal (Old Series), XIV., 4.*

Habitat.—A tall, robust, perennial herb, found in the North-West Panjáb plains, the Salt Range (according to **Stewart**), and the Western Himálaya from 6,000 to 9,000 feet.

MEDICINE. Root. 739 Seeds. 740 Leaves. 741 **Medicine.**—In the Panjáb most parts of this plant are given medicinally. The ROOT is prescribed in coughs, and the SEEDS are used as an emetic and in cases of hæmorrhoids. The LEAVES also are officinal. They are applied to the skin in cases of itch and as a poultice in boils and wounds. In Lahore, the seeds are given in colic and dysentery, and are applied to wounds.

SPECIAL OPINION.—"Is recommended in chronic affections of the skin" (*Civil Surgeon J. Anderson, M.B., Bijnor, N.-W. P.*).

FOOD. Stalks. 742 **Food.**—The STALKS are in some parts peeled and eaten. They have a mawkish sweet taste (*Stewart*).

Salvia officinalis, *Linn.* 743

GARDEN SAGE.

Vern.—*Salbia sefakuss* (Ainslie), HIND.

References.—*Voigt, Hort. Sub. Cal., 454; O'Shaughnessy, Beng. Dispens., 487; U. S. Dispens, 15th Ed., 1264; Firminger, Manual of Gardening for India, 159; Lisboa, U. Pl. Bomb., 168; Birdwood, Bomb. Pr., 66; Balfour, Cyclop., III., 510; Smith, Dic., 361; Journ. Agri.-Horti. Soc. (New Series), IV., 32; 1876-78, 44; Gaz., Mysore & Coorg, I., 64.*

Habitat.—The true Sage of European gardens is a native of the south of Europe, but has been introduced into England as a culinary herb. In the climate of India it is a very delicate plant, and can be kept alive through the hot and rainy seasons only with the greatest care (*Firminger*), so that its cultivation is practically confined to the gardens of the Europeans in India. The herb can easily be raised from seed in the cold weather, but Dr. Voigt states that plants introduced into the Calcutta Botanical Gardens in 1809, although they grew well, did not flower. The leaves are imported into India for culinary use, and Firminger recommends that no attempt should be made to allow the garden sage in India to flower, but that the plants should be raised annually from freshly-imported seeds, and the leaves plucked off, dried, and stored at the beginning of the hot season.

Medicine.—Sage is feebly tonic, astringent, and aromatic in its properties. By the ancients it was highly esteemed, but is now little used except as a condiment. It was at one time used in Europe as a substitute for tea (*U. S. Dispensatory*). MEDICINE. 744

Food.—The dried LEAVES, mostly imported from Europe, are used as a condiment by the Europeans in India. FOOD. Leaves. 745

S. plebeia, *R. Br.; Fl. Br. Ind., IV., 655.* 746

Syn.—S. BRACHIATA, *Roxb.*; (?) S. PARVIFLORA, *Roxb.*; OCIMUM FASTIGIATUM, *Roth.*; LUMNITZERA FASTIGIATA, *Spreng.*

Vern.—*Koka-buradi, bhú-tulsi*, BENG.; *Sathi, samúndar-sok*, PB.; *Kinro*, SIND; (The seeds—) *Kammar-kas*, BOMB.

References.—*DC., Prod., XII., 355; Roxb., Fl. Ind., Ed. C.B.C., 49; Voigt, Hort. Sub. Cal., 455; Dalz. & Gibs., Bomb. Fl., 209-210; Stewart, Pb. Pl., 172; Dymock, Mat. Med. W. Ind., 611; Atkinson, Him. Dist., 703; Drury, U. Pl., 380; Gazetteer, N.-W. P., I., 83; IV., lxxvi.; Ind. Forester, XII., App. 19.*

Habitat.—A stout annual herb, 6 to 18 inches high, met with throughout the plains of India, and ascending the hills to an altitude of 5,000 feet. It is distributed to China, the Malay islands, and Australia, but is absent from Ceylon.

Medicine.—The SEEDS are valued on account of their mucilaginous properties, and are given by Native practitioners in cases of gonorrhœa and menorrhagia. They are used in Bombay to increase the sexual powers (*Dymock*). MEDICINE. Seeds. 747

Domestic Uses.—[The mucilaginous SEEDS are employed to anoint women's hair and keep it glossy and in its place (*Trans., Med. & Phys. Soc., Bombay*). [The seeds are said to be used to kill vermin (*Dalzell*), but Dymock thinks this statement to be a mistake.—*Ed.*] DOMESTIC. Seeds. 748

SAMADERA, *Gærtn.; Gen. Pl., I., 310.*

Samadera indica, *Gærtn.; Fl. Br. Ind., I., 519; Wight, Ill., t. 68;* [SIMARUBEÆ. 749

This tree is the source of the NIEPA Bark of Commerce.

Syn.—NIOTA PENTA-PETALA, *Poir.*

Vern.—*Niepa*, TAM.; *Karinghota*, MALABAR; *Samadara*, SING.; *Kathai*-BURM.

References.—*DC., Prod., I., 592; Kurz, For. Fl. Burm., I., 200; Beddome, Fl. Sylv., 49; Gamble, Man. Timb., 64; Thwaites, En. Ceylon Pl., 70; Rheede, Hort. Mal., VI., t. 18; Grah., Cat. Bomb. Pl., 37; Dymock, Mat. Med. W. Ind., 2nd Ed., 147; Flück. & Hanb., Pharmacog., 133; U. S. Dispens., 15th Ed., 1744; Dymock, Warden, & Hooper, Pharm. Ind., I., 293; Year-Book Pharm. (1886), 196; Lisboa, U. Pl. Bomb., 36.*

Habitat.—A tree, 30 to 35 feet high, found in the Western Peninsula, South Konkan, and Malabar; also met with in Ceylon.

OIL. Seed. 750 **Oil.**—The SEED is brown, curved, and yields on expression an oil which is used medicinally.

MEDICINE. Bark. 751 Wood. 752 Oil. 753 Leaves. 754 Seeds. 755 Root. 756 **Medicine.**—The BARK is of a pale yellow colour; it contains a bitter principle called *Samaderin*, has a taste like quassia, and is used by the Natives on the Malabar coast as a febrifuge. An infusion of the WOOD is taken as a general tonic. Sandals made from the wood are supposed to keep off malaria and other diseases, but, probably, only from their protecting the feet and not from any medical property of the wood. The OIL extracted from the seeds are said to form a good local application in rheumatism. The bruised LEAVES are externally applied in erysipelas, and the SEEDS are worn round the neck as a preventive of asthma and chest affections (*Rheede*). The ROOT, as well as the seeds, is used medicinally by the Singalese. This drug may well be used as a substitute for quassia (*Dymock*).

CHEMISTRY 757 CHEMICAL COMPOSITION.—[**DeVrij** (1872) expressed from the seeds 33 per cent. of a light yellow bitter oil, which contains, according to **Oudemans**, 84 per cent. of olein and 16 per cent. of palmitin and stearin. The bitter principle, *samaderin*, was yellowish, and soluble in water and alcohol and amorphous; **Tonningen** (1858) had obtained it from the seed and bark in white scales, which became yellow with nitric or hydrochloric acid, and violet red with sulphuric acid. **Fluckiger** calls it quassiin. (*See Year-Book Pharm., 1886, p.196; Pharmacog. Indica.*)—*Ed.*]

TIMBER. 758 **Structure of the Wood.**—Light-yellow, soft, devoid of heartwood. Weight 26lb per cubic foot.

759 **Samadera lucida**, *Wall.; Fl. Br. Ind., I., 519.*

Syn.—NIOTA LUCIDA, *Wall.*

Vern.—*Ka-thay*, BURM.

Reference.—*Mason, Burma and Its People, 416, 764.*

Habitat.—A small tree, very nearly allied to **S. indica**, perhaps only a variety, found on the low grounds near the sea-coast in Burma and on the Andaman Islands.

MEDICINE. Leaves. 760 Plant. 761 **Medicine.**—The LEAVES are intensely bitter and taste like quassia. Perhaps this PLANT also possesses the virtues of the preceding (*Mason*).

Samara Ribes, *Kurz*; see **Embelia Ribes**, *Burm.*; MYRSINEÆ; Vol. [III., 242.

S. robusta, *Kurz*; see **Embelia robusta**, *Roxb.*; Vol. III., 243.

762 **SAMBUCUS**, *Linn.; Gen. Pl., II., 3.*

A genus of shrubs or small trees, comprising 10-12 species, found throughout all the temperate regions except in South Africa. Three species are natives of the Indian Peninsula, two of which are considered by the Natives to be of economic value.

763 **Sambucus Ebulus**, *Linn.; Fl. Br. Ind., III., 2*; CAPRIFOLIACEÆ.

THE DWARF ELDER, OR DANEWORT.

Vern.—*Richh kas, mushkiára, ganhúla gándal, gwándish, siske, tásar*, PB.

References.—*Stewart, Pb. Pl., 114; Brandis, For. Fl., 260; Gamble, Man. Timb., 213; Rovle, Ill. Him. Bot., 236; Treasury of Botany, II., 1013; Honigberger, Thirty-five years in the East, II., 340; Balfour, Cyclop. Ind., III., 513; Journal Agri.-Horti. Soc. Ind. (Old Series), XIV., 48; Ind. Forester, XIII., 68.*

Habitat.—A gregarious, herbaceous plant, growing from a perennial root-stock, found in Kashmír and some parts of the Western Himálaya, at altitudes between 6,000 and 10,000 feet. It is distributed to Europe, North Africa, and Western Asia.

Medicine.—The ROOTS and BERRIES have purgative properties, and are employed in Kashmír in the treatment of dropsy (*Honigberger*). MEDICINE. Roots. 764 Berries. 765

Domestic.—The smell of this plant, especially when bruised, is most fœtid like that of burnt flesh. Tinder is said to be made from its bark (?) on the Chenab (*Stewart*). DOMESTIC. 766

Sambucus javanica, *Blume; Fl. Br. Ind., III., 2.* 767

Syn.—S. WIGHTIANUS, *Wall.*; S. RUBRA, *Ham.*; S. THUNBERGIANA, *Miq.*; S. SP., *Griff.*

Vern.—*Galeni*, NEPAL.

References.—*Kurz, For. Fl. Burm., II., 3; Gamble, Man. Timb, 213.*

Habitat.—A small tree, with light-brown rather corky bark; met with in the Eastern Himálaya, from 4,000 to 8,000 feet and on the Khásia Hills. It is distributed to Java, China, and Japan.

Structure of the Wood.—White and soft, the pores small, and aggregated in groups. TIMBER 768

S. nigra, *Linn.* 769

THE COMMON ELDER OR BORE TREE; SUREAU, *Fr.*; FLIEDERBLUMEN, *Germ.*

Syn.—S. VULGARIS, *Lamk.*

Vern.—*Ukti khaman* (according to Ainslie) ARAB.

References.—*Voigt, Hort. Sub. Cal., 398; Ainslie, Mat. Ind., I., 118; Pharm. Ind., 109; Smith, Dict., 162; Treasury of Botany, II., 1013.*

Habitat.—A shrub or small tree reaching a height of 20 feet, native of all parts of Eupore and extending into Asia west of the Caucasus. It is cultivated in India to a small extent in botanical gardens, and is said to occur wild (?) in the Kangra District. The flowers are imported into India for medicinal use.

Medicine.—The FLOWERS are gently stimulant and sudorific. They are sometimes prescribed as a laxative to infants. In large doses they are said to cause nausea and diarrhœa. Elder flower water is used as a vehicle for other medicines, especially in lotions. It is mildly stimulant (*Watt, Calc. Exhib. Cat.*). MEDICINE. Flowers. 770

SPECIAL OPINION.—§ "Elder flower water is useful as an external application in urticaria" (*J. Parker, M.D., Deputy Sanitary Commissioner, Poona*).

SAND.

Sand, *Manual of Geology of India, Pt. I., 435.* 771

SABLE, *Fr.*; SAND, *Germ.*; ARENA, RENA, SAL BRUI, *Ital.*

Vern.—*Balu, reti*, HIND.; *Raml*, ARAB.; *Arena*, PERS.

References.—*W. W. Hunter, Statist. Acct. Beng., III., 372; Gazetteers:—Sind, 1, 22; Balfour, Cyclop. Ind., I., 917; III., 517.*

The following account of the occurrence of sand in India is an abstract of what will be found in the *Manual of the Geology of India* (*v.s. l.c.*).

SAND. **Occurrence of Sand.**

OCCURRENCE 772 OCCURRENCE.—Sand is found in most places along the Indian coast, and the rivers and streams of the Peninsula form immense deposits of it. Large tracts of blown sand, *i.e.*, sand drifted by the wind, form low hillocks on many parts of the Eastern coasts, *viz.*, north of Orissa in the Midnapore District, and southward at intervals along the whole of the East coast. The sand is derived from the sea-shore, and blown up into ridges at right angles to the prevailing wind. Small patches of it also are sometimes found on the banks of backwaters, and frequently the sand ridges extend inland for as far as two or three miles, the ground between the ridges being flat, and, in some cases, even marshy.

On the Malabar coast, sand dunes are equally common, and by accumulating on spits of sand, they contribute to the formation of lagoons or backwaters. In the northern portion of the west coast, about Bombay, no sand hills have been observed, probably because the detritus from the trap rocks there does not form a suitable material, but further north again, in Surat, Broach, and parts of Kathiawar and Cutch, there are considerable tracts of blown sand in the neighbourhood of the sea-shore.

Sand dunes, however, are not confined to the sea-shore, but occur to a greater or less extent along the beds of most of the rivers of India. By far the greatest and most important accumulation of blown sand, however, occurs in the tract of country known as the Great Indian Desert between Sind and Rájputána. This is a great sandy tract, which covers an area of upwards of 77.000 square miles. It is entirely destitute of streams of water, with but few hills of rock and, in the greater part of its extent, consists of dunes of blown sand called by the natives *thar, thur,* or *thul.* It is, however, neither absolutely barren, nor uninhabited; indeed, although the population is thin, villages are found throughout it, and immense herds of camels, cattle, sheep, and goats are pastured on the scanty vegetation. The general direction of the sand-drift is from south-west and south-south-west, the direction from which strong winds blow during the hot season, and it is evidently from this direction that the sand has accumulated. Many of the sand-hills are evidently of great antiquity; they often show marks of denudation from the action of rain, and in places are worn into ravines several feet in depth.

It is probable that this sand, which is indistinguishable from that of the sea-coast, has really been derived from the shore, and that in post-tertiary times the Ran of Cutch and the lower portion of the Indus valley were occupied by the sea. The form of the rocky ridges around Balmir and Jesalmir shows that they were shaped by subaerial not by marine denudation, and it is probable that the central portion of the desert was land, whilst the Indus valley, the Ran, and the Lúni valley, were occupied by sea.

Besides the Great Indian Desert, there are other large tracts in the Panjáb, repeating on a smaller scale the phenomena of the Thar and Rájputana desert. The most important of these is in the Sind-Sagar Doáb, between the Indus and Jhelum; but there is a barren tract in the Rachna Doáb between the Chenab and Ravi, and sand-hills occur in places also in the Bari Doáb between the Ravi and Sutlej.

USES. 773 **Uses.**—Except for mixing with clay to make pottery, etc., or with lime for mortar, sand is not much used in India for economic purposes. The clean sharp sand derived from the smaller streams is most employed, as sea-sand, owing to the presence of salt, is objectionable. Where *kunkar* lime is used, *surki* or pounded brick is preferred to sand, as it makes a stronger mortar. A valuable kind of fine sand, much used in Calcutta for making mortar, is dug up from the old bed of the Saraswati river at Magrah in the Hughli District.

Sandal wood, see **Santalum album**, *Linn.*, SANTALACEÆ.

SAND-BINDING PLANTS.

Sand-Binding Plants. 774

References.—*Baron F. von Mueller, Select Extra-tropical Plants, 465; Madden, Useful Native Plants of Australia, 85, 349, 637, 642, 643, 644; Man. Madras Adm., II., 27; R. & A. Dept. Correspondence regarding Sand-binding Plants, 1882-83; Agri.-Horti. Soc. Ind. Journal (Old Series), IX., 174; Balfour, Cyclop. Ind., III., 518.*

Cultivation of Sand-binding Plants.

CULTIVATION. 775

To a certain extent, sand-binding plants grow naturally along the sea shores of the Indian Peninsula, and on the margins of the inland sandy deserts, and some efforts have been made to cultivate such plants and so prevent the encroachments of sand blowing from the sea-shores, the deserts, and the beds of many of the great rivers on to the surrounding country. Much, however, remains to be done in this direction; but until the Natives are educated to an intelligent conception of the value of the efforts being made, little can be accomplished, as at present they are too apt, not appreciating the conservative design of the cultivation of these plants, to use them as fuel, and thus destroy their greatest protection from the advancing sand. Other countries have been more fortunate than India. Thus in Holland the great sea dyke owes its stability to these plants, which are carefully protected by Government; along the shores of Great Britain, as in Lincoln and Suffolk, etc., the quantity of dry land has been increased by the propagation of the Bent Star or Sand Carex (**Carex arenaria**) and the Lyme grass (**Elymus arenarius**), and in the Landes of Gascony, **Bremontier** recovered 100,000 acres of land from the blown sand by planting the cluster of pouch pine (**Pinus maritima**). In Australia, too, this class of plants has been largely utilised; indeed, **Baron F. von Mueller** enumerates no fewer than 80 genera, many species of which he recommends as useful to consolidate land on which blown sand has accumulated, and to prevent its further encroachments.

The following *précis* of the correspondence of the Government of India (Revenue and Agricultural Department) will perhaps best convey to the reader some idea of the efforts that have been made in India to cultivate this class of plants.

In August 1882, **Surgeon-General E. Balfour** wrote to the Private Secretary of His Excellency the Viceroy, suggesting that an effort might be made by the Agricultural Department to bind the loose sands blowing on and from the bed of the Indus, the Indian desert, as also the sands in the south of India, in the Tinnevelly and other districts, and that the seeds of the sand-binding plants referred to in his *Cyclopædia*, as well as those of other desert plants, might be collected and planted a little to windward of the sand tracts in double rows with a row on the edge of the sands. He also advocated that supplies of seed of sand-binding and sand-coast plants should be obtained from the Victorian Acclimatisation Society, from **Baron F von Mueller** of the Melbourne Botanic Gardens, and from the Cape Governments, and he added that the Bombay and Rájput Governments would also doubtless render assistance in the direction indicated. On receipt of this letter, the Revenue and Agricultural Department addressed the different Agri.-Horticultural Societies in India, the Governments of Madras, Bombay, the North-West Provinces, and Bengal, the Home Department, and the Acclimatization Society of Adelaide for any information that might be available on the subject of sand-binding plants. Replies from all these sources were received, abstracts of which are given below.

SAND-BINDING Plants.

Sand-binding Plants.

CULTIVATION

The *Director, Acclimatization Society, Melbourne,* merely referred the Government of India to **Baron F. von Mueller's** "Select plants for industrial culture and naturalisation" which appeared in 1880, and in which all the principal sand-binding plants are enumerated. The *Agri.-Horticultural Society of India* suggested the consultation of a paper by **Dr. Cleghorn**, published in 1855, in the ninth volume of the Journal of the Society, and reproduced by **Dr. Balfour** in his *Cyclopædia,* in which the cultivation of various plants indigenous to the sandy tracts in the Madras Districts, and their careful preservation under Government direction, is recommended for the purpose of consolidating the sands. The *Agri.-Horticultural Society of Madras* reported as to the protection afforded against blowing sands by the planting of **Casuarina** trees, and furnished a list of plants collected amongst the sand-hills to the south of Madras by the Superintendent of the Agri.-Horticultural Gardens, many of which, it was stated, might prove serviceable in fixing and binding blowing sands. The *Agri.-Horticultural Society of the Panjáb* replied that Hoshiarpur was the only place in the Province where attempts were made at resisting sand encroachments, and that there belts of the dense *múnj* grass (**Saccharum ciliare**) were planted with some success; but that it was not likely that any one plant would be found equally success ful throughout India. They recommended that the wild and easily grown weeds of each locality should be tried. The *Madras Government* forwarded a report from **Dr. Bidie**, in which he enumerated various sand-binding plants, all of which had been successful in varying circumstances in reclaiming sandy tracts. Protective measures against the encroachments of sand appear to have been undertaken at various periods in Madras, and with a fair amount of success, but as they were mainly due to the efforts of private individuals they were not continuous, nor did they extend over a wide tract of country. **Dr. Bidie** further noted that "at various parts along the Madras coast the phenomenon of natural drifted sand-heaps may be seen, and that these are mainly due to the presence of sand-binding plants, the locality frequented by these being that of the loose shifting sands. Among the varieties of sand-binding plants mentioned by **Dr. Bidie**, **Pandanus odoratissimus** is referred to as particularly useful when it is desirable to raise the sand-drift in large heaps, and at the same time to afford shelter from the sea-breeze. The Alexandrian laurel (**Calophyllum inophyllum**) and **Phænix sylvestris** are also said to be similarly valuable. **Casuarina muricata**, which thrives well near the coast, is mentioned as being a most important agent in the reclamation of waste sandy tracts. It is stated to be a most hardy plant, which will grow down to high-water mark, and even amongst loose sand. It acts also under certain conditions as a fructifier of the soil.

The *Superintendent of the Government Botanical Gardens, North-West Provinces and Oudh,* advised that sand-binding plants exclusively should not be used as an agent in the reclamation of waste sandy tracts. He referred to the action of nature in protecting desert-oases as pointing to the means that should be adopted in the case of sand-drifts. He urged that as these fertile patches are protected by forest vegetation which has sprung up spontaneously, the action of nature should be imitated by planting suitable trees and shrubs of quick growth in the sand-blown area to be protected. He recommended the planting of the trees in belts, facing more or less the direction from which the sand is usually blown. He added that in the event of these thriving and becoming established, a certain number of herbaceous plants of a binding character would soon of themselves make their appearance. *Dr. King, Superintendent of the Royal Botanic Gardens, Calcutta,* considered that the suitability of any particular

CULTIVATION

plant for sand-binding purposes was a question depending for its solution on local knowledge and local conditions, and said he had no doubt that suitable sand-binding plants could always be found in the immediate vicinity of the tract affected, but he doubted the feasibility of any one plant or set of plants being suitable for the widely varying climatic conditions of India. The *Bombay Government* reported that in Sind some fairly successful experiments had been made in the neighbourhood of Karachi, and also at Manora, the plant used in the latter place, where the experiments were most successful was **Ipomæa biloba.** The results of successful experiments made with the goats' foot creeper and the colocynth plant to stop the sand-drifts at Dumas near Surat were also referred to. This last experiment consisted in making a hedge (of dead *bávals* fixed with stakes driven into the ground) in the sandy soil near the border of the firm land. The sand blowing against the hedge formed a bank, and the belt of land behind was thus effectually protected and creepers grew freely over it. Another belt of land was then protected by another hedge, running parallel to the first, and so on till the margin of the sea was reached.

The above may be said to indicate the experiments that have been made in India with regard to the propagation of sand-binding plants, but it may be said that no experiments for this purpose can be successful, unless they are conducted over very considerable areas and under Government supervision, so that the plants grown may be allowed, to the fullest extent possible, to effect the purpose for which they are used, and not be interfered with on any pretext whatsoever. It would be well also, before introducing from abroad seeds of plants which might not thrive in India, to make extensive and long continued trial of the hardiest and most easily grown plants, indigenous to the various districts in which land protection is necessary. The appended list contains the more important and common sand-binding plants which occur in India :—

Acacia arabica, *Willd.*
A. eburnea, *Willd.*
A. Jacquemontii, *Benth.*
Agave americana, *Linn.*
Agrostis alba, *Linn.*
Alhagi maurorum, *Desv.*
Andropogon foveolatus, *Del.*
A. laniger, *Desf.*
Aristida depressa, *Retz.*
A. setacea, *Retz.*
Atriplex nummularia, *Lindley.*
Calotropis gigantea, *R.Br.*
C. procera, *R.Br.*
Canavalia obtusifolia, *DC.*
Capparis aphylla, *Roth.*
C. spinosa, *Linn.*
Casuarina equisetifolia, *Forst.*
Cenchrus catharticus, *Del.*
C. montanus, *Nees.*
Eleusine ægyptica, *Pers.*
E. flagellifera, *Nees.*
Eleusine scindica, *Duthie.*
Elionurus hirsutus, *Munro.*
Indigofera sp.
Hydrophylax maritima, *Linn.*
Ipomæa biloba, *Forsk.*
Jatropha Curcas, *Linn.*
J. glandulifera, *Roxb.*
Launæa pinnatifida, *Cass.*
Melanocenchris Royleana, *Nees.*
Opuntia Dillenii, *Haw.*
Pandanus odoratissimus, *Willd.*
Pennisetum cenchroides, *Rich.*
Perotis latifolia, *Ait*
Pupalia orbiculata, *Wight.*
Saccharum ciliare, *Anders.*
Salvadora oleoides, *Dcne.*
S. persica, *Linn.*
Spinifex squarrosus, *Linn.*
Sporobolus orientalis, *Kunth.*
Tamarisk gallica, *Linn.*
Zizyphus nummularia, *W. & A.*

SANDORICUM, *Cav.; Gen. Pl., I., 333.*

Sandoricum indicum, *Cav.; Fl. Br. Ind., I., 553;* MELIACEÆ. 776

Syn.—S. NERVOSUM, *Blume;* S. TERNATUM, *Blanco;* S. GLABERRIMUM, *Hassk.;* TRICHILIA NERVOSA, *Vahl.;* MELIA KŒTJAPE, *Burm.;* T. VENOSA, *Spreng.*

Vern.—*Thitto*, BURM.; *Santor*, MALAY.

References.—*Roxb., Fl. Ind., Ed. C.B.C., 368; Kurz, For. Fl. Burm., I., 217; Beddome, Fl. Sylv., 55; Gamble, Man. Timb., 72; Grah., Cat. Bomb. Pl., 31; Mason, Burma, 457, 525, 759; Rumphius, Amb., I., 167, t. 64; Lisboa, U. Pl. Bomb., 42; Gazetteer, Mysore & Coorg, I., 52; Ind Forester, I., 363; XIII., 134; Journ. Agri.-Horti. Soc. Ind. IX., Sel., 40, 57.*

Habitat.—A lofty, evergreen tree of Burma, introduced into Southern India. It is distributed to the islands of the Malayan Archipelago.

MEDICINE. Root. 777 **Medicine.**—Rumphius says that the ROOT, bruised with vinegar and water, was used by the Amboyans in his time as a carminative, and that some Native practitioners vaunted this preparation as an excellent medicine in cases of diarrhœa and dysentery.

FOOD. Fruit. 778 **Food.**—The FRUIT is of the size of an orange, and is occasionally called the wild mangosteen, from its resemblance to that fruit. It has a fleshy acid pulp and makes a very good jelly, but has a peculiar odour. The Natives eat the fruit raw, and esteem it excellent (*Mason*). Rumphius states that it was much used by the Amboyans, both raw and cooked with fish, in place of lemons.

TIMBER. 779 **Structure of the Wood.**—Sapwood grey; heartwood red, moderately hard, close grained, takes a beautiful polish. Weight 36℔ per cubic foot.

DOMESTIC. Wood. 780 **Domestic Uses.**—The WOOD is used for carts and boat-building. Rumphius states that the wood was employed by the Amboyans for house-building, and was thought by them a particularly good and durable timber for the purpose.

SANDSTONE.

781 SANDSTONE. **Sandstone,** *Manual, Geology of India, Pt. I., 52, 69, 96; Pt. II., 486, 503; III., 540.*

SANDSTONE, FREESTONE, MILLSTONE GRIT, *Eng.*; GRES, *Fr.*; SANDSTEIN, *Germ.*; PIETRA ARENARIA, *Ital.*

References.—*Mason, Burma and Its People, 587, 735; Baden Powell, Pb. Pr., 35, 36; Settle. Rep., Central Provinces (Chanda Dist.), 106; (Upper Godavery Dist.), 5; Gazetteers:—Bombay, VI., 11; Central Provinces (1870), 59; Balfour, Cyclop. Ind., III., 519.*

OCCURRENCE 782 OCCURRENCE.—The following account of the occurrence of sandstone in India has been mainly abridged from the *Manual of Geology* (*v. s., l.c.*).

The two great rock systems of India, the Vindhyan and the Gondwána, are essentially sandstone formations. The term—Vindhyan formation—was at first employed as a collective name for the beds in the great rock basin extending in an east and west direction from Sasseram, to Nimach, a distance of 600 miles, and for 300 miles north and south from Agra to Hoshangabad. Throughout the greater part of their extent, the Vindhyan sandstones are unconformably related to transition or gneissic rocks; but in the eastern part of their area, in Bundelkhand and the Son valley, they rest upon thick deposits nearly related to themselves, to which the name of Lower Vindhyan has been applied, while the original Vindhyan formation has now been distinguished as the Upper Vindhyans. The Lower Vindhyan is principally a limestone formation with subordinate bands of sandstone and shale, while the upper division of the Vindhyan system is, in the main, a sandstone formation, with distinct bands of shales, mostly coarse and flaggy.

The Gondwána system takes its name from the old term applied to the countries south of the Narbada valley (formerly the Gond kingdoms, and now forming the districts of Jabalpur, Nagpur, and Chhatisgarh in the Central Provinces). In this region, the most complete sequence of the formations constituting the present Gondwána system, is to be found, although it is, as a whole, very widely distributed in the Indian Peninsula.

OCCURRENCE

Representatives of this group occur in Sikkim, Bhután, and the Daffla hills, in Cutch, resting on marine jurassic rocks, and capped by Neocomian beds, in the desert between Sind and Rájputána, and on the east coast. The main occurrence, however, is south of a line formed by the valleys of the Narbada and Son, and north-east of a line drawn from the sea at Masulipatam through Khamamet and Warangal, north-east of Hyderabad. The main areas of the Gondwána rocks are in the Rájmahál hills and Damuda valley in Bengal, the Tributary Mahále of Orissa, Chhatisgarh, Chutia Nagpur, the Upper Son valley and Sátpura range south of the Narbada valley and the Godávari basin. Nearly the whole of the strata composing the Gondwána series is probably of fluveatile origin. This system also may be divided into an upper and lower series, the Tálchir, Dámuda, and Panchet groups, with their equivalents, being referred to the lower, while to the upper belong the Rájmahál, Mahadeva, and Jabalpur groups. The rocks of all these groups consist mainly of sandstones and shales of various kinds, intermixed with coal-bearing strata in the Damuda and Jabalpur groups, with ferruginous bands in the Mahadevi and Damuda groups and with lava or trap in the Rájmahál.

Besides the sandstone widely distributed in the two great systems above described, good sandstones are found in the Bagh beds which belong to the cretaceous period and occur to the south of Allirajpur and Bagh. Sandstones occur too among the rocks of the Siwalik and Nahan groups (which represent the upper and middle tertiary period of Europe), and at many Himálayan stations among the Eocene groups of the Sub-Himálayan series.

USES. 783

Uses of Sandstone.—The following note upon the 'Sandstones' of India has been kindly furnished by **Mr. H. B. Medlicott**, late Director of the Geological Survey:—"The rocks of the Vindhyan and Gondwána systems yield in many cases sandstones admirably adapted for building purposes. From Chunar, whence all the finer stone used in Calcutta is procured, to Ruphas in Bhartpur, whence the stone was taken for all the great buildings at Delhi, the rocks of the Vindhyan plateau have furnished sandstone to all the cities of the plains. In some places, especially in the southern basins of the lower Vindhyan rocks, this sandstone is often more or less altered into a quartzite and is no longer a freestone suitable for fine masonry. The Gondwána sandstone is generally coarser than the Vindhyan, but admirably fine sandstone from this formation has been quarried at several places along the Satpura range in the Narbada valley by the Great Indian Peninsula Railway, and is extensively used from Jabalpur to Khandwa. The sandstones of the jurassic beds in Kathiawar and Cutch have been much used locally as building stones.

The tertiary group in the central zone of the peninsula occasionally affords good stone, and has been extensively worked at Rajahmundry. Of the great tertiary series in Sind, the limestones are generally preferred to the sandstones. The molassic sandstones of the Sub-Himálayan zone are generally too soft for use, but the coarse hard sandstone of the Eocene series has been much used at the hill stations of Murree, Dharmsala, Dagshai, Sabathu, and Kasauli." "Besides being used for building purposes the sandstones of Bhartpur and of Chunar were largely used for telegraph posts; the facility with which some of the varieties split rendered it possible to obtain posts 16 feet long of material which would resist white ants and the action of the weather. These have of late years, however, been replaced by pillars of galvanised iron, as they were found liable to snap in two during strong hurricanes" (*Ball*). Sandstones suitable for millstones and for grindstones are found in many parts of India, and are utilised locally as such.

Flexible. 784

Sandstone, Flexible.—[The well-known flexible sandstones of India are obtained at Kariána, 60 miles due west from Delhi. It is a locally decomposed condition of a band of gneissose quartzite that is much quarried for quernstones (hand-mills) (*Man. Geology, India, I., 52*).—*Ed.*]

SANSEVIERIA, *Thumb.; Gen. Pl., III., 679.*

785 **Sansevieria zeylanica,** *Willd.; Fl. Br. Ind., VI., 270;* HÆMODORACEÆ.

THE BOW-STRING HEMP.

Syn.—It seems doubtful if the Indian plant (S. ROXBURGHIANA, *Schult.*) should be viewed as the same as that met with in Ceylon or distinct. [**Sir J. D. Hooker** (in the forthcoming volume of the *Flora of British India*) appears to regard it as distinct, and as possiby indigenous to India, but confined to the Western Peninsula and Ceylon. **Roxburgh** speaks of it as cultivated for its fibre. For all practical purposes both species may be regarded as one and the same, and as of equal value for their fibres.—*Ed.*]

Vern.—*Marúl, múrvá,* HIND.; *Murba, murahara, murgli, goráchakra, murgá, murgábi,* BENG.; *Murgali, morwa, ghannasaphan,* BOMB.; *Ghonasaphan, nagfan,* MAR.; *Murgali,* DEC.; *Marúl, marúl-kalung,* TAM.; *Mailai, mangi,* SALEM; *Tshama-cada, chága, sága,* TEL.; *Niyanda,* SING.; *Marura., muruvá,* SANS.

References.—*Roxb., Fl. Ind., Ed. C.B.C., 294; Dalz. & Gibs., Bomb. Fl., App., 91; Voigt, Hort. Sub. Cal., 656; Trimen, Sys. Cat. Cey. Pl., 93; Sir W. Jones, Treat. Pl. Ind., V., 108; Ainslie, Mat. Ind., II., 192; U. C. Dutt, Mat. Med. Hind., 310; Bidie, Cat. Raw Pr., Paris Exh., 118; Dymock, Mat Med. W. Ind., 2nd Ed., 842; Drury, U. Pl. Ind., 381; Useful Pl. Bomb. (XXV., Bomb. Gaz.), 236; Liotard, Mem. Paper-making Mat., 5, 6, 14, 18; Indian Fibres and Fibrous Substances, Cross, Bevan, King & Watt, 45; Christy, New Com. Pl., VI., 13, 43; Man. Madras Adm., I., 360; Nicholson, Man. Coimbatore, 40; Proceedings of the Government of Madras (Revenue Dept.) (1884), 187-188; Gazetteer, Mysore & Coorg, I., 67; Agri.-Horti. Soc. Ind., Trans., VIII. (Pro.), 381; Journal (Old Series), III., 23-26, 224, 226, (Sel.), 89-91, (Pro.), 41, 54, 62, 68; IX., 151; (Pro.) 149; X., 347; (New Series), IV., Pro., 1873, 1; VIII.5, 118, 121, 124; Ind. Forester, IX., 274; X., 89; XII., 40.*

Habitat.—A stemless bush, with perennial roots, and a rosette of six to eight succulent, radical leaves, the inner of which are often 4 feet long and end in a long straight spine, the scape, 1 to 2 feet long, rising from the centre of the leaves. Flowers racemose, greenish-white, erect, four to six together in clusters. It is found on the eastern coast of India from Bengal to Madras; common on the Coromandel coast, in Cumbum, and in the Dindigal district. [**Dalzell & Gibson** say that "in Malabar, of which country it is a native, it does not produce seed." In India it may be said to exist under cultivation mainly. It is distributed to Ceylon, Java, and the coasts of China and Africa.—*Ed.*]

FIBRE. Leaves. 786 **Fibre.**—From the succulent LEAVES is extracted a fibre, held in high esteem by the Natives on account of its elasticity and consequent suitableness for bow-strings. **Sir W. Jones** says: —"From the leaves of this plant the ancient Hindus extracted a very long elastic thread, called *Maurvi,* of which they made bow-strings, and which for that reason was ordained by **Menu** to form the sacrificial zone of the military classes." [**Roxburgh**, in his detailed account of this fibre, makes the following somewhat interesting remark: "I am inclined to think that the fine line, called China grass, which is employed for fishing lines, fiddle strings, etc., is made of these fibres."* **Roxburgh** thus would seem to have regarded China grassand Rhea as two widely distinct fibres. In his experiments 80℔ of the fresh leaves, yielded 1℔. of the clean dry fibre. He, therefore, concluded that the plant might be cultivated with advantage on account of its fibre.—*Ed.*] It is, in fact, easily cultivated. The fibre is used for the preparation of cordage

* Could this have been the Herba Bengalo alluded to by early travellers? *Conf. with Vol. IV., 223.—Ed.*

FIBRE.

and matting in the regions where it occurs, and is much valued in Europe for ropes used in deep-sea dredgings. In Trichinopoly it is employed for the manufacture of paper; but was reported on by **Messrs. Cross, Bevan, King & Watt** as too expensive for the paper-maker, except for very special qualities of paper.

The fibre is pliant, soft, and silky, and much resembles that of the pine-apple. It is usually prepared by taking the fresh leaves and placing one of them on a smooth board which is raised at one end. The lower end of the leaf is then pressed down by the toe of the workman, who squats on the plank, and with a blunt knife or piece of iron plate scrapes upwards along the surface of the leaf, and thus deprives it of its fleshy pulp by successive scrapings, turning the leaf over and over, as may be necessary. When the pulp is thoroughly removed, the fibre is washed for three or four minutes, and dried in the shade. Washing in brackish or salt water or continuous soaking in water is said to destroy the glossy white appearance of the fibre. With reference to the strength of *Murva* fibre **Dr. Royle** made some comparative experiments with this and Agave fibre, which showed that the two were about equal in strength. A series of comparative experiments was also made in 1838 by the Marine Board of the East India Company, who reported that "it was not equal in strength to the Europe or Manilla hemp, but that it seemed to take hot tar as well as the latter, and would answer generally for the same purposes as those to which the Europe and Manilla cordage is applied." In a further report it was stated that 40 maunds of the fresh plant produced one maund of fibre. The expenses, however, of the experiment were high, and the best methods of separating the fibre had not been followed.

The thread made from *Murva* fibre is sometimes woven into fine cloths, which readily take on various dyes.

MEDICINE. Root. 787

Medicine.—The fleshy ROOT is warm to the taste and of a not unpleasant odour. It is prescribed by Native practitioners in the form of an electuary, for consumptive complaints and coughs of long standing, to the quantity of a small teaspoonful twice daily (*Ainslie*). The JUICE of the tender shoots of the plant, is administered to children to clear their throats of phlegm (*Hort. Malab., II., 83*). The juice of the root and leaves is used as an antidote for snake bite, especially that of the Russell's viper (the *ghannas* snake, hence the Mahratta name of *Ghannasaphan*) (*Dymock*).

Juice. 788

SANTALUM, *Linn.; Gen. Pl., III., 224.* 789

The genus **Santalum** is described by **Bentham & Hooker** in the *Genera Plantarum* as composed of eight species, all closely allied to one another; they are indigenous to the East Indies, the Malaya Archipelago, Australia, and the islands of the Pacific Ocean. **Santalum album** is described in the *Flora of British India* as the only Indian representative of the genus; **S. myrtifolium**, *Roxb.*, being regarded as a synonym for **S. album** and not a distinct species, nor even a variety, as **DeCandolle** believed it to be.

Santalum album, *Linn.; Fl. Br. Ind., V., 231;* SANTALACEÆ. 790

THE WHITE SANDAL-WOOD TREE; BOIS DE SANTAL CITRIN, *Fr.;* WEISSES SANDELHOLZ, *Germ.*

Syn.—S. MYRTIFOLIUM, *Roxb.;* S. VERUM, *Linn.;* SIRIUM MYRTIFOLIUM, *Roxb.;* SANDALUM ALBUM, *Rumph.*

Vern.—*Chandal, sandal, chandan, chandoie* (=the tree), *safed chandan* (=the white wood), HIND.; *Chandan* (=the tree), *pitchandan* (=the yellow wood), *sufaid. chundun, srikhanda* (=the white wood), BENG; *Chandan* (=the tree), PB.; *Sukhad* (=the tree), SIND; *Chandan, sandal* (=the tree), *safed chandan* (=the white wood), BOMB.; *Gandhácha-koda, chandan,* MAR.; *Súkét, sukhud,* GUZ.; *Sundel, shandanak-kattai* (=the wood), TAM.; *Gandhapu-chekka, hari chandanam* (=the yellow wood), *chandanam, rakta krishna chandanam, tella chandanam, chanda napu chettu* (=the tree), *pita chandanam* (yellow wood), TEL.; *Gandha, gandada* (=the tree), *gandhaká-chekke* (=the wood), KAN.; *Chandana-mutti* (=the tree), *tsjandana-marum* (the wood), MALAY.; *San-ta-ku, ka-ra-mai, san-da-ku, nasaphiyu,* BURM.; *Rat-hihiri* (=the wood), SING.; *Chandana,* (=the tree), *krishna chandanam* (the dark-coloured heart-wood), *pitachandana* (yellow sandal), *srikhanda* (white sandal), *malayaja, bhadrasri, gandhasra, hari chandana* (fine old wood), SANS.; *Sandal-abiyaz* (=the wood), ARAB.; *Sandal supéd* (=the white wood), PERS.; *Tan-muh,* CHINESE.

References.—*DC., Prodr., XIV., 683; Roxb., Fl. Ind., Ed. C.B.C., 148; Voigt, Hort. Sub. Cal., 303; Brandis, For. Fl., 398; Kurz, For. Fl. Burm., II., 329; Beddome, Fl. Sylv., t. 256; Gamble, Man. Timb., 321; Dalz. & Gibs., Bomb. Fl., 224; Graham, Cat. Bomb. Pl., 177; Sir W. Elliot, Fl. Andhr., 34, 68, 100, 154, 162, 175; Sir W. Jones, Treat. Pl. Ind., V., 84; Rumphius, Amb., II., 42, t. 11; Pharm. Ind., 197, 461; British Pharm., 293; Flück. & Hanb., Pharmacog., 599; U. S. Dispens., 15th Ed., 1039, 1745; Fleming, Med. Pl. & Drugs (Asiatic Reser., XI.) 181; Ainslie, Mat. Ind., I., 376; O'Shaughnessy, Beng. Dispens., 532; Irvine, Mat. Med. Patna, 20; Medical Topog., 130; Moodeen Sheriff, Supp. Pharm. Ind., 223; U. C. Dutt, Mat. Med. Hind., 224, 295, 299; Sakharam Arjun, Cat. Bomb. Drugs, 119; Murray, Pl. & Drugs, Sind., 201; Bidie, Cat. Raw Pr., Paris Exh., 40, 62; Bent. & Trim., Med. Pl., IV., t. 252; Dymock, Mat. Med. W. Ind., 2nd Ed., 751; Year-Book Pharm., 1879, 467; 1882, 108; Birdwood, Bomb. Prod., 335; Baden Powell, Pb. Pr., 369; Drury, U. Pl. Ind., 383; Useful Pl. Bomb. (Vol. XXV., Bomb. Gaz.), 133, 204, 224, 395; Forbes Watson, Indian Prod., 141, 142, 170, 270; Kew Bulletin, 1888, 136; Piesse, Perfumery, 201; Ayeen Akbary (Gladwin's Trans.), Vol. I., 87, 91; II., 58; Ain-i-Akbari (Blochmann's Trans.), Vol. I., 81; Linschoten, Voyage to East Indies (Ed. Burnell, Tiele and Yule), Vol. II., 102, 103; Milburn, Oriental Commerce (1813), Vol. I., 290; (1825), 158, 159; Buchanan, Journey through Mysore and Canara, &c., Vol. I., 186, 202; II., 117, 132, 536; III., 192; Man. Madras Adm., Vol. I., 362; II., 72; Moore, Man. Trichinopoly, 79; Gribble, Man. Cuddapah, 71; Bombay Admin. Rep., 1872-73, 375; Gazetteers:—Bombay, IV., 23; VII., 40, 41; VIII., 261; XV. Pt. II., 70; XVIII., 44; Mysore & Coorg, I., 50, 66; III., 21; Agri.-Horti. Soc. Ind.:—Transactions, II., (App.) 314; Journal (New Series), I., 179; Ind. Forester, I., 28; II., 19; III., 271; VI., 321; VII., 1; VIII., 411; IX., 63, 75; X., 199, 247, 262, 318, 403, 550; XI., 273; Spons' Encycl., 1430, 1527; Balfour, Cyclop. Ind., III., 517.*

Habitat.—A small, evergreen tree, which rarely attains a height of over 40 feet. It grows naturally in the drier parts of Mysore, Coimbatore, and Salem districts and is extended south to Madura and north to Kolhapur; found generally at elevations of from 2,000 to 3,000 feet. It is a delicate tree, and suffers much from accidental injuries inflicted on the bark and stem, so that it flourishes most when protected by hedgerows and thorny jungles. It is cultivated and grows well in Bombay, Poona, and Gujerat and some parts of Northern India; but in regions out of its indigenous habitat, it usually loses to a great extent, if not altogether, the aromatic heartwood for which it is chiefly valued.

CULTIVATION.

CULTIVATION. 791

Before Mysore came under the protection of the East India Company no pains were taken in the cultivation of this valuable tree, although it was even then a Government monopoly, and very severe laws existed to prevent any person from cutting sandal-wood without permission. It would thus appear that fears were entertained that the supply of sandal-wood, at that time, seemed likely to become diminished; but better means of propagation and cultivation have since been instituted, and as it still continues a Government monopoly, the supply of the wood has gradually

CULTIVATION in Mysore.

increased, and its quality improved. In Madras cultivation of the wood is free, but the chief plantations are in the reserved forest. In Mysore, on the other hand (as already stated), the industry is a State monopoly. Formerly the seeds were chiefly spread through the agency of birds, but now, in most of the sandal-wood districts, immense plantations of the young trees exist, which are regularly transplanted into the open in the beginning of the second year of their growth.

The best methods of growing sandal-wood have been much discussed by Forest Officers, but the general conclusions arrived at seems to be as follows:—"The Sandal fruits are gathered while quite fresh, spread out to dry, and stored up in a dry place till before the beginning of the rains. The seed is then sown on *tilepot beds,* just covered with a mixture of sand and leaf manure. From then till the plants are transplanted, a year afterwards, the beds are kept constantly covered with old leaves, dead grass or any litter at hand. The leaves and litter, if properly watered, decay rapidly and require to be replenished frequently. The portion of each nursery under Sandal is shaded with boughs, so as to afford a broken half-shade similar to that in which sandal comes up naturally, in thickets and hedges. Transplanting begins the following year, as soon as the ground is thoroughly moistened by the first rains. The tilepots are lifted and carried away to pits, newly filled up with fresh mould. One tile is gently removed from the seedling, and the cylinder of earth and root resting on the other tile is slipped into the ground. The earth is filled in, and the remaining tile gently pulled out. The plants should be watered for a day or two after they are put out. *In situ* sowings of sandal seed were tried from 1868 to 1878 in Mysore, but the results were most unsuccessful, and it was not until the introduction of tilepot nurseries that regular planting was attended with any certainty of success. Sandal sowings *in situ* have been tried with gingelly, castor oil, and other plants as nurses, but the results were not sufficiently favourable to invite a repetition of the experiments. Doubtless the shade afforded by these plants is beneficial, but at the same time the drain on plant-food and on subsoil moisture is similar to that of very energetic weeds. Sandal cuttings cannot be propagated, even root-cuttings, which would seem likely to succeed, are in practice a failure" (*D. E. Hutchins, Ind. Forester*). The Sandal-wood tree is a slow grower and suffers much during the first years of its life from the depredations of animals. Thus, in the same paper, we find the following passage:—"During the first year, Sandal in nurseries should grow about 10 inches in height, but want of attention in keeping the plant supplied with leaf manure and water, will give sickly yellow plants not more than 4 or 5 inches in height at the end of the first year. On good soil in a yard cube pit, Sandal should be from 1 foot to 2 feet in height at the end of the second year from seed; and when planted side by side with **Casuarina** the Sandal will occasionally be higher than the **Casuarina**; but this rapid growth is not maintained beyond the first few years, and there are always considerable differences among Sandal trees of the same age, and growing side by side. When young, Sandal has to contend with many enemies, and at the end of two or three years there are more differences in the appearance of the growth than with **Casuarinas** or Blue gums of the same age. The smooth succulent character of the leaves of Sandal doubtless contributes to render them the favourite food for hares and deer. When planting Sandal, it is usually necessary to place thorns over each plant to keep off hares. If spotted deer are abundant in the locality, it becomes necessary to fence plots of Sandal planting, Self-sown seedlings of Sandal are rarely seen, except among clumps of thorns or other bushes, where they are naturally protected from browsing. The Sandal-wood tree attains its

CULTIVATION in Mysore. commercial maturity, *i.e.*, the age at which it pays best to cut it down at 27 to 30 years. At this period the heartwood is well developed (*i.e.*, at a general depth of about 2 inches below the surface), and the growth of this is so slow that it cannot in a year attain an increased value equal to the interest on its present selling price, *plus* the value of the space it would occupy. It is therefore found most profitable to cut it down between the age of 27 to 30 years." "It is felled at the end of the year, the largest roots which contain a very fine quality of the wood uprooted, and the trees are stripped of their bark and conveyed to various depôts, where they are cut into billets which are carefully dressed and sorted according to the quality of the wood. These billets form the Sandal-wood of commerce and are sold by weight at an annual auction where Native merchants congregate from all parts of India to make purchases. The pieces that are straight and have most heartwood fetch the highest prices, as the fragrance for which they are so much prized depends on the presence of essential oil which is chiefly situated in the dark central wood of the tree" (*Dr. G. Bidie, Memorandum on Sandalwood*).

OIL. Seeds. 792 Wood. 793 Roots. 794

Oil.—The SEEDS of the Sandal-wood tree yield by expression a thick and viscid oil, which is burnt by the poorer classes in lamps. Sandal-wood essential oil is distilled from the WOOD. The ROOTS yield the largest quantity and finest quality. The white or sap wood is rejected for distillation. The yield is about 2½ per cent. Sandal-wood oil is transparent, but of a pale yellow colour, and is one of the most favoured of Indian perfumes, especially among Muhammadan gentlemen.

Chemistry. 795

It has a resinous taste and a peculiar odour. Its specific gravity is about 0·980. According to P. Chapoteaut, in the *Year-Book of Pharmacy for 1882*, "100 parts of Sandalwood yield, upon distillation with steam, 1·25-2·8 parts of this essential oil, which is a thick liquid of ·945 specific gravity, and boils at 300-340°C. The oil consists of two substances boiling at 300 and 310°C. respectively, and answering to the formulæ $C_{15} H_{24}O$ and $C_{15} H_{26}O$. The latter of these bodies is an alcohol, and the former the corresponding aldehyde. Phosphoric anhydride absorbs water from both, converting them into hydrocarbons of the formulæ $C_{15} H_{22}$ and $C_{15} H_{24}$ respectively."

"The Mysore Government has long had establishments for extracting the oil, which is sold at the annual auction along with the wood, and chiefly bought up for exportation to China and Arabia. It is procured from the wood by distillation, the roots yielding the largest quantity and finest quality of oil. The body of the still is a large globular clay pot with a circular mouth, and is about 2½' deep by 6⅓' in circumference at the bilge. No capital is used, but the mouth of the still when charged is closed with a clay lid, having a small hole in its centre, through which a bent copper tube about 5⅓' long is passed for the escape of the vapour. The lower end of the tube is conveyed inside a copper receiver, placed in a large porous vessel containing cold water. When preparing the Sandal for distillation, the white or sapwood is rejected, and the heartwood is cut into small chips. Distillation is slowly carried on for ten days and nights, by which time the whole of the oil is extracted. As the water from time to time gets low in the still, fresh supplies are added from the refrigeratory. The quantity of oil yielded by wood of good quality is at the rate of 10 oz. per maund" (*Bidie, Memo. on Sandal-wood*). In **Spons'** *Encyclopædia* it is stated that "the yield from good wood is at the rate of 2½ per cent.; European distillers do not succeed in getting more than 30 oz. from 1 cwt."

MEDICINE. Wood. 796

Medicine.—Sandal-wood is described in Hindu medical works "as bitter, cooling, astringent, and useful in biliousness, vomiting, fever, thirst, and heat of the body. An emulsion of the WOOD is used as a cooling

MEDICINE.

application to the skin in erysipelas, prurigo, and sudamina" (*U. C. Dutt. Hind. Mat. Med.*). Ground up with water into a paste, it is commonly applied to local inflammations, to the temples in fevers, and to skin diseases, to allay heat and pruritus. It also acts as a diaphoretic. Sometimes a paste is made together with the juices of herbs, such as purslain, nightshade, etc (*Dymock*). The author of *Makhzan-ul-Adwiya* describes the wood as cold and dry, cardiac, tonic, astringent, alexipharmic, antiaphrodisiac, a resolvent of inflammatory swellings, etc. He recommends an emulsion in bilious fever on account of its cooling and protective influence over the heart, brain, stomach, etc. In the Konkans, Sandal-wood OIL with lime-juice and camphor is used as a cooling application to eruptions, etc., and a CONSERVE of sandal-wood made by boiling the wood cut up in small pieces until it is quite soft, with water in which a small quantity of impure carbonate of soda has been dissolved, and then preserving it in a thick syrup, is taken internally for the same purpose. In cases of morbid thirst the POWDER of the wood is recommended to be taken in cocoanut water, and in hot weather, after bathing, it is rubbed over the skin to cool it, allay the irritation of prickly heat, and check too copious perspiration (*Murray; U. C. Dutt; Lisboa*). From very ancient times Sandal-wood, in one or other of its forms, has been regarded throughout eastern countries as a very valuable remedy for gonorrhœa. Thus we find **Ainslie** stating that Sandal-wood in powder was given, in cow's milk, by the Vytians in this class of cases, and **Rumphius** also says that it was much esteemed by the Natives of Amboyna for the same purpose. **Dr. Henderson** of Glasgow was the first who directed the attention of European physicians to the use of the oil as a remedy for that disease, and since his time it has been employed internally in many cases where copaiba and cubebs had previously failed. Its action is similar to that of copaiba, but it does not communicate an unpleasant odour to the urine as the latter substance does.

Oil. 797

Conserve. 798

Powder. 799

It is usually given in doses of 10 to 30 minims, twice daily, either in capsules or as an emulsion.

SPECIAL OPINIONS.—§ "The oil is used internally as a remedy for gonorrhœa. The wood, rubbed up with water on a stone, is used as an application in prickly heat" (*Civil Surgeon J. H. Thornton, B.A., M.B., Monghyr*). "The oil, either alone or with copaiba, is a very good remedy for gonorrhœa. The wood, ground up into a paste, is used externally as noted in the above section" (*Surgeon-Major L. Beech, Cocanada*). "Demulcent, stimulant, doses 30 to 40 min., used in gonorrhœa, and gleet" (*1st Class Hospital Assistant Choonia Lall, City Branch Dispensary, Jubbulpore*). "Relieves headache and irritation in various skin diseases" (*Assistant Surgeon S. C. Bhattacharji, Chanda, Central Provinces*). "Used locally for headache" (*Assistant Surgeon Nehal Sing, Saharunpore*). "The oil, distilled from the wood, dissolved in spirit or mixed with sugar or mucilage, is tried in gonorrhœa in doses of 30 drops morning and evening, and has been found very useful" (*Assistant Surgeon N. R. Banerji, Etawah*). "The white sandal is used as a cooling application to the temple in case of headache as well as to the body in general in prickly heat" (*Assistant Surgeon R. Gupta, Bankipore*). "The wood ground up with water into a paste has been found very efficacious in pruritus, more particularly when due to heat, and also in sudamina" (*Civil Surgeon D. Basu, Faridpore, Bengal*). "A bolus of ground sandal checks hæmoptysis in its mild form, when taken twice a day for two or three days" (*Native Surgeon T. R. Moodelliar, Chingleput, Madras Presidency*).

TIMBER. 800

Structure of the Wood.—Sapwood white and scentless; heartwood yellowish brown, strongly scented, very hard, close-grained, and oily. Weight

TIMBER. about 61·5lb per cubic foot. According to the size and age of the tree the heartwood is of a light or dark yellow colour. In good specimens it has a smooth even contour, without a blemish or crack, and is surrounded by from 1 to 2 inches of sapwood. The spotted wood, known in Kanarese as *nágá* and *náwal kanu*—snake's and peacock's eye—for which Natives will often pay an enhanced price, is caused by the death of adventitious buds, the course of which from their origin upwards can be traced by a dark line of rich deposit appearing on longitudinal section as more or less annular spots. White ants will not touch perfect heartwood, but the white wood is often eaten extensively by them, and so through it the heartwood suffers (*Ind. Forester*).

DOMESTIC. Oil. 801 Paste. 802 **Domestic and Sacred.**—Throughout the East, Sandal-wood is much valued for a variety of domestic purposes. The essential OIL forms the basis of many of the ottos distilled in India, and alone has a peculiar fragrance much valued by the Natives for toilet purposes. The PASTE obtained by rubbing the wood on a stone with a little water is used for painting the body after bathing and is employed for making the *Shardana* or caste-marks of the Natives, especially in Southern India.

Sandal-wood carving is an established industry in some parts of the country. Richly carved boxes, cabinets, work tables, walking-sticks, etc., are made of the wood, and are much valued both by Natives and Europeans. The Kanara District is the chief home of the sandal-wood carving industry, and there, for upwards of a century, the handicraft has been well known, and the art handed down from father to son "Ground into powder it forms a favourite cosmetic with Burmese ladies" (*Kurz*).

SACRED. 803 The wood enters largely also into the religious ceremonies of the Hindus. Idols are carved in it. An emulsion of the wood is given as an offering to the gods and an incense made of the wood is burned before them. Large quantities are used by the Parsís in their fire temples. Rich Natives sometimes employ sandal-wood for cremating their dead relatives, and all, both rich and poor, add at least one piece of the wood to the funeral pile (*U. C. Dutt*).

TRADE.

TRADE IN SANDAL-WOOD.

A considerable trade in Sandal-wood has existed in India from an early date Thus, in 1825, **Milburn** wrote: "The produce of the (Malabar) coast is about 2,000 candies per annum and sometimes more. The Company used to send about 800 candies to China; all the remainder was sent by private traders to Bengal, Bombay, Cutch, and Muscat. The Company's Resident makes the purchase from the merchants on the sea-coast for ready money. These have always on hand a considerable stock, as sandal rather improves by keeping." [The figures of the modern trade are returned in value, not quantity, but such as they are, the trade statistics afford a tangible conception of the chief transactions. In the COASTWISE

Coastwise. 804 returns of India, Madras is shown as sending to Bombay a large quantity of "ornamental wood," but in the Statistics of Foreign trade "ornamented wood" is defined as Sandal, Ebony and other sorts. The returns of Foreign transactions moreover exhibit the exact shares taken by each of these sorts. It may be said that the foreign exports of sandal-wood from Bombay necessitate that almost the whole of the coastwise transactions from Madras to Bombay should be in sandal-wood. With these explanatory remarks the figures of the trade may be here exhibited. The total exports coastwise were in 1889-90 valued at R10,09,152, of which Madras furnished R7,70,791 worth, and of these Madras exports Bombay took R5,51,903 worth. But Bombay also exported coastwise R2,28,777 worth, the major portion of which went to ports within the presidency.

TRADE.

Bengal exported next to no 'ornamental woods," but it imported from Bombay R32,640 worth and from Madras R1,400 worth. Burma imported coastwise R15,288 worth from Bombay and R8,090 worth from Madras. The net result of these coastwise transactions may be said to demonstrate Bombay port as the great Indian (as it will also be shown to be the great foreign) emporium for sandal-wood. It obtains annually between 5 and 6 lakhs of rupees worth from the Madras presidency and about 1½ lakhs of rupees worth from Bombay presidency ports. From these sources, therefore, it may be accepted Bombay port town receives its supplies to meet its internal and external traffic in this article.

Foreign. 805

The FOREIGN trade may now be dealt with. The imports from foreign countries are not very important. They were valued at R16,404 in 1885-86, but the trade seems to be declining for, in 1889-90, they stood at R4,115. It may be regarded as somewhat remarkable that India should import any sandal-wood at all from foreign countries. Of last year's transactions the Straits Settlements supplied R3,780, mostly to Madras. Of these foreign imports about half are usually re-exported and mostly to France and Ceylon. The traffic in Indian sandal-wood is, however, the item of chief importance in the foreign trade. During the past five years the exports were valued at R4,44,241 in 1885-86; R4,75,038 in 1886-87; R6,51,316 in 1887-88; R5,05,013 in 1888-89; and R6,39,455 in 1889-90. The analysis of last year's transactions reveals the fact that Bombay furnished R4,36,397 worth and Madras the balance, *viz*, R2,03,058. The receiving countries were China (Hong-Kong R3,51,040 and Treaty Ports R57,239), France (R1,15,172), Germany (R75,202), the United Kingdom (R27,158), and other countries the remainder (say, R13,000). It would thus appear that England takes very little of the Indian sandal-wood and that China is the chief market.—*Ed., Dict. Econ. Prod.*] Dymock states that "Bombay alone imports annually from the Malabar coast from 700 to 800 tons of the wood and about 12,000lb of the oil. The wood is sold in Bombay at from R120 to R180 per kandy of 21 maunds (5¼ cwts.), while the oil is worth about R8½ per lb." "The revenue derived from the sale of sandal-wood forms the principal item of forest revenue in Mysore. In 1866-67, R74,598 were realised, the value of stock being R1,56,321. From the forests of South Canara and others on the western coast a large revenue is realised by the sale of the wood, this amounted to 3½ lakhs of rupees in eight years" (*Drury, Useful Plants of India*).

Santoninum, see Artemisia maritima, *Linn.*; I., 324; COMPOSITÆ.

SAPINDUS, *Linn.; Gen. Pl., I., 404.*

[*24, 25;* SAPINDACEÆ.

806 **Sapindus attenuatus,** *Wall.; Fl. Br. Ind., I., 684; Wight, Ic., t.*

Syn.—S. RUBER, *Kurz*; SCYTALIA RUBRA, *Roxb.*; NEPHELIUM RUBRUM, *G. Don, Wight*; EUPHORIA VERTICILLATA, *Wall.*; E. RUBRA, *Royle*.

Vern.—*Lal koi-púra* (Sylhet), ASSAM; *Achatta*, NEPAL; *Sirhútúngchir*, LEPCHA; *Tigroht*, MICHI.

References.—*Roxb., Fl. Ind., Ed. C.B.C., 329; Kurz, For. Fl. Burm., I., 298; Gamble, Man. Timb., 97; Trans Agri.-Horti. Soc. Ind., VII., 78.*

Habitat.—A shrub or small tree of the Eastern Himálaya, Assam, and Eastern Bengal, down to Chittagong.

FOOD. Fruit. 807

Food.—It produces a red or dark purple FRUIT of the size of an olive, which is eaten by the Natives in Sylhet.

808 **Sapindus Mukorossi,** *Gærtn.; Fl. Br. Ind., I., 683.*

THE SOAP-NUT TREE of NORTH INDIA.

Syn.—With reference to this species, **Mr. W. P. Hiern** in the *Flora of British India* writes:—"There are two forms of this tree, one with obtusely or shortly, and suddenly accuminate, leaflets (**S. detergens,** *Roxb.*), the other with more lanceolate acuminate leaflets (**S. acuminata,** *Wall; also Royle, Ill., 139*), sometimes as stated by **Dr. Royle** with the rachis of the leaves very narrowly bordered.

Vern.—*Ritha, aritha, dodan, kanmar,* HIND.; *Ritha,* BENG.; *Itá,* URIYA; *Dodan* (=the tree), *ritha, aritha, haritha* (=the fruit), PB.; *Kanmar,* N.-W. P.; *Kanmar, ritha,* BOMB.; *Phenila, urista,* SANS.

References.—*Roxb., Fl. Ind., Ed. C.B.C., 332; Voigt, Hort. Sub. Cal., 94; Brandis, For. Fl., 107; Gamble, Man. Timb., 96; Stewart, Pb. Pl., 32; Irvine, Mat. Med., Patna, 7; Med. Top. Ajmere, 125; Honigberger, Thirty-five years in the East, II., 341; Baden Powell, Pb. Pr., 330; Atkinson, Him. Dist., 307, 749; Drury, U. Pl., 384; Royle, Ill. Him. Bot. 137; Watson, Rep, 5; Balfour, Cyclop., III., 531; Treasury of Bot., 1017; Gazetteers:—Orissa, II., 179, App. vi.; N.-W. Provinces, IV., lxx.; Panjáb, Gurdaspur, 55; Ind. Forester:—II., 175; III., 45; VIII., 35; IX., 15; XII., 58; XIII., 58; Agri.-Horti. Soc. Ind. Trans, VII., 78; Journal, IV., 205; VIII. (Sel.), 179; IX., 410.*

Habitat.—A handsome, deciduous tree, with grey bark; cultivated throughout North-West India, Bengal, and Assam, and distributed to China, the Bonim Islands, and Japan. "On the Himálaya it ascends to an altitude of 4,000 feet. Royle speaks of it as wild in the valleys of the North-West Himálaya, but its orginal home requires further enquiries" (*Brandis*).

DYE. 809 **Dye.**—[Some of the tinctorial results for which the Indian dyer is famous can only be produced, it is said, if the fabric be first washed, the fruits of this tree being employed as the detergent.—*Ed.*]

GUM. 810 **Gum.**—A GUM (?) obtained from this tree was sent by the Madras Forest Department to the Amsterdam Exhibition.

OIL Fruit. 811 **Oil.**—The FRUIT contains a principle named *saponine,* and a fixed oil is derived from it by expression. It is to the presence of this principle that they owe their chief value as a substitute for soap.

MEDICINE. Fruits. 812 **Medicine.**—The FRUITS are used internally in cases of salivation, epilepsy, and as an expectorant. They are also recommended by Native practitioners for the cure of chlorosis. Seeds. 813 **Honigberger** states that "the SEEDS pounded with water are said often to put an end to an epileptic paroxysm, a small quantity being introduced into the patient's mouth."

FODDER. Leaves. 814 **Fodder.**—The LEAVES are given to cattle as fodder.

TIMBER. 815 **Structure of the Wood.**—Light yellow, moderately hard, compact, and close grained. Weight 44lb per cubic foot. It is not used.

DOMESTIC. Soap-Substitute. 816 **Domestic.**—[It is perhaps scarcely necessary to have to say that the fruits of this tree are largely used as a SOAP SUBSTITUTE, both for washing clothes and the hair. **Stewart** thinks that **Vigne** was in error when he said a decoction was employed to make the hair grow. The reader should consult the article **Detergents**, Vol. III., 84-85, for further particulars.—*Ed.*]

TRADE. 817 **Trade.**—Soap-nuts are exported from the Kumáon forest division to the extent of 20 tons per annum (*Atkinson*).

818 **S. trifoliatus,** *Linn.; Fl. Br. Ind., I., 682; Wight, Ill., t. 51.*

THE SOAP-NUT TREE of SOUTH INDIA.

Syn.—S. EMARGINATUS, *Vahl.;* S. LAURIFOLIA, *Vahl.;* S. ACUTUS, *Roxb;* S. ABSTERGENS *Roxb.*

Vern.—(The fruit=) *Ritha,* HIND.; *Bara-ritha, rithá,* BENG.; *Ud-rack,* BERAR; *Mukta maya, rettia,* URIYA; *Rithia,* C. P.; *Rhita, ritha,* BOMB.; *Rithe, ringin, rithá, rita,* MAR.; *Arithan, aritha,* GUZ.; *Rithá,* DEC.;

Pounanga, ponán-kottai, puvandi, púvanti, TAM.; *Kunkudu chettu, kúkudu, konkúdú, kunkudu-káyalu, kukudu-kayalu,* TEL.; *Antala, artala, thalay marathu, kúkate-káyi, kugate, antawala,* KAN.; *Urvanjik-kaya, ponnan-kotta, arita, rarak,* MALAY; *Miavmen-sue-khe-si, meavme-sue-khati,* BURM.; *Thalay marutha, anta-wála, puvella, punerai, gas-penela,* SING.; *Phenila, arishta phalam,* SANS.; *Rita, finduk-i-hindi* (Indian Filbert), ARAB.; *Bindake hindi* (Ainslie) *ratah,* PERS.

NOTE.—It seems likely that some of the above vern names refer to **S. mukorossi.**

Referenecs.—*DC., Prod., I., 608; Roxb., Fl. Ind., Ed. C.B.C., 331; Voigt, Hort. Sub. Cal., 94; Brandis, For. Fl., 106; Beddome, Fl. Sylv., t. 154; Gamble. Man. Timb., 96; Dals. & Gibs., Bomb. Fl., 34, 35; Graham, Cat. Bomb. Pl., 29; Mason, Burma and Its People, 517, 752; Sir W. Elliot, Fl. Andhr., 102, 103; Rheede, Hort. Mal., IV., 43, t. 19; Ainslie, Mat. Ind., II., 318; O'Shaughnessy, Beng. Dispens., 241; Pharm., Ind., 15; Irvine, Mat. Med., Patna, 81; Moodeen Sheriff, Supp. Pharm. Ind., 224; also Mat. Med. S. Ind. (in MSS.), Nos. 192, 193; U. C. Dutt, Mat. Med. Hind., 313; S. Arjun, Cat. Bomb. Drugs, 24; K. L. De, Indig. Drugs, Ind., 104; Murray, Pl. & Drugs, Sind, 67; Bidie, Cat. Raw Pr., Paris Exh., 51, 61, 104, 121; Dymock, Mat. Med. W. Ind., 2nd Ed., 188; Dymock, Warden & Hooper Pharmacog, Ind., I., 367; Year-Book Pharm., 1878, 291; Birdwood, Bomb. Prod., 15, 279, 341; Drury, U. Pl. Ind., 385; Atkinson, Him. Dist. (X., N.-W. P. Gaz.), 307; Useful Pl. Bomb. (XXV., Bomb. Gaz.) 51, 216, 252, 272; Man. Madras Adm., I., 314; II., 83; Nicholson, Man. Coimbatore, 407; Morris, Account Godavery, 18; Boswell, Man., Nellore, 98; Moore, Man., Trichinopoly, 80; Gribble, Man., Cuddapah Dist., 263; Settle. Rep., Central Provinces, Chanda, App., VI; Gazetteers:—Bombay, V., 285; VI., 14; XIII., 26; XV., 75; N.-W. P., I., 79; IV., lxx; Mysore & Coorg, I., 52, 59; III., 22; Ind. Forester:—II., 175; III., 201; IX., 413; X., 31, 546; XII., App. 10; Balfour, Cyclop. Ind., III, 531.*

Habitat.—A large, handsome tree, common about villages in South India and Ceylon. It is cultivated in Bengal, where it is doubtfully indigenous. There are said to be two forms—one with acuminate leaves, the other with emarginate pubescent leaves.

Gum.—It is said to yield a gum, but nothing is known regarding this product. GUM. 819

Oil.—A semi-solid oil is extracted from the KERNEL of the fruit. This is employed medicinally, but is regarded as too costly for general use. The FRUIT (*ritha*) is largely used in Southern India as a soap substitute; see **Detergents**, III., 84. OIL. Kernal. 820 Fruit. 821

Medicine.—The SOAP-NUT has been used medicinally by the Hindus from a very early period, and was afterwards adopted into the Muhammadan Pharmacopæia. The following account of its reputed action and uses in early times is taken from the *Makhzan-el-Adwiya*:—"The PULP of this fruit is at first sweetish, afterwards very bitter; it is hot and dry, tonic and alexipharmic; four grains in wine and *sherbet* cure colic; one *miskal* rubbed in water until it soaps and then strained, may be given to people who have been bitten by venomous reptiles, and to those suffering from diarrhœa or cholera. Three or four grains may be given by the nose, in all kinds of fits producing insensibility. Fumigations with it are useful in hysteria and melancholy; externally it may be applied, made into a plaster with vinegar, to the bites of reptiles, and to scrofulous swellings. The ROOT is said to be useful as an expectorant. Pessaries made of the KERNEL of the seed are used to stimulate the uterus in child-birth and amenorrhœa. One *miskal* of the pulp, with one-eighth of a *miskal* of Scammony, acts as a brisk purgative." Ainslie mentions its use by the Vytians as an expectorant in asthma. He also says: "I have been informed by my friend **Dr. Sherwood** that he has known several instances of the good effects of putting a MEDICINE. Soap-nut. 822 Pulp. 823 Root. 824 Kernel. 825

MEDICINE. Seeds.

little of the seeds formed by the soap-nut of this tree into the mouth of a person in an epileptic fit, by which means he was instantly brought to his recollection." It is used also by Native practitioners as an anthelmintic (*Dymock, Mat. Med. W. Ind.*). **Moodeen Sheriff**, in his *Materia Medica of Southern India*, gives a long account of the physiological actions and therapeutic uses of the Soap-nut. The following abstract may be here given :—The pericarp or pulp and kernel of the fruit, when given internally, is emetic, nauseant, and expectorant, and may be used successfully as a substitute for Ipecacuanha, to which it is equal, if not superior, in its effects. A thick watery solution of the drug is often dropped into the nostrils of patients suffering from hemicrania, hysteria, and epilepsy, and with good effect, **Moodeen Sheriff** is inclined to think, from his own experience of this method of administration. The same writer goes on to say that he has seen, in his own practice, the application of a poultice made with the soap-nut relieve the pain and swelling in parts bitten by scorpions and centipedes. He states, however, that it has no anthelmintic properties, as he has used it in very large doses for the removal of intestinal parasites, but without any effect further than emesis and slight purgation. The most convenient mode of administration is in the form of a draught, which is prepared by rubbing and bruising a nut in one or two ounces of water and straining the product through a cloth. From one drachm and a half to two drachms of the nut may be given as an emetic, from 20 to 40 grains as a nauseant, and from 10 to 18 grains as an expectorant (*Mat. Med. Southern India*).

Chemistry. 826

CHEMICAL COMPOSITION.—The following account of the chemical composition of the Soap-nut is extracted from **Dymock, Warden & Hooper's** *Pharmacographia Indica* :—"The saponin estimated by weighing the Sapogenin formed by boiling with dilute acid, amounts to 11·5 per cent.; this result is confirmed by determining the glucose before and after the treatment and calculating the increase of glucose into glucoside. The weight of the barium and lead precipitates points to a lower percentage of saponin. The fruits yield to water 40 per cent., and to alcohol 15 per cent. of extract. They contain in a ripe state over 10 per cent. of glucose, and a quantity of pectin, which renders the watery solution difficult of filtration. Submitted to distillation, the drug affords a small quantity of what appeared to be butyric acid. According to **Brannt**, no saponin is contained in the woody stone, seed, or husk. The thick cotyledons contain about 30 per cent. of a white fat, semi-fluid at 20°C. and melting to a clear oil at 30° C., which possesses a somewhat characteristic odour. The oil saponifies readily, and is employed medicinally and in the manufacture of soap."

SPECIAL OPINIONS.—§ "By Native doctors it is used as a very effectual remedy in one-sided headache. The pulp of one fruit is squeezed well with a little human or cow's milk, and the milk thus prepared is dropped into the nostril, and in coma and delirium into the eyes also. In this way it acts as a powerful derivative" (*Surgeon-Major D. R. Thomson, M.D., C.I.E., Madras*). "Used as an emetic when rubbed up with water" (*Surgeon J. McCloghey, Poona*). "The kernel of the seeds, mixed with equal parts of black pepper, and powdered is given in doses of grains 40 to 60 in epilepsy" (*Civil Surgeon J. Anderson, M.B., Bijnor, N.-W. P.*). "Said to act on the uterus and is used to bring on abortion" (*Surgeon H. W. Hill, Manbhúm*). It is an emetic and used as such in cases of poisoning. It is commonly employed for washing the hair and clothes (*Surgeon-Major Robb, Civil Surgeon, Ahmedabad*). "**Sapindus laurifolius** or **trifoliatus** and **S. emarginatus** are different plants. Both yield the Soapnut" (*Assistant Surgeon S. Arjun, L.M., Gorgaum, Bombay*).

Structure of the Wood.—Wood yellow and hard. Weight about 64lb per cubic foot. TIMBER. 827

Domestic Uses.—Besides the use of the fruit as a detergent, the WOOD is employed for house-building and in the construction of carts, and also for making a variety of small articles such as combs, boxes, etc. DOMESTIC. Wood. 828

SAPIUM, *Br.; Gen. Pl., III., 334.* 829

A genus of Euphorbiaceous plants containing twenty-five species, all of which are tropical. Six species are indigenous to the Indian Peninsula, Burma, or the Straits Settlements. The genus was joined by **Mueller** and more recently by **Baillon** to the allied EXCŒCARIÆ, but was restored in the *Genera Plantarum*. **Sir J. D. Hooker**, however, in the *Flora of British India*, remarks that the distinctive points are insufficient, and that it would be better again to unite the two.

[*1950;* EUPHORBIACEÆ.

Sapium baccatum, *Roxb.; Fl. Br. Ind., V., 470; Wight, Ic., t.* 830

Syn.—S. POPULIFOLIUM, *Wall.;* EXCŒCARIA BACCATA, *Muell.;* E. AFFINIS, *Griff.;* CARUMBIUM BACCATUM, *Kurz;* STILLINGIA PANICULATA, *Miq.*

Vern.—*Adamsali,* ASSAM; *Billa,* SYLHET; *Pudlikat, lal kainjal,* NEPAL; *Zinhlún, le-lun-pen,* BURM.

References.—*Roxb., Fl. Ind., Ed. C.B.C., 692; Brandis, For. Fl., 441; Kurz, For. Fl. Burm., II., 412; Gamble, Man. Timb., 367.*

Habitat.—A large, evergreen tree of Northern and Eastern Bengal, the Sikkim Himálaya, Assam, the Khásia Hills, Chittagong, and Burma. It is distributed to Malacca and Sumatra.

Food.—The BARK is chewed by the Natives of Assam. FOOD. Bark. 831

Structure of the Wood.—A soft, greyish-white light wood, coarsely fibrous, perishable (*Kurz*). **Roxburgh** mentions it as a large and useful timber tree, but does not say for what purposes it is utilized. TIMBER. 832

S. indicum, *Willd.; Fl. Br. Ind., V., 471; Wight, Ic., t. 1950.* 833

Syn.—S. HURMAIS, *Ham.;* S. BINGIRIUM, *Roxb.;* STILLINGIA INDICA, & BINGYRICA, *Baill.;* EXCŒCARIA INDICA, *Muell.*

Vern.—*Húruá, batúl, batan,* BENG.; *Hurná,* BOMB.; *Kirrimakálu,* SING.

References.—*DC., Prod., XV., ii., 1216; Roxb., Fl. Ind., Ed. C.B.C., 691; Brandis, For. Fl., 441; Kurz, For. Fl. Burm., II., 413; Gamble, Man. Timb., 367; Grah., Cat. Bomb. Pl., 181; Beddome, For. Man., 215; Rheede, Hort. Mal., IV., t. 51; Lisboa, U. Pl. Bomb., 125, 273; Balfour, Cyclop., III., 532; Jour. Agri.-Horti. Soc., X., 37.*

Habitat.—A small evergreen tree, found in the Sunderbans and the tidal forests of Tenasserim and Ceylon. It is said by **Graham** to occur in the Koncan.

Medicine.—The JUICE of the tree is very poisonous. MEDICINE. Juice. 834

Structure of the Wood.—Soft and white, with small brown heartwood. Weight 29lb per cubic foot. TIMBER. 835

Domestic Uses.—The WOOD is used for fuel in the regions where the tree occurs. The SEEDS are employed as a fish intoxicant by the Natives, see **Narcotics**, Vol. V., 332. DOMESTIC. Wood. 836 Seeds. 837

S. insigne, *Benth.; Fl. Br. Ind., V., 471.* 838

Syn.—FALCONERIA INSIGNIS, *Royle;* F. WALLICHIANA, *Royle;* CARUMBIUM INSIGNE, *Kurz.*

Var. MALABARICA.

Syn.—EXCŒCARIA INSIGNIS, *Bed., For. Man., 214, t. 22, f. 5; Wight, Ic., No. 1866.*

Vern.—*Khinna, khina lienda, lendwa,* HIND.; *Dúdla, bilodar, biloja, karálla, ledra,* PB.; *Dudla,* BOMB.; *Garpa shola,* ANAMALAIS.

References.—*Brandis, For. Fl., 442; Kurz, For. Fl. Burm., II., 412; Dalz. & Gibs., Bomb. Fl., 227; Beddome, For. Man., 214; Grah., Cat. Bomb. Pl., 367; Lisboa, U. Pl. Bomb., 125, 268; Ind. Forester, 1884, X., i, 30; Gazetteers:—N.-W. P., X., 317; Bombay, XV., 443.*

Habitat.—A robust, deciduous tree of the sub-Himálayan tract, from the Beas eastwards, ascending to 4,300 feet of Chittagong, Burma, and **Var.** MALABARICA on the Western Gháts as far north as Násik.

MEDICINE. Juice. 839 **Medicine.**—The whole plant is full of an acrid, milky JUICE, which, when applied to the skin, produces vesication (*Lisboa*).

TIMBER. 840 **Structure of the Wood.**—Greyish-white, very soft and spongy. Weight 23 to 29lb per cubic foot.

DOMESTIC. Wood. 841 **Domestic Uses.**—The WOOD is used for the cylinders of Native drums, and for making sandals (*Graham*; *Lisboa*).

842 **Sapium sebiferum,** *Roxb.; Fl. Br. Ind., V., 470.*

THE CHINESE TALLOW TREE.

Syn.—EXCŒCARIA SEBIFERA, *Muell.*; STILLINGIA SEBIFERA, *Michaux*; S. SINENSIS, *Baill.*; S. SEBIFERA, *Bojer*; CARUMBIUM SEBIFERUM, *Kurz*; CROTON SEBIFERUS, *Linn.*

Vern.—*Pippal-yang* (according to Ainslie), HIND.; *Mom-china*, BENG.; *Pista*, TRANS-INDUS; *Toyapippali* (according to Ainslie), SANS.; *Ya-ricou*, CHINESE.

References.—*DC., Prod., XV., ii., 1210; Roxb., Fl. Ind., Ed. C.B.C., 691; Voigt, Hort. Sub. Cal., 161; Brandis, For. Fl., 441; Kurz, For. Fl. Burm., II., 412; Gamble, Man. Timb., 366; Dalz. & Gibs., Bomb. Fl. Supp., 77; Grah., Cat. Bomb. Pl., 181; Ainslie, Mat. Ind., II., 433; Baden Powell, Pb. Pr., 423, 598; Atkinson, Him. Dist., 883; Drury, U. Pl., 209; Balfour, Cyclop., III., 740; Smith, Dic., 401; Journ. Agri.-Horti. Soc. Ind., VI., 164; Ind. Forester, II., 290, 394; V., 212; VI., 239; VIII., 124, 129; Report by Col. J. G. Macrae, Conservator of Forests, Sind Circle (6th July 1888); Report, Saharunpore Bot. Gardens (1863), N.-W. P. Sel. (1866), II., 108.*

Habitat.—A small, glabrous tree, indigenous to China, but introduced as a cultivated plant into various parts of India and elsewhere in warm countries. The Indian vernacular names given above are, therefore, mere adaptations.

DYE. Leaves. 843 **Dye.**—The LEAVES give a black dye.

OIL. Fruit. 844 Seed. 845 Shell. 846 **Oil and Oil-seed.**—The FRUIT is a three-celled capsule, each cell of which contains a single globose SEED, thickly coated with a white greasy substance—the so-called vegetable tallow. That substance is used in place of animal tallow in China, for the manufacture of candles, in soap-making, and also in dressing cloth. From the SHELL and seed an oil is extracted, which is burnt in lamps. In China, the tallow from these seeds almost entirely replaces animal tallow in the manufacture of candles. It is separated from the seeds by steaming them in tubs with convex open wicker bottoms, placed over cauldrons of boiling water. In China, this tallow is said to be hard and white, and to give a clear inodorous flame without smoke. In India it is found that it is not consistent enough, and when it has been used in candle-making, it is necessary to dip the candles made of it into wax, and so give them a hard external coating. The combustion, too, of these candles is described as imperfect, and they are said to yield a dim light and a thick smoke. A large cylindrical mass of the tallow, solid, and of a pure white colour very pure and inodorous, was exhibited at the Lahore Exhibition in 1864, and it was hoped then that the tallow tree might become an article of commercial importance in the Panjáb. Since that time, however, interest in it has declined, and for many years no attempts seem to have been made to utilise the tallow from this source. The labour and expense involved in collecting the seeds and extracting the tallow are said to have been far in excess of the value of the product. With reference to the oil **Roxburgh** remarks that cocoanut oil is much better for burning purposes. For further information see the article on **Oils,** Vol. V., 449. OIL.

Structure of the Wood.—White and even-grained, moderately hard. Weight 32lb per cubic foot. TIMBER. 847

Domestic Uses.—Besides the uses of the tallow and oil, as above described, the WOOD is made into bedsteads, tables, and toys, and is recommended by **Dr. Jameson** as well fitted for employment as printing blocks. It is a handsome tree, and is often planted for ornamental purposes. In Assam the *eri* silkworm is said to feed on its LEAVES. DOMESTIC. Wood. 848 Leaves. 849

SAPONARIA, *Linn.; Gen. Pl. I., 146.*

Saponaria Vaccaria, *Linn.; Fl. Br. Ind., I., 217;* CARYOPHYLLEÆ. 850

Syn.—S. PERFOLIATA, *Roxb.;* GYPSOPHILA VACCARIA, *W. & A.*

Vern.—*Musna,* SANTAL., HIND.; *Sabuni,* BENG.; *Guligafas,* PERS.

References.—*Roxb., Fl. Ind., Ed. C.B.C., 385; Boiss., Fl. Orient., I., 525; Aitchison, Bot. Afgh. Del. Com., 40; O'Shaughnessy, Beng. Dispens., 212; S. Arjun, Bomb. Drugs, 15; Murray, Pl. and Drugs, Sind, 95; Dymock, Warden, & Hooper, Pharm. Ind., I., 151; Atkinson, Him. Dist., 305; Gazetteers, N.-W. P., IV., lxviii.; X., 305; Ind. Forester, IV., 234; XII. (App.), 6.*

Habitat.—A tall, robust weed of cultivation, met with in wheat-fields throughout India and in Tibet.

Medicine.—The properties of this plant are in every respect identical with those of **S. officinalis,** the Soap-wort, Στρουθιον of **Dioscorides,** (*O'Shaughnessy*). It is considered by the Natives of India to have febrifugic and tonic properties in long continued fevers of a low type (*S. Arjun*) The mucilaginous SAP is said to be an efficacious remedy for itch (*Murray*). Preparations of this PLANT have emulsifying properties on account of the saponin it contains. MEDICINE. Sap. 851 Plant. 852

Domestic Uses.—The mucilaginous SAP is used as a soap substitute by the Natives of India (see **Detergents,** Vol. III., 85). DOMESTIC. Sap. 853

SAPOTA.

Sapota, see **Achras Sapota,** *Linn.;* Vol. I., 80; SAPOTACEÆ.

S. elengoides, *A. DC.*, and **S. tomentosa,** *A. DC.;* see **Sideroxylon tomentosum,** *Roxb.;* Vol. VI.

Sappan wood, see **Cæsalpinia Sappan,** *Linn.;* LEGUMINOSÆ, Vol. II., 10.

SAPPHIRE, *Ball, Man. Geol. Ind., 429.* 854

The Sapphire is classed along with the diamond, ruby, emerald, and pearl, etc., among "gems or precious stones," in contradistinction to the "inferior gems" amongst which are placed the carnelian, onyx, agate, etc. It is a blue transparent variety of corundum or native alumina, and is composed of oxide of alumina (Al_2O_3) together with a small quantity of oxide of cobalt, to which it owes its blue tint. It differs, therefore, from the oriental ruby merely in its colour. Crystals of Sapphire are usually double hexagonal pyramids with the basal plane sometimes well developed, but very often quite small or almost obliterated. Many of the crystals are very irregular, the corresponding angles, measured on different pairs of faces, frequently varying by several degrees.

Sapphire, *Mallet, Man. Geology of India, IV., 39.* 855

SAPHIR, *Fr.;* SAPPHIR, *Germ.;* SAFFIERSTIN, *Dut.;* ZAFFIRO, *It., Sp.;* JACHANT, *Rus.*

Vern.—*Nilam*, HIND., TAM., MAL.; *Nilam, kábúd, asmáni, surmai* (the colour of antimony—blue—grey), PB.; *Neela-hgnet-kha* (the yellow sapphire =), *ouk thapha-va* (the violet sapphire =), *Neela-khayan* (the green sapphire =), *Mya*, BURM.; *Nil*, SINGH.; *Sufir*, ARAB.

References.—*Encyclop. Brit., XXI., 302; Balfour, Cyclop. Ind., III., 532; Mason, Burma and Its People, 578, 579, 732; Baden Powell, Pb. Pr., 48-49; Tavernier, Travels in India, 399, 400; Calvert, Kulu, 54; Gazetteers, Central Provinces, 506; Settlement Rep., C. P. (Upper Godavery Dist.), 42; T. C. Plowden, Report to the Govt. of Ind., Foreign Department, on the Sapphires of Kashmir.*

Varieties. 856

VARIETIES.—Sapphires of various colours occur in India. Thus, there is the blue or true sapphire of popular language, the colour of which may be any shade of blue, from the palest to a deep indigo, the most esteemed tint being that of the blue cornflower. Violet sapphires (oriental amethysts) are also found in the same localities as those in which the true sapphire is met with. The most valuable sapphire found in the East Indies is the yellow sapphire or oriental topaz. A green gem, called by the Europeans in India an emerald, is often seen. It is, however, a green sapphire, and is much harder than the true emerald, which is a green beryl.

Occurrence. 857

Occurrence.—The following account of the occurrence of sapphires in India is mainly abstracted from **Mr. Mallet's** writings on the subject, in the *Manual of the Geology of India*. Sapphire is found along with many other varieties of corundum in the ruby mines of Upper Burma. According to **Mr. Spears**, the sapphires there are much rarer than rubies, although those found are of larger size. In Ceylon, rubies are of rare occurrence, while sapphires are found frequently, and are often of very large size. In the Salem District of Madras and in the valley of the Cauvery, **Captain Newbold** reports that sapphire is occasionally found, and, according to the *Central Provinces Gazetteer*, the same gem occurs in the Upper Godavari District. **Mr. J. Calvert** has stated that he found "sapphires worth R2,500 each besides other gems" in the Upper Raini valley near the head waters of the Beeas in Kúlu. In 1882 a remarkable discovery of sapphires was made in the Kashmír territory, and within a short time such quantities of gems were thrown on the market as to materially lower the value. Owing to the secrecy observed by the discoverers, very conflicting accounts were in circulation as to the place where the stones had been found, but this may be elucidated by the following account taken from the report by **Mr. T. C. Plowden**, late Resident in Kashmír, who himself visited the place and saw the sapphires being dug out there:—"Leaving the Chenab valley at Golabgarh in Padar, at the junction of the Bhutna stream with the Chenab and following the path towards Pzanskar, which runs up the Bhutna valley, two days' easy marching brings one to the village of Suniyan, the last village met with on this side of the passes into Pzanskar. To the north-east of this village rises a lofty range of mountains, known to geologists as the Pzanskar range which sends off many spurs into the Bhutna valley. It is near the southern end of one of these spurs between the villages of Machial and Suniyan that the sapphire deposits are found, about two miles in a north-westerly direction from Suniyan in a small valley. The greater portion of this valley is covered with masses of boulders and fragments of rock which have fallen from the surrounding cliff, and amongst these the sapphires are at present obtained. The rock of which the cliffs surrounding the valley are formed is of micaceous gneiss, interstratified with beds of crystalline siliceous limestone. The gneiss is also interstratified with veins of granite in portions of which corundum crystals have been developed which, when they are of sufficiently good colour, become sapphires."

Localities where found. (*W. R. Clark.*) | SARACA indica.

COLLECTION. 858

METHOD OF COLLECTION.--Sapphires like rubies occur in three ways—(1) *in situ*, imbedded in white crystalline limestone, (2) loose in the soil on the hillsides, and (3) in gem-bearing gravel. The second is the commonest method of occurrence, and the mode of collection now pursued in Kashmir is to dig up the soil, and then wash away the sand and mud, a rude adaptation of the methods employed by the diamond diggers of South Africa. Roughly estimated the average yield was 1 tola of sapphire to every 15 maunds of soil dug and washed. The stones, however, were of very uneven quality.

USES. 859

Uses.—Sapphires are much valued for ornamental purposes, and are made into brooches, rings, seals, etc. In consequence of its hardness the sapphire was usually mounted by the ancients in a partially rough state, the surface being polished but not cut. It has, however, occasionally been engraved as a gem.

TRADE. 860

Trade.—During the calendar year 1889, 36,010 carats of sapphire, valued at R15,000, were obtained in Kashmír. The outturn of Burmese sapphire during the same period cannot be ascertained, as both rubies and sapphires are included under one heading in the trade returns (*R. & A. Department Return of Outturn of Gems in India*).

SARACA, *Linn.; Gen. Pl., I., 583.*

[LEGUMINOSÆ.

Saraca indica, *Linn.; Fl. Br. Ind., II., 271; Wight, Ic., t. 206;* 861

Syn.—S. ARBORESCENS, *Burm.;* S. MINOR & ZOLLINGERIANA, *Miq.;* JONESIA ASOCA, *Roxb.;* J. PINNATA, *Willd.*

Vern.—*Asok,* HIND.; *Asok, asoka,* BENG.; *Aseka, ati,* CUTTACK; *Asok,* URIYA.; *Asoka,* MANIPUR; *Asok,* N.-W. P; & PB., *Asok, ashok, asoka, jassundi, jásúndi,* BOMB.; *Ashoka,* MAR.; *Ashopálava,* GUZ.; *Asek, kankéli,* TEL.; *Ahsunkar, asoka, asoge, ashoka,* KAN; *Thawgabo,* BURM.; *Diye ratembelá, diya-ratmal,* SING.; *Asoka, kankéli, vanjula,* SANS.

References.—*DC., Prod. II., 487; Roxb., Fl. Ind., Ed. C.B.C., 312; Voigt., Hort. Sub. Cal., 246; Brandis, For. Fl., 166; Kurz, For. Fl. Burm., I, 414; Beddome, Fl. Sylv., t. 57; Gamble, Man. Timb., 144; Dalz. & Gibs., Bomb. Fl., 82; Mason, Burma and its People, 403, 770; Trimen, Cat. Ceyl. Pl., 28; Elliot, Fl. Andhr., 82; Sir W. Jones, V., 111; Rheede, Hort. Mal., V., t. 59; Folkard, Plant-lore Legends & Lyrics, 229; U. C. Dutt, Mat. Med. Hind., 143, 292; Dymock, Mat. Med. W. Ind., 2nd Ed., 257; Lisboa, U. Pl. Bomb., 64, 279, 285; Kew Off. Guide to Bot. Gardens & Arboretum, 30; Bomb. Gar., VII., 40, 41; XV., 75; Gaz., Mysore & Coorg I., 52, 59; Settle. Rep., Fyzabad, 12; Ind. Forester, III., 203; XIV., 298.*

Habitat.—A low, erect tree of the Central and Eastern Himálaya, East Bengal, South India, Arracan, and Tenasserim. In Kumáon it occurs up to altitudes of 2,000 feet. It is found also in Ceylon and Malacca, and is distributed to the islands of the Malayan Archipelago. It is often cultivated in gardens for its handsome flowers, and very frequently so near temples (*Brandis*). "Bombay gardeners call the **Guatteria longifolia** (ANONACEÆ) "Asoka," and have an idea that it is the male of **Saraca indica**" (*Dymock*). Roxburgh remarks that "the plants and seed were probably brought originally from the Eastern frontier of Bengal, where it is indigenous. When the tree is in full blossom I do not think the whole Vegetable Kingdom, affords a more beautiful object."

MEDICINE. Bark. 862

Medicine.—The BARK is much used by Native physicians in uterine affections, and especially in menorrhagia. A decoction of the bark in milk is generally prescribed. A *ghrita* called *asoka ghrita* is also prepared, with a decoction of the bark and clarified butter, together with a number of aromatic herbs in the form of a paste (*U. C. Dutt, Hind. Mat. Med.*). In

MEDICINE. Orissa the bark is said to be used as an astringent in cases of internal hæmorrhoids (*W. W. Hunter*).

SPECIAL OPINIONS.—§ "The bark contains a large proportion of gallic acid. A decoction of it is chiefly used with dilute sulphuric acid in stopping uterine hœmorrhage" (*Civil Surgeon S. M. Shircore, Moorshedabad*). "FLOWERS pounded and mixed with water used in hœmorrhagic

Flowers. 863 dysentery" (*Surgeon A C. Mukerji, Noakhally*).

TIMBER. 864 **Structure of the Wood.**—Light reddish-brown, soft. Weight 50℔ per cubic foot. Brandis remarks that the heartwood is hard and dark coloured.

SACRED. 865 **Sacred Uses.**—The *Asoka* is one of the sacred trees of the Hindus, which they are ordered in the *Urapaj* to worship on the 13th day of the month *Chaitra, i.e.*, 27th December. Its flowers, probably on account of their beauty and the delicacy of their perfume, which, in the months of April and May, is exhaled throughout the night, are much used in temple decoration. "The tree is the Symbol of Love, and is dedicated to Kama, the Indian God of Love. Like the **Agnus castus** it is believed to have a certain charm in preserving chastity; thus Sita, the wife of Rama, when abducted by Ravana, escapes from the caresses of the demon and finds refuge in a grove of *Asokas*. In the legend of Buddha, when Maya is conscious of having conceived the Buddis-attva, she retires to a wood of *Asoka* trees and then sends for her husband The word Asoka signifies "that which is deprived of grief" (*Folkard, Plant-lore and Legends*), Mason (*Burma & Its People*) says the tree is held sacred among the Burmans, because under it, Gaudama was born, and immediately after his birth delivered his first address.

SARCOCEPHALUS, *Afzel.*; *Gen. Pl., II., 29.*

866 **Sarcocephalus cordatus,** *Miq.; Fl. Br. Ind., III., 22;* RUBIACEÆ.

Syn.—NAUCLEA CORDATA, *Roxb.;* N. COADUNATA, *Roxb.;* N. ROXBURGHII, *G. Don;* N. WALLICHIANA, *Br.;* N. PARVIFOLIA, *Wall.*

Vern. *Bakmi, vammi,* SING.; *Ma-u, ma-ulettanshe,* BURM.

References.—*Bedd., Fl. Sylv., t. 318; Kurz, For. Fl., II., 63; Gamble, Man. Timb., 218; Ind. Forester, X., 31; XII., 72, 73.*

Habitat.—A small tree, which occurs in the lower mixed forests of Burma, from Pegu and Martaban to Tenasserim It is found also in Ceylon, and is distributed to the Malay and Philippine Islands and North Australia.

TIMBER. 867 **Structure of the Wood**—Pale-coloured, rather light, coarse, and loose-grained. Weight 23-34℔ per cubic foot.

DOMESTIC. 868 **Domestic Uses.**—Beddome says it is used for making sandals, common furniture, doors, and for other purposes.

SARCOCHLAMYS, *Gaud.; Gen. Pl., III., 389.*

[URTICACEÆ.

869 **Sarcochlamys pulcherrima,** *Gaud.; Fl. Br. Ind., V., 588;*

Syn.—URTICA PULCHERRIMA, *Roxb.*

Vern.—*Dogal,* GARO HILLS; *Tsatya, shap-sha,* BURM.

References.—*DC., Prod., XVI., i., 235; Roxb., Fl. Ind., Ed. C.B.C., 656; Brandis, For. Fl., 405; Kurz, For. Fl., Burm., II., 426; Gamble, Man. Timb., 323; Darrah, Note on Cotton in Assam, 31; Liotard, Memo. on Indian Dyes, 127, viii.*

Habitat.—A bush or large shrub, with a stem often as thick as a man's leg, met with in Assam, the Khásia Hills, Sylhet, Chittagong, and Burma. It is distributed to Sumatra. It is found up to altitudes of 1,000 feet. Brandis says that in Pegu this plant, together with **Blumea grandis, Buddleia,** and other fast-growing large herbs and shrubs, forms the dense thicket which springs up in deserted clearings.

Dye.—In Assam a madder brown colour is produced with the bark of **Albizzia odoratissma,** in conjunction with the pounded LEAVES and TWIGS of the *dogal* tree (**Sarcochlamys plucherrima**), the yarn being boiled in the mixture (*Liotard, Memo. on Indian Dyes*). DYE. Leaves. 870 Twigs. 871

Fibre.—The LIBER gives a good fibre for ropes (*Kurz*). FIBRE. Liber. 872

Structure of the Wood.—Pale, reddish brown, rather light, of a fine silvery fibre, soft (*Kurz*). TIMBER. 873

SARCOCOCCA, *Lindl.; Gen. Pl., III., 266.*

[*t. 1877;* EUPHORBIACEÆ

Sarcococca pruniformis, *Lindl.; Fl. Br. Ind., V., 266; Wight, Ic.,* 874

Syn.—S. SALIGNA, *Muell.;* S. TRINERVIA, *Wight;* S. SUMATRANA, *Bl.* S. SALICIFOLIA, *Baill.;* BUXUS SALIGNA, *Don;* TRICERA NEPALENSIS, *Wall.*

Vern.—*Chilikat,* NEPAL; *Sukat sing,* KUMAON.

References.—*DC., Prod., XVI., i., 11; Brandis, For. Fl., 448; Beddome, Fl. Sylv., ccxvii; Gamble, Man. Timb., 371; Bedd., For. Man., 217; Atkinson, Him. Dist., 317; Ind. Forester, X., 35; Jour. Agri.-Horti. Soc. Ind., XIV., 11, 24.*

Habitat.—A small, evergreen shrub, met with in the Temperate Himálaya, at altitudes between 5,000 and 9,000 feet, in the Khásia Hills and Manipur, in the Deccan Peninsula, and in Ceylon. It is distributed to Afghánistán and to Sámatra.

Structure of the Wood.—White, moderately hard. TIMBER. 875

Domestic Uses.—The WOOD is sometimes used for making walking-sticks. DOMESTIC. Wood. 876

Sarcocolla, see **Astragalus? sp.**; LEGUMINOSÆ; Vol. I., 348.

SARCOSPERMA, *Hook. f.; Gen. Pl., II., 655.*

Sarcosperma arboreum, *Benth.; Fl. Br. Ind., III., 535;* 877

[SAPOTACEÆ

Syn.—SIDEROXYLON ARBOREUM, *Ham.*

Vern.—*Pahar lampati,* NEPAL; *Kulyatso,* LEPCHA.

References.—*Kurz, For. Fl. Burm., II., app. 575-76; Gamble, Man. Timb., 242; also Trees, Shrubs, etc., Darjeeling, 52.*

Habitat.—A large, bushy tree of the Eastern Himálaya. It is found in Sikkim, up to altitudes of 4,000 feet. It also occurs on the Khásia hills and Patkoye mountains in Assam, and in Lower Burma (Ava hills, *Kurz*).

Fodder.—The LEAVES are given to cattle (*Gamble*). FODDER. Leaves. 878

Structure of the Wood.—Pink, moderately hard, rather light. Weight 30·5lb per cubic foot. TIMBER. 879

Domestic Uses.—It is used in Sikkim to make canoes. DOMESTIC. 880

SARCOSTEMMA, *Br.; Gen. Pl., II., 763.* 881

With reference to the identification of plants of the genus **Sarcostemma** with the sacred *Homa* of the Parsís and the *Soma* of the early Sanskrit writers, the reader is referred to the generic note on **Ephedra,** Vol. III., 247. It is sufficient here to remark that **Sarcostemma,** although very probably not the original *Soma* of early writers, is at any rate one of the best known and most frequently used substitutes for the real article.

[*Wight, Ic., t. 595;* ASCLEPIADEÆ. 882

Sarcostemma brevistigma, *Wight & Arn.; Fl. Br. Ind., IV., 26;*

THE MOON PLANT.

Syn.—ASCLEPIAS ACIDA & APHYLLA, *Roxb.*

Vern.—*Somlatá,* HIND.; *Somlatá, soma,* BENG.; *Thorinjal,* SIND.; *Sóma, lama,* BOMB.; *Ran sher,* MAR.; *Tiga-tshumudú, konda pála, pulla tige, padma kâshtam, sóma latá, tige jemudu,* TEL.; *Muwa-kiriya,* SING.; *Soma,* SANS.

References.—*DC., Prod., VIII., 538; Roxb., Fl. Ind., Ed. C.B.C., 251; Voigt, Hort. Sub. Cal., 542; Dals. & Gibs., Bomb. Fl., 149; Elliot, Fl. Andhr., 96, 141, 159, 169, 181; Gibson, Cat. Bomb. Pl., 122; Ainslie, Mat. Ind., II., 378; U. C. Dutt, Mat. Med. Hind., 318; Drury, U. Pl., 385; Lisboa, U. Pl. Bomb., 165; Birdwood, Bomb. Pr., 53, 208; Gazetteers:—Bombay, V., 27; Mysore & Coorg, I., 62.*

Habitat.—A trailing, leafless, jointed shrub, not uncommonly met with in dry rocky places in the Deccan Peninsula. It occurs also in Bengal, but more rarely than on the western side of India.

DOMESTIC. 883 *Conf. with p. 126.* **Domestic Uses.**—**Dr. Gibson** (*Cat. Bomb. Pl.*) mentions that "it is often brought from a distance by farmers to extirpate white-ants from the sugarcane fields. A bundle of twigs is put into the trough of the well from which the field is watered, along with a bag of salt, hard packed, so that it may dissolve gradually. The water so impregnated destroys the ants without injuring the crop." The plant contains a large amount of milky sap, which **Roxburgh** says is "of a mild nature and acid taste, and is often used by Native travellers to allay their thirst." For its uses in

Liquor. 884 preparing an intoxicating LIQUOR, see **Ephedra,** Vol. III., 247; also **Dr. Watt's** numerous papers on the Soma of the Sanskrit authors.

885 886 887 The three other species of this genus, **S. Brunonianum,** *Wight & Arn.,* **S. intermedium,** *Dcne.,* and **S. Stocksii,** *Hook. f.,* are said by **Sir J. D. Hooker,** in the *Flora of British India,* to be indistinguishable in the dried state from this. They are used in a similar manner by the Natives of India, and have the same vernacular names.

SARCOSTIGMA, *W. & A.; Gen. Pl., I., 354.*

888 **Sarcostigma Kleinii,** *W. & A.; Fl. Br. Ind., I., 594; Wight, Ic.,* [*t. 1854;* OLACINEÆ.

Vern.—*Púvana, púvenagah* (=the plant), *adul, odul* (=the oil).

References.—*Dals. & Gibs., Bomb. Fl., 221; Drury, U. Pl. Ind., 386.*

Habitat.—A climbing, branched shrub, found in the Eastern and Western Peninsulas, Malacca (*Maingay*), Cochin, and Travancore (*Wight*), and the Koncan (*Stocks*).

MEDICINE. Oil. 889 **Medicine.**—This plant yields a medicinal OIL, highly esteemed in the treatment of rheumatism (*Drury*).

FOOD. Fruit. 890 **Food.**—The FRUIT of **S. edule,** which is said, in the *Flora of British India,* to be probably only a variety of **S. Kleinii,** is eaten by the Natives of the Andaman Islands (*Kurz*).

DOMESTIC. Oil. 891 **Domestic.**—The OIL is burnt in lamps.

Sarsaparilla, Indian, see **Hemidesmus indicus,** *R. Br.;* ASCLEPIADEÆ, [Vol. IV, 219.

S. Jamaica, see **Smilax officinalis,** *Hunb. Bonpl., & Kunth.*; LILIACEÆ, [Vol. VI.

Sarson Oil, see **Brassica campestris,** *Linn.*; CRUCIFERÆ; Vol. I., 522.

SASSAFRAS, *Nees; Gen. Pl., III., 160.*

892 **Sassafras officinale,** *Nees;* LAURINEÆ.

SASSAFRAS.

Syn.—LAURUS SASSAFRAS, *Linn.*

References.—*Watt, Calc. Exhib. Cat., V., 252; VII., 221; Pharm. Ind., 192; Gamble, Man. Timb, 313; Smith, Econ. Dict., 369.*

Habitat.—A tree, 20 to 40 feet high, native of the forests of North America, from Canada to Florida.

Medicine.—The dried ROOT is imported into India and used medicinally for its alterative, tonic, stimulant, and sudorific properties. It is useful in chronic rheumatism, secondary syphilis, scurvy, and skin diseases. The volatile OIL obtained by distillation of the wood is stimulant, carminative, and diaphoretic (*Pharm. Ind.*). In British practice it is only given in combination with sarsaparilla and guaiacum. MEDICINE. Root. 893 Oil. 894

Structure of the Wood.—Soft, porous, highly scented, preserving its odour a long time. TIMBER. 895

Satin-wood, see **Chloroxylon Swietenia,** *DC.*; MELIACEÆ; Vol. II., 270.

SAURAUJA, *Willd.; Gen. Pl., I., 184.* 896

Sauruja napaulensis, *DC.; Fl. Br. Ind., I., 286*; TERNSTRŒMIACEÆ.

Vern.—*Gogina, goganda,* HIND.; *Gogen,* NEPAL; *Kasúr,* LEPCHA; *Gogina, gogána,* KUMAON.

References.—*Brandis, For. Fl., 25; Gamble, Man. Timb., 29; List of Trees, Shrubs, & Climbers of Darjiling, 8; Atkinson, Him. Dist., 306; Ind. Forester, VIII., 405; XI., 2, 370; XIV., 343.*

Habitat.—A moderate-sized tree of the Himálaya from Bhután to Garhwál, also in the Temperate Khásia hills. It is found at altitudes between 2,500 and 5,000 feet.

Food & Fodder.—The FRUIT is succulent and palatable, and is eaten by the hill tribes. The LEAVES are lopped for cattle fodder (*Atkinson*). FOOD & FODDER. Fruit. 897 Leaves. 898

Structure of the Wood.—Pale-pink, very soft, spongy; shrinks much. Weight 25lb per cubic foot (*Gamble*). TIMBER. 899

S. punduana, *Wall.; Fl. Br. Ind., I., 287.* 900

Syn.—S. FASCICULATA, *var.* ABBREVIATA, *Choisy.*

Vern.—*Rata-gogén, sipha-rúng,* SIKKIM.

References.—*Kurz, For. Fl. Burm., I., 103; Gamble, Man. Timb., 29; List of Trees, Shrubs, & Climbers of Darjiling, 8*

Habitat.—A bush or small tree, found in the Sikkim Himálaya between the altitudes of 3,000 and 5,000 feet. It occurs also in the Khásia mountains, and in the tropical forests of Burma, at altitudes between 2,000 and 3,000 feet.

Structure of the Wood.—White, soft and even; finely fibrous (*Kurz*). TIMBER. 901

SAUROMATUM, *Schot.; Gen. Pl., III., 966.*

Sauromatum pedatum, *Schot.; DC., Monogr. Phanerogamiæ, II.,* [*569*; ARACEÆ. 902

Syn.—ARUM PEDATUM, *Willd.*; ARUM CLAVATUM, *Desf.*

Vern.—*Lot* (a name applied to several species of the genus), MAR.; *Bhasamkand,* C. P.

Reference.—*Dymock, Mat. Med. W. Ind., 817.*

Habitat.—A native of Western India and of the Central Provinces.

Medicine.—"The TUBERS are as large as small potatoes and of the same shape as those of *súran*" (**Amorphophallus campanulatus,** Vol. I., 225); "they are very acrid and poisonous, and are only used externally as a stimulating poultice by the Natives" (*Dymock*). MEDICINE. Tubers. 903

SAUSSUREA, *DC.; Gen. Pl., II., 471.*

Saussurea candicans, *Clarke; Fl. Br. Ind., III., 373*; COMPOSITÆ. 904

Syn.—S. BRAHUICA, *Boiss.*; APLOTAXIS CANDICANS, *DC.*; A. SCAPOSA, *Edgew*; CNICUS CANDICANS, *Wall*; CARDUUS HETEROMALLUS, *Don.*

Vern.—*Batula, kalí ziri,* PB.

References.—*DC., Prod., VI., 540; Stewart, Pb. Pl., 119; Boiss., Fl. Orient., III., 566; Atkinson, Him. Dist., 312; Gaz., N.-W. Prov., IV., lxxiii.*

Habitat.—A robust annual herb, found in sub-tropical and temperate Western India and the Himálaya, from the Salt Range, Hazára, and Kashmír to Bhután, altitude 2,000 to 7,000 feet. It is distributed to Afghánistán.

MEDICINE. Seeds. 905 **Medicine.**—The SEEDS are collected in the Panjáb for the drug-sellers (*Stewart*).

SPECIAL OPINION.—§"Carminative, used in *masálás* for horses" (*Surgeon-Major C. W. Calthrop, M.D., Morar*).

906 **Saussurea gossypiphora,** *Don; Fl. Br. Ind., III., 376.*

Syn.—S. GOSSYPINA, *Wall.*; APLOTAXIS GOSSYPINA, *DC.*; ERIOCORYNE NIDULARIS, *Wall, Mss.*

Vern.—*Kasbál, bút pesh*, PB.

References.—*Stewart, Pb. Pl., 119; Hooker, Him. Journals, I., 225; Balfour, Cyclop. Ind., III., 543.*

Habitat.—A herbaceous plant, with perennial root-stock, met with in the Alpine Himálaya, from Garhwal to Sikkim, at altitudes between 14,000 and 17,000 feet. **Sir J. D. Hooker** (*Him Jour.*) remarks of this very striking plant that it forms great clubs of the softest white wool, six inches to a foot high, its flowers and leaves seemingly uniformly clothed with the warmest fur that nature can devise.

SACRED. 907 **Sacred.**—**Madden** mentions that it is offered up at shrines on the Sutlej (*Stewart*).

908 **S. hypoleuca,** *Spreng.; Fl. Br. Ind., III., 374,*

Syn.—CARDUUS AURICULATUS, *Wall.*; APLOTAXIS AURICULATA, *DC.*

Vern.—For the vernacular names of this plant the reader is referred to **S. Lappa,** for both species seem to be known to the Natives by the same names.

References.—*DC., Prod., VI., 541; Dymock in Year-Book of Pharmacy (1878,) 242.*

MEDICINE. 909 **Habitat.**—A compositous herb, with decidedly nodding head; it occurs all over the Temperate Himálaya, from Kashmír to Sikkim, at altitudes between 7,000 and 13,000 feet.

Medicine.—It appears probable that part of the Costus used medicinally in India is derived from this species, although the true Costus is probably **S. Lappa.**

910 **S. Lappa,** *C. B. Clarke; Fl. Br. Ind., III., 376.*

THE COSTUS.

Syn.—AUCKLANDIA COSTUS, *Falc.*; APLOTAXIS LAPPA, *Dcne.*

Vern.—*Kút, kot, kust, kust-talk-putchuk, kur, pachak*, HIND.; *Pachak, kur*, BENG.; *Post-khai*, KASHMIR; *Rusta*, BHOTE; *Kút, kot, kúst talkh, kúth*, PB.; *Ouplate*, BOMB.; *Upaleta, kút*, GUZ.; *Kostum, putchuk, goshtam*, TAM.; *Changala, kustam*, TEL.; *Sepuddy*, MALAY.; *Goda mahanel*, SING.; *Kushtha, kashmirja* (according to **Stewart**), SANS.; *Kúst*, ARAB. & PERS.

References.—*Stewart, Pb. Pl., 121; J. H. Van Linschoten, Voyage to the East Indies, II., 129; Pharm. Ind., 127; Ainslie, Mat. Ind., II., 166; O'Shaughnessy, Beng. Dispens., 652; U. C. Dutt, Mat. Med. Hind., 180, 307; Dymock, Mat. Med. W. Ind., 2nd Ed., 449-456; Murray, Pl. & Drugs, Sind, 185; Irvine, Mat. Med. Patna, 52; Med. Top. Ajmere, 107, 142; Honigberger, Thirty-five years in the East, 262; Baden Powell, Pb. Pr., 356; Birdwood, Bomb. Pr., 48; Royle, Ill. Him. Bot., 360; Prod. Res., 223, 224; Aitchison, Trade Products of Le, 144, Davies, Trade & Resources of the N.-W. Frontier, lxxiii., cviii.*

ccclxxviii.; Balfour, Cyclop., I., 821; Smith, Econ. Dic., 134; Kew Off. Guide to the Mus. of Ec. Bot., 87; Gazetteer, Panjáb, Hazara Dist., 13; Journ. Agri.-Hort. Soc. Ind., XI., Part I.

Habitat.—A tall, very stout herb, with annual stem and thick perennial roots, indigenous to the moist, open slopes, surrounding the valley of Kashmír, at an elevation of 8,000 to 9,000 feet. It occurs also in parts of the basins of the Chenab and Jhelam, at elevations between 10,000 and 13,000 feet. The roots are dug up in the months of September and October, chopped up into pieces, 2 to 6 inches long, and exported without further prepararton.

History.—**Dr. Watt** has already written at some length on this subject and his remarks may therefore be reproduced:—"It would seem that for a long time Costus Root or *kust* was referred to a species of SCITAMINEÆ, most probably from the resemblance of the scent to that of Orris root. The genus to which it was attributed received the name of **Costus**, the perfume being said to be obtained from **C. arabicus**, *Willd.* The common and elegant plant of our jungles, **Costus speciosus**, *Sm.* (Vol. II., 579), was supposed to be nearly allied to the hypothetical species **C. arabicus**, but to be scentless. It is remarkable that while it has now been clearly proved that the plants which belong to the genus **Costus** have nothing to do with the Costus root of the ancients, the vernacular names *keo, kust* should in Bengali be given to **C. speciosus**, names which are also applied to the true **Costus**. The resemblance of the root to Orris or Iris, a plant nearly allied to **Costus speciosus**, is another remarkable coincidence. **Falconer**, in *Linn. Soc. Trans., Vol. XIX., Part I., 23 (1842)*, proved beyond doubt that the *kust* of Upper India was the root of what he called **Aucklandia Costus** (since reduced to **Saussurea Lappa**), and he concluded that this was the Costus of the ancients for the following reasons:— HISTORY. 911

"*1st*—It corresponds with the descriptions given by the ancient authors.

"*2nd*—Coincidence of names; in Kashmír the root is called *kût* and the Arabic vernacular is said to be *kust*, both being given as synonyms by the Persian *hakims*; they are also the names by which the medicine is known in all the bazárs of Hindustan Proper; in Bengal the Kashmír root is called *Patchak*, and it appears by a note in **Dr. Royle's** illustrations that **Garcia ab Horto** gives *Pucho* as the Malay synonym of **Costus arabicus**.

"*3rd*—*Koot* is used at the present day for the same purposes in China as Costus was formerly applied to by the Greeks and Romans.

"*4th*—The direct testimony of the Persians that *kust* comes from the borders of India, and that it was not a product of Arabia.

"*5th*—The commercial history of the root gathered in Kashmír under the name of *kust*" (*Dr. Dymock, Mat. Med.*).

"This root is collected in enormous quantities in the mountains of Kashmír, whence it is conveyed to Calcutta and Bombay, and thence shipped for China. The drug has a pungent, aromatic taste, and an odour resembling that of orris root. There is an excellent account of it, with a figure, in **Professor Guibourt's** *Histoire des Drogues* tome iii. p. 25 (*Science Papers by D. Hanbury, 257*). Costus root is remarkably similar to Elecampane both in external appearance and structure. Costus has been an important spice, incense, and medicine in the East from antiquity down to the present day; it would be of great interest to examine it chemically with regard to Elecampane" (*Pharmacographia*).

Kust is collected in large quantities in Kashmír and exported to the Panjáb, where it finds its way all over India and is shipped from Bombay

The Costus Root.

HISTORY.

and Calcutta to China and the Red Sea, a small quantity finding its way to Europe. Falconer describes two forms—*kust-i-talk* and *kust-i-shírín*, the latter being the chief article of commerce (*Watt, Calcutta Exhib. Cat.*, *V.*, *252*).

PERFUME. Roots. 912

Perfume.—As already indicated, the ROOTS of this plant are a valuable perfume. Stewart remarks that the loads of it when passing scent the air to some distance. Aitchison says that it is imported into Le from Kashmír and re-exported to Lhassa.

MEDICINE.

Medicine.—"*Kust* has been used in Hindu medicine from the earliest ages. It is said to be aphrodisiac and tonic, and useful in diseases arising from deranged air and phlegm, also in asthma and for resolving" tumours (*Meer Muhammad Husain*). It was formerly smoked as a substitute for opium. U. C. Dutt, in his *Hindu Materia Medica*, states that the "ROOT is described as aromatic, stimulant, and useful in cough, asthma, fever, dyspepsia, and skin disease." Mr. Baden Powell gives an interesting summary of the uses of *kut*: "the dried POWDER is the principal ingredient in a stimulating ointment for ulcers; it is used as an ingredient in a stimulating mixture for cholera. Stewart says it is officinal in the Panjáb, being applied in powder to ulcers, for worms in wounds, and for tooth-ache or rheumatism.

Root. 913

Powder. 914

CHEMISTRY 915

CHEMICAL COMPOSITION.—The ROOT contains much inulin, but no accurate chemical examination has been made. In the dry state, it is brown, very brittle, and apparently full of resin; it has a strong agreeable odour similar to that of orris root.

DOMESTIC. Root. 916

Domestic & Sacred.—The ROOT is used in perfumery, in powder or solution; it is largely employed for the hair, and as a protective of shawls and other valuable garments against the attacks of insects it has time out of count been valued. The Kashmír goods owe their peculiar odour to this root (*Watt*).

TRADE. 917

Trade.—The quantity collected is very large, amounting, according to Dr. Falconer, to about 2,000,000 pounds per annum. It is laden on bullocks and exported to the Panjáb, whence the greater portion is sent to Bombay and Calcutta. A great part of the imports into Bombay and Calcutta are exported to China and the Red Sea. In Kashmír, it is a Government monopoly; each village in the vicinity of the *kut* fields is assessed at a fixed amount which has to be delivered in the capital, and the surplus is bought up by the agents of the Maharajah and retailed again to dealers for export to Hindústan. In 1864 the revenue, obtained by the Kashmír State, from the sale of *kut*, was said to have amounted to nearly R1,90,000. According to Dr. Falconer, at the time he wrote, the cost of collection and transport to a depôt in Kashmír was 2*s.* 4*d.* per cwt.; on entering India its value was enhanced to from 16*s.* 9*d.* to 23*s.* 4*d.* per cwt., while its commercial value at Canton was 47*s.* 5*d.* per cwt.

As the drug is not mentioned separately in the trade returns of Bombay and Bengal, the amount of exports cannot be ascertained; but from the Consular reports we find that in the year 1875 the imports into two Chinese ports were, for Hankow 1,270 piculs, valued at £5,224 6*s.* 3*d.*, and Chefœ 277 piculs, valued at £1,197, so that it is a fairly important article of Chinese Commerce.

ADULTERANTS. 918

Substitutes and Adulterants.—In a communication made to the Agri.-Horticultural Society of India in 1860 by Mr. H. Cope of Amritsar, it is stated that "*kut* was adulterated not only with *tut* (the root of **Salvia lanata**), which is used, as a substitute for the genuine article, but that other foreign substances were used so that he had ascertained that unscrupulous dealers employed some 20 seers of *kut* to flavour 100 seers of trash." The principal substitutes or adulterants seem to be a species of **Ligularia**

Conf. with pp. 450, 501.

(*Stewart, Pb. Pl.*), and one of **Aconitum** (*Journal, Agri.-Horti. Soc. Ind.*), but many other plants are said to be used for the purpose. **Dymock**, however, states that there is no difficulty in obtaining the genuine article in Bombay, and that the adulterated root is probably specially prepared for the Chinese market. ADULTERANTS.

Saussurea obvallata, *Wall.; Fl. Br. Ind., III., 365.* 919

Syn.—CARDUUS OBVALLATUS & C. TECTUS, *Wall.*; APLOTAXIS OBVALLATA, *DC.*

Vern.—*Bergandu tongur*, N.-W. HIM., KARNAL & KUMAON; *Kanwal, birm-kanwal*, PB.

References.—*DC., Prod., VI., 541; Stewart, Pb. Pl., 129; Gazetteer, N.-W. P., X., 312.*

Habitat.—A herb met with in the Western Himálaya, from Kashmír to Sikkim, at altitudes between 10,000 and 15,000 feet.

Medicine.—The ROOT is used for application to bruises and cuts (*Duthie*). MEDICINE. Root. 920

Sacred.—Edgeworth mentions that this is one of the species offered up at the shrines of Budrinath (*Stewart*). SACRED. 921

Savoy, see **Brassica (oleracea) bulleata**, CRUCIFERÆ; I., 534.

SAW-MILLS.

Saw-Mills. 922

[The amount of timber cut up by hand is, in India, probably very considerably greater than the outturn ot the steam saw-mills. No data exists from which to frame, however, an estimate of the total production of timber, so that the present remarks must be restricted to exhibiting the published returns of steam saw-mills. And even of these full particulars are not forthcoming, since some of them are private concerns—not registered Companies. Mr. J. E. O'Conor (*Statistical Tables*) shows that, in 1889, there were, in India, sixty-two saw-mills, with a joint capital of R32,96,125, their outturn having been valued at R1,42,61,115. Of these mills, three are in Bombay, two in Madras, seven in Assam, forty-five in Lower Burma, and five in Upper Burma. These mills usually give employment to 5,294 permanent hands and 6,089 temporary. They turned out last year 216,737 tons of timber; 70,600 logs; 16,671 feet of sawn timber; 230,417 tea chests; and 22,539 shooks. The mills concerned in the production of tea chests were those of Assam. The largest and most important saw-mills are, however, those in Burma. Some idea of the importance of the saw mills of India may be learned from the returns of the traffic in timber, but more particularly of teak wood. The total exports coastwise were, in 1889-90, 146,109 tons, valued at R90,75,785; but these figures include of course hand-sawn as well as mills' timber. Of these exports Burma alone furnished 116,352 tons, 107,803 tons of which went to the other British provinces as follows: to Bengal 31,460 tons, to Bombay 55,092 tons, and to Madras 21,251 tons. After Burma, Bombay is the most important exporting province, but its exports came to only 24,269 tons of teak, the major portion of which went to ports within the presidency of Bombay or to the Native States of Kathiwar, Cambay, Cutch, etc. The foreign demand for Indian timber has, for some years past, manifested a tendency to expansion. Thus, for example, these exports were valued in 1885-86 at R1,08,655; in 1886-87 at R2,54,545; in 1887-88 at R2,78,130; in 1888-89 at R2,83,610; and in 1889-90 at R3,24,565. The analysis of these exports reveals the fact that Burma takes by far the largest share in the trade, and further that about one half of the total exports go to Ceylon, the next most important countries being Mauritius, East Coast of Africa, and the Straits Settlements.

Bengal and Bombay each export about the same amount of timber, and Madras a slightly lower amount, these three provinces furnishing between them about two-thirds of the total exports.

Although, having perhaps little bearing on the subject of Indian saw-mills, it may be here mentioned that India annually imports a very considerably larger amount of timber from foreign countries than she exports. The immense size of India doubtless renders this necessary, as railway charges within India would be heavier from the sources of timber than the steam freight from foreign countries. During the past five years the imports of timber from foreign countries were vulued as follows:—1885-86, R2,62,946; 1886-87, R2,98,289; 1887-88, R3,83,949; 1888-89, R4,63,776; and 1889-90, R5,19,308. Of the last year's imports Bengal took R1,82,085 worth; Madras R1,25,678 worth; Burma R1,00,796 worth; Bombay R64,170 worth, and Sind R46,579 worth. Great Britain heads the list of the supply countries with R1,52,605 worth; Ceylon usually follows, having last year furnished R1,14,411 worth; Hong-Kong R1,30,660 worth, and the Straits Settlements with R93,136 worth. All other countries furnished much smaller amounts. Austria and Japan, for example, having each contributed about R6,000 worth.—*Ed., Dict. Econ. Prod.*]

923 **SAXIFRAGA,** *Linn.; Gen. Pl., I., 635.*

A genus of Saxifragaceous plants, comprising 160 known species, 35 of which are found in India. They are herbs, most of them perennial, and inhabit cool, temperate, and especially Alpine, localities. Only one species is noted as being of economic value.

924 **Saxifraga ligulata,** *Wall.; Fl. Br. Ind., II., 398;* SAXIFRAGACEÆ.

Syn.—A variety of this plant—**ciliata**—described by **Royle** as occurring in Nepal and Kumaon at altitudes between 6,000 and 8,000 feet, is put to economic purposes where it occurs; but the vernacular names and uses for both species and variety seem to be similar.

Vern.—(The root=) *Pakhan-bed, silphora,* HIND.; *Bat piá, popal, wa, phúta, shaprochi, kúrgotar, dharposh, banpatrak, saprotri, til, kachálu, shibláeh makhán bed, dakachrú,* (the root=) *Pakhán bed, jintiána, maslún,* PB.; *Kamarghwal,* PUSHTU; *Pashanbheda,* BOMB.

References.—*Voigt, Hort. Sub. Cal., 267; Stewart, Pb. Pl., 103; Royle, Ill. Him. Bot., 226, 227; Murray, Pl. & Drugs, Sind, 143; Atkinson, Him. Dist., 309, 747; Balfour, Cyclop., III., 544.*

Habitat.—A small plant with leaves frequently a foot in diameter, which occurs on the Temperate Himálaya from Bhután to Kashmír at altitudes between 7,000 and 10,000 feet. It is met with also on the Khásia mountains at an altitude of 4,000 feet.

MEDICINE. Root. 925 **Medicine.**—The ROOT is used as a tonic in fevers, diarrhœa, and cough, and as an antiscorbutic. It is bruised and applied to boils and in ophthalmia. It is also considered absorbent and is given in dysentery (*Atkinson; Stewart*). In Sind the root is rubbed down and given with honey to children when teething.

DOMESTIC. Leaves. 926 **Domestic Uses.**—The large LEAVES are frequently employed as plates (*Stewart*).

SCÆVOLA, *Linn.; Gen. Pl., II., 539.*

927 **Scævola Kœnigii,** *Vahl.; Fl. Br. Ind., III., 421; Wight, Ill.,* [*t. 137;* GOODENOVIEÆ.

Syn.—S. SERICEA, *Forst.;* S. TACCADA, *Roxb.;* LOBELIA PLUMIERI, *Burm., non Linn.*

Vern.—*Bhadrak,* BOMB.; *Pinletan,* BURM.

References.—*Roxb. Fl. Ind., Ed. C.B.C., 177; Dalz. & Gibs., Bomb. Fl., 134; Burm., Fl. Ind., 186; Rumph., Amb., IV., 116, t. 54; Gamble, Man. Timb., 233; Kurz, For. Fl. Br. Burm., II., 84; Mur-*

ray, Pl. & Drugs of Sind, 180; Mason, Burma and Its People, 784; Birdwood, Bomb. Prod., 165; Lisboa, U. Pl. Bomb., 163; Gazetteer, Mysore & Coorg, I., 70.

Habitat.—A shrub with smooth succulent branches, found on the sea-shores of India, from Sind to Ceylon, and from thence to Burma and Malacca. It is distributed to Tropical Eastern Asia, Australia, and Polynesia.

Medicine.—In the time of **Rumphius** the JUICE of the BERRIES was instilled by the Amboyans into the eyes to clear off opacities and take away dimness of vision. MEDICINE. Juice. 928 Berries. 929

Food.—The LEAVES are eaten as a vegetable by the Natives. FOOD. Leaves. 930

Structure of the Wood.—It has a soft, spongy pith and coarse milky fibrous wood, which is useless for economic purposes. TIMBER. 931

Domestic.—"The Malays attach some superstitious qualities to its berries, and from the pith of the stem and thick branches they make artificial flowers" (*Lindley*). DOMESTIC. 932

[Vol. II., 519.

Scammony, see **Convolvulus Scammonia,** *Linn.*; CONVOLVULACEÆ;

SCHIMA, *Reinw.; Gen. Pl., I., 185.*

Schima crenata, *Korth.; Fl. Br. Ind., I., 289*; TERNSTRŒMIACEÆ. 933

Syn.—GORDONIA FLORIBUNDA, *Wall.*; G. OBLATA, *Roxb., Fl. Ind., Ed. C.B.C., 426.*

Vern.—*Theet-ya,* BURM.

References.—*Kurz, For. Fl. Br. Burm., I., 107; Griff., Notul, IV., 563; Mason, Burma & Its People, 408, 535, 752; Ind. Forester XIV., 341.*

Habitat.—An evergreen tree, 30 to 60 feet in height, found in the Eastern Peninsula from Tenasserim to Penang and distributed to Borneo and Sumatra.

Structure of the Wood.—Not of much value, but said to be hard and durable, though liable to warp and split. TIMBER. 934

Domestic.—The compact timber of this tree is used for house-posts and for rice mortars (*Mason*). DOMESTIC. 935

S. khasiana, *Dyer, in Fl. Br. Ind., I., 289.* 936

Habitat.—A tree of the Khásia mountains, found at altitudes between 4,000 and 6,000 feet.

Dye and Tan.—**Baillon** says that the BARK of this tree is used in dyeing and in the preparation of skins. DYE & TAN. Bark. 937

S. mollis, *Dyer, in Fl. Br. Ind., I., 288.* 938

Syn.—GORDONIA MOLLIS, *Wall.*

Vern.—This, like all the other members of the genus, is called *Theetya* by the Burmese, *i.e.*, itchwood, on account of the itching, which its chips or bark occasion when brought into contact with the skin.

References.—*Aplin, Report on Shan States (1887-88); Ind. Forester, XIV., 341.*

Habitat.—A large tree found on the Ava hills.

Domestic.—The WOOD is used for similar purposes to that of **S. crenata.** DOMESTIC. Wood. 939

S. Wallichii, *Choisy; Fl. Br. Ind., I., 289.* 940

Syn.—S. HYPOGLAUCA, *Miq.*; GORDONIA WALLICHII, *DC.*; G. INTEGRIFOLIA, *Roxb.*; G. CHILLAUNIA, *Ham.* With reference to the synonyms of **S. Wallichii, Dr. George Watt** (*Ind. Forester*) remarks that it is a very variable plant, but that a well marked form (**S. Noronhæ,** *Reinw.*) fully deserves, in his opinion, the independent position once assigned to it

Vern.—*Makusal, chilauni, makriya-chilauni,* HIND.; *Dingan* (Khasia), *Boldak* (Garo), *Makriah, chilauni, makusal,* ASSAM.; *Jam,* CACHAR; *Chilauni, goechassi,* NEPAL; *Sumbrong,* LEPCHA; *Gugera,* GOALPARA; *Theet-ya, a-nan-pho,* BURM.

References.—*Roxb., Fl. Ind., Ed. C.B.C., 426; Kurz, For. Fl. Burm., I., 106; Gamble, Man. Timb., 29; Mason, Burma & Its People, 535, 752; Pharm. Ind., I., 190; Dymock, Warden & Hooper, Pharmacog. Ind., I., 190; Ind. Forester, I., 85, 87; VII., 101; VIII., 403; XI., 252, 315, 355; XIV., 340, 341, 343.*

Habitat.—A large, evergreen tree of the Eastern Himálaya, from Nepál and Sikkim to Bhutan, found at altitudes between 2,000 and 5,000 feet. It occurs also in Assam, the Khásia hills, Chittagong, and Burma, and is distributed to Sumatra.

MEDICINE. Bark. 941 **Medicine.**—"The Hindi names for this tree signify 'that which causes itch,' 'that which causes monkeys' itch. The part of the tree which has this effect is the BARK, in which the liber cells appear like glistening white needles, which irritate the skin like cowhage, which drug it resembles in being a mechanical irritant" (*Pharmacog. Ind.*).

TIMBER. 942 **Structure of the Wood.**—Rough, red, moderately hard, close-grained, warps and shrinks much in seasoning. It is durable when well ventilated. Weight about 45℔ per cubic foot. **Gamble** remarks that the growth of this tree is moderately fast, and that, "as large quantities of the timber, well grown and straight, are available, it is to be hoped that it may be ere long in more extensive demand."

DOMESTIC. Wood. 943 **Domestic.**—The WOOD is used in Northern Bengal and Assam for many purposes, but chiefly for building. Many of the tea factories in Darjeeling have been built of it, and the Public Works Department have sometimes used it for bridges. **Mann** states that in Assam it is used for planks and ordinary building purposes, and for canoes. In 1875 several sleepers, made of it, were given over to the Northern Bengal State Railway for experiment, but the result does not appear to have been published (*Gamble*). **Hooker**, in his *Himálayan Journals*, says it is much prized by the Lepchas and Thibetans for ploughshares and other articles which need a hard wood.

SCHISMATOGLOTTIS, *Zoll. et Morr.; Gen. Pl., III., 984.*

[*Phanerog., II., 352;* AROIDEÆ.

944 **Schismatoglottis calyptrata,** *Zoll. et Morr.; DC., Monog.*

Syn.—CALLA CALYPTRATA, *Roxb., Fl. Ind., Ed. C.B.C., 631.*

Habitat.—A native of Amboyna, where it is used as an article of food and also medicinally. Introduced into Calcutta. See **Colocasia antiquorum,** *Schott.*; AROIDEÆ; Vol. II., 509.

945

SCHIST.

"The Schistose Rocks are those which have a 'schistose' *e.g.*, 'foliated texture.' Foliation is a term applied by **Professor Sedgwick** to those rocks which have had such a subsequent texture and structure given to them as to split into plates of different mineral matter, either with the bedding or across it" (*Geikie*). (*Conf.* with **Gneiss,** Vol. III., 517-18).

946 **Schist,** *Medlicott, in Man. Geol. Ind., I., 12.*

SCHIST, *Eng.;* SCHISTE, *Fr.;* SCHIEFER, *Germ.*

The following note has been kindly furnished by **Mr. H. B. Medlicott,** late Director of the Geological Survey:—

Schist is a foliated crystalline rock, the commonest varieties being mica-schist, talc-schist, chlorite-schist, hornblend-schist, etc. It is the commonest genus of metamorphic rock, and is always found in such regions of which there are many in India. It is quite too coarse a rock to rank as a building stone, being only available for rough rubble masonry, and, when obtainable in large slabs, for roofing, lintels, door-posts, flags, and the like.

SCHIZANDRA, *Mich.; Gen. Pl., I., 19.*

Schizandra grandiflora, *H. f. & T.; Fl. Br. Ind., I., 44;* MAGNOLIACEÆ. 947

Syn.—SPHÆROSTEMA GRANDIFLORUM, *H.f. & T.*; KADSURA GRANDIFLORA, *Wall.*

Vern.—*Singhata, taksielrik,* LEPCHA; *Sillangti,* KUMAON; *Klandru, kaljendru,* PB.

References.—*Brandis, For. Fl., 571; Gamble, Man. Timb., 4; Gazetteer, N.-W. P., X., 304; Ind. Forester, XI., 2; XIII., 68.*

Habitat.—A glabrous climbing shrub, of the Temperate Himálaya, from Simla to Bhután, found at altitudes between 6,000 and 10,000 feet.

Food.—The FRUIT is eaten. **FOOD. Fruit.**

Structure of the Wood.—Porous and with strong resinous smell. 948

TIMBER. 949

SCHLEICHERA, *Willd.; Gen. Pl., I., 404.*

Schleichera trijuga, *Willd.; Fl. Br. Ind., I., 681;* SAPINDACEÆ. 950

THE LAC TREE OF KOSUMBA; THE CEYLON OAK.

Syn.—S. PUBESCENS, *Roth.*; MELICOCCA TRIJUGA, *Juss.*; SCYTALIA TRIJUGA, *Roxb.*; STADMANNIA TRIJUGA, *Spreng.*; CUSSAMBIUM SPINOSUM, *Hamilt.*

Vern.—*Kosum, gausam, kusum,* HIND.; *Baru,* SANTAL; *Puvatti,* KADERS.; *Kassuma, koham, kocham,* PANCH MEHALS; *Rusam,* URIYA; *Baru,* KURKU; *Kússum, kojba,* C. P.; *Komur, púskú,* GOND.; *Gosam,* N.-W. P.; *Samma, jamoa, gausam, kússúmb,* PB.; *Peduman, gosam, kosam, kosamb, kocham, koshimb, assumar,* BOMB.; *Kusumb, peduman,* MAR.; *Pává, pú, púlachi, zolim-buriki, pumarum, puvu,* or *kúlá* in Ceylon) TAM.; *Púskú, may, roatanga, posuku, mayi, rotanga,* TEL.; *Chendala,* COORG; *Sagdi, sagade, chakota, akota,* KAN.; *Gyo, kyetmouk,* BURM.; *Kón,* SING.

References.—*DC., Prod., I., 615; Roxb., Fl. Ind., Ed. C.B.C., 351; Brandis, For. Fl., 105; Kurz, For. Fl. Burm., I., 289; Beddome, Fl. Sylv., t. 119; Gamble, Man. Timb., 95; Thwaites, En. Ceyl. Pl., 58; Dalz. & Gibs., Bomb. Fl., 35; Stewart, Pb. Pl., 32; Graham, Cat. Bomb. Pl., 29; Mason, Burma & Its People, 454, 752; Sir W. Elliot, Fl. Andhr., 114, 156; Trimen Cat. Ceylon Pl., 20; Rumphius, Amb., I., t. 57; O'Shaughnessy, Beng. Dispens., 242; S. Arjun, Cat. Bomb. Drugs, 25, 213; Dymock, Warden, & Hooper, Pharmacog. Ind., I., 370; Birdwood, Bomb. Prod., 259, 325; Baden Powell, Pb. Pr., 597; Atkinson, Him. Dist. (X., N.-W. P. Gaz.), 307, 814; Useful Pl. Bomb. (XXV., Bomb. Gaz.), 51, 150, 261, 278, 394; Liotard, Dyes, 33; Man, Madras Adm., II., 115; Settlement Reports:—Central Provinces, Chanda, App. vi.; Chhindwara, 28, 110; Baitool, 77, 127; Upper Godavery, 38; Nimar, 305; Raipore, 76, 77; Gazetteers:—Bombay, VII., 38, 39; XIII., 25; XV., 33; XVII., 25; N.-W. P., IV., lxx.; Mysore & Coorg, I., 48; Ind. Forester:—I., 120, 274; II., 18, 19; III., 23, 189, 201, 238; IV., 292, 322; VII., 277; VIII., 29, 103, 105, 127, 414, 438; IX., 128, 177, 487; X., 31, 33, 63, 325; XI., 357; XII., 188; XIII., 120; Balfour, Cyclop. Ind., II. 546.*

Habitat.—A large, deciduous tree, found in dry forests of the Sub-Himálayan tract, from the Sutlej eastwards, throughout Central and Southern India, Burma, and Ceylon. It is distributed to Java and Timor.

Resin.—It exudes a yellowish resin. The LAC produced upon this tree is known as *kusum* lac, and is the most highly prized quality (see **Coccus lacca**, Vol. II., 409). **RESIN. Lac.** 951

DYE. Flowers. 952 **Dye.**—A dye is said to be obtained from the FLOWERS. (*Settlement Rep., Chindwara Dist.*)

OIL. Seeds. 953 **Oil.**—The SEEDS yield an oil which is used in Malabar for culinary and lighting purposes. It is reputed to be the original Macassar oil, and has recently reappeared in German commerce under that name (*Pharmacog. Ind.*).

MEDICINE. Oil. 954 **Medicine.**—Messrs. Gehe & Co., in their trade report, state that the OIL is a valuable stimulating and cleansing application to the scalp, which promotes the growth of the hair. It has been long used by Native practitioners for the cure of itch and acne. Roxburgh remarks that the BARK rubbed up with oil is used for the cure of itch, but he does not mention the similar use made of the oil. Rev. A. Campbell tells us that the Santals use the bark in external application to relieve pains in the back and the loins.

Bark. 955

FOOD. Fruit. 956 **Food.**—The FRUIT contains a whitish pulp, which has a pleasant sub-acid taste and is often eaten during hot dry weather by the Natives, who ascribe to it cooling properties.

TIMBER. 957 **Structure of the Wood.**—Very hard, strong, durable, and takes a fine polish, sapwood whitish, heartwood light reddish brown. Weight about 70 per ℔ cubic foot.

DOMESTIC. Wood. 958 **Domestic.**—The WOOD is much used by Natives for the manufacture of articles where strength in small space is required. Thus it is employed for making pestles, the axles of wheels, and the teeth of harrows, and for the screw rollers of sugar mills, and of cotton and oil presses.

SCHREBERA, *Roxb.; Gen. Pl., II., 675.*

959 **Schrebera swietenioides,** *Roxb.; Fl. Br. Ind., III., 604; Wight, [Ill., t. 162;* OLEACEÆ.

Vern.—*Móka, góki, ghant, gantha, banpalás,* HIND.; *Ghantá párul* BENG.; *Jarjo, sandapsing,* KOL.; *Ghato,* ORAON; *Mokkak,* BHIL; *Jantia,* URIYA; *Jhán,* KURKU; *Moka,* C. P.; *Karindi, mokha, dhakha,* GOND.; *Patali, ghanta patali,* BUNDEL.; *Mokha, jháw,* RAJ.; *Moka gantha,* BOMB.; *Moga-linga,* TAM.; *Kalgante,* COORG; *Makkam mokob,* TEL.; *Thit-hswé-lwé,* BURM.; *Mushkaka, ghantápátali,* SANS.

References.—*DC., Prod., viii., 675; Roxb., Fl. Ind., Ed. C.B.C., 37; Brandis, For. Fl., 305; Kurz, For. Fl. Burm., II., 156; Beddome, Fl. Sylv., t. 248; Gamble, Man. Timb., 255; Dalz. & Gibs., Bomb. Fl., 138; U. C. Dutt, Mat. Med. Hind., 298, 310; Lisboa, U. Pl. Bomb., 97, 394; Gazetteer:—Bomb., xiii., 26; XV., 75; Ind. Forester:—III., 203; IV., 227; VIII., 417, 438; XI., 370, 371; XII., 311, 313; App. 16; XIII., 121.*

Habitat.—A deciduous tree, 40 to 50 feet high, found in the Tropical Himálaya of Kumáon, and in Central and Southern India and Burma. It is widely diffused, but nowhere abundant.

GUM. 960 **Gum.**—It yields a gum.

FOOD. Leaves. 961 **Food.**—The LEAVES were eaten as a vegetable in the Nasik District during the famine of 1877-78.

TIMBER. 962 **Structure of the Wood.**—Brownish grey, hard, polishes well, is durable does not warp or split. There is no heartwood proper, but irregular masses of purple or claret-coloured wood are scattered throughout the centre of the tree. Weight 56℔ per cubic foot.

DOMESTIC. Wood. 963 **Domestic.**—The WOOD is used for turning and for making combs and weaver's beams. It makes excellent cart wheels. Roxburgh says that he is inclined to think it would answer well in place of boxwood for scales to mathematical instruments, as it is not liable to warp.

SCHWEINFURTHIA, *A. Braun; Gen. Pl., II., 933.*

[*Wight, Ic., t. 1459;* SCROPHULARINEÆ.

Schweinfurthia sphærocarpa, *Braun.; Fl. Braun Ind., IV., 252;* 964

Syn.—S. PAPILIONACEA, *Boiss.;* ANTIRRHINUM PAPILIONACEUM, *Burm;* A. GLAUCUM, *Stocks;* LINARIA SPHÆROCARPA, *Benth.*

Vern.—(Bazar name, the drug=) *Sanipát,* HIND., BOMB., SANS.; *Sonpát,* SIND.

References.—*DC., Prod., X., 287; Boiss., Fl. Orient., IV., 387; Burm., Fl. Ind., 21, t. 39, f 2; Dymock, Mat. Med. W. Ind., 580.*

Habitat.—An annual or perennial branched glabrous herb found in rocky places in Sind and distributed to Baluchistan and Afghánistan.

Medicine.—The drug, which consists of the FRUIT broken up into small pieces, and the powdered LEAVES together with portions of the stem, has a slightly bitter, somewhat tea-like taste, and is prescribed by Native practitioners to patients suffering from typhoid symptoms. The POWDER it snuffed up for bleeding at the nose (*Dr. Stocks*).

MEDICINE. Fruit. 965 Leaves. 966 Powder. 967

SCILLA, *Linn.; Gen. Pl., III., 814.*

Scilla indica, *Roxb.;* see **Urginea indica,** *Kunth.;* LILIACEÆ; Vol. VI. Pt. IV.

S. indica, *Baker; Fl. Br. Ind., VI., 348.* 968

Syn.—[S. MACULATA, *Baker in Jour. Linn. Soc., XIII., 250;* LEDEBOURIA HYACINTHINA, *Roth.; Wight, Ic., t. 2040* L., MACULATA, *Dalz.;* BARNARIA INDICA, *Wight, Ic., t. 2041.*

Vern.—*Suphadie-khus,* BENG.; *Bhuikándá, pahárikánda, náni-janglikándo lahána ránakándá,* BOMB.; *Shiru-nari-vengayam,* TAM.

References.—*Grah., Cat. Bomb. Pl., 220; Dalz. & Gibs., Bomb. Fl., 251; O'Shaughnessy, Beng. Dispens., 663; Dymock, Mat. Med. W. Ind., 2nd Ed., 834; Irvine, Mat. Med. Patna, 106; Home Dept. Cor. rel. to Pharm. Ind., 230, 240, 291.*

Habitat.—Frequent in sandy places, especially near the sea, in the Deccan peninsula from the Concan and Nagpur southwards. **S. Hohenackeri,** *Fisch. et Mey.,* is a closely allied species, met with in the Panjáb.

Medicine.—The BULBS are scaly, about the size of a large nutmeg, composed of very smooth and fleshy scales, which are so imbricated that they might be mistaken for coats if not carefully examined; they are roundish or ovate in shape, sometimes slightly compressed on the sides; externally they are of a whitish-brown colour. They are usually found growing singly as if propagated by seeds, and not, as in the case of **Urginea indica,** in clusters, each of which contains in the centre a mother-bulb surrounded by many smaller ones (*Moodeen Sheriff, Mat. Med. Madras*). **Ainslie** notices the bulbs of **Scilla hyacinthina,** and says that they were employed in Southern India for the relief of strangury and fever in horses. **O'Shaughnessy,** after commenting on **Ainslie's** remark, says that he tried them as a substitute for the officinal squill (**Urginea Scilla**) of the British Pharmacopœia, but found that, although given in double and even treble the dose of the latter drug, they produced no marked effects. **Dymock,** however, states that "for many years they were issued from the Bombay Medical Stores in lieu of squills (*Ind. Jour. of Med. Phys. Sci., Jan. 18th, 1838 p. 9*), but that of late years **Urginea indica** has been in use; both appear to be equally satisfactory substitutes for squills." [The late **Moodeen Sheriff** (in his unfinished *Materia Medica of Madras*) describes them as more powerful than the bulbs of **Urginea indica** and quite equal to the officinal drug of the British Pharmacopœia. He says that they are particularly efficient if gathered soon after they have flowered, a fact which may have something to do with **O'Shaughnessy's** low estimate of their powers, since he remarks that the bulbs he made use of "had not flowered that season." The dose, according to **Moodeen Sheriff,** is from one to four grains. *Ed.*]

MEDICINE. Bulbs. 969

SCINDAPSUS, *Schott.; Gen. Pl., III., 992.*

Scindapsus pertusus, *Schott.*; a synonym for **Rhaphidophora pertusa,** *Schott.*; and **Pathos pertusa,** *Roxb., Fl. Ind., Ed. C.B.C., 146*; AROIDEÆ.

970 **S. officinalis,** *Schott.; DC., Mon. Phaner., II., 254; Wight, Ic., t. 778.*

Syn.—POTHOS OFFICINALIS, *Roxb., Fl. Ind., Ed. C. B. C., 145.*

Vern.—*Gajapipal, pippal-j'hanca, maidah, gaj-pipli, bari-pipli,* HIND.; *Gajapipal, gaj-pipul,* BENG.; *Dare jhapak',* SANTAL; *Gaj-pipali, háth, ungliya,* N.-W. P.; *Thora-pimpli,* MAR.; *Motto-piper,* GUZ.; *Hatti-pipli,* DEC.; *Atti-tippili,* TAM.; *Enuga-pippalu, gaja-pippallu,* TEL.; *Dodda-hipalli,* KAN.; *Atti-tippili, anait-tippili,* MALAY.; *Gaja-pippali, kari-pippali, kapi-balli, kola-balli, s'reyasi, vas'ira,* SANS.

References.—*Revd. A Campbell, Econ. Prod., Chutia Nagpore, No. 8435; Mason, Burma, 505, 816; Sir W. Elliot, Fl. Andhr., 56: Sir W. Jones, V., 151; Pharm. Ind., 250; Ainslie, Mat. Ind., II., 113; O'Shaughnessy, Beng. Dispens., 626; Moodeen Sheriff, Supp. Pharm. Ind., 226; U. C. Dutt, Mat. Med. Hind., 252, 297; Atkinson, Him. Dist., 318, 750; Birdwood, Bomb. Pr., 94; Home Dept. Cor., regarding new Pharm. Ind., 240; Gaz., Orissa (W. W. Hunter), II., 159, App. iv.*

Habitat.—A large climber, common in the tropical forests of many parts of India and Burma, adhering to the trees by thick adventitious roots.

MEDICINE. Fruit. 971 **Medicine.**—The dried sliced FRUIT is officinal in the Materia Medica of the Hindus. Stimulant, diaphoretic, and anthelmintic virtues are ascribed to it. It is also said to be aromatic and carminative, and useful in diarrhœa, asthma, and other affections caused by deranged phlegm. It is used principally as an aromatic adjunct to other medicines (*U. C. Dutt*). Among the Santals the fruit is applied externally for rheumatism (*Revd. A. Campbell*). [Dymock, writing apparently of **Rhaphidophora pertusa** (a nearly allied plant to that under consideration), says it is called *ghannaskúnda* in Marathi, because of its being used in the treatment of persons bitten by Russell's viper. For this purpose the JUICE, along with black pepper, is
Juice. 972 given internally, and the juice together with the juice of **Croton oblongifolius** and of the fruit of **Momordica Charantia,** is applied externally to the bitten part.—*Ed.*]

SCIRPUS, *Linn.; Gen. Pl., III., 1049.*

973 [A genus of Cyperaceous plants comprising about 200 species. These are widely scattered all over the world and are chiefly to be found in marshy places and stagnant water. Many species are indigenous to the East Indies, but of these only a few are described as being used for economic purposes. In the *Statistical Atlas* of Bombay the following remark occurs, which appears to relate to the edible sedges here dealt with:—"The tubers of these plants, found most largely in the Nál or inland lake to the west of Dholk and in marshy places in the Konkan, are eaten by the labouring classes in ordinary times, and very largely by the famine-stricken. These tubers materially support the latter in times of distress in the Nalkantha of Káthiáwár and Ahmadabad."—*Ed., Dict. Econ. Prdo.*]

974 **Scirpus dubius,** *Roxb.; Fl. Ind., Ed., C.B.C., 72;* CYPERACEÆ.

Vern.—*Alliki, gitti-gadda,* TEL.

Reference.—*Sir W. Elliot, Fl. Andhr., 13, 60.*

Habitat.—A native of wet, sandy, pasture land.

FOOD. Roots. 975 **Food.**—The tuberous ROOTS, which are said to be as good as yams (*Roxb.*), and the tender white SHOOTS which spring up after the monsoon
Shoots. 976 (*Elliot*), are eaten by the Natives of Chutia Nagpore and the other parts of India.

977 **S. Kysoor,** *Roxb.; Fl. Ind., Ed., C.B.C., 77.*

Vern.—*Kasuru, kesúr,* HIND., BENG.; *Kachera,* BOMB.; *Kaseru, dila,* PB.; *Gunda tunga gaddi,* TEL.; *Kaseruka,* SANS.

References.—*Sir. W. Elliot, Fl. Andhr., 65; U. C. Dutt, Mat. Med Hind., 304; Dymock, Mat. Med. W. Ind., 847; Atkinson, Econ. Prod., N.-W. P., Pt. V. (Foods), 101; Lisboa, U. Pl. Bomb., 184; Journ. Agri.-Horti. Soc. Ind., X., 356; XIII., Sel., 62.*

Habitat.—A weed common on the margins of tanks and rivers throughout India.

Medicine.—The tuberous ROOTS, which are about the size of a nutmeg, and of a black colour externally, have astringent properties and are given by Native practitioners in diarrhœa and vomiting. **MEDICINE. Roots. 978**

SPECIAL OPINIONS.—§ "*Kesur* is used to remove the taste of medicine from the mouth. It is chewed also for the purpose of checking sickness. I have often seen it used, but I cannot say whether it acts beneficially" (*Surgeon R. L. Dutt, M.D., Pabna*). "Astringent, given in diarrhœa" (*W. Dymock, Bombay*).

Food and Fodder.—The ROOTS are dug up in large quantities in the cold weather, sliced and eaten uncooked by the Natives of many parts of India. They are sweet and starchy, and are considered cooling and highly nutritious. **FOOD & FODDER. Roots. 979**

Sacred.—In Bengal, the TUBERS are given as offerings to the deities. **SACRED. Tubers. 980**

Scirpus maritimus, *Linn.; Boiss., Fl. Orient., V., 384.* **981**

Vern.—*Múrak, díla,* PB.; *Gurrapu sakatunga,* TEL.

References.—*Roxb., Fl. Ind., 75; Stewart, Pb. Pl., 265; Aitch., Bot. Afgh. Del. Com., 121; Sir W. Elliot, Fl. Andhr., 66; Gazetteer N.-W. P., IV., lxxix.*

Habitat.—Common in marshes and on the banks of streams in Northern India.

Fodder.—When fresh it forms good forage, but soon gets too dry (*Stewart*). **FODDER. 982**

SCLEROTIUM.

Sclerotium stipitatum, *Berk. & Curr.;* FUNGI. **983**

[In the brief review of the Fungi of India, given in Vol. III., 455, there will be found mentioned a fungus that inhabits the excavations produced by white-ants. By an unfortunate oversight its name has there been given as **Schrotium stipatum.** The substance here more especially alluded to cannot, however, be said to have been definitely determined. It is known in Tamil as *Patu manga,* and is reported to be highly valued as a medicine. The reader will find a detailed account of it in the Journal of the Agri.-Horticultural Society of India (*Vol. XIV., Old Series, Selections, 205-7*), and a scientific description of it in the Journal of the Linnean Society (*Vol. IX., 417*).—*Ed., Dict. Econ. Prod.*].

SCOLOPIA, *Schreb.; Gen. Pl., I., 127.*

Scolopia crenata, *Clos.; Fl. Br. Ind., I., 191;* BIXINEÆ. **984**

Syn.—PHOBEROS CRENATUS, *W. & A.;* P. ACUMINATUS, HOOKERIANUS, & ARNOTTIANUS, *Thwaites;* FLACOURTIA SAPIDA & CRENATA, *Wall.*

Vern.—*Hitterlu,* BOMB.

References.—*Dalz. & Gibs., Bomb. Fl., 11; Thwaites, Enum., 17, 400; Beddome, Fl. Sylv., t. 78; Gamble, Man. Timb., 17.*

Habitat.—A middle-sized tree found in hilly districts of Malabar, Kanara, Mysore, and Ceylon, and distributed to China and the Philippines.

Structure of the Wood.—White, very hard and dense, but liable to warp; said by **Beddome** to be used **for** planks. **TIMBER. 985**

986 **Scolopia rhinanthera,** *Clos.; Fl. Br. Ind., I., 190.*

Syn.—PHOBEROS RHINANTHERA, *Benn.;* P. MACROPHYLLA, *W. & A.;* FLACOURTIA INERMIS, *Wall.*

References.—*Gamble, Man. Timb., 17; Watt, Cal. Exhib. Cat., VII., 226.*

Habitat.—A tree of Malacca, Java, and Borneo. Found by **Kurz** in the Andaman Islands.

TIMBER. 987 **Structure of the Wood.**—Hard, red, close and even-grained. Weight 60lb per cubic foot.

SCOPOLIA, *Jacq.; Gen. Pl., II., 902.*

988 **Scopolia lurida,** *Dunal; Fl. Br. Ind, IV., 243;* SOLANACEÆ.

Syn.—ANISODUS LURIDANS, *Link. & Otto;* A. STRAMONIFOLIUS, *G. Don.*

References.—*DC., Prod., XIII., i, 555; Pharm. Ind., 181; Braithwaite Retrospect of Medicine, IX., 119.*

Habitat.—An erect, herbaceous plant of the Central Himálaya, found also in Nepál and Sikkim.

MEDICINE. Leaves. 989 **Medicine.**—The LEAVES, when bruised, emit a tobacco-like odour. A tincture, prepared from them and administered internally, was found to produce extreme dilatation of the pupil and, in two instances, to cause blindness during the use of the drug. (*Pharm. Ind.*)

S. præalta, *Dunal;* see **Physochlaina præalta,** *Hook. f.;* Vol. VI., [Part I., 226.

Scorodosma fœtidum, *Bunge;* see **Ferula fœtida,** *Regel;* UMBELLIFERÆ; Vol. III., 335.

990 ## SCORZONERA, *Linn.; Gen. Pl., II., 531.*

A genus of perennial, rarely annual, herbs, belonging to the Natural Order COMPOSITÆ. Three species are indigenous to the Panjáb and Western Himálaya, none of which seem to be of much economic value. In the adjacent countries of Balúchistán and Afghánistán, several other species are found. All these are used as vegetables by the inhabitants of those regions, but **S. mollis** and **S. divaricata,** *Turcz,* are the only Indian indigenous species that it has been thought necessary to deal with in this work.

991 **Scorzonera hispanica,** *Linn.; Boiss., Fl. Orient., III., 745;* COMPOSITÆ.

THE SPANISH SALSIFY; VIPER GRASS.

References.—*Firminger, Man. Gard. Ind., 163; DC., Origin Cult. Pl., 44; Smith, Econ. Dict., 371; Agri.-Horti. Soc. Ind., Trans., VII. (Pro., 116; Journal, IX. (Pro.), 97; (New Series), IV., 37.*

Habitat.—The SCORZONERA is wild in Europe, from Spain, where it abounds, the south of France and Germany, to the region of the Caucasus. In Sicily and Greece, this species is not found. It is cultivated in many parts of Europe for the sake of its ROOTS, which are used as a vegetable (*DC., Orig. Cult. Pl.*). In India it is occasionally cultivated in the gardens of Europeans.

FOOD. Roots. 992

993 **S. mollis,** *Bieb.; Boiss., Fl. Orient., III., 761;* also **S. divaricata,** *Turcz; Fl. Br. Ind.; III., 418.*

Vern.—*Kambul, jhag.,* AFGH.

Reference.—*Aitchison, Bot. Afgh. Del. Comm., 84.*

Habitat.—In the dry valleys of Afghánistán, Balúchistán, and Persia, extending thence to Eastern Europe.

FOOD. Roots. 994 Leaves. 995 **Food.**—Both the tuberous ROOTS and the LEAVES are collected, cooked, and eaten by the Natives of Afghánistán, and to some extent (**S. divaricata**) is similarly used by the hill tribes of the North-West Himalaya.

SCROPHULARIA, *Linn.; Gen. Pl., II., 937.*

Scrophularia dentata, *Royle; Fl. Br. Ind., IV., 256;* SCROPHULARINEÆ. 996

Syn.—S. KOTSCHYI, *H. f. & T.*

Vern.—*Shústi,* LADAK.

Reference.—*Stewart, Pb. Pl., 163.*

Habitat.—An annual herb with perennial root-stock, found on the Western Himálaya and in Western Thibet, at altitudes between 12,000 and 15,000 feet.

Fodder.—Browsed by goats but not by yaks (*Stewart*). FODDER. 997

SCUTELLARIA, *Linn.; Gen. Pl., II., 1201.*

Scutellaria linearis, *Benth.; Fl. Br. Ind., IV., 669;* LABIATÆ. 998

Vern.—*Mastiara,* PB.

References.—*Stewart, Pb. Pl., 173; Agri-Horti. Soc. Ind. Journal, XIV., 21; (New Series), I., 100; Gazetteer, N.-W. P., X., 315.*

Habitat.—Not uncommon on the Temperate Western Himálaya, from Kashmír to Kumáon, at altitudes between 3,000 and 8,000 feet. It is abundant on the Salt Range and is distributed to Afghánistán.

Food.—In the Salt Range this PLANT, although very bitter, is eaten by the Natives. FOOD. Plant. 999

SEA-SLUGS.

Sea-slugs. 1000

BÉCHE DE MER, TRIPANG, SEA LEECHES, HOLOTHURIA.

Vern.—*Hsen-hmyau,* BURM.; *Tripang-swala,* MALAY.; *Hoy-shun,* CHINA.

References.—*Forbes Watson, Indust. Survey, 370; Simmonds, Comm. Prod. of Sea, 105; Encyclop. Brit., III., 477; VII., 639; Balfour, Cyclop. Ind., I., 305; II., 96; III., 928; Royle, Production of Isinglass along the Coasts of India, 52, 54; Mason, Burma & Its People, 393, 728; Gazetteer of Burma, II., 415.*

Habitat.—Edible Holothurians are found on the coasts of the Mediterranean, the Eastern Archipelago, Australia, Mauritius, Ceylon, and Zanzibar, whence they are occasionally brought to Bombay for re-export to China. Several species are found on the Burmese coast, particularly that of the Mergui Archipelago, where they are captured in large quantities by the Natives, cured and sold to the Chinese. It is, however, from the coasts of New Caledonia, Tahiti, and the Fiji Islands that China is principally supplied with Sea-slugs. They are collected in large quantities throughout the Indian Archipelago, especially among the Eastern Islands.

Description.—The ordinary kind in size and appearance resembles a prickly cucumber, except that the colour is a whitish brown, another is perfectly black. One species is nearly 2 feet long, but they are generally much smaller and the average size may probably be taken as about 8 inches. The skin is in some species covered with spicules and prickles, in some it is quite smooth; and it may or may not be provided with "teats" or ambulacral feet disposed in rows. DESCRIPTION 1001

Four kinds are recognised in commerce as being of value. These are, according to the description of **Captain Andrew Cheyne,** who was for many years engaged in the fishery and preparation of these animals:—

(1) *Bankolungan,* in the fresh state, 11 to 13 inches long, of an oval shape, brown on the back, white and crusted with lime on the belly, with a row of teats on each side of the belly. They are hard, rigid, and with little power of locomotion, but able to expand and contract themselves at Bankolungan. 1002

DESCRIPTION. pleasure. This species is usually found on the inner edge of coral reefs in from 2 to 10 fathoms of water, and on a bottom of coral and sand. It can be obtained only by diving.

Keeskeesan. 1003 (2) *Keeskeesan,* 6 to 12 inches long, of an oval shape, quite black and smooth on the back, with a dark-greyish belly and one row of teats on each side. When contracted, it is similar in shape to a land tortoise. This species is found on the top of coral reefs in shallow water and on a bottom of coral and sand. It is more plentiful and, moreover, more easily caught than the preceding.

Talepan. 1004 (3) *Talepan,* 9 inches up to 2 feet long, of a dark red colour and narrower in proportion than the two above mentioned. The whole back is covered with large red prickles. They are found on all parts of the reefs, but chiefly in from 2 to 3 fathoms of water. They are softer and more difficult to cure than the others.

Munang. 1005 (4) *Munang,* seldom over 8 inches long, oval in shape, quite black and smooth, without teats or other excrescences, found in shallow water on the coral flats and often among turtle grass on the shore. These are chiefly procured from the Fiji Islands.

These are the kinds most esteemed in commerce. But the following are recognised as inferior in value :—

Sapatos China. 1006 (5) *Sapatos China,* about the same size as the *Munang,* of a reddish brown colour and with a wrinkled surface. It is found adhering to the coral rocks on the top of the reefs.

Lowlowan. 1007 (6) *Lowlowan,* of various lengths, black, wrinkled, and narrow. Found also on reefs.

Bilati blanco. 1008 (7) *Bilati blanco,* about 9 inches long, of an oval shape and a white and orange colour, and easily known by its voiding a white adhesive substance which sticks to the fingers when handled. Generally found on the inner edge of reefs, and on a sandy bottom. These generally bury themselves in the sand during the day and are most easily found on moonlight nights, when they emerge into the open water.

Matan. 1009 (8) *Matan,* differs from No. 7 only in colour, which is grey, brown and white, speckled.

Hangenan. 1010 (9) *Hangenan,* generally about a foot long, of a grey or greenish colour, wrinkled. Found on the lagoon side of coral reefs.

Sapatos grande. 1011 (10) *Sapatos grande,* 12 to 15 inches long, of a brown and white colour, wrinkled and very inferior.

PREPARATION. 1012 **Preparation.**—When caught they are first split and gutted, then boiled for a period varying according to the variety, but, speaking generally, from five to twenty minutes. When sufficiently cooked they ought, immediately on being taken out of the pot, to dry on the surface like a boiled egg. After this they are ready for drying. Drying operations are conducted on large platforms over a brisk fire. The platforms are erected one over the other, in large huts built for the purpose, and as the slugs become drier they are removed to higher platforms further away from the fire. When one batch appears dry, it is taken off the platforms, carefully examined, those not dry put up again, and the quantity thoroughly cured is sent on boardship where it is stowed away in bags. Should the ship be long in procuring a cargo they will require to be dried again every three months, but this may be done in the sun on platforms erected on deck (*Simmonds, Commercial Prod. of the Sea*).

FOOD. 1013 **Food.**—The Trepang is highly esteemed as an article of food by Chinese and Japanese epicures. It is minced down and made into a thick gelatinous soup, of which the Chinese are especially fond. It is seldom used by the Europeans in India, but is a favourite article of diet among the colonists of Manilla, and is said, when cooked by a Chinaman who understands the culinary art, to be a capital dish.

Trade.—A fairly extensive trade in Sea-slugs exists in India. From **Mr. O'Conor's** Trade Returns we find that, during the official year of 1889-90, the Straits Settlements sent 2,243℔ of this article, valued at R1,254, to Burma. During the same period 31,729℔ of Sea-slugs (Indian produce), valued at R4,450, were exported from Burma to the Straits Settlements, while a total of 43,287℔, valued at R9,530, of foreign Béche de mer was exported from Bengal and Madras to Ceylon, China, and the Straits Settlements. The total IMPORTS into China are large. Between 1868 and 1872, **Simmonds** states that they averaged 15,745 piculs of 133⅓℔ each The finest qualities of Sea-slug sell in China for as much as £100 per ton TRADE. 1014

Sebesten, see **Cordia Myxa,** *Linn.*, and **C. obliqua,** *Willd.*; BORAGINACEÆ; Vol. II., 563-565.

Sedges, see **Cyperus,** *Linn.*; Vol. II., 682; **Eriophorum,** *Linn.*; Vol. III, 266; **Fimbristylis,** *Vahl.*; Vol. III., 363; **Kyllinga,** *Rottb.*; Vol. IV., 569; **Scirpus,** *Linn.*; Vol. VI., etc.; CYPERACEÆ.

SECALE, *Linn.*; *Gen. Pl., III., 1203.*

Secale cereale, *Linn.*; *Boiss., Fl. Orient., V., 674*; GRAMINEÆ. 1015

COMMON RYE.

Vern.—*Gandam-dar, jow-thak-thak,* AFGH.

References.—*Stewart, Pb. Pl., 262; DC., Origin Cult. Pl., 370; Aitchison, Fl. Kuram Valley, 23, 110; Botany Afgh. Del. Comm., 126; Smith, Dic., 358; Kew Reports, 35, 79; Agri.-Horti. Soc. Ind. Journals, II., Sel., 178, 373; IV., Sel., 120; New Series, I., Sel. 19 t. V., Sel. 9.*

Habitat.—An annual corn-grass, cultivated in many parts of Europe. The indigenous habitat of wild rye is unknown, but **DeCandolle,** after reviewing the historical, philological, and botanical evidence, comes to the conclusion that most probably its "original area was in the region comprised between the Austrian Alps and the north of the Caspian Sea." It does not appear to exist in India to any extent either wild or cultivated, but Aitchison, in his *Kuram Valley Flora* and *Botany of the Afghan Delimitation Commission,* describes it as occurring abundantly as a weed in wheat-fields in these countries and to all appearance "perfectly wild."

Medicine.—Ergot of Rye, see **Claviceps purpurea,** *Tulsane,* Vol. II, 359, **Fungi.** MEDICINE. 1016

Food.—The GRAIN of rye is in Afghánistán, reaped with the wheat and ground up with it into flour. When a large proportion of rye is present, the flour is considered by the Natives to have injurious properties. FOOD. Grain. 1017

SECAMONE, *Br.*; *Gen. Pl., II., 746.*

[ASCLEPIADEÆ.

Secamone emetica, *Br.*; *Fl. Br. Ind., IV., 13; Wight, Ic., t. 1283*; 1018

Vern.—*Shada-bári,* BENG.

References.—*DC., Prod., VIII., 501; Kurz, For. Fl. Br. Burm., II., 195; Pharm. Ind., 142; O'Shaughnessy, Beng. Dispens., 451; Balfour, Cyclop. Ind., III., 559.*

Habitat.—A climbing shrub, common in jungles at the foot of mountains in the South Deccan Peninsula, and Ceylon.

Medicine.—The ROOT is acrid and is regarded by the Natives as possessing powerful emetic properties. **Dr. G. Bidie,** however, tried it in several cases and found it almost inert (*Pharm. Ind.*). MEDICINE. Root. 1019

SECHIUM, *Swartz; Gen. Pl.,*

1020 **Sechium edule,** *Swartz;* CUCURBITACEÆ.

THE CHAYOTE or CHOCO.

References.—*DC., Origin Cult. Pl., 273; Kew Bulletin (1887), No. 8, 7; (1889), No. 25, 28; Rep. on Govt. Bot. Gard. Saharunpore (1885), 5; (1886), 17; (1887), 20.*

Habitat—A cucurbitaceous plant, largely cultivated in Tropical America. **M. DeCandolle** believes it to be a native of Mexico and Central America. The cultivation of the Choco has, for some years, been successfully carried on in Ceylon, and more recently it has been introduced into the Darjíling and Saharanpore Botanical Gardens. It seems to thrive well, but is apparently "difficult to distribute, as the old plants do not bear moving about, whilst the large fleshy seeds must be sown directly they ripen" (*Saharunpore Report*).

FOOD. Fruit. 1021 Root. 1022 **Food.**—The FRUIT is pear-shaped, about 3 to 5 inches long, covered with soft prickles and either green or cream-coloured. When boiled it forms in the West Indies a favourite vegetable, and, with the addition of lime-juice and sugar, supplies an ingredient for tarts. The ROOT, when boiled or roasted, is farinaceous and wholesome (*Kew Bulletin.*)

1023

SECURINEGA, *Juss.; Gen. Pl., III., 275.*

[A genus of Euphorbiaceous plants, comprising eight species, widely distributed throughout tropical and temperate regions. None of the species now regarded as belonging to this genus are indigenous to the Indian Peninsula; but as several which were formerly looked on as belonging to it have been transferred to the genus **Flueggea** and have not been described in this work under that heading, it has been thought necessary to provide a place for them under **Securinega.** *Ed.*]

1024 **Securinega Leucopyrus,** *Muell.-Arg., in DC., Prodr., XV., ii., 451; [Wight, Ic., t. 1875;* EUPHORBIACEÆ.

Syn.—A synonym for FLUEGGIA LEUCOPYRUS, *Willd.; Fl. Br. Ind., V., 328;* F. VIROSA, *Dalz. & Gibs.;* CICCA LEUCOPYRUS, *Kurz;* PHYLLANTHUS LEUCOPYRUS, *Kon;* P. ALBICANS, *Wall.*

Vern.—*Achal,* NEPAL; *Hartho, aintha,* N.-W. P.; *Kakún, rithei, girthan, gargas, bháthi, bata, vanúthi, girk,* PB.; *Kiran,* SIND.; *Pera pastawane,* AFGH.; *Challa manta, sále manta,* C. P.; *Salepan, halepan,* RAJPUT.; *Parpo,* GOA.

References.—*Roxb., Fl. Ind., Ed. C.B.C., 679, 680; Beddome, For. Man., 197, t. 24, f. 4 & 6; Kurz, For. Fl. Br. Burm., II., 353; Gamble, Man. Timb., 354; Dalz. & Gibs., Bomb. Fl., 236; Thwaites, Enum., 281; Brandis, For. Fl., 456; Dymock, Mat. Med. W. Ind., 717; Gazetteers:—N.-W. P., X., 317; Bombay, XV., 442; Ind. Forester, XII., App. 21.*

Habitat.—A large, thorny shrub or small tree of the Panjáb plain and the Deccan Peninsula, from Kanara southwards. It is found also in Burma and Ceylon.

MEDICINE. Juice. 1025 Leaves. 1026 **Medicine.**—The JUICE of the LEAVES or the leaves made into a paste with tobacco are used to destroy worms in sores (*Dymock*).

FOOD. Fruit 1027 **Food.**—The FRUIT is eaten.

TIMBER. 1028 **Structure of the Wood.**—Pink, hard, close-grained. It is only used as fuel.

1029 **S. obovata,** *Muell.-Arg., in DC., Prodr., XV., ii., 449.*

Syn—FLUEGGIA MICROCARPA, *Blume; Fl. Br. Ind., V., 328;* F. LEUCOPYRUS, *Dalz. & Gibs. (not of Willd.);* F. LEUCOPHYLLA, *Wall.;* PHYLLANTHUS VIROSUS & RETUSUS, *Roxb.;* P. GRISEUS & GLAUCUS, *Wall.;* CICCA OBOVATA, *Kurz;* CHORIZANDRA PINNATA, *Wight, Ic., t. 1994.*

Vern.—*Dalme, dhani, bakarcha, ghari, gwala, darim,* HIND.; *Iktibi,* LEPCHA; *Ukieng, thaka,* MICHI; *Korchi,* GOND.; *Kandori, kodarsi,* BOMB.; *Yae-chinya,* BURM.

References.—*Roxb., Fl. Ind., Ed. C.B.C., 679, 680; Brandis, For. Fl., 455; Graham, Cat. Bomb. Pl., 180; Dalz. & Gibs., Bomb. Fl., 236; Kurz, For. Fl. Br. Burm., II., 354; Gamble, Man. Timb., 354; Lisboa, U. Pl. Bomb., 117, 171, 269, 273; Atkinson, Econ. Prod., N.-W. P., Part V., 44, 87; Gazetteers:—N.-W. P., IV., lxxvii.; X., 317, Bombay, XV., 442.*

Habitat.—A deciduous-leaved large shrub or small tree, found throughout India and in Burma and Ceylon. In the Himálaya it ascends to an altitude of 5,000 feet. It is distributed to China, the Malay Islands, Australia, and Tropical Africa.

Food.—It produces an abundance of small, round, pure white succulent BERRIES, which are, like those of the preceding species, said to be edible. FOOD. Berries. 1030

Structure of the Wood.—Reddish yellow, close-grained, durable Weight 52℔ per cubic foot. TIMBER. 1031

Domestic.—The WOOD is used for making agricultural implements. The BARK is very astringent and is said by **Roxburgh** to be employed to intoxicate fish. DOMESTIC. Wood. 1032 Bark. 1033

SEDUM, *Linn.; Gen. Pl, I., 659.* 1034

A genus of succulent herbs (the Stone crops) comprising about twenty species, indigenous to the Indian Peninsula. Few are of any economic importance, although several were included in the older systems of Materia Medica.

Sedum Rhodiola, *DC.; Fl. Br. Ind., II., 417;* CRASSULACEÆ. 1035

Syn.—S IMBRICATUM, *H. f. & T.*; RHODIOLA IMBRICATA, *Edgw.*; R. ROSEA, *Linn.*

Vern.—*Shrolo,* LADAK.

References.—*DC., Prod., III., 401; Stewart, Pb. Pl., 101; Aitchison in Journ. Linn. Soc., X., 74; Gazetteer, N.-W. P., X., 310.*

Habitat.—A herbaceous plant with perennial root-stock, common on the Western Alpine Himálaya from Kumáon to Kashmír, at altitudes between 12,000 and 17,000 feet. It is distributed to the arctic and alpine regions of Europe, Asia, and America.

Food.—The young LEAVES of the wild plant are eaten by the Natives of Lahoul (*Aitchison*). FOOD. Leaves. 1036

S. tibeticum, *H. f. & T.; Fl. Br. Ind., II., 418.* 1037

References.—*Stewart, Pb. Pl., 101; Aitchison in Journ. Linn. Soc., X., 74.*

Habitat.—A glabrous herb with perennial root-stock, found on the Western Alpine, Himálaya at altitudes between 12,000 and 16,000 feet. It is distributed to Afghánistán.

Food.—The LEAVES of this, as well as of the preceding species, are, according to Aitchison, eaten by the Natives of Lahoul. FOOD. Leaves. 1038

Selenite, see **Gypsum,** Vol. IV., 195.

SEMECARPUS, *Linn.; Gen. Pl., I., 424.*

[ANACARDIACEÆ.

Semecarpus albescens, *Kurz; Fl. Br. Ind., II., 35;* 1039

References.—*Kurz, For. Fl. Br. Burm., I., 313, and in Journ. Asiat. Soc. Beng. (1871), II., 51.*

Habitat.—A large evergreen tree "not unfrequent in the tropical forests of Martaban down to Tenasserim; rather rare in those of the Pegu Yomah up to 3,000 feet elevation" (*Kurz*).

Resin.—A black RESIN is said to be exuded by this tree (*Kurz*). RESIN. 1040

1041 **Semecarpus Anacardium,** *Linn. f.; Fl. Br. Ind., II., 30; Wight, Ic., t. 558.*
THE MARKING-NUT TREE.

Syn.—S. LATIFOLIUS, *Pers.;* ANACARDIUM LATIFOLIUM, *Lamk.;* A. OFFICINARUM, *Gærtn.*

Vern.—*Bhelá, bhiláwá, bilaran, bheyla, belatak,* HIND.; *Bhelá, bhelatuki,* BENG.; *Soso,* SANTAL; *Loso,* KOL.; *Bhallia,* URIYA; *Bawaræ,* GARO; *Bhola-guti,* ASSAM; *Bhalaiyo, bhalai,* NEPAL; *Kongki,* LEPCHA; *Cherun kuru, kampira,* MAL. (S.P.); *Kohka, biba,* GOND; *Bhiláwa, bhála, bhela, bhalian,* N.-W. P.; *Bhiláwa, bhela, bhiládar,* PB.; *Bhilawa, koka, bhallia* C. P.; *Biba, bhiba, bhilama, bilambi,* BOMB.; *Bibwa, bibu, bibha,* MAR.; *Bhilámu,* GUZ.; *Bhilavan, belatak,* DEC.; *Shén-kottai, sherán-kottai, shaing, shay- rang,* TAM.; *Jidi-vittulu, jiri, jidi, nella-jedi, nalla-jidi chettu, jidi chettu, tummeda mámidi,* TEL.; *Gerú, gheru, kari-gheru, ger,* KAN.; *Chyai-beng, clay ben, che, khi-si,* BURM.; *Kiri-badulla,* SING.; *Bhallátaká, arushkara, bhallátamu,* SANS.; *Beladin, habbul-fahm, hab-el-kalb,* ARAB.; *Biládur,* PERS.

References.—*Roxb., Fl. Ind., Ed. C.B.C., 269; Brandis, For. Fl., 124; Kurz, For. Fl. Burm., I., 312; Beddome, Fl. Sylv., t. 166; Gamble, Man. Timb., 111; Dalz. & Gibs., Bomb. Fl., 52; Stewart, Pb. Pl., 49; Rev. A Campbell, Rept. Econ. Pl., Chutia Nagpur, No. 7535; Graham, Cat. Bomb. Pl., 41; Sir W. Elliot, Fl. Andhr., 25, 74, 125, 184; Irvine, Mat. Med. Patna, 15; Medical Topog., 127; U. C. Dutt, Mat. Med. Hind., 141, 292, 293; Murray, Pl. & Drugs, Sind., 87; Dymock, Mat. Med. W. Ind., 2nd Ed., 203; Dymock, Warden, & Hooper, Pharmacog. Ind., Vol. I., 389; Year-Book Pharm., 1878, 291; Birdwood, Bomb. Prod., 20, 147, 261, 281; Baden Powell, Pb. Pr., 338, 597; Atkinson, Him. Dist. (Vol. X., N.-W. P. Gaz.), 308, 750, 780; Useful Pl. Bomb. (Vol. XXV., Bomb. Gaz.), 54, 151, 216, 242, 250, 264; Econ. Prod. N.-W. Prov., Pt. III. (Dyes and Tans), 85; Liotard, Dyes, 121, App. I.; McCann, Dyes and Tans, Beng., 137; Darrah, Note on Cotton in Assam, 34; Christy, New Com. Pl., VIII., 74; Aín-i-Akbari, Blochmann's Trans., I., 52; Man. Madras Adm., II., 82; Settlement Reports:—Central Provinces, Upper Godavery Dist., 38, 39; Chindwara, 110; Bhundara, 19; Mundlah, 88, 89; Chanda, App. VI.; Gazetteers:—Bombay, I., 137; XIII., 23; XV., 75; N.-W. P., I., 80; IV., lxx; Sind, 59; Mysore and Coorg, I., 50, 59; III., 22; Nellore Manual, 98, 116; Agri.-Horti. Soc. Ind., II., 1867, 80; I., Pt. IV, N. S., 398; Ind. Forester:—I., 362; II., 171, 175, 407; III., 24, 201; IV., 227; VIII., 106, 270, 412, 414; IX., 254, 255, 438; X., 222, 325; XI., 366; XII., App. 10; XIII., 120; Trans. Med. & Phys. Soc. Bombay (New Series), No. 12, 173; Smith, Dict. Econ. Pl., 268.*

Habitat.—A deciduous tree of the Sub-Himálayan tract, from the Sutlej eastward, ascending to an altitude of 3,500 feet, and found throughout the hotter parts of India as far east as Assam. It does not occur in Burma or Ceylon. It is distributed to the Eastern Archipelago and North Australia.

GUM. Juice. 1042 **Gum.**—The tree yields an acrid viscid JUICE, from which a varnish is made. A sample of gum prepared from this juice was sent by the Madras Forest Department to the Amsterdam Exhibition. It is usually described as a coarse black gum in amorphous carbonaceous masses with a shining coal-like fracture and having a dull brownish-black colour. It is said to be useless for commercial purposes (*Cooke*).

DYES & TANS Fruit. 1043 **Dyes and Tans.**—The pericarp of the FRUIT contains a bitter and powerfully astringent principle, which is universally used in India as a substitute for marking-ink. It gives a black colour to cotton fabrics, which is said to be insoluble in water, but soluble in alcohol. The JUICE of the pericarp is
Juice. 1044 mixed with lime water as a mordant before it is used to mark cloth. It is believed this substance is an ingredient in some of the marking-ink preparations sold at the present day in Europe.

In some parts of Bengal the fruits are regularly used as a dye for cotton cloths. They are employed either alone or with alum. The details of the process, as described in McCann's *Dyes and Tans of Bengal*, are as follows :—" In Balasore two jars are put on a brisk fire, one over the other; DYES & TANS. the upper one contains the *bhalia* fruit and has a hole in the bottom. The heat causes a black resinous juice to exude from the *bhalia* which runs into the lower jar. The cloth may either be dyed in this black liquid alone, or oil may be mixed with the liquid before the cloth is dipped in it. The cloth is then welt washed out with water. Lime water must be poured on the cloth to cause it to dry speedily." In Hazaribagh the method adopted is somewhat different. The *bhalia* fruit is soaked in water for three days, and then strained through a coarse cloth from the infusion thus obtained. The material to be dyed is washed well with water, and when half dry washed again in a solution of alum. When again half dry it is dipped in the *bhalia* infusion, worked well about till the required depth of colour is obtained, then removed and dried in the sun. When quite dry it is washed frequently in fresh water to get rid of the smell of the dye-stuff. The colour produced by the use of this dye-stuff is a dark grey or greyish black (*McCann*). Brandis says that the BARK is astringent, and is used as a Bark. 1045 dye.

"Pounded and boiled in rape oil, the fruit of this tree makes an excellent remedy for staying putrefaction when begun in a hide" (*Sir E. C. Buck, Dyes and Tans of the N.-W. P.*). "The NUTS of this tree are used by tan- Nuts. 1046 ners, especially in dressing the hides of the rhinoceros and buffalo to form targets" (*Buchanan, Stat. Dinagepur*).

Oil.—The KERNELS contain a small quantity of sweet oil. "The PERI- OIL. Kernels. 1047 CARP contains 32 per cent. of a vesicating oil of specific gravity ·991, easily soluble in ether and blackening on exposure to the air. It is similar to Pericarp. 1048 that of **Anacardium occidentale**, but Basiner found that it dissolves in potassa with a green colour, and its alcoholic solution turns black with basic lead acetate" (*Dymock*).

Medicine.—The acrid JUICE of the pericarp is a powerful vesicant, and MEDICINE. Juice. 1049 is often employed by Natives for producing fictitious marks of bruises. These, however, may be distinguished from the marks produced by blows by their deep bluish-black colour, and their presenting small vesicles or blisters on the surface. The ripe FRUIT is described in Hindu works of Fruit. 1050 medicine as having acrid, heating, stimulant, digestive, nervine, and escharotic properties. It is used in dyspepsia, skin diseases, piles, and nervous debility (*U. C. Dutt*). It is given internally by the Hindus of Southern India in small doses in scrofulous, venereal, and leprous affections, and externally an OIL prepared with the nut by boiling, is applied in rheumatism Oil. 1051 and sprains (*Ainslie*). In the Konkan, a single fruit is heated in the flame of a lamp; the oil from it is allowed to drop into a quarter of a seer of milk and this draught is given daily in cough caused by relaxation of the uvula and palate (*Dymock*).

In Muhammadan works on medicine the juice of the pericarp of the marking-nut is described as hot and dry, useful in all kinds of skin diseases, palsy, epilepsy, and other diseases of the nervous system. The dose prescribed is from ¼ to ½ a dirhem, and it is directed that when given internally it should always be mixed with oil or melted butter. Externally they apply it often in the form of a fumigation to cold swellings such as piles (*Dymock*). In the Panjáb the fruit is used to prepare a wash for cases of salivation, and its smoke is considered efficacious in impotency (*Stewart*). In its action, the oil of the marking-nut appears very closely to resemble that of the cashew-nut (**Anacardium occidentale**). It is a powerful vesicant, and when applied to the skin caused blistering within twelve hours. Extensive application of the oil produced painful micturition and hæmaturia. When administered internally in small doses, no physiological effect was observed. The marking-nut is occasionally used by Natives as a local

MEDICINE. irritant for the purpose of procuring abortion. It is also employed by malingerers to produce ophthalmia and skin eruptions.

SPECIAL OPINIONS.—§ "The acrid juice of the marking-nut is said to be applied to the *os uteri* by Natives when criminal abortion is intended" (*Joseph Parker, M.D., Deputy Sanitary Commissioner, Poona*). "The oil of the pericarp is used in marking linen and blackens under the influence of caustic lime" (*Deputy Surgeon-General R. F. Hutchinson, M.D., Morar*). "Is used by Sepoys to produce feigned disease. The oil is rubbed over a joint, the refuse is then burned, and the joint exposed to the smoke, a swelling of the joint is then produced and the impostor feigns rheumatism. The plant is well known" (*Surgeon-Major C. W. Calthrop, M.D., Morar*). "The juice of the marking-nuts or *bhéla* is frequently applied to the skin to simulate the marks of bruises by the people here" (*Assistant Surgeon Ram Chunder Gupta, Bankipore*). "Applied as a counter-irritant in rheumatism and to painful swellings; also to the gums in toothache" (*Shib Chunder Bhattacherji, Chanda, Central Provinces*). "Its chief use among native doctors is in the form of an electuary in syphilis" (*T. Ruthamn Moodelliar, Native Surgeon, Chingleput, Madras Presidency*). "It is applied externally for pains of rheumatism and sprains. It is considered aphrodisiac and is taken in the form of confection. It produces in some cases excessive itching and erysipelatous inflammation, for which the application of cocoanut oil or tamarind water is considered the best curative" (*Surgeon-Major Robb, Civil Surgeon, Ahmedabad*). "The juice is used as an escharotic in chronic rheumatic affections" (*Surgeon-Major A. S. G. Jayakar, Muskat*).

FOOD. Cup. 1052 Kernels. 1053

Food.—The yellow fleshy CUP on which the fruit rests is somewhat acrid in the fresh state, but when roasted in ashes it takes the flavour of a roasted apple and is eaten by Natives. The KERNELS of the nuts are also eaten. They are supposed to stimulate the mental powers, especially the memory.

TIMBER. 1054

Structure of the Wood.—Greyish-brown in colour, often with yellow streaks. It is full of an acrid juice which causes swelling and irritation of the skin when handled; timber-cutters for this reason object to felling it unless it has been ringed for some time previously. It cracks in seasoning and is not durable. Weight 42lb per cubic foot.

DOMESTIC. Seeds. 1055 Leaves. 1056 Wood. 1057

Domestic.—The oil from the SEEDS, mixed with the milk of a species of **Euphorbia**, is said by Brandis to be made into bird-lime by the wild tribes of the Satpura ranges. It is also used as a preventive against the attacks of white-ants, and as a lubricant to the wooden axles of native carts. The LEAVES are employed as plates. The WOOD is employed for making charcoal.

S. travancorica, *Bedd.; Fl. Sylv., t. 232; Fl. Br. Ind., II., 31.*

Vern.—*Natu sengote*, TEL.

Reference.—*Watt, Calcutta Exhib. Cat., Part I., 57.*

Habitat.—A very large tree, met with in the forests of the Tinnevelly and Travancore Hills.

RESIN. Juice. 1058

Resin.—It yields a caustic black JUICE similar to that of **S. Anacardium.**

SENECIO, *Linn.; Gen. Pl., II., 446.*

1059 **Senecio densiflorus,** *Wall.; Fl. Br. Ind., III., 355;* COMPOSITÆ.

Syn.—S. AUREUS and ANGULOSUS, *Wall.*; S. UNCINELLUS and DENSIFLORUS, *DC.*; SOLIDAGO DENSIFLORUS, *Wall.*

Vern.—*Chitawála*, PB.

References.—*DC., Prod., VI., 369; C. B. Clarke, Comp. Ind., 185; Stewart, Pb. Pl., 129; Gas., N.-W. P., X., 312.*

Habitat.—A tall shrubby plant, found in the Central and Eastern Himálaya, from Nepál to Bhután, at altitudes between 5,000 and 7,000 feet. It occurs also on the Khásia mountains between 4,000 and 6,000 feet, and in Burma.

MEDICINE. Leaves. 1060

Medicine.—In Hazara the LEAVES are applied to boils (*Stewart*).

Senecio Jacquemontianus, *Benth.; Fl. Br. Ind., III., 350.* 1061

Syn.—SENECILLIS JACQUEMONTIANA, *Dene.*

Vern.—*Poshkar*, KASMIR.

References.—*C. B. Clarke, Comp. Ind., 208; Stewart, Pb. Pl., 127.*

Habitat.—A tall yellow-flowered plant, found in the Western Himálaya at altitudes between 10,000 and 13,000 feet.

MEDICINE Root. 1062

Medicine.—**Stewart** states, on the authority of **Birdwood**, that the ROOT of a plant with this vernacular name is used for adulterating *kut* (see **Saussurea Lappa**, *C.B.C.*; page 480), and says that as the Kashmíris in Lahore make the same statement, there must be some foundation for it. He adds, however, that this may not be the plant used for that purpose, as in Kashmíri *poshkar* appears merely to signify a large herb with showy flowers.

S. quinquelobus, *Hook. f. & T.; Fl. Br. Ind., III., 353.* 1063

Syn.—PRENANTHES ? QUINQUELOBA, *Wall.*

Vern.—*Morta*, PB.

References.—*C B. Clarke, Comp. Ind., 209; Stewart, Pb. Pl., 129.*

Habitat.—A tall herbaceous plant, with perennial roots, found on the Temperate Himálaya, from Garhwál to Bhután, at altitudes between 10,000 and 12,000 feet.

MEDICINE Seeds. 1064

Medicine.—In Kanawar the SEEDS of what appears to be this species are given for colic.

S. tenuifolius, *Burm.; Fl. Br. Ind., III., 345.* 1065

Syn.—S. MULTIFIDUS, *Willd.*; S. LACINIOSUS, *Arn.*; DORONICUM TENUIFOLIUM, *Wight, Ic., t. 1129.*

Vern.—*Sanggye, mentog*, (Bazar flowers=) *nímbar*, PB.

References.—*DC., Prod., VI., 365; C. B. Clarke, Comp. Ind., 198; Stewart, Pb. Pl., 130; Gazetteer, N.-W. P., X., 312.*

Habitat.—A slender, much branched annual, met with in the Western Peninsula, and on the dry hills of the Western Ghat from the Konkan southward. It is distributed to Java.

MEDICINE. 1066

Medicine.—**Honigberger** states that it is officinal in Kashmír. The *nímbar* of the Lahore drug-sellers may probably be the produce of this plant (*Stewart*).

SACRED. 1067

Sacred.—In Lahoul it is held sacred to Buddha.

Senna, see the species of **Cassia,** *Linn.*; LEGUMINOSÆ; Vol. II., 210-226.

Serpentary root, see **Aristolochlia serpentaria,** *Linn.*; ARISTOLOCHIACEÆ; Vol. I., 317.

SERPENTINE.

Serpentine, *Ball, in Man. Geol. Ind., III., 446.* 1068

This mineral scientifically known as *ophite* is, when pure, a hydrous magnesium silicate, containing more water but less silica than talc. Iron peroxide is generally present in varying proportions, and there are traces of other colouring matters which give to it its varying and beautiful hues. Several varieties of SERPENTINE are distinguished: thus there are the noble or precious serpentine, which is partially translucent, and the fibrous, foliated, porcellanic, and resin-like, all of which receive special names. "Verd antique" marble consists of lime-stone with included serpentine.

The Gingelly Oil of Commerce.

Vern —*Kyouk-seing*, BURM.

References.—*Mem. G. S. I. (1872), V., 172; VIII., 282; X., 143; XVIII., 103; Journ. As. Soc. Beng. XXXIX, 237; I. Calvert, Kulu, 4; Mason, Burma and Its People, 586, 734; Madras Man. Admin., II., 39; Settlement Rept., Chanda Dist., C. P., 106; Forbes Watson, Ind. Surv. Ind., I., 413.*

OCCURRENCE.
Madras. **Occurrence.**—The following note has been kindly furnished by **H B.**
1069 **Medlicott, Esq.** late Director of the Geological Survey:—"In Madras ser-
Panjab. pentine, and more particularly serpentinous marbles, are found in the
1070 Kurnool and Cuddapah districts, also in parts of Salem, especially in the
Burma. neighbourhood of the magnesite deposits, the rock being more properly the
1071 mineral Baltimorite. In the Panjáb in the Puga and Haule valleys there
Andamans. is a dark green massive serpentine. In parts of British Burma serpentine
1072 is exceedingly abundant, and it also occurs in the rocks of the Andaman
Manipur. Islands and in the hills east of Manipur."
1073 Besides these localities it is described in the *Manual of the Geology of*
Bengal. *India* as occurring in Bengal in the Manbhum and Singbhum districts,
1074 and in the form of verd antique marble in Mirzapur. According to **Mr.**
Kulu. **Calvert** there is a serpentine quarry on the Rangal mountain in Kulu.
1075
MEDICINE. **Medicine.**—In Kulu serpentine is used medicinally for disease of the
Cups. liver. CUPS made of a serpentine called *zahr-muhra* are supposed in
1076 Ladak to split if poison is put into them.

For an account of its uses as a substitute for **Jade** see the account of that mineral in Vol. IV., 535.

[COMPOSITÆ; Vol. VI.

Serratula anthelmintica, *Roxb.*, see **Vernonia anthelmintica,** *Willd.*;

(*G. Watt.*)

1077 **SESAMUM**, *Linn.; Gen. Pl., II., 1058.*

There are referred to this genus some ten or twelve species of plants, the majority of which are natives of Africa. In India two, or perhaps only one, species occur wild, but **Sesamum indicum** is extensively cultivated and is often found, as an escape from cultivation, in the vicinity of human dwellings. It is sometimes also seen growing quite spontaneously in fields, even becoming a troublesome weed. **Blume** is reported to have observed it on the mountains of Java, in what he regarded as a truly wild state; but his description would lead to the supposition that the Javan plant might, with greater propriety, be regarded as an allied though distinct species. **DeCandolle** does not appear to have considered this explanation admissible, for, placing the greatest faith on **Blume's** observation, when taken in conjunction with the fact that **Rumphius** assigns to Sesamum Malayan names which are independent of any Sanskrit root, he has assumed that, from these facts, it was probable India obtained its stock of Sesamum from the Sunda Islands, some two or three thousand years ago. It will be found, however, from the remarks below (under the paragraph HISTORY) that the writer is more disposed to regard **S. indicum** as having been originally a native of India, or perhaps, rather of the upper northern tracts, its area of wild habitat having extended to Central Asia, but that its cultivation was probably first attempted in the Euphrates Valley and was extended to India by the Aryan conquerors.

[*163*; PEDALINEÆ.

1078 **Sesamum indicum,** *DC.; Fl. Br. Ind., V., 387; Wight, Ill., t.*

GINGELLY or SESAME OIL, *Eng.*; BENNÉ, HUILE DE SÉSAMÉ, *Fr.*; SESAMÖL, *Germ.*

Syn.—SESAMUM ORIENTALE, *Linn.*; S. LUTEUM, *Retz.*; S. OCCIDENTALE, *Heer & Regel.*

Vern.—*Til, tir, gingli, krishna-tél, barik-tel, míthá-tel, til-ká-tél* (oil), HIND.; *Tél, til, tilmi, rasi, sumsum, kala til, krishna til, bhadu til, kát til, rakta til, sánki til, khaslá til, khasa* (seed), BENG.; *Til*, KÓL.; *Rasi, khasa*, ORISSA; *Tilmin, kat tilmin*, SANTAL; *Til*, NEPAL; *Til, tili*, C. P.; *Til, tili, gingili* (*mitha tel*=sweet oil), N.-W. P.; *Bhun-*

guru, til, KUMAON; *Til, tili, kunjad,* PB.; *Til, kunjit,* AFG.; *Thirr, til,* SIND; *Til, mithá-tél* (oil), *bárik-tél* (seed), DECCAN; *Til, tal, krishna-til, barik-til, ashádi-tal* (white), *kala katwa* (black), *purbia* (red), BOMB.; *Til, silechá til, chokhóta-téla* (oil), *tila* (seed), MAR.; *Tal, til* (seed), *mithu tél,* GUZ.; *Nal-lenny* (oil), *yellú-cheddie, nuvvulu, ellu* (seed), TAM.; *Nuvvu, nuvvulu, manchi-núne* (oil), *polla nuvvulu* (seed), TEL.; *Wollelu, achchellu, ellu, valle-yanne* (oil), *yallu* (seed), KAN.; *Schit-elu, minak bijan, nallenna* (oil), *ellu, kárellu chitrallu* (seed), MALAY.; *Hnan, nahu-sí* (oil), BURM.; *Tun-pattala, tel-tala* (oil), *talla* or *talla-atta* (seed), SING.; *Tila, snehaphala, tila-taila* (oil), *tilaha* (seed), SANS.; *Duhn, djyldjylan, shiraj* (oil), *dhónul-hal* (oil), *sim-sim* (seed), *dhonu simsim* (oil), ARAB.; *Roghen, kunjed, kunjad* (seed), *roghane kunjad* (oil), *róghaneshírín* (seed), PERS.; *Semsem,* EGYPT; *Benjam,* SUMATRA.

Rumphius gives the following names to this plant: in Malabar and Hindustan, *gingelli* and *gingelin,* whence are descended the European names such as Spanish *Sorgelin;* Algerian and Sicilian, *Gingilena, gingulena, gurgulena, jugjolina* also *sanserlin.* In Arabic it is *Semsem* and the seed *golgolan* and, remarks **Rumphius,** who can doubt that the Arabic name is but a republication of the word *semen,* that is, fat or oil? In the Malayan tongue it is called *Widjin;* in the Ternatic *Widje* and among the Javanese and Balayans *Lenga;* with the Amboyans it is *Widjin;* in Banda, *Alalun;* and in China *Moa.* In concluding his notice of the various names known to him **Rumphius** while discussing the question whether it is the **Sesamum** of Latin writers makes the somewhat significant observation that the plant he has described differs widely from the account of it given by **Pliny, Theophrastus,** and **Dioscorides,** but thinks these differences may be accounted for by the great tendency to variation.

References.—*DC., Prodr., IX., 250; Boiss., Fl. Orient., IV., 81; Roxb., Fl. Ind., Ed. C.B.C., 491; Gamble, Man. Timb., 281; Thwaites, En. Ceyl. Pl., 209, 442; Trimen, Sys. Cat. Cey. Pl., 65; Dalz. & Gibs., Bomb. Fl., 161; Stewart, Pb. Pl., 149; DC., Orig. Cult. Pl., 419; Rev. A. Campbell, Rept. Econ. Pl., Chutia Nagpur, No. 8197, 9467; Graham, Cat. Bomb. Pl., 126; Mason, Burma and Its People, 504, 793; Sir W. Elliot, Fl. Andhr., 138, 155; Rheede, Hort. Mal., IX., 54, 55; Rumphius, Amb., t. 76, f. 1; Pharm. Ind., 151; Fluck. & Hanb., Pharmacog., 473-476; U. S. Dispens., 15th Ed., 1040; O'Shaughnessy, Beng. Dispens., 479; Irvine, Mat. Med. Patna, 108; U. C. Dutt, Mat. Med., Hindus, 216, 321; K. L. De, Indig. Drugs Ind., 106; Murray, Pl. & Drugs, Sind., 177; Waring, Bazar Med., 133; Bent. & Trim., Med. Pl., 198; Dymock, Mat. Med. W. Ind., 2nd Ed., 549; Year-Book Pharm., 1874, 105; Transactions of the Medical and Phy. Soc. Bombay (New Series), IV., 85, 155; Smith, Econ Dic., 193; Birdwood, Bomb. Prod., 127, 286; Baden Powell, Pb. Pr., 364, 420; Drury, U. Pl. Ind., 389; Atkinson, Him. Dist. (Vol. X., N.-W. P. Gaz.), 314, 750, 771; Duthie & Fuller, Field and Garden Crops, 35-36; Useful Pl. Bomb. (Vol. XXV., Bomb. Gaz.), 167, 219; Institutes of Manu, Burnell's Ed., 106; Gazetteers:—Bombay; Vol. II., 63, 269, 273, 277, 280, 284, 287, 295, 423, 536, 538, 541, 544, 547; Vol. III., 45, 145, 148, 151, 154, 158, 161, 164, 232, 234, 248, 294, 297, 298, 300, 302; Vol. IV., 53, 58, 232, 234, 237, 240, 243, 245, 247; Vol. V., 106, 119, 120, 193, 294, 299, 370, 377; Vol. VI., 39, 51, 247; Vol. VII., 78, 86, 94, 97, 149, 150, 554, 562, 570, 573, 575, 578, 580; Vol. VIII., 183, 189, 245, 260; Vol. X., 146, 148, 298, 300, 302, 305, 308, 310, 312, 314, 424; Vol. XI., 95, 97, 122, 128, 242, 244, 247, 249, 252, 424; Vol. XII., 52, 222, 348, 352, 358, 362, 370, 375, 379, 386, 391, 395, 400, 408, 413, 416, 420, 429; Vol. XIII., Part I., 286, 290, 336, Part II., 672, 675, 677, 680, 682, 689, 692, 694, 697; Vol. XV., Part II., 16, 19, 59; Vol. XVI., 91, 100, 348, 353, 363, 369, 394, 399, 407, 412; Vol. XVII., 246, 269, 343, 589, 594, 599, 604, 611, 617, 623, 629, 635, 642, 647; Vol. XVII., Part II., 34, 44, 46; Part III., 77, 80, 83, 87, 90, 93, 96, 100; Vol. XIX., 160, 164, 425, 429-446; Vol. XX., 229, 392, 395, 397, 399, 401, 403, 405; Vol. XXI., 244, 247, 497; Vol. XXII., 273, 365, 386, 627, 629, 632, 635, 637, 639, 641, 645, 647; Vol. XXIII., 319, 366, 530, 532, 534, 536, 538, 540, 542, 544; Vol. XXIV., 157, 171, 207, 216, 333. Bengal:—Vol. II., 64, 241; Vol. III., 80, 333; Vol. IV., 71, 246, 345; Vol. V., 84, 204, 308, 420; Vol. VI., 71, 293, 300; Vol. 273; VII, 242, 391; Vol. VIII, 60, 210; Vol. IX., 104, 302, 337, 338; Vol.*

X., 273; Vol. XI., 329, 330; Vol. XII., 89, 235; Vol. XIII, 83, 263, 29; Vol. XIV., 337; XVI., 103, 341; Vol. XVII., 313; Vol. XVIII., 104; Vol. XIX., 94, 302. Panjáb:—Delhi, 111, 139, 140; Gurgaon, 43; Hisar, 48; Rohtak, 93; Ludiana, 134; Simla, 55; Jullundur, 43; Hoshiarpur, 87, 94, 117; Kangra, Vol. I., 152, 153; Amritsar, 36; Gurdaspur, 50, 61; Sealkot, 67, 68; Lahore, 86, 90; Gujranwala, 52, 55; Ferozepore, 65, 69; Rawalpindi, 78, 81; Jhelum, 107, 108; Mooltan, 92, 93, 95, 100; Jhang, 105, 106, 107, 115; Montgomery, 88, 89, 102, 103, 104, 106, 107, 108, 111, 112; Muzuffargarh, 90, 93; Dera Ismail Khan, 119, 125, 128, 129, 131; Dera Ghazi Khan, 81, 84; Bannu, 139; Peshawar, 144, 159; Hazara, 129, 134, 136, 101; Kohat, 97; Jhang, 105. Central Provinces:—61, 223, 239, 365, 502. Sind:—9, 169, 170, 216, 217, 218, 492, 493, 534, 569, 571, 573, 574, 631, 632, 654, 670, 671, 851, 859, 860. N.-W. Provs.:—Vol. I., 82, 90, 93, 115, 119, 137, 152, 169, 225, 250, 252, 291, 317, 349, 493, 531, 571, 577, 589; Vol. III., 225, 463; Vol. IV., lxxiv.; Orissa:—Vol. II., 15 App. I., 180. Mysore & Coorg:—Vol. I., 63, 91; Vol. II., 11. Madras:—District Manuals, Salem, Vol. I., 147, 149; Vol. II., 9, 45, 67, 89, 105, 141, 159, 19[illegible], 214, 225, 237, 254, 268, 300, 306; Kistna, 366; Cuddapah, 47, 65, 68, 74; Madura, 105; North Arcot, 333, 334; Coimbatore, 224-225. Agricultural Reports:—Assam, 1888-89, 15; 1887-88, 17; 1886-87, 20; 1885-86, 17, Madras, 1878-79. Experimental Farm Reports, Cawnpore, 1886-87, 4, 18; 1885-86, 2, 5; 1884-85, 5; 1882-83, 8. Statistical Descriptions and Historical Accounts:—N.-W. Provs., Aligarh, 375, 479; Cawnpore, 27; Mainpuri, 50; Etawah, 35, 36; Eta, 19, 20, 30, 88; Muzaffarnagar, 28, 241; Meerut, 31, 38; Bulandshahr, 25. Ind. Forester:—X., 260; XIV., 370; Spons, Encycl., Encyclop. Brit.; Balfour, Cyclop. Ind., II., 583; Morton, Cycl. Agri.; Ure, Dic., Indus. Arts & Man.

Habitat.—An annual plant, which is commonly stated to be cultivated throughout the tropical regions of the globe. In India it would, perhaps, be more correctly described as a crop of the warm temperate or sub-tropical tracts, being grown as an autumn or even winter crop in the warmer parts of the country (the truly tropical areas), and as a summer one in the colder. Thus, for example, it is frequently stated that black sesamum is sown in February-March and reaped in May-June, and that white sesamum is sown in June and reaped in August and September. These dates are, however, applicable chiefly to the great table-land, the tarái, and lower hills of India. It is a *kharif* crop in the plains of the North-West Provinces, being sown in the middle or latter end of the rains and reaped in autumn. In the Panjáb it is essentially a rainy season crop. In the Central Provinces and a large part of Madras, two widely different crops are reared—one reaped in spring and the other in autumn. In Bombay generally and also in Sind, only one crop is grown, but the period of sowing is delayed till June or July and the harvest takes place from September to December. In Bengal, a perhaps even more direct adaptation to the periods of colder climatic influences occurs. The chief crop is sown in June and July and harvested in October, November, or December, while a less important crop is sown in January, February, or March, and reaped in June and July. In the moist tropical portions of Bengal, Assam, and Burma, the plant does not thrive so well as in the higher sandy-soiled tracts of Central and Northern India, regions subjected to a regular and not too excessive rainfall, or where the crop can be irrigated.

VARIETIES. 1079 **Races or Varieties of Sesamum.**—It will be found from the remarks below that the writer is disposed to think Sesamum may be viewed as indigenous in India; if, indeed, it cannot be regarded as occurring at the present day in a truly wild state. He has not, however, had the opportunity of extending his study of the plant to the numerous cultivated forms which exist in the country, and is, accordingly, unable to say whether these should be viewed as varieties or only cultivated races. Popular writers generally say that the white, black, red, and grey-seeded forms differ only

VARIETIES.

in the colour of the seed. But as opposed to this statement there stands out prominently one or two important considerations. In some parts of India the plant is a stunted herb which rarely exceeds 18 inches in height; in others it becomes a bush 3 or 4 feet high. In some fields all the flowers are white, leaves large, irregularly lobed; in other fields pink, or it may be dark red, and the leaves long, narrow, and almost quite entire. Added to these observations there is also the practical issue that certain forms can be grown under an environment of soil and climate quite unsuited to others. It would thus seem probable that careful study may reveal the existence not only of old and well differentiated races, but even of distinct varieties of **Sesamum indicum.** The information, such as it exists, leads forcibly to one conclusion, *viz.*, that Sesamum has been cultivated in India from as remote a period as rice, since its adaptations to climate and soil are quite as remarkable in the one case as in the other. Some of the Sesamum crops take only three months from sowing to harvest, others eight or ten. It may safely be said, therefore, that no subject of Indian agriculture would more richly reward careful study than that of Sesamum. Its ramifications not only extend into every phase of Indian agriculture, but into the early history of the human race, since there would seem no doubt but that *til* was the first oil-seed cultivated by man.

The tendency to variation under cultivation was one of the features specially dealt with by **Rumphius.** He says that on one stem the plant is frequently seen to have various forms of leaves, and, in fact, that the leaves appear to vary greatly according to the nature of climate and soil. The plant grown, for example (states **Rumphius**), in the Eastern islands, differs greatly from that of Hindustán. While discussing the differences between the black and white-seeded forms he says the white has broader leaves, of a more bright green colour and the flowers are also paler coloured than the black. The seed is either pure white or of a pale ashy colour. It has a sweeter taste and richer substance than the black. **Rumphius** thus appears to have regarded the white and black-seeded forms as distinct, and he states that the latter in the Malay peninsula sometimes attains the dimensions of a small bush five to six feet in height.

DYE. Oil. 1080

Dye.—The OIL is used in the process of dyeing silk, a pale orange colour (*Drury*). **Hawkes** (in the passage quoted below regarding sesamum oil in Madras) alludes to the oil being employed to brighten tinctorial results. Many oils are used by the Indian dyers, but it is not known whether they are supposed to themselves possess definite or specific tinctorial actions. The reader should consult the remarks on this subject in the paragraph under **Dyes** in the article **Ricinus communis,** *V., 509.*

Oil and Oil-cake.

OIL & OIL-CAKE. Oil. 1081

OIL.—Gingelly Oil is used in painting according to **Atkinson,** but its being a non-drying oil is opposed to its utility for that purpose. Sesamum is cultivated exclusively on account of its oil-yielding seed. It is, therefore, not necessary to do more than preserve in sequence order the paragraph which it is customary in this work to devote to the oils obtained from plants. The succeeding pages on sesamum seed and oil give the commercial phases of the subject, and naturally deal in detail with the methods of, and the extent to which, the oil is expressed in India. The medical properties and chemical composition of the oil will also be found below in the paragraphs devoted to these subjects. Suffice it therefore to give here a brief *resumé* of the leading properties of the oil by way of introduction to the more detailed discussion which follows.

There are, as already stated, at least two easily-recognised forms of this plant—one with white seeds (*safed til*) and the other with black seeds (*kala*

SESAMUM indicum.

Sesamum Oil and Oil-Cake.

OIL & OIL-CAKE. Oil.

til). The latter form is much more common, and yields a superior oil. It is sown in March and ripens in May, while the white form is sown in June and ripens in August. The oil is extracted by the same process as that for mustard-oil. Gingelly oil is clear and limpid, of colour varying from pale yellowish to dark amber; it has no smell, and is not liable to become rancid. It is composed essentially of oleine, which is often present to the extent of 75 per cent. But it is frequently adulterated with ground-nut oil. It is stated, however, that 10 per cent. of gingelly oil, mixed with other oils, may be detected by shaking one gramme of a cold mixture of sulphuric and nitric acids with one gramme of the mixed oils, when a fine green colour will be the result, a colour which no other oil produces. In India, gingelly oil is used for culinary purposes, in anointing the body, in soap-manufacture, and as a lamp oil. In England, it is chiefly employed in making soap and for burning in lamps. It resembles olive-oil in many of its properties, and is accordingly similarly used. The oil obtained from the black variety is generally stated to be more suitable for medicinal purposes than the white. It is also extensively employed in the manufacture of Indian perfumes.

Adulterants. 1082

ADULTERANTS AND SUBSTITUTES.—Gingelly oil is used in India to adulterate the oil of almonds (*Drury*) and *ghi* (*Duthie & Fuller*). After being kept for a time it becomes so mild that it may be used as a substitute for sweet oil in salads (*Drury*). Much of the imported olive oil into India is very probably only gingelly oil made in Europe (*Murray*). The test by which its presence may be detected has already been briefly alluded to above and will be also found discussed in greater detail in the paragraph which deals with the medicinal properties of the oil. A mixed oil, very extensively used in some parts of India, where it is known as *gora tel*, consists of ground-nut, sesamum, and safflower oils in varying proportions.

Perfumery. 1083

PERFUMERY.—Sesamum oil forms the basis of most of the fragrant or scented oils used by the Natives, either medicinally or for inunction before bathing. It is preferred for these purposes from the circumstance of its being little liable to turn rancid or thick, and from its possessing no strong taste or odour of its own (*U. C. Dutt*). It is sufficiently free from smell to admit of its being made the medium for extracting the perfume of the jasmine, the tuberose, the yellow rose, and narcissus, etc. This purpose is attained by adding one weight of flowers to three weights of oil in a bottle, which, being corked, is exposed to the rays of the sun for forty days; the oil is then supposed to be sufficiently impregnated for use (*Drury, Atkinson, &c.*). Another way of perfuming the oil is by keeping the seeds of sesamum between alternate layers of strong-scented flowers. By this means the scent becomes communicated to the oil-seed and fixed in the oil, which is pressed out in the ordinary manner (*Duthie & Fuller*). In the North-West Provinces the perfumed oil used to anoint the body is generally known as *phúlel*. Comparative experiments have not apparently been made between this process, and the system of *enfleurage* pursued in Europe, so that it cannot be said whether or not the practice might with advantage be adopted in other countries. In some respects it is more simple than the European method, and for a tropical country, where lard and other solid fats are liable to become rancid, has much to recommend it.

The reader should consult the article **Perfumes and Perfumery**, Vol. VI., Part I., 135.

Sesamum Oil-cake. 1084

SESAMUM OIL-CAKE.—This substance is frequently alluded to by Indian writers. Thus **Stocks**, speaking of Sind, says, it is called "*khur*" and is universally used for feeding oxen, camels, goats, and sheep. **Lisboa** remarks of Bombay that it is held that 'the cake, left after the expression of the oil, is very good fodder for fattening cattle.' **Stewart** writes of the Panjáb: 'The oil-cake is given to cattle, and sometimes

OIL & OIL-CAKE.

used by the poor as food when mixed with flour.' **Messrs. Duthie & Fuller** say of the North-West Provinces that "The oil-cake is used as cattle food, and in the western districts is much prized on this account, there being a considerable traffic in it. It is reported to be even occasionally used as human food by the poorer classes in times of distress'."

Many other similar passages might be quoted as exhibiting the use of this cake, from one end of India to the other, as an article of cattle food. It is thus somewhat significant that in Europe it would seem to hold an unfavoured position in the estimation of those interested in the rearing of cattle. The reader should consult the section OIL-CAKES of the article **Oils** for further information on this subject (Vol. V., 475), where he will find a comparative chemical analysis of this with the chief oil-cakes which are used in Europe as cattle food.

MEDICINE. Seeds. 1085

Medicine.—In Hindu medical works, three varieties of *til* SEEDS have been described,—black, white, and red. The black kind is the best suited for medicinal use. "Sesamum seeds are considered emollient, nourishing, tonic, diuretic, and lactagogue. They are said to be especially serviceable in piles, by regulating the bowels and removing constipation. Sesamum seeds ground to a paste with water are given with butter in bleeding piles. Sweetmeats made of the seeds are also beneficial in this disease. A poultice made of the seeds is applied to ulcers. Both the seeds and the OIL are used as demulcents in dysentery and urinary diseases in combination with other medicines of their class" (*Hindu Mat. Med.*) **O'Shaughnessy** (*Beng. Dispens.*, 479) regarded Jinjili Oil when carefully prepared as quite equal to Olive Oil for medicinal and pharmaceutical purposes. **Dr. A. Burn** (*Bombay Med. Phys. Trans., 1838, Vol. I.*) advocates a dressing of sesamum oil in the treatment of wounds, ulcers, etc. As a simple dressing he regards it as superior to any other, particularly during the hot season of the year. **Waring** (*Pharm. Ind.*) says that for many years he had employed it as a substitute for olive oil in the preparation of *Linimentum Calcis*, and found it answer well. **Drury**, compiling apparently from **Dr. Burn's** account, advocates the claims of this oil to greater consideration. **Baden Powell** says that in the Panjáb the oil is used in the treatment of rheumatism and boils. **Bentley & Trimen** say that "When of good quality Sesamé oil is quite equal to olive oil for use in medicine and pharmacy. It is largely used by the people of India for dietetical purposes, and it forms the vehicle for various drugs in the form of a medicated oil. For that purpose it is peculiarly serviceable, since it possesses little taste or smell and has no tendency to turn rancid on being kept. It is the basis also of most of the perfumed oils employed by the Natives before bathing.' In the *United States Dispensatory* it is stated that "this was known to the ancient Persians and Egyptians, and is esteemed by the modern Arabs and other people of the East, both as food and as an external application to promote softness of the skin. It is laxative in large doses."

Oil. 1086

The seeds (sometimes known as *benné* seeds) are said to be powerfully emmenagogue, and to be even capable of producing abortion. **Dr. Dymock** thinks, however, from the extent to which they are daily eaten by Hindu women that this statement must be incorrect. In the *Pharmacopœia India* the use of the seeds is recommended in amenorrhœa in the form of a warm sitz-bath containing a handful of the bruised seeds. **Waring** adds, however, that the alleged emmenagogue properties of these seeds deserve further investigation. A decoction of the seeds sweetened with sugar is prescribed in cough, and a compound decoction with linseed is employed as an aphrodisiac. A plaster of the ground seeds is also applied to burns, scalds, etc. From the fact of this oil remaining sweet it

SESAMUM indicum.

Medicinal Properties of Sesamum Oil.

MEDICINE.

seems worthy of enquiry whether it might not with advantage be substituted for linseed oil in the preparation of the emulsion known as "Carron Oil," now so largely employed in British practice in the treatment of scalds and burns. **Atkinson** mentions a somewhat curious medicinal property. "The dew," he says, "taken off the FLOWER in the early morning, is popularly supposed in Meerut district to be a panacea for all eye diseases."

Flower. 1087

Leaves. 1088

In the *United States Dispensatory* the LEAVES are placed among official drugs. They are said to "abound in a gummy matter, which they readily impart to water forming a rich, bland mucilage, much used in the Southern States as a drink in various complaints to which demulcents are applicable; as in cholera infantum, diarrhœa, dysentery, catarrh, and affections of the urinary passages." "One or two fresh leaves of full size, stirred about in half a pint of cool water, will soon render it sufficiently viscid. With dried leaves hot water is used. The leaves also serve for the preparation of emollient cataplasms." In India the leaves are to a small extent employed, but they do not enjoy the same favourable repute as in America. **Dr. Evers** (*Indian Medical Gazette, March 1875, p. 67*) gives, however, the results of his experiments with them and with the seeds to test their value as an emmenagogue "I have employed," he says, "the mucilage, obtained from the leaves of the Indian plant, in the treatment of sixteen cases of dysentery, and in all recovery followed. From six to seven days was the time necessary for such treatment. I confess, however, that my cases were not of the virulent type seen towards the end of the rainy season. The drug acts simply as a demulcent, and does not, in my opinion, exert any specific influence on the disease; furthermore, it is necessary to combine an opiate with it, to relieve the tenesmus, so that probably the opium added has as much to do in checking the disease as the mucilage itself." With regard to the value of the seeds as an emmenagogue **Dr. Evers** says: "In three cases of congestive dysmenorrhœa I administered the powder of the seeds in 10-grain doses, three or four times a day, with benefit. I have at the same time employed the hip-bath recommended by **Waring**. It is commonly believed in the south of India that the seeds, when eaten by pregnant women, are likely to induce abortion; but no instance of the kind has ever come under my notice, nor have I heard of any." The reader will find under the paragraph of SPECIAL OPINIONS below a brief note by a medical officer which supports the popular opinion that these seeds do possess the property attributed to them, of producing abortion. By the Natives of India a lotion made of the leaves and ROOTS is employed as a hair-wash (see *Vol. III., 86*) **Dr. Dymock** says that a powder made of the roasted and decorticated seeds is called *Ráhishi* in Arabic and *Arwah-i-Kunjad* in Persian: it is used as an emollient both externally and internally. Muhammadan writers generally speak of the seeds of this plant under the Arabic name *Simsim* or the Persian *Kunjad.* In Africa it is known as *Juljulan.*

SPECIAL OPINIONS.—§ "When the small fine thorns studding the prickly pear fruit penetrate the skin and cannot be removed by the forceps or other means, painting the affected part freely with gingelly oil effects their easy removal though not immediately. The thorns are softened under the action of the oil, or, rather, are almost dissolved and dislodged. A small vesicle appears at the site of each thorn, which bursts and eventually no trace of the thorn is to be found. An infant, illegitimately born, was thrown by its mother into a prickly pear bush, immediately after its birth. It was removed about three hours after and brought to the dispensary for treatment. Its body was studded with small thorns from head to foot, a very few were removed by the forceps. The removal of the rest

MEDICINE.

was effected in the above manner and with the above results. The child recovered perfectly." (*Surgeon-Major D. R. Thomson, M.D., C.I.E., Madras*). [**Dr. W. Dymock**, in a letter to the editor, says that he suspects some mistake in the statement that the seeds cause abortion, seeing that they are so extensively eaten by Hindus in sweetmeats, &c., *Ed.*] "I know of a case in which a large quantity of the seeds did actually produce abortion" (*Assistant Surgeon Bhagman Das (2nd), Rawal Pindi, Panjab*) "The seeds are useful in dysmenorrhœa attended with diminished menses, a hip-bath being found useful" (*Assistant Surgeon Bhagwan Das (2nd), Civil Hospital, Rawal Pindi, Panjab*). "The seed pulverised is taken internally (in doses of grs. viii) for amenorrhœa" (*Surgeon W. F. Thomas, 33rd M. N. I., Mangalore*). "The oil is used in our hospital in place of olive oil. Seeds are exported very largely to France and Italy, for the oil which is blended or substituted for the olive oil (*Surgeon-General W. R. Cornish, F.R.C.S., C.I.E., Madras*). "I have for a long time used the following in gonorrhœa and prefer it to copaiba or liquor potassa. ℞.—Oil Sesamé ♏. xx, Aqua Calcis ♏, xx, Aqua ℥j in mixture" (*Honorary Surgeon E. A. Morris, Tranquebar*).

CHEMISTRY. 1089

Chemical Composition.—The following statement of the chemistry of this substance is from **Flückiger & Hanbury's** *Pharmacographia*:—"The oil is a mixture of olein, stearin and other compounds of glycerin with acids of the fatty series. We prepared with it in the usual way a lead plaster, and treated the latter with ether in order to remove the oleate of lead. The solution was then decomposed by sulphuretted hydrogen, evaporated and exposed to hyponitric vapours. By this process we obtained 72·6 per cent. of *Elaïdic acid.* The specimen of sesamé oil prepared by ourselves consequently contained 76·0 per cent. of olein, inasmuch as it must be supposed to be present in the form of triolein. In commercial oils the amount of olein is certainly not constant.

"As to the solid part of the oil, we succeeded in removing fatty acids, freely melting, after repeated crystalizations, at 67° C., which may consist of stearic acid mixed with one or more of the allied homologous acids, as palmitic and myristic. By precipitating with acetate of magnesium, as proposed by **Heintz**, we finally isolated acids melting at 52·5 to 53°, 62 to 63°, and 69·2° C., which correspond to myristic, palmitic, and stearic acids.

"The small proportion of solid matter which separates from the oil on congealation cannot be removed by pressure, for even at many degrees below the freezing point it remains as a soft magma. In this respect sesamé oil differs from that of olive.

"Sesamé oil contains an extremely small quantity of a substance, perhaps resinoid, which has not yet been isolated. It may be obtained in solution by repeatedly shaking five volumes of the oil with one of glacial acetic acid. If a cold mixture of equal weights of sulphuric and nitric acids is added in like volume, the acetic solution acquires a greenish yellow hue. The same experiment being made with spirit of wine, substituted for acetic acid, the mixture assumes a blue colour, quickly changing to greenish yellow. The oil itself being gently shaken with sulphuric and nitric acids, takes a fine green hue, as shown in 1852 by **Behrens**, who at the same time pointed out that no other oil exhibits this reaction. It takes place even with the bleached and perfectly colourless oil. Sesamé oil added to other oils, if to a larger extent than 10 per cent., may be recognised by this test. The reaction ought to be observed with small quantities, say 1 gramme of the oil and 1 gramme of the acid mixture, previously cooled."

In the *United States Dispensatory* sesamum oil is said "to bear some resemblance to olive oil in its properties, and may be used for similar purposes. It is not a drying oil. At 12·7° C. (55°F.) it has the specific

SESAMUM indicum.

Sesamum Oil and Oil-Cake.

CHEMISTRY. gravity 0·919; and its point of congealation is −5°C. (23°F.) 'sp. gr. 0·914 to 0·923. When cooled to near 5°C (23°F.), it congeals to a yellowish white mass. Concentrated sulphuric acid converts it into a brownish-red jelly. If 10C. of the oil be agitated with 3 drops of a cold mixture of equal volumes of nitric and sulphuric acids, the oil will acquire a green colour, soon changing to brownish red' (*U. S.*). Its relation with nitric and sulphuric acids may serve for its detection when used as an adulterant, although it is said that if the other oil be in great excess, the oil of benné will not respond."

FOOD & FODDER. Oil. 1090 Seeds. 1091 **Food and Fodder.**—*Tili* OIL is not only used for human consumption like many other oils, but is also employed in sweetmeat-making and in adulteration of *ghí,* also occasionally for lighting and for anointing the body. For this last purpose it is sometimes scented by keeping the SEEDS between alternate layers of strongly-scented flowers, before the oil is pressed out : in its scented state the oil is called *phulel* and fetches R160 per maund. The seeds are also made into sweetmeats which are eaten by the Natives. They are toasted and ground into meal and made into cakes and other preparations eaten to some extent by the Hindu population of India. In the form of sweetmeat cakes prepared with sugar or molasses (*tilka-laddu* and *reori*) these seeds are offered for sale in every bazár of India. The oil bears a strong resemblance to olive oil, for which it is frequently substituted or used as an adulterant.

Oil-Cake. 1092 The OIL-CAKE (or residue remaining after the oil is extracted) is employed as cattle food, and in some parts of the country it is much prized as such. In times of drought and scarcity it is even used as human food by the poorer classes. As a fodder its stems and leaves are useless, but the empty

Capsules. 1093 CAPSULES are said to be eaten by cattle. For animals that have to perform hard labour a mixture of bruised sesamum and gram is regarded as useful (*Conf. with* Fodder, *Vol. III., 419*).

DOMESSTIC. Seeds. 1094 **Domestic and Sacred.**—Sesamum SEEDS "form an essential article of certain religious ceremonies of the Hindus, and have therefore received the names of *homadhánya* or the sacrificial grain, *pitritarpana* or the grain that is offered as an oblation to deceased ancestors" (*U. C. Dutt*). Atkinson, in his *Descriptive and Historical account of the Meerut District,* says that "At the festival of *Sakat,* held in the month of *Mágh,* the Hindus eat a composition of *gúr* and *til,* which they call *tilkut.*" "There is a proverb in frequent use :—'*tilon men til nahin kahna,*' that is to say, 'there is no oil in the seeds of sesamum,' which is thus equivalent to our proverb 'to swear black is white.' "

Oil. 1095 The OIL is occasionally used for purposes of illumination, and gives a clearer light than most other vegetable oils, but burns more rapidly (*Duthie & Fuller*).
Leaves. 1096 Anointing the body is another use to which the oil is applied either in the crude state, or scented, when it is termed *phulel.* The use of
Root. 1097 Sesamum by the sugar manufacturers is probably to regulate ebullition or to mechanically remove impurities (*Conf. with p. 234*). A lotion made
Stalks. 1098 from the LEAVES is used as a hair-wash and is supposed to promote the growth of the hair and make it black; a decoction of the ROOT is said to have the same properties. Sesamum STALKS when dry are used as fuel and as manure.

HISTORY.

HISTORY. 1099 Botanical evidence, if accepted by itself, might lead to the supposition that the sesamum of sub-tropical agriculture had originally been a native of Africa. In that great continent there are some eight or nine truly wild forms, out of a total of some ten or twelve species, referable to the genus. In Africa, too, the oil-yielding plant is known to have been cultivated from remote times. Indeed, it seems probable that had classic

HISTORY.

records of Africa existed similar to those of India and China, it might have been possible to trace its cultivation in that country nearly as far back as can be done for India, through the writings of the early Hindus. The word *sesame* or *sesamum* is common to Greek (σησάμη), Latin, and Arabic (*simsim*), with only slight variations. On this subject **DeCandolle** writes :—"**Theophrastus** and **Dioscorides** say that the Egyptians cultivated a plant called *sesame* for the oil contained in its seeds, and **Pliny** adds that it came from India. He also speaks of a sesame wild in Egypt from which oil was extracted, but this was probably the castor-oil plant. It is not proved that the ancient Egyptians, before the time of **Theophrastus**, cultivated sesame. No drawings or seeds have been found in the monuments. A drawing from the tomb of Rameses III. shows the custom of mixing small seeds with flour in making pastry, and in modern times this is done with sesame seeds, but others are also used, and it is not possible to recognise in the drawing those of the sesame in particular. If the Egyptians had known the species at the time of the exodus, eleven hundred years before **Theophrastus**, there would probably have been some mention of it in the Hebrew books, because of the various uses of the seed and especially of the oil. Yet commentators have found no trace of it in the Old* Testament. The name *semsem* or *simsim* is clearly Semitic, but only of the more recent epoch of the Talmud, and of the agricultural treatise of **Alawwam**, compiled after the Christian Era began. It was, perhaps, a Semitic people who introduced the plant and the name *semsem* (whence the sesam of the Greeks) into Egypt, after the epoch of the great monuments and of the exodus. They may have received it with the name from Babylonia, where **Herodotus** says that sesame was cultivated." **Flückiger & Hanbury**, however, hold that the Egyptian name for the plant *semsemat*, occurs in the Papyrus Ebers, is still existing in the Coptic *semsem* and in the Arabic *simsim*. These authors regard the plant as distinctly alluded to in "the most ancient documents of Egyptian, Hebrew, Sanskrit, Greek, and Latin literature."

But **DeCandolle** admits that an ancient cultivation in the Euphrates Valley, agrees with the existence of the Sanskrit name *Tila*, though he regards that name as "a word of which there are traces in several modern languages of India and particularly of Ceylon." He thus does not seem to have recognised that it *actually* has given origin to very nearly all the colloquial names for the plant which are in use in the various languages of modern India. From what he apparently takes to be the simple existence of a Sanskrit name he affirms: "We are thus carried back to India in accordance with the origin of which **Pliny** speaks, but it is possible that India itself may have received the species from the Sunda Isles, before the arrival of the Aryan conquerors. **Rumphius** gives three names for the sesame in three islands, very different one from the other, and from the Sanskrit word, which supports the theory of a more ancient existence in the Archipelago than on the continent of India." Perhaps the remark may be pardoned that **M. DeCandolle** generally tends to err on the side of putting, if anything, too great dependence on the mere presence in the classic literature of India, of names identified by modern writers as those referable to certain cultivated plants. In this particular instance, however, it would seem probable that he has been induced to set that evidence aside in favour of considerations of far less value, *viz.*, the existence of Malayan names not traceable to Sanskrit, and the observation of one botanist that a plant found in a wild state on the mountains of Java had been determined to be **Sesamum indicum.** From

* **Flückiger & Hanbury** cite Isaiah xxviii. 27 as a reference to this seed.

SESAMUM indicum.

History of the Sesamum Oil.

HISTORY.

these considerations he assumes that sesamum was probably brought to India from Sunda at a period prior to the Aryan invasion. But if this contention be accepted, it would, perhaps, be permissible to say that conversely we should be justified in looking for some trace of the Sunda name for the plant in the languages of India, if indeed the Sunda root might not also be expected to appear in the Sanskrit and other Aryan languages. Far from that being the case, however, there is a singular uniformity throughout the most diversified tongues of this country (a uniformity only very occasionally met with in the cultivated plants of India) in a name for the plant, its seed and oil which is clearly of Sanskrit and unmixed Sanskrit origin. That name too belongs to what might be called the earliest phase of the Aryan tongue. It enters into the most primitive conceptions of domestic life and religious ceremonial, and apparently assumes a generic from a specific significance, becoming "Oil" in more recent times on the discovery of other oil-yielding plants. And, indeed, most of the other Indian names given to sesamum come from the Arabic or Persian; few or none belong to the aboriginal languages of India. Of this nature may be mentioned (in addition to sesamum from *semsem*) the very general name *Gingeli* or *Gergelim* (the Indian commercial name), and *jinjalí* (the common Hind. and Mahr. name). Both these names **Dr. Rice** derives from the Arabic *chulchulan*, which denotes sesamum seed before being reaped. **Yule & Burnell** trace them from the Arabic *Al juljulán* pronounced in Spain *Al-jonjolín* whence the Spanish, *Al-jonjoli*; the Italian, *Ginggiolino* or *Zerzeline*; the Portuguese, *Girgelim, Zirzelim*; the French, *Jugeoline*; and the Philippine Island's name for sesamum of *Ajonjoli* (*Glossary*, 285). But the evidence deducible from Sanskrit literature is not in this case dependent upon the simple existence of a root from which the modern names appear to be derived. The early Sanskrit medical writers describe the various forms of the seed, assign to each the relative value maintained for it at the present day, and give nearly as complete an account of the oil, and of the medicinal and culinary uses of the seeds as can be found in modern works on the subject. Hence it may be said that there is no room for doubt that the *Tila* of Sanskrit authors is the *Til* of the present day, a position which cannot be very often upheld in the identification of modern with ancient names. In support of this statement the following passage may be given from **Dutt's** *Hindu Materia Medica* compiled from the Sanskrit authors:—"The word *Taila*, the Sanskrit for oil, is derived from *Tila*; it would therefore seem, that sesamum-oil was one of the first if not the first oil manufactured from an oil-seed by the Ancient Hindus. The *Bhávaprakása* describes three varieties of *Til* seeds, namely, black, white, and red. Of these, the black is regarded as the best suited for medicinal use. It yields also the largest quantity of oil, white *Til* is of intermediate quality. *Til* of red or other colours is said to be inferior and unfit for medicinal use. Sesamum seeds are used as an article of diet, being made into confectionery with sugar or ground into meal. They form an essential article of certain religious ceremonies of the Hindus, and have therefore received the names of *homadhánya* or the sacrificial-grain, *pitritarpana*, or the grain that is offered as an oblation to deceased ancestors, etc."

It will thus be seen that in this particular instance we possess abundant evidence that the *Tila* of Sanskrit authors is the *Til* of India, at the present day. But the important position which its seeds hold in the observances of Hinduism, secures for *Tila* an antiquity even greater than that of the Sanskrit medical writers. In addition to the synonyms mentioned by **Dutt** it is also known as *Sárála, Subandha, Taládhak*, and *Pútadhánya*. At the same time the word *Tila* has certain general meanings such as a "mole" or "spot," the comparison being doubtless to the

HISTORY.

size and colour of the seed. It also denotes a small "particle" or "portion," and occurs in certain proverbs or wise sayings of an ancient character. Numerous edible preparations made of the seeds, as also an extensive series of implements used in its culture or in the expression of the oil, have technical names in Sanskrit works in which the root *tila* is preserved. Such, for example, are *Tila-dhenu*, the special preparation of the seeds, made up in the form of a cow, which is used as an offering to the Brahmans; *Tila-piccata*, a sweetmeat of the seeds; *Tila-brishta*, fried sesamum; *Tilanna*, a mixture of *til* seed and rice. So, again, *Tila-homa*, a burnt-offering of sesamum seeds; *Tila-vratin*, eating only sesamum because of a vow; and *Tila-taila* or *tila-rasa*, sesamum-oil. The generic word *Taila*=oil, as derived from *Tila*, and the preservation of corresponding words directly taken from these (*til*, the sesamum plant, and *tel*, any oil) throughout the length and breadth of India and across the Himálaya into Afghánistán, is proof also of great antiquity for the original root of all these names. Indeed, even the Sanskrit redundancy of the root, to specially denote sesamum-oil, is very general in India, and in Ceylon it occurs as *Tel tala*. But as manifesting the gradual expansion of the meanings and associations of *Tila*, it may be pointed out that it is also the name of a chapter in the *Purána-sarva-sva; Tila-ganji-tírtha* is the name of a place mentioned in the *Kasika-ramana;* and *Tiladhenudána* is the title of a chapter in the *Váráha-Purána*. But it is, perhaps, needless to multiply examples of the extensive series of forms in which the word *Tila* occurs in Sanskrit literature. As exhibiting the important place which *Tila*, seeds and oil, took in ancient Hindu mythology, and, indeed, which they hold at the present day, it need only be necessary to cite the passages regarding it in the *Institutes of Manu*. In the third Lecture it is repeatedly mentioned. The peculiar form in which it should be offered to the Brahmans is dealt with. It is spoken of as one of the three things that purify at a *Çraddha*, and also as an offering that secures prosperity and confers offspring, while it delights the manes for a month. It is forbidden to eat anything mixed with sesamum seeds after sunset. The oil is alluded to as a hair-oil. The punishment of an unlearned man who accepts an offering of *Tila* is indicated, as also the peculiar transmigration that will fall to the lot of the thief of this seed. At the same time the simile of 'as sown so shall the harvest be' is illustrated by the remark that rice, sesame, beans, and barley will each bring forth according to its kind.

The *Institutes of Manu* were penned in India, and for the people who lived there, at the lowest computation, 2,000 years ago. But its rules of life and religion were framed on the time-honoured observances of the Vedas, hence the illustrations drawn from that great compilation of moral and material well-being may be accepted as showing that the presumption is greatly in favour of the idea that the plant was very probably known to the Aryans long prior to their invasion of India.

Though sesamum has not hitherto been recorded as found wild in any of the warmer tracts of Central Asia, it is cultivated everywhere on the Himálaya, in Afghánistán, Persia, Arabia, and Egypt. There would, therefore, seem very little evidence opposed to the statement that, if not originally a native of the warm temperate tracts of India (**Fluckiger & Hanbury** as well as **Bentley & Trimen** affirm without reservation that it is a native of India), it was probably brought to this country by the Aryans. On this supposition alone, as it would seem, can be accounted for, its Sanskrit, Persian, and Arabic names in use in India. That its cultivation may have originated independently, however, in more than one centre, seems quite likely. Indeed, if its cultivation sprang originally from one centre, it would

SESAMUM indicum.

History of the Sesamum Oil.

HISTORY,

seem justified, by the facts adduced, that we must look to Central Asia and Persia rather than to the Sunda Islands as its home. The writer would, in fact, venture the suggestion that it was probably first cultivated somewhere between the Euphrates valley and Bokhara south to Afghánistán and Upper India, and was very likely diffused into India proper and the Archipelago, before it found its way to Egypt and Europe. In part support of this idea, it may be stated that it has by no means been proved that India itself does not possess truly wild forms of the plant. The writer, some years ago, collected specimens on Parisnath hill in Behar, at an altitude of from 1,500 to 3,500 feet, and more recently others on the lower North-West Himálaya, which possess certain peculiarities suggestive at least of a degree of acclimatisation sufficient to arouse suspicion that they may in reality be indigenous. The Parisnath plants were found growing underneath the grassy vegetation with several miles of forest land intervening between them and cultivation. The Himálayan plants were also gathered in such situations as to suggest at least the doubt as to their being escapes from cultivation. What is curious, too, regarding these apparently wild states is the fact that they preserve certain recognisable structural features. They are erect, sparsely-branched herbs, generally 6 to 9 inches high; have long lanceolate almost entire leaves; small, remarkably dark-coloured flowers, instead of the white or pink flowers of the most prevalent form of the cultivated plant, and they possess two exceptionally large glands at the base of the short pedicels. These glands are rarely so well developed in the cultivated plant, but reappear in the neglected forms seen in the vicinity of cultivation. The subject, however, of the races or forms of sesamum met with in India is too imperfectly understood to justify more than the suggestion that their careful study may reveal the fact that the so-called acclimatised states manifest peculiarities that may be deemed by future investigators quite as much entitled to specific recognition as are the characters of the plant, found by **Blume**, and which **DeCandolle** accepts as proving that **Sesamum indicum** is a native of Java.

In addition to the fact of **Pliny's** having alluded to the oil as exported from Sind to Europe by way of the Red Sea, we have the subsequent reference in the Periplus (A.D. 80) to Guzerát as the country from which much sesamum oil was obtained. Passing over a gap of 1,500 years we next find various writers dealing with the subject. In 1510 it is mentioned by **Varthema** under the name of *zerzalino*; in 1552 by **Castanheda** as *gergelim*; in 1599 by **Fredericke** as *zezeline*; in 1606 by Gouvea as *gergelim*; in 1610 by **Mocquet** as *gerselin*; in 1661 by **Thevenot** as *telselin*; in 1673 by **Galland** as *georgeline*; in 1675 by **Heiden** as *jujoline*; in 1726 by **Valentijn** as the *gingeli* exported from Orissa; in 1727 by **Captain A. Hamilton** as *gingerly*; and in 1807 by **Dr. Buchanan-Hamilton** as *gingeli*. These brief historic records of sesamum have been taken from **Yule & Burnell's** *Anglo-Indian Colloquial Glossary* in order to exemplify both the gradual development of the modern knowledge in the seeds and certain mutations in the formation of its commercial name.

In the *Ain-i-Akbarí*, or the Administration Report for the year 1590 of the reign of the Emperor Akbar, frequent mention is made of sesamum, white and black, and what is somewhat significant, both kinds appear in the list of autumn crops. It is specially mentioned as grown in the *Subahs* of Agra, Allahabad, Oudh, Delhi, Lahore, Multan, and Malwa. There is, in fact, abundant evidence that to the people of India this oil-seed has been, from ancient times down to the present day, one of the most important agricultural crops of this nature. The expansion (within the past 30 or 40 years) which has taken place in the foreign exports manifests, however, an increased cultivation as the direct result of the benefits arising from the peaceful administration of India under British rule. The reader should, therefore, to complete the present brief historic sketch, consult the concluding section of this article which is devoted to THE TRADE IN SESAMUM SEED AND OIL.

CULTIVATION.

CULTIVATION

Area, Outturn, and Consumption.

Area. 1100

The chief facts regarding *til* cultivation in India (*e.g.*, influences of climate, season of sowing and reaping, area, yield, traffic, etc.), will be discovered from the following note, which was issued by the Government of India (Revenue and Agricultural Department) on the 12th February 1891, as a forecast of the season's crop :—

Third General Memo. on the Sesamum crop of the season 1890-91.

"The appended statement (A) tabulates the information available regarding the area and outturn of sesamum in the Provinces from which reports have been received. In most places there are two crops of Sesamum—a *kharif* and a *rabi* crop. This oil-seed is very generally sown mixed with other crops, and consequently it is difficult to estimate the acreage and yield correctly. The figures, therefore, must be accepted with more reserve.

"2. The information from Madras is incomplete; particulars of the late crop are not available. Up to the present the condition of the standing crops has been, on the whole, fair, but more rain is needed. The greater part of the area under Sesamum in the Bombay Presidency belongs to the Native State of Gujarat. The Baroda figures, however, have not yet been communicated. The condition and estimated outturn are nowhere large, principally the result of unfavourable weather, notwithstanding good sowing rain. In the North-Western Provinces and Oudh the crop was also affected by unseasonable rainfall—excessive at sowing and deficient afterwards. Most of the crop in the Central Provinces has been gathered, but there is a considerable area in the Nagpur country which is sown with cold weather Sesamum, and this will not come into the market for some time yet. The outturn has been affected by rain in November. In Berar the crop has suffered to a certain extent from want of rain.

"3. In the remaining British Provinces, the area under Sesamum is not considerable, with the sole exception perhaps of Bengal, where statistics are not at present available. The crop is probably grown extensively in Hyderabad, but statistics are not forthcoming for that Native State, nor for Central India and Rájputana, where Sesamum is grown for export on a small scale.

"4. For the current year the estimated area and outturn, as returned in Table A, are 2,032,000 * acres and 171,100 † tons, respectively.

"5. The annexed statement (B) shows the imports of Sesamum by rail and river into the chief seaport towns for the last five years. The average exports of the twelve blocks named for the past five years are about 92,000 tons. The order in which the Provinces stand as exporters is given below :—

Provinces.	Average exports in tons.	Provinces.	Average exports in tons.
Sind	20,000	Rajputána and Central India	4,000
Central Provinces . .	17,000	Bengal	2,000
Bombay . . .	15,000	Berar	2,000
Nizam's Territory . .	13,000	Mysore	1,100
Madras	6,000	Assam	900
Panjáb	5,000		
North-Western Provinces and Oudh . . .	4,000		

The Sind figures, however, include a large proportion of Panjáb exports.

"6. In 1889-90 the exports by sea were 83,777 tons, valued at R1,30,98,813, *i.e.*, 12·32 per cent. of the total value of seeds exported that year. The average weight and value of these exports during the last four quinquennial periods compare as follows :—

	Averages for five years ending 31st March			
	1874.	1879.	1884.	1889.
Tons (thousands) . .	38	51	106	108
Rupees (lakhs) . .	45·39	80·53	143·40	151·04

* Exclusive of the area of mixed Sesamum in the North-Western Provinces and Oudh.
† " " outturn " " " "

SESAMUM indicum.

Sesamum Cultivation in India.

CULTIVATION

Area.

"The bulk of the exports goes to France, as will be seen from the figures below of each of the years named:—

YEARS.	Total exports.	Consigned to France.
	Tons.	Tons.
1873-74	45,000	41,000
1878-79	51,000	42,000
1883-84	143,000	102,000
1888-89	77,000	56,000
1889-90	89,000	60,300

A

Area and Outturn of Sesamum in the Surveyed Provinces of India.

For further information see pages	PROVINCES.	Normal area.	Area in previous year, 1889-90.	Area in current year, 1890-91.	PERCENTAGE OF INCREASE (+) ABOVE OR DECREASE (—) BELOW Normal area.	Area in previous year.	Outturn in current year, 1890-91.	Yield per acre, seers.	Yield per acre, maunds.
		1	2	3	4	5	6	7*	8*
		Acres.	Acres.	Acres.			Tons.		
I. 519-522	Madras—								
	Farly crop .	315,000	370,000	339,000	+7·5	—8·3	18,161		
	Late crop .	216,000	238,000	189,000	—12·5	—20·6	10,125		
	Total Madras .	531,000	608,000	528,000	—·56	—13·1	28,286	60	1 1/3
III. 522-524	Bombay—								
	British Districts	335,000	296,000	233,000	—30·4	—21·3	27,363		
	Native States (*b*)	274,000	274,000	274,000	...	...	34,770		
	Total Bombay .	609,000	570,000	507,000	—16·75	—11·05	62,133	137	3 3/8
IV. 524	Sind—								
	British Districts	129,000	128,000	101,000	—21·89	—21·25	8,340		
	Native States .	900	900	900	...	...	114		
	Total Sind .	130,000	129,000	102,000	—21·54	—20·93	8,454	92·8	2 3/5
V. 525-526	N.-W. P. and Oudh	175,000	176,000	175,000(*a*)	*Nil*	—·57	15,675	100	2 1/3
VI. 526-528	Panjáb . . .	187,000	215,000	217,000	+15·48	+·93	25,000	129	3 1/4
VII. 529-531	Central Provinces .	337,932	448,000	409,000	+21·03	—8·7	27,150	74	1 7/8
VIII. 531	Berar . . .	144,000	91,000	94,000	—34·72	+3·29	4,433	53	1 13/40
	GRAND TOTAL	2,115,000	2,237,000	2,032,000	—3·92	—9·16	171,131	94·8	2 3/8

* The figures shown in columns 7 and 8 have been worked out from the figures given by the Government of India in columns 3 and 6.—*Ed., Dict. Econ. Prod.*

(*a*) This figure refers to unmixed *til*. The acreage of the mixed crop is estimated to be 4,000,000 and the outturn 35,000 tons.

(*b*) Less Baroda, for figures which have not been received.

CULTIVATION

Imports.

B

Statement showing the imports by rail and river into Bombay Town, Karachi, Calcutta, and Madras Seaports during the last five years (1885-86 to 1889-90) of Til or Jinjili.

Year and whither imported.		Whence exported.													Percentage to total imports into the four ports.
		Madras.	Bombay.	Sind.	N.-W. P. and Oudh.	Panjáb.	Central Provinces.	Berar.	Bengal.	Assam.	Rájputana and Central India.	Nizam's Territory.	Mysore.	Total.	
		1	2	3	4	5	6	7	8	9	10	11	12	13	14
		Tons.	Tons.	Tons.	Tons.	Tons.	Tons.	Tons.	Tons.	Tons.	Tons.	Tons.	Tons.	Tons.	Tons.
Bombay Town.	1885-86	461	19,212	...	2,242	371	16,733	5,147	42	...	5,127	7,217	...	56,552	64 70
	1886-87	2,405	20,267	...	3,467	435	24,460	3,195	...	...	6,214	13,327	...	74,360	69·69
	1887-88	1,932	16,201	...	4,014	694	12,075	1,990	...	...	3,900	22,739	...	63,545	52·43
	1888-89	60	7,192	...	1,474	1,147	14,561	756	...	...	1,950	11,875	...	39,015	55·68
	1889-90	42	11,400	...	2,841	65	17,766	679	...	...	3,190	7,022	30	43,035	56·29
Karachi	1885-86	...	...	18,119	...	4,596	...	...	...	..	4	...	...	22,719	25·99
	1886-87	...	...	17,205	5	2,138	...	...	...	...	...	...	...	19,348	18·13
	1887-88	...	...	32,413	409	8,779	...	...	..	...	...	...	...	41,601	34·33
	1888-89	...	...	15,833	81	5,857	...	...	...	...	...	...	...	21,771	31·07
	1889-90	...	...	15,395	...	4,065	...	...	...	...	...	...	...	19,460	25·45
Calcutta	1885-86	...	...	...	469	...	...	...	1,154	963	...	...	...	4,940§	5·65
	1886-87	...	...	...	720	...	242	...	1,106	387	39	...	...	3,668§	3·44
	1887-88	...	...	...	2,654	151	287	...	423	920	30	...	...	5,083§	4·19
	1888-89	...	...	...	532†	...	...	...	3,559‡	1,310	...	...	...	5,401§	7·71
	1889-90	...	...	...	477†	...	450	...	5,627‡	882	..	...	...	7,436§	9·73
Madras Seaports.	1885-86*	2,054	...	...	...	...	...	...	...	...	...	1,041	99	3,194	3·65
	1886-87*	7,525	...	...	...	..	...	...	...	...	...	264	1,541	9,330	8·74
	1887-88*	7,774	494	...	...	...	...	...	...	...	...	642	2,053	10,963	9·04
	1888-89†	3,039	...	...	...	...	...	...	...	...	...	463	380	3,882	5·54
	1889-90‡	4,358	...	...	...	...	...	...	...	...	...	497	1,670	6,525	8·53
Total	1885-86	2,515	19,212	18,119	2,711	4,967	16,733	5,147	1,196	963	5,131	8,258	99	87,405	The supplies from which the foreign exports were mainly drawn.
	1886-87	9,930	20,267	17,205	4,192	2,563	24,702	3,795	1,106	387	6,253	13,591	1,541	106,706	
	1887-88	9,706	16,695	32,413	7,077	9,624	12,362	1,990	423	920	3,930	23,381	2,054	121,192	
	1888-89	3,099	7,192	15,833	2,087	7,004	14,561	756	3,559	1,310	1,950	12,338	380	70,069	
	1889-90	4,400	11,400	15,395	3,318	4,130	18,216	679	5,627	882	3,190	7,519	1,700	76,456	

* Excluding Calicut, the figures for which are not shown separately.
† Including Calicut.
‡ These two years include river-borne trade.
§ Inclusive of river-borne trade.

Cultivation of Til.

CULTIVATION

It will be observed that the statement given in table A (p. 516) by no means represents the total area in India devoted to this crop, nor, accordingly, does it show the probable actual production. It gives, for example, no figures for Bengal (an important *til*-producing province), nor for Assam and Burma, nor the Native States of Central and South India and Rájputana. It also professedly excludes from consideration the large area in the North-West Provinces and Oudh devoted to mixed *til* and other crop cultivation. It is not expressly stated whether a similar exclusion applies to the other provinces, or whether the figures given express mixed production to the area of pure crops. The importance of these considerations will be seen from two facts:—(*a*) while the production of the North-West Provinces and Oudh is put down at 15,675 tons, a foot-note explains that in addition 35,000 tons were obtained from mixed *til* cultivation; (*b*) the yield worked out in columns 7 and 8 varies from 53 to 137 seers per acre. From this last consideration it seems probable that while mixed cultivation has been excluded, in the case mentioned, a calculation has been adopted in some of the other provinces to express mixed crops to the area of pure cultivation.

On the other hand, the figures given in table B (p. 517) refer to the actual movements of sesamum seed by rail and river, and, therefore, denote the surplus over local consumption from actual production (whether pure or mixed cultivation), from all the provinces. It should thus be carefully observed that the two tables exhibit an independent series of facts in the cultivation of, and trade in, sesamum. It has been customary in this work to resort to the figures of trade in the effort to arrive at some conception of production (when actual area of cultivation was not available), but with perhaps no other oil-seed would that system be more fallacious than the present. The extent to which the plant is grown for local consumption is probably greater than is the case with any of the other oil-seeds. The exports from India are, therefore, very likely only the surplus over and above a far larger quantity, much of which never moves beyond the district or village lands on which it was grown. Bearing these facts in view, it may be said the imports shown in table B, from the Presidencies into the port towns, represent the supply from which the foreign exports are drawn. But as the writer has had occasion to urge in connection with almost every article of Indian commerce, the distribution effected by road communications is often very serious, and before an exact statement of the receipts by any one port town can be arrived at, a balance sheet must be made out in which, in addition to rail, river, and road imports and exports, the effect of coastwise exchanges must be taken into consideration. The full force of these recommendations may be exhibited by the following analyses of the figures drawn from tables A and B:—

(1) Area returned in table A for	1889-90 . .	2,237,000 acres.
	1890-91 . .	2,032,000 ,,
(2) Forecast of outturn given in table A for 1890-91		171,131 tons.
(3) Imports into the port towns in table B for 1889-90		76,456 ,,
(4) Foreign exports for 1889-90		86,000 ,,

If now to the outturn in (2) be added the 35,000 tons mentioned above as produced in the North-West Provinces as a mixed crop, and if also a figure be accepted to represent the Bengal, Assam, Burma, and Native States' production, the actual outturn would appear to more likely exceed than fall short of 250,000 tons. Were 2 maunds to be accepted as the yield per acre, all over India, to obtain that supposed outturn, 3,500,000 acres of pure sesamum cultivation would have to be presumed,

CULTIVATION

or perhaps more than double that area were the crop, say, one half mixed. Indeed, in the memorandum quoted above, we are told that in the North-West Provinces 4,000,000 acres of mixed sesamum yielded only 35,000 tons or 0·24 maunds per acre. It is thus probable that the actual area, more or less under sesamum, in India, is over 10,000,000 acres, and that the local consumption is on an average two-thirds of the actual outturn. The imports into the port towns from the provinces are shown to have been only 76,456 tons, while the actual foreign exports were 89,000 tons. It need, therefore, be only added, in further support of the facts adduced, that the traffic in sesamum oil is by no means inconsiderable, and that to arrive at a possible conception of the area and outturn of this crop, it would be necessary to reduce from oil to the equivalent in seed the returns of the trade in sesamum oil. For example, the foreign exports of this oil were in 1889-90 returned as 91,120 gallons of pure oil, and 50,308 cwt. of dregs of gingelly oil. These figures to a large extent represent the oil obtained from seed pressed in the port towns, but there are rail, river, and road imports and exports of oil all over India, which should be also taken into account. At the same time it has been shown that the rail and river imports into the port towns are not equal to the foreign exports, so that a provision has to be made not only for the seed expressed in these towns, to meet the foreign trade in oil, but for the seed and oil used up by the city communities. It is thus evident that the area and outturn in Table A fall far short of the actual figures which will have to be determined in the future, a result which, it would appear likely, can only be obtained when both pure and mixed sesamum cultivation are taken into consideration. It is probably hopeless to expect that trade returns should tally with agricultural statistics when the latter deals with only half the crop. Hence, though unavoidable errors are, and must be, involved by all attempts at estimating the area and outturn of mixed crops, these errors have to be faced, if an approximation to accuracy be aimed at in the returns of a commercial product like that of sesamum.

As affording the most direct evidence on these issues, while furnishing at the same time certain particulars regarding the methods of cultivation, seasons of sowing and reaping, and yield, etc., etc., the following series of notes, from the Gazetteers, District Manuals, and Agricultural Department Reports, etc., may be here given, arranged provincially. It need only be remarked, in justification of this procedure, that the available material is too imperfect to allow of the more readable form of producing a compilation of the salient points, in place of a string of disjointed quotations. Regarding one feature of sesamum cultivation, in one province, much information exists, but nothing, or next to nothing, is said of it in another. This defect applies to every phase of the subject, and it can, therefore, be only hoped that the present admission of imperfect knowledge may lead to the publication of concise, though complete, reports, for each province, in place of the scattered notices that presently exist regarding the several districts. Indeed, it may be said that of some of the largest producing districts little or no information is available, while of the less important ones much has been written, according to the accidental interest taken in the subject by the authors of the District Manuals.

I.—MADRAS.

MADRAS. 1101

Before proceeding to give a few of the better passages that exist in works on South India regarding sesamum cultivation, the following facts from the forecast of the current crop (1890-91) may be furnished. It will be seen that the forecast not only furnishes definite information regarding

SESAMUM indicum.

Cultivation of Til

CULTIVATION in Madras.

Area.

the area under the crop, but exhibits the effect of rain in advancing or retarding the prospects.

"The total area returned under late-sown gingelly for 1890-91 is 189,000 acres, which is 12·5 per cent. below the normal and 20·6 per cent. below the revised figures (238,000 acres) reported for the previous year."

The following table exhibits the chief districts where sesamum is grown in Madras :—

	Districts.	Normal area.	Area in previous year.	Area in current year.	Percentage of increase (+) or decrease (—) over Normal area.	Percentage of increase (+) or decrease (—) over Area in previous year.	Remarks.
1	Godavari .	33,000	39,000	27,000	—18·2	—30·2	The decrease is extent is due to the fall of heavy rains during the sowing season, which encouraged wet cultivation, and also to the increase in the area under early crop. The crops are fair.
2	Bellary .	16,000	19,000	5,000	—68·7	—73·7	Decrease due to cotton being raised in lieu of gingelly, and also to want of rain.
3	North Arcot	32,000	50,000	29,000	—9·4	—42·0	The decrease in area is due partly to increase in the early crop area, and partly to falling off in the demand for the seed. The season also proved unfavourable, and the crops are reported to be suffering from want of rain in some taluks.
4	South Arcot	19,000	20,000	21,000	+10·5	+5·0	Increase due to cultivation of gingelly instead of paddy for want of timely rains for the latter crop. On the whole, the crops are in fair condition.
5	Salem .	14,000	11,000	12,000	—14·3	+9·1	Increase due to seasonable rains. Crops in fair condition.
6	Madura .	20,000	25,000	22,000	+10·0	—12·0	There was a decrease, as compared with the previous year, due to want of timely rains, and an increase as compared with the normal area, due to failure of other cultivation for want of rain. Crops on the whole good.
7	Tinnevelly .	9,000	11,000	8,000	—11·1	—27·3	Decrease due to want of rain. Crops fading in parts.
8	Malabar .	18,000	19,000	17,000	—5·6	—10·6	Decrease due to unfavourable season. Crops fair.
	Other Districts.	55,000	44,000	48,000	—12·8	+9·0	
	Total .	216,000	238,000	189,000	—12·5	—20·6	

The following quotations from the District Manuals of various localities may be accepted as affording a fairly good idea of the methods pursued in cultivating gingelly in this Presidency :— CULTIVATION in Madras.

GODAVERY.—"The mixed or sandy *regadæ* soil suits this plant best, but it also grows on very sandy soils, but is then inferior. The soil is prepared by ploughing about the month of April, and this operation is repeated two or three times. In May or June, when the ground is moist from recent rain, the seed, which should be of the very best description, is sown. Four seers are required for an acre of land. After sowing the ground should be again ploughed and bush-harrowed, and the seeds will spring up in eight or ten days. Fifteen days afterwards the field should be weeded; when two months' old, the plants will flower, and shortly after, the pods will appear, and in another month will be ripe. This crop is a very precarious one, being peculiarly liable to blight, and it invariably suffers if east winds prevail during its growth. The stalks, when cut, are stacked in a dry place, thatched with palmyra leaves, and allowed to remain eight days, after which the removal of the seeds may be commenced. This is effected by shaking the stacks, when about half the seeds will drop from the pods while the other half remains; the stalks are allowed to dry for a couple of days, when the remaining seeds are removed in a similar manner" (*Papers relating to the Survey and Settlement of the Western Delta taluqs of the Godavery District, 140*). Godavery. 1102

COIMBATORE.—"Universally grown both on wet, garden, and dry land; if the former it is usually before the regular crop as in Erode; or after it as in Dharapuram. In the latter it is grown often as a third crop in April and May; the usual succession is then *rágí* from June to September-October, paddy from October to February, gingelly from March to June. In some taluks it is also grown as a garden crop. On wet lands the moisture from the preceding crop is usually sufficient to start it, and the probable rains of April and May mature it; it is of course somewhat uncertain. It is less so when grown from June onward; on dry lands it is grown with the *Kár* rains in the Kangyam division of Dharapuram, sown broadcast with *cholam* and with *dholl* in lines; in Karur it is grown on the uplands with the rains of the south-west monsoon, and is also sown mixed with *Kkambu* and with cotton in July and is reaped in November-December; it is also grown as a separate crop. There are two sorts, *kár* and *tattu*, of which the former is the better and is grown in the hot weather; on garden lands, it is sown on land prepared as usual and watered; the young crop is not watered for about 15 or 20 days, and thereafter if there is no rain, only once in 10 or 15 days; water is stopped 15 days before pulling. The plants when pulled are stacked for a week, and the seed is then shaken from the pods and winnowed. The seed is about 10 measures and the yield from 150 to 350 or 400 measures. Its value is R15 to R35 or R40" (*Nicholson, Man. Dist., 224*). Coimbatore. 1103

NORTH ARCOT.—"There are two varieties of this very favourite dry crop the big and the small. It is an early crop, being sown, if the rains are sufficient, in April or May, and reaped four months later. It is sometimes irrigated, and is then sown as early as January or February. The plant is cut near the ground, and after being dried for a week or two is beaten with sticks. The oil is extracted in oil mills, a measure being yielded from four measures of seed. Gingelly oil is considered the very best, and is much used in native cookery, as well as for anointing the body; the cake is given to cattle. Much of the seed is exported from the district and sent to Europe, where a good deal of the so-called olive oil is extracted from it" (*Manual of the North Arcot District, 333 & 334*). North Arcot. 1104

CULTIVATION in Mysore & Coorg.

1105

II.—MYSORE AND COORG.

The cultivation of *til* is described as follows in the *Gazetteer of Mysore and Coorg*:—"The crop is known as *wollellu* or *phulagana-ellu*. It is raised exactly like the *kár-uddu*, cut down when ripe, and stacked for seven days. It is then exposed to the sun for three days, but at night is collected again into a heap; and, between every two days drying in the sun, it is kept a day in the heap. By this process the capsules burst of themselves, and the seed falls down on the ground. The cultivators sell the greater part of the seed to oil-makers. This oil is here in common use with the natives; both for the table and for unction. The seed is also made into flour, which is mixed with *jaggery*, and formed into a variety of sweet cakes. The straw is used for fuel and for manure. In Kolar it is more commonly called *achchellu*, and is cultivated as follows:—In Vaisákha plough twice, without manure, sow broadcast, and plough in the seed. In three months it ripens without further trouble, is cut down by the ground, and is afterwards managed exactly like the *uddu*. The seed is preserved in the same manner. The produce in a good crop is 20 seers, and in a middling one, 12 seers. The straw is used for fuel.

North of the Túmkúr District are cultivated two kinds of Sesamum, the *karu* or *wollellu*, and the *gur-ellu*. The last forms part of the watered crops; the *kar-ellu* is cultivated on dry field. The soil best fitted for it is *dare*, or stony land, which answers also for *sáme* and *hurali*. The ground on which *kar-ellu* has been cultivated will answer for the last-mentioned grain; but not so well as that which has been uncultivated. After it, even without dung, *sáme* thrives well. The same ground will every year produce a good crop of this *ellu*. If a crop of *ellu* is taken one year, and a crop of *sáme* the next, and so on successively, the crops of *ellu* will be poor, but those of *sáme* will be good. After the first rain in Vaisákha, which begins about the middle of April, plough three times. With the next rain sow broadcast, and plough in the seed. In between four and five months, it ripens without further trouble. On a *wokkala* land the seed is six seers, and the produce in a good crop is 5 *kolagas*, or eighty-fold. Which is to say that an acre sows ·55 peck, and produces 11·10 bushels. In the west the *kar-ellu* is sown on *rági* fields that consist of a red soil, and does not exhaust them. The field is ploughed as for *rági*, but it is not allowed manure. The seed is mixed with sand, sown broadcast, and harrowed with the rake drawn by oxen. It ripens in four months without further trouble. The seed is equal to half of the *rági* that would be sown on the same field, which is less than half a peck an acre. The produce is about 20 seers, or about 2½ bushels an acre. The straw is burned, and the ashes are used for manure" (*Gazetteer of Mysore & Coorg, Vol. I., 91*).

BOMBAY.

1106

III.—BOMBAY.

In the forecast for the crop of 1890-91, the Government of Bombay furnishes the following notes regarding the chief *til*-producing areas of Western India. These will be found to manifest the acreage devoted to the crop and to discuss the more important elements of uncertainty in sesamum cultivation.

Gujrat.

1107 "GUJARAT.—The final estimates (77,750 acres) are lower than those of the second report, but still are about 8 per cent. above the estimated average and the area of 1889-90. The later sowings in Surat have caused the reported decrease to disappear; but the revised figures show a decrease in Broach as compared with the early estimates. In Gujarat the sowing rain for both early and late kinds of sesamum was favourable. In Surat only was it excessive. The later rains were much less timely and the long

CULTIVATION in Bombay.

break resulted in withering and consequent diminution of yield. In Ahmedabad and Panch Mahals insects have done harm. The early crop has been harvested. The anna-estimates vary from as low as 6 annas in Panch Mahals to 14 annas in Surat. On the whole 8 annas may be taken as the anna-estimate of the province, *i.e.*, about half the average crop (16 annas).

Gujarat States. 1108

"GUJARAT STATES.—The area (258,000 acres) of sesamum is 3½ times that of the British districts of Gujarat, and, of the States, Kathiawar is the largest producer. The crop in Kathiawar was greatly reduced by deficiency of late rains, and in Halar it was further damaged by insects and cloudy weather. The yield is reported at 5 annas in Cutch to 9 annas in Kathiawar. [The reader might, in this connection, consult the concluding paragraph of the article **Sorghum** in which it is shown that Kathiawar prefers to cultivate **Sesamum** and to largely import its supplies of *juár Ed., Dict. Econ. Prod.*]

Deccan. 1109

"DECCAN.—Of the total (105,600 acres) 64 and 23 per cent., respectively, are credited to Khandesh and Nasik. This total signifies a diminution of nearly 24 per cent. as compared with last year, but twice as large if the comparison is made with the estimated average. No explanation other than that of unseasonableness of sowing rain can be offered. In Poona, however, and, to some extent elsewhere, the exclusion of niger seed hitherto erroneously included in the Agricultural Returns under sesamum accounts for some of the decrease. The crop was largely benefited by seasonable rain in August, but a long break lasting till the second week in September nullified the good prospects. The rain in September, moreover, was not heavy enough. Later on rain in November damaged the early crop, though it improved the late one, which is still in the field. These vicissitudes of season were more or less general. In Poona cloudy weather was experienced also. The outturn may be stated at from 7 annas in Poona to 10 or 11 annas in Khandesh. *Deccan States.*—The area is very small.

Karnatak. 1110

"KARNATAK.—The area (24,000 acres) is less than half that of last year and much below the average area. It is small even in a good year. The outturn is reported at from 6 annas in Dharwar to 8 annas in Belgaum and Bijapur. "*Karnatak States.*—The area is small. The estimated outturn may be slightly better than in the British collectorates.

Konkan. 1111

"KONKAN.—The area of 1890-91 is about equal to that of the Karnatak, but it much more nearly comes up to the average. Ratnagiri claims more than half the area. Kanara does not grow sesamum. The decrease in the Konkan is clearly due to excessive rain. The prospects of the crop were further lessened by a continuation of heavy rain during the growth of the crop, especially in Kolaba. On the whole, however, the crop is better than in the Deccan or Karnatak."

It is perhaps unnecessary to give, in great detail, the crop forecast of district cultivation in Bombay, similar to what has been done in the case of Madras, since the above notes on the divisions of the Presidency manifest the chief facts regarding the distribution of the crop. In the Statistical Atlas **Mr. Ozanne** gives the following brief sketch of the sesamum cultivation of Western India which may be usefully republished here:—"Is grown all over the Presidency and has the largest area under it in Khándesh. It is of three varieties—black, white, and grey, the last found only in Gujarat. These varieties differ only in colour. *Til* is grown unirrigated and unmanured in any soil, but has a preference for sandy loams. Sown in June and cut in November, sesame is grown generally with *bájri* and pulses either mixed or in separate furrows, and often by itself on land that has long lain fallow."

The following series of notes have been compiled from the Gazetteers. They convey some idea of the extent to which sesamum is grown in Western

CULTIVATION in Bombay. India, and exhibit a few of the more important ideas which prevail regarding the crop.

Kolhapur. 1112 KOLHAPUR.—*Til* is grown only in small quantities. It is of two kinds, black or brown and white. It is sown in June and harvested in September. The average acre outturn is 320lb.

Ahmednagar. 1113 AHMADNAGAR.—Two kinds are cultivated, the black and white, *gora* or *havra* and *kala*. *Til* is sown in June, usually with *bájrí*, either mixed or in separate furrows, and sometimes by itself on land that has long lain fallow; it is cut in November.

Kolaba. 1114 KOLABA.—Sesamum is raised mostly in Mángaon and Mahád, and grows best on fairly flat land. The soil does not require to have brushwood burnt on it (*ráb*) and is only ploughed twice after rain has fallen. No manure is used and the seed is sown broadcast, from the middle to the end of June. The crop does not require to be weeded and ripens about the beginning of November.

Nasik. 1115 NASIK.—Sesame is sown in June and July and reaped in October. It is grown almost entirely, north of the Satmálás.

Thana. 1116 THÁNA.—Two kinds of *til* are cultivated in Thána, black and white. Black *til* is generally grown after *harik*. It can also be grown after *náchni* or *vari*, but does not then yield so good an outturn. It is sown in June and ripens about November flourishing best on tolerably flat land. The white-seeded variety is grown after rice in the same way as the black *til*.

Khandesh. 1117 KHANDESH.—Sesamum is sown in June and harvested in September, and has an average acre yield of from 300 to 380 pounds. It has endless kinds known by their colour, the shades passing from dull black through brown to the purest white. In Khándesh all these forms sometimes grow together and yield seed known in trade as mixed *til*.

Poona. 1118 POONA.—Sesamum, of the black and white varieties, is grown throughout the district, but in considerable quantities only in Khed, Junnar, Mával and Haveli. It is sown in June usually with *bájrí*, either mixed in the same line or in separate lines and is cut in November. It springs up unsown in fallow lands.

Kathiawar. 1119 KATHIAWAR.—Gingelly seed is widely cultivated in this district. It grows in black soil, which requires to be thrice ploughed and twice hoed. There are three kinds, *ashádi tal* or white, *kála katwa* or black, and *purbia* or red. The white and black are usually sown in July and reaped in October, while *purbia* is sown in the *Purva Nakshatra*, in September, and reaped in December. Of the three, the white is the best tasted and the red the largest yielder. The oil obtained from the *ashádi* is sweeter and purer than that from *purbia*.

SIND. 1120

IV.—SIND.

The following note from the forecast of the crop of 1890-91 will be found instructive:—"The area (101,000 acres) is 21 per cent. below last year when it was up to average. The Indus inundation was low and rainfall unusually scanty. Furthermore the poor yields of the past two years have acted as a deterrent. The yield is fair, from 10 annas in Thar and Parkar to 13 annas in Karachi and the Upper Sind Frontier. It appears from the actual estimates reported (for there are no formulæ for Sind) that the average acre yield of sesamum is lower in Sind than in the Bombay Presidency."

According to the Gazetteer, sesamum forms one of the principal crops of Sind, and is cultivated more or less in all the districts. In Mahammad Khan good soil is said to be required for the crop and it is irrigated every eighteen days. It takes four-and-a-half months to mature; the average yield per *bígha* is 210lb. In Naushahro it is sown at the end of June in soft rich soil. It gets five to eight waterings, and takes about five months to mature. The details for the Larkhana District are very similar to the foregoing. The seed is sown in June and July, and the crop reaped in November and December. The average yield is said to be about ¾ maund per *bígha*; in Larkhana it is only twice watered.

CULTIVATION in N. W. Provinces & Oudh. 1121

V.—NORTH-WEST PROVINCES AND OUDH.

The Government of these provinces has published the following forecast regarding the crop of 1890-91. It may be said to manifest the importance of rain in all considerations regarding this crop:—"As stated in the second cotton forecast which is published simultaneously with this report, the rains from the middle of July to the end of September were exceptionally ill-distributed and ill-timed. The crop was injured in the beginning by heavy rains which obstructed proper and timely weeding and on the low lands left the crop to rot. The protracted drought in August and September prevented the plants from properly flowering and otherwise injured them. The area occupied by the crop this year is estimated at about 11 per cent. less than the normal area. The condition of the crop is reported to be 55, assuming that 100 represents a full outturn."

The details of the system of cultivation, seasons of sowing and reaping, etc., will, however, be perhaps best conveyed by giving here the chief passages from **Messrs. Duthie & Fuller's** account in the *Field and Garden Crops.* Indeed so completely do these concise passages cover the field that in the case of these provinces it is unnecessary to give any of the numerous scattered accounts which occur in the Gazetters and other such works.

"*Varieties.*—There are two forms, the black-seeded and the white-seeded; the former being generally known as *til,* and the latter as *tili.* *Til* ripens rather later than *tili,* and is more commonly grown mixed with high crops such as *juár,* while *tili* does best when mixed with cotton. *Tili* oil is preferred of the two for human consumption. Varieties. 1122

"*Area.*—Notwithstanding its economic importance the acreage under *til* is small, since it is very rarely grown as a sole crop in most districts of these Provinces. Fields of *til* are not uncommonly met with in the districts lying immediately under the Himálaya—Dehra returning 3,536 acres, Pilibhit 616 acres, Basti 1,301 acres, and Gorakhpur 857 acres. But the tract in which its cultivation as a sole crop is commonest is Bundelkhand, and the area under *til* in the five districts which are geographically included in this tract are shown below:— Area. 1123

Districts.	Acres.
Jalaun	6,000
Jhansi	21,400
Lalitpur	36,000
Hamirpur	49,000
Banda	35,700

"This amounts to no less than 8 per cent. of the total area under *kharif* crops in these five districts. The only other district in which *til* is largely grown alone is Allahabad (3,800 acres), and this is due to the fact that a large portion of the Allahabad district lies south of the Jumna, and is characterized by the same conditions as Bundelkhand. In no other district does its cultivation as a sole crop reach 300 acres. Judged by these returns *til* cultivation appears to be of insignificant importance over the greater part of the Provinces, but this is very far from being the case. Although not cultivated by itself, it is almost universally grown to a greater or less extent in fields of *juár, bajrá* and cotton, and it may be therefore said to

CULTIVATION in N.-W. Provinces & Oudh.

have a place on more than half the total area under ***kharif*** crops. It is, however, grown less commonly in the eastern than in the western districts, both because it does not thrive in a rice country, and because the ***mahua*** tree (**Bassia latifolia**) abounds in the eastern districts, and ***mahua*** oil is commonly consumed there.

Soil. 1124 "*Season and Soil.*—As has already been implied, ***til*** is a ***kharif*** crop and is sown at the commencement of the monsoon, and harvested in October and November. It prefers a light soil, and the wide extent of its cultivation in Bundelkhand is in great part limited to the light yellowish soil, locally known as ***ránkar***, which abounds in the raviny tracts near rivers. Indeed, a crop of ***til*** can be gathered from land which will yield no other crop but one of the inferior millets (***kodon*** *or* ***kutki***).

Method. 1125 "*Method of Cultivation.*—The method of its cultivation is the roughest possible. The seed is sown broadcast after two or three hurried ploughings and ploughed in. When grown with millet or cotton it gains the benefit of the care which these crops receive. It is in this case either sown broadcast, the seed being mixed with that of the principal crop before sowing, or it is disposed in parallel lines running across the field or along its margins. When mixed with other crops the amount of seed sown to the acre varies of course with the inclination of each individual cultivator. When grown alone from 8 to 12 seers of seed are used.

Harvesting 1126 "*Harvesting.*—When ripe the ***til*** plants are cut with a sickle to within 2 or 3 inches of the ground, and the stalks collected in shocks, heads uppermost, and allowed to dry. The seed capsules split open and the seed is extracted by beating the plant against the ground. The dry stalks called ***tilsota*** are used for fuel.

Injuries. 1127 "*Injuries.*—The ***til*** plant is very liable to damage from ill-timed rain, and this may explain the rarity of its cultivation as a sole crop in the thickly populated districts of the Ganges-Jumna Doáb, where risk must be reduced to the lowest minimum possible. Heavy rain, when the flowers are in process of fertilization, often ruins the crop, and hence, like ***bájra***, it is very liable to suffer if rain falls in October. Indeed, it is not uncommon for the crop to be an almost total failure.

Outturn. 1128 "*Outturn.*—Under the circumstances of its cultivation it is obviously impossible to frame any reliable estimate of its outturn per acre, which varies very greatly with the amount of seed sown. From 25 seers to 1½ maund are commonly gathered, when it is sown with ***juar*** or cotton. When grown alone from 4 to 6 maunds is the average return to the acre."

PANJAB. 1129

VI.—PANJAB.

Sesamum is said to be grown to a limited extent in almost every district of the Panjáb. The exports from the province find an outlet chiefly in Karachi, though smaller quantities are carried to Bombay. Some idea of the relative importance of the crop in the various districts and of the chief features of its cultivation may be gathered from the following forecast for the season 1890-91:—"This is the first separate forecast furnished on the sesamum (***til***) crop of this province. Special reports are received from sixteen districts, as in the remainder the crop is a very unimportant one. The total estimated area in these sixteen districts this year is 177,400 acres as compared with 181,400 acres last year; for the whole province the total area under this crop this year is estimated at 217,392 acres as compared with 215,117 acres last year, the increase being 2,275 acres or 1 per cent. In the districts of the Delhi Division in the South-East Panjáb the heavy rains of July and August, followed by an early cessation of the rains, did harm to this crop, and the result was a poor harvest. In

CULTIVATION in the Panjab.

the Lahore Division the crop was generally about an average one, but in the districts of the Rawalpindi Division lying near the hills, the crop was generally above the average, and in Gurdaspur and Sialkot it was a very good one. In these tracts the rains commenced early and the season throughout was favourable for this crop. The total estimated outturn is 24,610,400 seers or 439,471 cwt., the average outturn per acre being estimated at 278lb.

"On the whole, the crop of sesamum (*til*) for the year in the Panjáb must be classed as a good one."

Area. 1130

Statement of Area under Sesamum in hundreds of acres and estimated yield in hundreds of seers in each of the reporting Districts of the Pánjab for the kharif season of 1890.

No.	District.	Land irrigated by — Canals.		Wells.		Flooded and alluvial land, Sailaba, Bet khadir, etc.		Dry land dependent on rain.		Total.		Average produce per acre in seers.
		Area.	Total yield.	Area.	Total yield.	Area.	Total yield.	Area.	Total yield.	Area in acres.	Total yield in seers.	
1	2	3	4	5	6	7	8	9	10	11	12	13
1	Hissar	2	206	...	...	...	...	50	2,596	52	2,802	54
2	Gurgaon	3	1,116	...	...	3	312	44	3,700	50	5,128	103
3	Karnal	...	...	4	112	...	...	51	6,060	55	6,172	112
4	Kangra	1	100	...	...	...	...	89	7,071	90	7,171	80
5	Ferozepore	20	2,093	2	247	1	100	21	1,658	44	4,098	93
6	Multan	326	54,837	32	5,684	18	3,820	3	210	379	64,551	170
7	Montogomery	66	17,840	29	7,217	24	3,791	67	10,960	186	39,808	214
8	Lahore	11	1,063	2	222	3	204	66	5,151	82	6,640	81
9	Amritsar	3	525	5	788	3	450	51	5,396	62	7,159	115
10	Gurdaspur	2	396	4	736	5	515	258	37,738	269	39,385	146
11	Sialkot	...	...	11	1,640	8	1,475	51	6,856	70	9,971	142
12	Gujrat	...	...	6	617	4	567	86	5,739	96	6,923	72
13	Gujranwalla	2	272	5	536	2	78	81	7,384	90	8,270	92
14	Peshawar	15	2,340	1	200	1	150	74	5,779	91	8,469	93
15	Dera Ismail Khan	...	...	2	360	125	24,280	3	575	130	25,215	194
16	Muzaffargarh	11	2,231	...	...	17	2,111	...	...	28	4,342	155
	Total	462	83,019	103	18,359	214	37,853	995	106,873	177,400	2,46,10,400	139

Cultivation of Til

CULTIVATION in the Panjab.

Mr. Baden Powell says that in the Panjáb sesamum is generally cultivated, often being sown round the edges of fields, forming as it were a green hedge to the main crop. The brown or black forms are grown, but it is blanched by warming in hot water, the outer skin of the seed being rubbed off when the seed appears white. The yield of oil is about ⅜ths the weight of the seed employed.

The following notes from the District Gazetteers and Settlement Reports may be accepted as sufficiently manifesting the peculiarities of cultivation followed in the Panjáb :—

Jhang. 1131

"JHANG.—Sesamum is grown in small quantities on *sailab* lands and on rain lands in the upland. The writer has also seen it once or twice on the outskirts of a well, and such crops are sometimes irrigated. Very little is grown on the Chiniot *sailab* lands. *Til* loves a light soil, but requires much moisture. It will grow even on *rappar* lands—sand covered with only a thin layer of soil. The writer remembers being struck with the appearance of a very fine crop near the Trimmu ghât, and then seeing the land again later on, he found that it was nothing but a thin layer of mud on a substratum of sand. *Til* is much cultivated mixed with other crops—*jowár, másh,* and *múng*. The land is prepared by one or two ploughings. The seed is sown broadcast, mixed with sand, in August, and the early part of September. The amount used is about 7½℔. The flowers are liable to be nipped and to fall if the wind blows from the north. The root is also attacked by *múlá*" (*Settlement Report, Jhang District, 85 to 98*).

Montgomery. 1132

MONTGOMERY.—" *Til* is often sown with *moth* and *múng*, or *moth* alone; sometimes with *jowár*. It is essentially a rain crop, but is sometimes grown on canal-irrigated lands. After rain, plough, sow broadcast, mixing seed with earth if not sown with some other crop, and plough again. Sometimes the seed is simply thrown on the fallow ground and ploughed in. Two seers of seed go to the acre. *Til* plants should not be close together, according to the verse which may be translated :—" When barley grows scattered, *til* close together, the buffalo brings forth a male calf, and sons' wives give birth to daughters—all four are utterly bad.' Only one kind of *til*, the black, is known. The plant is affected by *tela* and lightning. When the crop is cut, the stalks are placed in a circle with their tops pointing inwards, and are left there for a fortnight with a weight upon them. This heatens and softens the pods. Then the stalks are placed on the ground with their tops pointing upwards, leaning against each other, or on a straw-rope. The action of the sun causes the pods to open, when the grain is shaken out on a cloth. Fifteen seers of *til* seed produce 6 seers of sweet oil. *Til* stalks, when dry, are used for fuel. They give forth a fierce flame" (*Gazetteer, Montgomery District, 111-112*).

KARNAL.—" No varieties of *til* are recognised. It must be grown in good stiff soil; and the soil must be new to give a good crop, which is probably the reason why it is chiefly cultivated in the Nardak whers virgin soil abounds. It is generally sown with *jawár* or *urad;* and the mode of cultivation is the same as that of the latter. When the plants are cut, they are put on end to dry. As they dry, the pods open, and the seed is then shaken out. The stems (*dánsra*) are of no use. The seed is taken to the oilman, who returns two-fifths of the weight in oil, keeping the oilcake (*khal*) which he sells. The oil is good for burning, and is the best of all oils for purposes of the kitchen. *Til* is very subject to attacks by carterpillars (*ál*). And if it once dries up, it never recovers. It is, however, never irrigated (*Gazetteer, 179*).

VII.—CENTRAL PROVINCES.

CULTIVATION in the CENTRAL PROVINCES. 1133

It will be found from the remarks under the paragraph TRADE that these provinces constitute the most important single area in the supply of the foreign demands through Bombay. Very little of a definite nature can, however, be found of the methods of cultivation pursued in the various districts. The following forecast of the crop of 1890-91, however, discusses the question of acreage, the influences that affect the crop, the yield, etc., while it exhibits at the same time the districts of greatest importance:—

"Last year's crop was an exceptionally good one; exports were large and high prices were realized by cultivators, so that they were stimulated to sow larger areas with *til* seed and such has been the case, except in the northern districts of Jabalpur, Damoh, and Mundla and the rice district of Bilaspur, where heavy rainfall at sowing time prevented cultivators from sowing as much land with *til* as they would otherwise have done. In this latter district, however, the crop is not of much commercial importance. In the three former districts the estimated decrease implies a diminished area of about 12,000 to 15,000 acres. This is, however, more than counterbalanced by the increased area sown with *til* in the districts of the Narbadda Valley and in the Nagpur country which export by far the largest quantities of *til*. The estimates are of course rough, but the net increase in the area under *til* can be but very little if at all below 50,000 acres.

"The weather, as was reported in the recent forecast of cotton for these provinces, has been irregular and the results to *til* have been very much the same as were reported of the cotton crop. That is to say, in the districts of the Jabalpur division it has been injured by rain; in Narsinghpur the rainfall has been timely, while in Hoshangabad and Nimar long breaks have done some damage. In the Nagpur country, the *til* has been somewhat injured by excessive rain in Nagpur and Chanda, while in Wardha the weather has just suited the crop. In spite, however, of heavy rain in some tracts and long breaks in others the crop has not suffered as much as might have been expected. In the districts of the Jabalpur division it is not far short of a full average crop ranging from 12 to 14 annas. In Narsinghpur a good average crop is expected, in Hoshangabad and Nimar a 10 annas crop. In Wardha a bumper crop is anticipated, and in Nagpur it is fair. It has to be remembered that in the Nagpur country about 70 per cent. of the *til* is the *rabí til* and has only recently been sown. Its prospects, as far as can be seen at present, are good, the weather having been favourable for sowing."

Forecast of the Sesamum (Til) crop in the Central Provinces for the season 1890-91.

DISTRICT.	Area under *Til* in preceding year (1889-90.)	Percentage by which area now under *Til* exceeds (+) or falls short of (—) that of previous year.	Estimated outturn in annas taking 16 annas to represent an average crop.	EXPLANATION.
	Acres.			
Saugor	30,000	+4	14	The heavy rains in the beginning of the season damaged the crops to a certain extent, but the long break in August and the subsequent timely rains improved matters to a great extent.

SESAMUM indicum.

Cultivation of Til

CULTIVATION in the Central Provinces.

Districts.	Area under *Til* in preceding year (1889-90.)	Percentage by which area now under *Til* exceeds (+) or falls short of (—) that of previous year.	Estimated outturn in annas taking 16 annas to represent an average crop.	Explanation.
	Acres.			
Damoh .	30,000	—6	12	The condition of the crop is not very satisfactory in both Tahsils. Owing to continuous moisture in the Damoh Tahsil the crops have suffered to a great extent.
Jabalpur .	40,000	—25	12	Sowings decreased owing to heavy rain. The plants look healthy notwithstanding the latter heavy rain.
Mandla .	23,000	— 10	14	Sowings decreased owing to heavy rain. The plants look healthy notwithstanding the latter heavy rain.
Seoni . .	22,000	+·05	12	The long break during the latter part of August and early part of September followed by unusually heavy rain now, has had an injurious effect.
Nursinghpur	16,000	+ 10	16	This year's rainfall was very satisfactory for cultivators, who availed themselves in clearing their fields in due time.
Hoshangabad.	26,000	+ 12½	10	The estimate is low and is due to want of rain for about a month subsequent to sowings.
Nimar .	34,000	+ 19	10	In Khandwa Tahsil where *til* is most grown the drought seriously affected the crop in August and subsequent rain has not caused much improvement.
Betul . .	17,000	+·43	12	The seeds germinated well, but the want of rain during the month of August and in the early part of September greatly injured the crops.
Chhindwara	10,500	+4	14	As a whole the average may safely be estimated at 12 annas as the crops in the Chhindwara Tahsil have not suffered and look promising.
Wardha .	54,000	+ 10·08	20	Bumper crops are expected this year on account of timely rainfall.
Nagpur .	18,000	+ 13·5	12	The excessive rainfall in the month of September has damaged the crop somewhat.
Chanda .	36,000	+7	10	Owing to too much rain the crop is said to suffer partially.
Raipur .	36,000	...	...	Report not received.
Bilaspur .	16,500	—25	10	The excessive rainfall interfered with weeding operations, and there being no break in the rains the young plants suffered from want of sunshine.
Sambalpur .	...	...	...	Report not received.

It is believed that in amplification of the facts contained in the above forecast, the following district account may be given as conveying the chief facts regarding the sesamum cultivation of these provinces :—

"The plant that yields the *gingelly* oil-seed of commerce has a sensible position among the products of the district. It is both a spring and autumn crop; the former being called *múghei* and the latter *howrí tillí*. The latter greatly preponderates in extent. It is essentially the crop of

CULTIVATION in the Central Provinces.

newly cleared land and of poor cultivators, as it pays inferior cultivation perhaps better than any other crop. The ground only requires to be partially cleared and scarcely turned with the plough; a mere handful of seed sows an acre; it is only once partially weeded; wild animals won't eat it till quite ripe, and it yields about 200lb of seed per acre on the poorest *kabrah* land, worth to the cultivator about R8. The total expense of cultivation may be R4 per acre, a considerable portion of which should be charged to the succeeding year's crops of *bajri* or *jowar*, as it consists chiefly of clearing the land of jungle. The oil is expressed for local consumption in the rudest of mills, holding at each operation about 18lb of seed, which results in 6lb of oil and 12 of oil-cake (*khull*). The oil-presser charges 6½ annas for this operation and thus makes about 7 annas a day for himself and the worn-out bullock that turns the mill. The mill has no exit for the oil at the bottom, both oil and cake coming out together at the top; water is freely used to facilitate the process, and thus the oil is of the worst possible quality (*Settlement Report, Nimar District, 195*)."

VIII.—BERAR.

BERAR. 1134

In the forecasts of the crop of 1890-91 the following brief note appears regarding this province:—"The total area under the sesamum crop is 94,516 acres. The probable outturn is estimated at 119,980 maunds or 1 maund and 10 seers per acre. Owing to a deficient rainfall, the crop has suffered to a certain extent, and in the Akola and Julgaon taluks it was partially destroyed by locusts and insects."

In the review of the trade of sesamum-seed below it will be found that Berar, but more particularly the Nizam's Dominions, afford a large portion of the seed exported from Bombay to foreign countries. The systems of cultivation, seasons of sowing and reaping, etc., are similar to those given in connection with Bombay Presidency and the Central Provinces.

IX.—BENGAL.

BENGAL. Dacca. 1135

DACCA.—Very little information is on record regarding the cultivation of sesamum in Bengal. In the Dacca District the plant is most extensively cultivated along the Lakhmia river and is frequently raised with a crop of rice. The following data, with regard to this method of mixed cultivation, are extracted from Mr. A. C. Sen's Report on the Agriculture of the Dacca District:—

Tillage. 1136

"*Tillage.*—The straw of the previous year's crop is collected in heaps and burnt, and the field is then ploughed. If the ground be sufficiently dry, the plough is followed by the ladder, otherwise a ploughing only is given. This is generally done in *Magh* (15th January to 15th February). After an interval of two to ten days the field is cross-ploughed, and the ladder is used twice. After three or four ploughings more have been given the land becomes ready for sowing.

Sowing.

"*Sowing.*—One-and-a-half seers of *til* and ten seers of *áman* paddy are mixed together and sown broadcast over a *bigha* of land. The sowing time extends from the middle of *Falgoon* (15th February to 15th March) to the end of *Chaitra* (15th March to 15th April). When the plants grow 4 to 5 inches high the field is hoed by a small *kodáli*. At the time of the hoeing the plants are thinned, if they come up too thickly. Eight or ten days after the first weeding is given. The second weeding comes in about a fortnight."

Reaping and Threshing. 1137

"*Reaping and Threshing.*—The *til* is cut in *Jeith* (June). After reaping the *til* is kept in heaps for a few days, and then threshed out by beating with a stick.

Yield. 1138

"*Yield.*—Two to three maunds per *bigha.*"

SESAMUM indicum.

Cultivation of Til

CULTIVATION in Bengal.

Dacca.

In some parts of the Dacca District, **Mr. Sen** informs us, *aus* paddy is also grown along with *áman* and *til* in the same field. "The tillage operation is similar to that described above. At the end of *Chaitra*, after a shower of rain, the field previously prepared is ploughed once more and broadcasted with 1½ seers of *til*, 10 seers of *aus*, and 6 seers of *áman* paddy. The seeds are well mixed together in a basket before sowing. When the plants appear, the field is first rolled with the ladder, then harrowed with the rake, and, lastly, two weedings are given at an interval of two to three weeks. The *til* is reaped in *Jeit*." "Very good crops," **Mr. Sen** adds, "are obtained in this way and the system of mixed cultivation is gaining favour with the cultivators."

As supplementing and amplifying the above short account, the following passages from **Sir W. W. Hunter's** *Statistical Account of Bengal* may be given :—

Midnapur. 1139

MIDNAPUR.—"Four varieties of *til* seed are grown, namely, *krishna til* and *sánkí til*, sown in jungle land in June and July and gathered in November and December; *khaslá til*, sown in sugarcane fields in March and April, and cut in June; and *bhadu til*, sown on jungle land in May and June, and cut in August and September" (vol. III., 80).

Hugli. 1140

HUGLI.—"It is stated that there are two varieties—*krishna til*, sown in June or July, and cut in September or October; and *kát til*, sown in January or February, and cut in July. Like *khesari* (**Lathyrus sativa**) *til* is often sown broadcast as a second product on rice fields, the first crop of which has been destroyed by inundation" (vol. III., 80).

Faridpur. 1141

FARIDPUR.—"*Til* is of two kinds: *til* sown on lowland in August or September and cut in November and December, and *kala til* sown on highland in February or March and cut in June or July. This plant is cultivated all over the district for the seed, as well as for the oil obtained from it, both of which are in much request" (vol. V., 308).

Rungpur. 1142

RUNGPUR.—"*Til* is of two varieties—*krishna til* and *raktá* or *áus til*. The first named variety is sown in August and September, and cut in November and December. It thrives best on high dry land, and is sown either singly or along with *thikri kalái*. The land requires to be ploughed four times and harrowed twice before sowing. In good years the produce varies from 1½ to 2 maunds per *bígha*, or from 3⅓ to 4½ cwts. per acre, the price being the same as for mustard. The second variety, *raktá* or *áus til*, is only cultivated on a very small scale in Rungpur. Sown in January and February and cut in May and June. The value of the crop is nearly equal to that of mustard" (vol. VII., 242.)

Rajshahye. 1143

RAJSHAHYE.—"*Til* sown on rice lands in March and reaped in July. Another variety of *til* known as *krishna til* is sown in April and cut in December, but is cultivated only to a very small extent in this district" (vol. VIII., 60).

Bogra. 1144

BOGRA.—A valuable oil-seed is the produce of the *til* plant, two or three varieties of which are found in Bográ. The best and most common kind is the *krishna* or black *til*, a crop that grows in the latter part of the rainy season, and matures in the beginning of the cold weather" (vol. VIII., 210).

Lohardugga. 1145

LOHARDUGGA.—"*Til, tilmí*, sown on high land in September or October and reaped in March, forms one of the staples of Palámau, and is largely grown throughout the southern portion of this sub-division. It is a hardy crop, grows on poor light soils, and does not require elaborate cultivation. The average yield of *til* is 4 maunds, or 3 cwt., per acre, and sells at R1-12 a maund, or 4*s.* 9*d.* a cwt." (vol. XVI., 341).

CULTIVATION in ASSAM. 1146

X.—ASSAM.

No information is available regarding the methods of cultivation pursued in this province. It may, however, be remarked that an export takes place yearly to Bengal, and that the crop must consequently be cultivated to some considerable extent.

BURMA. 1147

XI.—BURMA.

The cultivation of sesamum in Burma appears to be on a very small scale, if indeed it exists at all. It will be seen from the review of the trade that Burma affords the largest internal market, a very great amount, and one that is yearly increasing, being imported annually. The chief source of this supply is Madras. Notwithstanding the fact that practically no *til* cultivation exists in Burma, the consumption of the oil-seed must, to judge from the trade returns, be very general.

MANUFACTURE. 1148

MANUFACTURE AND USES OF THE OIL.

Most of the passages quoted above regarding the systems of cultivation which are pursued in the provinces of India will be found to allude to the methods of extraction and yield of the oil. These should, therefore, be read in connection with the more special passages given in this chapter on the subject of the manufacture and uses of sesamum oil.

The fact that there are widely different forms of sesamum cultivated in India; that these to the agriculturist have independent claims on his consideration, since they are grown at different seasons of the year, and that they yield varying proportions of oil with slightly diversified properties, does not seem to have attracted in Europe the attention which the subject deserves. It is, in fact, only necessary to add to these admissions of neglect, the further statement that with perhaps no other oil-seed do the practices of separation of oil vary to a greater extent than is the case with *til*, when it will be realised how obscure the traffic in this substance must be and to what extent its legitimate progression cannot help being retarded, when purchases are made blindly in a seed or oil one consignment of which may be highly valuable and another practically useless. Recognising that thus far there would seem no occasion to hesitate in stating clearly the urgency of more precise action, the writer regrets that the material before him is too imperfect to allow of a satisfactory account being drawn up on the botanical character of the plants grown, the methods of cultivation pursued, the merits of the seed placed on the market, and the systems of expression and quality of the oil made in India. The review which will be found below, of the internal trade of India in sesamum, indicates the areas from which Bombay port town draws its supplies, the supplies which constitute the major portion of the *til* exported to Europe, so that it would seem a more definite knowledge exists with regard to the gingelly seed of present European commerce than might be inferred from the disparaging remarks offered above. Whether that seed is uniformly of one quality and still more so, whether India might not furnish a quality of the seed better suited to European wants, are points regarding which no information can at present be furnished. As with the seed so with the oil expressed in India, the exports are shipped mainly from Bombay, but it is probably never a pure oil. It is obtained from two or more kinds of seeds which are mixed in the oil-mill and expressed together. Fortunately for the prospects of this trade, little or none of the Indian-made oil finds its way to Europe, the major portion being consigned to Arabia and the dregs to Ceylon. That there is a field in India for the preparation and export of a pure oil, would seem

MANUFACTURE and Uses.

a matter on which there can be but one opinion. So long, however, as a mixed oil is exported, it is probably undesirable that any effort should be made to divert the exports to Europe. Some thirty years ago, however, a very considerable export in this oil took place from India to the United Kingdom, a trade which appears to have been extinguished by the French sesamum oil-mills.

The series of notes below, arranged provincially, may serve to demonstrate the diversity that prevails as pursued in India in the systems of expression of the seed.

MADRAS. 1149

I.—MADRAS.

The following notes on the oil prepared in the Madras Presidency were originally published in the Madras Exhibition Jury Reports by **Lieutenant (now Colonel) Hawkes**, and no additional information appears to have been brought to light by subsequent writers on the subject:—

"The sesamum and its varieties are grown throughout the country. So universal is the use of this oil, that its name in almost all the vernacular languages signifies 'the oil.' The mode of extraction, sometimes adopted, is that of throwing the fresh seeds, without any cleansing process, into the common mill, and expressing in the usual way. The oil thus becomes mixed with a large portion of the colouring matter of the epidermis of the seed, and is neither so pleasant to the eye, nor so agreeable to the taste, as that obtained by first repeatedly washing the seeds in cold water, or by boiling them for a short time until the whole of the reddish-brown colouring matter is removed, and the seeds have become perfectly white; they are then dried in the sun, and the oil extracted as usual. In expressing this oil, the Natives of the Northern Division always add the bark of the *Tanghedi* (**Cassia auriculata**), or the *babúl* gum (**Acacia arabica**) to the seed to be pressed; this is probably done with a view of enhancing the value of the cake, which is used as an article of food for man and beast.

"The value of this oil in England was £47-10 per ton in January 1855, and £49 to £53-10 in January 1856. In different parts of the Presidency, the price of this oil varies from R1-5 to R6 per maund of 25lb. In South Arcot it is procurable at R27-12-5 per candy.

"The prices per maund at the undermentioned stations, for the quarter ending 31st October 1854, were as follows:—

	R	a.	p.		R	a.	p.
Arcot	3	8	0	Madura	5	8	3
Bangalore	3	7	3	Mangalore	4	1	8
Bellary	3	2	0	Nagpur	1	12	0
Berhampore	2	8	0	Palamcottah	4	12	0
Cannanore	6	0	0	Paulghaut	3	7	0
Cuddapah	2	13	0	Samulcottah	2	10	8
Jaulnah	2	6	0	Secunderabad	2	3	11
Jabbalpur	1	5	0	Trichinopoly	4	1	8
Madras	3	14	0	Vellore	3	14	0
Masulipatam	3	0	0	Vizagapatam	3	2	0

"In England this oil is chiefly used for the manufacture of soap, and for burning in table lamps, for which it is better suited than cocoanut oil, owing to the lower temperature at which it congeals, although the light it gives is not so bright. In India it is chiefly employed in cooking, for anointing the body, for making soap, for burning in lamps, &c., and by the dyer to brighten and fix his colours.

"The following tables will show the quantity and the destination of the exports of this oil:—

MANUFACTURE in Madras.

Year, 1847-48.		*Year, 1848-49.*	
Oil . . .	Gals. 19,520 R14,776	Oil . . .	Gals. 52,721 R36,294
Seed . . .	Qr. 17,518 R1,60,134	Seed . . .	Cwt. 144,125 R2,99,412
Year, 1852—54.		*Year, 1854-55.*	
Oil . . .	Gals. 119,180 R73,635	Oil . . .	Gals. 17,139 R12,720
Seed . . .	Cwt. 1,198,079 R6,93,760	Seed . . .	Cwt. 167,324 R4,31,726

Exported to the—	SEED. Cwt.	Gals.
United Kingdom	12,713	42,043
Ceylon	590	2,968
France	287,225	...
Pegu	741	19,698
Bombay	113	...
Malacca	33	3,593
Travancore	148	...
Mauritius and Bourbon	...	4,232

"The 'second sort gingelly' sometimes called 'bastard gingelly' is extracted from a variety of sesamum above mentioned. It differs but little from the true gingelly; the quantity of oil yielded by an equal amount of seed is somewhat less, but there appears to be no difference in the quality of the product.

"The following remarks upon the cultivation of the true gingelly, and its varieties in the Rajamundry District, have been furnished by F. Coplestone, Esq.—

Gingelly, or first sort Gingelly (the black seed).—This is the produce of the hill country called Reddyseema in the Rajamundry District. It is generally sown at the commencement of the monsoon (June), and ripens in four months, 160 seers of seed yields 50 seers of oil which is clear and sweet. The current value of the seed is R50 per candy of 500℔.

Bastard Gingelly, or second sort Gingelly, is the worst variety of this plant; the seeds are of mixed colours, white, red, and black. It is usually sown in the month of *Chytear*i (April) and ripens in three months,—160 seers of seed yield 35 seers of oil, which is of a brown colour and bitter. The current value of this seed is R35 per candy of 500℔.

White Gingelly is sown in the month of *Myglam* (January-February), and ripens in three months and a half. The oil is clean and sweet, 160 seers of seed yield 44 seers of oil, the current value of the seed is R44 per candy.

Pyrú Núvúlú is the red seed, sown generally on the islands called Lunkaloo. It ripens in three months, 160 seers of seed yield 45 seers of oil. The current price of the seed is R42 per candy. The term '*Pyrú*' is applied to the season after the general harvest in January, *viz.*, February, March, and April, and has no reference to these seeds except as indicating the time of their sowing.

The exports of this oil and seed are included in those of Gingelly."

II.—BOMBAY.

BOMBAY. 1150

In the Western Presidency it would appear that pure sesamum seed is rarely ground by itself. The oil is extracted chiefly by Lingayet Ganigrás from sesamum, linseed, safflower, and castor seed, grown in the district and bought by the oil pressers from the growers. From sesamum the oil is extracted by pressing the seeds in an oil-mill. The mortar of the oil-mill is a huge

MANUFACTURE in Bombay.

stone 8 feet long and about 12 feet round. The lower part is buried in the ground. The upper 3 feet are hollowed out and lined inside with wood, which has to be renewed once a year. None of the three grains, sesamum, linseed, or safflower, is put alone in the mortar. If any of these is pressed by itself it yields little oil, while of equal quantities of any two or more of these grains are pressed together, the outturn is greatly increased. After the stone mortar has been freshly lined with wood it does not hold more than 29℔ (8 seers) of seed. Afterwards when

Advantage claimed for mixing seeds.

1151 the roller or piston wears away the wood, the mortar holds daily a larger quantity of grain, till in the course of a year, it can hold 115℔ (32 seers) of seed. Before putting them into the mortar the seeds are slightly wetted. The roller is turned round and round in the mortar by means of bullocks yoked to a cross shaft which is attached to the roller from the outside. This process expresses and separates the oil from the seed. The oil is taken out for use, and the crushed seed is scraped out and used as cattle food. When a mortar holds only 29℔ (8 seers) of seeds, two bullocks take about two hours to express the oil. When the mortar begins to hold up to 115℔ (32 seers), the pressing takes about twice as long. So with a freshly repaired mill, oil is drawn out six times a day and only three times when the wooden lining gets worn.

N.-W. PROVINCES and OUDH.

III.—NORTH-WEST PROVINCES AND OUDH.

Messrs. Duthie & Fuller furnish the following facts regarding the
1152 system pursued in these provinces in the expression of the oil:—

"The oil is extracted by simple pressure in a mill, which is identical in form with the *kolhu* or pestle-mortar used for crushing sugarcane, but of a smaller size. The mill is worked by a single bullock, which has its eyes blind-folded to prevent, so it is said, giddiness. The animal is generally driven by a man or boy seated on the revolving beam, but a well-trained bullock may often be seen patiently going its round without any one to look after it. Oil-pressing is the peculiar occupation of a caste of men called *telis*, who are usually remunerated for the labour of pressing by receiving the oil-cake and a wage of grain equal in weight to the oil expressed. The oil-cake is used as cattle food, and in the western districts is much prized on this account, there being a considerable traffic in it. It is reported to be even occasionally used as human food by the poorer classes in times of distress. *Tili* oil is not only eaten raw after the manner of other oils, but is also commonly used in the manufacture of sweetmeats and in adulteration of *ghi*. It is occasionally used for lighting, and gives a clearer light than other vegetable oils, but burns more rapidly. Anointing the body is another use to which it is applied either in the crude state, or scented, when it is termed *phulel*" (*Field and Garden Crops, Pt. II.*, 37).

The method of preparing the oil in the Himálayan Districts has been described as follows by **Mr. Atkinson**:—"The mode of extracting the oil is usually the same in the hills and Bhabar. The seed is first sifted, cleaned, and dried, and then put into a *kolhu* or press worked by hand or by oxen. A little water is added, and after some time the oil runs out. The oil is then strained or allowed to stand in shallow vessels, when the impurities sink to the bottom. Every three parts of good seed yield one part of oil, which has risen in price much of late years and renders *til* a very valuable crop."

BENGAL.

IV.—BENGAL.

1153 Very little of a definite nature has been published regarding the methods pursued in the Lower Provinces in the preparation of this oil. The exports from Calcutta average only about 3,000 gallons, so that the subject

MANUFACTURE in Bengal.

is of small importance. The following abstract account of the process adopted in Behar may, however, be here given :—Mr. R. W. Bingham, in reporting of the resources of the Sasseram District, observed that of this seed there are two kinds, and that both are extensively sown in various parts. The first is sown in July, and is ready for reaping, say, in November; the second is sown in August; both crops come to maturity nearly at the same time. These plants are also sown as auxiliaries, but with the highland rain crops, such as *Ruhur, Motha*, etc. The seed has about the same value as *Sursún* in the bazárs, but the oil, being thinner and purer, and almost tasteless, while burning with little smoke, is extensively used in Indian perfumery. It is extracted from the seed in the same manner as mustard seed. The residue or cake is eaten by the poorer classes as an article of food, and is greedily devoured by cattle. It grows in sandy loams (*Agri.-Horti. Journ., XII.*, 339).

TRADE IN GINGELLY OR SESAMUM.

TRADE in the Oil. 1154

OIL AND DREGS.

Exports, Foreign. 1155

Foreign Exports.—It seems almost unnecessary to offer further comments on the subject of the Indian trade in this oil than has already been made. For some years past the exports to Europe have entirely disappeared, and the bulk of the oil which leaves India is now consigned to Arabia. The following table gives the volume and value of the oil exported, and shows at the same time the proportion of the total of the oil exports sent to Arabia and of the oil dregs shipped to Ceylon during the past ten years :—

Statement showing the Exports to Foreign Countries of Gingelly Oil and the Dregs of the Oil.

	Gingelly Oil.		Dregs of the Oil.		Total.		Countries to which greatest quantities were consigned.	
							Gingelly Oil.	Dregs of the Oil.
	Quantity.	Value.	Quantity.	Value.	Quantity.	Value.	Arabia.	Ceylon.
	Gallons.	R	Gallons.	R	Gallons.	R	Gallons.	Gallons.
1880-81 .	118,750	1,36,770	51,612	1,19,833	174,362	2,56,603	60,721	51,612
1881-82 .	111,701	1,20,182	49,380	1,07,298	161,081	2,27,480	79,381	48,450
1882-83 .	103,812	1,11,511	47,270	98,729	151,082	2,10,240	74,323	42,564
1883-84 .	89,012	1,03,309	70,115	1,38,375	159,127	2,41,684	53,169	57,779
1884-85 .	87,896	99,722	52,813	1,20,876	140,709	2,20,598	52,974	38,571
1885-86 .	77,625	89,636	51,289	1,00,966	128,914	1,90,602	52,001	36,085
1886-87 .	98,169	1,06,430	51,817	1,03,252	149,986	2,09,682	68,591	35,017
1887-88 .	76,311	84,820	41,196	81,713	117,507	1,76,533	42,899	33,804
1888-89 .	72,159	98,644	49,703	1,00,597	121,862	1,99,241	38,346	36,413
1889-90	91,120	1,31,216	50,308	1,03,588	141,428	2,34,804	50,333	37,235

TRADE in Seed.

Internal.
1156

SESAMUM SEED.

Internal and Coastwise Trade.—The section "CULTIVATION," of this article, deals with certain features of the trade in sesamum seed. It opens by republishing the forecast of the crop for 1890-91 which was issued by the Revenue and Agricultural Department of the Government of India, and after reviewing that forecast gives a series of notes on the systems of cultivation pursued in the various provinces. The statement there given incidentally manifests many features of the trade, and it shows the relative importance of the producing provinces by giving the areas devoted in each to the crop. Thus, for example, Table A (p. 516) shows the normal sesamum area in the provinces for which definite agricultural statistics are available. These are Bombay, 609,000; Madras, 531,000; the Central Provinces, 337,932; the Panjáb, 187,000; the North-West Provinces and Oudh, 175,000; Berar, 144,000; and Sind, 130,000 acres. While Sind would thus appear the least important of the provinces named, an inspection of Table B (p. 517), which exhibits the land trade by rail and river, will be found to reveal the fact that as a producing province, in the export trade, it is perhaps the most important, being followed, by the Central Provinces and then by Bombay. Madras, on the other hand, which has the largest area, in some of the annual returns of sesamum cultivation, participates to a very small extent in the *internal trade which is registered on rail and river routes.* But Table B has been drawn up with a view to show the gross imports by rail and river from the producing areas to the chief port towns—the towns from which the foreign exports are obtained. It takes, accordingly, no cognisance of the exports from these towns to the provinces of India, so that the figures given are not net imports. Further, since it does not embrace the coastwise transactions, the quantities shown against the port towns are not even the total gross imports. To these would have to be added the coastwise consignments, in order to furnish what, to the mercantile public, would seem required, namely, the grand total of all imports by whatever routes conveyed. It has already been remarked that returns of road traffic are not available. Relatively, however, while important in the local interchanges, from town to town and district to district, throughout the empire, in the case of the port towns the supplies drawn along the roads, are, it may be said, less important. The drain which takes place into the great commercial centres is largely the consequence of foreign demand, and the ultimate controlling power is the European dealer who naturally finds the largest and best markets in tracts of the country tapped by rail and river means of transport. The error in the returns of the port towns, due to defective registration of road traffic, may in this case, therefore, be set on one side. But in one province—Madras—a far more serious drawback to an accurate analysis of commercial transactions is dependent on the canal traffic. The writer has unfortunately been unable to overcome this difficulty, since the trade in oil seeds, borne on the canals to the Madras port towns, is not separately distinguished. The balance sheet which he has worked out, therefore, for the commercial coast towns of Madras shows on an average an excess of exports over imports, of some 20,000 tons. In only one other province, namely, Sind, is there a similar excess of published exports over imports which amounts, on an average, to some 3,000 tons. Doubtless this is due to the extensive boat traffic on the numerous mouths and affluents of the Indus, a proportion of which very probably escapes registration. It may, however, with these admissions of imperfection, serve a useful purpose to give here a balance sheet of the port towns so far as it has been possible to obtain the facts necessary for that purpose:—

TRADE in Seed.

Internal.

Analysis of the Sesamum Seed Trade with the Chief Port Towns of India for the year 1888-89.

		Imports.	Exports.
		Cwt.	Cwt.
BOMBAY	By Rail and River	758,585	13,390
	By Coastwise	160,239	74,497
	By Foreign	15,996	708,469
	TOTAL	934,820	796,356
	Less Exports	796,356	...
	Net balance available for local consumption	138,464	...
CALCUTTA	By Rail and River	105,031	2,144
	By Coastwise	29,271	88,644
	By Foreign	*Nil*	32,785
	TOTAL	134,302	123,571
	Less Exports	123,571	...
	Net balance available for local consumption	10,731	...
MADRAS PORTS (Madras, Pondicherry, Negapatam, Tuticorin, and Calicut).	By Rail and River	75,500	15,989
	By Coastwise	10,041	193,939
	By Foreign	18,128	336,613
	By Canal	Not available.	
	TOTAL	103,669	546,541
	Deduct Imports	...	103,669
	Excess not accounted for: no supplies for local consumption	...	442,872
KARACHI	By Rail and River	423,295	7
	By Coastwise	486	40,037
	By Foreign	4,593	459,549
	TOTAL	428,374	499,593
	Deduct Imports	...	428,374
	Excess not accounted for: no supplies for local consumption	...	71,219
BURMA (Rangoon and Moulmein).	By Coastwise	242,325	3,020
	By Foreign	29,128	30
	By Transfrontier routes to Upper Burma	10,868	18,745
	TOTAL	282,321	21,795
	Less Exports	21,795	...
	Net balance available for local consumption	260,526	...

SESAMUM indicum.

Trade in Gingelly

TRADE in Seeds.

Internal.

It will thus be seen that far from Madras being an unimportant province in the *til* supply, it is more important than Sind. The details of the figures given under the headings Rail and River, as also those for coastwise trade, show that practically the total Madras exports are drawn from Madras province, whereas about one-third of the Sind exports consists of Panjáb seed. The figure given under Bombay is the highest of all, but the share taken by the Presidency in the production of that export is remarkably low. Indeed, it may be said that the Central Provinces and the Nizam's Dominions produce the *til*-seed which appears in Bombay returns, since the Presidency furnishes on an average less than one-fifth of the total exports. In fact, Goa contributes as a rule larger quantities annually to Bombay port town than the Presidency does. This latter fact exhibits the importance of giving along side of the rail-borne trade to the port towns, the transactions which take place along the coast. Small quantities are drawn to Bombay by rail from Rájputana, the Panjáb, the North-West Provinces, Berar, and Madras. It will thus be seen that the great tableland which extends from the frontier of Rájputana to Madras and east and north-east of the Gangetic basin is the chief *til*-producing area of India. In the case of Sind, the high export, relatively to the small extent of land returned as under the crop, probably denotes a smaller local consumption than is the case with the other provinces of India.

Chief Til Seed Area. 1157

An important feature of the Madras trade in gingelly seed is the fact that that province furnishes Burma with the very large quantities which it requires annually. Bengal draws its supplies chiefly from the Lower Province and from Assam and its exports are comparatively unimportant. Indeed, Madras sends to Burma alone considerably more than the total exports from Bengal. The North-West Provinces and Oudh, while they furnish small quantities to Bombay and still smaller quantities to Bengal, may be said to grow *til* purely for local consumption. As has been stated, the Central Provinces furnish the largest quantities which appear in Bombay trade returns, but at present these provinces export little or no *til* to Bengal.

Foreign. 1158

Foreign Trade.—With regard to the Foreign trade in *til* seed it may be said, that while the exports have greatly expanded during the past 30 years, they might be described as having been practically stationary for the past ten years. This view of the trade may be demonstrated by the following averages of the quinquennial periods ending 31st March 1875, *viz.*, 580,943 cwt.; 1880, 1,317,279 cwt.; 1885, 2,324,028 cwt.; and 1890, 1,986,820 cwt. The highest recorded exports took place in the year 1883-84, when they stood at 2,843,382 cwt., since which date they have fluctuated at about 2,000,000 cwt. It has already been stated that the bulk of these exports go to France. The following table exhibits the total exports during the past twenty years as well as the share taken in these by France:—

Exports. 1159

Exports of Sesamum-seed from India.

Year.	Quantity.	Value.	Quantity consigned to France.
	Cwt.	R	Cwt.
1870-71	779,333	46,75,615	505,619
1871-72	565,854	33,95,224	495,414
1872-73	447,878	26,87,275	428,735
1873-74	908,430	54,49,184	828,578
1874-75	1,203,222	72,28,920	1,081,715
Average	580,943	46,87,243	668,012

TRADE in Seeds.

Foreign Exports.

Year.	Quantity.	Value.	Quantity consigned to France.
	Cwt.	R	Cwt.
1875-76	1,409,908	78,74,782	535,501
1876-77	1,307,815	86,82,937	1,172,219
1877-78	1,158,802	84,82,262	1,000,381
1878-79	1,039,687	79,96,210	834,502
1879-80	1,670,185	1,19,79,042	1,420,079
Average	1,317,279	90,03,046	992,556
1880-81	1,907,008	1,31,26,933	1,619,501
1881-82	1,917,854	1,21,77,307	1,493,429
1882-83	2,305,414	1,46,23,753	1,922,382
1883-84	2,843,382	1,97,97,536	2,045,140
1884-85	2,646,484	1,92,30,128	1,854,186
Average	2,324,028	1,57,91,131	1,784,927
1885-86	1,759,343	1,19,41,829	1,154,465
1886-87	2,114,484	1,41,08,994	1,509,506
1887-88	2,747,270	1,87,70,501	1,855,849
1888-89	1,537,444	1,14,70,019	1,121,999
1889-90	1,775,559	1,30,98,813	1,205,929
Average	1,986,820	1,38,78,031	1,369,549

French Til-seed Oil. 1160

It has already been pointed out that the bulk of the exports go to France where it is understood the oil is expressed and finds its way into European commerce as a substitute for, or adulterant with, olive oil.

Trans-frontier. 1161

Trans-frontier Trade. – In the table which gives above a balance sheet of the published items of the Burmese trade in gingelly-seed, the transactions between the Lower and the Upper Provinces have been shown as if across the frontier of British possessions. This was thought desirable in order to demonstrate the very large annual consumption of Madras seed which takes place in the Lower Province. The actual transfrontier trade, to and from the provinces of India, could not be given in the balance sheet (*p. 539*), since the quantities carried doubtless figure in the rail and river-borne trade. Of the transfrontier imports into British India the largest quantities are drawn from Kashmír. During the past two years these were, in 1888-89 2,965 cwt. and in 1889-90 7,048 cwt. Next in importance stands Nepál which in 1889-90 furnished India with 3,685 cwt.; then comes Khelat (the trade from which has, however, been declining) with 1,857 cwt., and last of all Hill Tipperah with 1,781 cwt. in 1889-90. The exports from British India across the frontier are very insignificant.

Market Seasons of the Black and White Til. 1162

In concluding this brief review of the sesamum-seed trade it may be of value to reiterate that there are two forms each of which possesses to a certain extent, properties peculiar to itself. These forms do not appear to have been separately recognised in European commerce. Were a preference to arise in the foreign demands, the one or the other might be separately grown to any required extent. They are sown at different seasons; the black form comes into market after the 1st of May and the white not till some short-time after the 1st of August. The so-called white gingelly seed varies in colour considerably from pure white to pink or red. The crop, which ripens in August, however, never becomes black, and in point of percentage of oil it yields less than the black or May crop. It will be observed from the

TRADE. remarks regarding cultivation above, that the periods of sowing and reaping these two crops, vary slightly in the provinces of India. Thus, for example, the black seed is in Bengal sown in June and the white in August, and the crops come into market about November and December instead of May and August. This fact may be accepted as corroborating the view of extensive adaptations, through antiquity of cultivation, or of considerable (almost specific) differences in the characteristics of the two crops.

(*W. R. Clark.*)

SESBANIA, *Pers.; Gen. Pl., I., 502.*

1163 **Sesbania aculeata,** *Pers.; Fl. Br. Ind., II., 114;* LEGUMINOSÆ.

Syn.—ÆSCHYNOMENE BISPINOSA, *Jacq.;* Æ. SPINULOSA, *Roxb.;* CORONILLA ACULEATA, *Willd.* Several varieties of this species are described in the *Flora of British India.*

1164 **Var. 1.—paludosa**=Æ. PALUDOSA & ULIGINOSA, *Roxb., Fl. Ind., Ed. C.B.C., 570.*

1165 **Var. 2.—sericea**=S. SERICEA, *DC., Prodr., II., 266.*

1166 **Var. 3.—cannabina**=S. CANNABINA, *Pers.,* ÆSCHYNOMENE CANNABINA, *Retz.;* CORONILLA CANNABINA, *Willd.;* S. AFFINIS, *Schrad.*

Vern.—*Jayanti, brihat-chakramed,* HIND.; *Jayanti, dhanicha, dhunchi, dhunsha,* BENG.; *Dhandain,* N.-W. P.; *Jhijan, jhanjhan, jaintar,* PB.; *Gadoreji,* SIND; *Rán-shewrá,* BOMB.; *Kán-sevari, bhuiavali, rán shevari,* MAR.; *Erra jíluga, erra-jilgua,* TEL.; *Nyaeh, pouk, najan ben,* BURM.; *Jayanti,* SANS.

References.—*DC., Prod. II., 265; Roxb. Fl., Ind., Ed. C.B.C., 570; Dalz. & Gibs., Bomb. Fl., 62; Stewart, Pb. Pl., 56; Sir W. Elliot, Fl. Andh., 52; U. C. Dutt, Mat. Med. Hind., 301; Baden Powell, Pb. Pr., 312, 509; Royle, Fib. Pl., 293; Cross, Bevan, & King, Rep. on Indian Fibres, 56; Balfour., Cyclop II., 584; Smith, Dic., 150; Kew Off. Guide to the Mus. of Ec. Bot., 40; Adm. Rep., Beng., 1882-83, 15; Agri. & Hort. Soc. Ind. Jour., IX., 415; (N. S.) 1885, Vol. VII., Pt. III. 224, 226, 228; Gazetteers:—Bomb., V., 25; VI., 14; XV., 432; N.-W. Prov., IV., lxx.; Panjáb, Montgomery, 19; Settle. Rept., Montgomery Dist., 19.*

Habitat.—A suffruticose annual, met with often in a state of cultivation, on the plains of India from the Western Himálaya to Ceylon and Siam. It has a cosmopolitan distribution throughout the tropics of the Old World.

CULTIVATION. 1167 CULTIVATION.—It is frequently grown on low wet ground which does not require much preparation as the plant is hardy. The time for sowing is after the soil has been moistened by the first showers of April or May. About 30lb of seed are allowed to the acre, and very little weeding is required. The crop is ready to be cut in September and October, but the fibre does not suffer if left standing till the seed is ripe in November. It is considered an ameliorating crop. The expense of cultivation, including land rent, is about R9 per acre (*Royle, Fibrous Plants*).

FIBRE. Stems. 1168 **Fibre.**—The STEMS of this plant have been long employed locally by the Natives of various parts of India to yield a strong and useful fibre, which they use as a substitute for hemp. Attention was directed to this plant by **Dr. Roxburgh**, who states, "It is deemed the coarsest, but not the least durable, of our Bengal substitutes for hemp. It is reckoned to be more durable, in the water or for purposes where it is often wet than *sun*, and is therefore universally employed for the drag-ropes and other cordage about fishing nets.

Royle (*Fibrous Plants*) observes that *dhunchi* forms "a very excellent fibre for common cord and twine purposes, and is certainly much superior in strength and durability to jute." He states that a rope made of *dhunchi* fibre was tried in the arsenal at Fort William and broke with not less than 75 cwt., though the Government Proof, required for such rope, was only

49 cwt. In 1887 *dhunchi* fibre was examined by **Messrs. Cross, Bevan, King & Watt**, who reported that it was "a strong fibre," "superior to jute in strength and durability, and best suited for the manufacture of cordage, for which purpose it should be preferred to either sunn-hemp or jute." **FIBRE.**

MODE OF PREPARATION.—The process of steeping and cleaning the fibre is similar to that required for *Sunn* (**Crotalaria juncea**, *q.v.*). **Preparation. 1169**

YIELD AND VALUE OF FIBRE.—In 1840 a sample was shown to the Agri.-Horticultura. Society of India by **M. Deneef**, who stated that a *bígáh* would yield 173lb of fibre and 92lb of seed, and that a woman could dress 4lb of fibre in a day. According to **Royle** the general produce of an acre is from 100 to 1,000lb of ill-cleaned fibre, and the current price of the fibre in the interior, was in his time about R1-8 a maund. The fibre was valued at the International Exhibition in 1851 at £30-35 a ton. **Yield. 1170** **Value. 1171**

Medicine.—The SEEDS of this species are mentioned by **Baden Powell** in his list of drugs; but no information is given as to the purpose for which they are used. **MEDICINE. Seeds. 1172**

Food.—The SEEDS were eaten at Poona during the famine of 1877-78. **FOOD. Seeds. 1173**

Sesbania ægyptiaca, *Pers.; Fl. Br. Ind., II., 114; Wight, Ic., t. 32.* **1174**

Syn.—ÆSCHYNOMENE SESBAN, *Linn.;* Æ. INDICA, *Burm.;* CORONILLA SESBAN, *Willd.*

Vern.—*Jaynti, jait, jhijan, janjhan, rásin, dhandiain, jet,* HIND. *Jayanti,* BENG.; *Saori, sewri, shewari,* BERAR; *Taitimúl, barjajanti,* URIYA; *Jaint,* N.-W. P.; *Jait, jant, jaintar* (Bazar seeds =) *riwás-an, jel,* PB.; *Saora,* C. P.; *Sewri, shevári, shewari, jait, janjan,* BOMB.; *Sevari,* MAR.; *Raysingani,* GUZ.; *Shewari, sheveri,* DEC.; *Champai, karumsembai,* TAM.; *Suiminta sominta,* TEL.; *Yethugyi,* BURM.; *Jaintjaintar, jyantika, jaya,* SANS.

References.—*DC., Prod., II., 264; Boiss., Fl. Orient., II., 193; Roxb., Fl. Ind., Ed. C.B.C., 569; Brandis, For. Fl., 137; Kurz, For. Fl. Burm., I., 362; Beddome, Fl. Sylv., 86; Anal. Gen., t. 12, f. 3; Gamble, Man. Timb., 118; Dalz. & Gibs., Bomb. Fl., Supp., 21; Stewart, Pb. Pl., 75; Mason, Burma and Its People, 504; Sir W. Elliot, Fl. Andhr., 169; Rheede, Hort. Mal., III., t, 127; Burmann, Fl. Ind., 169; Sakharam Arjun, Cat. Bomb. Drugs, 212; Murray, Pl. & Drugs, Sind, 118; Dymock, Mat. Med. W. Ind., 2nd Ed., 254; Dymock, Warden & Hooper, Pharmacog. Ind., I., 474; Baden Powell, Pb. Pr., 342, 597; Atkinson, Him. Dist. (X., N.-W. P. Gaz.), 750; Useful Pl. Bomb. (XXV., Bomb. Gaz.), 58, 197; Moore, Man., Trichinopoly, 80; Settlement Report, Central Provinces, Nimar, 280; Gazetteers:—Bombay, V., 25; Panjáb, (Delhi), 18; N.-W. P., I., 80; IV., lxx.; Mysore & Coorg., I., 59; Ind. Forester, XIII., 120; Balfour, Cyclop. Ind., II., 584.*

Habitat.—A soft-wooded shrub of short duration, found throughout India, from the Himálaya where it ascends to an altitude of 4,000 feet in the North-West to Ceylon and Siam. It is cosmopolitan in its distribution throughout the tropics of the Old World. **Brandis** says that "on the rich alluvial banks of the Kistna and Warna rivers in the Deccan, which are submerged during the annual floods, it is grown from seed as an annual, attaining 15 to 20 feet in one season."

Fibre.—The BARK is made into rope (*Brandis*). **FIBRE. Bark. 1175**

Medicine.—In Muhammadan medical works the SEEDS of this tree are described as stimulant, emmenagogue, and astringent. They are used to check diarrhœa and excessive menstrual flux, and to reduce enlargement of the spleen. The Hindus employ them in ointments for the cure of itch and various other cutaneous eruptions, and the JUICE of the bark administered internally is also given for these purposes. The LEAVES are much used in poultices to promote suppuration, and to resolve hydrocele and rheumatic swellings. In the Panjáb, the seeds are applied externally mixed with flour for itching of the skin. The Marathas have a superstition that the **MEDICINE. Seeds. 1176** **Juice. 1177** **Leaves. 1178**

MEDICINE. sight of the seeds will remove the pain of scorpion stings. In Dacca the juice of the fresh leaves is given as an anthelmintic in doses up to 2 ounces (*Taylor, Topography of Dacca*).

SPECIAL OPINIONS.—§"The leaves are sometimes ground up into a paste with turmeric and onions or garlic and used as a discutient. The leaves when thus treated act more as an escharotic than a counter-irritant.
Flowers. For persons suffering from coryza or nasal catarrh, the FLOWERS are boiled
1179 in gingelly oil and the oil used as a bath oil" (*Surgeon-Major D. R.*
Root. *Thomson, M.D., C.I.E., Madras*). The ROOT well bruised and made into
1180 a paste is an excellent application for scorpion stings. The leaves in the form of poultices applied to abscesses hasten suppuration and draw the pus towards the surface" (*Assistant Surgeon N. C. Dutt, Durbhanga*). "The fresh root is much praised by the officers of the Baroda State as a remedy for scorpion sting. Passes with the root in hand are made along the part from the point of extent of pain to the seat of sting. In one to two hours the pain subsides" (*Assistant Surgeon S. Arjun Ravat, L.M., Gorgaum, Bombay*). "Poultice of leaves useful in inflammatory swellings" (*Assistant Surgeon S. C. Bhattacherji, Chanda, Central Provinces*).

FODDER. **Fodder.**—The LEAVES and young BRANCHES are lopped for fodder.

Leaves. **Structure of the Wood.**—White, soft, fibrous, but rather close-grained.
1181 Weight 27lb per cubic foot.

Branches. **Domestic.**—It is grown in the Deccan to furnish poles as a substitute
1182 for the bamboo, and is often utilised while growing to shade and support
TIMBER. the piper vine and various cucurbitaceous plants. The WOOD is employ-
1183 DOMESTIC. ed to boil *jaggery* and is reduced to charcoal for gunpowder. In Burma
Wood. it is made into children's toys (*Kurz*). In Assam the soft pithy STEMS
1184 are platted into mats, portions of it being dyed black before being matted
Stems. so as to work out a bold pattern (*Cross, Bevan, & King*). It is in com-
1185 mon use in Bengal as a hedge plant, for which purpose its very quick growth renders it suitable (*Gamble*).

1186 **Sesbania grandiflora,** *Pers.; Fl. Br. Ind., II., 115.*

Syn.—ÆSCHYNOMENE GRANDIFLORA, *Linn.*; AGATI GRANDIFLORA, *Desv.* CORONILLA GRANDIFLORA, *Willd.*

Vern.—*Agust, agusta, bak, agasti, basna,* HIND.; *Agusta, bak, buka, agasti, bagfal, buko,* BENG.; *Hadga, heta,* BERAR; *Básna, bako,* N.-W. P.; *Augusta, basna, agásta,* BOMB.; *Agástá, agas'i, shevari, chopchini,* MAR. *Agathio,* GUZ.; *Agath-thi-nar, agáti,* TAM.; *Avasinana, avesi,* TEL.; *Agase,* KAN.; *Paukpan, paukhya,* BURM.; *Agati, agasti vaka vranári, buka,* SANS.

References.—*DC., Prod., II., 266; Roxb., Fl. Ind., Ed. C.B.C., 569; Brandis, For. Fl., 137; Kurz, For. Fl. Burm., I., 362; Beddome, Fl. Sylv., t. 86; Gamble, Man. Timb., 119; Dalz. & Gibs., Bomb. Fl. Supp., 22; Mason, Burma and Its People, 467, 767; Sir W. Jones, Treat. Pl. Ind., 143; Rheede, Hort. Mal., I., t. 51; Irvine, Mat. Med. Patna, 4; U. C. Dutt, Mat. Med. Hind., 92, 289, 322; Murray, Pl. & Drugs, Sind, 118; Dymock, Mat. Med. W. Ind., 2nd Ed., 253; Dymock, Warden, Hooper, Pharmacog. Ind., I., 472; Birdwood, Bomb. Prod., 26, 148; Useful Pl. Bomb. (XXV., Bomb. Gaz.), 58, 151; Econ. Prod., N.-W. Prov., Pt V. (Vegetables, Spices, and Fruits), 91, 93; Moore, Man. Trichinopoly, 75; W. W. Hunter, Orissa, II., 180, App. VI.; Settlement Report, Central Provinces, Chanda Dist., 82; Gazetteers:—Bombay, VI., 14; XV., 432; N.-W. P., IV., lxx.*

Habitat.—A short-lived, soft-wooded tree, which attains a height of 20 to 30 feet and is cultivated in Southern and Eastern India, the Ganges Doáb, and Burma. It is distributed, but usually in a state of cultivation, to Mauritius and North Australia.

GUM. **Gum.**—It yields a gum resembling Kino, of a garnet red colour when
1187 fresh, but becoming almost black by exposure to the air. This gum is

partially soluble in spirit and also in water, leaving a gelatinous residue of small bulk : it is very astringent (*Dymock*).

Fibre.—The inner BARK appears likely to yield good fibre (*Watt, Calc. Exhib. Cat.*). FIBRE. Bark. 1188

Medicine.—The BARK is very astringent and is given in infusion in the first stages of small-pox and other eruptive fevers. In many parts of India the JUICE of the LEAVES and FLOWERS is used as a popular remedy for nasal catarrh and headache; it is blown up the nostrils and causes a copious discharge of fluid relieving the pain and sense of weight in the frontal sinuses. The ROOT of the red-flowered variety, rubbed into a paste with water, is applied to rheumatic swellings. The leaves are said to be aperient. A poultice of the leaves is a popular remedy in Amboyna for bruises. The juice of the flowers is squeezed into the eyes to relieve dimness of vision (*Dymock; Arjun; Murray*). MEDICINE. Bark. 1189 Juice. 1190 Leaves. 1191 Flowers. 1192 Root. 1193

Food and Fodder.—The tender LEAVES, PODS, and FLOWERS are eaten by Natives as a vegetable and in curries. When taken very freely they are apt to produce diarrhœa (*Lisboa*). Cattle also eat the leaves and TENDER SHOOTS. FOOD & FODDER. Leaves. 1194 Pods. 1195 Flowers. 1196 Tender Shoots. 1197

Structure of the Wood.—White and soft, not durable. Weight 32 lb per cubic foot. TIMBER. 1198

Domestic and Sacred.—The WOOD is used in Bengal for the posts of Native houses and for firewood; in Berar and the Deccan the tree is grown as a substitute for bamboo. This species is also used as a support for the piper vine. The FLOWERS are sacred to Siva and are supposed to represent the male and female generative organ. DOMESTIC. Wood. 1199 Flowers. 1200

SESELI, *Linn.; Gen. Pl., I., 901.*

[UMBELLIFERÆ.

Seseli indicum, *W. & A.; Fl. Br. Ind., II., 693; Wight, Ic., t. 569;* 1201

Syn.—ATHAMANTHA DIFFUSA, *Wall.;* LIGUSTICUM INDICUM, *Wall.;* L. DIFFUSUM, *Roxb.*

Vern.—*Banjoán,* BENG.; *Kirminji-ajván,* MAR.; *Vanayamáni,* SANS.

References.—*DC., Prod., IV., 153; Roxb., Fl. Ind., Ed. C.B.C., 271; U. C. Dutt, Mat. Med Hind., 322.*

Habitat.—An annual diffuse herb, met with in the plains of India from the foot of the Siwaliks to Assam and Coromandel; frequent in Central Bengal.

Medicine.—The SEED is used as a medicine for cattle. It is also said to be carminative. MEDICINE. Seed. 1202

SPECIAL OPINION.—§ "I have found the seeds of **Seseli indicum** to act as a good anthelmintic for round worms, and they are also stimulant, carminative, and stomachic. Dose of simple powder, from twenty grains to a drachm (*Honorary Surgeon Moodeen Sheriff, Khan Bahadur, G.M., M.C, Triplicane, Madras*).

SESUVIUM, *Linn.; Gen. Pl., I., 855.*

[FICOIDEÆ.

Sesuvium Portulacastrum, *Linn.; Fl. Br. Ind., II., 659;* 1203

Syn.—S. REPENS, *Willd.;* PSAMMANTHE MARINA, *Hance.*

Vern.—*Dhápa,* BOMB.; *Vungaravasí,* TAM.; *Vangarreddi kúra,* TEL.

References.—*Roxb., Fl. Ind., Ed. C.B.C., 406; Dalz & Gibs., Bomb. Fl., 15; Sir W. Elliot, Fl. Andhr., 189; Murray, Pl. & Drugs, Sind, 108; Ind. Forester, III., 238; XII., 329.*

Habitat.—A succulent branching herb, met with on the sea-shores of India, from Sind to Calcutta and thence to Singapore. It is distributed to tropical and sub-tropical sea-shores.

FOOD. Seeds. 1204 Twigs. 1205 Leaves. 1206

Food.—In some parts of the coast it is cultivated as a substitute for spinach, and the SEEDS, TWIGS, and LEAVES were extensively eaten by the Natives during the famine of 1877-78.

SETARIA, *Beauv.; Gen. Pl., III., 1105.*

A genus of grasses containing about ten species, of which only four need be here specially dealt with.

1207 **Setaria glauca,** *Beauv.; Duthie, Fodder Grasses of N. Ind., 14;*
PIGEON OR BOTTLE GRASS. [GRAMINEÆ.

Syn.—PANICUM GLAUCUM, *Linn.*; PENNISETUM GLAUCUM, *R. Br.*

Vern.—*Bandra, bandri,* HIND.; *Pingi-natchi,* BENG.; *Kukra,* SANTAL; *Kaluku,* BERAR; *Dhusa, neori,* BUNDEL.; *Bindra,* N.-W. P.; *Bandra, bandri, dissi, kotu, ban kangni,* PB.; *Kutta choti, soma, kharkhura, billi, chhinchra,* RAJ.; *Pohwa, panhawa, thont wa,* C. P.; *Bhadli,* KHANDESH; *Bhádali,* MAR.; *Bhadli,* DEC.; *Nakakora,* TEL.

References.—*Rev. A. Campbell, Econ. Pl. Chutia Nagpur., No. 9210; Atkinson, Him. Dist., 320; Gazetteer, N.-W. P., I., 85; IV., lxxix.; Ind. Forester, XII., App., 23.*

Habitat.—An annual grass, very common all over the plains and up to moderate elevations on the hills. A variety is cultivated as a cereal in some of the Bombay Districts where it is known as *bhádh* (*Duthie*). It thrives best on rich or cultivated ground.

FOOD & FODDER. Grain. 1208 Grass. 1209

Food and Fodder.—In the Central Provinces and Chutia Nagpur, the GRAIN of the wild plant and in Bombay that of the cultivated variety is used as food. The GRASS in India is considered a fairly good fodder, but is, according to **Symonds,** unsuited for making hay. In Australia it is said to be highly valued as a fodder, and in the United States, where it is called "Pigeon" or "Bottle" grass, it is said to furnish fodder which is as nutritious as that from Hungarian grass (**S. italica**), but less productive.

1210 **S. intermedia,** *R. & S.; Duthie, Fodder Grasses of N. Ind., 14.*

Vern.—*Chiriya-chaina,* N.-W. P.; *Chota sarsata, undar punchha,* RAJ.; *Chota chikiya, noktowa, sawa,* C. P.; *Lundi,* BERAR.

Reference.—*Gazetteer, N.-W. P., IV., lxxix.*

Habitat.—This species occurs on the plains of Northern India, and at low elevations on the hills. In the Central Provinces, it occurs on both black and sandy soils.

FODDER. 1211

Fodder.—No information seems to be available as to the nutritive value of this species (*Duthie*).

1212 **S. italica,** *Beauv.; Duthie, Fodder Grasses of N. Ind., 15.*
THE ITALIAN MILLET.

Syn.—PANICUM ITALICUM, *Linn.*

Vern.—*Kángu, kángni, rala, rawla, bertia, kákun, kakni, kauni, kiranj, kirakang, chena, tángan, kora,* HIND., BENG., & DEC; *Kala kangni koní, kanghuní,* HIND.; *Erba,* SANTAL; *Tángun,* URIYA; *Rala, kungni,* C. P.; *Kákún,* BUNDEL.; *Kangní, tángun,* N.-W. P.; *Koni, kangní, mandira, shúngura, chína, gandra, mundúa, murhoa,* KUMAON; *Shálí, píngí,* KASHMIR; *Kangní, chiúrr, kher, khauní, shálí* (*a name given in Bombay to rabí crop of Sorghum, and in Madras to wild rice*), *sálan, shak, kusht, gal,* (*husket kangni called also chánwal kangni* (*lit. rice of kangani*) *according to* Baden Powell) *prál* (= the straw), PB.; *Gal,* PUSHTU; *Kirang,* SIND; *Kangni, kora-kang, kang, vavani,* BOMB.; *Kangu, káng, ríle, rála, chenna,* MAR.; *Kang, karáng,* GUZ.; *Tennai, tenai,* TAM.; *Koralú, kora,* TEL.; *Naoni, navani, vavani,* KAN.; *Tauna, navaria,* MALAY.; *Pyoung-lay-kouk, sami, puki,* BURM.; *Tana-hál,* ANDAMAN; *Tana-hál,* SING.; *Kangu, kanguni, priyangu,* SANS.; *Dukhn* (*according to* Ainslie), ARAB; *Gal, arzun,* PERS.; *Cay khe,* COCHIN CHINA.

References.—*Dalz. & Gibs., Bomb. Fl. Supp., 98; Stewart, Pb. Pl., 259; DC., Orig. Cult. Pl., 378; Ainslie, Mat. Ind., I., 226; Mason, Burma and Its People, 476, 816; Sir W. Elliot, Fl. Andhr., 82, 99; U. C. Dutt, Mat. Med. Hind., 303; Birdwood, Bomb. Prod., 11; Baden Powell, Pb. Pr., 237, 383; Drury, U. Pl. Ind., 326; Atkinson, Him. Dist. (X., N.-W. P. Gaz.), 320, 689; Duthie & Fuller, Field and Garden Crops, Part II., 5; Useful Pl. Bomb. (XXV., Bomb. Gaz.), 184, 276; Church, Food-Grains, Ind., 55; Man. Madras Adm., I., 288; Nicholson, Man. Coimbatore, 221; Morris, Account Godavery, 68; Moore, Man. Trichinopoly, 72; Man. Rev. Accts. Bombay, 101; Settlement Reports:—Panjáb, Montgomery, 107; Kangra, 25; Simla, xl., App. II., 4; Jhang, 85, 97; N.-W. P., Azamgarh, 115; Banda, 49; Central Provinces, Chanda, 81; Upper Godavery, 35; Nursingpur, 52; Baitool, 63; Hoshungabad, 286; Gazetteers:—Bombay, VIII., 182; XV., Pt. II., 18; Panjáb, Hoshiarpur, 94; N.-W. P., I., 85; III., 225; IV., lxxix.; Oudh, I., 419; Mysore & Coorg, II., 11.*

Habitat.—This millet is extensively cultivated in India, both in the plains and on the hills, up to 6,500 feet. Both **Dr. Watt** and **Mr. J. F. Duthie** state that it occurs wild in India on parts of the Himálayan region; but **DeCandolle** appears to doubt whether it has as yet been found truly wild anywhere. The last-named author, after reviewing the historical, philological, and botanical evidence, comes to the conclusion, however, that "the species existed before all cultivation, thousands of years ago, in China, Japan, and the Indian Archipelago." "Its cultivation must have early spread towards the West, since we know of Sanskrit names, but it does not seem to have been known in Syria, Arabia, and Greece, and it is probably, through Russia and Austria, that it early arrived among the lake-dwellers of the stone-age in Switzerland." **Mr. Duthie** says it is "both wild and cultivated in India and largely grown in other warm countries." "The Sanskrit name *Kangu* indicates its antiquity as a cultivated plant in India."

Cultivation in India.

CULTIVATION 1213

Kangni is pretty generally cultivated, although in comparatively small amounts, all over India. It is usually sown as a *kharif* crop; but the same land is often twice sown, once at the commencement of the rains and a second time between September and the end of January. Two principal varieties are cultivated, one straw yellow, the other reddish yellow.

Details as to the methods of cultivation and extent of the crop are not available for the whole of India, but the following extracts from District Manuals, Agricultural Reports and other publications will give, at any rate, some idea of its relative importance as a food stuff to the Natives of India.

N.-W. Provinces. 1214

NORTH-WEST PROVINCES.—**Messrs. Duthie & Fuller** (*Field and Garden Crops of the North-West Provinces*) give the following account:—"The area under *kakun* is even smaller than that under *chehna*. In each of the Meerut and Rohilkhand Divisions it amounts to about 1,200 acres. In the districts of the Agra Division it is somewhat larger (about 1,600 acres), and in the Allahabad Division it reaches the comparatively high figure of 8,000 acres. The area which it covers in the three districts of Azamgarh, Basti, and Gorakhpur is about the same as that in Rohilkhand. In the Jhansi Division it is reported to be grown on 2,600 acres. But it is far more commonly grown as a subordinate crop than by itself, and these figures greatly under-estimate its real agricultural importance. In the Doáb it is commonly sown on *juár* or *chari* fields on better class land, and in the Azamgarh District it is very generally mixed with *sawan*.

"It is sown with the commencement of the rains and reaped in September, being, as a rule, grown on the good land of the village, and often on the highly manured fields round the village site. As a general rule, it is followed by a spring crop. Its outturn is not so large as that of *sawan*,

SETARIA italica.

The Italian Millet—Kangni.

CULTIVATION averaging, when grown close, from 3½ to 5 maunds per acre. Great loss is suffered by the depredations of birds, who are particularly fond of the grain, and there is a common saying, "*kakun kheti, baj dharna*," *i.e.*, (the cultivation of the *kakun* is like keeping a hawk). The straw is no more nutritious as cattle fodder than rice straw, and is therefore not set much store by."

Panjab. 1215 PANJAB.—**Baden Powell** remarks: "This millet is cultivated in both harvests. The grain is much used in the Panjáb for feeding poultry, etc. It is very little used as food otherwise." The Report of the Settlement of the Simla District describes it as "Sown in *bakhil* lands, generally on the inferior fields. Sown in May, ripening in *Aser* (15th September to 15th October), it is not much sold, but is eaten boiled like rice. The straw is fed to cattle during the winter." In the Jhang District 116 acres are said to be under this millet, which is principally grown on the leased wells in the Government Bár to the east of Jhang. Stray patches are also to be seen on wells in villages, generally associated with cotton, rarely alone (*Settlement Report, Jhang District*).

Bombay. 1216 BOMBAY.—In Bombay, including Sind, the total area under this crop, during the year 1886-87, is said to have been 329,819 acres, and it is reported to be largely grown, especially in the Karnatak and Satara.

Madras. 1217 MADRAS.—In the Madras Manual of Administration it is described as one of the principal millets, but few details as to the area or method of cultivation are available. In the Saidapet Experimental Farm Manual it is said to have been "grown on more than one occasion on the farm with fair success producing a fair crop of grain as well as straw. The grain is difficult to thresh and does not command a ready sale." According to the Returns of Agricultural Statistics of British India, the price it commands in the Madras markets varies from R1-4-6 to R1-12-0 per maund of 80℔.

MEDICINE. 1218 **Medicine.**—Although much esteemed usually as an article of food, *Kangni* is sometimes objected to on account of its heating properties, and when taken as the sole food it is said to be sometimes apt to produce diarrhœa. Medicinally it is said to act as a diuretic and a stringent, and to be of use externally in rheumatism. It is a popular domestic remedy for alleviating the pains of parturition.

FOOD. Grain. 1219 **Food.**—The GRAIN is much esteemed as an article of human food in some parts of the country, and is eaten in the form of cakes and porridge in the North-West Provinces. In the Madras Presidency it is valued as an excellent material for making pastry. On this subject **Ainslie** wrote, "For the purpose of pastry it is little, if at all, inferior, to wheat, and when boiled with milk, it forms a light and pleasant meal for invalids." In the

Leaves. 1220 Chenab valley the LEAVES are used as a pot-herb. Boiled with milk, it forms a light and pleasant meal for invalids. The Brahmins specially esteem it. It is also grown as food for cage birds, and for feeding poultry in the Panjáb and North-West Provinces. When ripe the ears only are plucked, the straw being afterwards cut for fodder. **Baden Powell** says that the grain "is said to render beer more intoxicating."

Chemistry. 1221 The composition of Italian Millet, according to **Professor Church**, shows in 100 parts:—

"*Composition of Italian Millet* (*Husked*).

	In 100 parts.	In 1℔.
Water	10·2	1 oz. 277 grs.
Albuminoids	10·8	1 ,, 318 ,,
Starch	73·4	12 ,, 63 ,,
Oil	2·9	0 ,, 203 ,,
Fibre	1·5	0 ,, 105 ,,
Ash	1·2	0 ,, 84 ,,

The nutrient ratio according to this analysis is 1 : 7·4, the nutrient value 91. The percentage of flesh-forming material in the grain seems to vary a good deal from 9 to 13 (*Church, Food-Grains of India*).

FODDER. Straw. 1222 As fodder, the STRAW is not usually reckoned very nourishing and in many parts of the Panjáb Himálaya it is only used for feeding goats. In the Montgomery District, however, *bhusa* prepared with the straw is considered strengthening, and in some parts of Mysore it is thought to be next in quality as a fodder to that of *rágí*, while in other localities it is used only for bedding or thatching houses.

1223 **Setaria verticillata**, *Beauv.; Duthie, Fodder Grasses of N. Ind.*, *15*.

Syn.—PANICUM VERTICILLATUM, *Linn.*; PENNISETUM VERTICILLATUM, *R. Br.*

Vern.—*Dora-byara*, BENG.; *Bir-kauni*, SANTAL; *Chirchitta, bardanni, barti*, N.-W. P.; *Chirchira, barchitta, kutta*, PB.; *Kutta bari, gadar puchha, bandri, sarsata*, RAJ.; *Bandri, chakkarnitta-gadi, chikna bara, lapti, chilaya*, C. P.; *Jaljatang-jhara*, BERAR; *Chick-lenta*, TEL.

References.—*Rev. A. Campbell, Econ. Prod. Chutia Nagpur, No. 8705; Atkinson, Him. Dist. (X., Gaz. N.-W. P.), 320; Gazetteers, N.-W. P., I., 85; IV., lxxix.; Ind. Forester, XII., App., 23.*

Habitat.—A coarse, rank annual grass, common in shady places and in rich ground all over the plains of India and on the Himálaya up to an elevation of 6,000 feet.

FOOD & FODDER. Grain. 1224 **Food and Fodder.**—The GRAIN is eaten by poor people. Cattle eat it when it is young, *i.e.*, before the flowering spikes appear (*Duthie*).

(*G. Watt.*)

SHEEP, GOATS, AND ANTELOPES. 1225

The SHEEP, GOATS, and ANTELOPES belong to the Order Ungulata (or hoofed animals), an order which, by modern zoologists, is referred to two divisions —the **Subungulata** and **Ungulata vera.** The former is represented by the Sub-order PROBOSCIDEA or Elephants (*see Vol. III., 208-227*). The latter, by the Sub-order ARTIODACTYLA (the Ruminants, together with the Hippopotami and Pigs), which have an even number of digits or toes (either 2 or 4) on all feet. The ARTIODACTYLA are referred to four sections—the **Pecora, Tragulina, Tylopoda,** and **Suina.** The first of these (**Pecora**) is split into two *Families*—the *Bovidæ* (Oxen, *Vol. V., 659-674*, also the Sheep, Goats, and Antelopes) and the *Cervidæ* (the Deer, *Vol. III., 55-63*). The second, **Tragulina** (*Vol. III., 55*) includes the Chevrotains, or Mouse Deer. The third—the **Tylopoda**—is represented by the Camels and Llamas (*Vol. II, 50-64*). And lastly, **Suina,**—the Boar and Pig (see **Hog**, *Vol. IV., 253-254*).

The reader, from the references given above to other articles in this work, will be able to appreciate the group of animals which it is desired to specially deal with in this place. The isolation of the Oxen from the Sheep, Goats, and Antelopes was regarded as serving a useful economic purpose, since, although they blend into each other and form a continuous series of genera which constitutes the Family *Bovidæ*, the Oxen possess many features of interest and value to man quite distinct from the utility of the Sheep, Goats, and Antelopes.

Before dealing with the wider and more utilitarian characteristics of the domesticated animals of this series, it may perhaps be desirable to follow the course initiated with the Oxen, *viz.*, to give a few brief notes regarding the wild animals, taking these up in the alphabetical order of their scientific names.

1226 1. **Antilope cervicapra**, *Pallas; Blanf., Fauna, Br. Ind. (Mammalia), 521.*

THE INDIAN ANTELOPE OR BLACK BUCK.

Syn.—A. BEZOARTICA, *Gray.*

Vern.—*Haran, harna, harin, hirun* ♂, *harni* ♀, *kalwit* ♀, *mrig*, HIND.; *Kala* ♂, *goria* ♀, TIRHÚT; *Kalsar* ♂, *baoti* ♀, BEHAR.; *Bureta*, BHAGALPUR; *Báránt, sasin*, NEPAL; *Alali* ♂, *gandoli* ♀, BAORI;

INDIAN ANTELOPE or BLACK BUCK.

Bádú, HO KOL; *Bámani-haran*, URIYA; *Bamani-haran, bamuni-hiru, hiru, phandayat*, MAHR.; *Kutsar*, KORKU; *Irri* ♂, *sedi* ♀, *ledi, jinka*, TEL; *Chigri, húlékara*, KAN.; *Veli-man, ena* ♂, *harnia, mirga*, SANS.

References.—*Jerdon, Mam. Ind.*, 275-79; *Sterndale, Mam. Ind.*, 472-76.

Habitat.—Found, in open plains of short grass, in herds, occasionally, of several thousand animals. They are to be met with throughout India, especially in the North-Western Provinces, Rájputana, and parts of the Deccan, but are locally distributed and keep to particular tracts (*Blanford*). Are not found in Ceylon, nor to the east of the Bay of Bengal.

Characteristics 1227 CHARACTERISTICS.—Weight about 90 ℔; length of head and body 4 feet, tail 7 inches; height at shoulder about 32 inches. The colour varies from yellowish-fawn above, in the young, to blackish-brown above, or even almost black, in old animals. The horns are almost confined to the males. The speed and endurance of the antelope are amongst its chief characteristics, requiring a good horse to run down even a wounded animal. If captured young they are easily tamed. **Mr. Elliot** states "that the rutting-season commences about February or March, but fawns are seen of all ages at every season."

FOOD. Venison. 1228 **Food.**—The VENISON of the Indian antelope is excellent.

1229 2. **Boselaphus tragocamelus**, *W. Sclater; Blanford, Fauna Br. Ind.* [(*Mammalia*), 517.

THE NILGAI, OR BLUE BULL.

Syn.—ANTILOPE TRAGOCAMELUS, *Pallas*; PORTAX TRAGOCAMELUS, *Adams*; P. PICTA, *Horsfield*.

Vern.—*Nil, nilgao* ♂, *nilgai, silgai, silgao, rojra, rojh, rovi* ♀, *roz*, HIND.; *Guraya*, GOND.; *Murim* ♂, *susam* ♀, HO KOL.; *Manú-pota*, TAM.; *Mairu, maravi, kard-kadrai*, KAN.; *Manupotu*, TEL.

References.—*Jerdon, Mam. Ind.*, 272; *Sterndale, Mam. Ind.*, 476.

Habitat.—The Nilgai is found throughout India, from the base of the Himálayas to the south of Mysore. It is common in parts of the Eastern Panjáb, the North-West Provinces, Guzerat, and the Central Provinces. It is not found in Ceylon, Assam, nor Eastern Bengal.

Characteristics. 1230 CHARACTERISTICS.—General outline is horse-like, owing to the lean head, and long, compressed, deep neck. The colour of the male is iron grey,—varying from bluish to brownish grey; the female of a sandy or tawny colour. The length of male 6½-7 feet, tail 18-22 inches. The height at the shoulder 52-56 inches. It is generally found on level or undulating ground, or on hills, rarely in thick forests, preferring to keep much to the same ground. They are very tenacious of life, numerous instances being recorded of their reviving and making off after a supposed fatal wound.

FOOD. Flesh. 1231 **Food.**—**Sterndale** says that the FLESH is sometimes saturated with a bitter principle, owing to the Nilgai at times devouring large quantities of the intensely acrid berries of the *aonla* (**Phyllanthus Emblica**). The flesh, otherwise, is of fair quality, but inferior to most of the wild species of Indian *Bovidæ*.

DOMESTIC. 1232 **Domestic and Sacred.**—Nilgai are not difficult to tame, and may be taught to carry loads, draw light weights, and to be ridden, being thus superior to the Sambar stag, which will not bear the slightest burden. In some parts, the Hindus regard the animal as a kind of cow, and hence will not touch its flesh, the result being that the Nilgai becomes very tame.

1233 3. **Capra ægagrus**, *Gmelin; Blanford, Fauna Br. Ind.* (*Mammalia*), 502.

THE PERSIAN WILD GOAT; SIND IBEX.

Syn.—ÆGOCEROS ÆGAGRUS, *Kotschy*; C. CAUCASICA, *Gray*; C. BLYTHII, *Hume*.

PERSIAN WILD GOAT.

Vern.—*Pásang* (male), *boz*, *boz-pásang* (female), PERS.; *Kayik*, ASIA MINOR; *Thar*, PB.; *Borz*, *Afg.*; *Sair*, *sarah*, *phashin pachin* ♂ *borz-kuhi*, BALUCH.; *Chank* ♂ *hit*, *haraf* ♀, BRAHUI; *Ter*, *sarah*, SIND.

References.—*Jerdon, Mam. Ind.*, 292; *Hutton, Calc. Jour. N. H., II.*, 521; *id. I. As. Soc. Beng., XV.*, 161; *Blyth, Cat.*, 176; *Blanford, Jour As. Soc. Beng., XLIV, pt. 2*, 15; *Hume, Proc. As. Soc. Beng.*, 1874, *p.* 240; *Murray, Vert. Zoo., Sind*, 56; *Sterndale, Mam. Ind.*, 446.

Habitat.—Found throughout Asia Minor, Persia, Afghánistán, and Balúchistán, extending into Sind. It is met with on the barren hills of Balúchistán and Western Sind, but not east or north-east of the Bolan Pass and Quetta.

Characteristics. 1234
CHARACTERISTICS.—Male with a beard on the chin only; horns scimitar-shaped, curved backwards, greatly compressed. Good measure 40 inches round the curve, the extreme length known being 52·5 with a girth of 7 inches. A full-grown male stands 37 inches at the shoulder. It is a very active animal and leaps with wonderful precision. Found either solitary or in herds which sometimes number as many as 100.

MEDICINE. Bezoar-Stones. 1235
Medicine.—The concretions known as BEZOAR-STONES, which were formerly much used in medicine and as antidotes to poison, is a concretion found in the stomach of this goat. (See **Danford's**, also **Hutton's**, account.)

FOOD. Flesh. 1236
Food.—This interesting animal is very frequently hunted by Natives of the countries in which it is found for the sake of its FLESH. **Blanford** remarks that "there can be no doubt that **C. ægagrus** is one of the species, and probably the principal, from which tame goats are derived." **Hutton** did not think the Persian and Afghan goats could have been derived from either **C. ægagrus** or **C. falconeri**. (*Conf. with p. 560, also pp. 563 and 637.*)

DOMESTIC. Skins. 1237 Horns. 1238
Domestic Uses.—The SKINS are valued as water or flour bags. The HORNS are carried by mendicants as the insignia of their calling and as trumpeting horns.

1239
4. **Capra falconeri,** *Hügel; Blanford, Fauna Br. Ind.* (*Mammalia*), 505.

THE MARKHOR OR SNAKE-EATER.

Syn.—CARPA MEGACEROS, *Hutton*; C. JERDONI, *Hume.*

Vern.—*Markhor* (the snake-eater), KASHMIR; AFGHAN; *Rá-ché* (*rá-pho ché*, ♂ *and ráwa-che* ♀), LADAK; *Reskuh* (*matt* ♂ *hit*, *haraf* ♀) BRAHUI; *Pachin*, *sará* ♂ *buskuhi* ♀), BALUCH.

References.—*Jerdon, Mam. Ind.*, 291; *Hutton, Cal. Jour. Nat. Hist, II.*, 535, *pl. xx*; *Blyth, Cat.* 176; *Blanford, Jour. As. Soc. Beng., XLIV., pt. 2*, 17; *Sterndale, Mam. Ind.* 441; *Ward, Sportsman's Guide*, 20-24.

Habitat.—"This magnificent wild goat is found on the Pir Panjál range of the Himálaya, to the south of the valley of Kashmír, in the Hazara hills, and the hills on the north of the Jhelum and in the Hurdwar hills which separate the Jhelum from the Chenab river, not extending further east than the sources of the Beas. It is abundant on the hills to the west of the Indus, and extending north into Afghánistán. It is also met with in Ladak" (*Jerdon*).

Characteristics. 1240
CHARACTERISTICS.—**Blanford** remarks that "throughout the BOVIDÆ no species varies to so great an extent in the form of the horns as the Markhor." He, however, reduces all the conditions that have been described to four, *viz.*, (*a*) the true **C. falconeri** of Astor and Baltistan; (*b*) the Pir Panjál Markhor; (*c*) the Cabul Markhor or **C. megaceros,** *Hutton*; and (*d*) the Saliman race, the **C. jerdoni,** *Hume.* The Markhor is of a light greyish brown colour in summer, in winter becoming a dirty yellowish white with a bluish tinge. The adult male has a long black beard, and the neck and breast are covered with long black hair: the female has a short black beard but no mane. The horns are very long, massive, straight, and angular with spiral twists; they approximate closely at the base and thence diverge outwards and backwards. An old Gilgit male

SHEEP and Goats. **The Wild Sheep, Goats,**

MARKHOR. measured by **Colonel Biddulph** was 38·5 inches high, and 55 inches from between the horns to the root of the tail. **Blanford** adds that much larger dimensions have been recorded by other writers.

DOMESTIC. Horns. 1241 **Domestic Uses.**—The Markhor is much sought after by sportsmen, and its HORNS are considered a great trophy. **Blanford** says that it is in appearance by far the grandest of all wild goats. It has repeatedly bred in confinement with domestic goats, and it was at one time supposed that the tame races with spiral horns were derived from **C. falconeri.** It is not improbable (**Blanford** adds) that some are thus descended, but the spiral in the horns of tame goats is almost always in the reverse direction to that found in the Markhor—the first turn of which is outwards. On this subject **Henderson** (*Lahore to Yarkand, 137*) says that the few goats he saw had horns "with only one curve straight backwards, like the ibex, and not the spiral twist of the Markhor horns."

Conf. with p. 637.

1242 5. **Capra sibirica,** *Meyer; Blanford, Fauna Br. Ind.* (*Mammalia*), *503.*

THE HIMÁLAYAN IBEX.

Syn.—C. IBEX, *Hodgson;* C. SAKEEN, *Blyth;* C. HIMALAYANA, *Schinz.* IBEX & SAKIN SIBIRICA, *Hodgson.*

Vern.—*Skin,* ♂ *sakin, shyin, iskin,* and *dabmo* or *danmo* ♀ HIMALAYAN DISTRICTS & TIBET; *Búz, teringole, tangrol, skin* or *tin,* PB.; *Kyl,* KASHMIR.

References.—*Jerdon, Mam. Ind. 292; Blanford, Yark. Miss. Mam., 86, Aitchison, Tr. Linn. Soc. Zool., V., 64; Hodgson, Jour. As. Soc. Beng.; X., 913; XI., 283; XVI., 700; Sterndale, Mam. Ind., 444; Ward, Sportsman's Guide, 25-32.*

Habitat.—Found on the Western Himálaya from Kashmír to Nepál (*Hodgson*). In the west of Kashmír it is rare, and it is not found apparently to the west of the Jhelum river. It is abundant in Kanawar, on some of the ranges on both sides of the Sutlej, but rarer further east. It is much commoner on the north than on the south side of the great Himálayan range, and extends in its distribution throughout Central Asia to the Altai. It is chiefly found on or about precipitous cliffs at high elevations close to the snow. As the snow melts (May and June) the males forsake the females and retire to higher altitudes, descending in the early morning to feed.

Characteristics. 1243 CHARACTERISTICS.—Its general colour is a lightish brown, with a dark stripe running down the back in summer, dirty yellowish white in winter; the beard which is 6 to 8 inches long, is black; the horns are long, scimitar-shaped, curving over the neck, flattened at the sides, and strongly ridged in front, from 40 to 50 inches in length.

FIBRE. Horns. 1244 Undercoat. 1245 Hair. 1246 **Fibre (Fur & Wool).**—They are largely hunted by Europeans for their HORNS, and by Natives for the sake of a soft downy UNDERCOAT which in Kashmír is called *asali tús.* This is used as a lining for shawls, and for stockings and gloves, and is woven into the fine cloth called *túsi.* No wool is so rich, so soft, and so full. The HAIR is manufactured into coarse blanketing for tents or is twisted into ropes. In Ladak, large numbers are killed by the Natives during the winter, when they are forced to descend to the valleys. They are either snared at night or shot in the grey dawn of the morning, when they venture down to the streams to drink (*Jerdon*). **Baden Powell** says the hair of the Ibex makes the famous ibex-shawls. In another part of his work (quoting from **Cooper**) he gives the two first qualities of *pashm*—the white and grey—as obtained from the *shah-thosh,* which is probably this animal (*Conf.* with **Ovis vignei,** the *sha,* and **Pantholops hodgsoni** the *tsus, also p. 559*).

Conf. with pp. 559, 636.

6. Cemas goral, *Blanford, Fauna Br. Ind. (Mammalia), 516.* GORAL. 1247
THE GORAL.

Syn.—ANTILOPE GORAL, *Hardwick;* A. (NEMORHEDUS) GORAL, *Hodgson;* KEMAS GHORAL, *Ogilby;* NEMORHEDUS GORAL, *Horsfield.*

Vern.—*Deo chágal,* ASSAM; *Goral,* KUMAON; *Sáh, sár,* PB.; *Pij, pijur, rai, rom,* KASHMIR; *Suh-ging,* LEPCHA; *Ra-giyu,* BHUTIA.

References.—*Jerdon, Mam. Ind. 285; Sterndale, Mam. Ind. 457; Ward, Sportsman's Guide, 38-39.*

Habitat.—The whole range of the Himálaya, from Bhután to Kashmír, frequenting rocky places at altitudes between 3,000 and 8,000 feet, on grassy or mixed forest and grassy hills.

CHARACTERISTICS.—Of a dull rusty-brown colour, paler beneath, with a dark brown line from the vertex to the tail. The chest and front of fore-legs are of a deep brown colour. The ears externally are of a rusty brown with a large patch of pure white on the throat. The female is paler than the male, and the young are said to be redder in tint. The length of head and body is about 50 inches, height at the shoulder 28 to 30 inches, horns 8 inches. The horns spring from the crest of the frontals and incline backwards; they are ringed at the base and smooth for the apical half or third; in full-grown males they are usually 6 to 8 inches long. Characteristics. 1248

Food.—They are much hunted both by Europeans and Natives, and their FLESH is very palatable. FOOD. Flesh. 1249

7. Gazella bennetti, *Gray; Blanford, Fauna Br. Ind. (Mammalia), 526.* 1250
THE INDIAN GAZELLE; THE BALÚCHISTAN GAZELLE; GOAT-ANTELOPE in Bombay and Madras; RAVINE-DEER of sportsmen in Bengal.

Syn.—ANTILOPE BENNETTII, *Sykes;* A. ARABICA; *Elliot;* GAZELLA CHRISTII, *Gray;* G. FUSCIFRONS, *Blanford.*

Vern.—*Chinkára, chikára, kal-punch,* HIND.; *Phaskela,* N.-W. P.; *Ask,* or *ast, ahu,* BALUCH.; *Khasm,* BRAHUI; *Kalsipi* (*i.e.,* black tail), MAHR; *Tiska, budári, mudari,* KAN.; *Sank-húlé,* MYSORE; *Porsya* ♂, *chari* ♀, BAORI; *Burudu jinka,* TEL.

References.—*Jerdon, Mam. Ind., 280-81; Sterndale, Mam. Ind., 465.*

Habitat.—Found in Central India, extending throughout Balúchistan to the Persian Gulf, also in the desert parts of Rajputana, Hurriana, and Sind. It prefers the open bare plains, or low hills, and is never found in forests.

CHARACTERISTICS.—An adult buck 28·5 inches high at the croup, 26 at the shoulder; length $3\frac{1}{2}$ feet; tail $8\frac{1}{2}$ inches; horns 12 to 13 inches; weight about 50lb. The colour is a light chestnut above, with the breast and lower parts white, tail nearly black. The *Indian Gazelle* generally herds together in small parties of from two to six. It lives on grass and the leaves of bushes. When alarmed, it utters a sort of hiss by blowing through the nose, and stamps with the forefeet; whence its Kanarese name '*Tiska*' (*Elliot*). Characteristics. 1251

8. G. picticandata, *Brooke; Blanford, Fauna Br. Ind. (Mammalia), 529.* 1252
THE TIBETAN GAZELLE.

Syn.—PROCAPRA PICTICANDATA, *Hodgson.*

Vern.—*Goa, ragao,* TIBETAN.

Reference.—*Sterndale, Mam. Ind., 467.*

Habitat.—Commonly found in Ladak, and north of Nepal and Sikkim. According to **Kinloch**, its habitat is on the plateau to the south-east of the Tsomoriri lake, on the hills east of Hanle, and in the Indus Valley from Demchok, on the frontier of Ladak, as far down as Nyima.

CHARACTERISTICS.—Height of the male from 18 to 24 inches. Length from snout to rump 43, tail 0·75 inches. Colour in winter is a light sandy fawn above, the lower parts are white. They are not as a rule shy. Characteristics. 1253

GAZELLE. Fibre. 1254 **Fibre.**—This is probably the Antelope called *tsodkyi* in Tibet, which, Baden Powell says, affords a WOOL which is obtained from Lahaul (*Conf.* with **Pantholops hodgsoni** below).

1255 9. **Gazella subgutturosa,** *Blain; Blanford, Fauna Br. Ind.,* (*Mammalia*), *528.*

THE PERSIAN GAZELLE.

Syn.—ANTILOPE SUBGUTTUROSA, *Güldenstädt.*

Vern.—*Ahu,* PERS.

Reference.—*Sterndale, Mam. Ind., 466.*

Habitat.—Highlands of Persia, Central Asia, and on British territory in Pishin, north of Quetta.

Characteristics. 1256 CHARACTERISTICS.—The horns, which are confined to the males, are lyrate, annulate, with the points turned inwards. There is a well-marked lachrymal fossa, and infraorbital gland. Colour upper surface sandy, under surface white, as far as the tail. Tail blackish-rufous.

1257 10. **Hemitragus hylocrius,** *Blyth: Blanford, Fauna Br. Ind.,* (*Mammalia,*) *511.*

THE NILGHIRI WILD GOAT OR IBEX.

Syn—KEMAS HYLOCRIUS, *Ogilby;* CARPA WARRYATO, *Gray;* C. HYLOCRIUS, *Sclater.*

Vern.—*Warri-ádú, warri-átú,* TAM.; *Kard-ardu,* KAN.; *Mullá-átú,* MAL.

References.—*Jerdon, Mam. Ind., 283; Sterndale, Mam. Ind., 451.*

Habitat.—Found on the Western Ghâts (Nilgiri and Anaimalai hills) and southward towards Cape Comorin.

Characteristics. 1258 CHARACTERISTICS.—"The adult male, dark sepia brown, with a pale reddish-brown saddle, more or less marked and paler brown on the sides and beneath; legs somewhat grizzled with white, dark brown in front and paler posteriorly; the head is dark, grizzled with yellowish brown, and the eye is surrounded by a pale fawn-coloured spot; the horns are short, much curved, nearly in contact at the base, gradually diverging, strongly keeled internally, round externally, with numerous close rings not so prominent as in the last species. There is a large callous spot on the knees surrounded by a fringe of hair, and the male has a short stiff mane on the neck and withers. The hair is short, thick, and coarse" (*Jerdon*). The length of the adult male is, according to **Jerdon**, 4 feet 2 inches to 4 feet 8 inches, and the height at the shoulder 32 to 34 inches. **Sterndale** questions this latter measurement, which, he says, is much under the mark. The horns are occasionally 15 inches, rarely more than 12.

FOOD. Flesh. 1259 **Food.**—As an article of food, the FLESH when hung is said by the Rev. Mr. Baker, in a correspondence with Mr. Blyth, to be equal to Welsh mutton.

1260 11. **H. jemlaicus,** *Adams; Blanford, Fauna Br. Ind.* (*Mammalia*), *509.*

THE TEHR OR TAHR.

Syn.—CAPRA JEMLAHICA, *Ham.;* C. JHARAL, *Hodgson;* C. QUADRIMAMMIS, *Hodgson;* H. QUADRIMAMMIS, OR JHARAL, *Hodgson.*

Vern.—*Jharal,* NEPAL; *Jhula* ♂, *thar tahrni* ♀, KANAWAR; *Krás, jagla,* KASHMÍR; *Tehr, jehr, kart, esbu* ♂, *esbi* ♀, PB.

References.—*Hodgson, As. Res., xviii., pt. 2, p. 129; Jour. As. Soc. Beng., IV., 710; V., 254; Jerdon, Man. Ind., 286; Blanford, Jour. As. Soc. Beng., XLI., pt. 2, 40; Lydekker, Jour. As. Soc. Beng., XLVI., 286; Sterndale, Mam. Ind., 449; Ward, Sportsman's Guide, 33-34.*

Habitat.—Found throughout the entire range of the Himálaya at high elevations between the forest and snow limits.

Characteristics. 1261 CHARACTERISTICS.—The male is of various shades of brown, varying from dark to yellowish. There is no beard, the face being smooth and dark ashy, but on the fore quarters and neck, the hair lengthens into a magnificent mane, which sometimes reaches to the knees. The horns are triangu-

TEHR.

lar, the sharp edge to the front; they are 10 to 11 inches in circumference at the base where they touch, and taper to a fine point at a length of 12 to 14 inches. The height of a male is 36 to 40 inches, the length about 4 feet 8 inches to the root of the tail. The female is much smaller and of a reddish brown or fulvous drab above, with a dark streak down the back, whitish below; the horns also are much smaller.

Food.—The FLESH of the male is at certain seasons very rank and disagreeable to English tastes, but is in high favour with the Natives. That of the female is excellent. In autumn the *tahr* becomes immensely fat and heavy. **Dr. Falconer** (*Trans. Agri.-Hort. Soc. Ind., III., 76*) refers to this goat as affording a FLEECE very similar to that of the Chinese Tartary shawl fleece. This circumstance **Dr. Falconer** took as justifying the opinion that the Chinese goat, if domesticated on the higher ranges of the Indian side of the Himálaya, would continue to yield its much valued fleece. FOOD. Flesh. 1262 Fleece. 1263

12. Nemorhædus bubalinus, *W. Sclater; Blanford, Fauna Br. Ind. (Mammalia), 513.* 1264

THE HIMÁLAYAN GOAT-ANTELOPE or SEROW.

Syn.—ANTELOPE BUBALINA, *Hodgson;* and A. THAR, *Hodgson;* CAPRICORNIS THAR, *Ogliby;* C. BUBALINA, *Adams;* N. BUBALINA, *Jerdon.*

Vern.—*Serow, serowa,* N.-W. P.; *Saráo,* N. W. HIMÁLAYA; *Rámu, halj, sálábhir,* KASHMÍR; *Goa,* CHAMBA; *Aimu,* KUNÁWAR; *Yamu,* KULU; *Thar,* NEPAL; *Gya,* BHOTIA OF SIKKIM; *Sichi,* LEPTCHA; *Nga, paypa,* SHAN; *Shauli,* CHINESE.

References.—*Jerdon, Mam. Ind., 283; Blyth, Mam. and Birds of Burma, 46; Sterndale, Mam. Ind., 454; Ward, Sportsman's Guide, 35-37.*

Habitat.—The whole of the wooded ranges of the Himálaya from Kashmír to Bhután, and thence to the ranges dividing China from Burma; at elevations between 6,000 to 12,000 feet.

CHARACTERISTICS.—Black, more or less grizzled on the back, on the flanks mixed with deep clay colour it has a black dorsal stripe, the forearms and thighs anteriorly of a reddish brown colour. The rest of the limbs are hoary, beneath it is whitish in colour. The hair is scanty except on the neck on which there is a thick harsh rough mane. The horns are stout, roundish, ringed more than half way, tapering, much curved backwards, slightly divergent with the points inclining outwards; the average length is about 10 inches, but they are said to reach 14 occasionally. The length of the male is 5 to 5½ feet, the height at the shoulder about 3 feet 2 inches. Weight about 200lb. Characteristics. 1265

Food.—The FLESH is coarse. FOOD. Flesh. 1266

13. N. sumatrensis, *Cantor; Blanford, Fauna Br. Ind. (Mammalia), 514.* 1267

THE BURMESE GOAT-ANTELOPE.

Vern.—*Tau-tshiek (Tau-myin in Pegu)* BURMESE; *Kambing-utan,* MALAY.

Habitat.—According to **Blanford** this species differs so slightly from **N. bubalinus,** only in being more rufous and probably smaller in size, that he is inclined to regard them as one species. **Blyth** (*Cat. Mam. and Birds of Burma (1875), 46*) says that this animal is distributed from Arakan through Pegu to the extremity of the Malayan peninsula.

Medicine.—According to **Crawfurd** the HORNS are valued by the Chinese for certain alleged restorative properties. MEDICINE. Horns. 1268

41. Ovis hodgsoni, *Blyth; Blanford, Fauna Br. Ind. (Mammalia), 494.* 1269

THE GREAT TIBETAN SHEEP.

Syn.—O. AMMON, *Horsfield;* O. AMMONOIDES, *Hodgson.*

Vern.—*Nyan ♂, nyanmo ♀,* LADAK; *Nyang, nyand, hyan, nuan, niar, gnow,* TIBETAN.

GREAT TIBETAN SHEEP.

References.—*Hodgson, As. Res. (1833), xviii., pt. 2, 135; Blyth, Proc. Zoo. Soc. (1840), p. 65; Hodgson, Jour. As. Soc., X., 230, pl. 1, f. 1; Jerdon Mam. Ind., 298; Hooker, Himálayan Jour., I., 234; Blanford, Jour. As. Soc., Beng., xli., 40; Ward, Sportsman's Guide, 40-47.*

Habitat.—This magnificent wild sheep (probably the largest of the genus) does not usually occur on the Indian side of the great snowy ranges, but is said to be occasionally met with near the sources of the Ganges. In summer, it is seldom met with at a lower elevation than 15,000 feet, and it is often found much higher up amidst the snows.

Characteristics. 1270 CHARACTERISTICS.—Male usually stands 3½ to 4 feet in height, and length from nose to rump 6 to 6½ feet. Horns of an adult male 36 to 40 inches long, round the curve, and the girth 16 to 17 inches. The horns are said to be sometimes so enormous that the animal cannot feed on level ground as the horns reach below the level of the mouth. The longest horns on record are 53 inches, and girth 24 inches.

FOOD. Flesh. 1271 **Food.**—The FLESH is excellent; it is always tender even on the day it is killed, and of very good flavour (*Kinloch, Large Game Shooting in Tibet*).

DOMESTIC. Horns. 1272 **Domestic and Sacred.**—It is the shiest and wildest of all animals, and is very hard to kill. To shoot the **Ovis ammon** is the highest ambition of the sportsman on the Himálaya (*Jerdon*). **Cunningham** states that the HORNS along with those of the ibex and the *sha* (**O. vignei**) are placed on the religious piles of stones met with in Ladak and other Buddhist countries.

Blanford says that **O. brookei** has now been ascertained to be a wild hybrid between a male **O. hodgsoni** and female **O. vignei** (*Sterndale, Jour. Bomb. N. H. Soc., I., p. 35*),—a male of the great sheep in Zanskar having taken possession of a small flock of **O. vignei** ewes, and bred with them. The converse, a hybrid between the male **O. vignei** and female **O. hodgsoni**, has also been shot by **Major C. S Cumberland** (*Proc. Zoo. Soc., 1885, p 851*). The hybrid in the latter case was found with a flock of **O. hodgsoni**.

1273 15. **Ovis nahura**, *Gray*; *Blanford, Fauna Br. Ind. (Mammalia), 499.*

THE BHARAL OR BLUE WILD SHEEP.

Syn.—O. BURRHEL, *Blyth*; O. NAHOOR, NAYAUR, *Hodgson*; O. NAHURA, *Gray*; PSEUDOIS NAHOOR, *Hodgson.*

Vern.—*Bharal, bharar, bharut* (*males often menda*, a ram), HIND.; *Na, sna*, LADAK; *Wa, war*, PB.; *Nervati*, NEPAL; *Nao gnao*, BHOTIA.

References.—*Hodgson, As. Res. XVIII., pt. 2, p. 135; Blyth, Cat., p. 178; Jerdon, Mam. Ind., 296; Blanford, Jour. As. Soc. Beng., XLI., pt. 2, 40; id. Yark. Miss. Mam., p. 85, pl. XIV; Lydekker, Jour. As. Soc. Beng., XLIX., pt. 2, 13; Sterndale, Mam. Ind., 438; Ward, Sportsman's Guide, 52-56.*

Habitat.—Found on the Himálaya, from Sikkim and probably, according to **Jerdon**, Bhután, westwards to the valley of the Sutlej and, in **Sterndale's** opinion, even as far as Ladak and Western Tibet. **Blanford** mentions near Shigar in Bultistan, and near Sanju south-east of Yarkand to Moupin, and from the main Himálayan axis to the Kuenlun and Altyn Tágh. It is met with at great elevations, from the region of forest to the extreme limits of vegetation, or between 10,000 and 16,000 feet. In summer it generally keeps to the tops of hills and even in winter rarely descends below the forests.

Characteristics. 1274 CHARACTERISTICS.—The general colour of the animal is a dull slaty blue, slightly tinged with fawn; the belly, edge of the buttocks, and tail are white, a line along the flank dividing the darker tint from the belly. The edge of the hind limbs and tip of the tail is a deep black colour. The horns are moderately smooth with few wrinkles, rounded, nearly touching at the base, directed upwards, backwards, and outwards, the points being turned forwards and inwards. The female is smaller, the black marks are smaller

BHARAL.

and of less extent, the horns are small, straight, and slightly recurved, the nose is straighter. The young are darker and browner. The length of the head and body is 4½ to 5 feet; the height is 30 to 36 inches, the tail 7 inches, the horns 2 to 2½ feet round the curve, the circumference at the base is 12 to 13 inches (*Sterndale, Jerdon, etc*). (*Conf. with p. 567.*)

FOOD. Flesh. 1275

Food.—The FLESH in flavour is equal to the best Welsh mutton and is generally tender soon after the animal is killed. The *bharal* is fattest in September and October.

1276

16. Ovis poli, *Blyth; Blanford, Fauna Br. Ind. (Mammalia), 496.*

THE GREAT PAMIR SHEEP; MARCO POLO'S SHEEP.

Syn.—O. POLI and KARELINI, *Severtzoff.*

Vern.—*Kuchkár* ♂, *mesh* ♀, WAKHAN; *Rass, rush,* PAMIR; *Kulja* or *gulja* ♂, *Arkar* ♀, E. TURKISTAN.

References.—*Blyth, Proc. Zoo. Soc. (1840), 62; Stolicska, Proc. Zoo. Soc. (1874), 425; Jerdon, Mam. Ind., 299; Blanford, Proc. Zoo. Soc. (1884), 326; id. Yark. Miss. Mam. 80, 83; Sterndale, Mam. Ind., 424.*

Habitat.—"The high Pamir and the plateaus west and north of Eastern Turkestan, extending to the Altai. This sheep only comes within Indian limits in Hunza, north of Gilgit" (*Blanford*).

Characteristics. 1277

CHARACTERISTICS.—This magnificent wild sheep has immense horns, less massive, but more prolonged than those of **O. hodgsoni**. The horns of one specimen were 4 feet 8 inches in length round the curvature and 14½ inches in circumference at the base. **Blanford** says the extreme record measurements are 75 inches and 16·75 inches. **Severtzoff** estimates the weight of an old male at about 600lb.

Since this great sheep is not a native of India proper, it cannot receive more than the above passing mention. Its horns, like those of **O. hodgsoni** and **O. vignei,** are sometimes seen in the religious piles of stones met with in Ladak.

1278

17. O. vignei, *Blyth; Blanford, Fauna Br Ind. (Mammalia), 497.*

THE URIAL or SHÁ, by **Hutton** called the BEARDED SHEEP.

Syn.—O. CYCLOCEROS, *Hutton;* O. MONTANA, *Cunningham;* O. BLANFORDI, *Hume.*

Vern.—*Guch* ♂, *mish* ♀, PERS.; *Sha* (*shapo* ♂, *shamo* ♀, LADAK; *Urin,* ASTOR; *Koh-i-dumba,* AFGH.; *Kock gad* ♂, *garand* ♀, BALUCH & SIND; *Kar* ♂, *gad* ♀, BRAHUI; *Urial,* PB.

References.—*Blyth, Proc. Zool. Soc. (1840), p. 70; Hutton, Jour. As. Soc. Beng. XV., p. 152; id. Calc. Jour. Nat. Hist., II., p. 514 pl., XIX., (1842); Cunningham, Ladak, p. 199, pl. VII., (1854); Hume, Jour. As. Soc. Beng., XLVI., pt. 2, 327, pl. IV., (1877); Jerdon, Mam. Ind., 294; Sterndale, Mam. Ind., 435; Murray, Vertebate Zool. of Sind, 59; Ward, Sportsman's Guide, 48-51.*

Habitat.—Found over the whole Salt-range of the Panjáb, on the Sulaiman range across the Indus, the hills of Hazara, and those in the vicinity of Peshawar. According to the late **Mr. Dalgleish** it also occurs considerably farther east in Northern Tibet. It is also reported to be found at Astor and Gilgit to Afghanistan. It is met with at altitudes of from 800 to 2,000, rarely 3,000 feet (the *urial*) and up to 12,000 and 14,000 feet (the *sha*).

Characteristics. 1279

CHARACTERISTICS.—Of a general rufous brown colour, with a long thick black beard, mixed with white hairs from throat to breast, reaching to the knees; legs below the knees and feet white, belly white, the outside of the legs and a lateral line blackish in colour. The horns of the male are subtriangular, much compressed laterally and posteriorly, transversely sulcated, curving outwards and returning inward towards the face. The female is of a more uniform pale brown, with whitish belly, no beard, and short straight horns. The adult male is about 5 feet in length and 3 feet high;

URIAL. the horns measure from 25 to 30 inches long round the curve. **O. cycloceros** (*urial*) is regarded by some zoologists as a distinct species from **O. vignei** (*sha*) Blanford states that he cannot find any distinctive characters, those of colour being merely individual, while some of the skulls and horns appear indistinguishable.

FIBRE. Thosh. *Conf. with p. 636.* 1280 **Fibre.**—This animal apparently affords part of the wild *pashm* known as *thosh.*

FOOD. Flesh. 1281 **Food.**—The FLESH is good and well flavoured.

DOMESTIC. 1282 **Domestic Uses.**—Hutton says it possesses "a moderate sized lachcrymal sinus which appears to secrete, or, at all events, contains, a thick gummy substance of good consistency and a dull greyish colour. The Afghán and Baluch hunters make use of this gum, by spreading it over the pans of their matchlocks, to prevent the damp from injuring the priming. The *urial* has been bred freely with tame sheep. The occurrence of wild hybrids has been noted under **O. hodgsoni,** p. 556.

1283 18. **Pantholops hodgsoni,** *Hodgson; Blánford, Fauna Br. Ind. (Mammalia),* 524.
THE TIBETAN ANTELOPE: THE CHIRU.

Syn.—ANTILOPE (ORYX) KEMAS, *H. Smith;* A. CHIRU, *Lesson;* KEMAS HODGSONII, *Gray.*

Vern.—*Tsus* ♂, *chus* ♀, *chiru, chuhu, isos,* TIBETAN.

References.—*Jerdon, Mam. Ind., 282; Sterndale, Mam. Ind., 469; Ward, Sportsman's Guide, 72-73.*

Habitat.—Probably throughout the Tibetan plateau, from 12,000 to 18,000 feet elevation (*Blanford*).

Characteristics. 1284 CHARACTERISTICS.—Hodgson states that the male may measure 50 inches from nose to rump, tail with hair 9 inches, height at shoulder 32 inches. In colour pale fawn above, slaty grey towards the base, white below. The horns, ten in number, measure from 24-26 inches long, jet black, smooth and polished, encircled by a number of rings from 15 to 20 in number, extending from the base to within 6 inches of the top. Jerdon says that "it is probable this animal may have given rise to the belief in the unicorn; for at a little distance, when viewed laterally, there only appears to be one horn, there is so little divergence throughout their length."

FIBRE. Thosh. *Conf. with p. 636.* 1285 **Fibre.**—In some respects this answers to the animal that might at least afford some of the much prized form of wild *pashm* designated *thosh.*

DOMESTIC. Horns. 1286 **Domestic Uses.**—"The HORNS are beautifully adapted for knife handles" (*Kinloch*).

1287 19. **Tetracerus quadricornis,** *Gray; Blanford, Fauna Br. Ind. (Mammalia),* 519.
THE FOUR-HORNED ANTELOPE.

Syn.—TETRÁCEROS CHICKERA, *Blyth;* ANTILOPE (CERVICAPRÁ) QUADRICORNIS, *Blainville.*

Vern.—*Chousingha, chouka, doda,* HIND.; *Benkra,* MAHR.; *Bhokra phokra,* GUZ.; *Bhirki,* SAUGOR; *Bhirkura* ♂, *bhir* ♀, GONDI; *Bhirul,* BHEEL; *Kotari,* CHUTIA NAGPUR; *Kurus,* GONDS OF BASTAR; *Kondagori,* TEL.; *Kondguri, kaulla-kuri,* KAN.

References.—*Jerdon, Mam. Ind., 274; Sterndale, Mam. Ind., 479.*

Habitat.—In most parts of India, especially where the country is wooded and hilly. Throughout the Bombay Presidency it is common, also in the wooded parts of Rajputana and the Central Provinces (*Blanford*). It is not met with in Ceylon and Burma.

Characteristics. 1288 CHARACTERISTICS.—A small animal 40-42 inches in length, tail 5 inches, height at shoulder 24-26 inches, slightly higher at the croup than at the shoulder; weight about 43℔. Colour brownish-bay above, shading into white along the middle of the belly, but the colour varies somewhat according to locality. The anterior pair of horns are the shorter, measuring 1-1½ inches, the posterior 3-4 inches. In the Madras Presidency,

FOUR-HORNED ANTELOPE.

the anterior set of horns is said to be mostly absent, the skull of the adult animal showing only rudimentary projections. If taken young, it can be tamed.

FOOD. Flesh. 1289

Food.—The FLESH is not good eating, but can be made more palatable by being cooked with mutton fat.

DOMESTICATED GOATS. 1290

DOMESTICATED GOATS.

Hodgson gives the following as the chief distinctive features of Goats from Sheep :—

"Horns in both sexes : no mufle : no eye-pits : feet-pits in the fore-feet only, or none : no inguinal pores nor glands : no calcic tuft nor gland : mammæ two : odour intense in males : and a true beard in both sexes, or in males only."

"These animals are further distinguished by horns, directed rather upwards and backwards than circling sideways to the front, as in the sheep proper, by the obliquity of their insertion on the top of the head, their less volume, greater compression, less angularity, and, above all, by the keeled character of their sharp antral edge. The tail of the goats is shorter and flatter than in sheep; their chest or knees frequently, bare and callous; and their hairy pelage apt to be of great and unequal lengths." "It must be recollected" adds **Hodgson**, "that the so-called wild goats of the Himalaya (the *jharal* or *tehr*) are not goats at all; for they have four teats, a moist muzzle, and no interdigital pores or feet-pits." *(Conf. with Sheep, p. 567.)*

GOAT, *Eng.*; CHEVRE, *Fr.*; ZIEGE, *Germ.*; KAPROS, *Gr.*; BECCO, CAPRA, *It.*; CABRA, *Sp.*; KECHI, *Turk.*

Vern.—*Bakra* (he-goat), *bakri* (she-goat), HIND.; *Bakra*, (male), *bakri*, (female), N.-W. P.; *Gharsa* (wild-goat), *chhela* (male), *chheli* (female), PB.; *Bebek, kambing*, MALAY.; *Más, teys, tuyus*, ARAB.

References.—*Hodgson, Sheep and Goats of the Himalaya, Jour As. Soc., Bengal, XVI., 1003-1026; Hutton, Calcutta Jour. Nat. Hist., II., (1843) 514-542; Moorcroft; Vigne; Royle; Southey; Godron; Sterndale, Mam. Ind.; Wallace, India in 1887; Balfour Cyclopædia, India; Morton Cyclopædia, Agri.; Ure, Dict. Arts, etc., Encycl. Brit., etc., etc.*

Habitat.—The goat is now found in a state of domesticity over both Old and New Worlds, and various opinions have been expressed by naturalists as to the original stock from which it is descended. The prevalent and most probable opinion is that the various domestic breeds are descended from several wild species, some of which may be extinct. It was a domestic animal in Asia and Europe before the dawn of history, but was quite unknown in the New World before the advent of the Spaniards.

BREEDS.

Breeds of Goats in India.

Fleece-yielding. 1291

Writers on agriculture describe about twenty-five different breeds of goats in India; but as the distinctions between some of them are but little marked, and as they inter-breed freely with the village goat of the plains, it will be sufficient to describe here the commonest breeds and those in which the leading characteristics are most distinct. From the standpoint of the FLEECE there may be said to be four chief types :—

Shal-Wool. 1292

(*a*) The *pashm*-yielding goats of the upper alpine ranges of the Himalaya, more especially on the northern slopes and in Tibet. The wool or under-coat obtained from this breed is the "SHAWL WOOL" which is woven into *pashmina* fabrics and shawls. As already pointed out, in some respects, this animal resembles the Ibex (**C. sibirica**, pp. 552, 636). An idea of the importance of this fleece and of the manufactures therefrom, may drawn from the fact of the English name "shawl" having been derived from the Persian term *shal*. It would appear that the English word was not generally used till after the middle of the eighteenth century. The creation, in fact, of the British manufactures in shawls is almost solely attributable to **Mr. Moorcroft's** numerous reports on, and specimens of, the Kashmir manufactures of *pashm* and wool.

Pat. 1293

(*b*) The *pat*-yielding goats. The long, soft, mohair-like fleece of this

BREEDS of Goats.

Pat.

breed is made into the fabric known as *pattu*, which may be described as a coarse though durable tweed, largely used by the Natives of the Himálaya for clothing, and the better qualities of which are sometimes worn by Europeans, especially for shooting suits. The various breeds of this goat inhabit the southern slopes of the Himálaya, from the region of perpetual snow, down to the zone of oaks. By some writers the more alpine breeds are simply the *pashm* goat, altered through the moister nature of the southern, as compared with the northern, slopes of the Himálaya; by others all have been derived mainly from the same stock as the Sind and Baluchistan goats. They possess much in common with the "Wild goat," **C. ægagrus**, (*p. 551*) except that on the higher ranges they afford, in addition to the *pat* fleece, a winter coat of inferior *pashm*. Fuller practiculars will be found regarding these goats in the paragraph below on Himálayan goats.

Sind goat.

1294 **Hazara Goats.** *Conf. with p. 638.*

(*c*) *Sind, Rajputna, and Baluchistan goats'-hair-yielding goats.*—This group may be accepted as embracing all the hair-yielding goats of India. They are found on the lower hills up to altitudes of 6,000 feet. The breeds on the higher sections of this area approach the *pat*-yielding goats and those of the lower may be said to be scarcely separable from the ordinary village (non-hair-yielding) animal. The intermediate breeds (between these two extremes) afford the commercial (good) qualities of goats in hair. In many respects the goats of this section may be said to possess less evident derivation from the wild goat of Sind than do the *pat* goats.

Village goat.

1295 (*d*) *The village goat of the plains of India.*—The hair of this animal is too scanty, and withal too coarse, to be of any value. It is only utilizable in the manufacture of ropes, sacks, cheap floor-mats, etc., like the long, coarse hair combed out of the superior fleeces of the above breeds. The village goat of India is in fact reared more on account of its milk, flesh, and skin than its fleece; selection and development have accordingly, for centuries, been directed to these objects, with the not unnatural loss of any value as a fleece-yielder.

The development of the village goat of India and of the neighbouring Asiatic countries, where the objects named have been aimed at. has resulted in quite as diversified a series as can be shown under the fleece-yielding group. Some are tall with long legs, others short and well built. A very extensive range of colours is also met with, but white or black are the most prevalent. Others are grey, brown, chocolate, particoloured or blotched. In the character of the horn an equally diversified range exists. Some have no horns at all, others short, stout horns arching backwards, whilst straight horns may be seen, and horns spirally twisted, some even resembling those of the *markhor* (**C. falconeri**, *Conf. with p. 551*). In the matter of horns it may be here remarked that **Darwin** urged that a correlation exists between the horn and the hair or wool of both sheep and goats. The Angora white goat, with horns, has long curly hair, those without horns have a close coat. In general terms it may in fact be said that the more spirally twisted the horns the more curly the wool. The ears of the goat also afford useful characteristics as marking almost, degrees of domestication. It has been urged by many writers that dependence on man for protection has rendered the possession of erect, mobile ears (to catch every passing sound), unnecessary, and, accordingly, in most domestic breeds, the ears have become pendant. The degree to which this has been carried, as also the size and length of these largely disused organs, is characteristic of certain breeds both of village and fleece-yielding goats. The arching (romanizing) of the nose is another character of much interest. The peculiarities of the eyes have also been regarded as of value, but this is perhaps more in distinguishing sheep

BREEDS of Goats.

from goats than in separating the breeds of goats. No goat, for example, is known to possess the eye-pits which are so striking a peculiarity of sheep. Sheep also have feet-pits, but in goats these exist in the fore-feet only, or are entirely absent. Both **Hodgson** and **Blyth** urged the value of this character in distinguishing joints of meat, by the purchaser insisting that the hind quarter should be sold with the trotter attached—a practice now almost universal in India. It is thus possible to at once be sure if the joint be mutton or goats' flesh. The male goat is always intensely odoriferous, sheep never; but it is not known if there be distinctive characters in the odour of different goats. The mammæ or teats are always two in goats, but according to **Godron** these organs vary considerably in the different breeds. They are elongated in the common milch goat, hemispherical in the Angora race, bilobed and divergent in the goats of Syria and Nubia, etc. Some Indian goats possess teat-like formations on the neck, a peculiarity which **Mr. J. Thomson** observed in the Massai goats of Africa, but which **Prof. R. Wallace** informs the author he has seen in sheep and even in pigs. (*Conf. with pp. 569, 571.*)

ENUMERATION. 1296

Having thus briefly mentioned some of the more striking peculiarities of goats, an enumeration of the chief breeds of India may be now attempted, the remark being premised that with the exception of **Mr. B. H. Hodgson's** valuable paper on the Sheep and Goats of the Himálaya and of Tibet, and **Captain T. Hutton's** paper on the Sheep and Goats of Afghanistan, the subject has never been systematically studied. The writer is, therefore, unable to do more than allude by name to the forms referred to by various writers.

South Indian. 1297

I.—SOUTH INDIAN GOATS.—These are smaller than those found in the north, and have shorter and less abundant hair.

North Indian. 1298

II.—NORTH INDIAN GOATS.—The north Indian goat is a much finer animal than that from the south. Hfs build is more massive, and his hair longer and more flowing. The ears are long, large, and perfectly pendant. The colour of the hair is most frequently black, or black and tan, but some are white, black and white, or with a variety of black, white, and tan spots (*Wallace*). This appears to be the *Jamnapari* goat of **Hodgson** in the remarks below regarding the *dúgú* goats of the Himálaya.

Conf. with p. 565.

Nepal. 1299

III.—NEPAL GOATS.—These have long, flapping ears and rounded or Roman noses, whilst others have hollow or saddle backs. The colour of the hair is black, grey or white, with black blotches.

Bengal. 1300

IV.—BENGAL GOATS.—These are very much like the goats of Madras, but, if anything, they are smaller, more frequently black, and very often destitute of horns. They are never herded in flocks; each villager possesses one or two, and these are allowed to feed as best they can on the roads (picking up leaves, straw, etc.,), on wayside vegetation, hedges, etc. Like all other village goats they are most destructive, as they will eat almost anything and pull out at the roots the plants on which they browse. Their teeth tear rather than cut, hence the injury these animals do to plantations, if allowed access to young trees or shrubs. It will be seen that **Hodgson**, in the account given below of Himalayan goats, indentifies this animal with the *dúgú* goat of the Himálaya (*See p. 565*).

Hill Tracts. 1301

V.—THE GOATS OF HILL TRACTS, such as the central tableland, the Deccan, Sind, Rajputana and Baluchistan. These, in most of their characteristics, resemble the ordinary village goat, but their hair is more abundant and woolly. The horns often large, but only in exceptional or special breeds are they straight and twisted. They are herded with sheep and cattle in large flocks, and feed on the sub-arborescent vegetation of uncultured land, never taking grass, however good it may be, if leaves of bushes or young trees be available. These are the hair-yielding breeds of

SHEEP and Goats.

Domesticated Goats of India.

BREEDS of Goats.

India, but a very extensive series of animals has, by this classification, been lumped together. Until goats have been made the subject of a special study, this defect must continue to exist. The reader may be able to judge of the extent of our ignorance of the breeds of sheep and goats from the interest Hodgson (in 1847) was able to give to his study of the breeds of one tract of India, namely, the Himálaya.

Syrian. 1302 Angora Goats. 1303 Conf. with p. 638.

VI.—SYRIAN GOATS.—Hybrids of this breed, commonly designated Aden goats, are well known in India, whither they are imported by Arab traders and passed off under the name of Angora or Kashmír goats. They have long, flapping pendent ears and slender limbs, and are covered with long, shaggy hair, which, in the pure breed, is of a black colour. Their horns are somewhat erect and spiral, with an outward turn.

Himalayan. 1304

VII.—HIMALAYAN GOATS.—The remarks which the writer has to offer on this subject are simply an abstract of Hodgson's paper, which will be found in the *Journal of the Asiatic Society of Bengal, Vol. XVI, (1847), Pt. 2, pp. 1003-1026.* Mr. Hodgson was well qualified to deal with the subject he discusses, in the paper quoted above (and the original of the article will richly repay perusal), is illustrated with very good drawings of the Himalaya and Tibetan sheep and goats. Hodgson gives the following breeds of Goats :—

Changra. 1305

1ST CHANGRA.- This is the common domestic goat of Tibet—a breed of moderate size, which is distinguished by the uniform abundance of its long, flowing, straight hair, which descends below the knees and hocks, and covers pretty uniformly the whole animal. Even the legs are abundantly clothed, and the head, with its ample forelock and beard, shows the same tendency to copious development of hair. Underneath, especially in winter, the body is also covered with a sub-fleece of exceeding fine wool. This is not a very large animal as its mean height is only 2 feet, but its horns are long ($1\frac{1}{4}$ to $1\frac{1}{2}$ feet) and curved. The *chángrá* is wanton, capricious, restless, impatient of restraint, and in docility far inferior to that of Tibetan sheep, though he is better able to endure change of climate. An attempt to handle him evokes his impatience of all but lax control. He will not submit like his neighbour the *hunia* sheep to carry burdens. He may be bred and herded with facility, but he requires a large range and liberty to please himself whilst grazing.

In the dry cold plains of Tibet, the *chángrá* flourishes, and it is probable there are numerous sub-breeds, the more alpine ones originating the various qualities of *pashm* (see the remarks on this subject below); but although he may be reared in the Cis-Himalayan mountains and even in the lower or central tracts, the *chángrá* loses there his *pashm* fleece. He may also be kept alive in the southern divisions of the Himalaya or on the plains of India, but will not breed when removed to any great distance from his alpine habitat. Hodgson mentions the fact that a Kirghis breed allied to the *chángrá* had been conveyed in safety to Europe, and bred in the alpine parts of France. Bogle, also Turner, and later still Moorcroft, attempted, but failed to convey the Tibetan *pashm* goats to Europe. Even the Kirghis animal just mentioned did not produce *pashm* in Europe and the large sum of money spent in the effort to convey it to Europe and to acclimatise it was thus quite futile."

Hodgson somewhat significantly remarks that the *chángrá* is "closely allied to the celebrated *shawl* goat." It would thus appear that he accepted the true *pashm*-yielding animal as possibly distinct from the ordinary Tibetan goat. It would seem, however, probable, as already suggested, that the superior quality of *Tarfani pashm*" is obtained from at most but a spcial sub-breed of this animal. "The Natives of Tibet, says Hodgson, "manufacture ropes, caps, and coarse overalls out of the long hair, and a fine woollen cloth called *Tús*, out of the sub-fleece, mixed occasionally with the wool of the *silingia* sheep. The flesh of the *chángrd*, especially of the kids, is excellent, and is much eaten by the Tibetans and Cis-Himalayans, even the Hindus of the Central region, import large numbers for food and sacrifices, especially at the *Dasahara*, or great autumnal festival. But upon the whole, the Tibetans prefer the mutton of their sheep to that of their goats."

In general characteristics the *chángrá* is a medial-sized goat, with a fine small head, a spare and short neck, a long yet full body, short rigid limbs, and a short deer-like tail, rather shorter, more depressed, and more nearly nude below, than in the sheep, and frequently carried more or less elevated, especially in the males. The

BREEDS of Goats.

Himalayan.

Changra.

narrow, oblique muzzle is covered with hair: the longish face and nose quite straight, the short forehead, arched both lengthwise and across, and furnished with an ample forelock: and the small brownish yellow and saucy eye placed high up or near the base of the horns. The horns, which are inserted very obliquely on the top of the head, are in contact with their central sharp edges, but diverge towards their rounded posteal faces, and curve upwards, outwards and backwards, with much divergency and with one lax spiral twist, leaving the flat smooth points directed upwards and backwards. The compression of the horns is great, so that their basal section is elliptic or rather acute conoid, and the keel is neither very distinctly separated from the body of the horns, nor does it exhibit any salient knots, but is rather blended laterally into the surfaces, and chiefly indicated by the deflexion of the wrinckles of the horns, which are numerous and crowded but not heavy, and go pretty uniformly round the horns, but form a decided angle at the commencement of the keel. The ears are longish, narrow, obtusely pointed and pendant, with very little mobility. The short strong rigid limbs are supported on high vertical hoops, and have obtusely conic false hoofs, pretty amply developed behind them." Perhaps the most general colour of the *chángrá* is white, tinged with slaty blue. But the white is seldom unmixed, and the limbs and sides of the head are apt to be dark. There are frequently dark patches on the body, and often the whole body is black or tan, the limbs and face only being white." **Hodgson** recognises this animal as a strongly marked derivative from the "Wild goat," **C. ægagrus** (*See pp. 550-51*). He adds that "in *chángrá* there is, in fact, hardly any deviation from the wild type, except in the large and pendant ears; so that domestication would seem to have made less impression on these animals than on the sheep, though its effects on both groups have been less obliterative than is generally supposed."

Captain Hutton who, while resident at Kandahar, devoted much careful study to **C. ægagrus** in domestication and cross-bred it with the common goat, arrived at an emphatic opinion apposed to that advanced by **Hodgson**, namely, that the Persian and Afghan goats, at all events, were not derivable from **C. ægagrus. Hutton** wrote of **C. ægagrus** that it "is rendered interesting from its being now the prevalent opinion among naturalist, that from it have been derived our domestic breeds. The question, notwithstanding is far from being decided, and a few remarks on the subject may, therefore, be considered not unworthy of attention. If **ægagrus** be the stock from which our domestic goats have sprung, it should follow, that the differences which they now exhibit in general appearances, have been induced by domestication; and it is asserted, that the two breeds should be capable not only of freely producing offspring together, but that such offspring should likewise be, capable of breeding *inter se*; yet, on this point, there seems to being something more than a doubt for the offspring of the goats, which was formerly in the Paris Menagerie 'were either prematurely brought forth, or lived only a short time, in a sick or languishing condition.'" **Hutton** then proceeds to detail his experiments which may, by most readers, be regarded as establishing more than **Hutton** believed, the fact that a cross between **C. ægagrus** and the domestic goat is not only possible but that the progeny are likely to be fertile *inter se*. **Hutton** very wisely adds that all past experiments, including his own, have not been repeated sufficiently often to establish the point. He therefore passes to the consideration of other features which are of some interest. The female of **C. ægagrus,** he points out, is altogether destitute of a beard under the chin, whereas the domesticated breeds of Persia and Afghanistan uniformly possess a beard. **Hutton's,** half-bred animal followed its domesticated parent in the possession of a beard, from which circumstance he argues that the beard must have been a specific character in *both* male and female of the type from which the domestic goat had been descended. This contention, he maintained, was strengthened by the fact that the beard was retained even in the second and third generation of his cross-breeds. Further he contended that in **C ægagrus** the horns are very close together at the insertion, whereas those of the domesticated animal are far apart, and in all his crosses the horns were in this respect like those of the domesticated ancestor. The ears also are, he points out, small and erect in the wild animal, while in the offspring of the tame goat and in the tame goat itself, the ears are large and pendant. It may, however, be remarked that apparently all **Hutton's** experiments were from a female of **C. ægagrus** crossed with the domesticated male. The results might have been very different with a male wild animal on the tame female. His original half-bred female was crossed repeatedly by tame goats, and these again similarly crossed. In most of **Hutton's** half-breds the strongest strain was, and naturally, towards the domestic ancestor, except in their timidity and agility.

Conf. with p. 551

SHEEP and Goats.

Domesticated Goats of India.

BREEDS of Goats.

Himalayan.

In these characters alone they resembled their mother.* **Hutton's** observation of the use of the horns in the wild animal has been often alluded to by subsequent writers. Being struck with the immense size and strength of these appendages, he was disposed to think them next to useless, until on one occasion he witnessed one of his males of **C. ægagrus** miss its foothold and prepare for a fall that might have proved fatal. "No sooner did he feel himself falling than he bent his chin firmly down upon his breast, so as to bring his long recurved horns to the front, and *upon these* he received the shock of his fall, without sustaining the slightest injury." What wonder, therefore, that when domesticated and freed from the danger of falls, such as the wild animal must be constantly exposed to, the horns should become less and less necessary and alter materially in shape and form, and even disappear entirely in certain races.

Chapu. 1306

2ND, CHAPU.—This is the *Chyápú* or *Chápú* of the northern region of the sub-Himálaya. "This breed bears the same relation to *chángrá* as the *kágia* sheep do to the *barwál*, that is, it is invariably of much smaller size than the *chángrá*, and has a different habitat, with general similarity of structure and appearance; yet not wanting points of diversity. The ears of the *chyápú* are invariably smaller, and less pendant than those of the *chángrá*; and what is deserving of attention, the feet-pits are not constant in the *chyápú*, but are occasionally wanting, as in the *dúgú*, a species presently to be described." "The *chyápú* is further distinguished from the *chángrá* by the very various flexure of the horns of the former, which are sometimes erect, and sometimes curved backwards in the sickle style; sometimes spirally twisted, and sometimes not so; and, again, the ears of the *chyápú*, always short as compared with those of the *chángrá*, are occasionally so in the extre e, bearing the truncated appearance of the same organs in the *barwál* sheep. Lastly, the *chyápú* is neither so frequently, nor so much coloured as the *chángrá*. The *chyápú* is a small breed fully ⅓ less than the *chángrá*." "The long hair and fine sub-fleece, the ample fore-lock and beard common to both sexes, the sexes both horned, the invariable absence of the eye and groin-pits, the feet-pits present in fore-feet only, the long straight face, short arched forehead, keen and saucy eye, short spare neck, long full body, low rigid limbs, short high hoofs, conic obtuse false hoofs, and short depressed tail; and lastly, the invariable two teats, are marks alike of the *chángrá* and *chyápú*. But the gay and independent look of both is augmented in the lesser breed by the finer and more mobile ear, now erect, now forward, and anon backward, as each internal impulse or external signal prompt." "The females are not much less than the males, nor are their horns very materially less, nor different in form. The prevalent colour is white; but some are mottled or blotched with black or with tan; and the belly and limbs, and a lateral mark down the head from horns to nostrils, are often dark. So too are the ears; whilst the prevalent white colour is frequently flavescent and straw-tinged." "They are of strong constitutions and hardy habits, but love cold and short aromatic pastures, and as these can be found only in the Cachar region of the Cis-Himálayan mountains, to it the *chyápús* may be said to be confined, the immense numbers of them are imported into the central hilly region during the cold months to satisfy the flesh-loving habits of the people of that region, who also occasionally weave the long hair and fine wool of the *chyápú* into appropriate manufactures." "In economic point of view, I apprehend that the *chyápú*, not less than the *chángrá*, is an object well deserving the attention of all those who aspire to benefit their kind or themselves by multiplying the resources and materials of our stupendous manufacturing system." **Hodgson** adds that the *chyápú* would flourish wonderfully in the direst of our hilly countries, in Wales, England or Scotland.

Sinal. 1307

Angora Goats. 1308

Conf. with p 638.

3RD SINAL.—The *sinál* or *sinjál* breed is large and finely proportioned; the breed is, says **Hodgson**, the especial race of the Cachar, where the *chyápú*, though now abounding, is, no doubt, a not very remote immigrant from Tibet. "But the *sinál* now is, and has been for ages, proper to the more northern parts of the sub-Himálaya, including the whole of the northern region and a small part of the central. In these latitudes the *sinál* abounds from the Kali to the Tirsul or from Kumaon to Nepal proper; and probably beyond these limits, both west and east." "The Magars, Gúrúngs and Khas too, rear the *sinál*, whose ample hairy surcoat and fine sub-fleece, though both inferior to those of the *chángrá* and *chyápú* are yet capable of being, and actually are, applied to the manufacture of ropes and

* In connection with the subject of the persistency (or prepotency of the peculiarities of domesticated animals, the reader might consult the opening paragraphs of the chapter below on Domesticated Sheep, pp. 567-570, and again pp. 575-582. It is an accepted principle in breeding that the prepotency of the sire is likely to be stronger than that of the dame.

BREEDS of Goats.

Himalayan. Shal.

of blankets, serges, and caps, and only not more efficiently turned to economic use, because the Gurúngs alone of the above-named tribes are wise enough not to affect contempt of arts mechanical; for all arts, in short, but the glorious one of war!" The *sinál* stands to 2¼ to 2½ feet in height. It is a perfectly typical goat, even more so than the *chángrá*, having the horns less excessively compressed, and the keel more distinct. The long face is straight and the short forehead arched. The oblique small muzzle quite hairy and dry. The largish narrow and pointed ears quite pendant. The moderately compressed horns set on with the full usual obliquity on the top of the head, and in contact at their sharp keeled anterior edge, but separate and rounded behind, with an oval section and medial uniform wrinkling that is carried two-thirds towards the flat smooth tips. The direction of the horns is upwards and outwards with great divergency for a goat, and a single lax spiral turn having the points directed upwards and a little backwards. The neck is spare. The body long yet compact. The females smaller than the males but horned; the horns scarcely spirated. Colours white, or black, or brown, with white or fawn face and limbs; pure white being rarer than in any of the foregoing breeds.

The *sinál* is seldom seen out of his own district, being perhaps less patient of change than the *chángrá* or *chyápú*, and for foreign exportation is inferior to either of them, as well owing to their inferior hardihood, as to the smaller quantity and coarser quality of the fine sub-fleece. The mutton is good and the flesh of the kids greatly and justly prized, being far superior to that of lambs of any breed. The milk also is greatly and justly esteemed. The sub-fleece is frequently absent.

Dugu. 1309

4TH, DUGU.—The *dúgú* is the goat of the central region of the sub-Himálaya. **Hodgson** remarks that the central and lower regions of the sub-Himálaya are unsuited to goats or sheep owing to their rank pasture, excessive moisture, and enormous superabundance of leeches and other parasitic creatures generated by heat and moisture amid a luxuriant vegetation. This opinion is significant, as many subsequent writers have urged that the very region indicated (especially the lower basins of the Ravi, Beas and Sutlej) might, with great advantage, be thrown into immense sheep runs. It seems likely, however, that **Hodgson's** opinion was based more on experience in the central and eastern than the western extremity of the Himálaya. The curse of leeches certainly increases greatly in the more eastern sections of the lower Himálaya, where the rainfall is also much higher than to the west. But **Hodgson's** observation that the goat becomes more a concomitant of village life than an associate of pastoral avocation in the lower reaches of the Himálaya, is certainly true. He, therefore, speaks of the *dúgú* goat as bred only in small numbers by house-holders—and for home consumption of the milk and flesh, both of which are excellent and eagerly consumed by the higher castes. He further remarks that "the *dúgú* closely resembles, and is probably identical with, the ordinary domestic goat of the lower provinces, that of the upper provinces, *viz.*, the large gaunt roman-nosed, monstrous-eared *Jamnapari**, being unknown to these mountains, and unable to endure their climate in any part. The *Jamnapari* becomes in the mountains goitrous, casts its young prematurely, and hardly exists. But the little goat of moist Bengal does very well in the moist climate of the central and lower hills; and accordingly, I believe, that as the upper region of the hills is indebted to Tibet for its goats, so the central and lower regions are indebted to Bengal and Behar for theirs, and that the *dúgú* is at least in origin the common domestic goat of the Gangetic provinces from Allahabad to Calcutta."

Conf. with pp. 574, 617.

Conf. with p. 61.

Hodgson says of the *dúgú* goat that it is distinguished from all the breeds of Tibet and the higher Himálaya by the frequent absence, in the females particularly, of the long hair, and the nearly as frequent absence of the interdigital pits, belonging to these races or breeds. The males, however, of the *dúgú* breed are often as shaggy as the *chángrá* or *sinál*; whilst in the latter, as we have seen, the feet pits are not invariable. "The *dúgú* is of medial size and well proportioned, the male being much larger than the female, and frequently shaggy, whilst she is always smooth. There is no sub-fleece, and the hair is coarse and turned to no use, the skin only being of value when the flesh is disposed of. The muzzle of the *dúgú* is dry and hairy: the face unarched: the forehead considerably so: the ears largish and horizontal or pendant: the moderate horns turned up simply backwards, without spiral twist and with but a vague keel though it be traceable enough in the anteal sharp edge: the neck spare: the body longish yet full: the rigid limbs not short nor long, with high short hooffs and conic false hooffs: and, lastly, medial tail. depressed and nude below, and curvately raised in the males. The eye pits muffled and groin-pits are

* The tract of country bordering the Jamna river possesses the best Sheep and Goats in these provinces. *Conf. with p. 601.*

SHEEP and Goats.

Domesticated Goats of India.

as invariably absent as in the other breeds; and the feet pits more frequently wanting than in any. The beard is ample in both sexes, and the females always have horns and two teats and their hair is close and smooth."

FIBRE. 1310 **Fibre.**—Under the paragraph which is usually isolated by the heading **Fibre** should be described the hair, *pat*, and *pashm*, as also the manufactures from these, with, in addition, goats' skins. Since, however, these subjects are, by most writers, inseparably dealt with in conjunction with "sheep's wool" and "sheep's skin," etc., it has been found impossible to pursue the recognised course which it is customary to follow in this work. The special chapters below on "Pashm and Pashmina," on "Wool," and on "Skins" will, it is hoped, be found to possess sufficient details to meet the wants of the enquirer after the nature of the Indian products of these classes, as also the trade in them.

MEDICINE. Milk. 1311 Bile. 1312 Urine. 1313 Flesh. 1314 **Medicine.**—In Sanskrit systems of medicine, goats' MILK is described as sweet, cooling, and astringent. It is said to promote the digestive powers and to be useful in hœmorrhagic diseases, phthisis, and bowel complaints. The BILE of the goat is used in medicine either alone or in combination with those of the buffalo, wild boar, peacock, and *rohitaka* fish (**Labeo rohia**). Bile is considered laxative, and is chiefly used for soaking powders intended to be made into pill-masses. The URINE is used as a vehicle for the administration of the compound decoction of the root of **Nardostachys Jatamansi** (*q. v.* Vol. V., 338), while the FLESH is said to be easily digested and suited to the sick and convalescent. It enters also into the composition of a *grita*, the properties of which are highly extolled as a remedy for nervous diseases, and of an oil which is employed as an external application in convulsions, paralysis, masting of the limbs, and other diseases of the nervous system (*U. C. Dutt*). According to **Ainslie** "the Vytians have a notion that goats' flesh has virtues in incontinence of urine."

FOOD. Flesh. 1315 **Food.**—Goats' FLESH furnishes good nourishing food, and is often laid on the tables of Europeans in India without the difference between it and mutton being recognised. It is, however, comparatively hard and indigestible. Kids' flesh, on the other hand, is excellent eating and tastes like lamb or veal according to the manner of dressing. The flesh of suckling kids is best, as they have their milk-flesh and are plump and tender.
Milk. 1316 The MILK is rich, sweet, and nourishing, and is considered by some as superior to cows' milk. The globules in goats' milk are said to be smaller than in that of cows', and the milk is thence in a more perfect state of emulsion.

DOMESTIC. Hair. 1317 Undercoat. 1318 Skins. 1319 **Domestic and Sacred.**—In North-West India, the HAIR is used for textile fabrics, ropes, bags to contan grain, and mats. The UNDERCOAT of certain goats is the material of which the far-famed and costly Kashmír shawls are made. Goats' SKINS are tanned and sent to London, and, in recent years, to the United States of America also. They are bought by curriers, dyed and dressed, and are largely employed in book-binding, glove-making, and generally in fine leather work. Goats are largely employed to manure land by folding them during the night on certain areas, and their droppings during the day are, in the vicinity of coffee estates in Southern India, often collected by children for a similar purpose. The goat has a habit of shivering at intervals, and this is taken by the Hindus to be a kind of *afflatus divinus*. A similar notion was prevalent also among the ancient Greeks and Romans. In the North, one of these animals is often turned loose along a disputed boundary line, and where it shivers, there the mark is set up. The Thugs would only sacrifice a goat if their patroness Devi had signified acceptance by one of these tremors.

DOMESTICATED SHEEP.

BREEDS of Sheep. 1320

The following, taken from the *Fauna of British India*, are the distinctive features of sheep (*Conf. with Goats, p. 559*): —

"Tail short in all wild Asiatic forms. Suborbital gland and lachrymal fossa usually present (wanting in **O. nahura**). Interdigital glands present on all feet. Inguinal glands present. No muffle. No beard on chin, but frequently long hair on the neck. Mammæ two. Males non-odorous.

"The structural differences from the genus CAPRA (comprising the true goats) are very small, and one species, **O. nahura**, is absolutely intermediate. Both inhabit mountains and high plateaus, but the sheep keep more to open, undulating ground, the goat to crags and precipices. The flesh of all wild sheep is excellent, the males never having the rank odour that is characteristic of goats." ' The origin of tame sheep is quite unknown."

SHEEP, *Eng.*; BREBIS, MOUTON, *Fr.*; SCHAFE, *Germ.*; FAAR, *Dan.*; SCHAAP, *Dut.*; CASNEINRO, *Port.*; OWZI, *Russ.*; PECORA, OVEJAS, *Sp.*; FAR, *Swed.*; KOYUN, *Turk.*

Vern.—*Bhera, m'henda,* HIND.; *Luk,* TIBET; *Bhéra, méhndá,* N.-W. P. & OUDH; *Bher, chhatra* (male), *bhed* (female), *dúmba,* PB.; *Avi,* SANS.; *Gosfand,* PERS.

References.—*Hodgson, Jour. As. Soc. Beng. l.c.; Hutton, Cal. Jour. Nat. Hist. l.c.; Blyth, Jour. As.Soc. Bengal, l.c., also Zoo. Soc. London; Buchanan Hamilton, Account Kingdom Nepal, also Journey through Mysore, etc.; Royle, Prod. Res. Ind, articles Wool and Sheep; Baden Powell, Pb. Prod.; Shortt, Man. Ind. Cattle and Sheep; Wallace, India in 1887; Simonds (Pests), Jour. Royal Agri. Soc. I., new series, 1865; Burnes, Travels in Bokhara; Erman, Travels in Siberia; Youatt on Sheep; Southey, Colonial Sheep and Wool; Bischoff, Wool, Worsted and Sheep; Brown, British Sheep Farming; Coleman, Sheep and Pigs of Great Britain; Fream, Elements of Agriculture; Wallace, Farm Live Stock; Balfour, Cyclopædia India; Spons' Encyclopædia; Encyclopædia Britanica; Morton, Cycl. Agri.; Ure, Dict. Arts and Manufactures, etc., etc.*

Habitat.—Domesticated throughout the plains and lower hills of India, and up the Himálaya to sub-arctic zones. Although nothing definite has been published regarding the breeds of Indian sheep, it may safely be said that quite as extensive a diversity exists as in Europe, an assertion that will at once be realised when the immense size and the wide range (in climate, soil, and pasturage) of the vast empire is taken into consideration. There are breeds that are tropical (of which some thrive in swampy regions, others luxuriate on what may be called sandy deserts), also warm temperate, temperate, and arctic races. Indian zoologists are now agreed, however, that the notion held formerly cannot be accepted, that the Indian races of domesticated sheep are descended from the great wild sheep of the higher Himálaya and Tibet. Nothing, therefore, is known of the origin of the Indian sheep. **Blyth** was disposed to regard the fighting ram of India as possibly derived from **O. vignei**. **Hutton**, while repudiating any idea of the Afghan domesticated sheep having been derived from **O. vignei**, says that, however much pasturage and other agencies of domestication might fatten the tail, these could not add *several vertebræ* to it.* **Hodgson** speaks of the *barwál* sheep, "as the hero of a hundred fights," whose extraordinary massive horns show a normal approximation to the wild type. In fact **Hodgson** regarded all the Tibetan and alpine Himálayan sheep as descended from **Ovis hodgsoni**. **Blanford** (*Fauna British India*) says of **O. vignei** that it has been bred freely with tame sheep.

Fighting Rams. *Conf. with pp. 568, 571, 573, 583.* 1321

Breeds of Sheep in India.

BREEDS. 1322

As remarked regarding Goats, very little of a definite character has been written about the sheep of India. Indeed, **Hodgson's** paper on the Himálayan Sheep and Goats is the only scientific treatise on the

* *Conf. with p. 575.*

BREEDS of Sheep.

subject. **Buchanan-Hamilton** furnished certain particulars regarding the sheep of Mysore and of Nepal, and **Shortt's** *Manual of Indian Cattle and Sheep* has added a few particulars Most writers have contented themselves, however, by saying this and that regarding certain breeds, such as the Patna, Dumba, Meywar, Madras, Mysore, etc., etc., without apparently having considered it necessary to detail the characteristic features of the animals so designated. **Shortt's** account, if the illustrations can be viewed seriously, would seem to establish for South India certain well-marked breeds, and, doubtless, extensive diversities exist among the sheep of other provinces. But until an attempt has been made to study comparatively and to classify all the Indian breeds, the writings of isolated observers must be largely unintelligible to persons not intimately acquainted with the particular locality to which such special papers relate. Speaking generally, it may be said of, perhaps, more than half the breeds found on the plains of India, that they afford a kind of hair rather than of wool. They are reared chiefly on account of the mutton they afford, their fleece, like the hair of the village goat, being, comparatively speaking, valueless. In many respects, in fact, they approximate more nearly to the accepted type of the goat than of the sheep, and, as **Shortt** remarks of the Madras breed, they "resemble a greyhound with tucked-up belly, having some coarseness of form; the feet light, the limbs bony, the sides flat, and the tail short." In several of **Shortt's** pictures of the breeds of South India, the rams have a long mane extending almost to the knee, while the rest of the body is, comparatively speaking, naked. Indeed, the "hornless ram with mane" on plate 14 of **Shortt's** *Manual* would very probably be designated a peculiar diminutive bull rather than a ram, with, for the size of body, long antelope-like legs The ewes, in some parts of India, are often valued on account of their milking properties, and are thus to be found not in herds but as the solitary associates of village life, taking the place of the milch goat or associated with it. In the advanced agricultural doctrines of Europe, the presence of horns may be said to be regarded as proof of inferiority, except, in the case of special breeds, but in India, it is the rule, rather than the exception, for rams to have horns. The horned sheep of India are, at all events, not characteristic of, nor confined to, hilly country. Unless superiority in the tropics should be determined in the future therefore, as governed by altogether different principles than in temperate countries, the vast majority of the breeds of India would have to be accepted as manifesting the entire absence of culture and selection. But such a conclusion would only be in keeping with the accepted notions of Indian writers on this subject; indeed, the only selection that can be said to have taken place has been directed to perpetuating and developing the horns. The chief interest taken in sheep by the nobility of India has, for centuries, been in the possession of pets employed as fighting rams. The formation of a large head, massive horns, formidable mane, and long, powerful hind legs might be accepted as direct adaptations towards that purpose. In fighting the ram rushes at its adversary with great impetuosity, raises itself like the fighting goat on its hind legs, and falls with a crash that often destroys the horns and even fractures the skull of its adversary. And what is still more remarkable, it is sometimes seen to develop a propensity to bite and to strike down its antagonist by the fore-feet. The shepherd never isolates the rams from the ewes of his herd, and if he exercises any control over the progeny, it is in favouring rams with large heads and powerful horns. The romanising of the nose is a character less marked in plains than in hill sheep. The ears are often very large and pendent, and the tail exceptionally small, except in the sheep of arid tracts, which assume the condition of the so-called *dumba* breed. One other feature of the sheep of many parts of India must not be

Conf. with p. 635.

Fighting Rams.
Conf. with pp. 567, 571, 573.
1323

BREEDS of Sheep. *Conf. with pp. 561, 571.*

omitted, for, although never satisfactorily explained, it is too frequent to be devoid of significance. From the throat dangle "two long rounded pendulous lobules, from two to three inches in length," much after the same fashion as has been noticed regarding certain goats So far as the author can discover these (externally) goitre-like excrescences have never been investigated. They do not appear to be indicative of peculiar breeds, though they are more frequent in the sheep of certain tracts of country than of others.

But although many of the sheep of India yield a fleece of hair rather than of wool, certain breeds give fairly good wool. Of this class may be mentioned the black-headed sheep of Coimbatore, the woolly sheep of Mysore, the sheep of large portions of the Deccan, of Rajputana, of the Panjáb, and in Bengal and the North-West Provinces the so-called Patna sheep. Although the writer believes that there are possibly several very distinct breeds of large fat-tailed sheep (all designated *dumba*), these should be classed as wool-yielding breeds. They have been crossed with the Patna breed, with the merino and other imported sheep, and apparently with satisfactory results, though the improvement effected cannot be said to have been lasting. Some of the fine wools imported from Afghanistan and Persia are obtained from the breeds of dumba sheep, and this fact having been ascertained many years ago effort was put forth to secure stock of these sheep for breeding purposes. So far the result, however, has been unsatisfactory, for, when conveyed to the moister tracts of India, the fat-tail has been proved a source of danger. It is liable to disease, so that unless a breed could be produced in the natural habitat of this animal that would preserve its merit as a wool producer during successive crosses in which it was gradually developed into a condition suitable to the plains of India generally, it is not likely to be of much value to future breeders. It may, in fact, be said in conclusion that, so far as past experience goes, the breeds of most value as Indian stock for improvement are the Coimbatore, Mysore, Rájputana, and Patna. But it may be added that, perhaps, the majority of persons who have given this subject anything like careful consideration seem to incline to the view that except in certain tracts, there is very little hope of India as a whole, becoming of much greater moment than at present as a country of wool supply. Interest is far more keenly directed towards facilitating importation from the mountainous countries bordering on India, than in any material improvement of the wools of the plains. That these wools can be improved there is probably little doubt. Greater cleanliness in baling, more care in assorting, and the development of white in preference to parti-coloured or black stock would greatly improve the wool trade of India. But that India can ever hope to compete, say, with Australia in wool production, would seem a pure hallucination which could only be entertained by persons ignorant of the high temperature and extreme humidity of vast tracts of India. The questions, therefore that seem worthy of solution are—*1st*, the possibility of educating the shepherds in the notion of advantage from improvement within their power even now; and, *2nd*, when this has been attained, the desirability of extending the helping hand towards them in the supply of acclimatised and permanently improved stock. To expend large sums in the distribution of pedigree rams would seem the least hopeful course, for, unless these gifts are periodically repeated and for many years, the progeny are likely to acquire only a weakness of constitution calculated to operate in the ignorant mind more prejudicially than otherwise. In the writer's opinion a better experiment, and one that might not only, by example, lead the Native shepherds' towards self-help, but would afford the stock from which future advances might be made, would be for Government (in the absence of private enterprise) to own large herds of sheep in certain selected tracts.

Conf. with pp. 579, 617.

Domesticated Sheep of India.

BREEDS of Sheep.

Crossing.

1324 *Conf. with pp. 576, 580, 585, 588, 591, 594, 601, 609, 611, 613, 617, 618 637, 852.* The sheep in each case should be the local breed. By selection and elimination the flock might easily be brought to the condition of white wool yielding. The produce might then be sold in the open market and the sums realised freely published. After years of this experiment, and when others had been induced to follow in the new system, the time would arrive for the further step of crossing the various native breeds, and even for experimenting with foreign breeds. The course which has far too frequently been pursued in India might not inaptly be characterised as similar to an attempt to improve the breed of horses by crossing the costermonger's apology for that animal by the most expensive pedigree race-horse. No progression can be made till the defects of the indigenous stock are first eliminated. The records of breeding in Europe abundantly establish the necessity and utility of this dictum. It has been said of certain rearers, that they have taken a flock of the most depraved and mongrel character,

Prepotency.

1325 *Conf. with pp. 575-78, 637.* and within a very few years elaborated by "weeding" and careful crossing within the flock, a stock of high merit and robust constitution. No such experiment, so far as the records of Indian sheep-rearing testify, has ever been performed in this country. With sheep, as with tea, sugar-cane, and nearly every agricultural product that has secured recognition by the Europeans in India, the indigenous stock has been wastefully ignored. That such a course was justifiable may be admitted at first sight, when the great inferiority of the Indian stock, as compared with the European, American, and Australian triumphs of scientific agriculture are taken into consideration. But that acclimatization of exotics is the only, or indeed the most direct, way to improvement, surely no one will uphold who has given the study of animals and plants under domestication even the most casual consideration. Improvement to be lasting must work from indigenous towards exotic stock. This being so, the necessity for full particulars regarding the indigenous sheep of India will be recognised as the first and most natural step. All that can at present, however, be furnished is a few jottings under the names used to designate certain breeds.

Rajputana.

1326 I. RAJPUTANA (MEYWAR) SHEEP.—These are the finest and largest sheep in India, and many of them are annually sent in droves to different parts of Upper India for sale, so that the same breed is frequently designated Delhi, Hansi, and Tattyghar sheep. They have a poor wool, but the mutton is large and they get fat quickly. Their flesh is, however, somewhat coarse.

Bengal & Patna.

1327 II. BENGAL AND PATNA SHEEP.—The former is very inferior, but the latter is one of the best Indian breeds. Patna sheep are light-fleshed, but with wool of fair quality (*Conf. with pp. 617, 635*). They come early to maturity and are good and rapid feeders They are exported over most parts of Bengal, and even to some of the Madras districts, and the rams are much used for improving other breeds. *For crossing with the Dumba and other breeds Conf. with pp. 575, 580, 584, 586, 587, 589, 617, 618.*

Madras.

1328 III. MADRAS SHEEP —This breed is found in Chingleput, parts of Kistna, Godavery, Ganjam, Arcot, Salem, Trichinopoly, Tanjore, Madura, and Tinnevelly districts. These sheep seldom exceed 22 to 28 inches in height, and are covered with short coarse hair, the prevailing colour of which is red or brown. A number of them have black heads, legs, and bellies, and broken colours also appear. Many have, like the Nellore sheep, pendulous lobules hanging from the throat. A variety is sometimes met with in which the rams are hornless, and the throat and foreneck covered with a thick shaggy coat of hair extending like a frill from the throat to the breast and often reaching to the knees. Neither the Nellore nor the Madras breed furnishes wool or hair fit for textile purposes (*Shortt*).

Conf. with p. 569.

Nellore.

1329 IV. NELLORE SHEEP.—A breed of sheep of very large size is found in

BREEDS of Sheep.

Nellore. It, however, differs but very slightly from the red sheep of Madras generally. A good specimen may stand 30 to 36 inches in height, and if well fattened will scale when alive from 80 to 100lb. It is, however, rather tall and leggy. The prevailing colour is white or a light brownish white with black points; the body is well covered with short fur, and a light frill of hair frequently surrounds the throat and front of the chest in the males. Some are said to have two long rounded pendulous lobules, from 2 to 3 inches in length, hanging side by side from the throat. The tail is short. The ram has twisted horns of moderate size; the ewe has no horns (*Shortt*).

Conf. with p. 569.

Coimbatore. 1330

V. Coimbatore Sheep.—This is known as the *kurumba* breed. It is a wool-producing sheep. The animals belonging to it are small, the rams seldom exceeding 26 and the ewes 22 inches in height. The prevailing colour is white with a black head. They have very fair fleeces, the staple being from 4 to 5 inches long. The fleece usually weighs from 1 to 2lb, seldom over 3lb. The rams have long twisted horns; the ewes are hornless. They fatten well and the mutton of gram-fed animals is exceedingly rich and well tasted. The weight of the live animal ranges from 50 to 60lb, and is very seldom over 80lb. **Professor Wallace** says of this breed that he saw specimens with half or even the entire coat black. He observed stray specimens of the breed now and then far up in the Southern Maratha country. He adds, "Where the character of the land to the east changed abruptly to hard and poor soil, the breed of sheep changed with it to the inferior, but no doubt hardier, brick-brown variety."

Mysore. 1331

VI. Mysore Sheep.—This also is a woolly breed. The prevailing colour is from a light to a very dark grey or black. The rams stand about 25 inches and the ewes about 23 inches in height, and the ordinary live weight is from 40 to 60lb, but gram-fed wethers may scale up to 80lb. The fleece never exceeds 3 or 4lb in weight, and the staple averages 3 to 4 inches in length. The rams have large heavy horns, wrinkled and encircled outwards, and with the points directed inwards and forwards. The ewes are usually hornless, but some have light horns seldom exceeding 3 or 4 inches in length. This breed furnishes the best fighting rams of the plains of India and for this purpose they are much sought after by Rajahs and Chiefs. These rams with good feeding often attain a height of 30 inches and a weight of over 100lb.

Conf. with pp. 567, 573.

Bombay. 1332

VII. Bombay Sheep.—The reader will find a reference to the Deccan sheep and to the efforts that were put forth to improve the stock fifty years ago on page 579. One of the earliest notices of Deccan sheep, which the writer has discovered, is the brief mention of them by **Dr. Hove** in 1787. He says of the people of the Deccan: "Of sheep they had some, and they were the finest that I saw in India, with long wool, which was so soft and white as the finest Guzerat cotton. The inhabitants make their winter covering from this wool, and although they are made up together of a thick texture, yet remarkably light in proportion. I am rather surprised that nobody, either at Surat or Bombay, took notice of such a valuable article and introduced it into their settlements, which might, in time, become a great article of trade." **Professor Wallace** says that he found "that in about twelve hours' rail from Madras, in the direction of Bombay, large black sheep predominate and are numerous, especially in the neighbourhood of low rocky hills." "To the west, in the South Maratha country, sheep are mostly black, but white patches, and even white sheep, appear at times." (*See Trans. Agri.-Hort. Soc. Ind., VII., 114.*)

Nepal. 1333

VIII. Nepal Sheep.—These are of two kinds—the *ghorpalla* or village sheep, which are horned and a few of which are kept in each village. They are larger in size than the Tibetan sheep, but their flesh is coarse. They

BREEDS of Sheep.

fatten more readily than Tibetan sheep, and are imported largely into Darjiling and Jellapahar. The wool is of a coarse, hairy quality. The other breed in Nepal is known as the *ranbaria*. These run in flocks on mountains and forests, and are almost wild. They are smaller than the *ghorpalla*, with common looks, common wool, and coarse mutton (*Journ. Agri. Horti. Soc., Ind*). Dr. **Buchanan-Hamilton** figures the sheep used on the higher ranges of Nepal for carrying loads. It appears to be the same animal met with all along the Himálaya, and which, in Tibet, is often called the *bisa*. It has four horns, the middle pair erect and divergent, like those of some goats, and the lower reflexed, with the tips curving in towards the eyes. The face is exceptionally long, black, and the nose very much arched. This appears to be the *húniá* sheep, more fully described below under the section of Himalayan and Tibetan sheep.

Himalayan and Thibetan. 1334 *Conf. with pp. 608, 611.*

IX.—HIMÁLAYAN AND TIBETAN SHEEP.—Mr. **Hodgson** gives the following particulars regarding the breeds of this region:—

Hunia. 1335

1ST, HUNIA.—This is the *húniá* of Western and the *hálúk* of Eastern Tibet. It is a tall graceful animal, black-faced and polycerate (=many-horned), and is the universal beast of burden on the higher snowy ranges, being docile and sure-footed. It appears to be the *bisá* sheep already alluded to under Nepál. It is a rather large animal 4 to 4½ feet in length from the snout to the vent, and 2½ to 2¾ feet in height. The maximum length of the horns is 18 to 20 inches; they are present in both sexes, or, at all events, rarely absent from the female, always present in the male. They are much attenuated and consequently separated at the base, triangular, compressed, transversely wrinkled, and curve circularly to the sides, so as to describe two thirds of a circle with the smooth flat points again reverted outwards and sometimes backwards, so much so as to describe a second nearly perfect circle. But this perfect cork-screw twist is only seen in advanced age. The moderate-sized head of the *húniá* has great depth, moderate width, and considerable attenuation to the fine oblique muzzle which shows not the slighest sign of nudity or moisture, and has the narrow nostrils curving laterally upwards. "The nose is moderately arched but more so than in the wild race, and the forehead is less flat and less broad than in the argalis, being slightly arched both lengthwise and across." It will thus be seen that **Hodgson** compares this breed to the Great Tibetan Wild Sheep (**O. hodgsoni** or **O. ammon** as it was formerly known—the **Argali** of **Pallas**. "The longish, narrow, and pointed ears, **Hodgson** continues, differ from those of the wild race, only by being partially or wholly pendant, whereas in the wild race they are erect or horizontal, and much more mobile." The eyes of the *húnia*, are of good size and situated near to the base of the horns and remote from the muzzle. The neck is rather thin and short. The body moderately full, and elongate. The limbs long and fine, hardly less so than in the wild race. The hoofs compressed and high, the false hoofs small and obtuse. The feet pits are common to all four feet and provided with a distinct gland which yields a specific secretion, which is viscid and aqueous when fresh, candied when dry, and nearly void of odour. "Not so the secretion of the groin-glands, which in the *húniá* are conspicuous, and yield a greasy, fetid, sub-aqueous matter, which passes off constantly by a vaguely defined pore, quite similar to that of the axine deer but less definite in form than in the true antelopes" "The possession of these organs has been denied to the sheep by most writers." The tail of the *húniá*, continues **Hodgson**, is invariably short, though less remarkably so than in the argalis yet still retaining the same essentially deer-like character. "It is cylindrico-conic, and two-thirds nude below, differing little or not at all from the same organ in the several other tame races of these regions, where long tailed sheep are never seen till you reach the open plains of India; and, as upon those plains not only are *all* the sheep long-tailed, but *dumbas* or monstrous tailed sheep are common, whilst the latter also are totally unknown in the hills, it is a legitimate inference, that this caudal augmentation in most of its phases is an instance of degeneracy in these pre-eminently alpine animals." It is, therefore, he adds, "vain to look in the wild state for any prototype of the macropygean breeds, how great soever be the historic antiquity of the *dumbas*."

This fine breed, says **Hodgson**, is characteristic of extreme docility, superior size, graceful form, slender horns (often four or more), and by the almost invariable mark of a black face. They are nearly always white, a wholly black sheep in this breed being unknown. This genuinely Tibetan race cannot endure the rank pasture or high temperature of the sub-Himálaya south of the Cachar, or juxta-nivean region of these hills, where vegetation and temperature are European and quasi-Arctic. It does well in the

BREEDS of Sheep.

Himalayan. 1336

Cachar, and may be fattened or bred with care in the central region at altitudes not under 7,000 feet, where the maximum temperature in the shade is about 70°. "It is a hardy animal, feeding freely and fattening kindly. Its mutton and its fleece are both excellent in quality and very abundant in quantity, so that I should suppose the animal well worthy of the attention of sheep-rearers in cold climates. The wool is of the kind called long staple and has been valued at 8 pence per pound (*see Jour. Agri.-Hort. Soc., V., pt. IV., p. 205*).

Silingia. 1337 *Conf with pp. 585, 589-90.*

2ND, SILINGIA.—The *siling* sheep or *pelúk* of Eastern Tibet and of Siling. Eastern Tibet, says **Hodgson**, is the Kham of the Natives. It is a vast plateau, less elevated, less rugged, less cold, than the central section. Towards Assam, for example, in the valley of the Sánpú (or Brahmaputra) rice is grown. Siling or Tangut is a colder and loftier tract of country than the Kham generally, but the mean elevation of the Kham or the home of the *siling* sheep may be put at 7,000 to 8,000 feet. **Hodgson** regards Siling as identical with the *Serica regio* of classics. Serica or Sinica is, he says, Siling vel Sining vel Sering inclusive of Kham, a country of great celebrity, open to China by the Hoangho, and to India by the Sánpú, and to Western Asia and Europe by all the high plateau of high Asia. The reader might in this connection consult the remarks under **Silk** regarding the *Serica regio.* (*Conf. with Vol. VI, Pt. III, 22.*)

The *silingia* or sheep of Siling is nearly as common as the *húniá* in Kham, but less so in Utsang, and nearly or quite unknown in Nari, where the *húniá* most abounds. It is a delicate breed, both in structure and constitution, compared with the *húniá*, and though it will live and procreate in the Cachar or northern region of the sub-Himálaya, it is rare there and unknown south of it. In Nepal I procured my specimens from the Court, which imported them from Lassa: in Sikkim from the Barmúkh Rajah, who procured them from Kham. All parties with whom **Hodgson** had dealings extolled highly the unrivalled fineness of the fleece, from which the Chinese and the people of Siling manufacture *tús* and *málidah*, or the finest woollens known to these regions, save such as are the produce of European looms. The wool has been examined by competent authority, and is declared to be of shorter staple than that of the *húniá*, but suitable for combing, and worth in the market about the same price as the *húniá* fleece. Of the merits of the mutton, the Tibetans and people of Sikkim laud the flesh as highly as they do the fleece.

The animal is. very similar in general appearance to the *húniá*, but is somewhat smaller as well as of slighter make. Head moderate-sized with the nose considerably but not excessively arched, and somewhat slender, trigonal, compressed and wrinkled horns, curving circularly to the sides, but less tensely than in the *húniá*, and the flat smooth points reverted backwards and upwards. The ears are fairly lengthened (4 to 4½ inches) and pendant and the deer-like tail slightly elongated. The colour is usually white, but sometimes tinged with fawn especially upon the face and limbs; black is perhaps less rare in this than in the *húniá* breed. Hornless females are fairly frequent.

Barual. 1338

3RD, BARUAL.—The Barwal is a cis-Himálayan breed, and the ordinary sheep of the Cachar or northern region of the sub-Himálaya, where immense flocks are reared by the Gúrúng tribe, in all the tracts between the Júmla and Kirant. **Hodgson** says it extends in fact from Kumaon to Sikkim, or even still further beyond the western and eastern limits. It is especially the breed of the northern cis-Himálayan regions and although its strength enables it to live pretty well in the central region, yet it is seldom bred there, never in the southern region of the hills nor on the plains of India,. the heat of which it could probably not endure. The *barwal* (*barúál*) is the great fighting ram of the hill tribes of India. The hero, says **Hodgson**, of a hundred fights, it has great courage, vigorous frame, superior size, and enormous horns, covering and shielding the entire forehead. He is thus more than a match for any foreign or indigenons breed of sheep, and a terror even to bulls. In point of size it is slightly inferior to the *húniá*, but greatly superior to it in build, massiveness, and weight.

Fighting Rams. 1339 *Conf. with p. 567.*

The *barúál* is singularly remarkable for his massive horns, huge roman nose, and small horizontal truncated ears, pressed down by the horns in the adult male, and seeming as if the end had been cut off. The head is large, with a small golden brown eye; the nostrils narrow and oblique, showing faint symptoms of the nude muzzle, like "the wild argalis of Tibet;" neck is short and thick; the barrel compact and deep; and the limbs "supported on high short hoofs" are rather elevated, strong and perpendicular. Both sexes have horns, not a tithe of the females being void of them, and the males scarcely ever without them. The horns are inserted without obliquity, and in contact on the crest of the frontals or top of the head, which they entirely cover, and they are directed to the side with a more or less tense and perfect curve, which in old age is sometimes repeated on a smaller scale; but ordinarily the

SHEEP and Goats. **Domesticated Sheep of India.**

BREEDS of Sheep.

Himalayan.

smooth flat tips are directed outwards; "the cross furrows or wrinkles of the burwal's horns are as decided and heavy as in its wild prototype,'

"The flesh and fleece are both very abundant but coarse, well suited to the wants of the lusty, rude, and unshackled population of the Cachar, but not adapted probably for foreign exportation or exotic rearing. By far the largest number of the *ráhris* or coarse blankets and serges, manufactured in the sub-Himálaya, and extensively exported therefrom for Native use, in the plains of India are made from the wool of the *barúál*, which, likewise, entirely and exclusively clothes the tribes who rear it, and make the rearing of it their chief and almost sole occupation. The Gúrúngs, especially, are a truly shepherd, though not a nomadic, race, and they, it is principally, who breed the *barúál*, feeding their immense flocks near the snows in the hot weather, further off the snows in the cold weather, but never quitting their own proper habitat as well as that of their flocks, and which is the northern division of the sub-Himálaya. Coarse as is the wool of the *barúál* it is very superior to that of the Indian plains, and being of the long stapled kind, the animal might possibly prove a valuable addition to our European stores, either for the wool or for the flesh market, the *barúál* being of a hardy constitution, averse only to excessive heat, and feeding and fattening most kindly. The colour of the breed is almost invariably white; but reddish or tan legs and faces are sometimes found.

Cagia.
1340
Conf. with pp. 565, 617.

4TH, CÁGIA (KÁGIA OR KÁGYA) —This is the characteristic breed of the central region of the sub-Himálaya so far as that region can be said to have a breed for its rank pasture, and high temperature are very inimical to ovine animals. "There are few sheep in the central hilly region, and none in the lower, till you reach the open plains, and there is found a widely diffused breed, quite different in its superficial characters from any of the hill ones. What sheep are reared in the central region of the hills are of the *cágia* breed, but rather by householders than by shepherds, and for their flesh rather than for their wool. The *cágia* is a complete *barúál* in miniature: yet like as the two breeds are, each has its own region, nor does the great difference of size ever vary or disappear. Nor are there wanting other differential marks, such as the full-sized, pointed, and pendant ears of the *cágia* and its shorter stapled and finer wool. The *cágia* is a small, stout, and compact breed, possessed of great strength and soundness of constitution, impatient only of heat, and that much less so than the preceding breeds, eminently docile and tractable, affording mutton of unequaled quality, and wool not to be despised, yet to be praised with more qualification than the meat." This is the animal seen around Himálayan hill stations and which affords the better qualities of mutton eaten by the European residents; but the Tarai and even plains sheep are also driven up to the larger hill stations, where a large demand exists for mutton. The wool is of short staple but considerable fineness, though inferior very much to that of the *silingia* and somewhat to that of the *húniá*, but superior to the wool of the *barúál* in fineness, though not equal to it in length of fibre. The people of the central Himálaya, to which the *cágia* sheep is more especially restricted, dress almost entirely in cottons, and consequently do not much heed the fleece of their sheep. In Nepal, however, the Newars manufacture its wool into several stuffs and often mix it with cotton.

The *cágia* sheep is a handsome breed, but the head is too large and the legs too short. The eye is small and pale caloured; the ears longish, pointed, narrow and pendant; the body is full and deep; the tail short and deer-like; the nose only less romanised than in the *barúál*; and the horns, only inferior in thickness to that of its more alpine neighbour. The *cágia* is thus but slightly less armed than the *barúál*, but he is rarely used as a fighting ram. The colour is very generally white. Some few are black or ochreous yellow, and the young are apt to be of the last named hue, but turn white as they grow up. The males are almost invariably horned, and the females frequently so. Polycerate varieties seem unknown in the *cágia* as in the *barúál* breeds, though frequent in the *húniá* and by no means unusual in the *silingia*. **Hodgson** adds that while females, in all these breeds may be polled or not, they uniformly manifest a character that is remarkable, *viz*., the nose in none of the breeds is romanised in the females. The presence of two teats **Hodgson** gives as generic in the sheep and goats, but he states that in the *cágia* sheep four teats are by no means unusual.

Tarai.
1341

5TH, THE TARAI SHEEP, **Hodgson** says, is identical with that found all over the Gangetic provinces, and is characterised by medial size, black colour, a very coarse but true fleece, frequent absence of horns in one or both sexes, a nose romanised amply, very large drooping ears, and a long thick tail frequently passing into the monstrous *dúmba* "bussel."

In conclusion it may thus be repeated that **Hodgson's** trans- Himálayn sheep (the *hûniá* and the *silingia*) are, like his trans-Himálayan goats (the

INDIAN EXPERIMENTS in BREEDING.

chângrá or *chápu*), far superior to his cis-Himálayan (or Indian alpine) breeds of these animals. Efforts at breeding should mainly, therefore, be directed towards acclimatising on the southern or (Indian) slopes these Tibetan breeds, or of securing a larger supply of the Tibeten fleeces than as yet reach India. Any idea of expanding the Indian wool trade would, as it seems, be greatly retarded by ignorance of the breeds briefly indicated above If India cannot acclimatise and develop new Himálayan breeds from the *hunia* and *silingia* sheep, it should, by increased facilities of transport and friendly intercourse, endeavour to draw on the large supplies of Tibetan wool.

Hira & Dumba. 1342

X.—HIRA AND DUMBA SHEEP.—These are frequently brought into Northern India, and a few are sometimes imported into Madras as curiosities They are large sheep, but their chief peculiarity is the development of great masses of fat on either side of the tail, or at its root. From Afghánistán to Persia is said to be the habitat of these animals. Their tails form a reservoir, whence a store of nourishment is drawn during the winter months when fodder is scanty; but in the warmer climate of India they frequently are troublesome, as the tails, if not attended to and kept clean, are apt to ulcerate and become infested with maggots. It is said that in some parts of the country the tail grows to such a size that the animal is not able to carry it. A small wheeled carriage is, therefore, constructed to carry the weight which the animal drags after it wherever it goes (*Hutton*).

Cross the Dumba. 1343 *Conf. with pp.* 579, 585, 536, 587, 608, 611.

Hodgson calls this the *púchia* sheep (*púch* a tail), and says that its essential structure conforms entirely to his definition of the genus, whilst its deviations in subordinate points from the wild and tame sheep of the mountains, are due entirely to domestication. **Hutton**, on the other hand, believed the number of *vertebræ* in the *dumba* precluded it from being descended from **O. vignei** at least—the wild sheep of Afghánistán. The mutton from these animals is said to be very coarse.

Improvement. 1344 *Conf. with p.* 567.

Improvement of the Breeds in India.

In continuation of the remarks already made (*pp.* 567—570) on the improvement of sheep, it may be said that **Hodgson** briefly details the effects of domestication of sheep to be, "to augment exceedingly the size of the tail, in length and thickness, one or both, to increase the size and destroy the mobility of the ear, and to diminish the volume of the naturally massive horns, until they gradually disappear in one or both sexes; the romanising of the nose, out of all proportion to the 'modesty of nature,' as seen in the wild state, being a further and hardly less uniform consequence of domestication." It does not, however, follow that all these modifications have taken place in any one breed. Thus, for example, in a large number of the European breeds, the tail has by no means been either lengthened or thickened. The peculiarities of a breed when once acquired are, however, remarkably persistent or, as it is technically called, "prepotent." **Darwin** alludes to this fact repeatedly, and in connection with the Manchamp and Ancon breeds he says, had these "originated a century or two ago, we should have had no record of their birth; and many a naturalist would no doubt have insisted, especially in the case of the Manchamp race, that they had each descended from, or been crossed with, some unknown aboriginal form." It would thus appear that the racial characters of domesticated animals might almost be said to manifest very nearly as strong a persistency as most of the specific distinctions of wild species. And what is more remarkable the races of sheep, for example, when mixed together, exhibit a pronounced clanishness that is inimical to spontaneous crossing. The members of a particular race seek out and prefer each others company to that of any others of the herd. They also continue to

Prepotency. *Conf. with pp.* 569-70, 577, 637. 1345

Improvement of the Indian Breed of Sheep.

INDIAN EXPERIMENTS in BREEDING.

select the food most nearly akin to that on which the breed was first reared, so that a mixed herd of different races, if left free on the pasturage, breaks up naturally into sections which correspond to their racial distinctions. Crosses between breeds also manifest similar peculiarities to what may be seen in hybrids between wild species or between a wild and tame ancestor. The peculiarities thus produced would not, for example, be expected to be so persistent in the progeny, as the spontaneous characters that have appeared and been nurtured in the development of breeds. It may in fact be said that the peculiarities of crosses tend rapidly to return to those of the ancestor of greatest prepotency, and rarely assume racial or fixed proportions until after prolonged and repeated crossing and careful selection.

Selection.
1346
Conf. with pp. 570, 581, 591, 594, 611.

The birth of races is thus due mainly to selection from useful variations and to the crossing of such variations, within a breed, until the characters desired become fixed and developed into what, in time, is recognised as a new breed. Crosses between distinct breeds, which are intended for the butcher, are mostly repeated in each individual, instead of perpetuated. These and many such considerations have become the axioms of successful rearing of sheep in all parts of the globe.

Causes of Failure in India.
1347

Disregard of the crucial features of this subject has doubtless, largely caused the failures that have been experienced in Indian experiments to improve the breed of our sheep. It is impossible to disguise from a perusal of the reports of the experiments hitherto undertaken, that crossing of widely different breeds has alone been regarded as the criterion of success. It has apparently been thought enough to procure rams of a famed breed, without regard to the peculiarities either of the ewes, the climate, or the pasturage. Failure was in many cases thus inevitable. The early experimentors in Austraila very properly secured a hardy stock, before they proceeded to improvement. Similarity of climate and herbage to that of their country was recognised as the first consideration. Attention was thus turned to India (*see Youatt*; *also Royle, Prod. Res., 168*) for the first stock, but not to the superior breeds of the temperate Himaláya. The poorer races of the warmer and drier tracts were seen to be more appropriate. Once these had been acclimatised, an untiring energy by selection and crossing produced in time the final purpose, and the sheep runs of Australia can now claim a position in the world's supply of wool second to that of no other country.

Conf. with p. 610.

As with Indian rice, improved into the famed Carolina paddy, so it might almost be said with Indian sheep, developed into the highly prized stock of Australia, Indian agricultural reformers have thought it alone necessary to bring to India these much prized triumphs of scientific progression. The failure in both cases has been equally complete. And disappointed and disheartened the possibility of improvement has practically been abandoned. So uniformly has the writer urged the necessity of a more intimate and detailed knowledge of the actual condition of the products of Indian agriculture, that that theme may be said to pervade every chapter of this work. Until, however, we thoroughly understand the peculiarities of our Indian breeds of sheep, we are not in a position to try the experiment of improvement. We must know not only the external manifestations of our breeds, so as to be able to recognise them, one from the other, but we must fully appreciate their proclivities. And into this category must undoubtedly be placed the careful study of the relationship that exists between any given race and the climate (heat, humidity, etc.), and pasturage of the region in which it is at present most successfully reared. Nor must the purpose for which it is bred be forgotten. In many parts of India sheep cannot be produced as fleece-yielders; their value lies in the quality of the mutton and the utility of the skin. In other tracts the fleece is of primary importance, while in many sections of the Himalaya a race of

INDIAN EXPERIMENTS in BREEDING.

sheep (with mostly four horns and which possesses none of the good qualities which would be looked for by the European farmer), is valued as beasts of burden. **Dr. Buchanan-Hamilton** says of this breed, in Nepal, that a good wether has been known to carry a load of eighty pounds. In England it may be said two chief sections are recognised according to the character of fleece, *viz.*, long and short. Of the long wools the following are the chief breeds—Leicester, Border Leicester, Cotswold, Lincoln, Kentish, Devon Long Wool, South Devon, Wensleydale, and Roscommon. Of the short wool the following may be mentioned: Oxford Down, Southdown, Shropshire, Hampshire Down, Suffolk, Ryeland, Somerset, and Dorset Horned, and Clun Forest. But there are certain breeds peculiar to mountainous tracts, such as the Cheviot, Blackfaced Mountain, Herdwick, Lonk, Exmoor, Welsh Mountain, and Limestone. The true mountain breeds are nearly all horned, or the males only are so. The only horned plains' breeds are the Somerset and Dorset, in which both sexes possess horns. Hornless sheep (as in cattle) are spoken of as polled, but in certain hornless breeds, by reversion, horns sometimes appear, as, for example, in the Hampshire and Shropshire. This is generally taken as proof of descent from a horned stock; in other words, of the polled condition not having been rendered sufficiently prepotent. As **Mr. Spooner** expresses it (*Jour. Roy. Agri. Soc. Engl., XX.*) "rigorous weeding" is necessary to eradicate the tendency to horn and the other defects of which these structures may be taken as the truest index. But in rearing for the butcher, the conditions desired may be different and indeed opposed to those for good fleece-yielding. The English rearer has found it more profitable to forego a certain amount of quality of wool in the development of a good and quickly fattening sheep that also affords a fairly good wool. In Australia, on the other hand, the wool alone has been deemed worthy of consideration. Accordingly, several new breeds of Merino sheep have been developed apparently on Indian stock. The pure Merino sheep fattens very badly and yields a comparatively useless carcases, but it can live in a dry season on a scanty pasture, and thus is enabled to produce good fleece where other animals would starve. It is deficient in the principle of early maturity and general propensity to fatten and is, therefore, not a profitable breed where the meat market is a necessary consideration.

Conf. with p. 633.

It may thus be seen how essentially the first step towards the improvement of the Indian breeds is the thorough investigation of the character of existing stock. The great majority of Indian breeds are horned and, indeed, manifest most of the peculiarities which a European rearer would regard as indicative of poor quality. English or even Australian experience cannot, however, be rigorously followed. The conditions of India are dissimilar to those of almost any other sheep-rearing country. The tendency of the stock has, therefore, to be investigated, the prognostications of unfavourable departures fully understood, and the methods of selection and crossing, which are found best calculated to guard against these dangers, thoroughly established. In other words, we have to evolve a prepotency suited to the climate and forage, of sufficient strength at least to give a healthy stock on which to conduct the further experiments at cross-breeding with superior foreign races. In this direction nothing whatever has been done. **Royle** many years ago wrote: "Some amelioration might, no doubt, be effected in the wool-bearing flocks of North-Western India by judicious treatment, nutritious diet, and careful selection of the healthiest and most perfect specimens procurable in the country. Yet as the progress in this, though certain, would be slow, and, perhaps, not sufficiently great in degree, few are likely to attempt or to persevere in such an undertaking."

Conf. with p. 611.

Prepotency.
1348

Conf. with pp. 569-70, 575, 584, 637.
Conf. with p. 615.

Improvement of the Indian Breed of Sheep.

INDIAN EXPERIMENTS in BREEDING.

In all experiments hitherto performed in India, it must be admitted, the difficulty and danger has been found to lie in the chief responsibility devolving upon Government. The noblemen of India take little or no interest in sheep-breeding. There is no private enterprise to cope with the great problems that have to be solved. And in Government there is no guarantee of continuity. Experiments are often started under the supervision of an officer well qualified for the undertaking. He has scarcely commenced when, through the necessities of the public service, he is transferred to a remote part of the empire. His successor takes no interest in the subject, and the Government, in time, getting tired of a fruitless expenditure, abandon the experiment. Time passes by, and some officer more enlightened on this subject than his contemporaries, recognising the possibility of great improvements, proposes (perhaps in ignorance of past failures) to conduct the self-same experiment. If undertaken, the result is as before. This is no imaginary picture. The study of the records of the Government of India for the past half century or more reveals the fact that large sums have been spent in the attempt to improve the breed of sheep in India, and that whole flocks of rams and ewes have been periodically imported from England, the Cape of Good Hope, Australia, Spain, Germany, etc. Some writers, for example, in Madras, speak of the existing breeds having been greatly improved by these experiments and others, and probably with more chance of being correct, deny that any improvement has taken place. The attempt which was made to improve the Hazara sheep, though persisted in for some years, was ultimately pronounced a failure. This same experiment has only recently been proposed as worthy of fresh trial.

Conf. with pp. 608, 612, 615, 617, 633.

The Transactions and Journals of the Agri.-Horticultural Society of India contain numerous papers and reports on the improvement of Indian sheep, and on the wool of cross-breeds. The chief notices on these subjects, down to about 1842, will be found reviewed by **Dr. Royle** in the passages which may here be quoted from his *Productive Resources of India*:—

"The experiments already made in India," says **Royle**, "seem to have decided, for the present at least, that the Merino breed is the best fitted for introduction into that country, though the Southdown, and some other English breeds, may eventually be found eligible. The next subject for consideration is the country from which they should be imported into India; whether direct from Spain, from England, or from Saxony, or whether from the Cape of Good Hope or New Holland. Judging from the energetic zeal at present displayed, it is probable that some will be introduced from all these countries. But it is desirable, in the first instance, to import a breed from the climates most similar to that into which it is to be introduced. The sheep of the Cape and of New Holland being already much improved, and the climate of both being more like that of Northern India than is either that of England or of Germany, it would appear preferable to import chiefly from these two colonies, for introduction into the tableland or northern plains of India. But, as the pasturage of the Himalaya, as well as the temperature and moisture more nearly resemble those of England, it would appear, for the same reason, that some of the English breeds would be better suited to the mountains than the Merinos, which require both a warmer and a drier climate."

So, again, **Royle** says:—

Merinos imported into Bengal.

1349

"**Mr. H. Wood**, a Member of the Board of Superintendence for the Improvement of the Breeds of Cattle in Bengal, when at the Cape of Good Hope for the benefit of his health, conceived the idea that 'the introduction into India of the Spanish ram would probably produce wool worthy of mercantile notice, and thereby add an important article to the exports of the country.' The Bengal Government, in accordance with the proposition of the Board of Superintendence, and to give the experiment a fair trial, authorized, at an expense of R9,450, the provision, either from the Cape or New South Wales, of twenty Merino rams, and twenty Merino ewes; with 1,000 country ewes, to form the flock.

"After a trial of two years, the soil and climate of the North-Western Provinces not having proved so congenial as the Board had anticipated, the flock was divided and removed to the stations of Deyra Doon and of Sabathoo. These being within the

INDIAN EXPERIMENTS in BREEDING. Climate of N.-W. India not suited.

Himalayas, sanguine anticipations were entertained of the result, from the abundant and excellent pasturage, and the facility of changing the climate according to the season of the year. But these were never realized, as the Board of Superintendence were informed on the 3rd of August 1829, that the whole of the original stock of Merino sheep had died of old age, and that difficulty was experienced in rearing the produce, from the delicate nature of the animals and the exceeding moisture of the climate. To avoid this, it was suggested that the sheep of Sabathoo should be annually sent to the dry country beyond the snowy range, where little or no rain falls, or where, at least, there is no regular rainy season. 1350

Sent to Himalaya.

In September 1832, the Government finally reported, that in consequence of the large sums expended upon the Sheep without any apparent corresponding benefit, the Governor General (**Lord William Bentinck**) directed that the flock of Sabathoo should be gratuitously distributed among such of 1351

Experiment still unsuccessful.

the Hill Chiefs as might be disposed to receive them. Also, that the flock in the Deyra Doon should be transferred to **Mr. Vet.-Surgeon Hodgson**, of the Hauper Stud, free of charge, on condition of his furnishing periodical reports and specimens of the Wool. 1352

Flocks ordered to be distributed.

"**Mr. Moorcroft**, in his journies to Tibet, had also in view the improvement of the breed of sheep in India, as in his letter respecting the Prangos Hay Plant, from near Droz, he writes: 'I have purchased and made arrangements for the keep of upwards of a hundred head of a race of sheep, the smallest perhaps known, but which in fineness of fleece may vie with the Merino, under the advantages of a much hardier 1353

Moorcroft's Tibetan sheep.

constitution and of a better carcase.'

"Some little attention, we have seen, was early paid to the improvement of the breed of sheep in the Madras Presidency. 1354

Improvement in Madras Presidency; *Conf. with p. 635.*

In the beginning of 1838, the Government there sanctioned the purchase of Merino rams which had been recently imported from Australia by **Colonel Hazlewood**, of the Madras Army. These were of the Saxon breed, though imported from Sydney. **Colonel Hazlewood**, in a letter to **Captain Jacob** of the Bombay Artillery, mentions that experiments had been made in the Neilgherries by **Mr. Sullivan** with Merinos, and by **Sir William Rumbold** with South-downs, and that his own flock consisted of 700 white-wooled country ewes with 1355

in Mysore and Nilgiris.

Saxon rams. The ewes appear to have been obtained in Coimbatore and Baramahal; but Jalna and Beder are mentioned as the best places whence to obtain the white-wooled breed. 1356 The results obtained both at Bangalore and on the Neilgherry Hills, from crossing the white-wooled sheep of the country with Saxon, Merino, and South-down rams, are stated to be most satisfactory, both as to quantity and quality of wool and size of carcase.

Bombay Presidency.

"The most decisive results have been produced by **Major (now Colonel) Jervis** of the Bombay Military Service, who was first most active in urging the adoption of measures for the improvement of the breed of sheep in that Presidency, and has 1357

Proposed by Col. Jervis.

since submitted his improved fleeces to the judgment of competent persons in London. **Colonel Jervis** originally represented, in 1835, that many parts of the Deccan and of Gujerat are well adapted to sheep pastures, and stated that, 'if the wool 1358

Sheep from Cape of Good Hope and Afghanistan. *Conf. with pp. 575, 585, 586, 587.*

which is at present produced, and which, though of an interior sort, finds a ready market, were improved by means of a superior breed of sheep, there can be little doubt of the benefit which would ultimately result to this country. The Bombay Government, accordingly, ordered rams and ewes of the Saxon breed from the Cape of Good Hope, and as the best wool imported into Bombay was understood to be produced in Afghanistan and Cabool, **Colonel Pottinger**, as well as **Lieutenant** (now **Sir A. Burnes**,) were each requested to obtain three hundred ewes and eight 1359

A flock from England.

rams of a pure white colour from the pastoral districts in the vicinity of the Indus. A few were also ordered from Bussora, as the sheep there yield a very fine and lengthy fleece. The Court of Directors of the East India Company, likewise, on 1360

Sheep Farms established.

being applied to, sent out 120 rams and ewes of different breeds, including the South-down, Leicester—Cotswold, and Merino (**Lord Western's**) under the charge of the son of a respectable farmer, who delivered them, with but few losses, in good order 1361

Benefits of, participated in by Natives

in Bombay. (*Conf. with p. 629*).

"The Bombay Government subsequently reported that 'the sheep obtained from England, the Cape, and Cabool, have been distributed throughout the country; many 1362 of them having been entrusted to the care of gentlemen who understand the management of these animals, and take an interest in the undertaking. A sheep farm has been established at Ahmednugger, and another at the fort of Jooner, where the climate is good and pasturage plentiful; and these farms have been entrusted to the charge of **Mr. J. Webb**, of the Civil Service, who has a good practical acquaint-

SHEEP and Goats.

Improvement of the Indian Breed of Sheep.

INDIAN EXPERIMENTS in BREEDING. Results. 1363

ance with the management of sheep.* The natives in the interior who breed sheep are supplied from these farms with half-bred lambs, and are allowed to send their ewes to the Government farms to be kept with the rams. Many of the rams have been given to wealthy natives and Patells of villages (by whom they appear to be much prized), who have flocks of their own, and who breed sheep for the sake of the wool and not for the market.'

"The results of the experiments have led to the conclusion that the Cape-bred Merino sheep are far better adapted to the country than those imported from England; so much so that the Bombay Government have determined to import for the future only from the Cape of Good Hope. The Report of the Commerce of Bombay for 1836-37 states, 'from the active measures taken by Government to improve the fleeces of the Sheep in the extensive pastoral country of the Deccan, so well adapted for the carrying of such an improvement into effect, the export trade in wool promises, in a few years, to be one of the most important and valuable from Bombay.'

Col Jervis's Sheep Farm. *Conf. with p. 586-7.* 1364 Lord Westerns breed. 1365 Greatly improved fleeces 1366

Colonel Jervis, at whose recommendation the above experiments had been instituted, in the meantime, established a sheep farm on his own account in the Deccan; and imported for the native ewes of his flock a large number of the finest Merino rams from the celebrated breed of Lord Western. The results which he has obtained have been most satisfactory, as is evident from the following opinion of Messrs. Southey of Coleman-street, eminent wool-brokers. (1) The wool or rather hair of the native ewe of the Deccan may be set down as being of the value in London of 3*d.* per pound (2) The fleece of a yearling shorn near Poonah, in February 1839, the produce of the Deccan ewe crossed by the Merino rams imported from Europe, was pronounced 'a remarkable clean well prepared fleece of Wool, being fine in the hair, longer in staple, and of a better quality, than we have hitherto seen produced in the Indian Peninsula, and worth 15*d.* per pound. (3) A white fleece, inferior to the above, from having some dead hairs interspersed through the fleece, 12*d.* to 12½*d.* (4) A black fleece, with longer staple than the ordinary breed of Indian wool, 7*d.* to 7½*d.*, having become more valuable from the increased length of the staple."

The results of the Bombay sheep farm (alluded to above by Royle) were reported on in 1843 by Sir George Arthur, who visited the flocks for the purpose. He expressed himself much pleased with the manifest improvements in the condition of the sheep, and suggested that it would be necessary, if this improvement were to be kept up, to import annually a fresh stock of Merino rams till the improved stock had thoroughly superseded the country one on the farm. The writer has failed to discover any further direct record of the Bombay sheep farm, and if the experiment at improving the breed was persisted in, it must have been but for a very few years, when all interest in the subject was allowed to die out. In the

Improvement in Bengal. 1367 Sheep Farm in Madras. *Conf. with pp. 586, 629, 635.* 1368

Transactions of the Society we further learn that on a smaller scale than in Bombay improvement was attempted in Bengal. In 1836 ewes of the Patna stock were crossed by Southdown rams, and the wool of the progeny submitted to the Society for report. So, also, we learn that in the following year a similar cross was made in Baughalpur. In a like manner the subject was taken up in Madras, for, in addition to the experiments alluded to by Royle, we read that Sir Mark Cubbon had an experimental sheep farm at Heraganhalli, under the charge of a European Commissariat subordinate officer, and that Merino rams were imported yearly from Australia and the cross-breeds distributed all over the country. The breed of sheep throughout the province was thus, we are told, immensely improved both as to size, quality of mutton, and wool. In 1863, however, the farm

Cross between Saxon and Cutch breeds. 1369

* In the Proceedings of the Agricultural and Horticultural Society of Bombay 1838, p. 4, it is stated that at Faria Bagh, near Ahmednuggur, Major Byne is trying the cross betwixt the Saxon ram and Cutch ewe. "The wool of Cutch sheep is particularly long in the staple, though not fine; it is principally, exported to Persia for the making of carpets. A gentleman conversant with the wool trade in London has stated that wool of that sort is much wanted in this country; it will make Blankets, Carpets, and other coarse articles."—*Committee of House of Commons, Commerce and Finance, p. 467.*

INDIAN EXPERIMENTS in BREEDING

Cross of Indian Breeds Recommended. 1370

was given up, as it did not pay expenses, a fact which, Dr. Shortt says, was due to sheep-breeding alone having received attention; but it is affirmed of this experiment that the effects of the large importation of foreign sheep are still visible in the improved quality of some of the Madras breeds. In this connection it may, perhaps, be added that with regard to the improvement of Indian sheep, Dr. Shortt remarks that in his opinion the Patna, Coimbatore, and Mysore are the best breeds; further, that it is quite possible to improve both the mutton and wool of these animals, but that, at the same time, it must be remembered, that these sheep do not thrive in all districts. He suggests that more valuable results would be obtained by crossing these breeds with picked sheep of the Madras breed, than by the importation of expensive animals which are not so likely to withstand the vicissitudes of an Indian climate and the careless management of Indian shepherds. Professor R. Wallace apparently took a somewhat similar view, for he says: "It would be vain to try to improve Indian sheep by crossing with those from Europe." He offers this opinion, however, immediately after having mentioned the disappointing character of a cross-breed from Southdown rams on Leicester ewes which he saw near Darjeeling. The breed mentioned could scarcely be called an attempt to improve Indian sheep, but rather an experiment at rearing a European cross-breed on the Himálaya. He makes no mention of having seen, during his visit to India, any crosses between the indigenous and foreign breeds, so that his opinion, given above, was apparently based on general principles rather than individual cases. The writer would by no means be disposed to go to such an extreme, though he has urged selection and improvement with indigenous breeds as the first and rational step towards improvement. The Professor says of the Darjeeling cross-breed above, that "the surroundings did not seem to suit them, still they might have looked better had they not been subjected to 'in—and—in breeding' for such a length of time." "A number of the sheep had horns, and the wool was coarse and hairy, most unlike the qualitty of wool that a cross flock of the same description would produce in Europe. The tendency to hairiness of coat is a character common to mountain sheep. The assumption of it by fine woolled sheep indicates that local conditions have much to do with its production or development."

Selection the Rational course. *Conf. with pp. 570, 594.* 1371

As illustrative of the opinions that have from time to time been urged as to the possibility of improving the Indian supply of wool, the following letters may be quoted from the Journal of the Agri-Horticultural Society of India. It will be seen that apparently Bengal sheep had been taken to England and there cross-bred. The result having been highly satisfactory, it is assumed the same result could be attained by rams of high merit being taken to India. This opinion was in fact seriously advanced by the writer of the letters which follow, and he was at the time the highest authority in England on the subject of wool. To ignore the effect of climate and especially humidity, is of necessity misleading in the highest degree. This will at once be seen by reference to the case quoted above from Professor Wallace's *India in 1887*, where a cross of two superior fleece-yielding European sheep had produced what the Professor calls hair instead of wool.

Conf. with p. 594.

Conf. with pp. 585, 613.

To the Secretary to the Bengal Chamber of Commerce.

"Sir,—Having, on more than one occasion, been requested by the Honourable East India Company, and also by the East India and China Association, to offer an opinion on sundry samples of Indian sheep's wool, which they have at various times received from India, and to suggest such means as I conceive likely to improve the breed of Indian flocks and qualities of their fleeces, I take leave, at the recommendation of Mr. Stikeman, to forward you the accompanying samples, the produce of cross-

Improvement of the Indian Breed of Sheep.

INDIAN EXPERIMENTS in BREEDING.

breeds betwixt a Bengal ewe and an English Merino ram, the property of the **Right Honourable Lord Western**, from which may be seen the advantage likely to accrue to Indian flock-masters by due attention being paid to the cultivation of the breed of sheep in our Indian territories and dependencies, and I humbly conceive, that with increased attention, wool may become a source of wealth to our Indian possessions, at the same time prove an article of incalculable benefit to the manufacturers in the British nation.

"Being personally acquainted with many gentlemen who have long resided in India, I have heard, what appears to them, impediments in carrying my views into effect; to which I reply 'make it an object to the owner of sheep to accomplish the desired end, and you must succeed.'

"**Colonel Jervis** has already demonstrated, that with due care, a superior race of sheep may be reared, producing wool infinitely more valuable than the native stock; and I flatter myself the time is not distant when the wool of India will be so much improved in quality as to be classed in our wool reports amongst those of our other wool-growing colonies."

I remain, etc.,

London, 30th October 1844. THOS. SOUTHEY.

"*P.S.*—You will be pleased to lay before your Chamber of Commerce the accompanying, samples and give as much publicity to the subject as you may deem proper.

To the Honourable Court of Directors of the East India Company.

"Honourable Sirs,—In representing to your Honourable Court the cost and expenses attending the purchase of 12 Merino sheep from the flock of the **Right Honourable Lord Western**, which were brought by directions of your Honourable Court, I cannot refrain from offering the following observations on the wool of a Bengal ewe, which were shown me by His Lordship whilst on his domain, together with her progeny to the second generation; and His Lordship having kindly presented me with a fleece of each year's produce, I have much pleasure in representing to your Honourable Court specimens of the three years' growth, which, I presume, will clearly illustrate, that, by ordinary attention, the present breeds of Indian ewes may be converted into wool-bearing animals, producing wool that would realize double its present value—

No. 1. Native Indian ewe.
,, 2. First cross, with a Merino ram.
,, 3. Second cross, with an ewe of the above.
,, 4. Sample of Merino rams: wool shipped to Madras per *Lady Flora.*

"It affords me great gratification in thus demonstrating to your Honourable Court, a theory which I have long entertained, that the flocks in Your Honourable Court's territories in India and your dependencies is capable (with ordinary attention) of producing wool, both as to quality and quantity, that would become an article of vast importance both to the flock-owner in India, and the British empire."

I have, ete.,

Coleman Street, 31st October 1844. THOS. SOUTHEY.

A very extensive official correspondence has, within the past few years, taken place on the subject of the SHEEP and WOOL of India. A certain amount of that correspondence was published by **Mr. F. M. W. Schofield** in the form of a paper on wool, which appeared in the *Selections from the Records of the Government of India, 1888-89* (edited by the present writer), and the chief feature of that correspondence the reader will now find republished below under, (so far as has been found possible), provincial sections. The correspondence and other papers there brought together will be seen to occasionally refer to experiments at breeding sheep, and may, therefore, be read as an elaboration of the present chapter, while completing the available information regarding Indian wool. The present remarks on the improvement of the breeds of sheep in India may, therefore, be concluded by a reference to the most recent action taken by the Agri-Horticultural Society. In 1887 a special Committee of the Society was appointed, on the recommendation of Government, to consider the question of the improvement of the breeds of horses, cattle, and quality and pro-

Report of 1838. *Conf. with 584-585.* Report of 1845. *Conf. with p. 586 also 618.*

1372

INDIAN EXPERIMENTS in BREEDING.

duction of wool in Bengal. The Committee reported that (1) the climate of that Province was unfavourable to the production of good wool; (2) that considering this fact they were not prepared to recommend any direct Government aid in the direction of wool-raising in Bengal; (3) they suggested that it would be probably more beneficial to turn the attention of the *rayats* to improve the size and fattening qualities of the breed, and that sheep-breeding and raising on approved modern principles should be demonstrated to the *rayats*, who would thus be encouraged to go on themselves making experiments in that direction.

FIBRE. 1373

Fibre.—It is perhaps needless to say that sheep yield wool. This remark is, however, necessary to preserve the logical sequence of this article. The reader should, however, consult the chapter below on the subject of wool for information regarding the fibre or fleece of Indian sheep.

FOOD.

Prices of Mutton. 1374

Food.—There is a large meat-eating population in India, since, not only Europeans but all Muhammadans and the majority of Hindus are mutton-eaters; but among Natives, at any rate, goats' flesh is much more largely consumed than that of sheep. Sheep MUTTON is, however, procurable in most towns at 2 annas a seer for ordinary, and 3 to 4 annas a seer for fat mutton. In former years nearly every regiment in India supported a mutton club, and there was a station mutton club also in most districts, but those days are now gone by and a mutton club is almost a rarity. Still the ordinary gram-fed mutton of the bazárs is very palatable, and in hill stations some of the mutton, though small, equals in flavour the best Welsh mutton. The manufacture of TALLOW has become an important local industry in many parts of India, and for this purpose sheep and goats fat are largely utilized. The exports of tallow were in 1889-90 valued at R70,167 and the imports at R1,03,930, more than half of which came from the United Kingdom and about that quantity of the total imports was taken by Bombay. The imports taken by Bengal were valued at R22,773, and by Madras at R17,814. (*Conf. with Vol. V., 459*).

Tallow. 1375

DOMESTIC. Wool. 1376 Manure. 1377

Domestic.—Besides furnishing WOOL, sheep are much valued in agricultural districts for the MANURE which they supply. The shepherds of India are for the most part an itinerant class, who travel from place to place with their sheep. They camp in the fields under tents made of rough woollen cloths stretched over stakes, while the sheep are crowded into a pen, fenced with thorns. So highly is their manure prized that for one night of a flock of sheep the owner of a field will pay the shepherd some 12 to 16fb of rice or from 8 annas to R1 in cash.

Fighting Rams. *Conf. with pp. 567, 568, 571, 573.* 1378

Sheep-fights are a favourite amusement among the Natives of many parts of India, and the rams (especially of the *húniá* breed) used for fighting purposes are much prized. They are petted and pampered till they become quite savage, and will hit and strike with their fore-feet, and, in some instances, even evince a propensity to bite. They are pitted against one another, and large sums of money are often staked on the result. In fighting they first move backwards for a short distance to give impetus to their forward rush, and frequently in the fight have their heads or horns broken.

THE WOOL AND WOLLEN GOODS OF INDIA.

WOOL & WOOLLEN GOODS. 1379

In the review of the breeds of Indian sheep, given above, the writer has endeavoured to bring together the specific information available in Indian works, on the subject of the various qualities of wool obtainable in this country. The chapter on Pashm and Goats'-hair below will be found to similarly set forth the main lessons deducible from the study of the goats. The reader who may have perused these pages so far will, perhaps, readily admit how very difficult, and, indeed, how unsatisfactory (in

SHEEP: Wool.

Wool Production in Bengal.

BENGAL.

the present imperfect knowledge) would be any attempt at a complete statement of the woollen interests of India, province by province. The utmost that can, therefore, be accomplished is a review of the official correspondence and published records available on the subject of Indian wool. In this manner it is hoped the maximum amount of information may be conveniently exhibited to the general public, since many of the papers at the writer's disposal are not generally accessible. It need only be further explained that the official correspondence so overlaps itself that the communications, in some cases, cannot be arbitrarily arranged in provincial sections.

BENGAL.

I.—BENGAL.

1380 From what has been said regarding the sheep of this province, it may be inferred that only one—the Patna breed—is classed as a fleece-yielding animal. The experiments at crossing that breed, though first undertaken over half a century ago and spasmodically repeated from time to time ever since, have not as yet resulted in any improvement. Indeed, it may be said that the first and most natural step towards the improvement of this breed ought to be careful treatment, liberal feeding, and selection of valuable variations, together with the crossing of these until an improved quality had been rendered prepotent. It might, in fact, almost be held that it would be a needless waste of time to cross-breed with foreign races until defects are eliminated and a fairly superior indigenous stock secured, that would blend with foreign blood, while retaining the character of suitability to environment. In the more recent official correspondence, quoted below, it will be found that two gentlemen (**Messrs. Orrah** and **Abbott**) have recommended certain experiments at crossing with foreign stock. Some of **Mr. Orrah's** suggestions seem to the writer impracticable, however desirable they might be. To expect to be able to bring the alpine sheep of Tibet to the "tropical swamps" of Bengal is at variance with all past experience. **Mr. Abbott** proposes a more hopeful scheme, though a cross of the Patna sheep with the Australian Merino would probably result in a better fleece but inferior flesh-yielding animal. But here again the writer would anticipate little permanency, unless the Patna stock were to be first developed and fixed in a character likely to favour prepotency in the improvement effected by crossing. The mere distribution of a cross-breed of one generation does not seem likely to prove of much value. **Mr. Abbott's** liberal proposals should therefore be accepted as likely to prove beneficial only after many years' careful cultivation and selection, or until fixity of character had been attained. The offspring of an *established*, new breed, adapted to the conditions of the country, would be of lasting value; the progeny of a simple cross could have but a temporary effect, which, like the Bombay experiments, would disappear a few years after periodic renewal was discontinued. Indeed, it may be said that two mistakes have rendered all past experiments fruitless. These are the omission to eradicate the defects of the indigenous stock before attempting to cross: and impatience of the time necessary to effect improvement.

Improvement *Conf. with pp. 577, 637.* 1381

Conf. with pp. 570, 586, 587.

Conf. with pp. 576, 591.

It will be seen that, at the suggestion of the Bengal Government, a committee of the Agri-Horticultural Society was convened a few years ago with the object, among other things, to consider the possibility of improving the breed of Bengal sheep. The committee, after the most careful consideration, seem to have arrived at the opinion that the chief aim should be improvement for the meat market, not the supply of wool. This would appear a very justifiable conclusion so far as the greater portion of the province is concerned. Indeed, the conclusion arrived at was practically identical with that published by a similar committee that reported to the Society on the self-same subject nearly sixty years ago (*see opposite page*).

Conf. with p. 582.

BENGAL.

The high temperature and excessive humidity of large tracts of Bengal, however good the pasturage may be, was by both committees pronounced inimical to the formation of wool. If this view be accepted, it is doubtful how far it would be wise to cross with a breed like the Merino, which is, perhaps, the most inferior of all breeds from the butchers' standpoint. In the more northern and western portions of Bengal, however (especially in Behar and Chutia-Nagpur), there are large tracts of country for which **Mr. Abbott** might, after a few years' selection and crossing, be able to evolve a breed that would be of great value as a fleece-yielder.

Conf. with p 581.

One other section of this province seems, however, to call for investigation, *viz.*, the division of the Himálaya that geographically belongs to Bengal. The traffic from Tibet has, for many years, been regarded as of the utmost importance, and that portion which would be looked for across the snowy ranges along the Bengal frontier, has been viewed with great expectancy. The Mission that was on the eve, some few years ago, of starting for Tibet from Sikkim had, as one of its objects, the opening up of greater facilities for the importation of wool. The lower and outer ranges of the Eastern Himálaya are too moist and too severely infested with the pest of leeches for offering much prospects of future extended sheep-farming. The higher ranges of Sikkim, however, are above these influences and are known to support large heads of valuable sheep. These might be greatly extended and improved; but of far greater moment, doubtless, would be increased facilities for the extension of the import of Tibetan wool, which, within the past few years, has shown evidence of some capacity. It may be said that the passes into Bengal, from Sikkim East to Assam, should tap the region of the *silingia* sheep and might thus be naturally expected to afford a wool, perhaps superior to that obtainable in any other part of India. The more recent official and other correspondence furnished below will be found to deal very fully with the present position and future prospects of the Tibetan wool traffic into Bengal. But, before proceeding to quote from these papers, it may serve a useful purpose to republish a few passages illustrative of the interest taken in this subject fifty or sixty years ago. The Transactions of the Agri.-Horticultural Society of India contain many very instructive papers and reports. Thus we read of **Mr. C. R. Prinsep** having experimented with Bikanír sheep at Allipore near Calcutta: of Merino sheep bred and crossed with indigenous sheep in various parts of Bengal: of Jeypore sheep having attracted attention,—various reports were, in fact, furnished on the wool of that breed. Southdown sheep had also been imported and samples of the wool from the progeny (reared in Bengal) and from various crosses, were submitted for report. These and many other wools were made the subject of a special report issued by the Cattle Committee of the Society in 1838. As that report reviews the experiments that had, up to that date been made, it may be here reproduced:—

Tibetan Trade. *Conf. with pp. 573, 586, 590, 592, 597.* 1382

Crossing Breeds. *Conf. with pp. 570, 576, 580, 591, 594, 611, 613, 617, 618, 652.* 1383

"The committee have been required to give an opinion upon these specimens; but they would prefer to leave this preliminary measure to persons better informed on such matters, and recommend that the Secretary be directed to transmit these samples (excepting **Mr. Storm's** English Wool) to the Committee of Agriculture and Commerce of the Royal Asiatic Society of Great Britain, with a request that a report upon them be obtained from competent brokers; and that the committee be solicited to procure and send out to this Society samples of the most approved kind of wool as subjects for comparison.

"From the slight experience which some members of your committee have had in the breeding of sheep in this country, and the general premises deducible therefrom, they are disposed to think that the attempt to breed sheep in the plains of Bengal for the sake of the wool, would not be attended with success; although **Mr. Prinsep** states having 'found that the cross of the Dumba sheep with the Patna gives a lamb with a curly fur, precisely of the same nature as that of which the Persian

Wool Production in Bengal.

BENGAL.

caps are made' and is 'inclined to think this kind of fur might be largely produced even in Bengal' and, Mr. Prinsep adds, 'it has this recommendation that the *lamb* of the Dumba gives meat of the most estimable kind.' Mr. Gibbon observes a similar result in a cross between the Merino and Patna sheep; the wool of the lamb is very fine and curly until it has passed the second month.

"These samples warrant the committee in expressing a hope that, although, as a commercial speculation, they think the soil and climate of Bengal decidedly opposed to the successful rearing of sheep, there is still ground for the exercise of much useful experiment, and they would wish, all who do practically take an interest in the question, to submit their results to the Society, whether successful or not.

Conf. with p. 581.

"The site of, or below, Darjiling appears to your committee worthy the attention of the Society, in the prosecution of inquiry on the subject of improving Indian wool, so as to render it an article of commercial importance. "Colonel Lloyd has offered to forward the views of the Society, and Captain Bruce, who is on the eve of proceeding to Darjiling, states his intention of entering into the grazing of sheep extensively, and offers in like manner to be of service to this body.

Tibetan Trade.
1384

"The committee, with such able coadjutors, have no doubt but ere long they will be enabled to lay some interesting particulars before the Society; and suggest a copy of this report be forwarded to Colonel Lloyd and Captain Bruce, with a request that these gentlemen will afford them all the information they collect that on the experiments which Captain Bruce intends to make, and also that they will give the committee what information they can of the prospect of a wool trade with Tibet.

Bombay wool Exports.
1385

"The committee further recommend that the Secretary be directed to communicate with parties in such parts of the country as are known to be favourable to sheep, and to request samples of the several varieties, with as many particulars relating to the season of dropping their lambs, shearing, etc., as possible, and that he be desired to obtain from the Agricultural Societies of Bombay and Madras specimens of the varieties of wool, from the flocks which are understood to have been placed at their disposal by Government, and of the kind or kinds of wool which are now largely exported from Bombay especially of the Kerwan sheep, which are understood to furnish the bulk of that export."

It will thus be seen that three at least of the questions that may be regarded as occupying public attention at the present moment were discussed or put to practical test at least half a century ago, *viz.*, first, to cross Patna sheep with the Australian Merino breed; second, to cross Patna with *dumba* sheep; and, third, to open up a trade from Tibet through Darjiling. The possibility of improvement without the aid of foreign blood seems never once to have been even suggested.

Prizes at Cattle Shows Discontinued.
1386

Some few years later we learn that the Society had been holding cattle and sheep shows at which money awards and medals were given, but that the results had been so unsatisfactory that it was deemed necessary to discontinue those awards. A special committee was convened to consider this subject and their report which was adopted by the Society may be here quoted, the more so since it alludes to sheep-farming on a large scale as having been started in the Khasia Hills, at Meerut, and at Bhaugalpur:—

Report of the Cattle Committee.

Report of 1887. *Conf. with p. 582.*
1387

"With reference to the resolution passed at the last General Meeting of the Society, on 8th March 1843, 'that it be referred to the Cattle Committee to consider and report how far it would be advisable to withdraw the premium offered for imported cattle and sheep and the produce thereof, after the exhibition of 1844,'—your committee beg to state, that having duly considered the subject referred to them, they are of opinion, that the attempt to improve cattle and sheep by a money premium and medals has not held out sufficient encouragement, in the number of cattle brought forward at the shows, to induce a continuance of the annual exhibitions; and they, consequently, deem it advisable to recommend that such premiums for public competition be withdrawn, after the expiration of another year, to which period the engagements of the Society extend.

"Although not within the meaning of the resolution to which their attention has been more particularly directed, your committee do not consider it will be deemed out of place, if they bring to the notice of the Society a subject intimately connected with their department of labour, *viz.*, *the Improvement of Indian Wool.*

BENGAL.

Improvement of wool. 1388

"Your committee are aware that, with this object in view, several experiments, on an extneded scale are now in progress at Cherra Púnjí, at Meerut, at Bhaugalpur, and other parts of India, which, it is to be hoped, will, in the course of time, introduce a new and profitable article for exportation. Indeed, it may be mentioned, that a member of your committee has already sent from his own flocks several shipments of the staple to England, which have given a fair remuneration.

Sheep Farms. *Conf. with pp. 580, 629, 635, 638.* 1389

"With the view of attracting as much attention as possible to this important object, and of assisting to give a stimulus thereto, your committee would beg to suggest the propriety of offering a Schedule of Prizes for the best samples of wool from cross-breeds between the Merino and country sheep, as well as from other crosses. The parties sending such samples should possess, *bona-fide*, a certain number of sheep. The number required to make a candidate eligible to compete, together with other details connected with the Prizes, can be determined on hereafter, should the Society consider the suggestion worthy of adoption.

(Signed) C. K. Robinson, V.P. (Signed) Charles Huffnagle.
" Wm. Storm. " C. R. Prinsep.

"Resolved—That the Report of the committee be confirmed, and that with reference to the suggestion contained in the latter part thereof, they be requested to submit a detailed report, embodying a Schedule of Prizes, etc., for the information of the Society."

Prizes for wool. 1390

"In connection with the subject under consideration, the Secretary begged to submit the following extract of a letter from Major Napleton at Bhaugalpure, together with a report with which he had been favoured by Mr. Robert Smith, on the sample alluded to: 'I brought with me, says the Major, from Cabool some Kohistan ewes which have some very fine lambs, some Cabool *Dhumbas*, some Panjáb wethers, etc., and I wish to know if there are any prizes I could compete for at the Agricultural Shows. There is a remarkably fine breed of sheep to be procured near Monghyr, and I have some now in prime condition, having been 18 months on gram and *bhusa*, and if there is any prize for the finest and best fed country sheep, I feel confident I could carry it off. The fleece of the Kohistan ewe is large and fine, and I have several pieces of cloth called *burruck* made from the wool. I am thinking of sending you some wool, that its merits may be tested.'

Kohistan Dumba Sheep. *Conf. with pp. 575, 579.* 1391

"The Secretary stated, that he had submitted the sample of wool (the first cross between the Merino and the Patna sheep,) which was presented by Mr. Muller, at the last meeting, to Messrs. Adam F. Smith & Robert Smith, and both gentlemen had been so good as to favour the Society with their opinion.

"Mr. Adam F. Smith states, that he considers the sample a very creditable one for a *first* cross between a Merino ram and Patna ewe. The wool is of fair staple, and has a soft silky feel; it is, in short, not unlike the F.'s and S.'s of some of the best marks of Spain."

The following letters and remarks thereon deal with the subject of a cross of the Patna sheep with a Southdown ram:—

Cross with Patna and Merino. *Conf. with 570, 584, 586.* 1392

To the Secretary, Agricultural Society.

I beg to send you two fleeces, shorn from yearling lambs, the quality of which appears to be an improvement upon Bengali.

The lambs are the produce of Patna ewes, by an imported Southdown ram, and were bred by Mr. Ricketts of Chittagong.

Tipperah; Yours, etc.,
18th November 1845. F. Skipwith.

"*Report on the above Fleeces.*

"Mr. Speed having solicited me to give my opinion as to the quality of the two fleeces of wool, forwarded by Mr. Skipwith to the Agricultural Society of Calcutta, I do so with much pleasure.

"The wool is of decidedly good quality for the first cross, uniting length of staple, and (for the sort) softness, with great uniformity of quality throughout the fleece, which is much desired. The quality from its coarseness, will not admit of being used for other than blankets, and very coarse

SHEEP: Wool.

Wool Production in Bengal.

BENGAL.

cloth, its market value at present in the home markets, is about sixteen pence per ℔.

"I consider that Mr. Ricketts has acted correctly, in having crossed the Patna with the Southdown, and should strongly recommend him to carry out the improvement by crossing the production with the Merino, as he only requires now, the texture, he having procured the length of staple with carcass.

"India can, in my opinion, if sufficient care were displayed in the several crosses, produce as good a sample of wool as any of Her Majesty's Dominions, from the luxuriance of the food and the temperature* of the climate; as texture, with length of staple, is all that is necessary.

To produce a flock embracing strength, carcass, and fineness of wool, I recommend the Patna ewe crossed with the Southdown ram, then followed by a cross of the Merino; if the Southdown ram cannot be obtained, the Leicester ram may be used, but though it will produce carcass, the texture of the fleece will be wanting to a considerable degree, consequently requiring a second cross of the Merino before the necessary fineness is obtained" (*W. Stallard, Calcutta, 19th January 1846, in Jour. Agri.-Hort. Soc. Ind, V., Sel., pp. 10-11*).

Crossing with Chinese Breeds. *Conf. with p. 570.*

1393 The idea of improving Indian sheep by cross-breeding with certain valued Chinese breeds was entertained some few years ago and carefully tried, as will be seen from the following extracts from the publications of the Agri.-Horticultural Society of India:—

Report on Wool from Shanghai sheep, bred at Chittagong by James Cowell, Esq.

"In a letter, dated Chittagong, 13th March 1851, forwarding specimens of various products which he thought might be interesting to the Society, Mr. Archibald Scone, C S., writes:—

"One article is wool cut from sheep that Captain Marquard brought from Shanghai. The sheep are large bodied animals, tall and long, the wool seems to be long and fine. The sample in question was handed to Mr. James Cowell, merchant of this city, who has favoured the Society with the following report:—'This wool from Shanghai sheep, *viâ* Chittagong, is a very good specimen, being long, soft, and easily combed, which leads me to infer that sheep wool may form an article of export from the Northern Port of China, at no distant period. This sample is not assorted —head, belly, and back, being all mixed together, which it should not be. It is therefore difficult to affix a value to it, but I think, and my opinion is confirmed by a friend pratically acquainted with the article, that it would fetch at home in its present condition, from 10*d.* to 11*d.* per ℔. Sheep should be washed in a running stream before being sheared, the omission of which has caused the present sample to be dirtier and more discoloured than it would otherwise be. I fancy that a cross between a Merino ram and Chinese ewe would much improve the wool, and the experiment might probably be worth the time and attention of some of our countrymen in the Northern Ports of China." (*Jour. VIII., p. 38*), *Calcutta 1st April 1851.*

Memorandum respecting three fleeces of wool presented to the Society by J. A. Crawford, Esq.

"The fleece which I presented to the Society was shorn from a China ewe on the 28th ultimo. The ewe was one of three China sheep (a ram and two ewes) which were purchased for me by Captain Boltan of the steam ship *Reiver*. They are all three white sheep; one has no stain of colour, the other two (a ram and a ewe) have the slighest possible stain of a reddish tan colour about the ears. The ram, from the appearance of its fleece, when it arrived, must have been shorn shortly before it was purchased. All three sheep appeared to feel the heat very much on first arrival, 8th April. When, however, the very hot weather set in towards the close of the month, the ewes seemed to feel the heat more than the ram. This I attributed to.

* The temperature and humidity of India have demonstrated the unsatisfactory nature of all such high expectations.—*Ed., Dict. Econ. Prod.*

BENGAL.

the fact of their having their fleeces on, but I do not think that they had even then got their full fleeces. I immediately determined on shearing them. For this purpose I got a pair of English sheaιs. My goat-herd, who seems to know something about sheep, was set to work at once on one of the ewes. In reply to my questions he stated, in the most confident manner, that he knew all about sheep-shearing; of this I believed just as much as I thought proper to do. The manner in which he set about his work soon showed me that, though he could clip a sheep, he had no idea of shearing one. The consequence was that the wool was taken off in small quantities. This was on the 26th. On the 28th, I had the other ewe brought up before me, and having had it cast I showed him how to commence and set him to work. It was with some difficulty I could get him to work my way. First he tried to argue the question with me as to the use of following directions and then to complain of the trouble. I sat by him, however, and got half the fleece pretty fairly shorn. At this moment I was called away. On leaving I gave strict orders that he was not to go on shearing but merely to turn the sheep and await my return. I was not away more than a minute or so. On my return I found that my orders had been obeyed, but was horrified at catching the man in the act of separating the half fleece already shorn from that which was still on the sheep. Luckily I was in time to save the connection between the two at the neck, and eventually had the satisfaction of taking off the fleece in the state in which it now lies on the Society's table. Looking to the fact that the connection of the fleece has been preserved at the neck, the most difficult part of the sheep to shear, I have not the least doubt but that the whole fleece could have been shorn as one entire piece. Considering that it is now over twenty years since I was present at a sheep-shearing, and that the man had never shorn a fleece on this mode before, I think it will be admitted that it is very fairly shorn.

Clipping not Shearing. *Conf. with pp. 597, 624.* 1394

"These three sheep of mine are of the club-tailed or Dumba-breed. Their tails are, however, nothing like so large as those of the Cape sheep or the Dumbas from the Northern Provinces and Afghanistán. I am told that my sheep are of a breed which is remarkably prolific, twin lambs being the rule and three at a birth not an uncommon occurrence. Whether they will retain their prolific properties in this climate remains to be seen. From want of ground I am unable to experiment as largely as I could wish, but I intend to try on a small scale an experiment in crossing with the Patna and common Bengali sheep.

China Sheep of Dumba Breeds. 1395

"Doubts have been expressed to me as to whether it is possible to shear a Patna sheep, so as to preserve the fleece entire. I hope I shall be able to solve these doubts shortly, as I have the offer of some Patna sheep with fair fleeces to experiment on, and in the course of next month I shall be in a position to report the result to the Society. —*Calcutta, 24th May 1866.*"

"I send you herewith two packets of wool, one done up in plain paper, the other in a newspaper. The former is the wool of a Patna ewe, the latter is from a half-bred ram, bred from China and Patna stock. These two sheep were placed at my disposal by **Mr. J. Sheriff** of Hunter & Co., with a view to make a trial of shearing from them an entire fleece. The result has been a dead failure. I had both sheep carefully washed this morning, and both were shorn in my presence. From the first clip of the shears I saw it was hopeless to expect the fleece to come off entire from the ewe, and though the same care and pains were taken with it as my China sheep, the fleece of which is with the Society, it was impossible to prevent the wool coming away in bits. The same was the case with the ram, except that just along the ridge of the back, the wool seemed to be of closer texture and I had hoped from this that I should have got that piece of it off in a long strip; but the application of the shears showed at once that it was not to be done. The wool of the Patna ewe does not appear to have much of the character of English wool, but to partake more of the quality of goats'-hair. The wool from the ram, I think, will be acknowledged to be better than that from the ewe, and its superiority is doubtless attributable to the cross of the China sheep, at the same time, I think it is clear that one cross is not enough to the defect in the wool of the Patna sheep or to impart the good quality of the wool of the China sheep. In writing thus, I trust it will be borne in mind that I am merely comparing these two breeds, and that I do not, by any means, intend that the quality of the wool of the China sheep should be held to be good as compared with the quality of English wool. I sat and watched over the process of shearing both these sheep this morning, and I do not know it will be ever possible to get an entire fleece off a sheep of either breed.—*7th June 1866.*"

Cross of China and Patna. 1396

Impossible to Shear Patna. 1397

Patna wool a species of Hair. 1398

Note on Wool in Bengal by M. FINUCANE, Esq., *C.S., Director of Land Records and Agriculture, Bengal.*

"The Government of India having asked me, by demi-official communication, for

SHEEP: Wool.

Wool Production in Bengal.

BENGAL.

Tibetan wool Trade. *Conf. with pp. 573, 585, 586.* 1399

some information regarding the trade in wool in Bengal, and the possibility of improving the supply or quality of that article, I have made some enquiry on the subject. I now give the small amount of information on the subject which I have been able to gather. First, as regards the supply of wool from Tibet and the Northern Frontier, the following remarks occur in the Report of the External Trade of Bengal with Nepal, Sikkim, and Bhutan, published by the Government of Bengal, 1885:—"The quantity of wool available for export from Tibet is believed to be enormous. Between Kamba and Shigatse, within a march and a half of the Sikkim frontier at the head of the Lachen, sheep are killed, not for the sake of their hides or fleece, which are practically valueless for want of a market, but in order that their carcasses may be dried into jerket meat, and sold for 8 annas each. At Kamba itself carpets and rugs are manufactured of the finest quality, and of patterns evincing excellent taste and skill; but there is no outlet for these fabrics. Further north on the Great Chang Thang (or northern plateau), which begins just beyond the Sanpo, within five marches of the Kongra Lama, are prodigious flocks and herds which roam at will over the endless expanse. In noticing the improvements in the supplies of wool imported into Bengal from Tibet during 1883-84, it was remarked in the report for that year—'it is believed that this trade has dwindled during the current year (1884-85), partly owing to the difficulties placed in the way by Tibetan officials; the statistics recorded show that the belief was well-founded, for the quantity imported during 1884-85 was only one-tenth that imported during 1883-84, *viz*, 91 maunds against 911 maunds. With the exception of 19 maunds registered at Rungeet in 1883-84, and 5 maunds in 1882-83, the entire supply during the three years was brought through Pheydong. The value of manufactured woollen goods (chiefly blankets) during 1884-85 was R4,415 in excess of the figures of 1883-84, but R564 below those of 1882-83. By far the largest are supplies brought *viâ* Pheydong."

Sikkim and Tibet Wool Imports. *Conf. with pp. 585, 597.* 1400

The following statement shows the quantity of wool imported into British territory from Sikkim and Tibet during the past five years:—

	Maunds.		Maunds.
1882-83	168	1885-86	2,555
1883-84	911	1886-87	1,933
1884-85	91		

The falling off in imports of wool in 1884-85, as compared with the two previous years, has been attributed to the difficulties placed in the way of this trade by Tibetan officials; but though this may be one of the true causes of the decline in question, it is to be noted that the trade appears to have been at all times insignificant and irregular. At the same time that there was a decline in the imports of wool, it is to be observed that there was a very large increase in the imports of other articles, for example, musk and yak tails, which, however, may be accounted for by the greater facility with which, these less bulky articles may be smuggled.

Yaks Tails. 1401

Without, however, questioning the existence or the pernicious effects of restrictions placed by the Tibetan officials on the frontier trade—matters on which I have no knowledge and no special sources of information,—I may say that, having made some enquiry on the subject at Darjeeling, I have not seen or heard anything which would lead me to doubt that a considerable trade in Tibetan wool can be developed, even under existing conditions, by simply creating a steady demand and securing a steady sale for the article in Darjeeling. It will be seen from a letter from **Mr. Spencer Robinson**, which is annexed, that a merchant trading with Tibet has recently offered to deliver to that gentleman in Darjeeling ten thousand maunds of wool, provided he guaranteed the purchase of it at R16 per maund. The Tibetans, he adds, will not place any obstacles in the way of allowing the wool to come through. If the wool, as stated, can be delivered at Darjeeling at R16 a maund, or say, 3 to 3½ pence per pound, and the wool is worth in England 6½ to 7 pence per pound, as it is believed to be, there would appear to be little doubt that the existence of a steady demand at Darjeeling, or some other place nearer the frontier within British territory, would lead to a steady supply, so far as the resources of Tibet allow. I am not here arguing against the desirability of removing trade restriction—a question which does not come within my province, and on which I am not called upon to offer an opinion—but what I am arguing in favour of, is the creation of a steady demand for Tibetan wool in Darjeeling by establishing an agency, public or private, for the continuous purchase of it. The attempts beidg made by **Mr. Spencer Robinson**, who has, I am informed, much practical knowledge of the subject will, from this point of view, be watched with much interest. It will be seen from the annexed report, with which I have been favoured by the Chamber of Commerce, and from the Secretary of the Agri.-Horticultural Society, that Tibetan wool, as per sample, received from **Mr. Spencer Robinson** is supposed to be worth 6½ to 7

Tibet Wool Price of. 1402

BENGAL.

pence per pound in England, where the price is rapidly rising. If this estimate turns out to be correct, and wool is forthcoming from Tibet in large quantities, as stated to Mr. Spencer Robinson, the importation of wool from Tibet should be a highly remunerative business.

The following statement shows the exports or imports of wool from and to Calcutta according to the Custom House returns, and statistics of river and rail-borne trade in Bengal since 1881-82:—

Year.	Export.	Import.	Year.	Export.	Import.
	lb	lb		lb	lb
1881-82 . .	13,446	7,484	1884-85 . .	78,395	21,612
1882-83 . .	2,336	41,617	1885-86 . .	57,595	86,367
1883-84 .	32,684	19,613			

Trade in Bengal Wool. 1403

As regards the trade in wool produced in the plains of Bengal, there is very little information available in the records of the Bengal Government, and as the Government of India has called for very early report on the subject, I have not been able to ask local officers for further information regarding it. No attempts have ever been made in Bengal to improve the quantity or quality of wool produced in these provinces. The bare suggestion of the possibility of taking measures with this end in view has been made the subject of gibes and ridicule. It will, however, be seen from the papers annexed that practical men, who have given attention to the subject, are by no means of opinion that nothing can or that nothing ought to be done by Government in this matter. Mr. Abbott, a well-known indigo-planter of Tirhoot, is enthusiastically in favour of endeavouring to improve the breed of Behar sheep by judicious cross-breeding, and is willing himself to undertake the supervision of the experiment on an inexpensive plan, which will be found described in his letter annexed, and Mr. Orrah, Deputy Superintendent of the Bhagulpore Jail, is of the same opinion. I think that Mr. Abbott's plan is an eminently practical one, and that cross-breeding on the lines suggested by him and Mr. Orrah ought to be tried. The Revenue and Agricultural Department of the Government of India may perhaps be in a position to give some assistance and support, by supplying merino or other good rams and ewes. I am myself also in communication on the subject with Dr. Greenhill of Calcutta, who has been good enough to volunteer his assistance in the selection and importation of good rams from Australia.

Proposal to improve Bengal Stock. *Conf. with pp. 570, 576, 581, 594.* 1404

I annex some interesting papers with which, through the courtesy of Dr. Lethbridge, I have been favoured by Mr Orrah. This gentleman has given much attention to the subject of Indian wool, and probably knows more about it than any official in Bengal. Some of his suggestions for effecting improvements in the quality of Bengal wool, owing no doubt to his want of acquaintance with the actual conditions of Bengal peasant life, are, I think, impracticable; but he agrees with Mr. Abbott in thinking that much may be done by judicious cross-breeding. Mr. Orrah, being in charge of the manufacture of woollen articles in the Bhagulpore Jail, is in a position to offer an opinion of value on the subject, and he bears testimony to the fact that the attempts made to improve the breed of sheep in the North-Western Provinces have led to a marked improvement in the quality of the wool of that province, which, he says, is decidedly superior to that produced in Bengal. Further, he significantly notes that, since the system of cross-breeding sheep by continuous importation of fresh stock has ceased in the North-Western Provinces, the quality of the wool produced is there also deteriorating.

I regret that I am unable to furnish the Government of India at present with more accurate and detailed information on this subject, and have only two practical suggestions to make with a view to improvement in the supply and quality of wool from Bengal and the northern frontier. These are:—*1st.*—That arrangements may be made by which a steady demand and sale can be guaranteed for Tibetan wool in Darjeeling or elsewhere within British territory—such demand will create the supply, and will probably arise without the aid or interference of Government, when it is known that wool is forthcoming. If, in addition to this, the Government of Tibet can be induced to remove restrictions which must, as a matter of course, injuriously affect trade, all the better; but I am not in a position to offer an opinion on this point. Further, it would perhaps be at first desirable to allow specially favourable rates for carriage of wool from Darjeeling by rail to Calcutta, and orders to have this done have, I understand, been recently issued. It would also be well, if possible, to improve the means of communication by road with the Tibetan Frontier. *2nd.*—That as regards wool produced in the plains of Bengal, the suggestions made by Mr. Abbott for cross-breeding in his letter, which is annexed, be accepted as a tentative measure."

SHEEP: Wool.

Wool Production in Bengal.

BENGAL.

Letter from Spencer Robinson, *Esq., to* M. Finucane, *Esq., C.S., dated Teendaria, the 17th July 1887.*

Shipment of Tibet Wool to England. *Conf. with p. 597-600.*
1405

"I forward two samples of Tibet wool as received from that country. One is ewe wool, the other ram's wool. I have been selecting wool for a Calcutta merchant during the last week, who is sending it home to England. This is the first shipment of wool sent home. I received a valuation on this wool recently, which was $6\frac{1}{2}$ to 7 pence per pound in England. The wool trade with Tibet can be developed into a large business, and a merchant trading with Tibet offered to deliver me 10,000 maunds of wool in Darjeeling (provided I would guarantee to buy it) at R16 per maund. He states the Tibetans will not place any obstacles in the way of allowing the wool to come through the passes. He is sending me samples of cloth, etc., purchased by the Tibetans, and wishes me to forward them to English manufacturers and let him know the price of such cloth when landed in Calcutta."

Cloth.
1406

Letter from H. E. Abbott, *Esq., to* M. Finucane, *Esq., C.S., dated Jainpur, the 27th June 1887.*

Behar. *Conf. with p. 584.*
1407

"Suffice it then to say, as a commencement of correspondence, that as far as sheep-breeding in Behar is concerned, I live in the very centre of it, and roughly speaking, my tenants own at least a lakh of these useful little animals; but as far as wool or, quality of meat goes, they are of the most wretched description, though, like most things indigenous to the country, hardy to a degree. As I told you personally, I am convinced that judicious crossing would prove eminently beneficial (were Australian or British blood imported) to both wool and mutton. I do not think much outside trade is done in the wool line, the shepherds finding a sufficient market at present by weaving it into blankets, which they dispose of locally; but, with the railroad now at their doors, the industry only wants encouragement to develop into a very valuable adjunct to the commercial economy of the district. Be it always remembered that the grazing in Behar is for India the best *par excellence*. Sheep during the hot weather are driven to the northern *chours*, when for a mere nominal sum they fairly well keep themselves; while during the rains practical farming-planters are glad enough to let them have the magnificient grazing of the inevitably heavy undergrowth in their indigo lands, charging the value of one sheep in the hundred for the right. Therefore, Behar is eminently a country suited to sheep farming, and it behoves Government to improve the breed. Give me a dozen rams and double the number of ewes, allow me R20 a month to keep them up, and I will guarantee to make a present of 90 per cent. of every ram produced in the stock to sheep-farmers in the district, and 50 per cent. of the ewes. Your charge shall be no more than the above R20 as long as I am in the country, and when I leave, I will give back to Government the same amount of stock that they handed me, or at the end of five years I will ask no further aid from Government, but will return it in kind the amount of rams and ewes advanced me. I will, with pleasure, furnish statistics of the entire births, deaths, and distribution, of the stock: of course you will fully understand that, should the stock die, I have no claim whatever against you as far as the R20 per mensem goes. My mere wish is to show how Government aid, properly applied, can benefit the district at small expense, and be made in the long run almost self-supporting. After a bit, I believe we could make money by letting out the services of the rams. But this must be done gradually. I have already proved that, as far as horses, poultry and dogs are concerned one can, with imported blood and judicious local crossing, do very great things in Behar; and as I honestly believe, as far as Bengal is concerned, it could be made the nursery for grain, horses, oil-seeds, sheep, cattle, etc., for the whole of Southern India."

Wool locally consumed.
1408

Behar eminently suited for sheep farming.
1409

Proposal to improve the breed.
1410

Imported Rams.
1411

[Note.—Mr. Abbott has apparently been supplied with rams: see page 600.—*Ed.*]

Bhagulpore.
1412

Report by Mr. Orrah, *Deputy Superintendent, Central Jail, Bhaugulpore, dated 15th June 1887.*

Wool Characteristic.
1413

I have the honour to submit the following report upon Bengal wools, together with extracts.* Regarding the production and improvement of wool grown in Bengal, it will be necessary to enter somewhat minutely into the characteristics of wool, namely—

1st.—Softness.
2nd.—Elasticity.
3rd.—Length of staple.
4th.—Uniformity of staple.
5th.—Fineness.
6th.—Soundness.

Defects of Bengal wool.
1414

So far as my experience goes in the working of Bengal wools, I find them wanting in all the aforenamed characteristics. As regards want of softness and elasticity, I

* See foot note marked † to page 594.—*Ed., Dict. Econ. Prod.*

Wool Production in Bengal. (*G. Watt.*)

SHEEP: Wool.

BENGAL.

attribute this partly to the hardness of the water the sheep may have to drink, being of a limy character. The scarcity of yolk in the fibre, which is the natural reservoir from which the wool fibre gains its character for softness, etc., shows the sheep are not sufficiently and suitably fed. As regards length of staple, uniformity, fineness, and soundness, these deficiencies are more the result of want of care and attention to the breeding. Good breeding is undoubtedly the main cause by which sheep of all countries are improved, or for want of it are deteriorated. Whilst Bengal wools show the deteriorations all round, and contrast very unsatisfactorily with other provinces in the north-west of India, *there is one redeeming feature which is a very valuable one, possessed by the Bengal sheep, and that is they are very receptive, or in other words susceptible of being improved more rapidly than many other classes of sheep.* **Possible improvements. 1415** The Tibet sheep, for instance, will require much more time to eradicate the coarse hairs and kemps out of the fleece than would the Bengal sheep. To improve the quality and quantity of Bengal wool, I will mention a few principal methods of treatment necessary for this purpose, for part of which there is some evidence of its having been successful in India as well as other countries.

"*1st. Pasturage.*—I have spoken of the water as being hard: this seems to point out that land pasturage over Bengal is of a calcic or lime character,* and would undoubtedly not only account for hardness of water, but the pasturage thereof must contain in its grasses, herbs and trees and vegetation generally, a large proportion of calcium, therefore the food upon which the sheep live and graze must also contain **By attention to Pasturage. 1416** lime and we see, as a natural consequence, that the wool they grow is dry, natureless, stunted, and irregular in staple, harsh in feeling, coarse, brittle, inelastic, whilst the yolk of the fibre is scanty and poor. I would therefore suggest that the water which sheep drink should be tested, and if found to be hard, should be treated chemically with a small sprinkling of oxalic acid,† *viz.*, 1 oz. of the salt of this acid to 500 **By attention to the Water supplied to Sheep. 1417** gallons of water to deposit the lime of waters drunk by the sheep. The washing water for sheep should also have any lime that exists in it deposited before being used. Quantity and quality of wool, which is distinguished from the fleshy parts of sheep, by the large proportion of sulphur which it contains, is very much affected by the soil upon which the food grows: some soils growing poor grasses keep the sheep grazing thereon lean, and whilst giving finest of wool, yields only 1½℔, but a **Yield of Wool. 1418** merino fed upon good pasturage, of chemically treated soils, often gives a fleece weighing 10℔ to 11℔. It is calculated that 30,000,000 sheep‡ yield on an average 111,000,000℔ wool, or about 4℔ of wool to the fleece. This quantity of wool contains 5,000,000℔ of *sulphur*, which is, of course, all extracted from the soil. If we **Absorption of gypsum as a manure. 1419** suppose this sulphur to exist in, and to be extracted from, the soil in the form of gypsum, then the plants which the sheep live upon must take out from the soil to produce the wool alone 30,000,000℔ or 13,000 tons of gypsum. Though the proportion of this gypsum lost by any one sheep farm in a year is comparatively small, yet it is reasonable to believe that by the long growth of wool on land to which nothing is even added, either by art or from natural sources, those grasses must gradually cease to grow in which sulphur most largely abounds, and which therefore from growth of wool—in other words the produce of wool—is likely to diminish by lapse of time where sulphur has for centuries been yearly carried off the land: and again the produce is likely to be increased in amount when such land is dressed with gypsum or other manures in which sulphur naturally exists.§ This, I believe, could be obtained in a very cheap form from the gas works of the country; some of their waste products containing sulphur in a large degree, such, for instance, as the sulphate of lime, a waste product of these works. There probably also is a natural form of sulphur found in connection with the rock salt districts, as geologically it exists in some form in the same stratification. Iron pyrites also contain sulphur in large proportions. Though not acquainted with all the natural products of India, I am quite sure products could be obtained sufficiently low in rates for manuring purposes. No bones of **Bone manure. 1420** animals ought to leave India, for its land sadly needs it, being also a valuable manure for sheep farms, in addition to the various dungs of animals of all descriptions.

2nd, I would suggest as of first importance also, that a definite system of cross-breeding should be sustained with rams obtained from pure blood stock. There is, I believe, **Improvement by cross-breeding.** *Conf. with pp. 570, 576.* **1421** some difference of opinion amongst flock owners as to what countries the best rams

* This is doubtful.—*Direct., Land Records and Agri.*

† Impracticable, and if practicable, would be dangerous, oxalic acid being a poison.—*Director, Land Records and Agri.*

‡ As to the probable number of Sheep and Goats in India, see the article on "**Skins.**" *Vol. VI., Pt. III.*"—*Ed., Dict. Econ. Prod.*

§ Gypsum would no doubt be good manure for pasture land, but *beriwallahs* are not likely to use it.—*Director, Land Records and Agri.*

SHEEP: Wool.

Wool Production in Bengal.

BENGAL.

are obtainable from; but experience has always proved this fact, that the merino ram, whether of European, South African, or Australian breed, is decidedly the best for crossing with Indian sheep. As success has been obtained by crossing the merino rams of each one of these aforenamed countries with Indian sheep, the aim of flock-owners should be to obtain wools for clothing purposes of a finer quality. This could be accomplished, and some six different qualities be obtained, which would work in combination, or serve the special direct purposes of manufacture in general. With this end in view, I would suggest—

1st.—With South African breed. 1422
1st.—For producing finest fibres of a felty character, a cross of one South African merino ram, pure bred, with 100 Bengal ewes.

2nd.—English 1423
2nd.—For producing uniformity of lengths of staple, medium fineness, soundness, and elasticity, a cross of an English merino ram, pure bred, with 100 Bengal ewes.

3rd.—Australian. 1424
3rd.—For medium qualities and characteristics ranging between Nos. 1 and 2, a cross of one Australian merino ram, pure bred, with 100 Bengal ewes.

4th. Indian. 1425
4th.—For improvement from Indian stock rams for coarser wools and cloths, a cross with North-West part-bred merino rams, the result of crossings with European stock in years gone by. This crossing would give wool closely approaching the Agra wool, or wool used largely by the Cawnpore and Dhariwal Mills, and also exported largely to Europe.

5th.—Tibetan 1426
5th.—I would advise a trial being made of crossing some Tibet ewes with an English merino ram, pure bred, and the rams obtained from this crossing should then be crossed with Bengal ewes. The effect, I believe, would be that a Bengal wool would be produced, having a character distinctly its own, and a flannel and clothing wool also suitable for hosiery would be produced of an excellent character and high value.

6th.—Australian. 1427
6th.—To produce combing wool of very long, fine staple, soft in form, in length, soundness, and elasticity, of a high value, for European combing machinery, would, I believe, be obtained by crossing Australian fine merino ewes, pure bred, with Tibet rams.*

Selection of Cross-breeds from Stock. *Conf. with pp. 570, 576, 581.* 1428
In connection with this matter of breeding are several important factors, such as the ascertaining of defects, pursuing a good system of selection or rejection, and subdivision of sheep into classes. These should be done regularly by a yearly inspection, so as to form correct opinion of the nature and properties of the fleece borne by each, in order that the defective sheep may be removed and never again allowed to mix with those drafted and set apart for the production of fine wool. White-woolled sheep, free from grey or black, should be kept separate. Black-woolled sheep, free from white or grey, should be kept separate; also the rams of same should be similarly kept separate with their flocks, and not allowed to mix promiscuously. All part-coloured sheep should be extirpated.

Mixed wools. 1429
Kemp and hairy wool is very objectionable. Bengal wool is very kempy, that is full of white hairy coarse bristles or hairs which protrude and will not dye, or become amenable to any process or operation of improvement. This, however, is the result of deterioration in all its forms, and can only be eradicated by the carrying out of all such operations as are being suggested herein. If only a few kemps be seen in wool, it lowers its value immensely for clothing purposes. Good feeding, protection, and breeding will eradicate these objectionable features. Thus it is that changes in the fleeces of sheep are wrought by propagation or crossing of breeds possessing those qualities which it is wished to acquire. **Lord Western**, whose interest in the growth and improvement of wool production, years ago, records the effect of a union he made with one of his lordship's own merino tups and some East Indian ewes, on which a striking proof was exhibited of the influence of the male upon the progeny, the latter having a fleece infinitely superior to that of the dams. The ram was kept highly fed, and consequently their fleeces became long in fibre, heavy in weight, the breed of the ram being the merino, which is considered the best from which foreign stock can be improved. Purity of blood should be unquestionable, and the result will then be a stronger stamina capable of standing changes of climate better."

Selection of Stock. *Conf. with p. 581.* 1430
It is, however, impossible, in the brief space of a report like this, to enter more fully into the minute details of sheep husbandry. I subjoin, however, a few extracts obtained from my library, which may be of interest in showing what has been done in the past in the other Presidencies of India, and which I consider somewhat confirms by

* It is highly improbable that a Tibetan ram could exist, still less procreate, in Bengal. This result could only be secured by a long series of crosses bringing, after many years, a trace of the Tibetan blood from the alpine regions to the tropical plains. *Conf. with pp. 584, 613—Ed., Dict. Econ. Prod.*

BENGAL.

Superiority of N.-W. P. over Bengal Wool.

1431

facts some of the suggestions herein made.* Before concluding I might remark, *from my present workings of the North-West of India wools, as compared with Bengal wools, there is so decided a superiority of the North-West wools that I am obliged to use a large proportion to obtain more satisfaction from those who purchase our blankets and cloths, and I have no doubt in my own mind that this same superiority has been given to it by some such early action in the matter of breeding, etc., having been taken by the Government of the North-West of India in the years gone by.*

I am also convinced that Bengal could so improve its wools, and if the system of cross-breeding was kept up continuously might eventually supersede in character the wools of those provinces. I judge also from what I see of the North-West wools, at present there are signs in their wools now that the system of cross-breeding is not being kept up by continuous importations of fresh stock, a very desirable element to sustain and further improve their wools: this I would suggest Bengal should attend to continuously.

Further report by MR. ORRAH.

Since writing my report upon the wools of Bengal, I have received, by mail, a copy of the weekly paper of the 14th May 1887, called the "*Wool and Textile Fabrics*," which contains a report made by Mr. F. H. Bowman, D.Sc., F.R.S., an authority in England on technical matters upon Indian wools; I have, therefore, had a copy of the same written out, which I send to you with this letter:—

"We take the following paper on Colonial wools (by Mr. F. H. Bowman, D.Sc., F.R.S., etc., President of the Society of Dyers and Colourists) from the 'Report on the Colonial Sections of the Colonial and Indian Exhibition,' issued under the supervision of the Council of the Society of Arts, and edited by the Secretary to the Society, Sir H. Trueman Wood:

Colonial wool. 1432

"The whole of the wools exhibited from India, except one or two incidental specimens, were confined to those contained in a case within the Economic Court. In speaking of them as wools, the term is used in its widest sense so as to include all goat and sheep fibres. The samples were 23 in number, and no reference appears to be made to them specially in the official catalogue.† Few of these samples were named specifically, the largest portion being only distinguished by a number or letter without label. In character, they covered a wide range of quality from the very coarsest goat's hair down to the finest wool or pure pashmina, which is the undergrowth of the Thibetan shawl-goat, as well as the native Indian wools, of which there are at least eight varieties. These wools are interesting, as they contain almost every variation in the individual fibres which is to be found in all other races of sheep. Most of the hairs and wools exhibited in this section, however, are of comparatively small interest to European manufacturers, because the export is small and the quality such that they can only be used for the coarsest class of goods, and when worked by machinery they require to be mixed with other wools. They are, however, of considerable importance in India as forming the staple of the woollen industry in the mountain districts where the great bulk of the woollen goods are worn. Many of them are singular mixtures of coarse and fine fibres; so much so, that those who are only accustomed to the regular wools of cultivated sheep can hardly conceive it possible that many of the samples could be obtained from a single animal. The finest specimen in quality and regularity of fibre, and in all characteristics which are typical of the best wools, can scarcely be surpassed; but by far the largest numbers of samples are defaced by irregularities in the structure and quality of the fibre, which are only to be found in the most neglected sheep in the United Kingdom and the Colonies. To enumerate all the defects which are found in many of these wools when compared with the highest standards attainable in the Australasian Colonies would be to mention all the defects to be found in any wool, and, indeed, many of the samples probably resemble the covering of the primitive sheep from which all the truly domesticated varieties are originally derived. Without further knowledge in regard to the place of origin, a mere classification of these wools would be of little service, and specially since the wools of India scarcely come within the scope of this report, and will probably receive attention elsewhere.

* The extracts furnished by Mr. Orrah were chiefly from Dr. Royle's *Productive Resources of India and the Journals of the Agri.-Horti. Soc. of India.* As these publications have already been quoted above, it does not seem necessary to repeat them. *Conf. with pp. 578—580; 581—582, etc.—Ed., Dict. Econ. Prod.*

† In this particular case the Exhibition Catalogue had to be published before the specimens had been received.—*Ed., Dict., Econ. Prod. and Author of the Catalogue referred to.*

SHEEP: Wool.

Reports on Tibet Wool.

BENGAL.

"As already remarked, the wide range over which the growth of the wool extended, and the difference in climate and other conditions to which the sheep in the various colonies were subjected, rendered the present opportunity most valuable in making a comparative examination of the different wools. This survey brought home to the eye most forcibly the very wide range of conditions under which the sheep can be cultivated, and the high state of perfection to which it can attain in almost every part of the world, when due attention is paid to the culture and breed. It seems to indicate that special cases of sheep are more adapted to certain regions of the earth's surface than others, and that in many cases the environment of the sheep tends, in the course of generations under careful management, to produce a special character which becomes permanent, and may be retained as a pure breed. It also shows that certain characteristics of the wool, such as lustre in the long-woolled breeds, can only be retained permanently by the re-introduction of fresh blood from time to time; at any rate, in all the regions which lie nearest to the equator, a certain degree of equality of temperature and atmospheric moisture being necessary for its permanence. Thus it appears to be retained longest in New Zealand and the southern coast of the Australian Continent. The nature of the herbage also affects the quality of the wool in a marked degree, and probably one of the chief reasons why the Australian merinos deteriorate when introduced into Cape Colony is because the herbage is not fitted for the highest development of the sheep. One very marked lesson of the Exhibition is the fact that all the best wools exhibited show that whatever tends to improve the character of the sheep in any one direction, re-acts all round in a benefit to all the other characteristics. The same conditions which tend to increase the size of the sheep cause the wool to be better nourished, firmer, and more tenacious, without injury to the best qualities of the fibre, provided care is taken in the proper selection and purity of breed in the sheep. The question of difference in the lustre of the wool is an important one, and opens a wide field for investigation. It has been already noticed that the Victorian wools stand foremost in this respect amongst the merinos. When the fibres are examined by a microscope, it appears that while the fibres are equally fine when compared, say, with those from New South Wales, or South Australia, the development of the individual scales on the surface is larger, and they present fewer scales in the linear inch. On the other hand, as the fineness in diameter is maintained in the less lustrous fibres, and the development of the scales is greater in number and this gives a greater softness and pliability to the individual fibres with a large degree of serration, and therefore a higher felting power. It is for this reason probably that the wools of New South Wales are more adapted for fine clothing trade than the more lustrous Victorian, or the coarser fibre wools of South Australia. The judicious introduction of the best characters of certain classes of sheep into other breeds, as is clearly shown, may induce a permanent improvemeut of the new breed only under certain conditions, and it seems now beyond a doubt that it will always be necessary for the farmer to discover the special class for which his own climate and surroundings are the most advantageous if he is to attain the highest perfection in the production of wool. Those who are growers of the wool must remember that every year the demand for quality in the raw material is greater, and those only who aim at securing all the best properties which wool can possess will secure the markets of the future.

Crossing regarded as essential in Tropical Countries. 1433

Microscopic Peculiarities of Wool, 1434

Kemp. 1435

"In several instances 'kemps' were found associated with the wool. These kemps are fibres, usually shorter and thicker than the others, in which all traces of wool structure are absent. They are brittle, solid, and ivory like. This is the sure indication of want of trueness in breed, and is most objectionable, as these kempy fibres will neither felt nor take any dye. They cannot be removed from the fleece by any process, except picking them out, and hence they injure the quality of any goods for which the wool may be used. The defect was especially noticed in some of the cross-breds with the long-woolled sheep, and where it exists the value of the wool is most seriously deteriorated. Nothing can compensate for the want of condition in the wool when sheep are neglected, and it cannot be too strongly urged that every endeavour should be made to maintain in the bulk the high standard presented in the samples exhibited. Without this care and due attention to classification, the results of good breeding and cultivation may all be lost and rendered commercially unremunerative. An endeavour was made in preparing this report by each specimen exhibited in relation to the geographical position in which it was grown to determine, if possible, whether any general law with regard to characteristic properties could be drawn from this relation, but the differences in the breed of sheep, and in the care and attention bestowed on the wool, rendered any sound deduction impossible, and it therefore appears probable the selection of breed, good pasturage, and attention, have far more influence than mere geographical position within the range of the temperate zone."

Assortment essential. 1436 *Conf. with pp. 588, 597, 598, 610, 611, 633, 656.*

BENGAL.

From S. E. J. CLARKE, Esq., Secretary, Bengal Chamber of Commerce, to M. FINUCANE, Esq., C.S.,—dated Calcutta, the 28th July 1887.

Reports on Tibet Wool. *Conf. with pp.* 585, 609. 1437

I have only received your letters of the 21st and 27th instant, the former handing me two samples of ram's wool and ewe's wool from Tibet, of which you wish to know the value in Calcutta. The samples have been exmained by the Committee of the Chamber of Commerce, who direct me to send you the enclosed copies of letters written by **Messrs. Peel, Jacob & Co.** of this city, with reference to similar descriptions of wool sent to them from Darjeeling in the beginning of 1884. Such wool, say, 30 to 40 maunds, was then valued in Calcutta, at R18 per maund. It was, however, subsequently sold to the Elgin Mills Company of Cawnpore for R25 per maund.

Prices. *Conf. with* 590.

Through the courtesy of **Mr. J. L. Mackay, of Messrs. Mackinnon, Mackenzie & Co.** I am able to supplement the information given by **Messrs Peel, Jacob & Co.** by a London Valuation Report, dated 18th May of the current year, on the same samples of Tibetan wool sent home by **Mr. Mackay's** firm. The valuation given is from 6¾*d.* to 7*d.* per pound at the then current market rates. The *Economist* gives the price January—June 1884, of unwashed wool at 7*d.* per pound. The value of wool similar to the samples you have now sent would be probably at that time 5*d.* to 5½*d.* per pound. The higher quotation given by **Messrs. Buxton, Ronald & Co.** in May last is owing to the rise in the price of wool which has taken place during the last three years. Tibetan wool is not well-known in this market, so that it is difficult to say what the demand for it would be. The Committee of the Chamber of Commerce desire me to say that they are of opinion it would be advisable to send the samples to the Elgin Mills Company at Cawnpore, and also to the Egerton Woollen Mills Company, Limited, at Dhariwal, Amritsar, from both of which concerns you would be likely to receive valuable and practical reports as to the quality of the wool and its suitability to their requirements. They would probably also be in a position to say what place it would take in the home market.

Report on Tibet Wool. 1438

In conclusion, I am to say that, if you desire it, **Mr. Mackay** will be happy to send the samples home for valuation in London. An early reply to this suggestion will oblige.

Report on Wool Samples referred to in MR. CLARKE's letter.

Market for Indian Wool 50,000.000 lbs. annually. 1439

We have two samples of wool from you and value the first received at 5*d.* per pound, and that last received at 5½*d.* per pound. The wool is unwashed and unassorted; it is well grown, and is of a sound and healthy character. Such wool would sell in Europe in any quantity. Of similar wool from Bombay, Kurrachee, and Beyrout, we sell 50 million pounds annually. There is always a market for such wool at a price; at present the value of all carpet wools (that is, carpet wool) is remarkably low. Please refer to the figures in the enclosed circulars about East India wool, which will govern this wool also. Those figures are for washed wool, assorted into various colours and qualities; and it might be advisible to trace a small lot of the wool you have in view in this manner, and ship it to test this market. At the same time, we would certainly suggest that *five* or *ten* bales should be shipped in the natural state, and then we could report fully, and you would be prepared to act in the event of prices rising. It appears to us that the subject is one of great importance, for it is evident that the wool shown by your samples comes from a country *perfectly* adapted for the growth of a sound and healthy wool.

Assortment of Wool. *Conf. with p.* 596. 1440

In Bombay and Kurrachee it is customary to assort and wash the wool before it is shipped, and this plan commends itself to our buyers. We send you samples of a parcel of Candahar wool which was worth 5¾*d.* per pound in its original state as clipped from the sheep, and we give you the result in the assorted and marked washed state of this wool. We do not, however, know what the wool weighed before it was washed:—

	℔		℔
No.	1.—12,000	1st white	value 11*d.*
„	2.— 1,700	2nd „	„ 7½
„	3.— 6,000	Yellow	„ 9
„	4.— 2,500	„ pieces	„ 6½
„	5.— 1,200	Grey	„ 5½

SHEEP: Wool.

Reports on Tibet Wool.

TIBET WOOL.

From Messrs. Buxton, Ronald & Co. to Messrs. Duncan, Macneill & Co.,—dated London, the 18th May 1887.

With reference to the sample of Tibet wool submitted to us this day for valuation, we beg to say we consider the wool worth from $6\frac{1}{2}$ to *7d.* per pound at present market values.

From Messrs. Peel, Jacob & Co., to the Secretary to the Bengal Chamber of Commerce,—dated Calcutta, the 14th January, 1884.

The sample of wool referred to in your letter of 8th instant is to hand, and we will endeavour to send you a valuation for it in a few days. Our Liverpool correspondents, to whom we sent a sample of your wool, write us as follows:—

"We find the present value is about *5d.* for unwashed, and, say *9d.* per pound if washed. It is recommended to be washed before shipment, and if in addition the colours be assorted, each sort being of course packed separately, higher prices would be obtained. We understand there is usually a good demand for this article."

From Messrs. Peel, Jacob & Co., to the Secretary to the Bengal Chamber of Commerce,—dated Calcutta, the 25th January 1884.

We much regret we have been unable to send you any report on the last sample of wool sent us, as we have not received any communication so far from the Elgin Mills, Cawnpore, to whom we sent it. We have now the pleasure to enclose a report on your earlier samples which we have received from our home correspondents, who have gone to some trouble in the matter, and we have sent you by post the five samples referred to therein. We enclose, for your further information, a Liverpool Wool Circular, details in which may be of interest to you. We would ask your careful consideration of the report, and would recommend you to make a trial shipment as suggested, with the view of commencing a regular business.

Assortment essential. *Conf. with* *p. 596.*

1441

From the Agri-Horticultural Society of India, to the Director of Land Records and Agriculture, Bengal,—dated the 30th July 1887.

I am now in a position to reply to your demi-official No. 284 of the 21st instant regarding two samples of wool from Tibet, ram and ewe. As mentioned in my previous note on this subject, there is very little trade done in wool in Calcutta, and the dealers in Tibetan wool amount to probably less than half a dozen in number. I have obtained the opinion of two of these traders on the samples. They consider they are good raw wools, but are very dirty, and their value would depend on the washing and cleaning they should receive before being put upon the market. In the present state, the wool would be unsaleable here. The final market for wool of good quality is Amritsar, and the price there for staple of the quality of the samples would be from R1 to R5 per seer according to the cleaning to which it has been subjected; it would there meet in competition Australian and European wools which are imported *viâ* Bombay. Another market would be found at the mills in the North-Western Provinces, but prices are not good there, as best qualities of wool are not sought after.

Of the two samples the ewe's wool is the better; the brown spots in it would, however, probably depreciate its value. From a European point of view the samples would be much improved were the two qualities of wool of which they are each composed separated. The outer wool of the sheep is wiry and harsh as compared to the soft inner fleece which is the more valuable. Should you desire it, I can obtain a more precise valuation from Bombay in a few days.

Letter from the Agent, Elgin Mills Company, to the Director of Land Records and Agriculture, Bengal,—dated Cawnpore, the 16th August 1887.

With reference to your No. 327, dated 3rd instant, and the sample packets of ewe and ram's wool from Tibet, I have the pleasure to give the following particulars:—

Quality.—A very good combing wool, with about 33 per cent. natural grease.

Value.—Can be purchased in the Cawnpore market at from R23 to R25 per maund.

We might perhaps be able to relieve you of a small quantity, but our consumption is only about 10,000lb.

TIBET WOOL.

Letter from the Manager, Egerton Woollen Mills Company, to the Director, Land Records and Agriculture, Bengal,—No. F. 123 K., dated Dhariwal, Panjáb, the 17th August 1887.

I have received two samples of wool you have been good enough to send me.

I had a standing order with Mr. Prestage all last year, to purchase 500 maunds of this wool at Darjeeling, but he entirely failed to procure for us some 50 maunds. This year I made a contract with another gentleman, and he, with the greatest difficulty, has succeeded in getting me 250 maunds. I cannot say yet to what extent I should be likely to take this wool, as we get identically the same from Tibet through our part of the Himálayas, but it is certain that no single part of Tibet could possibly produce 10,000 maunds in a year, as that means the fleeces of over 200,000 sheep. Amount available in Tibet 1442

We gave R20 a maund for our each consignment delivered at Sealdah.

From the Agri-Horticultural Society of India, to the Government of India,—dated Calcutta, the 3rd January 1887.

I have the honour to acknowledge your No. 262—7-23 F. & S., dated the 19th ultimo, which I had the honour of placing before a General Meeting of the Society, and I am directed, in reply, to send you copy of a demi-official letter from Babu Protapa Chundra Ghosha reporting on the samples of wool which accompanied your letter.

Letter from Babu P. C. Ghosha, dated the 30th December 1887.

I have examined the six samples of Tibet wool sent by you, and I have got a Kashmíri shawl merchant to examine and value the same. He declares sample No. 5 to be the best of the batch. Of Black wool, the sample No. 3, though good in its way, is not so far superior to No. 6 as to justify the difference of R3 in the maund. Just as I stated in my previous letter, a great deal depends upon the kind of carding and cleaning each sample received before it is brought down to India for sale. As an extensive manufacturer of shawl and woollen goods, both at Amritsar and Kashmír, he is of opinion that in purchasing large quantities, the quality actually supplied is much inferior to the samples, and the difference of the quality of the sample, and that of the goods in bales varies indirectly as the quantity of the sample, and directly as that of the goods purchased. The smaller the quantity of sample, the better it becomes by handling. There, however, cannot be any question as to the quality of Tibet wool generally. They are superior in fineness and length of staple to any foreign wool brought to India. But one must not forget that the value of the wool, provided it be fine in texture, is regulated wholly and entirely by the degree of its cleanliness. Tibet wool is the only superior wool which is used by the shawl-makers, both in and out of India. As I said before, Rampur and Yarkand are the two principal marts. But the supply is so uncertain and precarious and the prices so varying, that it is quite difficult to form any idea on its export value. The wool-dealers of the frontiers are of opinion that India is a better market for such wool than any other country, and I think we ought to endorse the same.

From the notes on the samples sent it will be found that a very limited supply of the six different samples is available—

	Maunds.		Maunds.
I	100	IV	2,000
II	100	V	25
III	100	VI	200
			2,525

The total available supply of all sorts being about 2,500 maunds. Of this total, about 2,000 maunds are consumed annually by the people of India in the way of blankets and coarse cloth manufactured in Sikkim, South Bhutan, Garhwal, Nepal, Rumpur, and other hill places. The remainder, the better and finer quality, are all taken up by the shawl manufacturers. The prices paid by them for wool of average quality I have already quoted in my previous note on wool. It would, therefore, be much desirable to so far facilitate the import of that article to India as would make the material available to all classes of wool manufacturers of the country for home

SHEEP: Wool.

Wool Production in the N.-W. Provinces & Oudh.

TIBET WOOL.

consumption. Assam, Cachar, and the tea districts are the great places for woollen blankets which come from Tibet through Sikkim.

As for the value of the samples in Calcutta market, I regret to have to report that there are no Native purchasers for the same. I do not know whether the European manufacturers would care to get their wool from Tibet when they can get perhaps at less cost almost as good material from America, Australia, and Germany.

From the Secretary, Bengal Chamber of Commerce, to the Government of India,—No. 218-88, dated Calcutta, the 9th March 1888.

The Committee of the Chamber of Commerce direct me to acknowledge your Office No. 261—7-23 of 19th December, forwarding, for valuation, six samples of Tibetan wool received by your office from the Director of Land Records and Agriculture, North-Western Provinces and Oudh, and calling for any remarks the Chamber may have to make on the subject of a wool trade with Tibet.

In reply to the communication under acknowledgment, I am to say that the Committee can add nothing to what was said in my letter of 18th November last. So long as the Tibetans are allowed to occupy a position within the Sikkim border from which they can at will block the Jeylap-la the trade must necessarily depend upon their good will. This matter is, however, before the Government of India. On the general subject of improving the wool trade of Bengal, the Committee will be glad to receive any further communications from you showing how the suggestions of Mr. Abbott and Mr. Orrah have been dealt with.

Prices of wool. 1443

I now come to the valuation of the six samples of wool sent with your letter under reply, and returned herewith. For the London prices now given, the Committee are indebted to the courtesy of Messrs. Mackinnon Mackenzie & Co.

Sample No. I.—Price per maund at Josimatti R20, the seer weighing 80 tolas; about 100 maunds can be purchased. *London Report, No. 1.*—Grey, good quality (if greasy); probable value 5¼*d.* per pound.

No II.—White wool, supply available 100 maunds, the seer weighing 80 tolas; price at Josimatti R20. *London Report, No. 2.*—Cashmere, coarse white, 8¾*d.* per pound.

No. III.—Black wool, available supply 100 maunds. Local price R25 per maund. *London Report, No. 3.*—Black, good quality (if washed), 7½*d.* per pound.

No. IV.—White wool, available supply 2,000 maunds. Local price R20 per maund. *London Report, No. 4.*—White, good quality (if washed), 6½*d.* to 7*d.* per pound.

No. V.—Good white wool, available supply 25 maunds. Local price R40 per maund. *London Report.*—Cashmere, white, fine 10½*d.* per pound.

No. VI.—Black wool, available supply 200 maunds. Local price R22 per maund. *London Report, No. 6.*—Dark grey (if washed), 6½*d.* to 7*d.* per pound. The report goes on to say—"Our brokers say it is difficult to value on such small samples, but they assume the bulk to be fairly represented and *not loaded with sand*. This should be guarded against, as we are told the Natives put any rubbish up with their wool."

Conf. with pp 590-592.

NOTE.—Arrangements are now being made to supply Mr. H. E. Abbott, through the Director, Department of Land Records and Agriculture, Bengal, with a few merino ewes from the Doonagiri Estate in Kumaon. Mr. Abbott has already been supplied with two merino rams.

N. W. Provinces & Oudh. 1444

II.—NORTH-WEST PROVINCES AND OUDH.

Scattered here and there throughout the literature of Indian sheep and wool, repeated mention is made of the sheep of these provinces and of the experiments (persisted in for some years) to improve the quality of the wool. The writer has, however, failed to discover a paper, or series of papers, that contain the desired material to allow of a description being given of the sheep of these provinces, or of the nature of the experiments hitherto performed. The improvement that had been effected by crossing the indigenous breed will, however, be found alluded to by Mr. Orrah in his report regarding Bengal wools (*see p, 592*). From the recent official correspondence furnished below regarding the wool of the North-West Provinces, it will be learned that the extension of agricultural operations is

N.-W. P. & OUDH.

regarded as having curtailed the grazing land in Oudh and thus lessened the outturn of wool. The chief area of sheep farming seems to be the tract of country bordering the river Jamna (*Conf. with p. 565*). The traffic from Nepal and through Nepal from Tibet is spoken of as capable of great expansion with a regular and good demand in India. This possibility was, however, dealt with very fully by many early writers. Thus, for example, the following letter may be given, from the Transactions of the Agri-Horticultural Society of India (*Vol IV., p. 214 of 1837*), in which the suggestion is offered of the formation of sheep farms on the southern slopes of the Himálaya:—

"I have no doubt but a settlement of sheep might be formed in the Himálaya hills on the southern slope of the passes leading into Kanower. The rains which are supposed to militate against sheep are not of similar violence or continuation as those in the plains of Hindustan; and, as for pasture, I think from what I have seen and read, that multitudes might find ample and appropriate food.

"In the winter, when the snows are thick, the sheep could descend into the vallies, where shelter and dry food would be prepared. I am not aware that any particular disease or obstacle to an increase is likely to exist; of ravenous animals there are few or none, but bears. Eagles, however, would prove dangerous, without sufficient protection to the young lambs."

"The hill sheep is a strong, robust animal, and such as nature made him; nothing has been done, I imagine, to improve the fleece which is a strong and substantial substance, and of which I believe the coating of the mountaineer is made.

"I should be glad to learn if the Government would be likely to afford assistance to an individual, who entertained the project of introducing a better sort of sheep in the hills promising such benefit to the country. You are aware that though the British authority is paramount in the hills, it has not actually territorial possessions where I should suppose the preferable position for sheep. My idea is, that the southern face of the Himálaya, from the Sutlej, at the Borendo Gunnoss and other passes, at an elevation of from 11 to 13,000 feet, would be the spots best adapted; and if I remember right there is no timber save birch and juniper in the narrow valley where water runs, but the face of the country generally is undulating knolls, covered with wild strawberry plants to the knees. There is of course other pasturage underneath; in parts the grass is of a short herbaceous kind, resembling that upon our downs in England."

Improvement of Breed. *Conf. with p. 570.* 1445

As indicating some of the experiments which have been performed in these provinces with the object of improving the breed, the following may be given from the *Journal of the Agri-Horticultural Society, V., Sel., pp. 160-161*:—

Extract of a letter from H. HAMILTON BELL, Esq., dated Agra, the 20th October, 1846.

"I send you by this day's dâk banghy, as likely to be of some interest to your Society, a small parcel of wool, from the first cross of the Cape merino, and Jessulmere sheep, and I think you will consider it not unworthy attention and promising, as a really fine wool on the third or fourth cross. The merino rams suffer greatly from the extreme heat and require a good deal of attention; but they seem healthy. I am, however, a little doubtful of any reasult from my experiment. Little pasture land is now left in these zillahs, and after the rains have ceased for a month or two, it affords little sustenance. I have found a moderate feed of grain indispensable, and I scarcely anticipate a return from the wool that would repay the expense. Some of the ravines on the banks of the Jumna have a good deal of grazing land fit for sheep, and one of my villages is situated amongst them; but it is out of the way, and I cannot trust my flock at a distance under native care, at all events, till the breed has been brought as close to the merino as seems necessary.

Cross Bred Merino and Agra. 1446

["The above muster was referred to a party who has lately arrived from the woollen manufacturing districts of England, and whose practical knowledge of the article is great, and he reports it 'a very clean, useful article to the manufacturer, and, in the state of the sample, every way equal to it in quality and cleanliness: would be worth about 14 pence per ℔ in the English market some two or three months ago.' It is, however, worthy of remark, that to judge properly of the value of this wool, it would be necessary to have a *whole fleece* taken off and folded up unbroken, as the quality will vary much in different parts of the fleece.—*Ed. Jour. Agri.-Hort. Soc.*"]

SHEEP: Wool.

Production of Wool in Rajputana and Central India.

N.-W. P. & OUDH.

Demi-official from LIEUTENANT-COLONEL D. G. PITCHER, Officiating Director of Land Records and Agriculture, North-West Provinces and Oudh, to the Government of India,—dated Camp Chardah, Bahraich, the 17th March 1887.

OUDH.

Wool production diminished.

1447 Your demi-official, dated 4th February 1887, in regard to wool production in Oudh. I have been making enquiries while on tour through Oudh, and from all I can learn, production, at no time very extensive, has greatly decreased with the extension of cultivation and consequent diminished grazing area.

The whole of that produced is locally woven into coarse blankets. For 1885-86 Oudh exported by rail 36 maunds to Bengal and 71 maunds to Cawnpore. The two Woollen Mills at Cawnpore state that they get no wool from Oudh. I tried three or four years ago to induce the Cawnpore Woollen Mills to procure wool from the pargana in which I am now encamped, once famous for the quality of the blankets woven, but nothing came of it, and I now find that the *gareriyas*, or flock-masters, have nearly all cleared out since grazing in the forests has become so restricted, and since cultivation has extended. They have gone over to Nepal.

As with the increase of Indian woollen mills there must be a better price for wool than formerly, it will be interesting to notice whether the recent opening of the Bengal North-Western Railway to Nepalgunge near here, attracts any wool from Nepal. A branch line from Bahraich to Byramghat would very probably stimulate exports and imports to Nepal far more than the present alignment.

Extract from a demi-official letter from LIEUTENANT-COLONEL D. G. PITCHER, dated Srinagar, the 26th October 1887.

I have been making enquiries up here about the wool trade. At present but a small quantity is imported from Tibet; but if any firm would take the matter up systematically and make advances, any quantity, they say, can be procured at from R20 to R25 at Josimath, from whence to Najibabad it would cost in carriage another R4 to R5 per maund.

Demi-official from the Assistant Director of Land Records and Agriculture, North-West Provinces and Oudh, to the Government of India,—No. 4389, dated Cawnpore, the 10th November 1887.

On receipt of your demi-official, dated 20th May 1887, I consulted the principal correspondents of the Department in all the districts of the provinces about the production and disposal of wool in their districts. From their replies I see that the only part in these provinces which produces any wool worth consideration is the tract bordering the river Jamna, which enjoys large areas of grazing land for rearing sheep. Much of the wool that is produced there is used up locally in the manufacture of *kamals, lois, numdahs*, etc.

Jamna Wool. *Conf. with p. 565.*

1448 What is left is brought to the Mills at Cawnpore which work about 15,000 maunds annually, fully one-fourth of which is supplied from this tract. The chief source of the supply of wool is the Panjáb, which sends about two-thirds of the total quantity consumed at the Mills; a quantity is also received from Rájputana and Nepal. The several varieties command the following prices at Cawnpore:—

Prices of Wool.

1449

Desi (produced in the provinces) . .	R10 to R12 per maund.
Amritsar (Panjáb)	,, 14 to ,, 15 ,,
Narnal (Rájputana)	,, 15 to ,, 17 ,,
Nepal	,, 20 to ,, 22 ,,

RAJPUTANA.

1450

III.—RAJPUTANA AND CENTRAL INDIA.

The brief notices of the breeds of sheep in India will be found to embrace an allusion to the Meywar sheep. These animals are spoken of as the finest and largest sheep in India, but as having a poor wool, though high class mutton. The wool of Marwar and Bikanír is generally said to be the best found in these States. But the wool of Shekhawati (in Jeypore) is regarded as soft and superior to that of all the States. It very much resembles the Bikanír wool. The following communications from the Transactions of the Agri.-Horticultural Society of India (*Vol. II.*,

RAJPUTANA.

68-71) furnish perhaps the earliest commercial reference to the wool of Jeypore :—

Jeypore. 1451

TO CAPTAIN BENSON, Military Secretary to the Right Honourable the Governor General.

SIR,—Adverting to the notice of the 23rd February 1829, inviting suggestions from all classes on subjects connected with the commercial resources of India, I trust I shall be pardoned troubling you with a few observations on the sheep of the Jeypore district, the fleece of which, it appears to me, might form a profitable article of trade with England.

The expense ot the merino sheep in the hills, belonging to Government, being defrayed from this stud, led me to observe the fineness of the fleece of the droves of Jeypore sheep that occasionally pass through here; but not being a judge of wool myself, I deemed it advisable to obtain the opinion of some person in trade with regard to its value, etc., previous to my addressing you on the subject; and with this view I forwarded samples of the wool to Messrs. Mackintosh & Co. to transmit you a copy of my letter to them, accompanying the wool, and their reply.

"Of several hundred sheep (wethers) that I have seen brought from the direction of Jeypore, they have been invariably of a large size, white with, generall , black faces, and their fleece finer than that of any other sheep I have observed in India. The price at which they are usually sold here by the drovers, is one rupee a head, and I am informed by a native who is in the habit of bringing sheep from that part of the country, that they are still cheaper on the spot, and the wool obtainable at three seers per rupee, or about thirteen rupees per maund.

"That the fleece becomes finer on the sheep better pastured there can be no doubt, from the improvement in the wool of the few I purchased; the wool which Messrs. Mackintosh & Co. say is valued at R80 and that at R100 per maund, in Calcutta, being clipped from the same sheep within three months after they had been well fed.

I have, etc.,

HAWPER STUD DEPÔT, C. S. BARBERIE, *Lieut.*,
The 20th July 1831. *Sub-Asstt. of Stud.*

TO MESSRS. MACKINTOSH & CO., Calcutta.

"GENTLEMEN,—The wool of a breed of sheep of an adjacent district appearing to me to be finer than that of the lower provinces, I beg to transmit you a few packs, and shall feel obliged by your showing it to some person acquainted with the wool trade for the purpose of ascertaining if it be fit for the manufacture of blankets and the coarser sorts of broad cloth, and if so, what would be the probable price per ℔ in the market.

"The longer fleece was sheared from some sheep I bought from a drove passing through here about two months since, and the shorter I cut to-day from one of the same sheep; on neither occasion were the sheep washed, nor has the wool been cleaned since.

I have, etc.

HAWPER STUD DEPOT; C. S. BARBERIE.
The 17th May 1831.

TO LIEUTENANT C. S. Barberie, Hawper.

DEAR SIR,—Your letter of the 17th May reached us in due course, but in obtaining information respecting the samples of wool which accompanied it, we have necessarily delayed answering it. The blankets used here of native manufacture are prepared chiefly in the neighbourhood of Patna, and of materials very inferior to your samples; the prices of them vary from twelve annas to one rupee per blanket, and they weigh on an average 1½ seers. We believe the manufacture of blankets does not thrive here, and if wool be purchased in this market it will be speculatively, probably for foreign exportation. Your samples are valued here, the larger quantity at R80 per maund (of 82⅔℔), that tied with cotton yarn at R100 per maund. The following is the opinion of a Leeds' merchant well acquainted with the value of wool, to whom your samples have been submitted. 'The sample of wool sent is of a very low description, and to the best of my judgment would not sell for more than 5*d.* to 6*d.* per ℔, and the small muster tied with thread from 8*d.* to 9*d.* per ℔. If the wool can be laid in 25 to 30 per cent. below these prices, I should say they would sell readily, but

SHEEP: Wool.

Production of Wool and Woollen Goods

RAJPUTANA. Jeypore.

I would not recommend a large shipment, as the manufacturers have a prejudice against wool they cannot know.'

MACKINTOSH & Co."

CALCUTTA,
The 7th July 1831.

More recent information on the subject of Jeypore wool and woollen manufactures may, however, be here furnished. It does not appear from the records at the writer's disposal that any systematic effort has ever been put forth to improve the Jeypore breed, though several reports incidentally speak of merino rams, so that it is probable some of that breed may have been distributed to Jeypore from the Bombay and other sheep farms that existed in India well on to half a century ago:—

Note by SURGEON-MAJOR T. H. HENDLEY, Honorary Secretary, Jeypore Museum, on the production of Wool and Woollen Goods in the Jeypore State.

In Jeypore, sheep are principally reared by Gujars and Jats. In ancient times, Northern Jeypore, or at all events, that part of it which was included in Virata (Bairat), was famous as a sheep-producing country, and even now it is stated in the *Rajputana Gazetteer* that the principal export from Shekawati or North Jeypore, is wool. The best wool, however, is said to come from the western border, which is not indeed as good as that from Marwar and Bikaneer; good wool is also obtained at or near Malpura, south-west of Jeypore City, the seat of the *numdah* or felt industry. Sheep are kept in most villages and the wool is bought up by dealers and the *Namadgars* or felters, who may be either Musulmans or Hindus. *Khatiks* or butchers also sell the wool from dead or slaughtered animals, which being inferior in quality (*khos-ki-un*) is only used for making coarse felts. There are about 40 families of Namadgars, Musulmans, living in the Namadgar-ka-Mohalla, Jeypore City. About 15 families of Khatiks, who live near the old Kotwali, also deal and work in wool. The census return does not give particulars under this head for either the State or Capital.

Spring & Autumn wool. 1452

Sheep, and even lambs, are shorn twice a year, in the months of Chaitra (March-April) and Kartik (October-November). The wool obtained in the spring is white, that in the autumn yellowish in colour, owing to its having been worn in the rainy season. The Namadgars wash the coarse wool and carefully clean it in a cotton-cleaning machine. They tread the wool repeatedly in soap and water in large pans and squeeze it into balls which, after drying in the sun, are cleaned by the Pindara or cotton-cleaner with his bow or "tant," or with a machine.

The Superintendent of the Jeypore Jail, who until recently bought wool in the Jeypore bazar for carpet-making, has now purchased it from Sambhur through an agent who obtains it from the neighbouring villages; the best quality, he adds, comes from beyond the Jodhpore border, but it contains more *harut* or seeds of grass, thus making it difficult to work. Three qualities have been obtained; the last lot cost R14-12-0 per maund against R18 or R19 for wool of similar quality bought at Jeypore. This is the cheapest wool he ever bought in Jeypore. It contains less sand and dirt than usual. Some of the selected wool would be worth from R25 or R30 per maund in Jeypore. It comes in the form of a twisted rope tied in a knot, each piece weighing a little over 8 chittacks (Jeypore weight). The fleece from his own sheep, the wool being coarser, weighed about 8½ chittacks. From other enquiries, I find that in Jeypore the fleece of a full-grown sheep weighs from ¾ to 1 seer, or 2℔ 2 oz. In England the average is 4℔ to the fleece, varying from 10 and even 12℔ from the merino sheep to 1½℔ for an animal from Hereford, where wool is fine, as the sheep are kept very clean. In a warm climate it is natural for the fleece to weigh less. Sheep are not dipped before shearing in Jeypore.

I am informed that uncleaned wool may be purchased at rates varying from R10 to R12 or R13 per maund. The wool of dead sheep can be bought for R8 per maund. It is sent in large quantities to Bombay, Nagore in Marwar, and other places. Chaura-ka-Sarur, about 25 miles from Jeypore towards Shekhawati, is the best sheep-breeding ground in the State.

Uses of Wool in Jeypore. 1453 Felts. 1454

The principal use of wool in Jeypore is to make felt, the *numdahs* or *namads* of Persia, in which country carpets are generally placed in the centre of the room, set, as it were, in a frame-work of soft, thick *namad* or felt. I think there is little doubt that this art was introduced into Rajputana from Persia through Delhi. It is an industry of some importance. At Malpura, the principal seat of it, large numbers of saddle *namdahs* are made for Bengal cavalry regiments, also *gugis* or hooded cloaks for native horsemen and persons who are exposed to wet weather—*asans* or round prayer carpets used by Hindu devotees; the *Jai-nimaz* or Musulman prayer carpet, which is of oblong shape, marked out with a nich in coloured felt, gun covers, *chakmas* or square rugs, and *kamals* or coarse blankets. Some of these felts are remarkably fine and durable. The better kinds are beautifully white, the common ones yellow or dark grey. RAJPUTANA. Jeypore.

The *namdahs* are thick and fairly water-poof; many of them are tastefully ornamented with small pieces of coloured felt arranged in artistic designs, and most of the prayer carpets are made in different layers, of which the edges are cut into curves and tooth-like projections. Felt is also made at Tonk and in Jeypore. BLANKETS are manufactured at Hindon, Phagi, Madhorajpura, Choundlai, Chatson, Sambhur, Naraina, Bhandarej, Jobner, Chomu and Samodh. Fine woollen cloths (*lois*) are made in Nagore and Bikaneer not, I think, in Jeypore, though I am not quite sure whether coarse examples cannot be had from Losal on the west border. I cannot discover that any mixed cotton and wool cloths are made here. *Chagmas,* made in the autumn from yellow wool, are sold at R17 per maund; the best white ones made at Jeypore cost R20 per maund. Tonk *chakmas* of good quality are worth R24 per maund. Blankets cost from R0-8-0 to R1 each. Blankets. 1455 Lois. 1456

Demi-official from F. Henvey, Esq., Resident, Jeypore, to the Agent to the Governor General, Rajputana,—dated Jeypore, the 18th April 1887.

With reference to **Mr. Colvin's** demi-official letter of 9th March 1887, I enclose a statement showing the figures of the Jeypore wool trade. I asked **Babu Kanti Chunder** whether he could get me any interesting particulars about woollen fabrics, etc., and he has sent me a note, copy of which also is enclosed.

Statement showing particulars about the wool annually produced in, and exported from, the Jeypore State.

Average produce for the past three years.	Quantity of wool and woollen fabrics exported.	Route taken by produce.	Localities where fabrics are made.	Principal fabrics made.	Prices of fabrics,	Prices of wool per maund,	Present state of the wool trade.
Mds.	Mds.	Shekhawati to Delhi.			As. R		
4,362	3,468	Buswa .	Jeypore . .	Chakma	4 to 5	R10-8 in districts, and R12 8 in the Jeypore City.	It will appear from the following figures that there has been a gradual decrease, both of produce and export :—
		Mohwa .	Malpoora .	...	...		
		Parasoli .	Hindown .	Ghooghi	4 to 5		
		Sambhur .	Gangapore .	...	...		
		Nagar .	Sanganeer .	Dalli .	4 to 1-8		
		Daber .	Sikandra .	...	...		
		Lamba .	Chatsoo . .	Ausans .	2 to 1		
		Toda .	Chandloi .	...	...		
		Newai .	Chomoo . .	Blanket.	R1 to R2		
			Madhorajpoora				
			Phagi . .				
			Jhalli . .				
			Shekhawati .				

Sambat.	Produce.	Export.
	Mds.	Mds.
1940	5,210	4,363
1941	4,248	3,198
1942	3,628	2,844

NOTE.—This seems to be exclusive of Shekhawati, which is said to produce 1,728 maunds, valued at R32,560.

Explanatory note to accompany figures of wool trade.

Chakma.—A woollen sheet of oblong form, is made of different sizes. It is used for sheltering goods from rain and as a protection from cold, and also to spread Chakmas. 1457

SHEEP: Wool.

Production of Wool and Woollen Goods

RAJPUTANA. Jeypore. on the ground. Ordinary *chakmas* are 2 yards by 1 yard and are sold at about R1-8 each. *Chakmas* (prepared to order) are used for covering carts, ruths, etc., and can be made of any size as the purchaser may require, varying in prices according to the quantity and quality of wool and evenness of the fabric.

Ghooghi. 1458 "*Ghooghi.*—Is a warm covering, worn as a protection against rain and cold wind. It gradually spreads out from the top, which consists of a sort of a hood. *Ghooghis* are of two kinds: one for walking and the other for sowars or horsemen. The length of the one used by people on foot is just sufficient to cover the human body, that of the other used by sowar, serves to cover also the hip of the horse, and its length is always in proportion to its width. *Ghooghis* (made to order) can be prepared up to the value of R25. Ordinary *ghooghis* for walking purposes are sold at R1-8 or R2 each, those for sowars from R3 to R5 each, but the price generally varies according to the whiteness and softness of the wool, thickness and evenness of the fabric. Fabrics made of yellow, uncleaned and coarse wool, uneven in their make, and so thin that they are not water-tight, when turned towards the sun they let the light through and are not good, but those made of white and soft wool, even and thick, are costly.

Dallis. 1459 *Dalli.*—The length and breadth of the *dalli* are exactly according to the size of saddle or *kathi* under which it is placed to protect the back of the horse from injury, and its price varies according to its thickness, evenness, and the quality of wool of which it is made.

Ausans. 1460 *Ausans.*—Are of two kinds, square and circular. Square *ausans* are 2 feet square, varying in thickness like that of *chakma* and *ghooghi*. *Ausans* (prepared to order) are 1 inch to 1½ inch thick; circular *ausans* are of different diameters. Ordinary circular *ausans* prepared for sale are generally 2 feet in diameter and are worth about eight annas each. Circular *ausans* (prepared to order) are 1 yard to 1½ yard in diameter and are more costly. Hindus sit on *ausans* when they worship. These woollen fabrics are often ornamented with borders and flowers of layers of wool worked on the fabrics. Simple borders and flowers are made of white wool of which the fabrics are made, and sometimes pieces of broad-cloth of different colours are cut into flowers and pasted on the fabrics. The *ghooghis* and *chakmas* of Malpura owe their credit to the water of a certain well which gives a peculiar lustre to the fabrics.

The fabrics described above are made of white wool. Black wool is chiefly used for making blankets and grain-sacks.

Process of making Chakmas and Ghooghis.

Manufacture. 1461 First of all wool is cleaned of all foreign materials and then combed; when it is ready for use, a piece of cloth of the measurement of the fabric to be made is stretched out and the quantity of wool of which it is to be made is spread upon it evenly by a wooden instrument. Wool for preparing *chakmas, ghooghis, etc.*, is spread on low ground, so that the soap water which is required in the process may not uselessly flow away. After the wool has been spread on the measured cloth, water is sprinkled on the wool and then the wool is mixed with soap and water and pressed down with the hands and elbows. This process is technically called "Rudda" which makes the fabric lasting, even and exact to the measure. These fabrics are not made by any machine, but prepared by a peculiar process as follows: The cleaned wool is soaked in an infusion of soap, gum, alum, and water. It is then spread upon the floor over a piece of white cloth in flakes, saturated with the said fluid, forming any shape wanted and beat in with wooden handle until well set. The piece is then soaked again in the same solution and exposed to the sun, after which it is washed smoothly. These fabrics are remarkably tough and impervious to water. About a quarter seer of soap is required in preparing a *chakma* weighing 2 seers.

After the *chakma* and *ghooghi* have been pressed, the water is not wrung out, but they are spread over *pucca* walls or on beams of wood till dry. After the woollen fabrics have been prepared they are smoked with sulphur to make the colour bright. High, dry and sandy places are best suited for breeding of sheep, but clayey and damp soil is not good for them. In the villages of Malpura, Phagi, Chaksoo, and Dousa wool is produced plentifully, and in the first named village woollen fabrics are made. The sheep are large and good-looking, but wool of superior quality is produced in Shekhawati, and the mutton there is better and more nourishing, because the district is sandy, while Malpura, Dousa, and others are clayey and damp. There are places in Shekhawati, such as Bisas, Ramgarh, and Fatehpur conterminous with Bikaneer territory, which are remarkable for best sheep, yielding excellent wool. Young sheep yield soft and white wool, and as they

RAJPUTANA. Jeypore.

grow older their wool becomes coarse and yellowish. Superior fabrics are made of soft and white wool and are therefore costly. Inferior fabrics are made of coarse wool obtained from older sheep. Sheep are generally shorn twice a year. In Shekhawati sheep are shorn once at the end of Sawan (July) and again at the end of Phagan (March), because after three months wool is generally full of the thorns and prickles which grow there. In other districts of Jeypore territory wool is shorn at the end of Chait (March) and again at the beginning of Kartik (October). The average quantity of wool obtained from sheep a year weighs 2 chittacks.

Chakmas, Ghooghis, Ausans, etc., are not manufactured in Shekhawati, but the villagers, to meet their own wants, get blankets prepared by Chamars for their use and *dhablas* for their females. Wool is exported by merchants from Shekhawati to Delhi. The average quantity of wool produced in Shekhawati annually is 1,728 maunds, valued at R32,560. The average price of wool sold in Shekhawati district is 2 seers per rupee, while in other districts it is sold at 3 seers per rupee, and the reason for this difference is that Shekhawati yields soft wool and the other districts coarse wool."

Bikanir. 1462

Translation of a Kaifiat from the Council of Bikanír by A. P. THORNTON, Esq., Offg. Political Agent, Bikanír, dated 31st March 1887.

Production. 1463

Trade. 1464

We have received your Kaifiat, dated 16th March 1887, requesting information regarding production of wool. In reply, we beg to state that in Sumbat 1942, 15,811 maunds, and in Sumbat 1943, 19,073 maunds of wool were exported from Bikanír territory; this gives an average of 18,341½ maunds of wool exported per year; the amount of produce of wool cannot be approximately estimated, but in Sumbats 1942 and 1943, 1,624 and 1,429 maunds of woollen cloth were exported, and about 2,000 maunds of woollen cloth used in this State, therefore the average produce per year may be taken at 22,000 maunds, including the wool and woollen cloth exported to other territories. The average price is R20 per maund and R2 customs duty; it is exported to Bombay *viá* Bangla in the Panjáb; the state of trade is good because the quantity which remains over the quantity consumed in this State is all sold.

Prices. 1465

Jodhpore. 1466

Demi-official from MAJOR P. W. POWLETT, to MR. COLVIN,—dated Jodhpore, the 6th April 1887.

As desired in your demi-official of 9th March, I give the following information regarding the wool trade in Western Rajputana.

Marwar Wool. 1467

It is not possible to state with any accuracy the gross produce of wool, but for Marwar it may be roughly calculated at 50 to 60 thousand maunds. During the last four years wool has been exported from Marwar to Bombay chiefly by Railway as under:—

Production. 1468

1883-84.	1884-85.	1885-86.	Ten months of 1886-87.
Mds.	Mds.	Mds.	Mds.
39,180	32,100	43,400	43,150

The increase of export in 1885-86 and 1886-87 is attributed to the reduction in October 1884 of the customs duty on wool from R1-4 to half that amount, and the extension of the Jodhpore Railway line from Luni to Pachbhadra, which is to be opened in a few days, will, it is hoped, further stimulate the wool trade of Marwar.

Jeysalmir Sirohi Wool. 1469

The annual export from Jeysalmír and Sirohi may roughly be taken at 5,000 and 2,500* maunds, respectively. Sirohi wool is said to be much inferior in quality to that of Marwar or Jeysalmere, and consequently, while a maund of Sirohi wool seldom fetches more than R7 or R8, the price of same quantity of Marwar or Jeysalmere wool varies from R12 to R25. The best wool is obtained at the second shearing after the cold weather.

Prices. 1470

* To Bombay *viá* Madar and Pindwara.

Production of Wool and Woollen Goods

RAJPUTANA. Kishengarh. 1471 *Statement showing the estimated quantity of wool produced in the Kishengurh State.*

Estimated produce.	Exported.	Any facts regarding		
		Fabrics locally made.	Prices.	Present state of the trade.
Mds.	Mds.	(1) Ghooghi. (2) Chakma. (3) Kamal, etc.	Wool is sold at R18 per maund. Kamal, from R1 to R3 per piece. Ghooghi from 8 annas to R2; Ausan from 2 annas to R1; and Chakma from 8 annas to R1 and 4 annas per piece.	A little increase in the trade compared with last year.
1,000	700			

IV.—PANJAB

PANJAB. 1472 *Conf. with pp. 572-575.*

So much has been said regarding the Himálayan sheep that it is scarcely necessary to enlarge on that subject. **Hodgson's** description of the two great trans-Himálayan breeds and the two cis-Himálayan breeds is fully applicable to the Panjáb. What is more wanted in this place is some account of the breeds of the plains, for the Panjáb may fairly be characterised as the chief wool-producing province of India. It is all the more to be regretted therefore that absolutely nothing has been published regarding the various breeds of the province. The *dumba* sheep is fairly plentiful, but it seems probable the greatly fattened tail should scarcely be regarded as characteristic of a particular race, or, at all events it should be admitted that there are several distinct breeds which all possess that peculiarity. While unable to furnish a detailed account of the breeds of Panjáb sheep, it may safely be affirmed that there are several distinct forms, some that approach the *Cágia* (such, for example, as the Tarai or Siwalik sheep), others that blend towards the Afghan races, while still others approach the type of the Sind, Baluchistan, and Rájputana breeds. The Hazára race has for some time attracted considerable attention, and effort was made for many years to improve it by cross-breeding with the merino sheep.

Conf. with pp. 578, 612, 615, 617, 633.

A monograph of the woollen manufactures of the province was issued by the Government in 1884-85. The author of that useful report (**Mr. Johnstone**), while furnishing particulars of the manufactures and trade, is silent as to the nature of the wool produced, or the breeds of sheep in the province. A still older publication, **Mr. Baden Powell's** *Panjáb Products*, furnishes certain pieces of information, omitted by the subsequent compilers, who have dealt with the subject, and the Journals of the Panjáb Agricultural Society contain many still older and very useful papers. The writer considers a selection of brief notices from these sources of information, while lacking cohesion and continuity, may serve a useful purpose.

Himalayan Sheep. *Conf. with pp. 572-575.*

1473 HIMALAYAN SHEEP AND WOOL.—**Lord William Hay**, while Deputy Commissioner of Simla, wrote in 1853 the following letter (*Jour. Agri-Hort. Soc., Panjáb*), which deals with the subject of the Himálayan wool as also with *pashm*:—

"The wool consumed by the people of these States, as well as of Mundee and Sukeyt, is imported from the plains, or produced at the fairs, annually held at Rampur.

PANJAB Himalayan Sheep.

The wool and *pashm*, brought to these fairs for sale, are obtained from a peculiar breed of sheep and goats, found only in those elevated regions, lying north and north-east of the great Himálayan range, and known as Great and Little Tibet and Chinese Tartary. The sheep is called *Biangi*, and i s wool is remarkable for its softness and length of staple.

The *pashm* or shawl wool is the under-fleece of the goat, and is singularly soft and fine. There are two kinds of *pashm:* the black or rather grey, called *shabri*, and the white called *phum* or *pashm.* From the latter are manufactured the shawls for which Kashmír is so famous, and from the former various stuffs known as *patú* or *pushminas.*

The great marts for the sale of wool and *pashm* are Rodok and Garoo or Gartal, both places situated within the limits of the Chinese Empire. Rodok is a town of about 200 houses, on the right bank of the Indus, about half way between Leh, the capital of Ladak, and Garoo. At this place wool and *pashm* are always procurable, but no fair appears to be held at it. Garoo, on the other hand, is a spot, situated in the midst of a very elevated and rugged province of Tartary, and marked by only one or two houses belonging to the Gurpans or Chinese representatives, and a collection of black tents belonging to Tartar shepherds, who remain there during the summer months for the purpose of selling their wool, and pasturing their sheep; at this place, which is the most important wool mart in the Chinese Tartary, a fair is annually held during the month of Bhadon. When the wool is purchased, it is without loss of time packed on the backs of sheep and goats, and the merchants start off towards their various destinations. *Pashm* is chiefly purchased for the Kashmír and Rampur markets; some, however, finds its way to Sultanpur in Kulu, and a little to the plains, *viâ* Garhwal and Kumaon.

The principal fair at Rampur is held in the month of November, and lasts about three days. The *pashm*, is sold for ready money alone, and bought up by Kashmíris from Amritsir, Ludiana and Nurpur. Of white *pashm* not less than 600 maunds are annually disposed of. Of black not more than 100 maunds.

Difficulties of the Tibetan traffic 1474

The roads in these regions consist of narrow foot paths, barely passable for goats and sheep, traversing the wildest and most inhospitable regions, crossing. every now and then, rivers bridged by a fragile net-work of twigs, and traversing passes 15,000 feet high, and covered at all seasons of the year with deep snow. Owing to this want of roads, sheep and goats only are used for the conveyance of merchandize. For this purpose, the sheep and goats of Chumba are most highly prized, being remarkable for great strength and uncommon endurance. Thirty seers is the usual load, but a *pucka maund* is often carried by a stout sheep or goat. The sheep cost about R2-8 to R3-8, while goats average from R1-8 to R2-8. The traders in wool are for the most part inhabitants of Piti, Lahoul, and Kunawur. They resemble or rather are identical with the Tartars of Tibet. Their religion, language, manners, and customs are the same. They rarely, however, come lower than Rampur, being terribly afraid of the heat of the lower hills even during the winter months.

Supply Inexhaustible. 1475

The enquiries I have made leave no doubt in my mind that the supply of Tartar wool is inexhaustible. A wealthy and highly intelligent merchant of Rampur assured me that the supply would always equal the demand, however much it might increase, and he mentioned in proof of what he asserted, that before the annexation of the Panjáb wool and *pashm*, to the amount of a lakh and a half of rupees, were annually sold at Rampur. I do not feel competent to give a decided opinion as to the quality of the wool. Sufficient attention is not perhaps as yet paid to its cleanliness, but I doubt not that this will be rectified as soon as the Tartars discover that the price of wool depends very much on its cleanliness, and that by cleaning it carefully its bulk is reduced by nearly one-half." (*p. 180-184*).

Sheep Runs Recommended. *Conf. with pp. 610, 638.* 1476

Crossing with Merino. 1477

Captain T. Hutton, in a letter which shortly after appeared, reviewed **Lord W. Hay's** opinions, and added much additional information. It will also be found from the republication of his communication which follows, that he strongly recommended the formation of sheep runs on the higher Himálaya, and urged persistent crossing with the merino sheep, on the soft woolly indigenous stock, until, as in the Cape of Good Hope, a locally improved race had been developed. He was of opinion that a wool of great fineness would thus very likely be produced which would readily command a market. It may, however, be said that the general spirit of **Hodgson's** investigations revealed the existence of breeds of sheep on the Himálaya that without any crossing might in themselves be greatly improved and made a source of positive wealth to India. The chief

SHEEP: Wool.

Production of Wool and Woollen Goods

PANJAB. Himalayan. Sheep.

objections are the indolence of the people and the want of sufficient and cheap means of export. Hutton wrote:—

Biangi wool. 1478

"Samples of the wool, termed *biangi*, have more than once been transmitted to England for inspection, and the answer returned has always been to the effect that while, on the one hand, the wool in its present state is too fine and too deficient in felting properties to render it suitable for any of our existing manufactures, the expense of cleaning it from hairs with which it is intermixed, would, on the other hand, entirely exclude the article from the market."

Government wool agency. 1479

"In former years the Government of this country made enquiries as to the possibility of procuring supplies of this wool, and appointed an agent (the late Captain P. Gerard, if I mistake not) to receive it at Kotgarh beyond Simla; but although there was found to be no difficulty in obtaining the article, its dirty state rendered it so thoroughly useless to the European manufacturer, that the speculation was abandoned; that the *biangi*, or Tartar wool, might, therefore, be procured in quantities equal to any demand that might be made for it, is doubtless quite true; and Lord W. Hay, while admitting this, states at the same time that, 'sufficient attention is not perhaps as yet paid to its cleanliness, but he doubts not that this will be rectified as soon as the Tartars discover that the price of wool depends very much on its cleanliness, and that by cleaning it carefully its bulk is reduced by nearly one-half.' Here, then, is the very point upon which we are at issue, and which amounts to this, namely, that the wool which the British manufacturer has more than once pronounced to be unsuited to its wants, is procurable in any quantity!!"

Cleaning wool. *Conf. with pp. 589, 596, 633, 656.* 1480

"As to the question of cleaning it, Lord W. Hay is perhaps not aware that the dirty condition of the wool of which the manufacturer complains, is not owing to dust and animal filth, which might easily be washed out either before or after shearing, but to the intermixture of hair, all of which has to be thrown out before the wool can be rendered fit for use; and it is the expense attending upon this cleansing process that would raise the price so much as to exclude the wool from the market."

"That the Tartar either would or could cleanse it from hairs is not to be expected, and I doubt that he *would*, even if he could do so, because in his unenlightened eyes, reducing the bulk 'by nearly one-half,' would be regarded as tantamount to throwing away one-half of his produce; quantity, not quality, being his *motto*, and *dirt being an especial favourite!* As yet, therefore, the question does not appear to have been advanced one step beyond the position it held some *twenty years ago!* The course to be pursued is, as I formerly pointed out, to purchase a flock or flocks of Tartar ewes, and serve them, under European superintendence, with well-selected merino rams; by which means, in the course of two or three years, the hair would be entirely eradicated, and a pure and valuable fleece be ready for exportation to the mother country, and let me add, for successful competition with the very finest wools procurable in the markets of Europe. There is no locality, that I am aware of, better adapted for an experiment of this kind, than the heights beyond Cheen in Kunawur, and thence upwards through the Tartar district of Hungrung; in fact, the very country that will shortly be opened up by the completion of the new road into Tibet. There the runs would be entirely beyond the influence of the monsoon, and an abundant pasturage be found, if it were thought advisable to experiment upon them, for both the cis-and trans-Himálayan breeds of sheep."

Himalayan Sheep runs. *Conf. with p. 683.* 1481

"From Cheenee upwards to the head of Kunawur, the cis-Himálayan breed with coarser wool might be advantageously located and improved, and made to equal the Australian breed; while immediately across the intervening range at the back Soongnum, the Tartar district of Hungrung presents itself for the cultivation of the finer *biangi* wool amidst pastures of worm wood, and other plants which constitute the natural and appropriate food of the Tartar sheep, and upon which alone they appear to thrive. Thus, as the Australian wool was at length produced from repeated crosses of the merino and British sheep, in a dry climate peculiarly adapted to its growth, so in the trans-Himálayan portion of Kunawur may repeated crosses of merino and Himálayan sheep be productive of a wool in all respects as good; while in the Tartar regions a cross of the merino and the Tartar sheep may be expected to produce a fleece superior to any now known.* Both these experimental farms would be within easy distance of each other, and remain under the watchful eye of a Superintendent and suitable assistants, whose

Conf. with p. 576.

* All past experience would seem to strongly support the view that the improvement aimed at might be attained by the natural process of selection and only perfected where found necessary, by crosses of breeds. *Ed. Dict. Econ. Prod.*

PANJAB.

duty it would be to attend to the serving of the ewes, the lambing, shearing, health of the flock, and other duties belonging to such an undertaking, and who from his centrical position at Soongnum, or on the heights of Hungrung, might travel alternately to either side as occasion required. He would also take measures for the housing and feeding of the flocks during the winter, a duty which, if left to natives, would never be effectually performed. Tracts of land might be brought under cultivation for turnips, which grow to a goodly size at Soongnum; while a yellow flowering lucerne grows well on the higher tracts around Hungo and Chango in Hungrung."

"In short, the experiment is well worth the trial, and if carried out in the spirit of liberality, cannot fail to meet with brilliant success, for even should the supplies from Australia continue, our Indian fine wools would always command a sale even if not superior to the exports from the colonies; while in all probability, our Himálayan peaks would soon produce a fleece peculiar to those elevated tracts, and be the means of introducing still finer fabrics to the notice and the admiration of the European world." Great expectations of success. 1482

PANJÁB PLAINS WOOL.—The following passages from the older papers which appeared in the Journals of the Panjáb Agricultural Society, and which were afterwards republished in a volume of *Select Papers up to 1862*, may help to convey some idea of the wool of this province. It will be found that a committee of the Society urged the desirability of experiments at cross-breeding some of the races of Panjáb sheep with stock from Australia. (*Conf. with p. 613.*) The sheep of the Sind Sagor Doab was regarded as the best and that with which experiment might in the first instance be performed. It does not appear that the recommendation was ever seriously undertaken, however, but **Mr. Barnes** (in a report also furnished below, p. 613) will be seen to have deprecated the idea of any possible improvement through crossing with foreign blood. So far as the author has been able to learn, from the perusal of a very extensive literature on Indian sheep and wool, it would seem as if **Mr. Barnes** had been proved more nearly correct than the writers who urged that "it was a matter of immediate importance to import rams from Sydney." Selection (such as tried temporarily by **Mr. Robertson** at Madras, p. 635) so as to eliminate stock that produce kemp hairs and to obliterate parti-coloured animals, would seem a more likely step towards improvement (when taken in conjunction with the crossing of Indian breeds, till a degree at least of superiority had been fixed), rather than by the importation of widely distinct races, for the purpose of immediate crossing. Plains Wool. 1483 Cross Breeding considered unnecessary 1484

In FEROZPORE.—**Mr. E. Brandreth** wrote in 1852:—"The wool produced to the west of a line drawn from Mukoo through Zeera downwards to Fureedkote, is all white wool; to the eastward of this line the sheep are principally black. About 800 maunds of white wool are said to be produced in this district in the course of the year. About 400 maunds in Mumdote and about the same quantity in Fureedkote. The wool trade appears now to be the principal trade of Feerozpore; about forty thousand maunds are said to have been shipped for Bombay in the course of the last three years. The value of the export in wool may be set down as upwards of a lakh of rupees per annum." **Mr. W. Ford** wrote:—"To ensure a good supply of wool, attention should be paid to the breed of sheep. Wool should be carefully picked before packing; this may be done by putting it into a tank and stirring it about well, so as to get rid of all dirt (which materially injures its sale); it should then be carefully packed quite dry, as it is very liable to ignite." Ferozepore. 1485 Clearing and Assortment. 1486 *Conf. with pp. 596, 597, 610, 633, 656.*

In MOOLTAN.—**Mr. W. Ford** writing in 1852, says:—"To ensure a good supply of wool, attention should be paid to the breed of sheep; those of this district are longer than the Futtehgurh sheep, but have large heads, clumsy legs, long ears, long backs, long tails, and are under-bred in every way; their wool is also inferior. There is also a great difference between the sheep of the Mooltan district and those of Sind about Haiderabad; there the sheep are constantly crossed with the Kabul or *dhoomba*, are larger, and their wool more abundant and finer: I have, however, observed that sheep crossed with the *dhoomba* are very liable to disease at Lahore, Mooltan, and Umballa; probably they find the heat too great; they, however, appear to flourish about the lower range of the Rawul Pindi district. The wool of the Simla sheep is much finer than that of the Mooltan sheep, the consumption is greater than the supply, Mooltan. 1487 *Conf. with p. 575.* 1488

SHEEP : Wool.

Production of Wool and Woollen Goods

PANJAB. so a great deal of plain wool is imported. In 1851, the Mooltan district contained about 48,000 sheep, as far as I can ascertain. Each sheep is sheared twice a year, which gives ¾ of a seer of wool for each sheep; on 48,000 sheep this will give 36,000 seers; a seer is equal to 80 Company's rupees. Perhaps ¼ of this amount of wool ought to be deducted for lambs."

Jhung. 1489 In JHUNG.—Major G. W. Hamilton wrote in 1852:—"The sheep of this district are superior in appearance to those I have seen in other parts of India, black and pied sheep being less common than white, and the breed might doubtless be improved. The sheep in the district has been estimated at 150,000, but the pasturage is sufficient to maintain a much larger number, perhaps ten times as many. Sheep thrive well in this climate, and are a profitable stock, being bred not merely for wool, but for their milk, and are seldom sold to the butcher."

Goojerat. 1490 GOOJERAT.—"The Goojerat district does not seem to possess many sheep; Rawal Pindi has large flocks of the common black sort, but below the Salt-range, and in the
Shahpore. 1491 SHAHPORE district the animal is much larger than any I have seen elsewhere in India. The colour is white, with black on the ears, and spots of black on the body. The ears are peculiarly large and pendulous."

Jhelum. 1492 In JHELUM.—Captain Browne in 1852 wrote:—"If this wool be fit for the European market, it would be well worth the while of Government to endeavour to improve the indigenous breed by the introduction of some merino sheep. I am induced to make this recommendation from having observed the effects of this measure at the Cape of Good Hope, where a comparatively profitless breed of sheep has been exchanged for the well-paying merino. The climate, too, of the Salt-range approximates very nearly to that of the barren hilly country of South Africa, and the grass in our low hills seems very nearly described in the following extract from a work which touches on sheep farming at the Cape:

Conf. with pp. 578, 608.

'These animals thrive best (speaking of the merino) in those places which appear almost bare to the eye of the casual observer, but exhibit to the nice discriminator, short tufts of light curly vegetation sprinkled over the 'velt' among loose stones and undulations and underbushes On an acre of such ground a sheep will keep himself fat, free from disease, and in condition.'"

"The fleece of one sheep at the Cape is reckoned by the same author to weigh, on an average, about two and a half pounds, and is valued at one shilling and six pence per pound at the lowest rate. This calculation was made in 1837, but I don't know what the price of wool may be at this time. "In this district a sheep is shorn twice a year, and the fleece may average one seer, or two pounds, and with an improved breed it would, no doubt, be considerably more, as well as of greater value. The quantity of wool produced in the whole district is estimated at 2,500 maunds."

Googaira. 1493 In GOOGAIRA.—"About 1,600 maunds of wool are annually produced in the whole district, at least one-half of which is worked upon the spot. The remainder is sold to *Sowdagurs*, who come for the purpose from neighbouring towns. The average nerik is 5 seers per rupee."

Lahore. 1494 In LAHORE, etc.—Mr. J. Wedderburn wrote in 1852:—"From the enquiries I have instituted, it appears that native wool of all kinds can be procured in this district in the cold season to the extent of 3,200 maunds. Kusoor, 2,000 maunds; Choonean 1,000 maunds; Lahore, 200 maunds. It is of three sorts: 1st, white; 2nd, brown; 3rd, black. The white wool is the most plentiful, and is of two kinds. The finest is obtained from the back of the sheep, and is used in the manufacture of *looes*, or light blankets, the texture of which is like that of shawls, but very much coarser. The market rate is R10 per maund, while wool of inferior quality is collected from the neck, legs, and belly of the sheep. It is used in coarser fabrics, and sells at R8-8 per maund. The remarks referring to the white are applicable to the brown wool. The black wool, employed extensively in the manufacture of common blankets, is sold at R6-8 per maund.

Yield 1lb. 1495 "The sheep are sheared twice in the year; half a seer of wool is obtained from a sheep, at each shearing, or one seer annually. The hair of goats is employed only in making ropes and coarse blanket bags for placing on bullocks when they carry loads. There is a fine kind of wool known as *pashmina*, which is brought into the Panjáb from Bokhara and Tibet during the cold season. Amritsar is the only mart at which it can be purchased. *Dhossa*, a thick warm covering for the Natives, is made from the wool purchased at Bokhara, while of that brought down from Tibet the finest shawls are manufactured. The Bokhara wool is at R1-10 per seer. The Tibet wool at R2 per seer; 50 or 60 maunds of *pashmina* could be purchased at Amritsar in one season. The seer is 80 Company's rupees, and 40 seers make a maund."

Conf. with p. 611.

Mr. G. C. Barnes wrote:—"As to quality, the wool of this country is hard and

PANJAB.

wiry. The back of the animal furnishes the best sort; and young lambs of course have a finer wool than older sheep. But coarse as the material is, it is evidently rising in value, and becoming, by way of the Indus, an article of European demand. I do not anticipate much benefit even from crossing the breed of indigenous sheep with other species. The hill sheep will not stand the climate nor the rains. They are regularly driven across the high ranges to get out of the influence of the monsoon; and when the cross becomes acclimated, I fear that the wool will become as hard and wiry as the indigenous produce.

"Soft wool is produced only in temperate regions; as the power of the sun diminishes, the natural covering becomes warmer and of finer texture, until in the arctic wastes above the Himálaya, the wool becomes so soft as to change its nature and to afford material for shawls under the name of '*pashm*.'

Crossing. *Conf. with pp. 570, 581.*

"I do not mean to assert that the native breeds cannot be improved, but I am not sanguine that any great advantage will ensue from crosses. The climate will be 1496
sure to degenerate the quality as the produce of all temperate countries invariably degenerates, when transplanted to a tropical climate"

So much interest appears to have been taken some thirty years ago in the subject of Panjáb wool that a Committee of the Panjáb Agricultural Society was convened to investigate and report on the possible means of improvement. The report drawn up by the Committee is of such interest that no apology need be given for the still further republication from the Society's Journals of the following:—

Report of Committee on Wool, 1852.

"With reference to the improvement in quality of the Panjáb wool, your Committee observed that, although an increased demand shows the article to be of some value, yet its inferiority is admitted on all hands. But the European authorities, who have been consulted, unanimously state the wool to be capable of improvement, and that superiority can be best attained by crossing the breed of sheep. Your Committee will, therefore, turn their attention to the best mode of crossing. The fine races of Himálayan and Tibetan sheep, with their soft and downy fleeces, are of the utmost consideration. But it is to be feared that the animal, being a native of cold regions, would not bear the climate of the plains, and that the fleece would rapidly become coarse and hairy. If, in any of the plain districts of India, there should be found a breed of sheep, producing wool superior to that of the Panjáb, a crossing with such a breed would certainly succeed. There are excellent Australian breeds, which might, perhaps, easily become acclimatized in the Panjáb; and it is, in your Committee's opinion, a matter of immediate importance to import rams from Sydney. *Conf. with p. 611.*

Tibetan Breed. *Conf. with pp. 572-575, 594, 610, 611.* 1497

"With respect to the locality, in which it would be most expedient to rear flocks of an improved breed, your Committee have information before them, which leads to the conclusion that the best breeds of sheep below the hills, are to be found in the upper portion of the Sind-Sagar Doab. The climate there, too, is somewhat milder than in other plains districts."

It was also recognised that an important step towards improvement would be obtained by a report on as exhaustive a collection as possible of the existing qualities of Panjáb wool. It seems likely that the local information, from the districts (quoted above), was furnished in connection with the preparation of these collections. The samples brought together were sent to Europe and examined by an expert of high standing. The report was communicated to **Dr F. Royle**, and by him was transmitted to the Panjáb. The expressions of opinion there made are, however, as applicable to the wool of the present day as to that of 1854, and the report may, therefore, be here furnished:—

"SIR,—Having had the forty-two specimens of sheep and goats' wool, and two of silk forwarded by the Indian Government from Lahore, examined by those well acquainted with these products and their value in the market here, I beg to submit the following report with the appended tabular statement. In this I have arranged these wools according to their nature and the places from which they have been sent.

"**Mr. Southey**, so well known for the attention which he has for so many years paid to Colonial wool, has long been anxious about the wools imported of late years from Bombay, and which he has said would come largely into consumption and sell at fair prices, if they were sent in a clean state to market, especially if the long hairs were first picked out.

SHEEP: Wool.

Production of Wool and Woollen Goods

PANJAB.

White Wools.

1498

"Of the specimens sent the goat's wool, or hair, the produce of Tibet, which is so much valued for the manufacture of Kashmír shawls, is less esteemed here in the state in which it is sent, because long hairs are mixed with the fine wool, so that it cannot be worked in the machinery in use, until the hairs have been first picked out: this in England is an expensive operation. It is probable, however, that if these hairs were picked out in India, where labour is so much cheaper, the Tibet wool might sell at remunerating prices, as it has frequently been enquired after by those engaged in the manufacture of fine shawls.

"The black wool of the Panjáb, as well as of the Umballa district, which is so much used in North-West India for making blankets, is not much esteemed here, as there is a prejudice against black wools. But the Goojerat, No. 12 wool, being long in staple, would be useful for long purposes, and might sell at remunerating prices, and that of Simla is of fine quality. The wools placed under the head of mixed have been sent as white or black, but are mostly mixed. They are little approved of, and would probably not be much in demand; some of them **Messrs. Southey** have not priced as being unsaleable.

"*The white wools* from the district of *Simla and other parts of the Himálayas, are, on the contrary, thought so well of*, that there can be no doubt of their being largely consumed in this country, if they are sent to market. They are pronounced to be of very good quality and English in character, probably from the similarity of the Himálayan to an European climate. These wools are valued at from nine pence to thirteen pence a pound, and would probably sell at higher prices, if sent in a cleaner state to market.

"The white wools from the Panjáb are like those of the Himálaya, of a kind suited to the English market; but as they are generally mixed with long hairs, they are not priced so high as they otherwise would be, the better kinds ranging from seven pence to ten and a half pence; but if the hairs, dirt, etc., with which most of them are intermixed, could be picked out, where labour is so much cheaper, the value of the wool would be considerably increased. The wools, like those from the Himálaya, are pronounced to be of a very useful quality, and that there is no doubt large quantities would find a ready sale, and probably at higher prices than those quoted, if sent in a clean state."

"*Report on some Wools and Silk sent from Lahore, with observations and probable value in the English market, by* MESSRS. SOUTHEY:—

1.—Tibet wools, 1st sort, white, not in demand in English markets, and of little value.

2.—Tibet wools, 2nd sort, white, ditto.

3.—Ditto, 2nd sort, black (rather brown) ditto.

10.—*Pashm of Bokhara R65 per maund*, brought by way of Peshawar; goat's wool not in general demand, but clean varieties might sell for 2*s*. to 3*s*. per ℔.

19.—Black *pashm*, 1st sort, R1-4 per seer; these are all stated to be not in general demand: they are also objected to as being intermixed with long hairs. If these were picked out, some of the white kinds might be saleable, but the black kinds only at very low prices.

20.—Black *pashm*, 2nd sort, R1 per seer.

21.—Ditto, 3rd sort, 12 annas per seer, ditto.

22.—White *pashm*, 2nd sort, R2 per seer, ditto.

23.—Ditto, 3rd sort R2 per seer, ditto.

Sheep's wool, Himálayas.

9.—White lamb's wool from Simla, good quality, 11*d*.; good long, 13*d*.; but if picked clean, both would sell higher.

24.—White wool, 1st sort, Simla district, R2 for 2¼ seers, ditto.

32.—White wool, Spiti, within the Himálaya, good; but rather wanting of English character, 10*d*. to 11*d*.

39.—White wool, Jung, yellowish, of good quality, 9*d*.

Sheep's wool Panjáb and Amballa Districts.

5.—White Panjábi, 1st sort, Lahore, R10 per maund; white, yellowish rather Kempy, *i.e.*, with long hairs intermixed, 3*d*.

7.—Ditto ditto, Ferozepore, Kempy, yellowish, but a useful kind."

Mr. Baden Powell, who published some ten years later than the above reports, particulars of the Panjáb wool trade, in connection with the Lahore Exhibition, treats the subject under three sections: 1st *Pashm*; 2nd, the wool produced beyond the frontier, including the *dumba* wool of Peshawar; and 3rd, the Panjáb plains wool including the improved wool of the cross-bred Hazárá sheep. It does not seem necessary to say any-

PANJAB.

thing further on the subject of *Pashm* than will be found in the separate chapter below on that fibre. But of his second and third classes one or two particulars may be abstracted from **Mr. Baden Powell's** remarks. Speaking of the frontier and *dumba* wool he says: "The trade in these wools is now extensive, both by the Peshawar and other routes in the North-West Frontier. There is also a very considerable export to Karáchi and Bombay. It is a remark of **Barnes**, 'that our early commercial connection with the countries on the Indus was sought in order to find vent for British woollens, while the existing trade was almost confined to cottons; and this is the more singular, as there is good reason to believe that in return for these cottons we shall shortly receive raw wool from the countries of the Indus.'" This anticipation, adds **Mr. Baden Powell**, has now been completely fulfilled. Speaking of the third class—the wools of the Panjáb proper—**Mr. Baden Powell** says: "Wool being generally, in the Panjáb at least, produced without artifice or skill, there is but little to be said as to the origin and progress of its cultivation. The different kinds of wool are, and have been, localized for ages; the attempts to improve and cross different breeds have been few and insignificant; and there seems hitherto to have been no desire among the Natives, who rest abundantly satisfied with the breeds that exist, and neither know nor appreciate the benefits of improvement. Whatever has been done, such as the attempted introduction of merino wool into Hazárá, or the production of *pashm* in Spiti is due to European endeavours. Much remains to be done in improving and extending the produce, and still more in introducing good methods of cleaning, dressing and working up the wool." Speaking of the Hazárá experiments **Mr. Baden Powell** adds in a foot-note to the above allusion that "the Hazárá experiment must, I fear, be pronounced a failure; it has gone on for some time past always dying a slow death, although not yet extinct. The people, it is said, do not want and could not use the fine merino wool." A passage from the Revenue Report of Hazárá will be found below, which furnishes fuller particulars on the cross-breed that had been tried prior to the date of the report (*1862-63*), and it need only be added that apparently the subject has since been entirely forgotten. At all events, no recent reports have appeared, and it is not known whether the breed is still in existence or not.

Conf. with p. 577.

Conf. with pp. 578, 608, 612, 633.

But to conclude this abstract of the opinions held about the time of the Lahore Exhibition (1864), the following statement of the two chief classes of Panjáb wools may be furnished from **Mr. Baden Powell's** *Panjáb Products*:—

N.-W. Frontier. 1499

WOOLS OF THE N.-W. FRONTIER.—"We now come to the second class of wools, produced at or about Peshawar, Kábul, Kandahar, and Persia or Kirmán.

The most interesting varieties of wool are,—*1st.* that of the *dumba*, a large tailed sheep, at Peshawar and Kábul; from the latter place it obtains the name of '*Kabli pashm*;' it is used in the manufacture of '*chogas*' (cloaks with sleeves) as worn by the Afghans.

2nd.—Is '*pat*,' the hair of a goat common in and about Kábul; fabrics called '*pattu*' are made from this.

3rd—Is Kirmáni wool, a beautiful white, very soft wool, produced at Kirmán, it is called '*Wahab Shdhi.*'

4th—There is a Kandahari and Bukhára wool, among which we may include the Karakuli lamb *skins* * of Bukhára.

* The lamb skin (with the fleece on) of Karakul, a district about 20 *cos* distant to the south of Bukhára, is famous. About ten lakhs of rupees worth of these skins (the produce of Karakul and other districts of Bukhára all being called "*karakuli*)," is annually exported from Bukhára, to Persia, Turkistan, Russia, Kábul, and India. The greatest quantity goes to Persia, where the people make caps of "Karakul," called "*pupakh*." A piece of the best description of "*Karakuli*" sells from R25 to to R16 in Persia.

SHEEP: Wool.

Production of Wool and Woollen Goods

PANJAB.

N.-W. Frontier.

Wool obtained from the fat-tailed variety of sheep is used in the manufacture of clothes and carpets, and also exported to India. It is of wide distribution; the sheep abound at Peshawar, Kábul, Kandahar, Herat, and other places; Kelat and the surrounding country produce sheep's wool in great abundance. This sheep is apparently indigenous also to the Salt-range.

The following account of the trade in these wools from Kandahar is extracted from **Colonel Lumsden's** Report on Kandahar:—

At Birgand, Hazara, Herát, and Kandahar, when advances are made to the nomads on the future crop, the price on the spot is about 12 Company's annas per Kandahari maund of 4 Company's seers; but if purchased at the time of shearing it cost R1-4 for the same weight; and if taken on credit R1-8. A load of 48 maunds Kandahari or 192 Company's seers, is carried to Kandahar from any of the districts above-mentioned for Company's R12-8, and from this point to Karáchí for the same sum. The reduced rate for the latter distance is accounted for by the road being better, and below Dadar, perfectly safe. The gomashta or agent, proceeding with the investment, receives two-thirds of the profit, taking an equivalent share of risk; but if the arrangement with him is made on the Mahomedan principle (known as Mozarihat), when the agent runs no risk, one-fifth of the profit is absorbed in his pay.

The agent in Kandahar says that the tariff of boat-hire from Karáchí to Bombay varies so much that it is impossible to give even a fair approximation to the expenses of transit; the price in Bombay may be put down at R192 per kándi of sixty Kandahari maunds. Pure white wool is the most marketable; but brown and white are frequently mixed. The wool of Birgand and Herat is generally shorn twice a year; and if not exported, is manufactured into carpets, '*balásins*,' *masnadi namads*, and common felts. The fine wool known as kurak is procured from goats in the Herat, Guzak, and Hazárá districts.

Kirmán is a tract of country close by the Persian Gulf to the south of Persia. The wool finds its way into the Panjáb in considerable quantities. It is a soft delicate wool, but its principal use at present unfortunately appears to be the adulteration of genuine *pashm*. A table is annexed showing the imports of real *pashm* and Kirmáni wool into Amritsar, side by side: the increase of the latter is marked; the subject of the adulteration will be resumed when we come to speak of manufactured shawls.

Kirmani wool 1500 *Conf. with* *pp. 624, 636.*

Statement of Kirmāni Wool and real Pashm imports in the city of Amritsar from 1850-51 to 1861-62.

REAL PASHM.		KIRMANI WOOL.		REMARKS.
Year in which imported.	Quantity imported.	Year in which imported.	Quantity imported.	
	Mds.		Mds.	
1850-51	1,300	1850-51	40	There is no scarcity of pashm, but the agents from Amritsar no longer go up to Bashahr for it, owing to increased import of Kirmáni wool.
1851-52	1,250	1851-52	100	
1852-53	900	1852-53	250	
1853-54	950	1853-54	300	
1854-55	850	1854-55	400	
1855-56	700	1855-56	400	
1856-57	600	1856-57	500	
1857-58	600	1857-58	700	
1858-59	500	1858-59	700	This inferior wool has put the real *pashm* out of the market.
1859-60	400	1859-60	800	
1860-61	400	1860-61	1,000	
1861-62	500	1861-62	1,000	

Mr. Davies writes thus: it is evident that the quantity of shawl-goat's wool, imported into Amritsar has, for several years past, decreased. In its stead, sheep's wool from Kirmán in Persia has been largely introduced into the manufacture of shawls. This wool is fine of its kind, and long in the staple. It is much more easily and quickly worked than the more delicate goat wool. It is largely used in Persia in the fabrication of 'Jamewars,' which have superseded the use of Kashmír shawls in that country.

Plains wools 1501

WOOLS OF THE PLAINS.—We come now to the last class, representing the wool of plains. Among these I have included the wools of Hazárá, because they could not be included in any other. It is here that the first attempt at improving the breed was made, by the introduction of the merino sheep, but there does not seem any great prospect of success. Merino wool that was sent home in 1860; fetched 1*s.* 6*d.* a pound. At present, merino wool in Europe is chiefly produced—the best in Spain, the next best in Saxony. How unsuccessful the experiment may be, there are many other ways in which the breed of sheep might be improved and the wool trade stimulated: of pasturing grounds there is no lack.

PANJAB.

Plains wools.

'There can be no doubt,' wrote the Financial Commissioner in 1861, 'that the valleys of the Sutlej, Rávi, Chenab, Naincukh, and other tributaries of the Indus supply grazing grounds, not to be surpassed in any part of the world. The population inhabiting these are chiefly pastoral, but owing to sloth and ignorance, the wool they produce is small in quantity, full of dirt, and ill-cared for in every way.'

Conf. with pp. 565, 574.

1. Black and white sheep's wool, for blankets, etc.
2. Goat's hair, for grain bags, rope, etc., etc.
3. Camel's hair: the inner wool is used for *chogas* of a common kind, and is very soft. This wool is produced in the 'bar' and 'thal' tracts of Shahpúr, Rohtak, Jhang, and Gugaira, which are camel-feeding districts.

Hazara wool.

Conf. with pp. 578, 608, 612, 614, 615, 633.

1502

"In regard to wool, there is nothing now to communicate of an encouraging nature. When passing through Házárá in April, I saw the merino flock; and was greatly disappointed in it. It is under the care of a shepherd from Hindustan, who appears not at all to like the cold of the hills; and will not willingly venture into the localities best suited to the sheep. The latter appeared to me to be in a most wretched condition, offer a contrast to the half-breeds reared by the Sayads of Kaghan, who confess the wool obtained from these to be much softer and finer than that of their own sheep; though they do not appear greatly to appreciate these qualities. I consider, as I have heretofore stated, that there is no prospect whatever of merino wool being produced to any extent in our hills; or of any wool, least of all the finer kinds, being allowed to grow of sufficient length to be prized in the European markets; as the tangled and thorny woods and forests through which the sheep must pass, oblige the shepherds to shear them at least twice, and usually three times in the year. In the comparatively woodless tracts of trans-Himálaya regions, wool of the finest kinds, and I believe of considerable length, is produced almost everywhere, apart from the inner coat or *pashm*; and enquiries are being made in Spiti, with a view to ascertain how improvements may be effected; and a superior as well as a more abundant article obtained. It seems to me, however, to be doubtful whether any considerable increase of produce could be obtained from these regions, by any available means, at all events, not under existing circumstances"—(*Extract from Revenue Report of Hazár,á 1262-63*).

But that the experiments at cross-breeding were not in the Pánjab confined to the Házárá sheep, the following correspondence and report will show:—

Report on samples of wool from a cross-breed and from a Kohistan ewe, communicated in the following letters from MR. ROBERT SMITH, *of the Commissariat Department, to the Secretary,—Calcutta, 13th March 1843.*

Cross of Merino & Patna.

Conf. with pp. 570, 586, 587, 589.

1503

"I have examined the sample of wool (a first cross between a merino ram and Patna ewe) which you sent to me, and after carefully comparing it with numerous musters of commercial wools which I have by me, my opinion is as follows:—

"The wool is not so good as a sample of cross merino bred in this country which is in my possession, and which is worth 1*s.* 3*d.* per ℔ in the London market. The present sample is somewhat coarser in the fibre, but it has the advantage of a longer staple; its color is also not so pure as it might be, and the fibre has a shade of weakness in it. None of these disadvantages, however, are sufficient to disqualify the sample, *per se.* Wool of this kind, in large quantities, would find a market in England at about 1*s.* to. 13*d.* per ℔ and a little more breeding would bring it up to a much higher standard. Now that the Indus is open, fine woolled ewes from Mekram and Jhawar in Beloochistan might be readily procured, instead of breeding from the coarse woolled sheep of Patna; and with the Jeypoor sheep to give sire, a cross-breed might in a few years be established on this side of India, which would lay the foundation of much wealth to growers, and benefit the country and the revenue materially.

Desirable Cross of Indian Stock.

Conf. with pp. 570, 635.

1504

"While on this subject, could not the Society address the Government with reference particularly to Captain Postans's researches (page 434 of the Monthly Journal of Agricultural Society for December 1842) to procure samples of wool *

* "These samples were presented at the general meetings of the Society on the 8th March and 17th April, the first by Mr. C. J. Muller, Deputy Collector at Patna, the other by Major Napleton, at Bhaugulpore.

SHEEP: Wool.

Production of Wool and Woollen Goods

PANJAB. Plains.

from all parts of India, particularly from the North-West, that those who are desirous of entering into the trade in wool, might know the best sources of supply*:—

28th March 1843.

"I have the pleasure of replying to your note regarding the muster of Kohistan fleece-wool, which you sent to me some few days since.

Kohistan Sheep a Valuable Stock for Breeding.

Crossing with Patna and Merino. *Conf. with pp. 570, 586, 589, 617.*

1505 "From the matted structure of this fleece, it would not prove a marketable commodity in its unimproved state, since it could not be *combed*; but from its softness, length of staple, and fine fibre, the ewes of this breed, if of a tolerable large body, would be well adapted for laying the foundation of a valuable mixed stock, when crossed by merino rams, themselves of good blood. It is a great error that the few attempts which have been made in this respect in India, have been injudicious, in taking the dam from the hairy sheep of Bengal, and the sire from the merino; this is like breeding from the race-horse and the tattoo. The sire, under any circumstances, should be merino of the best blood, and the dam, the produce of merino and best soft woolled country sheep which can be obtained; that is, wool from the *second* generation. There can be no doubt that these Kohistan ewes crossed by the merino would at once yield a wool worth 1*s.* 3*d.* per ℔; but not having seen them, and consequently not knowing their size of carcase, I am unable to say if they would be profitable. If they would yield 4℔ of wool annually, they ought to be.

"In my opinion there is nothing to prevent the growth of good marketable wool in India, provided it be gone about in a proper manner. We have all kinds of climates, and the short sun-burnt grass of this country is precisely that on which the best wool is produced. Bhaugulpore, Mussourie, and the Dhoon, in fact, all hilly districts, with short scanty herbage, little jungle, and wide, sandy plains, with clumps of trees here and there would suit the sheep breeder; only he must commence with proper breeds, if he expects to succeed" (*Jour. Agri.-Hort. Soc. Ind., II., Part II., pp. 159-161*).

The following selection of passages may be given from the District Gazetteers (arranged alphabetically), and it need only be added that, although a few only are given, nearly every volume of the Panjáb Gazetteer says something about wool and woollen manufactures. Some of these deal more especially with *pashm* and *pashmina* goods, and have accordingly been republished in the special chapter on these subjects. It is, however, impossible to arbitrarily isolate the available information into sections that would literally correspond to "Wool," "Pashm," and "Hair." The passages which may now be given, it will be observed, carry our knowledge of this subject down to about the years 1881—84.

Amritsar.

1506 AMRITSAR—"The manufacture of *pashmina* or shawl wool into cloths of various textures and qualities, which is the leading trade of Amritsar, has been already noticed at some length. Opinions differ as to the prosperity or decadence of the shawl trade. But it must be a long time before the habit of shawl-wearing, common among the upper classes of natives, dies out entirely; and, although the European demand is variable, and foreign looms are quick to imitate Indian fabrics, the Amritsar dealers have displayed a facility in following changes of fashion which is very unusual among oriental products. The peculiarly soft and silky character of *pashmina* fabrics, even when the material is largely mixed with inferior wool, is unimitable by European power looms. A beautiful texture of fine shawl cloth, composed of equal parts of silk and *pashmina* is now made. The fabric is lustrous and exquistely soft, and is woven in self-colours. Modern taste inclines to plain surfaces, and the numerous sub-divisions of the trade dependent on the old style of coloured work, such as dyers, embroiderers, *rafúgars*, etc., have undoubtedly suffered a good deal from the changing fashion.

Carpet weaving.

1507 "The introduction of carpet-weaving promises to fill up to some extent the gap created by the falling off in the demand for elaborate shawls. The most important establishment employs about 300 persons who work on fifty looms. The greater part of these are boys, apprentices or *shágirds*, who are learning the trade. There are also several other smaller manufacturers. The Amritsar carpet, so far as can be judged from the products of the first years, promises to have a distinctive character.

* "The Cattle Committee have been requested to take this suggestion into consideration, and embody their opinion thereon in their next report, which will shortly be submitted to the Society." [*Conf. with various reports quoted on pp. 582, 586 Ed.*]

PANJAB.

The designs are mostly made by Kashmiris, and are based on shawl pattern motives. The colouring is very dark, sometimes rich, but inclining to gloom. The texture is much lighter than that turned out by the jails, and the carpets are softer and more pliant, but there is no reason to doubt their wearing qualities. In this respect they resemble, as might be expected, the carpets of Kashmír which are still softer and looser. Nearly all are sent to London or New York, and they appear to be unknown among Anglo-Indians. The Central Asian fabrics known in the market as *khoten* carpets are frequently brought into Amritsar. Many of these are admirable in colour and design and marked by an almost Chinese character. They have not, however, been used as models for imitation. A large number of Amritsar carpets were shown at the Calcutta Exhibition, 1883-84" (*Dist. Gazetteer, pp. 44-45*).

Dera Ghazi Khan. 1508

DERA GHAZI KHAN.—"In the border hills in this district there is an interesting domestic industry of woollen weaving, the products of which resemble the Arab or Semitic type of woven fabrics more than any other work found in India. The coarse and every-day forms of this pastoral craft are rough goat's hair ropes, the rude cloths on which grain is winnowed and cleaned, corn sacks, camel-bags and the like, which are used throughout this district and in the Deraját Division generally.

"More highly finished forms are camel-trappings, saddle-bags, *shatranjis* or rugs and similar articles woven by Biloch women in a somewhat harsh, worsted-like yarn, dyed in a few sober colours. The patterns are as simple as the material, but they are always good, and there is a quality of tone and colour in the stuff which more costly fabrics seldom possess.

"In addition to the woven pattern, saddle-bags are ornamented with tassels in which white cowries are strung, and with rosettes skilfully and ingeniously worked in floss silk of different colours, with *ghogis* (small oblong shells like seeds) sewn on the borders. The rugs have great wearing qualities, as warp and weft are both in hard wool; but being often crookedly woven, they do not always lie flat. The trade in Turkistán rugs and in some Algerian fabrics of a similar kind is supplied by a merely domestic industry, which finds employment for many hands. There are no signs that the Biloch weaving will grow to anything more than it is at present,—a household occupation for merely local use. The work is, however, interesting as an example of the instinctive 'rightness' and propriety of design and colour which seem to be invariable attributes of pastoral industries. It is curious that rugs almost identical in pattern and fabric, and similarly decorated with shells, are made in the Balkans, and sometimes sent to Paris for sale. The Banjáras of the Deccan weave a fabric identical in pattern with the Biloch work for women's petticoats and the peaks of bullock-saddles" (*Dist. Gazetteer, 89*).

Gurdaspur. Egerton Mills. 1509

GURDASPUR —"The Egerton Woollen Mills were started 1880, but manufacturing did not commence until the end of October 1882. The firm employs about 100 hands, who work for some eleven months in the year. There are three European supervisors, and work in the mills is very active. The cloth turned out is cheap and wonderfully good; and large contracts for the supply of regimental clothing have lately been secured. These mills are at Dháriwál, about seven miles from Gurdáspur on the banks of the canal. They are lit up at night with electric light. The range of building is very extensive.

"*Wool.*—Two sorts of wool are chiefly used—the *zer* and *gaddi*. The first comes from Sháhpúr and Siálkot, and the second from the country inhabited by the Gaddis, *i.e.*, Chambá and thereabouts. Women are employed in separating and cleaning the wool. A common industry in this district is the working of borders to *pashmina* shawls in different coloured wools. A man will work about one yard of this in a day. A yard of work is worth 4 annas, *i.e.*, 1½ annas woollen thread and 2½ annas as labour.

"Blankets or *lois* are also made. A good blanket, worth R20, will take about a month to weave, the cost being thus divisible: R14 stuff, and R6 labour. The blankets are made from district wool, and that which comes from Siálkot and Amritsar. The chief seats of this trade are Fatehgarh, Dharmkot, and Ikhláspur. There is some export of these blankets to Amritsar and Siálkot districts. The wool used is bought at R16 the maund, and the blankets sell at from R2 to R4 each. The manufacture of *pashmina* shawls may be divided into three heads—that of weaving shawls, weaving shawl borders (as before noted), and shawl embroidery. The shawl-work is carried on by Kashmiris at Sujánpur, Dinanagar, Dera Nanak, Pathankot, Kanjour, and Batálá, and the trade is apparently in the hands of a few men. Especially is this the case at Dera Nanak, where there are many shops full of workers, all seemingly employed by one master. The pay is wretchedly small and the workers have to supply their own materials. They sit working crowded together in small

Production of Wool and Woollen Goods

PANJAB.

shops, and their life must be a perfect slavery; yet they work at this unremunerative toil, the pay being but 2½ to 3 annas the day, when they could command from R5 to R6 a month as daily labourers at the neighbouring railway works. Of late years the trade has decreased in shawls, and prices now do not range high. There are three kinds of wool, used in the manufacture of shawls—Kashmiri, Rámpuri, and Wáhabsháhi. The Amritsar prices of these are—Kashmiri wool R5, Wáhabsháhi R3, Rámpuri R2 the seer. One-and-a-half seer of wool is calculated to make a length of 6 yards, at a cost of R18, namely, 3 seers of wool at R3 per seer, R9; spinning the thread, R4; wages of two persons for one month (one man and one woman), R5; total R18. The wool and silk used for shawl borders are obtained from Amritsar. The borders sell at from 2 to 8 annas per yard. The pattern resembles a thick flowerel ribbon. This is used in fringing the shawls. Embroidery work consists of working flowers and fancy work on shawls with worsted and silk thread. *Lois*, or wrappers, are also manufactured of wool, the first by Kashmiris, and the second by Juláhas. The *loi* manufacture was referred to under the head "Cotton."

"**Mr. Kipling** has kindly furnished the following note upon the manufactures of Gurdáspur:—It is customary to say of the woollen industries of the Gurdáspur district that they are dying out or falling off. But it seems doubtful whether they were ever really very prosperous. At Sujánpur, Dinanagar, Dera Nanak, Pathankot, Kanjour, and Batala, there are Kashmiri weavers and embroiderers who carry on their trades for a wretched pittance which would seem to be scarcely enough to keep body and soul together. They are, like so many more artizans of the province, practically enslaved to dealers, and earn but 2½ to 3 annas per diem. The masters in their turn find but a precarious sale for their goods, and the wonder is that so much good work is turned out under conditions so desperate. Fortunately, there are still large numbers of people in this country who wear coloured woollen shawls. A large crowd of the people of Bengal, such as was daily seen at the Calcutta Exhibition, shows at a glance that though Governments and Native Princes no longer encourage the manufacture of the best kind of shawls for their *tosha khans* and for gifts, there is still a market for ordinary woollen goods. Many of the native ladies of Calcutta insisted on

Trade in Kashmir Shawls.
1510 visiting the Exhibition, and it was seen that the wearing of shawls was by no means confined to the male sex. But the months during which a woollen shawl is comfortable in the North-Western Provinces, Bengal, and Bombay are but few, and in spite of the efforts of dealers who travel unceasingly, the consumption must be relatively small. There is not a town of any importance in India in which Panjáb woollen goods are not found awaiting sale. The adoption of a semi-Europeanized costume by many of the educated classes might perhaps be thought to tell heavily against the shawl trade. But against the number of educated natives who have adopted the closely fitting coat of English woollen cloth must be counted those of the uneducated classes, who, formerly wearing cotton alone, are now sufficiently prosperous to afford wool. And this would seem to be a large class. It seems clear that the Kashmiri shawl must for a long time to come be in some demand, but it is no less clear that there is an excessive supply. At the Panjáb Exhibition of 1881 the cheapness and good quality of the woollen goods from this district were commented upon by the jurors. A large *jámewár* (striped fabric suitable for a curtain) cost R6 only, and although somewhat coarse in texture, it was decidedly what English tradesmen called 'good value.' A speciality of the district is its *kenára báf*, woollen shawl-edgings or borders. Many of these are pretty in colour and capable of being utilised by European milliners and dress-makers. For furniture too, except in this country, the modern fanciful upholstery might find them a place. But the perpetual change in European fashions and the facility with which Western stream-driven loom can imitate and undersell any fabric that attracts public notice, forbid any hope of local industries receiving a payment benefit from European trade. At this moment the Rámpur *chaddar* and similar soft wool goods

Rampur Chaddars.
1511 are in some favour in England. It is true that a number of Panjáb *chaddars* are sent home and dyed in soft colours, which are supposed to be peculiarly Indian, but the greater part of the goods advertised as 'Amritsias' and under other oriental names are of French or English make. The narrow widths in which the cheaper cloths, such as *pattús, alwáns*, and *malidas* are made, render their adoption by Europeans almost impossible. But for this, which seems to be an insuperable difficulty to the ignorant hand-loom weaver, there might be a chance of employment for many weavers. There is no recognizable difference between the shawl-work of the Gurdáspur district and that of Amritsar and Kashmir. Much of the material used is brought from Amritsar, and some of the finished articles are thus disposed of.

"Mixed fabrics, English cotton thread, and country wool, are made at Pathankot, Sujánpur and Dinanagar. The *loi*, a coarse cold weather wrap in greyish white, is the usual article, and it is exported in some quantities to Amritsar, the North-

PANJAB.

Western Provinces, and Bengal. At Fatehgarh, Dharmkot, and Ikhláspur all-wool *lois* are made. *Pashmina* of course is not used in these goods, but the ordinary wool of the district.

"The establishment of a woollen cloth factory with English power-looms and English methods of dyeing and finishing cannot fail, if it proves successful, to have some influence on the production of self coloured woollen fabrics. The Egerton Woollen Mills Company, whose factory is at Dháriwál, 8 miles from Gurdáspur, produce blankets and all the coarser varieties of *lois* and *pattús* as well as more highly finished broad cloths, serges and other strong woollen goods. Their looms are driven by water-power supplied from the Bári Doáb Canal. For the coarser fabrics, country wool is used, but Australian wool is also imported and worked up in the finer goods. These cloths can be put in the market at rates relatively much cheaper than the ordinary hand-woven woollen goods, and seem likely in time to take their place to a large extent. But as the profits of such an enterprise must depend mainly on regular wholesale production and in contracts for military and police purposes, it may be long before the domestic blanket-weaver is driven to other occupations. The *súsis* of Batálá have a good reputation. They are striped like all *súsis,* but often have an admixture of silk. Colonel Harcourt, who has reported at length on the industries of the district, suggests that the fabric is very suitable for shirts, and there can be no doubt that it is a serviceable and agreeably coloured stuff. But the narrow width in which it is made would be a bar to its adoption for this or any other European purpose. Its chief use is for women's *pajáma*, each pair of which consumes a much larger quantity than the uninitiated would imagine. The *súsis* answer in some sort to the silk-bordered cotton goods of Bombay and the Central Provinces.

Lois. 1512

"*Lois* and wrappers of an inferior description, made of cotton and wool, in the proportion of two-thirds to one-third cotton, are largely manufactured in the towns of Sujánpur, Dinanagar, and Pathankot, and are exported to very distant parts of India,—Calcutta, Benares, and Lucknow. The total value of export may be fixed at R40,000. The usual time for export is November. During the Cabul war a good deal of this material was brought up for the use of syces in the expedition. The wool used in the manufacture of this article is imported from Sháhpur, and from the country inhabited by the Gaddis, *i.e.*, Chambá and thereabouts.

Blankets. 1513

"Blankets are also made in the towns of Fatehgarh, Dharmkot, and Ikhláspur from district, and that which comes from Siálkot and Amritsar. There are some export of these blankets to Amritsar and Siálkot districts. The amount of export is about R2,000. Besides the native manufacture of woollen articles, the Dharmkot woollen mills, which are situate on the Amritsar and Pathankot road, 7 miles from Gurdáspur, are now supplying the police and troops in the Panjáb with woollen fabrics of a very superior description. The amount of export is very great, though it cannot be stated with any degree of accuracy what it is, as no information on this head has been received from the Manager. There is, however, reason to believe that when the works, which are still under construction, are completed, this district will be the centre of trade in woollen goods" (*Dist. Gazetteer, pp. 56-59, 64, 74*).

Jhang. 1514 Yield 2 lb.

JHANG.—"Sheep are shorn twice a year, in September-October and April-May. About a seer of wool is given in the two shearings. Wool is now a very valuable commodity, and zamíndárs say that flockmasters in the Thal wear bracelets of gold. It mostly goes down to Karáchi. The figures below give the price of *Bar* wool and also of goat's hair at Maghiána for the last twenty years, in rupees per maund.

Prices. 1515

YEAR.		1861.	1862.	1863.	1864.	1865.	1866.	1867.	1868.	1869.	1870.	1871.	1872.	1873.	1874.	1875.	1876.	1877.	1878.	1879.	1880.
Rupees per maund.	Wool .	8	10	12	11	10	9	7	8⅓	9	8⅓	8⅓	10	12⅓	4⅓	10	10	9	10	12	18½
	Hair .	5½	5⅝	1	5⅝	5¼	5¼	5¼	4¾	5	5	5 5/16	5¼	5½	5⅓	5¾	9	6⅓	6½	8	9

Sheep skins are used for making women's shoes, covering saddles, etc. As far as the age at which put to the male, number of kids produced, and method of rearing, there is hardly ány difference between sheep and goats. A goat gives from 2 seers to ½ seer of milk a day; nothing is made from milk. A goat is usually killed when 5 or 6 years old. Sheep and goats produce about five times. Goat's hair is shorn every six months, and is made into pannier, bags, saddle bags, ropes, nose-

SHEEP : Wool.

Production of Wool and Woollen Goods

PANJAB.

bags, *salitas*, etc. It is called *jat*. The names of sheep and goats according to age are given below :—

	SHEEP.		GOATS.	
	Female.	Male.	Male.	Female.
To 6 months . .	Leli	Lela	Bakra Pathora	Pathori
To 1 year . .	Ghirapi	Ghirap	Chhilota	{ Kharapi Kharap
Afterwards . .	Bhed	Chhatra	Chhela	Chheli"

(*Dist. Gazetteer, pp. 126-27*).

Lahore. 1516 LAHORE.—"It is contended by some workmen that the fine *pashmina* woven at Lahore is superior to that of Amritsar. Whether this is true or not, there seems to be some reason for the belief that the trade has somewhat improved of late years. *Chadars, dhussas, patkas* and other articles are made. In attendance on the loom-embroiderers are always to be found Kashmíris, and there are many in Lahore. Besides fine goods, coarse woollen blankets (*lois*) are made. The greater part of this hand-weaving, both cotton and wool, is entirely unnoticed by Europeans, very few of whom venture into the city or tread the narrow alleys of such suburbs as Mozang. One slight indication of the extent of this domestic craft is afforded by the fact that the shuttle-maker's trade is, as such small trades go, a busy one. At every fair one or two stands will be found where weavers' shuttles are sold. A good shuttle lasts for many years, and is carefully handled and cherished. Perhaps it is fair to conclude that hand-loom weaving after all is scarcely so dead as might be expected from the large import of English piece goods."

Shahpur. 1517 Felt. 1518 SHAHPUR.—"Felt or *numda* rugs are made at Bhera and Khusháb, in both white and grey, unbleached or coloured wool, decorated with large barbaric patterns of red wool merely felted and beaten into the surface. The white felts bear no comparison with those of Kashmír and parts of Rájpútána, and the texture is so loose and imperfect that they seem to be always shedding the goat's hair with which they are intermixed. The wool is not perfectly cleaned, and they are peculiarly liable to the attacks of insects. But they are among the cheapest floor coverings produced in the Province.

Ghi. 1519 The chief animal products are wool, *ghí*, and hides. It is estimated that the shearings of the large flocks of the *Thal* and *Bár* yield annually not less than twelve thousand maunds, or upwards of four hundred tons of wool. Of this, probably two-thirds are exported, and the remainder consumed in the manufacture of blankets and felts. The fleece of the *Thal* sheep has the reputation of being the finest in the Panjáb. The sheep are sheared twice in the year, in the months of April and October, the average yield of each separate shearing, called a *pothi*, being about three-quarters of a *ser*. The wool is bought by the *pothi*, so that, in speaking of the market price, it is customary to quote the number of *pothis* obtainable for the rupee. Average selling price, four *pothis* per rupee, gives eight annas as the annual yield in cash per head of sheep to the owner. This will sufficiently account for the great rise in price of these animals of late years. The head-quarters of the trade in wool is Núrpur, in the *thal*, where a superior kind of blanket or *lúi* is made. A good deal of the wool which is produced in the *Bár* is made into felt at Bhera which supplies a large part of the Panjáb with this article (*Gaz., pp. 73-76*).

Woollen Manufactures of the Panjab.

MANUFACTURES OF THE PANJAB. 1520 Mr. D. C. Johnstone (*Monograph on Woollen Manufactures of the Panjáb in 1884-85*) furnishes much information of a practical nature more especially regarding the manufactures and the appliances. Space cannot be afforded to do more, however, than exhibit some of the passages from that report which deal with the raw materials :—

"The quantity and value of the woollen manufactures of this Province, though not insignificant, cannot for a moment be compared to the outturn of cotton goods: where

PANJAB.

in the one case the figures are units, in the other case they are tens. But, though this is so, when the figures for the whole Province are taken, there are, on the other hand, a few districts in which the manufacture of woollens is a really important industry.

Raw materials: 1521 Wool. 1522 Pashm. 1523 Goats' hair. 1524 Camel's hair. 1525

"The raw materials to be considered are four in number—sheep's WOOL, PASHM, or the wool of the Tibetan shawl goat, GOAT'S-HAIR, and CAMEL'S-HAIR; but it need hardly be remarked that the last two are not wool properly so-called. *All* wool has, in consequence of its structure, more or less of the property of felting, while goat's hair and camel's hair, having a different structure are *never* found to possess this property.

Estimated outturn. 1526

"The quantity of these materials produced in the Panjáb *per annum* can only be given very approximately, as some district reports contain no information on this point. But if for these districts a judicious conjectural addition be made for local outturn based on the relative size, situation and climate of each, the following figures are arrived at:—

	Sheep's wool.	Pashm.
	In Maunds.	*In Maunds.*
1. Produce of *Province*	63,000	*Nil.*
2. Imported by Sea	1,900	4,500
3. Imported from Afghánistán, etc., by land for use in Province	7,500	1,500
4. Imported from Kashmír, Ladákh, Chinese Tibet, etc., for use in Province . . .	3,000	1,500
5. Imported from Bikanír, Baháwalpur, Rajpút States, etc., for use in Province	2,000	*Nil.*
TOTAL .	95,400	7,500
Deduct export of local produce	13,000	*Nil.*
Balance worked up in Province	82,400	7,500

"*Pashm* is not grown within the Province at all, except to an inconsiderable amount in Spiti.

Wool-growing tracts. 1527

"As wool-growing tracts the important districts are Hissár, with 1,550 maunds; Fírozpúr, with 2,800 maunds; Lahore, with a local produce of over 4,000 maunds; Jhang, with about 3,600 maunds; Shahpur, with 8,000 maunds; Pesháwar nearly 3,000 maunds; Dera Ismail Khan, with 11,700 maunds; Amritsar, with 2,000 maunds; Mooltan, with 2,714 maunds; Ráwalpindi and Jhelum, with from 2,300 to 2,700 each. Of the wool of the plains that of the *Bar* country is deemed better than that of the *Thal*; but the plains wool is on the whole miserably poor, with the exception of that produced in the south-eastern part of the province. In the above figures Hissár is mentioned not so much for quantity as for quality.

Varieties of staples. 1528

"Of this product there are several different colours; in the plains black seems to be almost as common as white; and in the hills, and especially in Kángra and Kulu, sheep may be black, or white, or bluish-brown, or reddish-brown, or grey; while the staple varies in length from two inches in common breeds to six or even more in the case of certain hill breeds. Probably the softest and finest wool in the hills is the Lahauli wool, of four to five inches in length; but even this is said by the Manager of the Egerton Woollen Mills to be inferior to Australian wool. The long hill wool, however, takes dye badly.

Districts in which goat's hair is chiefly produced. 1529

In the production of goat's hair the Mooltan district is pre-eminent with the large figure of 1,403 maunds; and then come Shahpur with 1,100, and Gujrat with 600 maunds. Dera Gházi Khan is said to have an outturn of 1,820 maunds of goat's and camel's hair together, but no details are given. The total for the province is about 9,000 maunds, and the import small.

Districts where camel's hair is produced. 1530

Camels seem to be sheared chiefly in Shahpur and Hissár and Dera Ghází Khan. The total produce of the province is not more than 2,400 maunds, and import is small.

Imports of wool by sea. 1531

The wool imported by sea for use in the Panjáb is mainly Australian, and goes chiefly to the Egerton Woollen Mills; that imported from Afghánistán comes chiefly into Dera Ismail Khan and Dera Ghází Khan. Figures are not available, but most of this import is of the wool of the *dumba* sheep. That from Kashmír territory, Tibet

SHEEP: Wool.

Production of Wool and Woollen Goods

PANJAB.

and Central Asia, generally passes through Srinagar, or through Kulu and Kángra, and is used chiefly in Amritsar, Lahore, and Ludhiána, while a considerable amount has till lately been exported to Europe, a trade which has since the year under report suffered a severe check owing to the fall in prices in the London market. The wool imported from Bikanír and Baháwalpur finds its way chiefly to the river port of Fázilka, and the Bikanír article, which is remarkably fine, is taken to Karáchi for export to Europe. In fact, comparatively little of the wool, approaching to 2 lakhs of maunds, that went from or through the Punjab to Karáchi and Bombay, is Punjab wool: the mass of it being Rajputána wool (going *viá* Rewári and Fázilka and Mooltan), Himálayan, trans-Himálayan and Bikanírí.

Imports of pashm. 1532

The pashm imported comes from Persia by sea, from Bokhara, etc., *viá* Kabul (known as *Kabuli,*) from Kashmír, and from Tibet and Ladakh and countries beyond. The first mentioned, which is known as *Kirmáni* and is not so much esteemed as the others, finds its way to Amritsar and Lahore; the second to Lahore and Amritsar; the third, a small part of which is the product of the Kashmírí camel, to Ráwalpindi, Gujrát, Lahore and Amritsar; and the last to Nurpur in the Kángra district, to Amritsar, Ludhiána and Lahore. Of the *pashm* imported *viá* Kábul a small quantity is the product of the goat of Turfán which grows only pashm and no hair.

Wahab Shahi pashm. *Conf. with pp. 616—017, 636.* **1533**

In Amritsar, Nurpur, Ludhiána, and Lahore, a product called *Wahâb Shâhi pashm* is used. It comes from Persia and is passed off as pashm, being used by itself or mixed with real *pashm.* This is really only fine sheep's wool and not *pashm* at all. Most, too, of the stuff that goes to the Punjab *viá* Simla from Rámpur is not pashm though sold as such.

Wool locally produced, how utilized. 1534

It remains, with reference to this part of the subject, to give some idea of what becomes of the locally produced wool. Mooltan, Muzaffargarh, Jhang, Montgomery, Dera Ismail Khan, Gujranwála and Shahpur export, through the first-named district towards Karáchi, as a rule, something like 10,000 maunds of locally produced wool; but since the year under report all export has received a severe check, as stated above. Some of the wool of Rewári goes to Cawnpore and Meerut. And here we have stated the only considerable exports of Punjab-grown wool. It is not, however, the case that all the other districts keep their own wool for manufacture. Thus, while Rawálpindi, Hoshiárpur, Jullundur, Hazára, Peshâwar, Kohát and Bannu seem, except occasionally in small quantities, neither to import raw wool nor to export the locally grown article; Lahore, on the other hand, gets locally grown wool from Fírozpur, Gujrát, Gujránwála, and Siálkot; Ludhiána from Sirsa and Firozpur and Amritsar from the, districts to the north and from the hilly tracts of Kángra and Kulu.

Breeding. 1535

Except in Sirsa it does not appear that any efforts are made to improve stock and even there the only step taken is the castration of all but the finer rams: a good step, but one that should be supplemented by judicious introduction of fresh strains of blood.

Shearing. *Conf. with p. 589.* **1536**

Shearing is done twice a year, in spring and autumn, except in Hissár, Kángra and Kulu, where there is an intermediate shearing in June; and the usual wage is one-twentieth of the wool shorn.

Weight of fleeces. 1537

"The weight of a fleece varies from three chittaks, in the case of the inferior plains sheep, to one seer in Kángra hill-sheep. As these Kángra sheep are shorn three times, it may be taken that three seers is the very largest annual yield from any sheep in the Province. Excluding Kángra and Kulu sheep, the annual yield cannot exceed one seer *per* sheep, and the quality also is inferior, while average English sheep certainly give as much as five pounds and some breeds give seven and even eight pounds, a comparison which is eloquent of the poverty of the ordinary Panjáb breeds.

"It is only here and there that fleeces of dead sheep are used. They are plucked out and not shorn.

Clipping— (a) of goats; (b) of camels. 1538

"The clipping of goats is done once a year in nearly every district, and the yield is about half a seer at a clipping. Camels are clipped once a year. The yield at each clipping is one to two pounds for a male and two to four for a female, the cause of the difference being that the back and shoulders of the male camel are never clipped.

Selling price— (1) of wool; (2) of goat's hair; (3) of pashm. 1539

"Each sheep's fleece when cut is made up into a bundle, and in this state sells at an average rate of 2½ seers per rupee. The average price of goat's hair is about 13 seers per rupee; and of camel's hair 5 seers. The *pashm* imported is worth one rupee and eight annas a seer on an average and *Wahâb Shâhi pashm* (so-called) about 12 annas a seer.

Further processes for sheep's wool up to spinning. 1540

"At this stage it is necessary to divide the subject and to consider the further processes for the different materials separately. All the processes to be described now are not universally employed. The processes are sorting, washing, picking out foreign bodies by hand, teasing out matted portions by hand, carding or scutching with the bowstring, and combing.

PANJAB.

Sorting. 1541

"Wool sorting in the Panjáb is done in a very primitive style, and indeed in some parts it is not done at all. Where done only two qualities are recognised, the better wool and the worse wool.

Washing. 1542

"Washing of the wool is not common, nor is it very necessary except for wool loaded with sticky matter. Unless done carefully and with suitable soap it is very bad for the wool; and picking by hand or some other process is, in any case, still necessary for the removal of burrs, thorns, seeds, etc., entangled in the fibres.

Hand-picking. 1543

"Picking out of foreign bodies by hand is done everywhere. It is a very tedious process in the case of wool grown in tracts abounding in thorny bushes and undergrowth; but none of the other processes avail to remove burrs. The ordinary wages for this and the next process (teasing by hand) are certainly not more than one anna a day, the workers being nearly always women. The mere process of hand-picking involves a certain amount of teasing out of matted portions of wool; but where scutching and combing are uncommon, a state of affairs which appears to exist in the Jhang district and in Jullundur and Ludhiána, something more than this must be done; the wool must by hand be reduced to a homogeneous mass of fluff. But to effect this purpose in most districts either the *pinjan* (bowstring) or the comb is used.

Scutching or carding. 1544

The *pinjan* has been described in the Monograph on Cotton. A bow is suspended, string downwards, at such a height that the string passes through the wool to be operated on. The string is then made to vibrate violently either by twitching it or by striking it with a hammer, and the vibrating string catches up and scatters the wool about. Besides the opening out of the separate fibres, dust and all dirt, not viscous and not prickly, are shaken out. The instrument is used in nearly every district of the Panjáb; and nearly everywhere the work is done by men of some low caste. In most places there is a separate caste of *pinjas;* but it is also true that in many districts there is no such special caste. The wages vary from half an anna to one anna per seer scutched, which certainly does not give more than two annas per diem. The scutcher is generally a man, but in one or two of the south-eastern districts women scutch for themselves with a small *pinjan.*

"It must not be supposed that the bowstring and the comb are merely alternative instruments for effecting the same purpose. The former opens out the wool and loosens its mass, but leaves the fibres lying confusedly in all directions; while the latter tends to open out the wool and also to lay fibres side by side in parallel lines. The former is used when woollen thread is wanted, the latter when the spinning of worsted is the object.

Combs. 1545

The combs used in the Punjáb are of two sorts, single and double. The double are reported to be used only in Gujránwála, Amritsar, and Lahore; and the single comb is found in Siálkot and Fírozpúr.

"The double comb (*shána kanga*), which is the more effective of the two, consists of a piece of wood, laid on the ground, with two parallel rows of vertical iron teeth standing on it, there being 20 teeth about 4 inches high and the intervals between the two rows and between the teeth being 1 inch, and ½ an inch, respectively. The teeth are rigidly fixed to the platform which is kept steady by the operator's feet; and he does the combing by taking a flock of wool, striking it upon the teeth and drawing it gently downwards through the teeth at right angles to the rows. Before combing the wool is teased in the fingers, and sometimes, though not often, is scutched. It will now be clear that combing is an addition to, and is by no means a substitute for, the mutually similar processes of teasing and scutching. The single comb is a very primitive instrument and has very imperfect effects. In its rudest form it is a mere *panja* or claw which cleans rather than combs, though it does comb to some degree. The wages of a comber who combs out 4 seers *per diem* do not exceed 2 annas. Neither the single nor the double instrument is used for combing short-staple wool, nor could it be employed to any effect for such a purpose. The people do not seem to have discovered how much easier a comb is to work with when heated than when cold.

Wool when cleaned and prepared, made into balls. 1546

"The wool when teased or scutched or combed, as the case may be, is made up into balls (*punis*) and the next operation is spinning.

Spinning. The Charkhi. 1547

"The *charkhi*, with which wool as well as cotton-spinning is usually done, has been described in the *Cotton Monograph* as follows: 'It is formed of two parallel discs, the circumferences of which are connected by threads, and over the drum so formed passes a driving band also made of thread, which communicates a rapid motion to the axis of the spindle. The end of a 'puni' is presented to the point of the spindle, which seizes the fibre and spins a thread, the 'puni' being drawn away as the thread forms, as far as the spinner's arm will reach. Then the thread is slackened and allowed to coil itself on the body of the spindle until the spindle is full, when it is removed.' The process is the same for wool, and a spindle full of woollen or worsted thread is called *challi* or *mudha.*

The Dherna or Takli. 1548

But in some parts, notably Kulu, the *charkhi* is quite unknown and the instrument used is the *dherná* or *takli.* In Simla and a few

PANJAB.

other places both *charkhi* and *takli* are used for woollen spinning. A portion of a 'puni' is drawn out and held to the upper point of the instrument, and wound round it. The *dherná* is then spun round in the hand, and when it has got firm hold of the wool, it is allowed to hang in the air suspended by the thread it is spinning, the right hand of the operator keeping up the rotary motion, while the left hand regulates the draft of the wool. When the thread is getting so long as to put the *dherná* out of reach or to let it touch the ground, the draft of wool from the *puni* is stopped and the piece that has been spun is wound on the *dherná*. The *charkhi* is said to produce a more even and reliable thread than the *dherná;* and this can be readily understood to arise from the superior regularity of the rotary motion in the former machine.

Twisting. 1549 "When yarn has been spun it is generally found that it is too thin at places to bear the strain of weaving, or a coarse thick fabric is wanted. The yarn has therefore to be doubled or trebled, and sometimes more than three folds are given. For twisting, as this process is called, the *charkhi* can be used and also a form (called *masán*) of the *dherná* or *takli;* the difference being that the upper end of the spindle has a narrow, curved groove about half an inch long running from the point along and round the rod, and in this groove the threads twisted together are run.

Import of European yarn. 1550 "The import of European yarn is not very considerable. In the Lahore district it amounts to 500 maunds at least; but in other districts, judging by the reports sent in, it is used mainly in jails in small quantities. It is never used except for fine fabrics; or for knitting."

The loom used in weaving is the same as that for cotton, so that **Mr. Johnstone's** account of it may be omitted, but a few passages may be taken from the remainder of his monograph in illustration of the fabrics commonly turned out in the Panjáb:—

Felting after weaving. 1551 "The cloth after weaving is rough and threadbare in appearance and it has now to be felted. This is done by immersing it in water in which has been made a lather of soap or *ritha* (**Sapindus detergens** *p. 468*), and kneading the cloth with the hands or feet. If the cloth is then pegged out to dry, shrinking is avoided; if not pegged out, it shrinks considerably. Whether pegged out or not the surface becomes uniform and the separate threads are either not distinguishable at all or very little so. If the cloth has been made out of real worsted yarn no felting is attempted: such yarn is used when cloth like serge is made in which the threads are to remain visible; but such cloths of country make are uncommon. In all cases, too, washing after weaving has to be done to clean the cloth.

Process adopted in Kulu for raising the nap. 1552 "Finally, in some districts and specially in Kulu, a stiff brush (*thákárú*) is used to raise the nap. The bristles are made of small slivers of cane, which serve the purpose fairly well but are inferior in the requisite horny elasticity to the teazle (**Dipsacus Fullonum**), a plant that has been grown with success by a settler in Kulu and which could easily be grown anywhere in the Himálayas at moderate altitudes.

Namdas. 1553 "For articles made out of unspun wool the general name is *namda* or felt, and they are used for bed and floor rugs, for horse cloths, for lining ice boxes, and for other purposes.

"Though the details of manufacture differ, the principle is everywhere the same. The wool is scutched or hand-teased and washed. A layer of it is then spread out over a mat that can be rolled up like a door-chick. The thickness of the layer depends on the thickness of the *namda* wanted, and to produce a good article the thickness must be uniform. Then water is sprinkled well over the wool and the mat is carefully rolled up and subjected to pressure by the feet or hands and kneaded for a period varying from one to three hours. In some cases this finishes the process; but sometimes the mat is opened and the *namda* turned upside down and the process repeated. In very many districts, too, mere sprinkling of water is not deemed sufficient: the natural felting property of the wool being small, the wool has to be soaked in a solution of soap, which, drying, causes its fibres mutually to adhere; and in one district, *viz.*, Dehli, a mixture of chalk and gum has to be added. It is clear that the best *namda* must be those which are made from wool which felts merely with water, and that the use of any viscous substance to produce this effect is a sign that the wool is not really fit for *namda*-making. It stands to reason that a *namda* depending for its compactness on any substance soluble in water is at the mercy of the first heavy shower of rain or of the first more than momentary immersion in water. *Ritha* lather, used in some places, is not objectionable, as it assists real felting and is not sticky.

Namdas plain and ornamented. 1554 "*Namdas* are made of a single colour and also in patterns some of which are very pretty. As a *namda* is never intended to be washed, the dyes used in the pattern are seldom fast; for the use of a fixing ingredient would be an unnecessary expenditure.

PANJAB

To make a *namda* with pattern, the pattern is first laid out on the mat and the ground-work wool is spread over it; or the ground-work is spread out first. The patterns are sometimes geometrical, but sometimes contain conventional art foliage and flowers. In these latter the fundamental rule that where a curve springs from another curve or from a straight line, they should be mutually tangential, is ignored to the ruin of many fine combinations; and it is not unlikely that many other offences against true artistic principles are perpetrated. *Namda*-makers are not *Juláhas*, but belong to different castes. In Hazára 'telis' do the work. The usual wages are two to three annas *per diem*.

Blankets. 1555

"The *lois* of Sirsa and Fattahabad in the Hissár district and of Ludhiána are fine in texture and warm, while those of most of the Panjab districts have no special excellence.

"A description of the blankets (*i.e.*, *lois*, *pattú*, *kammals*, *bhurás*, etc.) of the province would occupy much space if it went into details. In point of intrinsic excellence we find the pattu of Waziri-Rupi, in Kulu, pre-eminent. In texture, in lightness, in warmth, and in simple artistic beauty, these blankets are very remarkable, and no mere description of the patterns could convey any adequate idea of them. And between these blankets and the wretched loosely-woven coarse *kammal* of the district of Hoshiárpur, there are fabrics of every intermediate degree of quality. Taken generally it may be said that the quality is poor; but it may be doubted whether any Western race, with the like appliances, would produce anything as good.

Kinds of blankets and woollen piece-goods. 1556

"There is no essential difference between *loi* and *bhúrá*, and *kammal*. *Lois* seem generally to be made white and *kammals* black (dyed or natural); while *bhúrá* are much the same as *lois*, and the two words may be taken almost to be synonyms. There is also no essential difference between the texture of *alwán* and that of *loi*. *Alwán* is cloth in long pieces for cutting up by the tailor; a *loi* may be of exactly the same make, but it has some sort of edging, *e.g.*, a single coloured line, and is for use as a blanket. In the Hissár district *dhablu* is made, cotton and wool being mixed in the manufacture. A similiar manufacture elsewhere is called *garbi* (or *garvi*) *patti*.

Quantity and value of woollen manufactures 1557

"I have not been able to make any estimate from the reports sent in of the quantity or value of the total woollen manufacture of the province for the year under report. But in 1880-81 it seems that the total annual outturn was estimated officially at R12,24,691; and the value has certainly not diminished. But a few reported details may be of interest. The Hissár district exports 18,000 yards of *loi* to Multan and Delhi, and imports 72,000 yards of blankets from Ludhiána and Firozpúr. The local outturn of Amballa is put at 9,713 blankets with an export of 9,795 to other districts and an import of R20,000 worth of *lois* from Ludhiána, Amritsar, Nurpur (Kángra), and Kashmír; and that of Ludhiána at over R24,000 worth of imitation shawls (half exported to Native States and Amritsar and elsewhere), and about 1½ lakh's worth of other things. Jullundur, with an outturn of 40,000 yards of blankets, nearly half of which is exported, imports 96,000 yards of European cloth and flannel and at least 32,000 yards of country woollens. The hill station of Murree imports R36,000 worth of European woollens and 10,000 pieces of pattú from Kashmír. Shahpur makes R30,000 worth of *lois* and blankets and 800 *ássans* and the export of these articles amounts to R15,000 in value. The people of Delhi and Gurgaon seem to prefer cotton-padded garments to woollen clothes for warmth.

Import of European piece goods. 1558

"The import of European woollens is said nowhere to compete very severely with the native industry generally. In Lahore it is thought that the manufacture of country-made medium stuffs is suffering and will suffer, and the same report comes from Hissár. In most other districts it is stated that the import of European goods has little or no effect on native manufactures.

Classes by whom woollens are used. 1559

"The European woollen fabrics imported are flannel, merino, broadcloth, knitted goods, etc., and are used by Europeans and a few of the wealthier natives. In the plains, while there is one tribe, the Odhs of Muzaffargarh and the Afghán frontier districts, that consider the wearing of woollen garments a religious obligation, and while the Khojahs of Lahore and elsewhere almost invariably wear the woollen bhúrá, cotton is much more worn than woollen; in fact, in many districts only woollen blankets for bedding are used and no woollen clothes. Even in such a district as Kángra cotton is largely worn and its use is becoming more and more common, though the Gaddi (shepherd) still wears nothing but woollen homespun.

Knitting. 1560

"Knitting of jerseys, gloves, and socks is done partly with European worsted but also and much more with native yarn. In Ludhiána R4,000 worth was knitted in the year under report, and in Amballa R2,000 worth. Knitting is sometimes done with two needles, in which case the sock or glove is made in two pieces which are afterwards sewn together. For seamless socks or gloves three to five needles are used."

Carpet-manufacture is briefly alluded to by **Mr. Johnstone**, but as his remarks are mainly taken from **Baden Powell**, the passage has been

PANJAB.

omitted. It would be beyond the scope of this work to deal with the designs and methods of manufacture of articles of a distinctly artistic nature. The raw products and primary manufactures are the features that mainly have to be dealt with. This explains the omission of direct treatment of articles of an artistic kind. **Mr. Johnstone's** remarks on the dyeing of wool and woollen goods are, however, sufficiently concise that they may be quoted here.

Dyeing. 1561 "Wool intended for coloured *namdas* is dyed, but, with this exception, dye is always applied to the yarn or the made fabric. The variety of dyes used is so great that a mere enumeration of them all would be too lengthy for such a report as this. The chief colours are red, blue, turquois blue (*firúsá*), yellow, green and black.

"(*a*) Red, made with cochineal, needs three tolas of cochineal for one yard of cloth or quarter seer of thread. The cochineal is put into boiling water in which the cloth is immersed and then sulphuric acid and saltpetre in certain proportions are added. The drying is done in the shade. If a deep red is wanted the cloth must then be put in boiling water to which four tolas of turmeric, one chittak pomegranate seeds, two chittaks of sulphuric acid and saltpetre have been added. The *gulanár* shade of red is made by doubling the turmeric.

"(*b*) A common red is also made from *lac*, got from the *ber* (**Zizyphus Jujuba**), the *kikar* of Sind (**Acacia arabica**), the *dhák* (**Butea frondosa**) of Hindustán (not of the Panjáb), the banyan (**Ficus indica**), and the pipal (**Ficus religiosa**).

"(*c*) Blue (*nilá*) is made in many shades, the basis being indigo and the fixing material chiefly sulphuric acid. Turquois blue (*firúsá*) is made from an imported dye with alum added during the process.

"(*d*) Yellow is got in many ways. One concoction used contains *akalbir* (**Datisca cannabina**), turmeric and alum; another *kesu*, or the flower of the *palás* or *dhák* (**Butea frondosa**), which, however, gives only a transient dye. Yellow may also be got out of the rind of pomegranate (*náspál*) with some fixing substance added.

"(*e*) Green can be got by dyeing first for blue and then for yellow as is done in Kulu; or by adding to the concoction for blue, turmeric, *akalbir* and alum.

"(*f*) Black is made in many ways. In the hills green walnut shells are used and the black colour produced is very intense and lasting. Another deep black is got from indigo added to a fermented compound of *gur, átá*, and the refuse after iron is smelted.

"In dying wool such substances as sulphuric acid and lime are added merely to help the wool to absorb the colour. *Akalbir* is used with the same object.

Woollen fabrics requiring bleaching are exposed to the fumes of burning sulphur. Aniline dyes as used in the province are never fast. I understand that their use is increasing. The *garbi chadar* (*pashmina*) can only be dyed *kháki*, but is generally undyed. Dyers are generally a separate caste and they are, as compared to spinners and weavers, well off."

ASSAM & BURMA. 1562

V.—ASSAM AND BURMA.

Absolutely no information is available regarding the sheep of these provinces. In one of the early volumes of the Transactions of the Agri-Horticultural Society of India, **Major Jenkins** is reported to have sent the fleeces of Tibetan sheep (imported into Upper Assam) for valuation. These were favourably reported on, but no trade to speak of appears to have developed in the article. In several papers allusion is made incidentally to sheep-farming under European management having been started on the Khasia hills, but the writer has failed to discover any particulars and need scarcely add that no such industry exists at the present day. In Manipur while sheep are to be had, they are by no means popular; but it is probable large portions of that little State would afford a rich field for future experiments since the pasturage is rich, the hills mostly low and grassy, and the rainfall by no means heavy, except on the western ranges bordering on Assam and Cachar.

BOMBAY.

VI.—BOMBAY.*

1563 It has already been stated that one of the earliest notices of the sheep of Western India is that given by **Dr. Hove**, a Polish Botanist, who visited India with the object of studying the cotton supply and manufactures of India. He was in the Deccan in 1787 and spoke in the highest terms of the sheep he there saw and the woollen manufactures of the Presidency The subject of Bombay wool next attracted attention about forty or fifty years after the date of **Hove's** visit, when Saxon and Merino rams were imported and an effort made to improve the breed. **Colonel Jervis** in 1835 represented to the Court of Directors of the Honourable the East India Company that the Deccan and Gujerat were well adapted to sheep. So statisfied was **Colonel Jervis** of success in the undertaking that he started a sheep farm on his own account. The Bombay Government also opened out two farms and placed these under a **Mr. J. Webb.** The farms were at Ahmednagar and near the fort of Júner. For this purpose large numbers of sheep were imported from Afghánistán, the Cape of Good Hope, England. The East India Company, for example, sent out 120 rams and ewes of the South-down, Leicester, Cotswold, and Merino breeds. Every thing was, in fact, done that could be thought of in the way of importation of stock. In 1843 we read of **Sir George Arthur** having reported on the farms in the most favourable terms, his previous colonial experience having highly qualified him to express an opinion. What came of all this enthusiasm and liberal expenditure of money is difficult to know. The subject seems to have fallen from public consideration even more rapidly than it had ascended The trade from Bombay in wool is by no means unimportant, but comparatively none of the exports are from the Presidency. The supply is drawn from Sind, Baluchistán, Afghánistán, Rájputana, and to a smaller extent from the Panjáb, the North-West Provinces, and the Central Provinces. The Presidency of Bombay by no means produces enough to meet the requirements of its local weavers, so that the once famed Deccan sheep might almost be said to have disappeared. With few subjects is a greater silence preserved in the valuable Bombay District Gazetteers than that of sheep and wool. The passages which are given below have been furnished more with the object of demonstrating this fact than from any great merit which they possess. We know practically nothing regarding Bombay wool. The returns of the traffic to and from the port town of Bombay by sea and rail manifest, however, an immense expansion. The total imports by sea to India, for example (mostly from Mekran, Sonmiani, and Persia), have increased during the past twenty years from 1,000,000℔ to 5,000,000℔; the re-exports of foreign wool (drawn largely by rail, road, and boat) have expanded from, say, 130,000℔ to 13,000,000℔ within that period; while the exports of Indian wool have fluctuated between 26,500,000℔ and 20,000,000℔. It will be seen from the tables furnished in the chapter on Wool Trade below that by far the major portion of this traffic, in raw wool, takes place with the port town of Bombay, the next most important sea-port being Karáchí. Karáchí has recently, however, sprung into great importance through having given birth to a large and yearly increasing re-export of foreign wool – wool that is drawn from across the Sind frontier. One other noticeable feature of the trade returns as bearing on the wool of Bombay may be here added, *viz.*, that nearly one-half of the Indian wool exported from Bombay is derived from Rájputana and Central India, less than one-sixth being usually obtained from the Bombay Presidency. It may thus be said that the wool that leaves Bombay comes mainly from

Sheep Farms. *Conf. with pp. 579, 587, 635, 638.*

1564

* *Conf. with Jour. Agri.-Hort. Soc. Ind., VII., 114.*

SHEEP : Wool.

Production of Wool and Woollen Goods

BOMBAY.

Rájputana, Central India, and the Panjáb, and that Karáchí wool is chiefly Baluchistan and Afghán, with lesser proportions of Sind and Panjáb wool. The chief local interest manifested in Bombay in wool may be said to be the manufacture of blankets. The following extracts from the Gazetteers will be seen to mainly deal with this branch of the trade:—

Dharwar. 1565

DHARWAR.—"White, black, or white and black striped blankets are woven by shepherds. Of 87,768 shepherds shown in the 1881 census, about one-tenth, or 8,700 are blanket-weavers. In the Ránebennur sub-division in the south-east large blankets, about sixteen feet by six, are woven; the blankets woven in the rest of the district are not larger than nine feet long and four broad for men, and seven and a half feet long and three broad for children. Generally the women spin the wool into thread, arrange and size the warp, and fill the shuttles; and the men weave. In Dhárwár, wool is not sold by the ordinary *sher* weight. Either the shearing of 100 sheep is bought in a lump for about £4 (R40), or the wool is bought by the *chitti* or four *sher* millet measure which costs about 16*s*. (R8), that is, at the rate of 14*d*. the pound. One *chitti* or fourteen pounds of wool works into four blankets, each nine feet long by four feet broad. Of these four blankets two are black, together worth 16*s*. (R8), and two are white, together worth 8*s*. (R4). To spin the wool and weave these four blankets take a man and woman about forty days, that is, after deducting 16*s*. (R8) as the cost of one *chitti* of wool, the men and women earn 8*s*. (R4) in forty days, or 6*s*. (R3) a month. At the rate of three blankets a month for each couple the 8,700 blanket-weavers, during the eight fair months, yield an estimated outturn of 104,400 blankets worth £31,320 (R3,13,200). This outturn is not enough to meet the local demand. Blankets are largely imported from Belari and Maisur, part of the imports being used locally and part being sent to the coast. Blanket-weavers generally sell their produce direct to the wearers on market days in local market towns. When not sold in the markets, blankets are sold to local blanket dealers who are generally rich shepherds and are sometimes Lingayat cloth-dealers. As white and white and black striped blankets fetch 4*s*. (R2) each and black blankets fetch 8*s*. (R4) each, most of the blankets woven are black" (*Gaz., XXII., pp. 380-1*).

Kolhapur. 1566 Yield 1lb.

KOLAPUR.—"The sheep are sheared twice a year in November and in June. The Dhangars cut the wool with a heavy pair of shearing scissors. An average fleece weighs half a pound which is worth 3*d*. to 3½*d*. (2 to 2¼ *as*.). Most of the local wool is woven into blankets and some is used for making felt or *burnus*, and native saddles. Very little raw wool leaves the State" (*Gaz., XXIV., 27*).

Kolaba. 1567

KOLABA.— Wool working is carried on at Mápgaon, Malgaon, and Alibag in the Alibag sub-division, and at Roha. The workers are Dhangars from the Deccan, of whom about 100 families earn their living by blanket-making. They have looms and weave coarse blankets, some with the wool of their own flocks and others with wool brought from the Deccan. The wool is bought either with their own or with borrowed money. The demand for their blankets is so great that, though they work for eight or nine hours a day during the whole year, they are unable to supply the demand and are forced to bring blankets from above the Sahyádris. Their average yearly earnings amount to about £12-10*s*. (R125). The craft is flourishing. In Mángaon and Mahád some Sangars or weaving Dhangars are engaged in making blankets which they sell to local merchants. The blankets vary in price from 1*s*. to 2*s*. (8 as. to R1) according to texture and the quality of the wool. Their average daily earnings vary from 6*d*. to 9*d*. (4 as. to 6 as.). Most of them have money on credit enough to buy the wool they use and keep some ready-made blankets in store" (*Gaz., XI., 132*).

Poona.

Yield 1lb. 1568

POONA.—"Sheep are sheared twice in a year, in June-July and in October-November. Each sheep on an average gives one pound of wool at each shearing worth 4½*d*. to 6*d*. (3 as. to 4 as.). The loss in carding, spinning, and weaving amounts to twenty-five per cent. Sometimes Dhangars are called to shear the sheep and are paid at the rate of 4*s*. (R2) the hundred. The wool is bought by the Dhangars whose women card it by means of a bamboo bowstring with gut twist, and spin it either fine with the help of the ordinary spinning wheel or coarse using the spindle. The threads are stiffened with a paste of tamarind stones pounded in the rough stone mortars which are generally to be seen outside of Dhangars' houses. The paste is applied with a large stiff brush. After the warp-threads have been placed and stretched the Dhangar takes two days to weave a blanket about eight feet long and two and half feet wide, the price of which varies from 2*s*. to 10*s*. (R1 to R5) according to the colour and fineness of the texture. White blankets and seats of *ásans* used while performing religious ceremonies have a special value, being considered more sacred" (*Gaz. XVIII., Pt. I., p. 67*).

BOMBAY. Sholapur. 1569

SHOLAPUR.—"Every two years they bear thrice, one lamb at a time. Sheep are reared more for their wool than for their milk. Twice every year, in March and again in July, their wool is cut. If black, it is sold to Sangars or blanket weavers at 6*d*. a pound (2 *shers* the rupee), and of mixed black and white at 5*d*. a pound (2½ *shers* the rupee). At each shearing 100 fleeces are worth about £1 (R10), that is, about £2 (R20) a year" (*Vol. XX., 17*).

Blankets.—"Almost all over the district blankets are woven by Dhangars and Sangars. Sangar weavers are chiefly found in the Bársi and Sángola sub-divisions. The wool is from their sheep, which are sheared twice a year. The wool is chiefly black with some dirty white threads. It has to be several times washed before it is ready for use. The blankets and seat cloths, or *ásans*, woven in the village of Gherdí in the Sángola sub-division have a local name. Blankets fetch 1*s*. to 10*s*. (R½ to 5) each. In some parts *burnus*, or coarse felt, is also made. Dhanga weavers earn 3*d*. to 6*d*. (2 as. to 4 as.) a day" (*Gaz., XX., 271*).

VII.—SIND AND BALUCHISTAN.

SIND and BALUCHISTAN. 1570

To a certain extent the subject of Sind and Baluchistan wool has already been indicated. The traffic in this article may, however, be described as a direct result of the prosperity of the port town of Karachí The following letter, addressed to the Chamber of Commerce, Bombay, in 1842, indicates the fact that the trade had then scarcely an existence:—

Account of the Wool produced in Upper Sind, Cutchi, and Baluchistan—by LIEUTENANT POSTANS.

"The following remarks are offered on the article of wool as produced in Upper Sind, Cutchi, and the higher country of Baluchistan, being the result of enquiries on the subject.

"Wool in Upper Sind is not a mercantile commodity, nor does its value as such appear to be known; the quantity produced is moreover unimportant and used by the natives entirely for purposes of home consumption, as mussuds, kumlies, rugs, etc., the sheep appear to be of a poor and inferior description, and are seen only in small flocks, though the whole of this track of country would seem to be well adapted in forage for feeding large quantities: the inundations, however, would probably, for a certain period of the year, render the soil too damp for this animal.

"In Cutchí the numerous large flocks of *dumba* sheep which are met with, particularly during the cold season (Zimistan) are principally those brought down by the Brahuí and other hill tribes for forage, and to avoid the inclement climate of the upper country. The flocks appertaining to the plains are not numerous, and the wool is used for the same purposes, as in Upper Sind before alluded to. The following statement from a Native Chief in Cutchí, respecting this article, may be relied upon, and it shows that the hill Beluchis manufactured the wool and brought the articles for sale to the lower country, proving the want as a supply in the plains:—

"'From the time of Meer Nusseer Khan of Kelat until now, the Sarapah tribe of Brahuís manufactured rugs, mussuds, carpets, etc., and traded with them of the Jhahwar tribes, the Neechari made woollen cloaks of various colours, ropes, etc., and took them to Shikarpore, Kyrpore, and Larkhana for sale; these are the articles made by the Brahuís of wool, and no one has yet purchased wool from Cutchí or taken it away for sale to various places. The Afghans in the neighbourhood of Candahar and Cabool make postins, shawls, etc., of value, and sell them in these countries. In the Boogtie and Murree hills on the eastern side of Cutchí, the valleys afford pasture to considerable flocks of the *Dumba* sheep, the wool from these parts is manufactured by the Beluchis themselves for their own use, the rest sold to the Hindoos in the small towns along the skirts of the hills, where it is used entirely for clothing or domestic purposes.'

"The mountainous division of Baluchistan, known as Jhahwar, is that in which wool is cultivated and forms the greater portion of the property of the Jhahwar tribes of Brahuís. The flocks as described to me over the Jhahwar province in the districts of Kozdor-kal-wadd, Zharee, Zedee, Pandran, etc., are extremely numerous, and if I am correctly informed, at least a lakh (1,00,000) of fleeces are produced annually therefrom. The following is a native statement on the subject:—

'Wool in the province of Jhahwar is produced in great quantity; formerly the Brahuís made the white into mussuds and the black wool into shawls, etc. Some was also taken to Kelat, Cutchí, and other places for sale, but this is the third or fourth

SHEEP: Wool.

Production of Wool and Woollen Goods

SIND and BALUCHISTAN.

Yield 1lb.

1571

year, that the Hindoos have become traders in wool; they pay the Brahuís in advance to secure the fleeces, and then send them to Bombay.'

"This information agrees with what I have elsewhere elicited, the sheep are sheared twice during the year, at the spring and autumn (March and October) the wool being sold by the fleece at an average of about 6 per rupee, each fleece weighing, it is said, something above half seer packa to one Bombay seer. The value of the article has of late become so well known to the Hindoo traders that they secure it by advancing money to the owners, and this in a country where there is little or no security; at the above rate, the profits must be considerable, thus Khorassan wool, under which denomination the above is, I believe, known in Bombay, appears to be worth about R140 to R145 per candy of 588lb, the same quantity could be purchased in Baluchistan for about R90, and the expense of transmission by way of Sonmeeanee and Kurachee does not greatly interfere with the profits.

"Independent of Jhahwar, wool is produced in various other places, in Baluchistan, in Sarawan, at Moostung, Khoran, Noskhey, etc., but not in the same quantity with that of the above district. In Afghanistan wool does not appear to be an article of export, finding its own value in the country, where it is in constant use for articles of clothing, etc., or of equal quality. Mekram furnishes a considerable supply of wool, but of an inferior quality to that from Beluchistan.

"From my enquiries I am led to believe that Sind (Upper or Lower) does not produce any of the wool at present exported to Bombay from the mouths of the Indus or Kurachee as a mercantile commodity, nor is it to be found in that country in sufficient quantity to form an article of trade, though there is apparently no reason why it should not do so as in the neighbouring country of Cutchí. The same may be said of Cutch-Gundava, but Mekram and the hilly tracts of Baluchistan furnish nearly all the article known in Bombay as Khorasan and Mekram wool. That Central Asia generally will be found to be rich in this staple commodity there can be no doubt, and as its value hereafter becomes known in these countries, it will doubtless be cultivated and become an important return in the trade of Bombay" (*Transactions of the Bombay Chamber of Commerce, 1842*).

In 1860 **Mr. P. M. Dalzell** (*Collector of Customs, Karáchí*) while in England was invited by the Bradford Chamber of Commerce to address the Chamber on the subject of Indian wool. He urged the necessity of Karáchí being made an "independent emporium of trade." He pointed out that eight-tenths of the wool of Karáchí came from Afghánistán and the country lying between that essentially pastoral region and the frontier of Sind. Seven years ago (1853) the exports of wool from Karáchí were valued at £20,000, last year (1860) they were returned at £400,000. The chief difficulty to a great expansion of that trade was the fact that Karáchí was not a sufficiently attractive market. The Kabulís could not purchase return goods of the kind they required and were, therefore, compelled to carry their wool to Bombay, and this entailed a loss of from 10 to 12 per cent. on the price obtained for the fleece. **Mr. Dalzell**, therefore, urged the establishment of mercantile firms in Karáchí with a fleet of steamers trading direct with England as the conditions necessary to expand the wool trade into proportions of the greatest value to England. All this (writing of 1892) may be said to have been attained; Karáchí has grown into an immense mercantile centre and has a yearly increasing supply of ships which trade direct with Europe, and a railway system that taps the whole of Northern India. Expressing the gross foreign exports of wool by the nominal pound sterling (*i.e.*, R10), they were valued in 1890-91 at £660,000 with, in addition, say, £200,000 exported coastwise mainly to Bombay. At the time at which **Mr. Dalzell** spoke the Karáchí traffic was entirely or nearly so with Bombay, so that these two items (foreign and coastwise, have to be taken conjointly, when it becomes apparent that the traffic in wool has been doubled within the period of thirty years that has transpired since the date of **Mr. Dalzell's** address to the Bradford Chamber of Commerce. This is a matter doubtless for congratulation, but the possibilities of Sind and of the countries from which it draws a large portion of its wool supplies have by no means been even now

SIND and BELUCHISTAN.

reached. The question of improving the breed of goats and sheep in that area has repeatedly been urged. It will be found in connection with the Panjáb that reference has been made to an experiment conducted many years ago to improve the fleece of the Hazára sheep. This was apparently fairly successfully accomplished, but the improved breed was not popular. The Natives had no use for a high class wool. The means of export were defective, at least considerably more so than at the present day; a market had not been created for the finer quality of wool. There was then comparatively no Karáchí demand, the improved sheep of Hazára fell into disfavour and soon died out. At the present moment the proposal to endeavour to improve the sheep and goats of Sind and Baluchistán has been urged once more and is being considered by the Government of India. **Major G. Gaisford**, Political Agent, Quetta and Peshin, in his original letter on this subject (*dated 6th January 1890*), suggested that from what he saw in Australia he was disposed to think the merino breed might do well in Quetta. The proposal contained in the letter quoted above was freely distributed and opinion invited from Local Governments and Chambers of Commerce. A concensus of opinion may be said to have been obtained in favour of great improvement being possible by more careful selection of the existing breed, the shearing of sheep of one colour and as near as possible of one quality of wool, together so as to avoid admixture, and the washing of the sheep before shearing. While most of the answers admit the possibility of improvement by merino crossing, the question may be said to have been more evaded than directly answered. The Chamber of Commerce, Karáchí, replied that while they considered the improvement of the breed of sheep was very desirable, the question of suitability of breed was one that an expert alone could decide. The Chamber concurred, however, in the desirability of coloured wools being kept separate and the proportion of white being preferentially increased. The Chamber of Commerce, Cawnpore, while endorsing the view that the various colours of wool should be kept distinct and the coarse locks from the legs kept apart from the general fleece, added that the value of a wool depends largely upon the absolute similarity of the fibre. A moderately coarse wool of uniform structure would fetch a much higher price than a finer wool having coarse hairs interspersed throughout it. **Mr. Hallen**, Superintendent of the Horse-Breeding Department in India, replied that "at the Hissar Cattle Farm English rams were, for several years, regularly imported. The results arrived at were unsatisfactory; indeed, it may be safely accepted as a fact that European sheep or their produce will not thrive on the plains of India. But with regard to **Major Gaisford's** suggestion to introduce merino sheep on the hills of Beluchistán, with the view of improving the wool and mutton of the districts, I am inclined to believe that, for the reasons noted, an experiment in this direction may be attempted, and would, therefore, suggest that merino rams be imported from the hotter districts of Australia where they have been found to thrive" **Mr. Orrah**, who has already been freely quoted in connection with the subject of Bengal sheep, was invited to favour the Government with his opinion on the proposal to endeavour, by cross-breeding, to produce a better quality of wool in the Beluchistán sheep, and he heartily concurred. His reply deals with the defects of Indian wool in general and then concludes with the following remarks :—

Hazara wool. *Conf. with pp. 577—578, 608, 612, 614, 615, 617.* 1572

Assortment of wool. *Conf. with. pp. 596, 599, 610, 611, 656.* 1573

"There are three descriptions of sheep or races in India—

Descriptions of Sheep: N.-W. Provinces. 1574

"The North-Western Provinces.—One, the common sheep of the plains of India with a very coarse fleece (but which have some very good qualities, such as strength and milling properties).

Dumba. 1575

"2nd, the *Dumba* or *Karoo* whose wool is mostly white of a very pure description.

SHEEP : Wool.

Production of Wool and Woollen Goods

SIND and BALUCHISTAN.

Description of Bhyangee.

1576 "3rd, the *Bhyangi* found in the Himalaya, the wool of which is soft and long in staple, but the milling properties of which are deficient, being more of the fur character than the true spirally wool fibre. It lays more like a hair, and has not the serrications which fine wool fibre should have. Of these three varieties I consider the pure white *Dombas* and *Karoo* sheep the most suitable for crossing with Australian merino rams. The rams should not be old, and from a good stock, South Down in build. This crossing should eventuate in a merino character of wool being obtained in the second or third generation, the pure white and gloss blending into fineness and length of staple and a lustre and soundness of the higher classes of combing and clothing wools which realise the best of values.

"In selecting rams or rams and ewes, as may be determined upon, I would advise that one-half be obtained from the combing wool breeds of stock and the other half from the clothing wool breeds, the first named being the long-stapled and the second the shorter-stapled wool, and should be kept strictly separate in the cross-breeding with Indian sheep, because two very different classes of wool will probably be produced and a different value and weight of production be the result. If they be allowed to mix, these points may be materially and detrimentally affected. The different crossings as formerly known between South Down, Leicesters, Cotswold, Forest-cheviots, Norfolks, etc., etc., etc., each and all comprehending innumerable breeds as represented in Australian, have now become more known in distinction as the stock of a certain district or noted flock owners and breeders, and it is not necessary now to say more than this until some experience has been gained. The hardier merino breeds of the South Down stamp should be obtained from the Australian stocks as a first trial, and if not successful, try another district of Australia or New South Wales; and if these do not succeed, next try New Zealand flocks, the climate of which I believe more closely resembles Beluchistan than the former-named countries, where almost as fine breeds of sheep exist in large quantities. It is only by *persistent* effort that success may be achieved. If, however, this be done, I have no doubt of India eventually becoming one of the most important wool-producing countries of the world."

It will thus be observed that in the above discussion regarding Baluchistán sheep (and this is the most recent on any aspect of the Indian wool trade) the opinions held differ in no respect from those advanced from year to year during the past 50 or 60 years. The one authority holds that cross-breeding with foreign races is unnecessary and that it had, indeed, proved useless, while the next deems crossing with superior breeds all that is required to raise India to a foremost place among the countries of the world's supply of wool.

MADRAS.

VIII.—MADRAS.

1577 In several of the reports regarding sheep-farming in Bengal and the efforts that were made to improve the wool of that province, mention has been incidentally made of the corresponding experiments in South India. Dr. Royle (*Productive Resources*), in a passage quoted above under the chapter on the *Improvement of the Breed of Sheep in India*, mentions the fact that about 1835 the Madras Government first took direct action in this matter. A letter from Colonel Haslewood to Captain Jacob of Bombay, dated April 1837, furnishes the chief data on which the various statements of the early action of the Madras Government has been based. The letter may, therefore, be here given in full :—

"I am happy to see that your Government have taken up my plan for improving the Indian wool. I have just received six more Saxon rams from the Raily flock at Sydney; the price there is ten, seven, and five guineas each according to the cross, or rather according to the size. Mr. Sullivan brought out two merino rams and two ewes, and I have seen the effect of crossing by these, and also by South-down rams imported by Sir William Rumbold on the Neelgherries. Even the red hairy sheep of India become South-down in size and wool in the second generation, and the white woolly sheep of India become merino and South-down in size and wool after one crossing. I have shorn Mr. Sullivan's merinos that have been two years in India. After twice washing and shearing, the day after the ewes gave five and four and a half pounds each. In fineness, length of staple, elasticity and *oiliness*, equal to any I ever saw in Tasmania, where two and a half pounds is the utmost ever got from a ewe of the merino kind, which seldom weigh more than fifty pounds per carcass when killed; and these ewes of **Mr. Sullivan's** had been shorn only seven months before.

"My flock I have removed here (Bangalore) from the hills,* as the rank grass there does not answer for sheep brought from below, although those bred there thrive exceedingly well. I am completing my flock here to 700 white woolly ewes, for which I have rams enough, pure Saxon. The rutting season begins here in June, and the lambing from November to January, and they may be shorn in February and in September. I do not know if you have any white woolly ewes indigenous to your provinces, although I know you have the black woolly; but you may get the white ones between Jalna and Beder, where I hear there are many flocks. We only have them in Coimbatore, and the Barramahal. After my flocks have given their first lambs, I shall turn them over to the Mysore Commissioner, and return to Tasmania" (*Trans. Agri.-Hort. Soc. Ind., VII., 128-129*).

Flock of 700 Ewes. *Conf. with pp. 579, 580, 587, 629, 638.* 1578

Dr. Shortt, in his *Manual of Indian Cattle and Sheep*, gives very little of a definite nature regarding the sheep of South India further than what has been discussed above under the chapter on BREEDS OF SHEEP IN INDIA. With the exception of the black-headed sheep of Coimbatore and the woolly sheep of Mysore, he speaks most disparagingly of the others, so far as wool production is concerned. In fact, he characterises them as yielding a poor quality of hair rather than wool. Curiously enough he makes little mention of the early experiments made towards improving the breeds of South India, so that he at least would not appear to have accepted the opinion as well founded, that the benefit thereby conferred can be traced in the present breeds. It would be interesting to know, however, whether early records exist of the Coimbatore and Mysore wool-yielding sheep that would establish these as purely indigenous breeds and not the descendants of the cross-breeding that was effected about the beginning of the present century. The sheep seen by **Dr. Hove** in the Deccan a hundred years ago was apparently very similar to the Coimbatore stock of the present day, and **Dr. Buchanan Hamilton's** Mysore sheep may be the same as the wool-yielding breed now to be found in that State. There is, however, no positive proof on these points, and Native opinion and tradition would be useful. So far as the writer can discover, the Coimbatore and Mysore sheep are purely indigenous and would afford perhaps better stock for experimenting with than the Patna in the production of a fleece sheep for the great central tableland of India and the Upper Gangetic basin Indeed, the much talked-of Patna breed affords an inferior fleece (*Conf. with p. 617*).

Conf. with pp. 568, 570, 580.

In the reports of the Experimental Farm at Sydapet frequent mention is made of sheep. Thus, for example, **Mr. Robertson** seems to have commenced about 1869 to endeavour by selection and crossing with Mysore, Coimbatore, Patna, Nellore, and Madras to evolve a useful prepotent stock. This he seems to have secured and designated the animal produced as the "Sydapet Breed." What has since come of this stock the records consulted by the writer do not say (*Conf. with p 611*).

The following brief notice of Madras sheep conveys some idea of the views currently held regarding the present position and prospects of sheep farming:—

"The wool produced in Southern India from the native breed of sheep is of very coarse quality and chiefly employed for making *cumblies*, a rough kind of blanket largely used by the Natives. Attempts have been made to improve the breed of the white-woolled country sheep by crossing with Australian, Merino and other rams. These crosses thrive best in the higher districts of the Peninsula, such as Coimbatore and the tableland of Mysore, where the temperature is somewhat cooler and forage more abundant than on the plains. In Mysore many of the sheep have foreign blood in them, and for a series of years the Madras Government endeavoured to improve the breed of sheep in the districts of Salem, Coimbatore, North Arcot, and Bellary by the distribution of superior rams. Although these efforts have improved to some extent the quality of the fleece, they cannot be said to have given any impetus to sheep-

* Nilghiris.

MADRAS.

breeding as in 1881-82 the total exports of wool from this Presidency only amounted to 868lb of the value of R220 In some districts, such as Ellore, the finer qualities of wool are used for making carpets of oriental patterns. These are mostly made on commission for European dealers, who secure them through their local agents. The exports for 1882-83 were 26,238lb valued at R5,173" (*Madras Man. Adm.*, *I.*, *p. 363*).

The people of Madras are essentially clad in cotton or silk. Wool enters but to a very small extent into their personal apparel, and hence the woollen industries of the Presidency are not very important. The following

Carpet-weaving. 1579

reference to the carpet-weaving will, however, be read with some interest:—

"As regards carpets, Mr. Havell considers that, though the ordinary Ellore and Masulipatam ones of small size prepared for the country bazaars are of inferior stuff and badly made, the best patterns in use made to order are not inferior to those of old South Indian carpets which are held up to the disparagement of modern productions. Aniline dyes are very rarely used, as they are at Warangal and other places in Hyderabad, and I have seen carpets from the Native looms at the three seats of the industry—Ellore, Masulipatam, and Ayyampet (Tanjore) which are in no respect inferior to old specimens in the hands of connoisseurs in London or in Native houses and palaces."

It is perhaps unnecessary to republish the various passages that occur in the district manuals of Madras. The information therein furnished does not add materially to what has already been indicated. The reader, should he desire further particulars, might, however, consult the Salem District Manual, Vol I., 142; North Arcot, p. 14; Coimbatore, p. 246; and Kurnool, p. 175. In the Coimbatore Manual it is stated that the special wool-yielding breed is generally designated *kurumba* because tended by shepherds of that tribe.

Pashm. 1580

PASHM & GOATS'-HAIR, & MANUFACTURES THEREFROM.

So much has been written on the shawl wool of Kashmír and Tibet that it may seem absurd to have to admit that the whole subject is still but very indifferently known. *Pashm* or *pam* is the under-coat of wool formed on certain goats. Speaking generally it may be said that there are two chief kinds of *pashm*—that of wild and that of domestic animals. By some writers the soft winter coat of the yak, the camel, and several antelopes is also classed as *pashm*, but these substances had better be regarded as at most inferior substitutes. Within recent years, however, a soft form of

Kirmani Pushm. *Conf. with pp. 615, 616, 624.* 1581

wool has begun to be largely imported into India from Persia which is known as *Kirmaní pashm*. This is taken to Amritsar, Lahore, Nurpur, Ludhiana, etc., and made into fabrics, shawls, etc., which are sold as *pashmina*, the Kirman wool being either mixed with a small proportion of *pashm* or used alone. But of the true *pashm* there are many kinds or qualities, the most highly prized of all being obtained from certain wild animals. No writer seems to have definitely determined the exact wild species that yield this substance. The ibex (**Carpa sibirica**) is often spoken

Asali Tus. 1582 *Conf. with p. 552, 559.*

of as affording a soft downy under-coat known as *asali tús*, of which it has been said no wool is so rich, so soft, and so full. Then again Mr. F. H. Cooper classifies Indian *pashm* into six kinds, his sixth kind being the down of a water-fowl. The first two which he characterises as the finest of all, are the white and grey forms of *shah-tush* or wild *pashm*. This, he says, is

Shah Tus. 1583 *Conf. with pp. 558, 639.*

"the inner winter coat or fine downy wool of a small species of wild goat, called *thosh*," in Tibet.

The long outer layer of hair (*pat*) found in both the wild and domesticated *pashm*-yielding goats, etc., is of a superior quality to ordinary goats'-hair. It is spun and woven into fabrics known as *pattú*. The ordinary or

PASHM GOATS.

coarser kinds of goats'-hair are usually made into ropes or sacking and the saddle-bags used by the carriers who trade across the Himálaya, their goods being packed into sacks borne by sheep, goats, asses, donkeys, ponies, camels, etc. As in *pashm* and wool so in goats'-hair, therefore, there are various qualities and colours depending upon the breed of animal or the nature of the country in which it lived. The goats'-hair of the plains of India generally is more hairy in character and often quite straight as compared with the more woolly and curled hair of the higher regions; the latter in fact gradually approaches in character that of the *pashm*-yielding species. As already explained, the Sind Ibex or Persian Wild Goat (**Capra ægagrus**) is by some writers supposed to be one of the species and probably the chief one from which the domesticated goats of India have developed (*p. 551*); but **Hinderson** (*p. 552*) notes that in the character of the horns the Yarkand goats approach much nearer to the Himálayan Ibex (**C. sibirica**), while other writers see a resemblance in some of the races of domesticated goats even to the *markhor* (**C. falconeri**). The late **Col. Sir O. B. St. John** found, near Quetta, a wild hybrid between **C. ægagrus** and **C. falconeri. Blanford** (*p. 552*) mentions the fact that the *markhor* has repeatedly been bred in captivity along with domestic goats, and that at one time it was supposed the tame races with spiral horns were derived from **C. falconeri**. He adds that "it is not improbable that some are thus descended." The objection to this conclusion rests mainly on the fact that the first turn of the spiral horn in domestic goats is mostly inwards, that of the *markhor* always outwards, but domestic goats with horns formed like those of the *markhor* have been recorded. The point of interest to which it is here desired to *more especially* directed attention, however, is the further observation, made by **Colonel Biddulph**, namely, that, while the Himálayan and Central Asiatic Ibex (**C. sibirica**) frequents localities at great altitudes, the *markhor* seeks the rocky ground within the limits of arborescent vegetation. This love for colder regions manifested by the Ibex seems to have been provided against by the development of an under-coat of downy wool (*pashm*) below the hair which the *markhor* does not possess. If therefore, the survival of certain characters such as the shape or rather conformation of the horn or the presence of *pashm* be accepted as denoting the origin of the cultivated races of goats, it might be inferred as **Hinderson** practically suggested of the Yarkand animal that the *pashm*-yielding domesticated goat has been developed from the Ibex. **Hodgson**, and following him, many other writers, regard the *pashm*-yielding goat as mainly derived from **C. ægagrus**. But it may safely be said that all authors admit the possibility of the Asiatic goats having been derived from more than one species, and the advisability of such a conclusion (apart from the diversified form, stature, colour, habit, etc., of the tame races) receives countenance from the admitted existence of hybrids between the wild species. The progression in characteristics from the village typical goat of the plains of India and the *dúgú* goat of the lower Himalaya to the Alpine *pashm*-yielding animal may, therefore, be due to more than selection and adaptation of one species to environment. It may mark the stages of adaptation and crossing (if it might not be called hybridization) of different species with the nearer approach in the extremes to the specific types. If there be any plausibility in this suggestion, the difficulty which the early writers foresaw in any attempt to breed the true *pashm* goat on the southern slopes of the Himálaya would at once assume a distinct position. There is sufficient justification at all events for the dictum that with few domesticated animals would there likely be experienced a greater difficulty in crossing the races of reputed merit than with goats, if it be desired to acclimatize and preserve the merit of the progeny in widely different regions. The suggestion has been offered

Ropes. 1584

Sack. 1585

Saddle-bags. 1586

Goats' hair. 1587

Crossing and Prepotency. *Conf. with pp. 570, 575, 577, 584.* 1588

SHEEP: Pashm.

Pashm and Goats'-hair,

PASHM GOATS.

Argora Goats. *Conf. with pp. 562, 564.*

1589 on more occasions than one, that the breeds of hill goats in India might be greatly improved by the introduction of some Angora blood. This might be so and, if the expenditure for such an experiment be of no serious moment, it might be tried. But it should be recollected that the Angora or Mohair goat is reared, not on account of an under-growth of woolly hair, but for its long ordinary hair. Further, that it is an inhabitant of a dry region at an elevation of about 2,500 feet, where its most favoured food is the leaves of the oak. It would, therefore, be very likely quite useless to attempt to cross such an animal with the *pashm*-yielding races of the higher Alpine regions of the Himálaya and trans-Himálaya, and probably equally futile to cross with it the Sind and Baluchistan low level goats. This latter suggestion has been recently made, however, but it is perhaps safe to say that if India be admitted, as in a condition of poverty in the character of its indigenous races of goats, the Angora goat should rather be crossed with the goats of the higher tracts of Baluchistán, such as those of Quetta and those along the Himálaya in the hotter valleys where the oak forms an important feature of goat and cattle fodder. That is to say it might be crossed with some local form of the *sinál* goat; see **Hodgson's** classification above. In 1869 **Mr. G. Landells** (*Jour Agri-Hort. Soc. Ind., N. S., Vol. I., Sel., 64*) suggested that the Angora goat might succeed in Hazara, but his suggestion was not apparently acted on. The Hazara goats yield *pat* not *pashm*, and may be thus regarded as allied in some respects to the Angora breed and to the *sinál* goat of the Himálaya. India may be said, however, to possess races of goats which under more careful treatment might develope into *pat* or hair-yielding stock quite as good as anything ever likely to be imported. The experiment proposed, well on to half a century ago, though never practically tested, to establish goat and sheep runs or farms on the higher southern or Indian slopes of the Himálaya, for the production of the better qualities of *pashm* and wool, seems to the writer well worthy of trial. The suggestion more recently offered seems less deserving of consideration, namely, to form herds of improved sheep in the lower basins of the Panjáb rivers, where the herbage is indifferent and leeches prevalent. **Dr. Falconer** seemed to think that the *pashm* goat might be acclimatized on the southern slopes of the Himálaya, but it must not be forgotten that the down of the ibex and of the *pashm* domesticated goat seems to be directly the result of the drier and ever so much colder nature of the northern as compared with the southern slopes. Indeed the *pashm* goat may be said to actually exist at Spiti and according to **Hodgson** the *chapú* is the acclimatised form of the Tibetan *chángrá*. If this be so, the goat, even if successfully reared on a more extended scale than at present on the southern slopes, would probably yield a far inferior *pashm* than the northern stock, if indeed it did not degenerate into a form of the *pat*-yielding (not *pashm*) goat. It may, however, be safely said that for present European commerce a *pashm* goat is not an indispensible necessity of success.

Sheep & Goat Runs *Conf. with pp. 580, 587, 609, 610, 629, 635.*

1590

Pashm. 1591 Pat. 1592 *Conf. with pp. 559, 636.*

This conclusion leads naturally therefore to the consideration of the chief recognised Indian qualities of *pashm* and *pat* or goats'-hair. One of the most detailed papers on *pashm* is that already alluded to, namely, **Mr. F. H. Cooper's** account, which appeared in the proceedings of the Agri-Horticultural Society of the Panjáb. In the Journals of that Society several other contributions amplified the knowledge of this subject, and it may be said that **Mr. Baden Powell's** excellent chapter (in the *Panjáb Products*) reviews all these papers, while the more modern articles add certain recent statistics, but contribute little or nothing not known well on to half a century ago. **Davies** (*Report on the Trade and Resources of the Countries on the North-western Boundary of British India*), published, in 1862, a very exhaustive statement of the traffic in this substance. **Aitchison** (*Handbook*

TRADE in Pashm.

of the Trade Products of Leh) carried the returns of traffic down to 1872, and subsequent writers have added particulars of certain branches of the trade in *pashm*. Thus, for example, the annual reports of the trans-frontier trade of India give the quantities annually carried into India, but it will be seen from the remarks that follow that the traffic into Kashmír is of greater importance than what actually reaches India. To return, therefore, to **Cooper's** classification of *pashm*, the following paragraphs incorporate the chief facts brought out by him as also by all subsequent writers:—

Shah-tus. 1593

1st, Shah-tus—white and grey. This is said to be derived from a wild goat. The princes in the localities where this is obtained and the "Magnates of Russian Siberia" are said to buy up all that can be got of it. It is valued not only as an extremely fine *pashm* but has attributed to it certain medicinal virtues. Shah-tus is, however, very scarce and is only brought to Kashmir when specially ordered. It is sold in balls of fine spun thread and very rarely as wool. Plain shawls without any ornament called (*Sádhá chádhar*, from 4 to 6 yards in length and 1 to 1½ in breadth), are known to have fetched from R80 to R200. The grey form which differs only in colour is valued at a much lower figure.

Sadha Chadhar. 1594

Tarfani Pashm. 1595

2nd, Tarfáni pashm.—This (the produce of the domesticated goat) is the article most prized in Kashmír; indeed, so high a value is placed upon it that for many years past the most stringent rules have been enforced against its exportation from Kashmír. "It is the production of the Tárfan Aksu, Kamal, and other hill districts, ranges east and north-east of Yarkand. It is brought down by the Argouns to Kashmír *viâ* Yarkand, in the form of coarse or uncleaned *pam* or *pashm*, mixed with the outer hair of the goat in various proportions, but separated at Yarkand or Ladákh from the *Tarfáni khudrang pam*, or coloured variety." "It sells, according to the fineness of the thread, at from half a rupee weight to 2½ rupees weight for one rupee (Chilkee rupee of 10 annas), while the value of shawls made of it, according to manufacture, may vary from 70 to 5,000 rupees."

Prior to the conquest of Ladákh by Kashmír the shawl-weavers of that State were mainly dependent on Changthán for their supply of *pashm*. The Tarfán *pashm* was, however, so much superior, that by bribery and by force, injunctions were soon established that secured for the sole use of Kashmír all the Tarfán *pashm* brought into Ladákh, and liberated the inferior forms of Changthán *pashm* for export to India. Alluding to the restrictions, on this trade, that existed even at the beginning of this century, **Dr. F. Royle** (*Productive Resources of India*) says "**Mr. Moorcroft**, who was deputed in 1814 to that part of Little Tibet, in Chinese Tartary, where the shawl goat is pastured, for the purpose of opening to Great Britain the means of obtaining the materials of the finest woollen fabrics, found that the Huneas were obliged to send all their best wool to Kashmír. In the year 1819, considerable advantage was anticipated from importing this wool into England; as a gentleman who was consulted and who professed to have a practical knowledge of the English wool market, valued it at eight shillings per pound." These high expectations were not, however, realized, and in the report of the transactions made by the Honourable the East India Company it is not quite clear whether they exported *pashm* or Himálayan wool. In the Asiatic Researches (*Vol. XII.*) the restrictions imposed by the rulers of Kashmír on the traffic in *pashm* is alluded to in the following terms:—"This is caused by strict injunctions to all owners of flocks, not to sell any shawl-wool except to the Cashmerians or their agents, in consequence of a representation having been made to the Government, that the Jonaree merchants had bought some last year, and that the Cashmerians would suffer if any of this kind

KASHMIR MANUFACTURES.

of wool were to pass into other hands (*Moorcroft's Journey to Lake Manosarovara*). So again Moorcroft (*Vol. I., 347*) wrote "by ancient custom and engagements, the export of the wool is exclusively confined to Kashmír, and all attempts to convey it to other countries are punished by confiscation. In a like manner it is considered illegal in Rodokh and Changthán to allow a trade in shawl-wool except through Ladákh; and in the latter country considerable impediments are opposed to the traffic in wool from Yarkand, although it is of superior quality and cheapness." So again "The wool of Changthán is sold to the Ladákhis alone, by virtue of an ancient agreement" (*p. 364*). It is not perhaps necessary to cite a greater number of authors who allude to this disability in the growth of a large trade in *pashm*. The fact has already been mentioned that the Panjáb manufacturers have had to seek in foreign wool, more especially that of Kirman in Persia, a substitute for an article that could be brought to them across the frontier and develope thereby a large return traffic from India. Dr. Jameson, in a paper communicated to the East India Company, made certain pertinent remarks which may be here quoted :—"In the Bari and Jetch Doabs, *viz.*, at Lahore and Amritsar in the former, and Jelalpore in the latter, and at Nurpur in the Kohistan, shawls are extensively manufactured but all of an inferior description, owing to the whole of the best shawl-wool being monopolised by Rajah Gulab Singh. This ought not to be the case, seeing that the great breeding country of the shawl-wool goat is in that tract of Chinese Tartary lying immediately to the north of the British passes in the Himálaya, and the wool-traders, in order to obtain a market, are obliged to carry their wool several hundred miles to Kashmír. Were a little encouragement given to them, wools in large quantity and of the finest quality would be imported into the British provinces by the Mana, Nítí, Onata, Dewra, and other passes. Several years ago the shawl-wool traders brought large quantities of wool to Srinugger (in Kashmír) through the Nití pass, but finding no demand for it, they were obliged to sell it at a great loss. Since then the attempt to get a market has never been developed." Davies more recently alludes to the restrictions imposed by Kashmír, more particularly to the complete appropriation of *Tarfáni pashm* and the liberation of the inferior qualities of Changthán. In this connection he mentions a fact lost sight of apparently by most persons, namely, that "The district of Spiti, geographically part of Ladakh, was purposely annexed to the British territory in 1846-47 in order to prevent the interposition of a foreign State between Rampur and the shawl-wool districts of Changthán." That action had not apparently the desired effect and Kashmír continued, and to the present day largely preserves, as a State monopoly, the right to all the finer shawl-wools that enter the geographical frontier of India. It thus restricts the supply to its own requirements, and has starved the Panjáb shawl manufacturers, as well as deprived Great Britain of a possible participation in an article the supply of which by all writers is affirmed to be practically inexhaustible.

It may, however, be contended that these restrictions no longer exist. Dr. Aitchison (*Hand book of the Trade of Leh, p. 188*) says, "A remarkable change has taken place in the trade of Leh, inasmuch as there are now no restrictions whatsoever upon any article that comes from Yarkand to Leh, and all Yarkand produce is free to be purchased by any one, hence there no longer exists the Kashmír monopoly of bygone days as regards *pashm* from Turkistán. The old monopoly, however, I regret to state, still exists in practice, although not in theory, as connected with the importation of *pashm* from Changthán. That this monopoly still exists is due to the fact that no strangers are allowed into the district of Chang-

KASHMIR TRADE IN PASHM.

thán from Leh, except certain Bhotes who are agents of the Kashmír Government. The whole trade between Leh, Changthán, and Lhassa being still carried on according to the system given by **Mr. Davies** in his report." But even were free egress allowed, the obstruction would still survive. The rule has been so long standing that the merest hint from the Kashmír officials not to render assistance in the transport of *pashm* owned by outside traders would at once be regarded as the modern phase of obstruction desired by the Rajah of Kashmír. The most hopeful view of this subject for the future may in fact be drawn from the near approach of railway communication into the heart of Kashmír. Formerly the *begar* system of transport existed. On this subject **Davies** explains that the custom "of trading by the Gyulpos (former Native rulers of Ladákh), with Changthán (Rudokh) the Maharajah's Government takes a prominent part in the trade of shawl-wool, tea, salt, sulphur, from Changthán, through Bustí Ram, as its commercial agent in Ladákh." Again "Tea is annually brought direct from Lhassa to Leh by a trader (who goes by the name of Chubba) on the part of the Lama of Lhassa. He takes saffron in return; he is allowed *begar* for the transport of his goods through the Maharajah's territories. The Maharajah's Government also sends a man every third year with Kashmír goods to Lhassa for the purpose of trade. He is in return allowed *begar* through the Lhassa territory. This man in Lhassa is called *Lúbchúk* or *vakíl*. The Changthán merchant sent by the Zong or Governor of Ghurdokh to Leh is entitled to *begar* in the Ladákh territory." In 1871 **Mr. Drew**, Joint Commissioner on the part of His Highness the Maharajah at Leh, put a stop to the system of forced State labour (*begar*) in carrying the goods to and from Kashmír and Lhassa, and arranged for the future that carriers should be paid in cash. Previous to this order, frequent complaints had been made by the people to the headmen of Rudokh and Gartok relative to the oppression of the officials in collecting wool for the Kashmír agent. In consequence of this new order difficulties arose between the Kashmír authorities and the traders, and *pashm* began to find a better outlet in being consigned to India. A fall in price took place in Amritsar from the sudden glutting of the market, and this was made the excuse for restoring the old arrangement of *begar* transport and literally, if not legally, to re-establish the monopoly by Kashmír in the traffic. **Dr. Aitchison**, from whom much of the above information has been derived, concludes his statement of the Kashmír grudge against the modern aspects of the *pashm* trade, by furnishing a table of the transactions for the six years ending 1872. He there shows the imports into Leh from Changthán to have been 3,450 maunds and from Turkistán 2,331 maunds, of which only one-twelfth part was allowed to find its way to the Panjáb. It will thus be seen how very true the remark, that has been offered above is, *viz.*, that the Indian transactions in *pashm* are unimportant compared with those of Kashmír. The great bulk of the *pashmina* goods of the Panjáb is therefore not made of *pashm*, but of Persian and other wools, and the possibilities of a future traffic in Tibetan *pashm* with England and other foreign countries, are rendered subservient to the selfish policy of Kashmír and the requirements of the half-starved weavers of that State.

Changthani. 1596

3rd, Changthaní.—This, as has already been explained, is regarded as inferior to the *Tarfání pashm*, but by many writers its inferiority is due more to careless preparation, such as the presence of particles of skin rather than to actual inferiority of staple. It is the produce of the domesticated goat of the Changthán province "and may be said to be produced along the northern base of the ranges, from about Rodokh in the west, or even from the banks of the Shegak, eastward to the Kailas ranges,

KASHMIR TRADE IN PASHM.

north of Man Thaloi or Mansarowar lakes; and even it is said as far as Lhassa. The wool is brought to Kashmír *viâ* Leh or Ladákh not only by the Argouns but also by numerous other traders." **Moorcroft** describes the source of this fleece as a tract of country that extends along the eastern frontier of Ládakh. Its more northern portion forms the separate province, called Rodokh, which lies along the northern border of the lake of Pangkok. The country is thinly inhabited; and the people are chiefly shepherds, who subsist by the sale of their wool to the merchants of Lé. The largest division of Chan-than, however, is called Garo and is in contact with Ladákh. Chan-than, **Mr. Moorcroft** adds, is the chief resort of the shawl-wool goat, and is also the pasturage of numerous flocks of sheep, whose wool is an article of trade. The breed of goat, says **Moorcroft**, that yields the shawl-wool is the same in Ladakh, as in Lhassa, Great Tibet, and Chinese Turkistan, but the wool is not so fine as in the districts on its eastern and northern frontier. The fleece is cut once a year, and the wool coarsely picked either in the place from whence it comes, or at Lé, is sold by the importers to the merchants at the city by whom it is sent to Kashmír. About 800 loads, adds **Moorcroft**, are exported to Kashmír. This appears to be the fleece of the *chángrá* goat of **Hodgson's** classification, although a remark of his, already alluded to, would seem to suggest the idea that the *Tarfani* was by **Hodgson** regarded as different—the produce of a distinct breed of goats.

An anonymous writer, in a little work on *Kashmir and Its Shawls*, offers some useful remarks on the subject of this fleece. Thus, for example, in allusion to the fact that the outer long hair is cut off with a knife instead of with a scissors, he says:—"The knife is too blunt to cut through the down as well as the hair, and so leaves it untouched, while a pair of scissors would cut off the down and hair together, and entail endless trouble and expense in subsequently separating the one from the other. It is not, however, to be supposed that the down subsequently combed off is altogether free from hair. There is, occasionally, a good deal, and the picking can only be done by hand; hence this work is very expensive." The hair is cut off by the knife in the direction of its growth, or from the head towards the tail. The comb used to remove the fine wool is made of five teeth of willow twigs bound together and is drawn through the new short coat (left after the removal of at least two-thirds of the hair) by combing towards the head. In cutting the hair, occasionally small bits of the skin are nipped off accidentally and these adhering to the fleece increase its impurity very seriously. It has already been suggested that probably the chief difference between the Tarfán and Changthan *pashm* is, in the former being cleaner and less adulterated than the lattter. No writer has definitely established the existence of a distinct breed in Tarfán from that of Tibet generally, although nearly every locality has some special property attributed to its *shál*. Thus, while it is admitted that the goat of Ladkáh is said to be identical to that of Changthán, the best fleece of the eastern *pashm*-yielding tract comes from a remote division even of Changthán, namely, Rodokh.

Tarfani Tus. 1597

4th, Tarfáni tús or khudrug.—This is a coloured *pashm* and may be designated an inferior quality or rejections from the selection of the superior qualities of Tarfání and Changthání wool. It is sold at from 12 to 18 rupees weight for one rupee (10 annas value). This is usually made into the cheap ordinary *pashmina* shawls valued at R20 to R80. It is the quality most largely permitted to pass into British territory for sale.

MANUFACTURES OF PASHM. 1598

Manufactures of Pashm or Pashmina.

So much has already been indicated on this subject that little more of a general nature remains to be told. **Mr. Moorcroft** devoted much patient

KASHMIR MANUFACTURES.

Pashmina.

study to this subject, and to him is due apparently the British trade in shawls and such traffic as exists in *shál-pashm*. Space cannot be afforded here to quote Mr. Moorcroft's technical account in full; the reader should consult the original (*Travels in the Himálayan Provinces of Hindustán, etc., Vol. II., 164-195*), where particulars will not only be found regarding the ordinary *pashmina* cloth but the shawl trade of Kashmír generally, the method of working out the patterns, of dyeing the wool, etc. In the anonymous publication, *Kashmír and Its Shawls*, a historic review of all that is known of this industry will be found, together with a reprint of Mr. Moorcroft's chief descriptive passages of the art as pursued in Kashmír. In the *Journál of Indian Art* several brief notices of Kashmír shawls will be found, and in a more recent publication a *Monograph on the Woollen Manufactures of the Panjáb* certain particulars are brought down to the returns for 1884-85. The writer must trust to the isolation of one or two striking peculiarities from these and other works as exhibiting the salient features of this industry, since to do the subject full justice a volume might easily be written on *pashm*, Kashmir shawls, and *pashmina* goods.

Kashmir Shawls. 1599

The author of the pamphlet on *Kashmír and Its Shawls*, thinks that the finer qualities were known in Europe at a very early date as may be inferred from the tradition that the light veil, fastened by a thin golden thread over the forehead, covering the back of the head, and falling on the shoulders, of Leonardo da Vinci's famous portrait of Mona Lisa, wife of Francesco of Giocondo, (a citizen of Florence), was in reality one of those earlier Kashmír fabrics. The painter of that picture died in 1519. The earliest authentic notice of these delicate and beautiful manufactures, however, is perhaps that given in the *Ain-i-Akbari* (1594). Abul Fuzl, the historian of the Emperor Akbar, records the various qualities which were most esteemed by the nobility of Delhi. These were first:—*Tús Assel*, incomparable for lightness, warmth, and softness; second, *Safed Alcheh*, also called *Terehdar*; third, *Zerdozy Gulabatum*, etc., which are of His Majesty's invention; and fourth, certain short pieces now by order of the Emperor made sufficiently long to be used for *Jamahs*. "Formerly shawls were but rarely brought from Kashmír, and those who had them, used to wear them over their shoulders in four-folds, so that they lasted for a long time." "His Majesty has introduced the custom of wearing two shawls, one above the other." "By the attention of His Majesty, the manufacture of shawls in Kashmír is in a very flourishing state, and, in Lahore, there were upwards of 1,000 manufacturers of this kind." There can thus be no doubt that the support of the Court of the Muhammadan Emperors did much to foster and advance the shawl industry of Kashmír. Bernier alludes to the futile attempts that were made in the time of Aurungzeb to introduce the industry into Patna, Agra, Lahore, etc., but adds that "the produce of foreign looms has never equalled that of Kashmír in its delicate softness."

A very extensive list of travellers might be quoted as having each contributed a little towards clearing up the mystery that for many years lingered around the Kashmír shawl industry—an industry that turned out goods which surpassed anything that could be produced elsewhere, as far as the softness and fineness of the material excelled the wool of Europe. Thus, in 1624, Father Antonio Andrada, a Portuguese Missionary in Tibet, crossed the mountains that separate Kashmír from Tibet, and in his journal deals with the cattle, sheep, etc., he had seen. Tavernier in his travels in India and the East (1636-1666) urged on the attention of France, the value of the Kirmani wool of Persia. Bernier, who journeyed in company with the Emperor Aurungzeb into Kashmír in 1663, tells us that the famous shawls

SHEEP: Pashm.

Pashm and Goats'-hair,

HISTORY OF KASHMIR MANUFACTURES.

of that little mountain State was made of two kinds of fleece, one, the fine wool of a certain breed of sheep, the other, the *tús* or down from a goat found in Tibet.

The authors of the *Lettres Edifiantes* asserted in 1712 that the shawl-wool came from Tibet. **Chardin** (a Frenchman who had settled in England and was employed in high diplomatic offices by **Charles II.**) spoke of camel's hair, locally known in Persia as *teftik* (called by Europeans *laine de chevron*), being used in the preparation of shawls. The **Rev. W. Tooke**, in his *History of the Reign of the Empress Catharine II. of Russia*, speaks of the under-down of goats being the material from which the high-priced shawls of Tibet and Kashmír is made. **M. Legoux de Flaix**, a gentleman of Pondicherry, investigated personally the shawl industry of Kashmír, and published in 1804 some useful information together with many misstatements, as, for example, the affirmation that the shawl-wool was the fleece of a sheep, not a goat. This so roused the aspirations of the Agricultural Society of Paris that it determined at all risks to obtain specimens of the breed. In the "Philosophical Transactions of Bengal," **Bogle** enlarged on the subject of the Tibetan goat fleece. **Craufurd** (*History of the Hindús*) attributes the wealth of Tibet to its goats and sheep. In 1803, **Khodja Yusuf** (an Armenian) was sent from Constantinople to Kashmír in order to have shawls made according to a new design furnished by himself. **Turner**, in the course of his embassy to Tibet (subsequent to **Bogle**), noticed large flocks of sheep with extremely fine and soft wool. Also large herds of goats which were considered superior to the Angora breed. Their hair, he says, is one of the richest products of Tibet. It reaches Kashmír *viá* Ladakh. He goes on to say of this fine under-coat of hair or wool that "It is covered by other long and harsh hair borne by the animal, and this protects the delicacy of the interior coat. These goats, no doubt, owe to the nature of the climate in which they live, this fine and warm covering, for all those that have been conveyed to Bengal have soon lost it, and been attacked by a kind of itch." "Those which I sent at different times to England fared no better. Some reached their destination alive, but in such a weak condition that they soon died. The sea is as dangerous to them as the heat of Bengal." During the exploration in 1808 of the sources of the Ganges by **Webb, Raper,** and **Hearsay**, so much attention was given to this subject that it was resolved to depute **Messrs. Moorcroft** and **Hearsay** to examine Lake Mansarovar, and, at the same time, to secure some of the animals that produce the shawl-wool. This expedition met with complete success. All doubt was removed as to the source of the shawl-wool. It was conclusively proved to be the under coat—the winter protection—of the Tibetan goat. What came of the specimens of this animal which they brought back with them has never transpired. **Mr. Moorcroft's** second expedition, the story of which he never lived to publish, narrates in the most minute detail, the methods of preparing the fleece and manufacture of it into the famed shawls of Kashmír. One of the earliest and to this day the most accurate writer on the breeds or races of Tibetan *shal*-goats was **Mr. B. H. Hodgson**, who, in 1847, published a most instructive paper on this subject. About the same time (*see Calcutta Journal, Natural History, Vol. II, pp. 514-542*) **Captain Thomas Hutton** issued his paper on the peculiarities of the Afghán wild sheep and goats, in which he discusses the probable origin of the *pashm*-yielding breed of the domesticated Tibetan goat. In a paper (*Jour. Asiatic Soc. Bengal, IX.*) he deals with the wool and woollen manufactures of Khorasán. **Captain Conolly** (*Jour. As. Society, 1840*) also devoted much careful study to the subject of goat's hair and under-fleece. **Southey** (*Colonial Wool*) furnished a special chapter in his work

S. 1599

TRADE IN PASHM MANUFACTURES.

to Indian Goats' Wool (*pp. 327-333*) in which he discusses the French experiments at cross-breeding the Kashmír and Tibetan goat. Shortly after (1865), **Mr. F. H. Cooper** contributed useful information on the qualities of the shawl-wool known to commerce.

But to return to the subject of the European demand for Kashmír shawls, it has been pointed out by the author of *Kashmír and Its Shawls* that, after the conquests of the Persian invader of India, **Nadir Shah**, in 1739, we are told of an ambassador having been despatched to Constantinople with fifteen elephant-loads of presents to the Sultān, amongst which were many Kashmír shawls. The author of *Kashmír and Its Shawls* thinks that the modern demand for these shawls dated from about that time, and he surmises that the wives of the ambassadors from the European courts, probably got presents of shawls from the Sultan. Be that as it may, the fashion first prevailed amongst French ladies to wear these "fine silky webs of wool," as **Larousse** describes them, and it then spread to England.

Trade in Pashm, Pashmina, & Shawls.

TRADE. 1600

The famine of 1819 drove many of the shawl-weavers from their homes in Kashmír to settle in the Panjáb, and the colonies of these skilled workmen that are now to be found in many parts of Upper India date from about that time, but they have been supplemented from year to year by successive waves of immigrants into British territory as the hand of oppression told more and more heavily upon them. In 1850 the trade in these expensive goods had grown to such proportions that French and British merchants established agencies both in Kashmír and in the Panjáb to purchase their annual supplies. It would be beyond the scope of this work to deal with the artistic designs worked out in the Kashmír shawls, but it may, in perfect fairness, be said that the effect of direct agencies was not an unmixed advantage to the weavers. These skilled workmen, with the increasing demand and great profits, became more and more the servants of middlemen and were urged to modify their patterns and the character of their goods, to meet the variations and fancies of the European market. Cheapness became an object, and degeneration followed to a large extent. In a table published as an appendix to *Kashmír and Its Shawls*, **Dr. Forbes Watson** gives the exports of all classes of *Páshmíná* goods from India as follows:—

	£		£
1850-51	171,709	1858-59	310,027
1851-52	146,270	1859-60	252,828
1852-53	215,659	1860-61	351,093
1853-54	170,153	1861-62	459,441
1854-55	197,890	1862-63	303,157
1855-56	209,279	1863-64	275,391
1856-57	290,640	1864-65	254,498
1857-58	227,618		

These figures do not of course mean the total exports from Kashmír, but the foreign exports from India. They therefore leave out of all consideration the consumption of this class of goods in India itself as also the traffic with Tibet and other Central Asiatic countries. Purchases, we are told, of *pashm* are largely made in Kashmír *pashmina* goods, so that in this way alone a very considerable export must yearly take place towards Tibet. It has already been shown that the coarsest *pashm* only finds its way to British India, and that even that article amounts to but about $\frac{1}{12}$th of the total quantity that reaches Kashmír. There is thus some foundation for the common statement that in Kashmír and in the capital town of the State alone, can the best qualities of shawls be procured. The most

SHEEP: Pashm.

Pashm and Goats'-hair,

TRADE IN PASHM MANUFACTURES.

skilled workmen appear to be those most closely watched by the State officials, By having secured to them the finest *pashm* and by various privileges, the art has been fostered and helped forward to its present proficiency. Indeed, it seems probable that were fashion to turn once more to the delicate and high class goods, the weavers of Kashmír would be able to show that they are not only equal to their forefathers but able to excell them, both in elegance of design and softness of colouration. For, be it remembered, that although the change in European taste and the after consequences of the Franco-Prussian War (scarcity of money in France) deprived the weavers of Kashmír and of the Panjáb of a large and profitable market for certain classes of their goods, the trade was not entirely ruined. India continued to demand a large supply, and the Central Asiatic traffic became, if anything greater than before. A far more serious injury was done to the local Kashmír industry by the establishment of colonies of their countrymen in British India, who found in Kirmani-wool, and who continue to find, a sufficient substitute for the cheaper requirements of modern trade.

In the *Reports of Trade and Navigation of British India,* shawls appear as exported under two headings, *viz.*, FOREIGN MERCHANDISE and INDIAN MANUFACTURE. Whether the former can be regarded as mainly Kashmír goods and the latter the shawls manufactured in the Panjáb, the writer is unable to say. Indeed, the former must include the re-exports of European shawls imported into India in the first instance. But even were the total value of these returns to be accepted as entirely Kashmír and Panjáb shawls, they would mark a serious decline from the values given by Dr. Forbes Watson:—

	VALUE OF THE EXPORTS OF SHAWLS FROM INDIA.				
	1885-86.	1886-87.	1887-88.	1888-89.	1889-90.
	R	R	R	R	R
Foreign Merchandise	14,487	10,624	11,233	20,864	21,322
Indian Manufacture	3,08,731	3,46,218	4,05,993	2,66,011	3,02,471
Total expressed at nominal . £	32,321	35,683	41,722	28,687	32,379

The following passages from modern publications may help to exemplify the present position of the *pashm* trade as well as to show the opinions held by various writers:—

In the opening paragraphs of the *Monograph on Woollen Manufactures of the Panjáb* it is estimated that the annual consumption of *pashm* in that province comes to 7,500 maunds, of which 4,500 maunds are imported by sea, 1,800 maunds obtained from Afghánistán, and 1,500 maunds obtained from Tibet. It seems probable that the imports from Tibet alone are pure *pashm*, the other supplies, particularly that by sea, being Persian (Kirmáni) wool, used as a substitute far *pashm*. *Wahab Sháhi pashm*, for example, is wool, not *pashm*; and, indeed, a large proportion of the stuff that comes to the Panjáb manufacturers *viâ* Rampur is wool, not *pashm*. The *pashm* imported is worth R1⅓ a seer; (2lb), the substitutes for it mentioned above fetch only 17 annas a seer. The Monograph then continues:—

Pashmina. 1601

The chief fabrics made of *pashm* are shawls, Rampur chadars, *pashmina alwán* ("a fine white serge-like stuff," as made in Simla), *jámawárs* (striped pattern, made also in wool), *rumáls* and *garbi chadars*. The last-named article is comparatively

PASHMINA: MANUFACTURES of PANJAB.

modern and probably has a future before it. In it the warp is of pashm and the wool of cotton. It is much stronger than the whole-*pashm* chadar; it is practically as warm and nearly as soft; and in delicacy of surface and attractiveness to the eye it may be said to surpass the older fabric. In 1880-81 the outturn of pashmína goods and of shawls was officially put at R11,04,642, and this figure may be taken to exceed somewhat the outturn for 1884-85.

Where made:

The use of pashm is practically confined to the districts of Ludhiána, Simla, Kángra Proper, Amritsar, Gujrát, and Lahore.

(1) *Ludhiana.*—Out of the 800 maunds of imported pashm (value R40,000) retained for local manufacture are made R40,000 worth of *alwán* (whereof R30,000 worth finds its way to Calcutta, Bombay, and elsewhere in India); R75,000 worth of Rampur chadars (whereof most are exported to Europe, Calcutta, Bombay and hill stations); R2,000 worth of *jamawárs*, exported to Europe for curtains at R200 per pair; and R4,000 worth of *garbi chadars*, of which half only are kept in the district. Ludhiana. 1602

(2) *Simla.*—The Rampur chadars made at Sabáthú, and, in fact, in almost any of the factories in the Panjáb, are much superior to the chadar imported from Rampur Bassahir itself. A "Rampur khás" chadar can be bought for R9 to R13, and the size being 9 feet by 4½, the rate is 3½ to 5 annas per square foot. A Sabáthú chadar, 12 feet by 6, goes from R25 to R45, or 5½ to 10 annas per square foot. There is only a small colony of pashm-weavers at Sabáthú and the outturn is 70 chadars and 80 pieces of *alwán*. Simla. 1603

(3) *Kángra Proper.*—Pashmina factories are only found at Nurpur and Triloknath. Thirty maunds of pashm are used in local manufacture. *Amlikár*, pashmína with silk embroidery work, is made at Nurpur, but the style and execution are very inferior to the genuine Kashmír *amlikár*. Kangra. 1604

(4) *Amritsar.*—This city imports all the different sorts of pashm including 3,000 maunds of *Wahab Sháhi* and 600 to 700 maunds of "Rampuri pashm," most of which is probably not real pashm. A large quantity of garbi chadars is made in the city. Amritsar. 1605

(5) *Gujrát.*—The pashm used is 2 maunds Kashmiri and 344 *Kirmáni* (*Wahább Sháhi*); and the value of the manufactures is R38,984, where of two-thirds are exported. The quality has deteriorated, and the cause can easily be understood to be the excessive use of the Persian material. Gujrat. 1606

There is no special information about the Lahore industry.

In shawl-making there is a comparatively minute division of labour; and the decline in the trade has made it impossible for all the different classes of workers to get steady occupation. The Kángra report states that the dyer would, if regularly employed, earn R2 *per diem*, but regular employment is not to be had. The designer (*nakkásh*) takes 10 days over a pattern and gets R2 to R15 for it. He draws the pattern with ink on white paper, the colours, etc., being indicated by technical marks. The *tarahband*, to whom the pattern is made over, then makes up the proper number of reels, and, at his direction, another man (*moharrir ta'lim*), whose pay is R3 to R4 per mensem, prepares the papers to guide the weavers. The *tarahband* gets 1¼ annas per 1,000 reels; but, the *moharrir ta'lim* being in his pay, his net earnings do not exceed annas 8 *per diem*. The pupil (*shágird*), who does the weaving, does not earn more than 1 anna *per diem* now-a-days; and the *rafúgar*, who sews pieces of fabric together with silk to make *doshálas* (double shawls), gets 4 annas a day when working; but his employment is not continuous; and as the single shawl is coming into favour both among Europeans and Natives, the *rafúgar's* position is gradually becoming more and more wretched. He, however, also gets 3 annas *per diem* for washing shawls when there are any to wash. There is also the *púrgar*, who picks out loose threads and gives the shawl a smooth appearance. Shawls are made chiefly in Nurpur and Triloknath in Kángra, where the number of factories has fallen from 80 to 20 in the last 20 years, and in Gujrát. Shawl manufactory. 1607

The causes given for the decline in the export of shawls and other pashmína goods are various. The persons concerned put it down to the check caused by the siege of Paris by the Germans, that city being formerly the chief customer, and to a subsequent change of fashion in Europe. I am inclined to attribute the decline rather to adulteration in the manufacture, to the success of the Rampur chadar industry in England, to the want of ingeunity in the production of new and artistic designs, and to the evil effect of the hard water of Nurpur on the material used. The change of fashion is a good deal the result of these causes; and, for the miserable wages now to be got, improved work is hardly to be expected in the future. Causes of the decline. 1608

Goat's hair (*jat*) is used in most districts for making ropes, nose-bags, sacking, *jhúls* for cattle, and matting for floors. The cleaning and opening up of the hair is done in some districts by laying the stuff out on the ground and beating it with a stick. In Hissár it is first washed and put into hot ashes to dry and then the beating Goats'-hair. 1609 Cleaning. 1610

SHEEP: Pashm.

Pashm and Goats'-hair,

PASHMINA MANUFACTURES of PANJAB.

is done. Lastly, in Muzaffargarh the bow-string (*pinjan*) is sometimes employed and the hair is scutched; and both in Muzaffargarh and in Gujránwála it seems that a modified form of the bow-string is also used. A peg is stuck into the ground and a string tied to it. The other end of the string is fastened to a stick in the operator's hand, and the goat's hair is deposited on the ground over the string, and the string is made to vibrate by being slackened and suddenly tightened up with a jerk. Sometimes there are two pegs at a short distance apart, in which case each peg has a string, the further end of which is attached to the operator's stick.

Spinning. 1611 Spinning of goat's hair is done with the *charkhi*, but more commonly with the *dherná*. Double threads are twisted just as woollen twist is made, and ropes are made by hand-twisting, generally of three strands.

Weaving. 1612 Weaving is done on a small loom without shuttle. The place of the shuttle is, as in carpet-weaving, taken by a stick, to which the woof thread is attached, and which is worked through the warp-threads alternately as it meets them.

Camel's hair 1613 Camel's hair (*milsee masal, mallas*) is twisted by hand and not spun. It is mixed with goat's hair to make sacking, and with cotton (which is used for the warp) to make *bhakla* cloth (Hissár); but it is mostly used for rope-making.

Neither goat's nor camel's hair are dyed.

In the *Journal of Indian Art* and in the District Gazetteers of the Panjáb, various reports have been published on the subject of pashm and of goats'-hair, as also on Pashmina and Pattú. The following may be given as fairly representing the opinions advanced by various writers as to the present position of the industry in these fibres:—

Pashmina. Amritsar. 1614 AMRITSAR.—"The most important among the numerous manufactures of Amritsar are those of *pashmina*, or shawl-wool, and silk. The *pashm* or wool used in the first-named kind is imported from Thibet *via* Rampur and Kashmir.

"The trade declined during 1866, owing, among other causes, to the adulteration of the wool with a fine but inferior sort imported *viá* Kabul, from the province of Kirman, whence the wool is known as *un Kirmani*. The trade is said to be now reviving. The *pashmina* fabrics are either plain uni-coloured cloth called *alwan, malida*, etc., which are made up into cloaks and articles of European apparel, either plain or embroidered with silk, or else are woven into shawls, the thread being previously dyed and wound-off expressly for the purpose. The shawls in which the pattern is produced in the loom are the most valuable; in others, the pattern is produced on a groundwork on plain-coloured *pashmina* by embroidery with the needle and fine *pashm* thread. Such shawls are called *amlikar*, as opposed to the *kannikar*, or loom-woven.

'The manufacture of *pashmina* work was first introduced some seventy years ago, about the time when Ranjit Singh was commencing to extend his rule over the whole Panjáb. It is almost exclusively conducted by Kashmiri Musalmans. It is calculated that soon after the manufacture was instituted, there were about 300 shops established in Amritsar in which *pashmina* work was carried on, and that shawls, etc., to the value of R30,000 were manufactured yearly in the city. Besides what was manufactured in the city itself, *pashmina* work was imported from Kashmir to the extent of some two lakhs of rupees in value yearly, and from other parts of the hills to the value of about R20,000. Part of this was sold in Amritsar, and part exported to Hindustan and Haidarabad in the Dakhan. The chief mart in Hindustan for export seems to have been Lucknow. In the year 1833 A.D. owing to a great famine in Kashmir, there was a large influx of Kashmiris into Amritsar. Shortly before the annexation of the Panjáb, the number of shops established in Amritsar had increased to 2,000 and the value of the *pashmina* work turned out yearly was as much as four lakhs of rupees. Also *pashmina* manufactures to the value of six lakhs of rupees were imported yearly from Kashmir, and to two lakhs from Nurpur, Bassáoli, and other parts of the hills. Now there are 4,000 looms in Amritsar, each worked by at least two men, and the value of the *pashmina* work manufactured yearly is estimated at eight lakhs of rupees, or £80,000. The manufacture, which requires the utmost skill and delicacy of manipulation, is learned by the workmen from the earliest childhood. Children are apprenticed (*shagird*) to master workmen, who after a time pay for their services, but usually to their relatives. The payment is made in advance, and if a *shagird* leaves his employer before his advances are worked off, the next employer is supposed to be responsible for the balance. The export of *pashmina* work from Amritsar to Europe commenced anout 40 years ago. The amount now exported yearly is estimated to be in value about twenty lakhs of rupees. This includes what is imported from Kahsmir and other places for re-export. Of this, sixteen lakhs' value is exported by European merchants settled in the Panjáb, and four lakhs' value by native merchants.

PASHMINA MANUFACTURES of PANJAB.

"The Amritsar long shawls of the first quality are sold at from R400 to R500 each; the same of the second, from R300 to R400; and of the third, from R200 to R300. Square shawls are sold, if of the first sort, from R250 to R300; of the second sort, from R175 to R250; and of the third sort, from R125 to R200. *Jamawars* (a kind of shawl distinguished by always having a stripe, flowered or plain, as the prevailing pattern) and *rumals* (square shawls) fetch from R25 to R50. The needlework *rumals* are sold at from R15 to R75. Shawls of the finest quality are made of the Changthani wool, which is imported *viá* Kulu and Sabathu, and is sold there at about R2 a seer. This pashm contains a large admixture of the coarser hair of the shawl-goat, and requires to be cleansed before spinning. This operation is performed with much difficulty. The second sort of shawls are made from a mixture (half and half) of Changthani and Kirmani wools, and it is very difficult to detect the admixture. The shawls of the third class, *viz., jamawar rumals* with straight lines and all other inferior sort of *pashmina* are made entirely from Kirmani wool. The price per seer of this wool is R1-10 annas; and as it contains only a small quantity of coarse hair, the weavers have less trouble and more profit in using it.

"The inferiority of Amritsar shawls to those of Kashmir has frequently been noticed, and is variously attributed to the air and climate of Kashmir, the quality of the water used in dyeing, etc. All these causes may to some extent be admitted. But the most prominent cause of the superiority of the Kashmir fabric is that the adulteration of the shawl-wool with that of Kirman is never practised. Indeed, the Kirmani wool is not allowed to be brought into Kashmir. Another reason is that in Kashmir the process of removing the coarse hair from the *pashm*, and spinning, are much more carefully performed. On the other hand, the scarlet colour of Amritsar is superior to that of Kashmir, the lac dye used being cheaper, and therefore less adulterated. The Amritsar blue and green are said to be also finer than the corresponding colours in Kashmir. Whatever may be accepted as the true cause of the difference there can be no doubt the real Kashmír shawls invariably command a higher price in the market than the Amritsar Fabrics. (*Journal, Indian, Art, July 1888.*)

Gurdaspur. 1615

GURDASPUR.—"For many years the shawl-weaving trade in the Gurdaspur district has been in a declining state. At Shujánpur, Dera Nának, Pathankot, and Batála in this district there are Kashmiri weavers and embroiderers who scarcely earn enough to keep body and soul together. They are practically the slaves of dealers, nor do they receive as wages more than two or three annas a day. The Englishman, seeing the squalor and misery of their lives in the midst of a thriving indigenous agricultural population, and knowing that, so far as Europe is concerned, the Kashmir shawl is dead, wonders that the manufacture survives at all, and that even tolerable work can be produced under such desperate conditions. It is not clear however, that the state of the Kashmiri weaver at its best was much better than it is now. The finest shawls ever exported from the "Happy Valley" were the work of half-starved artizans and the condition of Kashmíri immigrants in British India, though bad enough, is much better than in their own country. The persistent survival of the trade is worth a passing notice. It seems to be due to the fact that for nearly a century the shawl has been considered an essential article of the cold weather costume of the upper classes of Native society throughout the country, and that this fashion is but slowly giving way. It is quite true that in Bengal and the Presidency towns a *chapkan* of broad cloth of European make is found more comfortable and more convenient as a working dress than the costumes of the land. But there are very few among the coat-wearers who do not also possess shawls; and as a set-off to those who have become Europeanised, we must take into account the large number of the lower classes who formerly wore nothing but cotton, but whose prosperity is now marked by the wearing of shawls or scarves. In every town of importance there are shawl-dealers, while travelling hawkers visit the regions where a regular depôt is not established. During the time of the Franco-German war I have seen crowds of Kashmiri weavers listening with excitement to telegrams from the seat of war, and audibly cursing the cruel Germans for destroying the trade in the best shawls formerly taken by Paris. Yet, though the cessation of the French demand was a serious blow, and though Indian dealers have had cause to rue their connection with the London market, where their consignments have been disposed of by auction at ruinously low rates, and though European manufacturers have copied their fabrics and appropriated their names, there is still within the limits of the Indian Empire a larger market than is generally suspected. The unseen zenana absorbs a vast quantity, in addition to those worn by men. So, although the workmen fares but poorly, the dealer, on the whole, seems to hold his own.

"The work turned out in the Gurdaspur district, though resembling that of Kashmir and Amritsar, is not often of the best quality. Shawl edgings for attachment to plain fabrics, and *jamawars*, a shawl of coarse wool woven in broad stripes

SHEEP: Pashm.

Pashm and Goats'-hair,

PASHMINA MANUFACTURES of PANJAB.

of pattern, known in European trade as 'Turkish shawls,' and sometimes sold as veritable products of Stamboul, are the most characteristic things of local production. The woollen cloths known as *pattu, alwan, malida*, etc., are also made in small quantities. Much of the Gurdaspur work is disposed of at Amritsar. At Dera Nának, the birth-place of Geru Nának, the founder of the Sikh development of Hinduism, the cheapest Indian work—if not the cheapest woollen weaving in the world—is produced; for a coloured *jamewar* may here be had for R1-4 annas.

Mixed Fabrics. 1616

Mixed Fabrics.—*Lois* and wrappers of cotton thread and country wool in the proportion of two-thirds cotton are made at Pathankot, Sujánpur, and Dinanagar, and exported from these small towns to Amritsar, the North-West Provinces, Oudh and Bengal. The value of these goods is estimated at R40,000. At Fatehgarh and one or two other places all wool *lois* or blankets are made. *Pashmina*, or shawl wool, is not used in these coarse webs, but wool from the Shahpur district and the hills near Chamba." (*Journal, Indian Art, Oct. 1888.*)

Ludhiana. 1617

LUDHIANA.—"The production of fine shawls has now almost ceased. The Franco-Prussian war put a sudden stop to the manufacture of shawls above the value of R100. Incidentally it converted the population of Amritsar and Ludhiana to warm partizanship of France. Crowds of eager listeners used to collect at the railway station where the telegrams from the seat of war were read to them and received with loud expressions of satisfaction or disappointment as the French seemed to gain or lose.

"The decline in the French demand, however, was not the only cause of the falling off, which reduced the number of pashmína looms from 1,200 to 300. The Government and, dutifully following its example, the Durbar of Native States, no longer take goods shawls to be given as khíllats. The few shawls that appear on the trays laden with carriage clocks, gold-mounted rifles, musical boxes, epergnes and other knick-knacks presented at Durbars frequently travel back to the *toshakhana*, and might almost be considered as State theatrical properties. The management of the London sales to which goods are sent from Amritsar and Ludhiana has been more in the interest of the London buyer than in that of the distant maker.

"**Syud Ahsan Shah**, an Honorary Magistrate and shawl merchant of Ludhiana, writes that there is dishonesty and combination among the principal bidders, who purchase the lots at very low prices, and then divide the profits among themselves. A Parsi gentleman, named **Hormasjee**, who was present at the auction in London, says that bigger merchants make a combination and stop all competitions from petty traders. They purchase the lots themselves at very low prices, and then sell them to petty traders at a great profit. They also put the balance of their old stock to auction with a view to reduce the market value of fresh goods imported from India, and again purchase them to their advantage. Large stocks of goods for sale and want of competition thus reduce the market value of the goods, and traders are obliged to part with their stocks at a loss. The shawl trade has consequently declined, and out of 300 looms, only 200 now turn out superior stuff; the rest manufacture coarse stuff only.

"From this it would appear that 'the knock out auction' is not confined to Jew furniture brokers, and the lower ranks of commercial life in London. It must be admitted that the practice of consigning annually large quantities of goods to a limited number of dealers to fetch what may be given invites combination of the kind described by the worthy Magistrate of Ludhiana. The traders on this side are neither strong enough or united enough to combine to establish their own agency in London, nor would they be able to reach the limited and select market now commanded by a ring of dealers. On the part of these latter it is only fair to say that they complain at times of adulteration of goods professing to be pure pashmína with inferior wool. It is indisputable, however, that, on the whole, the Indian producer gets the worst of it, and it is heartily to be wished that he could find a direct means of reaching the purchasing public in Europe. In some years consigments have turned out more disastrously to the exporters than would be readily believed, and the fact that Indian goods may be sometimes purchased retail in Regent Street at a lower price than any dealer will part with them for in India, is one corroboration of their complaints. The merchant already quoted gives the following average prices of the pashmína goods now made:—

"Rampuri shawls, four yards by two, first quality, R60 each; second quality of the same size, R15. A shawl three yards by one and a half of good quality is worth R20, and the smaller size of inferior make R8. Good woollen stuff, double warp and woof, R8 to R20. Jamawárs, striped colour-woven fabrics, from R5 to R20. *Chadar joras* used by Natives from R20 to R50, and *rumals* from R10 to R50. **Syud Ahsan Shah** estimates the annual outturn of the Ludhiana manufactures as follows:—

PASHMINA MANUFACTURES of PANJAB.

	R
Rampuri shawls	70,000
Doshalás	20,000
Jamawárs	6,000
Rumáls	1,000
Small chadars	1,000
In all about	1,00,000

"The trade, it will be seen, though it is a comparatively recent one, dating from 1833 only, has undergone some vicissitudes, which began before the Franco-German war, with the extinction of the Native Government of Oudh. It does not seem likely to receive any great impetus in the future, but the consumption of shawls is so large among the upper classes of Natives that it must be long before it dies out altogether.

"It is for its weaving industry, however, that Ludhiana is principally famous; and this is of two sorts—woollen and cotton. The former of these, the manufacture of the cloths known as pashmína and Rampuri chadars from Tibetan and Rampur wool, is at present entirely in the hands of the Kashmírí colony, although some of the country weavers are said to be picking it up. The raw material is of two classes—*pasham* or the fine wool of the Thibetan goats; and *Rampurian*, or that of the nearer hills. The former is said to come from the *Barfani* country, which is rather indefinite geographically. Both wools are brought finally from Rampur, which appears to be the *entrepôt* of the trade, by the *gaddis* or hill men. These men used to take the direct route *viá* Rampur; but now generally reach Ludhiana from Umballa by rail. Within recent years (20 or 30) a third class of wool has begun to be imported from Kirman, in Persia, *viá* Karáchi and Lahore; and this is used as a substitute for Rampur wool. The wool from the hills is brought here in the months of October and November, and the annual amount of the sales is estimated at R50,000. The purchases are made, in the first instance, by Hindu merchants, who take large amount of it, and retail them to a second class of traders, or to the Kashmírís. The wool is spun into thread by women of all classes, Hindu and Muhammadan, rich and poor; and any woman can earn from rupees three to four a month by this. The Kashmírí gets a few rupees worth of wool or thread from the merchant (mahájan) and weaves it into *chadars* or piece 6 to 8 yards long and 1¼ to 1½ yards wide (Kashmírí measure). The cloth is of two descriptions—*pashminá* and *nagli pashminá*, the former entirely of *pashm*, and the latter a woof (*bâna*) of *pashm* on a warp (*tána*) of Rampur wool and sometimes of Kirmani. It is designated generally as *alwán*, and is white in colour when it comes off the loom, but may be dyed red, green, etc., according to taste. The *chadars* are purchased by well-to-do Natives for wearing over the shoulders like an ordinary cloak, the piece being cut into two lengths of about three yards each, which are joined at the corners and worn double. The shawl industry (*sál bafi*), or the weaving from *pashm* thread of Kashmír shawls was perhaps the most important branch of all, but it has never recovered from the complete stoppage of the trade in these articles on account of the Franco-Prussian war (1870). It is said that there were upwards of 1,000 Kashmírís engaged in it before that time, and an outturn of more than R1,00,000, worth of shawls; but France was the principal customer, and has ceased to take any since 1870; and there are now not more than 100 looms (single); the rest of the weavers having turned their hands to what they could, many being reduced to beggary. There appears now to be no demand anywhere for good shawls. Native States used to take them for dresses of honour, etc., but do not now to anything like the same extent. The only shawl work at present done is in coarse wool, what we know as *jámewárs*. These are worn by Natives as cloaks and are also exported towards Persia, where they are said to be used for waist cloths, or are cut into strips for borders of *chogas*, etc. A little fine work is still done in making borders for cloaks, the centre-piece being plain *alwán*. The coarse work turned out is not worth an hundredth part of what the fine shawl work was, a piece of *jámewár* selling for a few rupees, where a shawl would have sold for R200 to 300. An ordinary *chadar* of *pashmina* costs R20 to 30, and of *nagli pashmina*, R15 to 20. The looms are almost entirely single, and not more than two or three men ever work together, unless where apprentices learn the art from a master weaver. The district return gives 900 looms with 960 weavers, but **Ahsan Shah**, who is the representative of the body of Kashmírís, gives an estimate of 400 looms with 1,300 men and boys, weavers and apprentices. The Kashmírí population of the district is returned in the recent Census as 2,492, but a large proportion of these are in service or have other occupations. The *pashmina* and Rampur *chadars* of Ludhiana sell all over India; and the value exported is estimated at 1¾ lakhs, but the industry is said to have earned a bad reputation in recent years owing to the mixture of the inferior Kirmáni wool. The *pashmina* is mostly bought up from the weavers by large merchants,

PASHMINA MANUFACTURES in PANJAB.

either Hindus or Kashmíris. On the whole the *pashmina* industry appears to be on the decline, and **Ahsan Shah** says that the weavers are leaving the town, ast he cloth is becoming a drug in the market. The Kashmíris also knit stockings, gloves, etc. There are a good many looms at which common country blankets are woven by *Mazhbis* (Chúhras or Chamars converted to Sikhism). The miscellaneous looms of all these sorts are returned as 400.

WOOL TRADE. 1618

WOOL AND WOOLLEN GOODS.

So much of a historic nature has already been mentioned in the various provincial sections of this article, that it does not appear necessary to attempt to trace out the Indian Wool Trade from ancient times. It may, however, be said that wool was known to the earliest classic writers of India, and that the injunctions of the *Institutes of Manu* assign wool to be used for the sacrificial thread of the Vaisya. Many writers affirm that the art of weaving preceded that of spinning, and that the oldest woollen garments were plaited of cords of wool much after the fashion of reed and grass mats. The discovery of the art of felting appears also to have been early made and to have been practised by Asiatic craftsmen long anterior to the perfection of the industries of spinning and weaving wool. It has only been within recent times, however, that the explanation of this remarkable property has been made out. It was seen by **Youatt**, in 1835, that, when viewed under the microscope, wool was composed of very fine fibres more or less densely coated with minute scales, and that these fibres were curly and elastic. The scales or imbrications were further observed to differ in shape, arrangement, and number according to the breed of sheep. The long-stapled wools were found to have fewer imbrications, to be less curly, and to manifest the property of felting to a very much less extent than the short and very scaly wools. The shrinkage in felting is by some writers supposed to be also due to a certain amount of solution of the fibre taking place. It has, for example, been noted that the tendency to felting is enhanced if wool be washed or boiled in hot water, and still more so if acid be added. The volume and weight of the fabric is seen to diminish and the felting to become greater with each succeeding washing. The decomposition or washing out of the contents of the fibriles is thus supposed to facilitate interlacement, curling is increased, and the retention of that condition is viewed as at once intelligible when the agency of the scales is further taken into consideration. So very much does the successful utilization of wool depend on the nature of the fibre, that it becomes of paramount importance in selection and improvement of breeds of sheep, intended to be fleece-yielders, to ascertain the tendencies of this character. The wool of the major portion of the sheep of India is so deficient in scales that it has come to be regarded as hair rather than wool. Whether it be possible to improve this property, under tropical influences, is a point of the gravest doubt, but, at all events, the recent expansions of the woollen trade of India may, with perfect safety, be said to have been into regions known to produce woolly fleeces. It would be beside the scope of this article to deal with the origin and development of the British woollen manufactures; the reasons why Bradford naturally became the centre of the trade; the indebtedness to Flanders; or the discovery at Worstead in Norfolk of a peculiar method of carding and weaving that gave the name worsted to certain yarn. Suffice it to say that the modern European manufactures are, at the present day, regarded as far inferior to the articles formerly turned out, and largely so, through the fact that garments after a shower of rain are apt to shrink so seriously as to be unwearable. The manufacturer is accused of having so manipulated his goods that the natural felting takes place in the purchasers' hands. The discovery of a method of utilizing goats' and other hair along with wool and of working up old woollen mate-

Important consideration in Breeding. *Conf. with pp. 570.* 1619

PECULIARITIES OF WOOL.

rials into new fabrics (shoddy) have also greatly lowered the high esteem in which European woollen goods were formerly held. Indian wool appears to be mainly used in Europe in the manufacture of carpets, rugs, and blankets. Throughout the length and breadth of India wool-weaving may be met with, but mainly in the preparation of coarse blankets of a wool for which there is little or no market in Europe. Carpet-weaving, in spite of all that has been said to the contrary, still flourishes, and that, too, outside the precincts of the jails. But it is in the Panjáb and Kashmír that a high class indigenous industry exists in wool. In the Panjáb, owing to the conservative policy of the rulers of Kashmír, the quality of the goods turned out is very inferior to that of Kashmír itself. Some writers have tried to advance the idea that there is something in the climate or water of Kashmír favourable to wool-weaving. But the superiority of the goods of that State can easily be accounted for by the policy that has secured the better qualities of shawl-wool and thereby favoured the weavers of Kashmír. With a protective measure that retained to them the finest wools, it would have been strange indeed had the weavers of that State not attained to a higher proficiency than their fellow-countrymen colonised in British territory. So completely, in fact, has the shawl and *pashmina* industry of the Panjáb been starved of the better wools that for many years now the fleece used up by the looms in the Panjáb has been drawn mainly from Persia. The amount of true *pashm* woven in the Panjáb is, in fact, very small. Amritsar is the Bradford of India, but it has had to share the fate of competition in the production of cheap and inferior goods for a popular market. The industry is, however, a fairly flourishing one and of considerably greater importance than that of Kashmír. Within recent years also several large power-loom woollen mills have been established in India to produce goods in direct opposition to the imports from Europe. The degree of success that has attended this branch of the Indian woollen trade, the reader will have some means of judging of, by the statistical information furnished below. But enough has, perhaps, been said to illustrate some of the leading governing factors in the Indian wool and woollen goods trade. In fact the remarks that follow will be seen to be referred to two main sections: (*a*) RAW WOOL and (*b*) WOOLLEN GOODS. Each of these being discussed under two sub-sections, *viz.*, IMPORTS and EXPORTS, and the whole classified under the headings External Trade by Sea; Trans-frontier Land Trade; Coastwise Transaction; and Internal Trade. As manifesting the modern character of the Indian external transactions in RAW WOOL, it may be pointed out that in Milburn's *Oriental Commerce* (a work published in 1813) there is no separate article devoted to wool such as occurs on Sugar, Silk, Jute, Cotton, Indigo, etc. The Kashmír trade in shawls and fine woollen goods existed, however, and is briefly dealt with, but almost the only notice of wool is the occurrence of the name in a list of things that were not admissible at the Government Customs House at the Bombay Bunder, but which could be taken in at the "Muzjid Bunder." In 1805 the Bombay IMPORTS of woollen goods were valued at R3,45,299 (or, say, £34,529), last year (1890-91) the Bombay imports of woollen goods were valued at R74,18,526 (or, say, £741,852), and the total imports of woollen goods for all India at R1,81,82,126 (or, say, £1,818,212). The first record of EXPORTS of raw wool appears to have been in 1834 when the quantity that left India was given at 69,944℔. Once started, however, the traffic appears to have progressed rapidly. It stood at 486,528℔ in 1835; 1,196,664℔ in 1836; 2,444,019℔ in 1837, and passing over a gap of 35 years it became 24,122,562℔. Up to about that date the distinction did not seem to have been considered necessary into Indian wool and foreign wool re-exported from India. By the latter is mainly meant the wool imported

Chief Items of India's Wool Trade. 1620

CHIEF FEATURES of.

in the first instance by sea from the Persian Gulf or across the land frontier into Sind, the Panjáb, the North-West Provinces, and Bengal. Within recent years, partly through the establishment of direct communications with Europe from Karáchi and partly to the facilities that now exist, through the Sind-Pishín (Kandahar) Railway in carrying Baluchistán, Afghánistán, Kandahar, and other trans-frontier wools into Sind, the traffic in foreign wool has assumed very considerable proportions. The exports of Indian and foreign wools conjointly came last year (1890-91) to 34,133, 059℔ and the previous year they stood at 38,272,528℔. Thus in 57 years the exports of raw wools from India had increased from 69,944℔ to 34,133,059℔. The re-exports are, in fact, rapidly becoming a leading feature of India's wool traffic. Twenty years ago they stood at only 128,342℔. The returns of the Indian trade in wool are, in fact, replete with startling evidences of expansion, and these give very possibly but a foretaste of a still greater future. It will be seen in the special article on **Skins,** Vol. VI., Pt. III. pp. 244—250, that the writer has endeavoured to show that the Indian flock of sheep and goats cannot possibly be less than 50,000,000, since the skins supplied annually to the foreign and local markets come to well on to 40,000,000. But of that flock, perhaps, more than half are goats, and of the remainder a large percentage yield so inferior a fleece that, when clipped and sold, it is generally classed as hair instead of wool. It is the wool, however, of the village weavers of coarse blankets, rugs, and inferior carpets, but which, as a rule, escapes registration, since it is mainly used up locally. The wool of Indian commerce, to a very large extent, is imported across the land frontier or is derived from the Native States of Rájputana, Kattywar, Cutch, and of the Panjáb and the Himálaya. The only, strictly speaking, Indian wools that figure in trade returns are those of the Deccan and the mountainous tracts of South India. A writer quoted above (in the provincial sections) estimates that 30,000,000 sheep in India yield 111,000,000℔ of wool, or about 4℔ a head per annum. It is more than likely, however, that the sheep of the plains of India do not yield, on an average, much over 1½℔ a head. Indian sheep are generally sheared twice a year, in Spring and again in Autumn, and at each shearing it is certain the average yield does not exceed one pound. But, apart from this fact, it is very doubtful indeed if there be 30 million fleece-yielding sheep in India, if even so large a flock exists at all. **Mr. F. M. W. Schofield** (*Notes on the Wool Production and Wool Trade of India*) accepts the assumption of 30 million sheep, and, calculating 2℔ of wool per head with 20 per cent. added for the wool, hair, etc., of all other animals (used up along with, or as substitutes for, wool), he gives the total production at 72 millions or reduced to a population of 250 millions as equivalent to a consumption of ⅕th of a pound against 2⅓℔ in England. But such a calculation is obviously misleading, for, unless each district produces its own supply, the registration of transactions from district to district and province to province gives no indication of even a consumption of ⅕th of a pound. The modern features of the trade may be said in fact to point to a decline in the external demand for the so-called Indian wool, with a compensating increased consumption of the foreign wools of India. The exclusive location of the external traffic in a limited area argues against an extensive supply diffused all over India. This may be exemplified by the returns of 1890-91 :—

Probable Indian Flock of Sheep and Goats. 1621

Indian Consumption of wool. *Conf. with p. 671.* 1622

	℔
Indian wool exported from Bombay	15,318,510
Foreign do. do.	3,930,360
Indian wool exported from Sind	5,006,076
Foreign do. do.	8,857,856
TOTAL .	33,112,802

CHIEF FEATURES of.

This leaves, therefore, a little over one million pounds to have been exported from the whole of the rest of India. Before discussing the traffic in India generally (foreign and local), it may be as well to examine the sources from which Sind and Bombay obtained their supplies. During the year in question Sind imported 10,420,256℔ of wool, mostly by the Sind-Pishín (Kandahar) Railway. Of these imports Karáchi re-exported almost entirely to England 8,857,856℔ and consigned the balance by coastwise trade to Bombay from whence it was re-exported to Europe. Thus it may safely be said that of the modern exports of wool well on to one-third are carried by the ráilway that taps Baluchistán and Afghánistán. But of the Sind local wool a large proportion is drained from the Panjáb and Rájputana, though the sheep of Sind are also fleece-yielders. Turning now to Bombay it will be seen that during the past ten years the imports coastwise have greatly fallen off, with the establishment of direct trade from Karáchi to Europe. It may, however, be affirmed that the present traffic comes in round figures to 7 million pounds, drained from Sind, Kattywar, Cutch, and Madras. This leaves, therefore, some 12 or 13 million pounds to be accounted for. But an examination of the rail and road traffic reveals the fact that the imports of the port town of Bombay are derived mainly from Rájputana (6 to 8 million pounds); from the Panjáb (3 to 4 million pounds); and from Bombay Presidency (2 to 3 million pounds). Of the imports from the Panjáb a certain percentage is doubtless Rájputana, Afghán and Himálayan wool that merely finds its way *viá* the Panjáb to Bombay. In the Railway returns the distinction is not made into Indian and Foreign Wool, so that it is not possible to classify the returns. Enough has, however, been said to show that the wool exported from India to foreign countries comes from Baluchistán and Afghánistán, Sind, Rájputana, Kattywar, and Cutch, and to a much smaller extent from the Panjáb, Bombay, and Madras.

Sources of Supply. 1623

It is perhaps unnecessary to discuss this subject any further, since the tables that may now be furnished will fully exemplify the conclusions that have above been drawn, as also many other significant features of the Indian Wool Trade.

I.—FOREIGN TRADE IN RAW WOOL AND WOOLLEN GOODS.

FOREIGN.

Raw Wool. 1624

A.—RAW WOOL.—The Foreign is not only by far the most valuable section of the Indian Wool Trade, but the one of which the most precise information exists; it will be seen to be illustrated by tables I. to X below. Should future research prove the local trade in wool as co-extensive with the foreign, the facts regarding the latter must still remain of primary importance to the European manufacturer. Indeed, it may safely be said that India's interest in wool, so far as is presently known, is in the imports of British goods and the exports of raw wool. But the demands of the people of India for woollen fabrics are necessarily infinitely more restricted than for cotton or even silk. The local industries may, therefore, be regarded as engaged on the two extremes, the very coarsest blankets and the finest shawls and carpets. These local industries cannot, however, be viewed as holding a very high position from a national standpoint, though the luxurious manufactures of Kashmír and even of the Panjáb surpass in delicacy of texture and beauty of design anything turned out by the power-looms of Europe. The woollen goods of every-day use are mainly, however, supplied by Europe, and the woollen mills that exist in India cannot be regarded as having as yet anything like exercised comparatively the

SHEEP :
Wool Trade.

Trade in Wool

FOREIGN.

Raw Wool.

influence effected upon external supply, that has been attained by the cotton and jute manufacturers.

TABLE I.

Raw Wool Trade of India with Foreign Countries.

Years.	Imports.		Exports (Foreign wool).		Exports (Indian produce).	
	lb	R	lb	R	lb	R
1871-72 . .	1,511,411	4,23,419	128,342	37,986	24,122,562	90,28,997
1872-73 . .	1,733,884	5,27,053	426,634	2,35,844	20,394,718	83,80,418
1873-74 . .	1,254,900	3,85,624	647,826	2,84,966	20,333,372	93,83,357
1874-75 . .	1,542,767	4,27,717	152,353	60,181	21,290,782	95,99,009
1875-76 . .	1,749,188	4,55,007	370,944	1,55,401	23,767,692	1,09,42,002
1876-77 . .	2,145,584	5,32,116	531,364	2,55,415	24,056,767	1,07,73,720
1877-78 . .	2,340,135	5,86,454	537,660	2,31,999	23,075,323	94,36,448
1878-79 . .	2,722,041	6,27,721	1,223,166	5,12,449	26,568,518	1,05,84,574
1879-80 . .	3,564,939	8,72,729	2,298,058	9,18,270	26,368,794	1,09,59,723
1880-81 . .	2,775,554	7,23,434	3,145,431	15,64,873	22,602,690	1,01,41,371
1881-82 . .	2,990,077	7,54,350	5,176,734	22,76,945	21,580,618	81,45,513
1882-83 . .	2,781,257	6,89,313	4,819,024	21,24,270	21,561,303	79,04,058
1883-84 . .	2,526,942	6,51,368	5,198,984	22,71,611	20,036,196	75,58,409
1884-85 . .	2,591,421	6,18,212	6,602,000	28,02,932	18,928,173	71,35,760
1885-86 . .	3,095,026	7,77,217	8,179,584	33,37,918	23,148,763	87,23,211
1886-87 . .	3,170,582	8,07,573	10,540,478	44,32,555	23,208,643	89,95,517
1887-88 . .	3,475,085	9,34,096	11,207,112	52,24,909	23,877,031	97,23,462
1888-89 . .	4,500,219	12,48,829	13,156,968	61,95,273	21,960,848	96,87,529
1889-90 . .	5,100,556	13,67,440	14,402,296	69,35,273	23,870,232	1,08,56,367
1890-91 . .	4,236,826	11,56,154	12,788,216	62,46,808	21,344,843	96,83,223

The more striking features of the above table will be found alluded to in more than one place. But it may be here pointed out that the imports of foreign wool have increased from 1½ to, say, 5 million lb during the past twenty years. This result, it will be seen by Table II., has been mainly through the larger imports of Mekran and Persian wools. The re-exports have immensely expanded, *viz.*, from 128,342lb to 12,788,216lb. This has been entirely through the great facilities effected by Karáchi. Wool can now be shipped direct from Sind to Europe instead of having to bear the expènse of transit to Bombay. It is also cleaned, assorted, and baled in Karáchi and also in Bombay, by methods which are year by year made to conform more and more with the necessities of Europe. Although all writers are agreed that there is still great room for improvement in these directions, still the increasing prices paid for Indian and re-exported Indian wool, show that progress is being effected. It is scarcely necessary to comment on the exports of Indian wool. The trade appears to have been stationary for the past twenty years at least. Many writers, indeed, hold that the restriction of grazing lands, through extension of cultivation, and the establishment of forest reserves, preclude any material enhancement of the traffic in Indian wool. The requirements

Cleaning & Assorting. *Conf. with pp. 596, 597, 610, 611, 633.*
1625

FOREIGN.

Raw Wool.

Imports. 1626

of the country, it is urged, are already contending with foreign demand so that this side of the Indian wool trade may possibly decline rather than expand.

TABLE II.

Analysis of the Imports (by Sea) of Raw Wool from Foreign Countries.

Countries from whence Imported.	1875-76.		1880-81.		1885-86.		1890-91.	
	lb	R	lb	R	lb	R	lb	R
U. Kingdom	...	...	...	...	49,237	31,938	55,096	45,990
Germany	...	...	...	...	...	...	1	3
Aden	...	...	...	...	7,896	1,149	...	...
Arabia	52,304	7,278	31,864	6,425	47,566	9,451	60,228	15,358
Mekran and Sonmíani	735,746	1,33,058	1,294,228	2,50,847	1,018,871	1,53,719	789,672	1,21,595
Persia	945,909	3,10,747	1,370,460	4,43,598	1,873,946	5,44,843	3,201,212	9,28,310
Turkey in Asia	...	...	76,792	21,497	12,768	3,308	115,801	34,220
China (Hong-Kong)	11,200	2,550	...	...	...	...	...	...
Straits Settlements	...	...	...	...	5,880	400	672	70
Australia	...	...	...	...	78,862	32,409	14,144	10,608
Others	4,029	1,374	2,200	1,067	...	...	...	...
TOTAL	1,749,188	4,55,007	2,775,554	7,23,434	3,095,026	7,77,217	4,236,826	11,56,154
Provinces to which Imported—								
Bengal	319	148	...	...	85,174	32,917	50,196	42,313
Bombay	1,024,445	3,12,610	1,294,920	3,66,406	1,680,208	4,55,284	3,199,497	9,07,723
Sind	724,298	1,41,838	1,480,634	3,57,028	1,329,644	2,89,016	987,132	2,06,115
Madras	126	411	...	...	...	...	1	3
TOTAL	1,749,188	4,55,007	2,775,554	7,23,434	3,095,026	7,77,217	4,236,826	11,56,154

With regard to the foreign imports of raw wool it is perhaps only necessary to add to what has already been said that the growth of the transactions from Persia is the most noticeable. It is somewhat remarkable, that the *pashmína* manufacturers of Amritsar should prefer the Persian to the wools that are brought to India across the land frontier. This appears to be the case, however, since the rail-borne trade shows a very extensive export from Karáchi and Sind to the Panjáb. Thus in 1888-89, 5,062 maunds were imported by the Panjáb, of which Karáchi furnished 3,678 maunds (or, 301,596lb), and Sind 571 maunds. But Calcutta also sent 316 maunds to the Panjáb, and this was very possibly Australian wool.

TABLE III.

Exports. 1627

Analysis of the Foreign Wool Exported by Sea from India.

Countries to which Exported.	1875-76.		1880-81.		1885-86.		1890-91.	
	lb	R	lb	R	lb	R	lb	R
U. Kingdom	370,944	1,55,401	3,111,809	15,50,817	8,173,536	33,35,293	12,645,360	61,87,848
France	...	...	33,600	14,000	...	...	126,056	50,210
Germany	...	...	...	...	...	...	16,800	8,750
U. States	...	...	...	...	6,048	2,625	...	...
Others	...	...	22	56	...	...	...	...
TOTAL	370,944	1,55,401	3,145,431	15,64,873	8,179,584	33,37,918	12,788,216	62,46,808

SHEEP: Wool Trade.

Trade in Wool

FOREIGN. Exports.

Provinces from whence Exported.	1875-76.		1880-81.		1885-86.		1890-91.	
	lb	R	lb	R	lb	R	lb	R
Bombay	370,944	1,55,401	1,069,824	4,55,390	2,425,024	10,82,998	3,930,360	18,24,170
Sind .	...	...	2,075,585	11,09,427	5,754,560	22,54,920	8,857,856	44,22,638
Bengal .	...	...	22	56	...	...	...	...
TOTAL .	370,944	1,55,401	3,145,431	15,64,873	8,179,584	33,37,918	12,788,216	62,46,808

TABLE IV.

Analysis of the Indian Wool Exported by Sea from India.

Countries to which Exported.	1875-76.		1880-81.		1885-86.		1890-91.	
	lb	R	lb	R	lb	R	lb	R
U. Kingdom	23,717,020	1,09,24,836	22,374,128	1,00,35,598	23,031,102	86,78,713	20,945,851	94,53,962
France .	35,532	12,416	134,988	54,442	10,752	3,000	324,064	1,84,861
Italy .	...	...	...	...	...	...	1,344	600
Austria .	20	200	...	...	...	...	...	...
U. States .	15,120	4,550	24,590	13,631	42,000	9,907	...	...
Japan .	...	...	68,648	37,690	59,136	30,900	73,584	43,800
Mekran .	...	...	336	10	...	...	...	...
Others .	...	...	...	...	5,773	691	...	...
TOTAL .	23,767,692	1,09,42,002	22,602,690	1,01,41,371	23,148,763	87,23,211	21,344,843	96,83,223
Provinces from whence Exported—								
Bengal .	10,848	4,688	14,865	7,357	14,077	4,516	712,267	3,48,020
Bombay .	16,200,512	69,76,364	21,005,312	93,36,046	17,313,354	66,44,647	15,318,510	70,84,188
Sind .	7,556,332	39,60,950	1,576,485	7,96,458	5,792,232	20,71,085	5,006,076	22,06,680
Madras .	...	...	6,028	1,500	29,100	2,963	307,990	44,335
TOTAL .	23,767,692	1,09,42,002	22,602,690	1,01,41,371	23,148,763	87,23,211	21,344,843	96,83,223

Woollen Goods. 1628

B.—WOOLLEN GOODS.—Following the course pursued with the raw wool a statement may in the first place be furnished of the Imports, Re-exports and Exports, of Woollen Piece Goods, Shawls, Braids, etc., since 1871-72. The table No. V, of Imports shows a steady expansion of, say, from 5 million yards of woollen piece goods in 1871-72, to 13,110,184 yards in 1890-91. The traffic in shawls is no less remarkable, these having increased from 321,284 in 1875-76, to 1,085,727 shawls in 1890-91. This practically expresses the Bengal demand for English gaudily-coloured shawls which have become articles of a gentleman's dress.

FOREIGN.

Woollen Goods.

Imports. 1629

TABLE V.

Imports of Woollen Goods from Foreign Countries.

	Piece Goods.		Shawls.		Braids.		Other Sorts.	
	Yards.	R	No.	R	℔	R	℔	R
1871-72	4,456,140*	44,92,650	No returns.	...	Returned in ℔ or grosses.	1,99,029	No returns.	4,50,257
1872-73	6,872,570*	65,43,571				2,93,718		3,58,010
1873-74	6,068,538*	58,26,143				1,80,883		6,82,082
1874-75	5,043,281*	49,93,576				3,54,168		2,28,111
1875-76	7,233,629	72,50,456	321,284	10,14,939	137,119	2,60,113	369,229	1,72,095
1876-77	6,694,322	66,48,059	255,262	7,55,823	164,638	3,73,043	490,398	3,39,596
1877-78	7,069,693	62,88,600	271,460	7,09,037	236,546	4,93,405	424,345	3,36,767
1878-79	7,611,549	70,31,993	427,412	9,85,341	176,350	3,86,435	522,865	3,76,655
1879-80	7,672,043	74,70,554	446,582	9,23,554	190,745	3,50,417	592,258	5,34,239
1880-81	11,254,429	1,10,94,611	499,896	9,62,998	222,537	3,39,872	687,678	5,93,818
1881-82	8,850,816	92,22,304	395,622	7,65,278	231,427	4,65,407	678,024	7,59,331
1882-83	6,932,779	77,52,049	349,764	8,09,545	243,783	4,67,381	813,714	8,19,750
1883-84	9,316,192	95,70,474	591,425	13,21,642	158,303	3,04,851	957,588	9,73,564
1884-85	10,700,128	1,00,23,119	461,069	9,56,196	218,505	3,88,060	955,424	9,76,022
1885-86	11,223,258	1,12,56,264	616,782	12,18,054	263,348	4,70,726	1,039,134	9,73,572
1886-87	12,133,627	1,21,92,515	740,787	17,08,130	290,242	4,59,356	899,664	9,28,645
1887-88	13,806,388	1,40,37,181	515,372	10,73,689	284,150	4,94,104	1,406,454	15,52,578
1888-89	11,864,523	1,18,95,264	663,984	16,11,186	365,250	6,65,009	1,267,019	14,48,040
1889-90	10,215,322	1,03,98,967	901,511	23,42,540	274,629	5,15,266	1,172,321	12,95,575
1890-91	13,110,184	1,29,51,614	1,085,727	29,30,543	304,710	5,59,786	1,427,056	17,40,183

* During these years certain articles were returned as "pairs," "pieces," or "numbers" and were not, as in all subsequent years, expressed in yards. The total number of "pairs," "pieces," etc., which should be added to the figures given, averaged from 30,000 to 40,000. How many yards these were equivalent to is difficult to ascertain, but their value is given along with that of the yards of actual piece goods, so that the column of values denotes the progression of the total trade.

The re-export trade in Kashmír shawls, declined very seriously with the loss of the fashion in Europe for these articles. The industry has, however, changed its character. The goods now exported are not only very much cheaper than formerly but relatively so. The traffic in cheap shawls and *pashmina* goods may, in fact, be regarded as at present on the ascendant. To the figures here shown (to obtain any tangible conception of the total trade in *pashmina*), it becomes necessary to discover and add the share of the exports of Indian woollen goods derived from Amritsar. These are very possibly quite as much entitled to the name Kashmír shawls and *pashmina* as the goods returned in the table of re-exports. Indeed, there is reason for the belief that a large share of the re-exports are British shawls imported by India and re-exported to other countries. The isolation of Kashmír as a foreign country, along with Baluchistán, Afghánistán, and Persia, leads to considerable confusion in the returns both of the raw wool and of the woollen manufactures that in Europe would be classed as Kashmír, or Indian shawls, etc., whether made in Kashmír or in Amritsar.

Re-Exports. 1630

TABLE VI.

Re-Exports of Foreign Woollen Goods from India to Foreign Countries.

	Piece Goods.		Shawls.		Braids.		Other sorts.	
	Yds.	R	No.	R	℔	R	℔	R
1871-72	49,457	57,210	...	Returned conjointly with value of piece goods.	...	Returned conjointly with other sorts.	...	86,567
1872-73	98,294	1,68,689	...		...		...	28,206
1873-74	117,218	1,24,142	...		...		...	13,007
1874-75	92,697	1,23,195	...		...		...	107

SHEEP:
Wool Trade.

Trade in Wool

FOREIGN.

Woollen Goods.*

Re-exports.

	Piece Goods.		Shawls.		Brains.		Other Sorts.	
	Yds.	R	No.	R	℔	R	℔	R
1875-76 .	135,552	1,57,505	...	730	...	...	...	3,862
1876-77 .	131,087	1,28,471	2,375	11,167	...	...	33,620	21,916
1877-78 .	76,764	96,665	...	...	5	36	122,610	57,810
1878-79 .	136,998	1,29,138	2,201	7,688	4	19	75,363	37,865
1879-80 .	85,524	1,08,675	741	1,954	50	75	94,213	46,444
1880-81 .	135,077	1,49,897	1,128	4,390	250	425	37,944	48,539
1881-82 .	137,257	1,81,792	5,451	11,612	...	...	82,864	1,16,686
1882-83 .	142,206	1,57,288	2,882	13,885	...	...	56,420	78,803
1883-84 .	174,314	2,04,242	5,691	25,955	90	200	110,404	1,27,694
1884-85 .	208,094	2,22,333	2,958	5,871	252	384	79,335	96,213
1885-86 .	264,577	2,64,522	3,186	14,487	...	...	75,829	83,303
1886-87 .	282,824	2,90,648	3,104	10,624	...	...	118,284	1,58,852
1887-88 .	325,178	3,38,123	4,279	11,233	...	...	202,117	2,07,341
1888-89 .	460,891	4,28,175	10,961	20,864	295	835	231,116	3,24,058
1889-90 .	329,609	3,29,220	10,718	21,322	...	...	317,214	1,52,944
1890-91 .	430,000	4,49,347	25,722	30,517	...	...	227,555	1,71,334

Exports.
1631 The remarks already offered, regarding the Kashmír shawls that may be included in the registration of re-exports, are equally applicable to the statistical information that exists on the subject of the exports of Indian shawls and other woollen goods. Indeed, the value recorded for the shawls (an average of R53 each) would justify the opinion that many of them may have been fairly expensive articles, which in trade would be sold as Kashmír shawls, whether made in that State or not.

TABLE VII.

Exports of Indian Woollen Goods to Foreign Countries.

	Piece Goods.		Shawls.		Other Sorts.	
	Yds.	R	No.	R	℔	R
1871-72 . .	No returns.	*	17,391	16,77,191	No returns.	1,60,095
1872-73 . .			33,115	31,25,450		2,13,503
1873-74 . .			32,472	19,97,368		1,60,502 *
1874-75 . .			28,873	16,69,787		3,22,067
1875-76 . .	331,975		30,053	16,12,980	102,227	3,96,947
1876-77 . .	406,629	2,19,889	28,385	16,69,132	335,615	2,70,878
1877-78 . .	401,780	2,56,830	32,970	15,08,535	298,611	3,13,368
1878-79 . .	401,788	2,26,090	26,113	12,40,116	307,046	3,81,973
1879-80 . .	270,875	1,34,139	21,378	8,88,382	470,474	4,42,620
1880-81 . .	154,587	1,03,488	26,601	15,01,786	452,217	4,97,478
1881-82 . .	193,561	1,16,476	16,652	12,41,640	504,117	6,08,714
1882-83 . .	192,147	1,38,928	12,090	7,71,718	484,018	6,72,848
1883-84 . .	200,178	95,266	12,754	5,42,675	475,322	5,69,062
1884-85 . .	155,387	67,879	19,759	6,63,057	372,406	4,52,491
1885-86 . .	152,047	84,470	8,885	3,08,731	404,596	4,14,282
1886-87 . .	158,065	83,197	7,636	3,46,218	401,210	4,29,907
1887-88 . .	212,203	93,454	12,046	4,05,993	525,384	6,41,132
1888-89 . .	196,015	1,46,250	8,144	2,66,011	691,653	8,08,788
1889-90 . .	207,783	1,27,419	7,252	3,02,471	751,363	8,28,122
1890-91 . .	222,546	95,266	4,580	2,43,716	553,839	7,02,682

* During these years the values of piece goods, braids and other sorts, were given conjointly.

FOREIGN.

Woollen Goods.

Exports.

Having now shown the total Imports, Re-exports and Exports of woollen goods for a period of twenty years, it may be useful to analyse the returns for each fifth year, so as to show the countries from which or to which the transactions have been made, as also the degree of participation in the traffic taken by the provinces of India.

In the table which may now be given it will be seen that the British manufacturers practically enjoy a monopoly in the supply to India of the woollen piece goods, shawls, etc., which she requires. Indeed, the only competitors against Britain (of any moment) are Germany and Austria, but these countries together furnished last year only, say, £380,000 worth against £1,361,344 worth (nominal value) of the British goods imported by India.

Value. 1632

TABLE VIII.

Analysis of the value Returns of the Woollen Piece Goods, Shawals etc., Imported into India by Sea from Foreign Countries.

Years and Class of Goods.	Chief Countries from whence imported and value of these Imports.								
	United Kingdom.	Germany.	Austria.	France.	Italy.	Arabia	Persia.	Straits Settlements.	Grand totals, including balances not shown separately
	R	R	R	R	R	R	R	R	R
1875-76. Piece Goods.	69,77,526	74,868	...	49,411	41,331	9,332	9,663	32,279	72,50,456
1875-76. Shawls	10,14,017	...	...	...	...	...	...	...	10,14,939
1875-76. Braids	2,60,113	...	...	...	...	...	...	...	2,60,113
1875-76. Other Sorts.	1,58,322	1,872	...	...	...	...	4,568	2,529	1,72,095
TOTAL VALUE*.	84,09,978	76,740	...	49,495	42,200	9,447	14,437	35,328	86,97,603
1880-81. Piece Goods	1,04,96,165	1,35,915	88,431	84,893	2,27,747	1,536	2,094	45,927	1,10,94,611
1880-81. Shawls	9,48,643	...	...	8,477	2,984	..	1,059	...	9,62,998
1880-81. Braids	3,39,813	...	...	...	...	...	...	...	3,39,872
1880-81. Other Sorts.	5,57,128	...	2,486	..	1,545	...	13,720	3,234	5,93,818
TOTAL VALUE*.	1,23,41,749	1,36,055	91,156	93,712	2,32,276	1,646	16,873	49,338	1,29,91,299
1885-86. Piece Goods	1,03,22,216	4,95,951	2,09,232	26,583	72,383	1,530	29,926	43,928	1,12,56,264
1885-86. Shawls	11,43,217	5,218	39,889	...	2,552	...	23,252	...	12,18,054
1885-86. Braids	4,47,792	...	...	...	...	...	...	...	4,70,726
1885-86. Other Sorts.	8,75,550	11,599	4,058	...	...	...	52,773	11,117	9,73,572
TOTAL VALUE*.	1,27,88,775	5,12,768	2,53,179	28,070	74,935	2,510	1,05,951	55,416	1,39,18,616
1890-91. Piece Goods	94,88,138	16,70,652	12,47,036	2,29,909	9,562	...	11,945	2,21,895	1,29,51,614
1890-91. Shawls	21,97,446	1,820	2,79,805	4,38,946	120	...	484	3,896	29,30,543
1890-91. Braids	4,99,577	7,884	34,077	388	...	...	...	...	5,59,786
1890-91. Other So-ts.	14,28,284	1,22,930	21,689	3,122	870	...	1,13,899	28,688	17,40,183
TOTAL VALUE.	1,35,13,445	22,40,412	15,82,607	2,35,305	12,252	...	1,26,328	2,54,479	1,81,82,126

* The reason why the totals here shown are not what would be obtained by adding together the figures given, is due to the table exhibiting only the chief countries and principal articles: the totals (both vertically and horizontally) are the grand totals of all countries and in all classes of woollen goods. Of the European countries, not shown, Belgium may be stated to have within the past eight or ten years begun to be important. A fairly considerable amount of woollen goods is imported by post, and Government storesare also of some consideration. These appear in the grand total, but not in the other columns.

SHEEP:
Wool Trade.

Trade in Wool

FOREIGN.

Woollen Goods.

Value.

It is perhaps unnecessary to exhibit a detailed analysis of the shares taken by the receiving provinces in the various classes of woollen goods. It may safely be said that Bengal takes very nearly the whole of the shawls and braids, and that Bombay receives a little more than Bengal of the other classes. Thus, for example, in 1885-86 Bombay took "piece goods" to the value of R51,93,136 and Bengal R40,65,701; in 1890-91 Bombay took R64,97,992 and Bengal R39,46,510 worth. Of the other provinces Burma stands next in importance in the consumption of foreign piece goods. In the two years just named that province took R14,82,769 and R18,82,675 worth, respectively.

The following table may, however, be given of the total imports of all classes of woollen goods:—

TABLE IX.

Share of Imports taken by the Provinces.	1875-76.	1880-81.	1885-86.	1890-91.
	R	R	R	R
Bengal	47,51,205	59,87,736	58,40,274	75,10,334
Bombay	29,95,595	40,70,919	58,23,180	74,18,526
Sind	23,571	70,506	2,30,190	3,28,619
Madras	2,76,005	3,13,884	3,65,401	4,61,048
Burma	6,51,227	25,48,254	16,59,571	24,63,598
GRAND TOTAL .	86,97,603	1,29,91,299	1,39,18,616	1,81,82,126

The inferiority of Bengal wool, the backwardness of the woollen manufactures, and the growth of a demand among the well-to-do classes for winter garments of English woollen goods and shawls, are doubtless the reasons for the large share taken by Bengal in the imports of woollen goods. The climate of Bengal, as a whole, being more tropical than Bombay, it naturally would be looked to as a country that should require less woollen goods to head of population than the Western and Northern Provinces. Moreover the port town of Bombay is concerned not only in the supply of foreign woollen goods for that Presidency, but for a large part of Upper India. The demand for shawls by the immense population of Bengal is doubtless the expression of a modern popular taste manifested by the middle classes who consider a shawl as a necessity of a gentleman's apparel. Distance from the woollen manufacturers of Northern India, has doubtless operated also to favour imported goods rather than the more artistic, though more expensive shawls of the Panjáb and Kashmír.

The table of the re-export traffic, which may now be given, demonstrates the fact that a considerable section of this trade is very possibly in British, German, and Austrian goods sent to Persia, Arabia, the East Coast of Africa, etc.

FOREIGN Woollen Goods. Value.

Table X.

Analysis of the Value Returns of Woollen Goods Re-exported from India (e.g., first imported from Foreign Countries and thereafter exported to Foreign Countries).

Years and Classes of	Countries to which Re-exported shown by the Value of the Goods.									
	United Kingdom,	Turkey in Europe.	Ceylon.	Persia.	Abyssinia.	Arabia.	East Coast of Africa.	Aden.	Straits Settlements.	Grand total, including amounts to countries not separately shown.
	R	R	R	R	R	R	R	R	R	R
1875-76 Piece Goods	50,320	...	1,886	21,613	...	9,714	23,544	20,088	11,062	1,57,505
1875-76 Shawls	125	...	...	...	...	485	...	...	...	730
1875-76 Other Sorts	...	...	...	...	...	...	...	...	...	3,862
Total value	50,445	...	1,886	21,613	...	10,199	23,544	20,088	11,062	1,62,097
1880-81 Piece Goods	6,789	...	3,508	47,540	...	9,239	38,987	14,943	4,313	1,49,897
1880-81 Shawls	...	...	...	...	...	...	...	...	2,990	4,390
1880-81 Other Sorts	13,393	2,736	...	4,218	...	2,383	8,172	1,055	1,882	48,539
Total value	20,182	2,736	3,508	51,758	...	11,622	47,159	15,9 98	9,185	2,02,826
1885-86 Piece Goods	3,800	27,450	61,651	50,556	9,249	10,991	36,133	15,588	15,197	2,64,522
1885-86 Shawls	...	8,300	1,800	...	...	2,700	...	...	...	14,487
1885-86 Other Sorts	10,241	...	1,936	4,133	1,770	42,727	8,525	1,044	2,797	83,303
Total value	14,043	35,750	65,387	54,689	11,01	56,418	44,658	16,632	17,994	3,62,312
1890-91 Piece Goods	5,180	20	2,29,976	71,424	887	4,772	55,257	18,005	25,157	4,49,347
1890-91 Shawals	3,767	78	3,050	13,052	257	3,949	250	672	752	30,517
1890-91 Other Sorts	23,126	...	2,672	21,040	400	80,642	6,018	8,646	4,168	1,71,334
Total value	32,073	98	2,35,698	1,05,516	1,544	89,363	61,525	27,323	30,077	6,51,198

II.—TRANS-FRONTIER LAND TRAFFIC IN RAW WOOL AND WOOLLEN GOODS.

TRANS-FRONTIER TRADE. 1633

A review of this trade had best be dealt with under the two sections of (*a*) Raw Wool (including pashm) and (*b*) Piece Goods and Shawls.

Raw Wool and Pashm.—To exemplify the chief features of interest in the traffic in raw wool it is necessary to furnish one or two tables of the returns for the past ten or twelve years. This had best be done under the two sections (*a*) Imports and (*b*) Exports:—

SHEEP: Wool Trade.

Trade in Wool

TRANS-FRONTIER TRADE.

Raw Wool.

Imports.
1634

TABLE XI.

Statement of the Trans-frontier Imports of Raw Wool (including pashm) into India during the past twelve years.

Whence Imported.	1879-80.	1880-81.	1881-82.	1882-83.	1883-84.	1884-85.	1885-86.	1886-87.	1887-88.	1888-89.		1889-90.		1890-91.	
	Quantity.	Quantity.	Quantity.	Quantity.	Quantity.	Quantity.	Quantity.	Quantity.	Quantity.	Quantity.	Value.	Quantity.	Value.	Quantity.	Value.
	Cwt.	Cwt.	Cwt.	Cwt.	Cwt.	Cwt.	Cwt.	Cwt.	Cwt.	Cwt.	R	Cwt.	R	Cwt.	R
Lus Bela	3,515	6,594	6,803	6,086	6,844	4,778	7,257	4,504	7,283	8,935	1,31,436	8,305	1,30,026	6,251	93,123
Khelat	23,052	15,870	19,104	11,597	12,649	14,900	16,159	17,779	18,596	17,812	3,40,593	14,509	2,75,968	9,974	1,94,799
Kandahar	37,264	6,612	1,440	1,643	2,660	632	688	603	132	180	4,410	...	...	242	4,290
Sewestan	4,668	1,302	2,159	2,659	1,771	1,566	1,408	1,336	2,035	2,736	67,320	3,673	79,290	3,736	79,806
Girishk	...	...	...	2,228	1,451	165	505	733	414	110	2,250	508	9,938	...	...
Tirah	61	73	60	108	47	29	41	62	95	217	3,209	116	1,642	26	385
Kabul	11,036	8,065	5,323	4,284	4,169	3,093	3,711	6,453	5,952	4,112	71,176	3,263	56,361	4,707	78,092
Bajaur	525	730	532	683	735	588	506	1,178	666	614	9,159	406	6,013	249	3,740
Kashmir	587	415	137	115	246	367	373	292	244	203	4,880	826	17,602	1,400	50,795
Ladakh	1,678	1,014	2,573	2,281	2,000	3,067	4,383	1,742	2,071	3,183	93,962	2,879	91,870	2,173	70,384
Thibet	4,464	3,732	3,457	3,269	4,589	4,245	4,193	4,155	4,627	5,643	1,69,640	6,716	2,84,446	11,859	3,51,163
Bhutau	...	...	...	...	...	...	...	...	1,066	1,757	45,765	3,311	65,733	2,478	67,974
Nepal	50	26	74	529	22	151	446	22	56	320	8,411	199	3,764	872	17,227
Sikkim	..	4	...	123	669	67	1,877	1,129	70	32	833	6	141	1	15
Mekran	...	...	...	...	...	...	...	18	...	...	...	...	...	...	...
Herat	...	...	735	...	...	...	...	...	...	...	...	...	...	...	...
Pishin	22	...	...	...	...	...	...	...	...	...	...	...	...	...	...
Trans-frontier by Siud-Pishin Railway	...	10,176	14,525	26,727	34,526	30,759	40,141	50,165	65,697	98,486	13,40,500	1,01,784	13,85,410	76,152	10,36.520
TOTAL	86,922	54,613	56,920	62,332	72,378	64,417	81,688	90,172	1,09,004	1,43,910	22,56,244	1,47,132	24,22,697	1,21,540	20,57,431

TRANS-FRONTIER. Raw Wool. Imports.

It will thus be seen that the average weight of the imports during the first three years of the above series came to 66,152 cwt., and that the average during the last three amounted to 137,527 cwt., so that the traffic within the period was exactly doubled. But there are certain very instructive features of this prosperity. The Sind-Pishín (Kandahar) Railway began to carry wool in 1880-81. Prior to that date the Baluchistán and Afgánistán wool had to be carried to the Indian marts on camels and other beasts of burden. The imports from these countries in 1879-80 came to only 60,300 cwt., whereas the average amount during the past three years, carried by the Kandahar Railway alone, came to 92,140 cwt. It would thus appear that the opening up of that railway has had the immediate effect of immensely expanding the import from Kandahar. A similar, though much smaller, effect may be traced in the extension of railway communication to Darjeeling, the imports from Sikkim, Bhutan, and Eastern Tibet having been recently greatly augmented. During the first nine years of the series, exhibited by the table, the imports from Tibet may be said to have averaged 4,000 cwt., but in 1888-89, they became 5,643 cwt., in 1889-90, 6,716 cwt., and in 1890-91, 11,859 cwt. These and other illustrations of the recent expansion of the trans-frontier imports of wool may be taken as a foretaste of a possible still greater expansion, should the means of facilities of transport continue to be improved and extended. It may be added in this connection that the average value of the wool imported by the routes here dealt with, came to 2 annas 9 pie a pound, the most expensive being that from Tibet (7 annas 9 pie) and the cheapest from Bajaur (1 anna 7 pie). The Tibetan wools doubtless included pashm, hence the higher rate, but the superior wools of Kandahar, Kashmír, Nepal, Bhutan, and Sikkim may be said to have averaged during the past twelve years 3 annas 4 pie a pound.

The reader may have observed, in connection with the statistical information furnished above regarding the foreign trade in wool, that the Indian IMPORTS of *raw wool* from foreign countries (by sea) have steadily increased from 1,511,411℔ in 1871-72 to 5,100 556℔ in 1889-90 and 4,236,826℔ in 1890-91. Also that the RE-EXPORTS of foreign *raw wool* from India have expanded from 128,342℔ in 1871-72 to 14,402,296℔ in 1889-90 and 12,788,216℔ in 1890-91. An inquiry into the probable causes of these very significant features of India's enhanced trade in wool manifests the fact that the improvement in the import traffic by sea is due mainly to the expansion of the Bombay port town and the Karáchi receipts from Persia. The increased importance, on the other hand, of the re-exports of foreign wool from India is almost entirely due to the trans-frontier imports by Sind which find their way almost exclusively from Karáchi to Great Britain. To the British woollen manufacturer, therefore, (whatever political opinions may be held regarding Kandáhar), the Sind-Pishín Railway has become an essential condition of a supply of wool which has immensely expanded and is now yearly increasing in importance. The total receipts in raw wool from across the frontier were in 1890-91 121,540 cwt. (or, say, 13,600,000℔) and in 1889-90, 147,132 cwt. (or say, 16,500,000℔). The table given above has manifested the countries from which that very large supply was drawn, but it may be useful to display the shares taken by the provinces of India in these imports:—

SHEEP:
Wool Trade.

Trade in Wool

TRANS-FRONTIER.

Raw Wool.

Imports.

TABLE XII.

Provinces into which and countries from whence Trans-frontier Imports were made during the past four years.

Provinces into which imported.	Countries whence Exported.	1887-88.		1888-89.		1889-90.		1890-91.	
		Cwt.	R	Cwt.	R	Cwt.	R	Cwt.	R
Sind	Lus Bela .	7,283	1,07,830	8,395	1,31,436	8,305	1,30,026	6,251	93,123
	Khelat .	18,596	3,70,959	17,812	3,40,593	14,509	2,75,968	9,974	1,94,799
	Kandahar .	132	2,160	180	4,410	...	...	242	4,290
	Khorasan .	...	...	110	2,700	631	14,420	419	9,120
	Girishk .	414	81,750	110	2,250	508	9,938	...	...
	Mekran .	...	...	...	...	12	192	...	...
	Sind-Pishin Railway .	65,697	8,94,180	98,486	13,40,500	101,784	13,85,410	76,152	10,36,520
	Total .	92,122	13,83,879	125,093	18,21,889	125,749	18,15,954	93,038	13,37,852
Panjáb	Sewestan .	2,035	37,131	2,736	67,320	3,673	79,290	3,736	79,806
	Tirah .	95	1,011	217	3,209	116	1,642	26	383
	Kabul .	5,952	98,055	4,112	71,176	3,251	50,169	4,707	78,092
	Bajaur .	666	7,565	614	9,159	436	6,083	249	3,740
	Kashmír .	244	5,697	203	4,880	826	17,602	1,400	50,795
	Ladakh .	2,071	59,089	3,183	93,962	2,879	97,870	2,174	70,384
	Tibet .	322	22,410	422	27,668	2,361	1,76,387	2,087	1,08,954
	Total .	11,385	2,31,028	11,487	2,77,374	13,512	4,29,040	14,379	3,92,154
N.-W. P.	Tibet .	2,272	60,603	5,207	1,41,614	2,540	74,087	6,465	1,77,380
	Nepal .	5	91	4	152	6	160	11	280
	Total .	2,277	60,694	5,211	1,41,766	2,546	74,247	6,476	1,77,660
Bengal	Tibet .	2,033	51,154	14	358	1,815	33,978	3,307	64,829
	Nepal .	51	1,295	316	8,259	193	3,604	861	16,947
	Sikkim .	70	1,758	32	833	6	141	1	15
	Bhutan .	1,056	26,852	1,757	45,765	3,311	65,733	3,478	67,974
	Total .	3,220	81,059	2,119	55,215	5,325	1,03,456	7,647	1,49,756
GRAND TOTAL INTO BRITISH INDIA IN	Cwt.	109,004	17,56,781	143,910	22,96,244	147,132	24,22,697	121,540	20,57,431
	lb	12,208,448	...	16,117,920	...	16,478,784	...	13,612,480	...

It may thus be noted that by far the most important section of the trans-frontier traffic is that which finds its way into Sind, the major portion of which is now conveyed by the Sind-Pishin (Kandahar) Railway. Of the wool carried by that railway about three-fourths passes straight through the province to Karáchí and is exported direct to Europe. The balance may be said to be conveyed coastwise to Bombay and there shipped to Europe. In fact, with the exception of the Panjáb imports, it may safely be affirmed that the whole of the trans-frontier wool leaves India for Great Britain.

S. 1634

TRANS-FRONTIER TRADE. Exports.

EXPORTS.—The exports from India in raw wool across the land frontier are not very important. During the past three years they stood at 271 cwt. in 1888-89; 197 cwt. in 1889-90; and 177 cwt. in 1890-91. The exports were made almost entirely to Kashmír. 1635

Piece Goods & Shawls.

Piece Goods and Shawls.—Turning now to the trans-frontier traffic in manufactured woollen goods and shawls, considerable difficulty is experienced in dealing with quantity, since some of the entries are in pieces, others in cwt. Under these circumstances it is felt to be the safer course to exhibit the extent of the trade by the value of the goods. During the past three years the "piece goods" were valued at R7,37,675 for 1888-89; R8,77,389 for 1889-90; and R7,85,369 for 1890-91. It may be said that fully three-fourths of this traffic was in Kashmír goods. The value of the shawls stood at R6,17,660 for 1888-89; R7,14,600 for 1889-90; and R1,96,500 for 1890-91. Almost the entire supply was drawn from the capital (Kashmír) itself. The reader will, however, find the particulars dealt with in the chapter on **Pashm & Pashmina** to greatly amplify this section of the woollen trans-frontier traffic, and it need only be pointed out that since the re-exports from India are very small, the trade denoted by the imports from Kashmír must be viewed as largely denoting the Indian consumption of these goods. 1636

Woollen Piece Goods.

Exports of Woollen Piece Goods.—These were valued at R3,82,482 in 1888-89; R4,80,889 in 1889-90; and R6,19,241 in 1890-91. The three chief items in these exports are usually the traffic to Nepal, Zimme, and along the Sind-Pishin Railway. Last year, for example, the woollen piece goods consigned to Nepal were valued at R1,35,959; to Zimme R1,03,340, and by Sind-Pishin Railway R2,41,680. The other countries of importance are Tibet, Siam, and the Northern Shan States. The traffic in *shawls* is not very important. The following were the valuations of the exports: R27,685 in 1888-89; R22,700 in 1889-90; and R518 in 1890-91, and these exports were entirely to Nepal. It would thus appear that English and French imitations of Kashmir goods do not now at least penetrate to that State, although some writers affirm that it is difficult to feel satisfied that a shawl or dress cloth, purchased in Kashmir, has actually been manufactured there. The danger, however, is rather that an article sold as of pure *shál*-wool (the fleece of the Tibetan goat) is actually constructed of Persian or Afgánistán wool, cleverly designed and manipulated to be scarcely distinguishable from the genuine article. This is certainly true of the manufactures of Amritsar, but there is nothing to prove to the contrary that Kashmir does not obtain the so-called *Kirmáni* wool by its land routes; and the returns furnished above show that Kashmir even imports wool from India, probably Australian merino. 1637

III.—COASTWISE TRAFFIC IN WOOL AND WOOLLEN GOODS.

COASTWISE. 1638

The chief features of interest brought out by the study of the returns of the woollen trade carried along the coast of India may be said to be:—(*a*) the exemplification of the insignificance of the traffic in India as a whole; (*b*) the elaboration of the opinion formed from the study of the foreign trade, *viz.*, that the chief seats of the enterprise in raw wool are Karáchi and the port town of Bombay; and (*c*) the recent modifications in the traffic that appear to be directly the result of the commercial prosperity of Karáchí and of the province of Sind.

Raw Wool.

RAW WOOL.—Following the course adopted with the foreign trade, two tables may be here furnished of the total imports and exports of Indian and foreign wool for the past ten years:— 1639

SHEEP: Wool Trade.

Trade in Wool

COASTING. Raw Wool, Imports. 1640

TABLE XIII.

Statement of the Imports, Coastwise, of Indian and Foreign Raw Wool during the past ten years.

	Into Bengal.		Into Bombay.		Into Sind.		Into Madras.		Into Burma.		TOTAL IMPORTS.	
	lb	R	lb	R	lb	R	lb	R	lb	R	lb	R
Indian—												
1881-82	1,096	1,785	14,386,230	45,05,008	48,240	5,543	18	2	...	...	14,435,584	45,12,363
1882-83	21,746	7,516	12,770,763	38,90,212	48,248	6,362	...	...	...	...	12,840,757	39,04,090
1883-84	2,363	435	12,246,911	36,11,169	67,694	9,945	...	...	...	...	12,316,968	36,21,549
1884-85	1,747	460	10,808,789	32,17,047	56,772	9,320	...	...	...	...	10,867,308	32,26,827
1885-86	1,193	1,260	10,792,047	31,03,322	46,868	4,501	...	...	...	...	10,840,108	31,09,083
1886-87	1,225	130	6,931,513	20,02,295	27,786	2,524	...	...	9,694	2,560	6,970,218	20,97,509
1887-88	...	...	5,889,235	17,64,812	31,630	3,304	...	...	...	...	5,920,865	17,68,116
1888-89	2,478	490	6,622,452	20,74,718	80,198	10,265	...	...	...	...	6,705,128	20,85,473
1889-90	...	...	8,963,309	25,10,596	71,082	9,453	...	...	12,656	4,50	9,047,047	25,24,549
1890-91	8,762	1,414	5,365,707	16,76,569	39,380	4,467	...	...	...	...	5,413,849	16,82,450
Foreign—												
1881-82, 1882-83, 1883-84, 1884-85	Separ	ate	returns	do not a	ppear	to hav	e be	en p	ublish	ed for	these	years.
1885-86	32	55	1,593,605	3,87,731	13,440	4,800	30	66	...	...	1,607,107	3,92,652
1886-87	140	255	1,852,747	5,93,249	29,232	12,800	135	280	...	...	1,882,254	6,06,584
1887-88	60	30	2,281,552	7,56,579	38,752	18,200	223	474	...	...	2,320,587	7,75,283
1888-89	8	11	2,262,704	7,19,907	17,920	5,900	198	403	...	...	2,280,830	7,26,221
1889-90	20	50	3,292,790	10,66,752	...	...	50	100	...	...	3,292,860	10,66,902
1890-91	2	6	1,479,998	4,90,790	6,720	3,000	210	521	...	...	1,486,930	4,94,317

Exports. 1641

TABLE XIV.

Statement of the Exports, Coastwise, of Indian and Foreign Raw Wool during the past ten years.

	From Bengal.		From Bombay.		From Sind.		From Madras.		From Burma.		TOTAL EXPORTS.	
	lb	R	lb	R	lb	R	lb	R	lb	R	lb	R
Indian-												
1881-82	80	240	347,087	1,02,989	11,846,632	38,19,527	95,620	16,643	1,247	188	12,290,666	39,39,287
1882-83	336	30	343,521	1,04,130	9,552,107	30,71,690	141,605	25,328	...	...	10,037,569	32,01,178
1883-84	16,800	4,040	105,654	28,890	8,068,285	26,07,092	50,477	8,677	3,116	414	8,244,332	26,49,113
1884-85	35,814	9,872	92,634	20,833	6,988,201	22,64,827	14,897	2,575	2,365	316	7,133,911	22,98,423
1885-86	43,518	11,580	57,014	12,061	6,469,431	17,37,670	560	20	. .	...	6,570,523	17,61,340
1886-87	47,402	7,640	50,808	12,633	3,286,239	9,22,738	7,672	2,105	896	40	3,393,017	9,45,156
1887-88	51,126	12,555	46,052	11,663	2,133,074	6,04,934	39,190	6,020	...	...0	2,269,442	6,35,172
1888-89	68,928	18,339	27,040	4,953	2,969,853	9,08,072	60,037	5,715	984	8	3,126,842	9,37,159
1889-90	31,716	10,098	37,784	8,075	3,975,294	11,98,130	335,919	44,200	...	...	4,380,713	12,60,503
1890-91	29,744	12,924	9,536	1,746	1,896,064	5,38,141	190,129	22,441	3,812	340	2,129,285	5,75,592
Foreign-												
1881-82, 1882-83, 1883-84, 1884-85	Sep	arate	returns	do not	seem to	have bee	n prese	rved f	or th	ese	years.	
1885-86	...	...	6,720	1,800	2,180,528	5,29,055	...	...	...	...	2,187,248	5,30,855
1886-87	...	...	23,303	8,914	2,074,670	6,41,990	...	...	...	...	2,097,973	6,50,904
1887-88	...	...	25,088	11,540	2,108,400	6,85,875	...	...	...	...	2,133,488	6,97,415
1888-89	4,600	1,500	16,464	5,880	2,885,576	9,86,526	...	...	...	...	2,906,640	9,93,906
1889-90	...	...	...	...	3,191,708	10,41,093	...	...	...	...	3,191,708	10,41,093
1890 91	...	...	6,720	2,400	1,538,044	5,30,290	...	...	...	...	1,544,764	5,32,690

COASTWISE.

Raw Wool. Imports. 1642

It will be seen from the table of IMPORTS COASTWISE, that Bombay last year received 5,365,707lb of Indian wool and 1,479,998lb of foreign wool. The following may be given as the analysis of these imports:—

Indian Wool Imported into Bombay Coastwise in 1890-91.

	lb
From Sind	1,864,382
,, Madras	148,904
,, Ports within the Presidency	7,296
,, Goa	21,952
,, Cutch	841,176
,, Kattywar	2,477,181
,, Gaekwar's Territory	4,816
TOTAL	5,365,707

Foreign Wool Imported into Bombay Coastwise in 1890-91.

	lb
From Sind	1,479,998

It will thus be observed that of the Bombay coastwise imports in Indian wool, the Native State of Kattywar is by far the most important source of supply, but that Sind makes a by no means bad second. The foreign wool imported by Bombay coastwise is drawn entirely from Sind, so that the imports by Bombay from Sind usually constitute fully half its total coastwise receipts. But a very important feature has been exemplified by the table of coastwise imports to which it may be desirable to draw special attention. During the ten years dealt with, the Bombay imports have declined from 14,386,230lb to 5,365,707lb, and at the same time the exports from Sind which went almost exclusively to Bombay, have declined from 11,846,632lb to 1,896,064lb These very remarkable features of the modern trade in Indian wool find their solution in the birth and growth of a direct traffic from Karáchí to Europe. The exports from Karachi of both classes of wool came to 7,556,332lb in 1875-76; last year they were 13,863,932lb, of which 8,857,856lb were foreign wool brought mainly to Karáchí by the Sind-Pishin (Kandahar) Railway.

It is perhaps unnecessary to say anything further on the subject of the coastwise transactions. Bengal, Madras, and Burma take no part, it may almost be said, in the traffic, a fact indicative of the comparative unimportance of the wool of these provinces. Madras supplies annually from 100,000 to 300,000lb to Bombay—the wool apparently of its mountain breeds of sheep. Bombay is the chief importing province and Sind the exporting, from which circumstance it may be concluded that so far as can be learned by the study of the coastwise trade, Kattywar, Cutch, Rajputana, Sind, and the countries across the Sind and Baluchistan land frontiers, are the chief producing areas for the wool exported from Bombay to Europe.

Piece Goods & Shawls. 1643

PIECE GOODS AND SHAWLS.—Turning now to the subject of the traffic in piece goods and shawls carried from port to port along the coast of India. The total value of the imports of Indian woollen goods coastwise was last year R1,85,171; Madras took R1,60,885 worth, as follows: from Bengal R41,893 and from ports within the presidency R1,13,769 worth. Thus it may be said that the coastwise transactions in woollen goods are mainly in Madras manufactures, carried from port to port within that presidency. Of the imports coastwise of foreign woollen goods it may be said that last year these were valued at R15,91,647, of which R8,12,366 went from Bombay and R4,30,745 from Bengal. The receiving provinces were, Sind R4,05,295 worth, Madras R3,56,987 worth, and Burma R4,58,469 worth.

COASTWISE. Piece-Goods & Shawls. Imports.

The Sind and Madras supply was drawn mainly from Bombay, and the Burma from Bengal. The transactions in shawls are of no great moment. The total coastwise imports were valued last year at R47,807, of which R34,505 went from Bengal to Burma. Of braids and other sorts of woollen goods the coastwise imports came to only R1,08,960, of which R43,684 went from Bombay to Sind and R32,502 from Bengal to Burma.

These facts regarding the coastwise traffic in foreign woollen goods for the year 1890-91 may be taken as denoting the markets that are usually met by local maritime interchanges. The figures alter from year to year but not to such an extent as to necessitate elaborate tables. One point only need be here added, namely, that so far as can be inferred by the coastwise sea transactions, the Indian woollen mills do not appear to have begun to affect the markets supplied by the merchants who traffic along the coast. The woollen goods used up along the sea-board of India may, therefore, be said to be derived mainly from foreign countries.

INTERNAL. 1644

IV.—INTERNAL TRADE BY ROAD, RAIL, AND RIVER.

The study of nearly every branch of Indian commerce is beset with the difficulty of obtaining anything like satisfactory returns of the transactions carried by road, rail, and river. The Government has recently arranged that the traffic on the railways should be classified according to certain blocks. That is to say, each province is referred to a number of blocks and the traffic tabulated according to movement to, and from these blocks. Then again, other tables give the transactions to external blocks, that is to say, from a certain block in Bengal, for example, to a block in the Panjáb. Unfortunately, however, there is no imperial review of these provincial railway reports, so that, to give a statement of the total trade, as denoted by railway transations for any one year, necessitates an elaborate balance sheet being prepared, since the exports from certain provinces are the imports by others. But even when this labour has been gone through, the railway transactions by no means convey a conception of the total internal trade in any product. Large quantities of commerical commodities are carried by the steamers and Native boats that trade on the rivers and canals of India. A balance sheet has to be made out for these, but the total of the rail and river traffic would by no means be the grand total of the internal trade. Though rail and river facilities of transport are yearly draining an increasing share of the trade, the Native carts continue to largely participate in the transport of goods and their diffusion from district to district. Indeed, with wool, as with most natural products, extensive village manufactures are met by supplies of a purely local character that would entirely escape all registration. The road trade of India is a matter beyond the power of present administrative capabilities, though the imports and exports to and from certain large towns by road are recorded through the necessities of municipal and fiscal taxation.

A review of the chief features of the rail and river-borne trade of India was, however, published for the year 1888-89, and as the remarks already offered on the Indian wool trade exhibit the returns of foreign and constwise transactions for that year, the reader may find the following facts of interest as exemplifying more fully some of the opinions advanced regarding the value relatively of the provinces of India in the wool supply. The total imports of Indian and Foreign wools by all provinces and seaports came (in 1888-89) to 3,55,310 maunds (or, say, 28,135,423lb). But of that amount the following were the imports by Bombay port town and Karáchí together, *viz.*, 2,94,363 maunds (or 24,137,766lb), so that there was left therefore for consumption in the whole of the rest of India

INTERNAL.

(so far as the rail and river returns denote) about 5,000,000lb. These imports by Bombay and Karáchí must have been almost entirely exported, since the foreign transactions from India require about the amount here shown to meet the registered despatches. The share taken by Bombay in the rail-borne imports came to 9,958,737lb and by Karáchí 14,179,030lb. By reference to the tables already furnished, the reader will see (table No. XII) that the Sind-Pishin Railway carried to Karáchí, in the year 1888-89, 98,486 cwt. of raw foreign wool (or, say, 11,030,432lb), this would therefore have left a balance on the total imports of Karáchí of 3,148,598lb which may have been mainly Sind and Panjáb, Indian wools. It will be further seen from the study of the tables of foreign and coastwise trade that the view here arrived at of Karáchí transactions, approximates sufficiently closely to the other statistical returns, to justify the inference that the rail-borne statistics must have been fairly correct. The same may be said of the Bombay transactions, the recorded imports by the port town along the railways must have at least been all required to meet (with the coastwise and foreign imports) the net export. The imports by Calcutta by rail came to only 1,403 maunds (115,046lb), and by the Madras sea ports to 1,876 maunds (153,832lb), so that the share of the rail-borne imports taken by Karáchí, Bombay port town, Calcutta, and Madras came to 2,97,642 maunds, or, say, 24,406,644lb which thus left a balance for the provinces and Native States of India, of the wool recorded as carried by the railways, amounting to only 4,728,776lb. Unless, therefore, there be an extensive local production not registered, the consumption of wool in the plains of India may be said to be almost quite nominal, and certainly nothing like ⅛th of a pound to head of population, (the figure given by Mr. Schofield, *see p. 654*).

But of the imports received by the port town of Bombay by rail, 24,523 maunds were obtained from Bombay Presidency, 32,131 maunds from the Panjáb, and 64,500 maunds from Rájputana. Of the Karáchí imports 76,230 maunds are registered as obtained from Sind and 96,685 maunds from the Panjáb. Of these items, however, it may be said that the so-called Sind imports are in reality the traffic by the Sind-Pishin Railway, and of the Panjáb exports both to Bombay and Karáchí a very large slice of these is Rájputana wools. But of the receiving provinces Sind heads the list, having derived 33,708 maunds from the Panjáb and 657 back from Karáchí. The North-West Provinces and Oudh stand next in importance, having obtained in all 11,960 maunds, of which 10,046 maunds came from the Panjáb. None of the other provinces seem deserving of special notice, so far as the imports are concerned. It may, however, be remarked that of exporting provinces the Panjáb is by far the most important, having contributed 1,74,119 maunds to Karáchí, Bombay, the North-West Provinces, and Sind. Rájputana stands next in importance, having contributed 66,008 maunds to Bombay and the North-West Provinces. The opinions arrived at from the study of the foreign (sea and land) trade and coastwise traffic is thus abundantly confirmed by the railway returns, *viz.*, that the wool-producing provinces are the Panjáb and Rájputana, but that the wool brought by land routes is in the present state of the trade of greater interest than the indigenous produce. Further, that the production and cosumption of wool in the plains of India may be regarded as purely nominal.

V.—WOOLLEN MILLS OF INDIA.

WOOLLEN MILLS. 1645

Sufficient has, perhaps, been said already of the local manufactures of India. They are of very secondary consideration alongside of the value of the imports of foreign goods and the exports of raw wool. As remarked, the Panjáb, more especially Amritsar, is the seat of the Native woollen

Valuable Timber Trees.

WOOLLEN MILLS. manufactures of India. Carpet-weaving is practised here and there throughout the country, but it must be regaded as an industry to meet the requirements of the wealthy and mostly a foreign demand. Blanket-weaving is pursued throughout India, but it seems probable that the traffic is purely local, and that the wool so used up is locally produced.

Of the power-looms it may, however, be said that the requirements of the people of Upper India, more especially, and of the Native army, is sufficient to have encouraged the establishment of four or five large woollen-mills to compete with the foreign imported goods. Two of these mills are at Cawnpore in the North-West Provinces, another at Dhariwal in the Panjáb, and another at Bangalore in South India. In Bombay it is said one woollen mill has recently been constructed, and that several of the cotton mills have woollen departments. Interest has thus been created in the subject of woollen mills, and it is possible this branch of enterprise may in the future be considerably expanded.

(*W. R. Clark.*)

1646 **SHOREA**, *Roxb.; Gen. Pl., I., 193.*

A genus of valuable timber trees, all the species of which abound in various kinds of copalline resins. The members of the genus are indigenous to Tropical Asia and the Indian Archipelago. As all the species afford good timber, it is only where there is some special peculiarity about any particular one that it has been deemed necessary to give it a separate place in this work.

1647 **Shorea assamica**, *Dyer; Fl. Br. Ind., I., 307;* DIPTEROCARPEÆ.

Vern.—*Makai*, ASSAMESE.

References.—*Gamble, Man. Timb., 34, vi.; Ind. Forester, XI., 201.*

Habitat.—A gregarious tree, which attains a height of 90 to 100 feet. It was discovered originally by Mr. Mann as forming a large forest on the banks of the Dehing river in Upper Assam.

TIMBER. 1648 **Structure of the Wood.**—The timber is almost white when newly cut, but soon turns to a dark yellow or brown if exposed to the air. Its grain is very straight but not very close; it warps and splits when dried quickly, not otherwise. It is durable when kept in a well-ventilated place and free from damp, but is very liable to the attacks of white ants.

DOMESTIC. 1649 **Domestic Uses.**—A cheap timber for general purposes, as common deal is in England. It is easily worked and is not wasteful. It is therefore interesting to have to add that there is such a quantity as will supply the market for many years to come.

1650 **S. gratissima**, *Dyer; Fl. Br. Ind., I., 307.*

Syn.—HOPEA GRATISSIMA, *Wall.*

References.—*Kurz, For. Fl. Br. Burm., I., 121; Watt, Cal. Exhib. Cat., II., 228.*

Habitat.—A glabrous tree, found in the forests cf Tenasserim and Singapore.

TIMBER. 1651 **Structure of the Wood.**—Generally cross-grained. Heartwood brown, very hard.

1652 **S. obtusa**, *Wall; Fl. Br. Ind., I., 306.*

Syn.—S. LEUCOBOTRYA, *Miq.*; VATICA OBTUSA *Steud.*

Vern.—*Thit-ya* (=itchwood, a name derived from the itching which is caused when its chips or bark are brought into contact with the skin).

References.—*Kurz, For. Fl. Burm., I., 118; Gamble, Man. Timb., 39; Aplin, Report on Shan States (1887-88); Ind. Forester, I., 363; IV., 292; VIII., 416; X., 134; Gaz., Burma, I., 128, 132.*

Habitat.—A large tree, common in the *In* forests all over Burma, from Ava, Prome, and Martaban to Tenasserim and Siam; it ascends to an altitude of 2,000 feet.

Gum.—It exudes a white resin. GUM. 1653

Structure of the Wood.—Heartwood the colour of *sál* (**S. robusta**), very hard and durable; more even grained than that of either *sal* or *ingyin* (**S. siamensis**) (*Gamble*). Weight from 52 to 67lb per cubic foot, averaging 60lb. TIMBER. 1654

Domestic Uses.—The WOOD is much valued on account of its durability; it is used for canoes, in house-building, for tool handles, planes, etc. DOMEST Wood. 1655

Shorea robusta, *Gærtn.; Fl. Br. Ind., I., 306; Beddome, Fl. Sylv., t. 4.* 1656

THE SÁL TREE.

Syn. – VATICA ROBUSTA, *Steud.*

Vern.—*Sál, sála, salwa, sákhu, sakhua, sakher, sakoh,* (resin=) *rál, dhúná, dámar,* HIND.; *Sál, shal,* (resin=) *rál, dhúná,* BENG.; *Sarjum, sekura,* KOL; *Sarjom,* SANTAL; *Sargi,* BHUMIJ; *Sakwa,* KHARWA; *Sekwa,* ORAON; *Bolsal,* GARO; *Sakwa,* NEPAL; *Teturl,* LEPCHA; *Salwa, soringhi,* URIYA; *Sál, sarei, rinjal,* C. P.; *Sál, kándár, sákhu, koron,* N.-W. P.; *Koroh,* OUDH; *Sal, seral,* (resin=) *ral sard, r. sufed, r. kala, dhúna,* PB.; (resin=) *Ral,* DECCAN; *Sal,* (resin=) *ral, dhúná,* BOMB.; (resin=) *Rala, guggilu,* MAR.; (resin=) *Ral,* GUZ.; (resin=) *Kungiliyam,* TAM.; *Gúgal,* (resin=) *Guggilamu,* TEL.; *Kabbu,* (resin=) *guggala,* KAN.; *En-khyen,* BURM.; (resin=) *Dammala,* SING.; *Sala, asvakarna,* (resin=) *rála, guggilam, koushi-kaha,* SANS.; *Kaikahr,* ARAB.; *Lále-moáb bári,* PERS.

NOTE.—It is doubtful how far some of the above names denote the tree or the resin. **Moodeen Sheriff** says that in Persian *Sál* denotes **Tectona grandis.**

References.—*DC., Prod., XVI., ii., 628; Roxb., Fl. Ind., Ed. C.B.C., 440; Brandis, For. Fl., 26; Kurz, For. Fl. Burm., I., 119; Gamble, Man. Timb., 34; Stewart, Pb. Pl., 28; Rev. A. Campbell, Rept. Econ. Pl., Chutia Nagpur, No. 8402; Mason, Burma and Its People, 528, 737, 757; Pharm. Ind., I., 195; O'Shaughnessy, Beng. Dispens., 221; Irvine, Mat. Med. Patna, 91; Rankine, Med. Topog., Sarun, 71; Moodeen Sheriff, Supp. Pharm. Ind., 228; also Mat. Med. S. Ind. (in MSS.), 47; U. C. Dutt, Mat. Med. Hind., 120, 292, 316; S. Arjun, Cat. Bomb. Drugs, 20; Dymock, Mat. Med. W. Ind., 2nd Ed., 92; Dymock, Warden & Hooper, Pharmacog. Ind., I., 195; Birdwood, Bomb. Prod., 258; Baden Powell, Pb. Pr., 328, 411; Atkinson, Him. Dist. (X., N.-W. P. Gaz.), 306, 750, 779; Econ. Prod., N.-W. Prov., Pt. I. (Gums and Resins), 5; Pt. III. (Dyes and Tans), 82; Liotard, Dyes, 33; Wardle, Dye Report, 48; McCann, Dyes & Tans, Beng., 137; Church, Food-Grains, Ind., 174; Christy, New Com. Pl., V., 44; Man. Madras Adm., I., 313; Nicholson, Man. Coimbatore, 401; Rept. For. Adm., Chutia Nagpur (1885), 4, 6, 7, 9, 28; W. W. Hunter, Orissa, II., 5, app. I., 75, app. III., 120, app. IV., 179, app. VI.; Settlement Reports:—Panjáb, Kangra, 22; N.-W. P., Shahjehanpur, ix.; C. P., Mundlah Dist., 88; Bilaspore, 76; Chindwara Dist., 110; Gazetteers:—Panjáb, Hoshiarpur, 11; N.-W. P., III., 248; IV., lxviii.; C. P. (1870), 108, 123; Mysore and Coorg, II., 7; III., 28; Agri.-Horti. Soc., Ind.:—Trans., VIII., 109, (Pro.), 381; Journ., VI., 40; IX., (Sel.) 51; VIII., (Sel.) 179; XIII., 316, 322; (New Series), VII., 126-128; Ind. Forester:—I., 21, 74 76, 77, 78, 80, 81, 98, 196, 397, 398, 411; II., 93, 203, 292; III., 44, 200, 359; IV., 46, 99, 100, 292, 324, 387; V., 93, 212; VI., 125, 317, 345; VII., 42, 222; VIII., 114, 270, 301, 415; IX., 13, 177, 195, 218, 255, 349, 401, 413, 459, 475, 607; X., 60, 359, 403, 543, 545; XI., 252, 315, 436; XII., 188, 261, 397, 434; XIII., 296, 565; XIV., 299, 386; Indian Agriculturist, Aug. 14, 1886; Oct. 19, 1889; Spons, Encycl., 1644; Smith, Dict. Econ. Pl., 363.*

Habitat—A large, gregarious tree, often covering certain interrupted tracts—without the existence of connecting patches. It occurs along the base of the Tropical Himálaya from the Sutlej to Assam, in the Eastern districts of Central India, and on the Western Bengal hills. In Chutia Nagpur it is very abundant.

The Sal Tree

RESIN. 1657 **Resin.**—When tapped, the tree exudes large quantities of an aromatic RESIN—whitish at first, but becoming brown when dry. The method of tapping usually employed by the Natives is to cut out from three to five narrow strips of the bark, according to the size of the tree, about 3 or 4 feet from the ground. This is generally done in the month of July. In about twelve days these grooves fill up with resin which is gathered, and the grooves left to fill again. They give three yields, which, in the best trees, may amount to as much as 10℔. The first is the best in quality. A second yield in October and a third in January are also obtained from the same wounds, but small in quantity and inferior in quality. In some parts of the country, the Natives used to ring the trees and collect the resin daily as it exuded, and in this way large extents of forest, chiefly in Central India, were ruined before the protective operations of the Forest Department were brought into force. The resin usually occurs in small rough pieces, nearly opaque and very brittle; but "in some parts of the Upper Tista forests, large blocks, often 30 to 40 cubic inches in size, are found in the ground at the foot of the trees" (*Gamble*). The exudation has no taste or smell, a specific gravity of 1.097—1.123, is easily fusible, to a small extent soluble in alcohol, almost entirely so in ether, and perfectly in oil of turpentine and the fixed oils. Sulphuric acid dissolves it, imparting a red colour to the solution (*Mat. Med. W. Ind.*). The supply of *sál* resin in large quantities is rendered impossible by present forest conservancy; indeed, the writer is informed by **Dr. Dymock**, that the supply of this article now comes to Bombay almost entirely from Singapore and not from the extensive *sál* forests of India (*Watt*).

DYE & TAN. Ashes. 1658 Wood. 1659 Bark. 1660 **Dye and Tan.**—According to the **Rev. A Campbell** the ASHES of the WOOD are used in dyeing by the Santals. "**Dr. McCann**, in his *Report of Dyes and Tans*, compiled from the records of the Bengal Economic Museum, states that in Chutia Nagpur the BARK is used for the preparation of a red and a black dye. The bark has long been used locally as a tan, and it is to be feared that in dyeing it is more used as an auxiliary than as a dye-yielding stuff. As a tan it is much valued, being generally used along with **Terminalia, Mimusops,** and **Phyllanthus,** or with, in addition, the bark of **Ficus religiosa,** the *babul* (**Acacia arabica**), and the mango" (*Watt, Cal. Exhib. Cat.*).

In 1886, **Captain E. S. Wood**, Conservator of Forests, Oudh Circle, made some experiments with a view to ascertaining the value of *sál* extract as a tanning agent, and as this extract on analysis proved particularly rich in tannin, it may be well to give a fairly complete account as to the method by which it was obtained. For full particulars, as to the various experiments made with it, the reader is referred to **Dr. Watt's** *Selections from the Records of the Government of India, Revenue and Agricultural Department,* Vol. I., 95. The following is an abstract of **Captain Wood's** method of preparing the extract:—The extract is manufactured in a way similar to that practised by Catechu-makers in the Gonda forests. After choosing a good site, within easy reach of water, and of the bark used for boiling down, the huts of the men and the furnaces are erected. The furnace, built of clay with walls 6 inches thick, 10 feet to 15 feet long, an inner breadth of 2 feet 6 inches, and a similar height, has an arched top, pierced with holes on both sides along its length, to hold the earthen pots in which the bark is boiled, and has fireholes at each end and at one side, for introducing fuel and raking out ashes.

When the furnace is finished, the work of collecting and chipping the bark is begun. It is usually stripped off in pieces, 3 feet long, and conveyed to the encampment where it is cut into smaller pieces 3 inches by 4 inches. The earthen pots (*handis*) half-full of chips are filled with water

DYE & TAN.

till three-quarters full and then placed on the furnace to boil, after which they are allowed to simmer for one-and-a-half hours, till the liquor is a very deep red colour. It is then strained off into a fresh pot, the chips are thrown aside and when dry are used for the furnace as fuel. Three pots of the first boiled liquor fill one for the further boiling process, so a furnace holding 21 pots would, after boiling and straining, give seven pots of liquor, which are generally placed in the centre line, while the remainder, refilled with chips, are put on the sides of the furnace. The strained liquor is reduced to half its quantity by an hour's boiling, and half the pots are free to take the fresh outturn of liquor from the remaining 14.

The liquor gradually becoming stiff in about an hour, great care must be taken to keep it from burning, and as soon as it attains the consistency of treacle it is removed and poured into a trough. When the trough is full the contents are allowed to cool during the night, and next morning are poured into a circular pit 3 feet wide and 3 feet deep, and so on with each day's outturn till the pit is full and a new one has to be dug.

On a furnace of 21 pots in Ramgarh, three Khairaha women, engaged at R0-2-6 each per diem, did the boiling, and on an average turned out daily 16 seers of the extract. One coolie (on R0-2-0 per diem) was employed at the furnace for cutting bark, etc., two men for cutting poles and stripping bark, one man for cutting fuel, and a bullock cart for conveying fuel and bark to the furnace. During a period of 58 days (*viz.*, from 2nd February 1886 to March 31st, 1886), 291 maunds of *sál* bark were boiled and yielded 67 *ghurras* of extract, the average weight of a *ghurra*-full being 14 seers, the total weight of extract obtained was 23 maunds 18 seers, and the total expenditure R102-4-0, from which it will be seen that a maund of *sál* bark yielded 3·22 seers of extract.

Chemistry. 1661

The semi-fluid extract was analysed by **Dr. H. Warth** of the Central Forest School, who gives the following as its composition :—

Water	11.23	per cent.	
Insoluble in—			
Water	5.86	„	
Tannin	16.73	„	
Balance soluble in—			
Water	66.18	„	
TOTAL	100.00	„	
Of the whole	7.79	„	ashes were left.

"The colour of the iron precipitate is dingy green. The tannin is therefore of the same kind as that which is contained in **Acacia Catechu** extract. The tannin of gall apples, gall tannin, gives a bluish black precipitate with iron, and is therefore different. There was no Catechu acid present.

"For comparison I give you herewith the analysis of good genuine Burmese (Rangoon) cutch, which **Mr. Ribbentrop** sent me in January 1882 :—

CUTCH, RANGOON.

Catechu acid	0.0	per cent.
Catechu—tannic acid	7·35	„
Insoluble in water	14·35	„
Soluble in water	78·30	„
TOTAL	100·00	„
Ashes in the whole	18·89	„

SHOREA robusta.

The Sal—a good Tanning Material.

DYE & TAN. Chemistry.

"The method of analysis was the same in both cases—namely, precipitation with gelatine.

"Your extract is thus apparently twice as good as Burmese cutch. It would be of scientific interest to know how much bark was used in the preparation of the extract. Would you desire a publication of the results in the *Forester*, accompanied by a description to **Mr. Fisher**, the Editor? I would gladly make any suggestion I could on the possible improvement of the manufacture, if you would kindly let it be known how the extract has hitherto been made."

An analysis of the extract made by **W. N. Evans, Esq.**, Editor of *Leather*, the Journal of the Tanning Trade, shows that this extract contains 32.29 per cent. of tannin, and **Mr. Evans** goes on to state that "The colour of the solution made by the extract will be rather against its sale, but otherwise, judging from its strength, as about equivalent to chestnut wood extract of 30 degrees Bamin, which is selling from £12 to £15 per ton. No certainty can, however, be given to its real commercial value until it has been tested in some tannery, the quality of the leather it produces and the cost being the principal items involved."

Samples sent to the Cawnpore Harness and Saddlery Factory were reported on as follows by the foreman of that establishment:—"The concentrated essence of *sál* bark received has been tested and tried as far as practicable with the quantity received. The amount received, *viz.*, 1½ oz., was tried up to 21 ozs. of water, or fourteen times its own weight; and a rich brown liquor of 24° was the result. The quantity received was so very small, that its tanning properties could not be fairly tested; but judging from appearance and taste, I think it will be worth while to try it. It would be much better if it could be made solid, instead of a liquid, as by doing this there is not so much fear of evaporation, which will cause the extract to vary in strength." **Colonel Stewart**, the Superintendent of the factory, reported that "the liquor obtained from the extract of *sál* bark registered 24° of the barkometer; and it is probable that if a larger quantity had been sent, a tougher infusion could have been made." **Colonel Stewart** adds that he had not had "the means to test the extract chemically to discover the exact amount of tannin, and cannot make a practical test of such a small quantity; but is of opinion that the extract would give good results as a tanning agent, and it would be worth while to have sufficient extract to tan a given number of hides, and wishes to know if it would be possible to obtain 5 cwt. of this extract?"

OIL. Seeds. 1662

Oil.—The SEEDS yield an oil which is extracted by simple boiling (*Campbell*).

MEDICINE. Resin. 1663

Medicine.—"The RESIN is regarded as astringent and detergent, and is used in dysentery and for fumigations, plasters, etc. The resin thrown over the fire gives out thick volumes of fragrant smoke, and is much used for fumigating rooms occupied by the sick" (*Hindu Mat. Med.*). **Sakharam Arjun** states that he has seen good results follow the administration of *sál* with sugar in the treatment of dysentery (*Bombay Drugs, 20*).

Leaves. 1664

Mr. Campbell states that the LEAVES are used medicinally by the Santals.

SPECIAL OPINIONS.—§ "The resin is used by Native doctors for weak digestion, gonorrhœa, and as an aphrodisiac, and is prepared in the following manner:—A couple of ounces of the drug is powdered and fried in cows' *ghí* about 10 minutes and the whole thrown into a basin of cold water. The mixture which floats on the surface is gathered with the fingers and constantly squeezed within the water, in the course of a few minutes it gains consistence and assumes a white colour. Removing the water from time to time and thus continuing the above process from half an hour to one hour it gains the consistence and colour of butter. This is collected and preserved. A quantity, the size of a large nutmeg, is given internally twice a day" (*Surgeon-Major D R. Thompson, M.D., C.I.E., Madras*). "Twenty grains of the pulverised resin mixed with a pint of boiled milk taken every morning is considered a good aphrodisiac" (*Surgeon W. F. Thomas, 33rd Regiment, M. N. I., Mangalore*). MEDICINE.

Food.—The SEED ripens at the commencement of the rains and is collected and eaten especially in times of scarcity by the Santals and low caste tribes of Chutia Nagpur. For this purpose it is mixed with wood-ashes and boiled for two or three hours, well-washed to free it from the ashes, mixed with the flowers of the *mahua* tree (**Bassia latifolia**) and then reboiled or roasted. A sufficient quantity is cooked at one time to last the family for two or three days (*Journ. Agri.-Horti. Soc. Ind.*). Church, in his *Food-Grains of India*, describes its nutrient ratio as 1:12 and its nutrient value as 105; but remarks that the latter number is probably above the truth, for it is likely that a not inconsiderable part of the nutrients in these hard seeds of unappetising appearance, exists in an indigestible condition. FOOD. Seed. 1665

Structure of the Wood.—The *sál*, one of the most valuable timber trees in India, has a distinct sapwood which is small in amount, whitish and not durable. The heartwood is brown in colour, finely streaked with darker lines, coarse-grained, hard, strong, and tough, with a remarkably fibrous and cross-grained structure. The fibres of successive concentric strata do not run parallel, but at oblique angles, to each other, so that when the wood is dressed the fibres appear interlaced. It does not season well, but warps and splits in drying, and even when thoroughly seasoned, absorbs moisture with avidity in wet weather, increasing $\frac{1}{24}$th in bulk and correspondingly in weight. During the process of seasoning, it dries with great rapidity on the surface, while beneath it remains as wet as when first cut, and evaporation goes on afterwards with extreme slowness. The effect of this peculiarity is to cover the surface all over with superficial flaws from unequal shrinkage. With proper precautions, however, it can be made to dry slowly, and under these circumstances, it has been found by numerous experiments that the ratio of drying is $\frac{3}{4}$th of an inch annually all round the piece of wood. *Sál*, when once thoroughly seasoned, stands almost without a rival, as a timber, for strength, elasticity, and durability, which qualities it retains without being sensibly affected, for an immense length of time. Average weight of the seasoned *sál* about 55℔ per cubic foot (*Brandis; Gamble*). TIMBER. 1666

Its transverse strength has been tested by numerous experiments. Sir D. Brandis, after a long series of trials, found that the mean value of P. (the co-efficient of transverse strength) fluctuates between 708 and 916.

Domestic and Sacred.—The RESIN is used as an incense, and by boat builders instead of dammar for caulking boats. It is employed by the Santals to plug holes in earthen and even metal cooking vessels. Amongst many of the wild tribes of Central India the LEAVES are pinned into plates and cups, twisted into tobacco pipes, or formed into the wrappers for home-made cigars (*Rev. A. Campbell*). The day fixed for a hunting expedition is indicated by a branch of the *sál* tree being given with a certain number of leaves attached. A leaf is plucked off daily and the last one is removed on the morning of the hunt. DOMESTIC. Resin. 1667 Leaves. 1668

The TIMBER is the one most extensively used in Northern India. It is in constant request for piles, beams, planking, and railing of bridges, for beams, doors, and window posts of houses, for gun-carriages, for the bodies of carts, and, above all, for railway sleepers, the yearly consumption of which reaches some lakhs of cubic feet. In Assam it is the favourite wood for boat-building, and in the hills of Northern Bengal where it is found, Timber. 1669

DOMESTIC. perhaps, of the largest size now available, the trunks are hollowed out into canoes. Owing to the fact that when unseasoned it is not floatable, difficulty is experienced in most *sál* forests in getting the timber out of the forests in log. This is, however, overcome by floating the logs either with the assistance of boats or with floats of light wood or bamboos (*Gamble*). The CHARCOAL from the timber is said by the Kol iron smelters to be the best for their purpose and to produce superior iron. The local blacksmiths also prefer it for their work (*Rev. A. Campbell*).

Charcoal. 1670

SACRED. 1671 "The *sál* tree, called in Sanskrit *Sála* and *Asvakarna*, is of interest from a mythological point of view, since the mother of Buddha is represented as holding a branch of the tree in her hand when Buddha was born, and it was under the shade of a *Sála* tree that Buddha passed the last night of his life on earth. The small branches of the *sála* are used by Indian villagers to detect witches; they write upon branches the name of every woman over 12 years of age in the village; the branches are then placed in water and left for 4½ hours; if any woman's branch withers she is the witch" (*Pharmacog. Indica*).

1672 **Shorea siamensis**, *Miq.; Fl. Br. Ind., I., 304.*

Syn.—PENTACME SUAVIS, *A. DC.*; P. SIAMENSIS, *Kurz*; HOPEA (SHOREA?) SUAVA, *Wall.*

Vern.—*Ingyin, eng-kyn,* BURM.

References.—*DC., Prod., XVI., 2, 626, 631; Kurz, For. Fl. Br. Burm., I., 119; Gamble, Man. Timb., 39; Aplin, Rept. on Shan States, 6; Ind. Forester, IV., 292; VIII., 416; XIII., 134.*

Habitat.—A large, deciduous tree, very frequent in the *In* and dry forests of Burma, more especially those of the Ava and Prome districts, less frequent from Pegu and Martaban down to Tenasserim. It is distributed to Siam.

RESIN. 1673 **Resin.**—It yields a red resin.

TIMBER. 1674 **Structure of the Wood.**—Heartwood very hard, heavy and cross-grained; in this respect similar to that of *sál*, which it also resembles in colour. Weight about 55lb per cubic foot.

DOMESTIC. Wood. 1675 **Domestic Uses.**—The WOOD is much prized for its durability. It is used in house-building and for making canoes and bows and as planking.

1676 **S. stellata**, *Dyer; Fl. Br. Ind., I., 304.*

Syn—PARASHOREA STELLATA, *Kurz.*

Vern.—*Koung-mhoo,* BURM.

References.—*Kurz, For. Fl. Br. Burm., I., 117; Gamble, Man. Timb., 34.*

Habitat.—An evergreen tree frequent in the tropical forests of Martaban, rather rare along the eastern slopes of the Pegu Yomah, up to 1,500 feet elevation (*Kurz*).

TIMBER. 1677 **Structure of the Wood.**—White, hard, and rough. Weight about 50lb per cubic foot.

DOMESTIC. 1678 **Domestic.**—It is used for canoes and boat-building.

1679 **S. Talura**, *Roxb.; Fl. Br. Ind., I., 304; Wight, Ic., t. 164.*

Syn.—S. LACCIFERA, *Heyne*; VATICA LACCIFERA, *W. & A.*

Vern—*Talura, talari,* TAM.; *Jalari,* TEL.; *Jalaranda, jalari, jalada,* KAN.

References.—*DC., Prod, XVI., ii., 630; Roxb., Fl. Ind., Ed. C.B.C., 441; Brandis, For. Fl., 26; Beddome, Fl. Sylv., t. 6; Gamble, Man. Timb., 34; Liotard, Dyes, 33, 37; Watson, Rept., 18, 34; Gazetteers, Bombay, XV., 427; Madras Man. Adm., II., 76; Man. Cuddapah Dist., Madras, 263; Ind. Forester, X., 548; XII., 313.*

Habitat.—A large tree, met with in the forests of Mysore and the eastern districts of Madras.

Structure of the Wood.—Grey in colour, very hard, smooth, with small dark coloured irregularly shaped heartwood. Weight about 70lb per cubic foot. TIMBER. 1680

Domestic Uses.—It is much used for house-building and is largely sent down to Madras for that purpose. DOMESTIC. 1681

Shorea Tumbuggaia, *Roxb.; Fl. Br. Ind., I., 306; Wight, Ic., t. 27* 1682

Syn.—S. PENICILLATA, *A. DC.*; VATICA TUMBUGGAIA, *W. & A.*

Vern.—(Resin=) *Kálá-dámar,* HIND., BENG., & DEC.; *Cangu, congo, tumbugai,* (Resin=) *karuppu-dámar, tumb-ugai-pishin,* TAM.; *Thambá,* (Resin=) *nalla-dámar, nalla-rojan,* TEL.; *Vanbogu,* (Resin=) *kara-kundurukkan, tumbugai-pasha,* MALAY.

References.—*DC., Prod., XVI., ii., 630; Roxb., Fl. Ind., Ed. C.B.C., 440; Beddome, Fl. Sylv., 26; Gamble, Man. Timb., 39; Moodeen Sheriff, Mat. Med. Madras, 48; Birdwood, Bomb. Pr., 258; Watson, Rep., 6, 20, 32, 37; Madras Man. Adm., II., 76; Man. Cuddapah Dist., Madras, 262; Ind. Forester, X., 548.*

Habitat.—A large tree of the Western Peninsula, found in the dry forests of Cuddapah and Palghât in Mysore.

Resin.—It yields a dark coloured resin which is one of the common drugs in all the large markets of India. RESIN. 1683

Medicine.—The RESIN is recommended by **Moodeen Sheriff** as an external stimulant and a substitute for Abietis Resina and Pix Burgundica of European Pharmacopæias. MEDICINE. Resin. 1684

Structure of the Wood.—Smooth, harder than that of *sál,* but similar in appearance. Weight about 68lb per cubic foot. TIMBER. 1685

Domestic.—The RESIN is used as a substitute for pitch. The WOOD is employed in house-building, particularly for door frames and posts and for rafters. DOMESTIC. Resin. 1686 Wood. 1687

Shrews, see Rats, Mice & Marmots, Vol VI., p. 395.

SIDA, *Linn.; Gen. Pl, I., 203, 982.*

MALVACEÆ.

Sida carpinifolia, *Linn.; Fl. Br. Ind., I., 323; Wight, Ic., t. 95;* 1688
THE HORNBEAM-LEAVED SIDA.

Syn.—S. ACUTA, *Burm.*; S. LANCEOLATA, *Roxb.*; S. STIPULATA, *Cav.* S. STAUNTONIANA, *DC.*; S. SCOPARIA, *Lour.*

Vern.—*Bariára, kareta,* HIND.; *Pila-barélá-shikar, svt-berela koreta, bon-methi,* BENG.; *Isbadi, isarbadi,* DECCAN; *Bala, jangli-methi,* BOMB.; *Tupkaria, tukati, chikana, páta,* MAR.; *Jungli-methi,* GUZ.; *Vatta-tirippi, malaitangi, mayir-manikham, visha-boddi, chiti-mutti, mutu-va-pulogum,* TAM.; *Malatanni,* MALAY.; *Kát-say-nai, pyen-dan-gna-len,* BURM.; *Sirivadi-babila,* ANDAMAN; *Pata* or *balá* (generic names) SANS.

References.—*DC., Prod., I., 460, 461; Roxb., Fl. Ind., Ed. C.B.C., 515, 517; Thwaites, En. Ceyl. Pl., 27; Dalz. & Gibs., Bomb. Fl., 17; Mason, Burma and Its People, 519, 755; Rheede, Hort. Mal. X., t. 53; Pharm. Ind., 35; Ainslie Mat. Ind., II., 179; O'Shaughnessy, Beng. Dispens., 215; Moodeen Sheriff, Mat. Med. S. Ind. (in MSS.), 53; Sakharam Arjun, Cat. Bomb. Drugs, 18; Murray, Pl. & Drugs, Sind., 58; Dymock, Mat. Med. W. Ind., 2nd Ed., 99; Dymock, Warden & Hooper, Pharmacog. Ind., Vol. I., 206; Useful Pl. Bomb. (XXV., Bomb. Gaz.), 228; Liotard, Mem. Paper-making Mat., 31; Boswell, Man. Nellore, 142; Gazetteers:—Bombay, XV., 427; N.-W. P., IV., lxviii.; Ind. Forester, XIV., 273.*

Habitat.—A perennial under-shrub, generally distributed throughout the hotter parts of India.

Fibre.—A good FIBRE is obtained from the STEMS. FIBRE. Stems. 1689

Medicine.—The ROOT is medicinal and is described by **Moodeen Sheriff** in his forthcoming *Materia Medica of Madras* as "thin, long, cylindrical, MEDICINE. Roots. 1690

MEDICINE.

varying in length generally from 2 or 3 to 6 or more inches, and in thickness from that of a crow-quill to a goose-quill, very rough, knotty, contorted and often bent on itself once or twice. It is bitter in taste, and possesses no distinct smell, brown or dark brown externally and brownish-white internally." **Moodeen Sheriff** says it is very often confounded with another and larger root, which is described (under **Sida carpinifolia**) by **Ainslie** (in his *Materia Indica*) "as not unlike the common liquorice root." When administered in the form of a strong decoction the root of this plant has diaphoretic, antipyretic, stomachic and tonic properties, and has been found very useful in febrile affections and some forms of dyspepsia and also in mild cases of debility from previous illness (*Moodeen Sheriff*). **Sir W. O'Shaughnessy**, in a series of experiments made in Calcutta with this drug, found that given in the form of an infusion it promoted perspiration, increased the appetite, and was "in many respects a useful substitute for

Juice. 1691 more costly bitters." The expressed JUICE of the root in the form of an electuary is employed in the treatment of intestinal worms (*Beng. Dispens.*).

The roots of this and other species of **Sida** are largely used in Native medicine in the form of a weak infusion combined with ginger. Hindu practitioners regard them as tonic, astringent, and cooling, and prescribe

[R]oot-bark. 1692 them in nervous and urinary diseases, and in fever. The ROOT-BARK is often beaten up with milk and sugar, and aromatics and stimulants are sometimes added. In the Konkan the root of **S. carpinifolia** is applied with sparrows' dropping, to burst boils (*Dymock*). The Hindus of Southern

Leaves. 1693 India employ the LEAVES made warm and moistened with a little gingelly oil to hasten suppuration (*Ainslie*), and **Dymock** adds that in Konkan they are applied with other cooling leaves in ophthalmia. The Muhammadans of Western India believe this drug to have aphrodisiac properties. In Goa, the Portuguese value it as a diuretic, especially in rheumatic affections; they also use it as a demulcent in gonorrhœa.

1694 **Sida cordifolia,** *Linn.; Fl. Br. Ind., I., 324.*

[*Swartz*

Syn.—S. HERBACEA, MICANS, & ROTUNDIFOLIA, *Cav.;* S. ALTHÆIFOLIA

Vern.—*Kungyi, khareti, bariar,* HIND.; *Brelá, balá,* BENG.; *Kharent*) (Seeds=) *bíjband, chúka, hamás, kowár, símák,* PB.; *Burrayra,* (Seed= *bíjband,* SIND.; *Chikana,* MAR.; *Muttava, chiribenda, tetta gorra chettu tella antísa,* TEL.; *Balá, bátyálaka,* SANS.

References.—*DC., Prod., I., 464; Roxb., Fl. Ind., Ed. C.B.C., 517; Thwaites, En. Ceyl. Pl., 28; Dalz. & Gibs., Bomb. Fl., 17; Stewart, Pb. Pl., 23; Sir W. Elliot, Fl. Andhr., 41, 120, 174, 176; Rheede, Hort. Mal., X., t. 54; Fleming, Med. Pl. & Drugs (Asiatic Reser., XI.), 178; Ainslie, Mat. Ind., I., 205; Irvine Medical Topog., 126; U. C. Dutt, Mat. Med. Hind., 120, 293; Sakharam Arjun, Cat. Bomb. Drugs, 212; Murray, Pl. & Drugs, Sind., 59; Dymock, Mat. Med. W. Ind., 2nd Ed., 99; Baden Powell, Pb. Pr., 332; Atkinson, Him. (Dist., X., N.-W. P. Gaz.), 306; Gazetteers:—N.-W. P., I., 79; IV., lxviii.; Mysore and Coorg, I., 58; Ind. Forester, XII., App., 7; XIV., 273.*

Habitat.—A small annual or perennial weed, generally distributed in moist places, throughout tropical and sub-tropical India.

FIBRE. Plant. 1695 **Fibre.**—The PLANT yields a fine, white fibre.

MEDICINE. Seeds. 1696 **Medicine.**—The SEEDS are considered aphrodisiac and are administered in gonorrhœa. According to **Bellew**, they are given in the Panjáb for

Roots. 1697 colic and tenesmus. The ROOTS of this species also are regarded by Native practitioners as cooling, astringent, and tonic, and a decoction combined with ginger is given in intermittent fevers. In many nervous diseases such as hemiplegia and facial paralysis, the root of **Sida cordifolia** combined with asafœtida and rock-salt is administered internally and an oil called *balataila*, prepared from a strong decoction of this drug mixed with milk and sesamurn oil, is used as an external application in the same class of cases. A powder of the ROOT-BARK together with milk and sugar is given for the relief of frequent micturition and leucorrhœa (*U. C. Dutt*). MEDICINE. Root-bark. 1698

Sida humilis, *Willd; Fl. Br. Ind., I., 322.* 1699

Syn.—S. PILOSA, *Retz., not of Cav.;* S. RADICANS, *Wall.;* S. MULTICAULIS, *Cav.;* S. NERVOSA, *Wall., not of DC.*

Vern.—*Junka*, BENG.; *Bir, tandi, bariar, jokha sakam*, SANTAL; *Palampási*, TAM.; *Gáyapu áku*, TEL.

References.—*DC., Prod., I., 463; Roxb., Fl. Ind., Ed., C.B.C., 516; Thwaites, En. Ceyl. Pl., 23; Dalz. & Gibs., Bomb. Fl., 17; Sir W. Elliot Fl. Andhr., 59; Gazetteers.—Bomb., V., 24; XV., 427; N.-W P., I., 79; IV., lxviii.; Mysore and Coorg, I., 58; Man., Coimbatore Dist., Madras, 247.*

Habitat.—A very variable herbaceous plant, often procumbent, distributed generally throughout the hotter parts of India.

Medicine.—Among the Santals the LEAVES are pounded, and used as a local application to cuts and bruises. They are also given in the diarrhœa of pregnancy (*Rev. A. Campbell*). In the Coimbatore district, they are ground up with cummin seed, onions, and the succulent portion of aloe leaves, mixed with buffalo butter-milk, and given to cattle suffering from rinderpest (*Nicholson*). MEDICINE. Leaves. 1700

Food.—The LEAVES are eaten by the Santals as a potherb (*Rev. A. Campbell*). FOOD. Leaves. 1701

Domestic.—The LEAVES are employed to plug holes in iron or earthen cooking-pots. DOMESTIC. Leaves. 1702

S. rhombifolia, *Linn.; Fl. Br. Ind. I., 323.* 1703

Syn.—S. CANARIENSIS, *Willd.;* S. COMPRESSA, *Wall.*

Vern.—*Swet-berela, sahadebi*, HIND.; *Pitbalá, svet berelá*, BENG.; *Athiballa chettu*, TAM.; *Atibalá*, SANS.

References.—*DC., Prod., I., 462; Roxb., Fl. Ind., Ed. C.B.C., 517; Thwaites, En. Ceyl. Pl., 28; Dalz. & Gibs., Bomb. Fl., 17; Sir W. Elliot, Fl. Andhr., 17, 43, 44, 120; Sir W. Jones, Treat. Pl. Ind., 519, 755; Rheede, Hort. Mal., X., 18; Rumphius, Amb., V., t. 19; Fleming, Med. Pl. & Drugs (Asiatic Reser., XI.), 178; Irvine, Mat. Med. Patna, 125; U. C. Dutt, Mat. Med., Hind., 292; Murray, Pl. & Drugs, Sind., 59; Dymock, Mat. Med. W. Ind. Ed., 2nd., 99; Atkinson, Him. Dist. (X., N.-W. P. Gaz.), 306; Royle, Fibrous Pl., 262; Christy, New Com. Pl., VI., 36, 101; Rep. Agric. Dep., Bengal (1886-87), 21, 22; Gazetteers:—Bombay, XV., 427; N.-W. P., IV., lxviii.; Mysore & Coorg, I., 58 Agri.-Horti. Soc., Ind.:—Journ., VIII., 62, 63, 90, (Pro.) 101, 142; IX., 146-149, (Pro.) 69, 101; X., 61; XIV., 53, (Pro.) 69; (New Series), VI. (Sel.), 65; VIII., 117, 120, 124, 222; Ind. Forester, XIV., 269, 270, 273, 274, 276; Spons, Encycl., I., 996.*

Habitat.—A shrubby, very variable, perennial plant, widely distributed throughout the tropical regions of India.

Dr. Watt (in the *Selections from the Records of the Government of India, 1888-89*) has furnished so detailed a report on this fibrous plant that it does not seem necessary to do more here than to re-arrange the paragraphs of that paper according to the standard followed in this work. **Dr. Watt** writes:—The *Flora of British India* describes five varieties of this plant reducing to these the forms that were made species by the early writers. It seems probable that the fibres afforded by these varieties will not be found of equal merit; that being so, it would be desirable to have them separately dealt with. It is only by having all the varieties carefully cultivated and botanical specimens preserved (to allow of determin-

SIDA rhombifolia.

Varieties of the Sida Plant.

MEDICINE.

ation of the actual variety that afforded each particular fibre) that any real progress can be made towards organizing and developing a trade in **Sida** fibre. By this means alone, as it would appear, can the conflicting reports regarding **Sida** fibre be explained; one sample reported on by one author was the fibre from one variety and another sample from another. The following brief abstract taken, Dr. **Watt** explains, mainly from the *Flora of British India*, exhibits the varieties of **Sida rhombifolia**, *Linn.*, as accepted by modern botanists.

VARIETIES.

1704 "**Var. 1.—scabrida,** *Wight & Arnott, Prod.*, *57*.

Whole plant sprinkled with rigid, simple or 2—3 partite hairs; both sides of the leaves green, not tomentose below. Branches without tubercles under the leaves. Peduncles axillary solitary, a little more than half the length of the leaves, jointed at the base. Carpels bicuspidately awned.

This seems to be a form more particularly plentiful in South India.

1705 "**Var. 2.—retusa,** *Linn.*

Leaves obovate, retuse, hoary beneath, toothed towards the apex; stipules longer than the petiole. Peduncles solitary axillary, equalling the leaves, jointed near the middle. Carpels birostrate through the presence of short awns.

This form is fairly widely distributed, being the **S. retusa,** *Willd.*, as in *Roxburgh's Flora Indica* (*Ed. C.B.C., 517*), where it is said to be a native of Bengal; the **S. retusa** of *Dalzell & Gibson's Bombay Flora*, where it is said to be "very common;" the **S. chinensis,** *Retz.*, as in *Roxburgh's Flora Indica* (*l. c.*); and also the **S. philippica,** *DC.*, a form met with on the Coromandel Coast. It is the plant described by **Rheede,** *Hort. Mal., X., 18*; and by **Rumph.**, *Amb., V., t. 19*.

1706 "**Var. 3.—rhomboidea,** *Roxb.; Fl. Ind., Ed. C.B.C., 517*.

A shrubby plant without tubercles on the stem. Leaves rhomboid-lanceolate, serrate, hoary beneath; stipules longer than the petioles. Peduncles more than half the length of the leaves, jointed at the base, usually collected into leafy corymbs at the extremities of the branches. Awns of the capsules very short and inflected.

This is the *Mahábalá* of Sanskrit writers, the *Shwet-barjala* (white *barjala*) of Bengal, and the *Atibala chettu*, TELUGU. It is the **S. rhombifolia,** *Wall.*; the **S. rhomboidea,** *Roxb.*, as in *W. & A, Prod.;* and the **S. orientalis,** *Cav.* **Roxburgh** says it is a native of Bengal, where it blossoms during the cold season, the flowers opening at noon. It is met with in some parts of Madras, and seems also to be the plant reported on in the *Agri.-Horti. Soc. Journal* as affording the fibre which **Major Hannay** sent from Assam. In all probability this form is that which yields the best quality of fibre.

1707 "**Var. 4.—obovata,** *Wall.*

This is a large-leaved plant, the broadly obovate leaves measuring 1½ by 2 inches, hoary beneath, and having the apex coarsely toothed, and the base drawn out or cuneate. Peduncles longer than the petiole, but shorter than the blade.

1708 "**Var. 5.—microphylla,** *Cav.*

Leaves small, eliptic dentate, hoary beneath. Peduncle slightly exceeding the petiole. Carpels 5—7 awned.

This is the plant described by **Roxburgh** (*Fl. Ind., Ed. C.B.C., 515*), which, he says, is a native of Bengal, and which flowers the whole year round.

" **Var. 6.—rhombifolia.** VARIETIES. 1709

To these five forms must be added the condition which would answer to the type of the species—the **Sida rhombifolia**, *Willd.*, as in Roxburgh's *Fl. Ind., Ed. C.B.C*, *517*—the *Lal-bariala* (or red *bariala*) or *berela* of Bengal, and the *Atibula* of Sanskrit. This, Roxburgh states, is a native of Bengal, and flowers during the rainy season. **Aitchison** mentions it as met with in the Panjáb.

SIDA FIBRE.

FIBRE. 1710

The following is **Dr. Watt's** account of **Sida** fibre :—" It seems probable that the fibre experimented with in Bengal has been mostly obtained either from the last mentioned, or from the third, variety (**rhomboidea**). These, therefore, had better receive the first attention, as they will most likely be found to contain the best fibre. But the utilisation of **Sida** will mainly depend on the particular form that will produce the tallest stems with the fewest branches, and the highest percentage of fibre to weight of stems. It may readily be admitted that the feature, on which the industry will fail to be established, will be the yield as compared with jute. Hitherto the few Reports that have appeared exhibit the yield as considerably lower than that of jute. Experimenters should, however, not be too easily disheartened on this score, for it must be borne in mind that **Sida** has never been systematically cultivated, while everything has been done that is possible to improve the yield and quality of jute. A few years' cultivation may result in the production of a **Sida** stock that would give a nearer approximation than has as yet been attained to that of jute; and the fibre will most certainly fetch a considerably higher price than its rival. The claims of **Sida** fibre do not rest on the statements of one observer, more than its rejection should depend on the results of one experiment such as that recently reported on by the Agri-Horticultural Society of India, *viz*, that it possesses, like **Hibiscus, Abutilion,** and **Sansevieria** " no advantage over that of jute." That opinion is probably so far correct; but it is more to the verdict of the flax than the jute manufacturer that we have to look, since everything points towards the new fibre entering the higher textile markets for which jute is quite unsuited. To make gunny bags of **Sida** would, indeed, be an extravagance, and since jute serves that purpose sufficiently well, there would be nothing gained by disturbing the Bengal cultivators in order that **Sida** might be used as a jute substitute. **Sida** may, however, like **Hibiscus cannabinus,** be grown over a wide area where jute cultivation is impossible. The flax manufacturers of Europe admit that the time has come when they must seek for flax substitutes. In the opinion of many experts no fibre of modern times affords better hopes of success than **Sida**, and the matter may therefore be earnestly recommended to the attention of Government as one well worthy of the expenditure of a little money in the experiments here suggested, namely (*a*), to definitely determine which form of **Sida** yields the best fibre; (*b*) where that can be most successfully grown; (*c*) the present yield per acre; (*d*) the price at which the fibre could now be put into the market; and (*e*) to continue the experiments further in order to ascertain whether or not the fibre-yielding property of the selected stock could be improved.

" In order to justify these recommendations an abstract of existing literature on this subject may be here given. The first person who commercially drew attention to this fibre was **Major Hannay** of Assam. His action followed a few years after **Dr. Roxburgh** had described the various species of **Sida**, in which he remarked, under **S. rhombifolia,** that " the

SIDA rhombifolia.

The Sida Fibre.

FIBRE.

bark of this and the last (**rhomboidea**) yields abundance of very delicate flaxy fibres, and I think might be advantageously employed for many purposes. When the seed is sown thick on a good soil, the plants grow tall and slender, without branching, and in every way fit for such purposes." **Roxburgh** would thus appear to have cultivated these plants; but in the report of his fibre experiments he nowhere alludes to his results in testing the strength and endurance of the fibres obtained from them such as he published with most other Indian fibres. **Major Hannay's** observations seem to have been perfectly spontaneous, and it was only after his fibre had been communicated to the Agri.-Horticultural Society that it was discovered to be obtained from **Sida**. The following report on **Major Hannay's** Assam fibre will be read with interest and not unconnected with astonishment that the subject of it should have remained so many years in abeyance. "This fibre very much resembles our best dressed jute." "It is very attractive in its appearance. Its silvery bright and clear colour, its great cleanliness, and its excellent condition, are well exhibited—much better, indeed, than usually belongs to the great bulk of the jute which is exported to Europe, and hence might, for such reasons, obtain some preferable consideration." After testing the strength of this fibre, and ascertaining its indestructibility by water, I am convinced it is not jute, but I am not prepared to give it a name. But I think from the length of the staple, its similarity to silk, and its great strength, that it would fetch a high price in England. The line (only half an inch in circumference) sustained, after exposure to wet and sun for ten days 400℔" (*Journal, Agri.-Hort. Soc., Vol. VIII. (old series) 1854, p. 62*). In the subsequent volume, the members of the Agri.-Horticultural Society had their attention again directed to this fibre. **Major Hannay**, in 1853, sent seed of the plant from which his fibre was obtained. The plant was cultivated in the Society's garden and a report published. "I have now," **Mr. Joseph Willis** wrote, "to report on the specimen from our gardener **Mr. McMurray's** growth and preparation. I find it of excellent length, being about 10 feet. It is very completely freed from all ligneous adherents, and is in excellent condition, having its silvery brightness or lustre and colour in high perfection. The fibre is remarkably round; it is also fine, being somewhat coarser at the root end than in the upper parts, and near the top extremity it becomes exceedingly fine. The strength, although inferior to some of the best or *strong* fibres, which have been before us, is nevertheless excellent. I consider this fibre worthy of the best attention of those who may be engaged in vegetable fibrous productions, and more especially so, as it seems capable of being grown so well in Lower Bengal." **Mr. Haworth**, another high authority, recommended the fibre to be tried on jute machinery, and with this object a sample was sent to the Chamber of Commerce, Dundee, and another to **Messrs. Marshall & Co.** of Leeds. In the IXth Vol. of the Agri.-Horticultural Society's Journal (quoted above) **Mr. McMurray** gives particulars regarding the method of cultivation adopted by him. The seed was sown broadcast on the 16th May and the crop cut in September. The stems were covered over by grass rubbish to cause fermentation, which took place in four days, and they were afterwards steeped for twelve days. The fibre was then cleaned by the same method as is pursued with jute.

"Shortly after the date of these experiments, the Calcutta fibre merchants had their attention forcibly directed towards the establishment in India of jute mills in opposition to those in Dundee for which they had formerly been contented to supply the fibre. **Major Hannay's** discovery was thus lost sight of for nearly 40 years, until in 1880 the Bengal Gov-

FIBRE.

ernment drew the attention of the Society to the subject of the *burriala* fibre—the fibre of **Sida rhomboidea**. The samples of fibre then communicated by the Government to the Society had been furnished by **Rajah Kristendro Roy** of Balihar, in the district of Rajshaye. These were reported on by **Mr. Cogswell** (*Journal, Vol. VI., New Series, 224*) who stated that none of the samples had been steeped long enough. He recommended that long stems should be steeped for seven to ten days, and added: "A large sample should be prepared, and I will get it tested in one of the jute mills, to see what percentage of warp yarn can be spun from it, and a correct value shall be arrived at. There is much in these samples of a soft, bright, glossy, clean fibre, but it is very short in comparison with jute, barely half its length, the value being very materially reduced in consequence."

"As a result of this report fresh samples of the fibre were prepared by the Rajah, and these **Mr. Cogswell** considered had been oversteeped, and thereby considerably injured. But he states: "To a few, even experienced men, this fibre might be mistaken for that of fine jute, though not one-fourth of its average length, when deprived of the root ends, as this has been. Its colour is glossy bright in the extreme, and of a very high order. The fibre is strong, fine, round, and of excellent spinning properties, and is well suited for the finest yarns of jute manufacturers; some of it is so silky, as to render it in my opinion fit for higher purposes. I value it at about R4-8 or R4-12 per bazár maund. I think the flax manufacturers at home would be ready consumers of it."

"It is perhaps as well to point out in this connection that by length of fibre in the above Reports, **Mr. Cogswell** necessarily means length of the fibre ribbons and not of the ultimate filaments. Cotton, for example, would be a worthless fibre if fibres 10, 15, and 20 feet long were required by all manufacturers. The length of the ultimate filaments and their adhesions are in fact far more important points to the manufacturer of the higher textiles than the length of ribbons composed of such filaments, since the bleaching, carding, and spinning into fine yarn is consequent on the degree to which the ribbons can be broken up. It is noteworthy that **Mr. Cogswell** recognises that **Sida** fibre is "fit for higher purposes" than jute, and that he very properly recommends it to the consideration of flax not jute manufacturers. There are few higher practical authorities on fibres than **Mr. Cogswell**, and this testimony would seem therefore to require only to be published in Europe for encouraging demands to be made on the resources of the enlightened Native gentleman who is the modern pioneer of this much-neglected fibre. Recently, however, in order to test the fibre-extracting machines sent in competition for the Government reward, **Sida** fibre, along with several others, has been reported on, and the discouraging statement made that it possesses "no advantage over that of jute." To prevent this opinion which seems to have been formed on a provincial more than an imperial stand-point, in other words, a jute manufacturer's stand-point, from injuring the prospects of the **Sida** fibre industry, and thereby deterring flax spinners and others from giving their attention to this subject, it appears desirable to reproduce here the recent report published by **Messrs. Cross, Bevan, & King**, in which a chemical and microscopic comparison is drawn between **Sida** and **Jute**.

CHEMISTRY. 1711

"These distinguished chemists say: 'Although closely similar to jute in structure and general chemical characteristics, it is in appearance a superior fibre; it is softer to the touch, and in all respects more uniform. This superiority, moreover, is confirmed by comparative chemical investi-

SIDA spinosa.

The Sida Fibre.

CHEMISTRY.

gation, to show which we reproduce side by side analytical numbers obtained for the two fibres:—

	Jute.	Sida.
Moisture	10·3	10·7
Ash	1·2	0·6
Hydrolysis (*a*)	15·0	6·6
,, (*b*)	18·0	12·2
Cellulose	75·0	83·0
Mercerising	16·0	6·6
Nitration	125·0	137·0
Acid purification	1·0	0·3
Carbon percentage	46·5	45·0

"In the relatively high percentage of cellulose we have the most important factor of superiority. It is interesting to note the correlations with this cardinal point of difference of all other qualities determined. This fibre is undoubtedly worthy of extended investigation by practical men."

"In conclusion it may safely be said that if Chemistry can be trusted to indicate the properties of a fibre, **Sida** is unquestionably an infinitely superior fiber to jute. Under hydrolysis (for bleaching and cleaning with an alkali) it loses a very much smaller proportion of its weight, is therefore less easily disintegrated by the action of water, and is consequently much more durable. Similarly, it loses less under the acid purification, and by nitration obtains a considerably greater weight, while it possess a much larger percentage of cellulose. A fibre with such properties to recommend it is surely worthy of the time and expenditure necessary to ascertain whether or not all these advantages are financially counterbalanced by a less acreage yield."

"Although the only large sample of the fibre that has as yet been produced was raised in Eastern Bengal it seems highly probable that South India or Bombay would prove better suited than the damper regions of Bengal for the development of a **Sida** industry." (*Watt, Sel. Records Govt. of Ind.*)

MEDICINE.

Root. 1712 **Medicine.**—The medicinal properties of this resemble those of other species. The ROOT of the *var.* **retusa** is held in great repute by the Natives

Seeds. 1713 in the treatment of rheumatism. The STEMS abound in mucilage and are employed by the Natives as demulcents and emollients both for external and for internal use.

1714 **Sida spinosa,** *Linn.; Fl. Br. Ind., I., 323.*

Syn.—S. ALBA and ALNIFOLIA, *Linn.;* S. RETUSA, *Wight.;* S. GLANDULOSA, *Roxb., MSS.;* S. BORIARA, *Wall.*

Vern.—*Gulsakari, jangli-méthi,* HIND.; *Gorakchâuliâ, pila-barêla, bon-méthi,* BENG.; *Jangli-méthi,* DEC.; *Mayir-mâmkkam,* TAM.; *Mayilu-mânikyam, china muttama, muttava pulagum, ternalla benda,* TEL.; *Kadu-menthyá,* KAN; *Mayir-mânikkam, katta-ventiyam* MALAY.; *Koti-kâm-babila, man-manikam,* SING.; *Nâga-balâ,* SANS.; *Kulbahe-barri,* ARAB.; *Shanbalide-barri, sham lithe-dash-ti,* PERS.

References.—*DC., Prod., I., 460, 461; Roxb., Fl. Ind., Ed. C.B.C., 516; Thwaites, En. Ceyl. Pl., 28; Dalz. & Gibs., Bomb. Fl., 17; Sir W. Elliot, Fl. Andhr., 39, 120, 180; Moodeen Sheriff, Mat. Med. S. Ind. (in MSS.), 55; U. C. Dutt, Mat. Med. Hind., 310; Dymock, Mat. Med. W. Ind., 2nd Ed., 99; Gazetteers:—N.-W. P., I., 79; IV., lxviii.; Mysore and Coorg, I., 57.*

Habitat.—A small, shrubby, perennial plant, found throughout the hotter parts of India from the North-West Provinces to Ceylon, and distributed throughout the tropics generally.

Medicine.—The LEAVES are demulcent and refrigerant and are useful in some cases of gonorrhœa, gleet, and scalding urine. The ROOT acts as a gentle tonic and diaphoretic, and is employed in mild cases of debility and fever. The leaves are bruised in water, strained through cloth, and administered in the form of a draught; the root is used in decoction prepared in a similar manner to that of **S. carpinifolia** (*Moodeen Sheriff*). The roots are useful also in the treatment of some forms of cattle disease. MEDICINE. Leaves. 1715 Root. 1716

SIDEROXYLON, *Linn.; Gen. Pl., II., 655.* 1717

A genus of trees which takes its name (σιδερος iron, and χάιον wood) from the hardness of the wood of its different species. Seven species are indigenous to the East Indies, and the timber from all is more or less employed locally in the regions where they occur. As no further economic information is available regarding six of the species, it has been thought unnecessary to give them more than this passing notice.

[*Wight, Ic., t. 1218;* SAPOTACEÆ.

Sideroxylon tomentosum, *Roxb.; Fl. Br. Ind., III., 538;* 1718

Syn.—S. ARMATUM, *Roth.;* SAPOTA TOMENTOSA, ARMATA, & ELENGOIDES, *A.DC.;* ACHRAS TOMENTOSA & ELENGOIDES, *Bedd.*

Vern.—*Kanta bohul,* URIYA; *Pálá,* TAM., *Húdigolla,* KAN.; *Thit-cho,* BURM.

References.—*DC., Prod., VIII., 175; Roxb., Fl. Ind., Ed. C.B.C., 202; Kurz, For. Fl. Burm, II., 116; Beddome, Fl. Sylv., t. 235, also For. Man., 142; Gamble, Man. Timb., 241, also xxiv.; Dalz. & Gibs., Bomb. Fl., 139; Graham, Cat. Bomb. Pl., 105; Useful Pl. Bomb. (XXV., Bomb. Gaz.), 89; Aplin, Report on Shan States (1887-88); Gazetteer, Bomb., XV., 437.*

Habitat.—A small or moderately-sized tree of the Western Gháts, from the Konkan southwards; also of Ceylon and Burma.

Food.—The FRUIT, a smooth yellow berry, is used by the Natives in pickles and curries. It is eaten greedily by the *Sambre.* FOOD. Fruit. 1719

Structure of the Wood.—Light reddish-white, fibrous, rather heavy. TIMBER. 1720

Domestic.—The TIMBER is used for house-beams and carpenters' planes. DOMESTIC. Timber. 1721

SILENE, *Linn.; Gen. Pl., I., 147.*

Silene Griffithii, *Boiss.; Fl. Br. Ind., I., 220.* 1722

Syn.—S. WEBBIANA, *Wall.;* S. MULTIFIDA, *Edgew.;* S. VISCOSA, *Pers.*

References.—*Aitchison in Journ. Linn. Soc., X., 78; Gazetteer, N.-W. P., X., 305.*

Habitat.—A perennial herb, found on the Western Himálaya from Garhwál to Kabul and Kishtwar, at altitudes between 7,000 and 11,000 feet; distributed to Afghánistán.

Domestic.—In Lahoul, the ROOT and LEAVES of this species, mixed with a natural impure carbonate of soda, are added as a substitute for soap to the water used by the Natives (*Aitchison*). (*Conf. with Detergents, Vol. III., 85.*) DOMESTIC. Root. 1723 Leaves. 1724

SILICA.

Silica.—For an account of the various forms of Silica and the industrial uses made of them, the reader is referred to **Rock Crystals,** Vol. II., 170; **Clay,** Vol., II., 360, Flint, Vol. III., 404; **Glass,** Vol. III., 503; **Quartz,** Vol. V., 378; and **Sand,** Vol. VI., Part. II. 1725

Printed in the United States
By Bookmasters